Skills in
Mathematics for
**JEE Main &
Advanced**

Vectors & 3D Geometry

With Sessionwise Theory & Exercises...

Skills in
Mathematics for
JEE Main &
Advanced

Vectors & 3D Geometry

With Sessionwise Theory & Exercises...

Amit M. Agarwal

arihant
ARIHANT PRAKASHAN (Series), MEERUT

arihant

ARIHANT PRAKASHAN (Series), MEERUT

卐 Administrative & Production Offices

Regd. Office

'Ramchhaya' 4577/15, Agarwal Road, Darya Ganj, New Delhi -110002
Tele: 011- 47630600, 43518550

卐 Head Office

Kalindi, TP Nagar, Meerut (UP) - 250002 Tel: 0121-7156203, 7156204

卐 Sales & Support Offices

Agra, Ahmedabad, Bengaluru, Bareilly, Chennai, Delhi, Guwahati, Hyderabad, Jaipur, Jhansi, Kolkata, Lucknow, Nagpur & Pune.

卐 ISBN : 978-93-26191-64-7

PO No : TXT-XX-XXXXXXX-X-XX

Published by Arihant Publications (India) Ltd.

For further information about the books published by Arihant, log on to www.arihantbooks.com or e-mail at info@arihantbooks.com

Follow us on

PREFACE

"YOU CAN DO ANYTHING IF YOU SET YOUR MIND TO IT, I TEACH GEOMETRY TO JEE ASPIRANTS BUT BELIEVE THE MOST IMPORTANT FORMULA IS COURAGE + DREAMS = SUCCESS"

It is a matter of great pride and honour for me to have received such an overwhelming response to the previous editions of this book from the readers. In a way, this has inspired me to revise this book thoroughly as per the changed pattern of JEE Main & Advanced. I have tried to make the contents more relevant as per the needs of students, many topics have been re-written, a lot of new problems of new types have been added in etc. All possible efforts are made to remove all the printing errors that had crept in previous editions. The book is now in such a shape that the students would feel at ease while going through the problems, which will in turn clear their concepts too.

A Summary of changes that have been made in Revised & Enlarged Edition

- Theory has been completely updated so as to accommodate all the changes made in JEE Syllabus & Pattern in recent years.
- The most important point about this new edition is, now the whole text matter of each chapter has been divided into small sessions with exercise in each session. In this way the reader will be able to go through the whole chapter in a systematic way.
- Just after completion of theory, Solved Examples of all JEE types have been given, providing the students a complete understanding of all the formats of JEE questions & the level of difficulty of questions generally asked in JEE.
- Along with exercises given with each session, a complete cumulative exercises have been given at the end of each chapter so as to give the students complete practice for JEE along with the assessment of knowledge that they have gained with the study of the chapter.
- Previous Years questions asked in JEE Main & Adv, IIT-JEE & AIEEE have been covered in all the chapters.

However I have made the best efforts and put my all teaching experience in revising this book. Still I am looking forward to get the valuable suggestions and criticism from my own fraternity i.e. the fraternity of JEE teachers.

I would also like to motivate the students to send their suggestions or the changes that they want to be incorporated in this book. All the suggestions given by you all will be kept in prime focus at the time of next revision of the book.

Amit M. Agarwal

CONTENTS

SYLLABUS

JEE MAIN

Three Dimensional Geometry

Coordinates of a point in space, distance between two points, section formula, direction ratios and direction cosines, angle between two intersecting lines. Skew lines, the shortest distance between them and its equation. Equations of a line and a plane in different forms, intersection of a line and a plane, coplanar lines.

Vector Algebra

Vectors and scalars, addition of vectors, components of a vector in two dimensions and three dimensional space, scalar and vector products, scalar and vector triple product.

JEE ADVANCED

Three Dimensions

Distance between two points, direction cosines and direction ratios, equation of a straight line in space, skew lines, shortest distance between two lines, equation of a plane, distance of a point from a plane, angle between two lines, angle between two planes, angle between a line and the plane, coplanar lines.

Vectors

Addition of vectors, scalar multiplication, dot and cross products, scalar and vector triple products, and their geometrical interpretations.

CHAPTER

01

Vector Algebra

Learning Part

Session 1

- Scalar and Vector Quantities
- Representation of Vectors
- Position Vector of a Point in Space
- Direction Cosines
- Rectangular Resolution of a Vector in 2D and 3D Systems

Session 2

- Addition & Subtraction of Vectors
- Multiplication of Vector by Scalar
- Section Formula

Session 3

- Linear Combination of Vectors
- Theorem on Coplanar & Non-coplanar Vectors
- Linear Independence and Dependence of Vectors

Practice Part

- JEE Type Examples
- Chapter Exercises

Session 1

Scalar and Vector Quantities, Representation of Vectors, Position Vector of a Point in Space, Direction Cosines, Rectangular Resolution of a Vector in 2D and 3D Systems

Vectors represent one of the most important mathematical systems, which is used to handle certain types of problems in Geometry, Mechanics and other branches of Applied Mathematics, Physics and Engineering.

Scalar and Vector Quantities

Physical quantities are divided into two categories-Scalar quantities and Vector quantities. Those quantities which have only magnitude and which are not related to any fixed direction in space are called scalar quantities or briefly scalars. Examples of scalars are mass, volume, density, work, temperature etc.

A scalar quantity is represented by a real number along with a suitable unit. Second kind of quantities are those which have both magnitude and direction, such quantities are called vectors. Displacement, velocity, acceleration, momentum, weight, force etc., are examples of vector quantities.

Example 1. Classify the following measures as scalars and vectors

(i) 20 m north-west (ii) 10 Newton

(iii) 30 km/h

(iv) 50m/s towards north

(v) 10^{-19} coloumb

Sol. (i) Directed distance -Vector

(ii) Force-Vector

(iii) Speed-Scalar

(iv) Velocity-Vector

(v) Electric charge-Scalar

Representation of Vectors

Geometrically, a vector is represented by a directed line segment.

For example, $\mathbf{a} = \mathbf{AB}$. Here, A is called the initial point and B is called the terminal point or tip.

A directed line segment with initial point A and terminal point B is denoted by $\mathbf{AB}$ or $\overrightarrow{AB}$. Vectors are also denoted by small letters with an arrow above it or by small bold letters, e.g. **a, b, c** etc. or $\vec{a}, \vec{b}, \vec{c}$ etc.

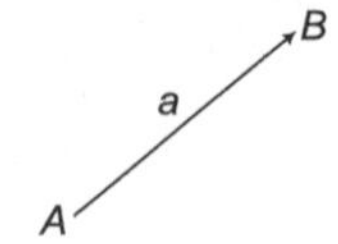

Here, in the figure $\mathbf{a} = \mathbf{AB}$ and magnitude or modulus of $\mathbf{a}$ is expressed as $|\mathbf{a}| = |\mathbf{AB}| = AB$ (Distance between initial and terminal points).

Remarks

1. The magnitude of a vector is always a non-negative real number.
2. Every vector **AB** has the following three characteristics

Length The length of **AB** will be denoted by $|\mathbf{AB}|$ or AB.

Support The line of unlimited length of which AB is a segment is called the support of the vector **AB**.

Sense The sense of **AB** is from A to B and that of **BA** is from B to A. Thus, the sense of a directed line segment is from its initial point to the terminal point.

Example 2. Represent graphically

(i) A displacement of 60 km, 40° east of north

(ii) A displacement of 50 km south-east

Sol. (i) The vector **OP** represent the required vector.

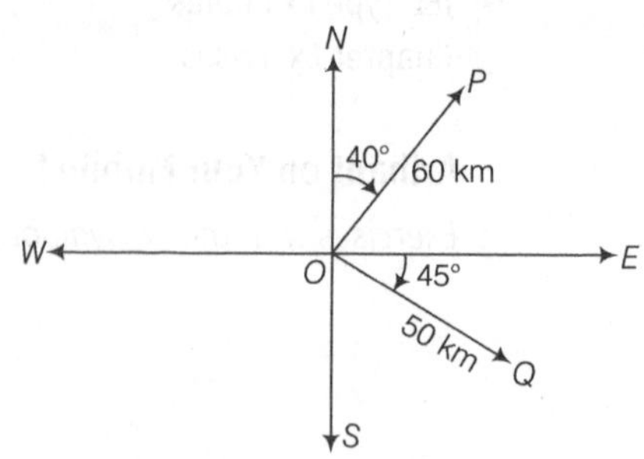

(ii) The vector **OQ** represent the required vector.

Types of Vectors

1. **Zero or null vector** A vector whose magnitude is zero is called zero or null vector and it is represented by 0.
   The initial and terminal points of the directed line segment representing zero vector are coincident and its direction is arbitrary.
2. **Unit vector** A vector whose modulus is unity, is called a unit vector. The unit vector in the direction of a vector **a** is denoted by $\hat{\mathbf{a}}$, read as "*a cap*". Thus, $|\hat{\mathbf{a}}|=1$.
   $$\hat{\mathbf{a}}=\frac{\mathbf{a}}{|\mathbf{a}|}=\frac{\text{Vector}}{\text{Magnitude of } a}$$
3. **Like and unlike vectors** Vectors are said to be like when they have the same sense of direction and unlike when they have opposite directions.
4. **Collinear or parallel vectors** Vectors having the same of parallel supports are called collinear vectors.
5. **Coinitial vectors** Vectors having the same initial point are called coinitial vectors.
6. **Coplanar vectors** A system of vectors is said to be coplanar, if they lie in the same plane or their supports are parallel to the same plane.
7. **Coterminous vectors** Vectors having the same terminal points are called coterminous vectors.
8. **Negative of a vector** The vector which has the same magnitude as the given vector **a** but opposite direction, is called the negative of **a** and is denoted by $-\mathbf{a}$. Thus, if $\mathbf{PQ}=\mathbf{a}$, then $\mathbf{QP}=-\mathbf{a}$.
9. **Reciprocal of a vector** A vector having the same direction as that of a given vector *a* but magnitude equal to the reciprocal of the given vector is known as the reciprocal of *a* and is denoted by $\mathbf{a}^{-1}$. Thus, if $|\mathbf{a}|=a$, then $|\mathbf{a}^{-1}|=1/a$.

Remark

A unit vector is self reciprocal.

10. **Localised vector** A vector which is drawn parallel to a given vector through a specified point in space is called a localised vector. For example, a force acting on a rigid body is a localised vector as its effect depends on the line of action of the force.
11. **Free vectors** If the value of a vector depends only on its length and direction and is independent of its position in the space, it is called a free vector.

Remark

Unless otherwise stated all vectors will be considered as free vectors.

12. **Equality of vectors** Two vectors **a** and **b** are said to be equal, if
    (i) $|\mathbf{a}|=|\mathbf{b}|$
    (ii) they have the same or parallel support.
    (iii) they have the same sense.
    Two unit vectors may not be equal unless they have the same direction.

Example 3. In the following figure, which of the vectors are:
(i) Collinear
(ii) Equal
(iii) Co-initial
(iv) Collinear but not equal

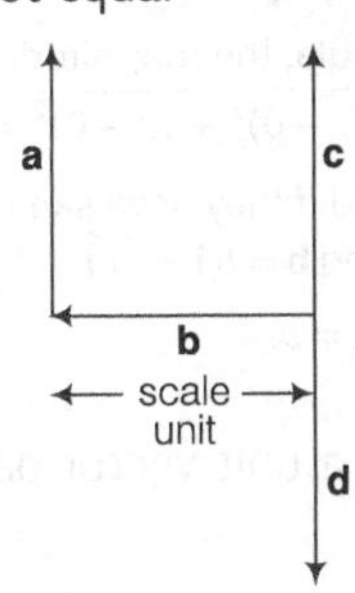

Sol. (i) ***a***,***c*** and ***d*** are collinear vectors.
(ii) ***a*** and ***c*** are equal vectors
(iii) ***b***,***c*** and ***d*** are co-initial vectors
(iv) ***a*** and ***d*** are collinear but they are not equal, as their directions are not same.

Position Vector of a Point in Space

Let *O* be the fixed point in space and $X'OX$, $Y'OY$ and $Z'OZ$ be three lines perpendicular to each other at *O*. Then, these three lines called *X*-axis, *Y*-axis and *Z*-axis which constitute the rectangular coordinate system. The planes *XOY*, *YOZ* and *ZOX*, called respectively, the *XY*-plane, the *YZ*-plane and the *ZX*-plane.

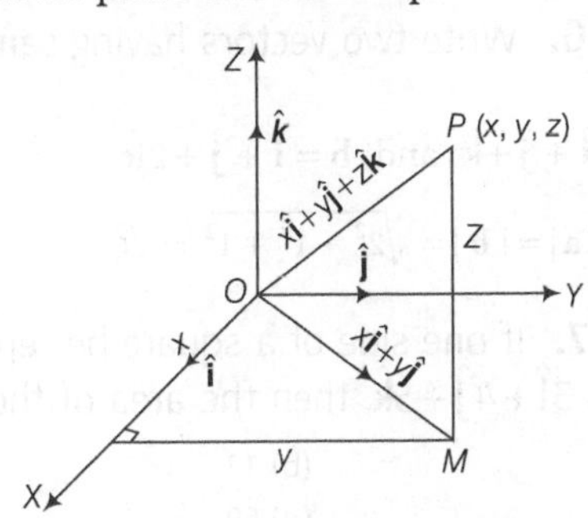

Now, let P be any point in space. Then, position of P is given by triad (x, y, z) where x, y, z are perpendicular distance from YZ-plane, ZX-plane and XY-plane respectively.

The vector **OP** is called the position vector of point P with respect to the origin O and written as

$$\mathbf{OP} = x\hat{\mathbf{i}} + y\hat{\mathbf{j}} + z\hat{\mathbf{k}}$$

where $\hat{\mathbf{i}}$, $\hat{\mathbf{j}}$ and $\hat{\mathbf{k}}$ are unit vectors parallel to X-axis, Y-axis and Z-axis. We usually denote position vector by $\mathbf{r}$.

Remarks

1. If A and B are any two points in space having coordinates (x_1, y_1, z_1) and (x_2, y_2, z_2) respectively, then distance between the points A and $B = \sqrt{(x_2 - x_1)^2 + (y_2 - y_1)^2 + (z_2 - z_1)^2}$.
2. Using distance formula, the magnitude of **OP** (or **r**) is given by $|\mathbf{OP}| = \sqrt{(x-0)^2 + (y-0)^2 + (z-0)^2} = \sqrt{x^2 + y^2 + z^2}$
3. Two vectors are equal if they have same components. i.e. if $\mathbf{a} = a_1\hat{\mathbf{i}} + a_2\hat{\mathbf{j}} + a_3\hat{\mathbf{k}}$ and $\mathbf{b} = b_1\hat{\mathbf{i}} + b_2\hat{\mathbf{j}} + b_3\hat{\mathbf{k}}$ are equal, then $a_1 = b_1$, $a_2 = b_2$ and $a_3 = b_3$.

Example 4. Find a unit vector parallel to the vector $-3\hat{\mathbf{i}} + 4\hat{\mathbf{j}}$.

Sol. Let $\mathbf{a} = -3\hat{\mathbf{i}} + 4\hat{\mathbf{j}}$

Then, $|\mathbf{a}| = \sqrt{(-3)^2 + (4)^2} = 5$

$\therefore$ Unit vector parallel to $\mathbf{a} = \hat{\mathbf{a}} = \dfrac{1}{|\mathbf{a}|} \cdot \mathbf{a}$

$$= \frac{-3\hat{\mathbf{i}} + 4\hat{\mathbf{j}}}{5} = \frac{-3}{5}\hat{\mathbf{i}} + \frac{4}{5}\hat{\mathbf{j}}$$

Example 5. Let $\mathbf{a} = 12\hat{\mathbf{i}} + n\hat{\mathbf{j}}$ and $|\mathbf{a}| = 13$, find the value of n.

Sol. Here, $\mathbf{a} = 12\hat{\mathbf{i}} + n\hat{\mathbf{j}}$

$\Rightarrow \quad |\mathbf{a}| = \sqrt{12^2 + n^2} = 13$

$\Rightarrow \quad 144 + n^2 = 169$

$\Rightarrow \quad n^2 = 25$ or $n = \pm 5$

Example 6. Write two vectors having same magnitude.

Sol. Let $\mathbf{a} = 2\hat{\mathbf{i}} + \hat{\mathbf{j}} + \hat{\mathbf{k}}$ and $\mathbf{b} = \hat{\mathbf{i}} + \hat{\mathbf{j}} + 2\hat{\mathbf{k}}$

Then, $|\mathbf{a}| = |\mathbf{b}| = \sqrt{2^2 + 1^2 + 1^2} = \sqrt{6}$

Example 7. If one side of a square be represented by the vectors $3\hat{\mathbf{i}} + 4\hat{\mathbf{j}} + 5\hat{\mathbf{k}}$, then the area of the square is

(a) 12 (b) 13
(c) 25 (d) 50

Sol. (d) Let $\mathbf{a} = 3\hat{\mathbf{i}} + 4\hat{\mathbf{j}} + 5\hat{\mathbf{k}}$ then $|\mathbf{a}|$

$$= \sqrt{3^2 + 4^2 + 5^2} = \sqrt{9 + 16 + 25} = 5\sqrt{2}$$

Thus, the length of a side of square $= 5\sqrt{2}$

Hence, area of square $= (5\sqrt{2})^2 = 25 \times 2 = 50$

Direction Cosines

Let $\mathbf{r}$ be the position vector of a point $P(x, y, z)$. Then, direction cosines of $\mathbf{r}$ are the cosines of angles α, β and γ that the vector $\mathbf{r}$ makes with the positive direction of X, Y and Z-axes respectively. We usually denote direction cosines by l, m and n respectively.

In the figure, we may note that ΔOAP is right angled triangle and in it we have

$$\cos\alpha = \frac{x}{r} \ (r \text{ stands for } |\mathbf{r}|)$$

Similarly, from the right angled triangles OBP and OCP, we get

$$\cos\beta = \frac{y}{r} \quad \text{and} \quad \cos\gamma = \frac{z}{r}$$

Thus, we have the following

$$\cos\alpha = l = \frac{x}{\sqrt{x^2 + y^2 + z^2}} = \frac{x}{|\mathbf{r}|} = \frac{x}{r}$$

$$\cos\beta = m = \frac{y}{\sqrt{x^2 + y^2 + z^2}} = \frac{y}{|\mathbf{r}|} = \frac{y}{r}$$

and $$\cos\gamma = n = \frac{z}{\sqrt{x^2 + y^2 + z^2}} = \frac{z}{|\mathbf{r}|} = \frac{z}{r}$$

Clearly, $\quad l^2 + m^2 + n^2 = 1.$

Here, $\alpha = \angle POX, \beta = \angle POY, \gamma = \angle POZ$ and $\hat{\mathbf{i}}, \hat{\mathbf{j}}$ and $\hat{\mathbf{k}}$ are the unit vectors along OX, OY and OZ respectively.

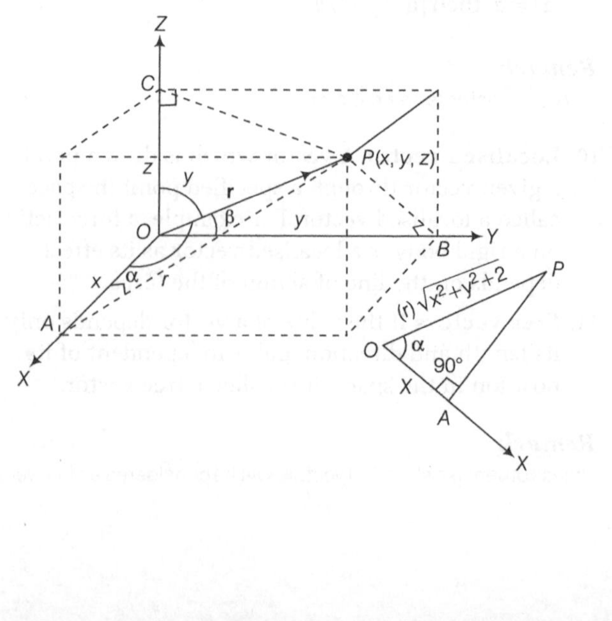

Remarks

1. The coordinates of point P may also be expressed as (lr, mr, nr).
2. The numbers lr, mr and nr, proportional to the direction cosines, are called the direction ratios of vector **r** and are denoted by a, b and c respectively.
3. If $\mathbf{r} = a\hat{\mathbf{i}} + b\hat{\mathbf{j}} + c\hat{\mathbf{k}}$, then a, b and c are direction ratios of the given vector.
   Also, if $a^2 + b^2 + c^2 = 1$, then a, b and c will be direction cosines of given vector.

Example 8. The direction cosines of the vector $3\hat{\mathbf{i}} - 4\hat{\mathbf{j}} + 5\hat{\mathbf{k}}$ are

(a) $\frac{3}{5}, \frac{-4}{5}, \frac{1}{5}$ (b) $\frac{3}{5\sqrt{2}}, \frac{-4}{5\sqrt{2}}, \frac{1}{\sqrt{2}}$

(c) $\frac{3}{\sqrt{2}}, \frac{-4}{\sqrt{2}}, \frac{1}{\sqrt{2}}$ (d) $\frac{3}{5\sqrt{2}}, \frac{4}{5\sqrt{2}}, \frac{1}{\sqrt{2}}$

Sol. (b) $\mathbf{r} = 3\hat{\mathbf{i}} - 4\hat{\mathbf{j}} + 5\hat{\mathbf{k}}$

$$\Rightarrow \quad |\mathbf{r}| = \sqrt{3^2 + (-4)^2 + 5^2} = 5\sqrt{2}$$

Hence, direction cosines are $\frac{3}{5\sqrt{2}}, \frac{-4}{5\sqrt{2}}, \frac{5}{5\sqrt{2}}$

i.e. $\frac{3}{5\sqrt{2}}, \frac{-4}{5\sqrt{2}}, \frac{1}{\sqrt{2}}$.

Example 9. Show that the vector $\hat{\mathbf{i}} + \hat{\mathbf{j}} + \hat{\mathbf{k}}$ is equally inclined to the axes OX, OY and OZ.

Sol. Let $\mathbf{a} = \hat{\mathbf{i}} + \hat{\mathbf{j}} + \hat{\mathbf{k}}$

If **a** makes angles α, β, γ with X, Y and Z-axes respectively, then

$$\cos\alpha = \frac{1}{\sqrt{1^2 + 1^2 + 1^2}} = \frac{1}{\sqrt{3}}$$

$$\cos\beta = \frac{1}{\sqrt{3}}$$

and $$\cos\gamma = \frac{1}{\sqrt{3}}$$

Thus, we have $\cos\alpha = \cos\beta = \cos\gamma$, i.e. $\alpha = \beta = \gamma$

Hence, **a** is equally inclined to the axes.

Rectangular Resolution of a Vector in 2D and 3D Systems

In Two Dimensional System

Any vector **r** in two dimensional system can be expressed as $\mathbf{r} = x\hat{\mathbf{i}} + y\hat{\mathbf{j}}$. The vectors $x\hat{\mathbf{i}}$ and $y\hat{\mathbf{j}}$ are called the perpendicular component vectors of **r**.

The scalars x and y are called the components or resolved parts of **r** in the directions of X-axis and Y-axis, respectively and the ordered pair (x, y) is known as coordinates of point whose position vector is **r**.

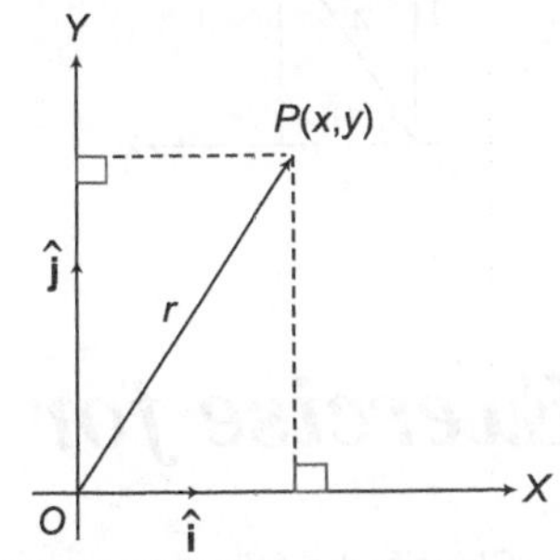

Also, the magnitude of $\mathbf{r} = \sqrt{x^2 + y^2}$ and if θ is the inclination of **r** with the X-axis, then $\theta = \tan^{-1}\left(\frac{y}{x}\right)$.

In three Dimensional System

Any vector **r** in three dimensional system can be expressed as

$$\mathbf{r} = x\hat{\mathbf{i}} + y\hat{\mathbf{j}} + z\hat{\mathbf{k}}$$

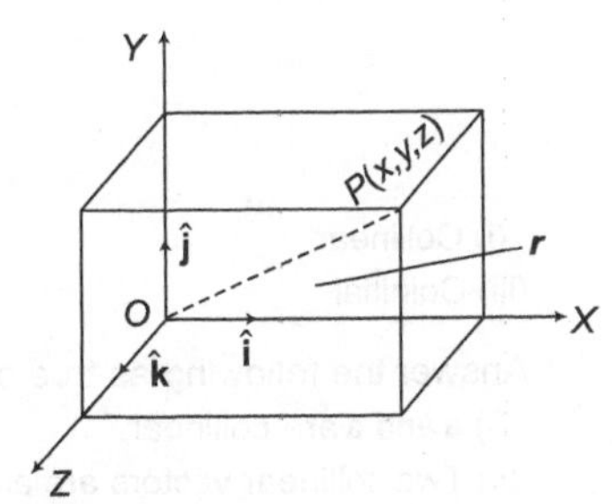

The vectors $x\hat{\mathbf{i}}$, $y\hat{\mathbf{j}}$ and $z\hat{\mathbf{k}}$ are called the right angled components of **r**.

The scalars x, y and z are called the components or resolved parts of **r** in the directions of X-axis, Y-axis and Z-axis, respectively and ordered triplet (x, y, z) is known as coordinates of P whose position vector is **r**. Also, the magnitude or modulus of

$$r = |\mathbf{r}| = \sqrt{x^2 + y^2 + z^2}.$$

Example 10. Let **AB** be a vector in two dimensional plane with the magnitude 4 units and making an angle of 30° with X-axis and lying in the first quadrant. Find the components of **AB** along the two axes of coordinates. Hence, represent **AB** in terms of unit vectors $\hat{\mathbf{i}}$ and $\hat{\mathbf{j}}$.

Sol. Let us consider A as origin. From the diagram, it can be seen that the component of **AB** along X-axis

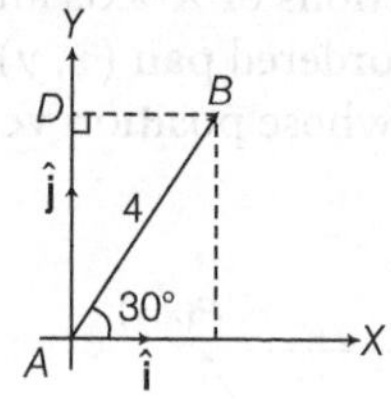

$$= AB \cos 30° = 4 \cos 30°$$

$$= 4 \times \frac{\sqrt{3}}{2} = 2\sqrt{3}$$

and the component of **AB** along Y-axis

$$= AB \sin 30° = 4 \times \frac{1}{2} = 2$$

Hence, $\mathbf{AB} = 2\sqrt{3}\hat{\mathbf{i}} + 2\hat{\mathbf{j}}$

Exercise for Session 1

1. Classify the following measures as scalars and vector :
   (i) 20 kg weight (ii) 45°
   (iii) 10 m south-east (iv) 50 m/sec^2
2. Represent the following graphically:
   (i) A displacement of 40 km, 30° west of south (ii) a displacement of 70 km, 40° north of west
3. In the given figure, $ABCDEF$ is a regular hexagon, which vectors are:

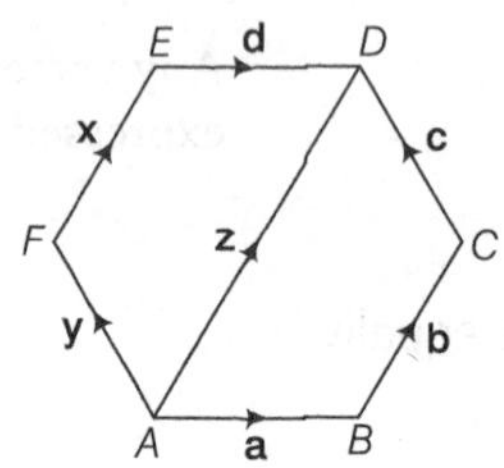

   (i) Collinear (ii) Equal
   (iii) Coinitial (iv) Collinear but not equal
4. Answer the following as true or false
   (i) **a** and **a** are collinear.
   (ii) Two collinear vectors are always equal in magnitude.
   (iii) Zero vector is unique.
   (iv) Two vectors having same magnitude are collinear.
5. Find the perimeter of a triangle with sides $3\hat{\mathbf{i}} + 4\hat{\mathbf{j}} + 5\hat{\mathbf{k}}$, $4\hat{\mathbf{i}} - 3\hat{\mathbf{j}} - 5\hat{\mathbf{k}}$ and $7\hat{\mathbf{i}} + \hat{\mathbf{j}}$.
6. Find the angle of vector $\mathbf{a} = 6\hat{\mathbf{i}} + 2\hat{\mathbf{j}} - 3\hat{\mathbf{k}}$ with X-axis.
7. Write the direction ratios of the vector $\mathbf{r} = \hat{\mathbf{i}} - \hat{\mathbf{j}} + 2\hat{\mathbf{k}}$ and hence calculate its direction cosines.

Session 2

Addition & Subtraction of Vectors, Multiplication of Vector by Scalar, Section Formula

Addition of Vectors

(Resultant of Vectors)

1. Triangle Law of Addition

If two vectors are represented by two consecutive sides of a triangle, then their sum is represented by the third side of the triangle, but in opposite direction. This is known as the triangle law of addition of vectors. Thus, if $\mathbf{AB} = \mathbf{a}$, $\mathbf{BC} = \mathbf{b}$ and $\mathbf{AC} = \mathbf{c}$, then $\mathbf{AB} + \mathbf{BC} = \mathbf{AC}$ i.e. $\mathbf{a} + \mathbf{b} = \mathbf{c}$.

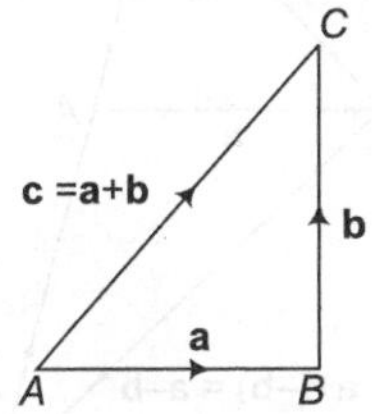

2. Parallelogram Law of Addition

If two vectors are represented by two adjacent sides of a parallelogram, then their sum is represented by the diagonal of the parallelogram whose initial point is the same as the initial point of the given vectors. This is known as parallelogram law of vector addition.

Thus, if $\mathbf{OA} = \mathbf{a}$, $\mathbf{OB} = \mathbf{b}$ and $\mathbf{OC} = \mathbf{c}$

Then, $\mathbf{OA} + \mathbf{OB} = \mathbf{OC}$ i.e. $\mathbf{a} + \mathbf{b} = \mathbf{c}$, where OC is a diagonal of the parallelogram $OACB$.

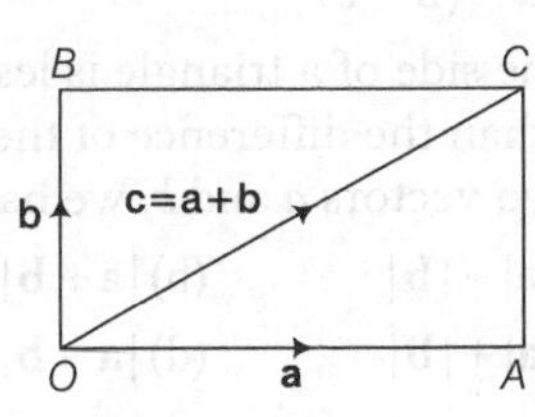

Remarks

1. The magnitude of $\mathbf{a} + \mathbf{b}$ is not equal to the sum of the magnitudes of $\mathbf{a}$ and $\mathbf{b}$.
2. From the figure, we have $\mathbf{OA} + \mathbf{AC} = \mathbf{OC}$ (By triangle law of vector addition)
   or $\mathbf{OA} + \mathbf{OB} = \mathbf{OC}$ ($\because \mathbf{AC} = \mathbf{OB}$), which is the parallelogram law. Thus, we may say that the two laws of vector addition are equivalent to each other.

3. Polygon law of addition

If the number of vectors are represented by the sides of a polygon taken in order, the resultant is represented by the closing side of the polygon taken in the reverse order.

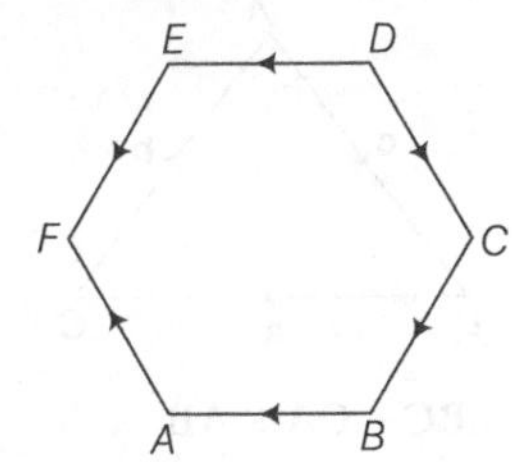

In the figure, $\mathbf{AB} + \mathbf{BC} + \mathbf{CD} + \mathbf{DE} + \mathbf{EF} = \mathbf{AF}$

4. Addition in Component Form

If the vectors are defined in terms of $\hat{\mathbf{i}}, \hat{\mathbf{j}}$ and $\hat{\mathbf{k}}$, i.e. if $\mathbf{a} = a_1\hat{\mathbf{i}} + a_2\hat{\mathbf{j}} + a_3\hat{\mathbf{k}}$ and $\mathbf{b} = b_1\hat{\mathbf{i}} + b_2\hat{\mathbf{j}} + b_3\hat{\mathbf{k}}$, then their sum is defined as

$\mathbf{a} + \mathbf{b} = (a_1 + b_1)\hat{\mathbf{i}} + (a_2 + b_2)\hat{\mathbf{j}} + (a_3 + b_3)\hat{\mathbf{k}}$.

Properties of Vector Addition

Vector addition has the following properties

(i) **Closure** The sum of two vectors is always a vector.

(ii) **Commutativity** For any two vectors $\mathbf{a}$ and $\mathbf{b}$,

$$\Rightarrow \quad \mathbf{a} + \mathbf{b} = \mathbf{b} + \mathbf{a}$$

(iii) **Associativity** For any three vectors $\mathbf{a}$, $\mathbf{b}$ and $\mathbf{c}$,

$$\Rightarrow \quad \mathbf{a} + (\mathbf{b} + \mathbf{c}) = (\mathbf{a} + \mathbf{b}) + \mathbf{c}$$

(iv) **Identity** Zero vector is the identity for addition. For any vector $\mathbf{a}$.

$$\Rightarrow \quad \mathbf{0} + \mathbf{a} = \mathbf{a} = \mathbf{a} + \mathbf{0}$$

(v) **Additive inverse** For every vector $\mathbf{a}$ its negative vector $-\mathbf{a}$ exists such that $\mathbf{a} + (-\mathbf{a}) = (-\mathbf{a}) + \mathbf{a} = \mathbf{0}$ i.e. $(-\mathbf{a})$ is the additive inverse of the vector $\mathbf{a}$.

Example 11. Find the unit vector parallel to the resultant vector of $2\hat{\mathbf{i}} + 4\hat{\mathbf{j}} - 5\hat{\mathbf{k}}$ and $\hat{\mathbf{i}} + 2\hat{\mathbf{j}} + 3\hat{\mathbf{k}}$.

Sol. Resultant vector, $\mathbf{r} = (2\hat{\mathbf{i}} + 4\hat{\mathbf{j}} - 5\hat{\mathbf{k}}) + (\hat{\mathbf{i}} + 2\hat{\mathbf{j}} + 3\hat{\mathbf{k}})$

$= 3\hat{\mathbf{i}} + 6\hat{\mathbf{j}} - 2\hat{\mathbf{k}}$

Unit vector parallel to $\mathbf{r} = \dfrac{1}{|\mathbf{r}|}\mathbf{r}$

$$= \frac{1}{\sqrt{3^2+6^2+(-2)^2}}(3\hat{\mathbf{i}}+6\hat{\mathbf{j}}-2\hat{\mathbf{k}})$$

$$= \frac{1}{7}(3\hat{\mathbf{i}}+6\hat{\mathbf{j}}-2\hat{\mathbf{k}})$$

Example 12. If **a**, **b** and **c** are the vectors represented by the sides of a triangle, taken in order, then prove that **a**+ **b**+ **c** = 0.

Sol. Let ABC be a triangle such that

$$\mathbf{BC} = \mathbf{a},\ \mathbf{CA} = \mathbf{b} \text{ and } \mathbf{AB} = \mathbf{c}$$

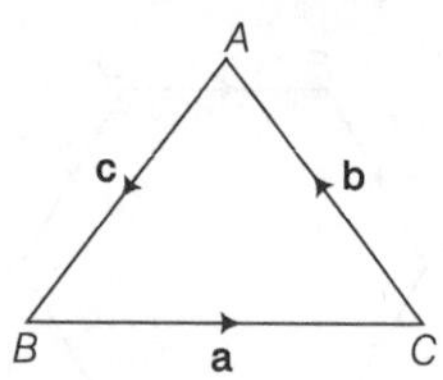

Then, $\mathbf{a}+\mathbf{b}+\mathbf{c} = \mathbf{BC}+\mathbf{CA}+\mathbf{AB}$

$= \mathbf{BA}+\mathbf{AB}$ $\quad(\because \mathbf{BC}+\mathbf{CA} = \mathbf{BA})$

$= -\mathbf{AB}+\mathbf{AB}$

$\mathbf{a}+\mathbf{b}+\mathbf{c} = 0$ **Hence proved.**

Example 13. If S is the mid-point of side QR of a ΔPQR, then prove that **PQ**+ **PR**= 2**PS**.

Sol. Clearly, by triangle law of addition, we have

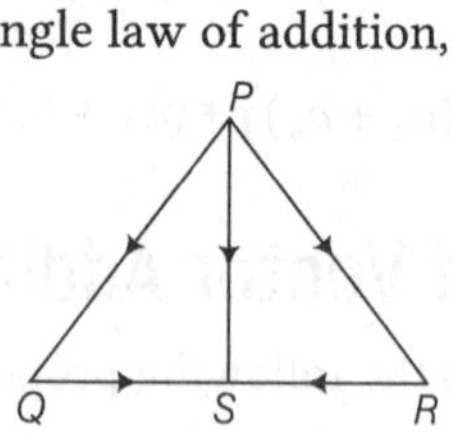

$\mathbf{PQ}+\mathbf{QS} = \mathbf{PS}$ ...(i)

and $\mathbf{PR}+\mathbf{RS} = \mathbf{PS}$ ... (ii)

On adding Eqs. (i) and (ii), we get

$(\mathbf{PQ}+\mathbf{QS})+(\mathbf{PR}+\mathbf{RS}) = 2\mathbf{PS}$

$\Rightarrow (\mathbf{PQ}+\mathbf{PR})+(\mathbf{QS}+\mathbf{RS}) = 2\mathbf{PS}$

$\Rightarrow \mathbf{PQ}+\mathbf{PR}+0 = 2\mathbf{PS}$

$[\because S$ is the mid-point of $\mathbf{QR} \therefore \mathbf{QS} = -\mathbf{RS}]$

Hence, $\mathbf{PQ}+\mathbf{PR} = 2\mathbf{PS}$ **Hence proved.**

Example 14. If $ABCDEF$ is a regular hexagon, prove that **AD**+ **EB**+ **FC** = 4**AB**.

Sol. We have,

$\mathbf{AD}+\mathbf{EB}+\mathbf{FC} = (\mathbf{AB}+\mathbf{BC}+\mathbf{CD})+(\mathbf{ED}+\mathbf{DC}+\mathbf{CB})+\mathbf{FC}$

$= \mathbf{AB}+(\mathbf{BC}+\mathbf{CB})+(\mathbf{CD}+\mathbf{DC})+\mathbf{ED}+\mathbf{FC}$

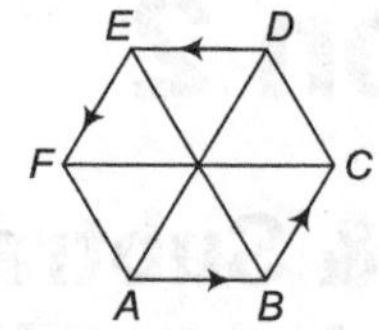

$= \mathbf{AB}+\mathbf{O}+\mathbf{O}+\mathbf{AB}+2\mathbf{AB} = 4\mathbf{AB}$

$(\because \mathbf{ED} = \mathbf{AB},\ \mathbf{FC} = 2\mathbf{AB})$

Hence proved.

Subtraction of Vectors

If **a** and **b** are two vectors, then their subtraction **a** − **b** is defined as **a** − **b** = **a** + (−**b**), where −**b** is the negative of **b** having magnitude equal to that of **b** and direction opposite to **b**.

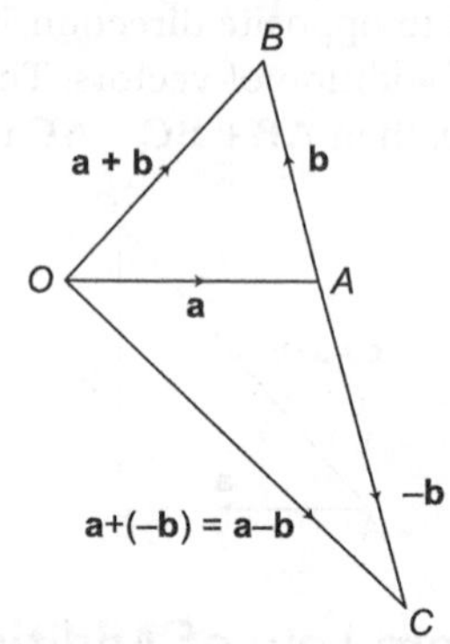

If $\mathbf{a} = a_1\hat{\mathbf{i}}+a_2\hat{\mathbf{j}}+a_3\hat{\mathbf{k}}$

and $\mathbf{b} = b_1\hat{\mathbf{i}}+b_2\hat{\mathbf{j}}+b_3\hat{\mathbf{k}}$

Then, $\mathbf{a}-\mathbf{b} = (a_1-b_1)\hat{\mathbf{i}}+(a_2-b_2)\hat{\mathbf{j}}+(a_3-b_3)\hat{\mathbf{k}}$

Properties of Vector Subtraction

(i) $\mathbf{a}-\mathbf{b} \neq \mathbf{b}-\mathbf{a}$

(ii) $(\mathbf{a}-\mathbf{b})-\mathbf{c} \neq \mathbf{a}-(\mathbf{b}-\mathbf{c})$

(iii) Since, any one side of a triangle is less than the sum and greater than the difference of the other two sides, so for any two vectors a and b, we have

(a) $|\mathbf{a}+\mathbf{b}| \le |\mathbf{a}|+|\mathbf{b}|$ (b) $|\mathbf{a}+\mathbf{b}| \ge |\mathbf{a}|-|\mathbf{b}|$

(c) $|\mathbf{a}-\mathbf{b}| \le |\mathbf{a}|+|\mathbf{b}|$ (d) $|\mathbf{a}-\mathbf{b}| \ge |\mathbf{a}|-|\mathbf{b}|$

Remark

If A and B are two points in space having coordinates (x_1, y_1, z_1) and (x_2, y_2, z_2), then

AB = Position Vector of B − Position Vector of A

$= (x_2\hat{\mathbf{i}}+y_2\hat{\mathbf{j}}+z_2\hat{\mathbf{k}})-(x_1\hat{\mathbf{i}}+y_1\hat{\mathbf{j}}+z_1\hat{\mathbf{k}})$

$= (x_2-x_1)\hat{\mathbf{i}}+(y_2-y_1)\hat{\mathbf{j}}+(z_2-z_1)\hat{\mathbf{k}}$

Example 15. If $A = (0, 1), B = (1, 0), C = (1, 2)$ and $D = (2, 1)$, prove that vector **AB** and **CD** are equal.

Sol. Here, $\mathbf{AB} = (1-0)\hat{\mathbf{i}} + (0-1)\hat{\mathbf{j}} = \hat{\mathbf{i}} - \hat{\mathbf{j}}$

and $\mathbf{CD} = (2-1)\hat{\mathbf{i}} + (1-2)\hat{\mathbf{j}} = \hat{\mathbf{i}} - \hat{\mathbf{j}}$

Clearly, $\mathbf{AB} = \mathbf{CD}$ **Hence proved.**

Example 16. If the position vectors of **A** and **B** respectively $\hat{\mathbf{i}} + 3\hat{\mathbf{j}} - 7\hat{\mathbf{k}}$ and $5\hat{\mathbf{i}} - 2\hat{\mathbf{j}} + 4\hat{\mathbf{k}}$, then find **AB**.

Sol. Let O be the origin, then we have

$$\mathbf{OA} = \hat{\mathbf{i}} + 3\hat{\mathbf{j}} - 7\hat{\mathbf{k}}$$

and $\mathbf{OB} = 5\hat{\mathbf{i}} - 2\hat{\mathbf{j}} + 4\hat{\mathbf{k}}$

Now, $\mathbf{AB} = \mathbf{OB} - \mathbf{OA} = (5\hat{\mathbf{i}} - 2\hat{\mathbf{j}} + 4\hat{\mathbf{k}}) - (\hat{\mathbf{i}} + 3\hat{\mathbf{j}} - 7\hat{\mathbf{k}})$

$$= 4\hat{\mathbf{i}} - 5\hat{\mathbf{j}} + 11\hat{\mathbf{k}}$$

Example 17. Vectors drawn from the origin O to the points A, B and C are respectively **a**, **b** and $4\mathbf{a} - 3\mathbf{b}$. Find **AC** and **BC**.

Sol. We have, $\mathbf{OA} = \mathbf{a}$, $\mathbf{OB} = \mathbf{b}$ and $\mathbf{OC} = 4\mathbf{a} - 3\mathbf{b}$

Clearly, $\mathbf{AC} = \mathbf{OC} - \mathbf{OA} = (4\mathbf{a} - 3\mathbf{b}) - (\mathbf{a})$

$$= 3\mathbf{a} - 3\mathbf{b}$$

and $\mathbf{BC} = \mathbf{OC} - \mathbf{OB} = (4\mathbf{a} - 3\mathbf{b}) - (\mathbf{b}) = 4\mathbf{a} - 4\mathbf{b}$

Example 18. Find the direction cosines of the vector joining the points $A(1, 2, -3)$ and $B(-1, -2, 1)$, directed from A to B.

Sol. Clearly, $\mathbf{AB} = (-1-1)\hat{\mathbf{i}} + (-2-2)\hat{\mathbf{j}} + (1+3)\hat{\mathbf{k}} = -2\hat{\mathbf{i}} - 4\hat{\mathbf{j}} + 4\hat{\mathbf{k}}$

Now, $|\mathbf{AB}| = \sqrt{(-2)^2 + (-4)^2 + (4)^2} = \sqrt{36} = 6$

$\therefore$ Unit vector along $\mathbf{AB} = \dfrac{\mathbf{AB}}{|\mathbf{AB}|} = \dfrac{-2\hat{\mathbf{i}} - 4\hat{\mathbf{j}} + 4\hat{\mathbf{k}}}{6}$

$$= -\frac{1}{3}\hat{\mathbf{i}} - \frac{2}{3}\hat{\mathbf{j}} + \frac{2}{3}\hat{\mathbf{k}}$$

Example 19. Let α, β and γ be distinct real numbers. The points with position vectors $\alpha\hat{\mathbf{i}} + \beta\hat{\mathbf{j}} + \gamma\hat{\mathbf{k}}$, $\beta\hat{\mathbf{i}} + \gamma\hat{\mathbf{j}} + \alpha\hat{\mathbf{k}}$ and $\gamma\hat{\mathbf{i}} + \alpha\hat{\mathbf{j}} + \beta\hat{\mathbf{k}}$

(a) are collinear

(b) form an equilateral triangle

(c) form a scalene triangle

(d) form a right angled triangle

Sol. (b) Let the given points be A, B and C with position vectors $\alpha\hat{\mathbf{i}} + \beta\hat{\mathbf{j}} + \gamma\hat{\mathbf{k}}$, $\beta\hat{\mathbf{i}} + \gamma\hat{\mathbf{j}} + \alpha\hat{\mathbf{k}}$ and $\gamma\hat{\mathbf{i}} + \alpha\hat{\mathbf{j}} + \beta\hat{\mathbf{k}}$.

As, α, β and γ are distinct real numbers, therefore ABC form a triangle.

Clearly, $\mathbf{AB} = \mathbf{OB} - \mathbf{OA} = (\beta\hat{\mathbf{i}} + \gamma\hat{\mathbf{j}} + \alpha\hat{\mathbf{k}}) - (\alpha\hat{\mathbf{i}} + \beta\hat{\mathbf{j}} + \gamma\hat{\mathbf{k}})$

$$= (\beta - \alpha)\hat{\mathbf{i}} + (\gamma - \beta)\hat{\mathbf{j}} + (\alpha - \gamma)\hat{\mathbf{k}}$$

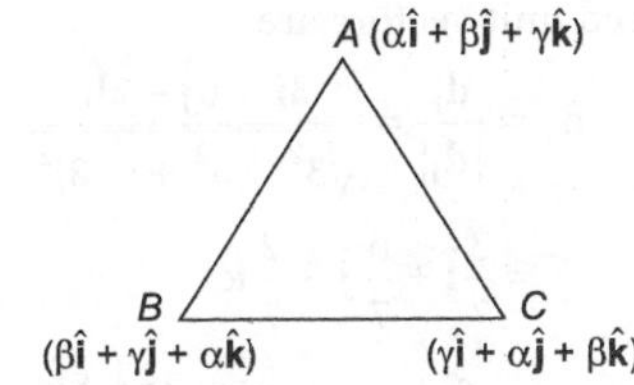

Now, $|\mathbf{AB}| = \sqrt{(\beta - \alpha)^2 + (\gamma - \beta)^2 + (\alpha - \gamma)^2}$

Similarly, $\mathbf{BC} = \mathbf{CA} = \sqrt{(\beta - \alpha)^2 + (\gamma - \beta)^2 + (\alpha - \gamma)^2}$

$\therefore \Delta ABC$ is an equilateral triangle.

Example 20. If the position vectors of the vertices of a triangle be $2\hat{\mathbf{i}} + 4\hat{\mathbf{j}} - \hat{\mathbf{k}}$, $4\hat{\mathbf{i}} + 5\hat{\mathbf{j}} + \hat{\mathbf{k}}$ and $3\hat{\mathbf{i}} + 6\hat{\mathbf{j}} - 3\hat{\mathbf{k}}$, then the triangle is

(a) right angled (b) isosceles

(c) equilateral (d) None of these

Sol. (a, b) Let A, B, C be the vertices of given triangle with position vectors, $2\hat{\mathbf{i}} + 4\hat{\mathbf{j}} - \hat{\mathbf{k}}$, $4\hat{\mathbf{i}} + 5\hat{\mathbf{j}} + \hat{\mathbf{k}}$ and $3\hat{\mathbf{i}} + 6\hat{\mathbf{j}} - 3\hat{\mathbf{k}}$ respectively.

Then, we have

$$\mathbf{OA} = 2\hat{\mathbf{i}} + 4\hat{\mathbf{j}} - \hat{\mathbf{k}}, \mathbf{OB} = 4\hat{\mathbf{i}} + 5\hat{\mathbf{j}} + \hat{\mathbf{k}}$$

and $\mathbf{OC} = 3\hat{\mathbf{i}} + 6\hat{\mathbf{j}} - 3\hat{\mathbf{k}}$

Clearly, $\mathbf{AB} = \mathbf{OB} - \mathbf{OA} = 2\hat{\mathbf{i}} + \hat{\mathbf{j}} + 2\hat{\mathbf{k}}$

$$\mathbf{BC} = -\hat{\mathbf{i}} + \hat{\mathbf{j}} - 4\hat{\mathbf{k}}$$

and $\mathbf{AC} = \hat{\mathbf{i}} + 2\hat{\mathbf{j}} - 2\hat{\mathbf{k}}$

Now, $\mathbf{AB} = |\mathbf{AB}| = \sqrt{2^2 + 1^2 + 2^2} = 3$

$$\mathbf{BC} = |\mathbf{BC}| = \sqrt{(-1)^2 + (1)^2 + (-4)^2} = 3\sqrt{2}$$

and $\mathbf{AC} = |\mathbf{AC}| = \sqrt{1^2 + 2^2 + (-2)^2} = 3$

$\because$ $AB = AC$ and $BC^2 = AB^2 + AC^2$

$\therefore$ The triangle is isosceles and right angled.

Example 21. The two adjacent sides of a parallelogram are $2\hat{\mathbf{i}} + 4\hat{\mathbf{j}} - 5\hat{\mathbf{k}}$ and $\hat{\mathbf{i}} + 2\hat{\mathbf{j}} + 3\hat{\mathbf{k}}$. Find the unit vectors along the diagonals of the parallelogram.

Sol. Let $OABC$ be the given parallelogram and let the adjacent sides OA and OB be represented by $a = 2\hat{\mathbf{i}} + 4\hat{\mathbf{j}} - 5\hat{\mathbf{k}}$ and $b = \hat{\mathbf{i}} + 2\hat{\mathbf{j}} + 3\hat{\mathbf{k}}$ respectively.

Now, the vectors along the two diagonals are

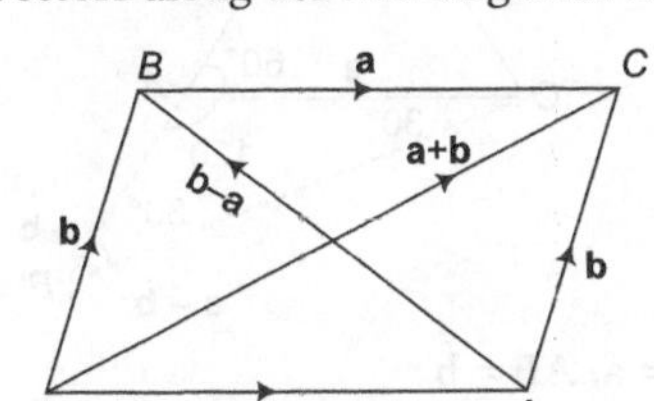

$$\mathbf{d}_1 = \mathbf{a} + \mathbf{b} = 3\hat{\mathbf{i}} + 6\hat{\mathbf{j}} - 2\hat{\mathbf{k}}$$

The required unit vectors are

$$\hat{\mathbf{n}}_1 = \frac{\mathbf{d}_1}{|\mathbf{d}_1|} = \frac{3\hat{\mathbf{i}} + 6\hat{\mathbf{j}} - 2\hat{\mathbf{k}}}{\sqrt{3^2 + 6^2 + (-2)^2}}$$

$$= \frac{3}{7}\hat{\mathbf{i}} + \frac{6}{7}\hat{\mathbf{j}} - \frac{2}{7}\hat{\mathbf{k}}$$

and $$\hat{\mathbf{n}}_2 = \frac{d_2}{|d_2|} = \frac{-\hat{\mathbf{i}} - 2\hat{\mathbf{j}} + 8\hat{\mathbf{k}}}{\sqrt{(-1)^2 + (-2)^2 + 8^2}}$$

$$= \frac{-1}{\sqrt{69}}\hat{\mathbf{i}} - \frac{2}{\sqrt{69}}\hat{\mathbf{j}} + \frac{8}{\sqrt{69}}\hat{\mathbf{k}}$$

Example 22. If **a** and **b** are any two vectors, then give the geometrical interpretation of the relation $|\mathbf{a} + \mathbf{b}| = |\mathbf{a} - \mathbf{b}|$.

Sol. Let **OA** = **a** and **AB** = **b**. Completing the parallelogram *OABC*.

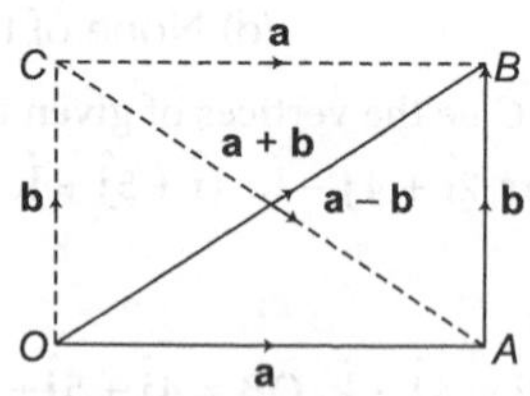

Then, **OC** = **b** and **CB** = **a**

From ΔOAB, we have

$$\mathbf{OA} + \mathbf{AB} = \mathbf{OB} \Rightarrow \mathbf{a} + \mathbf{b} = \mathbf{OB} \quad \text{...(i)}$$

From ΔOCA, we have

$$\mathbf{OC} + \mathbf{CA} = \mathbf{OA}$$

$$\Rightarrow \quad \mathbf{b} + \mathbf{CA} = \mathbf{a} \Rightarrow \mathbf{CA} = \mathbf{a} - \mathbf{b} \quad \text{...(ii)}$$

Clearly, $|\mathbf{a} + \mathbf{b}| = |\mathbf{a} - \mathbf{b}| \Rightarrow |\mathbf{OB}| = |\mathbf{CA}|$

Diagonals of parallelogram *OABC* are equal.

OABC is a rectangle.

$$\Rightarrow \quad \mathbf{OA} \perp \mathbf{OC} \Rightarrow \mathbf{a} \perp \mathbf{b}$$

Example 23. If the sum of two unit vectors is a unit vector, prove that the magnitude of their difference is $\sqrt{3}$.

Sol. Let $\hat{\mathbf{a}}$ and $\hat{\mathbf{b}}$ be two unit vectors represented by sides *OA* and *AB* of a ΔOAB.

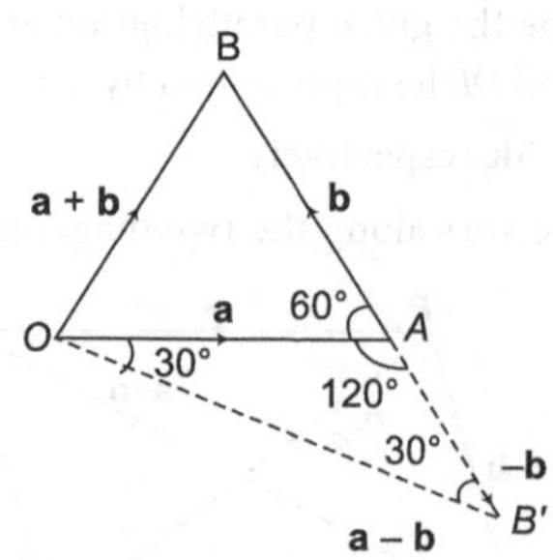

Then, $\mathbf{OA} = \hat{\mathbf{a}}$, $\mathbf{AB} = \hat{\mathbf{b}}$

$$\mathbf{OB} = \mathbf{OA} + \mathbf{AB} = \hat{\mathbf{a}} + \hat{\mathbf{b}}$$

(using triangle law of vector addition)

It is given that, $|\hat{\mathbf{a}}| = |\hat{\mathbf{b}}| = |\hat{\mathbf{a}} + \hat{\mathbf{b}}| = 1$

$$\Rightarrow \quad |\mathbf{OA}| + |\mathbf{AB}| = |\mathbf{OB}| = 1$$

ΔOAB is equilateral triangle.

Since, $|\mathbf{OA}| = |\hat{\mathbf{a}}| = 1 = |-\hat{\mathbf{b}}| = |\mathbf{AB}'|$

Therefore, $\Delta OAB'$ is an isosceles triangle.

$$\Rightarrow \quad \angle AB'O = \angle AOB' = 30^\circ$$

$$\Rightarrow \quad \angle BOB' = \angle BOA + \angle AOB' = 60^\circ + 30^\circ = 90^\circ$$

(since, $\Delta BOB'$ is right angled)

$\therefore$ In $\Delta BOB'$, we have

$$|BB'|^2 = |OB|^2 + |OB'|^2$$

$$= \left|\hat{\mathbf{a}} + \hat{\mathbf{b}}\right|^2 + \left|\hat{\mathbf{a}} - \hat{\mathbf{b}}\right|^2$$

$$2^2 = 1^2 + |\hat{\mathbf{a}} - \hat{\mathbf{b}}|^2$$

$$|\hat{\mathbf{a}} - \hat{\mathbf{b}}| = \sqrt{3}$$

Hence proved.

Multiplication of a Vector by a Scalar

If **a** is a vector and m is a scalar (i.e. a real number), then $m\,\mathbf{a}$ is a vector whose magnitude is m times that of **a** and whose direction is the same as that of **a**, if m is positive and opposite to that of **a**, if m is negative.

$\therefore$ Magnitude of $m\mathbf{a} = |m\mathbf{a}| \Rightarrow m$ (magnitude of **a**) $= m\,|\,\mathbf{a}\,|$

Again, if $\mathbf{a} = a_1\hat{\mathbf{i}} + a_2\hat{\mathbf{j}} + a_3\hat{\mathbf{k}}$,

then $m\,\mathbf{a} = (ma_1)\,\hat{\mathbf{i}} + (ma_2)\hat{\mathbf{j}} + (ma_3)\,\hat{\mathbf{k}}$

Properties of Multiplication of Vectors by a Scalar

The following are properties of multiplication of vectors by scalars, for vectors **a**, **b** and scalars m, n

(i) $m(-\mathbf{a}) = (-m)\,\mathbf{a} = -(m\mathbf{a})$

(ii) $(-m)(-\mathbf{a}) = m\,\mathbf{a}$

(iii) $m(n\mathbf{a}) = (mn)\,\mathbf{a} = n(m\mathbf{a})$

(iv) $(m + n)\,\mathbf{a} = m\mathbf{a} + n\mathbf{a}$

(v) $m(\mathbf{a} + \mathbf{b}) = m\,\mathbf{a} + m\,\mathbf{b}$

Example 24. If **a** is a non-zero vector of modulus a and m, is a non-zero scalar, then $m\mathbf{a}$ is a unit vector, if

(a) $m = \pm 1$ (b) $m = |\,\mathbf{a}\,|$

(c) $m = \dfrac{1}{|\,\mathbf{a}\,|}$ (d) $m = \pm 2$

Sol. (c) Since, $m\mathbf{a}$ is a unit vector, $|m\mathbf{a}| = 1$

$$\Rightarrow \quad |m|\,|\,\mathbf{a}\,| = 1$$

$$\Rightarrow \quad |m| = \frac{1}{|\mathbf{a}|} \Rightarrow m = \pm\frac{1}{|\,\mathbf{a}\,|}$$

Example 25. For a non-zero vector **a**, the set of real numbers, satisfying $|(5-x)\mathbf{a}|<|2\mathbf{a}|$ consists of all x such that

(a) $0 < x < 3$
(b) $3 < x < 7$
(c) $-7 < x < -3$
(d) $-7 < x < 3$

Sol. (b) We have, $|(5-x)\,\mathbf{a}|<|2\mathbf{a}|$

$$|5-x|\,|\mathbf{a}|<2|\,\mathbf{a}|$$
$$\Rightarrow |5-x|<2$$
$$\Rightarrow -2<5-x<2$$
$$\Rightarrow 3<x<7$$

Example 26. Find a vector of magnitude (5/2) units which is parallel to the vector $3\hat{\mathbf{i}}+4\hat{\mathbf{j}}$.

Sol. Here, $\mathbf{a}=3\hat{\mathbf{i}}+4\hat{\mathbf{j}}$

Then, $|\mathbf{a}|=\sqrt{3^2+4^2}=5$

∴ A unit vector parallel to

$$\mathbf{a}=\hat{\mathbf{a}}=\frac{a}{|\mathbf{a}|}\cdot=\frac{1}{5}(3\hat{\mathbf{i}}+4\hat{\mathbf{j}}) \quad \text{...(i)}$$

Hence, the required vector of magnitude (5/2) units and parallel to **a**

$$=\frac{5}{2}\cdot\hat{\mathbf{a}}=\frac{5}{2}\cdot\frac{1}{5}(3\hat{\mathbf{i}}+4\hat{\mathbf{j}})$$
$$=\frac{1}{2}(3\hat{\mathbf{i}}+4\hat{\mathbf{j}})$$

Section Formula

Let A and B be two points with position vectors **a** and **b** respectively. Let P be a point on AB dividing it is the ratio $m:n$.

Internal Division

If P divides AB internally in the ratio $m:n$. Then the position vector of P is given by

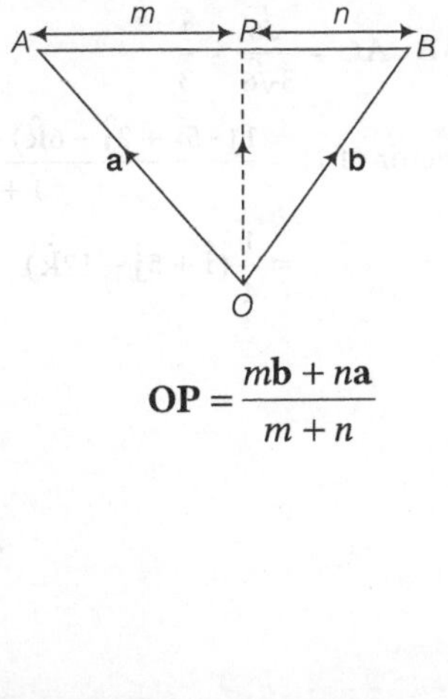

$$\mathbf{OP}=\frac{m\mathbf{b}+n\mathbf{a}}{m+n}$$

Proof

Let O be the origin. Then $\mathbf{OA}=\mathbf{a}$ and $\mathbf{OB}=\mathbf{b}$. Let r be the position vector of P which divides AB internally is the ratio $m:n$. Then

$$\frac{AP}{PB}=\frac{m}{n}$$

or $$n\mathbf{AP}=m\mathbf{PB}$$

or $$n(\text{PV of } P-\text{PV of } A)=m\,(\text{PV of } B-\text{PV of } P)$$

or $$n(\mathbf{r}-\mathbf{a})=m(\mathbf{b}-\mathbf{r})$$

or $$n\mathbf{r}-n\mathbf{a}=m\mathbf{b}-m\mathbf{r}$$

or $$\mathbf{r}(n+m)=m\mathbf{b}+n\mathbf{a}$$

or $$r=\frac{m\mathbf{b}+n\mathbf{a}}{m+n}$$

or $$\mathbf{OP}=\frac{m\mathbf{b}+n\mathbf{a}}{m+n}$$

External Division

If P divides AB externally in the ratio $m:n$. Then, the position vector of P is given by

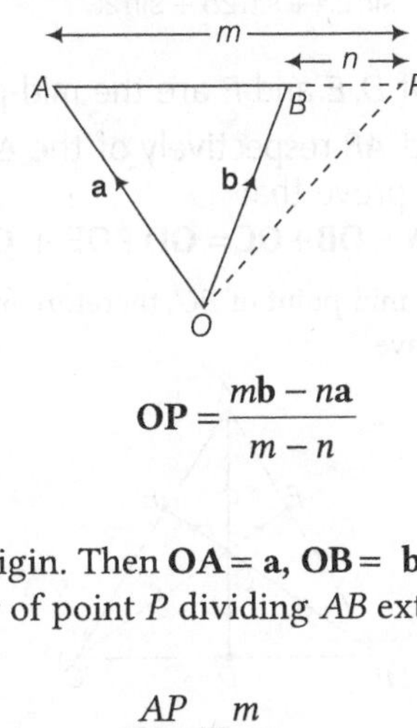

$$\mathbf{OP}=\frac{m\mathbf{b}-n\mathbf{a}}{m-n}$$

Proof

Let O be the origin. Then $\mathbf{OA}=\mathbf{a}$, $\mathbf{OB}=\mathbf{b}$. Let **r** be the position vector of point P dividing AB externally in the ratio $m:n$.

Then, $$\frac{AP}{BP}=\frac{m}{n}$$

or $$nAP=mBP$$

or $$n\mathbf{AP}=m\mathbf{BP}$$

or $$n\,(\text{PV of } P-\text{PV of } A)=m\,(\text{PV of } P-\text{PV of } B)$$

or $$n(\mathbf{r}-\mathbf{a})=m(\mathbf{r}-\mathbf{b})$$

or $$n\mathbf{r}-n\mathbf{a}=m\mathbf{r}-m\mathbf{b}$$

or $$r(m-n)=m\mathbf{b}-n\mathbf{a}$$

or $$\mathbf{r}=\frac{m\mathbf{b}-n\mathbf{a}}{m-n}$$

or $$\mathbf{OP}=\frac{m\mathbf{b}-n\mathbf{a}}{m-n}$$

Remarks

1. Position vector of mid-point of AB is $\frac{\mathbf{a}+\mathbf{b}}{2}$.
2. In ΔABC, having vertices $A(\mathbf{a})$, $B(\mathbf{b})$ and $C(\mathbf{c})$

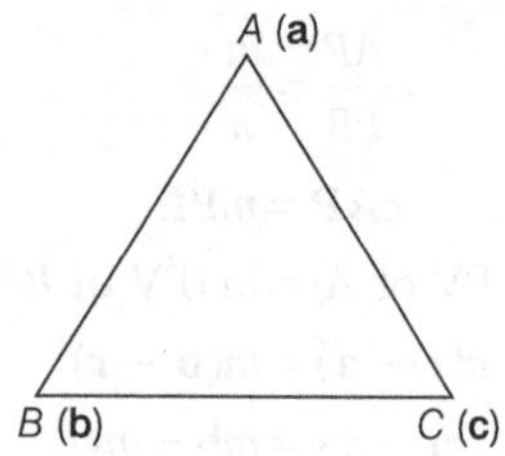

(i) Position vector of centroid is $\frac{\mathbf{a}+\mathbf{b}+\mathbf{c}}{3}$.

(ii) Position vector of incentre is $\frac{BC\mathbf{a} + AC\mathbf{b} + AB\mathbf{c}}{AB + BC + AC}$.

(iii) Position vector of orthocentre is $\frac{\tan A\,\mathbf{a} + \tan B\,\mathbf{b} + \tan C\,\mathbf{c}}{\tan A + \tan B + \tan C}$.

(iv) Position vector of circumcentre is $\frac{\sin 2A\,\mathbf{a} + \sin 2B\,\mathbf{b} + \sin 2C\,\mathbf{c}}{\sin 2A + \sin 2B + \sin 2C}$.

Example 27. If D, E and F are the mid-points of the sides BC, CA and AB respectively of the ΔABC and O be any point, then prove that

$$\mathbf{OA} + \mathbf{OB} + \mathbf{OC} = \mathbf{OD} + \mathbf{OE} + \mathbf{OF}$$

Sol. Since, D is the mid-point of BC, therefore by section formula, we have

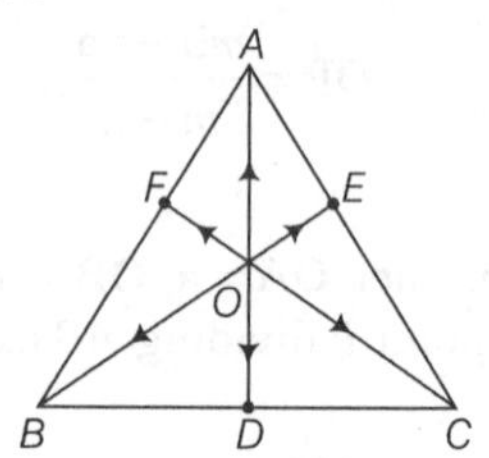

$$\mathbf{OD} = \frac{\mathbf{OB} + \mathbf{OC}}{2}$$

$\Rightarrow \quad \mathbf{OB} + \mathbf{OC} = 2\mathbf{OD}$ ...(i)

Similarly, $\quad \mathbf{OC} + \mathbf{OA} = 2\mathbf{OE}$ ...(ii)

and $\quad \mathbf{OB} + \mathbf{OA} = 2\mathbf{OF}$ ...(iii)

On adding Eqs. (i), (ii) and (iii), we get

$$2(\mathbf{OA} + \mathbf{OB} + \mathbf{OC}) = 2(\mathbf{OD} + \mathbf{OE} + \mathbf{OF})$$

$$\Rightarrow \quad \mathbf{OA} + \mathbf{OB} + \mathbf{OC} = \mathbf{OD} + \mathbf{OE} + \mathbf{OF}$$

Hence proved.

Example 28. Find the position vectors of the points which divide the join of points $A(2\mathbf{a} - 3\mathbf{b})$ and $B(3\mathbf{a} - 2\mathbf{b})$ internally and externally in the ratio 2 : 3.

Sol. Let P be a point which divide AB internally in the ratio 2 : 3. Then, by section formula, position vector of P is given by

$$\mathbf{OP} = \frac{2(3\mathbf{a} - 2\mathbf{b}) + 3(2\mathbf{a} - 3\mathbf{b})}{2+3}$$

$$= \frac{6\mathbf{a} - 4\mathbf{b} + 6\mathbf{a} - 9\mathbf{b}}{5} = \frac{12}{5}\mathbf{a} - \frac{13}{5}\mathbf{b}$$

Similarly, the position vector of the point (P') which divides AB externally in the ratio 2 : 3 is given by

$$\mathbf{OP}' = \frac{2(3\mathbf{a} - 2\mathbf{b}) - 3(2\mathbf{a} - 3\mathbf{b})}{2-3}$$

$$= \frac{6\mathbf{a} - 4\mathbf{b} - 6\mathbf{a} + 9\mathbf{b}}{-1} = \frac{5\mathbf{b}}{-1} = -5\mathbf{b}$$

Example 29. The position vectors of the vertices A, B and C of a triangle are $\hat{\mathbf{i}} - \hat{\mathbf{j}} - 3\hat{\mathbf{k}}$, $2\hat{\mathbf{i}} + \hat{\mathbf{j}} - 2\hat{\mathbf{k}}$ and $-5\hat{\mathbf{i}} + 2\hat{\mathbf{j}} - 6\hat{\mathbf{k}}$, respectively. The length of the bisector AD of the $\angle BAC$, where D is on the segment BC, is

(a) $\frac{3}{4}\sqrt{10}$ (b) $\frac{1}{4}$

(c) $\frac{11}{2}$ (d) None of these

Sol. (b)

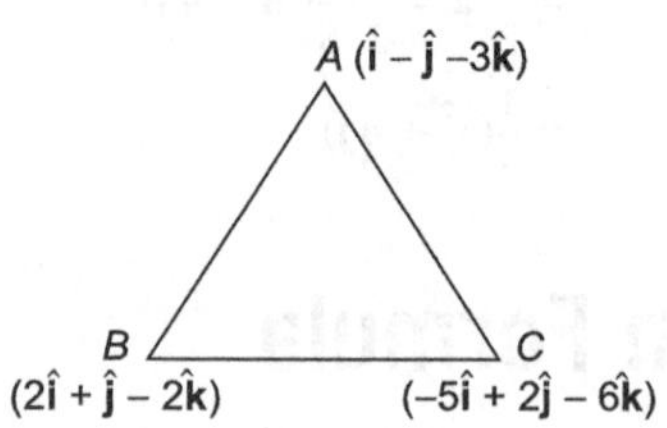

(a) $|\mathbf{AB}| = |(2\hat{\mathbf{i}} + \hat{\mathbf{j}} - 2\hat{\mathbf{k}}) - (\hat{\mathbf{i}} - \hat{\mathbf{j}} - 3\hat{\mathbf{k}})|$

$$= |\hat{\mathbf{i}} + 2\hat{\mathbf{j}} + \hat{\mathbf{k}}|$$

$$= \sqrt{1^2 + 2^2 + 1^2} = \sqrt{6}$$

$$|\mathbf{AC}| = |(-5\hat{\mathbf{i}} + 2\hat{\mathbf{j}} - 6\hat{\mathbf{k}}) - (\hat{\mathbf{i}} - \hat{\mathbf{j}} - 3\hat{\mathbf{k}})|$$

$$= |-6\hat{\mathbf{i}} + 3\hat{\mathbf{j}} - 3\hat{\mathbf{k}}|$$

$$= \sqrt{(-6)^2 + 3^2 + (-3)^2} = \sqrt{54} = 3\sqrt{6}$$

$$\mathbf{BD} : \mathbf{DC} = \mathbf{AB} : \mathbf{AC} = \frac{\sqrt{6}}{3\sqrt{6}} = \frac{1}{3}$$

$$\therefore \text{Position vector of } D = \frac{1(-5\hat{\mathbf{i}} + 2\hat{\mathbf{j}} - 6\hat{\mathbf{k}}) + 3(2\hat{\mathbf{i}} + \hat{\mathbf{j}} - 2\hat{\mathbf{k}})}{1+3}$$

$$= \frac{1}{4}(\hat{\mathbf{i}} + 5\hat{\mathbf{j}} - 12\hat{\mathbf{k}})$$

$\therefore$ $\quad \mathbf{AD}$ = Position vector of D –Position vector of A

$$\mathbf{AD} = \frac{1}{4}(\hat{\mathbf{i}} + 5\hat{\mathbf{j}} - 12\hat{\mathbf{k}}) - (\hat{\mathbf{i}} - \hat{\mathbf{j}} - 3\hat{\mathbf{k}}) = \frac{1}{4}(-3\hat{\mathbf{i}} + 9\hat{\mathbf{j}})$$

$$= \frac{3}{4}(-\hat{\mathbf{i}} + 3\hat{\mathbf{j}})$$

$$|\mathbf{AD}| = \frac{3}{4}\sqrt{(-1)^2 + 3^2} = \frac{3}{4}\sqrt{10}$$

Example 30. The median AD of the ΔABC is bisected at $E.BE$ meets AC in F. Then, $AF : AC$ is equal to

(a) 3/4 (b) 1/3

(c) 1/2 (d) 1/4

Sol. (b) Let position vector of A w.r.t. B is $\mathbf{a}$ and that of C w.r.t. B is $\mathbf{c}$.

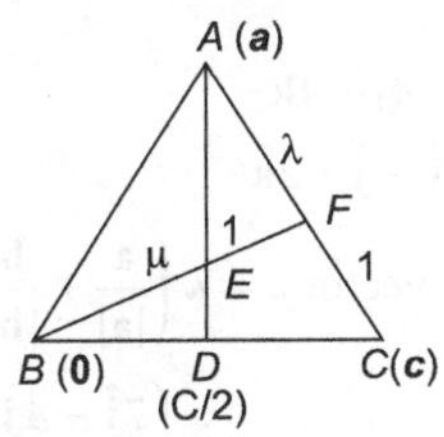

Position vector of D w.r.t.

$$B = \frac{0 + \mathbf{c}}{2} = \frac{\mathbf{c}}{2}$$

Position vector of

$$E = \frac{\mathbf{a} + \frac{\mathbf{c}}{2}}{2} = \frac{\mathbf{a}}{2} + \frac{\mathbf{c}}{4} \quad \text{...(i)}$$

Let $AF : FC = \lambda : 1$ and $BE : EF = \mu : 1$

Position vector of $F = \dfrac{\lambda\mathbf{c} + \mathbf{a}}{1 + \lambda}$

Now, position vector of

$$E = \frac{\mu\left(\frac{\lambda\mathbf{c} + \mathbf{a}}{1 + \lambda}\right) + 1 \cdot 0}{\mu + 1} \quad \text{...(ii)}$$

From Eqs. (i) and (ii), we get

$$\frac{\mathbf{a}}{2} + \frac{\mathbf{c}}{4} = \frac{\mu}{(1 + \lambda)(1 + \mu)}\mathbf{a} + \frac{\lambda\mu}{(1 + \lambda)(1 + \mu)}\mathbf{c}$$

$$\Rightarrow \quad \frac{1}{2} = \frac{\mu}{(1 + \lambda)(1 + \mu)}$$

and $$\frac{1}{4} = \frac{\lambda\mu}{(1 + \lambda)(1 + \mu)}$$

$$\Rightarrow \quad \lambda = \frac{1}{2}$$

$$\therefore \quad \frac{AF}{AC} = \frac{AF}{AF + FC} = \frac{\lambda}{1 + \lambda} = \frac{\frac{1}{2}}{\frac{3}{2}} = \frac{1}{3}$$

Magnitude of Resultant of Two Vectors

Let $\mathbf{R}$ be the resultant of two vectors $\mathbf{P}$ and $\mathbf{Q}$. Then, $\mathbf{R} = \mathbf{P} + \mathbf{Q}$

$$|\mathbf{R}| = R = \sqrt{P^2 + Q^2 + 2PQ\cos\theta}$$

where, $|\mathbf{P}| = P, |\mathbf{Q}| = Q$, $\tan\alpha = \dfrac{Q\sin\theta}{P + Q\cos\theta}$

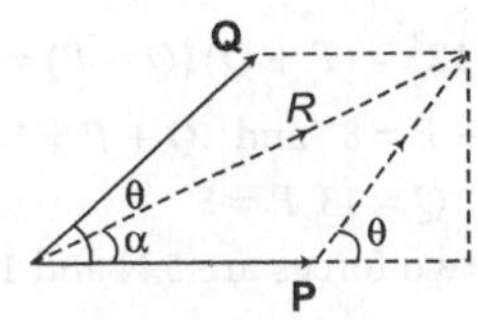

Deduction When $|\mathbf{P}| = |\mathbf{Q}|$, i.e. $P = Q$

$$\tan\alpha = \frac{P\sin\theta}{P + P\cos\theta}$$

$$= \frac{\sin\theta}{1 + \cos\theta} = \tan\frac{\theta}{2}$$

$$\therefore \quad \alpha = \frac{\theta}{2}$$

Hence, the angular bisector of two unit vectors $\mathbf{a}$ and $\mathbf{b}$ is along the vector sum $\mathbf{a} + \mathbf{b}$.

Remarks

1. The internal bisector of the angle between any two vectors is along the vector sum of the corresponding unit vectors.
2. The external bisector of the angle between two vectors is along the vector difference of the corresponding unit vectors.

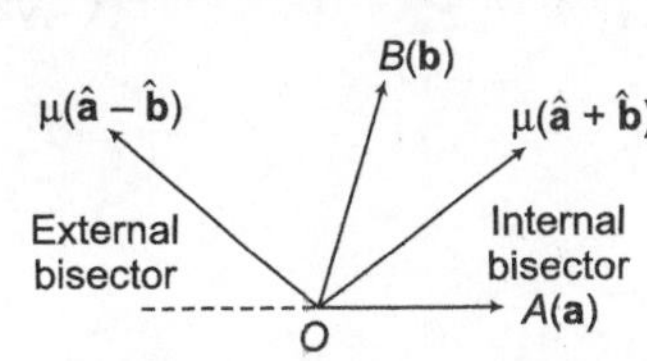

Example 31. The sum of two forces is 18 N and resultant whose direction is at right angles to the smaller force is 12 N. The magnitude of the two forces are

(a) 13, 5 (b) 12, 6

(c) 14, 4 (d) 11, 7

Sol. (a) We have, $|\mathbf{P}| + |\mathbf{Q}| = 18\text{N}; |\mathbf{R}| = |\mathbf{P} + \mathbf{Q}| + 12\,\text{N}$

$$\alpha = 90°$$

$$P + Q\cos\theta = 0$$

$$\Rightarrow \quad Q\cos\theta = -P$$

Now. $\quad R^2 = P^2 + Q^2 + 2PQ\cos\theta$

$$\Rightarrow \quad R^2 = P^2 + Q^2 + 2P(-P) = Q^2 - P^2$$

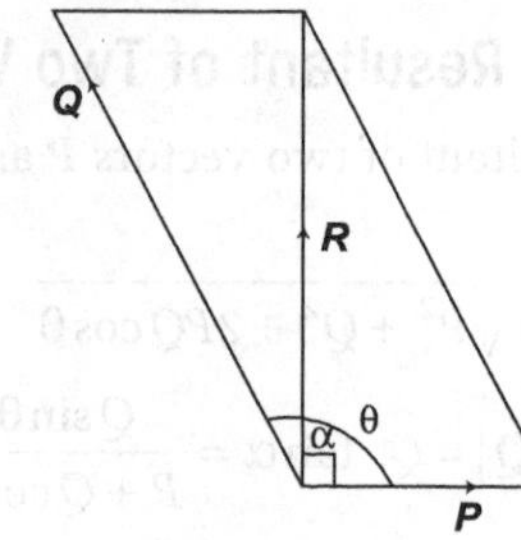

$$\Rightarrow \quad 12^2 = (P+Q)(Q-P) = 18(Q-P)$$
$$\Rightarrow \quad Q-P = 8 \text{ and } Q+P = 18$$
$$\Rightarrow \quad Q = 13, P = 5$$

∴ Magnitude of two forces are 5 N and 13 N.

Example 32. The length of longer diagonal of the parallelogram constructed on $5\mathbf{a}+2\mathbf{b}$ and $\mathbf{a}-3\mathbf{b}$, when it is given that $|\mathbf{a}| = 2\sqrt{2}, |\mathbf{b}| = 3$ and angle between $\mathbf{a}$ and $\mathbf{b}$ is $\frac{\pi}{4}$, is

(a) 15 (b) $\sqrt{113}$
(c) $\sqrt{593}$ (d) $\sqrt{369}$

Sol. (c) Length of the two diagonals will be

$$d_1 = |(5\mathbf{a}+2\mathbf{b}) + (\mathbf{a}-3\mathbf{b})|$$
and $$d_2 = |(5\mathbf{a}+2\mathbf{b}) - (\mathbf{a}-3\mathbf{b})|$$
$$\Rightarrow \quad d_1 = |6\mathbf{a} - \mathbf{b}|, d_2 = |4\mathbf{a}+5\mathbf{b}|$$

Thus,

$$d_1 = \sqrt{|6\mathbf{a}|^2 + |-\mathbf{b}|^2 + 2|6\mathbf{a}||-\mathbf{b}|\cos(\pi - \pi/4)}$$
$$= \sqrt{36(2\sqrt{2})^2 + 9 + 12\cdot 2\sqrt{2}\cdot 3\cdot\left(-\frac{1}{\sqrt{2}}\right)} = 15$$

$$d_2 = \sqrt{|4\mathbf{a}|^2 + |5\mathbf{b}|^2 + 2|4\mathbf{a}||5\mathbf{b}|\cos\frac{\pi}{4}}$$
$$= \sqrt{16\times 8 + 25\times 9 + 40\times 2\sqrt{2}\times 3\times\frac{1}{\sqrt{2}}}$$
$$= \sqrt{593}$$

∴ Length of the longer diagonal $= \sqrt{593}$

Example 33. The vector $\mathbf{c}$, directed along the internal bisector of the angle between the vectors $\mathbf{a} = 7\hat{\mathbf{i}} - 4\hat{\mathbf{j}} - 4\hat{\mathbf{k}}$ and $\mathbf{b} = -2\hat{\mathbf{i}} - \hat{\mathbf{j}} + 2\hat{\mathbf{k}}$ with $|\mathbf{c}| = 5\sqrt{6}$, is

(a) $\frac{5}{3}(\hat{\mathbf{i}} - 7\hat{\mathbf{j}} + 2\hat{\mathbf{k}})$ (b) $\frac{5}{3}(5\hat{\mathbf{i}} + 5\hat{\mathbf{j}} + 2\hat{\mathbf{k}})$
(c) $\frac{5}{3}(\hat{\mathbf{i}} + 7\hat{\mathbf{j}} + 2\hat{\mathbf{k}})$ (d) $\frac{5}{3}(-5\hat{\mathbf{i}} + 5\hat{\mathbf{j}} + 2\hat{\mathbf{k}})$

Sol. (a) Let $\mathbf{a} = 7\hat{\mathbf{i}} - 4\hat{\mathbf{j}} - 4\hat{\mathbf{k}}$

and $\mathbf{b} = -2\hat{\mathbf{i}} - \hat{\mathbf{j}} + 2\hat{\mathbf{k}}$

Now, required vector $\mathbf{c} = \lambda\left(\frac{\mathbf{a}}{|\mathbf{a}|} + \frac{\mathbf{b}}{|\mathbf{b}|}\right)$

$$= \lambda\left(\frac{7\hat{\mathbf{i}} - 4\hat{\mathbf{j}} - 4\hat{\mathbf{k}}}{9} + \frac{-2\hat{\mathbf{i}} - \hat{\mathbf{j}} + 2\hat{\mathbf{k}}}{3}\right)$$
$$= \frac{\lambda}{9}(\hat{\mathbf{i}} - 7\hat{\mathbf{j}} + 2\hat{\mathbf{k}})$$
$$|\mathbf{c}|^2 = \frac{\lambda^2}{81}\times 54 = 150$$
$$\Rightarrow \quad \lambda = \pm 15$$
$$\Rightarrow \quad \mathbf{c} = \pm\frac{5}{3}(\hat{\mathbf{i}} - 7\hat{\mathbf{j}} + 2\hat{\mathbf{k}})$$

Exercise for Session 2

1. If $\mathbf{a} = 2\hat{\mathbf{i}} - \hat{\mathbf{j}} + 2\hat{\mathbf{k}}$ and $\mathbf{b} = -\hat{\mathbf{i}} + \hat{\mathbf{j}} - \hat{\mathbf{k}}$, then find $\mathbf{a} + \mathbf{b}$. Also, find a unit vector along $\mathbf{a} + \mathbf{b}$.
2. Find a unit vector in the direction of the resultant of the vectors $\hat{\mathbf{i}} + 2\hat{\mathbf{j}} + 3\hat{\mathbf{k}}$, $-\hat{\mathbf{i}} + 2\hat{\mathbf{j}} + \hat{\mathbf{k}}$ and $3\hat{\mathbf{i}} + \hat{\mathbf{j}}$.
3. Find the direction cosines of the resultant of the vectors $(\hat{\mathbf{i}} + \hat{\mathbf{j}} + \hat{\mathbf{k}}), (-\hat{\mathbf{i}} + \hat{\mathbf{j}} + \hat{\mathbf{k}}), (\hat{\mathbf{i}} - \hat{\mathbf{j}} + \hat{\mathbf{k}})$ and $(\hat{\mathbf{i}} + \hat{\mathbf{j}} - \hat{\mathbf{k}})$.
4. In a regular hexagon *ABCDEF*, show that **AE** is equal to **AC** + **AF** − **AB**
5. Prove that 3**OD** + **DA** + **DB** + **DC** is equal to **OA** + **OB** − **OC**.
6. In a regular hexagon *ABCDEF*, prove that **AB** + **AC** + **AD** + **AE** + **AF** = 3**AD**.
7. *ABCDE* is a pentagon, prove that **AB** + **BC** + **CD** + **DE** + **EA** = 0.
8. The position vectors of *A*, *B*, *C*, *D* are **a**, **b**, 2**a** + 3**b** and **a** − 2**b**, respectively. Show that **DB** = 3**b** − **a** and **AC** = **a** + 3**b**.
9. If *P*(−1, 2) and *Q*(3, −7) are two points, express the vector **PQ** in terms of unit vectors $\hat{\mathbf{i}}$ and $\hat{\mathbf{j}}$. Also, find distance between point *P* and *Q*. What is the unit vector in the direction of **PQ**?
10. If $\mathbf{OP} = 2\hat{\mathbf{i}} + 3\hat{\mathbf{j}} - \hat{\mathbf{k}}$ and $\mathbf{OQ} = 3\hat{\mathbf{i}} - 4\hat{\mathbf{j}} + 2\hat{\mathbf{k}}$, find the modulus and direction cosines of **PQ**.
11. Show that the points *A*, *B* and *C* with position vectors $\mathbf{a} = 3\hat{\mathbf{j}} - 4\hat{\mathbf{j}} - 4\hat{\mathbf{k}}$, $\mathbf{b} = 2\hat{\mathbf{i}} - \hat{\mathbf{j}} + \hat{\mathbf{k}}$ and $\mathbf{c} = \hat{\mathbf{i}} - 3\hat{\mathbf{j}} - 5\hat{\mathbf{k}}$ respectively, form the vertices of a right angled triangle.
12. If $\mathbf{a} = 2\hat{\mathbf{i}} + 2\hat{\mathbf{j}} - \hat{\mathbf{k}}$ and $|x\mathbf{a}| = 1$, then find *x*.
13. If $\mathbf{p} = 7\hat{\mathbf{i}} - 2\hat{\mathbf{j}} + 3\hat{\mathbf{k}}$ and $\mathbf{q} = 3\hat{\mathbf{i}} + \hat{\mathbf{j}} + 5\hat{\mathbf{k}}$, then find the magnitude of $\mathbf{p} - 2\mathbf{q}$.
14. Find a vector in the direction of $5\hat{\mathbf{i}} - \hat{\mathbf{j}} + 2\hat{\mathbf{k}}$, which has magnitude 8 units.
15. If $\mathbf{a} = \hat{\mathbf{i}} + 2\hat{\mathbf{j}} + 2\hat{\mathbf{k}}$ and $\mathbf{b} = 3\hat{\mathbf{i}} + 6\hat{\mathbf{j}} + 2\hat{\mathbf{k}}$, then find a vector in the direction of **a** and having magnitude as $|\mathbf{b}|$.
16. Find the position vector of a point *P* which divides the line joining two points *A* and *B* whose position vectors are $\hat{\mathbf{i}} + 2\hat{\mathbf{j}} - \hat{\mathbf{k}}$ and $-\hat{\mathbf{i}} + \hat{\mathbf{j}} + \hat{\mathbf{k}}$ respectively, in the ratio 2 : 1.
    (i) internally (ii) externally
17. If the position vector of one end of the line segment *AB* be $2\hat{\mathbf{i}} + 3\hat{\mathbf{j}} - \hat{\mathbf{k}}$ and the position vector of its middle point be $3(\hat{\mathbf{i}} + \hat{\mathbf{j}} + \hat{\mathbf{k}})$, then find the position vector of the other end

Session 3

Linear Combination of Vectors, Theorem on Coplanar & Non-coplanar Vectors, Linear Independence and Dependence of Vectors

Linear Combination of Vectors

A vector **r** is said to be a linear combination of vectors **a**, **b** and **c**... etc., if there exist scalars x, y and z etc., such that $\mathbf{r} = x\mathbf{a} + y\mathbf{b} + z\mathbf{c} + \ldots$

For examples Vectors $\mathbf{r}_1 = 2\mathbf{a} + \mathbf{b} + 3\mathbf{c}$ and $\mathbf{r}_2 = \mathbf{a} + 3\mathbf{b} + \sqrt{2}\mathbf{c}$ are linear combinations of the vectors **a**, **b** and **c**.

Collinearity and Coplanarity of Vectors

Relation between Two Collinear Vectors (or Parallel Vectors)

Let **a** and **b** be two collinear vectors and let $\hat{\mathbf{x}}$ be the unit vector in the direction of **a**. Then, the unit vector in the direction of **b** is $\hat{\mathbf{x}}$ or $-\hat{\mathbf{x}}$ according as **a** and **b** are like or unlike parallel vectors. Now, $\mathbf{a} = |\mathbf{a}|\,\hat{\mathbf{x}}$ and $\mathbf{b} = \pm|\mathbf{b}|\,\hat{\mathbf{x}}$.

$$\therefore \quad \mathbf{a} = \left(\frac{|\mathbf{a}|}{|\mathbf{b}|}\right)|\mathbf{b}|\,\hat{\mathbf{x}} \Rightarrow \mathbf{a} = \pm\left(\frac{|\mathbf{a}|}{|\mathbf{b}|}\right)\mathbf{b}$$

$$\Rightarrow \quad \mathbf{a} = \lambda\mathbf{b}, \text{ where } \lambda = \pm\frac{|\mathbf{a}|}{|\mathbf{b}|}$$

Thus, if **a** and **b** are collinear vectors, then $\mathbf{a} = \lambda\mathbf{b}$ or $\mathbf{b} = \lambda\mathbf{a}$ for some scalar λ i.e, there exist two non-zero scalar quantities x and y so that $x\mathbf{a} + y\mathbf{b} = \mathbf{O}$

An Important Theorem

Theorem : Vectors **a** and **b** are two non-zero, non-collinear vectors and x, y are two scalars such that

$$x\mathbf{a} + y\mathbf{b} = 0$$

Then, $\quad x = 0, y = 0$

Proof It is given that $\quad x\mathbf{a} + y\mathbf{b} = 0 \quad$ ...(i)

Suppose that $x \neq 0$, then dividing both sides of (i) by the scalar x, we get

$$\mathbf{a} = -\frac{y}{x}\mathbf{b} \quad \text{...(ii)}$$

Now, $\frac{y}{x}$ is a scalar, because x and y are scalars.

Hence, Eq. (ii) expresses **a** as product of **b** by a scalar, so that **a** and **b** are collinear. Thus, we arrive at a contradiction because **a** and **b** are given to be non-collinear.

Thus our supposition that $x \neq 0$, is wrong.

Hence, $x = 0$. Similarly, $y = 0$

Remarks

1. $x\mathbf{a} + y\mathbf{b} = 0 \Rightarrow \begin{cases} \mathbf{a} = 0, \mathbf{b} = 0 \\ \text{or} \\ x = 0, y = 0 \\ \text{or} \\ \mathbf{a} \,||\, \mathbf{b} \end{cases}$

2. If **a** and **b** are two non-collinear (or non-parallel) vectors, then $x_1\mathbf{a} + y_1\mathbf{b} = x_2\mathbf{a} + y_2\mathbf{b}$

$\Rightarrow \quad x_1 = x_2$ and $y_1 = y_2$

Proof $x_1\mathbf{a} + y_1\mathbf{b} = x_2\mathbf{b} + y_2\mathbf{b}$

$\Rightarrow \quad (x_1 - x_2)\mathbf{a} + (y_1 - y_2)\mathbf{b} = 0$

$\Rightarrow \quad x_1 - x_2 = 0$ and $y_1 - y_2 = 0$

[$\because$ **a** and **b** are non-collinear]

$\Rightarrow \quad x_1 = x_2$ and $y_1 = y_2$

If $\mathbf{a} = a_1\hat{\mathbf{i}} + a_2\hat{\mathbf{j}} + a_3\hat{\mathbf{k}}$ and $b = b_1\hat{\mathbf{i}} + b_2\hat{\mathbf{j}} + b_3\hat{\mathbf{k}}$, then $\mathbf{a} \,||\, \mathbf{b}$

$\Rightarrow \quad \frac{a_1}{b_1} = \frac{a_2}{b_2} = \frac{a_3}{b_3}$

Test of Collinearity of Three Points

(i) Three points A, B and C are collinear, if $\mathbf{AB} = \lambda\mathbf{BC}$

(ii) Three points with position vectors **a**, **b** and **c** are collinear iff there exist scalars x, y and z not all zero such that $x\mathbf{a} + y\mathbf{b} + z\mathbf{c} = 0$, where $x + y + z = 0$

Proof Let us suppose that points A, B and C are collinear and their position vectors are **a**, **b** and **c** respectively. Let C divide the join of **a** and **b** in the ratio $y : x$. Then,

$$\mathbf{c} = \frac{x\mathbf{a} + y\mathbf{b}}{x + y}$$

or $\quad x\mathbf{a} + y\mathbf{b} - (x + y)\mathbf{c} = 0$

or $\quad x\mathbf{a} + y\mathbf{b} + z\mathbf{c} = 0$, where $z = -(x + y)$

Also, $x+y+z=x+y-(x+y)=0$

Conversely, let $x\mathbf{a}+y\mathbf{b}+z\mathbf{c}=0$, where $x+y+z=0$.
Therefore,

$$x\mathbf{a}+y\mathbf{b}=-z\mathbf{c}=(x+y)\mathbf{c} \quad (\because x+y=-z)$$

or $$\mathbf{c}=\frac{x\mathbf{a}+y\mathbf{b}}{x+y}$$

This relation shows that **c** divides the join of **a** and **b** in the ratio $y:x$. Hence, the three points A, B and C are collinear.

(iii) If $\mathbf{a}=a_1\hat{\mathbf{i}}+a_2\hat{\mathbf{j}}$, $\mathbf{b}=b_1\hat{\mathbf{i}}+b_2\hat{\mathbf{j}}$ and $\mathbf{c}=c_1\hat{\mathbf{i}}+c_2\hat{\mathbf{j}}$, then the points with position vector **a**, **b** and **c** will be collinear iff $\begin{vmatrix} a_1 & a_2 & 1 \\ b_1 & b_2 & 1 \\ c_1 & c_2 & 1 \end{vmatrix}=0$.

Proof The points with position vector **a**, **b** and **c** will be collinear iff there exist scalars x, y and z not all zero such that,

$x(a_1\hat{\mathbf{i}}+a_2\hat{\mathbf{j}})+y(b_1\hat{\mathbf{i}}+b_2\hat{\mathbf{j}})+z(c_1\hat{\mathbf{i}}+c_2\hat{\mathbf{j}})=0$ and $x+y+z=0$

$$\Rightarrow \quad xa_1+yb_1+zc_1=0$$
$$xa_2+yb_2+zc_2=0$$
$$x+y+z=0$$

Thus, the points will be collinear iff the above system of equation's have non-trivial solution

Hence, the points will be collinear

iff $\begin{vmatrix} a_1 & b_1 & c_1 \\ a_2 & b_2 & c_2 \\ 1 & 1 & 1 \end{vmatrix}=0$ or $\begin{vmatrix} a_1 & a_2 & 1 \\ b_1 & b_2 & 1 \\ c_1 & c_2 & 1 \end{vmatrix}=0$.

Example 34. Show that the vectors $2\hat{\mathbf{i}}-3\hat{\mathbf{j}}+4\hat{\mathbf{k}}$ and $-4\hat{\mathbf{i}}+6\hat{\mathbf{j}}-8\hat{\mathbf{k}}$ are collinear.

Sol. Let $\mathbf{a}=2\hat{\mathbf{i}}-3\hat{\mathbf{j}}+4\hat{\mathbf{k}}$ and $\mathbf{b}=-4\hat{\mathbf{i}}+6\hat{\mathbf{j}}-8\hat{\mathbf{k}}$

Consider, $\mathbf{b}=-4\hat{\mathbf{i}}+6\hat{\mathbf{j}}-8\hat{\mathbf{k}}=-2(2\hat{\mathbf{i}}-3\hat{\mathbf{j}}+4\hat{\mathbf{k}})=-2\mathbf{a}$

$\therefore$ The vectors **a** and **b** are collinear.

Example 35. Show that the points $A(1, 2, 3)$, $B(3, 4, 7)$ and $C(-3, -2, -5)$ are collinear. Find the ratio in which point C divides AB.

Sol. Clearly, $\mathbf{AB}=(3-1)\hat{\mathbf{i}}+(4-2)\hat{\mathbf{j}}+(7-3)\hat{\mathbf{k}}$

$$=2\hat{\mathbf{i}}+2\hat{\mathbf{j}}+4\hat{\mathbf{k}}$$

and $\mathbf{BC}=(-3-3)\hat{\mathbf{i}}+(-2-4)\hat{\mathbf{j}}+(-5-7)\hat{\mathbf{k}}$

$$=6\hat{\mathbf{i}}-6\hat{\mathbf{j}}-12\hat{\mathbf{k}}$$
$$=-3(2\hat{\mathbf{i}}+2\hat{\mathbf{j}}+4\hat{\mathbf{k}})=-3\mathbf{AB}$$

$\because \quad \mathbf{BC}=-3\mathbf{AB}$

$\therefore A$, B and C are collinear.

Now, let C divide AB in the ratio $k:1$, then

$$\mathbf{OC}=\frac{k\mathbf{OB}+1\cdot\mathbf{OA}}{k+1}$$

$$\Rightarrow \quad -3\hat{\mathbf{i}}-2\hat{\mathbf{j}}-5\hat{\mathbf{k}}=\frac{k(3\hat{\mathbf{i}}+4\hat{\mathbf{j}}+7\hat{\mathbf{k}})+(\hat{\mathbf{i}}+2\hat{\mathbf{j}}+3\hat{\mathbf{k}})}{k+1}$$

$$\Rightarrow \quad -3\hat{\mathbf{i}}-2\hat{\mathbf{j}}-5\hat{\mathbf{k}}=\left(\frac{3k+1}{k+1}\right)\hat{\mathbf{i}}+\left(\frac{4k+2}{k+1}\right)\hat{\mathbf{j}}+\left(\frac{7k+3}{k+1}\right)\hat{\mathbf{k}}$$

$$\Rightarrow \quad \frac{3k+1}{k+1}=-3; \frac{4k+2}{k+1}=-2 \text{ and } \frac{7k+3}{k+1}=-5$$

From, all relations, we get $k=\frac{-2}{3}$

Hence, C divides AB externally in the ratio $2:3$.

Example 36. If the position vectors of A, B, C and D are $2\hat{\mathbf{i}}+\hat{\mathbf{j}}$, $\hat{\mathbf{i}}-3\hat{\mathbf{j}}$, $3\hat{\mathbf{i}}+2\hat{\mathbf{j}}$ and $\hat{\mathbf{i}}+\lambda\hat{\mathbf{j}}$, respectively and $\mathbf{AB} \| \mathbf{CD}$, then λ will be

(a) –8 (b) –6

(c) 8 (d) 6

Sol. (b) $\mathbf{AB}=(\hat{\mathbf{i}}-3\hat{\mathbf{j}})-(2\hat{\mathbf{i}}+\hat{\mathbf{j}})=-\hat{\mathbf{i}}-4\hat{\mathbf{j}}$;

$$\mathbf{CD}=(\hat{\mathbf{i}}+\lambda\hat{\mathbf{j}})-(3\hat{\mathbf{i}}+2\hat{\mathbf{j}})=-2\hat{\mathbf{i}}+(\lambda-2)\hat{\mathbf{j}};$$

$\mathbf{AB} \| \mathbf{CD} \Rightarrow \mathbf{AB}=x\mathbf{CD}$

$$-\hat{\mathbf{i}}-4\hat{\mathbf{j}}=x\{-2\hat{\mathbf{i}}+(\lambda-2)\hat{\mathbf{j}}\}$$

$$\Rightarrow \quad -1=-2x, -4=(\lambda-2)x$$

$$\Rightarrow \quad x=\frac{1}{2} \text{ and } \lambda=-6$$

Example 37. The points with position vectors $60\hat{\mathbf{i}}+3\hat{\mathbf{j}}$, $40\hat{\mathbf{i}}-8\hat{\mathbf{j}}$ and $a\hat{\mathbf{i}}-52\hat{\mathbf{j}}$ are collinear, if a is equal to

(a) –40 (b) 40

(c) 20 (d) None of these

Sol (a) The three points are collinear if

$$\begin{vmatrix} 60 & 3 & 1 \\ 40 & -8 & 1 \\ a & -52 & 1 \end{vmatrix}=0$$

$$\Rightarrow \quad 60(-8+52)-3(40-a)+(-2080+8a)=0$$

$$\Rightarrow \quad 2640-120+3a-2080+8a=0$$

$$11a=-440$$

$$\Rightarrow \quad a=-40$$

Example 38. Let **a**, **b** and **c** be three non-zero vectors such that no two of these are collinear. If the vector $\mathbf{a}+2\mathbf{b}$ is collinear with **c** and $\mathbf{b}+3\mathbf{c}$ is collinear with **a** (λ being some non-zero scalar), then $\mathbf{a}+2\mathbf{b}+6\mathbf{c}$ is equal to

(a) **0** (b) $\lambda\mathbf{b}$

(c) $\lambda\mathbf{c}$ (d) $\lambda\mathbf{a}$

Sol. (a) As $\mathbf{a}+2\mathbf{b}$ and $\mathbf{c}$ are collinear $\mathbf{a}+2\mathbf{b}=\lambda\mathbf{c}$ ...(i)

Again, $\mathbf{b}+3\mathbf{c}$ is collinear with $\mathbf{a}$.

$\therefore \quad \mathbf{b}+3\mathbf{c}=\mu\mathbf{a}$ ...(ii)

Now, $\mathbf{a}+2\mathbf{b}+6\mathbf{c}=(\mathbf{a}+2\mathbf{b})+6\mathbf{c}=\lambda\mathbf{c}+6\mathbf{c}$

$=(\lambda+6)\mathbf{c}$ ...(iii)

Also, $\mathbf{a}+2\mathbf{b}+6\mathbf{c}=\mathbf{a}+2(\mathbf{b}+3\mathbf{c})=\mathbf{a}+2\mu\mathbf{a}$

$=(2\mu+1)\mathbf{a}$ ...(iv)

From Eqs. (iii) and (iv), we get

$(\lambda+6)\mathbf{c}=(2\mu+1)\mathbf{a}$

But $\mathbf{a}$ and $\mathbf{c}$ are non-zero, non-collinear vectors,

$\therefore \quad \lambda+6=0=2\mu+1$

Hence, $\quad \mathbf{a}+2\mathbf{b}+6\mathbf{c}=0$

Theorem of Coplanar Vectors

Let $\mathbf{a}$ and $\mathbf{b}$ be two non-zero, non-collinear vectors. Then any vector $\mathbf{r}$ coplanar with $\mathbf{a}$ and $\mathbf{b}$ can be uniquely expressed as a linear combination $x\mathbf{a}+y\mathbf{b}$; x and y being scalars.

Proof Let $\mathbf{a}$ and $\mathbf{b}$ be any two non-zero, non-collinear vectors and $\mathbf{r}$ be any vector coplanar with $\mathbf{a}$ and $\mathbf{b}$. We take any point O in the plane of $\mathbf{a}$ and $\mathbf{b}$

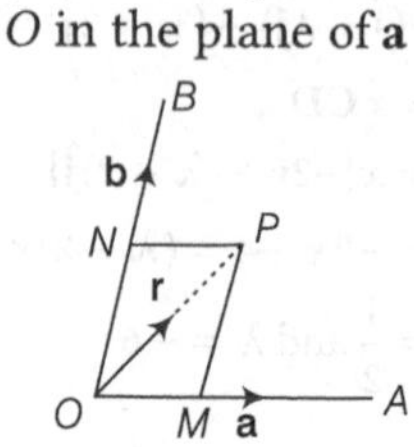

Let $\quad \mathbf{OA}=\mathbf{a}, \mathbf{OB}=\mathbf{b}$ and $\mathbf{OP}=\mathbf{r}$

Clearly, OA, OB and OP are coplanar.

Through P, we draw lines PM and PN, parallel to OB and OA respeetively meeting OA and OB at M and N respectively.

We have, $\mathbf{OP}=\mathbf{OM}+\mathbf{MP}$

$=\mathbf{OM}+\mathbf{ON}$ [$\because MP=ON$ and $MP \parallel ON$] ...(i)

Now, $\mathbf{OM}$ and $\mathbf{OA}$ are collinear vectors

$\mathbf{OM}=x\,\mathbf{OA}=x\mathbf{a}$, where x is scalar.

Similarly, $\mathbf{ON}=y\mathbf{OB}=y\mathbf{b}$, where y is a scalar.

Hence, from Eq. (i), $\mathbf{OP}=x\mathbf{a}+y\mathbf{b}$ or $\mathbf{r}=x'\mathbf{a}+y'\mathbf{b}$

Uniqueness: If possible, let $\mathbf{r}=x\mathbf{a}+y\mathbf{b}$ and $\mathbf{r}=x'\mathbf{a}+y'\mathbf{b}$ be two different ways of representing r.

Then, we have $\quad x\mathbf{a}+y\mathbf{b}=x'\mathbf{a}+y'\mathbf{b}$

$\Rightarrow \quad (x-x')\mathbf{a}+(y-y')\mathbf{b}=0$

But $\mathbf{a}$ and $\mathbf{b}$ are non-collinear vectors

$\therefore \quad x-x'=0$ and $y-y'=0$

$\Rightarrow \quad x'=x$ and $y'=y$

Thus, the uniqueness in established.

Test of Coplanarity of Three Vectors

(i) Three vectors $\mathbf{a}, \mathbf{b}, \mathbf{c}$ are coplanar iff any one of them is a linear combination of the remaining two, i.e. iff $\mathbf{a}=x\mathbf{b}+y\mathbf{c}$ where x and y are scalars.

(ii) If three points with position vectors $\mathbf{a}=a_1\hat{\mathbf{i}}+a_2\hat{\mathbf{j}}+a_3\hat{\mathbf{k}}$, $\mathbf{b}=b_1\hat{\mathbf{i}}+b_2\hat{\mathbf{j}}+b_3\hat{\mathbf{k}}$ and $\mathbf{c}=c_1\hat{\mathbf{i}}+c_2\hat{\mathbf{j}}+c_3\hat{\mathbf{k}}$ are coplanar,

then $\begin{vmatrix} a_1 & a_2 & a_3 \\ b_1 & b_2 & b_3 \\ c_1 & c_2 & c_3 \end{vmatrix}=0.$

If vectors $\mathbf{a}$,$\mathbf{b}$ and $\mathbf{c}$ are coplanar, then there exist scalars x and y such that $\mathbf{c}=x\mathbf{a}+y\mathbf{b}$.

Hence, $c_1\hat{\mathbf{i}}+c_2\hat{\mathbf{j}}+c_3\hat{\mathbf{k}}=x(a_1\hat{\mathbf{i}}+a_2\hat{\mathbf{j}}+a_3\hat{\mathbf{k}})$
$+y(b_1\hat{\mathbf{i}}+b_2\hat{\mathbf{j}}+b_3\hat{\mathbf{k}})$

Now, $\hat{\mathbf{i}}$, $\hat{\mathbf{j}}$ and $\hat{\mathbf{k}}$ are non-coplanar and hence independent.

Then, $\quad c_1=xa_1+yb_1, c_2=xa_2+yb_2$

and $\quad c_3=xa_3+yb_3$

The above system of equations in terms of x and y is consistent. Thus,

$$\begin{vmatrix} a_1 & b_1 & c_1 \\ a_2 & b_2 & c_2 \\ a_3 & b_3 & c_3 \end{vmatrix}=0 \text{ or } \begin{vmatrix} a_1 & a_2 & a_3 \\ b_1 & b_2 & b_3 \\ c_1 & c_2 & c_3 \end{vmatrix}=0.$$

Remark

If vectors $x_1\mathbf{a}+y_1\mathbf{b}+z_1\mathbf{c}$, $x_2\mathbf{a}+y_2\mathbf{b}+z_2\mathbf{c}$ and $x_3\mathbf{a}+y_3\mathbf{b}+z_3\mathbf{c}$ are coplanar(where $\mathbf{a}$, $\mathbf{b}$ and $\mathbf{c}$ are non-coplanar).

Then, $\begin{vmatrix} x_1 & y_1 & z_1 \\ x_2 & y_2 & z_2 \\ x_3 & y_3 & z_3 \end{vmatrix}=0$

Test of Coplanarity of Four Points

(i) To prove that four points $A(\mathbf{a})$, $B(\mathbf{b})$, $C(\mathbf{c})$ and $D(\mathbf{d})$ are coplanar, it is just sufficient to prove that vectors $\mathbf{AB}$, $\mathbf{AC}$ and $\mathbf{AD}$ and coplanar.

(ii) Four points with position vectors $\mathbf{a}$, $\mathbf{b}$, $\mathbf{c}$ and $\mathbf{d}$ are coplanar iff there exist scalars x, y, z and u not all zero such that $x\,\mathbf{a}+y\,\mathbf{b}+z\,\mathbf{c}+u\,\mathbf{d}=0$, where $x+y+z+u=0$.

(iii) Four points with position vectors

$\mathbf{a}=a_1\hat{\mathbf{i}}+a_2\hat{\mathbf{j}}+a_3\hat{\mathbf{k}}$,

$\mathbf{b}=b_1\hat{\mathbf{i}}+b_2\hat{\mathbf{j}}+b_3\hat{\mathbf{k}}$

$\mathbf{c}=c_1\hat{\mathbf{i}}+c_2\hat{\mathbf{j}}+c_3\hat{\mathbf{k}}$

and $\quad \mathbf{d}=d_1\hat{\mathbf{i}}+d_2\hat{\mathbf{j}}+d_3\hat{\mathbf{k}}$

will be coplanar, iff $\begin{vmatrix} a_1 & a_2 & a_3 & 1 \\ b_1 & b_2 & b_3 & 1 \\ c_1 & c_2 & c_3 & 1 \\ d_1 & d_2 & d_3 & 1 \end{vmatrix} = 0$

or $\begin{vmatrix} d_1 - a_1 & d_2 - a_2 & d_3 - a_3 \\ b_1 - a_1 & b_2 - a_2 & b_3 - a_3 \\ c_1 - a_1 & c_2 - a_2 & c_3 - a_3 \end{vmatrix} = 0$

Theorem on Non-coplanar Vectors

Theorem 1

If **a**, **b**, **c**, are three non-zero, non-coplanar vectors and x, y, z are three scalars such that

$$x\mathbf{a} + y\mathbf{b} + z\mathbf{c} = 0.$$

Then $\quad x = y = z = 0.$

Proof It is given that $x\mathbf{a} + y\mathbf{b} + z\mathbf{c} = 0$ ...(i)

Suppose that $x \neq 0$

Then Eq. (i) can be written as

$$x\mathbf{a} = -y\mathbf{b} - z\mathbf{c}$$

$$\Rightarrow \quad \mathbf{a} = -\frac{y}{x}\mathbf{b} - \frac{z}{x}\mathbf{c} \quad \text{...(ii)}$$

Now, $\frac{y}{x}$ and $\frac{z}{x}$ are scalars because x, y and z are scalars.

Thus, Eq. (ii) expresses **a** as a linear combination of **b** and **c**.

Hence, **a** is coplanar with **b** and **c** which is contrary to our hypothesis because **a**,**b** and **c** are given to be non-coplanar.

Thus, our supposition that $x \neq 0$ is wrong.

Hence, $x = 0$

Similarly, we can prove that $y = 0$ and $z = 0$

Theorem 2

If **a**,**b** and **c** are non-coplanar vectors, then any vector **r** can be uniquely expressed as a linear combination $x\mathbf{a} + y\mathbf{b} + z\mathbf{c}$; x, y and z being scalars.

or

Any vector in space can be expressed as a linear combination of three non-coplanar vectors.

Proof Take any point O.

Let **a**, **b**, **c** be any three non-coplanar vectors and **r** be any vector in space.

Let $\quad \mathbf{OA} = \mathbf{a}, \mathbf{OB} = \mathbf{b},$

$\mathbf{OC} = \mathbf{c}, \mathbf{OP} = \mathbf{r}$

Here, the three lines OA, OB, OC are not coplanar. Hence, they determine three different planes BOC, COA and AOB when taken in pairs.

Through P, draw planes parallel to these planes BOC, COA and AOB meeting OA, OB and OC in L, E and N respectively. Thus we obtain a parallelopiped with OP as diagonal and three coterminous edges OL, OE and ON along OA, OB and OC, respectively.

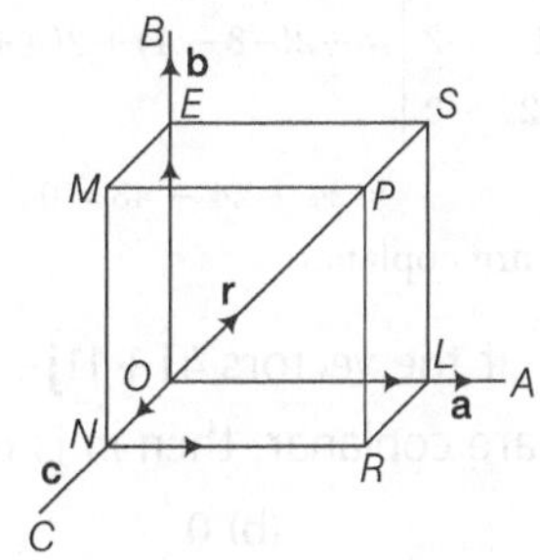

$\therefore$ **OL** is collinear with **OA**.

$\therefore$ $\mathbf{OL} = x\mathbf{OA} = x\mathbf{a}$, where x is a scalar.

Similarly, $\mathbf{OE} = y\,\mathbf{b}$ and $\mathbf{ON} = z\mathbf{c}$,

where y and z are scalars.

Now, $\mathbf{OP} = \mathbf{OR} + \mathbf{RP} = (\mathbf{ON} + \mathbf{NR}) + \mathbf{RP}$

$= \mathbf{ON} + \mathbf{OL} + \mathbf{OE} \quad [\because \mathbf{NR} = \mathbf{OL} \text{ and } \mathbf{RP} = \mathbf{OE}]$

$= \mathbf{OL} + \mathbf{OE} + \mathbf{ON} = x\mathbf{a} + y\mathbf{b} + z\mathbf{c}$

Thus, $\quad \mathbf{r} = x\mathbf{a} + y\mathbf{b} + z\mathbf{c}$

Hence, **r** can be expressed as a linear combination of **a**, **b** and **c**.

Uniqueness If possible let

$$\mathbf{r} = x\mathbf{a} + y\mathbf{b} + z\mathbf{c}$$

and $\quad \mathbf{r} = x'\mathbf{a} + y'\mathbf{b} + z'\mathbf{c}$

be two different ways of representing **r**, then we have

$$x\mathbf{a} + y\mathbf{b} + z\mathbf{c} = x'\mathbf{a} + y'\mathbf{b} + z'\mathbf{c}$$

$$\Rightarrow \quad (x - x')\mathbf{a} + (y - y')\mathbf{b} + (z - z')\mathbf{c} = 0$$

Now **a**, **b** and **c** are non-coplanar vectors

$$\therefore \quad x - x' = 0, y - y' = 0 \text{ and } z - z' = 0$$

$$\Rightarrow \quad x = x', y = y' \text{ and } z = z'$$

Hence, the uniqueness is established.

Remark

If **a**, **b**, **c** are any three non-coplanar vectors in space, then

$$x_1\mathbf{a} + y_1\mathbf{b} + z_1\mathbf{c} = x_2\mathbf{a} + y_2\mathbf{b} + z_2\mathbf{c}$$

$$\Rightarrow \quad x_1 = x_2, y_1 = y_2, z_1 = z_2$$

Proof $x_1\mathbf{a} + y_1\mathbf{b} + z_1\mathbf{c} = x_2\mathbf{a} + y_2\mathbf{b} + z_2\mathbf{c}$

$$\Rightarrow \quad (x_1 - x_2)\mathbf{a} + (y_1 - y_2)\mathbf{b} + (z_1 - z_2)\mathbf{c} = 0$$

$$\Rightarrow \quad x_1 - x_2 = 0, y_1 - y_2 = 0 \text{ and } z_1 - z_2 = 0$$

$$\Rightarrow \quad x_1 = x_2, y_1 = y_2 \text{ and } z_1 = z_2$$

Example 39. Check whether the given three vectors are coplanar or non-coplanar.
$-2\hat{i}-2\hat{j}+4\hat{k}, -2\hat{i}+4\hat{j}-2\hat{k}, 4\hat{i}-2\hat{j}-2\hat{k}$

Sol. Let $\mathbf{a} = -2\hat{i}-2\hat{j}+4\hat{k}$

$\mathbf{b} = -2\hat{i}+4\hat{j}-2\hat{k}$ and $\mathbf{c} = 4\hat{i}-2\hat{j}-2\hat{k}$

Now, consider

$$\begin{vmatrix} -2 & -2 & 4 \\ -2 & 4 & -2 \\ 4 & -2 & -2 \end{vmatrix} = -2(-8-4)+2(4+8)+4(4-16)$$

$$= 24 + 24 - 48 = 0$$

$\therefore$ The vectors are coplanar.

Example 40. If the vectors $4\hat{i}+11\hat{j}+m\hat{k}, 7\hat{i}+2\hat{j}+6\hat{k}$ and $\hat{i}+5\hat{j}+4\hat{k}$ are coplanar, then m is equal to

(a) 38 (b) 0
(c) 10 (d) −10

Sol. (c) Since the three vectors are coplanar, one will be a linear combination of the other two.

$\therefore \quad 4\hat{i}+11\hat{j}+m\hat{k} = x(7\hat{i}+2\hat{j}+6\hat{k}) + y(\hat{i}+5\hat{j}+4\hat{k})$

$\Rightarrow \quad 4 = 7x + y \quad$ ...(i)

$11 = 2x + 5y \quad$ ...(ii)

$m = 6x + 4y \quad$ ...(iii)

From Eqs. (i) and (ii), we get

$$x = \frac{3}{11} \text{ and } y = \frac{23}{11}$$

From Eq. (iii), we get

$$m = 6 \times \frac{3}{11} + 4 \times \frac{23}{11} = 10$$

Trick Since, vectors $4\hat{i}+11\hat{j}+m\hat{k}, 7\hat{i}+2\hat{j}+6\hat{k}$ and $\hat{i}+5\hat{j}+4\hat{k}$ are coplanar.

$$\therefore \quad \begin{vmatrix} 4 & 11 & m \\ 7 & 2 & 6 \\ 1 & 5 & 4 \end{vmatrix} = 0$$

$\Rightarrow \quad 4(8-30) - 11(28-6) + m(35-2) = 0$

$\Rightarrow \quad -88 - 11 \times 22 + 33m = 0$

$\Rightarrow \quad -8 - 22 + 3m = 0$

$\Rightarrow \quad 3m = 30 \Rightarrow m = 10$

Example 41. If **a**, **b** and **c** are non-coplanar vectors, prove that $3\mathbf{a}-7\mathbf{b}-4\mathbf{c}$, $3\mathbf{a}-2\mathbf{b}+\mathbf{c}$ and $\mathbf{a}+\mathbf{b}+2\mathbf{c}$ are coplanar.

Sol. Let $\alpha = 3\mathbf{a}-7\mathbf{b}-4\mathbf{c}, \beta = 3\mathbf{a}-2\mathbf{b}+\mathbf{c}$

and $\quad \gamma = \mathbf{a}+\mathbf{b}+2\mathbf{c}$

Also, let $\quad \alpha = x\beta + y - \gamma$

$\Rightarrow \quad 3\mathbf{a}-7\mathbf{b}-4\mathbf{c} = x(3\mathbf{a}-2\mathbf{b}+\mathbf{c}) + y(\mathbf{a}+\mathbf{b}+2\mathbf{c})$

$= (3x+y)\mathbf{a} + (-2x+y)\mathbf{b} + (x+2y)\mathbf{c}$

Since, **a**, **b** and **c** are non-coplanar vectors.
Therefore,

$$3x + y = 3, -2x + y = -7$$

and $\quad x + 2y = -4$

Solving first two, we find that $x = 2$ and $y = -3$. These values of x and y satisfy the third equation as well.

So, $x + 2$ and $y = -3$ is the unique solution for the above system of equation.

$\Rightarrow \quad \alpha = 2\beta - 3\gamma$

Hence, the vectors α, β and γ are coplanar, because α is uniquely written as linear combination of other two.

Trick For the vectors α, β, γ to be coplanar, we must have

$$\begin{vmatrix} 3 & -7 & -4 \\ 3 & -2 & 1 \\ 1 & 1 & 2 \end{vmatrix} = 0, \text{ which is true}$$

Hence, α, β, γ are coplanar.

Example 42. The value of λ for which the four points $2\hat{i}+3\hat{j}-\hat{k}, \hat{i}+2\hat{j}+3\hat{k}, 3\hat{i}+4\hat{j}-2\hat{k}$ and $\hat{i}-\lambda\hat{j}+6\hat{k}$ are coplanar

(a) 8 (b) 0
(c) −2 (d) 6

Sol. (c) The given four points are coplanar.

$$\begin{vmatrix} 2 & 1 & 3 & 1 \\ 3 & 2 & 4 & \lambda \\ -1 & 3 & -2 & 6 \\ 1 & 1 & 1 & 0 \end{vmatrix} = 0$$

$$\Rightarrow \quad \begin{vmatrix} 2 & 1 & 3 & 1 \\ 0 & 0 & 0 & -(\lambda+2) \\ -1 & 3 & -2 & 6 \\ 1 & 1 & 1 & 1 \end{vmatrix} = 0$$

Operating $(R_2 \to R_2 - R_1 - R_4)$

$$\Rightarrow \quad -(\lambda+2)\begin{vmatrix} 2 & 1 & 3 \\ -1 & 3 & -2 \\ 1 & 1 & 1 \end{vmatrix} = 0 \Rightarrow \lambda = -2$$

Example 43. Show that the points $P(\mathbf{a}+2\mathbf{b}+\mathbf{c})$, $Q(\mathbf{a}-\mathbf{b}-\mathbf{c})$, $R(3\mathbf{a}+\mathbf{b}+2\mathbf{c})$ and $S(5\mathbf{a}+3\mathbf{b}+5\mathbf{c})$ are coplanar given that **a**,**b** and **c** are non-coplanar.

Sol. To show that P, Q, R, S are coplanar, we will show that **PQ**, **PR**, **PS** are coplanar.

$$\mathbf{PQ} = -3\mathbf{b} - 2\mathbf{c}$$

$$\mathbf{PR} = 2\mathbf{a} - \mathbf{b} + \mathbf{c}$$

$$\mathbf{PS} = 4\mathbf{a} + \mathbf{b} + 4\mathbf{c}$$

Let $\quad \mathbf{PQ} = x\mathbf{PR} + y\mathbf{PS}$

$\Rightarrow \quad -3\mathbf{b} - 2\mathbf{c} = x(2\mathbf{a}-\mathbf{b}+\mathbf{c}) + y(4\mathbf{a}+\mathbf{b}+4\mathbf{c})$

$\Rightarrow \quad -3\mathbf{b} - 2\mathbf{c} = (2x+4y)\mathbf{a} + (-x+y)\mathbf{b} + (x+4y)\mathbf{c}$

As the vectors **a, b, c** are non-coplanar, we can equate their coefficients.

$$\Rightarrow \quad 0 = 2x + 4y$$
$$\Rightarrow \quad -3 = -x + y$$
$$\Rightarrow \quad -2 = x + 4y$$

$x = 2, y = -1$ is the unique solution for the above system of equations.

$$\Rightarrow \quad \mathbf{PQ} = 2\mathbf{PR} - \mathbf{PS}$$

PQ, **PR**, **PS** are coplanar because **PQ** is a linear combination of **PR** and **PS**

⇒ The points **P, Q, R, S** are also coplanar.

Trick For the vectors **PQ**, **PR** and **PS** to be coplanar, we must have $\begin{vmatrix} 0 & -3 & -2 \\ 2 & -1 & 1 \\ 4 & 1 & 4 \end{vmatrix} = 0$ which is true

∴ The **PQ**, **PR**, **PS** are coplanar.

Hence, the points P, Q, R, S are also coplanar.

Linear Independence and Dependence of Vectors

1. Linearly Independent Vectors

A set of non-zero vectors $\mathbf{a}_1, \mathbf{a}_2, \ldots, \mathbf{a}_n$ is said to be linearly independent, if

$$x_1\mathbf{a}_1 + x_2\mathbf{a}_2 + \ldots + x_n\mathbf{a}_n = 0$$
$$\Rightarrow \quad x_1 = x_2 = \ldots = x_n = 0.$$

2. Linearly Dependence Vectors

A set of vector $\mathbf{a}_1, \mathbf{a}_1, \ldots, \mathbf{a}_n$ is said to be linearly dependent, if there exist scalars $x_1, x_2, \ldots, x_n$ not all zero such that $x_1\mathbf{a}_1 + x_2\mathbf{a}_2 + \ldots + x_n\mathbf{a}_n = 0$

Properties of Linearly Independent and Dependent Vectors

(i) A super set of a linearly dependent set of vectors is linearly dependent.

(ii) A subset of a linearly independent set of vectors is linearly independent.

(iii) Two non-zero, non-collinear vectors are linearly independent.

(iv) Any two collinear vectors are linearly dependent.

(v) Any three non-coplanar vectors are linearly independent.

(vi) Any three coplanar vectors are linearly dependent.

(vii) Three vectors $\mathbf{a} = a_1\hat{\mathbf{i}} + a_2\hat{\mathbf{j}} + a_3\hat{\mathbf{k}}$, $\mathbf{b} = b_1\hat{\mathbf{i}} + b_2\hat{\mathbf{j}} + b_3\hat{\mathbf{k}}$ and $\mathbf{c} = c_1\hat{\mathbf{i}} + c_2\hat{\mathbf{j}} + c_3\hat{\mathbf{k}}$ will be linearly dependent vectors iff $\begin{vmatrix} a_1 & a_2 & a_3 \\ b_1 & b_2 & b_3 \\ c_1 & c_2 & c_3 \end{vmatrix} = 0.$

(viii) Any four vectors in 3-dimensional space are linearly dependent.

Example 44. Show that the vectors $\hat{\mathbf{i}} - 3\hat{\mathbf{j}} + 2\hat{\mathbf{k}}, 2\hat{\mathbf{i}} - 4\hat{\mathbf{j}} - \hat{\mathbf{k}}$ and $3\hat{\mathbf{i}} + 2\hat{\mathbf{j}} - \hat{\mathbf{k}}$ and linearly independent.

Sol. Let $\alpha = \hat{\mathbf{i}} - 3\hat{\mathbf{j}} + 2\hat{\mathbf{k}}$

$\beta = 2\hat{\mathbf{i}} - 4\hat{\mathbf{j}} - \hat{\mathbf{k}}$

and $\gamma = 3\hat{\mathbf{i}} + 2\hat{\mathbf{j}} - \hat{\mathbf{k}}$

Also, let $x\alpha + y\beta + z\gamma = 0$

$$\therefore \quad x(\hat{\mathbf{i}} - 3\hat{\mathbf{j}} + 2\hat{\mathbf{k}}) + y(2\hat{\mathbf{i}} - 4\hat{\mathbf{j}} - \hat{\mathbf{k}}) + z(3\hat{\mathbf{i}} + 2\hat{\mathbf{j}} - \hat{\mathbf{k}}) = 0$$

or $(x + 2y + 3z)\hat{\mathbf{i}} + (-3x - 4y + 2z)\hat{\mathbf{j}} + (2x - y - z)\hat{\mathbf{k}} = 0$

Equating the coefficient of $\hat{\mathbf{i}}$, $\hat{\mathbf{j}}$ and $\hat{\mathbf{k}}$, we get

$$x + 2y + 3z = 0$$
$$-3x - 4y + 2z = 0$$
$$2x - y - z = 0$$

Now, $\begin{vmatrix} 1 & 2 & 3 \\ -3 & -4 & 2 \\ 2 & -1 & -1 \end{vmatrix} = 1(4+2) - 2(3-4) + 3(3+8) = 41 \neq 0$

∴ The above system of equations have only trivial solution.

Thus, $x = y = z = 0$

Hence, the vectors α, β and γ are linearly independent.

Trick Consider the determinant of coefficients of $\hat{\mathbf{i}}$, $\hat{\mathbf{j}}$ and $\hat{\mathbf{k}}$

i.e. $\begin{vmatrix} 1 & -3 & 2 \\ 2 & -4 & -1 \\ 3 & 2 & -1 \end{vmatrix} = 1(4+2) + 3(-2+3) + 2(4+12)$

$$= 6 + 3 + 32 = 41 \neq 0$$

∴ The given vectors are non-coplanar. Hence, the vectors are linearly independent.

Example 45. If $\mathbf{a} = \hat{\mathbf{i}} + \hat{\mathbf{j}} + \hat{\mathbf{k}}$, $\mathbf{b} = 4\hat{\mathbf{i}} + 3\hat{\mathbf{j}} + 4\hat{\mathbf{k}}$ and $\mathbf{c} = \hat{\mathbf{i}} + \alpha\hat{\mathbf{j}} + \beta\hat{\mathbf{k}}$ are linearly dependent vectors and $|\mathbf{c}| = \sqrt{3}$, then

(a) $\alpha = 1, \beta = -1$ (b) $\alpha = 1, \beta = \pm 1$
(c) $\alpha = -1, \beta = \pm 1$ (d) $\alpha = \pm 1, \beta = 1$

Sol. (d) The given vectors are linearly dependent, hence there exist scalars x, y and z not all zero, such that

$$x\mathbf{a} + y\mathbf{b} + z\mathbf{c} = 0$$

i.e. $x(\hat{\mathbf{i}} + \hat{\mathbf{j}} + \hat{\mathbf{k}}) + y(4\hat{\mathbf{i}} + 3\hat{\mathbf{j}} + 4\hat{\mathbf{k}}) + z(\hat{\mathbf{i}} + \alpha\hat{\mathbf{j}} + \beta\hat{\mathbf{k}}) = 0$

i.e. $(x+4y+z)\hat{i}+(x+3y+\alpha z)\hat{j}+(x+4y+\beta z)\hat{k}=0$

$\Rightarrow \quad x+4y+z=0,\ x+3y+\alpha z=0,\ x+4y+\beta z=0$

For non-trivial solution $\begin{vmatrix} 1 & 4 & 1 \\ 1 & 3 & \alpha \\ 1 & 4 & \beta \end{vmatrix}=0 \Rightarrow \beta=1$

$|c|^2=3 \Rightarrow 1+\alpha^2+\beta^2=3$

$\Rightarrow \quad \alpha^2=2-\beta^2=2-1=1$

$\therefore \quad \alpha=\pm 1$

Trick $\quad |c|=\sqrt{1+\alpha^2+\beta^2}=\sqrt{3}$

$\Rightarrow \quad \alpha^2+\beta^2=2$

$\because$ **a**, **b** and **c** are linearly dependent, hence $\begin{vmatrix} 1 & 1 & 1 \\ 4 & 3 & 4 \\ 4 & \alpha & \beta \end{vmatrix}=0$

$\Rightarrow \quad \beta=1$

$\therefore \quad \alpha^2=1 \Rightarrow \alpha=\pm 1$

Exercise for Session 3

1. Show that the points $A(1, 3, 2), B(-2, 0, 1)$ and $C(4, 6, 3)$ are collinear.
2. If the position vectors of the points A, B and C be **a**, **b** and $3\mathbf{a}-2\mathbf{b}$ respectively, then prove that the points A, B and C are collinear.
3. The position vectors of four points P, Q, R and S are $2\mathbf{a}+4\mathbf{c}$, $5\mathbf{a}+3\sqrt{3}\mathbf{b}+4\mathbf{c}$, $-2\sqrt{3}\mathbf{b}+\mathbf{c}$ and $2\mathbf{a}+\mathbf{c}$ respectively, prove that PQ is parallel to RS.
4. If three points A, B and C have position vectors $(1, x, 3), (3, 4, 7)$ and $(y, -2, -5)$, respectively and if they are collinear, then find (x, y).
5. Find the condition that the three points whose position vectors, $\mathbf{a}=a\hat{i}+b\hat{j}+c\hat{k}$, $\mathbf{b}=\hat{i}+c\hat{j}$ and $\mathbf{c}=-\hat{i}-\hat{j}$ are collinear.
6. Vectors **a** and **b** are non-collinear. Find for what values of x vectors $\mathbf{c}=(x-2)\mathbf{a}+\mathbf{b}$ and $\mathbf{d}=(2x+1)\mathbf{a}-\mathbf{b}$ are collinear?
7. Let **a**, **b**, **c** are three vectors of which every pair is non-collinear. If the vectors $\mathbf{a}+\mathbf{b}$ and $\mathbf{b}+\mathbf{c}$ are collinear with **c** and **a** respectively, then find $\mathbf{a}+\mathbf{b}+\mathbf{c}$.
8. Show that the vectors $\hat{i}-\hat{j}-\hat{k}$, $2\hat{i}+3\hat{j}+\hat{k}$ and $7\hat{i}+3\hat{j}-4\hat{k}$ are coplanar.
9. If the vectors $2\hat{i}-\hat{j}+\hat{k}$, $\hat{i}+2\hat{j}-3\hat{k}$ and $3\hat{i}+a\hat{j}+5\hat{k}$ are coplanar, then prove that $a=4$.
10. Show that the vectors $\mathbf{a}-2\mathbf{b}+3\mathbf{c}$, $-2\mathbf{a}+3\mathbf{b}-4\mathbf{c}$ and $-\mathbf{b}+2\mathbf{c}$ are coplanar vector, where **a**, **b**, **c** are non-coplanar vectors.
11. If **a**, **b** and **c** are non-coplanar vectors, then prove that the four points $2\mathbf{a}+3\mathbf{b}-\mathbf{c}$, $\mathbf{a}-2\mathbf{b}+3\mathbf{c}$, $3\mathbf{a}+4\mathbf{b}-2\mathbf{c}$ and $\mathbf{a}-6\mathbf{b}+6\mathbf{c}$ are coplanar.

JEE Type Solved Examples :
Single Option Correct Type Questions

Ex. 1 *The non-zero vectors* $\mathbf{a}, \mathbf{b}$ *and* $\mathbf{c}$ *are related by* $\mathbf{a} = 8\mathbf{b}$ *and* $\mathbf{c} = -7\mathbf{b}$ *angle between* $\mathbf{a}$ *and* $\mathbf{c}$ *is*

(a) $\frac{\pi}{4}$ (b) $\frac{\pi}{2}$

(c) π (d) 0

Sol. (c) $\mathbf{a}$ and $\mathbf{b}$ vectors are in the same direction, $\mathbf{b}$ and $\mathbf{c}$ are in the opposite direction.

$\Rightarrow$ $\mathbf{a}$ and $\mathbf{c}$ are in opposite directions.

$\therefore$ Angle between $\mathbf{a}$ and $\mathbf{c}$ is π.

Ex. 2 *A unit vector* $\mathbf{a}$ *makes an angle* $\frac{\pi}{4}$ *with Z-axis. If* $\mathbf{a} + \hat{\mathbf{i}} + \hat{\mathbf{j}}$ *is a unit vector, then* $\mathbf{a}$ *is equal to*

(a) $\frac{\hat{\mathbf{i}}}{2} + \frac{\hat{\mathbf{j}}}{2} + \frac{\hat{\mathbf{k}}}{\sqrt{2}}$ (b) $\frac{\hat{\mathbf{i}}}{2} + \frac{\hat{\mathbf{j}}}{2} - \frac{\hat{\mathbf{k}}}{\sqrt{2}}$

(c) $-\frac{\hat{\mathbf{i}}}{2} - \frac{\hat{\mathbf{j}}}{2} + \frac{\hat{\mathbf{k}}}{\sqrt{2}}$ (d) None of these

Sol. (c) Let $\mathbf{a} = l\hat{\mathbf{i}} + m\hat{\mathbf{j}} + n\hat{\mathbf{k}}$, where $l^2 + m^2 + n^2 = 1$. $\mathbf{a}$ makes an angle $\frac{\pi}{4}$ with Z-axis.

$$\therefore \quad n = \frac{1}{\sqrt{2}}, l^2 + m^2 = \frac{1}{2} \quad \text{...(i)}$$

$$\therefore \quad \mathbf{a} = l\hat{\mathbf{i}} + m\hat{\mathbf{j}} + \frac{\hat{\mathbf{k}}}{\sqrt{2}}$$

$$\mathbf{a} + \hat{\mathbf{i}} + \hat{\mathbf{j}} = (l+1)\hat{\mathbf{i}} + (m+1)\hat{\mathbf{j}} + \frac{\hat{\mathbf{k}}}{\sqrt{2}}$$

Its magnitude is 1, hence $(l+1)^2 + (m+1)^2 = \frac{1}{2}$ ...(ii)

From Eqs. (i) and (ii), we get

$$2lm = \frac{1}{2} \Rightarrow l = m = -\frac{1}{2}$$

Hence, $\mathbf{a} = -\frac{\hat{\mathbf{i}}}{2} - \frac{\hat{\mathbf{j}}}{2} + \frac{\hat{\mathbf{k}}}{\sqrt{2}}$

Ex. 3 *If the resultant of two forces of magnitudes P and Q acting at a point at an angle of* $60°$ *is* $\sqrt{7}Q$*, then* P/Q *is*

(a) 1 (b) $\frac{3}{2}$

(c) 2 (d) 4

Sol. (c) $R^2 = P^2 + Q^2 + 2PQ\cos\theta$

$$\Rightarrow \quad (\sqrt{7}Q)^2 = P^2 + Q^2 + 2PQ\cos 60°$$

$$\Rightarrow \quad 7Q^2 = P^2 + Q^2 + PQ$$

$$\Rightarrow \quad P^2 + PQ - 6Q^2 = 0$$

$$\Rightarrow \quad P^2 + 3PQ - 2PQ - 6Q^2 = 0$$

$$\Rightarrow \quad P(P + 3Q) - 2Q(P + 3Q) = 0$$

$$\Rightarrow \quad (P - 2Q)(P + 3Q) = 0$$

$$\Rightarrow \quad P - 2Q = 0 \text{ or } P + 3Q = 0$$

From $P - 2Q = 0 \Rightarrow \frac{P}{Q} = 2$

Ex. 4 *A vector* $\mathbf{a}$ *has the components* $2p$ *and* 1 *w.r.t. a rectangular cartesian system. This system is rotated through a certain angle about the origin in the counter clockwise sense. If with respect to a new system,* $\mathbf{a}$ *has components* $(p+1)$ *and* 1*, then*

(a) $p = 0$ (b) $p = 1$ or $p = -\frac{1}{3}$

(c) $p = -1$ or $p = \frac{1}{3}$ (d) $p = 1$ or $p = -1$

Sol. (b) We have, $\mathbf{a} = 2p\hat{\mathbf{i}} + \hat{\mathbf{j}}$

On rotation, let $\mathbf{b}$ be the vector with components $(p+1)$ and 1 so that,

$$\mathbf{b} = (p+1)\hat{\mathbf{i}} + \hat{\mathbf{j}}$$

Now, $|\mathbf{a}| = |\mathbf{b}| \Rightarrow a^2 = b^2$

$$\Rightarrow \quad 4p^2 + 1 = (p+1)^2 + 1 \Rightarrow 4p^2 = (p+1)^2$$

$$\Rightarrow \quad 2p = \pm(p+1) \Rightarrow 3p = -1 \text{ or } p = 1$$

$$\therefore \quad p = -\frac{1}{3} \text{ or } p = 1$$

Ex. 5 *ABC is an isosceles triangle right angled at A. Forces of magnitude* $2\sqrt{2}$*, 5 and 6 act along* **BC, CA** *and* **AB** *respectively. The magnitude of their resultant force is*

(a) 4 (b) 5

(c) $11 + 2\sqrt{2}$ (d) 30

Sol. (b) $R\cos\theta = 6\cos 0° + 2\sqrt{2}\cos(180° - B) + 5\cos 270°$

$$R\cos\theta = 6 - 2\sqrt{2}\cos B \quad \text{...(i)}$$

$$R\sin\theta = 6\sin 0° + 2\sqrt{2}\sin(180° - B) + 5\sin 270°$$

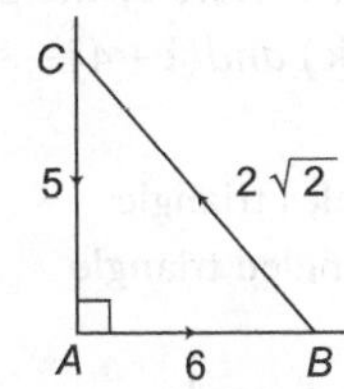

$$R\sin\theta = 2\sqrt{2}\sin B - 5 \quad \text{...(ii)}$$

From Eqs. (i) and (ii), we get

$$R^2 = 36 + 8\cos^2 B - 24\sqrt{2}\cos B + 8\sin^2 B + 25 - 20\sqrt{2}\sin B$$

$$= 61 + 8(\cos^2 B + \sin^2 B) - 24\sqrt{2}\cos B - 20\sqrt{2}\sin B$$

$\because$ ABC is a right angled isosceles triangle.

i.e. $\angle B = \angle C = 45°$

$\therefore$ $R^2 = 61 + 8(1) - 24\sqrt{2}\frac{1}{\sqrt{2}} - 20\sqrt{2}\cdot\frac{1}{\sqrt{2}} = 25$

$\therefore$ $R = 5$

Ex. 6 *A line segment has length* 63 *and direction ratios are* 3, − 2 *and* 6. *The components of line vector are*

(a) −27, 18, 54 (b) 27, − 18, 54

(c) 27, − 18, −54 (d) −27, − 18, − 54

Sol. (b) Let the components of line segment on axes are x, y and z.

So, $x^2 + y^2 + z^2 = 63^2$

Now, $\frac{x}{3} = \frac{y}{-2} = \frac{z}{6} = k$

$\because$ $(3k)^2 + (-2k)^2 + (6k)^2 = 63^2$

$k = \pm\frac{63}{7} = \pm 9$

$\therefore$ Components are (27, − 18, 54) or (−27, 18, −54).

Ex. 7 *If the vectors* $6\hat{\mathbf{i}} - 2\hat{\mathbf{j}} + 3\hat{\mathbf{k}}$, $2\hat{\mathbf{i}} + 3\hat{\mathbf{j}} - 6\hat{\mathbf{k}}$ *and* $3\hat{\mathbf{i}} + 6\hat{\mathbf{j}} - 2\hat{\mathbf{k}}$ *form a triangle, then it is*

(a) right angled (b) obtuse angled

(c) equilateral (d) isosceles

Sol. (b) $\mathbf{AB}$ = Position vectors of B Position vector of A

$= (2\hat{\mathbf{i}} + 3\hat{\mathbf{j}} - 6\hat{\mathbf{k}}) - (6\hat{\mathbf{i}} - 2\hat{\mathbf{j}} + 3\hat{\mathbf{k}}) = -4\hat{\mathbf{i}} + 5\hat{\mathbf{j}} - 9\hat{\mathbf{k}}$

$\Rightarrow |\mathbf{AB}| = \sqrt{16 + 25 + 81} = \sqrt{122}$

$\mathbf{BC} = \hat{\mathbf{i}} + 3\hat{\mathbf{j}} + 4\hat{\mathbf{k}}$

$\Rightarrow |\mathbf{BC}| = \sqrt{1 + 9 + 16} = \sqrt{26}$ and $\mathbf{AC} = -3\hat{\mathbf{i}} + 8\hat{\mathbf{j}} - 5\hat{\mathbf{k}}$

$\Rightarrow |\mathbf{AC}| = \sqrt{98}$

Therefore, $AB^2 = 122$, $BC^2 = 26$ and $AC^2 = 98$

$\Rightarrow$ $AB^2 + BC^2 = 26 + 122 = 148$

Since, $AC^2 < AB^2 + BC^2$, therefore ΔABC is an obtuse angled triangle.

Ex. 8 *The position vectors of the points A, B and C are* $(2\hat{\mathbf{i}} + \hat{\mathbf{j}} - \hat{\mathbf{k}})$, $(3\hat{\mathbf{i}} - 2\hat{\mathbf{j}} + \hat{\mathbf{k}})$ *and* $(\hat{\mathbf{i}} + 4\hat{\mathbf{j}} - 3\hat{\mathbf{k}})$ *respectively. These points.*

(a) form an isosceles triangle

(b) form a right angled triangle

(c) are collinear

(d) form a scalene triangle

Sol. (c) $\mathbf{AB} = (3-2)\hat{\mathbf{i}} + (-2-1)\hat{\mathbf{j}} + (1+1)\hat{\mathbf{k}}$

$= \hat{\mathbf{i}} - 3\hat{\mathbf{j}} + 2\hat{\mathbf{k}}$

$\mathbf{BC} = (1-3)\hat{\mathbf{i}} + (4+2)\hat{\mathbf{j}} + (-3-1)\hat{\mathbf{k}}$

$= -2\hat{\mathbf{i}} + 6\hat{\mathbf{j}} - 4\hat{\mathbf{k}}$

$\mathbf{CA} = (2-1)\hat{\mathbf{i}} + (1-4)\hat{\mathbf{j}} + (-1+3)\hat{\mathbf{k}}$

$= \hat{\mathbf{i}} - 3\hat{\mathbf{j}} - 2\hat{\mathbf{k}}$

$|\mathbf{AB}| = \sqrt{1 + 9 + 4} = \sqrt{14}$

$|\mathbf{BC}| = \sqrt{4 + 36 + 16} = \sqrt{56} = 2\sqrt{14}$

$|\mathbf{CA}| = \sqrt{1 + 9 + 4} = \sqrt{14}$

So, $|\mathbf{AB}| + |\mathbf{AC}| = |\mathbf{BC}|$ and angle between AB and BC is 180°.

So, points A, B and C cannot form an isosceles triangle.

Hence, A, B and C are collinear.

Ex. 9 *The position vector of a point C with respect to B is* $\hat{\mathbf{i}} + \hat{\mathbf{j}}$ *and that of B with respect to A is* $\hat{\mathbf{i}} - \hat{\mathbf{j}}$. *The position vector of C with respect to A is*

(a) $2\hat{\mathbf{i}}$ (b) $2\hat{\mathbf{j}}$

(c) $-2\hat{\mathbf{j}}$ (d) $-2\hat{\mathbf{i}}$

Sol. (a) Since, position vectors of a point C with respect to B is

$\mathbf{BC} = \hat{\mathbf{i}} + \hat{\mathbf{j}}$ ...(i)

Similarly, $\mathbf{AB} = \hat{\mathbf{i}} - \hat{\mathbf{j}}$ ...(ii)

Now, by Eqs. (i) and (ii),

$\mathbf{AC} = \mathbf{AB} + \mathbf{BC} = 2\hat{\mathbf{i}}$

Ex. 10 *In a* ΔABC, *if* $2\mathbf{AC} = 3\mathbf{CB}$, *then* $2\mathbf{OA} + 3\mathbf{OB}$ *is equal to*

(a) $5\mathbf{OC}$ (b) $-\mathbf{OC}$

(c) $\mathbf{OC}$ (d) None of these

Sol. (a) $2\mathbf{OA} + 3\mathbf{OB} = 2(\mathbf{OC} + \mathbf{CA}) + 3(\mathbf{OC} + \mathbf{CB})$

$= 5\mathbf{OC} + 2\mathbf{CA} + 3\mathbf{CB} = 5\mathbf{OC}$ $(\because 2\mathbf{CA} = -3\mathbf{CB})$

Ex. 11 *If* $\mathbf{a}, \mathbf{b}, \mathbf{c}$ *and* $\mathbf{d}$ *be the position vectors of the points A, B, C and D respectively, referred to same origin O such that no three of these points are collinear and* $\mathbf{a} + \mathbf{c} = \mathbf{b} + \mathbf{d}$, *then quadrilateral ABCD is a*

(a) square (b) rhombus

(c) rectangle (d) parallelogram

Sol. (d) Given, $\mathbf{a} + \mathbf{c} = \mathbf{b} + \mathbf{d}$

$\Rightarrow$ $\frac{1}{2}(\mathbf{a} + \mathbf{c}) = \frac{1}{2}(\mathbf{b} + \mathbf{d})$

Here, mid-points of **AC** and **BD** coincide, where **AC** and **BD** are diagonals. In addition, we know that, diagonals of a parallelogram bisect each other.

Hence, quadrilateral is parallelogram.

Ex. 12 *P is a point on the side BC of the* ΔABC *and Q is a point such that* **PQ** *is the resultant of* **AP**, **PB** *and* **PC**. *Then, ABQC is a*

(a) square

(b) rectangle

(c) parallelogram

(d) trapezium

Sol. (c) $\mathbf{AP}+\mathbf{PB}+\mathbf{PC}=\mathbf{PQ}$ or $\mathbf{AP}+\mathbf{PB}=\mathbf{PQ}+\mathbf{CP}$

$\Rightarrow \quad \mathbf{AB}=\mathbf{CQ}$

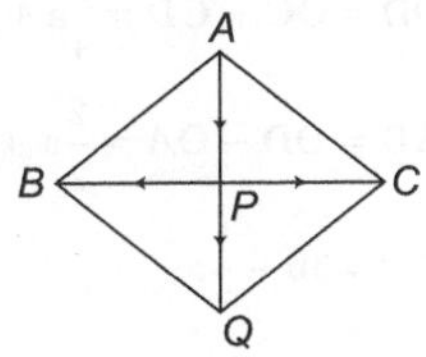

Hence, it is a parallelogram.

Ex. 13 *If ABCD is a parallelogram and the position vectors of A, B and C are $\hat{\mathbf{i}}+3\hat{\mathbf{j}}+5\hat{\mathbf{k}}$, $\hat{\mathbf{i}}+\hat{\mathbf{j}}+\hat{\mathbf{k}}$ and $7\hat{\mathbf{i}}+7\hat{\mathbf{j}}+7\hat{\mathbf{k}}$, then the position vector of D will be*

(a) $7\hat{\mathbf{i}}+5\hat{\mathbf{j}}+3\hat{\mathbf{k}}$ (b) $7\hat{\mathbf{i}}+9\hat{\mathbf{j}}+11\hat{\mathbf{k}}$

(c) $9\hat{\mathbf{i}}+11\hat{\mathbf{j}}+13\hat{\mathbf{k}}$ (d) $8\hat{\mathbf{i}}+8\hat{\mathbf{j}}+8\hat{\mathbf{k}}$

Sol. (b) Let position vector of D is $x\hat{\mathbf{i}}+y\hat{\mathbf{j}}+z\hat{\mathbf{k}}$, then $\mathbf{AB}=\mathbf{DC}$.

$\Rightarrow \quad -2\hat{\mathbf{j}}-4\hat{\mathbf{k}}=(7-x)\hat{\mathbf{i}}+(7-y)\hat{\mathbf{j}}+(7-z)\hat{\mathbf{k}}$

$\Rightarrow \quad x=7, y=9$ and $z=11$

Hence, position vector of D will be $7\hat{\mathbf{i}}+9\hat{\mathbf{j}}+11\hat{\mathbf{k}}$.

Ex. 14 *P is the point of intersection of the diagonals of the parallelogram ABCD. If O is any point, then* $\mathbf{OA}+\mathbf{OB}+\mathbf{OC}+\mathbf{OD}$ *is equal to*

(a) $\mathbf{OP}$ (b) $2\mathbf{OP}$

(c) $3\mathbf{OP}$ (d) $4\mathbf{OP}$

Sol. (d) We know that, P will be the mid-point of AC and BD.

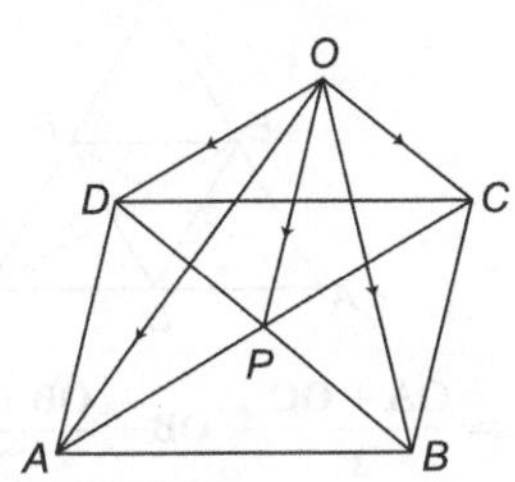

$\therefore \quad \mathbf{OA}+\mathbf{OC}=2\mathbf{OP}$ ...(i)

and $\quad \mathbf{OB}+\mathbf{OD}=2\mathbf{OP}$ ...(ii)

On adding Eqs. (i) and (ii), we get

$\mathbf{OA}+\mathbf{OB}+\mathbf{OC}+\mathbf{OD}=4\mathbf{OP}$

Ex. 15 *If C is the middle point of AB and P is any point outside AB, then*

(a) $\mathbf{PA}+\mathbf{PB}=\mathbf{PC}$

(b) $\mathbf{PA}+\mathbf{PB}=2\mathbf{PC}$

(c) $\mathbf{PA}+\mathbf{PB}+\mathbf{PC}=0$

(d) $\mathbf{PA}+\mathbf{PB}+2\mathbf{PC}=0$

Sol. (b) $\mathbf{PA}+\mathbf{PB}=(\mathbf{PA}+\mathbf{AC})+(\mathbf{PB}+\mathbf{BC})-(\mathbf{AC}+\mathbf{BC})$

$=\mathbf{PC}+\mathbf{PC}-(\mathbf{AC}-\mathbf{CB})=2\mathbf{PC}-0$

$(\because \mathbf{AC}=\mathbf{CB})$

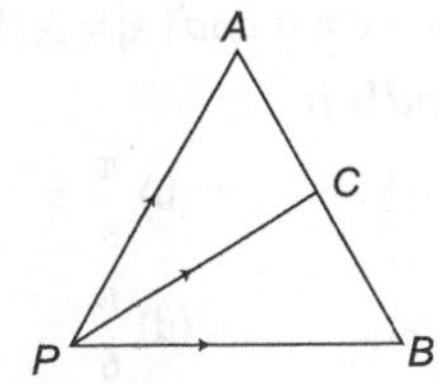

$\therefore \quad \mathbf{PA}+\mathbf{PB}=2\mathbf{PC}$

Ex. 16 *If O be the circumcentre and O′ be the orthocentre of the ΔABC, then* $\mathbf{O'A}+\mathbf{O'B}+\mathbf{O'C}$ *is equal to*

(a) $\mathbf{OO'}$ (b) $2\mathbf{O'O}$ (c) $2\mathbf{OO'}$ (d) 0

Sol. (b) $\quad \mathbf{O'A}=\mathbf{O'O}+\mathbf{OA}$

$\mathbf{O'B}=\mathbf{OO'}+\mathbf{OB}$

$\mathbf{O'C}=\mathbf{O'O}+\mathbf{OC}$

$\Rightarrow \mathbf{O'A}+\mathbf{O'B}+\mathbf{O'C}=3\mathbf{O'O}+\mathbf{OA}+\mathbf{OB}+\mathbf{OC}$

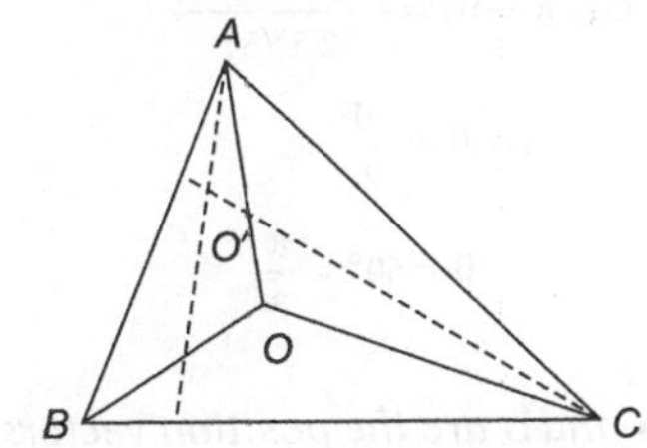

Since, $\quad \mathbf{OA}+\mathbf{OB}+\mathbf{OC}=\mathbf{OO'}=-\mathbf{O'O}$

$\therefore \quad \mathbf{O'A}+\mathbf{O'B}+\mathbf{O'C}=2\mathbf{O'O}$

Ex. 17 *Five points given by A, B, C, D and E are in a plane. Three forces* **AC**, **AD** *and* **AE** *act at A and three forces* **CB**, **DB** *and* **EB** *act B. Then, their resultant is*

(a) $2\mathbf{AC}$ (b) $3\mathbf{AB}$

(c) $3\mathbf{DB}$ (d) $2\mathbf{BC}$

Sol. (b) Points A, B, C, D and E are in a plane.

Resultant $=(\mathbf{AC}+\mathbf{AD}+\mathbf{AE})+\mathbf{CB}+\mathbf{BD}+\mathbf{EB})$

$=(\mathbf{AC}+\mathbf{CB})+(\mathbf{AD}+\mathbf{DB})+(\mathbf{AE}+\mathbf{EB})$

$=\mathbf{AB}+\mathbf{AB}+\mathbf{AB}=3\mathbf{AB}$

Ex. 18 *If the vectors represented by the sides AB and BC of the regular hexagon ABCDEF be* **a** *and* **b**, *then the vector represented by* **AE** *will be*

(a) $2\mathbf{b}-\mathbf{a}$ (b) $\mathbf{b}-\mathbf{a}$

(c) $2\mathbf{a}-\mathbf{b}$ (d) $\mathbf{a}+\mathbf{b}$

Sol. (a) As in figure, $\mathbf{AB}=\mathbf{a}$, $\mathbf{BC}=\mathbf{b}$,

So, $\quad \mathbf{AD}=2\mathbf{b}$ and $\mathbf{ED}=\mathbf{a}$

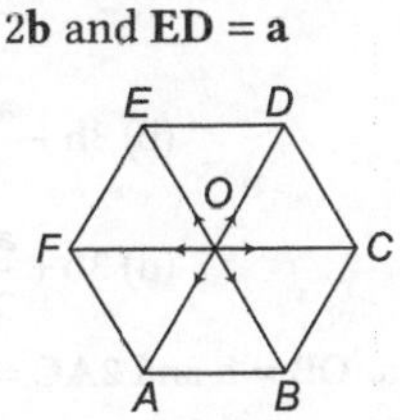

Now, $\quad \mathbf{AE}+\mathbf{ED}=\mathbf{AD}$

$\Rightarrow \quad \mathbf{AE}=\mathbf{AD}-\mathbf{ED}=2\mathbf{b}-\mathbf{a}$

Ex. 19 *If* $\mathbf{a}+\mathbf{b}+\mathbf{c}=0$ *and* $|\mathbf{a}|=3, |\mathbf{b}|=5, |\mathbf{c}|=7$, *then the angle between* **a** *and* **b** *is*

(a) $\frac{\pi}{2}$ (b) $\frac{\pi}{3}$

(c) $\frac{\pi}{4}$ (d) $\frac{\pi}{6}$

Sol. (b)

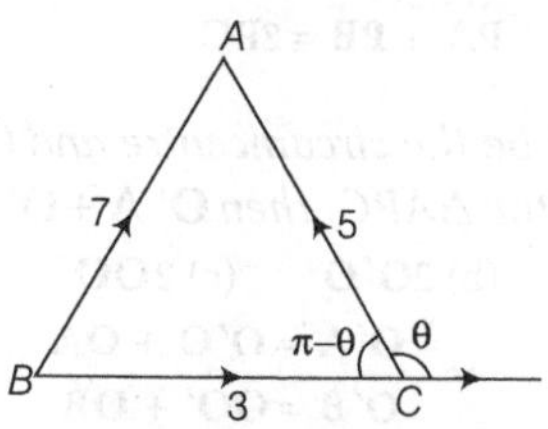

Let θ be the angle between **a** and **b**. Then, $\angle C = \pi - \theta$.

$$\therefore \quad \cos(\pi - \theta) = \frac{3^2 + 5^2 - 7^2}{2(3)(5)}$$

$$\therefore \quad -\cos\theta = \frac{-1}{2}$$

$$\therefore \quad \theta = 60^\circ = \frac{\pi}{3}$$

Ex. 20 *If* **a** *and* **b** *are the position vectors of A and B respectively, then the position vector of a point C on AB produced such that* $\mathbf{AC}=3\mathbf{AB}$ *is*

(a) $3\mathbf{a}-\mathbf{b}$ (b) $3\mathbf{b}-\mathbf{a}$

(c) $3\mathbf{a}-2\mathbf{b}$ (d) $3\mathbf{b}-2\mathbf{a}$

Sol. (d) Since, given that $\mathbf{AC}=3\mathbf{AB}$. It means that point *C* divides *AB* externally.

Thus, $\mathbf{AC}:\mathbf{BC}=3:2$

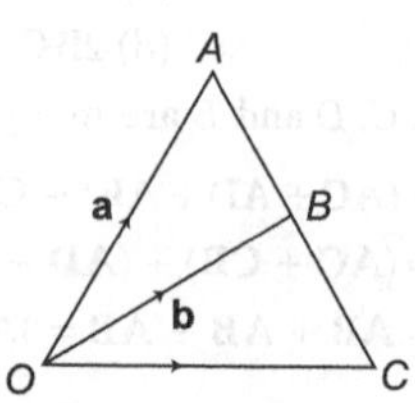

$$\text{Hence,} \quad \mathbf{OC} = \frac{3\cdot\mathbf{b}-2\cdot\mathbf{a}}{3-2} = 3\mathbf{b}-2\mathbf{a}$$

Ex. 21 *Let A and B be points with position vectors* **a** *and* **b** *with respect to the origin O. If the point C on OA is such that* $2AC=CO$, *CD is parallel to OB and* $|\mathbf{CD}|=3|\mathbf{OB}|$, *then* **AD** *is equal to*

(a) $3\mathbf{b}-\frac{\mathbf{a}}{2}$ (b) $3\mathbf{b}+\frac{\mathbf{a}}{2}$

(c) $3\mathbf{b}-\frac{\mathbf{a}}{3}$ (d) $3\mathbf{b}+\frac{\mathbf{a}}{3}$

Sol. (c) Since, $\mathbf{OA}=\mathbf{a}$, $\mathbf{OB}=\mathbf{b}$ and $2\mathbf{AC}=\mathbf{CO}$

By section formula, $\mathbf{OC}=\frac{2}{3}\mathbf{a}$

Therefore, $|\mathbf{CD}|=3|\mathbf{OB}|$

$\Rightarrow \quad \mathbf{CD}=3\mathbf{b}$

$\Rightarrow \quad \mathbf{OD}=\mathbf{OC}+\mathbf{CD}=\frac{2}{3}\mathbf{a}+3\mathbf{b}$

Hence, $\quad \mathbf{AD}=\mathbf{OD}-\mathbf{OA}=\frac{2}{2}\mathbf{a}+3\mathbf{b}-\mathbf{a}$

$$=3\mathbf{b}-\frac{1}{3}\mathbf{a}$$

Ex. 22 *If position vectors of a point A is* $\mathbf{a}+2\mathbf{b}$ *and* **a** *divides AB in the ratio* 2 : 3, *then the position vector of B is*

(a) $2\mathbf{a}-\mathbf{b}$ (b) $\mathbf{b}-2\mathbf{a}$

(c) $\mathbf{a}-3\mathbf{b}$ (d) $\mathbf{b}$

Sol. (c) If *x* be the position vector of *B*, then **a** divides *AB* in the ratio 2 : 3.

$$\therefore \quad \mathbf{a} = \frac{2x + 3(\mathbf{a}+2\mathbf{b})}{2+3}$$

$\Rightarrow \quad 5\mathbf{a}-3\mathbf{a}-6\mathbf{b}=2x$

$\Rightarrow \quad x=\mathbf{a}-3\mathbf{b}$

Ex. 23 *If D, E and F are respectively, the mid-points of AB, AC and BC in* ΔABC, *then* $\mathbf{BE}+\mathbf{AF}$ *is equal to*

(a) **DC** (b) $\frac{1}{2}\mathbf{BF}$

(c) $2\mathbf{BF}$ (d) $\frac{3}{2}\mathbf{BF}$

Sol. (a) $\mathbf{BE}+\mathbf{AF}=\mathbf{OE}-\mathbf{OB}+\mathbf{OF}-\mathbf{OA}$

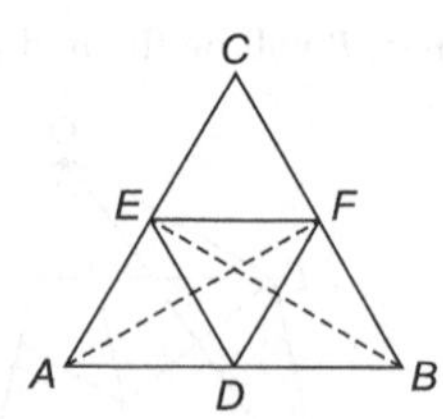

$$=\frac{\mathbf{OA}+\mathbf{OC}}{2}-\mathbf{OB}+\frac{\mathbf{OB}+\mathbf{OC}}{2}-\mathbf{OA}$$

$$=\mathbf{OC}-\frac{\mathbf{OA}+\mathbf{OB}}{2}=\mathbf{OC}-\mathbf{OD}=\mathbf{DC}$$

Ex. 24 *In a quadrilateral PQRS,* $\mathbf{PQ}=a$, $\mathbf{QR}=b$, $\mathbf{SP}=a-b$. *If M is the mid-point of QR and X is a point of SM such that,* $SX=\frac{4}{5}SM$, *then*

(a) $\mathbf{PX}=\frac{1}{5}\mathbf{PR}$

(b) $\mathbf{PX}=\frac{3}{5}\mathbf{PR}$

(c) $\mathbf{PX}=\frac{2}{5}\mathbf{PR}$

(d) None of the above

Sol. (b) If we take point P as the origin, the position vectors of Q and S are $\mathbf{a}$ and $\mathbf{b}-\mathbf{a}$ respectively.

In ΔPQR, we have

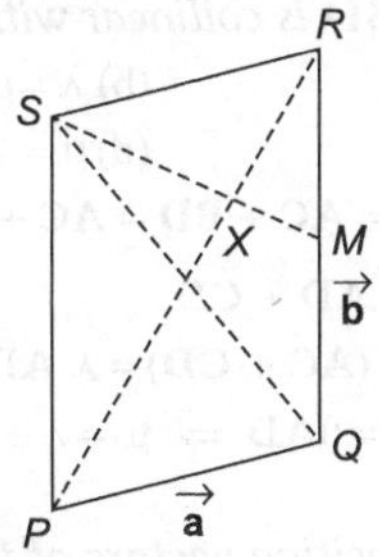

$$\mathbf{PR}=\mathbf{PQ}+\mathbf{QR} \Rightarrow \mathbf{PR}=\mathbf{a}+\mathbf{b}$$

$\therefore$ Position vector of $R = \mathbf{a}+\mathbf{b}$

$$\Rightarrow \quad \text{PV of } M=\frac{\mathbf{a}+(\mathbf{a}+\mathbf{b})}{2}=\left(\mathbf{a}+\frac{1}{2}\mathbf{b}\right)$$

Now, $\quad SX=\frac{4}{5}SM$

$$\Rightarrow \quad XM=SM-SX=SM-\frac{4}{5}SM=\frac{1}{5}SM$$

$$\therefore \quad SX:XM=4:1$$

$$\Rightarrow \quad \text{PV of } X=\frac{4\left(\mathbf{a}+\frac{1}{2}\mathbf{b}\right)+1(\mathbf{b}-\mathbf{a})}{4+1}$$

$$=\frac{3\mathbf{a}+2\mathbf{b}}{5} \Rightarrow \mathbf{PX}=\frac{3}{5}(\mathbf{a}+\mathbf{b})$$

$$\Rightarrow \quad \mathbf{PX}=\frac{3}{5}\mathbf{PR}$$

Ex. 25 *Orthocentre of an equilateral triangle ABC is the origin O. If* $\mathbf{OA}=\mathbf{a}, \mathbf{OB}=\mathbf{b}, \mathbf{OC}=\mathbf{c}$, *then* $\mathbf{AB}+2\mathbf{BC}+3\mathbf{CA}$ *is equal to*

(a) $3\mathbf{c}$ (b) $3\mathbf{a}$

(c) $\mathbf{0}$ (d) $3\mathbf{b}$

Sol. (b) For an equilateral triangle, centroid is the same as orthocentre

$$\therefore \quad \frac{\mathbf{OA}+\mathbf{OB}+\mathbf{OC}}{3}=\mathbf{0}$$

$$\therefore \quad \mathbf{OA}+\mathbf{OB}+\mathbf{OC}=\mathbf{0}$$

Now, $\quad \mathbf{AB}+2\mathbf{BC}+3\mathbf{CA}$

$$=\mathbf{OB}-\mathbf{OA}+2\mathbf{OC}-2\mathbf{OB}+3\mathbf{OA}-3\mathbf{OC}$$

$$=-\mathbf{OB}+2\mathbf{OA}-\mathbf{OC}$$

$$=-(\mathbf{OB}+\mathbf{OA}+\mathbf{OC})+3\mathbf{OA}=3\mathbf{OA}=3\mathbf{a}$$

Ex. 26 *If* $\mathbf{a},\mathbf{b}$, *and* $\mathbf{c}$ *are position vector of A, B and C respectively of* ΔABC *and if* $|\mathbf{a}-\mathbf{b}|=4, |\mathbf{b}-\mathbf{c}|=2, |\mathbf{c}-\mathbf{a}|=3$, *then the distance between the centroid and incentre of* ΔABC *is*

(a) 1 (b) $\frac{1}{2}$

(c) $\frac{1}{3}$ (d) $\frac{2}{3}$

Sol. (c) Let G be centroid and I be incenter.

$$|\mathbf{GI}|=|\mathbf{OI}-\mathbf{OG}|=\left|\frac{2\mathbf{a}+3\mathbf{b}+4\mathbf{c}}{9}-\frac{\mathbf{a}+\mathbf{b}+\mathbf{c}}{3}\right|$$

$$=\left|\frac{-\mathbf{a}+\mathbf{c}}{9}\right|=\frac{3}{9}=\frac{1}{3}$$

Ex. 27 *Let position vector of points A, B and C of triangle* ΔABC *respectively be* $\hat{\mathbf{i}}+\hat{\mathbf{j}}+2\hat{\mathbf{k}}, \hat{\mathbf{i}}+2\hat{\mathbf{j}}+\hat{\mathbf{k}}$ *and* $2\hat{\mathbf{i}}+\hat{\mathbf{j}}+\hat{\mathbf{k}}$. *Let* l_1, l_2 *and* l_3 *be the lengths of perpendiculars drawn from the orthocenter 'O' on the sides AB, BC and CA, then* $(l_1+l_2+l_3)$ *equals*

(a) $\frac{2}{\sqrt{6}}$ (b) $\frac{3}{\sqrt{6}}$

(c) $\frac{\sqrt{6}}{2}$ (d) $\frac{\sqrt{6}}{3}$

Sol. (c)

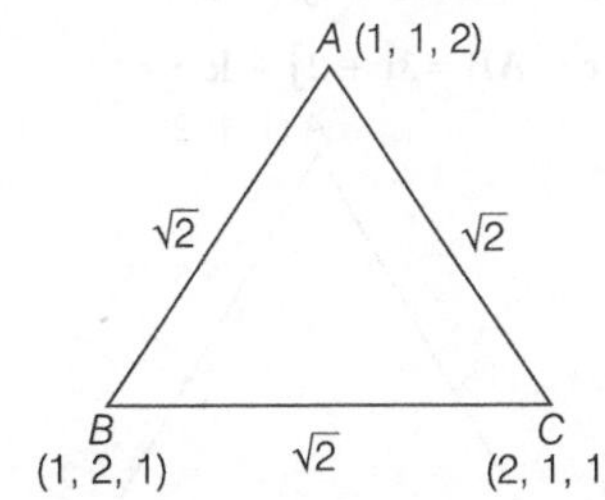

Clearly, triangle formed by the given points $\hat{\mathbf{i}}+\hat{\mathbf{j}}+2\hat{\mathbf{k}}$, $\hat{\mathbf{i}}+2\hat{\mathbf{j}}+\hat{\mathbf{k}}$ and $2\hat{\mathbf{i}}+\hat{\mathbf{j}}+\hat{\mathbf{k}}$ is equilateral as $AB=BC=AC=\sqrt{2}$.

$\therefore$ Distance of orthcentre 'O' from the sides is equal to inradius of the triangle.

$$\therefore l_1=l_2=l_3=\text{inradius}=r=\frac{\Delta}{s}=\frac{\frac{\sqrt{3}}{4}(\sqrt{2})^2}{\frac{3}{2}(\sqrt{2})}=\frac{1}{\sqrt{6}}$$

$$\Rightarrow \quad (l_1+l_2+l_3)=\frac{3}{\sqrt{6}}=\frac{\sqrt{6}}{2}$$

Ex. 28 *ABCDEF is a regular hexagon in the XY-plane with vertices in the anticlockwise direction. If* $\mathbf{AB}=2\hat{\mathbf{i}}$, *then* $\mathbf{CD}$ *is*

(a) $\hat{\mathbf{i}}+3\hat{\mathbf{j}}$ (b) $\hat{\mathbf{i}}+2\hat{\mathbf{j}}$

(c) $-\hat{\mathbf{i}}+3\hat{\mathbf{j}}$ (d) None of these

Sol.

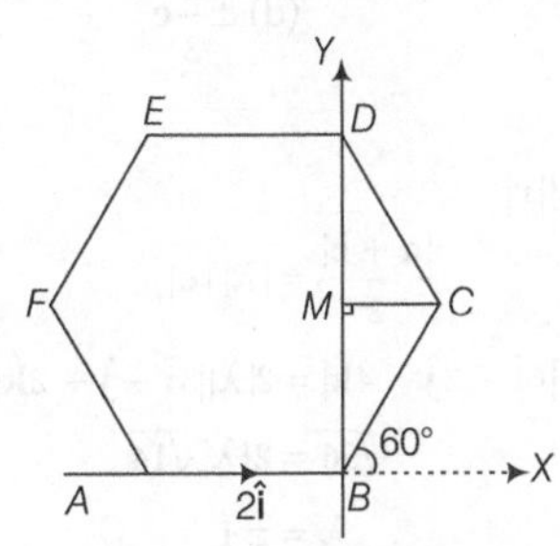

AB is along the X-axis and BD is along the Y-axis.

$\mathbf{AB} = 2\hat{\mathbf{i}} \Rightarrow AB = BC = CD = \ldots\ldots\ldots\ldots = 2$

From the figure, $BM = BC\sin 60° = 2\sin 60° = \sqrt{3}$

$$\therefore \quad \mathbf{BD} = 2\sqrt{3}\hat{\mathbf{j}}$$

$$\mathbf{BC} = BC\cos 60°\hat{\mathbf{i}} + BC\sin 60°\hat{\mathbf{j}} = \hat{\mathbf{i}} + \sqrt{3}\hat{\mathbf{j}}$$

$$\mathbf{CD} = \mathbf{BD} - \mathbf{BC} = 2\sqrt{3}\hat{\mathbf{j}} - (\hat{\mathbf{i}} + \sqrt{3}\hat{\mathbf{j}}) = -\hat{\mathbf{i}} + \sqrt{3}\hat{\mathbf{j}}$$

Ex. 29 *The vertices of triangle are $A(1, 1, 2)$, $B(4, 3, 1)$ and $C(2, 3, 5)$. A vector representing the internal bisector of the $\angle A$ is*

(a) $\hat{\mathbf{i}} + \hat{\mathbf{j}} + 2\hat{\mathbf{k}}$ (b) $2\hat{\mathbf{i}} - 2\hat{\mathbf{j}} + \hat{\mathbf{k}}$

(c) $2\hat{\mathbf{i}} + 2\hat{\mathbf{j}} + \hat{\mathbf{k}}$ (d) None of these

Sol. (c) From the figure, we have

$$\mathbf{b} = \mathbf{AC} = \hat{\mathbf{i}} + 2\hat{\mathbf{j}} + 3\hat{\mathbf{k}}$$

and $$\mathbf{c} = \mathbf{AB} = 3\hat{\mathbf{i}} + 2\hat{\mathbf{j}} + \hat{\mathbf{k}}$$

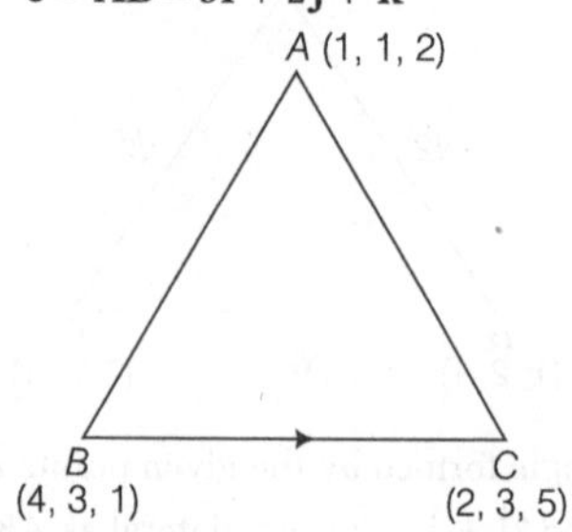

$\therefore$ Unit vector along the bisector of $\angle A$ is given by

$$= \frac{\mathbf{b}+\mathbf{c}}{2} = \frac{(\hat{\mathbf{i}} + 2\hat{\mathbf{j}} + 3\hat{\mathbf{k}}) + (3\hat{\mathbf{i}} + 2\hat{\mathbf{j}} - \hat{\mathbf{k}})}{\sqrt{14}}$$

$$= \frac{2\hat{\mathbf{i}} + 2\hat{\mathbf{j}} + \hat{\mathbf{k}}}{\sqrt{14}}$$

$\therefore$ Any vector along the angle bisector of

$$\angle A = 2\hat{\mathbf{i}} + 2\hat{\mathbf{j}} + \hat{\mathbf{k}}$$

Ex. 30 *Let $\mathbf{a} = (1, 1, -1)$, $\mathbf{b} = (5, -3, -3)$ and $\mathbf{c} = (3, -1, 2)$. If $\mathbf{r}$ is collinear with $\mathbf{c}$ and has length $\frac{|\mathbf{a}+\mathbf{b}|}{2}$, then $\mathbf{r}$ equals*

(a) $\pm 3\mathbf{c}$ (b) $\pm\frac{3}{2}\mathbf{c}$

(c) $\pm\mathbf{c}$ (d) $\pm\frac{2}{3}\mathbf{c}$

Sol. (c) Let $\mathbf{r} = \lambda\mathbf{c}$

Given $|\mathbf{r}| = |\lambda||\mathbf{c}|$

$$\therefore \quad \frac{|\mathbf{a}+\mathbf{b}|}{2} = |\lambda||\mathbf{c}|$$

$$\therefore \quad |6\hat{\mathbf{i}} - 2\hat{\mathbf{j}} - 4\hat{\mathbf{k}}| = 2|\lambda||3\hat{\mathbf{i}} - \hat{\mathbf{j}} + 2\hat{\mathbf{k}}|$$

$$\therefore \quad \sqrt{56} = 2|\lambda|\sqrt{14}$$

$$\therefore \quad \lambda = \pm 1$$

$$\therefore \quad \mathbf{r} = \pm\mathbf{c}$$

Ex. 31 *In a trapezium, the vector $\mathbf{BC} = \lambda\mathbf{AD}$. We will then find that $\mathbf{p} = \mathbf{AC} + \mathbf{BD}$ is collinear with $\mathbf{AD}$. If $\mathbf{p} = \mu\,\mathbf{AD}$, then*

(a) $\mu = \lambda + 1$ (b) $\lambda = \mu + 1$

(c) $\lambda + \mu = 1$ (d) $\mu = 2 + \lambda$

Sol. (a) We have, $\mathbf{p} = \mathbf{AC} + \mathbf{BD} = \mathbf{AC} + \mathbf{BC} + \mathbf{CD}$

$$= \mathbf{AC} + \lambda\mathbf{AD} + \mathbf{CD}$$

$$= \lambda\mathbf{AD} + (\mathbf{AC} + \mathbf{CD}) = \lambda\,\mathbf{AD} + \mathbf{AD} = (\lambda + 1)\mathbf{AD}$$

Therefore, $\mathbf{p} = \mu\mathbf{AD} \Rightarrow \mu = \lambda + 1$

Ex. 32 *If the position vectors of the points A, B and C be $\hat{\mathbf{i}} + \hat{\mathbf{j}}$, $\hat{\mathbf{i}} - \hat{\mathbf{j}}$ and $a\hat{\mathbf{i}} + b\hat{\mathbf{j}} + c\hat{\mathbf{k}}$ respectively, then the points A, B and C are collinear, if*

(a) $a = b = c = 1$

(b) $a = 1$, b and c are arbitrary scalars

(c) $a = b = c = 0$

(d) $c = 0$, $a = 1$ and b is arbitrary scalars

Sol. (d) Here, $\mathbf{AB} = -2\hat{\mathbf{j}}$, $\mathbf{BC} = (a-1)\hat{\mathbf{i}} + (b+1)\hat{\mathbf{j}} + c\hat{\mathbf{k}}$

The points are collinear, then $\mathbf{AB} = k(\mathbf{BC})$

$$-2\hat{\mathbf{j}} = k\{(a-1)\hat{\mathbf{i}} + (b+1)\hat{\mathbf{j}} + c\hat{\mathbf{k}}\}$$

On comparing, $k(a-1) = 0$, $k(b+1) = -2$, $kc = 0$

Hence, $c = 0$, $a = 1$ and b is arbitrary scalar.

Ex. 33 *Let a, b and c be distinct non-negative numbers and the vectors $a\hat{\mathbf{i}} + a\hat{\mathbf{j}} + c\hat{\mathbf{k}}$, $\hat{\mathbf{i}} + \hat{\mathbf{k}}$, $c\hat{\mathbf{i}} + c\hat{\mathbf{j}} + b\hat{\mathbf{k}}$ lie in a plane, then the quadratic equation $ax^2 + 2cx + b = 0$ has*

(a) real and equal roots

(b) real and unequal roots

(c) unreal roots

(d) both roots real and positive

Sol. (a) $a\hat{\mathbf{i}} + a\hat{\mathbf{j}} + c\hat{\mathbf{k}}$, $\hat{\mathbf{i}} + \hat{\mathbf{k}}$ and $c\hat{\mathbf{i}} + c\hat{\mathbf{j}} + b\hat{\mathbf{k}}$ are coplanar

$$\therefore \quad \begin{vmatrix} a & a & c \\ 1 & 0 & 1 \\ c & c & b \end{vmatrix} = 0 \Rightarrow c^2 - ab = 0$$

For, equation $ax^2 + 2cx + b = 0$

$$D = 4c^2 - 4ab = 0$$

So, roots are real and equal.

Ex. 34 *The number of distinct real values of λ for which the vectors $\lambda^3\hat{\mathbf{i}} + \hat{\mathbf{k}}$, $\hat{\mathbf{i}} - \lambda^3\hat{\mathbf{j}}$ and $\hat{\mathbf{i}} + (2\lambda - \sin\lambda)\hat{\mathbf{j}} - \lambda\hat{\mathbf{k}}$ are coplanar is*

(a) 0 (b) 1

(c) 2 (d) 3

Sol. (a) Put $\Delta = 0 \Rightarrow \lambda^7 + \lambda^3 + 2\lambda - \sin\lambda = 0$

Let $f(\lambda) = \lambda^7 + \lambda^3 + 2\lambda - \sin\lambda$

$\Rightarrow f'(\lambda) = (7\lambda^6 + 3\lambda^2 + 2 - \cos\lambda) > 0, \forall \in R$

$\therefore \quad f(\lambda) = 0$ has only one real solution $\lambda = 0$.

Ex. 35 *The points $A(2-x, 2, 2)$, $B(2, 2-y, 2)$, $C(2, 2, 2-z)$ and $D(1, 1, 1)$ are coplanar, then locus of $P(x, y, z)$ is*

(a) $\frac{1}{x}+\frac{1}{y}+\frac{1}{z}=1$ (b) $x+y+z=1$

(c) $\frac{1}{1-x}+\frac{1}{1-y}+\frac{1}{1-z}=1$ (d) None of these

Sol. (a) Here, $\mathbf{AB} = x\hat{\mathbf{i}} - y\hat{\mathbf{j}}$

$$\mathbf{AC} = x\hat{\mathbf{i}} - z\hat{\mathbf{k}};\ \mathbf{AD} = (x-1)\hat{\mathbf{i}} - \hat{\mathbf{j}} - \hat{\mathbf{k}}$$

As, these vectors are coplanar

$$\Rightarrow \begin{vmatrix} x & -y & 0 \\ x & 0 & -z \\ x-1 & -1 & -1 \end{vmatrix} = 0 \Rightarrow \frac{1}{x}+\frac{1}{y}+\frac{1}{z}=1$$

Ex. 36 $\mathbf{p}=2\mathbf{a}-3\mathbf{b}$, $\mathbf{q}=\mathbf{a}-2\mathbf{b}+\mathbf{c}$ *and* $\mathbf{r}=-3\mathbf{a}+\mathbf{b}+2\mathbf{c}$, *where* $\mathbf{a}, \mathbf{b}, \mathbf{c}$ *being non-zero non-coplanar vectors, then the vector* $-2\mathbf{a}+3\mathbf{b}-\mathbf{c}$ *is equal to*

(a) $\mathbf{p}-4\mathbf{q}$ (b) $\frac{-7\mathbf{q}+\mathbf{r}}{5}$

(c) $2\mathbf{p}-3\mathbf{q}+\mathbf{r}$ (d) $4\mathbf{p}-2\mathbf{r}$

Sol. (b) Let $-2\mathbf{a}+3\mathbf{b}-\mathbf{c} = x\mathbf{p}+y\mathbf{q}+z\mathbf{r}$

$$\Rightarrow -2\mathbf{a}+3\mathbf{b}-\mathbf{c} = (2x+y-3z)\mathbf{a} + (-3x-2y+z)\mathbf{b} + (y+2z)\mathbf{c}$$

$$\therefore\ 2x+y-3z=-2,\ -3x-2y+z=3$$

and $$y+2z=-1$$

On solving these, we get $x=0$, $y=-\frac{7}{5}$, $z=\frac{1}{5}$

$$\therefore\ -2\mathbf{a}+3\mathbf{b}-\mathbf{c} = \frac{(-7\mathbf{q}+\mathbf{r})}{5}$$

Trick Check alternates one-by-one

i.e. (a) $\mathbf{p}-4\mathbf{q} = -2\mathbf{a}+5\mathbf{b}-4\mathbf{c}$

(b) $\frac{-7\mathbf{q}+\mathbf{r}}{5} = -2\mathbf{a}+3\mathbf{b}-\mathbf{c}$

Ex. 37 *If a_1 and a_2 are two values of a for which the unit vector $a\hat{\mathbf{i}}+b\hat{\mathbf{j}}+\frac{1}{2}\hat{\mathbf{k}}$ is linearly dependent with $\hat{\mathbf{i}}+2\hat{\mathbf{j}}$ and $\hat{\mathbf{j}}-2\hat{\mathbf{k}}$, then $\frac{1}{a_1}+\frac{1}{a_2}$ is equal to*

(a) 1 (b) $\frac{1}{8}$ (c) $\frac{-16}{11}$ (d) $\frac{-11}{16}$

Sol. (c) $a\hat{\mathbf{i}}+b\hat{\mathbf{j}}+\frac{1}{2}\hat{\mathbf{k}} = l(\hat{\mathbf{i}}+2\hat{\mathbf{j}}) + m(\hat{\mathbf{j}}-2\hat{\mathbf{k}})$

$$\Rightarrow a=l,\ b=2l+m \text{ and } m=\frac{-1}{4}$$

$a\hat{\mathbf{i}}+b\hat{\mathbf{j}}+\frac{1}{2}\hat{\mathbf{k}}$ is unit vector

$$\therefore\ a^2+b^2=\frac{3}{4} \Rightarrow 5a^2-a-\frac{11}{16}=0$$

a_1 and a_2 are roots of above equation

$$\Rightarrow \frac{1}{a_1}+\frac{1}{a_2}=\frac{a_1+a_2}{a_1a_2}=-\frac{16}{11}$$

JEE Type Solved Examples : More than One Correct Option Type Questions

Ex. 38 *The vector $\hat{\mathbf{i}}+x\hat{\mathbf{j}}+3\hat{\mathbf{k}}$ is rotated through an angle θ and is doubled in magnitude. It now becomes $4\hat{\mathbf{i}}+(4x-2)\hat{\mathbf{j}}+2\hat{\mathbf{k}}$. The values of x are*

(a) 1 (b) $\frac{-2}{3}$

(c) 2 (d) $\frac{4}{3}$

Sol. (b,c) Let $\alpha = \hat{\mathbf{i}}+x\hat{\mathbf{j}}+3\hat{\mathbf{k}}$,

$$\beta = 4\hat{\mathbf{i}}+(4x-2)\hat{\mathbf{j}}+2\hat{\mathbf{k}}$$

Given, $2|\alpha| = |\beta|$

$$\Rightarrow 2\sqrt{10+x^2} = \sqrt{20+4(2x-1)^2}$$

$$\Rightarrow 10+x^2 = 5+(4x^2-4x+1)$$

$$\Rightarrow 3x^2-4x-4=0$$

$$\Rightarrow x=2, -\frac{2}{3}$$

Ex. 39 $\mathbf{a}, \mathbf{b}$ *and* $\mathbf{c}$ *are three coplanar unit vectors such that* $\mathbf{a}+\mathbf{b}+\mathbf{c}=0$. *If three vectors* $\mathbf{p}, \mathbf{q}$ *and* $\mathbf{r}$ *are parallel to* $\mathbf{a}, \mathbf{b}$ *and* $\mathbf{c}$ *respectively, and have integral but different magnitudes, then among the following options,* $|\mathbf{p}+\mathbf{q}+\mathbf{r}|$ *can take a value equal to*

(a) 1 (b) 0 (c) $\sqrt{3}$ (d) 2

Sol. (c,d) Let $\mathbf{a}, \mathbf{b}$ and $\mathbf{c}$ lie in the XY-plane.

Let $\mathbf{a}=\hat{\mathbf{i}}$, $\mathbf{b}=-\frac{1}{2}\hat{\mathbf{i}}+\frac{\sqrt{3}}{2}\hat{\mathbf{j}}$ and $\mathbf{c}=-\frac{1}{2}\hat{\mathbf{i}}-\frac{\sqrt{3}}{2}\hat{\mathbf{j}}$

Therefore, $|\mathbf{p}+\mathbf{q}+\mathbf{r}| = |\lambda\mathbf{a}+\mu\mathbf{b}+\nu\mathbf{c}|$

$$= \left|\lambda\hat{\mathbf{i}}+\mu\left(-\frac{1}{2}\hat{\mathbf{i}}+\frac{\sqrt{3}}{2}\hat{\mathbf{j}}\right)+\nu\left(-\frac{1}{2}\hat{\mathbf{i}}-\frac{\sqrt{3}}{2}\hat{\mathbf{j}}\right)\right|$$

$$= \left|\left(\lambda-\frac{\mu}{2}-\frac{\nu}{2}\right)\hat{\mathbf{i}}+\frac{\sqrt{3}}{2}(\mu-\nu)\hat{\mathbf{j}}\right|$$

$$= \sqrt{\left(\lambda-\frac{\mu}{2}-\frac{\nu}{2}\right)^2+\frac{3}{4}(\mu-\nu)^2}$$

$$= \sqrt{\lambda^2 + \mu^2 + \nu^2 - \lambda\mu - \lambda\nu - \mu\nu}$$

$$= \frac{1}{\sqrt{2}}\sqrt{(\lambda-\mu)^2 + (\mu-\nu)^2 + (\nu-\lambda)^2}$$

$$= \frac{1}{\sqrt{2}}\sqrt{1+1+4} = \sqrt{3}$$

$\Rightarrow$ $|\mathbf{p}+\mathbf{q}+\mathbf{r}|$ can take a value equal to $\sqrt{3}$ and 2.

Ex. 40 *A, B, C and D are four points such that* $\mathbf{AB} = m(2\hat{\mathbf{i}} - 6\hat{\mathbf{j}} + 2\hat{\mathbf{k}}), \mathbf{BC} = (\hat{\mathbf{i}} - 2\hat{\mathbf{j}})$ *and* $\mathbf{CD} = n(-6\hat{\mathbf{i}} + 15\hat{\mathbf{j}} - 3\hat{\mathbf{k}})$. *If CD intersects AB at some point E, then*

(a) $m \geq \frac{1}{2}$ (b) $n \geq \frac{1}{3}$ (c) $m = n$ (d) $m < n$

Sol. (a, b) Let $\mathbf{EB} = p\,\mathbf{AB}$ and $\mathbf{CE} = q\,\mathbf{CD}$

Then $0 < p$ and $q \leq 1$

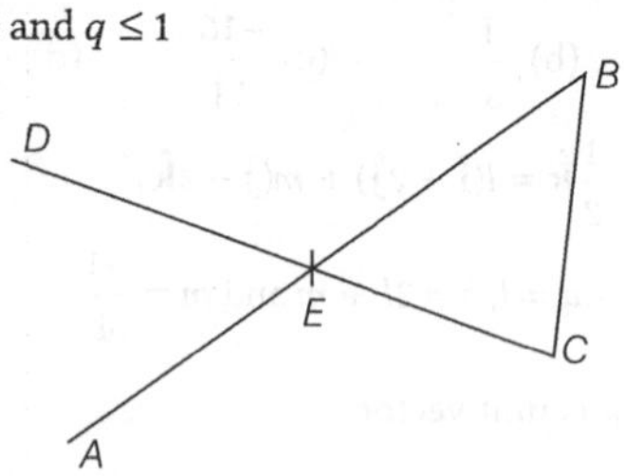

Since, $\mathbf{EB} + \mathbf{BC} + \mathbf{CE} = \mathbf{0}$

$$pm(2\hat{\mathbf{i}} - 6\hat{\mathbf{j}} + 2\hat{\mathbf{k}}) + (\hat{\mathbf{i}} - 2\hat{\mathbf{j}}) + qn(-6\hat{\mathbf{i}} + 15\hat{\mathbf{j}} - 3\hat{\mathbf{k}}) = \mathbf{0}$$

$$\Rightarrow (2pm + 1 - 6qn)\hat{\mathbf{i}} + (-6pm - 2 + 15qn)\hat{\mathbf{j}} + (2pm - 6qn)\hat{\mathbf{k}} = \mathbf{0}$$

$$\Rightarrow \quad 2pm - 6qn + 1 = 0,$$
$$-6pm - 2 + 15qn = 0$$
$$2pm - 6qn = 0$$

Solving these, we get

$$p = \frac{1}{(2m)} \quad \text{and} \quad q = \frac{1}{(3n)}$$

$$\therefore \quad 0 < \frac{1}{(2m)} \leq 1 \quad \text{and} \quad 0 < \frac{1}{(3n)} \leq 1$$

$$\Rightarrow \quad m \geq \frac{1}{2} \quad \text{and} \quad n \geq \frac{1}{3}$$

Ex. 41 *If non-zero vectors* **a** *and* **b** *are equally inclined to coplanar vector* **c**, *then* **c** *can be*

(a) $\frac{|\mathbf{a}|}{|\mathbf{a}| + 2|\mathbf{b}|}\mathbf{a} + \frac{|\mathbf{b}|}{|\mathbf{a}| + |\mathbf{b}|}\mathbf{b}$

(b) $\frac{|\mathbf{b}|}{|\mathbf{a}| + |\mathbf{b}|}\mathbf{a} + \frac{|\mathbf{a}|}{|\mathbf{a}| + |\mathbf{b}|}\mathbf{b}$

(c) $\frac{|\mathbf{a}|}{|\mathbf{a}| + 2|\mathbf{b}|}\mathbf{a} + \frac{|\mathbf{b}|}{|\mathbf{a}| + 2|\mathbf{b}|}\mathbf{b}$

(d) $\frac{|\mathbf{b}|}{2|\mathbf{a}| + |\mathbf{b}|}\mathbf{a} + \frac{|\mathbf{a}|}{2|\mathbf{a}| + |\mathbf{b}|}\mathbf{b}$

Sol. (b,d) Since, **a** and **b** are equally inclined to **c**, therefore **c** must be of the form $t\left(\frac{\mathbf{a}}{|\mathbf{a}|} + \frac{\mathbf{b}}{|\mathbf{b}|}\right)$

Now, $\frac{|\mathbf{b}|}{|\mathbf{a}| + |\mathbf{b}|}\mathbf{a} + \frac{|\mathbf{a}|}{|\mathbf{a}| + |\mathbf{b}|}\mathbf{b} = \frac{|\mathbf{a}|\,|\mathbf{a}|}{|\mathbf{a}| + |\mathbf{a}|}\left(\frac{\mathbf{a}}{|\mathbf{a}|} + \frac{\mathbf{b}}{|\mathbf{b}|}\right)$

Also, $\frac{|\mathbf{b}|}{2|\mathbf{a}| + |\mathbf{b}|}\mathbf{a} + \frac{|\mathbf{a}|}{2|\mathbf{a}| + |\mathbf{b}|}\mathbf{b} = \frac{|\mathbf{a}|\,|\mathbf{b}|}{2|\mathbf{a}| + |\mathbf{b}|}\left(\frac{\mathbf{a}}{|\mathbf{a}|} + \frac{\mathbf{b}}{|\mathbf{b}|}\right)$

Other two vectors cannot be written in the from $t\left(\frac{\mathbf{a}}{|\mathbf{a}|} + \frac{\mathbf{b}}{|\mathbf{b}|}\right)$

Ex. 42 *The vectors* $x\hat{\mathbf{i}} + (x+1)\hat{\mathbf{j}} + (x+2)\hat{\mathbf{k}}$, $(x+3)\hat{\mathbf{i}} + (x+4)\hat{\mathbf{j}} + (x+5)\hat{\mathbf{k}}$ *and* $(x+6)\hat{\mathbf{i}} + (x+7)\hat{\mathbf{j}} + (x+8)\hat{\mathbf{k}}$ *are coplanar if x is equal to*

(a) 1 (b) −3 (c) 4 (d) 0

Sol. (a, b, c, d)

$x\hat{\mathbf{i}} + (x+1)\hat{\mathbf{j}} + (x+2)\hat{\mathbf{k}}$, $(x+3)\hat{\mathbf{i}} + (x+4)\hat{\mathbf{j}} + (x+5)\hat{\mathbf{k}}$ and $(x+6)\hat{\mathbf{i}} + (x+7)\hat{\mathbf{j}} + (x+8)\hat{\mathbf{k}}$ are coplanar. We have determinant of their coefficients as $\begin{vmatrix} x & x+1 & x+2 \\ x+3 & x+4 & x+5 \\ x+6 & x+7 & x+8 \end{vmatrix}$

Applying $C_2 \to C_2 - C_1$ and $C_3 \to C_3 - C_1$, we have

$\begin{vmatrix} x & 1 & 2 \\ x+3 & 1 & 2 \\ x+6 & 1 & 2 \end{vmatrix} = 0$. Hence, $x \in R$.

Ex. 43 *Given three vectors* **a**, **b**, *and* **c** *are non-zero and non-coplanar vectors. Then which of the following are coplanar.*

(a) $\mathbf{a}+\mathbf{b}, \mathbf{b}+\mathbf{c}, \mathbf{c}+\mathbf{a}$ (b) $\mathbf{a}-\mathbf{b}, \mathbf{b}+\mathbf{c}, \mathbf{c}+\mathbf{a}$

(c) $\mathbf{a}+\mathbf{b}, \mathbf{b}-\mathbf{c}, \mathbf{c}+\mathbf{a}$ (d) $\mathbf{a}+\mathbf{b}, \mathbf{b}+\mathbf{c}, \mathbf{c}-\mathbf{a}$

Sol. (b, c, d) $\mathbf{c}+\mathbf{a} = (\mathbf{b}+\mathbf{c}) + (\mathbf{a}-\mathbf{b})$

$\mathbf{a}+\mathbf{b} = (\mathbf{b}-\mathbf{c}) + (\mathbf{c}+\mathbf{a})$

$\mathbf{a}+\mathbf{c} = (\mathbf{a}+\mathbf{b}) + (\mathbf{c}-\mathbf{a})$

So, vectors in options (b), (c) and (d) are coplanar.

Ex. 44 *In a four-dimensional space where unit vectors along the axes are* $\hat{\mathbf{i}}, \hat{\mathbf{j}}, \hat{\mathbf{k}}$ *and* $\hat{\mathbf{i}}$, *and* $\mathbf{a}_1, \mathbf{a}_2, \mathbf{a}_3, \mathbf{a}_4$ *are four non-zero vectors such that no vector can be expressed as a linear combination of others and* $(\lambda-1)(\mathbf{a}_1 - \mathbf{a}_2) + \mu(\mathbf{a}_2 + \mathbf{a}_3) + \gamma(\mathbf{a}_3 + \mathbf{a}_4 - 2\mathbf{a}_2) + \mathbf{a}_3 + \delta\mathbf{a}_4 = 0$, *then*

(a) $\lambda = 1$ (b) $\mu = -\frac{2}{3}$ (c) $\gamma = \frac{2}{3}$ (d) $\delta = \frac{1}{3}$

Sol. (a, b, d)

$(\lambda-1)(\mathbf{a}_1 - \mathbf{a}_2) + \mu(\mathbf{a}_2 + \mathbf{a}_3) + \gamma(\mathbf{a}_3 + \mathbf{a}_4 - 2\mathbf{a}_2) + \mathbf{a}_3 + \delta\mathbf{a}_4 = \mathbf{0}$

i.e. $(\lambda-1)\mathbf{a}_1 + (1-\lambda+\mu-2\gamma)\mathbf{a}_2 + (\mu+\gamma+1)\mathbf{a}_3 + (\gamma+\delta)\mathbf{a}_4 = \mathbf{0}$

Since, $\mathbf{a}_1, \mathbf{a}_2, \mathbf{a}_3$ and $\mathbf{a}_4$ are linearly independent, we have

$$\lambda - 1 = 0, 1 - \lambda + \mu - 2\gamma = 0,$$
$$\mu + \gamma + 1 = 0 \quad \text{and} \quad \gamma + \delta = 0$$

i.e. $\lambda = 1, \mu = 2\gamma, \mu + \gamma + 1 = 0, \gamma + \delta = 0$

Hence, $\lambda = 1, \mu = -\frac{2}{3}, \gamma = -\frac{1}{3}, \delta = \frac{1}{3}$

JEE Type Solved Examples : Statement I and II Type Questions

Directions (Ex. Nos. 45-51) *This section is based on* ***Statement*** *I and* ***Statement*** *II. Select the correct answer from the codes given below.*

(a) Both Statement I and Statement II are correct and Statement II is the correct explanation of Statement I
(b) Both Statement I and Statement II are correct but Statement II is not the correct explanation of Statement I
(c) Statement I is correct but Statement II is incorrect
(d) Statement II is correct but Statement I is incorrect

Ex. 45 **Statement I** *If* $|\mathbf{a}|=3, |\mathbf{b}|=4$ *and* $|\mathbf{a}+\mathbf{b}|=5$, *then* $|\mathbf{a}-\mathbf{b}|=5$.

Statement II *The length of the diagonals of a rectangle is the same.*

Sol. (a) We have, adjacent sides of triangle $|\mathbf{a}|=3, |\mathbf{b}|=4$

The length of the diagonal is $|\mathbf{a}+\mathbf{b}|=5$

Since, it satisfies the Pythagoras theorem, $\mathbf{a} \perp \mathbf{b}$

So, the parallelogram is a rectangle.

Hence, the length of the other diagonal is $|\mathbf{a}-\mathbf{b}|=5$.

Ex. 46 **Statement I** *If* $|\mathbf{a}+\mathbf{b}|=|\mathbf{a}-\mathbf{b}|$, *then* $\mathbf{a}$ *and* $\mathbf{b}$ *are perpendicular to each other.*

Statement II *If the diagonals of a parallelogram are equal in magnitude, then the parallelogram is a rectangle.*

Sol. (a) $\mathbf{a}+\mathbf{b}=\mathbf{a}-\mathbf{b}$ are the diagonals of a parallelogram whose sides are $\mathbf{a}$ and $\mathbf{b}$.

$$|\mathbf{a}+\mathbf{b}|=|\mathbf{a}-\mathbf{b}|$$

Thus, diagonals of the parallelogram have the same length.

So, the parallelogram is a rectangle, i.e. $\mathbf{a} \perp \mathbf{b}$.

Ex. 47 **Statement I** *If I is the incentre of* ΔABC, *then* $|\mathbf{BC}|\mathbf{IA}+|\mathbf{CA}|\mathbf{IB}+|\mathbf{AB}|\mathbf{IC}=0$

Statement II *The position vector of centroid of* ΔABC *is* $\dfrac{\mathbf{OA}+\mathbf{OB}+\mathbf{OC}}{3}$.

Sol. (b) We know that,

$$\mathbf{OI}=\frac{|\mathbf{CB}|\mathbf{OA}+|\mathbf{CA}|\mathbf{OB}+|\mathbf{AB}|\mathbf{OC}}{|\mathbf{BC}|+|\mathbf{CA}|+|\mathbf{AB}|}$$

and $$\mathbf{OG}=\frac{\mathbf{OA}+\mathbf{OB}+\mathbf{OC}}{3}$$

Ex. 48 **Statement I** *If* $\mathbf{u}$ *and* $\mathbf{v}$ *are unit vectors inclined at an angle* α *and* $\mathbf{x}$ *is a unit vector bisecting the angle between them, then* $\mathbf{x}=\dfrac{\mathbf{u}+\mathbf{v}}{2\sin\frac{\alpha}{2}}$.

Statement II *If ABC is an isosceles triangles with* $AB=AC=1$, *then vectors representing bisector of angle A is given by* $\mathbf{AB}=\dfrac{\mathbf{AB}+\mathbf{AC}}{2}$.

Sol. (d) We know that the unit vector along bisector of unit vectors $\mathbf{u}$ and $\mathbf{v}$ is $\dfrac{\mathbf{u}+\mathbf{v}}{2\cos\frac{\theta}{2}}$, where θ is the angle between vectors $\mathbf{u}$ and $\mathbf{v}$.

Also, in an isosceles ΔABC in which $AB=AC$, the median and bisector from A must be same line.

Ex. 49 **Statement I** *If* $\mathbf{a}=2\hat{\mathbf{i}}+\hat{\mathbf{k}}, \mathbf{b}=3\hat{\mathbf{j}}+4\hat{\mathbf{k}}$ *and* $\mathbf{c}=\lambda\mathbf{a}+\mu\mathbf{b}$ *are coplanar, then* $\mathbf{c}=4\mathbf{a}-\mathbf{b}$.

Statement II *A set vectors* $\mathbf{a}_1, \mathbf{a}_2, \mathbf{a}_3, \ldots, \mathbf{a}_n$ *is said to be linearly independent, if every relation of the form* $l_1\mathbf{a}_1+l_2\mathbf{a}_2+l_3\mathbf{a}_3+\ldots+l_n\mathbf{a}_n=0$ *implies that* $l_1=l_2=l_3=\ldots=l_n=0$ *(scalar).*

Sol. (b) $\mathbf{a}$, $\mathbf{b}$ and $\mathbf{c}$ are coplanar $\mathbf{c}=\lambda\mathbf{a}+\mu\mathbf{b} \Rightarrow \lambda=4$ and $\mu=-1$

Ex. 50 **Statement I** *Let A*($\mathbf{a}$), *B*($\mathbf{b}$) *and C*($\mathbf{c}$) *be three points such that* $\mathbf{a}=2\hat{\mathbf{i}}+\hat{\mathbf{k}}, \mathbf{b}=3\hat{\mathbf{i}}-\hat{\mathbf{j}}+3\hat{\mathbf{k}}$ *and* $\mathbf{c}=-\hat{\mathbf{i}}+7\hat{\mathbf{j}}-5\hat{\mathbf{k}}$. *Then, OABC is a tetrahedron.*

Statement II *Let A*($\mathbf{a}$), *B*($\mathbf{b}$) *and C*($\mathbf{c}$) *be three points such that vectors* $\mathbf{a}$, $\mathbf{b}$ *and* $\mathbf{c}$ *are non-coplanar. Then OABC is a tetrahedron.*

Sol. (a) Given vectors are non-coplanar.

Hence, the answer is (a).

Ex. 51 **Statement I** *Let* $\mathbf{a}, \mathbf{b}, \mathbf{c}$ *and* $\mathbf{a}$ *be the position vectors of four points A, B, C and D and* $3\mathbf{a}-2\mathbf{b}+5\mathbf{c}-6\mathbf{d}=0$. *Then points A, B, C and D are coplanar.*

Statement II *Three non-zero linearly dependent co-initial vectors* ($\mathbf{PQ}, \mathbf{PR}$ *and* $\mathbf{PS}$) *are coplanar. Then* $\mathbf{PQ}=\lambda\mathbf{PR}+\mu\mathbf{PS}$, *where* λ *and* μ *are scalars.*

Sol. (a)

$$3\mathbf{a}-2\mathbf{b}+5\mathbf{c}-6\mathbf{d}=(2\mathbf{a}-2\mathbf{b})+(-5\mathbf{a}+5\mathbf{c})+(6\mathbf{a}-6\mathbf{d})$$
$$=-2\mathbf{AB}+5\mathbf{AC}-6\mathbf{AD}=0$$

Therefore, $\mathbf{AB}$, $\mathbf{AC}$ and $\mathbf{AD}$ are linearly dependent.

Hence, by Statement II, Statement I is true.

JEE Type Solved Examples : Passage Based Questions

Passage I

(Ex. Nos. 52 to 54)

ABCD is a parallelogram. L is a point on BC which divides BC in the ratio 1:2. AL intersects BD at P. M is a point on DC which divides DC in the ratio 1:2 and AM intersects BD in Q.

Ex. 52 *Points P divides AL in the ratio*

(a) 1 : 2 (b) 1 : 3
(c) 3 : 1 (d) 2 : 1

Ex. 53 *Point Q divides DB in the ratio*

(a) 1 : 2 (b) 1 : 3
(c) 3 : 1 (d) 2 : 1

Ex. 54 *PQ : DB is equal to*

(a) $\frac{2}{3}$ (b) $\frac{1}{3}$
(c) $\frac{1}{2}$ (d) $\frac{3}{4}$

Sol. (Ex. Nos. 52-54)

52. (c)

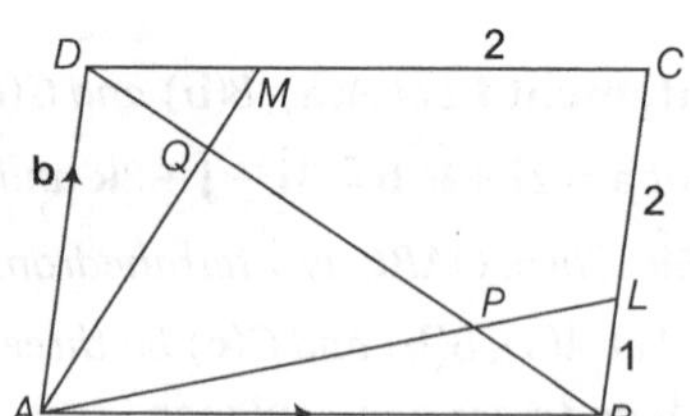

$$\mathbf{BL} = \frac{1}{3}\mathbf{b}$$

$\therefore$ $$\mathbf{AL} = \mathbf{a} + \frac{1}{3}\mathbf{b}$$

Let $\mathbf{AP} = \lambda \mathbf{AL}$ and P divides DB in the ratio $\mu : 1 - \mu$

Then, $$\mathbf{AP} = \lambda \mathbf{a} + \frac{\lambda}{3}\mathbf{b} \quad \text{...(i)}$$

Also, $$\mathbf{AP} = \mu \mathbf{a} + (1 - \mu)\mathbf{b} \quad \text{...(ii)}$$

From Eqs. (i) and (ii),

$$\lambda \mathbf{a} + \frac{\lambda}{3}\mathbf{b} = \mu \mathbf{a} + (1 - \mu)\mathbf{b}$$

$\therefore$ $$\lambda = \mu$$

and $$\frac{\lambda}{3} = 1 - \mu$$

$\therefore$ $$\lambda = \frac{3}{4}$$

53. (b) Hence, P divides AL in the ratio 3 : 1 and P divides DB in the ratio 1 : 3 Similarly Q divides DB in the ratio 1 : 3.

Thus, $$DQ = \frac{1}{4}DB$$

and $$PB = \frac{1}{4}DB$$

54. (b) $\therefore$ $$PQ = \frac{1}{2}DB,$$

i.e. $$PQ : DB = 1 : 2$$

Passage II

(Ex. Nos. 55 to 56)

Let A, B, C, D, E represent vertices of a regular pentagon ABCDE. Given the position vector of these vertices be **a, a + b, b,** λ**a** *and* λ**b** *respectively.*

Ex. 55 *The ratio* $\frac{AD}{BC}$ *is equal to*

(a) $1 - \cos\frac{3\pi}{5} : \cos\frac{3\pi}{5}$ (b) $1 + 2\cos\frac{2\pi}{5} : \cos\frac{\pi}{5}$

(c) $1 + 2\cos\frac{\pi}{5} : 2\cos\frac{\pi}{5}$ (d) None of these

Ex. 56 *AD divides EC in the ratio*

(a) $\cos\frac{2\pi}{5} : 1$ (b) $\cos\frac{3\pi}{5} : 1$

(c) $1 : 2\cos\frac{\pi}{5}$ (d) 1 : 2

Sol. (Ex. Nos. 55-56) Given $ABCDE$ is a regular pentagon

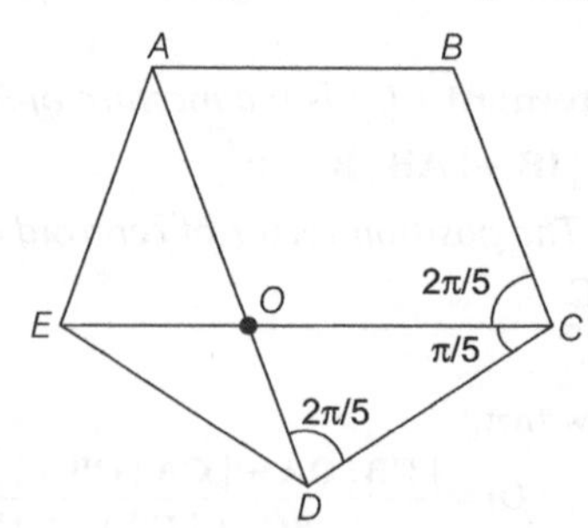

Let position vector point A and C be **a** and **b**, respectively.

AD is parallel to BC and AB is parallel to EC.

Therefore,

$AOCB$ is a parallelogram and position vector of B is **a** + **b**.

The position vectors of E and D are λ**b** and λ**a** respectively.

Also, $OA = BC = AB = OC = 1$ (let)

Therefore, $AOCB$ is rhombus.

$$\angle ABC = \angle AOC = \frac{3\pi}{5}$$

and $$\angle OAB = \angle BCO = \pi - \frac{3\pi}{5} = \frac{2\pi}{5}$$

Further, $OA = AE = 1$ and $OC = CD = 1$

Thus, ΔEAO and ΔOCD are isosceles.

In ΔOCD, using sine rule we get.

$$\frac{OC}{\sin\frac{2\pi}{5}} = \frac{OD}{\sin\frac{\pi}{5}}$$

$$\Rightarrow \quad OD = \frac{1}{2\cos\frac{\pi}{5}} = OE$$

$$\Rightarrow \quad AD = OA + OD = 1 + \frac{1}{2\cos\frac{\pi}{5}}$$

55. (c) $\frac{AD}{BC} = 1 + \frac{1}{2\cos\frac{\pi}{5}} = \frac{1 + 2\cos\frac{\pi}{5}}{2\cos\frac{\pi}{5}}$

56. (c) $\frac{OE}{OC} = \frac{1}{2\cos\frac{\pi}{5}}$

Passage III

(Ex. Nos. 57 to 58)

In a parallelogram OABC vectors **a**, **b**, **c** *respectively, the position vectors of vertices A, B, C with reference to O as origin. A point E is taken on the side BC which divides it in the ratio of* 2 : 1. *Also, the line segment AE intersects the line bisecting the angle ∠AOC internally at point P. If CP when extended meets AB in point F, then*

Ex. 57 *The position vector of point P is*

(a) $\frac{|\mathbf{a}||\mathbf{c}|}{3|\mathbf{c}| + 2|\mathbf{a}|}\left(\frac{\mathbf{a}}{|\mathbf{a}|} + \frac{\mathbf{c}}{|\mathbf{c}|}\right)$

(b) $\frac{3|\mathbf{a}||\mathbf{c}|}{3|\mathbf{c}| + 2|\mathbf{a}|}\left(\frac{\mathbf{a}}{|\mathbf{a}|} + \frac{\mathbf{c}}{|\mathbf{c}|}\right)$

(c) $\frac{2|\mathbf{a}||\mathbf{c}|}{3|\mathbf{c}| + 2|\mathbf{a}|}\left(\frac{\mathbf{a}}{|\mathbf{a}|} + \frac{\mathbf{c}}{|\mathbf{c}|}\right)$

(d) None of the above

Ex. 58 *The ratio in which F divides AB is*

(a) $\frac{2|\mathbf{a}|}{||\mathbf{a}| - 3|\mathbf{c}||}$ (b) $\frac{|\mathbf{a}|}{||\mathbf{a}| - 3|\mathbf{c}||}$

(c) $\frac{3|\mathbf{a}|}{||\mathbf{a}| - 3|\mathbf{c}||}$ (d) $\frac{3|\mathbf{c}|}{3||\mathbf{c}| - |\mathbf{a}||}$

Sol. (Ex. Nos. 57-58)

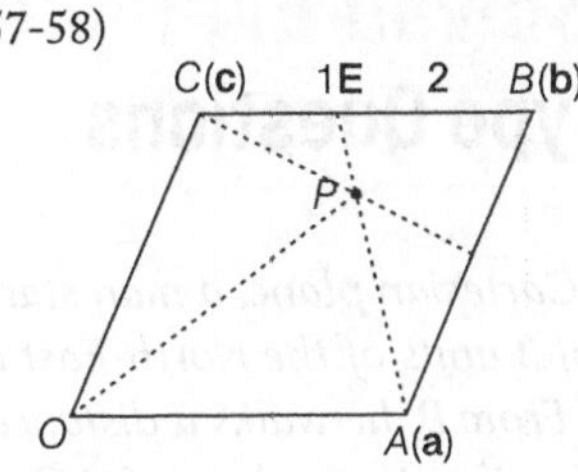

Let the position vector of A and C be **a** and **c** respectively. Therefore,

Position vector of

$$B = \mathbf{b} = \mathbf{a} + \mathbf{c} \quad \text{...(i)}$$

Also, position vector of

$$E = \frac{\mathbf{b} + 2\mathbf{c}}{3} = \frac{\mathbf{a} + 3\mathbf{c}}{3} \quad \text{...(ii)}$$

Now, point P lies on angle bisector of $\angle AOC$. Thus,

Position vector of point

$$P = \lambda\left(\frac{\mathbf{a}}{|\mathbf{a}|} + \frac{\mathbf{b}}{|\mathbf{b}|}\right) \quad \text{...(iii)}$$

Also, let P divides EA in ration $\mu : 1$. Therefore, Position vector of P

$$= \frac{\mu\mathbf{a} + \frac{\mathbf{a} + 3\mathbf{c}}{3}}{\mu + 1} = \frac{(3\mu + 1)\mathbf{a} + 3\mathbf{c}}{3(\mu + 1)} \quad \text{...(iv)}$$

Comparing Eqs. (iii) and (iv), we get

$$\lambda\left(\frac{\mathbf{a}}{|\mathbf{a}|} + \frac{\mathbf{c}}{|\mathbf{c}|}\right) = \frac{(3\mu + 1)\mathbf{a} + 3\mathbf{c}}{3(\mu + 1)}$$

$$\Rightarrow \quad \frac{\lambda}{|\mathbf{a}|} = \frac{3\mu + 1}{3(\mu + 1)} \text{ and } \frac{\lambda}{|\mathbf{c}|} = \frac{1}{\mu + 1}$$

$$\Rightarrow \quad \frac{3|\mathbf{c}| - |\mathbf{a}|}{3|\mathbf{a}|} = \mu$$

$$\Rightarrow \quad \frac{\lambda}{|\mathbf{c}|} = \frac{1}{\frac{3|\mathbf{c}| - |\mathbf{a}|}{3|\mathbf{a}|} + 1} \Rightarrow \lambda = \frac{3|\mathbf{a}||\mathbf{c}|}{3|\mathbf{c}| + 2|\mathbf{a}|}$$

57. (b) So, position vector of P is $\frac{3|\mathbf{a}||\mathbf{c}|}{3|\mathbf{c}| + 2|\mathbf{a}|}\left(\frac{\mathbf{a}}{|\mathbf{a}|} + \frac{\mathbf{c}}{|\mathbf{c}|}\right)$.

58. (d) Let F divides AB in ratio $t : 1$, then position vector of F is $\frac{t\mathbf{b} + \mathbf{a}}{t + 1}$

Now, points C, P, F are collinear, Then, $\mathbf{CF} = m\mathbf{CP}$

$$\Rightarrow \quad \frac{t(\mathbf{a} + \mathbf{c})}{t + 1} - \mathbf{c} = m\left\{\frac{3|\mathbf{a}||\mathbf{c}|}{3|\mathbf{c}| + 2|\mathbf{a}|}\left(\frac{\mathbf{a}}{|\mathbf{a}|} + \frac{\mathbf{c}}{|\mathbf{c}|}\right) - \mathbf{c}\right\}$$

Comparing coefficients, we get

$$\frac{t}{t + 1} = m\frac{3|\mathbf{c}|}{3|\mathbf{c}| + 2|\mathbf{a}|}$$

and $$\frac{-1}{t + 1} = m\frac{|\mathbf{a}| - 3|\mathbf{c}|}{3|\mathbf{c}| + 2|\mathbf{a}|}$$

$$t = \frac{3|\mathbf{c}|}{3|\mathbf{c}| - |\mathbf{a}|}$$

JEE Type Solved Examples : Matching Type Questions

Ex. 59 *In the Cartesian plane, a man starts at origin and walks a distance of 3 units of the North-East direction and reaches a point P. From P, he walks a distance of 4 units in the North-West direction to reach a point Q. Construct the parallelogram OPQR with PO and PQ as adjacent sides. Let M be the mid-point of PQ.*

Column I	Column II
A. The position vector of P is	(p) $\frac{3}{\sqrt{2}}(\hat{\mathbf{i}}+\hat{\mathbf{j}})$
B. The position vector of R is	(q) $\frac{1}{\sqrt{2}}(\hat{\mathbf{i}}+5\hat{\mathbf{j}})$
C. The position vector of M is	(r) $2\sqrt{2}(-\hat{\mathbf{i}}+\hat{\mathbf{j}})$
D. If the line OM meets the diagonal PR in the point T, then **OT** equals	(s) $\frac{\sqrt{2}}{3}(\hat{\mathbf{i}}+5\hat{\mathbf{j}})$

Sol. A → p, B → r, C → q, D → s

(A) Let $\hat{\mathbf{i}}$ and $\hat{\mathbf{j}}$ be the unit vectors along **OX** and **OY** respectively.

Now, $OP = 3$ and $\angle XOP = 45°$ implies that

$$\mathbf{OP} = (3\cos 45°)\hat{\mathbf{i}} + (3\sin 45°)\hat{\mathbf{j}} = \frac{3}{\sqrt{2}}(\hat{\mathbf{i}}+\hat{\mathbf{j}})$$

(B) Again, $\angle XOR = 135°$ and $OR = 4$ implies that

$$\mathbf{OR} = \frac{4}{\sqrt{2}}(-\hat{\mathbf{i}}+\hat{\mathbf{j}}) = 2\sqrt{2}(-\hat{\mathbf{i}}+\hat{\mathbf{j}})$$

(C) The position vector of Q is given by

$$\mathbf{OP}+\mathbf{PQ} = \mathbf{OP}+\mathbf{OR} = \frac{1}{\sqrt{2}}(-\hat{\mathbf{i}}+7\hat{\mathbf{j}})$$

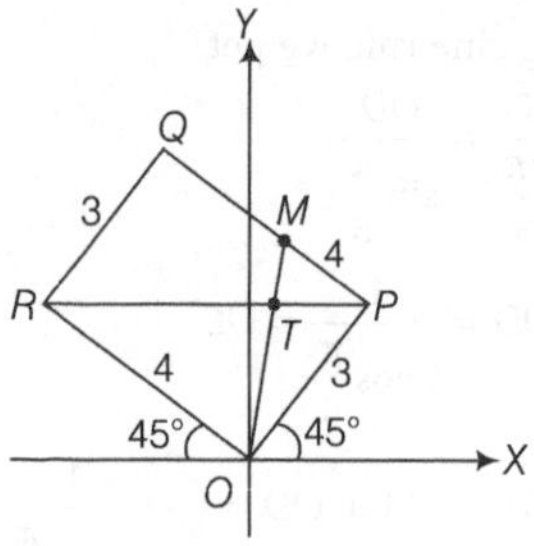

$$\therefore \quad \mathbf{OM} = \frac{\left(\frac{3}{\sqrt{2}}\right)(\hat{\mathbf{i}}+\hat{\mathbf{j}}) + \left(\frac{1}{\sqrt{2}}\right)(-\hat{\mathbf{i}}+7\hat{\mathbf{j}})}{2}$$

$$= \frac{2\hat{\mathbf{i}}+10\hat{\mathbf{j}}}{2\sqrt{2}} = \frac{\hat{\mathbf{i}}+5\hat{\mathbf{j}}}{\sqrt{2}}$$

(D) Now, $PT : RT = 1 : 2$

Therefore, $$\mathbf{OT} = \frac{1(\mathbf{OR})+2(OP)}{3}$$

$$= \frac{\left(\frac{4}{\sqrt{2}}\right)(-\hat{\mathbf{i}}+\hat{\mathbf{j}}) + 2\left[\left(\frac{3}{\sqrt{2}}\right)(\hat{\mathbf{i}}+\hat{\mathbf{j}})\right]}{3}$$

$$= \frac{\sqrt{2}}{3}(\hat{\mathbf{i}}+5\hat{\mathbf{j}})$$

JEE Type Solved Examples : Single Integer Answer Type Questions

Ex. 60 *P and Q have position vectors* **a** *and* **b** *relative to the origin O and X, Y divide* **PQ** *internally and externally respectively in the ratio 2 : 1. Vector* **XY** *is* $\lambda\mathbf{a}+\mu\mathbf{b}$*, then the value of* $|\lambda+\mu|$ *is*

Sol. (0) Since, X and Y divide **PQ** internally and externally in the ratio 2 : 1, then $X = \frac{2\mathbf{b}+\mathbf{a}}{3}$ and $y = 2\mathbf{b}-\mathbf{a}$

$\therefore$ **XY** = Position vector of y-Position vector of x

$$= 2\mathbf{b}-\mathbf{a}-\frac{2\mathbf{b}+\mathbf{a}}{3} = \frac{4\mathbf{b}}{3}-\frac{4\mathbf{a}}{3}$$

On comparing it with $\lambda\mathbf{a}+\mu\mathbf{b}$, we get

$$\lambda = -\frac{4}{3} \text{ and } \mu = \frac{4}{3}$$

$$\therefore \quad |\lambda+\mu| = \left|\frac{-4}{3}+\frac{4}{3}\right| = 0$$

Ex. 61 *If* $A(1,-1,-3)$, $B(2,1,-2)$ *and* $C(-5,2,-6)$ *are the position vectors of the vertices of* ΔABC. *The length of the bisector of its internal angle at A is* $\frac{\lambda\sqrt{10}}{4}$*, then value of* λ *is*

Sol. (3) We have, $\mathbf{AB} = \hat{\mathbf{i}}+2\hat{\mathbf{j}}+\hat{\mathbf{k}}$, $\mathbf{AC} = -6\hat{\mathbf{i}}+3\hat{\mathbf{j}}-3\hat{\mathbf{k}}$

$$\Rightarrow \quad |\mathbf{AB}| = \sqrt{6} \text{ and } |\mathbf{AC}| = 3\sqrt{6}$$

Clearly, point D divides BC in the ratio $AB : AC$, i.e. 1 : 3

$$\therefore \text{Position vector of } D = \frac{(-5\hat{\mathbf{i}}+2\hat{\mathbf{j}}-6\hat{\mathbf{k}})+3(2\hat{\mathbf{i}}+\hat{\mathbf{j}}-2\hat{\mathbf{k}})}{1+3}$$

$$= \frac{1}{4}(\hat{\mathbf{i}}+5\hat{\mathbf{j}}-12\hat{\mathbf{k}})$$

$$\therefore \quad \mathbf{AD} = \frac{1}{4}(\hat{\mathbf{i}}+5\hat{\mathbf{j}}-12\hat{\mathbf{k}}) - (\hat{\mathbf{i}}-\hat{\mathbf{j}}-3\hat{\mathbf{k}}) = \frac{3}{4}(-\hat{\mathbf{i}}+3\hat{\mathbf{j}})$$

$$\Rightarrow \quad |\mathbf{AD}| = AD = \frac{3}{4}\sqrt{10}$$

$$\therefore \quad \lambda = 3$$

Ex. 62 *Let ABC be a triangle whose centroid is G, orthocentre is H and circumcentre is the origin 'O'. If D is any point in the plane of the triangle such that no three of O, A, C and D are collinear satisfying the relation* $\mathbf{AD}+\mathbf{BD}+\mathbf{CH}+3\mathbf{HG}=\lambda\mathbf{HD}$, *then what is the value of the scalar 'Δ'*

Sol. LHS $= \mathbf{d}-\mathbf{a}+\mathbf{d}-\mathbf{b}+\mathbf{h}-\mathbf{c}+3(\mathbf{g}-\mathbf{h})$

$$= 2\mathbf{d}-(\mathbf{a}+\mathbf{b}+\mathbf{c})+3\frac{(\mathbf{a}+\mathbf{b}+\mathbf{c})}{3}-2\mathbf{h}$$

$$= 2\mathbf{d}-2\mathbf{h} = 2(\mathbf{d}-\mathbf{h}) = 2\mathbf{HD} \Rightarrow \lambda = 2$$

Ex. 63 *Let* $\mathbf{a}, \mathbf{b}$ *and* $\mathbf{c}$ *be unit vectors such that* $\mathbf{a}+\mathbf{b}-\mathbf{c}=0$. *If the area of triangle formed by vectors* $\mathbf{a}$ *and* $\mathbf{b}$ *is A, then what is the value of* $16A^2$?

Sol. (3) Given $\mathbf{a}+\mathbf{b}=\mathbf{c}$

Now, vector $\mathbf{c}$ is along the diagonal of the parallelogram which has adjacent side vectors $\mathbf{a}$ and $\mathbf{b}$. Since, $\mathbf{c}$ is also a unit vector, triangle formed by vectors $\mathbf{a}$ and $\mathbf{b}$ is an equilateral triangle.

Then, Area of triangle $= \frac{\sqrt{3}}{4} \Rightarrow A^2 = \frac{3}{10} \Rightarrow 16A^2 = 3$

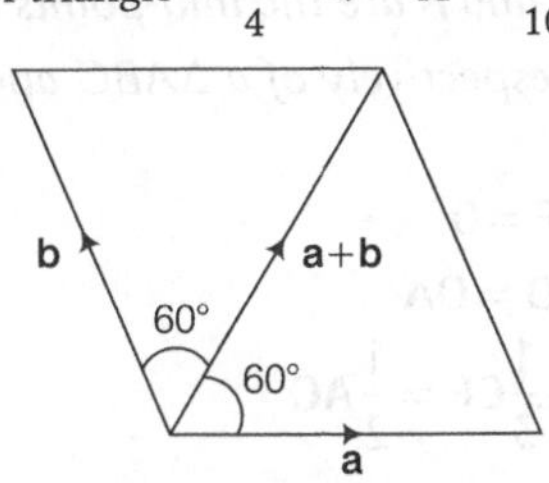

Ex. 64 *Find the least positive integral value of x for which the angle between vectors* $\mathbf{a}=x\hat{\mathbf{i}}-3\hat{\mathbf{j}}-\hat{\mathbf{k}}$ *and* $\mathbf{b}=2x\hat{\mathbf{i}}+x\hat{\mathbf{j}}-\hat{\mathbf{k}}$ *is acute.*

Sol. (2) Let $\mathbf{a}=x\hat{\mathbf{i}}-3\hat{\mathbf{j}}-\hat{\mathbf{k}}$ and $\mathbf{b}=2x\hat{\mathbf{i}}+x\hat{\mathbf{j}}-\hat{\mathbf{k}}$ be the adjacent sides of the parallelogram.

Now angle between $\mathbf{a}$ and $\mathbf{b}$ is acute, i.e. $|\mathbf{a}+\mathbf{b}|>|\mathbf{a}-\mathbf{b}|$

$\Rightarrow \quad |3x\hat{\mathbf{i}}+(x-3)\hat{\mathbf{j}}-2\hat{\mathbf{k}}|^2 > |-x\hat{\mathbf{i}}-(x+3)\hat{\mathbf{j}}|^2$

or $\quad 9x^2+(x-3)^2+4 > x^2+(x+3)^3$

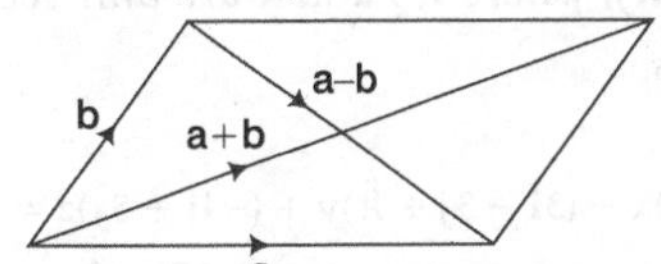

or $\quad 8x^2-12x+4>0$ or $2x^2-3x+1>0$

or $\quad (2x-1)(x-1)>0 \Rightarrow x<\frac{1}{2}$ or $x>1$

Hence, the least positive integral value is 2.

Ex. 65 *If the points* $a(\cos\alpha+\hat{\mathbf{i}}\sin\gamma)$, $b(\cos\beta+\hat{\mathbf{i}}\sin\beta)$ *and* $c(\cos\gamma+\hat{\mathbf{i}}\sin\gamma)$ *are collinear, then the value of* $|z|$ *is ... (where* $z=bc\sin(\beta-\gamma)+ca\sin(\gamma-\alpha)+ab\sin(\alpha+\beta)+3\hat{\mathbf{i}})$

Sol. (3) $\begin{vmatrix} a\cos\alpha & a\sin\alpha & 1 \\ b\cos\beta & b\sin\beta & 1 \\ c\cos\gamma & c\sin\gamma & 10 \end{vmatrix} = 0$

$\Rightarrow \quad bc\sin(\gamma-\beta)+a\sin(\alpha-\gamma)+ab\sin(\beta-\alpha)=0$

$\Rightarrow \quad |z|=3$

Subjective Type Questions

Ex. 66 *A particle in equilibrium is subjected to four forces*

$$\mathbf{F}_1=-10\hat{\mathbf{k}}, \qquad \mathbf{F}_2=u\left(\frac{4}{13}\hat{\mathbf{i}}-\frac{12}{13}\hat{\mathbf{j}}+\frac{3}{13}\hat{\mathbf{k}}\right)$$

$$\mathbf{F}_3=v\left(-\frac{4}{13}\hat{\mathbf{i}}-\frac{12}{13}\hat{\mathbf{j}}+\frac{3}{13}\hat{\mathbf{k}}\right)$$

and $\mathbf{F}_4=w(\cos\theta\,\hat{\mathbf{i}}+\sin\theta\,\hat{\mathbf{j}})$

Find the values of u, v and w in terms of θ.

Sol. Since, the particle is in equilibrium.

$$\mathbf{F}_1+\mathbf{F}_2+\mathbf{F}_3+\mathbf{F}_4=0$$

$$-10\hat{\mathbf{k}}+u\left(\frac{4}{13}\hat{\mathbf{i}}-\frac{12}{13}\hat{\mathbf{j}}+\frac{3}{13}\hat{\mathbf{k}}\right)+v\left(-\frac{4}{13}\hat{\mathbf{i}}-\frac{12}{13}\hat{\mathbf{j}}+\frac{3}{13}\hat{\mathbf{k}}\right)+w(\cos\theta\hat{\mathbf{i}}+\sin\theta\hat{\mathbf{j}})=0$$

$$\Rightarrow \left(\frac{4u}{13}-\frac{4v}{13}+w\cos\theta\right)\hat{\mathbf{i}}+\left(\frac{-12}{13}u-\frac{12}{13}v+w\sin\theta\right)\hat{\mathbf{j}}+\left(\frac{3}{13}u+\frac{3}{13}v-10\right)\hat{\mathbf{k}}=0$$

$$\Rightarrow \quad \frac{4u}{13}-\frac{4v}{13}+w\cos\theta=0 \quad \text{...(i)}$$

$$-\frac{12}{13}u-\frac{12}{13}v+w\sin\theta=0 \quad \text{...(ii)}$$

$$\frac{3}{13}u+\frac{3}{13}v-10=0 \quad \text{...(iii)}$$

From Eq. (iii), we get $\quad u+v=\frac{130}{3}$

From Eq. (ii), we get

$$-\frac{12}{13}(u+v)+w\sin\theta=0$$

$$\Rightarrow \quad -\frac{12}{13}\left(\frac{130}{3}\right)+w\sin\theta=0$$

$$\Rightarrow \quad w=\frac{40}{\sin\theta}=40\operatorname{cosec}\theta$$

On substituting the value of w in Eqs. (i) and (ii), we get

$$u-v=-130\cot\theta$$

and $\quad u+v=\frac{130}{3}$

On solving, we get

$$u + \frac{65}{3} - 65\cot\theta$$

$$v + \frac{65}{3} + 65\cot\theta \text{ and } w = 40 \text{ cosec } \theta$$

Ex. 67 *Find all values of 'λ' such that $x, y, z \neq (0, 0, 0)$ and $(\hat{\mathbf{i}} + \hat{\mathbf{j}} + 3\hat{\mathbf{k}})\, x + (3\hat{\mathbf{i}} - 3\hat{\mathbf{j}} + \hat{\mathbf{k}})\, y + (-4\hat{\mathbf{i}} + 5\hat{\mathbf{j}})z = \lambda(x\hat{\mathbf{i}} + y\hat{\mathbf{j}} + z\hat{\mathbf{k}})$, where $\hat{\mathbf{i}}, \hat{\mathbf{j}}$ and $\hat{\mathbf{k}}$ are unit vectors along the coordinate axes.*

Sol. Here,

$$(\hat{\mathbf{i}} + \hat{\mathbf{j}} + 3\hat{\mathbf{k}})x + (3\hat{\mathbf{i}} - 3\hat{\mathbf{j}} + \hat{\mathbf{k}})y + (-4\hat{\mathbf{i}} + 5\hat{\mathbf{j}})z = \lambda(x\hat{\mathbf{i}} + y\hat{\mathbf{j}} + z\hat{\mathbf{k}})$$

On comparing the coefficients of $\hat{\mathbf{i}}, \hat{\mathbf{j}}$ and $\hat{\mathbf{k}}$, we get

$$x + 3y - 4z = \lambda x$$

$$\Rightarrow \quad (1 - \lambda)x + 3y - 4z = 0 \quad \text{...(i)}$$

$$x - 3y + 5z = \lambda y$$

$$\Rightarrow \quad x - (3 + \lambda)y + 5z = 0 \quad \text{...(ii)}$$

$$3x + y = \lambda z$$

$$\Rightarrow \quad 3x + y - \lambda z = 0 \quad \text{...(iii)}$$

The Eqs. (i), (ii) and (iii) will have a non-trivial solution, if

$$\begin{vmatrix} 1-\lambda & 3 & -4 \\ 1 & -(3+\lambda) & 5 \\ 3 & 1 & -\lambda \end{vmatrix} = 0$$

$$[\because (x, y, z) \neq (0, 0, 0) \therefore \Delta = 0]$$

$$\Rightarrow (1 - \lambda)\{\lambda(3 + \lambda) - 5\} - 3\{-\lambda - 15\} - 4\{1 + 3(\lambda + 3)\} = 0$$

$$\Rightarrow \quad (1 - \lambda)\{\lambda^2 + 3\lambda - 5\} - 3\{-\lambda - 15\} - 4\{3\lambda + 10\} = 0$$

$$\Rightarrow \quad \lambda^3 + 2\lambda^2 + \lambda = 0$$

$$\Rightarrow \quad \lambda(\lambda^2 + 2\lambda + 1) = 0$$

$$\Rightarrow \quad \lambda(\lambda + 1)^2 = 0$$

$$\therefore \quad \lambda = 0 \text{ or } \lambda = -1$$

Ex. 68 *If G is the centroid of the ΔABC and if G' is the centroid of another $\Delta A'B'C'$, then prove that $AA' + BB' + CC' = 3GG'$.*

Sol. Here,

G is centroid of ΔABC and G' is centroid of $\Delta A'B'C'$, shown as in figure.

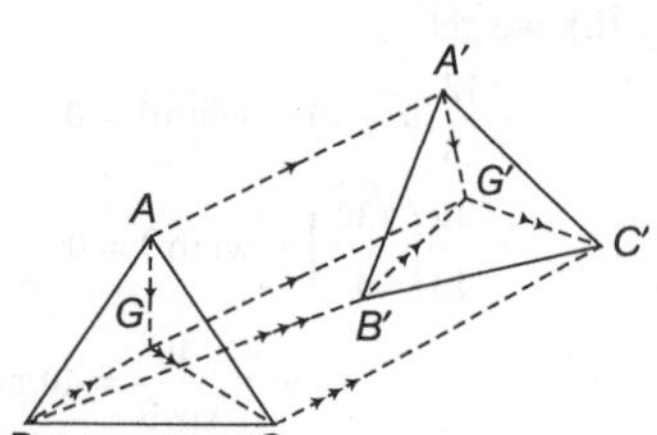

Clearly, $\quad \mathbf{AA'} = \mathbf{AG} + \mathbf{GG'} + \mathbf{G'A'} \quad$ (polygon law)

$$\mathbf{BB'} = \mathbf{BG} + \mathbf{GG'} + \mathbf{G'B'}$$

$$\mathbf{CC'} = \mathbf{CG} + \mathbf{CG'} + \mathbf{G'C'}$$

On adding these

$$\mathbf{AA'} + \mathbf{BB'} + \mathbf{CC'} = 3\mathbf{GG'} + (\mathbf{AG} + \mathbf{BG} + \mathbf{CG}) + (\mathbf{G'A'} + \mathbf{G'B'} + \mathbf{G'C'})$$

$$= 3\mathbf{GG'} + (\mathbf{AG} + 2\mathbf{DG}) + (\mathbf{G'A'} + 2\mathbf{G'D'})$$

(using **AD** and **A′D′** as the medians of ΔABC and $\Delta A'B'C'$, respectively)

$$= 3\mathbf{GG'} + (\mathbf{AG} + \mathbf{GA}) + \mathbf{G'A'} + \mathbf{A'G'}$$

$$= 3\mathbf{GG'} + O + O$$

$$\therefore \mathbf{AA'} + \mathbf{BB'} + \mathbf{CC'} = 3\mathbf{GG'}$$

Aliter

We know by triangle law

$$\mathbf{AA'} = \mathbf{OA'} - \mathbf{OA}$$

$$\mathbf{BB'} = \mathbf{OB'} - \mathbf{OB}$$

$$\mathbf{CC'} = \mathbf{OC'} - \mathbf{OC}$$

$$\Rightarrow \quad \mathbf{AA'} + \mathbf{BB'} + \mathbf{CC'} = (\mathbf{OA'} + \mathbf{OB'} + \mathbf{OC'}) - (\mathbf{OA} + \mathbf{OB} + \mathbf{OC})$$

$$= 3\mathbf{OG'} - 3\mathbf{OG'} = 3\mathbf{GG'}$$

Ex. 69 *If D, E and F are the mid-points of the sides BC, CA and AB, respectively of a ΔABC and O is any point, show that*

(i) $\mathbf{AD} + \mathbf{BE} + \mathbf{CF} = 0$

(ii) $\mathbf{OE} + \mathbf{OF} + \mathbf{DO} = \mathbf{OA}$

(iii) $\mathbf{AD} + \frac{2}{3}\mathbf{BE} + \frac{1}{3}\mathbf{CF} = \frac{1}{2}\mathbf{AC}$

Sol. Consider the point O as origin, we have,

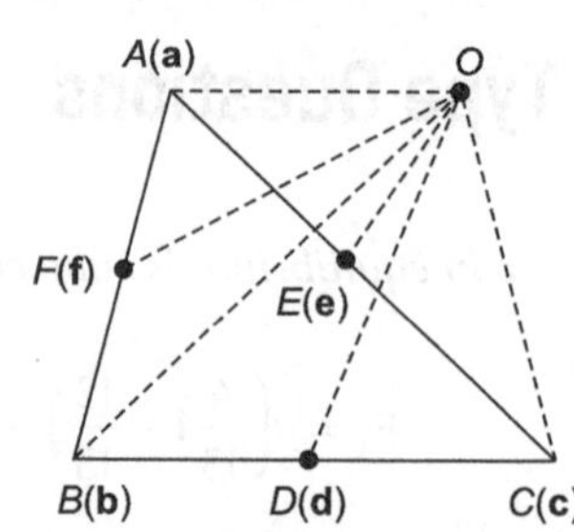

(i) $\mathbf{AD} + \mathbf{BE} + \mathbf{CF} = = (\mathbf{d} - \mathbf{a}) + (\mathbf{e} - \mathbf{b}) + (\mathbf{f} - \mathbf{c})$

$$= (\mathbf{d} + \mathbf{e} + \mathbf{f}) - (\mathbf{a} + \mathbf{b} + \mathbf{c}) = 0 \quad \text{[using Eq. (i)]}$$

$$\Rightarrow \quad \mathbf{AD} + \mathbf{BE} + \mathbf{CF} = 0$$

(ii) Here, $\mathbf{OE} + \mathbf{OF} + \mathbf{OD} = \mathbf{e} + \mathbf{f} - \mathbf{d}$

$$= \frac{\mathbf{c} + \mathbf{a}}{2} + \frac{\mathbf{a} + \mathbf{b}}{2} - \frac{\mathbf{b} + \mathbf{c}}{2} = \mathbf{a} = \mathbf{OA}$$

$$\therefore \quad \mathbf{OE} + \mathbf{OF} + \mathbf{OD} = \mathbf{OA}$$

(iii) Here, $\mathbf{AD} + \frac{2}{3}\mathbf{BE} + \frac{1}{3}\mathbf{CF} = (\mathbf{d} - \mathbf{a}) + \frac{2}{3}(\mathbf{e} - \mathbf{b}) + \frac{1}{3}(\mathbf{f} - \mathbf{c})$

$$= \frac{\mathbf{b} + \mathbf{c}}{2} - \mathbf{a} + \frac{2}{3}\left(\frac{\mathbf{c} + \mathbf{a}}{2} - \mathbf{b}\right) + \frac{1}{3}\left(\frac{\mathbf{a} + \mathbf{b}}{2} - \mathbf{c}\right)$$

$$= \mathbf{a}\left(-1 + \frac{1}{3} + \frac{1}{6}\right) + \mathbf{b}\left(\frac{1}{2} - \frac{2}{3} + \frac{1}{6}\right) + \mathbf{c}\left(\frac{1}{2} + \frac{1}{3} - \frac{1}{3}\right)$$

$$= -\frac{1}{2}\mathbf{a} + \frac{1}{2}\mathbf{c} = \frac{1}{2}(\mathbf{c} - \mathbf{a})$$

$$= \frac{1}{2}\mathbf{AC}$$

$$\mathbf{AD} + \frac{2}{3}\mathbf{BE} + \frac{1}{3}\mathbf{CF} = \frac{1}{2}\mathbf{AC}$$

Ex. 70 *If* **A** *and* **B** *be two vectors and k be any scalar quantity greater than zero, then prove that*

$$|\mathbf{A}+\mathbf{B}|^2 \le (1+k)|\mathbf{A}|^2\left(1+\frac{1}{k}\right)|\mathbf{B}|^2$$

Sol. We know, $(1 + k)|\mathbf{A}|^2 + \left(1+\frac{1}{k}\right)|\mathbf{B}|^2$

$$= |\mathbf{A}|^2 + k|\mathbf{A}|^2 + |\mathbf{B}|^2 + \frac{1}{k}|\mathbf{B}|^2 \quad \text{...(i)}$$

Also, $$k|\mathbf{A}|^2 + \frac{1}{k}|\mathbf{B}|^2 \ge 2\left(k|\mathbf{A}|^2 \cdot \frac{1}{k}|\mathbf{B}|^2\right)^{\frac{1}{2}} = 2|\mathbf{A}|\cdot|\mathbf{B}| \quad \text{...(ii)}$$

(since, Arithmetic mean ≥ Geometric mean)

So, $(1 + k)|\mathbf{A}|^2 + \left(1+\frac{1}{k}\right)|\mathbf{B}|^2 \ge |\mathbf{A}|^2 + |\mathbf{B}|^2 + 2|\mathbf{A}|\cdot|\mathbf{B}|$

$= (|\mathbf{A}| + |\mathbf{B}|)^2$ [using Eqs. (i) and (ii)]

And also, $|\mathbf{A}| + |\mathbf{B}| \ge |\mathbf{A} + \mathbf{B}|$

Hence, $(1 + k)|\mathbf{A}|^2 + \left(1 + \frac{1}{k}\right)|\mathbf{B}|^2 \ge |\mathbf{A} + \mathbf{B}|^2$

Ex. 71 *If O is the circumcentre and O' the orthocenter of* ΔABC *prove that*

(i) $\mathbf{SA}+\mathbf{SB}+\mathbf{SC} = 3\mathbf{SG}$, where S is any point in the plane of ΔABC.

(ii) $\mathbf{OA}+\mathbf{OB}+\mathbf{OC} = \mathbf{OO'}$

(iii) $\mathbf{O'A}+\mathbf{O'B}+\mathbf{O'C} = 2\mathbf{O'O}$

(iv) $\mathbf{AO'}+\mathbf{O'B}+\mathbf{O'C} = \mathbf{AP}$

where, **AP** is diameter of the circumcircle.

Sol. Let G be the centroid of ΔABC, first we shall show that circumcentre O, orthocenter O' and centroid G are collinear and $O'G = 2OG$.

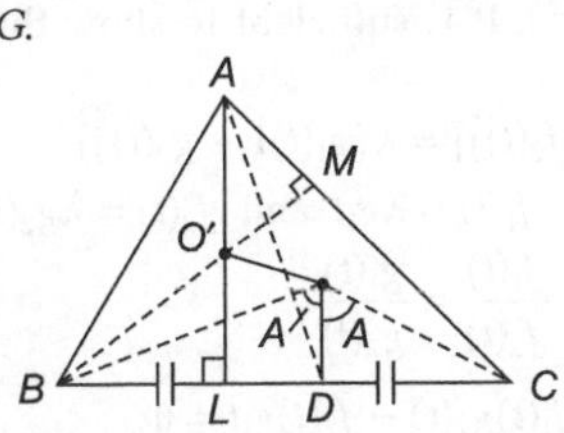

Let AL and BM be perpendiculars on the sides BC and CA, respectively. Let AD be the median and OD be the perpendicular from O on side BC. If R is the circumradius of circumcircle of ΔABC, then $OB = OC = R$.

In ΔOBD, we have $OD = R\cos A$ ...(i)

In ΔABM, $AM = AB\cos A = c\cos A$ ...(ii)

Form $\Delta AO'M$, $AO' = AM\sec(90° - C)$

$= c\cos A \operatorname{cosec} C$

$= \frac{c}{\sin C}\cdot\cos A = 2R\cos A$

$$\left(\because \frac{a}{\sin A} = \frac{b}{\sin B} = \frac{c}{\sin C} = 2R\right)$$

$$AO' = 2(OD) \quad \text{...(iii)}$$

Now, $\Delta AGO'$ and ΔOGD are similar.

$\therefore \quad \frac{OG}{O \cdot G} = \frac{GD}{GA} = \frac{OD}{AO'} = \frac{1}{2}$ [using Eq. (iii)]

$\Rightarrow \quad 2OG = O'G$

(i) We have, $\mathbf{SA} + \mathbf{SB} + \mathbf{SC} = \mathbf{SA} + (\mathbf{SB} + \mathbf{SC})$

$= \mathbf{SA} + 2\mathbf{SD}$ ($\because D$ is the mid-point of **BC**)

$= (1 + 2)\mathbf{SG} = 3\mathbf{SG}$

(ii) On replacing S by O in Eq. (i), we get

$\mathbf{OA} + \mathbf{OB} + \mathbf{OC} = 3\mathbf{OG}$

$= 2\mathbf{OG} + \mathbf{OG} = \mathbf{GO'} + \mathbf{OG}$

$= \mathbf{OG} + \mathbf{GO'} = \mathbf{OO'}$

(iii) $\mathbf{O'A} + \mathbf{O'B} + \mathbf{O'C} = 3\mathbf{O'G}$ [from Eq. (i)]

$= 2\mathbf{O'G} + \mathbf{O'G}$

$= 2\mathbf{O'G} + 2\mathbf{GO}$ ($\because 2\mathbf{OG} = \mathbf{O'G}$)

$= 2\mathbf{O'O}$

(iv) $\mathbf{AO} + \mathbf{O'B} + \mathbf{O'C} = 2\mathbf{AO'} + (\mathbf{O'A} + \mathbf{O'B} + \mathbf{O'C})$

$= 2\mathbf{AO'} + 2\mathbf{O'O}$ [From Eq. (iii)]

$= 2(\mathbf{AO'} + \mathbf{O'O}) = 2\mathbf{AO} = \mathbf{AP}$

($\because$ **AO** is the circumradius of ΔABC)

Ex. 72 *If* $\mathbf{c} = 3\mathbf{a} + 4\mathbf{b}$ *and* $2\mathbf{c} = \mathbf{a} - 3\mathbf{b}$, *show that,*

(i) **c** *and* **a** *have the same direction and* $|\mathbf{c}| > |\mathbf{a}|$.

(ii) **c** *and* **b** *have opposite direction and* $|\mathbf{c}| > |\mathbf{b}|$.

Sol. We have,

$$\mathbf{c} = 3\mathbf{a} + 4\mathbf{b} \text{ and } 2\mathbf{c} = \mathbf{a} - 3\mathbf{b}$$

$\Rightarrow \quad 2(3\mathbf{a} + 4\mathbf{b}) = \mathbf{a} - 3\mathbf{b}$

$\Rightarrow \quad 5\mathbf{a} = -11\mathbf{b}$

$\Rightarrow \quad \mathbf{a} = -\frac{11}{5}\mathbf{b}$ and $\mathbf{b} = -\frac{5}{11}\mathbf{a}$

(i) $\mathbf{c} = 3\mathbf{a} + 4\mathbf{b} = 3\mathbf{a} + 4\left(-\frac{5}{11}\mathbf{a}\right)$ $\left(\text{using } \mathbf{b} = -\frac{5}{11}\mathbf{a}\right)$

$$= 3\mathbf{a} - \frac{20}{11}\mathbf{a} = \frac{13}{11}\mathbf{a}$$

which shows that **c** and **a** have the same direction.

And $\mathbf{c} = \frac{13}{11}\mathbf{a}$

$\Rightarrow \quad |\mathbf{c}| = \frac{13}{11}|\mathbf{a}| \Rightarrow |\mathbf{c}| > |\mathbf{a}|$

(ii) We have, $\mathbf{c} = 3\mathbf{a} + 4\mathbf{b}$ and $\mathbf{a} = -\frac{11}{5}\mathbf{b}$

$$\mathbf{c} = 3\left(-\frac{11}{5}\mathbf{b}\right) + 4\mathbf{b} = -\frac{33}{5}\mathbf{b} + 4\mathbf{b}$$

$$\mathbf{c} = -\frac{13}{5}\mathbf{b}$$

This shows **c** and **b** have opposite directions.

Also, $\quad |\mathbf{c}| = \left| -\dfrac{13}{5}\mathbf{b} \right| = \dfrac{13}{5}|\mathbf{b}| \Rightarrow |\mathbf{c}| > |\mathbf{b}|$

Ex. 73 *A transversal cuts the sides OL, OM and diagonal ON of a parallelogram at A, B and C respectively. Prove that* $\dfrac{OL}{OA} + \dfrac{OM}{OB} = \dfrac{ON}{OC}$.

Sol. We have,

$$\mathbf{ON} = \mathbf{OL} + \mathbf{LN} = \mathbf{OL} + \mathbf{OM} \quad \text{...(i)}$$

Let $\quad \mathbf{OL} = x\mathbf{OA}, \mathbf{OM} = y\,\mathbf{OB} \quad$ ...(ii)

and $\quad \mathbf{ON} = z\mathbf{OC}$

So, $\quad |\mathbf{OL}| = x|\mathbf{OA}|, |\mathbf{OM}| = y|\mathbf{OB}|$ and $|\mathbf{ON}| = z|\mathbf{OC}|$

$$\therefore \quad x = \frac{OL}{OA}, y = \frac{OM}{OB} \text{ and } z = \frac{ON}{OC}$$

$\therefore$ From Eqs. (i) and (ii), we have

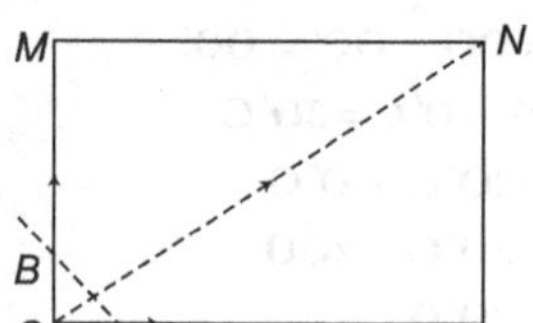

$$z\mathbf{OC} = x\mathbf{OA} + y\mathbf{OB}$$

$$\Rightarrow \quad x\,\mathbf{OA} + y\,\mathbf{OB} - z\,\mathbf{OC} = \mathbf{0}$$

$\therefore$ Points A, B and C are collinear, the sum of the coefficients of their PV must be zero.

$$\Rightarrow \quad x + y - z = 0$$

i.e. $\quad \dfrac{OL}{OA} + \dfrac{OM}{OB} = \dfrac{ON}{OC}$

Ex. 74 *If D, E and F be three points on the sides BC, CA and AB, respectively of a* ΔABC. *such that the points D, E and F are collinear then prove that* $\dfrac{\mathbf{BD}}{\mathbf{CD}} \cdot \dfrac{\mathbf{CE}}{\mathbf{AE}} \cdot \dfrac{\mathbf{AF}}{\mathbf{BF}} = 1$ *(Menelau's theorem)*

Sol. Here, D, E and F be the points on the sides **BC**, **CA** and **AB** respectively of ΔABC. Such that points D, E and F are collinear, be Shawn as the adjoining figuece.

Let B as the origin, $\mathbf{BA} = \mathbf{a}$ and $\mathbf{BC} = \mathbf{c}$

Then, $\mathbf{BF} = k\mathbf{a}$ and $\mathbf{BD} = l\mathbf{c}$

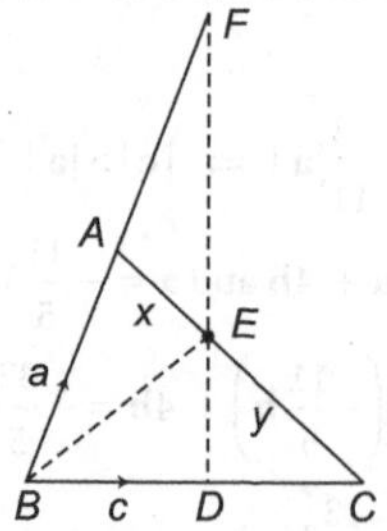

where, k and l are scalars.

$$\therefore \quad \frac{BD}{BC} = l \text{ and } \frac{BF}{BA} = k \quad \text{...(i)}$$

i.e. $\quad BC : BD = 1 : l$

$$\Rightarrow \quad \frac{BC}{BD} - 1 = \frac{1}{l} - 1 \Rightarrow \frac{DC}{BD} = \frac{1-l}{l}$$

$$\Rightarrow \quad \frac{BD}{DC} = \frac{l}{1-l} \text{ and } \frac{BA}{BF} = \frac{1}{k}$$

$$\Rightarrow \quad 1 - \frac{BA}{BF} = 1 - \frac{1}{k} \Rightarrow \frac{AF}{BF} = \frac{k-1}{k} \quad \text{...(ii)}$$

Now, let E divide the line AC in the ratio of $x : y$

$$\text{So, that } \mathbf{BE} = \frac{x\mathbf{c} + y\,\mathbf{a}}{x+y} = \frac{x \cdot \dfrac{\mathbf{BD}}{l} + y \cdot \dfrac{\mathbf{BF}}{k}}{x+y} \quad \text{...(iii)}$$

$$\Rightarrow \quad \mathbf{BE} - \frac{x}{l(x+y)}\mathbf{BD} - \frac{y}{k(x+y)}\mathbf{BF} = 0$$

Since, D, E and F are collinear.

Sum of coefficients must be zero.

$$\text{Hence, } 1 - \frac{x}{l(x+y)} - \frac{y}{k(x+y)} = 0$$

$$\Rightarrow \quad (x+y) - \frac{x}{l} - \frac{y}{k} = 0 \Rightarrow x + y = \frac{x}{l} + \frac{y}{k}$$

$$\Rightarrow \quad x\left(1 - \frac{1}{l}\right) = y\left(\frac{1}{k} - 1\right) \Rightarrow x\left(\frac{l-1}{l}\right) = y\left(\frac{1-k}{k}\right)$$

$$\Rightarrow \quad \frac{l}{l-1} \cdot \frac{y}{x} \cdot \frac{k-1}{k} = 1$$

$$\Rightarrow \quad \frac{BD}{DC} \cdot \frac{CE}{AE} \cdot \frac{AF}{BF} = 1 \quad \text{[using Eqs. (i), (ii) and (iii)]}$$

Ex. 75 *Let* $\mathbf{A}(t) = f_1(i)\hat{\mathbf{i}} + f_2(t)\hat{\mathbf{j}}$ *and* $\mathbf{B}(t) = g_1(t)\hat{\mathbf{i}} + g_2(t)\hat{\mathbf{j}}\ t \in [0, 1]$, *where* f_1, f_2, g_1 *and* g_2 *are continuous functions. Then show that* $\mathbf{A}(t)$ *and* $\mathbf{B}(t)$ *are parallel for some t.*

Sol. If $\mathbf{A}(t)$ and $\mathbf{B}(t)$ are non-zero vectors for all t and $\mathbf{A}(0) = 2\hat{\mathbf{i}} + 3\hat{\mathbf{j}}$, $\mathbf{A}(1) = 6\hat{\mathbf{i}} + 2\hat{\mathbf{j}}$, $\mathbf{B}(0) = 3\hat{\mathbf{i}} + 2\hat{\mathbf{j}}$, and $\mathbf{B}(1) = 2\hat{\mathbf{i}} + 6\hat{\mathbf{j}}$.

In order to prove that $\mathbf{A}(t)$ and $\mathbf{B}(t)$ are parallel vectors for some values of t. It is sufficient to show that $\mathbf{A}(t) = \lambda\mathbf{B}(t)$ for some λ.

$$\Leftrightarrow \quad \{f_1(t)\hat{\mathbf{i}} + f_2(t)\hat{\mathbf{j}}\} = \lambda\{g_1(t)\hat{\mathbf{i}} + g_2(t)\hat{\mathbf{j}}\}$$

$$\Leftrightarrow \quad f_1(t) = \lambda g_1 t \text{ and } f_2(t) = \lambda g_2(t)$$

$$\Leftrightarrow \quad \frac{f_1(t)}{f_2(t)} = \frac{g_1(t)}{g_2(t)}$$

$$\Leftrightarrow \quad f_1(t)g_2(t) - f_2(t)g_1 t = 0 \qquad \text{for some } t \in [0, 1]$$

Let $\quad f(t) = f_1(t)g_2(t) - f_2(t)g_1(t), t \in [0, 1]$

Since, f_1, f_2, g_1 and g_2 are continuous functions.

$\therefore F(t)$ is also a continuous function.

Also, $\quad f(0) = f_1(0)g_2(0) - g_1(0)f_2(0)$

$$= 2 \times 2 - 3 \times 3 = 4 - 9 = -5 < 0$$

and $\quad f(1) = f_1(1)g_2(1) - g_1(1)f_2(1)$

$$= 6 \times 6 - 2 \times 2 = 32 > 0$$

Thus, $F(t)$ is a continuous function on $[0, 1]$ such that $F(0) \cdot F(1) < 0$.

$\therefore$ By intermediate value theorem, there exists some $t \in (0, 1)$ such that

$$f(t) = 0$$

$$\Rightarrow \quad f_1(t)g_2(t) - f_2(t)g_1 t = 0$$

$$\Rightarrow \quad \mathbf{A}(t) = \lambda \mathbf{B}(t) \text{ for some } \lambda.$$

Hence, $\mathbf{A}(t)$ and $\mathbf{B}(t)$ are parallel vectors.

Ex. 76 *Prove that if* $\cos\alpha \neq 1, \cos\beta \neq 1$ *and* $\cos\gamma \neq 1$, *then the vectors* $\mathbf{a} = \hat{\mathbf{i}}\cos\alpha + \hat{\mathbf{j}} + \hat{\mathbf{k}}, \mathbf{b} = \hat{\mathbf{i}} + \hat{\mathbf{j}}\cos\beta + \hat{\mathbf{k}}$, $\mathbf{c} = \hat{\mathbf{i}} + \hat{\mathbf{j}} + \hat{\mathbf{k}}\cos\gamma$ *can never be coplanar.*

Sol. Suppose that, a, b and c are coplanar.

$$\Rightarrow \begin{vmatrix} \cos\alpha & 1 & 1 \\ 1 & \cos\beta & 1 \\ 1 & 1 & \cos\gamma \end{vmatrix} = 0$$

On applying $R_2 \to R_2 - R_1$ and $R_3 \to R_3 - R_1$

$$\Rightarrow \begin{vmatrix} \cos\alpha & 1 & 1 \\ 1-\cos\alpha & \cos\beta - 1 & 0 \\ 1-\cos\alpha & 0 & \cos\gamma - 1 \end{vmatrix} = 0$$

$$\Rightarrow \cos\alpha(\cos\beta - 1)(\cos\gamma - 1) - (1-\cos\alpha)(\cos\gamma - 1) - (1-\cos\alpha)(\cos\beta - 1) = 0$$

On dividing throughout by $(1-\cos\alpha)(1-\cos\beta)(1-\cos\gamma)$, we get

$$\frac{\cos\alpha}{1-\cos\alpha} + \frac{1}{1-\cos\beta} + \frac{1}{1-\cos\gamma} = 0$$

$$\Rightarrow \frac{-(1-\cos\alpha)+1}{1-\cos\alpha} + \frac{1}{1-\cos\beta} + \frac{1}{1-\cos\gamma} = 0$$

$$\Rightarrow -1 + \frac{1}{(1-\cos\alpha)} + \frac{1}{(1-\cos\beta)} + \frac{1}{(1-\cos\gamma)} = 0$$

$$\frac{1}{1-\cos\alpha} + \frac{1}{1-\cos\beta} + \frac{1}{1-\cos\gamma} = 1$$

$$\Rightarrow \operatorname{cosec}^2\frac{\alpha}{2} + \operatorname{cosec}^2\frac{\beta}{2} + \operatorname{cosec}^2\frac{\gamma}{2} = 2, \text{ which is not possible.}$$

As, $\quad \operatorname{cosec}^2\dfrac{\alpha}{2} \geq 1, \operatorname{cosec}^2\dfrac{\beta}{2} \geq 1$

and $\quad \operatorname{cosec}^2\dfrac{\gamma}{2} \geq 1$

$\because$ They cannot be coplanar.

Ex. 77 *If the vectors* $x\hat{\mathbf{i}} + \hat{\mathbf{j}} + \hat{\mathbf{k}}, \hat{\mathbf{i}} + y\hat{\mathbf{j}} + \hat{\mathbf{k}}$ *and* $\hat{\mathbf{i}} + \hat{\mathbf{j}} + z\hat{\mathbf{k}}$ *are coplanar where,* $x \neq 1, y \neq 1$ *and* $z \neq 1$, *then prove that* $\dfrac{1}{1-x} + \dfrac{1}{1-y} + \dfrac{1}{1-z} = 1$

Sol. The vectors are coplanar, if we can find two scalars λ and μ such that

$$(x\hat{\mathbf{i}} + \hat{\mathbf{j}} + \hat{\mathbf{k}}) = \lambda(\hat{\mathbf{i}} + y\hat{\mathbf{j}} + \hat{\mathbf{k}}) + \mu(\hat{\mathbf{i}} + \hat{\mathbf{j}} + z\hat{\mathbf{k}})$$

$$\Rightarrow \quad x = \lambda + \mu, 1 = \lambda y + \mu, 1 = \lambda + \mu z$$

$$\Rightarrow \quad x = \lambda + \mu, y = \frac{1-\mu}{\lambda}, z = \frac{1-\lambda}{\mu}$$

$$\Rightarrow \quad 1 - x = 1 - \lambda - \mu, 1 - y = \frac{\lambda - 1 + \mu}{\lambda},$$

$$1 - z = \frac{\mu - 1 + \lambda}{\mu}$$

$$\therefore \quad \frac{1}{1-x} + \frac{1}{1-y} + \frac{1}{1-z} = \frac{1}{1-\lambda-\mu} + \frac{\lambda}{\lambda+\mu-1} + \frac{\mu}{\lambda+\mu-1}$$

$$= \frac{-1+\lambda+\mu}{\lambda+\mu-1} = 1$$

$$\therefore \quad \frac{1}{1-x} + \frac{1}{1-y} + \frac{1}{1-z} = 1$$

Aliter

Thus, above problem could also be solved as

$$\begin{vmatrix} x & 1 & 1 \\ 1 & y & 1 \\ 1 & 1 & z \end{vmatrix} = 0 \Rightarrow \begin{vmatrix} x-1 & 0 & 1-z \\ 0 & y-1 & 1-z \\ 1 & 1 & z \end{vmatrix} = 0$$

(using $R_1 \to R_1 - R_3$ and $R_2 \to R_2 - R_3$)

$$\Rightarrow (x-1)(y-1)(z-1) \begin{vmatrix} 1 & 0 & -1 \\ 0 & 1 & -1 \\ \dfrac{-1}{1-x} & \dfrac{-1}{1-y} & \dfrac{-z}{1-z} \end{vmatrix} = 0$$

$$\left(\text{using } R_1 \to \frac{1}{x-1}R_1, R_2 \to \frac{1}{y-1}R_2, R_3 \to \frac{1}{z-1}R_3\right)$$

$$\Rightarrow \quad -\frac{1}{(1-x)}(1) + \frac{1}{(1-y)}(-1) - \frac{z}{(1-z)}(1) = 0$$

(expanding along R_3)

$$\Rightarrow \quad \frac{-1}{(1-x)} - \frac{1}{(1-y)} + \frac{(1-z)-1}{(1-z)} = 0$$

$$\Rightarrow \quad \frac{1}{1-x} + \frac{1}{1-y} + \frac{1}{1-z} = 1$$

Ex. 78 *If* $\mathbf{a}, \mathbf{b}$ *and* $\mathbf{c}$ *be any three non-coplanar vectors, then prove that the points* $l_1\mathbf{a} + m_1\mathbf{b} + n_1\mathbf{c}$, $l_2\mathbf{a} + m_2\mathbf{b} + n_2\mathbf{c}$, $l_3\mathbf{a} + m_3\mathbf{b} + n_3\mathbf{c}$ *and* $l_4\mathbf{a} + m_4\mathbf{b} + n_4\mathbf{c}$ *are coplanar, if*

$$\begin{vmatrix} l_1 & m_1 & n_1 & 1 \\ l_2 & m_2 & n_2 & 1 \\ l_3 & m_3 & n_3 & 1 \\ l_4 & m_4 & n_4 & 1 \end{vmatrix} = 0$$

Sol. We know that, four points having position vectors, $\mathbf{a}, \mathbf{b}, \mathbf{c}$ and $\mathbf{d}$ are coplanar, if there exists scalars x, y, z and t such that

$$x\mathbf{a} + y\mathbf{b} + z\mathbf{c} + t\mathbf{d} = 0 \quad \text{where, } x + y + z + t = 0$$

So, the given points will be coplanar, if there exists scalars x, y, z and t such that

$$x(l_1\mathbf{a} + m_1\mathbf{b} + n_1\mathbf{c}) + y(l_2\mathbf{a} + m_2\mathbf{b} + n_2\mathbf{c}) + z(l_3\mathbf{a} + m_3\mathbf{b} + n_3\mathbf{c}) + t(l_4\mathbf{a} + m_4\mathbf{b} + n_4\mathbf{c}) = 0$$

where, $x + y + z + t = 0$

$$\Rightarrow (l_1x + l_2y + l_3z + l_4t)\mathbf{a} + (m_1x + m_2y + m_3z + m_4t)\mathbf{b} + (n_1x + n_2y + n_3z + n_4t)\mathbf{c} = 0$$

where, $x + y + z + t = 0$

$$l_1x + l_2y + l_3z + l_4t = 0 \quad \text{...(i)}$$

$$m_1x + m_2y + m_3z + m_4t = 0 \quad \text{...(ii)}$$

$$n_1x + n_2y + n_3z + n_4t = 0 \quad \text{...(iii)}$$

and $$x + y + z + t = 0 \quad \text{...(iv)}$$

Eliminating x, y, z and t from above equations, we get

$$\begin{vmatrix} l_1 & l_2 & l_3 & l_4 \\ m_1 & m_2 & m_3 & m_4 \\ n_1 & n_2 & n_3 & n_4 \\ 1 & 1 & 1 & 1 \end{vmatrix} = 0$$

Ex. 79 *If $\mathbf{r}_1, \mathbf{r}_2$ and $\mathbf{r}_3$ are the position vectors of three collinear points and scalars l and m exists such that $\mathbf{r}_3 = l\,\mathbf{r}_1 + m\mathbf{r}_2$, then show that $l + m = 1$.*

Sol. Let A, B and C be the three points whose position vectors referred to O are $\mathbf{r}_1$, $\mathbf{r}_2$ and $\mathbf{r}_3$, respectively.

$$\mathbf{AB} = \mathbf{OB} - \mathbf{OA} = \mathbf{r}_2 - \mathbf{r}_1$$

$$\mathbf{BC} = \mathbf{OC} - \mathbf{OB} = \mathbf{r}_3 - \mathbf{r}_2$$

Now, if A, B and C are collinear points, then **AB** and **AC** are in the same line and $\mathbf{BC} = \lambda\,(\mathbf{AC})$

$$\Rightarrow (\mathbf{r}_3 - \mathbf{r}_2) = \lambda(\mathbf{r}_2 - \mathbf{r}_1)$$

$$\Rightarrow \mathbf{r}_3 = -\lambda\mathbf{r}_1 + (\lambda + 1)\,\mathbf{r}_2$$

$$\Rightarrow \mathbf{r}_3 = -\lambda\mathbf{r}_1 + m\mathbf{r}_2$$

where, $l = -\lambda$ and $m = \lambda + 1$

$$\Rightarrow l + m = -\lambda + (\lambda + 1) = 1$$

Ex. 80 *Show that points with position vectors $\mathbf{a} - 2\mathbf{b} + 3\mathbf{c}, -2\mathbf{a} + 3\mathbf{b} - \mathbf{c}$ and $4\mathbf{a} - 7\mathbf{b} + 7\mathbf{c}$ are collinear. It is given that vectors $\mathbf{a}$, $\mathbf{b}$ and $\mathbf{c}$ and non-coplanar.*

Sol. The three points are collinear, if we can find λ_1, λ_2 and λ_3, such that

$\lambda_1(\mathbf{a} - 2\mathbf{b} + 3\mathbf{c}) + \lambda_2(-2\mathbf{a} + 3\mathbf{b} - \mathbf{c}) + \lambda_3$
$(4\mathbf{a} - 7\mathbf{b} + 7\mathbf{c}) = 0$ with $\lambda_1 + \lambda_2 + \lambda_3 = 0$

On equating the coefficients **a**, **b** and **c** separately to zero, we get $\lambda_1 - 2\lambda_2 + 4\lambda_3 = 0$, $-2\lambda_1 + 3\lambda_2 - 7\lambda_3 = 0$ and $3\lambda_1 - \lambda_2 + 7\lambda_3 = 0$

On solving we get $\lambda_1 = -2, \lambda_2 = 1, \lambda_3 = 1$

So that, $\lambda_1 + \lambda_2 + \lambda_3 = 0$

Hence, the given vectors are collinear.

Vector Algebra Exercise 1 :
Single Option Correct Type Questions

1. If $\mathbf{a} = 3\hat{\mathbf{i}} - 2\hat{\mathbf{j}} + \hat{\mathbf{k}}, \mathbf{b} = 2\hat{\mathbf{i}} - 4\hat{\mathbf{j}} - 3\hat{\mathbf{k}}$ and $\mathbf{c} = -\hat{\mathbf{i}} + 2\hat{\mathbf{j}} + 2\hat{\mathbf{k}}$, then $\mathbf{a} + \mathbf{b} + \mathbf{c}$ is

(a) $3\hat{\mathbf{i}} - 4\hat{\mathbf{j}}$ (b) $3\hat{\mathbf{i}} + 4\hat{\mathbf{j}}$
(c) $4\hat{\mathbf{i}} - 4\hat{\mathbf{j}}$ (d) $4\hat{\mathbf{i}} + 4\hat{\mathbf{j}}$

2. What should be added in vector $\mathbf{a} = 3\hat{\mathbf{i}} + 4\hat{\mathbf{j}} - 2\hat{\mathbf{k}}$ to get its resultant a unit vector $\hat{\mathbf{i}}$?

(a) $-2\hat{\mathbf{i}} - 4\hat{\mathbf{j}} + 2\hat{\mathbf{k}}$ (b) $-2\hat{\mathbf{i}} + 4\hat{\mathbf{j}} - 2\hat{\mathbf{k}}$
(c) $2\hat{\mathbf{i}} + 4\hat{\mathbf{j}} - 2\hat{\mathbf{k}}$ (d) None of these

3. If $\mathbf{a} = 2\hat{\mathbf{i}} + \hat{\mathbf{j}} - 8\hat{\mathbf{k}}$ and $\mathbf{b} = \hat{\mathbf{i}} + 3\hat{\mathbf{j}} - 4\hat{\mathbf{k}}$, then the magnitude of $\mathbf{a} + \mathbf{b}$ is equal to

(a) 13 (b) $\frac{13}{3}$
(b) $\frac{3}{13}$ (d) $\frac{4}{13}$

4. If $\mathbf{a} = 2\hat{\mathbf{i}} + 5\hat{\mathbf{j}}$ and $\mathbf{b} = 2\hat{\mathbf{i}} - \hat{\mathbf{j}}$, then the unit vector along $\mathbf{a} + \mathbf{b}$ will be

(a) $\frac{\hat{\mathbf{i}} - \hat{\mathbf{j}}}{\sqrt{2}}$ (b) $\hat{\mathbf{i}} + \hat{\mathbf{j}}$
(c) $\sqrt{2}(\hat{\mathbf{i}} + \hat{\mathbf{j}})$ (d) $\frac{\hat{\mathbf{i}} + \hat{\mathbf{j}}}{\sqrt{2}}$

5. The unit vector parallel to the resultant vector of $2\hat{\mathbf{i}} + 4\hat{\mathbf{j}} - 5\hat{\mathbf{k}}$ and $\hat{\mathbf{i}} + 2\hat{\mathbf{j}} + 3\hat{\mathbf{k}}$ is

(a) $\frac{1}{7}(3\hat{\mathbf{i}} + 6\hat{\mathbf{j}} - 2\hat{\mathbf{k}})$
(b) $\frac{\hat{\mathbf{i}} + \hat{\mathbf{j}} + \hat{\mathbf{k}}}{\sqrt{3}}$
(c) $\frac{\hat{\mathbf{i}} + \hat{\mathbf{j}} + 2\hat{\mathbf{k}}}{\sqrt{6}}$
(d) $\frac{1}{\sqrt{69}}(-\hat{\mathbf{i}} - \hat{\mathbf{j}} + 8\hat{\mathbf{k}})$

6. If $\mathbf{a} = \hat{\mathbf{i}} + 2\hat{\mathbf{j}} + 3\hat{\mathbf{k}}, \mathbf{b} = -\hat{\mathbf{i}} + 2\hat{\mathbf{j}} + \hat{\mathbf{k}}$ and $\mathbf{c} = 3\hat{\mathbf{i}} + \hat{\mathbf{j}}$, then the unit vector along its resultant is

(a) $3\hat{\mathbf{i}} + 5\hat{\mathbf{j}} + 4\hat{\mathbf{k}}$ (b) $\frac{3\hat{\mathbf{i}} + 5\hat{\mathbf{j}} + 4\hat{\mathbf{k}}}{50}$
(c) $\frac{3\hat{\mathbf{i}} + 5\hat{\mathbf{j}} + 4\hat{\mathbf{k}}}{5\sqrt{2}}$ (d) None of these

7. If $\mathbf{a} = (2, 5)$ and $\mathbf{b} = (1, 4)$, then the vector parallel to $(\mathbf{a} + \mathbf{b})$ is

(a) (3, 5) (b) (1, 1)
(c) (1, 3) (d) (8, 5)

8. In the ΔABC, $\mathbf{AB} = \mathbf{a}$, $\mathbf{AC} = \mathbf{c}$ and $\mathbf{BC} = \mathbf{b}$, then

(a) $\mathbf{a} + \mathbf{b} + \mathbf{c} = 0$ (b) $\mathbf{a} + \mathbf{b} - \mathbf{c} = 0$
(c) $\mathbf{a} - \mathbf{b} + \mathbf{c} = 0$ (d) $-\mathbf{a} + \mathbf{b} + \mathbf{c} = 0$

9. If O is the origin and the position vector of A is $4\hat{\mathbf{i}} + 5\hat{\mathbf{j}}$, then a unit vector parallel to $\mathbf{OA}$ is

(a) $\frac{4}{\sqrt{41}}\hat{\mathbf{i}}$ (b) $\frac{5}{\sqrt{41}}\hat{\mathbf{i}}$
(c) $\frac{1}{\sqrt{41}}(4\hat{\mathbf{i}} + 5\hat{\mathbf{j}})$ (d) $\frac{1}{\sqrt{41}}(4\hat{\mathbf{i}} - 5\hat{\mathbf{j}})$

10. The position vectors of the points A, B and C are $\hat{\mathbf{i}} + 2\hat{\mathbf{j}} - \hat{\mathbf{k}}$, $\hat{\mathbf{i}} + \hat{\mathbf{j}} + \hat{\mathbf{k}}$ and $2\hat{\mathbf{i}} + 3\hat{\mathbf{j}} + 2\hat{\mathbf{k}}$, respectively. If A is chosen as the origin, then the position vectors of B and C are

(a) $\hat{\mathbf{i}} + 2\hat{\mathbf{k}}, \hat{\mathbf{i}} + \hat{\mathbf{j}} + 3\hat{\mathbf{k}}$ (b) $\hat{\mathbf{j}} + 2\hat{\mathbf{k}}, \hat{\mathbf{i}} + \hat{\mathbf{j}} + 3\hat{\mathbf{k}}$
(c) $-\hat{\mathbf{j}} + 2\hat{\mathbf{k}}, \hat{\mathbf{i}} - \hat{\mathbf{j}} + 3\hat{\mathbf{k}}$ (d) $-\hat{\mathbf{j}} + 2\hat{\mathbf{k}}, \hat{\mathbf{i}} + \hat{\mathbf{j}} + 3\hat{\mathbf{k}}$

11. The position vectors of P and Q are $5\hat{\mathbf{i}} + 4\hat{\mathbf{j}} + a\hat{\mathbf{k}}$ and $-\hat{\mathbf{i}} + 2\hat{\mathbf{j}} - 2\hat{\mathbf{k}}$, respectively. If the distance between them is 7, then the value of a will be

(a) –5, 1 (b) 5, 1
(c) 0, 5 (d) 1, 0

12. If position vector of points A, B and C are respectively $\hat{\mathbf{i}}, \hat{\mathbf{j}}$ and $\hat{\mathbf{k}}$ and $AB = CX$, then position vector of point X is

(a) $-\hat{\mathbf{i}} + \hat{\mathbf{j}} + \hat{\mathbf{k}}$ (b) $\hat{\mathbf{i}} - \hat{\mathbf{j}} + \hat{\mathbf{k}}$
(c) $\hat{\mathbf{i}} + \hat{\mathbf{j}} - \hat{\mathbf{k}}$ (d) $\hat{\mathbf{i}} + \hat{\mathbf{j}} + \hat{\mathbf{k}}$

13. The position vectors of A and B are $2\hat{\mathbf{i}} - 9\hat{\mathbf{j}} - 4\hat{\mathbf{k}}$ and $6\hat{\mathbf{i}} - 3\hat{\mathbf{j}} + 8\hat{\mathbf{k}}$ respectively, then the magnitude of $\mathbf{AB}$ is

(a) 11 (b) 12
(c) 13 (d) 14

14. If the position vectors of P and Q are $(\hat{\mathbf{i}} + 3\hat{\mathbf{j}} - 7\hat{\mathbf{k}})$ and $(5\hat{\mathbf{i}} - 2\hat{\mathbf{j}} + 4\hat{\mathbf{k}})$, then $|\mathbf{PQ}|$ is

(a) $\sqrt{158}$ (b) $\sqrt{160}$
(c) $\sqrt{161}$ (d) $\sqrt{162}$

15. If the position vectors of P and Q are $\hat{\mathbf{i}} + 2\hat{\mathbf{j}} - 7\hat{\mathbf{k}}$ and $5\hat{\mathbf{i}} - 3\hat{\mathbf{j}} + 4\hat{\mathbf{k}}$ respectively, the cosine of the angle between $\mathbf{PQ}$ and Z-axis is

(a) $\frac{4}{\sqrt{162}}$ (b) $\frac{11}{\sqrt{162}}$
(c) $\frac{5}{\sqrt{162}}$ (d) $\frac{-5}{\sqrt{162}}$

16. If the position vectors of A and B are $\hat{\mathbf{i}} + 3\hat{\mathbf{j}} - 7\hat{\mathbf{k}}$ and $5\hat{\mathbf{i}} - 2\hat{\mathbf{j}} + 4\hat{\mathbf{k}}$, then the direction cosine of $\mathbf{AB}$ along Y-axis is

(a) $\frac{4}{\sqrt{162}}$ (b) $-\frac{5}{\sqrt{162}}$
(c) – 5 (d) 11

17. The direction cosines of vector $\mathbf{a} = 3\hat{\mathbf{i}} + 4\hat{\mathbf{j}} + 5\hat{\mathbf{k}}$ in the direction of positive axis of X, is

(a) $\pm\frac{3}{\sqrt{50}}$ (b) $\frac{4}{\sqrt{50}}$

(c) $\frac{3}{\sqrt{50}}$ (d) $-\frac{4}{\sqrt{50}}$

18. The direction cosines of the vector $3\hat{\mathbf{i}} - 4\hat{\mathbf{j}} + 5\hat{\mathbf{k}}$ are

(a) $\frac{3}{5}, -\frac{4}{5}, \frac{1}{5}$ (b) $\frac{3}{5\sqrt{2}}, \frac{-4}{5\sqrt{2}}, \frac{1}{\sqrt{2}}$

(c) $\frac{3}{\sqrt{2}}, \frac{-4}{\sqrt{2}}, \frac{1}{\sqrt{2}}$ (d) $\frac{3}{5\sqrt{2}}, \frac{4}{5\sqrt{2}}, \frac{1}{\sqrt{2}}$

19. The point having position vectors $2\hat{\mathbf{i}} + 3\hat{\mathbf{j}} + 4\hat{\mathbf{k}}$, $3\hat{\mathbf{i}} + 4\hat{\mathbf{j}} + 2\hat{\mathbf{k}}$ and $4\hat{\mathbf{i}} + 2\hat{\mathbf{j}} + 3\hat{\mathbf{k}}$ are the vertices of

(a) right angled triangle
(b) isosceles triangle
(c) equilateral triangle
(d) collinear

20. If the position vectors of the vertices A, B and C of a ΔABC are $7\hat{\mathbf{j}} + 10\mathbf{k}$, $-\hat{\mathbf{i}} + 6\hat{\mathbf{j}} + 6\hat{\mathbf{k}}$ and $-4\hat{\mathbf{i}} + 9\hat{\mathbf{j}} + 6\hat{\mathbf{k}}$, respectively. The triangle is

(a) equilateral
(b) isosceles
(c) scalene
(d) right angled and isosceles also

21. If $\mathbf{a}$, $\mathbf{b}$ and $\mathbf{c}$ are the position vectors of the vertices A, B and C of the ΔABC, then the centroid of ΔABC is

(a) $\frac{\mathbf{a}+\mathbf{b}+\mathbf{c}}{3}$ (b) $\frac{1}{2}\left(\mathbf{a} + \frac{\mathbf{b}+\mathbf{c}}{2}\right)$

(c) $\mathbf{a} + \frac{\mathbf{b}+\mathbf{c}}{2}$ (d) $\frac{\mathbf{a}+\mathbf{b}+\mathbf{c}}{2}$

22. If in the given figure, $\mathbf{OA} = \mathbf{a}$, $\mathbf{OB} = \mathbf{b}$ and $AP : PB = m : n$, then $\mathbf{OP}$ is equal to

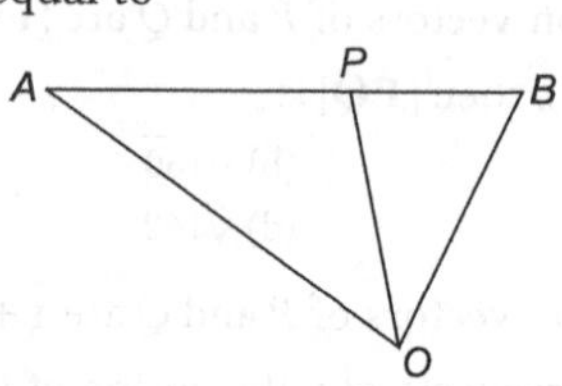

(a) $\frac{m\mathbf{a} + n\mathbf{b}}{m+n}$ (b) $\frac{n\mathbf{a} + m\mathbf{b}}{m+n}$

(c) $m\mathbf{a} - n\mathbf{b}$ (d) $\frac{m\mathbf{a} - n\mathbf{b}}{m-n}$

23. If $\mathbf{a}$ and $\mathbf{b}$ are position vector of two points A, B and C divides AB in ratio $2:1$, then position vector of C is

(a) $\frac{\mathbf{a}+2\mathbf{b}}{3}$ (b) $\frac{2\mathbf{a}+\mathbf{b}}{3}$

(c) $\frac{\mathbf{a}+2}{3}$ (d) $\frac{\mathbf{a}+\mathbf{b}}{2}$

24. The position vector of the points which divides internally in the ratio $2:3$ the join of the points $2\mathbf{a} - 3\mathbf{b}$ and $3\mathbf{a} - 2\mathbf{b}$, is

(a) $\frac{12}{5}\mathbf{a} + \frac{13}{5}\mathbf{b}$ (b) $\frac{12}{5}\mathbf{a} - \frac{13}{5}\mathbf{b}$

(c) $\frac{3}{5}\mathbf{a} - \frac{2}{5}\mathbf{b}$ (d) None of these

25. If O is origin and C is the mid-point of $A(2, -1)$ and $B(-4, 3)$. Then, value of $\mathbf{OC}$ is

(a) $\hat{\mathbf{i}} + \hat{\mathbf{j}}$ (b) $\hat{\mathbf{i}} - \hat{\mathbf{j}}$
(c) $-\hat{\mathbf{i}} + \hat{\mathbf{j}}$ (d) $-\hat{\mathbf{i}} - \hat{\mathbf{j}}$

26. If the position vectors of the points A and B are $\hat{\mathbf{i}} + 3\hat{\mathbf{j}} - \hat{\mathbf{k}}$ and $3\hat{\mathbf{i}} - \hat{\mathbf{j}} - 3\hat{\mathbf{k}}$, then what will be the position vector of the mid-point of AB

(a) $\hat{\mathbf{i}} + 2\hat{\mathbf{j}} - \hat{\mathbf{k}}$ (b) $2\hat{\mathbf{i}} + \hat{\mathbf{j}} - 2\hat{\mathbf{k}}$
(c) $2\hat{\mathbf{i}} + \hat{\mathbf{j}} - \hat{\mathbf{k}}$ (d) $\hat{\mathbf{i}} + \hat{\mathbf{j}} - 2\hat{\mathbf{k}}$

27. The position vectors of A and B are $\hat{\mathbf{j}} - \hat{\mathbf{j}} + 2\hat{\mathbf{k}}$ and $3\hat{\mathbf{i}} - \hat{\mathbf{j}} + 3\hat{\mathbf{k}}$. The position vector of the middle point of the line AB is

(a) $\frac{1}{2}\hat{\mathbf{i}} - \frac{1}{2}\hat{\mathbf{j}} + \hat{\mathbf{k}}$ (b) $2\hat{\mathbf{i}} - \hat{\mathbf{j}} + \frac{5}{2}\hat{\mathbf{k}}$

(c) $\frac{3}{2}\hat{\mathbf{i}} - \frac{1}{2}\hat{\mathbf{j}} + \frac{3}{2}\hat{\mathbf{k}}$ (d) None of these

28. If the vector $\mathbf{b}$ is collinear with the vector $\mathbf{a} = (2\sqrt{2}, -1, 4)$ and $|\mathbf{b}| = 10$, then

(a) $\mathbf{a} \pm \mathbf{b} = 0$ (b) $\mathbf{a} \pm 2\mathbf{b} = 0$
(c) $2\mathbf{a} \pm \mathbf{b} = 0$ (d) None of these

29. If $\mathbf{a} = (1, -1)$ and $\mathbf{b} = (-2, m)$ are two collinear vectors, then m is equal to

(a) 4 (b) 3
(c) 2 (d) 0

30. The points with position vectors $10\hat{\mathbf{i}} + 3\hat{\mathbf{j}}$, $12\hat{\mathbf{i}} - 5\hat{\mathbf{j}}$ and $a\hat{\mathbf{i}} + 11\hat{\mathbf{j}}$ are collinear, if a is equal to

(a) -8 (b) 4
(c) 8 (d) 12

31. The vectors $\hat{\mathbf{i}} + 2\hat{\mathbf{j}} + 3\hat{\mathbf{k}}$, $\lambda\hat{\mathbf{i}} + 4\hat{\mathbf{j}} + 7\hat{\mathbf{k}}$, $-3\hat{\mathbf{i}} - 2\hat{\mathbf{j}} - 5\hat{\mathbf{k}}$ are collinear, if λ is equal to

(a) 3 (b) 4
(c) 5 (d) 6

32. If the points $\mathbf{a} + \mathbf{b}$, $\mathbf{a} - \mathbf{b}$ and $\mathbf{a} + k\mathbf{b}$ be collinear, then k is equal to

(a) 0 (b) 2
(c) -2 (d) Any real number

33. If the position vectors of A, B, C and D are $2\hat{\mathbf{i}} + \hat{\mathbf{j}}$, $\hat{\mathbf{i}} - 3\hat{\mathbf{j}}$, $3\hat{\mathbf{i}} + 2\hat{\mathbf{j}}$ and $\hat{\mathbf{i}} + \lambda\hat{\mathbf{j}}$ respectively and $\mathbf{AB} \parallel \mathbf{CD}$, then λ will be

(a) -8 (b) -6
(c) 8 (d) 6

34. If the vectors $3\hat{i}+2\hat{j}-\hat{k}$ and $6\hat{i}-4x\hat{j}+y\hat{k}$ are parallel, then the value of x and y will be
(a) $-1, -2$ (b) $1, -2$
(c) $-1, 2$ (d) $1, 2$

35. If $\mathbf{a}$ and $\mathbf{b}$ are two non-collinear vectors and $x\mathbf{a}+y\mathbf{b}=0$
(a) $x=0$, but y is not necessarily zero
(b) $y=0$, but x is not necessarily zero
(c) $x=0$, $y=0$
(d) None of the above

36. Four non-zero vectors will always be
(a) linearly dependent
(b) linearly independent
(c) either (a) or (b)
(d) None of the above

37. The vectors $\mathbf{a}$, $\mathbf{b}$ and $\mathbf{a}+\mathbf{b}$ are
(a) collinear (b) coplanar
(c) non-coplanar (d) None of these

38. If $(x, y, z) \neq (0, 0, 0)$ and $(\hat{i}+\hat{j}+3\hat{k})x+(3\hat{i}-3\hat{j}+\hat{k})y$ $+(-4\hat{i}+5\hat{j})z=\lambda(x\hat{i}+y\hat{j}+z\hat{k})$, then the value of λ will be
(a) $-2, 0$ (b) $0, -2$
(c) $-1, 0$ (d) $0, -1$

39. The number of integral values of p for which $(p+1)\hat{i}-3\hat{j}+p\hat{i}, p\hat{i}+(p+1)\hat{j}-3\hat{k}$ and $-3\hat{i}+p\hat{j}+(p+1)\hat{k}$ are linearly dependent vectors is
(a) 0 (b) 1
(c) 2 (d) 3

40. The vectors $\mathbf{AB}=3\hat{i}+4\hat{k}$ and $\mathbf{AC}=5\hat{i}-2\hat{j}+4\hat{k}$ are the sides of a ΔABC. The length of the median through A is
(a) $\sqrt{18}$ (b) $\sqrt{72}$
(c) $\sqrt{33}$ (d) $\sqrt{288}$

41. In the figure, a vectors x satisfies the equation $x-w=v$. Then, x is equal to

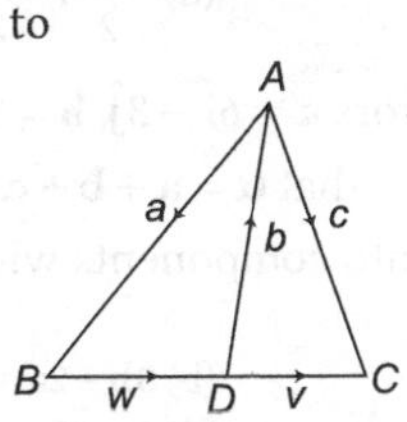

(a) $2\mathbf{a}+\mathbf{b}+\mathbf{c}$ (b) $\mathbf{a}+2\mathbf{b}+\mathbf{c}$
(c) $\mathbf{a}+\mathbf{b}+2\mathbf{c}$ (d) $\mathbf{a}+\mathbf{b}+\mathbf{c}$

42. Vectors $\mathbf{a}=\hat{i}+2\hat{j}+3\hat{k}$, $\mathbf{b}=2\hat{i}-\hat{j}+\hat{k}$ and $\mathbf{c}=3\hat{i}+\hat{j}+4\hat{k}$ are so placed that the end point of one vector is the starting point of the next vector. Then the vectors are
(a) not coplanar
(b) coplanar but cannot form a triangle
(c) coplanar and form a triangle
(d) coplanar and can form a right angled triangle

43. If $OP=8$ and $\mathbf{OP}$ makes angles 45° and 60° with OX-axis and OY-axis respectively, then $\mathbf{OP}$ is equal to
(a) $8(\sqrt{2}\hat{i}+\hat{j}\pm\hat{k})$ (b) $4(\sqrt{2}\hat{i}+\hat{j}\pm\hat{k})$
(c) $\frac{1}{4}(\sqrt{2}\hat{i}+\hat{j}\pm\hat{k})$ (d) $\frac{1}{8}(\sqrt{2}\hat{i}+\hat{j}\pm\hat{k})$

44. Let $\mathbf{a}$, $\mathbf{b}$ and $\mathbf{c}$ be three units vectors such that $3\mathbf{a}+4\mathbf{b}+5\mathbf{c}=0$. Then which of the following statements is true?
(a) $\mathbf{a}$ is parallel to $\mathbf{b}$
(b) $\mathbf{a}$ is perpendicular to $\mathbf{b}$
(c) $\mathbf{a}$ is neither parallel nor perpendicular to $\mathbf{b}$
(d) None of the above

45. A, B, C, D and E are five coplanar points, then $\mathbf{DA}+\mathbf{DB}+\mathbf{DC}+\mathbf{AE}+\mathbf{BE}+\mathbf{CE}$ is equal to
(a) $\mathbf{DE}$ (b) $3\mathbf{DE}$
(c) $2\mathbf{DE}$ (d) $4\mathbf{ED}$

46. If the vectors $\mathbf{a}$ and $\mathbf{b}$ are linearly independent satisfying $(\sqrt{3}\tan\theta+1)\,\mathbf{a}+(\sqrt{3}\sec\theta-2)\,\mathbf{b}=0$, then the most general value of θ are
(a) $n\pi-\frac{\pi}{6}, n\in Z$ (b) $2n\pi\pm\frac{11\pi}{6}, n\in Z$
(c) $n\pi\pm\frac{\pi}{6}, n\in Z$ (d) $2n\pi+\frac{11\pi}{6}, n\in Z$

47. The unit vector bisecting $\mathbf{OY}$ and $\mathbf{OZ}$ is
(a) $\frac{\hat{i}+\hat{j}+\hat{k}}{\sqrt{3}}$ (b) $\frac{\hat{j}-\hat{k}}{\sqrt{2}}$
(c) $\frac{\hat{j}+\hat{k}}{\sqrt{2}}$ (d) $\frac{-\hat{j}+\hat{k}}{\sqrt{2}}$

48. A line passes through the points whose position vectors are $\hat{i}+\hat{j}-2\hat{k}$ and $\hat{i}-3\hat{j}+\hat{k}$. The position vector of a point on it at unit distance from the first point is
(a) $\frac{1}{5}(5\hat{i}+\hat{j}-7\hat{k})$ (b) $\frac{1}{5}(4\hat{i}+9\hat{j}-15\hat{k})$
(c) $(\hat{i}-4\hat{j}+3\hat{k})$ (d) $\frac{1}{5}(\hat{i}-4\hat{j}+3\hat{k})$

49. If D, E and F are the middle points of the sides BC, CA and AB of the ΔABC, then $\mathbf{AD}+\mathbf{BE}+\mathbf{CF}$ is
(a) a zero vector (b) a unit vector
(c) 0 (d) None of these

50. If P and Q are the middle points of the sides BC and CD of the parallelogram $ABCD$, then $\mathbf{AP}+\mathbf{AQ}$ is equal to
(a) $\mathbf{AC}$ (b) $\frac{1}{2}\mathbf{AC}$
(b) $\frac{2}{3}\mathbf{AC}$ (d) $\frac{3}{2}\mathbf{AC}$

51. The figure formed by the four points $\hat{i}+\hat{j}-\hat{k}$, $2\hat{i}+3\hat{j}$, $3\hat{i}+5\hat{j}-2\hat{k}$ and $\hat{k}-\hat{j}$ is
(a) rectangle (b) parallelogram
(c) trapezium (d) None of these

52. A and B are two points. The position vector of $\mathbf{A}$ is $6\mathbf{b}-2\mathbf{a}$. A point P divides the line AB in the ratio $1:2$. If $\mathbf{a}-\mathbf{b}$ is the position vector of P, then the position vector of B is given by
(a) $7\mathbf{a}-15\mathbf{b}$
(b) $7\mathbf{a}+15\mathbf{b}$
(c) $15\mathbf{a}-7\mathbf{b}$
(d) $15\mathbf{a}+7\mathbf{b}$

53. If three points A, B and C are collinear, whose position vectors are $\hat{\mathbf{i}}-2\hat{\mathbf{j}}-8\hat{\mathbf{k}}$, $5\hat{\mathbf{i}}-2\hat{\mathbf{k}}$ and $11\hat{\mathbf{i}}+3\hat{\mathbf{j}}+7\hat{\mathbf{k}}$ respectively, then the ratio in which B divides AC is
(a) $1:2$ (b) $2:3$
(c) $2:1$ (d) $1:1$

54. If in a triangle, $\mathbf{AB}=\mathbf{a}$, $\mathbf{AC}=\mathbf{b}$ and D, E are the mid-points of AB and AC respectively, then $\mathbf{DE}$ is equal to
(a) $\frac{\mathbf{a}}{4}-\frac{\mathbf{b}}{4}$ (b) $\frac{\mathbf{a}}{2}-\frac{\mathbf{b}}{2}$
(c) $\frac{\mathbf{b}}{4}-\frac{\mathbf{a}}{4}$ (d) $\frac{\mathbf{b}}{2}-\frac{\mathbf{a}}{2}$

55. If $ABCD$ is parallelogram, $\mathbf{AB}=2\hat{\mathbf{i}}+4\hat{\mathbf{j}}-5\hat{\mathbf{k}}$ and $\mathbf{AD}=\hat{\mathbf{i}}+2\hat{\mathbf{j}}+3\hat{\mathbf{k}}$, then the unit vectors in the direction of BD is
(a) $\frac{1}{\sqrt{69}}(\hat{\mathbf{i}}+2\hat{\mathbf{j}}-8\hat{\mathbf{k}})$ (b) $\frac{1}{69}(\hat{\mathbf{i}}+2\hat{\mathbf{j}}-8\hat{\mathbf{k}})$
(c) $\frac{1}{\sqrt{69}}(-\hat{\mathbf{i}}-2\hat{\mathbf{j}}+8\hat{\mathbf{k}})$ (d) $\frac{1}{69}(-\hat{\mathbf{i}}-2\hat{\mathbf{j}}+8\hat{\mathbf{k}})$

56. If A, B and C are the vertices of a triangle whose position vectors are $\mathbf{a}$, $\mathbf{b}$ and $\mathbf{c}$ and G is the centroid of the ΔABC, then $\mathbf{GA}+\mathbf{GB}+\mathbf{GC}$ is
(a) 0 (b) $\mathbf{A}+\mathbf{B}+\mathbf{C}$
(c) $\frac{\mathbf{a}+\mathbf{b}+\mathbf{c}}{3}$ (d) $\frac{\mathbf{a}+\mathbf{b}-\mathbf{c}}{3}$

57. If $ABCDEF$ is regular hexagon, then $\mathbf{AD}+\mathbf{EB}+\mathbf{FC}$ is equal to
(a) 0 (b) $2\mathbf{AB}$
(c) $3\mathbf{AB}$ (d) $4\ \mathbf{AB}$

58. $ABCDE$ is a pentagon. Forces $\mathbf{AB}$, $\mathbf{AE}$, $\mathbf{DC}$ and $\mathbf{ED}$ act at a point. Which force should be added to this system to make the resultant $2\mathbf{AC}$?
(a) $\mathbf{AC}$ (b) $\mathbf{AD}$
(c) $\mathbf{BC}$ (d) $\mathbf{BD}$

59. If $ABCDEF$ is a regular hexagon and $\mathbf{AB}+\mathbf{AC}+\mathbf{AD}+\mathbf{AE}+\mathbf{AF}=\lambda\mathbf{AD}$, then λ is equal to
(a) 2 (b) 3
(c) 4 (d) 6

60. Let us define the length of a vector $a\hat{\mathbf{i}}+b\hat{\mathbf{j}}+c\hat{\mathbf{k}}$ as $|a|+|b|+|c|$. This definition coincides with the usual definition of length of a vector $a\hat{\mathbf{i}}+b\hat{\mathbf{j}}+c\hat{\mathbf{k}}$ if and only if
(a) $a=b=c=0$
(b) any two of a,b and c are zero
(c) any one of a,b and c is zero
(d) $a+b+c=0$

61. If $\mathbf{a}$ and $\mathbf{b}$ are two non-zero and non-collinear vectors, then $\mathbf{a}+\mathbf{b}$ and $\mathbf{a}-\mathbf{b}$ are
(a) linearly dependent vectors
(b) linearly independent vectors
(c) linearly dependent and independent vectors
(d) None of the above

62. If $|\mathbf{a}+\mathbf{b}|<|\mathbf{a}-\mathbf{b}|$, then the angle between $\mathbf{a}$ and $\mathbf{b}$ can lie in the interval.
(a) $(-\pi/2, \pi/2)$ (b) $(0, \pi)$
(c) $(\pi/2, 3\pi/2)$ (d) $(0, 2\pi)$

63. The magnitudes of mutually perpendicular forces $\mathbf{a}$, $\mathbf{b}$ and $\mathbf{c}$ are 2, 10 and 11 respectively. Then the magnitude of its resultant is
(a) 12 (b) 15
(c) 9 (d) None of these

64. If $\hat{\mathbf{i}}-3\hat{\mathbf{j}}+\hat{\mathbf{k}}$ bisects the angle between $\mathbf{a}$ and $-\hat{\mathbf{i}}+2\hat{\mathbf{j}}+2\hat{\mathbf{k}}$, where $\mathbf{a}$ is a unit vector, then
(a) $\mathbf{a}=\frac{1}{105}(41\hat{\mathbf{i}}+88\hat{\mathbf{j}}-40\hat{\mathbf{k}})$
(b) $\mathbf{a}=\frac{1}{105}(41\hat{\mathbf{i}}+88\hat{\mathbf{j}}+40\hat{\mathbf{k}})$
(c) $\mathbf{a}=\frac{1}{105}(-41\hat{\mathbf{i}}+88\hat{\mathbf{j}}-40\hat{\mathbf{k}})$
(d) $\mathbf{a}=\frac{1}{105}(41\hat{\mathbf{i}}-88\hat{\mathbf{j}}-40\hat{\mathbf{k}})$

65. Let $\mathbf{a}=\hat{\mathbf{i}}$ be a vector which makes an angle of 120° with a unit vector $\mathbf{b}$. Then, the unit vector $(\mathbf{a}+\mathbf{b})$ is
(a) $-\frac{1}{2}\hat{\mathbf{i}}+\frac{\sqrt{3}}{2}\hat{\mathbf{j}}$ (b) $-\frac{\sqrt{3}}{2}\hat{\mathbf{i}}+\frac{1}{2}\hat{\mathbf{j}}$
(c) $\frac{1}{2}\hat{\mathbf{i}}+\frac{\sqrt{3}}{2}\hat{\mathbf{j}}$ (d) $\frac{\sqrt{3}}{2}\hat{\mathbf{i}}-\frac{1}{2}\hat{\mathbf{j}}$

66. Given three vectors $\mathbf{a}=6\hat{\mathbf{i}}-3\hat{\mathbf{j}}$, $\mathbf{b}=2\hat{\mathbf{i}}-6\hat{\mathbf{j}}$ and $\mathbf{c}=-2\hat{\mathbf{i}}+21\hat{\mathbf{j}}$ such that $\alpha=\mathbf{a}+\mathbf{b}+\mathbf{c}$. Then, the resolution of the vector α into components with respect to $\mathbf{a}$ and $\mathbf{b}$ is given by
(a) $3\mathbf{a}-2\mathbf{b}$ (b) $3\mathbf{b}-2\mathbf{a}$
(c) $2\mathbf{a}-3\mathbf{b}$ (d) $\mathbf{a}-2\mathbf{b}$

67. 'I' is the incentre of ΔABC whose corresponding sides are a, b, c respectively. $a\,\mathbf{IA}+b\,\mathbf{IB}+c\,\mathbf{IC}$ is always equal to
(a) $\mathbf{0}$ (b) $(a+b+c)\mathbf{BC}$
(c) $(\mathbf{a}+\mathbf{b}+\mathbf{c})\mathbf{AC}$ (d) $(a+b+c)\mathbf{AB}$

68. If $\mathbf{x}$ and $\mathbf{y}$ are two non-collinear vectors and ABC is a triangle with side lengths a, b and c satisfying $(20a-15b)\mathbf{x}+(15b-12c)\mathbf{y}+(12c-20a)(\mathbf{x}\times\mathbf{y})=\mathbf{0}$, then ΔABC is
(a) an acute angled triangle (b) an obtuse angled triangle
(c) a right angled triangle (d) a scalene triangle

69. If **x** and **y** are two non-collinear vectors and a, b and c represent the sides of a ΔABC satisfying $(a-b)\mathbf{x}+(b-c)\mathbf{y}+(c-a)(\mathbf{x}\times\mathbf{y})=0$, then ΔABC is (where $\mathbf{x}\times\mathbf{y}$ is perpendicular to the plane of **x** and **y**)
(a) an acute angled triangle
(b) an obtuse angled triangle
(c) a right angled triangle
(d) a scalene triangle

70. If the resultant of two forces is of magnitude P and equal to one of them and perpendicular to it, then the other force is
(a) $P\sqrt{2}$ (b) P
(c) $P\sqrt{3}$ (d) None of these

71. If **b** is a vector whose initial point divides the join of $5\hat{\mathbf{i}}$ and $5\hat{\mathbf{j}}$ in the ratio $k:1$ and whose terminal point in the origin and $|\mathbf{b}|\le\sqrt{37}$, then k lies in the interval
(a) $[-6,-1/6]$ (b) $[-\infty,-6]\cup[-1/6,\infty]$
(c) $[0,6]$ (d) None of these

72. If $4\hat{\mathbf{j}}+7\hat{\mathbf{j}}+8\hat{\mathbf{k}}$, $2\hat{\mathbf{i}}+3\hat{\mathbf{j}}+4\hat{\mathbf{k}}$ and $2\hat{\mathbf{i}}+5\hat{\mathbf{j}}+7\hat{\mathbf{k}}$ are the position vectors of the vertices A, B and C respectively of ΔABC. The position vector of the point where the bisector of $\angle A$ meets BC is
(a) $\frac{1}{3}(6\hat{\mathbf{i}}+13\hat{\mathbf{j}}+18\hat{\mathbf{k}})$ (b) $\frac{2}{3}(6\hat{\mathbf{i}}+12\hat{\mathbf{j}}-8\hat{\mathbf{k}})$
(c) $\frac{1}{3}(-6\hat{\mathbf{i}}-8\hat{\mathbf{j}}-9\hat{\mathbf{k}})$ (d) $\frac{2}{3}(-6\hat{\mathbf{i}}-12\hat{\mathbf{j}}+8\hat{\mathbf{k}})$

73. If **a** and **b** are two unit vectors and θ is the angle between them, then the unit vector along the angular bisector of **a** and **b** will be given by
(a) $\frac{\mathbf{a}-\mathbf{b}}{2\cos(\theta/2)}$ (b) $\frac{\mathbf{a}+\mathbf{b}}{2\cos(\theta/2)}$
(c) $\frac{\mathbf{a}-\mathbf{b}}{\cos(\theta/2)}$ (d) None of these

74. A, B, C and D have position vectors **a**, **b**, **c** and **d**, respectively, such that $\mathbf{a}-\mathbf{b}=2(\mathbf{d}-\mathbf{c})$. Then,
(a) AB and CD bisect each other
(b) BD and AC bisect each other
(c) AB and CD trisect each other
(d) BD and AC trisect each other

75. On the xy plane where O is the origin, given points, $A(1,0)$, $B(0,1)$ and $C(1,1)$. Let P, Q and R be moving point on the line OA, OB, OC respectively such that $\mathbf{OP}=45t(\mathbf{OA})$, $\mathbf{OQ}=60t(\mathbf{OB})$, $\mathbf{OR}=(1-t)(\mathbf{OC})$ with $t>0$. If the three points P, Q and R are collinear, then the value of t is equal to
(a) $\frac{1}{106}$ (b) $\frac{7}{187}$
(c) $\frac{1}{100}$ (d) None of these

76. If **a**, **b** and **c** are three non-coplanar vectors such that $\mathbf{a}+\mathbf{b}+\mathbf{c}=\alpha\mathbf{d}$ and $\mathbf{b}+\mathbf{c}+\mathbf{d}=\beta\mathbf{a}$, then $\mathbf{a}+\mathbf{b}+\mathbf{c}+\mathbf{d}$ is equal to
(a) 0 (b) $\alpha\mathbf{a}$
(c) $\beta\mathbf{b}$ (d) $(\alpha+\beta)\mathbf{c}$

77. The position vectors of the points P and Q with respect to the origin O are $\mathbf{a}=\hat{\mathbf{i}}+3\hat{\mathbf{j}}-2\hat{\mathbf{k}}$ and $\mathbf{b}=3\hat{\mathbf{i}}-\hat{\mathbf{j}}-2\hat{\mathbf{k}}$, respectively. If M is a point on PQ, such that OM is the bisector of POQ, then OM is
(a) $2(\hat{\mathbf{i}}-\hat{\mathbf{j}}+\hat{\mathbf{k}})$ (b) $2\hat{\mathbf{i}}+\hat{\mathbf{j}}-2\hat{\mathbf{k}}$
(c) $2(-\hat{\mathbf{i}}+\hat{\mathbf{j}}-\hat{\mathbf{k}})$ (b) $2(\hat{\mathbf{i}}+\hat{\mathbf{j}}+\hat{\mathbf{k}})$

78. $ABCD$ is a quadrilateral. E is the point of intersection of the line joining the mid-points of the opposite sides. If O is any point and $\mathbf{OA}+\mathbf{OB}+\mathbf{OC}+\mathbf{OD}=x\mathbf{OE}$, then x is equal to
(a) 3 (b) 9
(c) 7 (d) 4

79. In the ΔOAB, M is the mid-point of AB, C is a point on OM, such that $2\mathbf{OC}=\mathbf{CM}$. X is a point on the side OB such that $\mathbf{OX}=2\mathbf{XB}$. The line XC is produced to meet OA in Y. Then, $\frac{OY}{YA}$ is equal to
(a) $\frac{1}{3}$ (b) $\frac{2}{7}$
(c) $\frac{3}{2}$ (d) $\frac{2}{5}$

80. Points X and Y are taken on the sides QR and RS, respectively of a parallelogram $PQRS$, so that $\mathbf{QX}=4\mathbf{XR}$ and $\mathbf{RY}=4\mathbf{YS}$. The line XY cuts the line PR at Z. Then, **PZ** is
(a) $\frac{21}{25}\mathbf{PR}$ (b) $\frac{16}{25}\mathbf{PR}$
(c) $\frac{17}{25}\mathbf{PR}$ (d) None of these

81. Find the value of λ so that the points P, Q, R and S on the sides OA, OB, OC and AB, respectively, of a regular tetrahedron $OABC$ are coplanar. It is given that $\frac{OP}{OA}=\frac{1}{3},\frac{OQ}{OB}=\frac{1}{2},\frac{OR}{OC}=\frac{1}{3}$ and $\frac{OS}{AB}=\lambda$.
(a) $\lambda=\frac{1}{2}$ (b) $\lambda=-1$
(c) $\lambda=0$ (d) for no value of λ

82. $OABCDE$ is a regular hexagon of side 2 units in the XY-plane. O being the origin and OA taken along the X-axis. A point P is taken on a line parallel to Z-axis through the centre of the hexagon at a distance of 3 units from O. Then, the vector **AP** is
(a) $-\hat{\mathbf{i}}+3\hat{\mathbf{j}}+\sqrt{5}\hat{\mathbf{k}}$ (b) $\hat{\mathbf{i}}-\sqrt{3}\hat{\mathbf{j}}+5\hat{\mathbf{k}}$
(c) $-\hat{\mathbf{i}}+\sqrt{3}\hat{\mathbf{j}}+\sqrt{5}\hat{\mathbf{k}}$ (d) $\hat{\mathbf{i}}+\sqrt{3}\hat{\mathbf{j}}+\sqrt{5}\hat{\mathbf{k}}$

Vector Algebra Exercise 2 : More than One Option Correct Type Questions

83. If the vectors $\hat{\mathbf{i}} - \hat{\mathbf{j}}, \hat{\mathbf{j}} + \hat{\mathbf{k}}$ and $\mathbf{a}$ form a triangle, then $\mathbf{a}$ may be

(a) $-\hat{\mathbf{i}} - \hat{\mathbf{k}}$ (b) $\hat{\mathbf{i}} - 2\hat{\mathbf{j}} - \hat{\mathbf{k}}$

(c) $2\hat{\mathbf{j}} + \hat{\mathbf{j}} + \hat{\mathbf{k}}$ (d) $\hat{\mathbf{i}} + \hat{\mathbf{k}}$

84. If the resultant of three forces
$\mathbf{F}_1 = p\hat{\mathbf{i}} + 3\hat{\mathbf{j}} - \hat{\mathbf{k}}, \mathbf{F}_2 = 6\hat{\mathbf{i}} - \hat{\mathbf{k}}$ and $\mathbf{F}_3 = -5\hat{\mathbf{i}} + \hat{\mathbf{j}} + 2\hat{\mathbf{k}}$
acting on a particle has a magnitude equal to 5 units, then the value of p is

(a) −6 (b) −4

(c) 2 (d) 4

85. Let ABC be a triangle, the position vectors of whose vertices are $7\hat{\mathbf{j}} + 10\hat{\mathbf{k}}, -\hat{\mathbf{i}} + 6\hat{\mathbf{j}} + 6\hat{\mathbf{k}}$ and $-4\hat{\mathbf{i}} + 9\hat{\mathbf{j}} + 6\hat{\mathbf{k}}$. Then, ΔABC is

(a) isosceles (b) equilateral

(c) right angled (d) None of these

86. The sides of a parallelogram are $2\hat{\mathbf{i}} + 4\hat{\mathbf{j}} - 5\hat{\mathbf{k}}$ and $\hat{\mathbf{i}} + 2\hat{\mathbf{j}} + 3\hat{\mathbf{k}}$. The unit vector parallel to one of the diagonals is

(a) $\frac{1}{7}(3\hat{\mathbf{i}} + 6\hat{\mathbf{j}} - 2\hat{\mathbf{k}})$ (b) $\frac{1}{7}(3\hat{\mathbf{i}} - 6\hat{\mathbf{j}} - 2\hat{\mathbf{k}})$

(c) $\frac{1}{\sqrt{69}}(\hat{\mathbf{i}} + 2\hat{\mathbf{j}} + 8\hat{\mathbf{k}})$ (d) $\frac{1}{\sqrt{69}}(-\hat{\mathbf{i}} - 2\hat{\mathbf{j}} + 8\hat{\mathbf{k}})$

87. If $A(-4, 0, 3)$ and $B(14, 2, -5)$, then which one of the following points lie on the bisector of the angle between **OA** and **OB** (O is the origin of reference)?

(a) (2, 2, 4) (b) (2, 11, 5)

(c) (−3, −3, −6) (d) (1, 1, 2)

88. If point $\hat{\mathbf{i}} + \hat{\mathbf{j}}, \hat{\mathbf{i}} - \hat{\mathbf{j}}$ and $p\hat{\mathbf{i}} + q\hat{\mathbf{j}} + r\hat{\mathbf{k}}$ are collinear, then

(a) $p = 1$ (b) $r = 0$

(c) $q \in R$ (d) $q \neq 1$

89. If $\mathbf{a}$, $\mathbf{b}$ and $\mathbf{c}$ are non-coplanar vectors and λ is a real number, then the vectors $\mathbf{a} + 2\mathbf{b} + 3\mathbf{c}$, $\lambda\mathbf{b} + \mu\mathbf{c}$ and $(2\lambda - 1)\mathbf{c}$ are coplanar when

(a) $\mu \in R$ (b) $\lambda = \frac{1}{2}$

(c) $\lambda = 0$ (d) no value of λ

Vector Algebra Exercise 3 : Statement I and II Type Questions

- **Directions** (Q. Nos. 90-92) This section is based on Statement I and Statement II. Select the correct answer from the codes given below.

(a) Both Statement I and Statement II are correct and Statement II is the correct explanation of Statement I

(b) Both Statement I and Statement II are correct but Statement II is not the correct explanation of Statement I

(c) Statement I is correct but Statement II is incorrect

(d) Statement II is correct but Statement I is incorrect

90. Statement I In ΔABC, $\mathbf{AB} + \mathbf{BC} + \mathbf{CA} = 0$

Statement II If $\mathbf{OA} = \mathbf{a}, \mathbf{OB} = \mathbf{b}$, then $\mathbf{AB} = \mathbf{a} + \mathbf{b}$

91. Statement I $\mathbf{a} = \hat{\mathbf{i}} + p\hat{\mathbf{j}} + 2\hat{\mathbf{k}}$ and $\mathbf{b} = 2\hat{\mathbf{i}} + 3\hat{\mathbf{j}} + q\hat{\mathbf{k}}$ are parallel vectors, if $p = \frac{3}{2}$ and $q = 4$.

Statement II $\mathbf{a} = a_1\hat{\mathbf{i}} + a_2\hat{\mathbf{j}} + a_3\hat{\mathbf{k}}$ and $\mathbf{b} = b_1\hat{\mathbf{i}} + b_2\hat{\mathbf{j}} + b_3\hat{\mathbf{k}}$ are parallel $\frac{a_1}{b_1} = \frac{a_2}{b_2} = \frac{a_3}{b_3}$.

92. Statement I If three points P, Q and R have position vectors $\mathbf{a}$, $\mathbf{b}$ and $\mathbf{c}$ respectively, and $2\mathbf{a} + 3\mathbf{b} - 5\mathbf{c} = 0$, then the points P, Q and R must be collinear.

Statement II If for three points A, B and C, $\mathbf{AB} = \lambda\,\mathbf{AC}$, then points A, B and C must be collinear.

Vector Algebra Exercise 4 : Passage based Type Questions

Passage I

(Q. Nos. 93 and 94)

- Let $OABCD$ be a pentagon in which the sides OA and CB are parallel and the sides OD and AB are parallel. Also, $OA : CB = 2 : 1$ and $OD : AB = 1 : 3$.

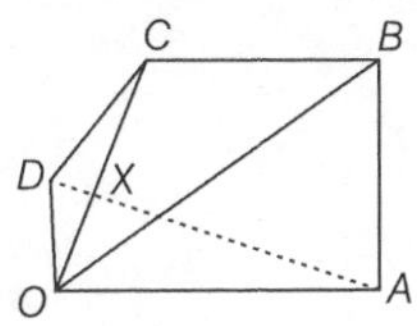

93. The ratio $\frac{OX}{XC}$ is

(a) 3/4 (b) 1/3
(c) 2/5 (d) 1/2

94. The ratio $\frac{AX}{XD}$ is

(a) 5/2 (b) 6
(c) 7/3 (d) 4

Passage II

(Q. Nos. 95 and 96)

- Consider the regular hexagon $ABCDEF$ with centre at O (origin).

95. $\mathbf{AD}+\mathbf{EB}+\mathbf{FC}$ is equal to

(a) $2\mathbf{AB}$ (b) $3\mathbf{AB}$
(c) $4\mathbf{AB}$ (d) None of these

96. Five forces **AB**, **AC**, **AD**, **AE**, **AF** act at the vertex A of a regular hexagon $ABCDEF$. Then, their resultant is

(a) $3\mathbf{AO}$ (b) $2\mathbf{AO}$
(c) $4\mathbf{AO}$ (d) $6\mathbf{AO}$

Passage III

(Q. Nos. 97 to 99)

- Three points A, B and C have position vectors $-2\mathbf{a}+3\mathbf{b}+5\mathbf{c}$, $\mathbf{a}+2\mathbf{b}+3\mathbf{c}$ and $7\mathbf{a}-\mathbf{c}$ with reference to an origin O. Answer the following questions.

97. Which of the following is true?

(a) $\mathbf{AC}=2\mathbf{AB}$ (b) $\mathbf{AC}=-3\mathbf{AB}$
(c) $\mathbf{AC}=3\mathbf{AB}$ (d) None of these

98. Which of the following is true?

(a) $2\mathbf{OA}-3\mathbf{OB}+\mathbf{OC}=\mathbf{0}$
(b) $2\mathbf{OA}+7\mathbf{OB}+9\mathbf{OC}=\mathbf{0}$
(c) $\mathbf{OA}+\mathbf{OB}+\mathbf{OC}=\mathbf{0}$
(d) None of the above

99. B divided AC in ratio

(a) 2 :1 (b) 2 : 3
(c) 2 : −3 (d) 1 : 2

Passage IV

(Q. Nos. 100 and 101)

- If two vectors **OA** and **OB** are there, then their resultant $\mathbf{OA}+\mathbf{OB}$ can be found by completing the parallelogram $OACB$ and $\mathbf{OC}=\mathbf{OA}+\mathbf{OB}$. Also, If $|\mathbf{OA}|=|\mathbf{OB}|$, then the resultant will bisect the angle between them.

100. A vector **C** directed along internal bisector of angle between vectors $\mathbf{A}=7\hat{\mathbf{i}}-4\hat{\mathbf{j}}-4\hat{\mathbf{k}}$ and $\mathbf{B}=-2\hat{\mathbf{i}}-\hat{\mathbf{j}}+2\hat{\mathbf{k}}$ with $|\mathbf{C}|=5\sqrt{6}$ is

(a) $\frac{5}{3}(\hat{\mathbf{i}}-\hat{\mathbf{j}}+\hat{\mathbf{k}})$ (b) $\frac{5}{3}(\hat{\mathbf{i}}-7\hat{\mathbf{j}}+2\hat{\mathbf{k}})$
(c) $\frac{5}{3}(5\hat{\mathbf{i}}+5\hat{\mathbf{j}}+2\hat{\mathbf{k}})$ (d) $\frac{5}{3}(-5\hat{\mathbf{i}}+5\hat{\mathbf{j}}+2\hat{\mathbf{k}})$

101. If internal and external bisectors of $\angle A$ of ΔABC meet the base BC at D and E respectively, then (D and E lie on same side of B)

(a) $BC=\frac{BD+BE}{4}$ (b) $BC^2=BD\times DE$
(c) $\frac{2}{BC}=\frac{1}{BD}+\frac{1}{BE}$ (d) None of these

Passage V

(Q. Nos. 102 and 103)

- Let $C : r(t)=x(t)\hat{\mathbf{i}}+y(t)\hat{\mathbf{j}}+z(t)\hat{\mathbf{k}}$ be a differentiable curve, i.e. $\lim_{x\to 0}\frac{r(t+h)-r(h)}{h}$ exist for all t,

$$\therefore \quad r'(t)=x'(t)\,\hat{\mathbf{i}}+y'(t)\,\hat{\mathbf{j}}+z'(t)\,\hat{\mathbf{k}}$$

If $r'(t)$, is tangent to the curve C at the point $P[x(t), y(t), z(t)]$ and $r'(t)$ points in the direction of increasing t.

102. The point P on the curve $r(t)=(1-2t)\hat{\mathbf{i}}+t^2\hat{\mathbf{j}}+2e^{2(t-1)}\hat{\mathbf{k}}$ at which the tangent vector $r'(t)$ is parallel to the radius vector $r(t)$ is

(a) (−1, 1, 2) (b) (1, −1, 2)
(c) (−1, 1, −2) (d) (1, 1, 2)

103. The tangent vector to $r(t)=2t^2\hat{\mathbf{i}}+(1-t)\hat{\mathbf{j}}+(3t^2+2)\hat{\mathbf{k}}$ at (2, 0, 5) is

(a) $4\hat{\mathbf{i}}+\hat{\mathbf{j}}-6\hat{\mathbf{k}}$ (b) $4\hat{\mathbf{i}}-\hat{\mathbf{j}}+6\hat{\mathbf{k}}$
(c) $2\hat{\mathbf{i}}-\hat{\mathbf{j}}+6\hat{\mathbf{k}}$ (d) $2\hat{\mathbf{i}}+\hat{\mathbf{j}}-6\hat{\mathbf{k}}$

Vector Algebra Exercise 5 : Matching Type Question

104. **a** and **b** form the consecutive sides of a regular hexagon *ABCDEF*.

Column I	Column II
a. If $\mathbf{CD} = x\mathbf{a} + y\mathbf{b}$, then	p. $x = -2$
b. If $\mathbf{CE} = x\mathbf{a} + y\mathbf{b}$, then	q. $x = -1$
c. If $\mathbf{AE} = x\mathbf{a} + y\mathbf{b}$, then	r. $y = 1$
d. If $\mathbf{AD} = -x\mathbf{b}$, then	s. $y = 2$

Vector Algebra Exercise 6 : Single Integer Answer Type Questions

105. If the resultant of three forces $\mathbf{F}_1 = p\hat{\mathbf{i}} + 3\hat{\mathbf{j}} - \hat{\mathbf{k}}$, $\mathbf{F}_2 = -5\hat{\mathbf{i}} + \hat{\mathbf{j}} + 2\hat{\mathbf{k}}$ and $\mathbf{F}_3 = 6\hat{\mathbf{i}} - \hat{\mathbf{k}}$ acting on a particle has a magnitude equal to 5 units. Then, what is difference in the values of p?

106. Vectors along the adjacent sides of parallelogram are $\mathbf{a} = \hat{\mathbf{i}} + 2\hat{\mathbf{j}} + \hat{\mathbf{k}}$ and $\mathbf{b} = 2\hat{\mathbf{i}} + 4\hat{\mathbf{j}} + \hat{\mathbf{k}}$. Find the length of the longer diagonal of the parallelogram.

107. If vectors $\mathbf{a} = \hat{\mathbf{i}} + 2\hat{\mathbf{j}} - \hat{\mathbf{k}}, \mathbf{b} = 2\hat{\mathbf{i}} - \hat{\mathbf{j}} + \hat{\mathbf{k}}$ and $\mathbf{c} = \lambda\hat{\mathbf{i}} + \hat{\mathbf{j}} + 2\hat{\mathbf{k}}$ are coplanar, then find the value of $(\lambda - 4)$.

108. If $\mathbf{a} + \mathbf{b}$ is along the angle bisector of **a** and **b**, where $|\mathbf{a}| = \lambda|\mathbf{b}|$, then the number of digits in value of λ is

109. Let **p** be the position vector of orthocentre and **g** is the position vector of the centroid of ΔABC, where circumcentre is the origin. If $\mathbf{p} = k\mathbf{g}$, then the value of k is

110. In a ΔABC, a line is drawn passing through centroid dividing AB internally in ratio 2 : 1 and AC in $\lambda : 1$ (internally). The value of λ is

111. A vector **a** has component $2p$ and 1 with respect to a rectangular cartesian system. The system is rotated through a certain angle about the origin in the counter clockwise sense. If with to the new system, **a** has components $(p+1)$ and 1, where p take the values p_1 and p_2. Then, the value of $3|p_1 + p_2|$ is

Vector Algebra Exercise 7 : Subjective Type Questions

112. A vector **a** has components a_1, a_2 and a_3 in a right handed rectangular cartesian system $OXYZ$. The coordinate system is rotated about Z-axis through angle $\frac{\pi}{2}$. Find components of **a** in the new system.

113. Find the magnitude and direction of $\mathbf{r}_1 - \mathbf{r}_2$ when $|\mathbf{r}_1| = 5$ and points North-East while $|\mathbf{r}_2| = 5$ but points North-West.

114. Let $OACB$ be a parallelogram with O at the origin and **OC** a diagonal. Let D be the mid-point of **OA**. Using vector methods prove that **BD** and **CO** intersects in the same ratio. Determine this ratio.

115. ΔABC is a triangle with the point P on side BC such that $3\mathbf{BP} = 2\mathbf{PC}$, the point Q is on the line **CA** such that $4\mathbf{CO} = \mathbf{QA}$. Find the ratio in which the line joining the common point R of **AP** and **BQ** and the point S divides **AB**.

116. In a ΔABC internal angle bisectors AI, BI and CI are produced to meet opposite sides in A', B' and C', respectively. Prove that the maximum value of $\dfrac{AI \cdot BI \cdot CI}{AA' \cdot BB' \cdot CC'}$ is $\dfrac{8}{27}$.

117. Let $\mathbf{r}_1, \mathbf{r}_2, \mathbf{r}_3, \ldots, \mathbf{r}_n$ be the position vectors of points $P_1, P_2, P_3, \ldots, P_n$ relative to an origin O. Show that if the vectors equation $a_1\mathbf{r}_1 + a_2\mathbf{r}_2 + \ldots + a_n\mathbf{r}_n = 0$ holds, then a similar equation will also hold good with respect to any other origin O', if $a_1 + a_2 + a_3 + \ldots + a_n = 0$.

118. Let $OABCD$ be a pentagon in which the sides OA and CB are parallel and the sides OD and AB are parallel as shown in figure. Also, $OA : CB = 2 : 1$ and $OD : AB = 1 : 3$. If the diagonals OC and AD meet at x, find $OX : OC$.

119. If $\mathbf{u}$, $\mathbf{v}$ and $\mathbf{w}$ is a linearly independent system of vectors, examine the system $\mathbf{p}, \mathbf{q}$ and $\mathbf{r}$, where
$\mathbf{p} = (\cos a)\mathbf{u} + (\cos b)\mathbf{v} + (\cos c)\mathbf{w}$
$\mathbf{q} = (\sin a)\mathbf{u} + (\sin b)\mathbf{v} + (\sin c)\mathbf{w}$
$\mathbf{r} = \sin(x + a)\mathbf{u} + \sin(x + b)\mathbf{v} + \sin(x + c)\mathbf{w}$ for linearly dependent.

Vector Algebra Exercise 8 : Questions Asked in Previous Years Exam

120. If the vectors $\mathbf{AB} = 3\hat{\mathbf{i}} + 4\hat{\mathbf{k}}$ and $\mathbf{AC} = 5\hat{\mathbf{i}} - 2\hat{\mathbf{j}} + 4\hat{\mathbf{k}}$ are the sides of a ΔABC, then the length of the median through A is [JEE Main 2013, 2003]

(a) $\sqrt{18}$ (b) $\sqrt{72}$
(c) $\sqrt{33}$ (d) $\sqrt{45}$

121. Let $\mathbf{a}, \mathbf{b}$ and $\mathbf{c}$ be three non-zero vectors which are pairwise non-collinear. If $\mathbf{a} + 3\mathbf{b}$ is collinear with $\mathbf{c}$ and $\mathbf{b} + 2\mathbf{c}$ is collinear with $\mathbf{a}$, then $\mathbf{a} + 3\mathbf{b} + 6\mathbf{c}$ is [AIEEE 2011]

(a) $\mathbf{a} + \mathbf{c}$ (b) $\mathbf{a}$
(c) $\mathbf{c}$ (d) 0

122. The non-zero vectors $\mathbf{a}$, $\mathbf{b}$ and $\mathbf{c}$ are related by $\mathbf{a} = 8\mathbf{b}$ and $\mathbf{c} = -7\mathbf{b}$. Then, the angle between a and c is [AIEEE 2008]

(a) π (b) 0
(c) $\dfrac{\pi}{4}$ (d) $\dfrac{\pi}{2}$

123. If C is the mid-point of AB and P is any point outside AB, then [AIEEE 2005]

(a) $\mathbf{PA} + \mathbf{PB} + \mathbf{PC} = 0$ (b) $\mathbf{PA} + \mathbf{PB} + 2\mathbf{PC} = 0$
(c) $\mathbf{PA} + \mathbf{PB} = \mathbf{PC}$ (d) $\mathbf{PA} + \mathbf{PB} = 2\mathbf{PC}$

124. If $\mathbf{a}, \mathbf{b}$ and $\mathbf{c}$ are three non-zero vectors such that no two of these are collinear. If the vector $\mathbf{a} + 2\mathbf{b}$ is collinear with $\mathbf{c}$ and $\mathbf{b} + 3\mathbf{c}$ is collinear with $\mathbf{a}$ (λ being some non-zero scalar), then $\mathbf{a} + 2\mathbf{b} + 6\mathbf{c}$ is equal to [AIEEE 2004]

(a) $\lambda\mathbf{a}$ (b) $\lambda\mathbf{b}$
(c) $\lambda\mathbf{c}$ (d) 0

125. If $\mathbf{a}$, $\mathbf{b}$, $\mathbf{c}$ are non-coplanar vectors and λ is a real number, then the vectors $\mathbf{a} + 2\mathbf{b} + 3\mathbf{c}$, $\lambda\mathbf{b} + 4\mathbf{c}$ and $(2\lambda - 1)\mathbf{c}$ are non-coplanar for [AIEEE 2004]

(a) all values of λ
(b) all except one value of λ
(c) all except two values of λ
(d) no value of λ

126. Consider points A, B, C and D with position vectors $7\hat{\mathbf{i}} - 4\hat{\mathbf{j}} + 7\hat{\mathbf{k}}$, $\hat{\mathbf{i}} - 6\hat{\mathbf{j}} + 10\hat{\mathbf{k}}$, $-\hat{\mathbf{i}} - 3\hat{\mathbf{j}} + 4\hat{\mathbf{k}}$ and $5\hat{\mathbf{i}} - \hat{\mathbf{j}} + 5\hat{\mathbf{k}}$, respectively. Then, $ABCD$ is a [AIEEE 2003]

(a) square (b) rhombus
(c) rectangle (d) None of these

127. If $\begin{vmatrix} a & a^2 & 1+a^3 \\ b & b^2 & 1+b^3 \\ c & c^2 & 1+c^3 \end{vmatrix} = 0$ and vectors $(1, a, a^2)$, $(1, b, b^2)$ and $(1, c, c^2)$ are non-coplanar, then the product abc equal to [AIEEE 2003]

(a) 2 (b) -1
(c) 1 (d) 0

128. The vector $\hat{\mathbf{i}} + x\hat{\mathbf{j}} + 3\hat{\mathbf{k}}$ is rotated through an angle θ and doubled in magnitude, then it becomes $4\hat{\mathbf{i}} + (4x - 2)\hat{\mathbf{j}} + 2\hat{\mathbf{k}}$. The value of x are [AIEEE 2002]

(a) $\left\{-\dfrac{2}{3}, 2\right\}$ (b) $\left(\dfrac{1}{3}, 2\right)$
(c) $\left\{\dfrac{2}{3}, 0\right\}$ (d) $\{2, 7\}$

Answers

Exercise for Session 1

1. (i) vector (ii) scalar (iii) vector (iv) scalar

2.

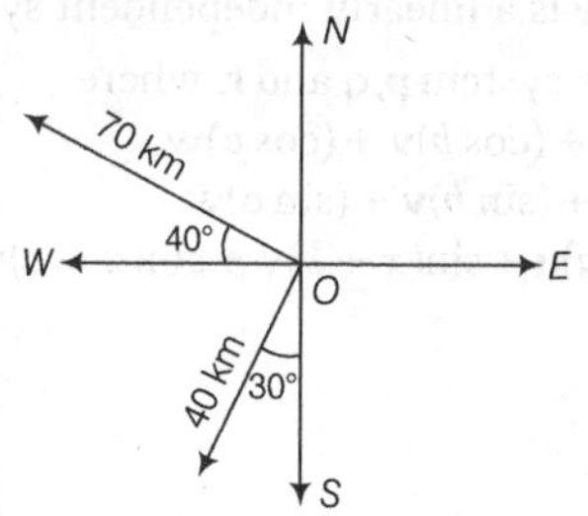

3. (i) **a, d; b, x, z; c, y** (ii) **b, x; a, d; c, y**
(iii) **a, y, z** (iv) **b, z; x, z**

4. (i) True (ii) False
(iii) False (iv) False

5. $\sqrt{450}$ **6.** $\cos^{-1}\frac{6}{7}$

7. Direction ratios are 1, – 1, 2 and Direction cosines are $\frac{1}{\sqrt{6}}, \frac{-1}{\sqrt{6}}, \frac{2}{\sqrt{6}}$.

Exercise for Session 2

1. $\hat{\mathbf{i}} + \hat{\mathbf{k}}; \frac{1}{\sqrt{2}}\hat{\mathbf{i}} + \frac{1}{\sqrt{2}}\hat{\mathbf{k}}$ **2.** $\frac{3}{5\sqrt{2}}\hat{\mathbf{i}} + \frac{5}{5\sqrt{2}}\hat{\mathbf{j}} + \frac{4}{5\sqrt{2}}\hat{\mathbf{k}}$

3. $\left(\frac{1}{\sqrt{3}}, \frac{1}{\sqrt{3}}, \frac{1}{\sqrt{3}}\right)$

9. $4\hat{\mathbf{i}} - 9\hat{\mathbf{j}}, \sqrt{97}, \frac{4}{\sqrt{97}}\hat{\mathbf{i}} - \frac{9}{\sqrt{97}}\hat{\mathbf{j}}$

10. $\sqrt{59}; \frac{1}{\sqrt{59}}\hat{\mathbf{i}} - \frac{7}{\sqrt{59}}\hat{\mathbf{j}} + \frac{3}{\sqrt{59}}\hat{\mathbf{k}}$ **12.** $\pm\frac{1}{3}$

13. $\sqrt{66}$ **14.** $\frac{8}{\sqrt{30}}(5\hat{\mathbf{i}} - \hat{\mathbf{j}} + 2\hat{\mathbf{k}})$

15. $7/3, (\hat{\mathbf{i}} + 2\hat{\mathbf{j}} + 2\hat{\mathbf{k}})$

16. (i) $-\frac{1}{3}\hat{\mathbf{i}} + \frac{4}{3}\hat{\mathbf{j}} + \frac{1}{3}\hat{\mathbf{k}}$ (ii) $-3\hat{\mathbf{i}} + 3\hat{\mathbf{k}}$

17. $4\hat{\mathbf{i}} + 3\hat{\mathbf{j}} + 7\hat{\mathbf{k}}$

Exercise for Session 3

4. (2,–3) **5.** $a - 2b = 1$
6. $x = \frac{1}{3}$ **7.** 0

Chapter Exercises

1. (c) **2.** (a) **3.** (a) **4.** (d) **5.** (a) **6.** (c)
7. (c) **8.** (b) **9.** (c) **10.** (d) **11.** (a) **12.** (a)
13. (d) **14.** (d) **15.** (b) **16.** (b) **17.** (c) **18.** (b)
19. (c) **20.** (d) **21.** (a) **22.** (b) **23.** (a) **24.** (b)
25. (c) **26.** (b) **27.** (b) **28.** (c) **29.** (c) **30.** (c)
31. (a) **32.** (d) **33.** (b) **34.** (a) **35.** (c) **36.** (a)
37. (b) **38.** (d) **39.** (b) **40.** (c) **41.** (b) **42.** (b)
43. (b) **44.** (d) **45.** (b) **46.** (d) **47.** (c) **48.** (a)
49. (a) **50.** (d) **51.** (c) **52.** (a) **53.** (b) **54.** (d)
55. (c) **56.** (a) **57.** (d) **58.** (c) **59.** (b) **60.** (b)
61. (b) **62.** (c) **63.** (b) **64.** (d) **65.** (c) **66.** (c)
67. (a) **68.** (c) **69.** (a) **70.** (a) **71.** (b) **72.** (a)
73. (b) **74.** (d) **75.** (b) **76.** (a) **77.** (b) **78.** (d)
79. (b) **80.** (a) **81.** (b) **82.** (c) **83.** (a,b,d) **84.** (b,c)
85. (a,c) **86.** (a,d) **87.** (a,c,d) **88.** (a,b,d) **89.** (a,b,c) **90.** (c)
91. (a) **92.** (a) **93.** (c) **94.** (b) **95.** (c) **96.** (d)
97. (c) **98.** (a) **99.** (d) **100.** (b) **101.** (c) **102.** (a)
103. (b) **104.** a → q, r ; b → p, r
105. (2, – 4) **106.** (7) **107.** (2)
108. (1) **109.** (3) **110.** (2) **111.** (2)
112. $(a_2, -a_1, a_3)$ **113.** $5\sqrt{5}$, West to East
114. 2 :1 **115.** 6 : 1 **118.** 2 : 5
120. (c) **121.** (d) **122.** (a) **123.** (d) **124.** (d) **125.** (c)
126. (d) **127.** (b) **128.** (a)

Solutions

1. $\mathbf{a}+\mathbf{b}+\mathbf{c}=(3+2-1)\hat{\mathbf{i}}+(-2-4+2)\hat{\mathbf{j}}+(1-3+2)\hat{\mathbf{k}}$
$=4\hat{\mathbf{i}}-4\hat{\mathbf{j}}$

2. Let $\mathbf{b}$ should be added, then $\mathbf{a}+\mathbf{b}=\hat{\mathbf{i}}$

$\Rightarrow \quad \mathbf{b}=\hat{\mathbf{i}}-\mathbf{a}=\hat{\mathbf{i}}-(3\hat{\mathbf{i}}+4\hat{\mathbf{j}}-2\hat{\mathbf{k}})$
$=-2\hat{\mathbf{i}}-4\hat{\mathbf{j}}+2\hat{\mathbf{k}}$

3. $|\mathbf{a}+\mathbf{b}|=|3\hat{\mathbf{i}}+4\hat{\mathbf{j}}-12\hat{\mathbf{k}}|$
$=|\sqrt{3^2+4^2+(12)^2}|=13$

4. $\mathbf{a}+\mathbf{b}=4\hat{\mathbf{i}}+4\hat{\mathbf{j}}$,

Therefore, unit vector $=\dfrac{4(\hat{\mathbf{i}}+\hat{\mathbf{j}})}{\sqrt{32}}=\dfrac{\hat{\mathbf{i}}+\hat{\mathbf{j}}}{\sqrt{2}}$

5. Resultant vector $=(2\hat{\mathbf{i}}+4\hat{\mathbf{j}}-5\hat{\mathbf{k}})+(\hat{\mathbf{i}}+2\hat{\mathbf{j}}+3\hat{\mathbf{k}})$
$=3\hat{\mathbf{i}}+6\hat{\mathbf{j}}-2\hat{\mathbf{k}}$

Unit vector $=\dfrac{3\hat{\mathbf{i}}+6\hat{\mathbf{j}}-2\hat{\mathbf{k}}}{\sqrt{9+36+4}}=\dfrac{1}{7}(3\hat{\mathbf{i}}+6\hat{\mathbf{j}}-2\hat{\mathbf{k}})$

6. $\mathbf{R}=3\hat{\mathbf{i}}+5\hat{\mathbf{j}}+4\hat{\mathbf{k}}$

$\Rightarrow \quad \mathbf{R}=\dfrac{3\hat{\mathbf{i}}+5\hat{\mathbf{j}}+4\hat{\mathbf{k}}}{5\sqrt{2}}$

7. $\mathbf{a}+\mathbf{b}=3\hat{\mathbf{i}}+9\hat{\mathbf{j}}=3(\hat{\mathbf{i}}+3\hat{\mathbf{j}})$. Hence, it is parallel to (1, 3).

8. $\mathbf{AB}+\mathbf{BC}+\mathbf{CA}=0$

$\Rightarrow \quad \mathbf{a}+\mathbf{b}-\mathbf{c}=0$

9. Unit vector parallel to $\mathbf{OA}=\dfrac{4\hat{\mathbf{i}}+5\hat{\mathbf{j}}}{\sqrt{16+25}}=\dfrac{1}{\sqrt{41}}(4\hat{\mathbf{i}}+5\hat{\mathbf{j}})$

10. $\mathbf{OA}=\hat{\mathbf{i}}+2\hat{\mathbf{j}}-\hat{\mathbf{k}}, \mathbf{OB}=\hat{\mathbf{i}}+\hat{\mathbf{j}}+\hat{\mathbf{k}}$

and $\quad \mathbf{OC}=2\hat{\mathbf{i}}+3\hat{\mathbf{j}}+2\hat{\mathbf{k}}$

Position vector of B w.r.t origin at A at

$\mathbf{AB}=\mathbf{OB}-\mathbf{OA}=-\hat{\mathbf{j}}+2\hat{\mathbf{k}}$

Position vector of C w.r.t. origin at A is

$\mathbf{AC}=\mathbf{OC}-\mathbf{OA}=\hat{\mathbf{i}}+\hat{\mathbf{j}}+3\hat{\mathbf{k}}$

11. $7=\sqrt{(5+1)^2+(4-2)^2+(a+2)^2}$

$\Rightarrow \quad a+2=\pm 3 \Rightarrow a=-5, 1$

12. $\mathbf{AB}=\mathbf{CX} \Rightarrow \hat{\mathbf{j}}-\hat{\mathbf{i}}=$ position vector of point $X-\hat{\mathbf{k}}$

$\therefore$ Position vector of point $X=-\hat{\mathbf{i}}+\hat{\mathbf{j}}+\hat{\mathbf{k}}$

13. $\mathbf{AB}=(6-2)\hat{\mathbf{i}}+(-3+9)\hat{\mathbf{j}}+(8+4)\hat{\mathbf{k}}$
$=4\hat{\mathbf{i}}+6\hat{\mathbf{j}}+12\hat{\mathbf{k}}$

$|\mathbf{AB}|=\sqrt{16+25+144}=14$

14. $\mathbf{PQ}=(5-1)\hat{\mathbf{i}}+(-2-3)\hat{\mathbf{j}}+(4+7)\hat{\mathbf{k}}$
$=4\hat{\mathbf{i}}-5\hat{\mathbf{j}}+11\hat{\mathbf{k}}$

$|\mathbf{PQ}|=\sqrt{16+25+121}=\sqrt{162}$

15. $\mathbf{PQ}=\mathbf{OQ}-\mathbf{OP}=4\hat{\mathbf{i}}-5\hat{\mathbf{j}}+11\hat{\mathbf{k}}$

$\therefore \quad \dfrac{\mathbf{PQ}}{|\mathbf{PQ}|}=\dfrac{4}{\sqrt{162}}\hat{\mathbf{i}}-\dfrac{5}{\sqrt{162}}\hat{\mathbf{j}}+\dfrac{11}{\sqrt{162}}\hat{\mathbf{k}}$

$\therefore \quad \cos\gamma=\dfrac{11}{\sqrt{162}}$, where γ is the angle of $\mathbf{PQ}$ with Z-axis.

16. $\mathbf{AB}=4\hat{\mathbf{i}}-5\hat{\mathbf{j}}+11\hat{\mathbf{k}}$

Direction cosine along Y-axis $=\dfrac{-5}{\sqrt{16+25+121}}=\dfrac{-5}{\sqrt{162}}$

17. $\dfrac{3}{\sqrt{3^2+4^2+5^2}}=\dfrac{3}{\sqrt{50}}$

18. Vector $\mathbf{A}=3\hat{\mathbf{i}}-4\hat{\mathbf{j}}+5\hat{\mathbf{k}}$. We know that, direction cosines of

$\mathbf{A}=\dfrac{3}{\sqrt{3^2+4^2+5^2}}, \dfrac{-4}{\sqrt{3^2+4^2+5^2}}, \dfrac{5}{\sqrt{3^2+4^2+5^2}}$

$=\dfrac{3}{5\sqrt{2}}, \dfrac{-4}{5\sqrt{2}}, \dfrac{1}{\sqrt{2}}$

19. Here, $\quad \mathbf{OA}=2\hat{\mathbf{i}}+3\hat{\mathbf{j}}+4\hat{\mathbf{k}}$

$\mathbf{OB}=3\hat{\mathbf{i}}+4\hat{\mathbf{j}}+2\hat{\mathbf{k}}$

and $\quad \mathbf{OC}=4\hat{\mathbf{i}}+2\hat{\mathbf{j}}+3\hat{\mathbf{k}}$

So, $\mathbf{AB}=\hat{\mathbf{i}}+\hat{\mathbf{j}}-2\hat{\mathbf{k}}, \mathbf{BC}=\hat{\mathbf{i}}-2\hat{\mathbf{j}}+\hat{\mathbf{k}}, \mathbf{CA}=2\hat{\mathbf{i}}-\hat{\mathbf{j}}-\hat{\mathbf{k}}$

Clearly, $|AB|=|BC|=|CA|=\sqrt{6}$

So, these points are vertices of an equilateral triangle.

20. Given, position vectors of A, B and C are $7\hat{\mathbf{j}}+10\hat{\mathbf{k}}$, $-\hat{\mathbf{i}}+6\hat{\mathbf{j}}+6\hat{\mathbf{k}}$ and $-4\hat{\mathbf{i}}+9\hat{\mathbf{j}}+6\hat{\mathbf{k}}$, respectively.

$\therefore \quad |\mathbf{AB}|=|-\hat{\mathbf{i}}-\hat{\mathbf{j}}-4\hat{\mathbf{k}}|=\sqrt{18}$

$|\mathbf{BC}|=|-3\hat{\mathbf{i}}+3\hat{\mathbf{j}}|=\sqrt{18}$

$|\mathbf{AC}|=|-4\hat{\mathbf{i}}+2\hat{\mathbf{j}}-4\hat{\mathbf{k}}|=\sqrt{36}$

Clearly, $AB=BC$ and $(AC)^2=(AB)^2+(BC)^2$

Hence, triangle is right angled isosceles.

24. Position vectors of the points which divides internally is

$\dfrac{3(2\mathbf{a}-3\mathbf{b})+2(3\mathbf{a}-2\mathbf{b})}{5}=\dfrac{12\mathbf{a}-13\mathbf{b}}{5}$

25. Coordinate of C is $\left(\dfrac{2-4}{2}, \dfrac{-1+3}{2}\right)=(-1, 1)$

$\therefore \quad \mathbf{OC}=-\hat{\mathbf{i}}+\hat{\mathbf{j}}$

26. $\dfrac{3\hat{\mathbf{i}}-\hat{\mathbf{j}}-3\hat{\mathbf{k}}+\hat{\mathbf{j}}+3\hat{\mathbf{j}}-\hat{\mathbf{k}}}{2}=2\hat{\mathbf{i}}+\hat{\mathbf{j}}-2\hat{\mathbf{k}}$

27. $\dfrac{\mathbf{a}+\mathbf{b}}{2}=2\hat{\mathbf{i}}-\hat{\mathbf{j}}+\dfrac{5}{2}\hat{\mathbf{k}}$

28. It is given that $\mathbf{b}$ is collinear with the vector $\mathbf{a}$.

$\therefore \quad \mathbf{b}=\lambda\mathbf{a}$ ...(i)

$=2\sqrt{2}\lambda\hat{\mathbf{i}}-\lambda\hat{\mathbf{j}}+4\lambda\hat{\mathbf{k}}$

Also, $\quad |\mathbf{b}|=10$

$\Rightarrow \quad \sqrt{(2\sqrt{2}\lambda)^2+(-\lambda)^2+(4\lambda)^2}=10$

$\Rightarrow \quad 25\lambda^2=100$

$\Rightarrow \quad \lambda=\pm 2$ ... (ii)

From Eqs. (i) and (ii), we have

$\mathbf{b}=\pm 2\mathbf{a} \Rightarrow 2\mathbf{a}+\mathbf{b}=0$

29. Condition for collinearity, $\mathbf{b} = \lambda\ \mathbf{a}$

$\Rightarrow \quad (-2\hat{\mathbf{i}} + m\hat{\mathbf{j}}) = \lambda\,(\hat{\mathbf{i}} - \hat{\mathbf{j}})$

Comparison of coefficient, we get

$\Rightarrow \quad \lambda = -2$ and $-\lambda = m$

So, $\quad m = 2$

30. If given points be A, B and C, then $\mathbf{AB} = k(\mathbf{BC})$ or

$$2\hat{\mathbf{i}} - 8\hat{\mathbf{j}} = k[(a-12)\hat{\mathbf{i}} + 16\hat{\mathbf{j}}]$$

$\Rightarrow \quad k = -\dfrac{1}{2}$

Also, $\quad 2 = k\,(a-12)$

$\Rightarrow \quad a = 8$

31. $\begin{vmatrix} 1 & 2 & 3 \\ \lambda & 4 & 7 \\ -3 & -2 & -5 \end{vmatrix} = 0$

$\Rightarrow \quad \lambda = 3$

32. $(\mathbf{a} - \mathbf{b}) - (\mathbf{a} + \mathbf{b}) = [(\mathbf{a} + k\mathbf{b}) - (\mathbf{a} - \mathbf{b})]$

$\Rightarrow \quad -2\mathbf{b} = (k+1)\mathbf{b}$

Hence, $k \in R$

33. $\mathbf{AB} = -\hat{\mathbf{i}} - 4\hat{\mathbf{j}}, \mathbf{CD} = -2\hat{\mathbf{i}} + (\lambda - 2)\hat{\mathbf{j}}$

$\because \quad \mathbf{AB} \parallel \mathbf{CD}$

So, $\quad \dfrac{-1}{-2} = \dfrac{-4}{\lambda - 2} \Rightarrow \lambda - 2 = -8$

$\Rightarrow \quad \lambda = -6$

34. Obviously, $\dfrac{3}{6} = \dfrac{2}{-4x} = -\dfrac{1}{y}$

$\Rightarrow \quad x = -1$ and $y = -2$

35. If **a** and **b** are two non-zero, non-collinear vectors and x and y are two scalars such that $x\mathbf{a} + y\mathbf{b} = 0$ then $x = 0$ and $y = 0$ because one will be a scalar multiple of the other and hence collinear which is a contradiction.

36. Four or more than four non-zero vectors are always linearly dependent.

37. These are coplanar because $1\,(\mathbf{a}) + 1\,(\mathbf{b}) = \mathbf{a} + \mathbf{b}$

38. Comparing the coefficients of $\hat{\mathbf{i}}$, $\hat{\mathbf{j}}$ and $\hat{\mathbf{k}}$, and the corresponding equations are

$$x + 3y - 4z = \lambda x \quad \text{or} \quad (1-\lambda)x + 3y - 4z = 0 \qquad \text{...(i)}$$

$$x - (\lambda + 3)y + 5z = 0 \qquad \text{... (ii)}$$

$$3x + y - \lambda z = 0 \qquad \text{... (iii)}$$

These Eqs. (i), (ii) and (iii) have a non-trivial solution, if

$$\begin{vmatrix} (1-\lambda) & 3 & -4 \\ 1 & -(\lambda+3) & 5 \\ 3 & 1 & -\lambda \end{vmatrix} = 0 \Rightarrow \lambda = 0, -1$$

39. The vectors are linearly dependent

$\Rightarrow \quad \begin{vmatrix} p+1 & -3 & p \\ p & p+1 & -3 \\ -3 & p & p+1 \end{vmatrix} = 0$

$\Rightarrow \quad (2p-2)\begin{vmatrix} 1 & -3 & p \\ 1 & p+1 & -3 \\ 1 & p & P+1 \end{vmatrix} = 0$

$\Rightarrow \quad 2(p-1)\begin{vmatrix} 1 & -3 & p \\ 0 & p+4 & -3-p \\ 0 & p+3 & 1 \end{vmatrix} = 0$

$\Rightarrow \quad 2(p-1)(p+4+(p+3)^2) = 0$

$\Rightarrow \quad (p-1)(p^2 + 7p + 13) = 0$

Roots of $p^2 + 7p + 13 = 0$ are (imaginary)

$\therefore \quad p = 1$

Only integral value of p is 1.

40. PV of $\mathbf{AD} = \dfrac{(3+5)\hat{\mathbf{i}} + (0-2)\hat{\mathbf{j}} + (4+4)\hat{\mathbf{k}}}{2}$

$= 4\hat{\mathbf{i}} - \hat{\mathbf{j}} + 4\hat{\mathbf{k}}$

$|\mathbf{AD}| = \sqrt{16+16+1} = \sqrt{33}$

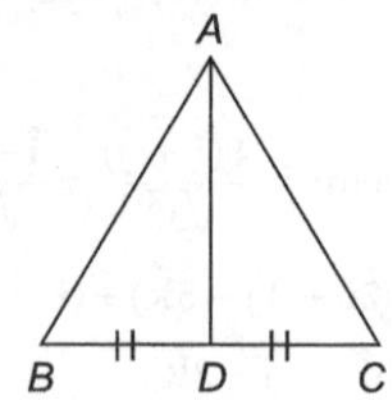

41. $\mathbf{v} = \mathbf{b} + \mathbf{c}$

$\mathbf{w} = \mathbf{b} + \mathbf{a}$

We have, $\quad \mathbf{x} = \mathbf{v} + \mathbf{w} = \mathbf{a} + 2\mathbf{b} + \mathbf{c}$

42. Note that $\mathbf{a} + \mathbf{b} = \mathbf{c}$

43. Here is the only vector $4(\sqrt{2}\hat{\mathbf{i}} + \hat{\mathbf{j}} \pm \hat{\mathbf{k}})$, whose length is 8.

44. $3\mathbf{a} + 4\mathbf{b} + 5\mathbf{c} = 0$

Hence, **a**, **b** and **c** are coplanar.

No other conclusion can be derived from it.

45. A, B, C, D and E are five coplanar points.

$\mathbf{DA} + \mathbf{DB} + \mathbf{AE} + \mathbf{BE} + \mathbf{CE}$

$= (\mathbf{DA} + \mathbf{AE}) + (\mathbf{DB} + \mathbf{BE}) + (\mathbf{DC} + \mathbf{CE})$

$= \mathbf{DE} + \mathbf{DE} + \mathbf{DE} = 3\mathbf{DE}$

46. $\sqrt{3}\tan\theta + 1 = 0$ and $\sqrt{3}\sec\theta - 2 = 0$

$\Rightarrow \quad \theta = \dfrac{11\pi}{6}$

$\Rightarrow \quad \theta = 2n\pi + \dfrac{11\pi}{6}, n \in Z$

47. $\hat{\mathbf{j}}$ and $\hat{\mathbf{k}}$ are unit vectors along Y and Z-axes, then unit vector bisecting **OY** and **OZ** is $\dfrac{\hat{\mathbf{j}} + \hat{\mathbf{k}}}{\sqrt{2}}$.

48.

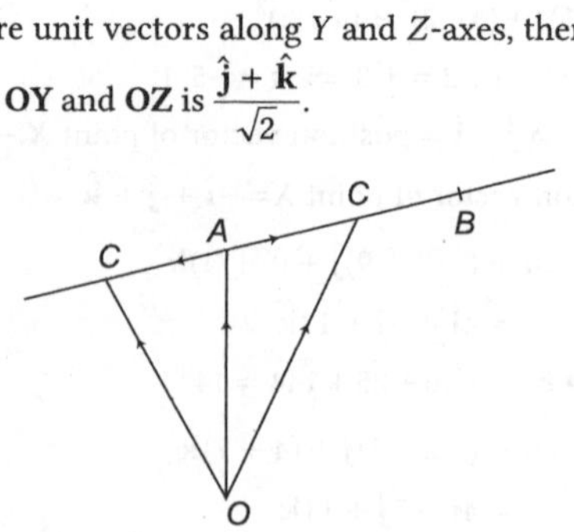

We have, $A(\hat{\mathbf{i}} + \hat{\mathbf{j}} - 2\hat{\mathbf{k}})$ and $B(\hat{\mathbf{i}} - 3\hat{\mathbf{j}} + \hat{\mathbf{k}})$

On line AB points C and C' are at distance 1 unit from A.

$\mathbf{OC} = \mathbf{OA} + \mathbf{AC}$, where $\mathbf{AC}$ is unit vector in direction of $\mathbf{AB}$

$$\therefore \quad \mathbf{OC} = \mathbf{OA} + \frac{\mathbf{AB}}{|\mathbf{AB}|}$$

Similarly, $$\mathbf{OC'} = \mathbf{OA} - \frac{\mathbf{AB}}{|\mathbf{AB}|}$$

49. $$\mathbf{AD} = \mathbf{OD} - \mathbf{OA} = \frac{\mathbf{b}+\mathbf{c}}{2} - \mathbf{a} = \frac{\mathbf{b}+\mathbf{c}-2\mathbf{a}}{2}$$

[where, O is the origin for reference]

Similarly, $$\mathbf{BE} = \mathbf{OE} - \mathbf{OB} = \frac{\mathbf{c}+\mathbf{a}}{2} - \mathbf{b} = \frac{\mathbf{c}+\mathbf{a}-2\mathbf{b}}{2}$$

and $$\mathbf{CF} = \frac{\mathbf{a}+\mathbf{b}-2\mathbf{b}}{2}$$

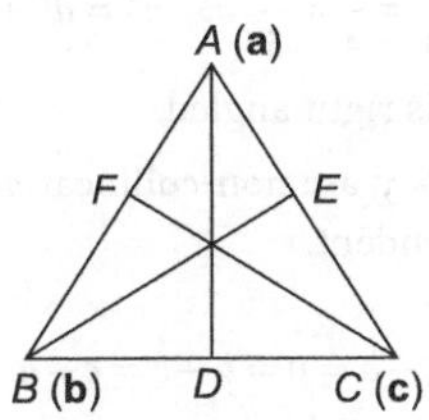

Now, $$\mathbf{AD} + \mathbf{BE} + \mathbf{CF} = \frac{\mathbf{b}+\mathbf{c}-2\mathbf{a}}{2} + \frac{\mathbf{c}+\mathbf{a}-2\mathbf{b}}{2} + \frac{\mathbf{a}+\mathbf{b}-2\mathbf{c}}{2} = \mathbf{0}$$

50. $$\mathbf{AP} = \mathbf{AB} + \mathbf{BP} = \mathbf{AB} + \frac{1}{2}\mathbf{BC} = \mathbf{AB} + \frac{1}{2}\mathbf{AD} \quad \text{...(i)}$$

$$\mathbf{AQ} = \mathbf{AD} + \mathbf{DQ} = \mathbf{AD} + \frac{1}{2}\mathbf{DC} = \mathbf{AD} + \frac{1}{2}\mathbf{AB} \quad \text{... (ii)}$$

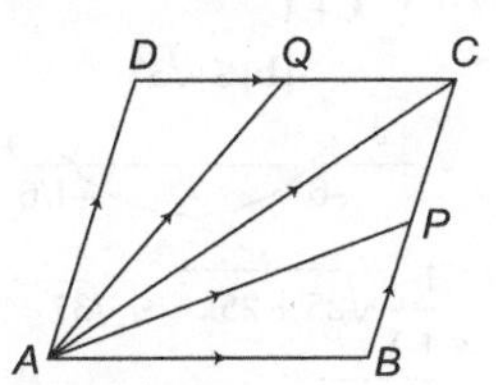

By Eqs. (i) and (ii), we get

$$\mathbf{AP} + \mathbf{AQ} = \frac{3}{2}(\mathbf{AB} + \mathbf{AD})$$

$$= \frac{3}{2}(\mathbf{AB} + \mathbf{BC}) = \frac{3}{2}\mathbf{AC}$$

51. Let $A \equiv (1, 1, -1)$, $B \equiv (2, 3, 0)$, $C \equiv (3, 5, -2)$ and $D \equiv (0, -1, 1)$.

So, $\mathbf{AB} = (1, 2, 1)$, $\mathbf{BC} = (1, 2, -2)$, $\mathbf{CD} = (-3, -6, 3)$ and $\mathbf{DA} = (1, 2, -2)$.

Clearly, $\mathbf{BC} \parallel \mathbf{DA}$ but AB in not parallel to CD.

So, it is a trapezium.

52. $$\mathbf{OP} = \frac{1\,(\mathbf{OB}) + 2\,(6\mathbf{b} - 2\mathbf{a})}{1+2}$$

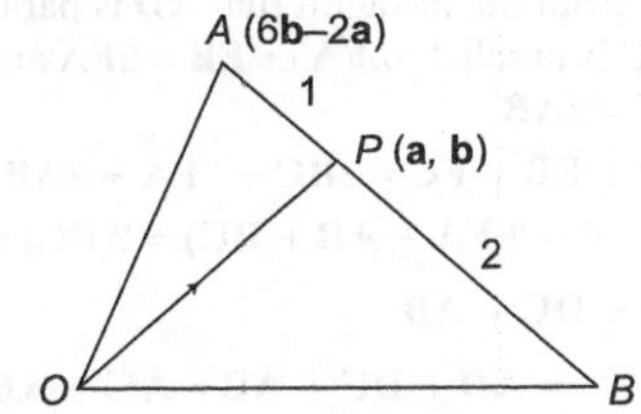

$$\Rightarrow \quad 3(\mathbf{a} - \mathbf{b}) = \mathbf{OB} + 12\mathbf{b} - 4\mathbf{a}$$

$$\Rightarrow \quad \mathbf{OB} = 7\mathbf{a} - 15\mathbf{b}$$

53. Let the B divide AC in ratio $\lambda : 1$, then

$$5\hat{\mathbf{i}} - 2\hat{\mathbf{k}} = \frac{\lambda(11\hat{\mathbf{i}} + 3\hat{\mathbf{j}} + 7\hat{\mathbf{k}}) + \hat{\mathbf{i}} - 2\hat{\mathbf{j}} - 8\hat{\mathbf{k}}}{\lambda + 1}$$

$$\Rightarrow \quad 3\lambda - 2 = 0$$

$$\Rightarrow \quad \lambda = \frac{2}{3}, \text{ i.e. ratio} = 2 : 3$$

54. We know by fundamental theorem of proportionality that, $\mathbf{DE} = \frac{1}{2}\mathbf{BC}$

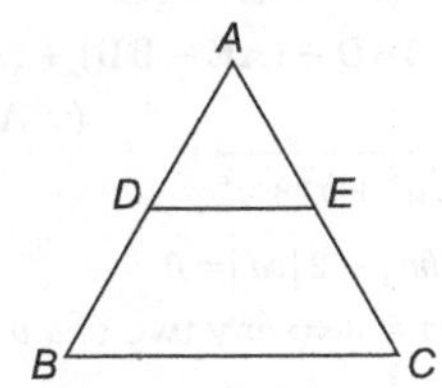

In triangle, $\mathbf{BC} = \mathbf{b} - \mathbf{a}$

Hence, $$\mathbf{DE} = \frac{1}{2}(\mathbf{b} - \mathbf{a})$$

55. Since, $\mathbf{AB} + \mathbf{BD} = \mathbf{AD}$

$$\mathbf{BD} = \mathbf{AD} - \mathbf{AB}$$

$$\Rightarrow \quad = (\hat{\mathbf{i}} + 2\hat{\mathbf{j}} + 3\hat{\mathbf{k}}) - (2\hat{\mathbf{i}} + 4\hat{\mathbf{j}} - 5\hat{\mathbf{k}})$$

$$= -\hat{\mathbf{i}} - 2\hat{\mathbf{j}} + 8\hat{\mathbf{k}}$$

Hence, unit vector in the direction of $\mathbf{BD}$ is

$$\frac{-\hat{\mathbf{i}} - 2\hat{\mathbf{j}} + 8\hat{\mathbf{k}}}{|-\hat{\mathbf{i}} - 2\hat{\mathbf{j}} + 8\hat{\mathbf{k}}|} = \frac{-\hat{\mathbf{i}} - 2\hat{\mathbf{j}} + 8\hat{\mathbf{k}}}{\sqrt{69}}$$

56. Position vectors of vertices A, B and C of the ΔABC = $\mathbf{a}$, $\mathbf{b}$ and $\mathbf{c}$. We know that, position vector of centroid of the triangle, $G = \dfrac{\mathbf{a}+\mathbf{b}+\mathbf{c}}{3}$.

Therefore, $\mathbf{GA} + \mathbf{GB} + \mathbf{GC}$

$$= \left(\mathbf{a} - \frac{\mathbf{a}+\mathbf{b}+\mathbf{c}}{3}\right) + \left(\mathbf{b} - \frac{\mathbf{a}+\mathbf{b}+\mathbf{c}}{3}\right) + \left(\mathbf{c} - \frac{\mathbf{a}+\mathbf{b}+\mathbf{c}}{3}\right)$$

$$= \frac{1}{3}(2\mathbf{a} - \mathbf{b} - \mathbf{c} + 2\mathbf{b} - \mathbf{a} - \mathbf{c} + 2\mathbf{c} - \mathbf{a} - \mathbf{b}) = 0$$

57. A regular hexagon $ABCDEF$.

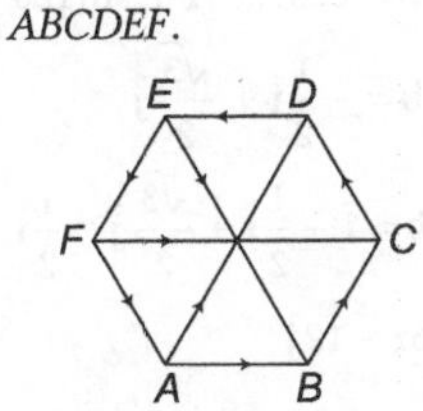

We know from the hexagon that **AD** is parallel to **BC** or **AD** = 2**BC** is parallel to **FA** or **EB** = 2**FA** and **FC** is parallel to **AB** or **FC** = 2**AB**.

Thus, **AD** + **EB** + **FC** = 2**BC** + 2**FA** + 2**AB**

$= 2(\mathbf{FA} + \mathbf{AB} + \mathbf{BC}) = 2(\mathbf{FC}) = 2(2\mathbf{AB}) = 4\mathbf{AB}$

58. $\mathbf{AE} + \mathbf{ED} + \mathbf{DC} + \mathbf{AB}$

$= \mathbf{AD} + \mathbf{DC} + \mathbf{AB} = \mathbf{AC} + \mathbf{AB}$

Obviously, if **BC** is added to this system, then it will be $\mathbf{AC} + \mathbf{AB} + \mathbf{BC} = \mathbf{AC} + \mathbf{AC} = 2\mathbf{AC}$.

59. By triangle law, $\mathbf{AB} = \mathbf{AD} - \mathbf{BD}, \mathbf{AC} = \mathbf{AD} - \mathbf{CD}$

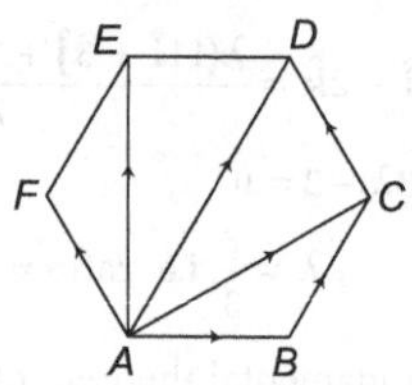

Therefore,

$$\mathbf{AB} + \mathbf{AC} + \mathbf{AD} + \mathbf{AE} + \mathbf{AF}$$

$$= 3\mathbf{AD} + (\mathbf{AE} - \mathbf{BD}) + (\mathbf{AF} - \mathbf{CD}) = 3\mathbf{AD}$$

Hence, $\lambda = 3$ $\quad (\because \mathbf{AE} = \mathbf{BD}, \mathbf{AF} = \mathbf{CD})$

60. $|a| + |b| + |c| = \sqrt{a^2 + b^2 + c^2}$

$\Leftrightarrow 2|ab| + 2|bc| + 2|ca| = 0$

$\Leftrightarrow ab = bc = ca = 0 \Leftrightarrow$ any two of a, b and c are zero

61. Since, **a** and **b** are non-collinear, so **a** + **b** and **a** − **b** will also be non-collinear.

Hence, **a** + **b** and **a** − **b** are linearly independent vectors.

62. $|\mathbf{a} + \mathbf{b}| < |\mathbf{a} - \mathbf{b}|$

$$\Rightarrow \frac{\pi}{2} < \theta < \frac{3\pi}{2}$$

63. $R = \sqrt{4 + 100 + 121} = 15$

64. We must have $\lambda(\hat{\mathbf{i}} - 3\hat{\mathbf{j}} + 5\hat{\mathbf{k}}) = \mathbf{a} + \dfrac{2\hat{\mathbf{k}} + 2\hat{\mathbf{j}} - \hat{\mathbf{i}}}{3}$

Therefore, $3\mathbf{a} = 3\lambda(\hat{\mathbf{i}} - 3\hat{\mathbf{j}} + 5\hat{\mathbf{k}}) - (2\hat{\mathbf{k}} + 2\hat{\mathbf{j}} - \hat{\mathbf{i}})$

$= \hat{\mathbf{i}}(3\lambda + 1) - \hat{\mathbf{j}}(2 + 9\lambda) + \hat{\mathbf{k}}(15\lambda - 2)$

or $3|\mathbf{a}| = \sqrt{(3\lambda + 1)^2 + (2 + 9\lambda)^2 + (15\lambda - 2)^2}$

or $9 = (3\lambda + 1)^2 + (2 + 9\lambda)^2 + (15\lambda - 2)^2$

or $315\lambda^2 - 18\lambda = 0 \Rightarrow \lambda = 0, \dfrac{2}{35}$

If $\lambda = 0$, $\mathbf{a} = \hat{\mathbf{i}} - 2\hat{\mathbf{j}} - 2\hat{\mathbf{k}}$ (not acceptable)

For $\lambda = \dfrac{2}{35}, \mathbf{a} = \dfrac{41}{105}\hat{\mathbf{i}} - \dfrac{88}{105}\hat{\mathbf{j}} - \dfrac{40}{105}\hat{\mathbf{k}}$

65. $\mathbf{b} = \cos 120° \hat{\mathbf{i}} + \sin 120° \hat{\mathbf{j}}$

or $\mathbf{b} = -\dfrac{1}{2}\hat{\mathbf{i}} + \dfrac{\sqrt{3}}{2}\hat{\mathbf{j}}$

Therefore, $\mathbf{a} + \mathbf{b} = \hat{\mathbf{i}} - \dfrac{1}{2}\hat{\mathbf{i}} + \dfrac{\sqrt{3}}{2}\hat{\mathbf{j}} = \dfrac{1}{2}\hat{\mathbf{i}} + \dfrac{\sqrt{3}}{2}\hat{\mathbf{j}}$

66. $\alpha = \mathbf{a} + \mathbf{b} + \mathbf{c} = 6\hat{\mathbf{i}} + 12\hat{\mathbf{j}}$

Let $\alpha = x\mathbf{a} + y\mathbf{b} \Rightarrow 6x + 2y = 6$

and $-3x - 6y = 12$

$\therefore \quad x = 2, y = -3$

$\therefore \quad \alpha = 2\mathbf{a} - 3\mathbf{b}$

67. Let the incentre be at the origin and be $A(\mathbf{p}), B(\mathbf{q})$ and $C(\mathbf{r})$. Then,

$$\mathbf{IA} = \mathbf{p}, \mathbf{IB} = \mathbf{q} \text{ and } \mathbf{IC} = \mathbf{r}$$

Incentre I is $\dfrac{ap + bq + cr}{a + b + c}$, where $p = BC$, $q = AC$ and $r = AB$

Incentre is at the origin. Therefore,

$$\frac{a\mathbf{p} + b\mathbf{q} + c\mathbf{r}}{a + b + c} = \mathbf{0},$$

or $a\mathbf{b} + b\mathbf{q} + c\mathbf{r} = \mathbf{0}$

$\Rightarrow \quad a\mathbf{IA} + b\mathbf{IB} + c\mathbf{IC} = \mathbf{0}$

68. Since **x**, **y** and **x** × **y** are linearly independent, we have

$$20a - 15b = 15b - 12c = 12c - 20a = 0$$

$$\Rightarrow \frac{a}{3} = \frac{b}{4} = \frac{c}{5} \Rightarrow c^2 = a^2 + b^2$$

Hence, ΔABC is right angled.

69. As **x**, **y** and **x** × **y** are non-collinear vectors, vectors are linearly independent.

Hence,

$$a - b = 0 = b - c = c - a$$

or $a = b = c$

Therefore, the triangle is equilateral.

70. $|\mathbf{AB}| = |Q| = \sqrt{P^2 + P^2} = P\sqrt{2}$

71. The point the divides $5\hat{\mathbf{i}}$ and $5\hat{\mathbf{j}}$ in th ratio of

$$k : 1 \text{ is } \frac{(5\hat{\mathbf{j}})k + (5\hat{\mathbf{i}})1}{k + 1}$$

$$\therefore \quad \mathbf{b} = \frac{5\hat{\mathbf{i}} + 5k\hat{\mathbf{j}}}{k + 1}$$

Also $|\mathbf{b}| \le \sqrt{37}$

$$\Rightarrow \frac{1}{k + 1}\sqrt{25 + 25k^2} \le \sqrt{37}$$

or $5\sqrt{1 + k^2} \le \sqrt{37}(k + 1)$

On Squaring both sides, we get

$$25(1 + k^2) \le 37(k^2 + 2k + 1)$$

or $6k^2 + 37k + 6 \ge 0$ or $(6k + 1)(k + 6) \ge 0$

$$k \in (-\infty, -6) \cup \left[-\frac{1}{6}, \infty\right]$$

72. Let the bisector of $\angle A$ meets BC at D, then AD divides BC in the ratio $AB : AC$.

$\therefore$ Position vectors of D

$$= \frac{|\mathbf{AB}|(2\hat{\mathbf{i}} + 5\hat{\mathbf{j}} + 7\hat{\mathbf{k}}) + |\mathbf{AC}|(2\hat{\mathbf{i}} + 3\hat{\mathbf{j}} + 4\hat{\mathbf{k}})}{|\mathbf{AB}| + |\mathbf{AC}|}$$

Here, $|\mathbf{AB}| = |-2\hat{\mathbf{i}} - 4\hat{\mathbf{j}} - 4\hat{\mathbf{k}}| = 6$

and $\quad |\mathbf{AC}| = |-2\hat{\mathbf{i}} - 2\hat{\mathbf{j}} - \hat{\mathbf{k}}| = 3$

$\therefore$ Position vector of D

$$= \frac{6(2\hat{\mathbf{i}} + 5\hat{\mathbf{j}} + 7\hat{\mathbf{k}}) + 3(2\hat{\mathbf{i}} + 3\hat{\mathbf{j}} + 4\hat{\mathbf{k}})}{6+3}$$

$$= \frac{18\hat{\mathbf{i}} + 39\hat{\mathbf{j}} + 54\hat{\mathbf{k}}}{9}$$

$$= \frac{1}{3}(6\hat{\mathbf{i}} + 13\hat{\mathbf{j}} + 18\hat{\mathbf{k}})$$

73. Vector in the direction of angular bisector of $\mathbf{a}$ and $\mathbf{b}$ is $\frac{\mathbf{a}+\mathbf{b}}{2}$.

Unit vector in this direction is $\frac{\mathbf{a}+\mathbf{b}}{|\mathbf{a}+\mathbf{b}|}$

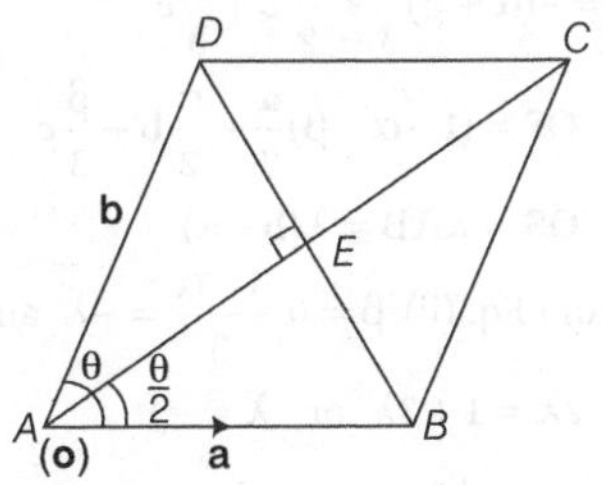

From the figure, position vector of E is $\frac{\mathbf{a}+\mathbf{b}}{2}$.

Now in triangle AEB, $AE = AB\cos\frac{\theta}{2}$

$$\Rightarrow \quad \left|\frac{\mathbf{a}+\mathbf{b}}{2}\right| = \cos\frac{\theta}{2}$$

Hence, unit vector along the bisector is $\frac{\mathbf{a}+\mathbf{b}}{2\cos(\theta/2)}$.

74. $\quad \mathbf{a} - \mathbf{b} = 2(\mathbf{d} - \mathbf{c})$

$$\therefore \quad \frac{\mathbf{a}+2\mathbf{c}}{2+1} = \frac{\mathbf{b}+2\mathbf{d}}{2+1}$$

Hence, AC and BD trisect each other as LHS is the position vector of a point trisecting A an C, and RHS that of B and D.

75. Again, it is given that the point P, Q and R are collinear.

$$\Rightarrow \quad \mathbf{PQ} = \lambda\mathbf{QR}$$

$$\Rightarrow \quad 15t(4\hat{\mathbf{j}} - 3\hat{\mathbf{i}}) = \lambda[(1-t)(\hat{\mathbf{i}}+\hat{\mathbf{j}}) - 60t\hat{\mathbf{j}}]$$

$$\Rightarrow \quad = \lambda[(1-t)\hat{\mathbf{i}} + (1-61t)\hat{\mathbf{j}}]$$

$$\Rightarrow \quad \frac{45t}{t-1} = \frac{60t}{1-61t}$$

$$\Rightarrow \quad \frac{3t}{t-1} = \frac{4t}{1-61t}$$

$$\Rightarrow \quad 3(1-61t) = 4(t-1)$$

$$\Rightarrow \quad 3 - 183t = 4t - 4 \Rightarrow 187t = 7$$

$$\therefore \quad t = \frac{7}{187}$$

76. We have, $\mathbf{a} + \mathbf{b} + \mathbf{c} = \alpha\mathbf{d}$

and $\quad \mathbf{b} + \mathbf{c} + \mathbf{d} = \beta\mathbf{a}$

$\therefore \quad \mathbf{a} + \mathbf{b} + \mathbf{c} + \mathbf{d} = (\alpha+1)\mathbf{d}$

and $\quad \mathbf{a} + \mathbf{b} + \mathbf{c} + \mathbf{d} = (\beta+1)\mathbf{a}$

$\Rightarrow \quad (\alpha+1)\mathbf{d} = (\beta+1)\mathbf{a}$

If $\alpha \neq -1$, then $(\alpha+1)\mathbf{d} = (\beta+1)\mathbf{a}$

$$\Rightarrow \quad \mathbf{d} = \frac{\beta+1}{\alpha+1}\mathbf{a}$$

$$\Rightarrow \quad \mathbf{a} + \mathbf{b} + \mathbf{c} = \alpha\mathbf{d}$$

$$\Rightarrow \quad \mathbf{a} + \mathbf{b} + \mathbf{c} = \alpha\left(\frac{\beta+1}{\alpha+1}\right)\mathbf{a}$$

$$\Rightarrow \quad \left[1 - \frac{\alpha(\beta+1)}{\alpha+1}\right]\mathbf{a} + \mathbf{b} + \mathbf{c} = 0$$

a, b and **c** are coplanar which is contradiction to the given condition.

$$\therefore \quad \alpha = -1$$

and so $\mathbf{a} + \mathbf{b} + \mathbf{c} + \mathbf{d} = 0$

77. Since $|\mathbf{OP}| = |\mathbf{OQ}| = \sqrt{14}$, ΔOPQ is an isosceles.

Hence, the internal bisector OM is perpendicular to PQ and M is the mid-point of P and Q. Therefore,

$$\mathbf{OM} = \frac{1}{2}(\mathbf{OP} + \mathbf{OQ}) = 2\hat{\mathbf{i}} + \hat{\mathbf{j}} - 2\hat{\mathbf{k}}$$

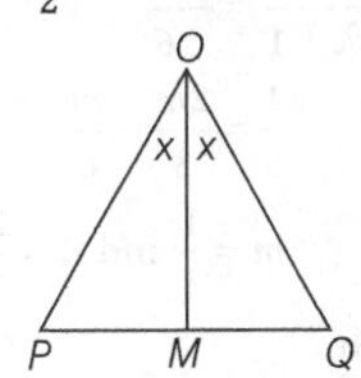

78.

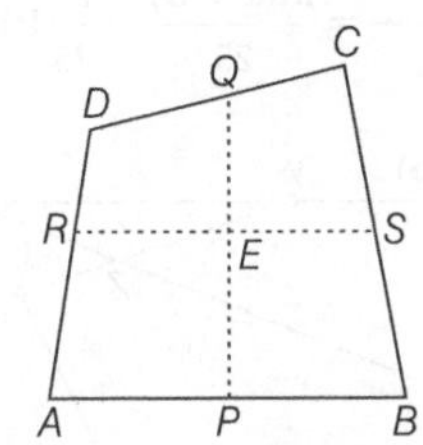

Let $\quad \mathbf{OA} = \mathbf{a}, \mathbf{OB} = \mathbf{b}, \mathbf{OC} = \mathbf{c}$

and $\quad \mathbf{OD} = \mathbf{d}$

Therefore, $\quad \mathbf{OA} + \mathbf{OB} + \mathbf{OC} + \mathbf{OD}$

$$= \mathbf{a} + \mathbf{b} + \mathbf{c} + \mathbf{d}$$

P, the mid-point of **AB**, is $\frac{\mathbf{a}+\mathbf{b}}{2}$.

Q, the mid-point of **CD**, is $\frac{\mathbf{c}+\mathbf{d}}{2}$.

Therefore, the mid-point of **PQ** is $\frac{\mathbf{a}+\mathbf{b}+\mathbf{c}+\mathbf{d}}{4}$.

Similarly, the mid-point of **RS** is $\frac{\mathbf{a}+\mathbf{b}+\mathbf{c}+\mathbf{d}}{4}$

i.e. $\quad \mathbf{OE} = \frac{\mathbf{a}+\mathbf{b}+\mathbf{c}+\mathbf{d}}{4}$

$$\Rightarrow \quad x = 4$$

79.

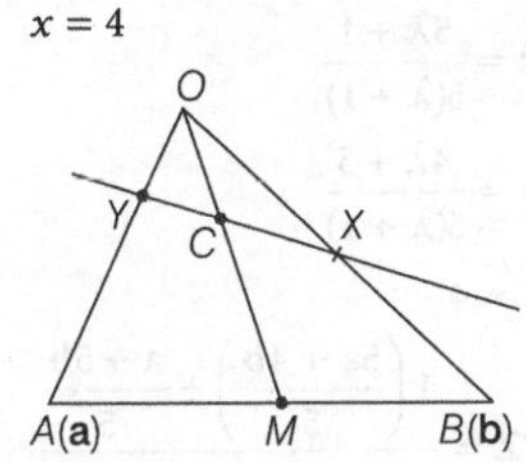

$\mathbf{OA} = \mathbf{a}$, $\mathbf{OB} = \mathbf{b}$

$\therefore \quad \mathbf{OM} = \dfrac{\mathbf{a}+\mathbf{b}}{2}$

$\therefore \quad \mathbf{OC} = \dfrac{\mathbf{a}+\mathbf{b}}{6}$

$\mathbf{OX} = \dfrac{2}{3}\mathbf{b}$

Let $\quad \dfrac{OY}{YA} = \lambda$

$\therefore \quad \mathbf{OY} = \dfrac{\lambda}{\lambda+1}\mathbf{a}$

Now points Y, C and X are collinear.

$\therefore \quad \mathbf{YC} = m\mathbf{CX}$

$\therefore \quad \dfrac{\mathbf{a}+\mathbf{b}}{6} - \dfrac{\lambda}{\lambda+1}\mathbf{a} = m\dfrac{2\mathbf{b}}{3} - m\dfrac{\mathbf{a}+\mathbf{b}}{6}$

Comparing coefficients of **a** and **b**

$\therefore \quad \dfrac{1}{6} - \dfrac{\lambda}{\lambda+1} = -\dfrac{m}{6}$

and $\quad \dfrac{1}{6} = \dfrac{2m}{3} - \dfrac{m}{6}$

$\therefore \quad m = \dfrac{1}{3}$ and $\lambda = \dfrac{2}{7}$

80. $\dfrac{4\left(\dfrac{5\mathbf{a}+4\mathbf{b}}{3}\right) + \dfrac{\mathbf{a}+5\mathbf{b}}{5}}{4+1} = \dfrac{21(\mathbf{a}+\mathbf{b})}{25} = \dfrac{21}{25}\mathbf{PR}$

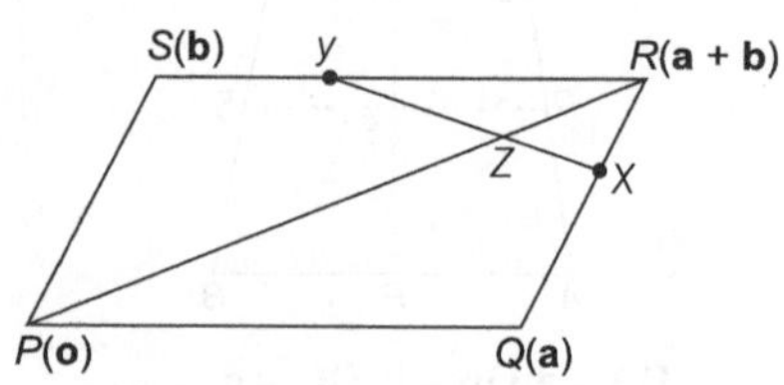

PV of X is $\dfrac{4(\mathbf{a}+\mathbf{b})+\mathbf{a}}{5} = \dfrac{5\mathbf{a}+4\mathbf{b}}{5}$

PV of Y is $\dfrac{4\mathbf{b}+\mathbf{a}+\mathbf{b}}{5} = \dfrac{\mathbf{a}+5\mathbf{b}}{5}$

Now, $\quad \mathbf{PZ} = m\mathbf{PR}$

$\mathbf{PZ} = m(\mathbf{a}+\mathbf{b})$

Let Z divided YX in the ratio $\lambda:1$

PV of $Z = \dfrac{\lambda\mathbf{OX}+\mathbf{OY}}{\lambda+1}$

$\therefore \quad \mathbf{PZ} = \dfrac{\lambda\left(\dfrac{5\mathbf{a}+4\mathbf{b}}{5}\right) + \dfrac{\mathbf{a}+5\mathbf{b}}{5}}{\lambda+1} = m(\mathbf{a}+\mathbf{b})$

Comparing coefficients of **a** and **b**

$m = \dfrac{5\lambda+1}{5(\lambda+1)}$

and $\quad m = \dfrac{4\lambda+5}{5(\lambda+1)}$

$\therefore \quad \lambda = 4$

$\therefore \quad \mathbf{PZ} = \dfrac{4\left(\dfrac{5\mathbf{a}+4\mathbf{b}}{5}\right) + \dfrac{\mathbf{a}+5\mathbf{b}}{5}}{4+1}$

$= \dfrac{21(\mathbf{a}+\mathbf{b})}{25} = \dfrac{21}{25}\mathbf{PR}$

81. Let $\mathbf{OA} = \mathbf{a}, \mathbf{OB} = \mathbf{b}$ and $\mathbf{OC} = \mathbf{c}$,

then $\mathbf{AB} = \mathbf{b} - \mathbf{a}$ and $\mathbf{OP} = \dfrac{1}{3}\mathbf{a}$.

$OQ = \dfrac{1}{2}\mathbf{b}, \mathbf{OR} = \dfrac{1}{3}\mathbf{c}.$

Since P, Q, R and S are coplanar, then

$\mathbf{PS} = \alpha\mathbf{PQ} + \beta\mathbf{PR}$

(**PS** can be written as a linear combination of **PQ** and **PR**)

$= \alpha(\mathbf{OQ} - \mathbf{OP}) + \beta(\mathbf{OR} - \mathbf{OP})$

i.e. $\mathbf{OS} - \mathbf{OP} = -(\alpha+\beta)\dfrac{\mathbf{a}}{3} + \dfrac{\alpha}{2}\mathbf{b} + \dfrac{\beta}{3}\mathbf{c}$

$\Rightarrow \quad \mathbf{OS} = (1-\alpha-\beta)\dfrac{\mathbf{a}}{3} + \dfrac{\alpha}{2}\mathbf{b} + \dfrac{\beta}{3}\mathbf{c} \quad \text{... (i)}$

Given $\quad \mathbf{OS} = \lambda\mathbf{AB} = \lambda(\mathbf{b}-\mathbf{a}) \quad \text{...(ii)}$

From Eq. (i) and Eq. (ii), $\beta = 0$, $\dfrac{1-\alpha}{3} = -\lambda$ and $\dfrac{\alpha}{2} = \lambda$

$\Rightarrow \quad 2\lambda = 1 + 3\lambda$ or $\lambda = -1$

82.

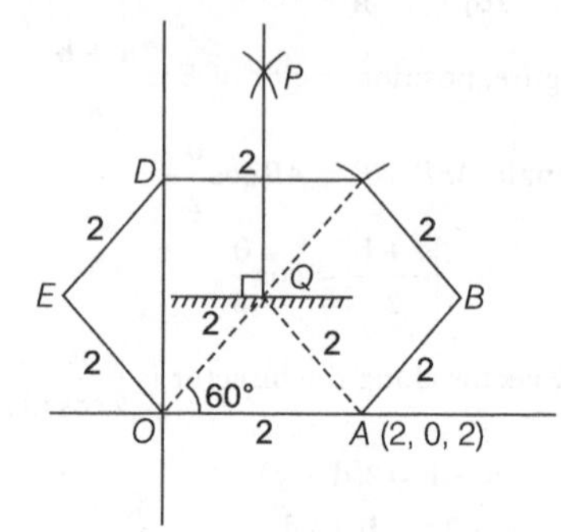

Here, coordinate of Q are $(2\cos 60°, 2\sin 60°)$

$\Rightarrow \quad Q(1, \sqrt{3}, 0)$

$\therefore \quad P(1, \sqrt{3}, z)$

$OP = 3$

$\Rightarrow \quad \sqrt{1+3+z^2} = 3$ or $z^2 = 5$

$z = \sqrt{5}$

$\therefore \quad P(1, \sqrt{3}, \sqrt{5}) \Rightarrow \mathbf{OP} = \hat{\mathbf{i}} + \sqrt{3}\hat{\mathbf{j}} + \sqrt{5}\hat{\mathbf{k}}$

Now, $\mathbf{AP} = \mathbf{OP} - \mathbf{OA} = \hat{\mathbf{i}} + \sqrt{3}\hat{\mathbf{j}} + \sqrt{5}\hat{\mathbf{k}} - 2\hat{\mathbf{i}}$

$= -\hat{\mathbf{i}} + \sqrt{3}\hat{\mathbf{j}} + \sqrt{5}\hat{\mathbf{k}}$

83. $\mathbf{a} = [\pm(\hat{\mathbf{i}} - \hat{\mathbf{j}}) \pm (\hat{\mathbf{j}} + \hat{\mathbf{k}})]$

$= \pm(\hat{\mathbf{i}} + \hat{\mathbf{k}}), \pm(\hat{\mathbf{i}} - 2\hat{\mathbf{j}} - \hat{\mathbf{k}})$

84. Let **R** be the resultant. Then

$\mathbf{R} = \mathbf{F}_1 + \mathbf{F}_2 + \mathbf{F}_3 = (p+1)\hat{\mathbf{i}} + 4\hat{\mathbf{j}}$

Given, $|\mathbf{R}| = 5$. Therefore, $(p+1)^2 + 16 = 25$

or $\quad p+1 = \pm 3$ or $p = 2, -4$

85. We have, $\quad \mathbf{AB} = -\hat{\mathbf{i}} - \hat{\mathbf{j}} - 4\hat{\mathbf{k}}, \mathbf{BC} = -3\hat{\mathbf{i}} + 3\hat{\mathbf{j}}$

and $\quad \mathbf{CA} = 4\hat{\mathbf{i}} - 2\hat{\mathbf{j}} + 4\hat{\mathbf{k}}.$

Therefore, $|\mathbf{AB}| = |\mathbf{BC}| = 3\sqrt{2}$ and $|\mathbf{CA}| = 6$

Clearly, $|\mathbf{AB}|^2 + |\mathbf{BC}|^2 = |\mathbf{AC}|^2$

Hence, the triangle is right angled isosceles triangle.

86. Let $\mathbf{a} = 2\hat{i} + 4\hat{j} - 5\hat{k}$ and $\mathbf{b} = \hat{i} + 2\hat{j} + 3\hat{k}$.

Then, the diagonals of the parallelogram are

$$\mathbf{p} = \mathbf{a} + \mathbf{b}$$

and $\mathbf{q} = \mathbf{b} - \mathbf{a}$,

i.e. $\mathbf{p} = 3\hat{i} + 6\hat{j} - 2\hat{k},\ \mathbf{q} = -\hat{i} - 2\hat{j} + 8\hat{k}$

So, unit vectors along the diagonals are

$$\frac{1}{7}(3\hat{i} + 6\hat{j} - 2\hat{k}) \text{ and } \frac{1}{\sqrt{69}}(-\hat{i} - 2\hat{j} + 8\hat{k})$$

87. $\mathbf{OA} = -4\hat{i} + 3\hat{k};\ \mathbf{OB} = 14\hat{i} + 2\hat{j} - 5\hat{k}$

$$\mathbf{a} = \frac{-4\hat{i} + 3\hat{k}}{5};\ \mathbf{b} = \frac{14\hat{i} + 2\hat{j} - 5\hat{k}}{15}$$

$$\mathbf{r} = \frac{\lambda}{15}[-12\hat{i} + 9\hat{j} + 14\hat{i} + 2\hat{j} - 5\hat{k}]$$

$$= \frac{\lambda}{15}[2\hat{i} + 2\hat{j} + 4\hat{k}] = \frac{2\lambda}{15}[\hat{i} + \hat{j} + 2\hat{k}]$$

88. Points $A(\hat{i} + \hat{j})$, $B(\hat{i} - \hat{j})$ and $C(p\hat{i} + q\hat{j} + r\hat{k})$ are collinear

Now, $\mathbf{AB} = -2\hat{j}$

and $\mathbf{BC} = (p-1)\hat{i} + (q+1)\hat{j} + r\hat{k}$

Vectors **AB** and **BC** must be collinear

$\Rightarrow$ $p = 1, r = 0$ and $q \neq -1$

89. For coplanar vectors, $\begin{vmatrix} 1 & 2 & 3 \\ 0 & \lambda & \mu \\ 0 & 0 & 2\lambda - 1 \end{vmatrix} = 0$

or $(2\lambda - 1)\lambda = 0$ or $\lambda = 0, \frac{1}{2}$

90. In ΔABC, $\mathbf{AB} + \mathbf{BC} = \mathbf{AC} = -\mathbf{CA}$

or $\mathbf{AB} + \mathbf{BC} + \mathbf{CA} = \mathbf{0}$

$\mathbf{OA} + \mathbf{AB} = \mathbf{OB}$ is the triangle law of addition.

Hence, Statement 1 is true and Statement 2 is is false.

91. $\frac{1}{2} = \frac{p}{3} = \frac{2}{q} \Rightarrow p = \frac{3}{2}$ and $q = 4$

92.

$$2\mathbf{a} + 3\mathbf{b} - 5\mathbf{c} = 0$$

$$3(\mathbf{b} - \mathbf{a}) = 5(\mathbf{c} - \mathbf{a})$$

$$\Rightarrow \quad \mathbf{AB} = \frac{5}{3}\mathbf{AC}$$

Hence, **AB** and **AC** must be parallel since there is a common point A. The points A, B and C must be collinear.

Solutions (Q.Nos. 93-94)

Let the position vectors of A, B, C and D be $\mathbf{a}, \mathbf{b}, \mathbf{c}$ and $\mathbf{d}$, respectively. Then,

$$OA : CB = 2 : 1$$

$\Rightarrow$ $\mathbf{OA} = 2\mathbf{CB} \Rightarrow \mathbf{a} = 2(\mathbf{b} - \mathbf{c})$ ... (i)

and $OD : AB = 1 : 3$

$$3\mathbf{OD} = \mathbf{AB}$$

$\Rightarrow$ $3\mathbf{d} = (\mathbf{b} - \mathbf{a}) = \mathbf{b} - 2(\mathbf{b} - \mathbf{c})$ [using Eq. (i)]

$= -\mathbf{b} + 2\mathbf{c}$ ... (ii)

Let $OX : XC = \lambda : 1$ and $AX : XD = \mu : 1$

Now, X divides OC in the ratio $\lambda : 1$. Therefore,

$$\text{PV of } X = \frac{\lambda \mathbf{c}}{\lambda + 1} \quad \text{...(iii)}$$

X also divided AD in the ratio $\mu : 1$. Therefore,

$$\text{PV of } X = \frac{\mu \mathbf{d} + \mathbf{a}}{\mu + 1} \quad \text{...(iv)}$$

From Eqs. (iii) and (iv), we get

$$\frac{\lambda \mathbf{c}}{\lambda + 1} = \frac{\mu \mathbf{d} + \mathbf{a}}{\mu + 1}$$

or $\left(\frac{\lambda}{\lambda+1}\right)\mathbf{c} = \left(\frac{\mu}{\mu+1}\right)\mathbf{d} + \left(\frac{1}{\mu+1}\right)\mathbf{a}$

or $\left(\frac{\lambda}{\lambda+1}\right)\mathbf{c} = \left(\frac{\mu}{\mu+1}\right)\left(\frac{-\mathbf{b} + 2\mathbf{c}}{3}\right) + \left(\frac{1}{\mu+1}\right)2(\mathbf{b} - \mathbf{c})$

[using Eqs. (i) and (ii)]

or $\left(\frac{\lambda}{\lambda+1}\right)\mathbf{c} = \left(\frac{6-\mu}{3(\mu+1)}\right)\mathbf{b} + \left(\frac{2\mu}{3(\mu+1)} - \frac{2}{\mu+1}\right)\mathbf{c}$

or $\left(\frac{\lambda}{\lambda+1}\right)\mathbf{c} = \left(\frac{6-\mu}{3(\mu+1)}\right)\mathbf{b} + \left(\frac{2\mu - 6}{3(\mu+1)}\right)\mathbf{c}$

or $\left(\frac{6-\mu}{3(\mu+1)}\right)\mathbf{b} + \left(\frac{2\mu-6}{3(\mu+1)} - \frac{\lambda}{\lambda+1}\right)\mathbf{c} = 0$

or $\frac{6-\mu}{3(\mu+1)} = 0$

and $\frac{2\mu - 6}{3(\mu+1)} - \frac{\lambda}{\lambda+1} = 0$

(as **b** and **c** are non-collinear)

or $\mu = 6, \lambda = \frac{2}{5}$

Hence, $OX : XC = 2 : 5$ and $AX / XD = \frac{\mu}{1} = \frac{6}{1}$

93. (c) **94.** (b)

Solutions (Q.Nos. 95-96)

Consider the regular hexagon $ABCDEF$ with centre at O (origin).

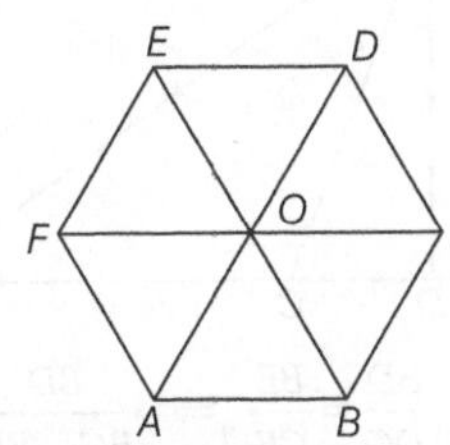

$$\mathbf{AD} + \mathbf{EB} + \mathbf{FC} = 2\mathbf{AO} + 2\mathbf{OB} + 2\mathbf{OC}$$

$= 2(\mathbf{AO} + \mathbf{OB}) + 2\mathbf{OC}$

$= 2\mathbf{AB} + 2\mathbf{AB}$ [$\because \mathbf{OC} = \mathbf{AB}$]

$= 4\mathbf{AB}$

$\mathbf{R} = \mathbf{AB} + \mathbf{AC} + \mathbf{AD} + \mathbf{AE} + \mathbf{AF}$

$= \mathbf{ED} + \mathbf{AC} + \mathbf{AD} + \mathbf{AE} + \mathbf{CD}$

[$\because \mathbf{AB} = \mathbf{ED}$ and $\mathbf{AF} = \mathbf{CD}$]

$= (\mathbf{AC} + \mathbf{CD}) + (\mathbf{AE} + \mathbf{ED}) + \mathbf{AD}$

$= \mathbf{AD} + \mathbf{AD} + \mathbf{AD} = 3\mathbf{AD} = 6\mathbf{AO}$

95. (c) **96.** (d)

Solutions (Q.Nos. 97-99)

97. $\mathbf{AB} = \mathbf{OB} - \mathbf{OA} = 3\mathbf{a} - \mathbf{b} - 2\mathbf{c}$

$\mathbf{AC} = \mathbf{OC} - \mathbf{OA} = 9\mathbf{a} - 3\mathbf{b} - 6\mathbf{c} = 3\mathbf{AB}$

98. $2\mathbf{OA} - 3\mathbf{OB} + \mathbf{OC}$

$$= 2(-2\mathbf{a} + 3\mathbf{b} + 5\mathbf{c}) - 3(\mathbf{a} + 2\mathbf{b} + 3\mathbf{c}) + (7\mathbf{a} - \mathbf{c}) = 0$$

99. $\because\ 2\mathbf{OA} - 3\mathbf{OB} + \mathbf{OC} = 0$

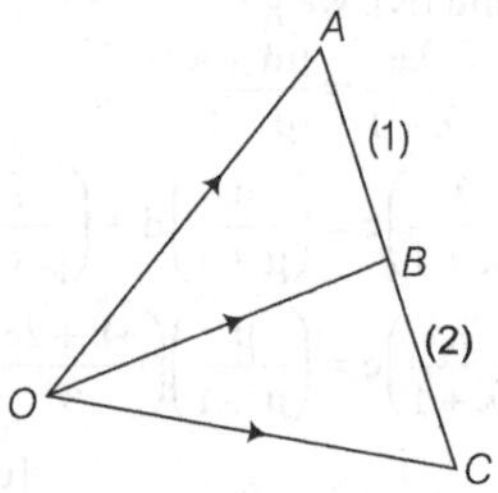

$$\Rightarrow \quad \mathbf{OB} = \frac{2\mathbf{OA} + \mathbf{OC}}{3}$$

$\Rightarrow$ B Divides AC in 1 : 2.

Solutions (Q.Nos. 100-101)

100. Here, $\mathbf{c} = t(\hat{\mathbf{a}} + \hat{\mathbf{b}}) = t\left(\frac{7\hat{\mathbf{i}} - 4\hat{\mathbf{j}} - 4\hat{\mathbf{k}}}{9} + \frac{(-2\hat{\mathbf{i}} - \hat{\mathbf{j}} + 2\hat{\mathbf{k}})}{3}\right)$

$$\Rightarrow \quad \mathbf{c} = t\left(\frac{\hat{\mathbf{i}} - 7\hat{\mathbf{j}} + 2\hat{\mathbf{k}}}{9}\right)$$

Also, $\quad |\mathbf{c}| = 5\sqrt{6} \Rightarrow \frac{t}{9}.\sqrt{1 + 49 + 4} = 5\sqrt{6}$

$\therefore \quad t = 15 \Rightarrow \mathbf{c} = \frac{15}{9}(\hat{\mathbf{i}} - 7\hat{\mathbf{j}} + 2\hat{\mathbf{k}})$

or $\quad = \frac{5}{3}(\hat{\mathbf{i}} - 7\hat{\mathbf{j}} + 2\hat{\mathbf{k}})$

101. Here, $\frac{AB}{AC} = \frac{BD}{DC}$ and $\frac{AB}{AC} = \frac{BE}{CE}$

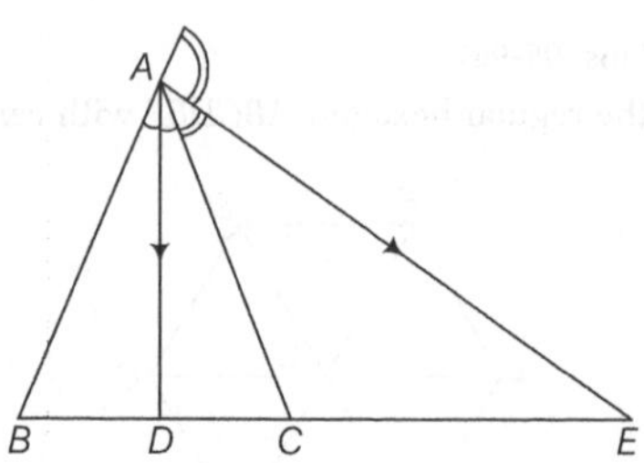

$$\Rightarrow \quad \frac{BD}{DC} = \frac{BE}{CE} \Rightarrow \frac{BD}{BC - BD} = \frac{BE}{BE - BC}$$

$$\Rightarrow \quad BD \cdot BE - BD \cdot BC = BC \cdot BE - BD \cdot BE$$

$$\Rightarrow \quad 2BD \cdot BE = (BD + BE) \cdot BC$$

or $\quad \frac{2}{BC} = \frac{1}{BD} + \frac{1}{BE}$

Solutions (Q.Nos. 102-103)

102. $r'(t) = -2\hat{\mathbf{i}} + 2t\hat{\mathbf{j}} + 4e^{2(t-1)}\hat{\mathbf{k}}$

Since, $r'(t)$ is parallel to $r(t)$,

so $\quad r(t) = \alpha r'(t)$

$$1 - 2t = -2\alpha, t^2 = \alpha t, 2e^{2(t-1)}$$

$$= 4\alpha e^{2(t-1)}, \alpha = \frac{1}{2}$$

The only value of t which satisfies all three equations is $t = 1$. So, $r(1)$ is the required point $(-1, 1, 2)$.

103. $(2, 0, 5)$ corresponding to $r(1)$ and $r'(t) = 4t\hat{\mathbf{i}} - \hat{\mathbf{j}} + 6t\hat{\mathbf{k}}$

So, the required tangent vector is $r'(1) = 4\hat{\mathbf{i}} - \hat{\mathbf{j}} + 6\hat{\mathbf{k}}$.

104.

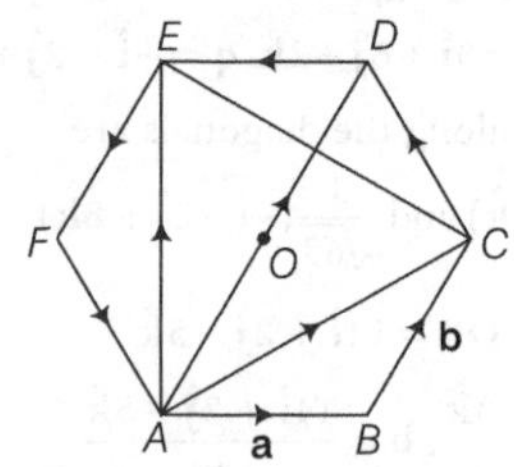

$\mathbf{AB} = \mathbf{a}, \mathbf{BC} = \mathbf{b}$

$\therefore \quad \mathbf{AC} = \mathbf{AB} + \mathbf{BC} = \mathbf{a} + \mathbf{b}$ ... (i)

$\mathbf{AD} = 2\mathbf{BC} = 2\mathbf{b}$ ... (ii)

(because AD is parallel to BC and twice is length)

$\mathbf{CD} = \mathbf{AD} - \mathbf{AC} = 2\mathbf{b} - (\mathbf{a} + \mathbf{b}) = \mathbf{b} - \mathbf{a}$

$\mathbf{FA} = -\mathbf{CD} = \mathbf{a} - \mathbf{b}$ ... (iii)

$\mathbf{DE} = -\mathbf{AB} = -\mathbf{a}$ ... (iv)

$\mathbf{EF} = -\mathbf{BC} = -\mathbf{b}$ ... (v)

$\mathbf{AE} = \mathbf{AD} + \mathbf{DE} = 2\mathbf{b} - \mathbf{a}$ ... (vi)

$\mathbf{CE} = \mathbf{CD} + \mathbf{DE} = \mathbf{b} - \mathbf{a} - \mathbf{a} = \mathbf{b} - 2\mathbf{a}$ ... (vii)

105. Let $\mathbf{R}$ be the resultant. Then,

$$\mathbf{R} = \mathbf{F}_1 + \mathbf{F}_2 + \mathbf{F}_3 = (p + 1)\hat{\mathbf{i}} + 4\hat{\mathbf{j}}$$

Given, $|\mathbf{R}| = 5$, Therefore $R^2 = 25$

$\because \quad (p + 1)^2 + 16 = 25$ or $p + 1 = \pm 3$ or $p = 2, -4$

106. Vectors along to sides are $\mathbf{a} = \hat{\mathbf{i}} + 2\hat{\mathbf{j}} + \hat{\mathbf{k}}$ and $\mathbf{b} = 2\hat{\mathbf{i}} + 4\hat{\mathbf{j}} + \hat{\mathbf{k}}$

Clearly the vector along the longer diagonal is

$$\mathbf{a} + \mathbf{b} = 3\hat{\mathbf{i}} + 6\hat{\mathbf{j}} + 2\hat{\mathbf{k}}$$

Hence, length of the longer diagonal is

$$|\mathbf{a} + \mathbf{b}| = |3\hat{\mathbf{i}} + 6\hat{\mathbf{j}} + 2\hat{\mathbf{k}}| = 7$$

107. Vector $\mathbf{a} = \hat{\mathbf{i}} + 2\hat{\mathbf{j}} - \hat{\mathbf{k}}$, $\mathbf{b} = 2\hat{\mathbf{i}} - \hat{\mathbf{j}} + \hat{\mathbf{k}}$, $\mathbf{c} = \lambda\hat{\mathbf{i}} + \hat{\mathbf{j}} + 2\hat{\mathbf{k}}$ are coplanar.

$$\Rightarrow \quad \begin{vmatrix} 1 & 2 & -1 \\ 2 & -1 & 1 \\ \lambda & 1 & 2 \end{vmatrix} = 0 \text{ or } \lambda - 3 + 2(-5) = 0 \text{ or } \lambda = 13$$

Number of digits in value of λ is 2.

108. Since, angle bisector of $\mathbf{a}$ and $\mathbf{b}$

$$\Rightarrow \quad h(\hat{\mathbf{a}} + \hat{\mathbf{b}}) = h\left(\frac{\mathbf{a}}{|\mathbf{a}|} + \frac{\mathbf{b}}{|\mathbf{b}|}\right) \quad \text{... (i)}$$

given, $\mathbf{a} + \mathbf{b}$ is along angle bisector.

$$\Rightarrow \quad \mu\left(\frac{\mathbf{a}}{|\mathbf{a}|} + \frac{\mathbf{b}}{|\mathbf{b}|}\right) = \mathbf{a} + \mathbf{b}$$

only, when $|\mathbf{a}| = |\mathbf{b}| = \mu$

$\therefore \quad |\mathbf{a}| = |\mathbf{b}| \Rightarrow \lambda = 1$

109.

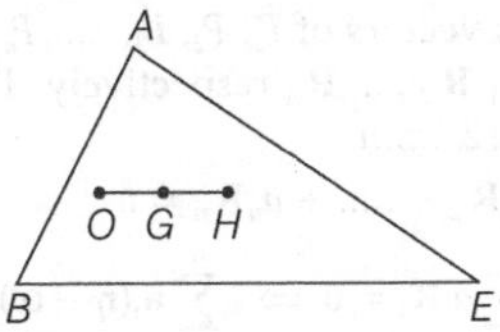

Here, O is circum centre $= \mathbf{0}$, G is centroid $= \mathbf{g}$

H is orthocentre $= \mathbf{p}$

Since, $$\frac{\mathbf{OG}}{\mathbf{GH}} = \frac{1}{2}$$

$$\Rightarrow \quad \frac{\mathbf{g}}{\mathbf{p}-\mathbf{g}} = \frac{1}{2} \Rightarrow 2\mathbf{g} = \mathbf{p} - \mathbf{g}$$

or $\quad \mathbf{p} = 3\mathbf{g}$

$\therefore \quad k = 3$

110. $\mathbf{XG} = k\mathbf{GY}$

$$\frac{\mathbf{b}+\mathbf{c}}{3} - \frac{2\mathbf{b}}{3} = k\left(\frac{\lambda\mathbf{c}}{1+\lambda} - \frac{\mathbf{b}+\mathbf{c}}{3}\right), k = 1$$

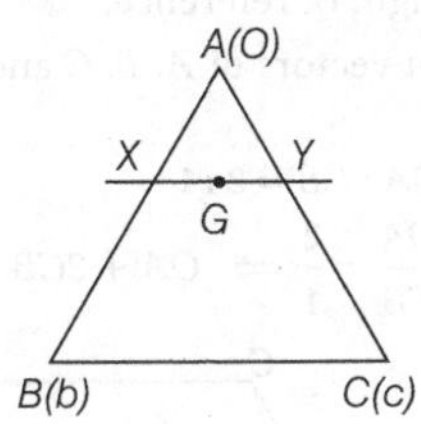

$$\frac{2\mathbf{c}}{3} = \frac{\lambda}{1+\lambda}\mathbf{c}$$

$$\Rightarrow \quad \frac{\lambda}{1+\lambda} = \frac{2}{3}$$

$$\Rightarrow \quad \lambda = 2$$

111. We have, $\quad \mathbf{a} = 2p\hat{\mathbf{i}} + \hat{\mathbf{j}}$

Let $\mathbf{b}$ be the vector obtained from $\mathbf{a}$ by rotating the axes. Then, the components of $\mathbf{b}$ are $p+1$ and 1. Therefore,

$\mathbf{b} = (p+1)\,\hat{\alpha} + \hat{\beta}$

where $\hat{\alpha}$ and $\hat{\beta}$ are unit vectors along the new axes.

But $|\mathbf{b}| = |\mathbf{a}|$

$$\Rightarrow \quad 4p^2 + 1 = (p+1)^2 + 1$$

$$\Rightarrow \quad 3p^2 - 2p - 1 = 0 \Rightarrow p = 1, -\frac{1}{3}$$

$$\Rightarrow \quad p_1 = 1 \text{ and } p_2 = -\frac{1}{3}$$

$$\therefore \quad 3\,|p_1 + p_2| = 3\left|1 - \frac{1}{3}\right| = 2$$

112. Here, $\mathbf{a}$ is rotated about Z-axis, the Z-component of $\mathbf{a}$ will remain unchanged namely a_3

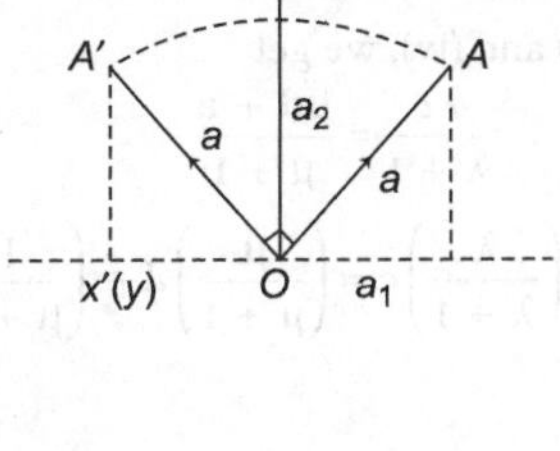

Now, if it is turned through an angle $\frac{\pi}{2}$. As shown in adjoining figure.

$\therefore$ Now components are $(a_2, -a_1, a_3)$.

113. Here, $\mathbf{r}_1 = \mathbf{OA}$ pointing North-East and $\mathbf{r}_2 = \mathbf{OB}$ pointing North-West. Where $|\mathbf{OA}| = |\mathbf{OB}| = 5$.

As shown in figure,

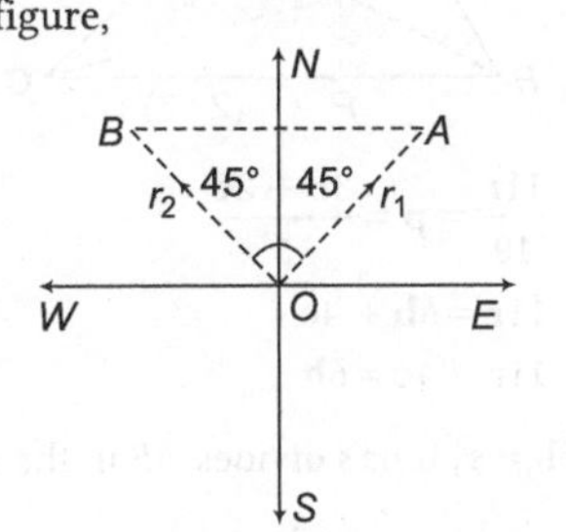

$$\therefore \quad \angle BOA = 90°$$

$$\Rightarrow \quad \mathbf{r}_1 - \mathbf{r}_2 = \mathbf{BA} \quad \text{(using triangle law)}$$

Clearly, $\angle BOA$ is right angled at O.

$$\therefore \quad BA^2 = OA^2 + OB^2 = 5^2 + 5^2 = 50$$

$$\Rightarrow \quad |BA| = 5\sqrt{5}$$

or $\quad |\mathbf{r}_1 - \mathbf{r}_2| = 5\sqrt{5}$

i.e. $\mathbf{r}_1 - \mathbf{r}_2$ has magnitude $5\sqrt{5}$ and points from West to East.

114. Let **OACB** be a parallelogram shown as

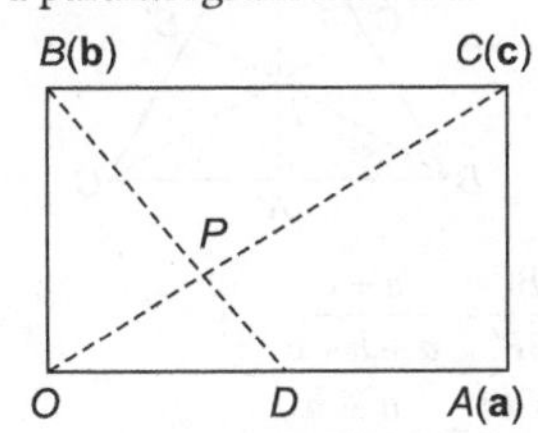

Here, $$\mathbf{OD} = \frac{1}{2}\mathbf{BC}$$

$$\Rightarrow \quad \mathbf{OP} + \mathbf{PD} = \frac{1}{2}(\mathbf{BP} + \mathbf{PC}) \quad [\text{using } \Delta \text{ law}]$$

$$\Rightarrow \quad 2\mathbf{OP} + 2\mathbf{PD} = \mathbf{BP} + \mathbf{PC}$$

$$-2\mathbf{PO} + 2\mathbf{PD} = -\mathbf{PB} + \mathbf{PC}$$

$$\Rightarrow \quad \mathbf{PB} + 2\mathbf{PD} = \mathbf{PC} + 2\mathbf{PO}$$

$$\Rightarrow \quad \frac{\mathbf{PB} + 2\mathbf{PD}}{1+2} = \frac{\mathbf{PC} + 2\mathbf{PO}}{1+2}$$

The common point P of BD and CO divides each in the ratio 2 : 1.

115. Let S be the point of intersection of **AB** and **CR**. Let A be the origin and the position vectors of the points B, C, P, Q, R and S be $\mathbf{b}, \mathbf{c}, \mathbf{p}, \mathbf{q}, \mathbf{r}$ and $\mathbf{s}$ respectively.

$$\therefore \quad \mathbf{p} = \frac{3\mathbf{b} + 2\mathbf{c}}{5}$$

and $$\mathbf{q} = \frac{4\mathbf{c}}{5} \quad \text{...(i)}$$

$$\Rightarrow \quad \frac{5\mathbf{p} - 3\mathbf{b}}{2} = \frac{5\mathbf{q}}{4} \Rightarrow 10\mathbf{p} - 6\mathbf{b} = 5\mathbf{q}$$

i.e. $\quad 10\mathbf{p} = 5\mathbf{q} + 6\mathbf{b} \Rightarrow \dfrac{10\mathbf{p}}{11} = \dfrac{5\mathbf{q}+6\mathbf{b}}{11} = \mathbf{r}$

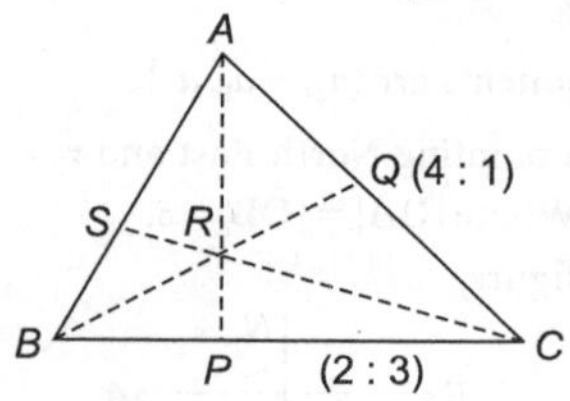

$\Rightarrow \quad \dfrac{11\mathbf{r}}{10} = \mathbf{p} = \dfrac{3\mathbf{b}+2\mathbf{c}}{5}$ [using Eq. (ii)]

$11\mathbf{r} = 6\mathbf{b} + 4\mathbf{c}$

$11\mathbf{r} - 4\mathbf{c} = 6\mathbf{b}$

$\dfrac{11\mathbf{r}-4\mathbf{c}}{7} = \dfrac{6}{7}\mathbf{b} = \mathbf{s}$, thus s divides AB in the ratio 6 : 1.

116. Since, angle bisectors divides opposite side in the ratio of sides containing the angle.

$\Rightarrow \quad BA' = \dfrac{ac}{b+c}$ and $CA' = \dfrac{ab}{a+c}$

Now, BI is also angle bisector of $\angle B$ for $\Delta ABA'$.

$\Rightarrow \quad \dfrac{AI}{AI'} = \dfrac{b+c}{a} \Rightarrow \dfrac{AI}{AA'} = \dfrac{b+c}{a+b+c}$

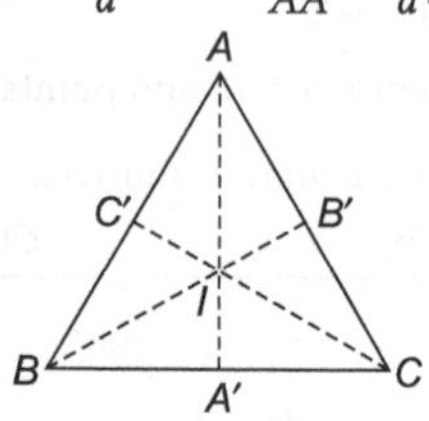

Similarly, $\quad \dfrac{BI}{BB'} = \dfrac{a+c}{a+b+c}$

and $\quad \dfrac{CI}{CC'} = \dfrac{a+b}{a+b+c}$

$\Rightarrow \quad \dfrac{\mathbf{AI}\cdot\mathbf{BI}\cdot\mathbf{CI}}{\mathbf{AA'}\cdot\mathbf{BB'}\cdot\mathbf{CC'}} = \dfrac{(b+c)(a+c)(a+b)}{(a+b+c)(a+b+c)(a+b+c)}$ ...(i)

As we know $\mathbf{AM} \geq \mathbf{GM}$, we get

$$\frac{\dfrac{b+c}{a+b+c} + \dfrac{c+a}{a+b+c} + \dfrac{a+b}{a+b+c}}{3} \geq \left[\frac{(a+b)(b+c)(c+a)}{(a+b+c)^3}\right]^{\frac{1}{3}}$$

$\Rightarrow \quad \dfrac{2(a+b+c)}{3(a+b+c)} \geq \dfrac{[(a+b)(b+c)(c+a)]^{1/3}}{a+b+c}$

$\Rightarrow \quad \dfrac{(a+b)(b+c)(c+a)}{(a+b+c)} \leq \dfrac{8}{27}$ ...(ii)

From Eqs. (i) and (ii), we get

$$\frac{AI\cdot BI\cdot CI}{AA'\cdot BB'\cdot CC'} \leq \frac{8}{27}$$

117. Let the position vector of O' with reference to O as the origin be α.

Then, $\quad \mathbf{OO'} = \alpha$

Now, $\mathbf{O'}P_i$ = Position vector

or P_i – Position vector of $O' = \mathbf{r}_i - \alpha$

$i = 1, 2, \ldots\ldots, n$

Let the position vectors of $P_1, P_2, P_3, \ldots\ldots, P_n$ with respect to O' as the origin be $\mathbf{R}_1, \mathbf{R}_2, \ldots, \mathbf{R}_n$ respectively. Then, $\mathbf{R}_i = \mathbf{O'}$

$P_i = \mathbf{r}_i - \alpha, i = 1, 2, \ldots, n$ [using Eq. (i)]

Now, $a_1\mathbf{R}_1 + a_2\mathbf{R}_2 + \ldots + a_n\mathbf{R}_n = 0$

$\Rightarrow \quad \displaystyle\sum_{i=1}^{n} a_i\mathbf{R}_i = 0 \Rightarrow \sum_{i=1}^{n} \mathbf{a}_i(\mathbf{r}_i - \alpha) = 0$

$\Rightarrow \quad \displaystyle\sum_{i=1}^{n} a_i\mathbf{r}_i - \sum_{i=1}^{n} a_i\,\alpha = 0$

$\Rightarrow \quad \mathbf{0} - \alpha\left(\displaystyle\sum_{i=1}^{n} a_i\right) = 0 \qquad \left[\because \displaystyle\sum_{i=1}^{n} a_i\mathbf{r}_i = 0 \text{ (given)}\right]$

$\Rightarrow \quad \displaystyle\sum_{i=1}^{n} a_i = 0$

Thus, $\displaystyle\sum_{i=1}^{n} a_i\mathbf{R}_i = 0$ will hold good, if $\displaystyle\sum_{i=1}^{n} a_i = 0$.

118. Let O be the origin of reference.

Let the position vectors of A, B, C and D be $\mathbf{a}$, $\mathbf{b}$, $\mathbf{c}$ and $\mathbf{d}$, respectively.

Then, $\quad OA : CB = 2 : 1$

$\Rightarrow \quad \dfrac{OA}{CB} = \dfrac{2}{1} \Rightarrow OA + 2CB$

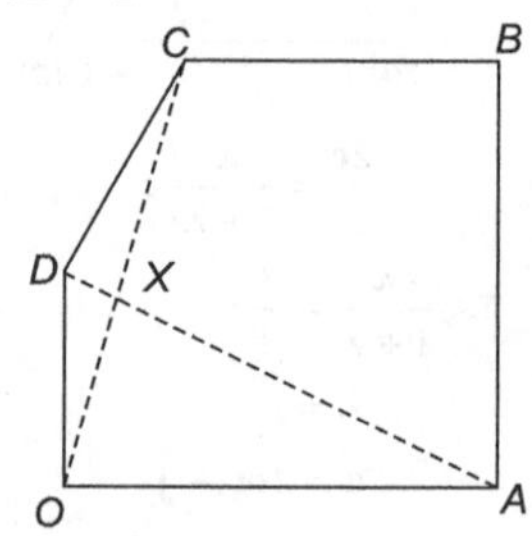

$\Rightarrow \quad \mathbf{OA} = 2\mathbf{CB}$

$\Rightarrow \quad \mathbf{a} = 2(\mathbf{b} - \mathbf{c})$ ...(i)

and $\quad OD : AB = 1 : 3$

$\Rightarrow \quad \dfrac{OD}{AB} = \dfrac{1}{3} \Rightarrow 3OD = AB$

$\Rightarrow \quad 3\mathbf{OD} = \mathbf{AB}$

$\Rightarrow \quad 3\mathbf{d} = (\mathbf{b} - \mathbf{a}) = \mathbf{b} - 2(\mathbf{b} - \mathbf{c})$ [using Eq. (i)]

$\Rightarrow \quad 3\mathbf{d} = -\mathbf{b} + 2\mathbf{c}$ ...(ii)

Let $OX : XC = \lambda : 1$ and $AX : XD = \mu : 1$

Now, X divides OC in the ratio $\lambda : 1$. Therefore,

PV of $X = \dfrac{\lambda c}{\lambda + 1}$ ...(iii)

X also divides AD in the ratio $\mu : 1$

PV of $X = \dfrac{\mu\mathbf{d} + \mathbf{a}}{\mu + 1}$

From Eqs. (iii) and (iv), we get

$$\frac{\lambda c}{\lambda + 1} = \frac{\mu\mathbf{d} + \mathbf{a}}{\mu + 1}$$

$\Rightarrow \quad \left(\dfrac{\lambda}{\lambda+1}\right)\mathbf{c} = \left(\dfrac{\mu}{\mu+1}\right)\mathbf{d} + \left(\dfrac{1}{\mu+1}\right)\mathbf{a}$

$$\Rightarrow \left(\frac{\lambda}{\lambda+1}\right)\mathbf{c} = \left(\frac{\mu}{\mu+1}\right)\left(\frac{-\mathbf{b}+2\mathbf{c}}{3}\right) + \left(\frac{1}{\mu+1}\right)2(\mathbf{b}-\mathbf{c})$$

[using Eqs. (i) and (iv)]

$$\Rightarrow \left(\frac{\lambda}{\lambda+1}\right)\mathbf{c} = \left(\frac{2}{\mu+1} - \frac{\mu}{3(\mu+1)}\right)\mathbf{b} + \left(\frac{2\mu}{3(\mu+1)} - \frac{2}{\mu+1}\right)\mathbf{c}$$

$$\Rightarrow \left(\frac{\lambda}{\lambda+1}\right)\mathbf{c} = \left(\frac{6-\mu}{3(\mu+1)}\right)\mathbf{b} + \left(\frac{2\mu-6}{3(\mu+1)}\right)\mathbf{c}$$

$$\Rightarrow \left(\frac{6-\mu}{3(\mu+1)}\right)\mathbf{b} + \left(\frac{2\mu-6}{3(\mu+1)} - \frac{\lambda}{\lambda+1}\right)\mathbf{c} = 0$$

$$\Rightarrow \frac{6-\mu}{3(\mu+1)} = 0 \text{ and } \frac{2\mu-6}{3(\mu+1)} - \frac{\lambda}{\lambda+1} = 0$$

(since, **b** and **c** are non-collinear)

$$\Rightarrow \mu = 6 \text{ and } \lambda = \frac{2}{5}$$

Hence, $OX : XC = 2:5$

119. let l, m and n be scalars such that

$$l\mathbf{p} + m\mathbf{q} + n\mathbf{r} = 0$$

$$\Rightarrow \{l(\cos a)\mathbf{u} + (\cos b)\mathbf{v} + (\cos c)\mathbf{w}\} + m\{(\sin a)\mathbf{u} + (\sin b)\mathbf{v} + (\sin c)\mathbf{w}\} + n\{\sin(x+a)\mathbf{u} + \sin(x+b)\mathbf{v} + \sin(x+c)\mathbf{w}\} = 0$$

$$\Rightarrow \{l\cos a + m\sin a + n\sin(x+a)\}\mathbf{u} + \{l\sin b + m\sin(x+b)\}\mathbf{v} + \{l\cos c + m\sin c + n\sin(x+c)\}\mathbf{w} = 0$$

$$\Rightarrow l\cos a + m\sin a + n\sin(x+a) = 0 \quad \text{...(i)}$$

$$l\cos b + m\sin b + n\sin(x+b) = 0 \quad \text{...(ii)}$$

$$l\cos c + m\sin c + n\sin(x+c) = 0 \quad \text{...(iii)}$$

This is a homogeneous system of linear equations in l, m and n. The determinant of the coefficient matrix is

$$\Delta = \begin{vmatrix} \cos a & \sin a & \sin(x+a) \\ \cos b & \sin b & \sin(x+b) \\ \cos c & \sin c & \sin(x+c) \end{vmatrix} = \begin{vmatrix} \cos a & \sin a & 0 \\ \cos b & \sin b & 0 \\ \cos c & \sin c & 0 \end{vmatrix} = 0$$

(using $C_3 \to C_3 - \sin x\, C_1 - \cos x\, C_2$)

⇒ So, the above system of equations has non-trivial solutions also. This means that l, m and n may attain non-zero values also.

Hence, the given system of vectors is a linearly dependent system of vectors.

120. We know that, the sum of three vectors of a triangle is zero.

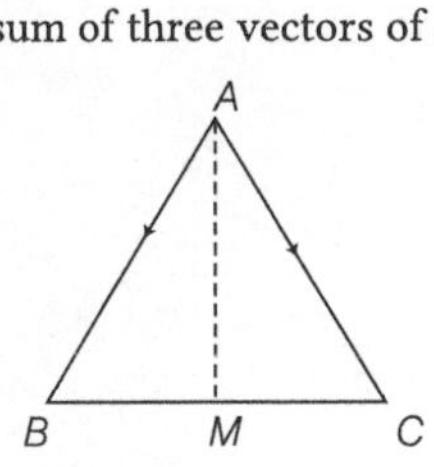

$$\therefore \quad \mathbf{AB} + \mathbf{BC} + \mathbf{CA} = 0$$

$$\Rightarrow \quad \mathbf{BC} = \mathbf{AC} - \mathbf{AB} \quad [\because \mathbf{AC} = -\mathbf{CA}]$$

$$\Rightarrow \quad \mathbf{AB} = \frac{\mathbf{AC} - \mathbf{AB}}{2} \quad [\because M \text{ is a mid-point of } BC]$$

Also, $\mathbf{AB} + \mathbf{BM} + \mathbf{MA} = 0$

[by properties of a triangle]

$$\Rightarrow \quad \mathbf{AB} + \frac{\mathbf{AC} - \mathbf{AB}}{2} = \mathbf{AM} \quad [\because \mathbf{AM} = -\mathbf{MA}]$$

$$\Rightarrow \quad \mathbf{AM} = \frac{\mathbf{AB} + \mathbf{AC}}{2}$$

$$= \frac{3\hat{\mathbf{i}} + 4\hat{\mathbf{k}} + 5\hat{\mathbf{i}} - 2\hat{\mathbf{j}} + 4\hat{\mathbf{k}}}{2}$$

$$= 4\hat{\mathbf{i}} - \hat{\mathbf{j}} + 4\hat{\mathbf{k}}$$

$$\Rightarrow \quad |\mathbf{AM}| = \sqrt{4^2 + 1^2 + 4^2} = \sqrt{33}$$

121. As, $\mathbf{a} + 3\mathbf{b}$ is collinear with **c**.

$$\therefore \quad \mathbf{a} + 3\mathbf{b} = \lambda\mathbf{c} \quad \text{...(i)}$$

Also, $\mathbf{b} + 2\mathbf{c}$ is collinear with **a**.

$$\Rightarrow \quad \mathbf{b} + 2\mathbf{c} = \mu\mathbf{a} \quad \text{...(ii)}$$

From Eq. (ii), we get

$$\mathbf{a} + 3\mathbf{b} + 6\mathbf{c} = (\lambda + 6)\mathbf{c} \quad \text{...(iii)}$$

From Eq. (ii), we get

$$\mathbf{a} + 3\mathbf{b} + 6\mathbf{c} = (1 + 3\mu)\mathbf{a} \quad \text{...(iv)}$$

From Eqs. (iii) and (iv), we get

$$\therefore \quad (\lambda + 6)\mathbf{c} = (1 + 3\mu)\mathbf{a}$$

Since, **a** is not collinear with **c**.

$$\Rightarrow \quad \lambda + 6 = 1 + 3\mu = 0$$

From Eq. (iv), we get

$$\mathbf{a} + 3\mathbf{b} + 6\mathbf{c} = 0$$

122. Since, $\mathbf{a} = 8\mathbf{b}$ and $\mathbf{c} = -7\mathbf{b}$

So, **a** is parallel to **b** and **c** is anti-parallel to **b**.

⇒ **a** and **c** are anti-parallel.

So, the angle between **a** and **c** is π.

123. Let P be the origin outside of AB and C is mid-point of AB, then

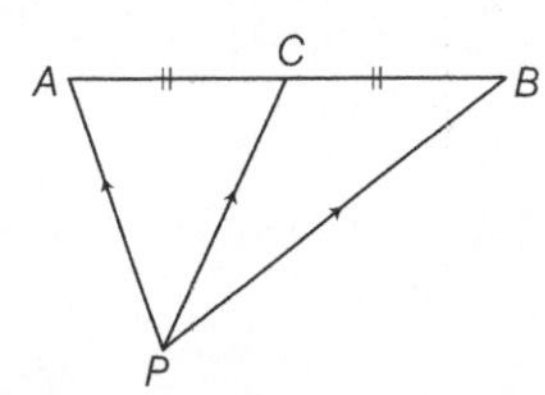

$$\mathbf{PC} = \frac{\mathbf{PA} + \mathbf{PB}}{2} \Rightarrow 2\mathbf{PC} = \mathbf{PA} + \mathbf{PB}$$

124. If $\mathbf{a} + 2\mathbf{b}$ is collinear with **c**, then $\mathbf{a} + 2\mathbf{b} = t\mathbf{c}$ ...(i)

Also, $\mathbf{b} + 3\mathbf{c}$ is collinear with **a**, then

$$\mathbf{b} + 3\mathbf{c} = \lambda\mathbf{a}$$

$$\Rightarrow \quad \mathbf{b} = \lambda\mathbf{a} - 3\mathbf{c} \quad \text{...(ii)}$$

From Eqs. (i) and (ii), we get

$$\mathbf{a} + 2(\lambda\mathbf{a} - 3\mathbf{c}) = t\mathbf{c}$$

$$\Rightarrow \quad (\mathbf{a} - 6\mathbf{c}) = t\mathbf{c} - 2\lambda a$$

On comparing the coefficients of **a** and **c**, we get

$$1 = -2\lambda \Rightarrow \lambda = -\frac{1}{2}$$

and $\quad -6 = t \Rightarrow t = -6$

From Eq. (i), we get

$$\mathbf{a} + 2\mathbf{b} = -6\mathbf{c}$$

$$\Rightarrow \quad \mathbf{a} + 2\mathbf{b} + 6\mathbf{c} = 0$$

125. The three vectors $(\mathbf{a} + 2\mathbf{b} + 3\mathbf{c})$, $(\lambda\mathbf{b} + 4\mathbf{c})$ and $(2\lambda - 1)\mathbf{c}$ are non-coplanar, if

$$\begin{vmatrix} 1 & 2 & 3 \\ 0 & \lambda & 4 \\ 0 & 0 & 2\lambda - 1 \end{vmatrix} \neq 0$$

$$\Rightarrow \quad (2\lambda - 1)(\lambda) \neq 0$$

$$\Rightarrow \quad \lambda \neq 0, \frac{1}{2}$$

So, these three vectors are non-coplanar for all except two values of λ.

126. Given that, $\mathbf{OA} = 7\hat{\mathbf{i}} - 4\hat{\mathbf{j}} + 7\hat{\mathbf{k}}$

$$\mathbf{OB} = \hat{\mathbf{i}} - 6\hat{\mathbf{j}} + 10\hat{\mathbf{k}}$$

$$\mathbf{OC} = -\hat{\mathbf{i}} - 3\hat{\mathbf{j}} + 4\hat{\mathbf{k}}$$

$$\mathbf{OD} = 5\hat{\mathbf{i}} - \hat{\mathbf{j}} + 5\hat{\mathbf{k}}$$

Now, $AB = \sqrt{(7-1)^2 + (-4+6)^2 + (7-10)^2}$

$$= \sqrt{36 + 4 + 9}$$

$$= \sqrt{49} = 7$$

$$BC = \sqrt{(1+1)^2 + (-6+3)^2 + (10-4)^2}$$

$$= \sqrt{4 + 9 + 36}$$

$$= \sqrt{49} = 7$$

$$CD = \sqrt{(-1-5)^2 + (-3+1)^2 + (4-5)^2}$$

$$= \sqrt{36 + 4 + 1}$$

$$= \sqrt{41}$$

and $DA = \sqrt{(5-7)^2 + (-1+4)^2 + (5-7)^2}$

$$= \sqrt{4 + 9 + 4}$$

$$= \sqrt{17}$$

Hence, option (d) is correct.

127. Since, $\begin{vmatrix} a & a^2 & 1 + a^3 \\ b & b^2 & 1 + b^3 \\ c & c^2 & 1 + c^3 \end{vmatrix} = \begin{vmatrix} a & a^2 & 1 \\ b & b^2 & 1 \\ c & c^2 & 1 \end{vmatrix} + \begin{vmatrix} a & a^2 & a^3 \\ b & b^2 & b^3 \\ c & c^2 & c^3 \end{vmatrix} = 0$

$$\Rightarrow \quad \begin{vmatrix} a & a^2 & 1 \\ b & b^2 & 1 \\ c & c^2 & 1 \end{vmatrix} = abc \begin{vmatrix} a & a^2 & 1 \\ b & b^2 & 1 \\ c & c^2 & 1 \end{vmatrix} = 0$$

$$\Rightarrow \quad (1 + abc) \begin{vmatrix} a & a^2 & 1 \\ b & b^2 & 1 \\ c & c^2 & 1 \end{vmatrix} = 0 \quad \left[\because \begin{vmatrix} a & a^2 & 1 \\ b & b^2 & 1 \\ c & c^2 & 1 \end{vmatrix} \neq 0 \right]$$

$$\Rightarrow \quad 1 + abc = 0$$

$$\Rightarrow \quad abc = -1$$

128. Since, the vector $\hat{\mathbf{i}} + x\hat{\mathbf{j}} + 3\hat{\mathbf{k}}$ is doubled in magnitude, then it becomes

$$4\hat{\mathbf{i}} + (4x - 2)\hat{\mathbf{j}} + 2\hat{\mathbf{k}}$$

$$\therefore \quad 2|\hat{\mathbf{i}} + x\hat{\mathbf{j}} + 3\hat{\mathbf{k}}| = |4\hat{\mathbf{i}} + (4x - 2)\hat{\mathbf{j}} + 2\hat{\mathbf{k}}|$$

$$\Rightarrow \quad 2\sqrt{1 + x^2 + 9} = \sqrt{16 + (4x - 2)^2 + 4}$$

$$\Rightarrow \quad 40 + 4x^2 = 20 + (4x - 2)^2$$

$$\Rightarrow \quad 3x^2 - 4x - 4 = 0$$

$$\Rightarrow \quad (x - 2)(3x + 2) = 0$$

$$\Rightarrow \quad x = 2, -\frac{2}{3}$$

CHAPTER

02

Product of Vectors

Learning Part

Session 1
- Product of Two Vectors
- Components of a Vector Along and Perpendicular to Another Vector
- Application of Dot Product in Mechanics

Session 2
- Vector or Cross Product of Two Vectors
- Area of Parallelogram and Triangle
- Moment of a Force and Couple
- Rotation About an Axis

Session 3
- Scalar Triple Product

Session 4
- Vector Triple Product

Practice Part

- JEE Type Examples
- Chapter Exercises

Arihant on Your Mobile !

Exercises with the symbol can be practised on your mobile. See inside cover page to activate for free.

Session 1

Product of Two Vectors, Components of a Vector Along and Perpendicular to Another Vector, Application of Dot Product in Mechanics

Product of Two Vectors

Product of two vectors is processed by two methods. When the product of two vectors results in a scalar quantity, then it is called **scalar product**. It is also known as **dot product** because we denote it by putting a **dot** (.) between two vectors.

When the product of two vectors results in a vector quantity, then this product is called **vector product**. It is also known as **cross product** because we denote it by putting a **cross** (×) between two vectors.

Scalar or Dot Product of Two Vectors

If **a** and **b** are two non-zero vectors and θ be the angle between them, then their scalar product (or dot product) is denoted by $\mathbf{a} \cdot \mathbf{b}$ and is defined as the scalar $|\mathbf{a}||\mathbf{b}| \cos \theta$, where $|\mathbf{a}|$ and $|\mathbf{b}|$ are modulii of a and b respectively and $0 \le \theta \le \pi$.

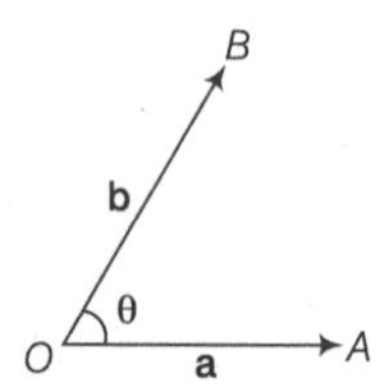

Remarks

1. $\mathbf{a} \cdot \mathbf{b} \in R$
2. $\mathbf{a} \cdot \mathbf{b} \le |\mathbf{a}||\mathbf{b}|$
3. $\mathbf{a} \cdot \mathbf{b} > 0 \Rightarrow$ Angle between **a** and **b** is acute.
4. $\mathbf{a} \cdot \mathbf{b} < 0 \Rightarrow$ Angle between **a** and **b** is obtuse.
5. The dot product of a zero and non-zero vector is a scalar zero.

Geometrical interpretation of scalar product

Let **a** and **b** be two vectors represented by **OA** and **OB** respectively. Let θ be the angle between **OA** and **OB**. Draw $BL \perp OA$ and $AM \perp OB$.

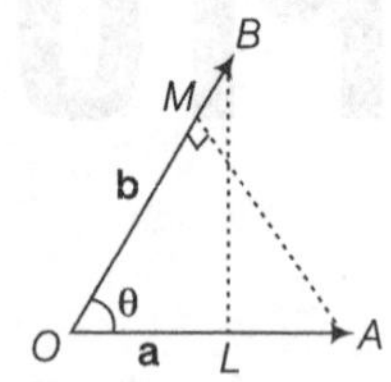

From triangles *OBL* and *OAM*, we have $OL = OB \cos \theta$ and $OM = OA \cos \theta$. Here, *OL* and *OA* are known as projection of **b** on **a** and **a** on **b** respectively.

Now, $\mathbf{a} \cdot \mathbf{b} = |\mathbf{a}||\mathbf{b}| \cos \theta$

$= |\mathbf{a}| (OB \cos \theta) = |\mathbf{a}| (OL)$

= (Magnitude of **a**) (Projection of **b** on **a**) ...(i)

Again, $\mathbf{a} \cdot \mathbf{b} = |\mathbf{a}||\mathbf{b}| \cos \theta$

$= |\mathbf{b}| (|\mathbf{a}| \cos \theta) = |\mathbf{b}| (OA \cos \theta) = |\mathbf{b}| (OM)$

$\mathbf{a} \cdot \mathbf{b}$ = (Magnitude of **b**)

(Projection of **a** on **b**) ...(ii)

Thus, geometrically interpreted, **the scalar product of two vectors is the product of modulus of either vector and the projection of the other in its direction.**

Remarks

1. Projection of **a** on $\mathbf{b} = \dfrac{\mathbf{a} \cdot \mathbf{b}}{|\mathbf{b}|} = \mathbf{a} \cdot \mathbf{b}$
2. Projection of **b** on $\mathbf{a} = \dfrac{\mathbf{a} \cdot \mathbf{b}}{|\mathbf{a}|} = \mathbf{a} \cdot \mathbf{b}$
3. Angle between two vectors it **a** and **b** be two vectors inclined at an angle θ, then $\mathbf{a}, \mathbf{b} = |\mathbf{a}| \cdot |\mathbf{b}| \cos \theta$

$$\Rightarrow \quad \cos\theta = \frac{\mathbf{a} \cdot \mathbf{b}}{|\mathbf{a}| \cdot |\mathbf{b}|}$$

$$\Rightarrow \quad \theta = \cos^{-1}\left(\frac{\mathbf{a} \cdot \mathbf{b}}{|\mathbf{a}| \cdot |\mathbf{b}|}\right)$$

If $\mathbf{a} = a_1\hat{\mathbf{i}} + a_2\hat{\mathbf{j}} + a_3\hat{\mathbf{k}}$ and $\mathbf{b} = b_1\hat{\mathbf{i}} + b_2\hat{\mathbf{j}} + b_3\hat{\mathbf{k}}$

$$\theta = \cos^{-1}\left(\frac{a_1b_1 + a_2b_2 + a_3b_3}{\sqrt{a_1^2 + a_2^2 + a_3^2}\sqrt{b_1^2 + b_2^2 + b_3^2}}\right)$$

Properties of Scalar Product

(i) **Commutativity** The scalar product of two vector is commutative i.e., $\mathbf{a} \cdot \mathbf{b} = \mathbf{b} \cdot \mathbf{a}$.

(ii) **Distributivity of scalar product over vector addition** The scalar product of vectors is distributive over vector addition i.e.,

(a) $\mathbf{a} \cdot (\mathbf{b} + \mathbf{c}) = \mathbf{a} \cdot \mathbf{b} + \mathbf{a} \cdot \mathbf{c}$ (Left distributivity)

(b) $(\mathbf{b} + \mathbf{c}) \cdot \mathbf{a} = \mathbf{b} \cdot \mathbf{a} + \mathbf{c} \cdot \mathbf{a}$ (right distributivity)

(iii) Let $\mathbf{a}$ and $\mathbf{b}$ be two non-zero vectors $\mathbf{a} \cdot \mathbf{b} = 0 \Leftrightarrow \mathbf{a} \perp \mathbf{b}$.

As $\hat{\mathbf{i}}, \hat{\mathbf{j}}$ and $\hat{\mathbf{k}}$ are mutually perpendicular unit vectors along the coordinate axes, therefore

$\hat{\mathbf{i}} \cdot \hat{\mathbf{j}} = \hat{\mathbf{j}} \cdot \hat{\mathbf{i}} = 0, \hat{\mathbf{j}} \cdot \hat{\mathbf{k}} = \hat{\mathbf{k}} \cdot \hat{\mathbf{j}} = 0; \hat{\mathbf{k}} \cdot \hat{\mathbf{i}} = \hat{\mathbf{i}} \cdot \hat{\mathbf{k}} = 0$

(iv) For any vector $\mathbf{a}, \mathbf{a} \cdot \mathbf{a} = |\mathbf{a}|^2$.

As $\hat{\mathbf{i}}, \hat{\mathbf{j}}$ and $\hat{\mathbf{k}}$ are unit vectors along the coordinate axes, therefore $\hat{\mathbf{i}} \cdot \hat{\mathbf{i}} = |\hat{\mathbf{i}}|^2 = 1, \hat{\mathbf{j}} \cdot \hat{\mathbf{j}} = |\hat{\mathbf{j}}|^2 = 1$ and $\hat{\mathbf{k}} \cdot \hat{\mathbf{k}} = |\hat{\mathbf{k}}|^2 = 1$.

(v) If m is a scalar and $\mathbf{a}$ and $\mathbf{b}$ be any two vectors then $(m\mathbf{a}) \cdot \mathbf{b} = m(\mathbf{a} \cdot \mathbf{b}) = \mathbf{a} \cdot (m\mathbf{b})$.

(vi) If m and n are scalars and $\mathbf{a}$ and $\mathbf{b}$ be two vectors, then $(m\mathbf{a}) \cdot n\mathbf{b} = mn(\mathbf{a} \cdot \mathbf{b}) = (mn\mathbf{a}) \cdot \mathbf{b} = \mathbf{a} \cdot (mn\mathbf{b})$

(vii) For any vectors $\mathbf{a}$ and $\mathbf{b}$, we have

(a) $\mathbf{a} \cdot (-\mathbf{b}) = -(\mathbf{a} \cdot \mathbf{b}) = (-\mathbf{a}) \cdot \mathbf{b}$

(b) $(-\mathbf{a}) \cdot (-\mathbf{b}) = \mathbf{a} \cdot \mathbf{b}$

(viii) For any two vectors $\mathbf{a}$ and $\mathbf{b}$, we have

(a) $|\mathbf{a} + \mathbf{b}|^2 = |\mathbf{a}|^2 + |\mathbf{b}|^2 + 2\mathbf{a} \cdot \mathbf{b}$

(b) $|\mathbf{a} - \mathbf{b}| = |\mathbf{a}|^2 + |\mathbf{b}|^2 - 2\mathbf{a} \cdot \mathbf{b}$

(c) $(\mathbf{a} + \mathbf{b}) \cdot (\mathbf{a} - \mathbf{b}) = |\mathbf{a}|^2 - |\mathbf{b}|^2$

(d) $|\mathbf{a} + \mathbf{b}| = |\mathbf{a}| + |\mathbf{b}| \Rightarrow \mathbf{a} \,||\, \mathbf{b}$

(e) $|\mathbf{a} + \mathbf{b}|^2 = |\mathbf{a}|^2 + |\mathbf{b}|^2 \Rightarrow \mathbf{a} \perp \mathbf{b}$

(f) $|\mathbf{a} + \mathbf{b}| = |\mathbf{a} - \mathbf{b}| \Rightarrow \mathbf{a} \perp \mathbf{b}$

Scalar Product in Terms of Components

If $\mathbf{a} = a_1 \hat{\mathbf{i}} + a_2 \hat{\mathbf{j}} + a_3 \hat{\mathbf{k}}$ and $\mathbf{b} = b_1 \hat{\mathbf{i}} + b_2 \hat{\mathbf{j}} + b_3 \hat{\mathbf{k}}$, then $\mathbf{a} \cdot \mathbf{b} = a_1 b_1 + a_2 b_2 + a_3 b_3$.

Thus, scalar product of two vectors is equal to the sum of the products of their corresponding components. In particular, $\mathbf{a} \cdot \mathbf{a} = |\mathbf{a}|^2 = a_1^2 + a_2^2 + a_3^2$.

Example 1. If θ is the angle between the vectors $\mathbf{a} = 2\hat{\mathbf{i}} + 2\hat{\mathbf{j}} - \hat{\mathbf{k}}$ and $\mathbf{b} = 6\hat{\mathbf{i}} - 3\hat{\mathbf{j}} + 2\hat{\mathbf{k}}$, then

(a) $\cos\theta = \dfrac{4}{21}$ (b) $\cos\theta = \dfrac{3}{19}$

(c) $\cos\theta = \dfrac{2}{19}$ (d) $\cos\theta = \dfrac{5}{21}$

Sol. (a) Angle between $\mathbf{a}$ and $\mathbf{b}$ is given by

$$\cos\theta = \frac{\mathbf{a} \cdot \mathbf{b}}{|\mathbf{a}||\mathbf{b}|} = \frac{(2\hat{\mathbf{i}} + 2\hat{\mathbf{j}} - \hat{\mathbf{k}}) \cdot (6\hat{\mathbf{i}} - 3\hat{\mathbf{j}} + 2\hat{\mathbf{k}})}{\sqrt{2^2 + 2^2 + (-1)^2} \cdot \sqrt{6^2 + (-3)^2 + 2^2}} = \frac{12 - 6 - 2}{3 \cdot 7} = \frac{4}{21}$$

Example 2. $(\mathbf{a} \cdot \hat{\mathbf{i}})\hat{\mathbf{i}} + (\mathbf{a} \cdot \hat{\mathbf{j}})\hat{\mathbf{j}} + (\mathbf{a} \cdot \hat{\mathbf{k}})\hat{\mathbf{k}}$ is equal to

(a) $\mathbf{a}$ (b) $2\mathbf{a}$

(c) $3\mathbf{a}$ (d) $\mathbf{0}$

Sol. (a) Let $\mathbf{a} = a_1 \hat{\mathbf{i}} + a_2 \hat{\mathbf{j}} + a_3 \hat{\mathbf{k}}$

$\therefore \quad \mathbf{a} \cdot \hat{\mathbf{i}} = (a_1 \hat{\mathbf{i}} + a_2 \hat{\mathbf{j}} + a_3 \hat{\mathbf{k}}) \hat{\mathbf{i}} = a_1,$

$\mathbf{a} \cdot \hat{\mathbf{j}} = a_2, \mathbf{a} \cdot \hat{\mathbf{k}} = a_3$

So, $(\mathbf{a} \cdot \hat{\mathbf{i}})\hat{\mathbf{i}} + (\mathbf{a} \cdot \hat{\mathbf{j}})\hat{\mathbf{j}} + (\mathbf{a} \cdot \hat{\mathbf{k}})\hat{\mathbf{k}}$

$= a_1 \hat{\mathbf{i}} + a_2 \hat{\mathbf{j}} + a_3 \hat{\mathbf{k}} = \mathbf{a}$

Example 3. If $|\mathbf{a}| = 3, |\mathbf{b}| = 4$, then a value of λ for which $\mathbf{a} + \lambda\mathbf{b}$ is perpendicular to $\mathbf{a} - \lambda\mathbf{b}$.

(a) 9 / 16 (b) 3 / 4

(c) 3 / 2 (d) 4 / 3

Sol. (b) $\mathbf{a} + \lambda\mathbf{b}$ is perpendicular to $\mathbf{a} - \lambda\mathbf{b}$.

$$\therefore \quad (\mathbf{a} + \lambda\mathbf{b}) \cdot (\mathbf{a} - \lambda\mathbf{b}) = 0$$

$$\Rightarrow \quad |\mathbf{a}^2| - \lambda(\mathbf{a} \cdot \mathbf{b}) + \lambda(\mathbf{b} \cdot \mathbf{a}) - \lambda^2 |\mathbf{b}|^2 = 0$$

$$\Rightarrow \quad |\mathbf{a}|^2 - \lambda^2 |\mathbf{b}|^2 = 0$$

$$\Rightarrow \quad \lambda = \pm \frac{|\mathbf{a}|}{|\mathbf{b}|} = \pm \frac{3}{4}$$

Example 4. The projection of $\mathbf{a} = 2\hat{\mathbf{i}} + 3\hat{\mathbf{j}} - 2\hat{\mathbf{k}}$ on $\mathbf{b} = \hat{\mathbf{i}} + 2\hat{\mathbf{j}} + 3\hat{\mathbf{k}}$ is

(a) $\dfrac{1}{\sqrt{14}}$ (b) $\dfrac{2}{\sqrt{14}}$

(c) $\sqrt{14}$ (d) $\dfrac{-2}{\sqrt{14}}$

Sol. (b) Projection of $\mathbf{a}$ on $\mathbf{b}$

$$= \mathbf{a} \cdot \frac{\mathbf{b}}{|\mathbf{b}|} = \frac{(2\hat{\mathbf{i}} + 3\hat{\mathbf{j}} - 2\hat{\mathbf{k}}) \cdot (\hat{\mathbf{i}} + 2\hat{\mathbf{j}} + 3\hat{\mathbf{k}})}{|\hat{\mathbf{i}} + 2\hat{\mathbf{j}} + 3\hat{\mathbf{k}}|}$$

$$= \frac{2 + 6 - 6}{\sqrt{14}} = \frac{2}{\sqrt{14}}$$

Example 5. If $\mathbf{a} = 5\hat{\mathbf{i}} - \hat{\mathbf{j}} + 7\hat{\mathbf{k}}$ and $\mathbf{b} = \hat{\mathbf{i}} - \hat{\mathbf{j}} + \lambda\hat{\mathbf{k}}$, then find λ such that $\mathbf{a} + \mathbf{b}$ and $\mathbf{a} - \mathbf{b}$ are orthogonal.

Sol. Clearly, $\mathbf{a} + \mathbf{b} = (5\hat{\mathbf{i}} - \hat{\mathbf{j}} + 7\hat{\mathbf{k}}) + (\hat{\mathbf{i}} - \hat{\mathbf{j}} + \lambda\hat{\mathbf{k}})$

$= 6\hat{\mathbf{i}} - 2\hat{\mathbf{j}} + (7 + \lambda)\hat{\mathbf{k}}$

and $\mathbf{a}-\mathbf{b}=(5\hat{\mathbf{i}}-\hat{\mathbf{j}}+7\hat{\mathbf{k}})-(\hat{\mathbf{i}}-\hat{\mathbf{j}}+\lambda\hat{\mathbf{k}})$

$$=4\hat{\mathbf{i}}+(7-\lambda)\hat{\mathbf{k}}$$

Since, $\mathbf{a}+\mathbf{b}$ and $\mathbf{a}-\mathbf{b}$ are orthogonal

$$\therefore \quad (\mathbf{a}+\mathbf{b})\cdot(\mathbf{a}-\mathbf{b})=0$$

$$\Rightarrow \quad [6\hat{\mathbf{i}}-2\hat{\mathbf{j}}+(7+\lambda)\hat{\mathbf{k}}](4\hat{\mathbf{i}}+(7-\lambda)\mathbf{k})=0$$

$$\Rightarrow \quad 6\times 4-2\times 0+(7+\lambda)(7-\lambda)=0$$

$$\Rightarrow \quad 24+49-\lambda^2=0$$

$$\Rightarrow \quad \lambda^2=73$$

$$\Rightarrow \quad \lambda=\pm\sqrt{73}$$

Example 6. If $\mathbf{a}, \mathbf{b}$ and $\mathbf{c}$ are unit vectors such that $\mathbf{a}+\mathbf{b}+\mathbf{c}=0$, then find the value of $\mathbf{a}\cdot\mathbf{b}+\mathbf{b}\cdot\mathbf{c}+\mathbf{c}\cdot\mathbf{a}$.

Sol. Consider, $\mathbf{a}+\mathbf{b}+\mathbf{c}=0$

On squaring both sides, we get

$$(\mathbf{a}+\mathbf{b}+\mathbf{c})^2=0^2$$

$$\Rightarrow \quad (\mathbf{a}+\mathbf{b}+\mathbf{c})\cdot(\mathbf{a}+\mathbf{b}+\mathbf{c})=0\cdot 0$$

$$\Rightarrow \quad |\mathbf{a}|^2+|\mathbf{b}|^2+|\mathbf{c}|^2+2\mathbf{a}\cdot\mathbf{b}+2\mathbf{b}\cdot\mathbf{c}+2\mathbf{c}\cdot\mathbf{a}=0$$

$$\Rightarrow \quad 1+1+1+2(\mathbf{a}\cdot\mathbf{b}+\mathbf{b}\cdot\mathbf{c}+\mathbf{c}\cdot\mathbf{a})=0$$

$$\Rightarrow \quad \mathbf{a}\cdot\mathbf{b}+\mathbf{b}\cdot\mathbf{c}+\mathbf{c}\cdot\mathbf{a}=-\frac{3}{2}$$

Example 7. If $\mathbf{a}, \mathbf{b}$ and $\mathbf{c}$ are mutually perpendicular vectors of equal magnitudes, then find the angle between the vectors $\mathbf{a}$ and $\mathbf{a}+\mathbf{b}+\mathbf{c}$.

Sol. Let θ be the angle between the vectors $\mathbf{a}$ and $\mathbf{a}+\mathbf{b}+\mathbf{c}$.

Then, $\cos\theta=\dfrac{\mathbf{a}\cdot(\mathbf{a}+\mathbf{b}+\mathbf{c})}{|\mathbf{a}|\,|\mathbf{a}+\mathbf{b}+\mathbf{c}|}$

$$=\frac{\mathbf{a}\cdot\mathbf{a}+\mathbf{a}\cdot\mathbf{b}+\mathbf{a}\cdot\mathbf{c}}{|\mathbf{a}|\,|\mathbf{a}+\mathbf{b}+\mathbf{c}|}$$

$$=\frac{|\mathbf{a}|^2}{|\mathbf{a}|\,|\mathbf{a}+\mathbf{b}+\mathbf{c}|}$$

[$\because \mathbf{a}\cdot\mathbf{b}=\mathbf{a}\cdot\mathbf{c}=0$ as $\mathbf{a},\mathbf{b},\mathbf{c}$ are mutually perpendicular vectors]

$$=\frac{|\mathbf{a}|}{|\mathbf{a}+\mathbf{b}+\mathbf{c}|} \quad \text{...(i)}$$

Now consider,

$$|\mathbf{a}+\mathbf{b}+\mathbf{c}|^2=(\mathbf{a}+\mathbf{b}+\mathbf{c})\cdot(\mathbf{a}+\mathbf{b}+\mathbf{c})$$

$$=|\mathbf{a}|^2+|\mathbf{b}|^2+|\mathbf{c}|^2+2(\mathbf{a}\cdot\mathbf{b}+\mathbf{b}\cdot\mathbf{c}+\mathbf{c}\cdot\mathbf{a})$$

$$=3|\mathbf{a}|^2+2(0)=3|\mathbf{a}|^2$$

[$\because |\mathbf{a}|=|\mathbf{b}|=|\mathbf{c}|$ and $\mathbf{a}\cdot\mathbf{b}=\mathbf{b}\cdot\mathbf{c}=\mathbf{c}\cdot\mathbf{a}=0$]

$$|\mathbf{a}+\mathbf{b}+\mathbf{c}|=\sqrt{3}\,|\mathbf{a}|$$

From Eq. (i), we get

$$\cos\theta=\frac{1}{\sqrt{3}}$$

$$\Rightarrow \quad \theta=\cos^{-1}\frac{1}{\sqrt{3}}$$

Example 8. If the vectors $\mathbf{a}=(c\log_2 x)\hat{\mathbf{i}}-6\hat{\mathbf{j}}+3\hat{\mathbf{k}}$ and $\mathbf{b}=(\log_2 x)\hat{\mathbf{i}}+2\hat{\mathbf{j}}+(2c\log_2 x)\hat{\mathbf{k}}$ make an obtuse angle for any $x\in(0,\infty)$. Then, determine the interval to which c belongs.

Sol. For the vectors $\mathbf{a}$ and $\mathbf{b}$ to be inclined at an obtuse angle, we must have

$$\mathbf{a}\cdot\mathbf{b}<0, \forall\ x\in(0,\infty)$$

$$\Rightarrow \quad c(\log_2 x)^2-12+6c\log_2 x<0, \forall\ x\in(0,\infty)$$

$$\Rightarrow \quad cy^2+6cy-12<0, \forall\ y\in R, \text{ where } y=\log_2 x$$

$$\Rightarrow \quad c<0 \text{ and } 36c^2+48c<0,$$

(using $ax^2+bx+c<0$, $\forall\ x\in R$ iff $a<0$ and $D<0$)

$$\Rightarrow \quad c<0 \text{ and } c\,(3c+4)<0 \Rightarrow c\in\left(-\frac{4}{3},0\right)$$

Example 9. If $a+2b+3c=4$, then find the least value of $a^2+b^2+c^2$.

Sol. Consider vectors $\mathbf{p}=a\hat{\mathbf{i}}+b\hat{\mathbf{j}}+c\hat{\mathbf{k}}$ and $\mathbf{q}=\hat{\mathbf{i}}+2\hat{\mathbf{j}}+3\hat{\mathbf{k}}$

Now, $\cos\theta=\dfrac{a+2b+3c}{\sqrt{a^2+b^2+c^2}\sqrt{1^2+2^2+3^2}}$

$$\Rightarrow \quad \cos^2\theta=\frac{(a+2b+3c)^2}{14(a^2+b^2+c^2)}\le 1 \quad [\because \cos^2\theta\le 1]$$

$$\Rightarrow \quad a^2+b^2+c^2\ge\frac{(a+2b+3c)^2}{14}=\frac{16}{14}=\frac{8}{7}$$

Hence, least value of $a^2+b^2+c^2$ is $\frac{8}{7}$.

Example 10. Find the unit vector which makes an angle of 45° with the vector $2\hat{\mathbf{i}}+2\hat{\mathbf{j}}-\hat{\mathbf{k}}$ and an angle of 60° with the vector $\hat{\mathbf{j}}-\hat{\mathbf{k}}$.

Sol. Let the unit vector be $\mathbf{c}=c_1\hat{\mathbf{i}}+c_2\hat{\mathbf{j}}+c_3\hat{\mathbf{k}}$ so that; it makes an angle of 45° with $2\hat{\mathbf{i}}+2\hat{\mathbf{j}}-\hat{\mathbf{k}}$.

$$\Rightarrow \quad \frac{2c_1+2c_2-c_3}{3}=\cos 45^\circ \quad (\because |\hat{\mathbf{c}}|=1)$$

$$\Rightarrow \quad 2c_1+2c_2-c_3=\frac{3}{\sqrt{2}} \quad \text{...(i)}$$

Also, it makes an angle of 60° with $\hat{\mathbf{j}}-\hat{\mathbf{k}}$.

$$\Rightarrow \quad \frac{c_2-c_3}{\sqrt{2}}=\cos 60^\circ \quad (\because |\hat{\mathbf{j}}-\hat{\mathbf{k}}|=\sqrt{2} \text{ and } |\hat{\mathbf{c}}|=1)$$

$$\Rightarrow \quad c_2-c_3=\frac{\sqrt{2}}{2} \quad \text{...(ii)}$$

$$\Rightarrow \quad c_1^2+c_2^2+c_3^2=1 \quad \text{...(iii)}$$

(using $|\mathbf{c}|=|c_1\hat{\mathbf{i}}+c_2\hat{\mathbf{j}}+c_3\hat{\mathbf{k}}|=1$)

From Eq. (ii), $c_2=\dfrac{1}{\sqrt{2}}+c_3$ and from Eq. (i), $c_1=\dfrac{-c_3}{2}+\dfrac{1}{2\sqrt{2}}$

On substituting in Eq. (iii), we get

$$\frac{c_3^2}{4}+\frac{1}{2}+\frac{2c_3}{\sqrt{2}}+\frac{1}{8}-\frac{c_3}{2\sqrt{2}}+c_3^2+c_3^2=1$$

$$\Rightarrow \quad \frac{9}{4}c_3^2+\frac{3c_3}{2\sqrt{2}}+\frac{5}{8}=1 \Rightarrow c_3=-\frac{1}{\sqrt{2}},\frac{1}{3\sqrt{2}}$$

Hence, the required vectors are $\left(\frac{1}{\sqrt{2}},0,-\frac{1}{\sqrt{2}}\right)$ and $\left(\frac{1}{3\sqrt{2}},\frac{4}{3\sqrt{2}},\frac{1}{3\sqrt{2}}\right)$.

Example 11. Show that the median to the base of an isosceles triangle is perpendicular to base.

Sol. Let ABC be an isosceles triangle in which $AB = AC$.

Let A be the origin of reference and let

$$\mathbf{AB}=\mathbf{b}, \mathbf{AC}=\mathbf{c}$$

Let D be the middle point of **BC**.

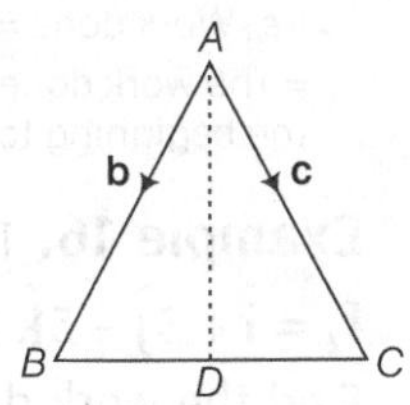

Then, $\quad \mathbf{AD}=\frac{\mathbf{b}+\mathbf{c}}{2}$

Now, $\quad \mathbf{BC}=\mathbf{c}-\mathbf{b}$

$$\therefore \quad \mathbf{AD}\cdot\mathbf{BC}=\left(\frac{\mathbf{b}+\mathbf{c}}{2}\right)\cdot(\mathbf{c}-\mathbf{b})$$

$$=\frac{1}{2}(|\mathbf{c}|^2-|\mathbf{b}|^2)=\frac{1}{2}(|\mathbf{AC}|^2-|\mathbf{AB}|^2)$$

$$=\frac{1}{2}(0)=0$$

Hence, median to the base of an isosceles triangle is perpendicular to base.

Example 12. Using vector method, prove that in a triangle, $a^2=b^2+c^2-2bc\cos A$ (cosine law).

Sol. In a ΔABC,

Let $\quad \mathbf{AB}=\mathbf{c}, \mathbf{BC}=\mathbf{a}, \mathbf{CA}=\mathbf{b}$

$\because \quad \mathbf{a}+\mathbf{b}+\mathbf{c}=0$

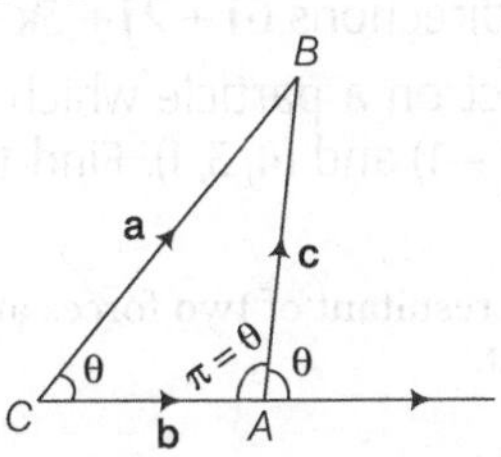

We have, $\quad \mathbf{a}=-(\mathbf{b}+\mathbf{c})$

$\therefore \quad |\mathbf{a}|=|-(\mathbf{b}+\mathbf{c})| \Rightarrow |\mathbf{a}|^2=|\mathbf{b}+\mathbf{c}|^2$

$\Rightarrow \quad |\mathbf{a}|^2=|\mathbf{b}|^2+|\mathbf{c}|^2+2\mathbf{b}\cdot\mathbf{c}$

$\Rightarrow \quad |\mathbf{a}^2|=|\mathbf{b}|^2+|\mathbf{c}|^2+2|\mathbf{b}||\mathbf{c}|\cos(\pi-A)|$

Since, angle between **b** and **c** = The angle between CA produced and AB

$\therefore \quad a^2=b^2+c^2-2bc\cos A$

Example 13. Using vector method, prove that in a triangle, $a=b\cos C+c\cos B$ (projection formula).

Sol. In a ΔABC,

Let $\quad \mathbf{AB}=\mathbf{c}, \mathbf{BC}=\mathbf{a}, \mathbf{CA}=\mathbf{b}$

$\because \quad \mathbf{a}+\mathbf{b}+\mathbf{c}=0$

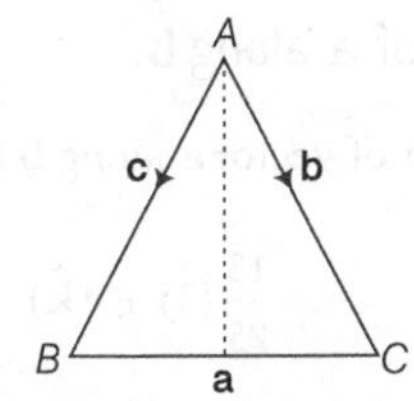

We have, $\quad |\mathbf{a}|=|-(\mathbf{b}+\mathbf{c})|$

$\therefore \quad \mathbf{a}\cdot\mathbf{a}=-(\mathbf{b}+\mathbf{c})\cdot\mathbf{a}$

or $\quad |\mathbf{a}|^2=-\mathbf{b}\cdot\mathbf{a}-\mathbf{c}\cdot\mathbf{a}$

$$=-|\mathbf{b}||\mathbf{a}|\cos(\pi-C)-|\mathbf{c}||\mathbf{a}|\cos(\pi-B)$$

Since, the angle between **b** and **a** $=(\pi-C)$ and angle between **c** and **a** $=(\pi-\mathbf{B})$

$\therefore \quad a^2=ab\cos C+ac\cos B$

$\Rightarrow \quad a=b\cos C+c\cos B$

Components of a Vector Along and Perpendicular to Another Vector

Let **a** and **b** be two vectors represented by **OA** and **OB** and let θ be the angle between **a** and **b**. Draw $MB\perp OA$, shown as

$\therefore \quad \mathbf{b}=\mathbf{OM}+\mathbf{MB}$

$\Rightarrow \quad \mathbf{OM}=(OM)\mathbf{a}=(OB\cos\theta)\,\mathbf{a}=(|\mathbf{b}|\cos\theta)\,\mathbf{a}$

$$=\left(|\mathbf{b}|\frac{(\mathbf{a}\cdot\mathbf{b})}{|\mathbf{a}|^2}\right)\mathbf{a}$$

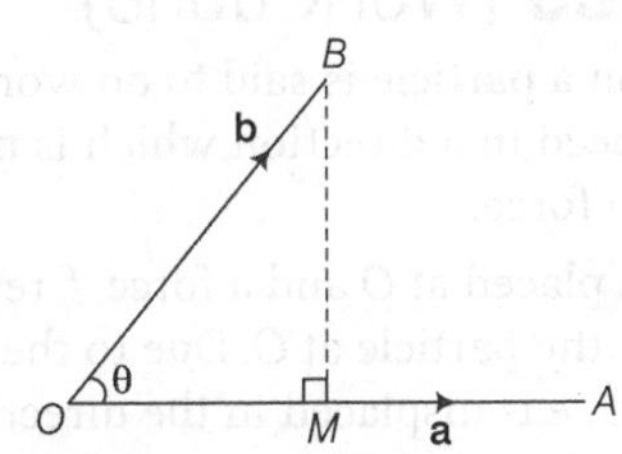

$$=\left(\frac{\mathbf{a}\cdot\mathbf{b}}{|\mathbf{a}|}\right)\mathbf{a}=\frac{\mathbf{a}\cdot\mathbf{b}\,\mathbf{a}}{|\mathbf{a}||\mathbf{a}|}=\left(\frac{\mathbf{a}-\mathbf{b}}{|\mathbf{a}|^2}\right)\mathbf{a}$$

$$\mathbf{b}=\mathbf{OM}+\mathbf{MB}$$

$$\mathbf{MB}=\mathbf{b}-\mathbf{OM}=\mathbf{b}-\left(\frac{\mathbf{a}\cdot\mathbf{b}}{|\mathbf{a}|^2}\right)\mathbf{a}$$

Thus, the components of **b** along and perpendicular to **a** are $\left(\frac{\mathbf{a}\cdot\mathbf{b}}{|\mathbf{a}|^2}\right)\mathbf{a}$ and $\mathbf{a}-\left(\frac{\mathbf{a}\cdot\mathbf{b}}{|\mathbf{a}|^2}\right)\mathbf{a}$ respectively.

Example 14. If $\mathbf{a}=4\hat{\mathbf{i}}+6\hat{\mathbf{j}}$ and $\mathbf{b}=3\hat{\mathbf{j}}+4\hat{\mathbf{k}}$, then find the component of **a** along **b**.

Sol. The component of vector **a** along **b** is $\frac{(\mathbf{a}\cdot\mathbf{b})}{|\mathbf{b}|^2}$

$$=\frac{18}{25}(3\hat{\mathbf{j}}+4\hat{\mathbf{k}})$$

Example 15. Express the vector $\mathbf{a}=5\hat{\mathbf{i}}-2\hat{\mathbf{j}}+5\hat{\mathbf{k}}$ as sum of two vectors such that one is parallel to the vector $\mathbf{b}=3\hat{\mathbf{i}}+\hat{\mathbf{k}}$ and the other is perpendicular to **b**.

Sol. Required vectors are

$$\left(\frac{\mathbf{a}\cdot\mathbf{b}}{|\mathbf{b}|^2}\right)\mathbf{b} \text{ and } \mathbf{a}-\left(\frac{\mathbf{a}\cdot\mathbf{b}}{|\mathbf{b}|^2}\right)\mathbf{b}$$

Clearly, $\left(\frac{\mathbf{a}\cdot\mathbf{b}}{|\mathbf{b}|^2}\right)\mathbf{b}=2(3\hat{\mathbf{i}}+\hat{\mathbf{k}})=6\hat{\mathbf{i}}+2\hat{\mathbf{k}}$ and so,

$$\mathbf{a}-\left(\frac{\mathbf{a}-\mathbf{b}}{|\mathbf{b}|^2}\right)\mathbf{b}=(5\hat{\mathbf{i}}-2\hat{\mathbf{j}}+5\hat{\mathbf{k}})-(6\hat{\mathbf{i}}+2\hat{\mathbf{k}})$$

$$=-\hat{\mathbf{i}}-2\hat{\mathbf{j}}+3\hat{\mathbf{k}}$$

Note that $(6\hat{\mathbf{i}}+2\hat{\mathbf{k}})+(-\hat{\mathbf{i}}-2\hat{\mathbf{j}}+3\hat{\mathbf{k}})$

$$=5\hat{\mathbf{i}}-2\hat{\mathbf{j}}+5\hat{\mathbf{k}}=\mathbf{a}$$

Application of Dot Product in Mechanics (Work done)

A force acting on a particle is said to do work, if the particle is displaced in a direction which is not perpendicular to force.

Let a particle be placed at O and a force f represented by **OB** be acting on the particle at O. Due to the application of force **f**, the particle is displaced in the direction of **OA**. Let **OA** be the displacement.

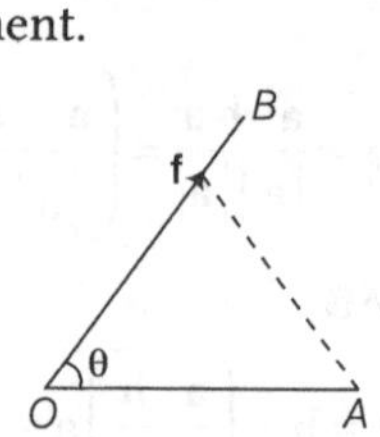

Then, the component of **OA** in the the direction of force **f** is, $|\mathbf{OA}|\cos\theta$

$$\text{Work done}=|\mathbf{f}||\mathbf{OA}|\cos\theta=\mathbf{f}\cdot\mathbf{OA}$$

$$=\mathbf{f}\cdot\mathbf{d}, \text{ where } \mathbf{d}=\mathbf{OA}$$

$\therefore$ Work done = (Force) $\cdot$ (Displacement)

Remarks

1. The work done by the resultant of a number of forces $\mathbf{f}_1\mathbf{f}_2, \mathbf{f}_3, \ldots \mathbf{f}_n$ in a displacement **d** of a particle is equal to the sum of work done by the forces separately
   i.e. Work done $=\mathbf{f}_1\cdot\mathbf{d}+\mathbf{f}_2\cdot\mathbf{d}+\ldots+\mathbf{f}_n\cdot\mathbf{d}$
   $=(\mathbf{f}_1+\mathbf{f}_2+\ldots+\mathbf{f}_n)\cdot\mathbf{d}$
   $=\mathbf{R}\cdot\mathbf{d}$ where, $\mathbf{R}=\mathbf{f}_1+\mathbf{f}_2+\ldots+\mathbf{f}_n$
2. The work done by a force **f** when its point of application experiences a number of consecutive displacements $\mathbf{d}_1, \mathbf{d}_2, \mathbf{d}_3 \ldots, \mathbf{d}_n$, is equal to the work done by the forces in single displacement from the beginning to end.
   i.e., Work done $=\mathbf{f}\cdot(\mathbf{d}_1+\mathbf{d}_2+\ldots+\mathbf{d}_n)$
   = The work done by the force **f** in the single displacement from the beginning to end

Example 16. Two forces $\mathbf{f}_1=3\hat{\mathbf{i}}-2\hat{\mathbf{j}}+\hat{\mathbf{k}}$ and $\mathbf{f}_2=\hat{\mathbf{i}}+3\hat{\mathbf{j}}-5\hat{\mathbf{k}}$ acting on a particle at A move it to B. Find the work done if the position vector of A and B are $-2\hat{\mathbf{i}}+5\hat{\mathbf{k}}$ and $3\hat{\mathbf{i}}-7\hat{\mathbf{j}}+2\hat{\mathbf{k}}$.

Sol. Let **R** be the resultant of two forces $\mathbf{f}_1$ and $\mathbf{f}_2$ and **d** be the displacement.

Then, $\mathbf{R}=(3\hat{\mathbf{i}}-2\hat{\mathbf{j}}+\hat{\mathbf{k}})+(\hat{\mathbf{i}}+3\hat{\mathbf{j}}-5\hat{\mathbf{k}})$

$$=4\hat{\mathbf{i}}+\hat{\mathbf{j}}-4\hat{\mathbf{k}}$$

and $\mathbf{d}=(3\hat{\mathbf{i}}-7\hat{\mathbf{j}}+2\hat{\mathbf{k}})-(-2\hat{\mathbf{i}}+5\hat{\mathbf{k}})=5\hat{\mathbf{i}}-7\hat{\mathbf{j}}-3\hat{\mathbf{k}}$

$\therefore$ The total work done = The work done by resultant

$$=\mathbf{R}\cdot\mathbf{d}=(4\hat{\mathbf{i}}+\hat{\mathbf{j}}-4\hat{\mathbf{k}})\cdot(5\hat{\mathbf{i}}-7\hat{\mathbf{j}}-3\hat{\mathbf{k}})$$

$$=20-7+12=25 \text{ units}$$

Example 17. Forces of magnitudes 5 and 3 units acting in the directions $6\hat{\mathbf{i}}+2\hat{\mathbf{j}}+3\hat{\mathbf{k}}$ and $3\hat{\mathbf{i}}-2\hat{\mathbf{j}}+6\hat{\mathbf{k}}$, respectively act on a particle which is displaced from the point $(2, 2, -1)$ and $(4, 3, 1)$. Find the work done by the forces.

Sol. Let **R** be the resultant of two forces and **d** be the displacement.

Then,

$$\mathbf{R}=5\frac{(6\hat{\mathbf{i}}+2\hat{\mathbf{j}}+3\hat{\mathbf{k}})}{\sqrt{36+4+9}}+3\frac{(3\hat{\mathbf{i}}-2\hat{\mathbf{j}}+6\hat{\mathbf{k}})}{\sqrt{9+4+36}}$$

$$=\frac{1}{7}(39\hat{\mathbf{i}}+4\hat{\mathbf{j}}+33\hat{\mathbf{k}})$$

and $\mathbf{d}=(4\hat{\mathbf{i}}+3\hat{\mathbf{j}}+\hat{\mathbf{k}})-(2\hat{\mathbf{i}}+2\hat{\mathbf{j}}-\hat{\mathbf{k}})=2\hat{\mathbf{i}}+\hat{\mathbf{j}}+2\hat{\mathbf{k}}$

$\therefore$ Total work done $=\mathbf{R}\cdot\mathbf{d}=\frac{1}{7}(78+4+66)$

$$=\frac{148}{7} \text{ units}$$

Exercise for Session 1

1. Find the angle between the vectors $\hat{\mathbf{i}} - 2\hat{\mathbf{j}} + 3\hat{\mathbf{k}}$ and $3\hat{\mathbf{i}} - 2\hat{\mathbf{j}} + \hat{\mathbf{k}}$.
2. Find and angle between two vectors **a** and **b** with magnitudes $\sqrt{3}$ and 2 respectively such that $\mathbf{a} \cdot \mathbf{b} = \sqrt{6}$
3. Show that the vectors $2\hat{\mathbf{i}} - \hat{\mathbf{j}} + \hat{\mathbf{k}}$ and $\hat{\mathbf{i}} - 3\hat{\mathbf{j}} - 5\hat{\mathbf{k}}$ are at right angles.
4. If $\mathbf{r} \cdot \hat{\mathbf{i}} = \mathbf{r} \cdot \hat{\mathbf{j}} = \mathbf{r} \cdot \hat{\mathbf{k}}$ and $|\mathbf{r}| = 3$, then find vector **r**.
5. Find the angle between the vectors $\mathbf{a} + \mathbf{b}$ and $\mathbf{a} - \mathbf{b}$, if $\mathbf{a} = 2\hat{\mathbf{i}} - \hat{\mathbf{j}} + 3\hat{\mathbf{k}}$ and $\mathbf{b} = 3\hat{\mathbf{i}} + \hat{\mathbf{j}} - 2\hat{\mathbf{k}}$.
6. Find the projection of the vector $\hat{\mathbf{i}} + 3\hat{\mathbf{j}} + 7\hat{\mathbf{k}}$ on the vector $7\hat{\mathbf{i}} - \hat{\mathbf{j}} + 8\hat{\mathbf{k}}$.
7. If the projection of vector $x\hat{\mathbf{i}} - \hat{\mathbf{j}} + \hat{\mathbf{k}}$ on vector $2\hat{\mathbf{i}} - \hat{\mathbf{j}} + 5\hat{\mathbf{k}}$ is $\frac{1}{\sqrt{30}}$, then find the value of x.
8. If $|\mathbf{a}| + |\mathbf{b}| = |\mathbf{c}|$ and $\mathbf{a} + \mathbf{b} = \mathbf{c}$, then find the angle between **a** and **b**.
9. If three unit vectors **a**, **b** and **c** satisfy $\mathbf{a} + \mathbf{b} + \mathbf{c} = 0$, then find the angle between **a** and **b**.
10. If $\mathbf{a} = x\hat{\mathbf{i}} + (x - 1)\hat{\mathbf{j}} + \hat{\mathbf{k}}$ and $\mathbf{b} = (x + 1)\hat{\mathbf{i}} + \hat{\mathbf{j}} + a\hat{\mathbf{k}}$ make an acute angle, $\forall x \in R$, then find the values of **a**.
11. Find the component of $\hat{\mathbf{i}}$ in the direction of the vector $\hat{\mathbf{i}} + \hat{\mathbf{j}} + 2\hat{\mathbf{k}}$.
12. Find the vector components of a vector $2\hat{\mathbf{i}} + 3\hat{\mathbf{j}} + 6\hat{\mathbf{k}}$ along and perpendicular to the non-zero vector $2\hat{\mathbf{i}} + \hat{\mathbf{j}} + 2\hat{\mathbf{k}}$.
13. A particle is acted upon by constant forces $4\hat{\mathbf{i}} + \hat{\mathbf{j}} - 3\hat{\mathbf{k}}$ and $3\hat{\mathbf{i}} + \hat{\mathbf{j}} - \hat{\mathbf{k}}$ which displace it from a point $\hat{\mathbf{i}} + 2\hat{\mathbf{j}} + 3\hat{\mathbf{k}}$ to the point $5\hat{\mathbf{i}} + 4\hat{\mathbf{j}} + \hat{\mathbf{k}}$. Find the work done by the forces in standard units.

Session 2

Vector or Cross Product of Two Vectors, Area of Parallelogram and Triangle, Moment of a Force and Couple, Rotation About an Axis

Vector or Cross Product of Two Vectors

Let **a** and **b** be two non-zero, non- parallel vectors. Then the vector product $\mathbf{a} \times \mathbf{b}$, in that order, is defined as a vector whose magnitude is

$$|\mathbf{a}||\mathbf{b}|\sin\theta$$

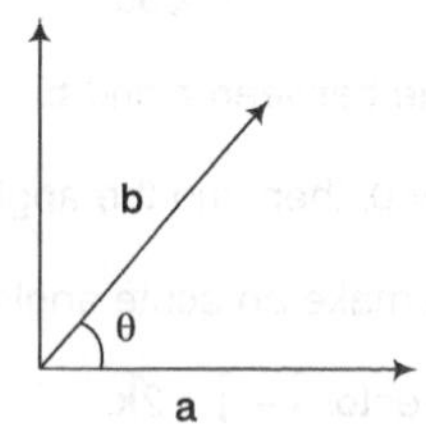

where θ is the angle between **a** and **b**, whose direction is perpendicular to the plane of **a** and **b** in such a way that **a**, **b** and this direction constitute a right handed system.

In other words, $\mathbf{a} \times \mathbf{b} = |\mathbf{a}||\mathbf{b}|\sin\theta\, n$, where θ is the angle between **a** and **b**, $\hat{\mathbf{n}}$ is a unit vector perpendicular to the plane of **a** and **b** such that **a**, **b** and n form a right handed system.

Properties of Vector Product

(i) Vector product is not commutative i.e., if **a** and **b** are any two vectors, then $\mathbf{a} \times \mathbf{b} \neq \mathbf{b} \times \mathbf{a}$, however $\mathbf{a} \times \mathbf{b} = -(\mathbf{b} \times \mathbf{a})$.

(ii) Vector product is not associative, i.e. $\mathbf{a} \times (\mathbf{b} \times \mathbf{c}) \neq (\mathbf{a} \times \mathbf{b}) \times \mathbf{c}$

(iii) If **a** and **b** are two vectors and m is a scalar, then

$$m\mathbf{a} \times \mathbf{b} = m(\mathbf{a} \times \mathbf{b}) = \mathbf{a} \times m\mathbf{b}$$

(iv) If **a** and **b** are two vectors and m and n are scalars, then $m\mathbf{a} \times n\mathbf{b} = mn(\mathbf{a} \times \mathbf{b}) = m(\mathbf{a} \times \mathbf{b}) = n(m\mathbf{a} \times \mathbf{b})$

(v) Distributivity of vector product over vector addition. Let **a**, **b** and **c** be any three vectors. Then,

(a) $\mathbf{a} \times (\mathbf{b} + \mathbf{c}) = \mathbf{a} \times \mathbf{b} + \mathbf{a} \times \mathbf{c}$ (left distributivity)

(b) $(\mathbf{b} + \mathbf{c}) \times \mathbf{a} = \mathbf{b} \times \mathbf{a} + \mathbf{c} \times \mathbf{a}$ (right distributivity)

(vi) For any three vectors **a**, **b** and **c**, we have $\mathbf{a} \times (\mathbf{b} - \mathbf{c}) = \mathbf{a} \times \mathbf{b} - \mathbf{a} \times \mathbf{c}$.

(vii) The vector product of any vector (zero or non-zero) with zero vector is a zero vector i.e.

$$\mathbf{a} \times \mathbf{0} = \mathbf{0} \times \mathbf{a} = \mathbf{0}$$

(viii) The vector product of two non-zero vectors is zero vector iff they are parallel (collinear) i.e. $\mathbf{a} \times \mathbf{b} = \mathbf{0} \Leftrightarrow \mathbf{a} \,||\, \mathbf{b}$, **a** and **b** are non-zero vectors.

It follows from the above property that $\mathbf{a} \times \mathbf{a} = \mathbf{0}$ for every non-zero vector **a**, which in turn implies that $\hat{\mathbf{i}} \times \hat{\mathbf{i}} = \hat{\mathbf{j}} \times \hat{\mathbf{j}} = \hat{\mathbf{k}} \times \hat{\mathbf{k}} = \mathbf{0}$.

(ix) Vector product of orthonormal triad of unit vectors $\hat{\mathbf{i}}, \hat{\mathbf{j}}$ and $\hat{\mathbf{k}}$ using the definition of the vector product obtain

$$\hat{\mathbf{i}} \times \hat{\mathbf{j}} = \hat{\mathbf{k}},\ \hat{\mathbf{j}} \times \hat{\mathbf{k}} = \hat{\mathbf{i}},\ \hat{\mathbf{k}} \times \hat{\mathbf{i}} = \hat{\mathbf{j}},$$
$$\hat{\mathbf{j}} \times \hat{\mathbf{i}} = -\hat{\mathbf{k}},\ \hat{\mathbf{k}} \times \hat{\mathbf{j}} = -\hat{\mathbf{i}},\ \hat{\mathbf{i}} \times \hat{\mathbf{k}} = -\hat{\mathbf{j}}$$

(x) **Lagrange's identity** If **a**, **b** are any two vectors, then

$$|\mathbf{a} \times \mathbf{b}|^2 = |\mathbf{a}|^2\,|\mathbf{b}|^2 - (\mathbf{a} \cdot \mathbf{b})^2$$
$$\text{or}\quad |\mathbf{a} \times \mathbf{b}|^2 + (\mathbf{a} \cdot \mathbf{b})^2 = |\mathbf{a}|^2\,|\mathbf{b}|^2$$

Vector Product in Terms of Components

If $\mathbf{a} = a_1\hat{\mathbf{i}} + a_2\hat{\mathbf{j}} + a_3\hat{\mathbf{k}}$ and $\mathbf{b} = b_1\hat{\mathbf{i}} + b_2\hat{\mathbf{j}} + b_3\hat{\mathbf{k}}$

Then, $a \times b = (a_2b_3 - a_3b_2)\hat{\mathbf{i}} - (a_1b_3 - a_3b_1)\hat{\mathbf{j}}$

$$+ (a_1b_2 - a_2b_1)\hat{\mathbf{k}} = \begin{vmatrix} \hat{\mathbf{i}} & \hat{\mathbf{j}} & \hat{\mathbf{k}} \\ a_1 & a_2 & a_3 \\ b_1 & b_2 & b_3 \end{vmatrix}$$

Example 18. If $\mathbf{a} = 2\hat{\mathbf{i}} + 3\hat{\mathbf{j}} - 5\hat{\mathbf{k}}$ and $\mathbf{b} = m\hat{\mathbf{i}} + n\hat{\mathbf{j}} + 12\hat{\mathbf{k}}$ and $\mathbf{a} \times \mathbf{b} = \mathbf{0}$. Then, find the values of m and n.

Sol. Clearly, $\mathbf{a} \times \mathbf{b} = \begin{vmatrix} \hat{\mathbf{i}} & \hat{\mathbf{j}} & \hat{\mathbf{k}} \\ 2 & 3 & -5 \\ m & n & 12 \end{vmatrix}$

$$= \hat{\mathbf{i}}(36 + 5n) - \hat{\mathbf{j}}(24 + 5m) + \hat{\mathbf{k}}(2n - 3m)$$

Since, $\mathbf{a} \times \mathbf{b} = \mathbf{0}$

$\therefore (36 + 5n)\hat{\mathbf{i}} - (24 + 5m)\hat{\mathbf{j}} + (2n - 3m)\hat{\mathbf{k}} = 0\hat{\mathbf{i}} + 0\hat{\mathbf{j}} + 0\hat{\mathbf{k}}$

On comparing the coefficients of $\hat{\mathbf{i}}, \hat{\mathbf{j}}$ and $\hat{\mathbf{k}}$, we get

$$36 + 5n = 0,\ -(24 + 5m) = 0 \text{ and } 2n - 3m = 0$$

$$\Rightarrow \quad n = -\frac{36}{5} \text{ and } m = -\frac{24}{5}$$

Example 19. Show that $(\mathbf{a}-\mathbf{b})\times(\mathbf{a}+\mathbf{b})=2(\mathbf{a}\times\mathbf{b})$

Sol. Consider, $(\mathbf{a}-\mathbf{b})\times(\mathbf{a}+\mathbf{b})=(\mathbf{a}-\mathbf{b})\times\mathbf{a}+(\mathbf{a}-\mathbf{b})\times\mathbf{b}$

[By distributivity of vector product over vector addition]

$=\mathbf{a}\times\mathbf{a}-\mathbf{b}\times\mathbf{a}+\mathbf{a}\times\mathbf{b}-\mathbf{b}\times\mathbf{b}$

[Again, by distributivity of vector product over vector addition]

$=0+\mathbf{a}\times\mathbf{b}+\mathbf{a}\times\mathbf{b}-0$ $[\because \mathbf{a}\times\mathbf{b}=-(\mathbf{b}\times\mathbf{a})]$

$=2(\mathbf{a}\times\mathbf{b})$ **Hence Proved**

Example 20. If a is any vector, then $(\mathbf{a}\times\hat{\mathbf{i}})^2+(\mathbf{a}\times\hat{\mathbf{j}})^2+(\mathbf{a}\times\hat{\mathbf{k}})^2$ is equal to

(a) $|\mathbf{a}|^2$ (b) 0

(c) $3|\mathbf{a}|^2$ (d) $2|\mathbf{a}|^2$

Sol. (d) Let $\mathbf{a}=a_1\hat{\mathbf{i}}+a_2\hat{\mathbf{j}}+a_3\hat{\mathbf{k}}$

$\therefore\ \mathbf{a}\times\hat{\mathbf{i}}=(a_1\hat{\mathbf{i}}+a_2\hat{\mathbf{j}}+a_3\hat{\mathbf{k}})\times\hat{\mathbf{i}}=-a_2\hat{\mathbf{k}}+a_3\hat{\mathbf{j}}$

$(\mathbf{a}\times\hat{\mathbf{i}})^2=(\mathbf{a}\times\hat{\mathbf{i}})\cdot(\mathbf{a}\times\hat{\mathbf{i}})$

$=(-a_2\hat{\mathbf{k}}+a_3\hat{\mathbf{j}})\cdot(-a_2\hat{\mathbf{k}}+a_3\hat{\mathbf{j}})=a_2^2+a_3^2$

Similarly, $(\mathbf{a}\times\hat{\mathbf{j}})^2=a_3^2+a_1^2$

and $(\mathbf{a}\times\hat{\mathbf{k}})^2=a_1^2+a_2^2$

$\therefore\ (\mathbf{a}\times\hat{\mathbf{i}})^2+(\mathbf{a}\times\hat{\mathbf{j}})^2+(\mathbf{a}\times\hat{\mathbf{k}})^2$

$=2(a_1^2+a_2^2+a_3^2)=2|\mathbf{a}|^2$

Example 21. If $\mathbf{a}\cdot\mathbf{b}=0$ and $\mathbf{a}\times\mathbf{b}=0$, prove that $\mathbf{a}=0$ or $\mathbf{b}=0$.

Sol. Given, $\mathbf{a}\cdot\mathbf{b}=0$ and $\mathbf{a}\times\mathbf{b}=0$

Now, $\mathbf{a}\cdot\mathbf{b}=0\Rightarrow\mathbf{a}=0$ or $\mathbf{b}=0$ or $\mathbf{a}\perp\mathbf{b}$

and $\mathbf{a}\times\mathbf{b}=0\Rightarrow\mathbf{a}=0$ or $\mathbf{b}=0$ or $\mathbf{a}\,||\,\mathbf{b}$

Since, $\mathbf{a}\perp\mathbf{b}$ and $\mathbf{a}\,||\,\mathbf{b}$ can never hold simultaneously.

$\therefore$ $\mathbf{a}\cdot\mathbf{b}=0$ and $\mathbf{a}\times\mathbf{b}=0$

$\Rightarrow$ $\mathbf{a}=0$ or $\mathbf{b}=0$

Example 22. If **a**, **b** and **c** are vectors such that $\mathbf{a}\cdot\mathbf{b}=\mathbf{a}\cdot\mathbf{c}$, $\mathbf{a}\times\mathbf{b}=\mathbf{a}\times\mathbf{c}$ and $\mathbf{a}\neq 0$, then show that $\mathbf{b}=\mathbf{c}$

Sol. $\mathbf{a}\cdot\mathbf{b}=\mathbf{a}\cdot\mathbf{c}$ and $\mathbf{a}\neq 0$

$\Rightarrow$ $\mathbf{a}\cdot\mathbf{b}-\mathbf{a}\cdot\mathbf{c}=0$ and $\mathbf{a}\neq 0$

$\Rightarrow$ $\mathbf{a}\cdot(\mathbf{b}-\mathbf{c})=0$ and $\mathbf{a}\neq 0$

$\Rightarrow$ $\mathbf{a}\perp(\mathbf{b}-\mathbf{c})$ or $\mathbf{b}=\mathbf{c}$ ...(i)

Again, $\mathbf{a}\times\mathbf{b}=\mathbf{a}\times\mathbf{c}$ and $\mathbf{a}\neq 0$

$\Rightarrow$ $\mathbf{a}\times(\mathbf{b}-\mathbf{c})=0$ and $\mathbf{a}\neq 0$

$\Rightarrow$ $\mathbf{a}\,||\,(\mathbf{b}-\mathbf{c})=0$ and $\mathbf{b}=\mathbf{c}$ ...(ii)

$\therefore$ From Eqs. (i) and (ii), we have

$\mathbf{b}=\mathbf{c}$

[as **a** cannot be both parallel and perpendicular to $(\mathbf{b}-\mathbf{c})$]

Example 23. If **a**,**b** and **c** are three non-zero vectors such that $\mathbf{a}\cdot(\mathbf{b}\times\mathbf{c})=0$ and **b** and **c** are not parallel vectors, prove that $\mathbf{a}=\lambda\mathbf{b}+\mu\mathbf{c}$ where λ and μ are scalar.

Sol. We have, $\mathbf{a}\cdot(\mathbf{b}\times\mathbf{c})=0$

$\Rightarrow$ $\mathbf{a}=0$ or $\mathbf{b}\times\mathbf{c}=0$ or $\mathbf{a}\perp(\mathbf{b}\times\mathbf{c})$

$\Rightarrow$ $\mathbf{a}=0$ or $\mathbf{b}\,||\,\mathbf{c}$ or $\mathbf{a}\perp(\mathbf{b}\times\mathbf{c})$

But $\mathbf{a}\neq 0$ and $\mathbf{b}\neq\mathbf{c}$

$\therefore$ $\mathbf{a}\perp(\mathbf{b}\times\mathbf{c})$

$\Rightarrow$ **a** lies in the plane of **b** and **c**.

$\Rightarrow$ **a**, **b** and **c** are coplanar.

$\Rightarrow$ $a=\lambda b+\mu c$

Example 24. If $\mathbf{a}\times\mathbf{b}=\mathbf{a}\times\mathbf{c}$, $\mathbf{a}\neq 0$, show that $\mathbf{b}=\mathbf{c}+t\mathbf{a}$ for some scalar t.

Sol. We have, $\mathbf{a}\times\mathbf{b}=\mathbf{a}\times\mathbf{c}$

$\Rightarrow$ $\mathbf{a}\times\mathbf{b}-\mathbf{a}\times\mathbf{c}=0$

$\Rightarrow$ $\mathbf{a}\times(\mathbf{b}-\mathbf{c})=0$

$\Rightarrow$ $\mathbf{a}=0$ or $(\mathbf{b}-\mathbf{c})=0$ or $\mathbf{a}\,||\,(\mathbf{b}-\mathbf{c})$

$\Rightarrow$ $\mathbf{a}\,||\,(\mathbf{b}-\mathbf{c})$ $(\because \mathbf{a}\neq 0$ and $\mathbf{b}\neq 0)$

$\Rightarrow$ $\mathbf{b}-\mathbf{c}=t\mathbf{a}$ (for some scalar t)

$\Rightarrow$ $\mathbf{b}=\mathbf{c}+t\mathbf{a}$

Example 25. For any two vector **u** and **v**, prove that

(i) $(\mathbf{u}\cdot\mathbf{v})^2+|\mathbf{u}\times\mathbf{v}|^2=|\mathbf{u}|^2|\mathbf{v}|^2$

(ii) $(1+|\mathbf{u}|^2)(1+|\mathbf{v}|^2)=(1-\mathbf{u}\cdot\mathbf{v})^2+|\mathbf{u}+\mathbf{v}+(\mathbf{u}\times\mathbf{v})|^2$

Sol. (i) To show $(\mathbf{u}\cdot\mathbf{v})^2+|\mathbf{u}\times\mathbf{v}^2|=|\mathbf{u}|^2|\mathbf{v}|^2$

Let θ be the angle between **u** and **v**.

$\Rightarrow$ $\mathbf{u}\cdot\mathbf{v}=uv\cos\theta$

and $|\mathbf{u}\times\mathbf{v}|=uv\sin\theta$

$\Rightarrow$ $(\mathbf{u}\cdot\mathbf{v})^2+|\mathbf{u}\times\mathbf{v}|^2=u^2v^2\cos^2\theta+u^2v^2\sin^2\theta$

$\Rightarrow$ $(\mathbf{u}\cdot\mathbf{v})^2+|\mathbf{u}\times\mathbf{v}|^2=u^2v^2$

$\Rightarrow$ $(\mathbf{u}\cdot\mathbf{v})^2+|\mathbf{u}\times\mathbf{v}|^2=|\mathbf{u}|^2|\mathbf{v}|^2$

(ii) Taking RHS $(1-\mathbf{u}\cdot\mathbf{v})^2+|\mathbf{u}+\mathbf{v}+(\mathbf{u}\times\mathbf{v})|^2$

$\Rightarrow 1+(\mathbf{u}\cdot\mathbf{v})^2-2\mathbf{u}\cdot\mathbf{v}+|[\mathbf{u}+\mathbf{v}+(\mathbf{u}\times\mathbf{v})]\cdot[\mathbf{u}+\mathbf{v}+(\mathbf{u}\times\mathbf{v})]|$

$\Rightarrow 1+|\mathbf{u}|^2|\mathbf{v}|^2\cos^2\theta-2|\mathbf{u}||\mathbf{v}|\cos\theta+\mathbf{u}\cdot\mathbf{u}+\mathbf{u}\cdot\mathbf{v}+\mathbf{u}(\mathbf{u}\times\mathbf{v})+\mathbf{v}\cdot\mathbf{u}+\mathbf{v}\cdot\mathbf{v}+\mathbf{v}\cdot(\mathbf{u}\times\mathbf{v})+(\mathbf{u}\times\mathbf{v})+(\mathbf{u}\times\mathbf{v})\cdot\mathbf{u}+(\mathbf{u}\times\mathbf{v})\cdot\mathbf{v}+(\mathbf{u}\times\mathbf{v})\cdot(\mathbf{u}\times\mathbf{v})$

$\Rightarrow 1+|\mathbf{u}|^2|\mathbf{v}|^2\cos^2\theta-2|\mathbf{u}||\mathbf{v}|\cos\theta+|\mathbf{u}|^2+|\mathbf{u}||\mathbf{v}|\cos\theta+0+|\mathbf{u}||\mathbf{v}|\cos\theta+|\mathbf{v}^2+0+0+0+|\mathbf{u}+\mathbf{v}|^2$

$\Rightarrow 1+|\mathbf{u}|^2|\mathbf{v}|^2\cos^2\theta+|\mathbf{u}|^2+|\mathbf{v}|^2+|\mathbf{u}|^2|\mathbf{v}|^2\sin^2\theta$

$\Rightarrow 1+|\mathbf{u}|^2|\mathbf{v}|^2+|\mathbf{u}|^2+|\mathbf{v}|^2$

$=(1+|\mathbf{u}|^2)(1+|\mathbf{v}|^2)=$ LHS

Angle between Two Vectors

If θ is the angle between **a** and **b**, then $\sin\theta = \frac{|\mathbf{a}\times\mathbf{b}|}{|\mathbf{a}||\mathbf{b}|}$.

Expression for sin θ

If $\mathbf{a} = a_1\hat{\mathbf{i}} + a_2\hat{\mathbf{j}} + a_3\hat{\mathbf{k}}, \mathbf{b} = b_1\hat{\mathbf{i}} + b_2\hat{\mathbf{j}} + b_3\hat{\mathbf{k}}$ and θ be angle between **a** and **b**, then

$$\sin^2\theta = \frac{(a_2b_3 - a_3b_2)^2 + (a_1b_3 - a_3b_1)^2 + (a_1b_2 - a_2b_1)^2}{(a_1^2 + a_2^2 + a_3^2)(b_1^2 + b_2^2 + b_3^2)}$$

Example 26. The sine of the angle between the vector $\mathbf{a} = 3\hat{\mathbf{i}} + \hat{\mathbf{j}} + \hat{\mathbf{k}}$ and $\mathbf{b} = 2\hat{\mathbf{i}} - 2\hat{\mathbf{j}} + \hat{\mathbf{k}}$ is

(a) $\sqrt{\frac{74}{99}}$ (b) $\sqrt{\frac{25}{99}}$

(c) $\sqrt{\frac{37}{99}}$ (d) $\frac{5}{\sqrt{41}}$

Sol. (a) $\mathbf{a}\times\mathbf{b} = \begin{vmatrix} \hat{\mathbf{i}} & \hat{\mathbf{j}} & \hat{\mathbf{k}} \\ 3 & 1 & 1 \\ 2 & -2 & 1 \end{vmatrix} = 3\hat{\mathbf{i}} - \hat{\mathbf{j}} - 8\hat{\mathbf{k}}$

$$\sin\theta = \frac{|\mathbf{a}\times\mathbf{b}|}{|\mathbf{a}||\mathbf{b}|} = \frac{\sqrt{74}}{\sqrt{11}\cdot\sqrt{9}} = \sqrt{\frac{74}{99}}$$

Example 27. If $|\mathbf{a}| = 2, |\mathbf{b}| = 5$ and $|\mathbf{a}\times\mathbf{b}| = 8$, then find the value of $\mathbf{a}\cdot\mathbf{b}$.

Sol. We have, $|\mathbf{a}| = 2, |\mathbf{b}| = 5$

and $|\mathbf{a}\times\mathbf{b}| = 8.$

Let θ be the angle between **a** and **b**.

Then, $\sin\theta = \frac{|\mathbf{a}\times\mathbf{b}|}{|\mathbf{a}||\mathbf{b}|} = \frac{8}{2\times 5} = \frac{4}{5}$

Now, $\cos\theta = \pm\sqrt{1-\sin^2\theta}$

$$= \pm\sqrt{1 - \frac{16}{25}} = \pm\frac{3}{5}$$

$$\mathbf{a}\cdot\mathbf{b} = |\mathbf{a}|\mathbf{b}|\cos\theta = \pm\left(2\cdot 5\cdot\frac{3}{5}\right) = \pm 6$$

Vector Normal to the Plane of Two Given Vectors

If **a** and **b** are two non-zero, non-parallel vectors and let θ be the angle between them. $\mathbf{a}\times\mathbf{b} = |\mathbf{a}||\mathbf{b}|\sin\theta\,\hat{\mathbf{n}}$, where $\hat{\mathbf{n}}$ is a unit vector perpendicular to the plane of **a** and **b** such that **a**, **b**, $\hat{\mathbf{n}}$ form a right handed system.

$$\Rightarrow \quad (\mathbf{a}\times\mathbf{b}) = |\mathbf{a}\times\mathbf{b}|\,\hat{\mathbf{n}}$$

$$\Rightarrow \quad \hat{\mathbf{n}} = \frac{\mathbf{a}\times\mathbf{b}}{|\mathbf{a}\times\mathbf{b}|}$$

Thus, $\hat{\mathbf{n}} = \frac{\mathbf{a}\times\mathbf{b}}{|\mathbf{a}\times\mathbf{b}|}$ is a unit vector perpendicular to the plane of **a** and **b**. Note that $-\frac{\mathbf{a}\times\mathbf{b}}{|\mathbf{a}\times\mathbf{b}|}$ is also a unit vector perpendicular to the plane of **a** and **b**. Vectors of magnitude 'λ' normal to the plane of **a** and **b** are given by

$$\pm\frac{\lambda(\mathbf{a}\times\mathbf{b})}{|\mathbf{b}\times\mathbf{b}|}$$

Example 28. The unit vector perpendicular to the vectors $6\hat{\mathbf{i}} + 2\hat{\mathbf{j}} + 3\hat{\mathbf{k}}$ and $3\hat{\mathbf{i}} - 6\hat{\mathbf{j}} - 2\hat{\mathbf{k}}$, is

(a) $\frac{2\hat{\mathbf{i}} - 3\hat{\mathbf{j}} + 6\hat{\mathbf{k}}}{7}$ (b) $\frac{2\hat{\mathbf{i}} - 3\hat{\mathbf{j}} - 6\hat{\mathbf{k}}}{7}$

(c) $\frac{2\hat{\mathbf{i}} + 3\hat{\mathbf{j}} - 6\hat{\mathbf{k}}}{7}$ (d) $\frac{2\hat{\mathbf{i}} + 3\hat{\mathbf{j}} + 6\hat{\mathbf{k}}}{7}$

Sol. (c) Let $\mathbf{a} = 6\hat{\mathbf{i}} + 2\hat{\mathbf{j}} + 3\hat{\mathbf{k}}$ and $\mathbf{b} = 3\hat{\mathbf{i}} - 6\hat{\mathbf{j}} - 2\hat{\mathbf{k}}$

$$\mathbf{a}\times\mathbf{b} = \begin{vmatrix} \hat{\mathbf{i}} & \hat{\mathbf{j}} & \hat{\mathbf{k}} \\ 6 & 2 & 3 \\ 3 & -6 & -2 \end{vmatrix}$$

$$= 14\hat{\mathbf{i}} + 21\hat{\mathbf{j}} - 42\hat{\mathbf{k}} = 7(2\hat{\mathbf{i}} + 3\hat{\mathbf{j}} - 6\hat{\mathbf{k}})$$

$$|\mathbf{a}\times\mathbf{b}| = 7|2\hat{\mathbf{i}} + 3\hat{\mathbf{j}} - 6\hat{\mathbf{k}}| = 7\cdot 7$$

$$\therefore \quad \frac{\mathbf{a}\times\mathbf{b}}{|\mathbf{a}\times\mathbf{b}|} = \frac{1}{7}(2\hat{\mathbf{i}} + 3\hat{\mathbf{j}} - 6\hat{\mathbf{k}})$$

which is a unit vector perpendicular to **a** and **b**.

Example 29. Find unit vectors perpendicular to the plane determined by the points

$P(1, -1, 2), Q(2, 0, -1)$ and $R(0, 2, 1)$

Sol. Clearly, required unit vector is a unit vector perpendicular to the plane of **PQ** and **PR**.

Now,

$$\mathbf{PQ} = (2\hat{\mathbf{i}} - \hat{\mathbf{k}}) - (\hat{\mathbf{i}} - \hat{\mathbf{j}} + 2\hat{\mathbf{k}})$$
$$= \hat{\mathbf{i}} + \hat{\mathbf{j}} - 3\hat{\mathbf{k}}$$
$$\mathbf{PR} = (2\hat{\mathbf{j}} + \hat{\mathbf{k}}) - (\hat{\mathbf{i}} - \hat{\mathbf{j}} + 2\hat{\mathbf{k}})$$
$$= -\hat{\mathbf{i}} + 3\hat{\mathbf{j}} - \hat{\mathbf{k}}$$

and $\mathbf{PQ}\times\mathbf{PR} = \begin{vmatrix} \hat{\mathbf{i}} & \hat{\mathbf{j}} & \hat{\mathbf{k}} \\ 1 & 1 & -3 \\ -1 & 3 & -1 \end{vmatrix} = 8\hat{\mathbf{i}} + 4\hat{\mathbf{j}} + 4\hat{\mathbf{k}}$

∴ Required unit vectors

$$= \pm\frac{\mathbf{PQ}\times\mathbf{PR}}{|\mathbf{PQ}\times\mathbf{PR}|} = \pm\frac{(8\hat{\mathbf{i}} + 4\hat{\mathbf{j}} + 4\hat{\mathbf{k}})}{4\sqrt{6}}$$

$$= \pm\frac{1}{\sqrt{6}}(2\hat{\mathbf{i}} + \hat{\mathbf{j}} + \hat{\mathbf{k}})$$

Example 30. Let **A**, **B** and **C** be unit vectors. Suppose $\mathbf{A}\cdot\mathbf{B}=\mathbf{A}\cdot\mathbf{C}=0$ and the angle between **B** and **C** is $\frac{\pi}{4}$.

Then,

(a) $\mathbf{A}=\pm 2\,(\mathbf{B}\times\mathbf{C})$ (b) $\mathbf{A}=\pm\sqrt{2}\,(\mathbf{B}\times\mathbf{C})$

(c) $\mathbf{A}=\pm 3\,(\mathbf{B}+\mathbf{C})$ (d) $\mathbf{A}=\pm\sqrt{3}\,(\mathbf{B}\times\mathbf{C})$

Sol. (b) Since, $\mathbf{A}\cdot\mathbf{B}=0$

$$\Rightarrow \quad \mathbf{A}\perp\mathbf{B} \text{ and } \mathbf{A}\cdot\mathbf{C}=0$$

$$\Rightarrow \quad \mathbf{A}\perp\mathbf{C}$$

$$\therefore \quad \mathbf{A}=\pm\frac{\mathbf{B}\times\mathbf{C}}{|\mathbf{B}\times\mathbf{C}|}$$

[$\because$ **A** is a unit vector perpendicular to both **B** and **C**]

Here, $$|\mathbf{B}\times\mathbf{C}|=|\mathbf{B}||\mathbf{C}|\sin\frac{\pi}{4}$$

$$=1\cdot 1\cdot\frac{1}{\sqrt{2}}=\frac{1}{\sqrt{2}}$$

So, $$\mathbf{A}=\pm\frac{(\mathbf{B}\times\mathbf{C})}{\frac{1}{\sqrt{2}}}=\pm\sqrt{2}\,(\mathbf{B}\times\mathbf{C})$$

Right Handed System and Left Handed System of Vectors

(i) **Right handed system of vectors** Three mutually perpendicular vectors **a**, **b** and **c** from a right handed system of vector iff $\mathbf{a}\times\mathbf{b}=\mathbf{c}, \mathbf{b}\times\mathbf{c}=\mathbf{a}, \mathbf{c}\times\mathbf{a}=\mathbf{b}$.

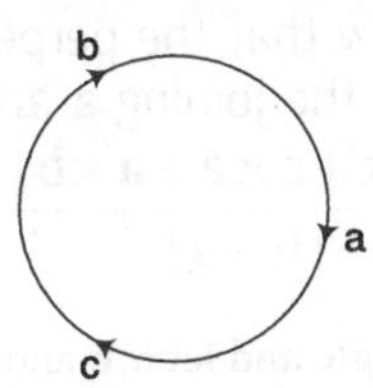

For example, the unit vectors $\hat{\mathbf{i}}$, $\hat{\mathbf{j}}$ and $\hat{\mathbf{k}}$ form a right handed system, $\hat{\mathbf{i}}\times\hat{\mathbf{j}}=\hat{\mathbf{k}}, \hat{\mathbf{j}}\times\hat{\mathbf{k}}=\hat{\mathbf{i}}, \hat{\mathbf{k}}\times\hat{\mathbf{i}}=\hat{\mathbf{j}}$

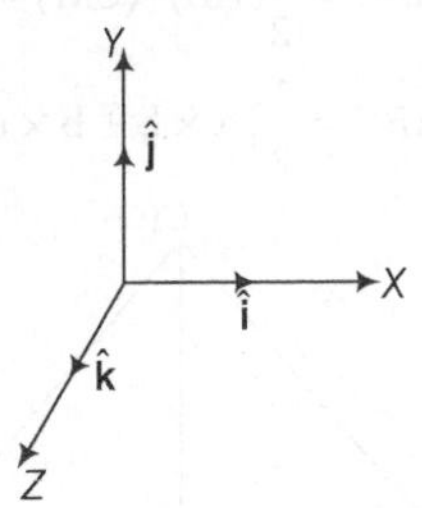

(ii) **Left handed system of vectors** The vectors **a**, **b** and **c** mutually perpendicular to one another form a left handed system of vectors iff

$$\mathbf{c}\times\mathbf{b}=\mathbf{a}, \mathbf{a}\times\mathbf{c}=\mathbf{b}, \mathbf{b}\times\mathbf{a}=\mathbf{c}.$$

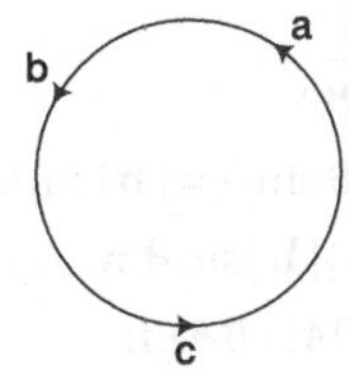

Example 31. The vectors **c**, $\mathbf{a}=x\hat{\mathbf{i}}+y\hat{\mathbf{j}}+z\hat{\mathbf{k}}$ and $\mathbf{b}=\hat{\mathbf{j}}$ are such that **a**, **c** and **b** form a right handed system, then **c** is

(a) $z\hat{\mathbf{i}}-x\hat{\mathbf{k}}$ (b) 0

(c) $y\hat{\mathbf{j}}$ (d) $-z\hat{\mathbf{i}}+x\hat{\mathbf{k}}$

Sol. (a) **a**, **c** and **b** form a right handed system.

Hence, $$\mathbf{b}\times\mathbf{a}=\mathbf{c}$$

$$\Rightarrow \quad \mathbf{c}=\hat{\mathbf{j}}\times(x\hat{\mathbf{i}}+y\hat{\mathbf{j}}+z\hat{\mathbf{k}})$$

$$=-x\hat{\mathbf{k}}+z\hat{\mathbf{i}}=z\hat{\mathbf{i}}-x\hat{\mathbf{k}}$$

Example 32. If **a**, **b** and **c** are three non-zero vectors such that $\mathbf{a}\times\mathbf{b}=\mathbf{c}$ and $\mathbf{b}\times\mathbf{c}=\mathbf{a}$, prove that **a**, **b** and **c** are mutually at right angles and $|\mathbf{b}|=1$ and $|\mathbf{c}|=|\mathbf{a}|$.

Sol. $\mathbf{a}\times\mathbf{b}=\mathbf{c}$ and $\mathbf{a}=\mathbf{b}\times\mathbf{c}$

$\Rightarrow$ $\mathbf{c}\perp\mathbf{a}$, $\mathbf{c}\perp\mathbf{b}$ and $\mathbf{a}\perp\mathbf{b}$, $\mathbf{a}\perp\mathbf{c}$ $\Rightarrow$ $\mathbf{a}\perp\mathbf{b}$, $\mathbf{b}\perp\mathbf{c}$ and $\mathbf{c}\perp\mathbf{a}$

$\Rightarrow$ **a**, **b** and **c** are mutually perpendicular vectors.

Again, $\mathbf{a}\times\mathbf{b}=\mathbf{c}$ and $\mathbf{b}\times\mathbf{c}=\mathbf{a}$

$$\Rightarrow \quad |\mathbf{a}\times\mathbf{b}|=|\mathbf{c}| \text{ and } |\mathbf{b}\times\mathbf{c}|=|\mathbf{a}|$$

$$\Rightarrow \quad |\mathbf{a}||\mathbf{b}|\sin\frac{\pi}{2}=|\mathbf{c}| \text{ and } |\mathbf{b}||\mathbf{c}|\sin\frac{\pi}{2}=|\mathbf{a}|$$

$$\Rightarrow \quad |\mathbf{a}||\mathbf{b}|=|\mathbf{c}|$$

and $$|\mathbf{b}||\mathbf{c}|=|\mathbf{a}| \qquad (\because \mathbf{a}\perp\mathbf{b} \text{ and } \mathbf{b}\perp\mathbf{c})$$

$$\Rightarrow \quad |\mathbf{b}|^2|\mathbf{c}|=|\mathbf{c}|$$

$$\Rightarrow \quad |\mathbf{b}|^2=1 \Rightarrow |\mathbf{b}|=1$$

On putting in $|\mathbf{a}||\mathbf{b}|=|\mathbf{c}| \Rightarrow |\mathbf{a}|=|\mathbf{c}|$

Geometrical Interpretation of Vector Product

If **a** and **b** are two non-zero, non-parallel vectors represented by **OA** and **OB** respectively and let θ be the angle between them. Complete the parallelogram *OACB*. Draw $BL\perp OA$.

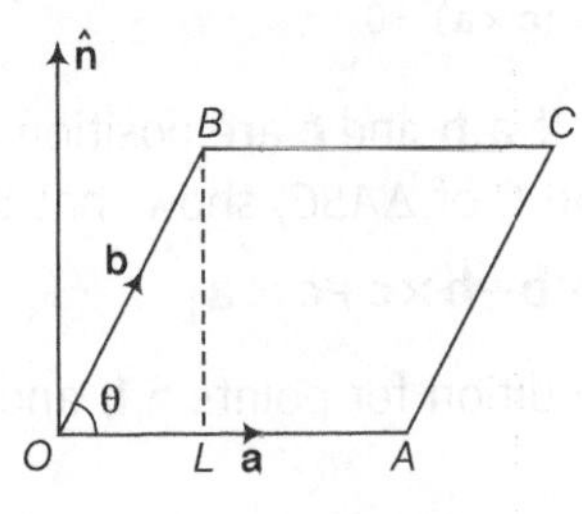

In ΔOBL, $\sin\theta = \dfrac{BL}{OB}$

$\Rightarrow$ $BL = OB\sin\theta = |\mathbf{b}|\sin\theta$

Now, $\mathbf{a}\times\mathbf{b} = |\mathbf{a}||\mathbf{b}|\sin\theta\,\mathbf{n}$

$= (OA)(BA)\,\mathbf{n}$

$=$ (Base $\times$ Height) $\mathbf{n}$ = (Area of parallelogram $OACB$) $\mathbf{n}$

$=$ Vector area of the parallelogram $OACB$

Thus, $\mathbf{a}\times\mathbf{b}$ is a vector whose magnitude is equal to the area of the parallelogram having $\mathbf{a}$ and $\mathbf{b}$ as its adjacent sides and whose direction $\mathbf{n}$ is perpendicular to the plane of $\mathbf{a}$ and $\mathbf{b}$ such that $\mathbf{a}$, $\mathbf{b}$ and $\mathbf{n}$ form a right handed system. Hence, $\mathbf{a}\times\mathbf{b}$ represents the vector area of the parallelogram having adjacent sides along $\mathbf{a}$ and $\mathbf{b}$.

Area of Parallelogram and Triangle

(i) The area of a parallelogram with adjacent sides $\mathbf{a}$ and is $|\mathbf{a}\times\mathbf{b}|$.

(ii) The area of a parallelogram with diagonals $\mathbf{d}_1$ and $\mathbf{d}_2$ is $\frac{1}{2}|\mathbf{d}_1\times\mathbf{d}_2|$.

(iii) The area of a plane quadrilateral $ABCD$ is $\frac{1}{2}|\mathbf{AC}\times\mathbf{BD}|$, where AC and BD are its diagonals.

(iv) The area of a triangle with adjacent sides $\mathbf{a}$ and $\mathbf{b}$ is $\frac{1}{2}|\mathbf{a}\times\mathbf{b}|$.

(v) The area of a ΔABC is $\frac{1}{2}|\mathbf{AB}\times\mathbf{AC}|$

or $\frac{1}{2}|\mathbf{BC}\times\mathbf{BA}|$

or $\frac{1}{2}|\mathbf{CB}\times\mathbf{CA}|$

(vi) If $\mathbf{a}$, $\mathbf{b}$ and $\mathbf{c}$ are position vectors of a ΔABC, then its area $= \frac{1}{2}|(\mathbf{a}\times\mathbf{b})+(\mathbf{b}\times\mathbf{c})+(\mathbf{c}\times\mathbf{a})|$

Remark

Three points with position vectors $\mathbf{a}$, $\mathbf{b}$ and $\mathbf{c}$ are collinear, if $(\mathbf{a}\times\mathbf{b})+(\mathbf{b}\times\mathbf{c})+(\mathbf{c}\times\mathbf{a})=0$

Example 33. If $\mathbf{a},\mathbf{b}$ and $\mathbf{c}$ are position vectors of the vertices A, B and C of ΔABC, show that the area of ΔABC is $\frac{1}{2}|\mathbf{a}\times\mathbf{b}+\mathbf{b}\times\mathbf{c}+\mathbf{c}\times\mathbf{a}|$.

Deduce the condition for points $\mathbf{a},\mathbf{b}$ and $\mathbf{c}$ to be collinear.

Sol. Area of $\Delta ABC = \frac{1}{2}|\mathbf{AB}\times\mathbf{AC}|$

Now, $\mathbf{AB}$ = Position vector of B – Position vector of A

$\mathbf{AB} = \mathbf{b}-\mathbf{a}$

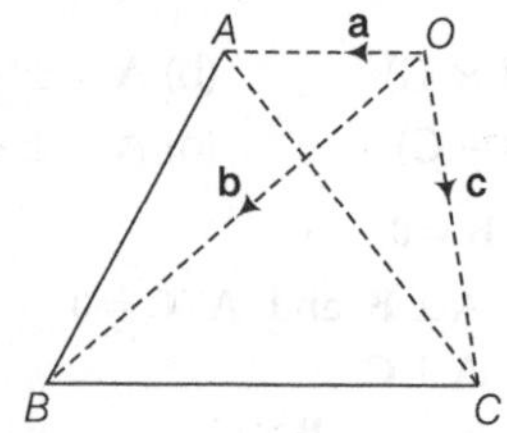

$\mathbf{AC}$ = Position vector of C – Position vector of A

$\mathbf{AC} = \mathbf{c}-\mathbf{a}$

$\mathbf{AB}\times\mathbf{AC} = (\mathbf{b}-\mathbf{a})\times(\mathbf{c}-\mathbf{a})$

$= \mathbf{b}\times\mathbf{c}-\mathbf{b}\times\mathbf{a}-\mathbf{a}\times\mathbf{c}+\mathbf{a}\times\mathbf{a}$ $(\because \mathbf{a}\times\mathbf{a}=0)$

$= \mathbf{a}\times\mathbf{b}+\mathbf{b}\times\mathbf{c}+\mathbf{c}\times\mathbf{a}$

Hence, area of $\Delta ABC = \frac{1}{2}|\mathbf{AB}\times\mathbf{AC}|$

$= \frac{1}{2}|\mathbf{a}\times\mathbf{b}+\mathbf{b}\times\mathbf{c}+\mathbf{c}\times\mathbf{a}| = 0$

If the points A, B and C are collinear, then area of $\Delta ABC = 0$

$\Rightarrow$ $\frac{1}{2}|\mathbf{a}\times\mathbf{b}+\mathbf{b}\times\mathbf{c}+\mathbf{c}\times\mathbf{a}| = 0$

$\Rightarrow$ $|\mathbf{a}\times\mathbf{b}+\mathbf{b}\times\mathbf{c}+\mathbf{c}\times\mathbf{a}| = 0$

$\Rightarrow$ $\mathbf{a}\times\mathbf{b}+\mathbf{b}\times\mathbf{c}+\mathbf{c}\times\mathbf{a} = 0$

Thus, $\mathbf{a}\times\mathbf{b}+\mathbf{b}\times\mathbf{c}+\mathbf{c}\times\mathbf{a} = 0$

is the required condition of collinearity of three points $\mathbf{a},\mathbf{b}$ and $\mathbf{c}$.

Example 34. Show that the perpendicular distance of the point $\mathbf{c}$ from to the joining $\mathbf{a}$ and $\mathbf{b}$ is

$$\frac{|\mathbf{b}\times\mathbf{c}+\mathbf{c}\times\mathbf{a}+\mathbf{a}\times\mathbf{b}|}{|\mathbf{b}-\mathbf{a}|}$$

Sol. Let ABC be a triangle and let $\mathbf{a},\mathbf{b}$ and $\mathbf{c}$ be the position of its vertices A, B and C respectively. Let CM be the perpendicular from C on AB.

Then, area of $\Delta ABC = \frac{1}{2}(AB)\cdot(CM) = \frac{1}{2}|\mathbf{AB}||\mathbf{CM}|$

Also, area of $\Delta ABC = \frac{1}{2}|\mathbf{a}\times\mathbf{b}+\mathbf{b}\times\mathbf{c}+\mathbf{c}\times\mathbf{a}|$

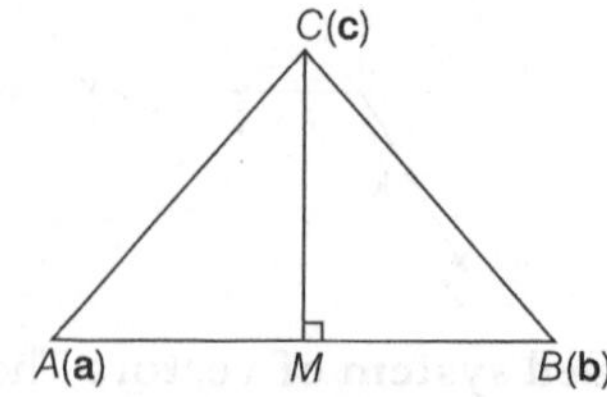

$\therefore$ $\frac{1}{2}|\mathbf{AB}||\mathbf{CM}| = \frac{1}{2}|\mathbf{a}\times\mathbf{b}+\mathbf{b}\times\mathbf{c}+\mathbf{c}\times\mathbf{a}|$

$\Rightarrow$ $\mathbf{CM} = \dfrac{|\mathbf{b}\times\mathbf{c}+\mathbf{c}\times\mathbf{a}+\mathbf{a}\times\mathbf{b}|}{|\mathbf{b}-\mathbf{a}|}$

Example 35.

(i) Find the area of the quadrilateral whose diagonals are given by

$$3\hat{\mathbf{i}}+\hat{\mathbf{j}}-2\hat{\mathbf{k}}, \hat{\mathbf{i}}-3\hat{\mathbf{j}}+4\hat{\mathbf{k}}$$

(ii) $A_1, A_2, \ldots, A_n$ are the vertices of a regular plane polygon with n sides. O is the centre. Show that

$$\sum_{i=1}^{n-1} (\mathbf{OA}_i \times \mathbf{OA}_{i+1}) = (1-n)(\mathbf{OA}_2 \times \mathbf{OA}_1)$$

Sol. (i) Area of the quadrilateral $= \frac{1}{2} | d_1 \times d_2 |$

$$= \frac{1}{2} \begin{vmatrix} \hat{\mathbf{i}} & \hat{\mathbf{j}} & \hat{\mathbf{k}} \\ 3 & 1 & -2 \\ 1 & -3 & 4 \end{vmatrix} = \frac{1}{2} | -2\hat{\mathbf{i}} - 14\hat{\mathbf{j}} - 10\hat{\mathbf{k}} |$$

$$= \frac{1}{2}\sqrt{4+196+100} = \frac{10\sqrt{3}}{2} = 5\sqrt{3}$$

(ii) $A_1, A_2, \ldots, A_n$ are the vertices of a regular plane polygon of n sides and centre O.

Let $| \mathbf{OA}_i | = k, \forall\, i = 1 < 2, 3, \ldots, n$

Let $\hat{\mathbf{e}}_i$ be the unit vector along $\mathbf{OA}_i$

$$\mathbf{OA}_i = k\hat{\mathbf{e}}_i$$

$$\mathbf{OA}_i \times \mathbf{OA}_{i+1} = k\hat{\mathbf{e}}_i \times k\hat{\mathbf{e}}_{i+1} = k^2 \hat{\mathbf{x}}_i$$

where $\hat{\mathbf{x}}_i$ is a unit vector in the direction perpendicular to the plane of the polygon and $\hat{\mathbf{x}}_i = \hat{\mathbf{x}}_{i+1}$ for $i = 1, 2, 3, \ldots, n-1$

$$\therefore \quad \text{LHS} = \sum_{i=1}^{n-1} (\mathbf{OA}_i \times \mathbf{OA}_{i+1}) = k^2 \sum_{r=1}^{n-1} \hat{\mathbf{x}}_i$$

$$= k^2 (n-1)\hat{\mathbf{x}}_i = (n-1)(\mathbf{OA}_i \times \mathbf{OA}_2)$$

$$= (1-n)(\mathbf{OA}_2 \times \mathbf{OA}_1)$$

Moment of a Force and Couple

Moment of a Force

(i) **About a point** Let a force $\mathbf{F}$ be applied at a point P of a rigid body then, the moment of $\mathbf{F}$ about a point O measures the tendency of $\mathbf{F}$ to turn the body about point O. If this tendency of rotation about O is in anti-clockwise direction, the moment is positive, otherwise it is negative.

Let $\mathbf{r}$ be the position vector of P relative to O. Then, the moment or torque of $\mathbf{F}$ about the point O is defined as the vector $\mathbf{M} = \mathbf{r} \times \mathbf{F}$.

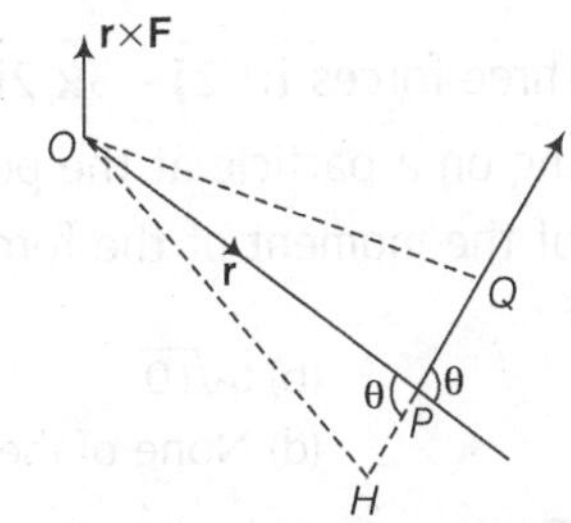

If several forces are acting through the same point P then the vector sum of the moment of the separate forces about O is equal to the moment of their resultant force about O.

Remark

Moment of a force $\mathbf{F}$ about a point $\mathbf{A} = \mathbf{AB} \times \mathbf{F}$, where $\mathbf{B}$ is any point on $\mathbf{F}$.

(ii) **About a line** The moment of a force $\mathbf{F}$ acting at a point P about a line L is a scalar given by $(\mathbf{r} \times \mathbf{F}) \cdot \hat{\mathbf{a}}$, where $\mathbf{a}$ is a unit vector in the direction of the line and $\mathbf{OP} = \mathbf{r}$, where O is any point on the line.

Thus, the moment of a force $\mathbf{F}$ about a line is the resolved part (component) along this line, of the moment of $\mathbf{F}$ about any point on the line.

Remark

The moment of a force about a point is a vector while the moment about a straight line is a scalar quantity.

Example 36. Find the moment about $(1, -1, -1)$ of the force $3\hat{\mathbf{i}} + 4\hat{\mathbf{j}} - 5\hat{\mathbf{k}}$ acting at $(1, 0, -2)$.

Sol. Let $A \equiv (1, -1, -1), B \equiv (1, 0, -2)$

and $\mathbf{F} = 3\hat{\mathbf{i}} + 4\hat{\mathbf{j}} - 5\hat{\mathbf{k}}$

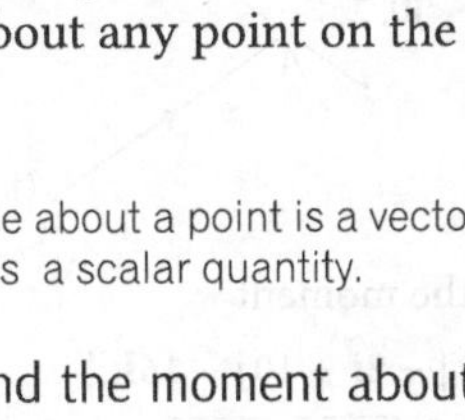

Then, moment of force $\mathbf{F}$ about A is given by $\mathbf{AB} \times \mathbf{F}$.

Here, $\quad \mathbf{AB} = (\hat{\mathbf{i}} - 2\hat{\mathbf{k}}) - (\hat{\mathbf{i}} - \hat{\mathbf{j}} - \hat{\mathbf{k}}) = \hat{\mathbf{j}} - \hat{\mathbf{k}}$

$$\therefore \quad \mathbf{AB} \times \mathbf{F} = \begin{vmatrix} \hat{\mathbf{i}} & \hat{\mathbf{j}} & \hat{\mathbf{k}} \\ 0 & 1 & -1 \\ 3 & 4 & -5 \end{vmatrix}$$

$$= \hat{\mathbf{i}}(-5+4) - \hat{\mathbf{j}}(0+3) + \hat{\mathbf{k}}(0-3)$$

$$= -\hat{\mathbf{i}} - 3\hat{\mathbf{j}} - 3\hat{\mathbf{k}}$$

Example 37. Three forces $\hat{\mathbf{i}}+2\hat{\mathbf{j}}-3\hat{\mathbf{k}}, 2\hat{\mathbf{i}}+3\hat{\mathbf{j}}+4\hat{\mathbf{k}}$ and $\hat{\mathbf{i}}-\hat{\mathbf{j}}+\hat{\mathbf{k}}$ acting on a particle at the point (0, 1, 2). The magnitude of the moment of the forces about the point (1, – 2, 0) is

(a) $2\sqrt{35}$ (b) $6\sqrt{10}$

(c) $4\sqrt{7}$ (d) None of these

Sol. (b) Total force $\mathbf{F} = (\hat{\mathbf{i}}+2\hat{\mathbf{j}}-3\hat{\mathbf{k}}) + (2\hat{\mathbf{i}}+3\hat{\mathbf{j}}+4\hat{\mathbf{k}}) + (\hat{\mathbf{i}}-\hat{\mathbf{j}}+\hat{\mathbf{k}}) = 4\hat{\mathbf{i}}+4\hat{\mathbf{j}}+2\hat{\mathbf{k}}$

Moment of the forces about

$$P = \mathbf{r}\times\mathbf{F} = \mathbf{PA}\times\mathbf{F}$$

$$\mathbf{PA} = (0-1)\,\hat{\mathbf{i}} + (1+2)\,\hat{\mathbf{j}} + (2-0)\,\hat{\mathbf{k}} = -\hat{\mathbf{i}}+3\hat{\mathbf{j}}+2\hat{\mathbf{k}}$$

$\therefore$ Moment about $P = (-\hat{\mathbf{i}}+3\hat{\mathbf{j}}+2\hat{\mathbf{k}})\times(4\hat{\mathbf{i}}+4\hat{\mathbf{j}}+2\hat{\mathbf{k}})$

$$= \begin{vmatrix} \hat{\mathbf{i}} & \hat{\mathbf{j}} & \hat{\mathbf{k}} \\ -1 & 3 & 2 \\ 4 & 4 & 2 \end{vmatrix} = -2\hat{\mathbf{i}}+10\hat{\mathbf{j}}-16\hat{\mathbf{k}}$$

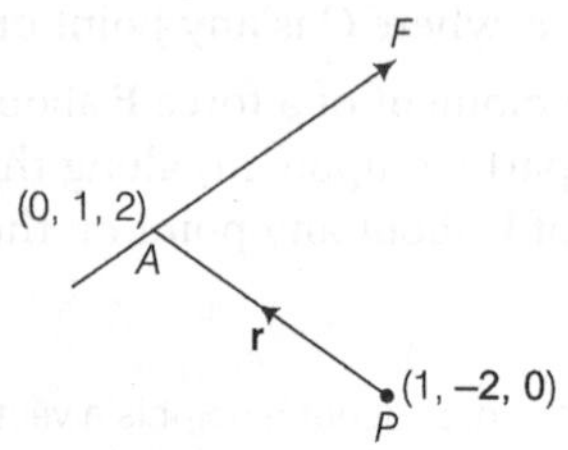

Magnitude of the moment

$$= |-2\hat{\mathbf{i}}+10\hat{\mathbf{j}}-16\hat{\mathbf{k}}|$$

$$= 2\sqrt{1^2+5^2+8^2} = 2\sqrt{90} = 6\sqrt{10}$$

Example 38. Find the moment about a line through the origin having the direction of $2\hat{\mathbf{i}}-2\hat{\mathbf{j}}+\hat{\mathbf{k}}$ due to 30 kg force acting at a point (– 4, 2, 5) in the direction of $12\hat{\mathbf{i}}-4\hat{\mathbf{j}}-3\hat{\mathbf{k}}$.

Sol. Let **F** be the force. Then,

$$\mathbf{F} = \frac{30\,(12\hat{\mathbf{i}}-4\hat{\mathbf{j}}-3\hat{\mathbf{k}})}{\sqrt{144+16+9}} = \frac{30}{13}(12\hat{\mathbf{i}}-4\hat{\mathbf{j}}-3\hat{\mathbf{k}})$$

Suppose the force **F** acts at point $P(-4, 2, 5)$ the moment of **F** acting at P about a line in the direction $2\hat{\mathbf{i}}-2\hat{\mathbf{j}}+\hat{\mathbf{k}}$ is equal to the resolve part along the line of moment of **F** about a point on the line.

$$\therefore \quad \mathbf{r} = \mathbf{OP} = (-4\hat{\mathbf{i}}+2\hat{\mathbf{j}}+5\hat{\mathbf{k}}) - (0) = -4\hat{\mathbf{i}}+2\hat{\mathbf{j}}+5\hat{\mathbf{k}}$$

Let **M** be the moment **F** about O. Then,

$$\mathbf{M} = \mathbf{r}\times\mathbf{F} = \frac{30}{13}\begin{vmatrix} \hat{\mathbf{i}} & \hat{\mathbf{j}} & \hat{\mathbf{k}} \\ -4 & 2 & 5 \\ 12 & -4 & -3 \end{vmatrix}$$

$$= \frac{30}{13}(4\hat{\mathbf{i}}+48\hat{\mathbf{j}}-8\hat{\mathbf{k}})$$

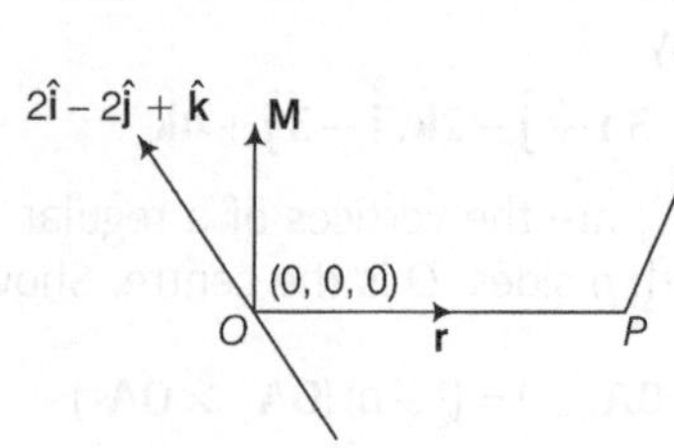

Let **a** be unit vector in the direction of $2\hat{\mathbf{i}}-2\hat{\mathbf{j}}+\hat{\mathbf{k}}$. Then,

$$\mathbf{a} = \frac{2\hat{\mathbf{i}}-2\hat{\mathbf{j}}+\hat{\mathbf{k}}}{\sqrt{4+4+1}} = \frac{1}{3}(2\hat{\mathbf{i}}-2\hat{\mathbf{j}}+\hat{\mathbf{k}})$$

Thus, the moment of **F** about the given line

$$= \mathbf{M}\cdot\mathbf{a} = \frac{30}{13}(14\hat{\mathbf{i}}+48\hat{\mathbf{j}}-8\hat{\mathbf{k}}) \cdot \frac{1}{3}(2\hat{\mathbf{i}}-2\hat{\mathbf{j}}+\hat{\mathbf{k}}) = -\frac{760}{13}$$

Moment of a Couple

A system consisting of a pair of equal unlike parallel forces is called a couple. The vector sum to two forces of a couple is always zero vector.

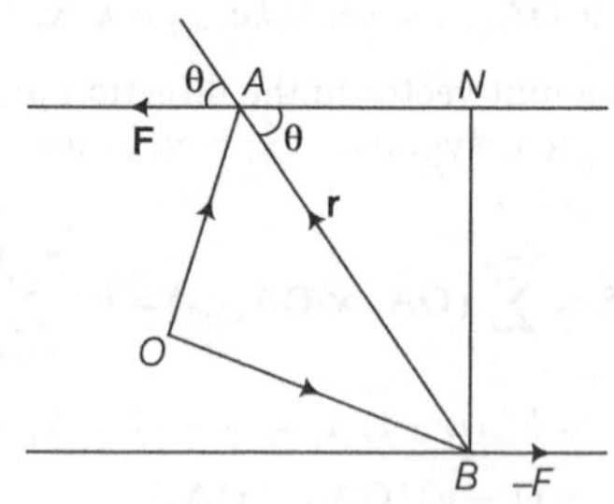

The moment of a couple is a vector perpendicular to the plane of couple and its magnitude is the product of the magnitude of either force with perpendicular distance between the lines of the forces.

$$\mathbf{M} = \mathbf{r}\times\mathbf{F}, \text{ where } \mathbf{r} = \mathbf{BA}$$

$$|\mathbf{M}| = |\mathbf{BA}\times\mathbf{F}| = |\mathbf{F}||\mathbf{BA}|\sin\theta$$

where θ is the angle between **BA** and **F**

$$= |\mathbf{F}|(BN) = |\mathbf{F}|\alpha$$

where, $\alpha = BN$ is the arm of the couple and + *ve* or – *ve* sign is to be taken accordingly as the forces indicate a counter clockwise rotation or clockwise rotation.

Example 39. The moment of the couple formed by the forces $5\hat{\mathbf{i}}+\hat{\mathbf{k}}$ and $-5\hat{\mathbf{i}}-\hat{\mathbf{k}}$ acting at the points (9, – 1, 2) and (3, – 2, 1) respectively, is

(a) $-\hat{\mathbf{i}}+\hat{\mathbf{j}}+5\hat{\mathbf{k}}$ (b) $\hat{\mathbf{i}}-\hat{\mathbf{j}}-5\hat{\mathbf{k}}$

(c) $2\hat{\mathbf{i}}-2\hat{\mathbf{j}}-10\hat{\mathbf{k}}$ (d) $-2\hat{\mathbf{i}}+2\hat{\mathbf{j}}+10\hat{\mathbf{k}}$

Sol. (b) Moment of the couple,

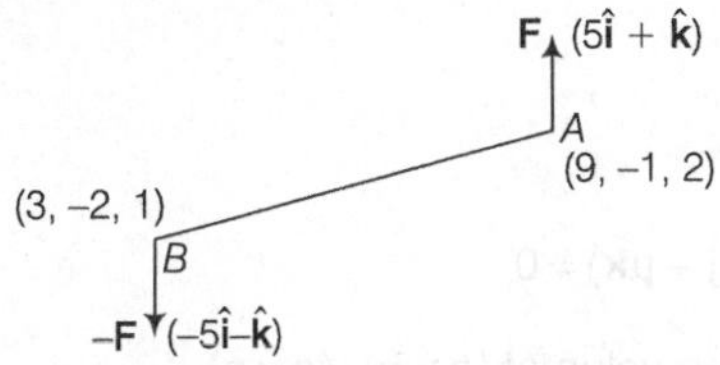

$$= \mathbf{BA} \times \mathbf{F} = \{(9-3)\,\hat{\mathbf{i}} + (-1+2)\,\hat{\mathbf{j}} + (2-1)\,\hat{\mathbf{k}}\} \times (5\hat{\mathbf{i}} + \hat{\mathbf{k}})$$

$$= (6\hat{\mathbf{i}} + \hat{\mathbf{j}} + \hat{\mathbf{k}}) \times (5\hat{\mathbf{i}} + \hat{\mathbf{k}}) = \begin{vmatrix} \hat{\mathbf{i}} & \hat{\mathbf{j}} & \hat{\mathbf{k}} \\ 6 & 1 & 1 \\ 5 & 0 & 1 \end{vmatrix}$$

$$= \hat{\mathbf{i}} - \hat{\mathbf{j}} - 5\hat{\mathbf{k}}$$

Rotation About an Axis

When a rigid body rotates about a fixed axis ON with an angular velocity ω, then velocity v of a particle P is given by

$$v = \omega \times \mathbf{r},$$

were, $\mathbf{r} = \mathbf{OP}$

and $\omega = |\omega|$ (unit vector along ON)

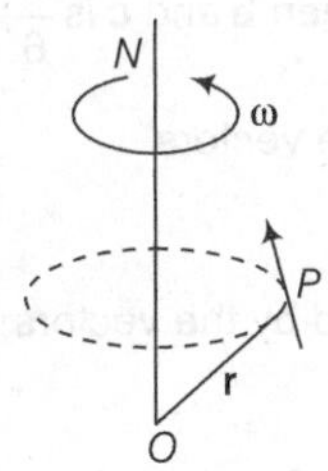

Example 40. A particle has an angular speed of 3 rad/s and the axis of rotation passes through the points (1, 1, 2) and (1, 2, –2). Find the velocity of the particle at point P (3, 6, 4).

Sol. Clearly, $\mathbf{OA} = \hat{\mathbf{i}} + \hat{\mathbf{j}} + 2\hat{\mathbf{k}}$

$$\mathbf{OB} = \hat{\mathbf{i}} + 2\hat{\mathbf{j}} - 2\hat{\mathbf{k}}$$

$$\mathbf{AB} = \hat{\mathbf{j}} - 4\hat{\mathbf{k}} = |\mathbf{AB}| = \sqrt{17}$$

and $\mathbf{AP} = (3\hat{\mathbf{i}} + 6\hat{\mathbf{j}} + 4\hat{\mathbf{k}}) - (\hat{\mathbf{i}} + \hat{\mathbf{j}} + 2\hat{\mathbf{k}})$

$$= 2\hat{\mathbf{i}} + 5\hat{\mathbf{j}} + 2\hat{\mathbf{k}}$$

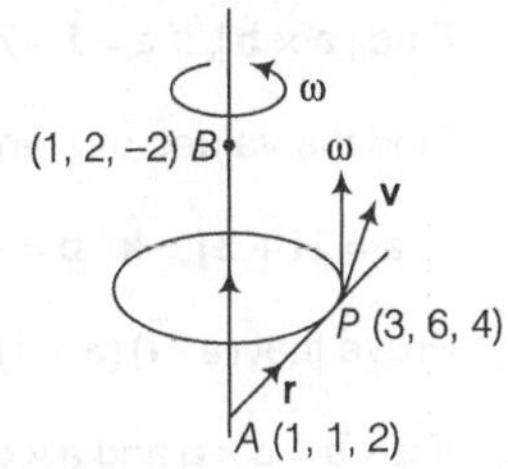

$\therefore$ $$\omega = \frac{3}{\sqrt{17}}(\hat{\mathbf{j}} - 4\hat{\mathbf{k}})2 \text{ and } \mathbf{r}$$

Now, $$\mathbf{v} = \omega \times \mathbf{r} = \frac{3}{\sqrt{17}}(\hat{\mathbf{j}} - 4\hat{\mathbf{k}}) \times (2\hat{\mathbf{i}} + 5\hat{\mathbf{j}} + 2\hat{\mathbf{k}})$$

$$= \frac{3}{\sqrt{17}}(22\hat{\mathbf{i}} - 8\hat{\mathbf{j}} - 2\hat{\mathbf{k}})$$

Example 41. A rigid body is spinning about a fixed point (3, –2, –1) with an angular velocity of 4 rad/s, the axis of rotation being in the direction of (1, 2, – 2). Find the velocity of the particle at the point (4, 1, 1).

Sol. $$\omega = 4\left(\frac{\hat{\mathbf{i}} + 2\hat{\mathbf{j}} - 2\hat{\mathbf{k}}}{\sqrt{1+4+4}}\right) = \frac{4}{3}(\hat{\mathbf{i}} + 2\hat{\mathbf{j}} - 2\hat{\mathbf{k}})$$

$$\mathbf{r} = \mathbf{OP} - \mathbf{OA}$$

$$= (4\hat{\mathbf{i}} + \hat{\mathbf{j}} + \hat{\mathbf{k}}) - (3\hat{\mathbf{i}} - 2\hat{\mathbf{j}} - \hat{\mathbf{k}})$$

$$= \hat{\mathbf{i}} + 3\hat{\mathbf{i}} + 2\hat{\mathbf{k}}$$

$$\mathbf{v} = \omega \times \mathbf{r} = \frac{4}{3}(\hat{\mathbf{i}} + 2\hat{\mathbf{j}} - 2\hat{\mathbf{k}}) \times (\hat{\mathbf{i}} + 3\hat{\mathbf{j}} + 2\hat{\mathbf{k}})$$

$$\Rightarrow \quad \frac{4}{3}(10\hat{\mathbf{i}} - 4\hat{\mathbf{j}} + \hat{\mathbf{k}})$$

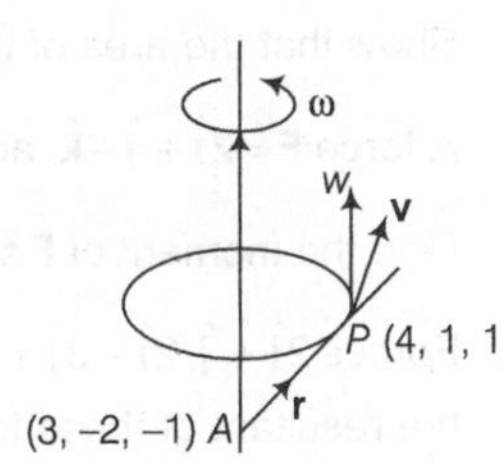

Exercise for Session 2

1. Find $|\mathbf{a}\times\mathbf{b}|$, if $\mathbf{a}=\hat{\mathbf{i}}-7\hat{\mathbf{j}}+7\hat{\mathbf{k}}$ and $\mathbf{b}=3\hat{\mathbf{i}}-2\hat{\mathbf{j}}-2\hat{\mathbf{k}}$.
2. Find the values of λ and μ for which $(2\hat{\mathbf{i}}+6\hat{\mathbf{j}}+27\hat{\mathbf{k}})\times(\hat{\mathbf{i}}+\lambda\hat{\mathbf{j}}+\mu\hat{\mathbf{k}})=0$
3. If $\mathbf{a}=2\hat{\mathbf{i}}+3\hat{\mathbf{j}}-\hat{\mathbf{k}}$, $\mathbf{b}=-\hat{\mathbf{i}}+2\hat{\mathbf{j}}-4\hat{\mathbf{k}}$, $\mathbf{c}=\hat{\mathbf{i}}+\hat{\mathbf{j}}+\hat{\mathbf{k}}$, then find the value of $(\mathbf{a}\times\mathbf{b})\cdot(\mathbf{a}\times\mathbf{c})$.
4. Prove that $(\mathbf{a}\cdot\hat{\mathbf{i}})(\mathbf{a}\times\hat{\mathbf{i}})+(\mathbf{a}\cdot\hat{\mathbf{j}})(\mathbf{a}\times\hat{\mathbf{j}})+(\mathbf{a}\times\hat{\mathbf{k}})(\mathbf{a}\times\hat{\mathbf{k}})=0$.
5. If $\mathbf{a}\times\mathbf{b}=\mathbf{c}\times\mathbf{d}$ and $\mathbf{a}\times\mathbf{c}=\mathbf{b}\times\mathbf{d}$, then show that $\mathbf{a}-\mathbf{d}$ is parallel to $\mathbf{b}-\mathbf{c}$.
6. If $(\mathbf{a}\times\mathbf{b})^2+(\mathbf{a}\cdot\mathbf{b})^2=144$ and $|\mathbf{a}|=4$, then find the value of $|\mathbf{b}|$.
7. If $|\mathbf{a}|=2$, $|\mathbf{b}|=7$ and $(\mathbf{a}\times\mathbf{b})=3\hat{\mathbf{i}}+2\hat{\mathbf{j}}+6\hat{\mathbf{k}}$, find the angle between $\mathbf{a}$ and $\mathbf{b}$.
8. Let the vectors $\mathbf{a}$ and $\mathbf{b}$ be such that $|\mathbf{a}|=3$, $|\mathbf{b}|=\frac{\sqrt{2}}{3}$ and $\mathbf{a}\times\mathbf{b}$ is a unit vector, then find the angle between $\mathbf{a}$ and $\mathbf{b}$.
9. If $|\mathbf{a}|=\sqrt{26}$, $|\mathbf{b}|=7$, and $|\mathbf{a}\times\mathbf{b}|=35$, find $\mathbf{a}\cdot\mathbf{b}$.
10. Find a unit vector perpendicular to the plane of two vectors $\mathbf{a}=\hat{\mathbf{i}}-\hat{\mathbf{j}}+2\hat{\mathbf{k}}$ and $\mathbf{b}=2\hat{\mathbf{i}}+3\hat{\mathbf{j}}-\hat{\mathbf{k}}$.
11. Find a vector of magnitude 15, which is perpendicular to both the vectors $4\hat{\mathbf{i}}-\hat{\mathbf{j}}+8\hat{\mathbf{k}}$ and $-\hat{\mathbf{j}}+\hat{\mathbf{k}}$.
12. Let $\mathbf{a}=\hat{\mathbf{i}}+4\hat{\mathbf{j}}+2\hat{\mathbf{k}}$, $\mathbf{b}=3\hat{\mathbf{i}}-2\hat{\mathbf{j}}+7\hat{\mathbf{k}}$ and $\mathbf{c}=2\hat{\mathbf{i}}-\hat{\mathbf{j}}+4\hat{\mathbf{k}}$.
    Find a vector $\mathbf{d}$ which is perpendicular to both $\mathbf{a}$ and $\mathbf{b}$ and $\mathbf{c}\cdot\mathbf{d}=15$.
13. Let $\mathbf{a}$, $\mathbf{b}$ and $\mathbf{c}$ be unit vectors such that $\mathbf{a}\cdot\mathbf{b}=0=\mathbf{a}\cdot\mathbf{c}$. If the angle between $\mathbf{b}$ and $\mathbf{c}$ is $\frac{\pi}{6}$, then find $\mathbf{a}$.
14. Find the area of the triangle whose adjacent sides are determined by the vectors
    $\mathbf{a}=-2\hat{\mathbf{i}}-5\hat{\mathbf{k}}$ and $\mathbf{b}=\hat{\mathbf{i}}-2\hat{\mathbf{j}}-\hat{\mathbf{k}}$.
15. Find the area of the parallelogram whose adjacent sides are represented by the vectors
    $3\hat{\mathbf{i}}+\hat{\mathbf{j}}-2\hat{\mathbf{k}}$ and $\hat{\mathbf{i}}-3\hat{\mathbf{j}}+4\hat{\mathbf{k}}$.
16. Show that the area of the parallelogram having diagonals $3\hat{\mathbf{i}}+\hat{\mathbf{j}}-2\hat{\mathbf{k}}$ and $\hat{\mathbf{i}}-3\hat{\mathbf{j}}+4\hat{\mathbf{k}}$ is $5\sqrt{3}$.
17. A force $\mathbf{F}=2\hat{\mathbf{i}}+\hat{\mathbf{j}}-\hat{\mathbf{k}}$ acts at point A whose position vector is $2\hat{\mathbf{i}}-\hat{\mathbf{j}}$. Find the moment of force $\mathbf{F}$ about the origin.
18. Find the moment of $\mathbf{F}$ about point $(2,-1,3)$, when force $\mathbf{F}=3\hat{\mathbf{i}}+2\hat{\mathbf{j}}-4\hat{\mathbf{k}}$ is acting on point $(1,-1,2)$.
19. Forces $2\hat{\mathbf{i}}+\hat{\mathbf{j}}$, $2\hat{\mathbf{i}}-3\hat{\mathbf{j}}+6\hat{\mathbf{k}}$ and $-\hat{\mathbf{i}}+2\hat{\mathbf{j}}-\hat{\mathbf{k}}$ act at a point P, with position vector $4\hat{\mathbf{i}}-3\hat{\mathbf{j}}-\hat{\mathbf{k}}$. Find the moment of the resultant of these force about the point Q whose position vector is $6\hat{\mathbf{i}}+\hat{\mathbf{j}}-3\hat{\mathbf{k}}$.

Session 3

Scalar Triple Product

Scalar Triple Product

The scalar triple product is defined for three vectors and it is defined as the dot product of one of the vectors with the cross product of the other two.

If $\mathbf{a}, \mathbf{b}, \mathbf{c}$ are any three vector, then their scalar product is defined as $(\mathbf{a} \times \mathbf{b}) \cdot \mathbf{c}$.

We denote it by $[\mathbf{a}, \mathbf{b}, \mathbf{c}]$.

It is also called the mixed or box product.

Remark

Result of scalar triple product is always a scalar.

Geometrical Interpretation of Scalar Triple Product

Let $\mathbf{a}$, $\mathbf{b}$ and $\mathbf{c}$ be three vectors. Consider a parallelopiped having coterminous edges **OA**, **OB** and **OC** such that $\mathbf{OA} = a$, $\mathbf{OB} = b$ and $\mathbf{OC} = c$. Then, $\mathbf{a} \times \mathbf{b}$ is a vector perpendicular to the plane of $\mathbf{a}$ and $\mathbf{b}$. Let ϕ be the angle between $\mathbf{c}$ and $\mathbf{a} \times \mathbf{b}$.

If $\mathbf{n}$ is a unit vector along $\mathbf{a} \times \mathbf{b}$, then ϕ is the angle between $\mathbf{n}$ and $\mathbf{c}$.

Now, $[\mathbf{a}\ \mathbf{b}\ \mathbf{c}] = (\mathbf{a} \times \mathbf{b}) \cdot \mathbf{c}$

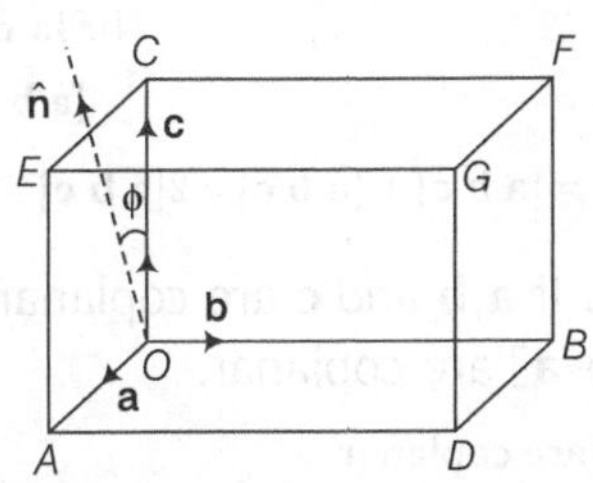

$= $ (Area of parallelogram $OADB$) $\mathbf{n} \cdot \mathbf{c}$

$= $ (Area of parallelogram $OADB$) $(\mathbf{n} \cdot \mathbf{c})$

$= $ (Area of parallelogram $OADB$) $(|\mathbf{c}||\mathbf{n}|\cos\phi)$

$= $ (Area of parallelogram $OADB$) $(|\mathbf{c}|\cos\phi)$

$= $ (Area of parallelogram $OADB$) (OL)

$= $ Area of base $\times$ height

$= $ Volume of parallelopiped

Height of parallelopiped

$$= \frac{\text{Volume of parallelopiped}}{\text{Area of base}}$$

Properties of Scalar Triple Product

(i) If $\mathbf{a}$, $\mathbf{b}$ and $\mathbf{c}$ are given by

$$\mathbf{a} = a_1\hat{\mathbf{i}} + a_2\hat{\mathbf{j}} + a_3\hat{\mathbf{k}}$$
$$\mathbf{b} = b_1\hat{\mathbf{i}} + b_2\hat{\mathbf{j}} + b_3\hat{\mathbf{k}}$$
$$\mathbf{c} = c_1\hat{\mathbf{i}} + c_2\hat{\mathbf{j}} + c_3\hat{\mathbf{k}}$$

Then, $(\mathbf{a} \times \mathbf{b}) \cdot \mathbf{c} = \begin{vmatrix} a_1 & a_2 & a_3 \\ b_1 & b_2 & b_3 \\ c_1 & c_2 & c_3 \end{vmatrix}$

(ii) $(\mathbf{a} \times \mathbf{b}) \cdot \mathbf{c} = \mathbf{a} \cdot (\mathbf{b} \times \mathbf{c})$ i.e. position of dot and cross can be interchanged without altering product. Hence, it is also represented by $[\mathbf{a}\ \mathbf{b}\ \mathbf{c}]$.

(iii) $[\mathbf{a}\ \mathbf{b}\ \mathbf{c}] = [\mathbf{b}\ \mathbf{c}\ \mathbf{a}] = [\mathbf{c}\ \mathbf{a}\ \mathbf{b}]$

(iv) $[\mathbf{a}\ \mathbf{b}\ \mathbf{c}] = -[\mathbf{b}\ \mathbf{a}\ \mathbf{c}]$

(v) $[k\mathbf{a}\ \mathbf{b}\ \mathbf{c}] = k[\mathbf{a}\ \mathbf{b}\ \mathbf{c}]$, $[k_1\mathbf{a}\ k_2\mathbf{b}\ k_3\mathbf{c}] = k_1k_2k_3\ [\mathbf{a}\ \mathbf{b}\ \mathbf{c}]$

(vi) $[\mathbf{a} + \mathbf{b}\ \mathbf{c}\ \mathbf{d}] = [\mathbf{a}\ \mathbf{c}\ \mathbf{d}] + [\mathbf{b}\ \mathbf{c}\ \mathbf{d}]$

(vii) $\mathbf{a}$, $\mathbf{b}$ and $\mathbf{c}$ in that order form a right handed system, if $[\mathbf{a}\ \mathbf{b}\ \mathbf{c}] > 0$.

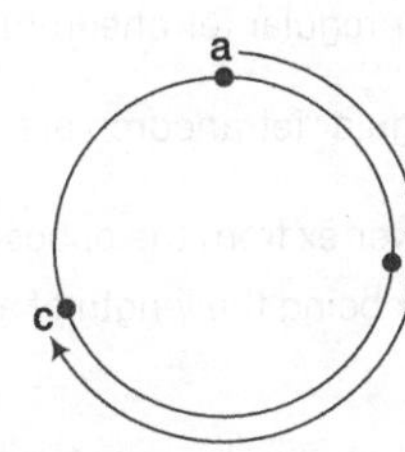

(viii) The necessary and sufficient condition for three non-zero, non-collinear vector $\mathbf{a}$, $\mathbf{b}$ and $\mathbf{c}$ to be coplanar is that $[\mathbf{a}\ \mathbf{b}\ \mathbf{c}] = 0$ i.e., $\mathbf{a}$, $\mathbf{b}$ and $\mathbf{c}$ are coplanar $\Leftrightarrow [\mathbf{a}\ \mathbf{b}\ \mathbf{c}] = 0$.

(ix) $[x_1\mathbf{a} + y_1\mathbf{b} + z_1\mathbf{c},\ x_2\mathbf{a} + y_2\mathbf{b} + z_2\mathbf{c},\ x_3\mathbf{a} + y_3\mathbf{b} + z_3\mathbf{c}]$

$$= \begin{vmatrix} x_1 & y_1 & z_1 \\ x_2 & y_2 & z_2 \\ x_3 & y_3 & z_3 \end{vmatrix} [\mathbf{a}\,\mathbf{b}\,\mathbf{c}]$$

Remarks

1. Four points **A**, **B**, **C**, **D** are coplanar if [**AB**, **AC**, **AD**] = 0
2. Four points **a**, **b**, **c** and **d** are coplanar, if
   [**d b c**] + [**d c a**] + [**d a b**] = [**a b c**]
   or [**a b c**] + [**a c d**] + [**a d b**] = [**d b c**]
3. [**a a b**] = [**b b a**] = [**c c b**] = 0
   i.e., if any two vectors are same, then vectors are coplanar.

Volume of Tetrahedron (A pyramid having a triangular base)

If $OABC$ is a tetrahedron as shown in figure, where $\mathbf{OA} = \mathbf{a}$, $\mathbf{OB} = \mathbf{b}$, and $\mathbf{OC} = \mathbf{c}$, then volume of tetrahedron $= \frac{1}{6}[\mathbf{a\ b\ c}]$

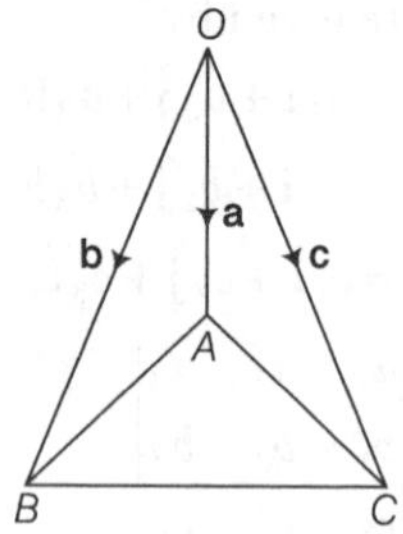

Remarks

1. The six mid-points of the six edges of a tetrahedron lie in a sphere, if the pair of opposite edges are perpendicular to each other.
2. Centre of the sphere is the centroid of the tetrahedron.
3. $GA^2 + GB^2 + GC^2 + GO^2 = 12r^2$, G being the centroid.
4. The angle between any two plane faces of a regular tetrahedron is $\cos^{-1}\frac{1}{3}$.
5. Angle between the any edge and a face not containing the angle is $\cos^{-1}\sqrt{\frac{1}{3}}$ (for regular tetrahedron).
6. Any two edges of regular tetrahedron are perpendicular to each other.
7. The distance of any vertex from the opposite face of regular tetrahedron is $\sqrt{\frac{2}{3}}k$, k being the length of any edge.

Example 42. Find the volume of the parallelopiped whose edges are represented by $\mathbf{a} = 2\hat{\mathbf{i}} - 3\hat{\mathbf{j}} + 4\hat{\mathbf{k}}$, $\mathbf{b} = \hat{\mathbf{i}} + 2\hat{\mathbf{j}} - \hat{\mathbf{k}}$ and $\mathbf{c} = 3\hat{\mathbf{i}} - \hat{\mathbf{j}} + 2\hat{\mathbf{k}}$.

Sol. Here, $[\mathbf{a\ b\ c}] = \begin{vmatrix} 2 & -3 & 4 \\ 1 & 2 & -1 \\ 3 & -1 & 2 \end{vmatrix} = 2(4-1) + 3(2+3) + 4(-1-6)$

$$= 6 + 15 - 28 = -7$$

$\therefore$ The volume of the parallelopiped $= |[\mathbf{a\ b\ c}]| = 7$.

Example 43. Let $\mathbf{a} = x\hat{\mathbf{i}} + 12\hat{\mathbf{j}} - \hat{\mathbf{k}}$, $\mathbf{b} = 2\hat{\mathbf{i}} + 2x\hat{\mathbf{j}} + \hat{\mathbf{k}}$ and $\mathbf{c} = \hat{\mathbf{i}} + \hat{\mathbf{k}}$. If $\mathbf{b}, \mathbf{c}, \mathbf{a}$ in that order form a left handed system, then find the value of x.

$$[x_1\mathbf{a} + y_1\mathbf{b} + z_1\mathbf{c}, x_2\mathbf{a} + y_2\mathbf{b} + z_2\mathbf{c}, x_3\mathbf{a} + y_3\mathbf{b} + z_3\mathbf{c}] = \begin{vmatrix} x_1 & y_1 & z_1 \\ x_2 & y_2 & z_2 \\ x_3 & y_3 & z_3 \end{vmatrix} [\mathbf{a\ b\ c}]$$

Sol. Since, $\mathbf{b}, \mathbf{c}, \mathbf{a}$ form a left handed system, three fore $[\mathbf{b}, \mathbf{c}, \mathbf{a}] < 0$

$$\Rightarrow \begin{vmatrix} 2 & 2x & -1 \\ 1 & 0 & 1 \\ x & 12 & -1 \end{vmatrix} < 0$$

$$\Rightarrow 2(0-12) - 2x(-1-x) + 1(12-0) < 0$$

$$\Rightarrow -24 + 2x + 2x^2 + 12 < 0$$

$$\Rightarrow 2x^2 + 2x - 12 < 0 \Rightarrow x^2 + x - 6 < 0$$

$$\Rightarrow (x-2)(x+3) < 0 \Rightarrow x \in (-3, 2)$$

Example 44. For any three vectors $\mathbf{a}, \mathbf{b}$ and $\mathbf{c}$ prove that $[\mathbf{a}+\mathbf{b}\ \mathbf{b}+\mathbf{c}\ \mathbf{c}+\mathbf{a}] = 2[\mathbf{a\ b\ c}]$

Sol. We have, $[\mathbf{a}+\mathbf{b}\ \mathbf{b}+\mathbf{c}\ \mathbf{c}+\mathbf{a}]$

$$= \{(\mathbf{a}+\mathbf{b}) \times (\mathbf{b} \times \mathbf{c})\} \cdot (\mathbf{c}+\mathbf{a})$$

$$= \{\mathbf{a} \times \mathbf{b} + \mathbf{a} \times \mathbf{c} + \mathbf{b} \times \mathbf{b} + \mathbf{b} \times \mathbf{c}\} \cdot (\mathbf{c}+\mathbf{a})$$

$$\{\because \mathbf{b} \times \mathbf{b} = 0\}$$

$$= \{\mathbf{a} \times \mathbf{b} + \mathbf{a} \times \mathbf{c} + \mathbf{b} \times \mathbf{c}\} \cdot (\mathbf{c}+\mathbf{a})$$

$$= (\mathbf{a} \times \mathbf{b}) \cdot \mathbf{c} + (\mathbf{a} \times \mathbf{c}) \cdot \mathbf{c} + (\mathbf{b} \times \mathbf{c}) \cdot \mathbf{c} + (\mathbf{a} \times \mathbf{b}) \cdot \mathbf{a} + (\mathbf{a} \times \mathbf{c}) \cdot \mathbf{a} + (\mathbf{b} \times \mathbf{c}) \cdot \mathbf{a}$$

$$= [\mathbf{a\ b\ c}] + 0 + 0 + 0 + 0 + [\mathbf{b\ c\ a}]$$

$$[\because [\mathbf{a\ c\ c}] = 0, [\mathbf{b\ c\ c}] = 0, [\mathbf{a\ b\ a}] = 0, [\mathbf{a\ c\ a}] = 0]$$

$$= [\mathbf{a\ b\ c}] + [\mathbf{a\ b\ c}] = 2[\mathbf{a\ b\ c}]$$

Example 45. If $\mathbf{a}, \mathbf{b}$ and $\mathbf{c}$ are coplanar show $[\mathbf{a}+\mathbf{b}\ \mathbf{b}+\mathbf{c}\ \mathbf{c}+\mathbf{a}]$ are coplanar.

Sol. Since, $[\mathbf{a\ b\ c}]$ are coplanar

$$[\mathbf{a\ b\ c}] = 0 \quad \text{...(i)}$$

and shown in above example

$$[\mathbf{a}+\mathbf{b}\ \mathbf{b}+\mathbf{c}\ \mathbf{c}+\mathbf{a}] = 2[\mathbf{a\ b\ c}] = 0 \quad \text{[using Eq. (i)]}$$

which shows $[\mathbf{a}+\mathbf{b}\ \mathbf{b}+\mathbf{c}\ \mathbf{c}+\mathbf{a}]$ are coplanar, if $[\mathbf{a\ b\ c}]$ are coplanar.

Example 46. For any three vectors $\mathbf{a}, \mathbf{b}$ and $\mathbf{c}$ prove that $[\mathbf{a\ b\ c}]^2 = \begin{vmatrix} \mathbf{a}\cdot\mathbf{a} & \mathbf{a}\cdot\mathbf{b} & \mathbf{a}\cdot\mathbf{c} \\ \mathbf{b}\cdot\mathbf{a} & \mathbf{b}\cdot\mathbf{b} & \mathbf{b}\cdot\mathbf{c} \\ \mathbf{c}\cdot\mathbf{a} & \mathbf{c}\cdot\mathbf{b} & \mathbf{c}\cdot\mathbf{c} \end{vmatrix}$

Sol. Let $\mathbf{a} = a_1\hat{\mathbf{i}} + a_2\hat{\mathbf{j}} + a_3\hat{\mathbf{k}}$

$\mathbf{b} = b_1\hat{\mathbf{i}} + b_2\hat{\mathbf{j}} + b_3\hat{\mathbf{k}}$

$\mathbf{c} = c_1\hat{\mathbf{i}} + c_2\hat{\mathbf{j}} + c_3\hat{\mathbf{k}}$

Then, $[\mathbf{a}\ \mathbf{b}\ \mathbf{c}][\mathbf{a}\ \mathbf{b}\ \mathbf{c}] = \begin{vmatrix} a_1 & a_2 & a_3 \\ b_1 & b_2 & b_3 \\ c_1 & c_2 & c_3 \end{vmatrix}\begin{vmatrix} a_1 & a_2 & a_3 \\ b_1 & b_2 & b_3 \\ c_1 & c_2 & c_3 \end{vmatrix}$

On multiplying row-by-row, we get

$[\mathbf{a}\ \mathbf{b}\ \mathbf{c}]^2 =$

$$\begin{vmatrix} a_1a_1 + a_2a_2 + a_3a_3 & a_1b_1 + a_2b_2 + a_3b_3 & a_1c_1 + a_2c_2 + a_3c_3 \\ b_1a_1 + b_2a_2 + b_3a_3 & b_1b_1 + b_2b_2 + b_3b_3 & b_1c_1 + b_2c_2 + b_3c_3 \\ c_1a_1 + c_2a_2 + c_3a_3 & c_1b_1 + c_2b_2 + c_3b_3 & c_1c_1 + c_2c_2 + c_3c_3 \end{vmatrix}$$

$$= \begin{vmatrix} \mathbf{a}\cdot\mathbf{a} & \mathbf{a}\cdot\mathbf{b} & \mathbf{a}\cdot\mathbf{c} \\ \mathbf{b}\cdot\mathbf{a} & \mathbf{b}\cdot\mathbf{b} & \mathbf{b}\cdot\mathbf{c} \\ \mathbf{c}\cdot\mathbf{a} & \mathbf{c}\cdot\mathbf{b} & \mathbf{c}\cdot\mathbf{c} \end{vmatrix}$$

Example 47. If $\mathbf{a}, \mathbf{b}, \mathbf{c}, \mathbf{l}$ and $\mathbf{m}$ are vectors, prove that

$$[\mathbf{a}\ \mathbf{b}\ \mathbf{c}]\ (\mathbf{l} \times \mathbf{m}) = \begin{vmatrix} \mathbf{a} & \mathbf{b} & \mathbf{c} \\ \mathbf{a}\cdot\mathbf{l} & \mathbf{b}\cdot\mathbf{l} & \mathbf{c}\cdot\mathbf{l} \\ \mathbf{a}\cdot\mathbf{m} & \mathbf{b}\cdot\mathbf{m} & \mathbf{c}\cdot\mathbf{m} \end{vmatrix}$$

Sol. Let $\mathbf{a} = a_1\hat{\mathbf{i}} + a_2\hat{\mathbf{j}} + a_3\hat{\mathbf{k}}$, $\mathbf{b} = b_1\hat{\mathbf{i}} + b_2\hat{\mathbf{j}} + b_3\hat{\mathbf{k}}$,

$\mathbf{c} = c_1\hat{\mathbf{i}} + c_2\hat{\mathbf{j}} + c_3\hat{\mathbf{k}}$, $\mathbf{l} = l_1\hat{\mathbf{i}} + l_2\hat{\mathbf{j}} + l_3\hat{\mathbf{k}}$

and $\mathbf{m} = m_1\hat{\mathbf{i}} + m_2\hat{\mathbf{j}} + m_3\hat{\mathbf{k}}$.

$$\Rightarrow [\mathbf{a}\ \mathbf{b}\ \mathbf{c}](\mathbf{l} \times \mathbf{m}) = \begin{vmatrix} a_1 & a_2 & a_3 \\ b_1 & b_2 & b_3 \\ c_1 & c_2 & c_3 \end{vmatrix}\begin{vmatrix} \hat{\mathbf{i}} & \hat{\mathbf{j}} & \hat{\mathbf{k}} \\ l_1 & l_2 & l_3 \\ m_1 & m_2 & m_3 \end{vmatrix}$$

On multiplying row-by-row, we get

$$= \begin{vmatrix} a_1\hat{\mathbf{i}} + a_2\hat{\mathbf{j}} + a_3\hat{\mathbf{k}} & a_1l_1 + a_2l_2 + a_3l_3 & a_1m_1 + a_2m_2 + a_3m_3 \\ b_1\hat{\mathbf{i}} + b_2\hat{\mathbf{j}} + b_3\hat{\mathbf{k}} & b_1l_1 + b_2l_2 + b_3l_3 & b_1m_1 + b_2m_2 + b_3m_3 \\ c_1\hat{\mathbf{i}} + c_2\hat{\mathbf{j}} + c_3\hat{\mathbf{k}} & b_1l_1 + b_2l_2 + b_3l_3 & c_1m_1 + c_2m_2 + c_3m_3 \end{vmatrix}$$

$$= \begin{vmatrix} \mathbf{a} & \mathbf{a}\cdot\mathbf{l} & \mathbf{a}\cdot\mathbf{m} \\ \mathbf{b} & \mathbf{b}\cdot\mathbf{l} & \mathbf{b}\cdot\mathbf{m} \\ \mathbf{c} & \mathbf{c}\cdot\mathbf{l} & \mathbf{c}\cdot\mathbf{m} \end{vmatrix} = \begin{vmatrix} \mathbf{a} & \mathbf{b} & \mathbf{c} \\ \mathbf{a}\cdot\mathbf{l} & \mathbf{b}\cdot\mathbf{l} & \mathbf{c}\cdot\mathbf{l} \\ \mathbf{a}\cdot\mathbf{m} & \mathbf{b}\cdot\mathbf{m} & \mathbf{c}\cdot\mathbf{m} \end{vmatrix}$$

Example 48. If $\mathbf{a}$ and $\mathbf{b}$ are non-zero and non-collinear vectors, then show that
$\mathbf{a} \times \mathbf{b} = [\mathbf{a}\ \mathbf{b}\ \mathbf{i}]\ \hat{\mathbf{i}} + [\mathbf{a}\ \mathbf{b}\ \mathbf{j}]\ \hat{\mathbf{j}} + [\mathbf{a}\ \mathbf{b}\ \mathbf{k}]\ \hat{\mathbf{k}}$

Sol. Let $\mathbf{a} \times \mathbf{b} = x\hat{\mathbf{i}} + y\hat{\mathbf{j}} + z\hat{\mathbf{k}}$ ...(i)

$(\mathbf{a} \times \mathbf{b})\cdot\hat{\mathbf{i}} = (x\hat{\mathbf{i}} + y\hat{\mathbf{j}} + z\hat{\mathbf{k}})\cdot\hat{\mathbf{i}}$

$(\mathbf{a} \times \mathbf{b})\cdot\hat{\mathbf{i}} = x$

Also, $(\mathbf{a} \times \mathbf{b})\cdot\hat{\mathbf{j}} = y$ ...(ii)

$(\mathbf{a} \times \mathbf{b})\cdot\hat{\mathbf{k}} = z$

$\mathbf{a} \times \mathbf{b} = [\mathbf{a}\ \mathbf{b}\ \mathbf{i}]\hat{\mathbf{i}} + [\mathbf{a}\ \mathbf{b}\ \mathbf{j}]\hat{\mathbf{j}} + [\mathbf{a}\ \mathbf{b}\ \mathbf{k}]\ \hat{\mathbf{k}}$

Example 49. If $\mathbf{a}, \mathbf{b}$ and $\mathbf{c}$ are any three vectors in space, then show that
$(\mathbf{c} + \mathbf{b}) \times (\mathbf{c} + \mathbf{a})\cdot(\mathbf{c} + \mathbf{b} + \mathbf{a}) = [\mathbf{a}\ \mathbf{b}\ \mathbf{c}]$

Sol. Here, $(\mathbf{c} + \mathbf{b}) \times (\mathbf{c} + \mathbf{a})\cdot(\mathbf{c} + \mathbf{b} + \mathbf{a})$

$\Rightarrow (\mathbf{c}\times\mathbf{c} + \mathbf{c}\times\mathbf{a} + \mathbf{b}\times\mathbf{c} + \mathbf{b}\times\mathbf{a})\cdot(\mathbf{c} + \mathbf{b} + \mathbf{a})$

$\Rightarrow (\mathbf{c}\times\mathbf{a} + \mathbf{b}\times\mathbf{c} + \mathbf{b}\times\mathbf{a})\cdot(\mathbf{c} + \mathbf{b} + \mathbf{a})$ $(\because \mathbf{c}\times\mathbf{c} = 0)$

$\Rightarrow (\mathbf{c}\times\mathbf{a})\cdot\mathbf{c} + (\mathbf{c}\times\mathbf{a})\cdot\mathbf{b} + (\mathbf{c}\times\mathbf{a})\cdot\mathbf{a} + (\mathbf{b}\times\mathbf{c})\cdot\mathbf{c} + (\mathbf{b}\times\mathbf{c})\cdot\mathbf{b}$
$+ (\mathbf{b}\times\mathbf{c})\cdot\mathbf{a} + (\mathbf{b}\times\mathbf{a})\cdot\mathbf{c} + (\mathbf{b}\times\mathbf{a})\cdot\mathbf{b} + (\mathbf{b}\times\mathbf{a})\cdot\mathbf{a}$

$\Rightarrow 0 + [\mathbf{c}\ \mathbf{a}\ \mathbf{b}] + 0 + 0 + 0 + [\mathbf{b}\ \mathbf{c}\ \mathbf{a}] + [\mathbf{b}\ \mathbf{a}\ \mathbf{c}] + 0 + 0$

$\Rightarrow [\mathbf{a}\ \mathbf{b}\ \mathbf{c}] + [\mathbf{a}\ \mathbf{b}\ \mathbf{c}] - [\mathbf{a}\ \mathbf{b}\ \mathbf{c}]$ $(\because \mathbf{c}\ \mathbf{a}\ \mathbf{b} = \mathbf{a}\ \mathbf{b}\ \mathbf{c})$

$\Rightarrow [\mathbf{a}\ \mathbf{b}\ \mathbf{c}]$ $(\because [\mathbf{b}\ \mathbf{a}\ \mathbf{c}] = -[\mathbf{a}\ \mathbf{b}\ \mathbf{c}])$

Example 50. If $\mathbf{u}, \mathbf{v}$ and $\mathbf{w}$ are three non-coplanar vectors, then $(\mathbf{u} + \mathbf{v} - \mathbf{w})\cdot[(\mathbf{u} - \mathbf{v}) \times (\mathbf{v} - \mathbf{w})]$ is equal to

(a) 0 (b) $\mathbf{u}\cdot(\mathbf{v}\times\mathbf{w})$

(c) $\mathbf{u}\cdot(\mathbf{w}\times\mathbf{v})$ (d) $3\mathbf{u}\cdot(\mathbf{v}\times\mathbf{w})$

Sol. (b) $(\mathbf{u} + \mathbf{v} - \mathbf{w})\cdot[\mathbf{u} - \mathbf{v} \times (\mathbf{v} - \mathbf{w})]$

$\Rightarrow (\mathbf{u} + \mathbf{v} - \mathbf{w})\cdot[(\mathbf{u}\times\mathbf{v}) \times (\mathbf{u}\times\mathbf{w}) - 0 + (\mathbf{v}\times\mathbf{w})]$
$= [\mathbf{u}\ \mathbf{u}\ \mathbf{v}] + [\mathbf{v}\ \mathbf{u}\ \mathbf{v}] - [\mathbf{w}\ \mathbf{u}\ \mathbf{v}] - [\mathbf{u}\ \mathbf{u}\ \mathbf{w}] - [\mathbf{v}\ \mathbf{u}\ \mathbf{w}]$
$+ [\mathbf{w}\ \mathbf{u}\ \mathbf{w}] + [\mathbf{u}\ \mathbf{v}\ \mathbf{w}] + [\mathbf{v}\ \mathbf{v}\ \mathbf{w}] - [\mathbf{w}\ \mathbf{v}\ \mathbf{w}]$
$= 0 + 0 - [\mathbf{u}\ \mathbf{v}\ \mathbf{w}] - 0 + [\mathbf{u}\ \mathbf{v}\ \mathbf{w}] + 0 + [\mathbf{u}\ \mathbf{v}\ \mathbf{w}] + 0 - 0$
$= [\mathbf{u}\ \mathbf{v}\ \mathbf{w}] = \mathbf{u}\cdot(\mathbf{v}\times\mathbf{w})$

Example 51. If $\mathbf{a}$, $\mathbf{b}$ and $\mathbf{c}$ are non-coplanar vectors and λ is a real number, then the vector $\mathbf{a} + 2\mathbf{b} + 3\mathbf{c}$, $\lambda\mathbf{b} + 4\mathbf{c}$ and $(2\lambda - 1)\mathbf{c}$ are non-coplanar for

(a) no value of λ

(b) all except one value of λ

(c) all except two values of λ

(d) all values of λ

Sol. (c) Since, $\mathbf{a}$, $\mathbf{b}$ and $\mathbf{c}$ are non-coplanar vectors.

$\therefore$ $[\mathbf{a}\ \mathbf{b}\ \mathbf{c}] \neq 0$

Now, $\mathbf{a} + 2\mathbf{b} + 3\mathbf{c}$, $\lambda\mathbf{b} + 4\mathbf{c}$ and $(2\lambda - 1)\,\mathbf{c}$ will be non-coplanare iff

$(\mathbf{a} + 2\mathbf{b} + 3\mathbf{c})\cdot\{(\lambda\mathbf{b} + 4\mathbf{c}) \times (2\lambda - 1)\,\mathbf{c}\} \neq 0$

i.e., $(\mathbf{a} + 2\mathbf{b} + 3\mathbf{c})\cdot\{\lambda(2\lambda - 1)(\mathbf{b}\times\mathbf{c}) \neq 0$

i.e., $\lambda(2\lambda - 1)\,[\mathbf{a}\ \mathbf{b}\ \mathbf{c}]\} \neq 0$

$\therefore$ $\lambda \neq 0, \frac{1}{2}$

Thus, given vectors will be non-coplanar for all values of λ except two values 0 and $\frac{1}{2}$.

Example 52. If x, y and z are distinct scalars such that $[x\mathbf{a} + y\mathbf{b} + z\mathbf{c}, x\mathbf{b} + y\mathbf{c} + z\mathbf{a}, x\mathbf{c} + y\mathbf{a} + z\mathbf{b}] = 0]$ where $\mathbf{a}$, $\mathbf{b}$ and $\mathbf{c}$ are non-coplanar vectors, then

(a) $x + y + z = 0$ (b) $xy + yz + zx = 0$

(c) $x^3 + y^3 + z^3 = 0$ (d) $x^2 + y^2 + z^2 = 0$

Sol. (a) $\mathbf{a}$, $\mathbf{b}$ and $\mathbf{c}$ are non-coplanar.

$\therefore$ $[\mathbf{a}\ \mathbf{b}\ \mathbf{c}] \neq 0$

Now, consider $[x\mathbf{a} + y\mathbf{b} + z\mathbf{c}, x\mathbf{b} + y\mathbf{c} + z\mathbf{a}, x\mathbf{c} + y\mathbf{a} + z\mathbf{b}] = 0$

$$\Rightarrow \begin{vmatrix} x & y & z \\ z & x & y \\ y & z & x \end{vmatrix} [\mathbf{a}\ \mathbf{b}\ \mathbf{c}] = 0 \Rightarrow \begin{vmatrix} x & y & z \\ z & x & y \\ y & z & x \end{vmatrix} = 0 \quad [\because [\mathbf{a}\ \mathbf{b}\ \mathbf{c}] \neq 0]$$

$$\Rightarrow (x + y + z)(x^2 + y^2 + z^2 - xy - yz - zx) = 0$$

$$\Rightarrow \frac{1}{2}(x + y + z)\{(x - y)^2\} + (y - z)^2 + (z - x)^2\} = 0$$

$$\Rightarrow x + y + z = 0 \text{ or } x = y = z$$

But x, y and z are distinct

$$\therefore x + y + z = 0.$$

Example 53. If $\mathbf{a}, \mathbf{b}$ and $\mathbf{c}$ are three non-coplanar uni-modular vectors, each inclined with other at an angle 30°, then volume of tetrahedron whose edges are $\mathbf{a}, \mathbf{b}$ and $\mathbf{c}$ is

(a) $\dfrac{\sqrt{3\sqrt{3} - 5}}{12}$ (b) $\dfrac{3\sqrt{3} - 5}{12}$

(c) $\dfrac{5\sqrt{2} + 3}{12}$ (d) None of these

Sol. (a) Since, the volume of tetrahedron with edges $\mathbf{a}$, $\mathbf{b}$ and $\mathbf{c}$ is $[\mathbf{a}\ \mathbf{b}\ \mathbf{c}]$.

Where, $\mathbf{a} \cdot \mathbf{a} = \mathbf{b} \cdot \mathbf{b} = \mathbf{c} \cdot \mathbf{c} = 1$

and $\mathbf{a} \cdot \mathbf{c} = \mathbf{b} \cdot \mathbf{c} = \mathbf{a} \cdot \mathbf{c} = \dfrac{\sqrt{3}}{2}$ (given)

$$\therefore V = \frac{1}{6}[\mathbf{a}\ \mathbf{b}\ \mathbf{c}]$$

$$\Rightarrow V^2 = \frac{1}{36}[\mathbf{a}\ \mathbf{b}\ \mathbf{c}]^2 = \frac{1}{36}\begin{vmatrix} \mathbf{a}\cdot\mathbf{a} & \mathbf{a}\cdot\mathbf{b} & \mathbf{a}\cdot\mathbf{c} \\ \mathbf{b}\cdot\mathbf{a} & \mathbf{b}\cdot\mathbf{b} & \mathbf{b}\cdot\mathbf{c} \\ \mathbf{c}\cdot\mathbf{a} & \mathbf{c}\cdot\mathbf{b} & \mathbf{c}\cdot\mathbf{c} \end{vmatrix}$$

$$= \frac{1}{36}\begin{vmatrix} 1 & \frac{\sqrt{3}}{2} & \frac{\sqrt{3}}{2} \\ \frac{\sqrt{3}}{2} & 1 & \frac{\sqrt{3}}{2} \\ \frac{\sqrt{3}}{2} & \frac{\sqrt{3}}{2} & 1 \end{vmatrix} = \frac{1}{36}\left(\frac{3\sqrt{3}}{4} - \frac{5}{4}\right)$$

$$\therefore V = \frac{1}{12}\sqrt{3\sqrt{3} - 5}$$

Exercise for Session 3

1. If $\mathbf{a}$ and $\mathbf{b}$ are two vectors such that $|\mathbf{a} \times \mathbf{b}| = 2$, then find the value of $[\mathbf{a}\ \mathbf{b}\ \mathbf{a} \times \mathbf{b}]$.
2. If the vectors $2\hat{\mathbf{i}} - 3\hat{\mathbf{j}}$, $\hat{\mathbf{i}} + \hat{\mathbf{j}} - \hat{\mathbf{k}}$ and $3\hat{\mathbf{i}} - \hat{\mathbf{k}}$ form three concurrent edges of a parallelopiped , the find the volume of the parallelopiped.
3. If the volume of a parallelopiped whose adjacent edges are $\mathbf{a} = 2\hat{\mathbf{i}} + 3\hat{\mathbf{j}} + 4\hat{\mathbf{k}}$, $\mathbf{b} = \hat{\mathbf{i}} + \alpha\hat{\mathbf{j}} + 2\hat{\mathbf{k}}$, $\mathbf{c} = \hat{\mathbf{i}} + 2\hat{\mathbf{j}} + \alpha\hat{\mathbf{k}}$ is 15, then find the value of α, where $\alpha > 0$.
4. The position vector of the four angular points of a tetrahedron are $\mathbf{A}(\hat{\mathbf{j}} + 2\hat{\mathbf{k}})$, $\mathbf{B}(3\hat{\mathbf{i}} + \hat{\mathbf{k}})$, $\mathbf{C}(4\hat{\mathbf{i}} + 3\hat{\mathbf{j}} + 6\hat{\mathbf{k}})$ and $\mathbf{D} = (2\hat{\mathbf{i}} + 3\hat{\mathbf{j}} + 2\hat{\mathbf{k}})$. Find the volume of the tetrahedron *ABCD*.
5. Find the altitude of a parallelopiped whose three coterminous edges are vectors $\mathbf{A} = \hat{\mathbf{i}} + \hat{\mathbf{j}} + \hat{\mathbf{k}}$, $\mathbf{B} = 2\hat{\mathbf{i}} + 4\hat{\mathbf{j}} - \hat{\mathbf{k}}$ and $\mathbf{C} = \hat{\mathbf{i}} + \hat{\mathbf{j}} + 3\hat{\mathbf{k}}$ with $\mathbf{A}$ and $\mathbf{B}$ as the sides of the base of the parallelopiped.
6. Examine whether the vectors $\mathbf{a} = 2\hat{\mathbf{i}} + 3\hat{\mathbf{j}} + 2\hat{\mathbf{k}}$, $\mathbf{b} = \hat{\mathbf{i}} - \hat{\mathbf{j}} + 2\hat{\mathbf{k}}$ and $\mathbf{c} = 3\hat{\mathbf{i}} + 2\hat{\mathbf{j}} + 4\hat{\mathbf{k}}$ from a left handed or a right handed system.
7. Prove that the four points $4\hat{\mathbf{i}} + 5\hat{\mathbf{j}} + \hat{\mathbf{k}}$, $-(\hat{\mathbf{j}} + \hat{\mathbf{k}})$, $(3\hat{\mathbf{i}} + 9\hat{\mathbf{j}} + 4\hat{\mathbf{k}})$ and $4(-\hat{\mathbf{i}} + \hat{\mathbf{j}} + \hat{\mathbf{k}})$ are coplanar.
8. Prove that $[\mathbf{a}\ \mathbf{b}\ \mathbf{c}][\mathbf{u}\ \mathbf{v}\ \mathbf{w}] = \begin{vmatrix} \mathbf{a}\cdot\mathbf{u} & \mathbf{b}\cdot\mathbf{u} & \mathbf{c}\cdot\mathbf{u} \\ \mathbf{a}\cdot\mathbf{v} & \mathbf{b}\cdot\mathbf{v} & \mathbf{c}\cdot\mathbf{v} \\ \mathbf{a}\cdot\mathbf{w} & \mathbf{b}\cdot\mathbf{w} & \mathbf{c}\cdot\mathbf{w} \end{vmatrix}$
9. If $[\mathbf{a}\ \mathbf{b}\ \mathbf{c}] = 2$, then find the value of $[(\mathbf{a} + 2\mathbf{b} - \mathbf{c})\ (\mathbf{a} - \mathbf{b})\ (\mathbf{a} - \mathbf{b} - \mathbf{c})]$.
10. If $\mathbf{a}$, $\mathbf{b}$ and $\mathbf{c}$ are three non-coplanar vectors, then find the value of

$$\frac{\mathbf{a}\cdot(\mathbf{b}\times\mathbf{c})}{\mathbf{b}\cdot(\mathbf{c}\times\mathbf{a})} + \frac{\mathbf{b}\cdot(\mathbf{c}\times\mathbf{a})}{\mathbf{c}\cdot(\mathbf{a}\times\mathbf{b})} + \frac{\mathbf{c}\cdot(\mathbf{a}\times\mathbf{b})}{\mathbf{a}\cdot(\mathbf{b}\times\mathbf{c})}$$

Session 4
Vector Triple Product

Vector Triple Product

It is defined for three vectors **a**, **b** and **c** as the vector $\mathbf{a}\times(\mathbf{b}\times\mathbf{c})$.

This vector being perpendicular to **a** and $\mathbf{b}\times\mathbf{c}$. But $\mathbf{b}\times\mathbf{c}$ is a vector perpendicular to the plane of **b** and **c**.

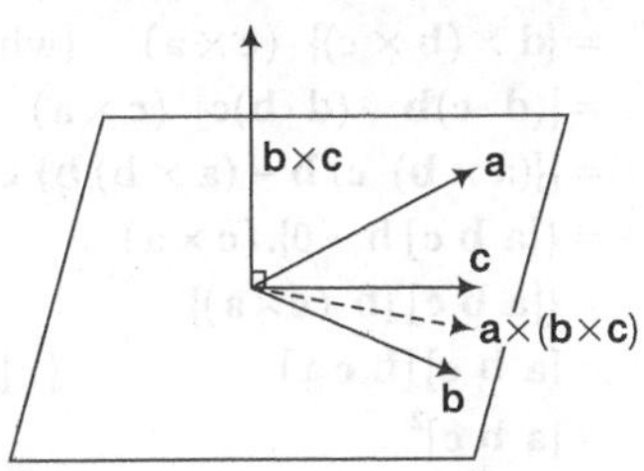

$\therefore$ $\mathbf{a}\times(\mathbf{b}\times\mathbf{c})$ lie in the plane of **b** and **c**, i.e., it is coplanar with **b** and **c**.

i.e., $$\mathbf{a}\times(\mathbf{b}\times\mathbf{c}) = l\mathbf{b} + m\mathbf{c} \quad \text{...(i)}$$

Taking the scalar product of this equation with **a**, we get

$$0 = l(\mathbf{a}\cdot\mathbf{b}) + m(\mathbf{a}\cdot\mathbf{c}) \quad \left[\begin{array}{l}\because \mathbf{a}\times(\mathbf{b}\times\mathbf{c}) \text{ is } \perp \text{ to } \mathbf{a}\\ \therefore (\mathbf{a}\times(\mathbf{b}\times\mathbf{c})\cdot\mathbf{a} = 0\end{array}\right]$$

$$\Rightarrow \quad l(\mathbf{a}\cdot\mathbf{b}) = -m(\mathbf{a}\cdot\mathbf{c})$$

$$\Rightarrow \quad \frac{l}{\mathbf{a}\cdot\mathbf{c}} = -\frac{m}{\mathbf{a}\cdot\mathbf{b}} = \lambda \quad \text{(say)}$$

$$\Rightarrow \quad l = \lambda(\mathbf{a}\cdot\mathbf{c})$$

and $$m = -\lambda(\mathbf{a}\cdot\mathbf{b})$$

Substituting the value of l and m in Eq. (i), we get

$$\mathbf{a}\times(\mathbf{b}\times\mathbf{c}) = \lambda\,[(\mathbf{a}\cdot\mathbf{c})\mathbf{b} - (\mathbf{a}\cdot\mathbf{b})\,\mathbf{c}]$$

Here, the value of λ can be determined by taking specific values of **a**, **b** and **c**.

If we choose the coordinate axes in such a way that,

$$\mathbf{a} = a_1\hat{\mathbf{i}},\ \mathbf{b} = b_1\hat{\mathbf{i}} + b_2\hat{\mathbf{j}}$$

and $$\mathbf{c} = c_1\hat{\mathbf{i}} + c_2\hat{\mathbf{j}} + c_3\hat{\mathbf{k}},$$

it is easy to show that $\lambda = 1$.

Hence, $$\mathbf{a}\times(\mathbf{b}\times\mathbf{c}) = (\mathbf{a}\cdot\mathbf{c})\,\mathbf{b} - (\mathbf{a}\cdot\mathbf{b})\,\mathbf{c}$$

and $\mathbf{a}\times(\mathbf{b}\times\mathbf{c}) = (\mathbf{a}\times\mathbf{b})\times\mathbf{c}$, if some of all **a**, **b** and **c** are zero vectors or **a** and **c** are collinear.

Remarks

1. $\mathbf{a}\times(\mathbf{b}\times\mathbf{c})$ is a linear combination of those two vectors which are with in brackets.
2. If $\mathbf{r} = \mathbf{a}\times(\mathbf{b}\times\mathbf{c})$, then **r** is perpendicular to **a** and lie in the plane of **b** and **c**.
3. $\mathbf{a}\times(\mathbf{b}\times\mathbf{c}) = (\mathbf{a}\cdot\mathbf{c})\,\mathbf{b} - (\mathbf{a}\cdot\mathbf{b})\,\mathbf{c}$
   $(\mathbf{a}\times\mathbf{b})\times\mathbf{c} = (\mathbf{c}\cdot\mathbf{a})\,\mathbf{b} - (\mathbf{c}\cdot\mathbf{b})\mathbf{a}$
   Aid to memory
   $$\text{I}\times(\text{II}\times\text{III}) = (\text{I}\cdot\text{III})\,\text{II} - (\text{I}\cdot\text{II})\,\text{III}$$
4. $(\mathbf{a}\times\mathbf{b})\times(\mathbf{c}\times\mathbf{d}) = ((\mathbf{a}\times\mathbf{b})\cdot\mathbf{d})\,\mathbf{c} - ((\mathbf{a}\times\mathbf{b})\cdot\mathbf{c})\,\mathbf{d}$
   $= [\mathbf{a}\ \mathbf{b}\ \mathbf{d}]\,\mathbf{c} - [\mathbf{a}\ \mathbf{b}\ \mathbf{c}]\,\mathbf{d}$
   $\Rightarrow$ The vector $(\mathbf{a}\times\mathbf{b})\times(\mathbf{c}\times\mathbf{d})$ lies in the plane of **c** and **d**.
   Also, $(\mathbf{a}\times\mathbf{b})\times(\mathbf{c}\times\mathbf{d}) = -(\mathbf{c}\times\mathbf{d})\times(\mathbf{a}\times\mathbf{b})$
   $= -\{((\mathbf{c}\times\mathbf{d})\cdot\mathbf{b})\,\mathbf{a} - ((\mathbf{c}\times\mathbf{d})\cdot(\mathbf{a}\ \mathbf{b})\}$
   $= -[\mathbf{c}\ \mathbf{d}\ \mathbf{b}]\,\mathbf{a} + [\mathbf{c}\ \mathbf{d}\ \mathbf{a}]\mathbf{b}$
   Which shows that $(\mathbf{a}\times\mathbf{b})\times(\mathbf{c}\times\mathbf{d})$ lies in the plane of **a** and **b**.
   Thus, the vector $(\mathbf{a}\times\mathbf{b})\times(\mathbf{c}\times\mathbf{d})$ lies along the common section of the plane of **c**,**d** and that of the plane of **a**, **b**.

Lagrange's Identity

$$(\mathbf{a}\times\mathbf{b})\cdot(\mathbf{c}\times\mathbf{d}) = \begin{vmatrix}\mathbf{a}\cdot\mathbf{c} & \mathbf{a}\cdot\mathbf{d}\\ \mathbf{b}\cdot\mathbf{c} & \mathbf{b}\cdot\mathbf{d}\end{vmatrix}$$

Proof $\quad \text{LHS} = (\mathbf{a}\times\mathbf{b})\cdot(\mathbf{c}\times\mathbf{d}) = \mathbf{u}\cdot(\mathbf{c}\times\mathbf{d})$

where $\quad \mathbf{u} = \mathbf{a}\times\mathbf{b} = (\mathbf{u}\times\mathbf{c})\cdot\mathbf{d}$

$$= ((\mathbf{a}\times\mathbf{b})\times\mathbf{c})\cdot\mathbf{d}$$
$$= ((\mathbf{c}\cdot\mathbf{a})\cdot\mathbf{b} - (\mathbf{c}\cdot\mathbf{a})\cdot\mathbf{a})\cdot\mathbf{d}$$
$$= (\mathbf{c}\cdot\mathbf{a})\,(\mathbf{b}\cdot\mathbf{d}) - (\mathbf{c}\cdot\mathbf{b})\,(\mathbf{a}\cdot\mathbf{d})$$
$$= (\mathbf{a}\cdot\mathbf{c})\,(\mathbf{b}\cdot\mathbf{d}) - (\mathbf{b}\cdot\mathbf{c})\,(\mathbf{a}\cdot\mathbf{d})$$
$$= \begin{vmatrix}\mathbf{a}\cdot\mathbf{c} & \mathbf{a}\cdot\mathbf{d}\\ \mathbf{b}\cdot\mathbf{c} & \mathbf{b}\cdot\mathbf{d}\end{vmatrix}$$

| Example 54. If $\mathbf{a} = \hat{\mathbf{i}} + \hat{\mathbf{j}} + \hat{\mathbf{k}}, \mathbf{b} = \hat{\mathbf{i}} + \hat{\mathbf{j}}, \mathbf{c} = \hat{\mathbf{i}}$ and $(\mathbf{a}\times\mathbf{b})\times\mathbf{c} = \lambda\mathbf{a} + \mu\mathbf{b}$, then $\lambda + \mu$ is equal to

(a) 0 (b) 1
(c) 2 (d) 3

Sol. (a) $(\mathbf{a}\times\mathbf{b})\times\mathbf{c} = \lambda\mathbf{a} + \mu\mathbf{b}$

$$\Rightarrow \quad (\mathbf{a}\cdot\mathbf{c})\,\mathbf{b} - (\mathbf{b}\cdot\mathbf{c})\,\mathbf{a} = \lambda\mathbf{a} + \mu\mathbf{b}$$

$$\Rightarrow \quad \lambda = -\mathbf{b}\cdot\mathbf{c}, \mu = \mathbf{a}\cdot\mathbf{c}$$

$$\therefore \quad \lambda + \mu = \mathbf{a}\cdot\mathbf{c} - \mathbf{b}\cdot\mathbf{c} = (\mathbf{a} - \mathbf{b})\cdot\mathbf{c}$$

$$= \{(\hat{\mathbf{i}} + \hat{\mathbf{j}} + \hat{\mathbf{k}}) - (\hat{\mathbf{i}} + \hat{\mathbf{j}})\} = \hat{\mathbf{k}}\cdot\hat{\mathbf{i}} = 0$$

Example 55. If $\mathbf{a}$, $\mathbf{b}$ and $\mathbf{c}$ are three non-parallel unit vectors such that $\mathbf{a} \times (\mathbf{b} \times \mathbf{c}) = \frac{1}{2}\mathbf{b}$, then find the angles which $\mathbf{a}$ makes with $\mathbf{b}$ and $\mathbf{c}$.

Sol. We have, $\mathbf{a} \times (\mathbf{b} \times \mathbf{c}) = \frac{1}{2}\mathbf{b}$

$\Rightarrow \quad (\mathbf{a} \cdot \mathbf{c})\mathbf{b} - (\mathbf{a} \cdot \mathbf{b})\mathbf{c} = \frac{1}{2}\mathbf{b}$

$\Rightarrow \quad \mathbf{a} \cdot \mathbf{c} = \frac{1}{2}$ and $\mathbf{a} \cdot \mathbf{b} = 0$ (comparing $\mathbf{c}$ and $\mathbf{b}$)

$\Rightarrow \quad \mathbf{a} \cdot \mathbf{c} = \frac{1}{2}$ and $\mathbf{a} \perp \mathbf{b}$

Suppose $\mathbf{a}$ makes angle θ with $\mathbf{c}$. Then, $\mathbf{a} \cdot \mathbf{c} = \frac{1}{2}$

$\Rightarrow \quad |\mathbf{a}||\mathbf{c}|\cos\theta = \frac{1}{2} \Rightarrow \cos\theta = \frac{1}{2}$ $(\because |\mathbf{a}||\mathbf{c}| \neq 0)$

$\Rightarrow \quad \theta = \frac{\pi}{3}$

Thus, a is perpendicular to $\mathbf{b}$ and makes an angle $\frac{\pi}{3}$ with $\mathbf{c}$.

Example 56. If $\mathbf{a} = -\hat{\mathbf{i}} + \hat{\mathbf{j}} + \hat{\mathbf{k}}$ and $\mathbf{b} = 2\hat{\mathbf{i}} + \hat{\mathbf{k}}$, then find the vector x satisfying the conditions.

(i) that it is coplanar with $\mathbf{a}$ and $\mathbf{b}$.

(ii) it is perpendicular to $\mathbf{b}$.

(iii) $\mathbf{a} \cdot \mathbf{x} = 7$

Sol. Since, x is in the plane of $\mathbf{a}$ and $\mathbf{b}$ and is perpendicular to $\mathbf{b}$.

$\therefore \quad x = \lambda\{\mathbf{b} \times (\mathbf{a} \times \mathbf{b})\}$

$\Rightarrow \quad x = \lambda\{(\mathbf{b} \cdot \mathbf{b})\mathbf{a} - (\mathbf{b} \cdot \mathbf{a})\mathbf{b}\}$

$= \lambda\{5(-\hat{\mathbf{i}} + \hat{\mathbf{j}} + \hat{\mathbf{k}}) - (-1)(2\hat{\mathbf{i}} + \hat{\mathbf{k}})\}$

$= \lambda\{-5\hat{\mathbf{i}} + 5\hat{\mathbf{j}} + 5\hat{\mathbf{k}} + 2\hat{\mathbf{i}} + \hat{\mathbf{k}}\}$

$= \lambda\{-3\hat{\mathbf{i}} + 5\hat{\mathbf{j}} + 6\hat{\mathbf{k}}\}$

Now, $\mathbf{a} \cdot \mathbf{x} = 7$

$\Rightarrow \quad -3\lambda + 5\lambda + 6\lambda = 7$

$\Rightarrow \quad 8\lambda = 7 \Rightarrow \lambda = \frac{7}{8}$

Hence, $\quad x = \frac{7}{8}(-3\hat{\mathbf{i}} + 5\hat{\mathbf{j}} + 6\hat{\mathbf{k}})$

Example 57. Prove that $\mathbf{a} \times (\mathbf{b} \times \mathbf{c}) + \mathbf{b} \times (\mathbf{c} \times \mathbf{a}) + \mathbf{c} \times (\mathbf{a} \times \mathbf{b}) = 0$

Sol. We have, $\mathbf{a} \times (\mathbf{b} \times \mathbf{c}) + \mathbf{b} \times (\mathbf{c} \times \mathbf{a}) + \mathbf{c} \times (\mathbf{a} \times \mathbf{b})$

$= \{(\mathbf{a} \cdot \mathbf{c})\mathbf{b} - (\mathbf{a} \cdot \mathbf{b})\mathbf{c}\} + \{(\mathbf{b} \cdot \mathbf{a})\mathbf{c} - (\mathbf{b} \cdot \mathbf{c})\mathbf{a}\} + \{(\mathbf{c} \cdot \mathbf{b})\mathbf{a} - (\mathbf{c} \cdot \mathbf{a})\mathbf{b}\}$

$= [(\mathbf{a} \cdot \mathbf{c})\mathbf{b} - (\mathbf{a} \cdot \mathbf{b})\mathbf{c} + (\mathbf{a} \cdot \mathbf{b})\mathbf{c} - (\mathbf{b} \cdot \mathbf{c})\mathbf{a} + (\mathbf{b} \cdot \mathbf{c})\mathbf{a} - (\mathbf{a} \cdot \mathbf{c})\mathbf{b}] = 0$

Example 58. Show that the vectors $\mathbf{a} \times (\mathbf{b} \times \mathbf{c})$, $\mathbf{b} \times (\mathbf{c} \times \mathbf{a})$ and $\mathbf{c} \times ((\mathbf{a} \times \mathbf{b})$ are coplanar.

Sol. Let $\mathbf{p} = \mathbf{a} \times (\mathbf{b} \times \mathbf{c})$, $\mathbf{q} = \mathbf{b} \times (\mathbf{c} \times \mathbf{a})$ and $\mathbf{r} = \mathbf{c} \times (\mathbf{a} \times \mathbf{b})$, then

$\mathbf{p} + \mathbf{q} + \mathbf{r} = \mathbf{a} \times (\mathbf{b} \times \mathbf{c}) + \mathbf{b} \times (\mathbf{c} \times \mathbf{a}) + \mathbf{c} \times (\mathbf{a} \times \mathbf{b}) = 0$

$\Rightarrow \quad \mathbf{p} = (-1), \mathbf{q} = (-1)\,\mathbf{r}$

which shows $\mathbf{p}$ is linear combination of $\mathbf{q}$ and $\mathbf{r}$.

So, $\mathbf{p}$, $\mathbf{q}$ are coplanar.

Hence, $\mathbf{a} \times (\mathbf{b} \times \mathbf{c})$, $\mathbf{b} \times (\mathbf{c} \times \mathbf{a})$ and $\mathbf{c} \times (\mathbf{a} \times \mathbf{b})$ are coplanar.

Example 59. Prove that $[\mathbf{a} \times \mathbf{b}\ \mathbf{b} \times \mathbf{c}\ \mathbf{c} \times \mathbf{a}] = [\mathbf{a}\ \mathbf{b}\ \mathbf{c}]^2$

Sol. We have, $[\mathbf{a} \times \mathbf{b}\ \mathbf{b} \times \mathbf{c}\ \mathbf{c} \times \mathbf{a}]$

$= \{(\mathbf{a} \times \mathbf{b}) \times (\mathbf{b} \times \mathbf{c})\} \cdot (\mathbf{c} \times \mathbf{a})$

$= \{\mathbf{d} \times (\mathbf{b} \times \mathbf{c})\} \cdot (\mathbf{c} \times \mathbf{a})$ [where, $\mathbf{d} = (\mathbf{a} \times \mathbf{b})$]

$= [(\mathbf{d} \cdot \mathbf{c})\mathbf{b} - (\mathbf{d} \cdot \mathbf{b})\mathbf{c}] \cdot (\mathbf{c} \times \mathbf{a})$

$= [\{(\mathbf{a} \times \mathbf{b}) \cdot \mathbf{c}\}\,\mathbf{b} - (\mathbf{a} \times \mathbf{b})\,\mathbf{b}\}\,\mathbf{c}] \cdot (\mathbf{c} \times \mathbf{a})$

$= \{[\mathbf{a}\ \mathbf{b}\ \mathbf{c}]\,\mathbf{b} - 0\}.(\mathbf{c} \times \mathbf{a})$ $[\because (\mathbf{a}\ \mathbf{b}\ \mathbf{b}) = 0]$

$= \{[\mathbf{a}\ \mathbf{b}\ \mathbf{c}]\,\{\mathbf{b} \cdot (\mathbf{c} \times \mathbf{a})\}$

$= [\mathbf{a}\ \mathbf{b}\ \mathbf{c}]\,[\mathbf{b}\ \mathbf{c}\ \mathbf{a}]$ $\{\because [\mathbf{a}\ \mathbf{b}\ \mathbf{c}] = [\mathbf{b}\ \mathbf{c}\ \mathbf{a}]\}$

$= [\mathbf{a}\ \mathbf{b}\ \mathbf{c}]^2$

Example 60. If $\mathbf{a}$, $\mathbf{b}$ and $\mathbf{c}$ are coplanar show $[\mathbf{a} \times \mathbf{b}\ \mathbf{b} \times \mathbf{c}\ \mathbf{c} \times \mathbf{a}]$ are coplanar.

Sol. Since, $[\mathbf{a}\ \mathbf{b}\ \mathbf{c}]$ are coplanar.

$\Rightarrow \quad [\mathbf{a}\ \mathbf{b}\ \mathbf{c}] = 0$

and $[\mathbf{a} \times \mathbf{b}\ \mathbf{b} \times \mathbf{c}\ \mathbf{c} \times \mathbf{a}] = [\mathbf{a}\ \mathbf{b}\ \mathbf{c}]^2 = 0$

$\therefore [\mathbf{a} \times \mathbf{b}\ \mathbf{b} \times \mathbf{c}\ \mathbf{c} \times \mathbf{a}]$ are coplanar, if $\mathbf{a}$, $\mathbf{b}$ and $\mathbf{c}$ are coplanar.

Example 61. If $\mathbf{A}$, $\mathbf{B}$ and $\mathbf{C}$ are vectors such that $|\mathbf{B}| = |\mathbf{C}|$, prove that $\{(\mathbf{A} + \mathbf{B}) \times (\mathbf{A} + \mathbf{C})\} \times (\mathbf{B} \times \mathbf{C}) \cdot (\mathbf{B} + \mathbf{C}) = 0$.

Sol. Let $\mathbf{R}_1 = \mathbf{A} + \mathbf{B}$, $\mathbf{R}_2 = \mathbf{A} + \mathbf{C}$, $\mathbf{R}_3 = \mathbf{B} + \mathbf{C}$

$\therefore$ LHS $= \{(\mathbf{A} + \mathbf{B}) \times (\mathbf{A} + \mathbf{C})\} \times (\mathbf{B} \times \mathbf{C})\} \times (\mathbf{B} \times \mathbf{C}) \cdot (\mathbf{B} + \mathbf{C})$

$\Rightarrow \quad \{(\mathbf{R}_1 \times \mathbf{R}_2) \times (\mathbf{B} \times \mathbf{C})\} \cdot \mathbf{R}_3$

$\Rightarrow \quad [\{\mathbf{R}_1 \cdot (\mathbf{B} \times \mathbf{C})\}\,\mathbf{R}_2 - \{\mathbf{R}_2 \cdot (\mathbf{B} \times \mathbf{C})\,\mathbf{R}_1\}] \cdot \mathbf{R}_3$

$\Rightarrow \quad [\mathbf{A} + \mathbf{B}\ \mathbf{B}\ \mathbf{C}]\,[\mathbf{R}_2 \cdot \mathbf{R}_3] - [\mathbf{A} + \mathbf{C}\ \mathbf{B}\ \mathbf{C}]\,(\mathbf{R}_1 \cdot \mathbf{R}_3)$

$\Rightarrow \quad \{[\mathbf{A}\ \mathbf{B}\ \mathbf{C}] + [\mathbf{B}\ \mathbf{B}\ \mathbf{C}]\}\,[(\mathbf{A} + \mathbf{C}) \cdot (\mathbf{B} + \mathbf{C})] - \{[\mathbf{A}\ \mathbf{B}\ \mathbf{C}] + [\mathbf{C}\ \mathbf{B}\ \mathbf{C}]\}\,[(\mathbf{A} + \mathbf{B}) \cdot (\mathbf{B} + \mathbf{C})]$

$\Rightarrow \quad [\mathbf{A}\ \mathbf{B}\ \mathbf{C}]\,(\mathbf{A} \cdot \mathbf{B} + \mathbf{A} \cdot \mathbf{C} + \mathbf{C} \cdot \mathbf{B} + \mathbf{C} \cdot \mathbf{C}) - [\mathbf{A}\ \mathbf{B}\ \mathbf{C}]\,(\mathbf{A} \cdot \mathbf{B} + \mathbf{A} \cdot \mathbf{C} + \mathbf{B} \cdot \mathbf{B} + \mathbf{B} \cdot \mathbf{C})$

$\Rightarrow \quad [\mathbf{A}\ \mathbf{B}\ \mathbf{C}]\,(\mathbf{A} \cdot \mathbf{B} + \mathbf{A} \cdot \mathbf{C} + \mathbf{C} \cdot \mathbf{B} + |\mathbf{C}|^2 - \mathbf{A} \cdot \mathbf{B} - \mathbf{A} \cdot \mathbf{C} + \mathbf{B} \cdot \mathbf{B} + \mathbf{B} \cdot \mathbf{C})$

$\Rightarrow \quad [\mathbf{A}\ \mathbf{B}\ \mathbf{C}]\,(|\mathbf{C}|^2 - |\mathbf{B}|^2)$

$\Rightarrow \quad [\mathbf{A}\ \mathbf{B}\ \mathbf{C}]\,(0)$ $(\because |\mathbf{B}| = |\mathbf{C}|)$

$\Rightarrow \quad 0 = \text{RHS}$

Example 62. If $\mathbf{b}$ and $\mathbf{c}$ are two non-collinear vector such that $\mathbf{a} \parallel (\mathbf{b} \times \mathbf{c})$, then prove that $(\mathbf{a} \times \mathbf{b}) \cdot (\mathbf{a} \times \mathbf{c})$ is equal to $|\mathbf{a}|^2 (\mathbf{b} \cdot \mathbf{c})$.

Sol. Since, $\mathbf{a} \parallel (\mathbf{b} \times \mathbf{c})$, therefore $\mathbf{a} \perp \mathbf{b}$ and $\mathbf{a} \perp \mathbf{c}$

$\Rightarrow$ $\mathbf{a} \cdot \mathbf{b} = 0$ and $\mathbf{a} \cdot \mathbf{c} = 0$

Now, consider $(\mathbf{a} \times \mathbf{b}) \cdot (\mathbf{a} \times \mathbf{c}) = \begin{vmatrix} \mathbf{a} \cdot \mathbf{a} & \mathbf{a} \cdot \mathbf{c} \\ \mathbf{b} \cdot \mathbf{a} & \mathbf{b} \cdot \mathbf{c} \end{vmatrix} = \begin{vmatrix} \mathbf{a} \cdot \mathbf{a} & 0 \\ 0 & \mathbf{b} \cdot \mathbf{c} \end{vmatrix}$

$= (\mathbf{a} \cdot \mathbf{a})(\mathbf{b} \cdot \mathbf{c}) = |\mathbf{a}^2|(\mathbf{b} \cdot \mathbf{c})$.

Reciprocal System of Vectors

The two system of vectors are called reciprocal system of vectors if by taking dot product, we get unity.

Thus, if $\mathbf{a}$, $\mathbf{b}$ and $\mathbf{c}$ be three non-coplanar vectors and if.

$$\mathbf{a}' = \frac{\mathbf{b} \times \mathbf{c}}{|\mathbf{a}\ \mathbf{b}\ \mathbf{c}|}, \mathbf{b}' = \frac{\mathbf{c} \times \mathbf{a}}{|\mathbf{a}\ \mathbf{b}\ \mathbf{c}|} \text{ and } \mathbf{c}' = \frac{\mathbf{a} \times \mathbf{b}}{|\mathbf{a}\ \mathbf{b}\ \mathbf{c}|}$$

Then $\mathbf{a}'$, $\mathbf{b}'$ and $\mathbf{c}'$ are said to be reciprocal system of vectors for the vector $\mathbf{a}$, $\mathbf{b}$ and $\mathbf{c}$.

Remarks

1. If $\mathbf{a}, \mathbf{b}, \mathbf{c}$ and $\mathbf{a}', \mathbf{b}', \mathbf{c}'$ are reciprocal system of vectors, then

$\mathbf{a} \cdot \mathbf{a}' = \dfrac{\mathbf{a} \cdot (\mathbf{b} \times \mathbf{c})}{(\mathbf{a}\ \mathbf{b}\ \mathbf{c})} = \dfrac{[\mathbf{a}\ \mathbf{b}\ \mathbf{c}]}{[\mathbf{a}\ \mathbf{b}\ \mathbf{c}]} = 1$

Similarly, $\mathbf{b} \cdot \mathbf{b}' = \mathbf{c} \cdot \mathbf{c}' = 1$

2. $\mathbf{a} \cdot \mathbf{b}' = \mathbf{a} \cdot \mathbf{c}' = \mathbf{b} \cdot \mathbf{a}' = \mathbf{b} \cdot \mathbf{a}' = \mathbf{c} \cdot \mathbf{a}' = \mathbf{c} \cdot \mathbf{b}' = 0$

3. $[\mathbf{a}\ \mathbf{b}\ \mathbf{c}] \cdot [\mathbf{a}'\ \mathbf{b}'\ \mathbf{c}'] = 1$

Proof : We have,

$[\mathbf{a}'\mathbf{b}'\mathbf{c}'] = \left[\dfrac{\mathbf{b} \times \mathbf{c}}{[\mathbf{a}\ \mathbf{b}\ \mathbf{c}]} = \dfrac{\mathbf{c} \times \mathbf{a}}{[\mathbf{a}\ \mathbf{b}\ \mathbf{c}]} = \dfrac{\mathbf{a} \times \mathbf{b}}{[\mathbf{a}\ \mathbf{b}\ \mathbf{c}]}\right]$

$= \dfrac{1}{[\mathbf{a}\ \mathbf{b}\ \mathbf{c}]^3} [\mathbf{b} \times \mathbf{c}\ \mathbf{c} \times \mathbf{a}\ \mathbf{a} \times \mathbf{b}] = \dfrac{1}{[\mathbf{a}\ \mathbf{b}\ \mathbf{c}]^3} [\mathbf{a}\ \mathbf{b}\ \mathbf{c}]^2$

$= \dfrac{1}{[\mathbf{a}\ \mathbf{b}\ \mathbf{c}]}$

$\therefore [\mathbf{a}'\mathbf{b}'\mathbf{c}'] \cdot [\mathbf{a}\ \mathbf{b}\ \mathbf{c}] = 1$

4. The orthogonal triad of vectors $\hat{\mathbf{i}}$, $\hat{\mathbf{j}}$ and $\hat{\mathbf{k}}$ is self reciprocal.

Let $\hat{\mathbf{i}}$, $\hat{\mathbf{j}}$ and $\hat{\mathbf{k}}$ be the system of vectors reciprocal to the system $\hat{\mathbf{i}}$, $\hat{\mathbf{j}}$ and $\hat{\mathbf{k}}$ then, we have,

$\hat{\mathbf{i}} = \dfrac{\hat{\mathbf{j}} \times \hat{\mathbf{k}}}{[\hat{\mathbf{i}}\ \hat{\mathbf{j}}\ \hat{\mathbf{k}}]} = \hat{\mathbf{j}}$

Similarly, $\hat{\mathbf{j}}' = \hat{\mathbf{j}}$ and $\hat{\mathbf{k}}' = \hat{\mathbf{k}}$

5. $\mathbf{a}$, $\mathbf{b}$ and $\mathbf{c}$ are non-coplanar iff $\mathbf{a}'$, $\mathbf{b}'$, $\mathbf{c}'$ are non-coplanar.

$\because$ $[\mathbf{a}\ \mathbf{b}\ \mathbf{c}]\ [\mathbf{a}'\ \mathbf{b}'\ \mathbf{c}'] = 1$

$\therefore$ $[\mathbf{a}'\ \mathbf{b}'\ \mathbf{c}'] = \dfrac{1}{[\mathbf{a}\ \mathbf{b}\ \mathbf{c}]}$

So, $[\mathbf{a}\ \mathbf{b}\ \mathbf{c}] \neq 0 \Leftrightarrow [\mathbf{a}'\ \mathbf{b}'\ \mathbf{c}'] \neq 0$

Thus, $\mathbf{a}$, $\mathbf{b}$ and $\mathbf{c}$ are non-coplanar iff $[\mathbf{a}'\ \mathbf{b}'\ \mathbf{c}']$ are non-coplanar.

6. If $\mathbf{a}, \mathbf{b}, \mathbf{c}$ are non-coplanar vectors, then

$$\mathbf{r} = \frac{[\mathbf{x}\ \mathbf{b}\ \mathbf{c}]\ \mathbf{a} + [\mathbf{x}\ \mathbf{c}\ \mathbf{a}]\ \mathbf{b} + [\mathbf{x}\ \mathbf{a}\ \mathbf{b}]\ \mathbf{c}}{[\mathbf{a}\ \mathbf{b}\ \mathbf{c}]}$$

Example 63. Find the set of vectors reciprocal to the set of vectors $2\hat{\mathbf{i}} + 3\hat{\mathbf{j}} - \hat{\mathbf{k}}$, $\hat{\mathbf{i}} - \hat{\mathbf{j}} - 2\hat{\mathbf{k}}$, $-\hat{\mathbf{i}} + 2\hat{\mathbf{j}} + 2\hat{\mathbf{k}}$

Sol. Let the given vector be $\mathbf{a}$, $\mathbf{b}$, $\mathbf{c}$.

Now, $[\mathbf{a}\ \mathbf{b}\ \mathbf{c}] = \begin{vmatrix} 2 & 3 & -1 \\ 1 & -1 & -2 \\ -1 & 2 & 2 \end{vmatrix}$

$= 2(-2+4) - 3(2-2) - 1(2-1)$

$= 4 - 1 = 3$

$\mathbf{b} \times \mathbf{c} = \begin{vmatrix} \hat{\mathbf{i}} & \hat{\mathbf{j}} & \hat{\mathbf{k}} \\ 1 & -1 & -2 \\ -1 & 2 & 2 \end{vmatrix} = 2\hat{\mathbf{i}} + \hat{\mathbf{k}}$

$\mathbf{c} \times \mathbf{a} = \begin{vmatrix} \hat{\mathbf{i}} & \hat{\mathbf{j}} & \hat{\mathbf{k}} \\ -1 & 2 & 2 \\ 2 & 3 & -1 \end{vmatrix} = -8\hat{\mathbf{i}} + 3\hat{\mathbf{j}} - 7\hat{\mathbf{k}}$

$\mathbf{a} \times \mathbf{b} = \begin{vmatrix} \hat{\mathbf{i}} & \hat{\mathbf{j}} & \hat{\mathbf{k}} \\ 2 & 3 & -1 \\ 1 & -1 & -2 \end{vmatrix} = -7\hat{\mathbf{i}} + 3\hat{\mathbf{j}} - 5\hat{\mathbf{k}}$

Hence, $\mathbf{a}' = \dfrac{\mathbf{b} \times \mathbf{c}}{[\mathbf{a}\ \mathbf{b}\ \mathbf{c}]} = \dfrac{2\hat{\mathbf{i}} + \hat{\mathbf{k}}}{3}$

$\mathbf{b}' = \dfrac{\mathbf{c} \times \mathbf{a}}{[\mathbf{a}\ \mathbf{b}\ \mathbf{c}]} = \dfrac{-8\hat{\mathbf{i}} + 3\hat{\mathbf{j}} - 7\hat{\mathbf{k}}}{3}$

and $\mathbf{c}' = \dfrac{\mathbf{a} \times \mathbf{b}}{[\mathbf{a}\ \mathbf{b}\ \mathbf{c}]} = \dfrac{-7\hat{\mathbf{i}} + 3\hat{\mathbf{j}} - 5\hat{\mathbf{k}}}{3}$

Example 64. Find a set of vector reciprocal to the vectors $\mathbf{a}$, $\mathbf{b}$ and $\mathbf{a} \times \mathbf{b}$.

Sol. Let the given vectors be denoted by $\mathbf{a}$, $\mathbf{b}$ and $\mathbf{c}$ where $\mathbf{c} = \mathbf{a} \times \mathbf{b}$.

$\therefore$ $[\mathbf{a}\ \mathbf{b}\ \mathbf{c}] = (\mathbf{a} \times \mathbf{b}) \cdot \mathbf{c} = (\mathbf{a} \times \mathbf{b}) = (\mathbf{a} \times \mathbf{b})^2$ ...(i)

and let the reciprocal system of vector be $\mathbf{a}'$, $\mathbf{b}'$ and $\mathbf{c}'$.

$\therefore$ $\mathbf{a}' = \dfrac{\mathbf{b} \times \mathbf{c}}{[\mathbf{a}\ \mathbf{b}\ \mathbf{c}]} = \dfrac{\mathbf{b} \times (\mathbf{a} \times \mathbf{b})}{[\mathbf{a} \times \mathbf{b}]^2}$

$\mathbf{b}' = \dfrac{\mathbf{c} \times \mathbf{a}}{[\mathbf{a}\ \mathbf{b}\ \mathbf{c}]} = \dfrac{(\mathbf{a} \times \mathbf{b}) \times \mathbf{a}}{[\mathbf{a} \times \mathbf{b}]^2}$

$\mathbf{c}' = \dfrac{\mathbf{a} \times \mathbf{b}}{[\mathbf{a}\ \mathbf{b}\ \mathbf{c}]} = \dfrac{\mathbf{a} \times \mathbf{b}}{[\mathbf{a} \times \mathbf{b}]^2}$

$\therefore \mathbf{a}'$, $\mathbf{b}'$ and $\mathbf{c}'$ are required reciprocal system of vectors for $\mathbf{a}$, $\mathbf{b}$ and $\mathbf{a} \times \mathbf{b}$.

Example 65. If $\mathbf{a}' = \dfrac{\mathbf{b} \times \mathbf{c}}{[\mathbf{a}\ \mathbf{b}\ \mathbf{c}]}$, $\mathbf{b}' = \dfrac{\mathbf{c} \times \mathbf{a}}{[\mathbf{a}\ \mathbf{b}\ \mathbf{c}]}$, $\mathbf{c}' = \dfrac{\mathbf{a} \times \mathbf{b}}{[\mathbf{a}\ \mathbf{b}\ \mathbf{c}]}$, then show that $\mathbf{a} \times \mathbf{a}' + \mathbf{b} \times \mathbf{b}' + \mathbf{c} \times \mathbf{c}' = 0$, where $\mathbf{a}$, $\mathbf{b}$ and $\mathbf{c}$ are non-coplanar.

Sol. Here, $\mathbf{a} \times \mathbf{a}' = \dfrac{\mathbf{a} \times (\mathbf{b} \times \mathbf{c})}{[\mathbf{a}\ \mathbf{b}\ \mathbf{c}]}$

$$\mathbf{a} \times \mathbf{a}' = \frac{(\mathbf{a} \cdot \mathbf{c})\mathbf{b} - (\mathbf{a} \cdot \mathbf{b})\,\mathbf{c}}{[\mathbf{a}\ \mathbf{b}\ \mathbf{c}]} \quad \text{...(i)}$$

Similarly,

$$\mathbf{b} \times \mathbf{b}' = \frac{(\mathbf{b} \cdot \mathbf{a})\,\mathbf{c} - (\mathbf{b} \cdot \mathbf{c})\,\mathbf{a}}{[\mathbf{a}\ \mathbf{b}\ \mathbf{c}]} \quad \text{...(ii)}$$

$$\mathbf{c} \times \mathbf{c}' = \frac{(\mathbf{c} \cdot \mathbf{b})\mathbf{a} - (\mathbf{c} \cdot \mathbf{a})\,\mathbf{b}}{[\mathbf{a}\ \mathbf{b}\ \mathbf{c}]} \quad \text{...(iii)}$$

$$\mathbf{a} \times \mathbf{a}' + \mathbf{b} \times \mathbf{b}' + \mathbf{c} \times \mathbf{c}'$$

$$= \frac{(\mathbf{a} \cdot \mathbf{c})\mathbf{b} - (\mathbf{a} \cdot \mathbf{b})\,\mathbf{c} + (\mathbf{b} \cdot \mathbf{a})\mathbf{c} - (\mathbf{b} \cdot \mathbf{c})\,\mathbf{a} + (\mathbf{c} \cdot \mathbf{b})\mathbf{a} - (\mathbf{c} \cdot \mathbf{a})\,\mathbf{b}}{[\mathbf{a}\ \mathbf{b}\ \mathbf{c}]} \quad (\because \mathbf{a} \cdot \mathbf{b} = \mathbf{b} \cdot \mathbf{a})$$

$$= 0$$

Example 66. If (e_1, e_2, e_3) and (e_1', e_2', e_3') are two sets of non-coplanar vectors such that $\hat{\mathbf{i}} = 1, 2, 3,$

we have $e_{\mathbf{i}} \cdot e_{\mathbf{j}}' = \begin{cases} 1, \text{ if } \hat{\mathbf{i}} = \hat{\mathbf{j}} \\ 0, \text{ if } \hat{\mathbf{i}} \neq \hat{\mathbf{j}} \end{cases}$, then show that

$[e_1, e_2, e_3], [e_1'\ e_2'\ e_3'] = 1.$

Sol. We have, $e_1 \cdot e_2' = 0, e_1 \cdot e_3' = 0$

$$\Rightarrow \quad e_1 \perp e_2' \text{ and } e_1 \perp e_3'$$

$$\therefore \quad e_1 \,||\, (e_2' \times e_3')$$

$$\therefore \quad e_1 = \lambda\,(e_2' \times e_3') \quad \text{...(i)}$$

$$e_1 \cdot e_1' = \lambda (e_2' \times e_3') \cdot e_1'$$

$$1 = \lambda\,[e_1'\ e_2'\ e_3'] \quad (\because e_1 e_1' = 1, \text{ given})$$

$$\lambda = \frac{1}{[e_1'\ e_2'\ e_3']}$$

From Eq. (i),

$$e_1 = \frac{e_2' \times e_3'}{[e_1'\ e_2'\ e_3']}$$

Similarly, $e_2 = \dfrac{e_3' \times e_1'}{[e_1'\ e_2'\ e_3']}$

and $e_3 = \dfrac{e_1' \times e_2'}{[e_1'\ e_2'\ e_3']}$

$$\therefore \quad [e_1\ e_2\ e_3] = \frac{[e_2' \times e_3'\ e_3' \times e_1'\ e_1' \times e_2']}{[e_1'\ e_2'\ e_3']^3}$$

$$\Rightarrow \quad [e_1, e_2, e_3][e_1'\ e_2'\ e_3']^3 = [e_2' \times e_3'\ e_3' \times e_1'\ e_1' \times e_2'] \quad \text{...(ii)}$$

Now, $[e_2' \times e_3'\ e_3' \times e_1'\ e_1' \times e_2'] = [e_1'\ e_2'\ e_3']^2$ ...(iii)

$\therefore$ From Eqs. (ii) and (iii), we get

$$[e_1\ e_2\ e_3][e_1'\ e_2'\ e_3']^3 = [e_1'\ e_2'\ e_3']^2$$

$$[e_1\ e_2\ e_3][e_1'\ e_2'\ e_3'] = 1$$

Solving of Vector Equations

Solving a vector equation means determining an unknown vector (or a number of vectors satisfying the given conditions)

Generally, to solve vector equations we express the unknown as the linear combination of three non-coplanar vector as; $\mathbf{r} = x\mathbf{a} + y\mathbf{b} + z(\mathbf{a} \times \mathbf{b})$; as $\mathbf{a}$, $\mathbf{b}$ and $\mathbf{a} \times \mathbf{b}$ are non-coplanar and find x, y and z using given conditions.

Sometimes, we can directly solve the given condition it would be more clear from some examples.

Example 67. Solve the vector equation $\mathbf{r} \times \mathbf{b} = \mathbf{a} \times \mathbf{b}$, $\mathbf{r} \cdot \mathbf{c} = 0$ provided that $\mathbf{c}$ is not perpendicular to $\mathbf{b}$.

Sol. We are given,

$$\mathbf{r} \times \mathbf{b} = \mathbf{a} \times \mathbf{b}$$

$$\Rightarrow \quad (\mathbf{r} - \mathbf{a}) \times \mathbf{b} = 0$$

Hence, $(\mathbf{r} - \mathbf{a})$ and $\mathbf{b}$ are parallel.

$$\Rightarrow \quad \mathbf{r} - \mathbf{a} = t\mathbf{b}$$

and we known $\mathbf{r} \cdot \mathbf{c} = 0,$ ...(i)

$\therefore$ Taking dot product of Eq. (i) by $\mathbf{c}$, we get

$$\mathbf{r} \cdot \mathbf{a} - \mathbf{a} \cdot \mathbf{c} = t(\mathbf{b} \cdot \mathbf{c})$$

$$\Rightarrow \quad 0 - \mathbf{a} \cdot \mathbf{c} = t(\mathbf{b} \cdot \mathbf{c})$$

$$\Rightarrow \quad t = -\left(\frac{\mathbf{a} \cdot \mathbf{c}}{\mathbf{b} \cdot \mathbf{c}}\right) \quad \text{...(ii)}$$

$\therefore$ From Eqs. (i) and (ii) solution of $\mathbf{r}$ is

$$\mathbf{r} = \mathbf{a} - \left(\frac{\mathbf{a} \cdot \mathbf{c}}{\mathbf{b} \cdot \mathbf{c}}\right)\mathbf{b}$$

Example 68. Solve for $\mathbf{x}$, such that $\mathbf{A} \cdot \mathbf{X} = \mathbf{C}$ and $\mathbf{A} \times \mathbf{X} = \mathbf{B}$ with $C \neq 0$.

Sol. We have, $\mathbf{A} \times \mathbf{X} = \mathbf{B}$

Taking vector product of both sides with $\mathbf{A}$, we get

$$\mathbf{A} \times \mathbf{B} = \mathbf{A} \times (\mathbf{A} \times \mathbf{X})$$

$$= (\mathbf{A} \cdot \mathbf{X})\,\mathbf{A} - (\mathbf{A} \cdot \mathbf{A})\,\mathbf{X}$$

$$= \mathbf{C}\mathbf{A} - |\,\mathbf{A}\,|^2\,\mathbf{X}$$

(using $\mathbf{A} \cdot \mathbf{X} = \mathbf{C}$ and $\mathbf{A} \cdot t\mathbf{A} = |\,\mathbf{A}\,|^2$)

$$\Rightarrow \quad |\,\mathbf{A}\,|^2\,\mathbf{X} = \mathbf{C}\mathbf{A} - \mathbf{A} \times \mathbf{B}$$

$$\text{or} \quad \mathbf{X} = \frac{\mathbf{C}\mathbf{A} + \mathbf{B} \times \mathbf{A}}{|\,\mathbf{A}\,|^2}$$

Example 69. Solve the vector equation $\mathbf{r} \times \mathbf{a} + k\mathbf{r} = \mathbf{b}$, where $\mathbf{a}$ and $\mathbf{b}$ are two given vector and k is any scalar.

Sol. Since, $\mathbf{a}$, $\mathbf{b}$ and $\mathbf{a} \times \mathbf{b}$ are two non-coplanar vectors.

$$\mathbf{r} = x\mathbf{a} + y\mathbf{b} + z(\mathbf{a} \times \mathbf{b}) \quad \text{...(i)}$$

(where, x, y and z scalars)

On putting $\mathbf{r}$ in $\mathbf{r} \times \mathbf{a} + k\mathbf{r} = \mathbf{b}$, we get

$$\{x\mathbf{a} + y\mathbf{b} + z(\mathbf{a} \times \mathbf{b})\} \times \mathbf{a} + k\{x\mathbf{a} + y\mathbf{b} + z(\mathbf{a} \times \mathbf{b})\} = \mathbf{b}$$

$$\Rightarrow y(\mathbf{b} \times \mathbf{a}) + z\{(\mathbf{a}\cdot\mathbf{a})\mathbf{b} - (\mathbf{a}\cdot\mathbf{b})\mathbf{a}\} + k\{x\mathbf{a} + y\mathbf{b} + z(\mathbf{a} \times \mathbf{b})\} = \mathbf{b}$$

$$\Rightarrow \{kx - z(\mathbf{a}\cdot\mathbf{b})\}\mathbf{a} + \{ky + z(\mathbf{a}\cdot\mathbf{a})\}\mathbf{b} + \{(y + zk)\}(\mathbf{a} \times \mathbf{b}) = \mathbf{b}$$

$$\Rightarrow kx - z(\mathbf{a}\cdot\mathbf{b}) = 0,\ ky + z(\mathbf{a}\cdot\mathbf{a}) = 1$$

$$\Rightarrow -y + zk = 0$$

On solving these equations, we get

$$z = \frac{1}{k^2 + |\mathbf{a}|^2},$$

$$x = \frac{\mathbf{a}\cdot\mathbf{b}}{k(|\mathbf{a}|^2 + k^2)}$$

and $$y = \frac{k}{k^2 + |\mathbf{a}|^2}$$

On putting theses value in Eq. (i), we get the solution,

$$\mathbf{r} = \frac{(\mathbf{a}\cdot\mathbf{b})}{k(k^2 + |\mathbf{a}|^2)} + \frac{k}{k^2 + |\mathbf{a}|^2}(\mathbf{b}) + \frac{1}{k^2 + |\mathbf{a}|^2}(\mathbf{a} \times \mathbf{b})$$

$$\mathbf{r} = \frac{1}{k^2 + |\mathbf{a}|^2}\left[\frac{(\mathbf{a}\cdot\mathbf{b})\,\mathbf{a}}{k} + (k)\,\mathbf{b} + (\mathbf{a} \times \mathbf{b})\right]$$ is required solution.

Example 70. Solve for vectors **A** and **B**, where
$\mathbf{A} + \mathbf{B} = \mathbf{a}$, $\mathbf{A} \times \mathbf{B} = \mathbf{b}$, $\mathbf{A}\cdot\mathbf{a} = 1$

Sol. We have, $\mathbf{A} + \mathbf{B} = \mathbf{a}$

$$\Rightarrow \mathbf{A}\cdot\mathbf{a} + \mathbf{B}\cdot\mathbf{a} = \mathbf{a}\cdot\mathbf{a}$$

$$\Rightarrow 1 + \mathbf{B}\cdot\mathbf{a} = \mathbf{a}^2 \quad \text{(given } \mathbf{A}\cdot\mathbf{a} = 1)$$

$$\Rightarrow \mathbf{B}\cdot\mathbf{a} = \mathbf{a}^2 - 1 \quad \text{...(i)}$$

Also, $\mathbf{A} \times \mathbf{B} = \mathbf{b}$

$$\Rightarrow \mathbf{a} \times (\mathbf{A} \times \mathbf{B}) = \mathbf{a} \times \mathbf{b}$$

$$\Rightarrow (\mathbf{a}\cdot\mathbf{B})\mathbf{A} - (\mathbf{a}\cdot\mathbf{A})\mathbf{B} = \mathbf{a} \times \mathbf{b}$$

$$\Rightarrow (\mathbf{a}^2 - 1)\mathbf{A} - \mathbf{B} = \mathbf{a} \times \mathbf{b}$$

[using Eq. (i) and $\mathbf{a}\cdot\mathbf{A} = 1$] ...(ii)

and $\mathbf{A} + \mathbf{B} = \mathbf{a}$

From Eqs. (i) and (ii), we get

$$\mathbf{A} = \frac{(\mathbf{a} \times \mathbf{b}) + \mathbf{a}}{\mathbf{a}^2} \text{ and } \mathbf{B} = \mathbf{a} - \left\{\frac{(\mathbf{a} \times \mathbf{b}) + \mathbf{a}}{\mathbf{a}^2}\right\}$$

$$\Rightarrow \mathbf{B} = \frac{(\mathbf{b} \times \mathbf{a}) + \mathbf{a}(\mathbf{a}^2 - 1)}{\mathbf{a}^2}$$

Thus, $$\mathbf{A} = \frac{(\mathbf{a} \times \mathbf{b}) + \mathbf{a}}{\mathbf{a}^2} \text{ and } \mathbf{B} = \frac{(\mathbf{b} \times \mathbf{a})\ \mathbf{a}(\mathbf{a}^2 - 1)}{\mathbf{a}^2}$$

Exercise for Session 4

1. Find the value of $\alpha \times (\beta \times \gamma)$, where $\alpha = 2\hat{i} - 10\hat{j} + 2\hat{k}$, $\beta = 3\hat{i} + \hat{j} + 2\hat{k}$, $\gamma = 2\hat{i} + \hat{j} + 3\hat{k}$.
2. Find the vector of length 3 unit which is perpendicular to $\hat{i} + \hat{j} + \hat{k}$ and lies in the plane of $\hat{i} + \hat{j} + \hat{k}$ and $2\hat{i} - 3\hat{j}$.
3. Show that $(\mathbf{b} \times \mathbf{c}).\ \mathbf{a} \times \mathbf{d}) + (\mathbf{c} \times \mathbf{a})(\mathbf{b} \times \mathbf{d}) + (\mathbf{a} \times \mathbf{b})(\mathbf{c} \times \mathbf{d}) = 0$
4. Prove that $\hat{i} \times (\mathbf{a} + \hat{i}) + \hat{j} \times (\mathbf{a} \times \hat{j}) + \hat{k} \times (\mathbf{a} \times \hat{k}) = 2\mathbf{a}$.
5. Prove that $[\mathbf{a} \times \mathbf{b}\ \ \mathbf{a} \times \mathbf{c}\ \ \mathbf{d}] = (\mathbf{a}\cdot\mathbf{d})[\mathbf{a}\ \mathbf{b}\ \mathbf{c}]$
6. If **a**, **b** and **c** are non-coplanar unit vector such that $\mathbf{a} \times (\mathbf{b} \times \mathbf{c}) = \frac{\mathbf{b} + \mathbf{c}}{\sqrt{2}}$, **b** and **c** are non-parallel, then prove that the angle between **a** and **b** is $\frac{3\pi}{4}$.
7. Find a set of vectors reciprocal to the set of vectors $-\hat{i} + \hat{j} + \hat{k}$, $\hat{i} - \hat{j} + \hat{k}$, $\hat{i} + \hat{j} + \hat{k}$.
8. If **a**, **b**, **c** and **a′**, **b′**, **c′** are reciprocal system of vectors, then prove that
   $$\mathbf{a}' \times \mathbf{b}' + \mathbf{b}' \times \mathbf{c}' + \mathbf{c}' \times \mathbf{a}' = \frac{\mathbf{a} + \mathbf{b} + \mathbf{c}}{[\mathbf{a}\ \mathbf{b}\ \mathbf{c}]}.$$
9. Solve $\mathbf{r} \times \mathbf{b} = \mathbf{a}$, where **a** and **b** are given vectors such that $\mathbf{a}\cdot\mathbf{b} = 0$.
10. Find vector **r**, if $\mathbf{r}\cdot\mathbf{a} = \mathbf{m}$ and $\mathbf{r} \times \mathbf{b} = \mathbf{c}$, where $\mathbf{a}\cdot\mathbf{b} \neq 0$.

JEE Type Solved Examples :
Only One Option Correct Type Questions

Ex. 1 *If* $|\mathbf{a}|=5, |\mathbf{a}-\mathbf{b}|=8$ *and* $|\mathbf{a}+\mathbf{b}|=10$, *then* $|\mathbf{b}|$ *is equal to*

(a) 1 (b) $\sqrt{57}$
(c) 3 (d) None of these

Sol. (b) We know that for any two vectors

$$|\mathbf{a}+\mathbf{b}|^2+|\mathbf{a}-\mathbf{b}|^2=2(|\mathbf{a}|^2+|\mathbf{b}|^2)$$

$$\Rightarrow \quad (10)^2+(8)^2=2[(5)^2+|\mathbf{b}|^2]$$

$$\Rightarrow \quad 100+64=50+2|\mathbf{b}|^2 \Rightarrow |\mathbf{b}|^2=57$$

$$\therefore \quad |\mathbf{b}|=\sqrt{57}$$

Ex. 2 *Angle between diagonals of a parallelogram whose sides are represented by* $\mathbf{a}=2\hat{\mathbf{i}}+\hat{\mathbf{j}}+\hat{\mathbf{k}}$ *and* $\mathbf{b}=\hat{\mathbf{i}}-\hat{\mathbf{j}}-\hat{\mathbf{k}}$

(a) $\cos^{-1}\left(\frac{1}{3}\right)$ (b) $\cos^{-1}\left(\frac{1}{2}\right)$
(c) $\cos^{-1}\left(\frac{4}{9}\right)$ (d) $\cos^{-1}\left(\frac{5}{9}\right)$

Sol. (a) Let $\mathbf{c}$ and $\mathbf{d}$ be the diagonals of parallelogram.

Then, $\mathbf{c}=\mathbf{a}+\mathbf{b}$ and $\mathbf{d}=\mathbf{a}-\mathbf{b}$

$\Rightarrow \quad \mathbf{c}=3\hat{\mathbf{i}}$ and $\mathbf{d}=\hat{\mathbf{i}}+2\hat{\mathbf{j}}+2\hat{\mathbf{k}}$

Let θ be the angle between $\mathbf{c}$ and $\mathbf{d}$.

$$\text{Then,} \quad \cos\theta=\frac{\mathbf{c}\cdot\mathbf{d}}{|\mathbf{c}||\mathbf{d}|}=\frac{3\hat{\mathbf{i}}\cdot(\hat{\mathbf{i}}+2\hat{\mathbf{j}}+2\hat{\mathbf{k}})}{\sqrt{(3)^2}\sqrt{1^2+2^2+2^2}}$$

$$=\frac{3}{3\times 3}=\frac{1}{3}$$

$$\therefore \quad \theta=\cos^{-1}\left(\frac{1}{3}\right)$$

Ex. 3 *Let* $\mathbf{a},\mathbf{b},\mathbf{c}$, *be vectors of length* 3, 4, 5 *respectively and* $\mathbf{a}$ *be perpendicular to* $(\mathbf{b}+\mathbf{c})$, $\mathbf{b}$ *to* $(\mathbf{c}+\mathbf{a})$ *and* $\mathbf{c}$ *to* $(\mathbf{a}+\mathbf{b})$, *then the value of* $(\mathbf{a}+\mathbf{b}+\mathbf{c})$ *is*

(a) $2\sqrt{5}$ (b) $2\sqrt{2}$
(c) $10\sqrt{5}$ (d) $5\sqrt{2}$

Sol. (d) We have, $|\mathbf{a}|=3$, $|\mathbf{b}|=4$ and $|\mathbf{c}|=5$. It is given that

$\mathbf{a}\perp(\mathbf{b}+\mathbf{c})$, $\mathbf{b}\perp(\mathbf{c}+\mathbf{a})$ and $\mathbf{c}\perp(\mathbf{a}+\mathbf{b})$

$$\Rightarrow \quad \mathbf{a}\cdot(\mathbf{b}+\mathbf{c})=0, \mathbf{b}\cdot(\mathbf{c}+\mathbf{a})=0 \text{ and } \mathbf{c}\cdot(\mathbf{a}+\mathbf{b})=0$$

$$\Rightarrow \quad \mathbf{a}\cdot\mathbf{b}+\mathbf{a}\cdot\mathbf{c}=\mathbf{b}\cdot\mathbf{c}+\mathbf{b}\cdot\mathbf{a}=\mathbf{c}\cdot\mathbf{a}+\mathbf{c}\cdot\mathbf{b}=0$$

or $\quad \mathbf{a}\cdot\mathbf{b}+\mathbf{b}\cdot\mathbf{c}+\mathbf{c}\cdot\mathbf{a}=0$ (adding all the above equations)

$$\text{Now,} \quad |\mathbf{a}+\mathbf{b}+\mathbf{c}|^2=|\mathbf{a}|^2+|\mathbf{b}|^2+|\mathbf{c}|^2+2(\mathbf{a}\cdot\mathbf{b}+\mathbf{b}\cdot\mathbf{c}+\mathbf{c}\cdot\mathbf{a})$$

$$=3^2+4^2+5^2=50$$

$$\therefore \quad |\mathbf{a}+\mathbf{b}+\mathbf{c}|=5\sqrt{2}$$

Ex. 4 *Let* $a, b>0$ *and* $\alpha=\frac{\hat{\mathbf{i}}}{a}+\frac{4\hat{\mathbf{j}}}{b}+b\hat{\mathbf{k}}$ *and* $\beta=b\hat{\mathbf{i}}+a\hat{\mathbf{j}}+\frac{1}{b}\hat{\mathbf{k}}$, *then the maximum value of* $\frac{10}{5+\alpha\cdot\beta}$ *is*

(a) 1 (b) 2
(c) 4 (d) 8

Sol. (a) $$\alpha\cdot\beta=\frac{b}{a}+\frac{4a}{b}+1\geq 5 \quad (\because \text{AM} \geq \text{GM})$$

So, $$\left(\frac{10}{5+\alpha\cdot\beta}\right)_{\max}=1.$$

Ex. 5 *If the unit vectors* $\mathbf{e}_1$ *and* $\mathbf{e}_2$ *are inclined at an angle* 2θ *and* $|\mathbf{e}_1-\mathbf{e}_2|<1$, *then for* $\theta\in[0,\pi]$, θ *may lie in the interval*

(a) $\left[0,\frac{\pi}{6}\right)$ (b) $\left[\frac{\pi}{6},\frac{\pi}{2}\right]$
(c) $\left(\frac{5\pi}{6},\pi\right]$ (d) $\left[\frac{\pi}{2},\frac{5\pi}{6}\right]$

Sol. (a) It is given that $\mathbf{e}_1$ and $\mathbf{e}_2$ are two unit vectors inclined at an angle 2θ and $|\mathbf{e}_1-\mathbf{e}_2|<1$.

$$\therefore \quad |\mathbf{e}_1-\mathbf{e}_2|<1 \Rightarrow |\mathbf{e}_1-\mathbf{e}_2|^2<1$$

$$\Rightarrow \quad 4\sin^2\theta<1 \quad [\because |\mathbf{e}_1-\mathbf{e}_2|^2=4\sin^2\theta]$$

$$\Rightarrow \quad \sin^2\theta<\frac{1}{4}$$

$$\therefore \quad \theta\in\left[0,\frac{\pi}{6}\right)$$

Ex. 6 *If* $\mathbf{a}=3\hat{\mathbf{i}}-\hat{\mathbf{j}}+5\hat{\mathbf{k}}$ *and* $\mathbf{b}=\hat{\mathbf{i}}+2\hat{\mathbf{j}}-3\hat{\mathbf{k}}$ *are given vector. A vector* $\mathbf{c}$ *which is perpendicular to Z-axis satisfying* $\mathbf{c}\cdot\mathbf{a}=9$ *and* $\mathbf{c}\cdot\mathbf{b}=-4$. *If inclination of* $\mathbf{c}$ *with X-axis and Y-axis is* α *and* β *respectively, then which of the following is not true?*

(a) $\alpha>\frac{\pi}{4}$ (b) $\beta>\frac{\pi}{2}$
(c) $\alpha>\frac{\pi}{2}$ (d) $\beta<\frac{\pi}{2}$

Sol. (c) $\mathbf{c}$ lies in XY-plane

$$\therefore \quad \mathbf{c}=x\hat{\mathbf{i}}+y\hat{\mathbf{j}}$$

From the given conditions $3x-y=9$

and $\quad x+2y=-4$

Solving, we get $\mathbf{c}=2\hat{\mathbf{i}}-3\hat{\mathbf{j}}$

$$\therefore \quad \alpha=\cos^{-1}\left(\frac{2}{\sqrt{13}}\right), \beta=\cos^{-1}\left(\frac{-3}{\sqrt{13}}\right)$$

Ex. 7 *If A is 3×3 matrix and **u** is a vector. If A**u** and **u** are orthogonal for all real **u**, then matrix A is a*

(a) singular (b) non-singular
(c) symmetric (d) skew-symmetric

Sol. (a) $A\mathbf{u} \cdot \mathbf{u} = 0$

$$\Rightarrow \quad |A||\mathbf{u}|^2 = 0, \text{ Since } |\mathbf{u}| \ne 0 \Rightarrow |A| = 0$$

$\therefore$ A is singular.

Ex. 8 *Let the cosine of angle between the vectors **p** and **q** be λ such that $2\mathbf{p} + \mathbf{q} = \hat{\mathbf{i}} + \hat{\mathbf{j}}$ and $\mathbf{p} + 2\mathbf{q} = \hat{\mathbf{i}} - \hat{\mathbf{j}}$, then λ is equal to*

(a) $\frac{5}{9}$ (b) $-\frac{4}{5}$
(c) $\frac{3}{9}$ (d) $\frac{7}{9}$

Sol. (b) It is given that $2\mathbf{p} + \mathbf{q} = \hat{\mathbf{i}} + \hat{\mathbf{j}}$

and $\mathbf{p} + 2\mathbf{q} = \hat{\mathbf{i}} - \hat{\mathbf{j}}$

$$\Rightarrow \quad \mathbf{p} = \frac{1}{3}\hat{\mathbf{i}} + \hat{\mathbf{j}} \text{ and } \mathbf{q} = \frac{1}{3}\hat{\mathbf{i}} - \hat{\mathbf{j}}$$

Let θ be the angle between **p** and **q**. Then

$$\cos\theta = \frac{\mathbf{p} \cdot \mathbf{q}}{|\mathbf{p}||\mathbf{q}|} = \frac{\left(\frac{1}{3}\hat{\mathbf{i}} + \hat{\mathbf{j}}\right) \cdot \left(\frac{1}{3}\hat{\mathbf{i}} - \hat{\mathbf{j}}\right)}{\sqrt{\left(\frac{1}{3}\right)^2 + (1)^2}\sqrt{\left(\frac{1}{3}\right)^2 + (-1)^2}}$$

$$= \frac{\frac{1}{9} - 1}{\sqrt{\frac{1}{9} + 1}\sqrt{\frac{1}{9} + 1}} = \frac{-\frac{8}{9}}{\frac{10}{9}} = -\frac{8}{10} = -\frac{4}{5}$$

$$\therefore \quad \lambda = \cos\theta = -\frac{4}{5}$$

Ex. 9 *Let **a**, **b** and **c** be vectors with magnitudes 3, 4 and 5, respectively and $\mathbf{a} + \mathbf{b} + \mathbf{c} = 0$, then the values of $\mathbf{a} \cdot \mathbf{b} + \mathbf{b} \cdot \mathbf{c} + \mathbf{c} \cdot \mathbf{a}$ is*

(a) 47 (b) 25
(c) 50 (d) – 25

Sol. (d) We observe, $|\mathbf{a}|^2 + |\mathbf{b}|^2 = 3^2 + 4^2 = 5^2 = |\mathbf{c}|^2$

$$\mathbf{a} \cdot \mathbf{b} = 0$$

$$\mathbf{b} \cdot \mathbf{c} = |\mathbf{b}||\mathbf{c}| \cdot \cos\left(\pi - \cos^{-1}\frac{4}{5}\right)$$

$$= 4 \times 5\left\{-\cos\left(\cos^{-1}\frac{4}{5}\right)\right\}$$

$$= 4 \times 5 \times \left(-\frac{4}{5}\right) = -16$$

$$\mathbf{c} \cdot \mathbf{a} = |\mathbf{c}||\mathbf{a}| \cdot \cos\left(\pi - \cos^{-1}\frac{3}{5}\right)$$

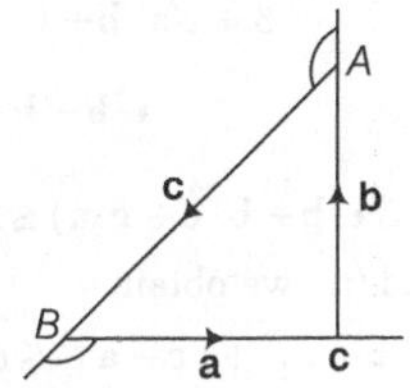

$$= 5 \cdot 3\left\{-\cos\left(\cos^{-1}\frac{3}{5}\right)\right\} = 5 \cdot 3 \cdot \left(-\frac{3}{5}\right) = -9$$

$$\therefore \quad \mathbf{a} \cdot \mathbf{b} + \mathbf{b} \cdot \mathbf{c} + \mathbf{c} \cdot \mathbf{a} = 0 - 16 - 9 = -25$$

Trick $\because \mathbf{a} + \mathbf{b} + \mathbf{c} = 0$

On squaring both the sides, we get $|\mathbf{a} + \mathbf{b} + \mathbf{c}|^2 = 0$

$$\Rightarrow \quad |\mathbf{a}|^2 + |\mathbf{b}|^2 + |\mathbf{c}|^2 + 2(\mathbf{a} \cdot \mathbf{b} + \mathbf{b} \cdot \mathbf{c} + \mathbf{c} \cdot \mathbf{a}) = 0$$

$$\Rightarrow \quad 2(\mathbf{a} \cdot \mathbf{b} + \mathbf{b} \cdot \mathbf{c} + \mathbf{c} \cdot \mathbf{a}) = -(9 + 16 + 25)$$

$$\Rightarrow \quad \mathbf{a} \cdot \mathbf{b} + \mathbf{b} \cdot \mathbf{c} + \mathbf{c} \cdot \mathbf{a} = -25$$

Ex. 10 *Let **u**, **v** and **w** be such that $|\mathbf{u}| = 1$, $|\mathbf{v}| = 2$, $|\mathbf{w}| = 3$. If the projection **v** along **u** is equal to that of **w** along **u** and **v**, **w** are perpendicular to each other, then $|\mathbf{u} - \mathbf{v} + \mathbf{w}|$ equals.*

(a) $\sqrt{14}$ (b) $\sqrt{7}$
(c) 2 (d) 14

Sol. (a) We have,

Projection of **v** and **u** = Projection of **w** along **u**

$$\Rightarrow \quad \frac{\mathbf{v} \cdot \mathbf{u}}{|\mathbf{u}|} = \frac{\mathbf{w} \cdot \mathbf{u}}{|\mathbf{u}|} \Rightarrow \mathbf{v} \cdot \mathbf{u} = \mathbf{w} \cdot \mathbf{u} \quad \text{... (i)}$$

Also, **v** and **w** are perpendicular to each other

$$\therefore \quad \mathbf{v} \cdot \mathbf{w} = 0 \quad \text{... (ii)}$$

Now, $|\mathbf{u} - \mathbf{v} + \mathbf{w}|^2 = |\mathbf{u}|^2 + |\mathbf{v}|^2 + |\mathbf{w}|^2 - 2(\mathbf{u} \cdot \mathbf{v}) - 2(\mathbf{v} \cdot \mathbf{w}) + 2(\mathbf{u} \cdot \mathbf{w})$

$$\Rightarrow \quad |\mathbf{u} - \mathbf{v} + \mathbf{w}|^2 = 1 + 4 + 9$$

$$\Rightarrow \quad |\mathbf{u} - \mathbf{v} + \mathbf{w}| = \sqrt{14}$$

Ex. 11 *If **a**, **b** and **c** are unit vectors, then $|\mathbf{a} - \mathbf{b}|^2 + |\mathbf{b} - \mathbf{c}|^2 + |\mathbf{c} - \mathbf{a}|^2$ does not exceed*

(a) 4 (b) 9
(c) 8 (d) 6

Sol. (b) We have, $|\mathbf{a} - \mathbf{b}|^2 + |\mathbf{b} - \mathbf{c}|^2 + |\mathbf{c} - \mathbf{a}|^2$

$$= |\mathbf{a}| + |\mathbf{b}|^2 - 2(\mathbf{a} \cdot \mathbf{b}) + |\mathbf{b}|^2 + |\mathbf{c}|^2 - 2(\mathbf{b} \cdot \mathbf{c}) + |\mathbf{c}|^2 + |\mathbf{a}|^2 - 2(\mathbf{c} \cdot \mathbf{a})$$

$$= 2[|\mathbf{a}|^2 + |\mathbf{b}|^2 + |\mathbf{c}|^2 - (\mathbf{a} \cdot \mathbf{b} + \mathbf{b} \cdot \mathbf{c} + \mathbf{c} \cdot \mathbf{a})]$$

$$= 2[3 - (\mathbf{a} \cdot \mathbf{b} + \mathbf{b} \cdot \mathbf{c} + \mathbf{c} \cdot \mathbf{a})]$$

$$= 6 - 2(\mathbf{a} \cdot \mathbf{b} + \mathbf{b} \cdot \mathbf{c} + \mathbf{c} \cdot \mathbf{a}) \quad \text{...(i)}$$

Now, $|\mathbf{a} + \mathbf{b} + \mathbf{c}|^2 \ge 0$

$$\Rightarrow |\mathbf{a}|^2 + |\mathbf{b}|^2 + |\mathbf{c}|^2 + 2(\mathbf{a} \cdot \mathbf{b} + \mathbf{b} \cdot \mathbf{c} + \mathbf{c} \cdot \mathbf{a}) \ge 0$$

$\Rightarrow \quad 3 + 2(\mathbf{a}\cdot\mathbf{b} + \mathbf{b}\cdot\mathbf{c} + \mathbf{c}\cdot\mathbf{a}) \geq 0$

$\Rightarrow \quad \mathbf{a}\cdot\mathbf{b} + \mathbf{b}\cdot\mathbf{c} + \mathbf{c}\cdot\mathbf{a} \geq -\frac{3}{2}$

$\Rightarrow \quad -2(\mathbf{a}\cdot\mathbf{b} + \mathbf{b}\cdot\mathbf{c} + \mathbf{c}\cdot\mathbf{a}) \leq 3 \quad$ ...(ii)

From Eqs. (i) and (ii), we obtain

$|\mathbf{a}-\mathbf{b}|^2 + |\mathbf{b}-\mathbf{c}|^2 + |\mathbf{c}-\mathbf{a}|^2 \leq 6 + 3$

$\Rightarrow |\mathbf{a}-\mathbf{b}|^2 + |\mathbf{b}-\mathbf{c}|^2 + |\mathbf{c}-\mathbf{a}|^2 \leq 9$

Ex. 12 *The vectors* $\mathbf{a} = 2\lambda^2\hat{\mathbf{i}} + 4\lambda\hat{\mathbf{j}} + \hat{\mathbf{k}}$ *and* $\mathbf{b} = 7\hat{\mathbf{i}} - 2\hat{\mathbf{j}} + \lambda\hat{\mathbf{k}}$ *make an obtuse angle whereas the angle between* $\mathbf{b}$ *and* $\hat{\mathbf{k}}$ *is acute and less than* $\pi/6$,

(a) $0 < \lambda < \frac{1}{2}$ (b) $\lambda > \sqrt{159}$

(c) $-\frac{1}{2} < \lambda < 0$ (d) null set

Sol. (d) As angle between $\mathbf{a}$ and $\mathbf{b}$ is obtuse, $\mathbf{a}\cdot\mathbf{b} < 0$

$\Rightarrow (2\lambda^2\hat{\mathbf{i}} + 4\lambda\hat{\mathbf{j}} + \hat{\mathbf{k}})\cdot(7\hat{\mathbf{i}} - 2\hat{\mathbf{j}} + \lambda\hat{\mathbf{k}}) < 0$

$\Rightarrow \quad 14\lambda^2 - 8\lambda + \lambda < 0$

$\Rightarrow \quad \lambda(2\lambda - 1) < 0$

$\Rightarrow \quad 0 < \lambda < \frac{1}{2} \quad$ ...(i)

Angle between $\mathbf{b}$ and $\mathbf{k}$ is acute and less than $\frac{\pi}{6}$.

$\mathbf{b}\cdot\mathbf{k} = |\mathbf{b}|\cdot|\mathbf{k}|\cos\theta$

$\Rightarrow \quad \lambda = \sqrt{53 + \lambda^2}\,1\cdot\cos\theta$

$\Rightarrow \quad \cos\theta = \frac{\lambda}{\sqrt{53 + \lambda^2}}$

$\theta < \frac{\pi}{6} \Rightarrow \cos\theta > \cos\frac{\pi}{6}$

$\Rightarrow \quad \cos\theta > \frac{\sqrt{3}}{2} \Rightarrow \frac{\lambda}{\sqrt{53 + \lambda^2}} > \frac{\sqrt{3}}{2}$

$\Rightarrow \quad 4\lambda^2 - 3(53 + \lambda^2) > 0$

$\Rightarrow \quad \lambda^2 > 159$

$\Rightarrow \quad \lambda < -\sqrt{159} \quad$ ...(ii)

From Eqs. (i) and (ii), $\lambda = \phi$

$\therefore$ Domain of λ is null set.

Ex. 13 *The locus of a point equidistant from two given points whose position vectors are* $\mathbf{a}$ *and* $\mathbf{b}$ *is equal to*

(a) $\left[\mathbf{r} - \frac{1}{2}(\mathbf{a}+\mathbf{b})\right]\cdot(\mathbf{a}+\mathbf{b}) = 0$

(b) $\left[\mathbf{r} - \frac{1}{2}(\mathbf{a}+\mathbf{b})\right]\cdot(\mathbf{a}-\mathbf{b}) = 0$

(c) $\left[\mathbf{r} - \frac{1}{2}(\mathbf{a}+\mathbf{b})\right]\cdot\mathbf{a} = 0$

(d) $[\mathbf{r} - (\mathbf{a}+\mathbf{b})]\cdot\mathbf{b} = 0$

Sol. (b) Let $P(\mathbf{r})$ be a point on the locus.

$\therefore \quad AP = BP$

$\Rightarrow \quad |\mathbf{r}-\mathbf{a}| = |\mathbf{r}-\mathbf{b}| \Rightarrow |\mathbf{r}-\mathbf{a}|^2 = |\mathbf{r}-\mathbf{b}|^2$

$\Rightarrow \quad (\mathbf{r}-\mathbf{a})\cdot(\mathbf{r}-\mathbf{a}) = (\mathbf{r}-\mathbf{b})\cdot(\mathbf{r}-\mathbf{b})$

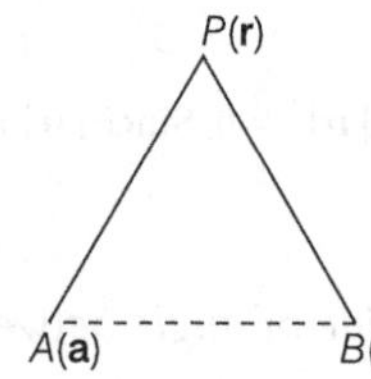

$\Rightarrow \quad 2\mathbf{r}\cdot(\mathbf{a}-\mathbf{b}) = \mathbf{a}\cdot\mathbf{a} - \mathbf{b}\cdot\mathbf{b}$

$\Rightarrow \quad \mathbf{r}\cdot(\mathbf{a}-\mathbf{b}) = \frac{1}{2}(\mathbf{a}+\mathbf{b})\cdot(\mathbf{a}-\mathbf{b})$

$\therefore \quad \left[\mathbf{r} - \frac{1}{2}(\mathbf{a}+\mathbf{b})\right]\cdot(\mathbf{a}-\mathbf{b}) = 0$

This is the locus of P.

Ex. 14 *In cartesian coordinates the point A is* (x_1, y_1), *where* $x_1 = 1$ *on the curve* $y = x^2 + x + 10$. *Then tangent at A cuts the X-axis at B. The value of the dot product* $\mathbf{OA}\cdot\mathbf{AB}$ *is*

(a) $-\frac{520}{3}$ (b) -148

(c) 140 (d) 12

Sol. (b) Given curve is $y = x^2 + x + 10$ ...(i)

When $x = 1$, $y = 1^2 + 1 + 10 = 12$

$\therefore \quad A = (1, 12)$

$\therefore \quad \mathbf{OA} = \hat{\mathbf{i}} + 12\hat{\mathbf{j}}$

From Eq. (i), $\frac{dy}{dx} = 2x + 1$

Equation of tangent at A is $y - 12 = \left(\frac{dy}{dx}\right)_{(1,12)}(x-1)$

$\Rightarrow \quad y - 12 = (2\times 1 + 1)(x-1)$

$\Rightarrow \quad y - 12 = 3x - 3$

$\therefore \quad y = 3(x+3)$

This tangent cuts X-axis (i.e. $y = 0$) at $(-3, 0)$

$\therefore \quad B = (-3, 0)$

$\mathbf{OB} = -3\hat{\mathbf{i}} + 0\cdot\hat{\mathbf{j}} = -3\hat{\mathbf{i}}$, $\mathbf{OA}\cdot\mathbf{AB} = \mathbf{OA}\cdot(\mathbf{OB} - \mathbf{OA})$

$(\hat{\mathbf{i}} + 12\hat{\mathbf{j}})\cdot(-3\hat{\mathbf{i}} - \hat{\mathbf{i}} - 12\hat{\mathbf{j}}) = (\hat{\mathbf{i}} + 12\hat{\mathbf{j}})\cdot(-4\hat{\mathbf{i}} - 12\hat{\mathbf{j}})$

$= -4 - 144 = -148$

Ex. 15 *In a tetrahedron OABC, the edges are of lengths,* $|OA| = |BC| = a$, $|OB| = |AC| = b$, $|OC| = |AB| = c$. *Let* G_1 *and* G_2 *be the centroids of the triangle ABC and AOC such that* $OG_1 \perp BG_2$, *then the value of* $\frac{a^2 + c^2}{b^2}$ *is*

(a) 2 (b) 3

(c) 6 (d) 9

Sol. (b) $OG_1 \cdot BG_2 = 0$.

$$\Rightarrow \frac{a+b+c}{3} \cdot \frac{a+c-3b}{3} = 0$$

$$\Rightarrow a^2 + c^2 - 3b^2 + 2a \cdot c - 2b \cdot c - 2a \cdot b = 0$$

Now, $|c-a|^2 = b^2, |c-b|^2 = a^2$ and $|a-b|^2 = c^2$

$$\therefore 2a \cdot c = a^2 + c^2 - b^2,\ 2b \cdot c = b^2 + c^2 - a^2,$$

$$2a \cdot b = a^2 + b^2 - c^2$$

Putting in the above result, we get $2a^2 + 2c^2 - 6b^2 = 0$

$$\Rightarrow \frac{a^2 + c^2}{b^2} = 3$$

Ex. 16 *If OABC is a tetrahedron such that* $\mathbf{OA}^2 + \mathbf{BC}^2 = \mathbf{OB}^2 + \mathbf{CA}^2 = \mathbf{OC}^2 + \mathbf{AB}^2$, *then which of the following is not true?*

(a) $OA \perp BC$ (b) $OB \perp AC$

(c) $OC \perp AB$ (d) $AB \perp AC$

Sol. (d) Let $\mathbf{OA} = \mathbf{a}$, $\mathbf{OB} = \mathbf{b}$, $\mathbf{OC} = \mathbf{c}$

Then from the given conditions

$$\mathbf{a} \cdot \mathbf{a} + (\mathbf{b} - \mathbf{c}) \cdot (\mathbf{b} - \mathbf{c}) = \mathbf{b} \cdot \mathbf{b} + (\mathbf{c} - \mathbf{a}) \cdot (\mathbf{c} - \mathbf{a})$$

$$\Rightarrow -2\mathbf{b} \cdot \mathbf{c} = -2\mathbf{c} \cdot \mathbf{a}$$

$$\Rightarrow \mathbf{c} \cdot (\mathbf{b} - \mathbf{a}) = 0 \Rightarrow \mathbf{BA} \cdot \mathbf{OC} = 0$$

Hence $AB \perp OC$. Similarly,

$$BC \perp OA \text{ and } CA \perp OB$$

Ex. 17 *If a, b, c and A, B, C* $\in R - \{0\}$ *such that* $aA + bB + cC + \sqrt{(a^2 + b^2 + c^2)(A^2 + B^2 + C^2)} = 0$, *then value of* $\frac{aB}{bA} = \frac{bC}{cB} + \frac{cA}{aC}$ *is*

(a) 3 (b) 4

(c) 5 (d) 6

Sol. (a) Let $\mathbf{r}_1 = a\hat{\mathbf{i}} + b\hat{\mathbf{j}} + c\hat{\mathbf{k}}$ and $\mathbf{r}_2 = A\hat{\mathbf{i}} + B\hat{\mathbf{j}} + C\hat{\mathbf{k}}$

$$\Rightarrow \mathbf{r}_1 \cdot \mathbf{r}_2 = aA + bB + cC$$

$$\Rightarrow |\mathbf{r}_1||\mathbf{r}_2| = \sqrt{(a^2 + b^2 + c^2)(A^2 + B^2 + C^2)}$$

$$\therefore \mathbf{r}_1 \mathbf{r}_2 = -|\mathbf{r}_1||\mathbf{r}_2|$$

$\Rightarrow$ $\mathbf{r}_1$ and $\mathbf{r}_2$ are anti-parallel

$$\Rightarrow \frac{a}{A} = \frac{b}{B} = \frac{c}{C} = k, \text{ where } k \text{ is any constant}$$

$$\Rightarrow \frac{aB}{bA} + \frac{bC}{cB} + \frac{cA}{aC} = 3$$

Ex. 18 *The unit vector in ZOX plane making angles* 45° *and* 60° *respectively, with* $a = 2\hat{\mathbf{i}} + 2\hat{\mathbf{j}} - \hat{\mathbf{k}}$ *and* $\mathbf{b} = \hat{\mathbf{j}} - \hat{\mathbf{k}}$, *is*

(a) $\frac{1}{\sqrt{2}}(-\hat{\mathbf{i}} + \hat{\mathbf{k}})$ (b) $\frac{1}{\sqrt{2}}(\hat{\mathbf{i}} - \hat{\mathbf{k}})$

(c) $\frac{\sqrt{3}}{2}(\hat{\mathbf{i}} + \hat{\mathbf{k}})$ (d) None of these

Sol. (b) Let the required vector be $\mathbf{r} = x\hat{\mathbf{i}} + z\hat{\mathbf{k}}$, since $\mathbf{r}$ is a unit vector.

$$\therefore x^2 + y^2 = 1$$

It is given that $\mathbf{r}$ makes 45° and 60° angles with $\mathbf{a}$ and $\mathbf{b}$ respectively.

$$\therefore \cos 45° = \frac{\mathbf{r} \cdot \mathbf{a}}{|\mathbf{a}||\mathbf{r}|} \text{ and } \cos 60° = \frac{\mathbf{r} \cdot \mathbf{b}}{|\mathbf{r}||\mathbf{b}|}$$

$$\Rightarrow \frac{1}{\sqrt{2}} = \frac{2x - y}{3} \text{ and } \frac{1}{2} = -\frac{-y}{\sqrt{2}}$$

$$\Rightarrow 2x - y = \frac{3}{\sqrt{2}} \text{ and } y = \frac{1}{\sqrt{2}}$$

$$\Rightarrow x = \frac{1}{\sqrt{2}}$$

$$\therefore \mathbf{r} = \frac{1}{\sqrt{2}}(\hat{\mathbf{i}} - \hat{\mathbf{k}})$$

Ex. 19 *A unit vector perpendicular to the vector* $-\hat{\mathbf{i}} + 2\hat{\mathbf{j}} + 2\hat{\mathbf{k}}$ *and making equal angles with X and Y-axes can be*

(a) $\frac{1}{3}(2\hat{\mathbf{i}} + 2\hat{\mathbf{j}} - \hat{\mathbf{k}})$ (b) $\frac{1}{3}(2\hat{\mathbf{i}} - 2\hat{\mathbf{j}} - \hat{\mathbf{k}})$

(c) $\frac{1}{3}(2\hat{\mathbf{i}} + 2\hat{\mathbf{j}} + \hat{\mathbf{k}})$ (d) $\frac{1}{3}(2\hat{\mathbf{i}} - 2\hat{\mathbf{j}} + \hat{\mathbf{k}})$

Sol. (a) Let the required vector be $\mathbf{r} = l\hat{i} + m\hat{\mathbf{j}} + n\hat{\mathbf{k}}$, where l, m, n are the direction cosines of $\mathbf{r}$ such that $l = m$.

It is given that $\mathbf{r}$ is perpendicular to $-\hat{\mathbf{i}} + 2\hat{\mathbf{j}} + 2\hat{\mathbf{k}}$. Therefore,

$$\mathbf{r} \cdot (-\hat{\mathbf{i}} + 2\hat{\mathbf{j}} + 2\hat{\mathbf{k}}) = 0$$

$$\Rightarrow -l + 2m + 2n = 0$$

$$\Rightarrow l + 2n = 0 \quad [\because l = m]$$

$$\Rightarrow l = -2n$$

Now, $l^2 + m^2 + n^2 = 1$

$$\Rightarrow 4n^2 + 4n^2 + n^2 = 1$$

$$\Rightarrow n = \pm \frac{1}{3}$$

$$\therefore l = \mp \frac{2}{3}, m = \mp \frac{2}{3}, n = \mp \frac{1}{3}$$

Hence, $\mathbf{r} = \mp \frac{1}{3}(2\hat{\mathbf{i}} + 2\hat{\mathbf{j}} - \hat{\mathbf{k}})$

Ex. 20 *If* $(\mathbf{a} + 3\mathbf{b}) \cdot (7\mathbf{a} - 5\mathbf{b}) = 0$ *and* $(\mathbf{a} - 4\mathbf{b}) \cdot (7\mathbf{a} - 2\mathbf{b}) = 0$. *Then, the angle between* **a** *and* **b** *is*

(a) 60° (b) 30°

(c) 90° (d) None of these

Sol. (a) We have, $(\mathbf{a} + 3\mathbf{b}) \cdot (7\mathbf{a} - 5\mathbf{b}) = 0$

$$\Rightarrow 7|\mathbf{a}|^2 + 16\mathbf{a} \cdot \mathbf{b} - 15|\mathbf{b}|^2 = 0 \quad \text{...(i)}$$

and $(\mathbf{a} - 4\mathbf{b}) \cdot (7\mathbf{a} - 2\mathbf{b}) = 0$

$$\Rightarrow 7|\mathbf{a}|^2 - 30\mathbf{a} \cdot \mathbf{b} + 8|\mathbf{b}|^2 = 0 \quad \text{...(ii)}$$

From Eqs. (i) and (ii), we get

$$\mathbf{a}\cdot\mathbf{b}=\frac{15|\mathbf{b}|^2-7|\mathbf{a}|^2}{16}$$

$$\mathbf{a}\cdot\mathbf{b}=\frac{1}{30}(7|\mathbf{a}|^2+8|\mathbf{b}|^2)$$

$$\Rightarrow \quad \frac{15|\mathbf{b}|^2-7|\mathbf{a}|^2}{8}=\frac{1}{15}(7|\mathbf{a}|^2+8|\mathbf{b}|^2)$$

$$\Rightarrow \quad 15(15|\mathbf{b}|^2-7|\mathbf{a}|^2)=8(7|\mathbf{a}|^2+8|\mathbf{b}|^2)$$

$$\Rightarrow \quad 225|\mathbf{b}|^2-105|\mathbf{a}|^2=56|\mathbf{a}|^2+64|\mathbf{b}|^2$$

$$\Rightarrow \quad 161|\mathbf{b}|^2=161|\mathbf{a}|^2$$

$$\Rightarrow \quad |\mathbf{b}|^2=|\mathbf{a}|^2$$

From Eq. (i), we get

$$16\mathbf{a}\cdot\mathbf{b}=15|\mathbf{b}|^2-7|\mathbf{a}|^2=15|\mathbf{b}|^2-7|\mathbf{b}|^2$$

$$\Rightarrow \quad 16\mathbf{a}\cdot\mathbf{b}=8|\mathbf{b}|^2$$

$$\Rightarrow \quad \mathbf{a}\cdot\mathbf{b}=\frac{1}{2}|\mathbf{b}|^2$$

$$\Rightarrow \quad |\mathbf{a}||\mathbf{b}|\cos\theta=\frac{1}{2}|\mathbf{b}|^2$$

$$\Rightarrow \quad \cos\theta=\frac{1}{2}$$

$$\therefore \quad \theta=60^\circ$$

Ex. 21 *Let two non-collinear vectors* **a** *and* **b** *inclined at an angle* $\frac{2\pi}{3}$ *be such that* $|\mathbf{a}|=3$ *and* $|\mathbf{b}|=2$. *If a point P moves so that at any time t its position vector* **OP** *(where O is the origin) is given as* $\mathbf{OP}=\left(t+\frac{1}{t}\right)\mathbf{a}+\left(t-\frac{1}{t}\right)\mathbf{b}$, *then least distance of P from the origin is*

(a) $\sqrt{2\sqrt{133}-10}$ (b) $\sqrt{2\sqrt{133}+10}$

(c) $\sqrt{5+\sqrt{133}}$ (d) None of these

Sol. (b) We have, $|\mathbf{OP}|^2=\left(t+\frac{1}{t}\right)^2|\mathbf{a}|^2+\left(t-\frac{1}{t}\right)^2|\mathbf{b}|^2+2\left(t^2-\frac{1}{t^2}\right)|\mathbf{a}||\mathbf{b}|\cos\left(\frac{2\pi}{3}\right)$

$$\therefore \quad |\mathbf{OP}|^2=9\left(t+\frac{1}{t}\right)^2+4\left(t-\frac{1}{t}\right)^2+2\left(t^2-\frac{1}{t^2}\right)3\cdot 2\cdot\left(\frac{-1}{2}\right)$$

$$=9\left(t^2+\frac{1}{t^2}+2\right)+4\left(t^2+\frac{1}{t^2}-2\right)-6\left(t^2-\frac{1}{t^2}\right)$$

$$=7t^2+\frac{19}{t^2}+10$$

$$\Rightarrow \quad |\mathbf{OP}|^2\geq 2\cdot\sqrt{7t^2\cdot\frac{19}{t^2}}+10 \quad (\because \text{AM} \geq \text{GM})$$

$\therefore$ Minimum value of $|\mathbf{OP}|=\sqrt{10+2\sqrt{133}}$

Ex. 22 *If* **a, b, c** *be non-zero vectors such that* **a** *is perpendicular to* **b** *and* **c** *and* $|\mathbf{a}|=1, |\mathbf{b}|=2, |\mathbf{c}|=1, \mathbf{b}\cdot\mathbf{c}=1$ *and there is a non-zero vector* **d** *coplanar with* $\mathbf{a}+\mathbf{b}$ *and* $2\mathbf{b}-\mathbf{c}$ *and* $\mathbf{d}\cdot\mathbf{a}=1$, *then minimum value of* $|\mathbf{d}|$ *is*

(a) $\frac{2}{\sqrt{13}}$ (b) $\frac{3}{\sqrt{13}}$

(c) $\frac{4}{\sqrt{5}}$ (d) $\frac{4}{\sqrt{13}}$

Sol. (d) $\mathbf{a}\cdot\mathbf{b}=\mathbf{a}\cdot\mathbf{c}=0, |\mathbf{a}|=|\mathbf{c}|=1, |\mathbf{b}|=2$ and $\mathbf{b}\cdot\mathbf{c}=1$

let $\quad \mathbf{d}=x(\mathbf{a}+\mathbf{b})+y(2\mathbf{b}-\mathbf{c})$

But $\quad \mathbf{d}\cdot\mathbf{a}=1$

$$\Rightarrow \quad x(1+0)+0=1$$

$$\Rightarrow \quad x=1$$

$$\Rightarrow \quad \mathbf{d}=\mathbf{a}+\mathbf{b}+y(2\mathbf{b}-\mathbf{c})$$

$$\Rightarrow \quad |\mathbf{d}|^2=|\mathbf{a}|^2+|\mathbf{b}|+2\mathbf{a}\cdot\mathbf{b}+y^2(2\mathbf{b}-\mathbf{c})^2+2y(\mathbf{a}+\mathbf{b})\cdot(2\mathbf{b}-\mathbf{c})$$

$$\Rightarrow \quad |\mathbf{d}|^2=1+4+y^2(16+1-4)+2y(8-1)$$

$$=13y^2-14y+5$$

$$\therefore \quad |\mathbf{d}|_{\min}=\sqrt{\frac{4\times 13\times 5-14\times 14}{4\times 13}}=\frac{4}{\sqrt{13}}$$

Ex. 23 *A groove is in the form of a broken line ABC and the position vectors of the three points are respectively* $2\hat{\mathbf{i}}-3\hat{\mathbf{j}}+2\hat{\mathbf{k}}, 3\hat{\mathbf{i}}+2\hat{\mathbf{j}}-\hat{\mathbf{k}}$ *and* $\hat{\mathbf{i}}+\hat{\mathbf{j}}+\hat{\mathbf{k}}$. *A force of magnitude* $24\sqrt{3}$ *acts on a particle of unit mass kept at the point A and moves it along the groove to the point C. If the line of action of the force is parallel to the vector* $\hat{\mathbf{i}}+2\hat{\mathbf{j}}+\hat{\mathbf{k}}$ *all along, the number of units of work done by the force is*

(a) $144\sqrt{2}$ (b) $144\sqrt{3}$

(c) $72\sqrt{2}$ (d) $72\sqrt{3}$

Sol. (c) $\mathbf{F}=(24\sqrt{3})\dfrac{\hat{\mathbf{i}}+2\hat{\mathbf{j}}+\hat{\mathbf{k}}}{|\hat{\mathbf{i}}+2\hat{\mathbf{j}}+\hat{\mathbf{k}}|}=\dfrac{24\sqrt{3}}{\sqrt{6}}(\hat{\mathbf{i}}+2\hat{\mathbf{j}}+\hat{\mathbf{k}})$

$$=12\sqrt{2}(\hat{\mathbf{i}}+2\hat{\mathbf{j}}+\hat{\mathbf{k}})$$

Displacement, $\mathbf{r}=$ Position Vector of $C-$ Position Vector of A

$$=(\hat{\mathbf{i}}+\hat{\mathbf{j}}+\hat{\mathbf{k}})-(2\hat{\mathbf{i}}-3\hat{\mathbf{j}}+2\hat{\mathbf{k}})$$

$$=(-\hat{\mathbf{i}}+4\hat{\mathbf{j}}-\hat{\mathbf{k}})$$

Work done by the force

$$W=\mathbf{r}\cdot\mathbf{F}=(-\hat{\mathbf{i}}+4\hat{\mathbf{j}}-\hat{\mathbf{k}})\cdot 12\sqrt{2}(\hat{\mathbf{i}}+2\hat{\mathbf{j}}+\hat{\mathbf{k}})$$

$$=12\sqrt{2}(-1+8-1)=72\sqrt{2}$$

Ex. 24 *For any vectors* **a, b**; $|\mathbf{a}\times\mathbf{b}|^2+(\mathbf{a}\cdot\mathbf{b})^2$ *is equal to*

(a) $|\mathbf{a}|^2|\mathbf{b}|^2$ (b) $|\mathbf{a}+\mathbf{b}|$

(c) $|\mathbf{a}|^2-|\mathbf{b}|^2$ (d) 0

Sol. (a) We have, $|\mathbf{a}\times\mathbf{b}| = |\mathbf{a}||\mathbf{b}|\sin\theta$

$$\Rightarrow \quad |\mathbf{a}\times\mathbf{b}|^2 = |\mathbf{a}|^2|\mathbf{b}|^2\sin^2\theta$$
$$= |\mathbf{a}|^2|\mathbf{b}|^2(1-\cos^2\theta)$$
$$= |\mathbf{a}|^2|\mathbf{b}|^2 - |\mathbf{a}|^2|\mathbf{b}|^2\cos^2\theta$$
$$= |\mathbf{a}|^2|\mathbf{b}|^2 - (|\mathbf{a}||\mathbf{b}|\cos\theta)^2$$
$$= |\mathbf{a}|^2|\mathbf{b}|^2 - (\mathbf{a}\cdot\mathbf{b})^2$$
$$\therefore |\mathbf{a}\times\mathbf{b}|^2 + (\mathbf{a}\cdot\mathbf{b})^2 = |\mathbf{a}|^2|\mathbf{b}|^2$$

Ex. 25 *If* $\mathbf{a} = \hat{\mathbf{i}}+\hat{\mathbf{j}}+\hat{\mathbf{k}}, \mathbf{b} = \hat{\mathbf{i}}+\hat{\mathbf{j}}-\hat{\mathbf{k}}$, *then vectors perpendicular to* **a** *and* **b** *is/are*

(a) $\lambda(\hat{\mathbf{i}}+\hat{\mathbf{j}})$ (b) $\lambda(\hat{\mathbf{i}}+\hat{\mathbf{j}}+\hat{\mathbf{k}})$
(c) $\lambda(\hat{\mathbf{i}}-\hat{\mathbf{j}})$ (d) None of these

Sol. (c) Any vector perpendicular to both **a** and $\mathbf{b} = \lambda(\mathbf{a}\times\mathbf{b})$

$$\text{Now, } \mathbf{a}\times\mathbf{b} = \begin{vmatrix} \hat{\mathbf{i}} & \hat{\mathbf{j}} & \hat{\mathbf{k}} \\ 1 & 1 & 1 \\ 1 & 1 & -1 \end{vmatrix}$$
$$= -2\hat{\mathbf{i}} + 2\hat{\mathbf{j}} = -2(\hat{\mathbf{i}}-\hat{\mathbf{j}})$$
$$= -2(\hat{\mathbf{i}}-\hat{\mathbf{j}})$$

$\therefore$Required vector $= \lambda(\hat{\mathbf{i}}-\hat{\mathbf{j}})$

Ex. 26 *If* $\mathbf{a}\times\mathbf{b} = \mathbf{b}\times\mathbf{c} \neq 0$, *then the correct statement is*

(a) $\mathbf{b}\,||\,\mathbf{c}$ (b) $\mathbf{a}\,||\,\mathbf{b}$
(c) $(\mathbf{a}+\mathbf{c})\,||\,\mathbf{b}$ (d) None of these

Sol. (c) We have,

$$\mathbf{a}\times\mathbf{b} = \mathbf{b}\times\mathbf{c}$$
$$\Rightarrow \quad \mathbf{a}\times\mathbf{b} - \mathbf{b}\times\mathbf{c} = 0$$
$$\Rightarrow \quad \mathbf{a}\times\mathbf{b} + \mathbf{c}\times\mathbf{b} = 0$$
$$\Rightarrow \quad (\mathbf{a}+\mathbf{c})\times\mathbf{b} = 0$$
$$\therefore \quad (\mathbf{a}+\mathbf{c})\,||\,\mathbf{b}$$

[$\because$ if vector product of two vectors is zero, then both vectors are parallel to each other]

Ex. 27 *If* $\mathbf{a} = \hat{\mathbf{i}}+2\hat{\mathbf{j}}+3\hat{\mathbf{k}}, \mathbf{b} = -\hat{\mathbf{i}}+2\hat{\mathbf{j}}+\hat{\mathbf{k}}$ *and* $\mathbf{c} = 3\hat{\mathbf{i}}+\hat{\mathbf{j}}$. *If* $(\mathbf{a}+\mathbf{b})\perp\mathbf{c}$, *then t is equal to*

(a) 5 (b) 4
(c) 3 (d) 2

Sol. (a) We have, $\mathbf{a} = \hat{\mathbf{i}}+2\hat{\mathbf{j}}+3\hat{\mathbf{k}}$

$$\mathbf{b} = -\hat{\mathbf{i}}+2\hat{\mathbf{j}}+\hat{\mathbf{k}}$$

and $\mathbf{c} = 3\hat{\mathbf{i}}+\hat{\mathbf{j}}$

Since, $(\mathbf{a}+t\mathbf{b})$ is perpendicular to **c**

$$\therefore \quad (\mathbf{a}+t\mathbf{b})\cdot\mathbf{c} = 0$$
$$[(1-t)\hat{\mathbf{i}} + (2+2t)\hat{\mathbf{j}} + (3+t)\hat{\mathbf{k}}]\cdot(3\hat{\mathbf{i}}+\hat{\mathbf{j}}) = 0$$
$$\Rightarrow \quad 3(1-t) + (2+2t) = 0$$
$$\Rightarrow \quad t = 5$$

Ex. 28 *If* $\mathbf{a} = 2\hat{\mathbf{i}}-3\hat{\mathbf{j}}+\hat{\mathbf{k}}, \mathbf{b} = -\hat{\mathbf{i}}+\hat{\mathbf{k}}, \mathbf{c} = 2\hat{\mathbf{j}}-\hat{\mathbf{k}}$, *then the area (in sq units) of parallelogram with diagonals* $\mathbf{a}+\mathbf{b}$ *and* $\mathbf{b}+\mathbf{c}$ *will be*

(a) $\sqrt{21}$ (b) $2\sqrt{21}$
(c) $\frac{1}{2}\sqrt{21}$ (d) None of these

Sol. (c) We have, $\mathbf{a} = 2\hat{\mathbf{i}}-3\hat{\mathbf{j}}+\hat{\mathbf{k}}$

$$\mathbf{b} = -\hat{\mathbf{i}}+\hat{\mathbf{k}} \text{ and } \mathbf{c} = 2\hat{\mathbf{j}}-\hat{\mathbf{k}}$$

Since, $(\mathbf{a}+\mathbf{b})$ and $(\mathbf{b}+\mathbf{c})$ are the diagonals of the parallelogram

Now, $\mathbf{a}+\mathbf{b} = \hat{\mathbf{i}}-3\hat{\mathbf{j}}+2\hat{\mathbf{k}}$ and $\mathbf{b}+\mathbf{c} = -\hat{\mathbf{i}}+2\hat{\mathbf{j}}$

$$\therefore \text{ Area of parallelogram} = \frac{1}{2}|(\mathbf{a}+\mathbf{b})\times(\mathbf{b}+\mathbf{c})|$$
$$= \frac{1}{2}|(\hat{\mathbf{i}}-3\hat{\mathbf{j}}+2\hat{\mathbf{k}})\times(-\hat{\mathbf{i}}+2\hat{\mathbf{j}})|$$
$$= \left|\frac{1}{2}\begin{vmatrix} \hat{\mathbf{i}} & \hat{\mathbf{j}} & \hat{\mathbf{k}} \\ 1 & -3 & 2 \\ -1 & 2 & 0 \end{vmatrix}\right| = \left|\frac{1}{2}(-4\hat{\mathbf{i}}-2\hat{\mathbf{j}}-\hat{\mathbf{k}})\right|$$
$$= \frac{1}{2}\sqrt{(-4)^2(-2)^2+(-1)^2} = \frac{\sqrt{21}}{2} \text{ sq units.}$$

Ex. 29 *The coordinates of the mid-points of the sides of* ΔPQR *are* $(3a, 0, 0), (0, 3b, 0)$ *and* $(0, 0, 3c)$ *respectively, then the area of* ΔPQR *is*

(a) $18\sqrt{b^2c^2+c^2a^2+a^2b^2}$ (b) $9\sqrt{b^2c^2+c^2a^2+a^2b^2}$
(c) $\frac{9}{2}\sqrt{b^2c^2+c^2a^2+a^2b^2}$ (d) $18\sqrt{ab+bc+ca}$

Sol. (a) Let *L*, *M*, *N* be the mid-points of the sides of ΔPQR.

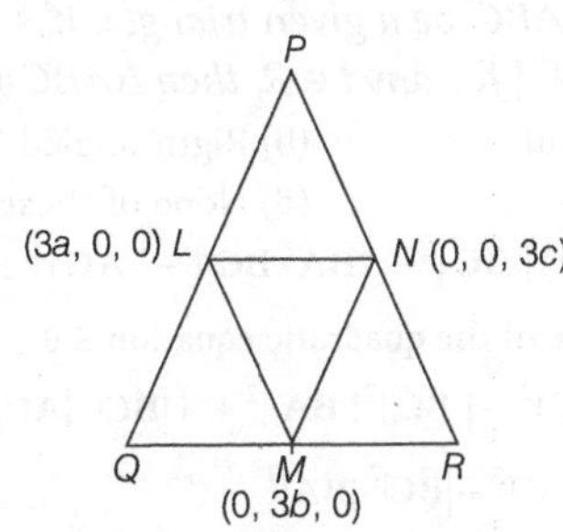

$$\text{Area of } \Delta LMN = \frac{1}{2}|\mathbf{MN}\times\mathbf{ML}|$$
$$= \frac{1}{2}|(-3b\hat{\mathbf{j}}+3c\hat{\mathbf{k}})\times(3a\hat{\mathbf{i}}-3b\hat{\mathbf{j}})|$$
$$= \frac{1}{2}|9(bc\hat{\mathbf{i}}+ca\hat{\mathbf{j}}+ab\hat{\mathbf{k}})|$$
$$= \frac{9}{2}\sqrt{(bc)^2+(ca)^2+(ab)^2}$$

Now, area of $\Delta PQR = 4\times(\text{Area of } \Delta LMN)$

$$= 4\times\frac{9}{2}\sqrt{b^2c^2+c^2a^2+a^2b^2}$$
$$= 18\sqrt{b^2c^2+c^2a^2+a^2b^2}$$

Ex. 30 *In a parallelogram ABCD,* $\mathbf{AB} = \hat{\mathbf{i}} + \hat{\mathbf{j}} + \hat{\mathbf{k}}$ *and diagonal* $\mathbf{AC} = \hat{\mathbf{i}} - \hat{\mathbf{j}} + \hat{\mathbf{k}}$ *and area of parallelogram is* $\sqrt{8}$ *sq units, then* $\angle BAC$ *is equal to*

(a) $\frac{\pi}{6}$ (b) $\frac{\pi}{3}$

(c) $\sin^{-1}\left(\frac{\sqrt{8}}{3}\right)$ (d) $\cos^{-1}\left(\frac{\sqrt{8}}{3}\right)$

Sol. (c) We have, $\mathbf{AB} = \hat{\mathbf{i}} + \hat{\mathbf{j}} + \hat{\mathbf{k}}$ and $\mathbf{AC} = \hat{\mathbf{i}} - \hat{\mathbf{j}} + \hat{\mathbf{k}}$

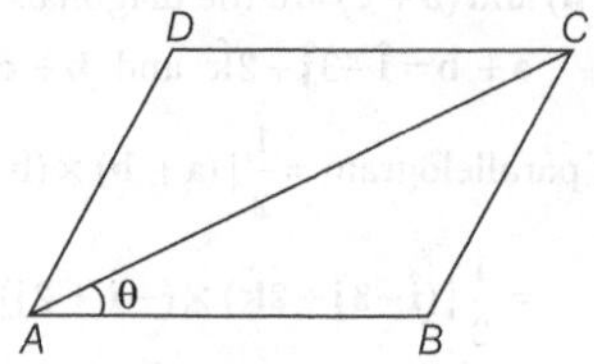

Let θ be the $\angle BAC$. Then,

$$\sin\theta = \frac{|\mathbf{AB} \times \mathbf{AC}|}{|\mathbf{AB}||\mathbf{AC}|}$$

Now, $$\mathbf{AB} \times \mathbf{AC} = \begin{vmatrix} \hat{\mathbf{i}} & \hat{\mathbf{j}} & \hat{\mathbf{k}} \\ 1 & 1 & 1 \\ 1 & -1 & 1 \end{vmatrix} = 2\hat{\mathbf{i}} - 2\hat{\mathbf{k}}$$

∴ $$|\mathbf{AB} \times \mathbf{AC}| = \sqrt{8}$$

Hence, $$\sin\theta = \frac{\sqrt{8}}{\sqrt{3} \times \sqrt{3}}$$

$$\Rightarrow \quad \theta = \sin^{-1}\left(\frac{\sqrt{8}}{3}\right)$$

Ex. 31 *Let* ΔABC *be a given triangle. If* $|\mathbf{BA} - t\mathbf{BC}| \geq |\mathbf{AC}|$ *for any* $t \in R$, *then* ΔABC *is*

(a) Equilateral (b) Right angled

(c) Isosceles (d) None of these

Sol. (b) $|\mathbf{BA}|^2 + t^2|\mathbf{BC}|^2 - 2\mathbf{BA} \cdot \mathbf{BC}\, t - |\mathbf{AC}|^2 \geq 0, \forall\, t \in R$ ...(i)

Discriminante of the quadratic equation ≤ 0

$\Rightarrow 4(\mathbf{BA} \cdot \mathbf{BC})^2 - |\mathbf{BC}|^2|\mathbf{BA}|^2 + 4|\mathbf{BC}|^2|\mathbf{AC}|^2 \leq 0$ ...(ii)

Using $(\mathbf{BA} \cdot \mathbf{BC})^2 - |\mathbf{BC}|^2|\mathbf{BA}|^2$

$$= -|\mathbf{BA} \times \mathbf{BC}|^2$$
$$= -|(\mathbf{BC} + \mathbf{CA}) \times \mathbf{BC}|^2$$
$$= -|\mathbf{CA} \times \mathbf{BC}|^2$$

Using Eq. (ii) in Eq. (i).

$$|\mathbf{BC}|^2|\mathbf{AC}|^2 \leq |\mathbf{AC} \times \mathbf{BC}|^2$$

But $$|\mathbf{AC} \times \mathbf{BC}| = |\mathbf{AC}||\mathbf{BC}|\sin C$$

$$\Rightarrow \quad \sin^2 C \geq 1$$

$$\Rightarrow \quad \sin C = \pm 1$$

$$\Rightarrow \quad \angle C = \frac{\pi}{2}$$

Ex. 32 *If* $a^2 + b^2 + c^2 = 1$ *where* $a, b, c \in R$, *then the maximum value of* $(4a - 3b)^2 + (5b - 4c)^2 + (3c - 5a)^2$ *is*

(a) 25 (b) 50

(c) 144 (d) None of these

Sol. (b) Let $\mathbf{r}_1 = a\hat{\mathbf{i}} + b\hat{\mathbf{j}} + c\hat{\mathbf{k}}$, $\mathbf{r}_2 = 3\hat{\mathbf{i}} + 4\hat{\mathbf{j}} + 5\hat{\mathbf{k}}$ ...(i)

$$|\mathbf{r}_1 \times \mathbf{r}_2| \leq |\mathbf{r}_1|^2|\mathbf{r}_2|^2$$

Now, $$\mathbf{r}_1 \times \mathbf{r}_2 = \begin{vmatrix} \hat{\mathbf{i}} & \hat{\mathbf{j}} & \hat{\mathbf{k}} \\ a & b & c \\ 3 & 4 & 5 \end{vmatrix}$$

$$= \hat{\mathbf{i}}(5b - 4c) + \hat{\mathbf{j}}(3c - 5a) + \hat{\mathbf{k}}(4a - 3b)$$

So, from Eq. (i), we get

$$(5b - 4c)^2 + (3c - 5a)^2 + (4a - 3b)^2 \leq 50$$

Ex. 33 *If* **a**, **b** *and* **c** *are pth, qth, rth terms of HP and* $\mathbf{u} = (q - r)\hat{\mathbf{i}} + (r - p)\hat{\mathbf{j}} + (p - q)\hat{\mathbf{k}}$, $\mathbf{v} = \frac{\hat{\mathbf{i}}}{a} + \frac{\hat{\mathbf{j}}}{b} + \frac{\hat{\mathbf{k}}}{c}$, *then*

(a) **u** and **v** are parallel vectors

(b) **u** and **v** are orthogonal vectors

(c) $\mathbf{u} \cdot \mathbf{v} = 1$

(d) $\mathbf{u} \times \mathbf{v} = \hat{\mathbf{i}} + \hat{\mathbf{j}} + \hat{\mathbf{k}}$

Sol. (b) Let A be the first term and D be the common difference of the corresponding AP. Then,

$$\frac{1}{a} = A + (p - 1)D, \frac{1}{b} = A + (q - 1)D, \frac{1}{c} = A + (r - 1)D$$

$$\Rightarrow a^{-1}(q - r) + b^{-1}(r - p) + c^{-1}(p - q) = 0$$

$$\Rightarrow \quad \mathbf{v} \cdot \mathbf{u} = 0$$

$$\Rightarrow \quad \mathbf{u} \perp \mathbf{v}$$

Hence, **u** and **v** are orthogonal vectors.

Ex. 34 *If the vector product of a constant vector* **OA** *with variable vector* **OB** *in a fixed plane OAB be a constant vector, then the locus of B is*

(a) a straight line perpendicular to **OA**

(b) a circle with centre O radius equal to $|\mathbf{OA}|$

(c) a straight line parallel to **OA**

(d) None of the above

Sol. (c) Let $A(\alpha, \beta)$ point be given and O be taken as the origin

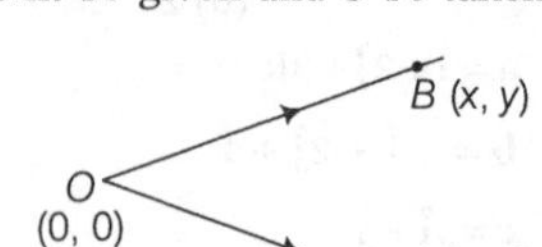

We have, $\mathbf{OA} = \alpha\hat{\mathbf{i}} + \beta\hat{\mathbf{j}}$

and $\mathbf{OB} = x\hat{\mathbf{i}} + y\hat{\mathbf{j}}$

Now, $|\mathbf{OA} \times \mathbf{OB}| = |(\alpha y - \beta x)\hat{\mathbf{k}}| = \text{constant}$

$\Rightarrow \quad \alpha y - \beta x = \text{constant}$

$\therefore$ Locus of $B(x, y)$ is a line parallel to OA because slope of $OA = \dfrac{\beta}{\alpha}$

Ex. 35 *Unit vector perpendicular to the plane of ΔABC with position vectors $\mathbf{a}, \mathbf{b}, \mathbf{c}$ of the vertices A, B, C is*

(a) $\dfrac{\mathbf{a}\times\mathbf{b}+\mathbf{b}\times\mathbf{c}+\mathbf{c}\times\mathbf{a}}{\Delta}$

(b) $\dfrac{\mathbf{a}\times\mathbf{b}+\mathbf{b}\times\mathbf{c}+\mathbf{c}\times\mathbf{a}}{2\Delta}$

(c) $\dfrac{\mathbf{a}\times\mathbf{b}+\mathbf{b}\times\mathbf{c}+\mathbf{c}\times\mathbf{a}}{4\Delta}$

(d) None of the above

Sol. (b) The required vector is given by

$$\hat{\mathbf{n}} = \frac{\mathbf{AB}\times\mathbf{AC}}{|\mathbf{AB}\times\mathbf{AC}|}$$

$$\begin{aligned}\mathbf{AB}\times\mathbf{AC} &= (\mathbf{b}\times\mathbf{a})\times(\mathbf{c}-\mathbf{a})\\ &= \mathbf{b}\times\mathbf{c}-\mathbf{b}\times\mathbf{a}-\mathbf{a}\times\mathbf{c}+\mathbf{a}\times\mathbf{a}\\ &= \mathbf{b}\times\mathbf{c}+\mathbf{a}\times\mathbf{b}+\mathbf{c}\times\mathbf{a} \quad [\because \mathbf{a}\times\mathbf{a}=0]\end{aligned}$$

We also know that,

Area of $\Delta ABC = \dfrac{1}{2}|\mathbf{a}\times\mathbf{b}+\mathbf{b}\times\mathbf{c}+\mathbf{c}\times\mathbf{a}|$

$$\therefore \quad \hat{\mathbf{n}} = \frac{\mathbf{a}\times\mathbf{b}+\mathbf{b}\times\mathbf{c}+\mathbf{c}\times\mathbf{a}}{|\mathbf{a}\times\mathbf{b}+\mathbf{b}\times\mathbf{c}+\mathbf{c}\times\mathbf{a}|} = \frac{\mathbf{a}\times\mathbf{b}+\mathbf{b}\times\mathbf{c}+\mathbf{c}\times\mathbf{a}}{2\Delta}$$

$$\left[\because \Delta = \frac{1}{2}|\mathbf{a}\times\mathbf{b}+\mathbf{b}\times\mathbf{c}+\mathbf{c}\times\mathbf{a}|\right]$$

Ex. 36 *The vector $\mathbf{r}$ satisfying the conditions that*

I. *it is perpendicular to $3\hat{\mathbf{i}}+2\hat{\mathbf{j}}+2\hat{\mathbf{k}}$ and $18\hat{\mathbf{i}}-22\hat{\mathbf{j}}-5\hat{\mathbf{k}}$*

II. *it makes an obtuse angle with Y-axis.*

III. $|\mathbf{r}| = 14$

(a) $2(-2\hat{\mathbf{i}}-3\hat{\mathbf{j}}+6\hat{\mathbf{k}})$

(b) $2(2\hat{\mathbf{i}}-3\hat{\mathbf{j}}+6\hat{\mathbf{k}})$

(c) $4\hat{\mathbf{i}}+6\hat{\mathbf{j}}-12\hat{\mathbf{k}}$

(d) None of the above

Sol. (a) Let $\mathbf{a} = 3\hat{\mathbf{i}}+2\hat{\mathbf{j}}+2\hat{\mathbf{k}}$ and $\mathbf{b} = 18\hat{\mathbf{i}}-22\hat{\mathbf{j}}-5\hat{\mathbf{k}}$

Then, the required vector $\mathbf{r}$ is given by

$$\begin{aligned} \mathbf{r} &= \lambda(\mathbf{a}\times\mathbf{b})\\ \Rightarrow \quad \mathbf{r} &= \lambda(34\hat{\mathbf{i}}+51\hat{\mathbf{j}}-102\hat{\mathbf{k}})\\ &= 17\lambda(2\hat{\mathbf{i}}+3\hat{\mathbf{j}}-6\hat{\mathbf{k}})\end{aligned}$$

Now, $\quad |\mathbf{r}| = 14 \Rightarrow 119|\lambda| = 14$

$\Rightarrow \quad |\lambda| = \dfrac{2}{17}$

Since, $\mathbf{r}$ makes an obtuse angle with Y-axis. Therefore,

$$\lambda = -\frac{2}{17}$$

Hence, $\quad \mathbf{r} = -2(2\hat{\mathbf{i}}+3\hat{\mathbf{j}}-6\hat{\mathbf{k}})$

or $\quad \mathbf{r} = 2(-2\hat{\mathbf{i}}-3\hat{\mathbf{j}}+6\hat{\mathbf{k}})$

Ex. 37 *Let a, b, c denote the lengths of the sides of a triangle such that*

$$(a-b)\,\mathbf{u}+(b-c)\,\mathbf{v}+(c-a)\,(\mathbf{u}\times\mathbf{v}) = 0$$

for any two non-collinear vectors $\mathbf{u}$ and $\mathbf{v}$, then the triangle is

(a) right angled (b) equilateral

(c) isosceles (d) scalene

Sol. (b) Since, $\mathbf{u}, \mathbf{v}$ and $\mathbf{u}\times\mathbf{b}$ are non-coplanar vectors.

$\therefore \quad (a-b)\,\mathbf{u}+(b-c)\mathbf{v}+(c-a)\,(\mathbf{u}\times\mathbf{v}) = 0$

$\Rightarrow \quad a-b = 0 = b-c = c-a$

$\Rightarrow \quad a = b = c$

So, the triangle is equilateral.

Ex. 38 *The value of $\hat{\mathbf{i}}\cdot(\hat{\mathbf{j}}\times\hat{\mathbf{k}})+\hat{\mathbf{j}}\cdot(\hat{\mathbf{k}}\times\hat{\mathbf{i}})+\hat{\mathbf{k}}\cdot(\hat{\mathbf{i}}\times\hat{\mathbf{j}})$ is*

(a) 3 (b) 2

(c) 1 (d) 0

Sol. (a) We have, $\hat{\mathbf{i}}\cdot(\hat{\mathbf{j}}\times\hat{\mathbf{k}})+\hat{\mathbf{j}}\cdot(\hat{\mathbf{k}}\times\hat{\mathbf{i}})+\hat{\mathbf{k}}\cdot(\hat{\mathbf{i}}\times\hat{\mathbf{j}})$

$= \hat{\mathbf{i}}\cdot\hat{\mathbf{i}}+\hat{\mathbf{j}}\cdot\hat{\mathbf{j}}+\hat{\mathbf{k}}\cdot\hat{\mathbf{k}} \quad [\because \hat{\mathbf{i}}\times\hat{\mathbf{j}} = \hat{\mathbf{k}}, \hat{\mathbf{j}}\times\hat{\mathbf{k}} = \hat{\mathbf{i}}, \hat{\mathbf{k}}\times\hat{\mathbf{i}} = \hat{\mathbf{j}}]$

$= |\hat{\mathbf{i}}|^2+|\hat{\mathbf{j}}|^2+|\hat{\mathbf{k}}|^2 = 1+1+1 = 3 \quad [\because \hat{\mathbf{i}}, \hat{\mathbf{j}}, \hat{\mathbf{k}}$ are unit vectors$]$

Ex. 39 *For non-zero vectors $\mathbf{a}, \mathbf{b}, \mathbf{c}$; $|(\mathbf{a}\times\mathbf{b})\cdot\mathbf{c}| = |\mathbf{a}||\mathbf{b}||\mathbf{c}|$ holds if and only if*

(a) $\mathbf{a}\cdot\mathbf{b} = 0, \mathbf{b}\cdot\mathbf{c} = 0$ (b) $\mathbf{b}\cdot\mathbf{c} = 0, \mathbf{c}\cdot\mathbf{a} = 0$

(c) $\mathbf{c}\cdot\mathbf{a} = 0, \mathbf{a}\cdot\mathbf{b} = 0$ (d) $\mathbf{a}\cdot\mathbf{b} = \mathbf{b}\cdot\mathbf{c} = \mathbf{c}\cdot\mathbf{a} = 0$

Sol. (d) We have, $|(\mathbf{a}\times\mathbf{b})\cdot\mathbf{c}| = |\mathbf{a}||\mathbf{b}||\mathbf{c}|$

$\Rightarrow \quad ||\mathbf{a}||\mathbf{b}||\mathbf{c}|\sin\theta\cos\alpha| = |\mathbf{a}||\mathbf{b}||\mathbf{c}|$

$\Rightarrow \quad |\sin\theta||\cos\alpha| = 1$

$\Rightarrow \quad \theta = \dfrac{\pi}{2}$ and $\alpha = 0$

$\Rightarrow \quad \mathbf{a}\perp\mathbf{b}$ and $\mathbf{c}\,||\,\hat{\mathbf{n}}$

$\Rightarrow \quad \mathbf{a}\perp\mathbf{c}$ and $\mathbf{c}\perp$ both $\mathbf{a}$ and $\mathbf{b}$

$\Rightarrow$ **a, b, c** are mutually perpendicular.

$\Rightarrow \quad \mathbf{a}\cdot\mathbf{c} = \mathbf{b}\cdot\mathbf{c} = \mathbf{c}\cdot\mathbf{a} = 0$

Ex. 40 *The position vectors of three vertices A, B, C of a tetrahedron $OABC$ with respect to its vertex O are $\hat{\mathbf{i}}, 6\hat{\mathbf{j}}, \hat{\mathbf{k}}$, then its volume (in cu units) is*

(a) 3 (b) $\dfrac{1}{3}$

(c) $\dfrac{1}{6}$ (d) 6

Sol. (d) We have, $A(6\hat{i}+0\hat{j}+0\hat{k})$, $B(0\hat{i}+6\hat{j}+0\hat{k})$, $C(0\hat{i}+0\hat{j}+\hat{k})$ and $O(0\hat{i}+0\hat{j}+0\hat{k})$

$\therefore$ $\mathbf{OA}=6\hat{i}+0\hat{j}+0\hat{k}$

$\mathbf{OB}=0\hat{i}+6\hat{j}+0\hat{k}$

and $\mathbf{OC}=0\hat{i}+0\hat{j}+\hat{k}$

$\therefore$ Volume of tetrahedron

$$=\frac{1}{6}[\mathbf{OA}\ \mathbf{OB}\ \mathbf{OC}]=\frac{1}{6}[6\hat{i}\ 6\hat{j}\ \hat{k}]$$

$$=\frac{1}{6}\begin{vmatrix}6&0&0\\0&6&0\\0&0&1\end{vmatrix}=\frac{1}{6}[6(6-0)]=6 \text{ cu units}$$

Ex. 41 *A parallelopiped is formed by planes drawn parallel to coordinate axes through the points $A=(1,2,3)$ and $B=(9,8,5)$. The volume of that parallelopiped is equal to (in cubic units)*

(a) 192 (b) 48
(c) 32 (d) 96

Sol. (d) Translating the axes through $A(1,2,3)$.

A changes to $(0,0,0)$, B changes to $(8,6,2)$.

$\therefore$ Coterminous edges are of lengths 8, 6, 2.

Volume of parallelopiped $=8\cdot6\cdot2=96$ cu units.

Ex. 42 *If $|\mathbf{a}|=1$, $|\mathbf{b}|=3$ and $|\mathbf{c}|=5$, then the value of $[\mathbf{a}-\mathbf{b}\ \mathbf{b}-\mathbf{c}\ \mathbf{c}-\mathbf{a}]$ is*

(a) 0 (b) 1
(c) -1 (d) None of these

Sol. (a) $[\mathbf{a}-\mathbf{b}\ \mathbf{b}-\mathbf{c}\ \mathbf{c}-\mathbf{a}]+\begin{bmatrix}1&-1&0\\0&1&-1\\-1&0&1\end{bmatrix}[\mathbf{abc}]$

$$=[1(1-0)+1(0-1)+0(0+1)][\mathbf{abc}]$$
$$=0\times[\mathbf{abc}]=0$$

Ex. 43 *If $\mathbf{a}, \mathbf{b}, \mathbf{c}$ are three non-coplanar vectors, then $3\mathbf{a}-7\mathbf{b}-4\mathbf{c}$, $3\mathbf{a}-2\mathbf{b}+\mathbf{c}$ and $\mathbf{a}+\mathbf{b}+\lambda\mathbf{c}$ will be coplanar, if λ is*

(a) -1 (b) 1
(c) 3 (d) 2

Sol. (d) Let $\alpha=3\mathbf{a}-7\mathbf{b}-4\mathbf{c}$, $\beta=3\mathbf{a}-2\mathbf{b}+\mathbf{c}$ and $\gamma=\mathbf{a}+\mathbf{b}+\lambda\mathbf{c}$

For α, β and γ to be coplanar $[\alpha\ \beta\ \gamma]=0$

$$\Rightarrow \begin{vmatrix}3&-7&-4\\3&-2&1\\1&1&\lambda\end{vmatrix}[\mathbf{a}\ \mathbf{b}\ \mathbf{c}]=0$$

$$[3(-2\lambda-1)+7(3\lambda-1)-4(3+2)(abc)]=0$$

$$\Rightarrow (15\lambda-30)[\mathbf{a}\ \mathbf{b}\ \mathbf{c}]=0$$

Since, $\mathbf{a}, \mathbf{b}, \mathbf{c}$ are non-coplanar

$$\therefore 15\lambda-30=0 \Rightarrow \lambda=2$$

Ex. 44. *Let $\mathbf{r}=(\mathbf{a}\times\mathbf{b})\sin x+(\mathbf{b}\times\mathbf{c})\cos y+(\mathbf{c}\times\mathbf{a})$, where $\mathbf{a}$, $\mathbf{b}$ and $\mathbf{c}$ are non-zero non-coplanar vectors. If $\mathbf{r}$ is orthogonal to $3\mathbf{a}+5\mathbf{b}+2\mathbf{c}$, then the value of $\sec^2 y+\operatorname{cosec}^2 x+\sec y\operatorname{cosec} x$ is*

(a) 3 (b) 4
(c) 5 (d) 6

Sol. (a) $\mathbf{r}\cdot(3\mathbf{a}+5\mathbf{b}+2\mathbf{c})=0$

$$\Rightarrow \mathbf{a}\cdot(\mathbf{b}\times\mathbf{c})[2\sin x+3\cos y+5]=0$$
$$\Rightarrow 2\sin x+3\cos y+5=0$$
$$\Rightarrow 2\sin x+3\cos y=-5 \quad [\because \mathbf{a}\cdot(\mathbf{b}\times\mathbf{c})\neq0]$$
$$\Rightarrow \sin x=-1,\ \cos y=1$$
$$\Rightarrow \operatorname{cosec} x=-1,\ \sec y=-1$$

Ex. 45 *Let a, b, c be distinct non-negative numbers. If the vectors $a\hat{i}+a\hat{j}+c\hat{k}$, $\hat{i}+\hat{k}$ and $c\hat{i}+c\hat{j}+b\hat{k}$ lie in a plane, then c is*

(a) HM of a and b (b) 0
(c) AM of a and b (d) GM of a and b

Sol. (d) Since, the given points lie in a plane

$$\therefore \begin{vmatrix}a&a&c\\1&0&1\\c&c&b\end{vmatrix}=0$$

On applying $C_1\to C_1-C_2$

$$\Rightarrow \begin{vmatrix}0&a&c\\1&0&1\\0&c&b\end{vmatrix}=0$$

$$\Rightarrow -1(ab-c^2)=0$$

$$\Rightarrow c^2=ab$$

Hence, c is GM of a and b.

Ex. 46 *If $\mathbf{a}, \mathbf{b}$ and $\mathbf{c}$ are non-coplanar vectors and λ is a real number, then $[\lambda(\mathbf{a}+\mathbf{b})\,|\,\lambda^2\mathbf{b}|\,\lambda\mathbf{c}]=[a\ \mathbf{a}+\mathbf{c}\ \mathbf{b}]$ for*

(a) exactly two values of λ
(b) exactly one value of λ
(c) no value of λ
(d) exactly three values of λ

Sol. (c) Given, $[\lambda(\mathbf{a}+\mathbf{b})\lambda^2\mathbf{b}\lambda\mathbf{c}]=[\mathbf{a}\,\mathbf{b}+\mathbf{c}\,\mathbf{b}]$

$$\Rightarrow \begin{vmatrix}\lambda(a_1+b_1)&\lambda(a_2+b_2)&\lambda(a_3+b_3)\\\lambda^2b_1&\lambda^2b_2&\lambda^2b_3\\\lambda c_1&\lambda c_2&\lambda c_3\end{vmatrix}$$

$$=\begin{vmatrix}a_1&a_2&a_3\\b_1+c_1&b_2+c_2&b_3+c_3\\b_1&b_2&b_3\end{vmatrix}$$

$$\Rightarrow \lambda^4\begin{vmatrix}a_1+b_1&a_2+b_2&a_3+b_3\\b_1&b_2&b_3\\c_1&c_2&c_3\end{vmatrix}$$

$$= \begin{vmatrix} a_1 & a_2 & a_3 \\ b_1 + c_1 & b_2 + c_2 & b_3 + c_3 \\ b_1 & b_2 & b_3 \end{vmatrix}$$

[applying $R_1 \to R_1 - R_2$ in LHS and $R_2 \to R_2 - R_3$ in RHS]

$$\Rightarrow \quad \lambda^4 \begin{vmatrix} a_1 & a_2 & a_3 \\ b_1 & b_2 & b_3 \\ c_1 & c_2 & c_3 \end{vmatrix} = - \begin{vmatrix} a_1 & a_2 & a_3 \\ b_1 & b_2 & b_3 \\ c_1 & c_2 & c_3 \end{vmatrix}$$

$$\Rightarrow \quad \lambda^4 = -1$$

Hence, no real value of λ exists.

Ex. 47 *In a regular tetrahedron, let θ be angle between any edge and a face not containing the edge. The value of $\cos^2 \theta$ is*

(a) 1/6 (b) 1/9

(c) 1/3 (d) None of these

Sol. (c) Let $OABC$ be the tetrahedron. Let G be the centroid of the face OAB, then $GA = \frac{1}{\sqrt{3}} AC$.

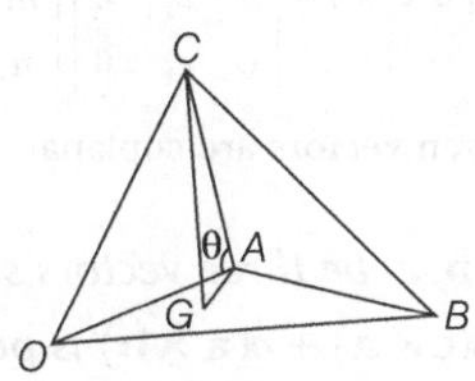

Then, $\cos\theta = \frac{GA}{CA} = \frac{1}{\sqrt{3}}$

$\therefore \quad \cos^2\theta = \frac{1}{3}$

Ex. 48 *DABC be a tetrahedron such that AD is perpendicular to the base ABC and $\angle ABC = 30°$. The volume of tetrahedron is 18. If value of $AB + BC + AD$ is minimum, then the length of AC is*

(a) $6\sqrt{2-\sqrt{3}}$ (b) $3(\sqrt{6}-\sqrt{2})$

(c) $6\sqrt{2+\sqrt{3}}$ (d) $3(\sqrt{6}+\sqrt{2})$

Sol. (a) $\quad \text{Volume} = \frac{1}{3} AD\left(\frac{1}{2} AB \cdot BC \sin 30°\right)$

$$\Rightarrow \quad 18 = \frac{1}{12}(AD \cdot AB \cdot BC)$$

$$\Rightarrow \quad AD \cdot AB \cdot BC = 216$$

Now, $AB + BC + AD \geq 3(AD \cdot AB \cdot BC)^{1/3}$

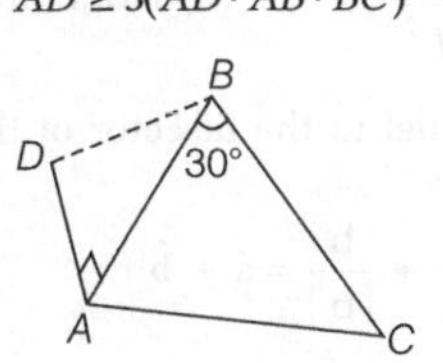

$$\Rightarrow \quad AB + BC + AD \geq 18$$

Minimum value occurs when $AB = BC = AD = 6$

Hence, $\quad AC = \sqrt{AB^2 + BC^2 - 2AB \cdot BC \cdot \cos 30°}$

$$= 6\sqrt{2-\sqrt{3}}$$

Ex. 49 *If* $\mathbf{a} = \hat{\mathbf{i}} + \hat{\mathbf{j}} + \hat{\mathbf{k}}, \mathbf{b} = \hat{\mathbf{i}} - \hat{\mathbf{j}} + \hat{\mathbf{k}}, \mathbf{c} = \hat{\mathbf{i}} + 2\hat{\mathbf{j}} - \hat{\mathbf{k}}$, *then the value of* $\begin{vmatrix} \mathbf{a}\cdot\mathbf{a} & \mathbf{a}\cdot\mathbf{b} & \mathbf{a}\cdot\mathbf{c} \\ \mathbf{b}\cdot\mathbf{a} & \mathbf{b}\cdot\mathbf{b} & \mathbf{b}\cdot\mathbf{c} \\ \mathbf{c}\cdot\mathbf{a} & \mathbf{c}\cdot\mathbf{b} & \mathbf{c}\cdot\mathbf{c} \end{vmatrix}$ *is*

(a) 2 (b) 4

(c) 16 (d) 64

Sol. (c) We have, $\mathbf{a} = \hat{\mathbf{i}} + \hat{\mathbf{j}} + \hat{\mathbf{k}}$

$$\mathbf{b} = \hat{\mathbf{i}} - \hat{\mathbf{j}} + \hat{\mathbf{k}} \text{ and } \mathbf{c} = \hat{\mathbf{i}} + 2\hat{\mathbf{j}} - \hat{\mathbf{k}}$$

We know that, $\begin{vmatrix} \mathbf{a}\cdot\mathbf{a} & \mathbf{a}\cdot\mathbf{b} & \mathbf{a}\cdot\mathbf{c} \\ \mathbf{b}\cdot\mathbf{a} & \mathbf{b}\cdot\mathbf{b} & \mathbf{b}\cdot\mathbf{c} \\ \mathbf{c}\cdot\mathbf{a} & \mathbf{c}\cdot\mathbf{b} & \mathbf{c}\cdot\mathbf{c} \end{vmatrix} = [\mathbf{a}\ \mathbf{b}\ \mathbf{c}]^2$

$$= \begin{vmatrix} 1 & 1 & 1 \\ 1 & -1 & 1 \\ 1 & 2 & -1 \end{vmatrix}^2$$

$$= [1(1-2) - 1(-1-1) + 1(2+1)]^2$$

$$= [-1+2+3]^2 = [4]^2 = 16$$

Ex. 50 *The value of a so that the volume of parallelopiped formed by* $\hat{\mathbf{i}} + a\hat{\mathbf{j}} + \hat{\mathbf{k}}, \hat{\mathbf{j}} + a\hat{\mathbf{k}}$ *and* $a\hat{\mathbf{i}} + \hat{\mathbf{k}}$ *becomes minimum is*

(a) -3 (b) 3

(c) $1/\sqrt{3}$ (d) $\sqrt{3}$

Sol. (c) Volume of the parallelopiped

$$V = [\hat{\mathbf{i}} + a\hat{\mathbf{j}} + \hat{\mathbf{k}}\hat{\mathbf{j}} + a\hat{\mathbf{k}}\ a\hat{\mathbf{i}} + \hat{\mathbf{k}}]$$

$$= (\hat{\mathbf{i}} + a\hat{\mathbf{j}} + \hat{\mathbf{k}}) \cdot \{(\hat{\mathbf{j}} + a\hat{\mathbf{k}}) \times (a\hat{\mathbf{i}} + \hat{\mathbf{k}})\}$$

$$= (\hat{\mathbf{i}} + a\hat{\mathbf{j}} + \hat{\mathbf{k}}) \cdot (\hat{\mathbf{i}} + a^2\hat{\mathbf{j}} - a\hat{\mathbf{k}}) = 1 + a^3 - a$$

$$\frac{dV}{da} = 3a^2 - 1, \frac{d^2V}{da^2} = 6a, \frac{dV}{da} = 0$$

$$\Rightarrow \quad 3a^2 - 1 = 0 \Rightarrow a = \pm \frac{1}{\sqrt{3}}$$

At $\quad a = \frac{1}{\sqrt{3}}, \frac{d^2V}{da^2} = \frac{6}{\sqrt{3}} > 0$

$\therefore V$ is minimum at $a = \frac{1}{\sqrt{3}}$

Ex. 51 *If* $\mathbf{a}, \mathbf{b}$ *and* $\mathbf{c}$ *be any three non-zero and non-coplanar vectors, then any vector* $\mathbf{r}$ *is equal to*

(a) $z\mathbf{a} + x\mathbf{b} + y\mathbf{c}$ (b) $x\mathbf{a} + y\mathbf{b} + z\mathbf{c}$

(c) $y\mathbf{a} + z\mathbf{b} + x\mathbf{c}$ (d) None of these

where, $x = \frac{[\mathbf{r}\ \mathbf{b}\ \mathbf{c}]}{[\mathbf{a}\ \mathbf{b}\ \mathbf{c}]}, y = \frac{[\mathbf{r}\ \mathbf{c}\ \mathbf{a}]}{[\mathbf{a}\ \mathbf{b}\ \mathbf{c}]}, z = \frac{[\mathbf{r}\ \mathbf{a}\ \mathbf{b}]}{[\mathbf{a}\ \mathbf{b}\ \mathbf{c}]}$

Sol. (b) Since, $\mathbf{a}$, $\mathbf{b}$ and $\mathbf{c}$ are three non-coplanar vectors, we may assume $\mathbf{r} = \alpha\mathbf{a} + \beta\mathbf{b} + \gamma\mathbf{c}$

$$[\mathbf{r}\ \mathbf{b}\ \mathbf{c}] = (\alpha\mathbf{a} + \beta\mathbf{b} + \gamma\mathbf{c})\cdot(\mathbf{b}\times\mathbf{c}) = \alpha\{\mathbf{a}\cdot(\mathbf{b}\times\mathbf{c})\}$$
$$= \alpha[\mathbf{a}\ \mathbf{b}\ \mathbf{c}]$$
$$\Rightarrow \quad \alpha = \frac{[\mathbf{r}\ \mathbf{b}\ \mathbf{c}]}{[\mathbf{a}\ \mathbf{b}\ \mathbf{c}]}$$
But $$x = \frac{[\mathbf{r}\ \mathbf{b}\ \mathbf{c}]}{[\mathbf{a}\ \mathbf{b}\ \mathbf{c}]}$$
$$\therefore \quad \alpha = x$$
Similarly, $\beta = y, \gamma = z$
$$\therefore \quad \mathbf{r} = x\mathbf{a} + y\mathbf{b} + z\mathbf{c}$$

Ex. 52 *The position vectors of vertices of ΔABC are $\mathbf{a}$, $\mathbf{b}$, $\mathbf{c}$ and $\mathbf{a}\cdot\mathbf{a} = \mathbf{b}\cdot\mathbf{b} = \mathbf{c}\cdot\mathbf{c} = 3$. If $[\mathbf{a}\ \mathbf{b}\ \mathbf{c}] = 0$, then the position vectors of the orthocentre of ΔABC is*

(a) $\mathbf{a}+\mathbf{b}+\mathbf{c}$ (b) $\frac{1}{3}(\mathbf{a}+\mathbf{b}+\mathbf{c})$

(c) 0 (d) None of these

Sol. (a) Hence, $[\mathbf{a}\ \mathbf{b}\ \mathbf{c}] = 0$

So, the points O, A, B and C are coplanar. Also, $OA = OB = OC = \sqrt{3}$, hence origin O is the circumcentre.

Position vector of the centroid G is $\frac{\mathbf{a}+\mathbf{b}+\mathbf{c}}{3}$.

Now, orthocentre divides OG in the ratio of 3 : 2 externally. So, position vectors of orthocentre is $\mathbf{a}+\mathbf{b}+\mathbf{c}$.

Ex. 53 *If α and β are two mutually perpendicular unit vectors $\{r\alpha + r\beta + s(\alpha\times\beta)\}$, $[\alpha + (\alpha\times\beta)]$ and $\{s\alpha + s\beta + t(\alpha\times\beta)\}$ are coplanar, then s is equal to*

(a) AM of r and t (b) GM of r and t

(c) HM of r and t (d) None of these

Sol. Since, α and β are two mutually perpendicular vectors and $(r\alpha + r\beta + s(\alpha\times\beta)$, $[\alpha + (\alpha\times\beta)]$, $\{s\alpha + s\beta + t(\alpha\times\beta)\}$ are coplanar

$$\Rightarrow \quad \begin{vmatrix} r & r & s \\ 1 & 0 & 1 \\ s & s & t \end{vmatrix} = 0$$
$$\Rightarrow \quad s^2 = rt$$

Ex. 54 *Let $\mathbf{b} = -\hat{\mathbf{i}} + 4\hat{\mathbf{j}} + 6\hat{\mathbf{k}}$ and $\mathbf{c} = 2\hat{\mathbf{i}} - 7\hat{\mathbf{j}} - 10\hat{\mathbf{k}}$. If $\mathbf{a}$ be a unit vector and the scalar triple product $[\mathbf{a}\ \mathbf{b}\ \mathbf{c}]$ has the greatest value, then $\mathbf{a}$ is equal to*

(a) $\frac{1}{\sqrt{3}}(\hat{\mathbf{i}} + \hat{\mathbf{j}} + \hat{\mathbf{k}})$ (b) $\frac{1}{\sqrt{5}}(\sqrt{2}\hat{\mathbf{i}} - \hat{\mathbf{j}} - \sqrt{2}\hat{\mathbf{k}})$

(c) $\frac{1}{3}(2\hat{\mathbf{i}} + 2\hat{\mathbf{j}} - \hat{\mathbf{k}})$ (d) $\frac{1}{\sqrt{59}}(3\hat{\mathbf{i}} - 7\hat{\mathbf{j}} - \hat{\mathbf{k}})$

Sol. (c) $\mathbf{b}\times\mathbf{c} = 2\hat{\mathbf{i}} + 2\hat{\mathbf{j}} - \hat{\mathbf{k}}$

$$[\mathbf{a}\ \mathbf{b}\ \mathbf{c}] = \mathbf{a}\cdot(2\hat{\mathbf{i}} + 2\hat{\mathbf{j}} - \hat{\mathbf{k}}) = 1\cdot 3\cdot\cos\theta \le 3$$

The greatest value of $[\mathbf{a}\ \mathbf{b}\ \mathbf{c}] = 3$, which is obtained when $\theta = 0$.

So, $$\mathbf{a} = \frac{\mathbf{b}\times\mathbf{c}}{|\mathbf{b}\times\mathbf{c}|} = \frac{2}{3}\hat{\mathbf{i}} + \frac{2}{3}\hat{\mathbf{j}} - \frac{1}{3}\hat{\mathbf{k}}$$

Ex. 55 *The vectors*
$$\mathbf{u} = (al + a_1l_1)\hat{\mathbf{i}} + (am + a_1m_1)\hat{\mathbf{j}} + (an + a_1n_1)\hat{\mathbf{k}}$$
$$\mathbf{v} = (bl + b_1l_1)\hat{\mathbf{i}} + (bm + b_1m_1)\hat{\mathbf{j}} + (bn + b_1n_1)\hat{\mathbf{k}} \text{ and}$$
$$\mathbf{w} = (cl + c_1l_1)\hat{\mathbf{i}} + (cm + c_1m_1)\hat{\mathbf{j}} + (cn + c_1n_1)\hat{\mathbf{k}}$$

(a) form an equilateral triangle

(b) are coplanar

(c) are collinear

(d) are mutually perpendicular

Sol. (b) We have,

$$[\mathbf{u}\ \mathbf{v}\ \mathbf{w}] = \begin{vmatrix} al + a_1l_1 & am + a_1m_1 & an + a_1n_1 \\ bl + b_1l_1 & bm + b_1m_1 & bn + b_1n_1 \\ cl + c_1l_1 & cm + c_1m_1 & cn + c_1n_1 \end{vmatrix}$$
$$\Rightarrow \quad [\mathbf{u}\ \mathbf{v}\ \mathbf{w}] = \begin{vmatrix} a & a_1 & 0 \\ b & b_1 & 0 \\ c & c_1 & 0 \end{vmatrix} \begin{vmatrix} l & l_1 & 0 \\ m & m_1 & 0 \\ n & n_1 & 0 \end{vmatrix} = 0$$

Hence, the given vectors are coplanar.

Ex. 56 *Let $\mathbf{a}, \mathbf{b}, \mathbf{c}$ be three vectors such that $[\mathbf{a}\ \mathbf{b}\ \mathbf{c}] = 2$. If $\mathbf{r} = l(\mathbf{b}\times\mathbf{c}) + m(\mathbf{c}\times\mathbf{a}) + n(\mathbf{a}\times\mathbf{b})$ is perpendicular to $\mathbf{a}+\mathbf{b}+\mathbf{c}$, then the value of $(l+m+n)$ is*

(a) 2 (b) 1

(c) 0 (d) None of these

Sol. (c) It is given that $\mathbf{r}$ perpendicular $(\mathbf{a}+\mathbf{b}+\mathbf{c})$

$$\therefore \quad \mathbf{r}\cdot(\mathbf{a}+\mathbf{b}+\mathbf{c}) = 0$$
$$\Rightarrow \quad l[\mathbf{a}\ \mathbf{b}\ \mathbf{c}] + m[\mathbf{c}\ \mathbf{a}\ \mathbf{b}] + n[\mathbf{a}\ \mathbf{b}\ \mathbf{c}] = 0$$
$$\Rightarrow \quad 2(l + m + n) = 0 \qquad [\because [\mathbf{a}\ \mathbf{b}\ \mathbf{c}] = 2]$$
$$\Rightarrow \quad l + m + n = 0$$

Ex. 57 *If $\mathbf{a}$, $\mathbf{b}$ and $\mathbf{c}$ are three mutually perpendicular vectors, then the projection of the vectors $l\frac{\mathbf{a}}{|\mathbf{a}|} + m\frac{\mathbf{b}}{|\mathbf{b}|} + n\frac{(\mathbf{a}\times\mathbf{b})}{|\mathbf{a}\times\mathbf{b}|}$ along the angle bisector of the vectors $\mathbf{a}$ and $\mathbf{b}$ is*

(a) $\frac{l+m}{\sqrt{2}}$ (b) $\sqrt{l^2 + m^2 + n^2}$

(c) $\frac{\sqrt{l^2 + m^2}}{\sqrt{l^2 + m^2 + b^2}}$ (d) None of these

Sol. (a) A vector parallel to the bisector of the angle between the vectors $\mathbf{a}$ and $\mathbf{b}$ is

$$\frac{\mathbf{a}}{|\mathbf{a}|} + \frac{\mathbf{b}}{|\mathbf{b}|} = \hat{\mathbf{a}} + \hat{\mathbf{b}}$$

$\therefore$Units vector along the bisector

$$= \frac{\hat{\mathbf{a}} + \hat{\mathbf{b}}}{|\hat{\mathbf{a}} + \hat{\mathbf{b}}|} = \frac{1}{\sqrt{2}}(\hat{\mathbf{a}} + \hat{\mathbf{b}})$$

$$(\hat{\mathbf{a}} + \hat{\mathbf{b}})^2 = |\mathbf{a}|^2 + |\hat{\mathbf{b}}| + 2\hat{\mathbf{a}} \cdot \hat{\mathbf{b}} = 1 + 1 + 0 = 2$$

$\therefore$ Required projection

$$= \left[l\frac{\mathbf{a}}{|\mathbf{a}|} + m\frac{\mathbf{b}}{|\mathbf{b}|} + n\frac{\mathbf{a} \times \mathbf{b}}{|\mathbf{a} \times \mathbf{b}|} \right] \cdot \frac{1}{\sqrt{2}}(\hat{\mathbf{a}} + \hat{\mathbf{b}})$$

$$= \frac{1}{\sqrt{2}}(l + m)$$

$$\left[\because \frac{\mathbf{a} \cdot \hat{\mathbf{a}}}{|\mathbf{a}|} = \frac{\mathbf{b} \cdot \hat{\mathbf{b}}}{|\mathbf{b}|} = 1 \text{ and } \hat{\mathbf{a}} \cdot \hat{\mathbf{b}} = \hat{\mathbf{a}} \cdot (\mathbf{a} \times \mathbf{b}) = \hat{\mathbf{b}} \cdot (\hat{\mathbf{a}} \times \mathbf{b}) = 0 \right]$$

Ex. 58 *If the volume of parallelopiped formed by the vectors* **a**, **b**, **c** *as three coterminous edges is 27 cu units, then the volume of the parallelopiped have* $\alpha = \mathbf{a} + 2\mathbf{b} - \mathbf{c}$, $\beta = \mathbf{a} - \mathbf{b}$ *and* $\gamma = \mathbf{a} - \mathbf{b} - \mathbf{c}$ *as three coterminous edges is*

(a) 27 (b) 9
(c) 81 (d) None of these

Sol. (c) We have, $|[\mathbf{a}\ \mathbf{b}\ \mathbf{c}]| = 27$ cu units

Now, $$[\alpha\ \beta\ \gamma] = \begin{vmatrix} 1 & 2 & -1 \\ 1 & -1 & 0 \\ 1 & -1 & -1 \end{vmatrix} [\mathbf{a}\ \mathbf{b}\ \mathbf{c}] = 3[\mathbf{a}\ \mathbf{b}\ \mathbf{c}]$$

$\therefore$Required volume $= |[\alpha\ \beta\ \gamma]|$

$= 3|[\mathbf{a}\ \mathbf{b}\ \mathbf{c}]|$

$= 3 \times 27 = 81$ cu units

Ex. 59 *If V is the volume of the parallelopiped having three coterminous edges as* **a**, **b** *and* **c**, *then the volume of parallelopiped having three coterminous edges as* $\alpha = (\mathbf{a} \cdot \mathbf{a})\mathbf{a} + (\mathbf{a} \cdot \mathbf{b})\mathbf{b} + (\mathbf{a} \cdot \mathbf{c})\mathbf{c}$, $\beta = (\mathbf{a} \cdot \mathbf{b})\mathbf{a} + (\mathbf{b} \cdot \mathbf{b})\mathbf{b} + (\mathbf{b} \cdot \mathbf{c})\mathbf{c}$ *and* $\gamma = (\mathbf{a} \cdot \mathbf{c})\,\mathbf{a} + (\mathbf{b} \cdot \mathbf{c})\,\mathbf{b} + (\mathbf{c} \cdot \mathbf{c})\,\mathbf{c}$ *is*

(a) V^3 (b) $3V$
(c) V^2 (d) $2V$

Sol. (a) We have, $|[\mathbf{a}\ \mathbf{b}\ \mathbf{c}]| = V$

Let V_1 be the volume of the parallelopiped formed by the vectors α, β and γ. Then,

$$V_1 = |[\alpha\ \beta\ \gamma]|$$

Now, $$[\alpha\ \beta\ \gamma] = \begin{vmatrix} \mathbf{a} \cdot \mathbf{a} & \mathbf{a} \cdot \mathbf{b} & \mathbf{a} \cdot \mathbf{c} \\ \mathbf{a} \cdot \mathbf{b} & \mathbf{b} \cdot \mathbf{b} & \mathbf{b} \cdot \mathbf{c} \\ \mathbf{a} \cdot \mathbf{c} & \mathbf{b} \cdot \mathbf{c} & \mathbf{c} \cdot \mathbf{c} \end{vmatrix} [\mathbf{a}\ \mathbf{b}\ \mathbf{c}]$$

$\Rightarrow$ $[\alpha\ \beta\ \gamma] = [\mathbf{a}\ \mathbf{b}\ \mathbf{c}]^2\,[\mathbf{a}\ \mathbf{b}\ \mathbf{c}]$

$\Rightarrow$ $[\alpha\ \beta\ \gamma] = [\mathbf{a}\ \mathbf{b}\ \mathbf{c}]^3$

$\because$ $V_1 = |[\alpha\ \beta\ \gamma]$

$= |[\mathbf{a}\ \mathbf{b}\ \mathbf{c}]^3| = V^3$

Ex. 60 *Let* **r**, **a**, **b** *and* **c** *be four non-zero vectors such that* $\mathbf{r} \cdot \mathbf{a} = 0$,

$|\mathbf{r} \times \mathbf{b}| = |\mathbf{r}||\mathbf{b}|, |\mathbf{r} \times \mathbf{c}| = |\mathbf{r}||\mathbf{c}|$, *then* $[\mathbf{a}\ \mathbf{b}\ \mathbf{c}]$ *is*

(a) $|\mathbf{a}||\mathbf{b}||\mathbf{c}|$ (b) $-|\mathbf{a}||\mathbf{b}||\mathbf{c}|$
(c) 0 (d) None of these

Sol. (c) Given, $\mathbf{r} \cdot \mathbf{a} = 0$

$$|\mathbf{r} \times \mathbf{b}| = |\mathbf{r}||\mathbf{b}|$$

and $$|\mathbf{r} \times \mathbf{c}| = |\mathbf{r}||\mathbf{c}|$$

This shows **r** is perpendicular to both **b** and **c**.

$\Rightarrow$ **r** is perpendicular to **a**, **b** and **c**

$\therefore$ $[\mathbf{a}\ \mathbf{b}\ \mathbf{c}] = 0$

Ex. 61 *If* **a**, **b** *and* **c** *are any three vector forming a linearly independent system, then* $\theta \in R$,

$$\left[\mathbf{a}\cos\theta + \mathbf{b}\sin\theta + \mathbf{c}\cos 2\theta,\ \mathbf{a}\cos\left(\frac{2\pi}{3} + \theta\right) + \mathbf{b}\sin\left(\frac{2\pi}{3} + \theta\right) + \mathbf{c}\cos 2\left(\frac{2\pi}{3} + \theta\right)\right.$$

$$\left.\mathbf{a}\cos\left(\theta - \frac{2\pi}{3}\right) + \mathbf{b}\sin\left(\theta - \frac{2\pi}{3}\right) + \mathbf{c}\cos 2\left(\theta - \frac{2\pi}{3}\right)\right] \textit{ is}$$

(a) $[\mathbf{a}\ \mathbf{b}\ \mathbf{c}]\cos\theta$
(b) $[\mathbf{a}\ \mathbf{b}\ \mathbf{c}]\cos 2\theta$
(c) $[\mathbf{a}\ \mathbf{b}\ \mathbf{c}]\cos 3\theta$
(d) None of the above

Sol. (d) Since **a**, **b** and **c** are linearly independent,

$$[\mathbf{a}\ \mathbf{b}\ \mathbf{c}] \neq 0$$

We know that,

$$[a x_1 + \mathbf{b} x_2 + \mathbf{c} x_3 \mathbf{a} y_1 + \mathbf{b} y_2 + \mathbf{c} y_3\ \mathbf{a} z_1 + \mathbf{b} z_2 + \mathbf{c} z_3]$$

$$= [\mathbf{a}\ \mathbf{b}\ \mathbf{c}] \cdot \begin{vmatrix} x_1 & x_2 & x_3 \\ y_1 & y_2 & y_3 \\ z_1 & z_2 & z_3 \end{vmatrix}$$

Hence, the given system can be written as

$$[\mathbf{a}\ \mathbf{b}\ \mathbf{c}] \begin{vmatrix} \cos\theta & \sin\theta & \cos 2\theta \\ \cos\left(\frac{2\pi}{3} + \theta\right) & \sin\left(\frac{2\pi}{3} + \theta\right) & \cos 2\left(\theta + \frac{2\pi}{3}\right) \\ \cos\left(\theta - \frac{2\pi}{3}\right) & \sin\left(\theta - \frac{2\pi}{3}\right) & \cos 2\left(\theta - \frac{2\pi}{3}\right) \end{vmatrix}$$

On applying $R_3 \to R_3 + R_2 + R_1$, we get

$$[\mathbf{a}\ \mathbf{b}\ \mathbf{c}] \begin{vmatrix} \cos\theta & \sin\theta & \cos 2\theta \\ \cos\left(\frac{2\pi}{3} + \theta\right) & \sin\left(\frac{2\pi}{3} + \theta\right) & \cos 2\left(\theta + \frac{2\pi}{3}\right) \\ 0 & 0 & 0 \end{vmatrix}$$

$\Rightarrow [\mathbf{a}\ \mathbf{b}\ \mathbf{c}]\,(0) = 0$

Ex. 62 *Let* $\mathbf{a}, \mathbf{b}, \mathbf{c}$ *be three non-coplanar vectors and* $\mathbf{d}$ *be a non-zero vector, which is perpendicular to* $\mathbf{a}+\mathbf{b}+\mathbf{c}$*. Now, if* $\mathbf{d}=(\sin x)(\mathbf{a}\times\mathbf{b})+(\cos y)(\mathbf{b}\times\mathbf{c})+2(\mathbf{c}\times\mathbf{a})$*, then the minimum value of* (x^2+y^2) *is*

(a) π^2 (b) $\frac{\pi^2}{2}$

(c) $\frac{\pi^2}{4}$ (d) $\frac{5\pi^2}{4}$

Sol. (d) Given, $\mathbf{d}\cdot(\mathbf{a}+\mathbf{b}+\mathbf{c})=0$

and $\mathbf{d}=\sin x\,(\mathbf{a}\times\mathbf{b})+\cos y\,(\mathbf{b}\times\mathbf{c})+2\,(\mathbf{c}\times\mathbf{a})$ ...(i)

$\because$ $\mathbf{a}\cdot\mathbf{d}=\cos y[\mathbf{a}\ \mathbf{b}\ \mathbf{c}]$ ... (ii)

$\mathbf{b}\cdot\mathbf{d}=2\,[\mathbf{b}\ \mathbf{c}\ \mathbf{a}]$ ... (iii)

$\mathbf{c}\cdot\mathbf{d}=\sin x\,[\mathbf{a}\ \mathbf{b}\ \mathbf{c}]$ ... (iv)

On adding Eqs. (ii), (iii) and (iv), we get

$$\mathbf{a}\cdot\mathbf{d}+\mathbf{b}\cdot\mathbf{d}+\mathbf{c}\cdot\mathbf{d}=(\cos y+2+\sin x)[\mathbf{a}\ \mathbf{b}\ \mathbf{c}]$$

$\because$ $\sin x+\cos y+2=0$

$\Rightarrow$ $\sin x+\cos y=-2$

$\Rightarrow$ $\sin x=-1$

$[\because -1<\sin x\le 1 \text{ and } -1\le\cos y\le 1]$

and $\cos y=-1$

Since, we have to find the minimum value of

$$x^2+y^2;\ x=-\frac{\pi}{2}, y=\pi$$

$\therefore$ $$x^2+y^2=\frac{\pi^2}{4}+\pi^2=\frac{5\pi^2}{4}$$

Ex. 63 *Let* $\mathbf{a}, \mathbf{b}, \mathbf{c}$ *be three vectors of magnitude 1, 1 and 2 respectively. If* $\mathbf{a}\times(\mathbf{a}\times\mathbf{c})+\mathbf{b}=0$*, then the acute angle between* $\mathbf{a}$ *and* $\mathbf{c}$ *is*

(a) $\frac{\pi}{3}$ (b) $\frac{\pi}{4}$

(c) $\frac{\pi}{6}$ (d) None of these

Sol. (c) Given, $|\mathbf{a}|=1, |\mathbf{b}|=1,$ and $|\mathbf{c}|=2$

Also, $\mathbf{a}\times(\mathbf{a}\times\mathbf{c})+\mathbf{b}=0$

$\Rightarrow$ $(\mathbf{a}\cdot\mathbf{c})\,\mathbf{a}-(\mathbf{a}\cdot\mathbf{a})\mathbf{c}+\mathbf{b}=0$

$\Rightarrow$ $(\mathbf{a}\cdot\mathbf{c})\mathbf{a}-\mathbf{c}+\mathbf{b}=0$ $[\because \mathbf{a}\cdot\mathbf{a}=|\mathbf{a}|^2=1]$

$\Rightarrow$ $(\mathbf{a}\cdot\mathbf{c})\mathbf{a}-\mathbf{c}=-\mathbf{b}$

$\Rightarrow$ $|(\mathbf{a}\cdot\mathbf{c})\mathbf{a}-\mathbf{c}|=|-\mathbf{b}|$

$\Rightarrow$ $|(\mathbf{a}\cdot\mathbf{c})\mathbf{a}-\mathbf{c}|^2=|\mathbf{b}|^2$

$\Rightarrow$ $|(\mathbf{a}\cdot\mathbf{c})\mathbf{a}|^2+|\mathbf{c}|^2-2\{(\mathbf{a}\cdot\mathbf{c})\mathbf{a}\cdot\mathbf{c}\}=|\mathbf{b}|^2$

$\Rightarrow$ $(\mathbf{a}\cdot\mathbf{c})\,|\mathbf{a}|^2+|\mathbf{c}|^2-2(\mathbf{a}\cdot\mathbf{c})\,(\mathbf{a}\cdot\mathbf{c})=|\mathbf{b}|^2$

$\Rightarrow$ $(\mathbf{a}\cdot\mathbf{c})^2\,\{|\mathbf{a}|^2-2\}+|\mathbf{c}|^2=|\mathbf{b}|^2$

$\Rightarrow$ $-(\mathbf{a}\cdot\mathbf{c})^2+4=1$

$\Rightarrow$ $\mathbf{a}\cdot\mathbf{c}=\pm\sqrt{3}$

$\Rightarrow$ $|\mathbf{a}||\mathbf{c}|\cos\theta=\sqrt{3}$

where, θ is an acute angle between $\mathbf{a}$ and $\mathbf{c}$.

$\Rightarrow$ $\cos\theta=\frac{\sqrt{3}}{2}$

$\therefore$ $\theta=\frac{\pi}{6}$

Ex. 64 *Let* $\mathbf{a}=2\hat{\mathbf{i}}+\hat{\mathbf{j}}+\hat{\mathbf{k}}, \mathbf{b}=\hat{\mathbf{i}}+2\hat{\mathbf{j}}-\hat{\mathbf{k}}$ *and* $\mathbf{c}$ *is a unit vector coplanar to them. If* $\mathbf{c}$ *is perpendicular to* $\mathbf{a}$*, then* $\mathbf{c}$ *is equal to*

(a) $\frac{1}{\sqrt{2}}(-\hat{\mathbf{j}}+\hat{\mathbf{k}})$ (b) $-\frac{1}{\sqrt{3}}(\hat{\mathbf{i}}+\hat{\mathbf{j}}+\hat{\mathbf{k}})$

(c) $\frac{1}{\sqrt{5}}(\hat{\mathbf{i}}-2\hat{\mathbf{j}})$ (d) $\frac{1}{\sqrt{3}}(\hat{\mathbf{i}}+\hat{\mathbf{j}}+\hat{\mathbf{k}})$

Sol. (a) $\mathbf{a}\times(\mathbf{a}\times\mathbf{b})=(\mathbf{a}\cdot\mathbf{c})\,\mathbf{a}-(\mathbf{a}\cdot\mathbf{a})\mathbf{b}$

$$=3(\hat{\mathbf{i}}+\hat{\mathbf{j}}+\hat{\mathbf{k}})-6(\hat{\mathbf{i}}+2\hat{\mathbf{j}}-\hat{\mathbf{k}})=-9\hat{\mathbf{j}}+9\hat{\mathbf{k}}$$

$\therefore$Required unit vector $=\frac{\mathbf{a}\times(\mathbf{a}\times\mathbf{b})}{|\mathbf{a}\times(\mathbf{a}\times\mathbf{b})|}$

$$=\pm\frac{1}{\sqrt{2}}(-\hat{\mathbf{j}}+\hat{\mathbf{k}})$$

Ex. 65 *Let* $\mathbf{a}=2\hat{\mathbf{i}}+\hat{\mathbf{j}}-2\hat{\mathbf{k}}$ *and* $\mathbf{b}=\hat{\mathbf{i}}+\hat{\mathbf{j}}$*. If* $\mathbf{c}$ *is a vector such that* $\mathbf{a}\cdot\mathbf{c}=|\mathbf{c}|, |\mathbf{c}-\mathbf{a}|=2\sqrt{2}$ *and the angle between* $\mathbf{a}\times\mathbf{b}$ *and* $\mathbf{c}$ *is 30°, then* $|(\mathbf{a}\times\mathbf{b})\times\mathbf{c}|$ *is equal to*

(a) $\frac{2}{3}$ (b) $\frac{3}{2}$

(c) 2 (d) 3

Sol. (b) $|\mathbf{c}-\mathbf{a}|=2\sqrt{2}$

$\Rightarrow$ $|\mathbf{c}|^2+|\mathbf{a}|^2-2\mathbf{a}\cdot\mathbf{c}=8$

$\Rightarrow$ $|\mathbf{c}|^2+(\sqrt{9})^2-2|\mathbf{c}|=8$

$\Rightarrow$ $|\mathbf{c}|^2-2|\mathbf{c}|+1=0$

$\Rightarrow$ $(|\mathbf{c}|-1)^2=0\Rightarrow|\mathbf{c}|=1$

Now, $\mathbf{a}\times\mathbf{b}=2\hat{\mathbf{i}}+2\hat{\mathbf{j}}+\hat{\mathbf{k}}$

$\therefore$ $|\mathbf{a}\times\mathbf{b}|=\sqrt{4+4+1}=3$

$\therefore$ $|(\mathbf{a}\times\mathbf{b})\times\mathbf{c}|=|\mathbf{a}\times\mathbf{b}||\mathbf{c}|\sin 30°$

$$=3\times1\times\frac{1}{2}=\frac{3}{2}$$

Ex. 66 *Let* $\hat{\mathbf{a}}$ *and* $\hat{\mathbf{b}}$ *be two unit vectors such that* $\hat{\mathbf{a}}\cdot\hat{\mathbf{b}}=\frac{1}{3}$ *and* $\hat{\mathbf{a}}\times\hat{\mathbf{b}}=\hat{\mathbf{c}}$*. Also* $\bar{F}=\alpha\hat{\mathbf{a}}+\beta\hat{\mathbf{b}}+\gamma\hat{\mathbf{c}}$*, where* α, β, γ *are scalars. If* $\alpha=k_1(\hat{\mathbf{F}}\cdot\hat{\mathbf{a}})-k_2(\hat{\mathbf{F}}\cdot\hat{\mathbf{b}})$*, then the value of* $2(k_1+k_2)$ *is*

(a) $2\sqrt{3}$ (b) $\sqrt{3}$

(c) 3 (d) 1

Sol. (c) $\bar{\mathbf{F}}=\alpha\hat{\mathbf{a}}+\beta\hat{\mathbf{b}}+\gamma\hat{\mathbf{c}}$

$\therefore$ $\bar{F}\cdot(\hat{\mathbf{b}}\times\hat{\mathbf{c}})=\alpha\,[\hat{\mathbf{a}}\ \hat{\mathbf{b}}\ \hat{\mathbf{c}}]$

$\therefore \quad \alpha = \dfrac{\hat{\mathbf{F}} \cdot (\hat{\mathbf{b}} \times \hat{\mathbf{c}})}{[\hat{\mathbf{a}}\,\hat{\mathbf{b}}\,\hat{\mathbf{c}}]}$

Now, $\quad \hat{\mathbf{b}} \times \hat{\mathbf{c}} = \hat{\mathbf{b}} \times (\hat{\mathbf{a}} \times \hat{\mathbf{b}}) = (\hat{\mathbf{b}} \cdot \hat{\mathbf{b}})\hat{\mathbf{a}} - (\hat{\mathbf{b}} \cdot \hat{\mathbf{a}})\hat{\mathbf{b}}$

$= \hat{\mathbf{a}} - \dfrac{1}{3}\hat{\mathbf{b}}$

$[\hat{\mathbf{a}}\hat{\mathbf{b}}\hat{\mathbf{c}}] = (\hat{\mathbf{a}} \times \hat{\mathbf{b}})^2 = |\hat{\mathbf{a}}|^2 |\hat{\mathbf{b}}|^2 - (\hat{\mathbf{a}} \cdot \hat{\mathbf{b}})^2$

$= 1 - \dfrac{1}{9} = \dfrac{8}{9}$

$\therefore \quad \alpha = \dfrac{9}{8}\left\{\hat{\mathbf{F}} \cdot \hat{\mathbf{a}} - \dfrac{1}{3}\hat{\mathbf{F}} \cdot \hat{\mathbf{b}}\right\}$

$\therefore \quad k_1 = \dfrac{9}{8}, k_2 = \dfrac{3}{8}$

$\Rightarrow \quad k_1 + k_2 = \dfrac{12}{8} = \dfrac{3}{2}$

$\therefore \quad 2(k_1 + k_2) = 3$

Ex. 67 *Let* $\mathbf{a} = \hat{\mathbf{i}} - \hat{\mathbf{j}}, \mathbf{b} = \hat{\mathbf{j}} - \hat{\mathbf{k}}$ *and* $\mathbf{c} = \hat{\mathbf{k}} - \hat{\mathbf{i}}$. *If* $\mathbf{d}$ *is a unit vector such that* $\mathbf{a} \cdot \mathbf{d} = 0 = (\mathbf{b}\ \mathbf{c}\ \mathbf{d})$, *then* $\mathbf{d}$ *is equal to*

(a) $\pm \dfrac{(\hat{\mathbf{i}} + \hat{\mathbf{j}} - 2\hat{\mathbf{k}})}{\sqrt{6}}$ (b) $\pm \dfrac{(\hat{\mathbf{i}} + \hat{\mathbf{j}} - \hat{\mathbf{k}})}{\sqrt{3}}$

(c) $\pm \dfrac{(\hat{\mathbf{i}} + \hat{\mathbf{j}} + \hat{\mathbf{k}})}{\sqrt{3}}$ (d) $\pm \hat{\mathbf{k}}$

Sol. (a) We have, $\mathbf{a} \cdot \hat{\mathbf{d}} = 0$ and $[\mathbf{b}\ \mathbf{c}\ \hat{\mathbf{d}}] = 0$

$\Rightarrow \mathbf{a} \perp \hat{\mathbf{d}}$ and $\mathbf{b}, \mathbf{c}, \hat{\mathbf{d}}$ are coplanar.

$\Rightarrow \hat{\mathbf{d}} \perp \mathbf{a}$ and $\hat{\mathbf{d}}$ lies in the plane of $\mathbf{b}$ and $\mathbf{c}$, we know that the vector $\mathbf{r} = \mathbf{a} \times (\mathbf{b} \times \mathbf{c})$ is perpendicular to $\mathbf{a}$ and lies in the plane of $\mathbf{b}$ and $\mathbf{c}$

$\therefore \quad \hat{\mathbf{d}} = \pm \dfrac{\mathbf{r}}{|\mathbf{r}|}$

Now, $\quad \mathbf{r} = \mathbf{a} \times (\mathbf{b} \times \mathbf{c})$

$\Rightarrow \quad \mathbf{r} = (\mathbf{a} \cdot \mathbf{c})\,\mathbf{b} - (\mathbf{a} \cdot \mathbf{b})\mathbf{c}$

$\Rightarrow \quad \mathbf{r} = -(\hat{\mathbf{j}} - \hat{\mathbf{k}}) + (\hat{\mathbf{k}} - \hat{\mathbf{i}})$

$= -\hat{\mathbf{i}} - \hat{\mathbf{j}} + 2\hat{\mathbf{k}}$

$\therefore \quad \hat{\mathbf{d}} = \pm \dfrac{\mathbf{r}}{|\mathbf{r}|} = \pm \dfrac{(-\hat{\mathbf{i}} - \hat{\mathbf{j}} + 2\hat{\mathbf{k}})}{\sqrt{1+1+4}}$

$= \pm \dfrac{\hat{\mathbf{i}} + \hat{\mathbf{j}} - 2\hat{\mathbf{k}}}{\sqrt{6}}$

Ex. 68 *If* $\mathbf{a}, \mathbf{b}$ *and* $\mathbf{c}$ *are non-coplanar unit vectors such that* $\mathbf{a} \times (\mathbf{b} \times \mathbf{c}) = \dfrac{1}{\sqrt{2}}(\mathbf{b} + \mathbf{c})$, *then the angle between* $\mathbf{a}$ *and* $\mathbf{b}$ *is*

(a) $\dfrac{3\pi}{4}$ (b) $\dfrac{\pi}{4}$

(c) $\dfrac{\pi}{2}$ (d) π

Sol. (a) We have, $\mathbf{a} \times (\mathbf{b} \times \mathbf{c}) = \dfrac{\mathbf{b} \times \mathbf{c}}{\sqrt{2}}$

$\Rightarrow \quad (\mathbf{a} \cdot \mathbf{c})\mathbf{b} - (\mathbf{a} \cdot \mathbf{b})\mathbf{c} = \dfrac{1}{\sqrt{2}}\mathbf{b} + \dfrac{1}{\sqrt{2}}\mathbf{c}$

$\Rightarrow \quad (\mathbf{a} \cdot \mathbf{c}) - \dfrac{1}{\sqrt{2}} = 0 \quad [\because \mathbf{a}, \mathbf{b}, \mathbf{c}$ are non-coplanar]

and $\quad \mathbf{a} \cdot \mathbf{b} = -\dfrac{1}{\sqrt{2}}$

$\Rightarrow \quad |\mathbf{a}||\mathbf{b}|\cos\theta = -\dfrac{1}{\sqrt{2}} \Rightarrow \cos\theta = -\dfrac{1}{\sqrt{2}}$

$\therefore \quad \theta = \dfrac{3\pi}{4}$

Ex. 69 *The unit vector which is orthogonal to the vector* $3\hat{\mathbf{i}} + 2\hat{\mathbf{j}} + 6\hat{\mathbf{k}}$ *and is coplanar with the vectors* $2\hat{\mathbf{i}} + \hat{\mathbf{j}} + \hat{\mathbf{k}}$ *and* $\hat{\mathbf{i}} - \hat{\mathbf{j}} + \hat{\mathbf{k}}$ *is*

(a) $\dfrac{2\hat{\mathbf{i}} - 6\hat{\mathbf{j}} + \hat{\mathbf{k}}}{\sqrt{41}}$ (b) $\dfrac{2\hat{\mathbf{i}} - 3\hat{\mathbf{j}}}{\sqrt{13}}$

(c) $\dfrac{3\hat{\mathbf{j}} - \hat{\mathbf{k}}}{\sqrt{10}}$ (d) $\dfrac{4\hat{\mathbf{i}} + 3\hat{\mathbf{j}} - 3\hat{\mathbf{k}}}{\sqrt{34}}$

Sol. (c) Let $\mathbf{a} = 3\hat{\mathbf{i}} + 2\hat{\mathbf{j}} + 6\hat{\mathbf{k}}, \mathbf{b} = 2\hat{\mathbf{i}} + \hat{\mathbf{j}} + \hat{\mathbf{k}}, \mathbf{c} = \hat{\mathbf{i}} - \hat{\mathbf{j}} + \hat{\mathbf{k}}$

Then, by definition, a vector orthogonal to $\mathbf{a}$ and coplanar to $\mathbf{b}$ and $\mathbf{c}$ is given by

$\Rightarrow \quad \mathbf{a} \times (\mathbf{b} \times \mathbf{c})$

$\Rightarrow \mathbf{a} \times (\mathbf{b} \times \mathbf{c}) = (\mathbf{a} \cdot \mathbf{c})\mathbf{b} - (\mathbf{a} \cdot \mathbf{b})\mathbf{c}$

$= 7(2\hat{\mathbf{i}} + \hat{\mathbf{j}} + \hat{\mathbf{k}}) - (14)(\hat{\mathbf{i}} - \hat{\mathbf{j}} + \hat{\mathbf{k}}) = 21\hat{\mathbf{j}} - 7\hat{\mathbf{k}}$

Hence, a unit vector $= \dfrac{\mathbf{a} \times (\mathbf{b} \times \mathbf{c})}{|\mathbf{a} \times (\mathbf{b} \times \mathbf{c})|} = \dfrac{3\hat{\mathbf{j}} - \hat{\mathbf{k}}}{\sqrt{10}}$

Ex. 70 *Let* $\mathbf{a}, \mathbf{b}$ *and* $\mathbf{c}$ *be non-zero vectors such that* $(\mathbf{a} \times \mathbf{b}) \times \mathbf{c} = \dfrac{1}{3}|\mathbf{b}||\mathbf{c}|\,\mathbf{a}$. *If* θ *is the acute angle between the vector* $\mathbf{b}$ *and* $\mathbf{c}$, *then* $\sin\theta$ *is equal to*

(a) $\dfrac{2\sqrt{2}}{3}$ (b) $\dfrac{\sqrt{2}}{3}$

(c) $\dfrac{2}{3}$ (d) $\dfrac{1}{3}$

Sol. (a) $(\mathbf{a} \times \mathbf{b}) \times \mathbf{c} = \dfrac{1}{3}|\mathbf{b}||\mathbf{c}|\mathbf{a}$

$\Rightarrow \quad (\mathbf{a} \cdot \mathbf{c})\mathbf{b} - (\mathbf{b} \cdot \mathbf{c})\,\mathbf{a} = \dfrac{1}{3}|\mathbf{b}||\mathbf{c}|\mathbf{a}$

$\Rightarrow \quad (\mathbf{a} \cdot \mathbf{c})\mathbf{b} = \{(\mathbf{b} \cdot \mathbf{c}) + \dfrac{1}{3}|\mathbf{b}||\mathbf{c}|\}\mathbf{a}$

$\Rightarrow \quad (\mathbf{a} \cdot \mathbf{c})\mathbf{b} = |\mathbf{b}||\mathbf{c}|\left\{\cos\theta + \dfrac{1}{3}\right\}\mathbf{a}$

As $\mathbf{a}$ and $\mathbf{b}$ are not parallel, $\mathbf{a} \cdot \mathbf{c} = 0$ and $\cos\theta + \dfrac{1}{3} = 0$

$\Rightarrow \quad \cos\theta = -\dfrac{1}{3} \Rightarrow \sin\theta = \dfrac{2\sqrt{2}}{3}$

Ex. 71 *The value for* $[\mathbf{a}\times(\mathbf{b}+\mathbf{c}), \mathbf{b}\times(\mathbf{c}-2\mathbf{a}), \mathbf{c}\times(\mathbf{a}+3\mathbf{b})]$ *is equal to*

(a) $[\mathbf{a}\ \mathbf{b}\ \mathbf{c}]^2$

(b) $7[\mathbf{a}\ \mathbf{b}\ \mathbf{c}]^2$

(c) $-5[\mathbf{a}\times\mathbf{b}\ \mathbf{b}\times\mathbf{c}\ \mathbf{c}\times\mathbf{a}]$

(d) None of the above

Sol. (b) Let $\mathbf{a}\times\mathbf{b}=l$, $\mathbf{b}\times\mathbf{c}=m$ and $\mathbf{c}\times\mathbf{a}=n$,

$$\therefore\quad [\mathbf{a}\times(\mathbf{b}+\mathbf{c}), \mathbf{b}\times(\mathbf{c}-2\mathbf{a}), \mathbf{c}\times(\mathbf{a}+3\mathbf{b})]$$
$$=[l-n, m+2l, n-3m]$$
$$=\begin{vmatrix} 1 & 0 & -1 \\ 2 & 1 & 1 \\ 0 & -3 & 1 \end{vmatrix}[lmn]$$
$$=7[lmn]=7[\mathbf{a}\times\mathbf{b}\ \mathbf{b}\times\mathbf{c}\ \mathbf{c}\times\mathbf{a}]$$
$$=7[\mathbf{a}\ \mathbf{b}\ \mathbf{c}]^2$$

Ex. 72 *If* **a**, **b**, **c** *and* **p**, **q**, **r** *are reciprocal system of vectors, then* $\mathbf{a}\times\mathbf{p}+\mathbf{b}\times\mathbf{q}+\mathbf{c}\times\mathbf{r}$ *is equal to*

(a) $[\mathbf{a}\ \mathbf{b}\ \mathbf{c}]$ (b) $[\mathbf{p}+\mathbf{q}+\mathbf{r}]$

(c) 0 (d) $\mathbf{a}+\mathbf{b}+\mathbf{c}$

Sol. (c) $\mathbf{p}=\dfrac{\mathbf{b}\times\mathbf{c}}{[\mathbf{a}\ \mathbf{b}\ \mathbf{c}]}$, $\mathbf{q}=\dfrac{\mathbf{c}\times\mathbf{a}}{[\mathbf{a}\ \mathbf{b}\ \mathbf{c}]}$, $\mathbf{r}=\dfrac{\mathbf{a}\times\mathbf{b}}{[\mathbf{a}\ \mathbf{b}\ \mathbf{c}]}$

$$\mathbf{a}\times\mathbf{p}=\mathbf{a}\times\frac{(\mathbf{b}\times\mathbf{c})}{[\mathbf{a}\ \mathbf{b}\ \mathbf{c}]}=\frac{(\mathbf{a}\cdot\mathbf{c})\mathbf{b}-(\mathbf{a}\cdot\mathbf{b})\mathbf{c}}{[\mathbf{a}\ \mathbf{b}\ \mathbf{c}]}$$

Similarly, $\mathbf{b}\times\mathbf{q}=\dfrac{(\mathbf{a}\cdot\mathbf{b})\mathbf{c}-(\mathbf{b}\cdot\mathbf{a})\mathbf{a}}{[\mathbf{a}\ \mathbf{b}\ \mathbf{c}]}$

and $\mathbf{c}\times\mathbf{r}=\dfrac{(\mathbf{b}\cdot\mathbf{c})\mathbf{a}-(\mathbf{a}\cdot\mathbf{c})\mathbf{b}}{[\mathbf{a}\ \mathbf{b}\ \mathbf{c}]}$

$$\therefore\quad \mathbf{a}\times\mathbf{p}+\mathbf{b}\times\mathbf{q}+\mathbf{c}\times\mathbf{r}$$
$$=\frac{1}{[\mathbf{a}\ \mathbf{b}\ \mathbf{c}]}\{(\mathbf{a}\cdot\mathbf{c})\mathbf{b}-(\mathbf{a}\cdot\mathbf{b})\mathbf{c}+(\mathbf{a}\cdot\mathbf{b})\mathbf{c}-(\mathbf{b}\cdot\mathbf{c})\mathbf{a}+(\mathbf{b}\cdot\mathbf{c})\mathbf{a}-(\mathbf{a}\cdot\mathbf{c})\mathbf{b}\}$$
$$=\frac{1}{[\mathbf{a}\ \mathbf{b}\ \mathbf{c}]}\times 0=0$$

Ex. 73 *Solve* $\mathbf{a}\cdot\mathbf{r}=x$, $\mathbf{b}\cdot\mathbf{r}=y$, $\mathbf{c}\cdot\mathbf{r}=z$, *where* **a**, **b**, **c** *are given non-coplanar vectors.*

Sol. Given $\mathbf{a}\cdot\mathbf{r}=x$, $\mathbf{b}\cdot\mathbf{r}=y$, $\mathbf{c}\cdot\mathbf{r}=z$

Let $\mathbf{a}'$, $\mathbf{b}'$, $\mathbf{c}'$, be the reciprocal vectors of **a**, **b**, **c**, respectively.

Then, $\mathbf{a}=\dfrac{\mathbf{b}\times\mathbf{c}}{[\mathbf{a}\ \mathbf{b}\ \mathbf{c}]}$, $\mathbf{b}'=\dfrac{\mathbf{c}\times\mathbf{a}}{[\mathbf{a}\ \mathbf{b}\ \mathbf{c}]}$, $\mathbf{c}=\dfrac{\mathbf{a}\times\mathbf{b}}{[\mathbf{a}\ \mathbf{b}\ \mathbf{c}]}$

Now, $\mathbf{r}=(\mathbf{r}\cdot\mathbf{a})\mathbf{a}'+(\mathbf{r}\cdot\mathbf{b})\mathbf{b}'+(\mathbf{r}\cdot\mathbf{c})\mathbf{c}'$
$=x\mathbf{a}'+y\mathbf{b}'+z\mathbf{c}'$

JEE Type Solved Examples : More than One Option Correct Type Questions

Ex. 74 *If* $\mathbf{z}_1=a\hat{\mathbf{i}}+b\hat{\mathbf{j}}$ *and* $\mathbf{z}_2=c\hat{\mathbf{i}}+d\hat{\mathbf{j}}$ *are two vectors in* $\hat{\mathbf{i}}$ *and* $\hat{\mathbf{j}}$ *system, where* $|\mathbf{z}_1|=|\mathbf{z}_2|=r$ *and* $\mathbf{z}_1\cdot\mathbf{z}_2=0$, *then* $\mathbf{w}_1=a\hat{\mathbf{i}}+c\hat{\mathbf{j}}$ *and* $\mathbf{w}_2=b\hat{\mathbf{i}}+d\hat{\mathbf{j}}$ *satisfy*

(a) $|\mathbf{w}_1|=r$

(b) $|\mathbf{w}_2|=r$

(c) $\mathbf{w}_1\cdot\mathbf{w}_2=0$

(d) None of the above

Sol. (a, b, c) $|z_1|=|z_2|=r$ and $z_1, z_2=0$

$$\Rightarrow\quad a^2+b^2=c^2+d^2=r \quad \text{...(i)}$$

and $ac+bd=0$

as, $ac=-bd$

$$\Rightarrow\quad \frac{a}{b}=\frac{c}{-d}=\lambda \quad \text{...(ii)}$$

From Eqs. (i) and (ii),

$$a^2(1+\lambda^2)=d^2(1+\lambda^2)$$
$$\Rightarrow\quad a^2=d^2 \text{ and } b^2=c^2$$

Now, $|w_1|=a^2+c^2=a^2+b^2=h=|w_2|$

$$w_1\cdot w_2=ab+bd=0$$

Ex. 75 *If unit vectors* $\hat{\mathbf{i}}$ *and* $\hat{\mathbf{j}}$ *are at right angles to each other and* $\mathbf{p}=3\hat{\mathbf{i}}+4\hat{\mathbf{j}}$, $\mathbf{q}=5\mathbf{i}$, $4\mathbf{r}=\mathbf{p}+\mathbf{q}$ *and* $2\mathbf{s}=\mathbf{p}-\mathbf{q}$, *then*

(a) $|\mathbf{r}+k\mathbf{s}|=|\mathbf{r}-k\mathbf{s}|$ for all real k

(b) **r** is perpendicular to **s**

(c) $\mathbf{r}+\mathbf{s}$ is perpendicular to $\mathbf{r}-\mathbf{s}$

(d) $|\mathbf{r}|=|\mathbf{s}|=|\mathbf{p}|=|\mathbf{q}|$

Sol. (a, b, d) We have, $\mathbf{p}=3\hat{\mathbf{i}}+4\hat{\mathbf{j}}$ and $\mathbf{p}=5\hat{\mathbf{i}}$

Also, $4\mathbf{r}=\mathbf{p}+\mathbf{q}=3\hat{\mathbf{i}}+4\hat{\mathbf{j}}+5\hat{\mathbf{i}}$
$=8\hat{\mathbf{i}}+4\hat{\mathbf{j}}\Rightarrow\mathbf{r}=2\hat{\mathbf{i}}+\hat{\mathbf{j}}$

and $2\mathbf{s}=\mathbf{p}-\mathbf{q}=3\hat{\mathbf{i}}+4\hat{\mathbf{j}}+5\hat{\mathbf{i}}=-2\hat{\mathbf{i}}+4\hat{\mathbf{j}}$

$\Rightarrow\quad \mathbf{s}=-\hat{\mathbf{i}}+2\hat{\mathbf{j}}$

Now, $|\mathbf{r}+k\mathbf{s}|=|\mathbf{r}-k\mathbf{s}|$

$$\Rightarrow\quad |2\hat{\mathbf{i}}+\hat{\mathbf{j}}-k\hat{\mathbf{i}}+2k\hat{\mathbf{j}}|^2=|2\hat{\mathbf{i}}+\hat{\mathbf{j}}+k\hat{\mathbf{i}}-2k\hat{\mathbf{j}}|^2$$
$$\Rightarrow\quad (2-k)^2+(1+2k)^2=(2+k)^2+(1-2k)^2$$

Which is true for all values of k.

Now, $\mathbf{r}\cdot\hat{\mathbf{s}}=(2\hat{\mathbf{i}}+\hat{\mathbf{j}})(-\hat{\mathbf{i}}+2\hat{\mathbf{j}})$
$=-2+2=0$

$\therefore\quad \mathbf{r}\perp\mathbf{s}$

Also, $(\mathbf{r}+\mathbf{s})\cdot(\mathbf{r}-\mathbf{s}) = (\hat{\mathbf{i}}+3\hat{\mathbf{j}})(3\hat{\mathbf{i}}-\hat{\mathbf{j}}) = 3-3=0$

$\therefore \quad (\mathbf{r}+\mathbf{s}) \perp (\mathbf{r}-\mathbf{s})$

Also, $\quad |\mathbf{r}| = \sqrt{(2)^2+1} = \sqrt{5}$

$|\mathbf{s}| = \sqrt{(-1)^2+(2)^2} = \sqrt{5}$

$|\mathbf{p}| = \sqrt{3^2+4^2} = 5$

$|\mathbf{q}| = \sqrt{5^2} = 5$

$\therefore \quad |\mathbf{r}| = |\mathbf{s}|$ and $|\mathbf{p}| = |\mathbf{q}|$

Ex. 76 *$\mathbf{a}, \mathbf{b}$ and $\mathbf{c}$ are three vectors such that $\mathbf{a}\cdot\mathbf{a} = \mathbf{b}\cdot\mathbf{b} = \mathbf{c}\cdot\mathbf{c} = 3$ and $|\mathbf{a}-\mathbf{b}|^2 + |\mathbf{b}-\mathbf{c}|^2 + |\mathbf{c}-\mathbf{a}|^2 = 27$, then*

(a) $\mathbf{a}, \mathbf{b}$ and $\mathbf{c}$ are necessarily coplanar.

(b) $\mathbf{a}, \mathbf{b}$ and $\mathbf{c}$ represent sides of a triangle in magnitude and direction

(c) $\mathbf{a}\cdot\mathbf{b}+\mathbf{b}\cdot\mathbf{c}+\mathbf{c}\cdot\mathbf{a}$ has the least value $-9/2$

(d) $\mathbf{a}, \mathbf{b}$ and $\mathbf{c}$ represent orthogonal triad of vectors

Sol. (a, b, c) Here,

$|\mathbf{a}-\mathbf{b}|^2 + |\mathbf{b}-\mathbf{c}|^2 + |\mathbf{c}-\mathbf{a}|^2$

$= 2(|\mathbf{a}|^2 + |\mathbf{b}|^2 + |\mathbf{c}|^2 - \mathbf{a}\cdot\mathbf{b} - \mathbf{b}-\mathbf{b}\cdot\mathbf{c} - \mathbf{c}\cdot\mathbf{c})$

$\Rightarrow \quad \mathbf{a}\cdot\mathbf{b}+\mathbf{b}\cdot\mathbf{c}+\mathbf{c}\cdot\mathbf{a} = -\frac{9}{2}$

Now, $|\mathbf{a}+\mathbf{b}+\mathbf{c}|^2 = |\mathbf{a}|^2 + |\mathbf{b}|^2 + 2(\mathbf{a}\cdot\mathbf{b}+\mathbf{b}\cdot\mathbf{c}+\mathbf{c}\cdot\mathbf{a})$

$= 3+3+3-2\left(\frac{9}{2}\right) = 0$

$\therefore \quad \mathbf{a}+\mathbf{b}+\mathbf{c} = 0 \quad \text{...(i)}$

Also, $\quad |\mathbf{a}+\mathbf{b}+\mathbf{c}|^2 \ge 0$

$\Rightarrow \quad \mathbf{a}\cdot\mathbf{b}+\mathbf{b}\cdot\mathbf{c}+\mathbf{c}\cdot\mathbf{a} \ge -\frac{9}{2} \quad \text{...(ii)}$

Thus, least value is $-9/2$

Ex. 77 *If $\mathbf{a}$ and $\mathbf{b}$ are non-zero vectors such that $|\mathbf{a}+\mathbf{b}| = |\mathbf{a}-2\mathbf{b}|$, then*

(a) $2\,\mathbf{a}\cdot\mathbf{b} = |\mathbf{b}|^2$

(b) $\mathbf{a}\cdot\mathbf{b} = |\mathbf{b}|^2$

(c) least value of $\mathbf{a}\cdot\mathbf{b} + \frac{1}{|\mathbf{b}|^2+2}$ is $\sqrt{2}$

(d) least value of $\mathbf{a}\cdot\mathbf{b} + \frac{1}{|\mathbf{b}|+2}$ is $\sqrt{2}-1$

Sol. (a, d) $\quad |\mathbf{a}+\mathbf{b}| = |\mathbf{a}-2\mathbf{b}|$

$\Rightarrow \quad \mathbf{a}\cdot\mathbf{b} = \frac{|\mathbf{b}|^2}{2}$

Also, $\quad \mathbf{a}\cdot\mathbf{b} + \frac{1}{|\mathbf{b}|^2+2} = \frac{|\mathbf{b}|^2+2}{2} + \frac{1}{|\mathbf{b}|^2+2} - 1$

$\ge \sqrt{2}-1$ (using AM $\ge$ GM)

Ex. 78 *If vectors $\mathbf{b} = (\tan\alpha, -1, 2\sqrt{\sin\alpha/2})$ and $\mathbf{c} = \left(\tan\alpha, \tan\alpha, -\frac{3}{\sqrt{\sin\alpha/2}}\right)$ are orthogonal and vectors $\mathbf{a} = (1, 3, \sin 2\alpha)$ makes an obtuse angle with the Z-axis, then the value of α is*

(a) $\alpha = (4n+1)\pi + \tan^{-1}2$

(b) $\alpha = (4n+1)\pi - \tan^{-1}2$

(c) $\alpha = (4n+2)\pi + \tan^{-1}2$

(d) $\alpha = (4n+2)\pi - \tan^{-1}2$

Sol. (b, d), Since, $\mathbf{a} = (1, 3, \sin 2\alpha)$ makes an obtuse angle with the Z-axis, its z-component is negative.

Thus, $\quad -1 \le \sin 2\alpha < 0 \quad \text{...(i)}$

But $\mathbf{b}\cdot\mathbf{c} = 0 \quad (\because \text{orthogonal})$

$\tan^2\alpha - \tan\alpha - 6 = 0$

$\therefore \quad (\tan\alpha - 3)(\tan\alpha + 2) = 0$

$\Rightarrow \quad \tan\alpha = 3, -2$

Now, $\quad \tan\alpha = 3.$

Therefore, $\quad \sin 2\alpha = \frac{2\tan\alpha}{1+\tan^2\alpha} = \frac{6}{1+9} = \frac{3}{5}$

(not possible as $\sin 2\alpha < 0$)

Now, if $\tan\alpha = -2$,

$\Rightarrow \quad \sin 2\alpha = \frac{2\tan\alpha}{1+\tan^2\alpha} = \frac{-4}{1+4} = \frac{-4}{5}$

$\Rightarrow \quad \tan 2\alpha > 0$

Hence, 2α is the third quadrant. Also, $\sqrt{\sin\alpha/2}$ is meaningful.

If $0 < \sin\alpha/2 < 1$, the

$\alpha = (4n+1)\pi - \tan^{-1}2$

and $\quad \alpha = (4n+2)\pi - \tan^{-1}2$

Ex. 79 *If $\mathbf{a}$ and $\mathbf{b}$ are any two unit vectors, then the possible integers in the range of $\frac{3|\mathbf{a}+\mathbf{b}|}{2} + 2|\mathbf{a}-\mathbf{b}|$, is/are*

(a) 2 (b) 3

(c) 4 (d) 5

Sol. (b, c, d) We have, $|\mathbf{a}| = |\mathbf{b}| = 1$

Let θ be the angle between $\mathbf{a}$ and $\mathbf{b}$.

$|\mathbf{a}+\mathbf{b}| = 2\cos\frac{\theta}{2}$

and $\quad |\mathbf{a}-\mathbf{b}| = 2\sin\frac{\theta}{2}$

$\because \quad 3\cos\frac{\theta}{2} + 4\sin\frac{\theta}{2} : \theta \in [0, \pi]$

$-5 \le 3\cos\frac{\theta}{2} + 4\sin\frac{\theta}{2} \le 5$

The possible range are 3, 4 or 5.

Ex. 80 *Which of the following expressions are meaningful?*

(a) $\mathbf{u}\cdot(\mathbf{v}\times\mathbf{w})$ (b) $(\mathbf{u}\cdot\mathbf{v})\cdot\mathbf{w}$

(c) $(\mathbf{u}\cdot\mathbf{v})\,\mathbf{w}$ (d) $\mathbf{u}\times(\mathbf{v}\cdot\mathbf{w})$

Sol. (a, c) (i) Since, $\mathbf{v}\times\mathbf{w}$ is a vector, therefore, $\mathbf{u}\cdot(\mathbf{v}\times\mathbf{w})$ is a scalar quantity.

$\therefore$ (a) is meaningful.

(ii) $(\mathbf{u}\cdot\mathbf{v})$ is scalar.

$\therefore(\mathbf{u}\cdot\mathbf{v})\cdot\mathbf{w}$ is not meaningful.

(iii) $(\mathbf{u}\cdot\mathbf{v})$ is a scalar.

So, $(\mathbf{u}\cdot\mathbf{v})\mathbf{w}$ is a scalar multiple of $\mathbf{w}$.

$\therefore(\mathbf{u}\cdot\mathbf{v})\mathbf{w}$ is meaningful.

(iv) $(\mathbf{v}\cdot\mathbf{w})$ is a scalar.

So, $\mathbf{u}\times(\mathbf{v}\cdot\mathbf{w})$ is not meaningful as cross product is taken for two vector quantity and not for a vector and scalar.

Ex. 81 *If* $\mathbf{a}+2\mathbf{b}+3\mathbf{c}=0$, *then* $\mathbf{a}\times\mathbf{b}+\mathbf{b}\times\mathbf{c}+\mathbf{c}\times\mathbf{a}=$

(a) $2(\mathbf{a}\times\mathbf{b})$ (b) $6(\mathbf{b}\times\mathbf{c})$

(c) $3(\mathbf{c}\times\mathbf{a})$ (d) 0

Sol. (a, b, c) $\mathbf{a}=-(2\mathbf{b}+3\mathbf{c})$

$$\mathbf{a}\times\mathbf{b}+\mathbf{b}\times\mathbf{c}+\mathbf{c}\times\mathbf{a}$$
$$=-(2\mathbf{b}+3\mathbf{c})\times\mathbf{b}+\mathbf{b}\times\mathbf{c}+\mathbf{c}\times\{-(2\mathbf{b}+3\mathbf{c})\}$$
$$=-3\mathbf{c}\times\mathbf{b}+\mathbf{b}\times\mathbf{c}-2\mathbf{c}\times\mathbf{b}=6(\mathbf{b}\times\mathbf{c})$$

Similarly, putting the values of $\mathbf{b}$ and $\mathbf{c}$ in terms of $\mathbf{a}$ and $\mathbf{a}$, $\mathbf{b}$ respectively in $\mathbf{a}\times\mathbf{b}+\mathbf{b}\times\mathbf{c}+\mathbf{c}\times\mathbf{a}$, we get the desired results.

Ex. 82 *Let* $\alpha=a\hat{\mathbf{i}}+b\hat{\mathbf{j}}+c\hat{\mathbf{k}}$, $\beta=b\hat{\mathbf{i}}+c\hat{\mathbf{j}}+a\hat{\mathbf{k}}$ *and* $\gamma=c\hat{\mathbf{i}}+a\hat{\mathbf{j}}+b\hat{\mathbf{k}}$ *be three coplanar vectors with* $a\neq b$ *and* $\mathbf{v}=\hat{\mathbf{i}}+\hat{\mathbf{j}}+\hat{\mathbf{k}}$. *Then* $\mathbf{v}$ *is perpendicular to*

(a) α (b) β

(c) γ (d) None of these

Sol. (a, b, c) It is given that α,β and γ are coplanar vectors.

Therefore, $$[\alpha\ \beta\ \gamma]=0\Rightarrow\begin{vmatrix}a&b&c\\b&c&a\\c&a&b\end{vmatrix}=0$$

or $3abc-a^3-b^3-c^3=0$

or $a^3+b^3+c^3-3abc=0$

or $(a+b+c)(a^2+b^2+c^2-ab-bc-ca)=0$

or $a+b+c=0$ $[\because a^2+b^2+c^2-ab-bc-ca\neq 0]$

$\Rightarrow$ $\mathbf{v}\cdot\alpha=\mathbf{v}\cdot\beta=\mathbf{v}\cdot\gamma=0$

Hence, $\mathbf{v}$ is perpendicular to α, β and γ

Ex. 83 *If* $\mathbf{a}$ *is perpendicular to* $\mathbf{b}$ *and* p *is non-zero scalar such that* $p\mathbf{r}+(\mathbf{r}\cdot\mathbf{b})\mathbf{a}=\mathbf{c}$, *then* $\mathbf{r}$ *satisfy*

(a) $[\mathbf{r}\ \mathbf{a}\ \mathbf{c}]=0$ (b) $p^2\mathbf{r}=p\mathbf{a}-(\mathbf{c}\cdot\mathbf{a})\,\mathbf{b}$

(c) $p^2\mathbf{r}=p\mathbf{b}-(\mathbf{a}\cdot\mathbf{b})\mathbf{c}$ (d) $p^2\mathbf{r}=p\mathbf{c}-(\mathbf{b}\cdot\mathbf{c})\mathbf{a}$

Sol. (a, d) Given, $\mathbf{a}\cdot\mathbf{b}=0$

and $p\mathbf{r}+(\mathbf{r}\cdot\mathbf{b})\,\mathbf{a}=\mathbf{c}$ ... (i)

On taking dot product by $\mathbf{b}$, we get

$$p(\mathbf{r}\cdot\mathbf{b})+(\mathbf{r}\cdot\mathbf{b})\mathbf{a}\cdot\mathbf{b}=\mathbf{b}\cdot\mathbf{c}$$
$$\Rightarrow\quad p(\mathbf{r}\cdot\mathbf{b})=\mathbf{b}\cdot\mathbf{c}$$
$$\Rightarrow\quad p\left(\frac{\mathbf{c}-p\mathbf{r}}{\mathbf{a}}\right)=\mathbf{b}\cdot\mathbf{c}$$
$$\Rightarrow\quad p\mathbf{c}-p^2\mathbf{r}=(\mathbf{b}\cdot\mathbf{c})\mathbf{a}$$
$$\therefore\quad p^2\mathbf{r}=p\mathbf{c}-(\mathbf{b}\cdot\mathbf{c})\mathbf{a}$$
$$\Rightarrow\quad \mathbf{r}=\frac{\mathbf{c}}{p}-\frac{(\mathbf{b}\cdot\mathbf{c})}{p^2}\mathbf{a}$$
$$\Rightarrow\quad [\mathbf{r}\ \mathbf{a}\ \mathbf{c}]=0$$

Ex. 84 *If* $\alpha(\mathbf{a}\times\mathbf{b})+\beta(\mathbf{b}\times\mathbf{c})+\gamma(\mathbf{c}\times\mathbf{a})=0$, *then*

(a) $\mathbf{a}, \mathbf{b}, \mathbf{c}$ are coplanar if all of $\alpha,\beta,\gamma\neq 0$

(b) $\mathbf{a}, \mathbf{b}, \mathbf{c}$ are coplanar if any one of $\alpha,\beta,\gamma\neq 0$

(c) $\mathbf{a}, \mathbf{b}, \mathbf{c}$ are non-coplanar for any $\alpha,\beta,\gamma\neq 0$

(d) None of the above

Sol. (a, b) We have, $\alpha(\mathbf{a}\times\mathbf{b})+\beta(\mathbf{b}\times\mathbf{c})+\gamma(\mathbf{c}\times\mathbf{a})=0$

Taking dot product with $\mathbf{c}$, we have

$$\alpha[\mathbf{a}\ \mathbf{b}\ \mathbf{c}]=0$$

Similarly, taking dot product with b and c, we have

$$\gamma[\mathbf{a}\ \mathbf{b}\ \mathbf{c}]=0,\ \beta[\mathbf{a}\ \mathbf{b}\ \mathbf{c}]=0$$

Now, even if one of $\alpha,\beta,\gamma\neq 0$, then we have $[\mathbf{a}\ \mathbf{b}\ \mathbf{c}]=0$

$\Rightarrow$ $\mathbf{a}, \mathbf{b}, \mathbf{c}$ are coplanar.

Ex. 85 *If* $\mathbf{a}=\hat{\mathbf{i}}+\hat{\mathbf{j}}+\hat{\mathbf{k}}$ *and* $\mathbf{b}=\hat{\mathbf{i}}-\hat{\mathbf{j}}$, *then the vectors* $(\mathbf{a}\cdot\hat{\mathbf{i}})\hat{\mathbf{i}}+(\mathbf{a}\cdot\hat{\mathbf{j}}+(\mathbf{a}\cdot\hat{\mathbf{k}})\hat{\mathbf{k}}$, $(\mathbf{b}\cdot\hat{\mathbf{i}})\hat{\mathbf{i}}+(\mathbf{b}\cdot\hat{\mathbf{j}})\hat{\mathbf{j}}+(\mathbf{b}\cdot\hat{\mathbf{k}})\hat{\mathbf{k}}$ *and* $\hat{\mathbf{i}}+\hat{\mathbf{j}}-2\hat{\mathbf{k}}$

(a) are mutually perpendicular

(b) are coplanar

(c) form a parallelopiped of volume 6 units

(d) form a parallelopiped of volume 3 units

Sol. (a, c) Given $\mathbf{a}=\hat{\mathbf{i}}+\hat{\mathbf{j}}+\hat{\mathbf{k}}$ and $\mathbf{b}=\hat{\mathbf{i}}-\hat{\mathbf{j}}$

$(\mathbf{a}\cdot\hat{\mathbf{i}})\hat{\mathbf{i}}+(\mathbf{a}\cdot\hat{\mathbf{j}})\hat{\mathbf{j}}+(\mathbf{a}\cdot\hat{\mathbf{k}})\hat{\mathbf{k}}=\hat{\mathbf{i}}+\hat{\mathbf{j}}+\hat{\mathbf{k}}=\mathbf{x}$ (say)

and $(b\cdot\hat{\mathbf{i}})\hat{\mathbf{i}}+(b\cdot\hat{\mathbf{j}})\hat{\mathbf{j}}+(b\cdot\hat{\mathbf{k}})\hat{\mathbf{k}}=\hat{\mathbf{i}}-\hat{\mathbf{j}}=\mathbf{y}$ (say)

and $\hat{\mathbf{i}}+\hat{\mathbf{j}}-2\hat{\mathbf{k}}=\hat{\mathbf{z}}$ (say)

Clearly, $\mathbf{x}\cdot\mathbf{y}=\mathbf{yz}=\mathbf{z}\cdot\mathbf{x}=0$

$\therefore$ $\mathbf{x}$, $\mathbf{y}$ and $\mathbf{z}$ are mutually perpendicular

$$\text{Volume of parallelopiped}=\begin{vmatrix}1&1&1\\1&-1&0\\1&1&-2\end{vmatrix}$$
$$=1(2-0)-1(-2-0)+1(1+1)$$
$$=2+2+2=6$$

$\because$ $\mathbf{x}$, $\mathbf{y}$ and $\mathbf{x}$ are not coplanar, i.e., $[\mathbf{x}\ \mathbf{y}\ \mathbf{z}]\neq 0$

$\therefore$Volume of parallelopiped formed by $\mathbf{x}$, $\mathbf{y}$ and $\mathbf{z}$ is 6 cu units.

Ex. 86 *The volume of the parallelopiped whose coterminous edges are represented by the vectors* $2\mathbf{b}\times\mathbf{c}$, $3\mathbf{c}\times\mathbf{a}$ *and* $4\mathbf{a}\times\mathbf{b}$ *where* $\mathbf{a}=(1+\sin\theta)\hat{\mathbf{i}}+\cos\theta\hat{\mathbf{j}}+\sin 2\theta\hat{\mathbf{k}}$

$$\mathbf{b}=\sin\left(\theta+\frac{2\pi}{3}\right)\hat{\mathbf{i}}+\cos\left(\theta+\frac{2\pi}{3}\right)\hat{\mathbf{j}}+\sin\left(2\theta+\frac{4\pi}{3}\right)\hat{\mathbf{k}},$$

$$\mathbf{c}=\sin\left(\theta-\frac{2\pi}{3}\right)\hat{\mathbf{i}}+\cos\left(\theta-\frac{2\pi}{3}\right)\hat{\mathbf{j}}+\sin\left(2\theta-\frac{4\pi}{3}\right)\hat{\mathbf{k}} \text{ is 18}$$

cubic units, then the value of θ *in the interval* $\left(0,\frac{\pi}{2}\right)$, *is/are*

(a) $\frac{\pi}{9}$ (b) $\frac{2\pi}{9}$

(c) $\frac{\pi}{3}$ (d) $\frac{4\pi}{9}$

Sol. (a, b, d) Volume $=|[2\mathbf{b}\times\mathbf{c}\ \ 3\mathbf{c}\times\mathbf{a}\ \ 4\mathbf{a}\times\mathbf{b}]|=18$

$$\Rightarrow \quad 24[\mathbf{a}\ \mathbf{b}\ \mathbf{c}]^2=18$$

$$\Rightarrow \quad |[\mathbf{a}\ \mathbf{b}\ \mathbf{c}]|=\frac{\sqrt{3}}{2}$$

$$\text{Now, } [\mathbf{a}\ \mathbf{b}\ \mathbf{c}]=\begin{vmatrix}(1+\sin\theta) & \cos\theta & \sin 2\theta\\ \sin\left(\theta+\frac{2\pi}{3}\right) & \cos\left(\theta+\frac{2\pi}{3}\right) & \sin\left(\theta+\frac{4\pi}{3}\right)\\ \sin\left(\theta-\frac{2\pi}{3}\right) & \cos\left(\theta-\frac{2\pi}{3}\right) & \sin\left(2\theta-\frac{4\pi}{3}\right)\end{vmatrix}$$

Applying $R_1\to R_1+R_2+R_3$ and expanding

$$|[\mathbf{a}\ \mathbf{b}\ \mathbf{c}]|=\sqrt{3}\,|\cos 3\theta|=\frac{\sqrt{3}}{2}$$

$$\Rightarrow \quad \cos 3\theta=\pm\frac{1}{2}\Rightarrow 3\theta=\frac{\pi}{3},\frac{2\pi}{3},\frac{4\pi}{3}$$

$$\Rightarrow \quad \theta=\frac{\pi}{9},\frac{2\pi}{9},\frac{4\pi}{9}$$

Ex. 87 *If* $\mathbf{a}=x\hat{\mathbf{i}}+y\hat{\mathbf{j}}+z\hat{\mathbf{k}}$, $\mathbf{b}=y\hat{\mathbf{i}}+z\hat{\mathbf{j}}+x\hat{\mathbf{k}}$ *and* $\mathbf{c}=z\hat{\mathbf{i}}+x\hat{\mathbf{j}}+y\hat{\mathbf{k}}$, *then* $\mathbf{a}\times(\mathbf{b}\times\mathbf{c})$ *is/are*

(a) parallel to $(y-z)\hat{\mathbf{i}}+(z-x)\hat{\mathbf{j}}+(x-y)\hat{\mathbf{k}}$

(b) orthogonal to $\mathbf{i}+\mathbf{j}+\mathbf{k}$

(c) orthogonal to $(y+z)\hat{\mathbf{i}}+(z+x)\hat{\mathbf{j}}+(x+y)\hat{\mathbf{k}}$

(d) parallel to $\hat{\mathbf{i}}+\hat{\mathbf{j}}+\hat{\mathbf{k}}$

Sol. (a, b, c) $\mathbf{a}\times(\mathbf{b}\times\mathbf{c})=(\mathbf{a}\cdot\mathbf{c})\,\mathbf{b}-(\mathbf{a}\cdot\mathbf{b})\mathbf{c}$

$$=(xz+yx+yz)\,(y\hat{\mathbf{i}}+z\hat{\mathbf{j}}+x\hat{\mathbf{k}})$$

$$-(xy+yz+zx)\,(z\hat{\mathbf{i}}+x\hat{\mathbf{j}}+y\hat{\mathbf{k}})$$

$$=(xy+yz+zx)\,[(y-z)\hat{\mathbf{i}}+(z-x)\hat{\mathbf{j}}+(x-y)\hat{\mathbf{k}}]$$

Clearly, 1 to $\hat{\mathbf{i}}+\hat{\mathbf{j}}+\hat{\mathbf{k}}$ and also to $(y+z)\hat{\mathbf{i}}+(z+x)\hat{\mathbf{j}}+(x+y)\hat{\mathbf{k}}$, as dot product are zero, clearly parallel to

$$(y-z)\hat{\mathbf{i}}+(z-x)\hat{\mathbf{j}}+(x-y)\hat{\mathbf{k}}$$

Ex. 88 *If* **a**, **b** *and* **c** *are three non-zero vectors, then which of the following statement(s) is/are true?*

(a) $\mathbf{a}\times(\mathbf{b}\times\mathbf{c})$, $\mathbf{b}\times(\mathbf{c}\times\mathbf{a})$, $\mathbf{c}\times(\mathbf{a}\times\mathbf{b})$ from a right handed system.

(b) $\mathbf{c}$, $(\mathbf{a}\times\mathbf{b})\times\mathbf{c}$, $\mathbf{a}\times\mathbf{b}$ from a right handed system

(c) $\mathbf{a}\cdot\mathbf{b}+\mathbf{b}\cdot\mathbf{c}+\mathbf{c}\cdot\mathbf{a}<0$, if $\mathbf{a}+\mathbf{b}+\mathbf{c}=0$

(d) $\frac{(\mathbf{a}\times\mathbf{b})\cdot(\mathbf{b}\times\mathbf{c})}{(\mathbf{b}\times\mathbf{c})\cdot(\mathbf{a}\times\mathbf{c})}=-1$, if $\mathbf{a}+\mathbf{b}+\mathbf{c}=0$

Sol. (b, c, d)

(a) $\mathbf{a}\times(\mathbf{b}\times\mathbf{c})+\mathbf{b}\times(\mathbf{c}\times\mathbf{a})+\mathbf{c}\times(\mathbf{a}\times\mathbf{b})=0$

$\Rightarrow$ vector are coplanar, so do not form RHS.

(b) $(\mathbf{a}\times\mathbf{b})\times\mathbf{c}$, $\mathbf{a}\times\mathbf{b}$, $\mathbf{c}$ in that order form RHS.

$\Rightarrow$ $\mathbf{c}$, $(\mathbf{a}\times\mathbf{b})\times\mathbf{c}$, $\mathbf{a}\times\mathbf{b}$ also form RHS as they are in same cyclic order.

(c) $\mathbf{a}+\mathbf{b}+\mathbf{c}=0\Rightarrow|\mathbf{a}+\mathbf{b}+\mathbf{c}|^2=0$

$\Rightarrow a^2+b^2+c^2=-2(\mathbf{a}\cdot\mathbf{b}+\mathbf{b}\cdot\mathbf{c}+\mathbf{c}\cdot\mathbf{a})$

$\Rightarrow \quad \mathbf{a}\cdot\mathbf{b}+\mathbf{b}\cdot\mathbf{c}+\mathbf{c}\cdot\mathbf{a}\cdot<0$

(d) $a+b+c=0$

$\Rightarrow \quad \mathbf{a}\times\mathbf{b}=\mathbf{b}\times\mathbf{c}=\mathbf{c}\times\mathbf{a}$

Ex. 89 *Let the unit vectors* **a** *and* **b** *be perpendicular and unit vector* **c** *is inclined at angle* α *to* **a** *and* **b**. *If* $\mathbf{c}=l\mathbf{a}+m\mathbf{b}+n(\mathbf{a}\times\mathbf{b})$, *then*

(a) $l=m$ (b) $n^2=1-2l^2$

(c) $n^2=-\cos 2\alpha$ (d) $m^2=\frac{1+\cos 2\alpha}{2}$

Sol. (a, b, c, d)

$$\mathbf{a}\cdot\mathbf{b}=0,\ \mathbf{c}\cdot\mathbf{a}=\mathbf{c}\cdot\mathbf{b}=\cos\alpha$$

Take dot products with **a** , **b** and **c** respectively.

$$\Rightarrow \quad l=m,\ l^2+m^2+n^2=1$$

$$n^2=-\cos 2\alpha,$$

$$m^2=\frac{1+\cos 2\alpha}{2}$$

Ex. 90 *If* $\mathbf{a}\times(\mathbf{b}\times\mathbf{c})$ *is perpendicular to* $(\mathbf{a}\times\mathbf{b})\times\mathbf{c}$, *we may have*

(a) $(\mathbf{a}\cdot\mathbf{c})\,|\mathbf{b}|^2=(\mathbf{a}\cdot\mathbf{b})\,(\mathbf{b}\cdot\mathbf{c})$ (b) $\mathbf{a}\cdot\mathbf{b}=0$

(c) $\mathbf{a}\cdot\mathbf{c}=0$ (d) $\mathbf{b}\cdot\mathbf{c}=0$

Sol. (a, c)

$$\mathbf{a}\times(\mathbf{b}\times\mathbf{c})=(\mathbf{a}\cdot\mathbf{c})\mathbf{b}-(\mathbf{a}\cdot\mathbf{b})\mathbf{c}$$

and $\quad(\mathbf{a}\times\mathbf{b})\times\mathbf{c}=-(\mathbf{c}\cdot\mathbf{b})\mathbf{a}+(\mathbf{a}\cdot\mathbf{c})\mathbf{b}$

We have been given

$$(\mathbf{a}\times(\mathbf{b}\times\mathbf{c}))\cdot((\mathbf{a}\times\mathbf{b})\times\mathbf{c})=0$$

$\therefore \quad ((\mathbf{a}\cdot\mathbf{c})\mathbf{b}-(\mathbf{a}\cdot\mathbf{b})\mathbf{c})\cdot((\mathbf{a}\cdot\mathbf{c})\mathbf{b}-(\mathbf{c}\cdot\mathbf{b})\mathbf{a})=0$

or $\quad(\mathbf{a}\cdot\mathbf{c})^2|\mathbf{b}|^2-(\mathbf{a}\cdot\mathbf{c})(\mathbf{b}\cdot\mathbf{c})(\mathbf{a}\cdot\mathbf{b})$

$$-(\mathbf{a}\cdot\mathbf{b})(\mathbf{a}\cdot\mathbf{c})(\mathbf{b}\cdot\mathbf{c})+(\mathbf{a}\cdot\mathbf{b})(\mathbf{b}\cdot\mathbf{c})(\mathbf{c}\cdot\mathbf{a})=0$$

or $\quad(\mathbf{a}\cdot\mathbf{c})^2|\mathbf{b}|^2=(\mathbf{a}\cdot\mathbf{c})(\mathbf{a}\cdot\mathbf{b})(\mathbf{b}\cdot\mathbf{c})$

or $(\mathbf{a}\cdot\mathbf{c})((\mathbf{a}\cdot\mathbf{c})(\mathbf{b}\cdot\mathbf{b})-(\mathbf{a}\cdot\mathbf{b})(\mathbf{b}\cdot\mathbf{c}))=0$

$$\mathbf{a}\cdot\mathbf{c}=0$$

or $(\mathbf{a}\cdot\mathbf{c})|\mathbf{b}|^2=(\mathbf{a}\cdot\mathbf{b})(\mathbf{b}\cdot\mathbf{c})$

Ex. 91 *If* $(\mathbf{a}\times\mathbf{b})\times(\mathbf{c}\times\mathbf{d})\cdot(\mathbf{a}\times\mathbf{d})=0$, *then which of the following may be true?*

(a) **a**, **b**, **c** and **d** are necessarily coplanar

(b) **a** lies in the plane of **c** and **d**

(c) **b** lies in the plane of **a** and **d**

(d) **c** lies in the plane of **a** and **d**

Sol. (b, c, d) $(\mathbf{a}\times\mathbf{b})\times(\mathbf{c}\times\mathbf{d})\cdot(\mathbf{a}\times\mathbf{d})=0$

or $([\mathbf{a}\ \mathbf{c}\ \mathbf{d}]\,\mathbf{b}-[\mathbf{b}\ \mathbf{c}\ \mathbf{d}]\mathbf{a})\cdot(\mathbf{a}\times\mathbf{d})=0$

or $[\mathbf{a}\ \mathbf{c}\ \mathbf{d}][\mathbf{d}\ \mathbf{a}\ \mathbf{d}]=0$

Hence, either **c** or **b** must lie in the plane of **a** and **d**.

Ex. 92 *The angles of a triangle, two of whose sides are represented by vectors* $\sqrt{3}(\hat{\mathbf{a}}\times\hat{\mathbf{b}})$ *and* $\hat{\mathbf{b}}-(\mathbf{a}\cdot\hat{\mathbf{b}})\hat{\mathbf{a}}$, *where* $\hat{\mathbf{b}}$ *is a non-zero vector and* $\hat{\mathbf{a}}$ *is a unit vector in the direction of* $\hat{\mathbf{a}}$, *are*

(a) $\tan^{-1}(\sqrt{3})$ (b) $\tan^{-1}(1/\sqrt{3})$

(c) $\cot^{-1}(0)$ (d) $\tan^{-1}(1)$

Sol. (a, b, c) Consider $\mathbf{V}_1\cdot\mathbf{V}_2=0$

$$\Rightarrow \quad A=90^\circ$$

Using the sine law, $\left|\dfrac{\mathbf{b}-(\hat{\mathbf{a}}\cdot\mathbf{b})\hat{\mathbf{a}}}{\sin\theta}\right|=\dfrac{\sqrt{3}\,|\hat{\mathbf{a}}\times\hat{\mathbf{b}}|}{\cos\theta}$

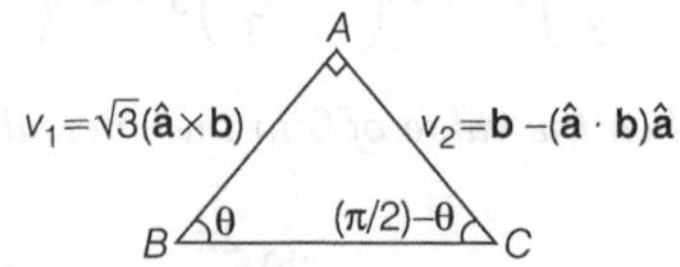

or $\tan\theta=\dfrac{1}{\sqrt{3}}\dfrac{|\mathbf{b}-(\hat{\mathbf{a}}\cdot\hat{\mathbf{b}})\hat{\mathbf{a}}|}{|\hat{\mathbf{a}}\times\hat{\mathbf{b}}|}$

$$=\frac{1}{\sqrt{3}}\frac{|(\hat{\mathbf{a}}\times\hat{\mathbf{b}})\times\hat{\mathbf{a}}|}{|\hat{\mathbf{a}}\times\hat{\mathbf{b}}|}$$

$$=\frac{1}{\sqrt{3}}\frac{|\hat{\mathbf{a}}\times\hat{\mathbf{b}}||\hat{\mathbf{a}}|\sin 90^\circ}{|\hat{\mathbf{a}}\times\hat{\mathbf{b}}|}=\frac{1}{\sqrt{3}}$$

or $\theta=\dfrac{\pi}{6}$

JEE Type Solved Examples : Statement Type I & II Questions

Directions (Q. Nos. 93-96) This section is based on Statement I and Statement II. Select the correct answer from the codes given below.

(a) Both Statement I and Statement II are correct and Statement II is the correct explanation of Statement I

(b) Both Statement I and Statement II are correct but Statement II is not the correct explanation of Statement I

(c) Statement I is correct but Statement II is incorrect

(d) Statement II is correct but Statement I is incorrect

Ex. 93 *Let the vectors* **PQ**, **OR**, **RS**, **ST**, **TU** *and* **UP** *represent the sides of a regular hexagon.*

Statement I $\mathbf{PQ}\times(\mathbf{RS}+\mathbf{ST})\neq 0$

Statement II $\mathbf{PQ}\times\mathbf{RS}=0$ *and* $\mathbf{PQ}\times\mathbf{ST}\neq 0$

Sol. (c) Clearly, $\mathbf{RS}+\mathbf{ST}=\mathbf{RT}$, which is not parallel to **PQ**.

$$\therefore \quad \mathbf{PQ}\times(\mathbf{RS}+\mathbf{ST})\neq 0$$

So, Statement I is correct.

Also, **PQ** is not parallel to **RS**.

$$\therefore \quad \mathbf{PQ}\times\mathbf{RS}\neq 0$$

So, Statement II is not correct.

Ex. 94 **p**, **q** *and* **r** *are three vectors defined by*

$\mathbf{p}=\mathbf{a}\times(\mathbf{b}+\mathbf{c})$, $\mathbf{q}=\mathbf{b}\times(\mathbf{c}+\mathbf{a})$ *and* $\mathbf{r}=\mathbf{c}\times(\mathbf{a}+\mathbf{b})$

Statement I **p**, **q** *and* **r** *are coplanar.*

Statement II *Vectors* **p**, **q**, **r** *are linearly independent.*

Sol. (c) **Statement I** $\mathbf{p}+\mathbf{q}+\mathbf{r}=\mathbf{a}\times(\mathbf{b}+\mathbf{c})+\mathbf{b}\times(\mathbf{c}+\mathbf{a})+\mathbf{c}\times(\mathbf{a}+\mathbf{b})$

$$=\mathbf{a}\times\mathbf{b}+\mathbf{a}\times\mathbf{c}+\mathbf{b}\times\mathbf{c}+\mathbf{b}\times\mathbf{a}+\mathbf{c}\times\mathbf{a}+\mathbf{c}\times\mathbf{b}$$

$$=\mathbf{a}\times\mathbf{b}+\mathbf{a}\times\mathbf{c}+\mathbf{b}\times\mathbf{c}-\mathbf{a}\times\mathbf{b}-\mathbf{a}\times\mathbf{c}-\mathbf{b}\times\mathbf{c}$$

$\therefore \mathbf{p}=-\mathbf{q}-\mathbf{r}$ (a linear combination of **q** and **r**)

Therefore, **p**, **q**, **r** are coplanar and hence statement I is true.

Statement II $\mathbf{p}+\mathbf{q}+\mathbf{r}=0\Rightarrow\mathbf{p},\mathbf{q},\mathbf{r}$ are not linearly independent.

Therefore, statement II is not true.

Ex. 95 **Statement I** *If in a* ΔABC, $\mathbf{BC}=\dfrac{\mathbf{p}}{|\mathbf{p}|}-\dfrac{\mathbf{q}}{|\mathbf{q}|}$ *and* $\mathbf{AC}=\dfrac{2\mathbf{p}}{|\mathbf{p}|}$; $|\mathbf{p}|\neq|\mathbf{q}|$, *then the value of* $\cos 2A+\cos 2B+\cos 2C$ *is* -1.

Statement II *If in* ΔABC, $\angle C=90^\circ$, *then* $\cos 2A+\cos 2B+\cos 2C=-1$

Sol. (b) **Statement II** In $\Delta ABC, \angle C = 90°$

$$\therefore \quad \cos 2A + \cos 2B + \cos 2C$$
$$= 2\cos(A+B)\cos(A-B) + \cos 180°$$
$$= \cos(180° - C)\cos(A-B) - 1$$
$$= -2\cos C\cos(A-B) - 1$$
$$= 0 - 1 = -1 \quad [\because \cos C = \cos 90° = 0]$$

Therefore, Statement II is true.

Statement I $\mathbf{BC} = \hat{p} - \hat{q}, \mathbf{AC} = 2\hat{p}$

$$\therefore \quad \mathbf{AB} = \mathbf{AC} + \mathbf{CB} = 2\hat{p} - (\hat{p} - \hat{q}) = \hat{p} + \hat{q}$$

Now, $\mathbf{AB} \cdot \mathbf{BC} = (\hat{p} + \hat{q}).(\hat{p} - \hat{q}) = |\hat{p}|^2 - |\hat{q}|^2 = 1 - 1 = 0$

$$\therefore \quad \angle B = 90°$$

Now, $\cos 2A + \cos 2B + \cos 2C$

$$= \cos 2A + \cos 2C + \cos 2B$$
$$= 2\cos(A+C)\cos(A-C) + \cos 180°$$
$$= 2\cos(180° - B)\cos(A-C) - 1$$
$$= -2\cos B\cos(A-C) - 1 = -1 \quad (\because B = 90°)$$

Therefore, Statement I is also true.

Thus both Statements are true but Statement II is not the correct explanation of Statement I.

● ***Ex. 96*** **Statement I** *If* **a** *is perpendicular to* **b** *and* **c**, *then* $\mathbf{a} \times (\mathbf{b} \times \mathbf{c}) = 0$

Statement II *If* **b** *is perpendicular to* **c**, *then* $\mathbf{b} \times \mathbf{c} = 0$

Sol. (c) If **a** is perpendicular to **b** and **c**, then $\mathbf{a} \parallel (\mathbf{b} \times \mathbf{c})$

$$\therefore \quad \mathbf{a} \times (\mathbf{b} \times \mathbf{c}) = 0$$

Therefore, Statement I is true.

But, if $\mathbf{b} \perp \mathbf{c}$, then $\mathbf{b} \times \mathbf{c} \neq 0$

Therefore, Statement II is not true.

JEE Type Solved Examples : Passage Based Type Questions

Passage I
(Ex. Nos. 97-99)

Let $\mathbf{a} = 2\hat{\mathbf{i}} + 3\hat{\mathbf{j}} - 6\hat{\mathbf{k}}$, $\mathbf{b} = 2\hat{\mathbf{i}} - 3\hat{\mathbf{j}} + 6\hat{\mathbf{k}}$ and $\mathbf{c} = -2\hat{\mathbf{i}} + 3\hat{\mathbf{j}} + 6\hat{\mathbf{k}}$. Let $\mathbf{a}_1$ be the projection of **a** on **b** and $\mathbf{a}_2$ be the projection of $\mathbf{a}_1$ and **c**. Then

● ***Ex. 97*** $\mathbf{a}_2$ *is equal to*

(a) $\frac{943}{49}(2\hat{\mathbf{i}} - 3\hat{\mathbf{j}} - 6\hat{\mathbf{k}})$ (b) $\frac{943}{49^2}(2\hat{\mathbf{i}} - 3\hat{\mathbf{j}} - 6\hat{\mathbf{k}})$

(c) $\frac{943}{49}(-2\hat{\mathbf{i}} + 3\hat{\mathbf{j}} + 6\hat{\mathbf{k}})$ (d) $\frac{943}{49^2}(-2\hat{\mathbf{i}} + 3\hat{\mathbf{j}} + 6\hat{\mathbf{k}})$

Sol. (b)

$$\mathbf{a}_1 = \left[(2\hat{\mathbf{i}} + 3\hat{\mathbf{j}} - 6\hat{\mathbf{k}}).\frac{(2\hat{\mathbf{i}} - 3\hat{\mathbf{j}} + 6\hat{\mathbf{k}})}{7}\right]\frac{2\hat{\mathbf{i}} - 3\hat{\mathbf{j}} + 6\hat{\mathbf{k}}}{7}$$
$$= \frac{-41}{49}(2\hat{\mathbf{i}} - 3\hat{\mathbf{j}} + 6\hat{\mathbf{k}})$$
$$\mathbf{a}_2 = \frac{-41}{49}\left((2\hat{\mathbf{i}} - 3\hat{\mathbf{j}} + 6\hat{\mathbf{k}}).\frac{(-2\hat{\mathbf{i}} + 3\hat{\mathbf{j}} + 6\hat{\mathbf{k}})}{7}\right) \times \frac{(-2\hat{\mathbf{i}} + 3\hat{\mathbf{j}} + 6\hat{\mathbf{k}})}{7}$$
$$= \frac{-41}{(49)^2}(-4 - 9 + 36)(-2\hat{\mathbf{i}} + 3\hat{\mathbf{j}} + 6\hat{\mathbf{k}})$$
$$= \frac{943}{49^2}(2\hat{\mathbf{i}} - 3\hat{\mathbf{j}} - 6\hat{\mathbf{k}})$$

● ***Ex. 98*** $\mathbf{a}_1 . \mathbf{b}$ *is equal to*

(a) -41 (b) $-\frac{41}{7}$

(c) 41 (d) 287

Sol. (a) $\mathbf{a}_1 \cdot \mathbf{b} = -\frac{41}{49}(2\hat{\mathbf{i}} - 3\hat{\mathbf{j}} + 6\hat{\mathbf{k}}) \cdot (2\hat{\mathbf{i}} - 3\hat{\mathbf{j}} + 6\hat{\mathbf{k}}) = -41$

● ***Ex. 99.*** *Which of the following is true?*

(a) **a** and $\mathbf{a}_2$ are collinear

(b) $\mathbf{a}_1$ and **c** are collinear

(c) **a**, $\mathbf{a}_1$ and **b** are coplanar

(d) **a**, $\mathbf{a}_1$ and $\mathbf{a}_2$ are coplanar

Sol. (c) **a**, $\mathbf{a}_1$ and **b** are coplanar because $\mathbf{a}_1$ and **b** are collinear.

Passage II
(Ex. Nos. 100-102)

Let **a**, **b** be two vectors perpendicular to each other and $|\mathbf{a}| = 2, |\mathbf{b}| = 3$ and $\mathbf{c} \times \mathbf{a} = \mathbf{b}$.

● ***Ex. 100*** *The least value of* $|\mathbf{c} - \mathbf{a}|$ *is*

(a) 1 (b) $\frac{1}{2}$

(c) $\frac{1}{4}$ (d) $\frac{3}{2}$

● ***Ex. 101*** *When* $|\mathbf{c} - \mathbf{a}|$ *is least the value of* α *(when* α *is angle between* **a** *and* **c***) equals*

(a) $\tan^{-1}\left(\frac{1}{2}\right)$ (b) $\tan^{-1}\left(\frac{3}{4}\right)$

(c) $\cos^{-1}\left(\frac{2}{3}\right)$ (d) None of these

Ex. 102 *When $|\mathbf{c}-\mathbf{a}|$ attains least value, then the value of $|\mathbf{c}|$ is*

(a) $\frac{1}{2}$ (b) $\frac{7}{2}$

(c) $\frac{5}{2}$ (d) 4

Sol. For (Ex. Nos. 100-102)

Here, $|\mathbf{a}|=2, |\mathbf{b}|=3$

$$\mathbf{c}\times\mathbf{a}=\mathbf{b} \Rightarrow |\mathbf{c}|\,|\mathbf{a}|\sin\alpha=|\mathbf{b}|$$

$$\Rightarrow \quad |\mathbf{c}|=\frac{3}{2}\operatorname{cosec}\alpha$$

Consider, $|\mathbf{c}-\mathbf{a}|^2=|\mathbf{c}|^2-2(\mathbf{a}\cdot\mathbf{c})+|\mathbf{a}|^2$

$$=\frac{9}{4}\operatorname{cosec}^2\alpha+4-2(2)\cdot\left(\frac{3}{2}\right)\cdot\operatorname{cosec}\alpha\cdot\cos\alpha$$

$$=\frac{25}{4}+\frac{9}{4}\cdot\cot^2\alpha-6\cot\alpha$$

$$=\frac{25}{4}+\left(\frac{3}{2}\cot\alpha-2\right)^2-4$$

$$=\frac{9}{4}+\left(\frac{3}{2}\cot\alpha-2\right)^2\geq\frac{9}{4}$$

$|\mathbf{c}-\mathbf{a}|\geq\frac{3}{2}$ and least possible when

$$\frac{3}{2}\cot\alpha=2 \Rightarrow \tan\alpha=\frac{3}{4} \Rightarrow \alpha=\tan^{-1}\frac{3}{4}$$

Also, $\quad |\mathbf{c}|=\frac{3}{2\sin\alpha}=\frac{3}{2\left(\frac{3}{5}\right)}=\frac{5}{2}$

100. (d) ***101.*** (b) ***102.*** (c)

Passage III

(Ex. Nos. 103-105)

Consider a triangular pyramid *ABCD* the position vectors of whose angular points are *A* (3, 0, 1), *B* (− 1, 4, 1), *C* (5, 2, 3) and *D*(0, − 5, 4). Let *G* be the point of intersection of the medians of triangle *BCD*.

Ex. 103 *The length of vector* **AG** *is*

(a) $\sqrt{17}$ (b) $\frac{\sqrt{51}}{3}$

(c) $\frac{3}{\sqrt{6}}$ (d) $\frac{\sqrt{59}}{4}$

Ex. 104 *Area of triangle ABC in sq units is*

(a) 24 (b) $8\sqrt{6}$

(c) $4\sqrt{6}$ (d) None of these

Ex. 105 *The length of the perpendicular from vertex D on the opposite face is*

(a) $\frac{14}{\sqrt{6}}$ (b) $\frac{2}{\sqrt{6}}$

(c) $\frac{3}{\sqrt{6}}$ (d) None of these

Sol. For (Ex. Nos. 103-105)

Point G is $\left(\frac{4}{3},\frac{1}{3},\frac{8}{3}\right)$. Therefore,

$$|\mathbf{AG}|^2=\left(\frac{5}{3}\right)^2+\frac{1}{9}+\left(\frac{5}{3}\right)^2=\frac{51}{9}$$

or $\quad |\mathbf{AG}|=\frac{\sqrt{51}}{3}$

$$\mathbf{AB}=-4\hat{\mathbf{i}}+4\hat{\mathbf{j}}+0\hat{\mathbf{k}}$$

$$\mathbf{AC}=2\hat{\mathbf{i}}+2\hat{\mathbf{j}}+2\hat{\mathbf{k}}$$

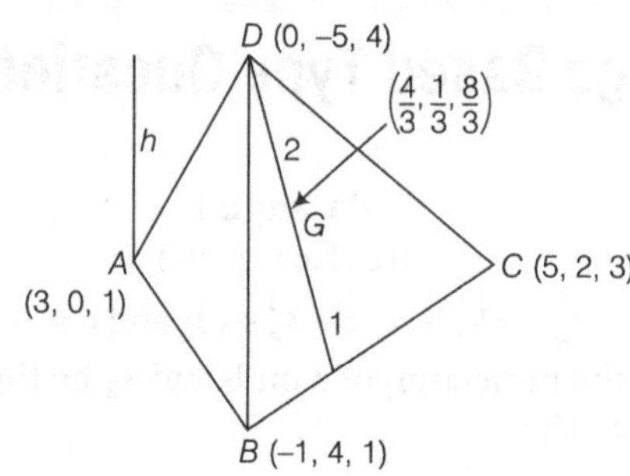

$$\therefore \quad \mathbf{AB}\times\mathbf{AC}=-8\begin{vmatrix}\hat{\mathbf{i}} & \hat{\mathbf{j}} & \hat{\mathbf{k}}\\ 1 & -1 & 0\\ 1 & 1 & 1\end{vmatrix}$$

$$=8(\hat{\mathbf{i}}+\hat{\mathbf{j}}-2\hat{\mathbf{k}})$$

Area of $\Delta ABC=\frac{1}{2}|\mathbf{AB}\times\mathbf{AC}|=4\sqrt{6}$

$$\mathbf{AD}=-3\hat{\mathbf{i}}-5\hat{\mathbf{j}}+3\hat{\mathbf{k}}$$

The length of the perpendicular from the vertex *D* on the opposite face

$$=|\text{Projection of } \mathbf{AD} \text{ on } \mathbf{AB}\times\mathbf{AC}|$$

$$=\left|\frac{(-3\hat{\mathbf{i}}-5\hat{\mathbf{j}}+3\hat{\mathbf{k}})\cdot(\hat{\mathbf{i}}+\hat{\mathbf{j}}-2\hat{\mathbf{k}})}{\sqrt{6}}\right|$$

$$=\left|\frac{-3-5-6}{\sqrt{6}}\right|$$

$$=\frac{14}{\sqrt{6}}$$

103. (b) ***104.*** (c) ***105.*** (a)

Passage IV

(Ex. Nos. 106-108)

Let A, B, C represent the vertices of a triangle, where A is the origin and B and C have position vectors **b** and **c** respectively. Points **M, N** and **P** are taken on sides **AB, BC** and **CA** respectively, such that $\frac{\mathbf{AM}}{\mathbf{AB}}=\frac{\mathbf{BN}}{\mathbf{BC}}=\frac{\mathbf{CP}}{\mathbf{CA}}=\alpha$

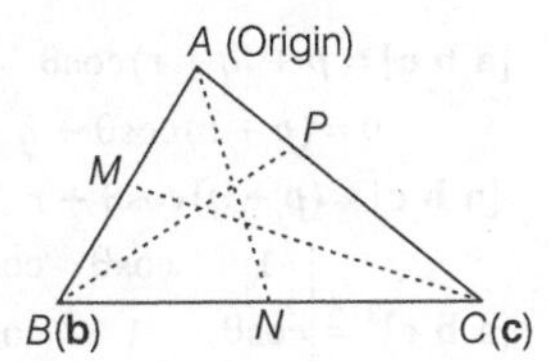

[Now answer the following questions]

Ex. 106. $\mathbf{AN}+\mathbf{BP}+\mathbf{CM}$ *is*

(a) $3\alpha(\mathbf{b}+\mathbf{c})$ (b) $\alpha(\mathbf{b}+\mathbf{c})$
(c) $(1-\alpha)(\mathbf{b}+\mathbf{c})$ (d) 0

Sol. (d) Since $\frac{\mathbf{AM}}{\mathbf{AB}}=\alpha$,

$\therefore$ P.V. of $\mathbf{M}=\alpha\,\mathbf{b}$

Since, $\frac{\mathbf{BN}}{\mathbf{BC}}=\alpha$

$\therefore$ $\frac{\mathbf{BN}}{\mathbf{NC}}=\frac{\alpha}{1-\alpha}$

$\therefore$Position vector of $\mathbf{N}=(1-\alpha)\mathbf{b}+\alpha\mathbf{c}$

Since, $\frac{\mathbf{CP}}{\mathbf{CA}}=\alpha$

$\therefore$ $\frac{\mathbf{AP}}{\mathbf{PC}}=\frac{1-\alpha}{\alpha}$

$\therefore$ P.V. of $\mathbf{p}=(1-\alpha)\mathbf{c}$

Now, $\mathbf{AN}=(1-\alpha)\mathbf{b}+\alpha\mathbf{c}$

$\mathbf{BP}=(1-\alpha)\mathbf{c}-\mathbf{b}$

$\mathbf{CM}=\alpha\mathbf{b}-\mathbf{c}$

$\therefore$ $\mathbf{AN}+\mathbf{BP}+\mathbf{CM}=\mathbf{0}$

Ex. 107. *The vectors* **AN, BP** *and* **CM** *are*

(a) concurrent (b) sides of a triangle
(c) non-coplanar (d) None of these

Sol. (b) Since $\mathbf{AN}+\mathbf{BP}+\mathbf{CM}=\mathbf{0}$

Hence, **AN, BP** and **CM** from the sides of a triangle.

Ex. 108. *If* Δ *represents the area enclosed by the three vectors* **AN, BP** *and* **CM**, *then the value of* α *for which* Δ *is least*

(a) does not exist (b) $\frac{1}{2}$
(c) $\frac{1}{4}$ (d) None of these

Sol. (b) $\Delta=\frac{1}{2}|\mathbf{AN}\times\mathbf{BP}|=\frac{1}{2}|\{(1-\alpha)\mathbf{b}+\alpha\mathbf{c}\}\times\{(1-\alpha)\mathbf{c}-\mathbf{b}\}|$

$$=\frac{1}{2}|(1-\alpha)^2(\mathbf{b}\times\mathbf{c})+\alpha(\mathbf{b}\times\mathbf{c})|$$

$$=\frac{1}{2}|\mathbf{b}\times\mathbf{c}|(\alpha^2-\alpha+1)$$

$$=\frac{1}{2}|\mathbf{b}\times\mathbf{c}|\left\{\left(\alpha-\frac{1}{2}\right)^2+\frac{3}{4}\right\}$$

$\therefore\Delta$ is least, if $\alpha=\frac{1}{2}$

Passage V

(Ex. Nos. 109-110)

If AP, BQ and CR are the altitudes of acute ΔABC and $9AP+4BQ+7CR=0$.

Ex. 109. $\angle ACB$ *is equal to*

(a) $\frac{\pi}{4}$ (b) $\frac{\pi}{3}$
(c) $\cos^{-1}\left(\frac{1}{3\sqrt{7}}\right)$ (d) $\cos^{-1}\left(\frac{1}{\sqrt{7}}\right)$

Ex. 110. $\angle ABC$ *is equal to*

(a) $\cos^{-1}\left(\frac{2}{\sqrt{7}}\right)$ (b) $\frac{\pi}{2}$
(c) $\cos^{-1}\left(\frac{\sqrt{7}}{3}\right)$ (d) $\frac{\pi}{3}$

Sol. For (Ex. Nos. 109-110)

Since, sum of three vectors $9\mathbf{AP}$, $4\mathbf{BQ}$ and $7\mathbf{CR}$ is zero, there is a Δ whose sides have lengths $9|\mathbf{AP}|$, $4|\mathbf{BQ}|$, $7|\mathbf{CR}|$ and are parallel to the corresponding vectors.

H is orthocentre.

$\Rightarrow \angle BHP=90°-\angle QBC=\angle ACB$

$\Rightarrow$ Angle between **AP** and **BQ** equal to $\angle ACB$, similarly angle between **BQ** and **CR** be $\angle BAC$.

$\Rightarrow$ $\frac{AB}{7|\mathbf{CR}|}=\frac{BC}{9|\mathbf{AP}|}=\frac{AC}{4|\mathbf{BQ}|}$

2. ar. $(\Delta ABC)=|\mathbf{AB}\times\mathbf{CR}|=|\mathbf{BC}\times\mathbf{AP}|=|\mathbf{CA}\times\mathbf{BQ}|$

$\Rightarrow$ $\frac{c^2}{7}=\frac{a^2}{9}=\frac{b^2}{4}$

$\therefore$ $a:b:c=3:2:\sqrt{7}$

$\Rightarrow$ $\cos C=\frac{a^2+b^2-c^2}{2ab}=\frac{1}{2}$

$\Rightarrow$ $\angle C=60°=\frac{\pi}{3}$

and $\cos B=\frac{a^2+b^2-c^2}{2ac}=\frac{2}{\sqrt{7}}$

109. (b) **110.** (a)

Passage VI

(Ex. Nos. 111 to 113)

Let $\mathbf{a}, \mathbf{b}, \mathbf{c}$ are non-zero unit vectors inclined pairwise with the same angle θ. p, q, r are non-zero scalars satisfying $\mathbf{a} \times \mathbf{b} + \mathbf{b} \times \mathbf{c} = p\mathbf{a} + q\mathbf{b} + r\mathbf{c}$

Ex. 111 *Volume of parallelopiped with edges* $\mathbf{a}, \mathbf{b}$ *and* $\mathbf{c}$ *is*

(a) $p + (q + r)\cos\theta$ (b) $(p + q + r)\cos\theta$
(c) $2p - (q + r)\cos\theta$ (d) None of these

Ex. 112 *The value of* $\left(\dfrac{q}{p} + 2\cos\theta\right)$ *is*

(a) 1 (b) 0
(c) $2[\mathbf{a}\ \mathbf{b}\ \mathbf{c}]$ (d) None of these

Ex. 113 *The value of* $|(p + q)\cos\theta + r|$ *is*

(a) $(1 + \cos\theta)\sqrt{1 - 2\cos\theta}$

(b) $2\sin^2\dfrac{\theta}{2}|\sqrt{1 + 2\cos\theta}|$

(c) $(1 - \sin\theta)\sqrt{1 + 2\cos\theta}$

(d) None of the above

Sol. For (Ex. Nos. 111-113)

Volume of parallelopiped $= [\mathbf{a}\ \mathbf{b}\ \mathbf{c}] = \mathbf{a} \cdot (\mathbf{b} \times \mathbf{c})$

Now, we have

$$\mathbf{a} \times \mathbf{b} + \mathbf{b} \times \mathbf{c} = p\mathbf{a} + q\mathbf{b} + r\mathbf{c}$$

$\Rightarrow \mathbf{a} \cdot (\mathbf{a} \times \mathbf{b}) + \mathbf{a} \cdot (\mathbf{b} \times \mathbf{c}) = p(\mathbf{a} \cdot \mathbf{a}) + q(\mathbf{a} \cdot \mathbf{b}) + r(\mathbf{a} \cdot \mathbf{c})$

$\Rightarrow [\mathbf{a}\ \mathbf{a}\ \mathbf{b}] + [\mathbf{a}\ \mathbf{b}\ \mathbf{c}] = p|\mathbf{a}|^2 + q|\mathbf{a}||\mathbf{b}|\cos\theta + r|\mathbf{a}||\mathbf{c}|\cos\theta$

$\Rightarrow [\mathbf{a}\ \mathbf{b}\ \mathbf{c}] = p + q\cos\theta + r\cos\theta = p + (q + r)\cos\theta$

$\therefore$ Volume of parallelopiped $= p + (q + r)\cos\theta$

Taking dot products with $\mathbf{a}, \mathbf{b}, \mathbf{c}$ respectively with given equation

$$[\mathbf{a}\ \mathbf{b}\ \mathbf{c}] = p + (q + r)\cos\theta \quad \text{...(i)}$$

$$0 = (p + r)\cos\theta + q \quad \text{...(ii)}$$

$$[\mathbf{a}\ \mathbf{b}\ \mathbf{c}] = (p + q)\cos\theta + r \quad \text{...(iii)}$$

Also, $$[\mathbf{a}\ \mathbf{b}\ \mathbf{c}]^2 = \begin{vmatrix} 1 & \cos\theta & \cos\theta \\ \cos\theta & 1 & \cos\theta \\ \cos\theta & \cos\theta & 1 \end{vmatrix}$$

$\Rightarrow$ $[\mathbf{a}\ \mathbf{b}\ \mathbf{c}]^2 = (1 - \cos\theta)^2(1 + 2\cos\theta)$

$\therefore$ $v = |[\mathbf{a}\ \mathbf{b}\ \mathbf{c}]| = |1 - \cos\theta||\sqrt{1 + 2\cos\theta}|$

$$= 2\sin^2\frac{\theta}{2} \cdot |\sqrt{1 + 2\cos\theta}|$$

From Eqs. (i) and (iii), $p = r$ substituting in Eq. (ii), we get

$$2p\cos\theta + q = 0$$

$\Rightarrow$ $\dfrac{q}{p} + 2\cos\theta = 0$

111. (a) ***112.*** (c) ***113.*** (b)

JEE Type Solved Examples : Matching Type Questions

Ex. 114 *Match the items of Column I with items of Column II.*

Column I	Column II
A. If $|\mathbf{a} + \mathbf{b}| = |\mathbf{a} + 2\mathbf{b}|$, then angle between $\mathbf{a}$ and $\mathbf{b}$ is	p. 90°
B. If $|\mathbf{a} + \mathbf{b}| = |\mathbf{a} - 2\mathbf{b}|$, then angle between $\mathbf{a}$ and $\mathbf{b}$ is	q. obtuse
C. If $|\mathbf{a} + \mathbf{b}| = |\mathbf{a} - \mathbf{b}|$ then angle between $\mathbf{a}$ and $\mathbf{b}$ is	r. 0°
D. Angle between $\mathbf{a} \times \mathbf{b}$ and a vector perpendicular to the vector $\mathbf{c} \times (\mathbf{a} \times \mathbf{b})$ is	s. acute

Sol. A $\to$ q; B $\to$ s; C $\to$ p; D $\to$ r

(A) $|\mathbf{a} + \mathbf{b}| = |\mathbf{a} + 2\mathbf{b}|$

$$a^2 + b^2 + 2\mathbf{a} \cdot \mathbf{b} = a^2 + 4b^2 + 4\mathbf{a} \cdot \mathbf{b}$$

or $$2\mathbf{a} \cdot \mathbf{b} = -3b^2 < 0$$

Hence, angle between $\mathbf{a}$ and $\mathbf{b}$ is obtuse.

(B) $|\mathbf{a} + \mathbf{b}| = |\mathbf{a} - 2\mathbf{b}|$

or $a^2 + b^2 + 2\mathbf{a} \cdot \mathbf{b} = a^2 + 4b^2 - 4\mathbf{a} \cdot \mathbf{b}$

or $$6\mathbf{a} \cdot \mathbf{b} = 3b^2 > 0$$

Hence, angle between $\mathbf{a}$ and $\mathbf{b}$ is acute.

(C) $|\mathbf{a} + \mathbf{b}| = |\mathbf{a} - \mathbf{b}|$

$\Rightarrow$ $\mathbf{a} \cdot \mathbf{b} = 0$

Hence, $\mathbf{a}$ is perpendicular to $\mathbf{b}$.

(D) $\mathbf{c} \times (\mathbf{a} \times \mathbf{b})$ lies in the plane of vectors $\mathbf{a}$ and $\mathbf{b}$.

A vector perpendicular to this plane is parallel to $\mathbf{a} \times \mathbf{b}$

Hence, angle is 0°.

Ex. 115 *Match the items of Column I with items of Column II.*

Column I	Column II
A. Let $|\mathbf{a}| = |\mathbf{b}| = 2$, $\mathbf{x} = \mathbf{a} + \mathbf{b}$, $\mathbf{y} = \mathbf{a} - \mathbf{b}$. If $|\mathbf{x} \times \mathbf{y}| = 2\{\lambda - (\mathbf{a} \cdot \mathbf{b})^2\}^{\frac{1}{2}}$, then the value of λ is	p. 4
B. The non-zero value of λ for which angle between $\lambda\hat{\mathbf{i}} + \hat{\mathbf{j}} + \hat{\mathbf{k}}$ and $\hat{\mathbf{i}} + \lambda\hat{\mathbf{j}} + \hat{\mathbf{k}}$ is $\frac{\pi}{3}$, is	q. 42
C. If $|\mathbf{a}| = |\mathbf{b}| = 1$ and $|\mathbf{c}| = 2$, then the maximum value of $|\mathbf{a} - 2\mathbf{b}|^2 + |\mathbf{b} - 2\mathbf{c}|^2 + |\mathbf{c} - 2\mathbf{a}|^2$ is	r. 16
	s. 7

Sol. A → (r), B → (p), C → (q)

(A) $\mathbf{x} \times \mathbf{y} = (\mathbf{a} \times \mathbf{b}) \times (\mathbf{a} - \mathbf{b})$

$= \mathbf{a} \times \mathbf{a} - \mathbf{a} \times \mathbf{b} + \mathbf{b} \times \mathbf{a} - \mathbf{b} \times \mathbf{b} = -2(\mathbf{a} \times \mathbf{b})$

LHS $= |\mathbf{x} \times \mathbf{y}| = |-2(\mathbf{a} \times \mathbf{b})| = 2|\mathbf{a}||\mathbf{b}|\sin\theta$

RHS $= 2(\lambda - (\mathbf{a} \cdot \mathbf{b})^2)^{\frac{1}{2}} = 2(\lambda - |\mathbf{a}|^2|\mathbf{b}|^2\cos^2\theta)^{\frac{1}{2}}$

$\because \quad 8\sin\theta = 2(\lambda - 16\cos^2\theta)^{\frac{1}{2}}$

$\Rightarrow \quad 16\sin^2\theta = \lambda - 16\cos^2\theta$

$\Rightarrow \quad 16(\sin^2\theta + \cos^2\theta) = \lambda$

$\therefore \quad \lambda = 16$

(B) We have, $\cos\frac{\pi}{3} = \frac{1}{2}$

$$= \frac{(\lambda\hat{\mathbf{i}} + \hat{\mathbf{j}} + \hat{\mathbf{k}}) \cdot (\hat{\mathbf{i}} + \lambda\hat{\mathbf{j}} + \hat{\mathbf{k}})}{\sqrt{\lambda^2 + 2}\sqrt{\lambda^2 + 2}}$$

$\because \quad \frac{1}{2} = \frac{\lambda + \lambda + 1}{\lambda^2 + 2}$

$\Rightarrow \quad \lambda^2 + 2 = 2(2\lambda + 1)$

$\Rightarrow \quad \lambda^2 - 4\lambda = 0$

$\Rightarrow \quad \lambda(\lambda - 4) = 0$

$\Rightarrow \quad \lambda = 0$ and 4

$\therefore \quad \lambda = 4$

(C) We have, $|\mathbf{a} - 2\mathbf{b}|^2 + |\mathbf{b} - 2\mathbf{c}|^2 + |\mathbf{c} - 2\mathbf{a}|^2$

$= |\mathbf{a}|^2 + 4|\mathbf{b}|^2 - 4\mathbf{a} \cdot \mathbf{b} + |\mathbf{b}|^2 + 4|\mathbf{c}|^2 - 4\mathbf{b} \cdot \mathbf{c} + |\mathbf{c}|^2 + 4|\mathbf{a}|^2 - 4\mathbf{c} \cdot \mathbf{a}$

$\Rightarrow \quad 1 + 4 + 1 + 16 + 4 + 4 - 4(\mathbf{a} \cdot \mathbf{b} + \mathbf{b} \cdot \mathbf{c} + \mathbf{c} \cdot \mathbf{a})$

$\Rightarrow \quad 30 - 4(\mathbf{a} \cdot \mathbf{b} + \mathbf{b} \cdot \mathbf{c} + \mathbf{c} \cdot \mathbf{a})$

We know that,

$$(\mathbf{a} + \mathbf{b} + \mathbf{c})^2 = \Sigma \mathbf{a}^2 + 2\Sigma \mathbf{a} \cdot \mathbf{b} \geq 0$$

i.e. $1 + 1 + 4 + 2(\mathbf{a} \cdot \mathbf{b} + \mathbf{b} \cdot \mathbf{c} + \mathbf{c} \cdot \mathbf{a}) \geq 0$

$\Rightarrow \quad \mathbf{a} \cdot \mathbf{b} + \mathbf{b} \cdot \mathbf{c} + \mathbf{c} \cdot \mathbf{a} \geq -3$

$\Rightarrow \quad \geq 30 - 4(-3) \geq 30 + 12 \geq 42$

Ex. 116 *Match the items of Column I with items of Column II.*

Column I	Column II
A. Given two vectors $\mathbf{a} = 2\hat{\mathbf{i}} - 3\hat{\mathbf{j}} + 6\hat{\mathbf{k}}$, $\mathbf{b} = -2\hat{\mathbf{i}} + 2\hat{\mathbf{j}} - \hat{\mathbf{k}}$ and $\lambda = \frac{\text{Projection of } \mathbf{a} \text{ on } \mathbf{b}}{\text{Projection of } \mathbf{b} \text{ on } \mathbf{a}}$, then the value of 3λ is	p. 0
B. If $\mathbf{a} = \hat{\mathbf{i}} + 2\hat{\mathbf{j}} + 3\hat{\mathbf{k}}$, $\mathbf{b} = -\hat{\mathbf{i}} + 2\hat{\mathbf{j}} + \hat{\mathbf{k}}$, $\mathbf{c} = 3\hat{\mathbf{i}} + \hat{\mathbf{j}}$ and $\mathbf{a} + p\mathbf{b}$ is normal to $\mathbf{c}$, then p is	q. 7
C. Let $\mathbf{a}$, $\mathbf{b}$, $\mathbf{c}$ be three non-zero vectors such that $\mathbf{a} + \mathbf{b} + \mathbf{c} = 0$, then $\lambda(\mathbf{a} \times \mathbf{b}) + \mathbf{c} \times \mathbf{b} + \mathbf{a} \times \mathbf{c} = 0$, where λ is equal to	r. 5
D. The points whose position vectors are $p\hat{\mathbf{i}} + q\hat{\mathbf{j}} + r\hat{\mathbf{k}}$, $q\hat{\mathbf{i}} + r\hat{\mathbf{j}} + p\hat{\mathbf{k}}$ and $r\hat{\mathbf{i}} + p\hat{\mathbf{j}} + q\hat{\mathbf{k}}$ are collinear, then the value of $(p^2 + q^2 + r^2 - pq - qr - rp)$ is	s. 2

Sol. A → (q), B → (r), C → (s), D → (p)

(A) Given, $\quad \mathbf{a} = 2\hat{\mathbf{i}} - 3\hat{\mathbf{j}} + 6\hat{\mathbf{k}}$

and $\quad \mathbf{b} = -2\hat{\mathbf{i}} + 2\hat{\mathbf{j}} - \hat{\mathbf{k}}$

Now, $\lambda = \frac{\text{Projection of } \mathbf{a} \text{ on } \mathbf{b}}{\text{Projection of } \mathbf{b} \text{ on } \mathbf{a}} = \frac{\left(\frac{\mathbf{a} \cdot \mathbf{b}}{|\mathbf{b}|}\right)}{\left(\frac{\mathbf{b} \cdot \mathbf{a}}{|\mathbf{a}|}\right)} = \frac{|\mathbf{a}|}{|\mathbf{b}|}$

$$= \frac{\sqrt{2^2 + (-3)^2 + 6^2}}{\sqrt{(-2)^2 + 2^2 + (-1)^2}} = \frac{\sqrt{4 + 9 + 36}}{\sqrt{4 + 4 + 1}} = \frac{7}{3}$$

$\therefore \quad 3\lambda = 7$

(B) Given, $\mathbf{a} = \hat{\mathbf{i}} + 2\hat{\mathbf{j}} + 3\hat{\mathbf{k}}$,

$\mathbf{b} = -\hat{\mathbf{i}} + 2\hat{\mathbf{j}} + \hat{\mathbf{k}}$ and $\mathbf{c} = 3\hat{\mathbf{i}} + \hat{\mathbf{j}}$

Since, $\mathbf{a} + p\mathbf{b}$ is normal to $\mathbf{c}$

$\Rightarrow \quad (\mathbf{a} + p\mathbf{b}) \cdot \mathbf{c} = 0$

$\Rightarrow \quad [(1 - p)\hat{\mathbf{i}} + (2 + 2p)\hat{\mathbf{j}} + (3 + p)\hat{\mathbf{k}}] \cdot (3\hat{\mathbf{i}} + \hat{\mathbf{j}}) = 0$

$\Rightarrow \quad 3(1 - p) + 2 + 2p = 0$

$\Rightarrow \quad 3 - 3p + 2 + 2p = 0$

$\therefore \quad p = 5$

(C) Given, $\mathbf{a} + \mathbf{b} + \mathbf{c} = 0$

$\Rightarrow \quad \mathbf{a} \times \mathbf{a} + \mathbf{a} \times \mathbf{b} + \mathbf{a} \times \mathbf{c} = \mathbf{a} \times 0$

$\Rightarrow \quad \mathbf{a} \times \mathbf{b} + \mathbf{a} \times \mathbf{c} = 0 \quad$ ...(i)

Also, $\quad \mathbf{b} \times \mathbf{a} + \mathbf{b} \times \mathbf{b} + \mathbf{b} \times \mathbf{c} = \mathbf{b} \times 0$

$\Rightarrow \quad \mathbf{b} \times \mathbf{a} + \mathbf{b} \times \mathbf{c} = 0$

$\Rightarrow \quad -\mathbf{a} \times \mathbf{b} - \mathbf{c} \times \mathbf{b} = 0$

$\Rightarrow \quad \mathbf{a} \times \mathbf{b} + \mathbf{c} \times \mathbf{b} = 0 \quad$ ...(ii)

On adding Eqs. (i) and (ii), we get

$$2(\mathbf{a} \times \mathbf{b}) + \mathbf{c} \times \mathbf{b} + \mathbf{a} \times \mathbf{c} = 0$$

On comparing, we get $\lambda = 2$

(D) $\begin{vmatrix} p & q & r \\ q & r & p \\ r & p & q \end{vmatrix} = 0$

$\Rightarrow (p+q+r)(p^2+q^2+r^2-pq-qr-rp)=0$

$\therefore \quad p^2+q^2+r^2-pq-qr-rp=0$

Ex. 117 *Match the items of Column I with items of Column II.*

Column I		Column II
A. If **a** and **b** are two unit vectors inclined at $\frac{\pi}{3}$, then $16[\mathbf{a}\ \mathbf{b}+\mathbf{a}\times\mathbf{b}\ \mathbf{b}]$ is	p.	-12
B. If **b** and **c** are orthogonal unit vectors and $\mathbf{b}\times\mathbf{c}=\mathbf{a}$, then $[\mathbf{a}+\mathbf{b}+\mathbf{c}\ \ \mathbf{a}+\mathbf{b}\ \ \mathbf{b}+\mathbf{c}]$ is	q.	1
C. If $\lvert\mathbf{a}\rvert=\lvert\mathbf{b}\rvert=\lvert\mathbf{c}\rvert=2$ and $\mathbf{a}\cdot\mathbf{b}=\mathbf{b}\cdot\mathbf{c}=\mathbf{c}\cdot\mathbf{a}=2$, then $[\mathbf{a}\ \mathbf{b}\ \mathbf{c}]\cos 45^\circ$ is equal to	r.	3
D. $\mathbf{a}=2\hat{\mathbf{i}}+3\hat{\mathbf{j}}-\hat{\mathbf{k}}$, $\mathbf{b}=-\hat{\mathbf{i}}+2\hat{\mathbf{j}}-4\hat{\mathbf{k}}$, $\mathbf{c}=\hat{\mathbf{i}}+\hat{\mathbf{j}}+\hat{\mathbf{k}}$ and $\mathbf{d}=3\hat{\mathbf{i}}+2\hat{\mathbf{j}}+\hat{\mathbf{k}}$, then $\frac{1}{7}(\mathbf{a}\times\mathbf{b})\cdot(\mathbf{c}\times\mathbf{d})$ is equal to	s.	4

Sol. A → (p); B → (q); C → (s); D → (r)

(A) Given **a** and **b** are two unit vectors, i.e., $|\mathbf{a}|=|\mathbf{b}|=1$ and angle between them is $\frac{\pi}{3}$.

$$\sin\theta=\frac{|\mathbf{a}\times\mathbf{b}|}{|\mathbf{a}|\,|\mathbf{b}|} \Rightarrow \sin\frac{\pi}{3}=|\mathbf{a}\times\mathbf{b}|$$

$$\frac{\sqrt{3}}{2}=|\mathbf{a}\times\mathbf{b}|$$

Now, $[\mathbf{a}\ \mathbf{b}+\mathbf{a}\times\mathbf{b}\ \mathbf{b}]=[\mathbf{a}\ \mathbf{b}\ \mathbf{b}]+[\mathbf{a}\ \mathbf{a}\times\mathbf{b}\ \mathbf{b}]$

$=0+[\mathbf{a}\ \mathbf{a}\times\mathbf{b}\ \mathbf{b}]$

$=(\mathbf{a}\times\mathbf{b})\cdot(\mathbf{b}\times\mathbf{a})=-(\mathbf{a}\times\mathbf{b})\cdot(\mathbf{a}\times\mathbf{b})$

$=-|\mathbf{a}\times\mathbf{b}|^2=-\frac{3}{4}$

(B) If **b** and **c** are orthogonal $\mathbf{b}\cdot\mathbf{c}=0$

Also, it is given that $\mathbf{b}\times\mathbf{c}=\mathbf{a}$.

Now $[\mathbf{a}+\mathbf{b}+\mathbf{c}\ \mathbf{a}+\mathbf{b}\ \mathbf{b}+\mathbf{c}]$

$=[\mathbf{a}\ \mathbf{a}+\mathbf{b}\ \mathbf{b}+\mathbf{c}]+[\mathbf{b}+\mathbf{c}\ \mathbf{a}+\mathbf{b}\ \mathbf{b}+\mathbf{c}]$

$=[\mathbf{a}\ \mathbf{b}\ \mathbf{c}]=\mathbf{a}\cdot(\mathbf{b}\times\mathbf{c})$

$=\mathbf{a}\cdot\mathbf{a}=|\mathbf{a}|^2=1$ (because **a** is a unit vector)

(C) We know that, $[\mathbf{a}\times\mathbf{b}\ \mathbf{b}\times\mathbf{c}\ \mathbf{c}\times\mathbf{a}]=[\mathbf{a}\ \mathbf{b}\ \mathbf{c}]^2$

and $=[\mathbf{a}\ \mathbf{b}\ \mathbf{c}]^2=\begin{vmatrix}\mathbf{a}\cdot\mathbf{a} & \mathbf{a}\cdot\mathbf{b} & \mathbf{a}\cdot\mathbf{c}\\ \mathbf{b}\cdot\mathbf{a} & \mathbf{b}\cdot\mathbf{b} & \mathbf{b}\cdot\mathbf{c}\\ \mathbf{c}\cdot\mathbf{a} & \mathbf{c}\cdot\mathbf{b} & \mathbf{c}\cdot\mathbf{c}\end{vmatrix}=\begin{vmatrix}4 & 2 & 1\\ 2 & 4 & 2\\ 2 & 2 & 4\end{vmatrix}=32$

$\therefore \quad [\mathbf{a}\ \mathbf{b}\ \mathbf{c}]=4\sqrt{2}$

(D) $(\mathbf{a}\cdot\mathbf{c})(\mathbf{b}\cdot\mathbf{d})-(\mathbf{b}\cdot\mathbf{c})(\mathbf{a}\cdot\mathbf{d})=21$

JEE Type Solved Examples : Single Integer Answer Type Questions

Ex. 118 *Given that* $\mathbf{u}=\hat{\mathbf{i}}-2\hat{\mathbf{j}}+3\hat{\mathbf{k}}$; $\mathbf{v}=2\hat{\mathbf{i}}+\hat{\mathbf{j}}+4\hat{\mathbf{k}}$; $\mathbf{w}=\hat{\mathbf{i}}+3\hat{\mathbf{j}}+3\hat{\mathbf{k}}$ *and* $(\mathbf{u}\cdot\mathbf{R}-15)\hat{\mathbf{i}}+(\mathbf{v}\cdot\mathbf{R}-30)\hat{\mathbf{j}}+(\mathbf{w}\cdot\mathbf{R}-20)\hat{\mathbf{k}}=\mathbf{0}$. *Then, the greatest integer less than or equal to* $|\mathbf{R}|$ *is*

Sol. (6) Let $\mathbf{R}=x\hat{\mathbf{i}}+y\hat{\mathbf{j}}+z\hat{\mathbf{k}}$

$\mathbf{u}=\hat{\mathbf{i}}-2\hat{\mathbf{j}}+3\hat{\mathbf{k}}$; $\mathbf{v}=2\hat{\mathbf{i}}+\hat{\mathbf{j}}+4\hat{\mathbf{k}}$; $\mathbf{w}=\hat{\mathbf{i}}+3\hat{\mathbf{j}}+3\hat{\mathbf{k}}$

$(\mathbf{u}\cdot\mathbf{R}-15)\hat{\mathbf{i}}+(\mathbf{v}\cdot\mathbf{R}-30)\hat{\mathbf{j}}+(\mathbf{w}\cdot\mathbf{R}-20)\hat{\mathbf{k}}=\mathbf{0}$ (given)

So, $\mathbf{u}\cdot\mathbf{R}=15 \Rightarrow x-2y+3z=15$ ...(i)

$\mathbf{v}\cdot\mathbf{R}=30 \Rightarrow 2x+y+4z=30$ ...(ii)

$\mathbf{w}\cdot\mathbf{R}=25 \Rightarrow x+3y+3z=25$ ...(iii)

Solving, we get

$x=4$

$y=2$

$z=5$

Now, $|\mathbf{R}|=\sqrt{4^2+2^2+5^2}=\sqrt{45}$

$[|\mathbf{R}|]=[\sqrt{45}]=6$

Ex. 119 *The position vector of a point P is* $\mathbf{r}=x\hat{\mathbf{i}}+y\hat{\mathbf{j}}+z\hat{\mathbf{k}}$, *where* $x, y, z\in N$ *and* $\mathbf{a}=\hat{\mathbf{i}}+2\hat{\mathbf{j}}+\hat{\mathbf{k}}$. *If* $\mathbf{r}\cdot\mathbf{a}=20$ *and the number of possible of P is* 9λ, *then the value of* λ *is*

Sol. (9) $\mathbf{r}\cdot\mathbf{a}=20 \Rightarrow x+2y+z=20$, $x, y, z\in N$

$\therefore$ Number of non-negative integer solution are

$\Rightarrow {}^{17}C_1+{}^{15}C_1+{}^{13}C_1+\ldots+{}^{1}C_1$

$=81=9\lambda \Rightarrow \lambda=9$

Ex. 120 *Let* **u** *be a vector on rectangular coordinate system with sloping angle 60°. Suppose that* $|\mathbf{u}-\hat{\mathbf{i}}|$ *is geometric mean of* $|\mathbf{u}|$ *and* $|\mathbf{u}-2\hat{\mathbf{i}}|$, *where* $\hat{\mathbf{i}}$ *is the unit vector along the X-axis. Then, the value of* $(\sqrt{2}+1)|\mathbf{u}|$ *is*

Sol. (1) Since, angle between **u** and $\hat{\mathbf{i}}$ is 60°, we have

$$\mathbf{u}\cdot\hat{\mathbf{i}}=|\mathbf{u}||\hat{\mathbf{i}}|\cos 60^\circ=\frac{|\mathbf{u}|}{2}$$

Given that, $|\mathbf{u}|, |\mathbf{u}-\hat{\mathbf{i}}|, |\mathbf{u}-2\hat{\mathbf{i}}|$ are in GP.

So, $|\mathbf{u}-\hat{\mathbf{i}}|^2=|\mathbf{u}|\,|\mathbf{u}-2\hat{\mathbf{i}}|$

Squaring both sides,

$$[|\mathbf{u}|^2 + |\hat{\mathbf{i}}|^2 - 2\mathbf{u}\cdot\hat{\mathbf{i}}]^2 = |\mathbf{u}|^2[|\mathbf{u}|^2 + 4|\hat{\mathbf{i}}|^2 - 4\mathbf{u}\cdot\hat{\mathbf{i}}]$$

$$\left(|\mathbf{u}|^2 + 1 - \frac{2|\mathbf{u}|}{2}\right)^2 = |\mathbf{u}|^2\left(|\mathbf{u}|^2 + 4 - \frac{|\mathbf{u}|}{2}\right)$$

or $\quad |\mathbf{u}| + 2|\mathbf{u}| - 1 = 0$

$\Rightarrow \quad |\mathbf{u}|^2 = -\dfrac{2 \pm 2\sqrt{2}}{2}$

or $\quad |\mathbf{u}| = \sqrt{2} - 1$

$\Rightarrow \quad (\sqrt{2}+1)|\mathbf{u}| = (\sqrt{2}+1)(\sqrt{2}-1) = 2 - 1 = 1$

Ex. 121 *Let $A(2\hat{\mathbf{i}} + 3\hat{\mathbf{j}} + 5\hat{\mathbf{k}})$, $B(-\hat{\mathbf{i}} + 3\hat{\mathbf{j}} + 2\hat{\mathbf{k}})$ and $C(\lambda\hat{\mathbf{i}} + 5\hat{\mathbf{j}} + \mu\hat{\mathbf{k}})$ are vertices of a triangle and its median through A is equally inclined to the positive directions of the axes, the value of $2\lambda - \mu$ is*

Sol. (4) Median through A is

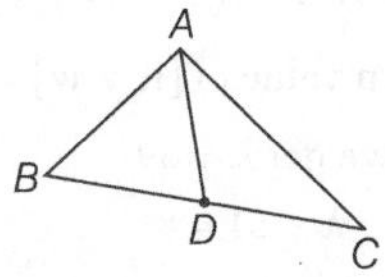

$$\mathbf{AD} = \frac{\mathbf{AB} + \mathbf{AC}}{2}$$

$$\mathbf{AB} = -3\hat{\mathbf{i}} - 3\hat{\mathbf{k}}$$

$$\mathbf{AC} = (\lambda - 2)\hat{\mathbf{i}} + 2\hat{\mathbf{j}} + (\mu - 5)\hat{\mathbf{k}}$$

$$\mathbf{AD} = \frac{1}{2}[(\lambda - 5)\hat{\mathbf{i}} + 2\hat{\mathbf{j}} + (\mu - 8)\hat{\mathbf{k}}]$$

We have, **AD** is equally inclined to the positive direction of axes

$$\cos\theta = \frac{\frac{1}{2}[(\lambda-5)\hat{\mathbf{i}} + 2\hat{\mathbf{j}} + (\mu-8)\hat{\mathbf{k}}]\cdot\hat{\mathbf{i}}}{|\mathbf{AD}||\hat{\mathbf{i}}|}$$

$$= \frac{\frac{1}{2}[(\lambda-5)\hat{\mathbf{i}} + 2\hat{\mathbf{j}} + (\mu-8)\hat{\mathbf{k}}]\cdot\hat{\mathbf{j}}}{|\mathbf{AD}||\hat{\mathbf{j}}|}$$

$$= \frac{\frac{1}{2}[(\lambda-5)\hat{\mathbf{i}} + 2\hat{\mathbf{j}} + (\mu-8)\hat{\mathbf{k}}]\cdot\hat{\mathbf{k}}}{|\mathbf{AD}||\hat{\mathbf{k}}|}$$

$\Rightarrow \quad \dfrac{\lambda - 5}{2} = 1 = \dfrac{\mu - 8}{2}$

$\Rightarrow \quad \lambda = 7$ and $\mu = 10$

$\therefore \quad 2\lambda - \mu = 14 - 10 = 4$

Ex. 122 *Three vectors $\mathbf{a}(|\mathbf{a}| \neq 0)$, $\mathbf{b}$ and $\mathbf{c}$ are such that $\mathbf{a} \times \mathbf{b} = 3\mathbf{a} \times \mathbf{c}$, also $|\mathbf{a}| = |\mathbf{b}| = 1$ and $|\mathbf{c}| = \frac{1}{3}$. If the angle between $\mathbf{b}$ and $\mathbf{c}$ is $60°$ and $|\mathbf{b} - 3\mathbf{c}| = \lambda|\mathbf{a}|$, then the value of λ is*

Sol. (1) We have, $\quad \mathbf{a} \times \mathbf{b} = 3\mathbf{a} \times \mathbf{c}$

$\Rightarrow \quad \mathbf{a} \times (\mathbf{b} - 3\mathbf{c}) = 0$

$\Rightarrow$ **a** is parallel to $\mathbf{b} - 3\mathbf{c}$.

Now, $\quad \mathbf{b} - 3\mathbf{c} = \lambda\mathbf{a}$

$\Rightarrow \quad |\mathbf{b} - 3\mathbf{c}|^2 = \lambda^2|\mathbf{a}|^2$

$\Rightarrow \quad |\mathbf{b}|^2 + 9|\mathbf{c}|^2 - 6(\mathbf{b}\cdot\mathbf{c}) = \lambda^2|\mathbf{a}|^2$

$\Rightarrow \quad 1 + 9 \times \dfrac{1}{9} - 6 \times 1 \times \dfrac{1}{3} \times \dfrac{1}{2} = \lambda^2(1)^2$

$\Rightarrow \quad 1 + 1 - 1 = \lambda^2 \Rightarrow \lambda^2 = 1$

$\therefore \quad \lambda = \pm 1$

Ex. 123 *If $\mathbf{a}, \mathbf{b}, \mathbf{c}$ are unit vectors such that $\mathbf{a}\cdot\mathbf{b} = 0 = \mathbf{a}\cdot\mathbf{c}$ and the angle between $\mathbf{b}$ and $\mathbf{c}$ is $\frac{\pi}{3}$, then the value of $|\mathbf{a} \times \mathbf{b} - \mathbf{a} \times \mathbf{c}|$.*

Sol. $\mathbf{a}\,.\,\mathbf{b} = 0 \Rightarrow \mathbf{a} \perp \mathbf{b}$

$\mathbf{a}\,.\,\mathbf{c} = 0 \Rightarrow \mathbf{a} \perp \mathbf{c}$

$\Rightarrow \quad \mathbf{a} \perp \mathbf{b} - \mathbf{c}$

$\therefore \quad |\mathbf{a} \times \mathbf{b} - \mathbf{a} \times \mathbf{c}| = |\mathbf{a} \times (\mathbf{b} - \mathbf{c})|$

$= |\mathbf{a}||\mathbf{b} - \mathbf{c}| = |\mathbf{b} - \mathbf{c}|$

Now, $\quad |\mathbf{b} - \mathbf{c}|^2 = |\mathbf{b}|^2 + |\mathbf{c}|^2 - 2|\mathbf{b}||\mathbf{c}|\cos\dfrac{\pi}{3}$

$= 2 - 2 \times \dfrac{1}{2} = 1$

$|\mathbf{b} - \mathbf{c}| = 1$

Ex. 124 *If the area of the triangle whose vertices are $A(-1, 1, 2)$; $B(1, 2, 3)$ and $C(t, 1, 1)$ is minimum, then the absolute value of parameter t is*

Sol. $\mathbf{AB} = 2\hat{\mathbf{i}} + \hat{\mathbf{j}} + \hat{\mathbf{k}}$, $\mathbf{AC} = (t+1)\hat{\mathbf{i}} + 0\hat{\mathbf{j}} - \hat{\mathbf{k}}$

$$\mathbf{AB} \times \mathbf{AC} = \begin{vmatrix} \hat{\mathbf{i}} & \hat{\mathbf{j}} & \hat{\mathbf{k}} \\ 2 & 1 & 1 \\ t+1 & 0 & -1 \end{vmatrix}$$

$$= -\hat{\mathbf{i}} + (t+3)\hat{\mathbf{j}} - (t+1)\hat{\mathbf{k}}$$

$$= \sqrt{1 + (t+3)^2 + (t+1)^2}$$

$$= \sqrt{2t^2 + 8t + 11}$$

Area of $\Delta ABC = \dfrac{1}{2}|\mathbf{AB} \times \mathbf{AC}|$

$= \dfrac{1}{2}\sqrt{2t^2 + 8t + 1}$

Let $f(t) = \Delta^2 = \dfrac{1}{4}(2t^2 + 8t + 1)$

$f(t) = 0 \Rightarrow t = -2$

At $\quad t = -2$, $f''(t) > 0$

So, Δ is minimum at $t = -2$

Ex. 125 *Let* $\mathbf{OA}=\mathbf{a}, \mathbf{OB}=10\mathbf{a}+2\mathbf{b}$ *and* $\mathbf{OC}=\mathbf{b}$, *where O, A and C are non-collinear points. Let p denote the area of quadrilateral OACB, and let q denote the area of parallelogram with OA and OC as adjacent sides. If* $p=kq$, *then k is equal to*

Sol. Here, $\mathbf{OA}=\mathbf{a}$, $\mathbf{OB}=10\mathbf{a}+2\mathbf{b}$, $\mathbf{OC}=\mathbf{b}$

q = Area of parallelogram with OA and OC as adjacent side.

$$\therefore \quad q=|\mathbf{a}\times\mathbf{b}| \quad ...(i)$$

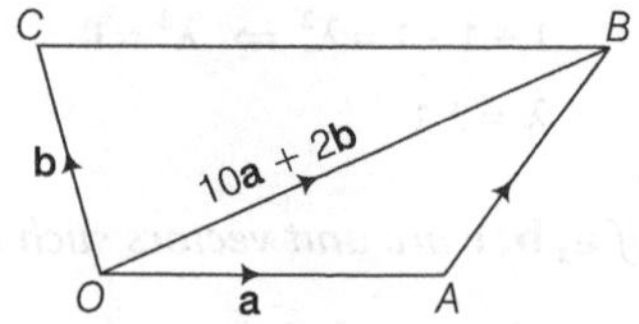

p = Area of quadrilateral $OABC$

$= \text{Area of } \Delta OAB + \text{Area of } \Delta OBC$

$$=\frac{1}{2}|\mathbf{a}\times(10\mathbf{a}+2\mathbf{b})|+\frac{1}{2}|(10\mathbf{a}+2\mathbf{b})\times\mathbf{b}|$$

$$=|\mathbf{a}\times\mathbf{b}|+5|\mathbf{a}\times\mathbf{b}|$$

$$\therefore \quad p=6|\mathbf{a}\times\mathbf{b}|$$

$$\text{or} \quad p=6q \quad \text{[From Eq. (i)]}$$

$$\therefore \quad k=6$$

Ex. 126. *If* $\mathbf{x}, \mathbf{y}$ *are two non-zero and non-collinear vectors satisfying* $[(a-2)\alpha^2+(b-3)\alpha+c]\mathbf{x}+[(a-2)\beta^2+(b-3)\beta+c]\mathbf{y}+[(a-2)\gamma^2+(b-3)\gamma+c](\mathbf{x}\times\mathbf{y})=0$ *where* α, β, γ *are three distinct real numbers, then find the value of* $(a^2+b^2+c^2-4)$.

Sol. (9) Since, $\mathbf{x}$ and $\mathbf{y}$ are non-collinear vectors, therefore $\mathbf{x}$, $\mathbf{y}$ and $\mathbf{x}\times\mathbf{y}$ are non-coplanar vectors.

So, $[(a-2)\alpha^2+(b-3)\alpha+c]\mathbf{x}+[(a-2)\beta^2+(b-3)\beta+c]\mathbf{y}+[(a-2)\gamma^2+(b-3)\gamma+c](\mathbf{x}\times\mathbf{y})=0$

$\Rightarrow$ Coefficient of each vector $\mathbf{x}$, $\mathbf{y}$ and $\mathbf{x}\times\mathbf{y}$ is zero.

$$(a-2)\alpha^2+(b-3)\alpha+c=0$$

$$(a-2)\beta^2+(b-3)\beta+c=0$$

$$(a-2)\gamma^2+(b-3)\gamma+c=0$$

The above three equations will satisfy if the coefficients of α, β and γ are zero because α, β and γ are three distinct real numbers

$$a-2=0 \text{ or } a=2,$$

$$b-3=0 \text{ or } b=3 \text{ and } c=0$$

$$\therefore \quad a^2+b^2+c^2=2^2+3^2+0^2=4+9=13$$

Ex. 127 *Let* $\mathbf{v}=2\hat{\mathbf{i}}+\hat{\mathbf{j}}-\hat{\mathbf{k}}$ *and* $\mathbf{w}=\hat{\mathbf{i}}+3\hat{\mathbf{k}}$. *If* $\hat{\mathbf{u}}$ *is unit vector and the maximum value of* $[\mathbf{u}\ \mathbf{v}\ \mathbf{w}]=\sqrt{\lambda}$, *then the value of* $(\lambda-51)$ *is*

Sol. (8) We have, $[\mathbf{u}\ \mathbf{v}\ \mathbf{w}]=\mathbf{u}\cdot(\mathbf{v}\times\mathbf{w})$

$$\Rightarrow \quad [\mathbf{u}\ \mathbf{v}\ \mathbf{w}]\le|\mathbf{u}||\mathbf{v}\times\mathbf{w}| \quad [\because \mathbf{a}\cdot\mathbf{b}\le|\mathbf{a}||\mathbf{b}|]$$

$$\Rightarrow \quad [\mathbf{u}\ \mathbf{v}\ \mathbf{w}]\le|\mathbf{v}\times\mathbf{w}| \quad [\because |\mathbf{u}|=1]$$

$$\text{Now,} \quad \mathbf{v}\times\mathbf{w}=\begin{vmatrix}\hat{\mathbf{i}} & \hat{\mathbf{j}} & \hat{\mathbf{k}}\\ 2 & 1 & -1\\ 1 & 0 & 3\end{vmatrix}=3\hat{\mathbf{i}}-7\hat{\mathbf{j}}-\hat{\mathbf{k}}$$

$$\therefore \quad |\mathbf{v}\times\mathbf{w}|=\sqrt{9+49+1}=\sqrt{59}$$

Hence, maximum value of $[\mathbf{u}\ \mathbf{v}\ \mathbf{w}]=\sqrt{59}$

On comparing, we get $\lambda=59$

$$\therefore \quad \lambda-51=8$$

Ex. 128 *Let* $\mathbf{a}=\alpha\hat{\mathbf{i}}+2\hat{\mathbf{j}}-3\hat{\mathbf{k}}, \mathbf{b}=\hat{\mathbf{i}}+2\alpha\hat{\mathbf{j}}-2\hat{\mathbf{k}}$ *and* $\mathbf{c}=2\hat{\mathbf{i}}-\alpha\hat{\mathbf{j}}+\hat{\mathbf{k}}$. *Then the value of* 6α, *such that* $\{(\mathbf{a}\times\mathbf{b})\times(\mathbf{b}\times\mathbf{c})\}\times(\mathbf{c}\times\mathbf{a})=0$, *is*

Sol. (4) $\mathbf{a}=\alpha\hat{\mathbf{i}}+2\hat{\mathbf{j}}-3\hat{\mathbf{k}}$, $\mathbf{b}=\hat{\mathbf{i}}+2\alpha\hat{\mathbf{j}}-2\hat{\mathbf{k}}$, $\mathbf{c}=2\hat{\mathbf{i}}-\alpha\hat{\mathbf{j}}+\hat{\mathbf{k}}$

$$\{(\mathbf{a}\times\mathbf{b})\times(\mathbf{b}\times\mathbf{c})\}\times(\mathbf{c}\times\mathbf{a})=\mathbf{0}$$

$$\text{or} \quad \{[\mathbf{a}\ \mathbf{b}\ \mathbf{c}]\mathbf{b}-[\mathbf{a}\ \mathbf{b}\ \mathbf{b}]\mathbf{c}\}\times(\mathbf{c}\times\mathbf{a})=\mathbf{0}$$

$$\text{or} \quad [\mathbf{a}\ \mathbf{b}\ \mathbf{c}]\mathbf{b}\times(\mathbf{c}\times\mathbf{a})=\mathbf{0}$$

$$\text{or} \quad [\mathbf{a}\ \mathbf{b}\ \mathbf{c}]((\mathbf{a}\cdot\mathbf{b})\mathbf{c}-(\mathbf{b}\cdot\mathbf{c})\mathbf{a})=\mathbf{0}$$

$$\text{or} \quad [\mathbf{a}\ \mathbf{b}\ \mathbf{c}]=0 \quad (\because \mathbf{a} \text{ and } \mathbf{c} \text{ are not collinear})$$

$$\Rightarrow \quad \begin{vmatrix}\alpha & 2 & -3\\ 1 & 2\alpha & -2\\ 2 & -\alpha & 1\end{vmatrix}=0$$

or $\alpha(2\alpha-2\alpha)-2(1+4)-3(-\alpha-4\alpha)=0$ or $10-15\alpha=0$

$$\therefore \quad \alpha=\frac{2}{3}$$

$$6\alpha=\frac{6\times2}{3}=4$$

Subjective Type Questions

Ex. 129 *Let* **a** *and* **b** *be two unit vectors such that* $\hat{\mathbf{a}} + \hat{\mathbf{b}}$ *is also a unit vector. Then, find the angle between* $\hat{\mathbf{a}}$ *and* $\hat{\mathbf{b}}$.

Sol. Since, $\hat{\mathbf{a}} + \hat{\mathbf{b}}$ is a unit vector.

$$\Rightarrow \quad |\hat{\mathbf{a}} + \hat{\mathbf{b}}| = 1 \Rightarrow |\hat{\mathbf{a}} + \hat{\mathbf{b}}|^2 = 1$$

$$\Rightarrow \quad (\hat{\mathbf{a}} + \hat{\mathbf{b}}) \cdot (\hat{\mathbf{a}} + \hat{\mathbf{b}}) = 1$$

$$\Rightarrow \quad \hat{\mathbf{a}} \cdot \hat{\mathbf{a}} + \hat{\mathbf{b}} \cdot \hat{\mathbf{b}} + 2\hat{\mathbf{a}} \cdot \hat{\mathbf{b}} = 1 \Rightarrow 1 + 1 + 2\hat{\mathbf{a}} \cdot \hat{\mathbf{b}} = 1$$

$$\Rightarrow \quad \hat{\mathbf{a}} \cdot \hat{\mathbf{b}} = -\frac{1}{2} \Rightarrow |\hat{\mathbf{a}}| |\hat{\mathbf{b}}| \cos\theta = -\frac{1}{2}$$

$$\Rightarrow \quad \cos\theta = -\frac{1}{2} \quad (\because |\hat{\mathbf{a}}| = |\hat{\mathbf{b}}| = 1)$$

$$\Rightarrow \quad \theta = 120^\circ$$

Hence, the angle between $\hat{\mathbf{a}}$ and $\hat{\mathbf{b}} = 120^\circ$.

Ex. 130 *Determine the values of c so that for all real x, the vectors* $cx\hat{\mathbf{i}} - 6\hat{\mathbf{j}} + 3\hat{\mathbf{k}}$ *and* $x\hat{\mathbf{i}} + 2\hat{\mathbf{j}} + 2cx\hat{\mathbf{k}}$ *make an obtuse angle with each other.*

Sol. Given, $\mathbf{a} = cx\hat{\mathbf{i}} - 6\hat{\mathbf{j}} + 3\hat{\mathbf{k}}$ and $\mathbf{b} = x\hat{\mathbf{i}} + 2\hat{\mathbf{j}} + 2cx\hat{\mathbf{k}}$

$\because$ **a** and **b** make an obtuse angle with each other.

$$\therefore \quad \cos\theta = \frac{\mathbf{a} \cdot \mathbf{b}}{|\mathbf{a}||\mathbf{b}|} < 0$$

$$\text{i.e.,} \quad \frac{cx^2 - 12 + 6cx}{\sqrt{c^2x^2 + 36 + 9}\sqrt{x^2 + 4 + 4c^2x^2}} < 0$$

$$\Rightarrow \quad cx^2 + 6cx - 12 < 0 \quad \text{...(i)}$$

Now, two cases are possible.

Case I $\quad c \neq 0$

$\Rightarrow cx^2 + 6cx - 12$ is a quadratic equation which has real solution "iff $A < 0$ and $B^2 - 4AC < 0$"

i.e. if $c < 0$ and $36c^2 + 48c < 0$

i.e. if $c < 0$ and $12c(3c + 4) < 0$

$$\Rightarrow \quad 3c + 4 > 0 \quad [\because c > 0]$$

$$\Rightarrow \quad -\frac{4}{3} < c < 0 \quad \text{...(ii)}$$

Case II $\quad c = 0$

$\Rightarrow -12 < 0$ which is an identity.

$\therefore \quad c = 0$ satisfy Eq. (i) ...(iii)

$\therefore$ From Eqs. (ii) and (iii), we get $-\frac{4}{3} < c \leq 0$

Ex. 131 *A, B, C and D are four points in space. Using vector methods, prove that* $AC^2 + BD^2 + AD^2 + BC^2 \geq AB^2 + CD^2$ *what is the implication of the sign of equality.*

Sol. Let the position vector of *A, B, C* and *D* be **a**, **b**, **c** and **d**, respectively.

Then, $AC^2 + BD^2 + AD^2 + BC^2$

$$= (\mathbf{c} - \mathbf{a}) \cdot (\mathbf{c} - \mathbf{a}) + (\mathbf{d} - \mathbf{b}) \cdot (\mathbf{d} - \mathbf{b}) + (\mathbf{d} - \mathbf{a}) \cdot (\mathbf{d} - \mathbf{a}) + (\mathbf{c} - \mathbf{b}) \cdot (\mathbf{c} - \mathbf{b})$$

$$= |\mathbf{c}|^2 + |\mathbf{a}|^2 - 2\mathbf{a} \cdot \mathbf{c} + |\mathbf{d}|^2 + |\mathbf{b}|^2 - 2\mathbf{d} \cdot \mathbf{b} + |\mathbf{d}|^2 + |\mathbf{a}|^2 - 2\mathbf{a} \cdot \mathbf{d} + |\mathbf{c}|^2 + |\mathbf{b}|^2 - 2\mathbf{b} \cdot \mathbf{c}$$

$$= |\mathbf{a}|^2 + |\mathbf{b}|^2 - 2\mathbf{a} \cdot \mathbf{b} + |\mathbf{c}|^2 + |\mathbf{d}|^2 - 2\mathbf{c} \cdot \mathbf{d} + |\mathbf{a}|^2 + |\mathbf{b}|^2 + |\mathbf{c}|^2 + |\mathbf{d}|^2 - 2\mathbf{a} \cdot \mathbf{b} + 2\mathbf{c} \cdot \mathbf{d} - 2\mathbf{a} \cdot \mathbf{c} - 2\mathbf{b} \cdot \mathbf{d} - 2\mathbf{a} \cdot \mathbf{d} - 2\mathbf{b} \cdot \mathbf{c}$$

$$= (\mathbf{a} - \mathbf{b}) \cdot (\mathbf{a} - \mathbf{b}) + (\mathbf{c} - \mathbf{d}) \cdot (\mathbf{c} \cdot \mathbf{d}) + (\mathbf{a} + \mathbf{b} - \mathbf{c} - \mathbf{d}) \cdot (\mathbf{a} + \mathbf{b} - \mathbf{c} - \mathbf{d})$$

$$= AB^2 + CD^2 + (\mathbf{a} + \mathbf{b} - \mathbf{c} - \mathbf{d}) \cdot (\mathbf{a} + \mathbf{b} - \mathbf{c} - \mathbf{d}) \geq AB^2 + CD^2$$

$$\therefore \quad AC^2 + BD^2 + AD^2 + BC^2 \geq AB^2 + CD^2$$

Ex. 132 *Using vector method, prove that the altitudes of a triangle are concurrent.*

Sol. Let the point of intersection *O* of two altitudes *BQ* and *CR* be taken as origin and the position vectors of the vertices *A, B, C* be **a**, **b**, **c** respectively. Let *AO* produced meet *BC* at *P*. We will show that *AP* is perpendicular to *BC* , showing there by that the three altitudes are concurrent.

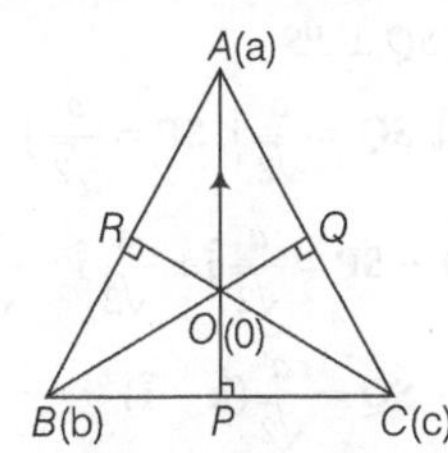

$$\therefore \quad \mathbf{OB} = \mathbf{b}, \mathbf{BQ} = \mu\mathbf{b}$$

as its collinear with **OB**.

Similarly, since $\quad \mathbf{OC} = \mathbf{C}$

$$\therefore \quad \mathbf{CR} = \nu\mathbf{C}$$

Now, $\quad \mathbf{AC} = \mathbf{c} - \mathbf{a}$ and $\mathbf{AB} = \mathbf{b} - \mathbf{a}$

Since, $BQ \perp AC$, we have $\mu\mathbf{b} \cdot (\mathbf{c} - \mathbf{a})$ and so $\mathbf{a} \cdot \mathbf{b} = \mathbf{b} \cdot \mathbf{c}$

Again, since $CR \perp AB$, $\nu\mathbf{c} \cdot (\mathbf{b} - \mathbf{a}) = 0$

$$\therefore \quad \mathbf{b} \cdot \mathbf{c} = \mathbf{c} \cdot \mathbf{a}$$

$$\therefore \quad \mathbf{a} \cdot \mathbf{b} = \mathbf{b} \cdot \mathbf{c} = \mathbf{c} \cdot \mathbf{a}$$

$$\text{or} \quad \mathbf{a} \cdot (\mathbf{c} - \mathbf{b}) = 0$$

$$\Rightarrow \quad \lambda\mathbf{a} \cdot (\mathbf{c} - \mathbf{b}) = 0$$

$$\therefore \quad \mathbf{AP} \cdot \mathbf{BC} = 0 \Rightarrow AP \perp BC$$

Ex. 133 *Using vector method, prove that the angle in a semi-circle is a right angle.*

Sol. Take the centre *O* as origin and *AB* is the diameter, so that $OA = OB$.

If the point *A* is *a*, then *B* is $-a$ and $|a| = r =$ radius.

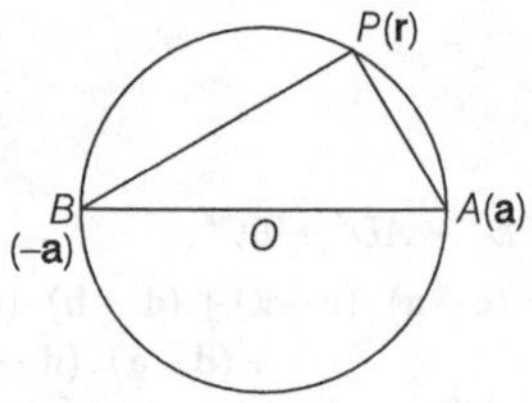

Let P be any point r on the circumference, so that $|r| = OP = r$

Then, $\mathbf{AP}$ = Position vector of P – Position vector of $A = r - a$

and $\mathbf{BP}$ = Position vector of P – Position vector of $B = r + a$

$\therefore \quad \mathbf{AP} \cdot \mathbf{BP} = (r - a) \cdot (r + a) = r^2 - a^2 = r^2 - r^2 = 0$

Ex. 134 *The corner P of the square OPQR is folded up so that the plane OPQ is perpendicular to the plane OQR, find the angle between OP and QR.*

Sol. After folding OPQ, $PS \perp SR$.

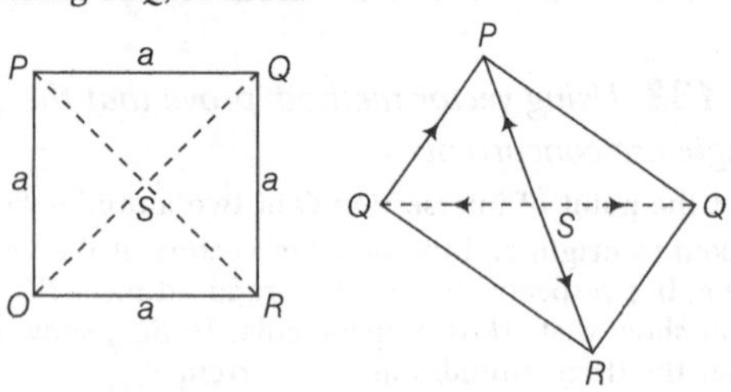

Here, $SQ \perp SR$, $SQ \perp PS$

Let $\mathbf{SR} = \frac{a}{\sqrt{2}}\hat{\mathbf{j}}$, $\mathbf{SQ} = \frac{a}{\sqrt{2}}\hat{\mathbf{i}}$, $\mathbf{SP} = \frac{a}{\sqrt{2}}\hat{\mathbf{j}}$

$$\mathbf{OP} = -\mathbf{SO} + \mathbf{SP} = \frac{a}{\sqrt{2}}\hat{\mathbf{j}} + \frac{a}{\sqrt{2}}\hat{\mathbf{i}} = \frac{a}{\sqrt{2}}(\hat{\mathbf{j}} + \hat{\mathbf{i}})$$

$$\mathbf{QR} = \mathbf{SR} - \mathbf{SQ} = \frac{a}{\sqrt{2}}(\hat{\mathbf{k}} - \hat{\mathbf{i}})$$

$$|\mathbf{OP}| = \frac{a}{\sqrt{2}}\sqrt{2} = a \Rightarrow |\mathbf{QR}| = a$$

Cosine of angle between $\mathbf{OP}$ and

$$\mathbf{QR} = \frac{\mathbf{OP} \cdot \mathbf{QR}}{|\mathbf{OP}||\mathbf{QR}|} = \frac{a^2(\hat{\mathbf{i}} + \hat{\mathbf{j}}) \cdot (\hat{\mathbf{k}} - \hat{\mathbf{i}})}{2 \qquad a^2}$$

$$\Rightarrow \quad \cos\theta = \frac{1}{2}(-1) = -\frac{1}{2} \Rightarrow \theta = \frac{2\pi}{3}$$

Ex. 135 *In a ΔABC, prove by vector method that* $\cos 2A + \cos 2B + \cos 2C \ge -\frac{3}{2}$

Sol. As we know, $(\mathbf{OA} + \mathbf{OB} + \mathbf{OC})^2 \ge 0$ ...(i)

and $\quad |\mathbf{OA}|^2 = |\mathbf{OB}|^2 = |\mathbf{OC}|^2 = R^2$ ...(ii)

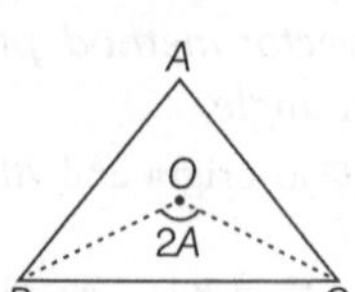

$\therefore$ Using Eq. (i),

$|\mathbf{OA}|^2 + |\mathbf{OB}|^2 + |\mathbf{OC}|^2 + 2(\mathbf{OA} \cdot \mathbf{OB} + \mathbf{OB} \cdot \mathbf{OC} + \mathbf{OC} \cdot \mathbf{OA}) \ge 0$

$\Rightarrow 3R^2 + 2R^2(\cos 2A + \cos 2B + \cos 2C) \ge 0$

$\Rightarrow \quad \cos 2A + \cos 2B + \cos 2C \ge -\frac{3}{2}$

Ex. 136 *Let $\beta = 4\hat{\mathbf{i}} + 3\hat{\mathbf{j}}$ and γ be two vectors perpendicular to each other in the XY-plane. Find all the vectors in the same plane having the projections 1, 2 along β and γ, respectively.*

Sol. Here, $\beta = 4\hat{\mathbf{i}} + 3\hat{\mathbf{j}}$

Since, γ is perpendicular to β i.e. $\beta \cdot \gamma = 0$

$\therefore$ We can choose $\gamma = 3\lambda\hat{\mathbf{i}} - 4\lambda\hat{\mathbf{j}}$ for all values of λ.

Let the required vector be $\alpha = l\hat{\mathbf{i}} + m\hat{\mathbf{j}}$.

Now, projection of α along $\beta = \frac{\alpha \cdot \beta}{|\beta|}$

$$1 = \frac{4l + 3m}{5} \Rightarrow 4l + 3m = 5 \qquad \text{...(i)}$$

Similarly, projection of α along $\gamma = \frac{\alpha \cdot \gamma}{|\gamma|}$

$$\Rightarrow \quad 2 = \frac{3\lambda l - 4\lambda m}{5\lambda}$$

$$\Rightarrow \quad 3l - 4m = 10 \qquad \text{...(ii)}$$

On solving Eqs. (i) and (ii), we get

$l = 2$ and $m = -1$

$\alpha = 2\hat{\mathbf{i}} - \hat{\mathbf{j}}$

Ex. 137 *If* **a**, **b** *and* **c** *are three coplanar vectors. If* **a** *is not parallel to* **b**, *show that*

$$\mathbf{c} = \frac{\begin{vmatrix} \mathbf{c}\cdot\mathbf{a} & \mathbf{a}\cdot\mathbf{b} \\ \mathbf{c}\cdot\mathbf{b} & \mathbf{b}\cdot\mathbf{b} \end{vmatrix}\mathbf{a} + \begin{vmatrix} \mathbf{a}\cdot\mathbf{a} & \mathbf{c}\cdot\mathbf{a} \\ \mathbf{a}\cdot\mathbf{b} & \mathbf{c}\cdot\mathbf{b} \end{vmatrix}\mathbf{b}}{\begin{vmatrix} \mathbf{a}\cdot\mathbf{a} & \mathbf{a}\cdot\mathbf{b} \\ \mathbf{a}\cdot\mathbf{b} & \mathbf{b}\cdot\mathbf{b} \end{vmatrix}}$$

Sol. Since, **a**, **b** and **c** are coplanar, we may write

$$\mathbf{c} = \lambda_1\mathbf{a} + \lambda_2\mathbf{b}$$

$$\Rightarrow \quad \mathbf{a}\cdot\mathbf{c} = \lambda_1\mathbf{a}\cdot\mathbf{a} + \lambda_2\mathbf{a}\cdot\mathbf{b} \qquad \text{...(i)}$$

$$\text{and} \quad \mathbf{b}\cdot\mathbf{c} = \lambda_1\mathbf{b}\cdot\mathbf{a} + \lambda_2\mathbf{b}\cdot\mathbf{b} \qquad \text{...(ii)}$$

On solving Eqs. (i) and (ii), by Cramer's rule, we find that

$$\lambda_1 = \frac{\begin{vmatrix} \mathbf{a}\cdot\mathbf{c} & \mathbf{a}\cdot\mathbf{b} \\ \mathbf{b}\cdot\mathbf{c} & \mathbf{b}\cdot\mathbf{b} \end{vmatrix}}{\begin{vmatrix} \mathbf{a}\cdot\mathbf{a} & \mathbf{a}\cdot\mathbf{b} \\ \mathbf{a}\cdot\mathbf{b} & \mathbf{b}\cdot\mathbf{b} \end{vmatrix}} \text{ and } \lambda_2 = \frac{\begin{vmatrix} \mathbf{a}\cdot\mathbf{a} & \mathbf{a}\cdot\mathbf{c} \\ \mathbf{b}\cdot\mathbf{a} & \mathbf{b}\cdot\mathbf{c} \end{vmatrix}}{\begin{vmatrix} \mathbf{a}\cdot\mathbf{a} & \mathbf{a}\cdot\mathbf{b} \\ \mathbf{a}\cdot\mathbf{b} & \mathbf{b}\cdot\mathbf{b} \end{vmatrix}}$$

On substituting λ_1 and λ_2, we get

$$\mathbf{c} = \frac{\begin{vmatrix} \mathbf{c}\cdot\mathbf{a} & \mathbf{a}\cdot\mathbf{b} \\ \mathbf{c}\cdot\mathbf{b} & \mathbf{b}\cdot\mathbf{b} \end{vmatrix}\mathbf{a} + \begin{vmatrix} \mathbf{a}\cdot\mathbf{a} & \mathbf{c}\cdot\mathbf{a} \\ \mathbf{a}\cdot\mathbf{b} & \mathbf{c}\cdot\mathbf{b} \end{vmatrix}\mathbf{b}}{\begin{vmatrix} \mathbf{a}\cdot\mathbf{a} & \mathbf{a}\cdot\mathbf{b} \\ \mathbf{a}\cdot\mathbf{b} & \mathbf{b}\cdot\mathbf{b} \end{vmatrix}}$$

Ex. 138 *In ΔABC, D is the mid-point of side AB and E is the centroid of ΔCDA. If $\mathbf{OE}\cdot\mathbf{CD}=0$, where O is the circumcentre of ΔABC, using vectors prove that $AB = AC$.*

Sol. Let us take O to be the origin and position vector of the vertices A, B and C be $\mathbf{a}$, $\mathbf{b}$ and $\mathbf{c}$, respectively.

We have,

$$|\mathbf{a}| = |\mathbf{b}| = |\mathbf{c}|$$

Now, $$\mathbf{D} = \frac{\mathbf{a}+\mathbf{b}}{2} \quad (\because \text{ mid-point of } AB)$$

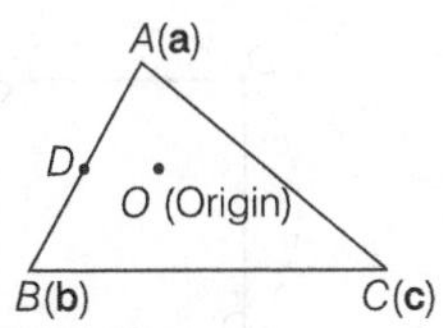

$$\mathbf{E} = \frac{3\mathbf{a}+\mathbf{b}+2\mathbf{c}}{2}$$

$\therefore$ $$\mathbf{CD} = \frac{\mathbf{a}+\mathbf{b}}{2} = \frac{\mathbf{a}+\mathbf{b}-2\mathbf{c}}{2}$$

and $$\mathbf{OE} = \frac{3\mathbf{a}+\mathbf{b}+2\mathbf{c}}{2}$$

$\therefore$ $$\mathbf{OE}\cdot\mathbf{CD} = 0$$

$\Rightarrow$ $$\frac{1}{4}(3\mathbf{a}+\mathbf{b}+2\mathbf{c})\cdot(\mathbf{a}+\mathbf{b}-2\mathbf{c}) = 0$$

$$3|\mathbf{a}|^2 + |\mathbf{b}|^2 - 4|\mathbf{c}|^2 + 4\mathbf{a}\cdot\mathbf{b} - 4\mathbf{a}\cdot\mathbf{c} = 0$$

$\Rightarrow$ $$\mathbf{a}\cdot\mathbf{b} = \mathbf{a}\cdot\mathbf{c}$$

$\Rightarrow$ $$3|\mathbf{a}|^2 + |\mathbf{b}|^2 - 4|\mathbf{c}|^2 = 0$$

$\Rightarrow$ $$|\mathbf{a}|^2 + |\mathbf{b}|^2 - 2\mathbf{a}\cdot\mathbf{b} = |\mathbf{a}|^2 + |\mathbf{c}| - 2\mathbf{a}\cdot\mathbf{c} \quad (\because |\mathbf{a}| = |\mathbf{b}| = |\mathbf{c}|)$$

$\Rightarrow$ $$|\mathbf{a}-\mathbf{b}|^2 = |\mathbf{a}-\mathbf{c}|^2$$

$\Rightarrow$ $$|\mathbf{AB}|^2 = |\mathbf{AC}|^2$$

$\Rightarrow$ $$|\mathbf{AB}| = |\mathbf{AC}|$$

Ex. 139 *Let I be the incentre of ΔABC. Using vectors prove that for any point P $a(PA)^2 + b(PB)^2 + c(PC)^2 = a(IA)^2 + b(IB)^2 + c(IC)^2 + (a+b+c)(IP)^2$ where a, b and c have usual meanings.*

Sol. We have, $\mathbf{IP} + \mathbf{IA} = \mathbf{PA}$

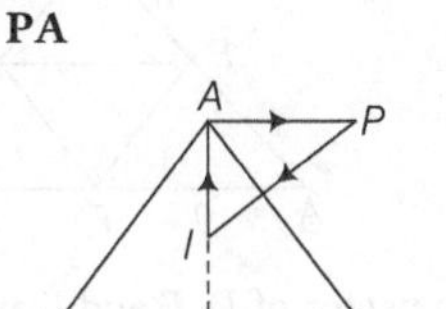

$\Rightarrow$ $$|\mathbf{PA}|^2 = |\mathbf{PI}|^2 + |\mathbf{IA}|^2 + 2\mathbf{PI}\cdot\mathbf{IA}$$

$$a|\mathbf{PA}|^2 = a|\mathbf{PI}|^2 + a|\mathbf{IA}|^2 + 2\mathbf{PI}\cdot(a\mathbf{IA}) \quad \text{...(i)}$$

Similarly, $$b|\mathbf{PB}|^2 = b|\mathbf{PI}|^2 + b|\mathbf{IB}|^2 + 2\mathbf{PI}\cdot(b\,\mathbf{IB}) \quad \text{...(ii)}$$

$$c|\mathbf{PC}|^2 = c|\mathbf{PI}|^2 + c|\mathbf{IC}|^2 + 2\mathbf{PI}\cdot(c\,\mathbf{IC})$$

On adding Eqs. (i), (ii) and (iii), we get

$$a|\mathbf{PA}|^2 + b|\mathbf{PB}|^2 + c|\mathbf{PC}|^2 = (a+b+c)|\mathbf{PI}|^2 + a|\mathbf{IA}|^2 + b|\mathbf{IB}|^2 + c|\mathbf{IC}|^2 + 2\mathbf{PI}\cdot|a\mathbf{IA} + b\mathbf{IB} + c\mathbf{IC}|^2$$

$\Rightarrow$ $$a|\mathbf{PA}|^2 + b|\mathbf{PB}|^2 + c|\mathbf{PC}|^2 = (a+b+c)|\mathbf{PI}|^2 + a|\mathbf{IA}|^2 + b|\mathbf{IB}|^2 + c|\mathbf{IC}|^2$$

($\because a\,\mathbf{IA} + b\,\mathbf{IB} + c\,\mathbf{IC} = 0$) shown as, since D be point of intersection of AI with side BC, we have $BD : DC = c : b$ and

$$\mathbf{AI} : \mathbf{ID} = b + c : a$$

$\Rightarrow$ $$\mathbf{ID} = \frac{c\,\mathbf{IC} + b\,\mathbf{IB}}{b+c} \text{ and } a\,\mathbf{AI} = (\mathbf{b}+\mathbf{c})\,\mathbf{ID}$$

$$a\,\mathbf{AI} = c\,\mathbf{IC} + b\,\mathbf{IB} \Rightarrow a\,\mathbf{IA} + b\,\mathbf{IB} + c\,\mathbf{IC} = 0$$

Ex. 140 *If two circles intersect, prove by using vector method, that the line joining their centres is perpendicular to their common chord.*

Sol. Let O be the centre of the first circle and C be the centre of second. Let a and b be the radii of the two circles. Position vector of C is $\mathbf{c}$ and AB be point of intersection of two circles.

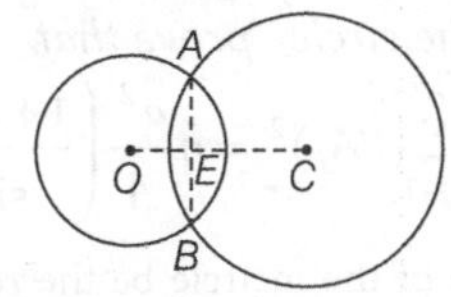

If $\mathbf{r}$ is the position vector of A.

$\Rightarrow$ $$\mathbf{CA} = \mathbf{OA} - \mathbf{OC} = \mathbf{r} - \mathbf{c} \quad \text{...(i)}$$

$$(\because \mathbf{OA} = \mathbf{r} \text{ and } \mathbf{OC} = \mathbf{c})$$

Also, $$\mathbf{r}\cdot\mathbf{r} = a^2 \text{ and } (\mathbf{r}-\mathbf{c})\cdot(\mathbf{r}-\mathbf{c}) = b^2 \quad \text{...(ii)}$$

Hence, at the point of intersection of two circle

$$a^2 - 2\mathbf{r}\cdot\mathbf{c} + |\mathbf{c}|^2 = b^2 \Rightarrow \mathbf{r}\cdot\mathbf{c} = \frac{1}{2}[b^2 - a^2 - |\mathbf{c}|^2]$$

If E is the point of intersection of OC and AB, then

$$\mathbf{OA} = \mathbf{OE} + \mathbf{EA} = \lambda\mathbf{c} + k_1\mathbf{AB}$$

$$\mathbf{OB} = \mathbf{OE} + \mathbf{EB} = \lambda\mathbf{c} + k_2\mathbf{AB}$$

$\Rightarrow$ $$2\mathbf{OA}\cdot\mathbf{c} = 2\mathbf{r}\cdot\mathbf{c} = 2\,[\lambda\mathbf{c} + k_1\mathbf{AB}]\cdot\mathbf{c} = \mathbf{a}^2 - \mathbf{b}^2 + |\mathbf{c}|^2$$

and $$2\mathbf{OB}\cdot\mathbf{c} = 2\mathbf{r}\cdot\mathbf{c} = 2\{\lambda\mathbf{c} + k_2\mathbf{AB}\}\cdot\mathbf{c} = \mathbf{a}^2 - \mathbf{b}^2 + |\mathbf{c}|^2$$

$\Rightarrow$ $$2[\lambda\mathbf{c} - k_1\mathbf{AB}]\cdot\mathbf{c} = 2\,[\lambda\mathbf{c} + k_2\mathbf{AB}]\,\mathbf{c} \Rightarrow \mathbf{AB}\cdot\mathbf{c} = 0$$

Hence, AB is perpendicular to OC.

Ex. 141 *Using vector method prove that $\cos(A-B) = \cos A\cos B + \sin A\sin B$.*

Sol. Let OX and OY be two lines perpendicular to each other and $\angle POX = A$, $\angle QOX = B$. So that, $\angle POQ = A - B$ shown as,

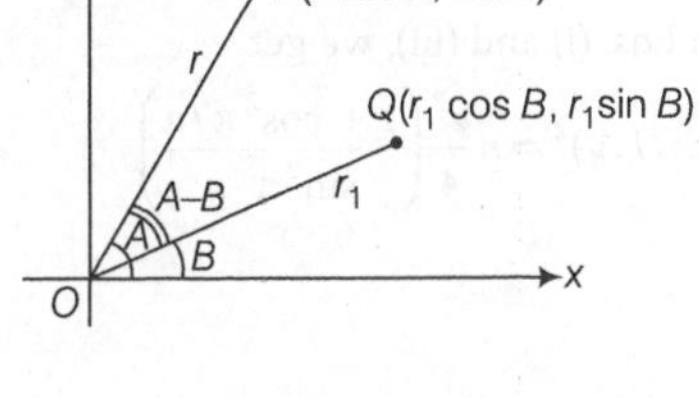

Let $\hat{\mathbf{i}}$ and $\hat{\mathbf{j}}$ denote unit vectors along OX and OY so that,

$$\hat{\mathbf{i}}\cdot\hat{\mathbf{i}} = \hat{\mathbf{j}}\cdot\hat{\mathbf{j}} = 1 \text{ and } \hat{\mathbf{i}}\cdot\hat{\mathbf{j}} = \hat{\mathbf{j}}\cdot\hat{\mathbf{i}} = 0$$

Also, let $\quad OP = r$ and $OQ = r_1$

$\therefore P\,(r\cos A,\, r\sin A)$ and $Q\,(r_1\cos B,\, r_1\sin B)$

$$\therefore \quad \mathbf{OP} = (r\cos A)\,\hat{\mathbf{i}} + (r\sin A)\,\hat{\mathbf{j}} \quad \text{...(i)}$$

and $\mathbf{OQ} = (r_1\cos B)\,\hat{\mathbf{i}} + (r_1\sin B)\,\hat{\mathbf{j}}$

By definition

$$\mathbf{OP}\cdot\mathbf{OQ} = |\mathbf{OP}|\,|\mathbf{OQ}|\,\cos\angle POQ = r_1 r\cos(A-B)$$

$$\therefore \quad \mathbf{OP}\cdot\mathbf{OQ} = rr_1\cos(A-B) \quad \text{...(ii)}$$

Also, from Eq. (i)

$$\begin{aligned}\mathbf{OP}\cdot\mathbf{OQ} &= rr_1\cos A\cos B + rr_1\sin A\sin B\\ &= rr_1(\cos A\cos B + \sin A\sin B) \quad \text{...(iii)}\end{aligned}$$

From Eqs. (ii) and (iii), we get

$rr_1\cos(A-B) = rr_1(\cos A\cos B + \sin A\sin B)$

$\Rightarrow \cos(A-B) = \cos A\cos B + \sin A\sin B$

Ex. 142 *A circle is inscribed in an n-sided regular polygon $A_1, A_2 \ldots, A_n$ having each side a unit for any arbitrary point P on the circle, prove that*

$$\sum_{i=1}^{n}(PA_i)^2 = n\frac{a^2}{4}\left(\frac{1+\cos^2\pi/n}{\sin^2\pi/n}\right)$$

Sol. Let the centre of the incircle be the reference point.

Then, $\mathbf{PA}_i = \mathbf{OA}_i - \mathbf{OP}$

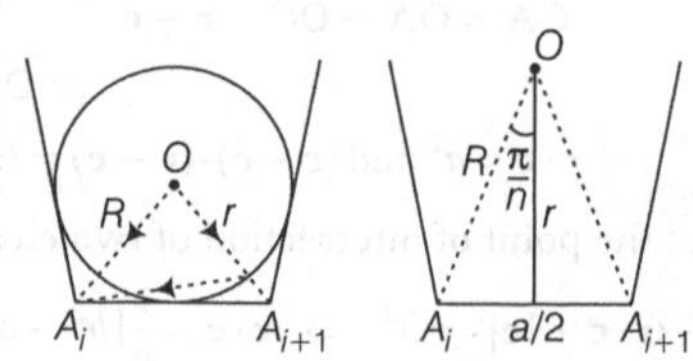

$$\mathbf{PA}_i\cdot\mathbf{PA}_i = (\mathbf{OA}_i - \mathbf{OP})\cdot(\mathbf{OA}_i - \mathbf{OP})$$

$$(\mathbf{PA}_i)^2 = (|\mathbf{OA}_i|)^2 + (|\mathbf{OP}|)^2 - 2\mathbf{OA}_i\cdot\mathbf{OP}$$

$$\begin{aligned}\sum_{i=1}^{n}(\mathbf{PA}_i)^2 &= \sum_{i=1}^{n}(|\mathbf{OA}_i|)^2 + (|\mathbf{OP}|^2) - 2\mathbf{OA}_i\cdot\mathbf{OP}\\ &= nR^2 + nr^2 - 2\mathbf{OP}\cdot\sum_{i=1}^{n}\mathbf{OA}_i \quad \text{...(i)}\\ &= n(R^2+r^2) - 2\,OP\cdot(0)\end{aligned}$$

Now, $\quad R = \frac{a}{2}\operatorname{cosec}\frac{\pi}{n},\, r = \frac{a}{2}\cot\frac{\pi}{n} \quad$ ...(ii)

$$\begin{aligned}\therefore \quad R^2 + r^2 &= \frac{a^2}{4}\left(\operatorname{cosec}^2\frac{\pi}{n} + \cot^2\frac{\pi}{n}\right)\\ &= \frac{a^2}{4}\left(\frac{1+\cos^2\pi/n}{\sin^2\pi/n}\right) \quad \text{...(iii)}\end{aligned}$$

$\therefore$ From Eqs. (i) and (iii), we get

$$\Rightarrow \quad \sum_{i=1}^{n}(PA_i)^2 = n\frac{a^2}{4}\left(\frac{1+\cos^2\pi/n}{\sin^2\pi/n}\right)$$

Ex. 143 *If* $\mathbf{a}, \mathbf{b}, \mathbf{c}$ *and* $\mathbf{d}$ *are the position vector of the vertices of a cyclic quadrilateral ABCD, prove that*

$$\frac{|\mathbf{a}\times\mathbf{b}+\mathbf{b}\times\mathbf{d}+\mathbf{d}\times\mathbf{a}|}{(\mathbf{b}-\mathbf{a})\cdot(\mathbf{d}-\mathbf{a})} + \frac{|\mathbf{b}\times\mathbf{c}+\mathbf{c}\times\mathbf{d}+\mathbf{d}\times\mathbf{b}|}{(\mathbf{b}-\mathbf{c})\cdot(\mathbf{d}\cdot\mathbf{a})} = 0$$

Sol. Consider, $\dfrac{|\mathbf{a}\times\mathbf{b}+\mathbf{b}\times\mathbf{d}+\mathbf{d}\times\mathbf{a}|}{(\mathbf{b}-\mathbf{a})\cdot(\mathbf{d}-\mathbf{a})}$

$$\begin{aligned}&= \frac{(\mathbf{a}-\mathbf{d})\times(\mathbf{b}-\mathbf{a})}{(\mathbf{b}-\mathbf{a})\cdot(\mathbf{d}-\mathbf{a})} = \frac{|\mathbf{a}-\mathbf{d}|\,|\mathbf{b}-\mathbf{a}|\sin A}{|\mathbf{b}-\mathbf{c}|\,|\mathbf{d}-\mathbf{c}|\cos A}\\ &= \tan A \quad \text{...(i)}\end{aligned}$$

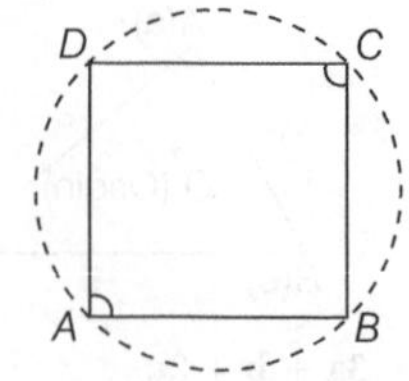

Again, $\dfrac{|\mathbf{b}\times\mathbf{c}+\mathbf{c}\times\mathbf{d}+\mathbf{d}\times\mathbf{b}|}{(\mathbf{b}-\mathbf{c})\cdot(\mathbf{d}-\mathbf{c})} = \dfrac{|(\mathbf{b}-\mathbf{c})\times(\mathbf{c}-\mathbf{d})|}{(\mathbf{b}-\mathbf{c})\cdot(\mathbf{d}-\mathbf{c})}$

$$= \frac{|\mathbf{b}-\mathbf{c}|\,|\mathbf{c}-\mathbf{d}|\sin C}{|\mathbf{b}-\mathbf{c}|\,|\mathbf{d}-\mathbf{c}|\cos C} = \tan C$$

As cyclic quadrilateral

$$A = 180^\circ - C$$

$$\Rightarrow \quad \tan A = \tan(180^\circ - C)$$

$$\Rightarrow \quad \tan A + \tan C = 0$$

$$\Rightarrow \quad \frac{|\mathbf{a}\times\mathbf{b}+\mathbf{b}\times\mathbf{d}+\mathbf{d}\times\mathbf{a}|}{(\mathbf{b}-\mathbf{a})\cdot(\mathbf{d}-\mathbf{a})} + \frac{|\mathbf{b}\times\mathbf{c}+\mathbf{c}\times\mathbf{d}+\mathbf{d}\times\mathbf{b}|}{(\mathbf{b}-\mathbf{c})\cdot(\mathbf{d}-\mathbf{c})} = 0$$

Ex. 144 *In ΔABC, points D, E and F are taken on the sides BC, CA and AB, respectively such that* $\dfrac{BD}{DC} = \dfrac{CE}{EA} = \dfrac{AF}{AB} = n.$

Prove that, $\Delta DEF = \dfrac{n^2-n+1}{(n+1)^2}\Delta ABC.$

Sol. Take A is the origin and let the position vectors of the points B and C be $\mathbf{b}$ and $\mathbf{c}$, respectively.

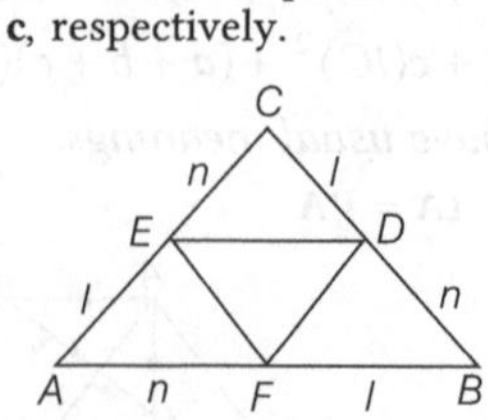

$\therefore$ The position vector of D, E and F are

$$\frac{n\mathbf{c}+\mathbf{b}}{\mathbf{n}+1}, \frac{\mathbf{c}}{\mathbf{n}+1}, \frac{n\mathbf{b}}{\mathbf{n}+1}$$

$$\mathbf{FD} = \mathbf{AD} - \mathbf{AF} = \frac{n\mathbf{c}+\mathbf{b}-n\mathbf{b}}{n+1} = \frac{n\mathbf{c}+(1-n)\,\mathbf{b}}{n+1}$$

and $\quad \mathbf{EF} = \mathbf{AF} - \mathbf{AE} = \dfrac{n\mathbf{b}-\mathbf{c}}{n+1}$

Now, vector area of $\Delta ABC = \frac{1}{2}(\mathbf{b}\times\mathbf{c})$ and vector area of ΔDEF

$$= \frac{1}{2}(\mathbf{FD}\times\mathbf{FE})$$

$$= \frac{1}{2(n+1)^2}\{(n\mathbf{b}-\mathbf{c})\times n\mathbf{c}+(1-n)\mathbf{b}\}$$

$$= \frac{1}{2(n+1)^2}\{n^2\mathbf{b}\times\mathbf{c}+(1-n)\,\mathbf{b}\times\mathbf{c}\}$$

$$= \frac{1}{2(n+1)^2}[(n^2-n+1)(\mathbf{b}\times\mathbf{c})]$$

$$= \frac{n^2-n+1}{2(n+1)^2}\,\Delta ABC$$

$\therefore$ Area of $\Delta\ DEF = \frac{n^2-n+1}{2(n+1)^2}$ area of $\Delta\ ABC$

Ex. 145 *Let the area of a given $\Delta\ ABC$ be Δ. Points A_1, B_1 and C_1 are the mid-points of the sides BC, CA and AB, respectively. Point A_2 is the mid-point of CA, lines C_1A_1 and AA_2 meet the median BB_1 at E and D, respectively. If Δ is the area of the quadrilateral A_1A_2DE, using vectors prove that $\frac{\Delta_1}{\Delta} = \frac{11}{56}$.*

Sol. Let the position verctor of A, B and C be $\mathbf{a}$, $\mathbf{b}$ and $\mathbf{c}$, respectively.

We have, $\mathbf{AC}_1 = \frac{b}{2}$, $\mathbf{AB}_1 = \frac{c}{2}$

$$\mathbf{AA}_2 = \frac{\mathbf{b}+\mathbf{c}}{2},\ \mathbf{AA}_2 = \frac{3\mathbf{c}+\mathbf{b}}{4}$$

Equation of the lines BB_1, AA_2 and C_1A_1 are

$$\mathbf{r} = \mathbf{b}+\lambda_1\left(\frac{\mathbf{c}}{2}-\mathbf{b}\right)$$

$$\mathbf{r} = \lambda_2\frac{3\mathbf{c}+\mathbf{b}}{4} \text{ and } \mathbf{r} = \frac{\mathbf{b}}{2}+\lambda_3\left(\frac{\mathbf{c}}{2}\right)$$

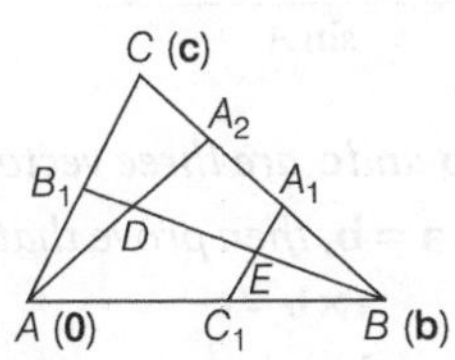

For the point D, we have

$$\mathbf{b}+\lambda_1\left(\frac{\mathbf{c}}{2}-\mathbf{b}\right) = \lambda_2\left(\frac{3\mathbf{c}+\mathbf{b}}{4}\right)$$

$$\mathbf{b}\left(1-\lambda_1-\frac{\lambda_2}{4}\right)+\frac{\mathbf{c}}{4}(2\lambda_1-3\lambda_2) = 0$$

$$\Rightarrow \quad \lambda_1 = \frac{6}{7},\ \lambda_2 = \frac{4}{7}$$

$$\therefore \quad \mathbf{AD} = \frac{3\mathbf{c}+\mathbf{b}}{7}$$

For the point E, we have $\mathbf{b}+\lambda_1\left(\frac{\mathbf{c}}{2}-\mathbf{b}\right) = \frac{\mathbf{b}}{2}+\frac{\lambda_3}{2}\mathbf{c}$

$$\Rightarrow \quad \mathbf{b}\left(\frac{1}{2}-\lambda_1\right)+\frac{\mathbf{c}}{2}(\lambda_1-\lambda_3) = 0$$

$$\Rightarrow \quad \lambda_1 = \lambda_3 = \frac{1}{2}$$

$$\therefore \quad \mathbf{AE} = \frac{2\mathbf{b}+\mathbf{c}}{4}$$

Now, $\mathbf{EA}_2 = \frac{3\mathbf{c}+\mathbf{b}-2\mathbf{b}-\mathbf{c}}{4} = \frac{2\mathbf{c}-\mathbf{b}}{7}$

$$\mathbf{DA}_1 = \frac{\mathbf{b}+\mathbf{c}}{2}-\frac{3\mathbf{c}+\mathbf{b}}{7} = \frac{5\mathbf{b}+\mathbf{c}}{14}$$

Area of quadrilateral $EA_1A_2\ D = \frac{1}{2}|\mathbf{EA}_2\times\mathbf{DA}_1|$

$$= \frac{1}{112}|(2\mathbf{c}-\mathbf{b})\times(5\mathbf{b}+\mathbf{c})|$$

$$= \frac{1}{112}|10\mathbf{c}\times\mathbf{b}-\mathbf{b}\times\mathbf{c}|$$

$$= \frac{11}{112}|\mathbf{c}\times\mathbf{b}| = \frac{11}{56}\cdot\frac{1}{2}|\mathbf{c}\times\mathbf{b}| = \frac{11}{56}$$

Thus, required ratio is $\frac{11}{56}$.

Ex. 146 *Let ABC be an acute angled triangle with centriod G and the internal bisectors of angles A, B and C meets BC, CA and AB in M, N and K respectively using vectors, prove that if G lies on one of the sides of $\Delta\ MNK$, then one of the altitudes of $\Delta\ ABC$ equals the sum of other two.*

Sol. Let G be on MK.

Let the position vectors of B and C with reference to origin A be $\mathbf{b}$ and $\mathbf{c}$, respectively.

$$\therefore \quad BC = a,\ CA = b \text{ and} AB = c$$

Now, $\frac{BM}{MC} = \frac{c}{b}$

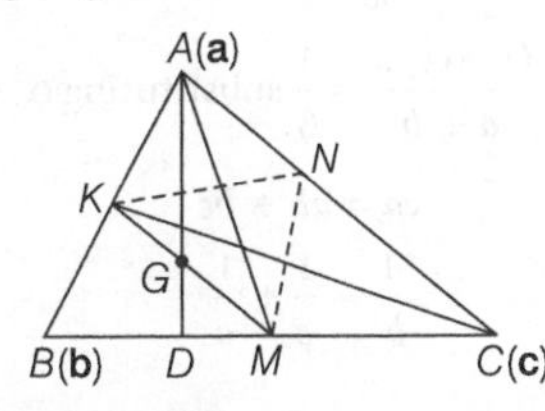

$\therefore$ Position vector of $M = \frac{b\mathbf{b}+c\mathbf{c}}{b+c}$

Similarly, position vector of $K = \frac{b\,\mathbf{b}}{a+b}$

$$\mathbf{MK} = \mathbf{PV} \text{ of } K - \mathbf{PV} \text{ of } M = \frac{b\mathbf{b}}{a+b}-\frac{b\mathbf{b}+c\mathbf{c}}{b+c}$$

PV of $G = \frac{\mathbf{b}+\mathbf{c}}{3}$

$$\mathbf{GK} = \mathbf{PV} \text{ of } \mathbf{K} - \mathbf{PV} \text{ of } G = \frac{b\mathbf{b}}{a+b}-\frac{\mathbf{b}+\mathbf{c}}{3}$$

Since, G lies on MK, $\mathbf{MK} \times \mathbf{GK} = 0$

$$\Rightarrow \left(\frac{\mathbf{bb}}{a+b} - \frac{\mathbf{bb}+\mathbf{cc}}{b+c}\right) \times \left(\frac{\mathbf{bb}}{a+b} - \frac{\mathbf{b}+\mathbf{c}}{3}\right) = 0$$

$$\Rightarrow \quad -\frac{\mathbf{bb}}{a+b} \times \frac{\mathbf{b}+\mathbf{c}}{3} - \frac{\mathbf{bb} \times \mathbf{cc}}{a+c}$$

$$\times \frac{\mathbf{bb}}{a+b} + \frac{\mathbf{bb}+\alpha\mathbf{c}}{b+c} \times \frac{\mathbf{b}+\mathbf{c}}{3} = 0$$

$$\Rightarrow -\frac{b(\mathbf{b} \times \mathbf{c})}{3(a+b)} - \frac{\mathbf{bc}\,(\mathbf{c} \times \mathbf{b})}{(b+c)\,(a+b)} + \frac{\mathbf{b}\,(\mathbf{b} \times \mathbf{c})}{3\,(b+c)} + \frac{\mathbf{b}\,(\mathbf{c} \times \mathbf{b})}{3} = 0$$

$$\left[\frac{b}{3\,(a+b)} - \frac{bc}{(b+c)\,(a+b)} + \frac{b}{3(b+c)} + \frac{c}{3(b+c)}\right] (\mathbf{c} \times \mathbf{b}) = 0$$

$$\Rightarrow \frac{b}{3(a+b)} - \frac{bc}{(b+c)\,(a+b)} - \frac{b}{3(b+c)} + \frac{c}{3(b+c)} = 0$$

$$\Rightarrow b(b+c) - 3bc - b(a+b) + c(a+b) = 0$$

$$\Rightarrow b^2 + bc - 3bc - ab - b^2 - ac + bc$$

$$\Rightarrow \quad ac = ab + bc$$

$$\Rightarrow \quad \frac{1}{b} = \frac{1}{c} + \frac{1}{a}$$

$$\Rightarrow \quad \frac{2\Delta}{b} = \frac{2\Delta}{c} + \frac{2\Delta}{a} \text{ (where, } \Delta \text{ denotes area of } \Delta\, ABC)$$

$\Rightarrow P_b = P_a + P_c$ denotes the altitudes drawn through A, B and C, respectively.

Aliter

$$g = \alpha\, \mathbf{m} + (1-\alpha)\,\mathbf{k}$$

$$\therefore \quad \frac{\mathbf{b}+\mathbf{c}}{3} = \frac{\alpha\,(b\,\mathbf{b} + \alpha\mathbf{c})}{b+c} + \frac{(1-\alpha)\, b\, \mathbf{b}}{a+b}$$

On comparing coefficients of **b** and **c**, we get

$$\frac{1}{3} = \frac{\alpha c}{b+c}$$

$$\Rightarrow \quad \alpha = \frac{b+c}{3c}$$

and $\dfrac{\alpha b}{b+c} + \dfrac{(1-\alpha)\, b}{a+b} = \dfrac{1}{3}$ substituting α, we get

$$ca = ab + bc$$

$$\Rightarrow \quad \frac{1}{b} = \frac{1}{c} + \frac{1}{a}$$

Ex. 147 *Three poles of height x, $x+y$ and $x+z$ are posted at the vertices A, B and C of a triangular park of sides a, b and c, respectively. A plane sheet is mounted at the tops of the poles. If the plane of the sheet is inclined at an angle θ to the horizontal plane, prove using vector*

$$\theta = \tan^{-1}\left\{\frac{\sqrt{\dfrac{y^2}{c^2} + \dfrac{z^2}{b^2} - \dfrac{2yz}{bc}\cos A}}{\sin A}\right\}$$

Sol. Let A', B' and C' be the tops of the poles at A, B and C, respectively. Through A' draw a $\Delta A'\, B_1C_1$ congruent to $\Delta\, ABC$ and parallel to the horizontal plane of the park. Take $A'\, B_1$ as the X-axis and a line perpendicular to it as the Y-axis (in the plane of $\Delta\, A'\, B_1C_1$) and a line through A' and perpendicular to the plane $A'\, B_1C_1$ as the Z-axis.

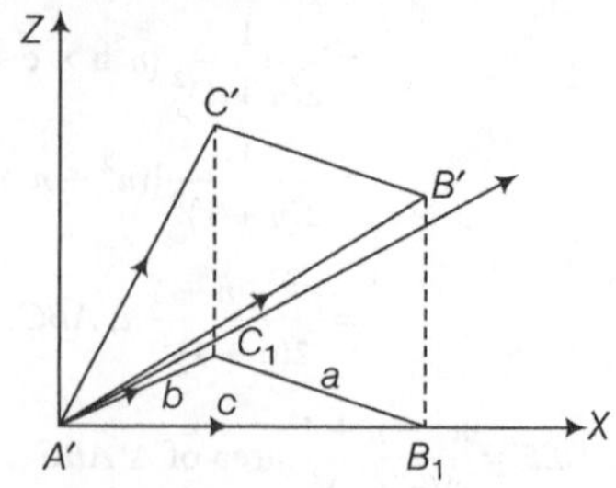

If $\hat{\mathbf{i}}$, $\hat{\mathbf{j}}$ and $\hat{\mathbf{k}}$ are the unit vectors along these axes, then

$$\mathbf{A'\,B_1} = c\,\hat{\mathbf{i}}$$
$$\mathbf{A'\,C_1} = (b\cos A)\hat{\mathbf{i}} + (b\sin A)\hat{\mathbf{j}}$$
$$\mathbf{A'B'} = c\hat{\mathbf{i}} + y\,\hat{\mathbf{k}}$$
$$\mathbf{A'\,C'} = (b\cos A)\hat{\mathbf{i}} + (b\sin A)\hat{\mathbf{j}} + z\hat{\mathbf{k}}$$

Since, the planes $A'\,B'\,C'$ is inclined at an angle θ to the plane $A'\,B_1C_1$, angle between the normals to the planes is $(\pi - \theta)$.

Obviously, the unit vector normal to the plane $A'\,B_1C_1$ is **k** and the normal vector to $A'\,B'\,C'$ is

$$[(b\cos A)\,\hat{\mathbf{i}} + (b\sin A)\,\hat{\mathbf{j}} + (z)\,\hat{\mathbf{k}}] \times (c\hat{\mathbf{i}} + y\,\hat{\mathbf{k}})$$
$$= (yb\sin A)\,\hat{\mathbf{i}} - (yb\cos A - zc)\hat{\mathbf{j}} - (bc\sin A)\hat{\mathbf{k}}$$

$$\cos(\pi - \theta) = \frac{\{(yb\sin A)\hat{\mathbf{i}} - (yb\cos A - zc)\hat{\mathbf{j}} - (bc\sin A)\hat{\mathbf{k}}\} \cdot \hat{\mathbf{k}}}{\sqrt{y^2b^2\sin^2 A + y^2b^2\cos^2 A + z^2c^2 - 2yzbc\cos A + b^2c^2\sin^2 A}}$$

$$\Rightarrow \cos\theta = \frac{\sin A}{\sqrt{\dfrac{y^2}{c^2} + \dfrac{z^2}{b^2} - \dfrac{2yz}{bc}\cos A + \sin^2 A}}$$

$$\Rightarrow \tan\theta = \frac{\sqrt{\dfrac{y^2}{c^2} + \dfrac{z^2}{b^2} - \dfrac{2yz}{bc}\cos A}}{\sin A}$$

Ex. 148 *If **a**, **b** and **c** are three vectors such that $\mathbf{a} \times \mathbf{b} = \mathbf{c}$, $\mathbf{b} \times \mathbf{c} = \mathbf{a}$ and $\mathbf{c} \times \mathbf{a} = \mathbf{b}$, then prove that $|\mathbf{a}| = |\mathbf{b}| = |\mathbf{c}|$*

Sol. Here, $\quad \mathbf{a} \times \mathbf{b} = \mathbf{c}$ (given)

$$\Rightarrow \quad (\mathbf{a} \times \mathbf{b}) \cdot \mathbf{c} = \mathbf{c} \cdot \mathbf{c}$$

$$\Rightarrow \quad [\mathbf{a}\,\mathbf{b}\,\mathbf{c}] = |\mathbf{c}|^2 \quad \text{...(i)}$$

Also, $\quad \mathbf{b} \times \mathbf{c} = \mathbf{a}$ (given)

$$(\mathbf{b} \times \mathbf{c}) \cdot \mathbf{a} = \mathbf{a} \cdot \mathbf{a}$$

$$\Rightarrow \quad [\mathbf{b}\ \mathbf{c}\ \mathbf{a}] = |\mathbf{a}|^2 \quad \text{...(ii)}$$

and $\quad \mathbf{c} \times \mathbf{a} = \mathbf{b}$

$$\Rightarrow \quad (\mathbf{c} \times \mathbf{a}) \cdot \mathbf{b} = \mathbf{b} \cdot \mathbf{b} \quad \text{(given)}$$

$$\Rightarrow \quad [\mathbf{c}\ \mathbf{a}\ \mathbf{b}] = |\mathbf{b}|^2 \quad \text{(given)}$$

Since, $\quad [\mathbf{a}\ \mathbf{b}\ \mathbf{c}] = [\mathbf{b}\ \mathbf{c}\ \mathbf{a}] = [\mathbf{c}\ \mathbf{a}\ \mathbf{b}]$ ...(iii)

$\therefore$ From Eqs. (i), (ii) and (iii), we get

$$|\mathbf{c}|^2 = |\mathbf{a}|^2 = |\mathbf{b}|^2 \Rightarrow |\mathbf{c}| = |\mathbf{a}| = |\mathbf{b}|$$

Ex. 149 *If* **a**, **b**, **c** *and* **d** *are four coplanar points, then show that* $[\mathbf{a}\ \mathbf{b}\ \mathbf{c}]=[\mathbf{b}\ \mathbf{c}\ \mathbf{d}]+[\mathbf{c}\ \mathbf{a}\ \mathbf{d}]+[\mathbf{a}\ \mathbf{b}\ \mathbf{d}]$

Sol. Since, **a**, **b**, **c** and **d** are coplanar points.

We have, $\mathbf{b}-\mathbf{a}$, $\mathbf{c}-\mathbf{a}$ and $\mathbf{d}-\mathbf{a}$ are coplanar.

$$\Rightarrow \quad [\mathbf{b}-\mathbf{a}\ \mathbf{c}-\mathbf{a}\ \mathbf{d}-\mathbf{a}]=0$$

$$\Rightarrow \quad \{(\mathbf{b}-\mathbf{a})\times(\mathbf{c}-\mathbf{a})\}\cdot(\mathbf{d}-\mathbf{a})=0$$

$$\Rightarrow \quad (\mathbf{b}\times\mathbf{c})\cdot\mathbf{d}-(\mathbf{b}\times\mathbf{c})\cdot\mathbf{a}-(\mathbf{b}\times\mathbf{c})\cdot\mathbf{a}=0$$

$$-(\mathbf{a}\times\mathbf{c})\cdot\mathbf{d}+(\mathbf{a}\times\mathbf{c})\cdot\mathbf{a}=0$$

$$\Rightarrow \quad [\mathbf{b}\ \mathbf{c}\ \mathbf{d}]-[\mathbf{b}\ \mathbf{c}\ \mathbf{a}]-[\mathbf{b}\ \mathbf{a}\ \mathbf{d}]-[\mathbf{a}\ \mathbf{c}\ \mathbf{d}]=0$$

$$\Rightarrow \quad [\mathbf{a}\ \mathbf{b}\ \mathbf{c}]=[\mathbf{b}\ \mathbf{c}\ \mathbf{d}]+[\mathbf{a}\ \mathbf{b}\ \mathbf{d}]+[\mathbf{c}\ \mathbf{a}\ \mathbf{d}]$$

Ex. 150 *Let* $\hat{\mathbf{u}}$ *and* $\hat{\mathbf{v}}$ *be unit vectors. If* **w** *is a vector such that* $\mathbf{w}+(\mathbf{w}\times\mathbf{u})=\mathbf{v}$, *then prove that* $|(\mathbf{u}\times\mathbf{v})\cdot\mathbf{w}|\le\frac{1}{2}$ *and that the equality holds if and only if* **u** *is perpendicular to* **v**.

Sol.
$$\mathbf{w}+(\mathbf{w}\times\mathbf{u})=\mathbf{v}$$

$$\Rightarrow \quad \mathbf{w}\times\mathbf{u}=v-\mathbf{w} \qquad \text{...(i)}$$

$$\Rightarrow \quad (\mathbf{w}\times\mathbf{u})^2=v^2+w^2-2\mathbf{v}\cdot\mathbf{w}$$

$$\Rightarrow \quad 2\mathbf{v}\cdot\mathbf{w}=1+w^2-(\mathbf{u}\times\mathbf{w})^2 \qquad \text{...(ii)}$$

Also, taking dot product of Eq. (i) with **v**, we get

$$\mathbf{w}\cdot\mathbf{v}+(\mathbf{w}\times\mathbf{u})\cdot\mathbf{v}=\mathbf{v}\cdot\mathbf{v}$$

$$\Rightarrow \quad \mathbf{v}\cdot(\mathbf{w}\times\mathbf{u})=1-\mathbf{w}\cdot\mathbf{v} \qquad \text{...(iii)}\ (\because \mathbf{v}\cdot\mathbf{v}=|\mathbf{v}|^2=1)$$

Now,
$$\mathbf{v}\cdot(\mathbf{w}\times\mathbf{u})=1-\frac{1}{2}[1+w^2-(\mathbf{u}\times\mathbf{w})^2]$$

[using Eqs. (ii) and (iii)]

$$=\frac{1}{2}-\frac{w^2}{2}+\frac{(\mathbf{u}\times\mathbf{w})^2}{2} \qquad (\because 0\le\cos^2\theta\le1)$$

$$=\frac{1}{2}(1-w^2+w^2\sin^2\theta) \qquad \text{...(iv)}$$

As we know, $0\le w^2\cos^2\theta\le w^2$

$$\therefore \quad \frac{1}{2}\ge\frac{1-w^2\cos^2\theta}{2}\ge\frac{1-w^2}{2}$$

$$\Rightarrow \quad \frac{1-w^2\cos^2\theta}{2}\le\frac{1}{2} \qquad \text{...(v)}$$

From Eqs. (iv) and (v), we get $|\mathbf{v}\cdot(\mathbf{w}\times\mathbf{u})|\le\frac{1}{2}$

Equality holds only when $\cos^2\theta=0\Rightarrow\theta=\frac{\pi}{2}$

i.e., $\mathbf{u}\perp\mathbf{w}=0\Rightarrow\mathbf{u}\cdot\mathbf{w}=0$

$$\mathbf{w}+(\mathbf{w}\times\mathbf{u})=\mathbf{v}$$

$$\mathbf{u}\cdot\mathbf{w}+\mathbf{u}\cdot(\mathbf{w}\times\mathbf{u})=\mathbf{u}\cdot\mathbf{v}$$

$$0+0=\mathbf{u}\cdot\mathbf{v} \ \Rightarrow\ \mathbf{u}\cdot\mathbf{v}=0\Rightarrow\mathbf{u}\perp\mathbf{v}$$

Ex. 151 *Prove that*

$$\mathbf{R}+\frac{[\mathbf{R}.(\beta\times(\beta\times\alpha)]\alpha}{|\alpha\times\beta|^2}+\frac{\mathbf{R}.(\alpha\times(\alpha\times\beta))]\beta}{|\alpha\times\beta|^2}=\frac{[\mathbf{R}\,\alpha\,\beta\,](\alpha\times\beta)}{|\alpha\times\beta\,|^2}$$

Sol. α, β and $\alpha\times\beta$ are three non-coplanar vectors. Any vector **R** can be represented as a linear combination of these vectors.

$$\Rightarrow \quad \mathbf{R}=k_1\alpha+k_2\beta+k_3(\alpha\times\beta) \qquad \text{...(i)}$$

$$\Rightarrow \quad \mathbf{R}.(\alpha\times\beta)=k_3(\alpha\times\beta).(\alpha\times\beta)=k_3(\alpha\times\beta)^2$$

$$\Rightarrow \quad k_3=\frac{\mathbf{R}.(\alpha\times\beta)}{|\alpha\times\beta\,|^2}=\frac{[\mathbf{R}\alpha\beta]}{|\alpha\times\beta|^2}$$

On taking dot product of Eq. (i) with $\alpha\times(\alpha\times\beta)$

$$\Rightarrow \quad \mathbf{R}.\alpha\times(\alpha\times\beta)=k_2(\alpha\times(\alpha\times\beta)).\beta$$

$$k_2[(\alpha.\beta)\alpha-(\alpha.\alpha)\beta].\beta=k_2[(\alpha.\beta)^2-\alpha^2\beta^2]$$

$$=-k_2|\alpha\times\beta\,|^2$$

$$\Rightarrow \quad k_2=\frac{-[\mathbf{R}.(\alpha\times(\alpha\times\beta))]}{|\alpha\times\beta\,|^2}$$

Similarly,
$$k_1=-\frac{[\mathbf{R}.(\beta\times(\beta\times\alpha))]}{|\alpha\times\beta\,|^2}$$

$$\Rightarrow \quad \mathbf{R}=\frac{-[\mathbf{R}.(\beta\times(\beta\times\alpha))]\,\alpha}{|\alpha\times\beta\,|^2}-\frac{[\mathbf{R}.(\alpha\times(\alpha\times\beta))]\,\beta}{|\alpha\times\beta\,|^2}+\frac{[\mathbf{R}.(\alpha\times\beta)](\alpha\times\beta)}{|\alpha\times\beta\,|^2}$$

$$\Rightarrow \quad \mathbf{R}+\frac{[\mathbf{R}.(\beta\times(\beta\times\alpha))]\alpha}{|\alpha\times\beta\,|^2}+\frac{[\mathbf{R}.(\alpha\times(\alpha\times\beta))]\,\beta}{|\alpha\times\beta\,|^2}=\frac{[\mathbf{R}.(\alpha\times\beta)]\,(\alpha\times\beta)}{|\alpha\times\beta\,|^2}$$

Ex. 152 *If* **a**, **b** *and* **c** *represents the sides of tetrahedron and* θ *be an angle between* **a** *and* **b**, ϕ *be an angle between* **a** *and* **c**, ψ *be an angle between* **b** *and* **c**, *then prove that the volume of the tetrahedron is given by*

$$v^2=\frac{a^2b^2c^2}{36}\begin{vmatrix}1&\cos\theta&\cos\theta\\\cos\theta&1&\cos\psi\\\cos\phi&\cos\psi&1\end{vmatrix}$$

Sol. *OABC* represent a tetrahedron, where

$$\mathbf{OA}=\mathbf{a},\mathbf{OB}=\mathbf{b},\mathbf{OC}=\mathbf{c} \text{ relative to } O$$

Volume of tetrahedron $(v)=\frac{1}{6}[\mathbf{a}\cdot(\mathbf{b}\times\mathbf{c})]$

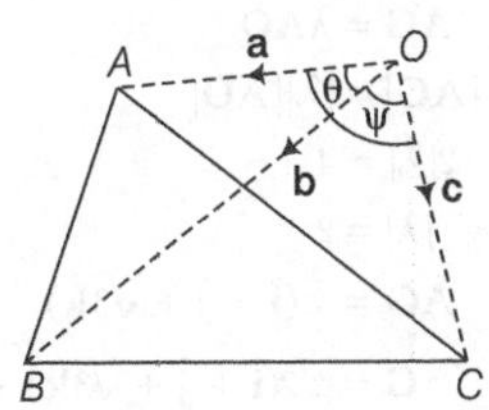

Also,
$$v^2=\frac{1}{36}[\mathbf{a}\cdot(\mathbf{b}\times\mathbf{c})]^2=\frac{1}{36}\begin{vmatrix}\mathbf{a}\cdot\mathbf{a}&\mathbf{a}\cdot\mathbf{b}&\mathbf{a}\cdot\mathbf{c}\\\mathbf{b}\cdot\mathbf{a}&\mathbf{b}\cdot\mathbf{b}&\mathbf{b}\cdot\mathbf{c}\\\mathbf{c}\cdot\mathbf{a}&\mathbf{c}\cdot\mathbf{b}&\mathbf{c}\cdot\mathbf{c}\end{vmatrix}$$

$$=\frac{1}{36}\begin{vmatrix}a^2&ab\cos\theta&ac\cos\phi\\ab\cos\theta&b^2&bc\cos\psi\\ac\cos\phi&bc\cos\psi&c^2\end{vmatrix}$$

$$=\frac{1}{36}a^2b^2c^2\begin{vmatrix}1&\cos\theta&\cos\phi\\\cos\theta&1&\cos\psi\\\cos\phi&\cos\psi&1\end{vmatrix}$$

Ex. 153 *A pyramid with vertex at the point P, whose position vector is $4\hat{\mathbf{i}}+2\hat{\mathbf{j}}+2\sqrt{3}\,\hat{\mathbf{k}}$ has a regular hexagonal base ABCDEF. Position vectors of points A and B are $\hat{\mathbf{i}}$ and $\hat{\mathbf{i}}+2\hat{\mathbf{j}}$, respectively. Centre of the base has the position vector $\hat{\mathbf{i}}+\hat{\mathbf{j}}+\sqrt{3}\hat{\mathbf{k}}$. Altitude drawn from P on the base meets the diagonal AD at point G. Find all possible position vectors of G. It is given that volume of the pyramid is $6\sqrt{3}$ cu units.*

Sol. Let the centre of base be (0).

$$\mathbf{AB}=2\hat{\mathbf{j}} \Rightarrow |\mathbf{AB}|=2$$

$$\Delta OAB=\frac{1}{4}\cdot 4\sqrt{3}=\sqrt{3}$$

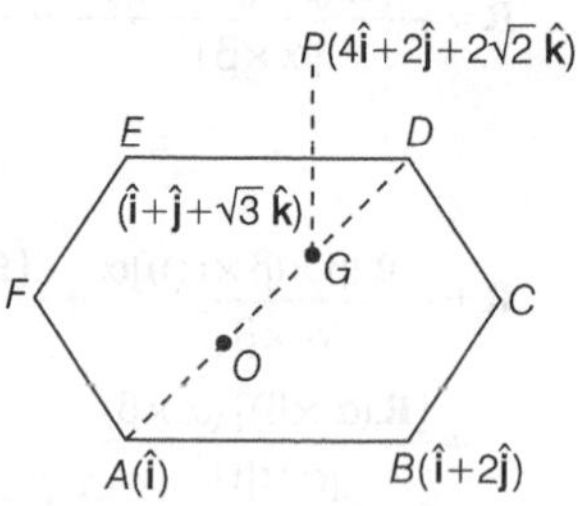

$\Rightarrow$ Base are $a=6\sqrt{3}$ sq unit.

Let height of the pyramid be h.

$$\Rightarrow \quad \frac{1}{3}\cdot 6\sqrt{3}h=6\sqrt{3}$$

$$\Rightarrow \quad h=3 \text{ units}$$

$$\mathbf{AP}=3\hat{\mathbf{i}}+2\hat{\mathbf{j}}+2\sqrt{3}\hat{\mathbf{k}}$$

$$\Rightarrow \quad |\mathbf{AP}|=\sqrt{9+4+12}=5 \text{ units}$$

$$\Rightarrow \quad \mathbf{AP}=\sqrt{9+4+12}=5 \text{ units}$$

$$\Rightarrow \quad \mathbf{AG}=\sqrt{25-9}=4 \text{ units}$$

$$|\mathbf{AG}|=4 \text{ units}$$

Now, **AQ** and **AO** are collinear.

$$\Rightarrow \quad \mathbf{AG}=\lambda\mathbf{AO}$$

$$\Rightarrow \quad |\mathbf{AG}|=|\lambda||\mathbf{AO}|$$

$$\Rightarrow \quad 2|\lambda|=4$$

$$\Rightarrow \quad |\lambda|=2$$

$$\Rightarrow \quad \mathbf{AG}=\pm(\hat{\mathbf{i}}+\hat{\mathbf{j}}+\sqrt{3}\hat{\mathbf{k}})$$

$$\Rightarrow \quad \mathbf{G}=\pm 2(\hat{\mathbf{i}}+\hat{\mathbf{j}}+\sqrt{3}\hat{\mathbf{k}})+\hat{\mathbf{i}}$$

$$=-(\hat{\mathbf{i}}+2\hat{\mathbf{j}}+2\sqrt{3}\hat{\mathbf{k}}),\ -3\hat{\mathbf{i}}+2\hat{\mathbf{j}}+2\sqrt{3}\hat{\mathbf{k}}$$

Ex. 154 *Let $\hat{\mathbf{a}}$, $\hat{\mathbf{b}}$ and $\hat{\mathbf{c}}$ be the non-coplanar unit vectors. The angle between $\hat{\mathbf{b}}$ and $\hat{\mathbf{c}}$ be α and angle between $\hat{\mathbf{c}}$ and $\hat{\mathbf{a}}$ be β and between $\hat{\mathbf{a}}$ and $\hat{\mathbf{b}}$ be γ. If $A(\hat{\mathbf{a}}\cos\alpha, 0)$, $B(\hat{\mathbf{b}}\cos\beta, 0)$ and $C(\hat{\mathbf{c}}\cos\gamma, 0)$, then show that in ΔABC.*

$$\frac{|\hat{\mathbf{a}}\times(\hat{\mathbf{b}}\times\hat{\mathbf{c}})|}{\sin A}=\frac{|\hat{\mathbf{b}}\times(\hat{\mathbf{c}}\times\hat{\mathbf{a}})|}{\sin B}=\frac{|\hat{\mathbf{c}}\times(\hat{\mathbf{a}}\times\hat{\mathbf{b}})|}{\sin C}$$

$$=\frac{\pi|\hat{\mathbf{a}}(\hat{\mathbf{b}}\times\hat{\mathbf{c}})|}{|\Sigma \sin\alpha\cos\beta\cos\gamma\,\hat{\eta}|}$$

where, $\hat{\eta}_1=\dfrac{\hat{\mathbf{b}}\times\hat{\mathbf{c}}}{|\hat{\mathbf{b}}\times\hat{\mathbf{c}}|}$, $\hat{\eta}_2=\dfrac{\hat{\mathbf{c}}\times\hat{\mathbf{a}}}{|\hat{\mathbf{c}}\times\hat{\mathbf{a}}|}$ *and* $\hat{\eta}_3=\dfrac{\hat{\mathbf{a}}\times\hat{\mathbf{b}}}{|\hat{\mathbf{a}}\times\hat{\mathbf{b}}|}$

Sol. We know from sine rule,

$$\frac{AB}{\sin C}=\frac{AC}{\sin B}=\frac{BC}{\sin A}$$

$$=2R=\frac{(AB)(BC)(CA)}{2(\Delta ABC)} \qquad \text{...(i)}$$

$$BC=|\mathbf{BC}||=|\hat{\mathbf{c}}\cos\gamma-\hat{\mathbf{b}}\cos\beta|$$

$$=|(\hat{\mathbf{a}}\cdot\hat{\mathbf{b}})\hat{\mathbf{c}}\cdot(\hat{\mathbf{c}}\cdot\hat{\mathbf{a}})\hat{\mathbf{b}}|=|\hat{\mathbf{a}}\times(\hat{\mathbf{b}}\times\hat{\mathbf{c}})|$$

Similarly, $\quad AC=|\mathbf{AC}|=|\hat{\mathbf{b}}\times(\hat{\mathbf{c}}\times\hat{\mathbf{a}})|$

and $\quad AB=|\mathbf{AB}|=|\hat{\mathbf{c}}\times(\hat{\mathbf{a}}\times\hat{\mathbf{b}})|$

Also, $\Delta ABC=\dfrac{1}{2}|\mathbf{BC}\times\mathbf{BA}|$

$$=\frac{1}{2}|(\hat{\mathbf{c}}\cos\gamma-\hat{\mathbf{b}}\cos\beta)\times(\hat{\mathbf{a}}\cos\alpha-\hat{\mathbf{b}}\cos\beta)|$$

$$=\frac{1}{2}|(\hat{\mathbf{c}}\times\hat{\mathbf{a}})\cos\alpha\cos\gamma+(\hat{\mathbf{b}}\times\hat{\mathbf{c}})\cos\gamma\cos\beta+(\hat{\mathbf{a}}\times\hat{\mathbf{b}})\cos\beta\cos\alpha|$$

$$=\frac{1}{2}|\hat{\eta}_1\sin\alpha\cos\beta\cos\gamma+\hat{\eta}_2\sin\beta\cos\alpha\cos\gamma+\hat{\eta}_3\sin\gamma\cos\alpha\cos\beta|$$

$$\Rightarrow \quad 2\Delta ABC=|\Sigma\hat{\eta}_1\sin\alpha\cos\beta\cos\gamma|$$

$\therefore$ Eq. (i) reduces to

$$\frac{|\hat{\mathbf{a}}\times(\hat{\mathbf{b}}\times\hat{\mathbf{c}})|}{\sin A}=\frac{|\hat{\mathbf{b}}\times(\hat{\mathbf{c}}\times\hat{\mathbf{a}})|}{\sin B}=\frac{|\hat{\mathbf{c}}\times(\hat{\mathbf{a}}\times\hat{\mathbf{b}})|}{\sin C}$$

$$=\frac{\pi|\hat{\mathbf{a}}\times(\hat{\mathbf{b}}\times\hat{\mathbf{c}})|}{|\Sigma\sin\alpha\cos\beta\cos\gamma\,\hat{\eta}|}$$

Ex. 155 *Let a and b be given non-zero and non-collinear vectors, such that $\mathbf{c}\times\mathbf{a}=\mathbf{b}-\mathbf{c}$. Express c in terms of a, b and* $\mathbf{a}\times\mathbf{b}$

Sol. Let $\mathbf{c}=x_1\mathbf{a}+x_2\mathbf{b}+x_3(\mathbf{a}\times\mathbf{b})$

$$\Rightarrow \mathbf{c}\times\mathbf{a}=x_2(\mathbf{b}\times\mathbf{a})-x_3\mathbf{a}\times(\mathbf{a}\times\mathbf{b})$$

$$=x_2(\mathbf{b}\times\mathbf{a})-x_3(\mathbf{a}\cdot\mathbf{b})\mathbf{a}+x_3|\mathbf{a}|^2\mathbf{b}$$

We have been given, $\mathbf{c}\times\mathbf{a}=\mathbf{b}-\mathbf{c}$

$$\Rightarrow \mathbf{b}-x_1\mathbf{a}-x_2\mathbf{b}-x_3(\mathbf{a}\times\mathbf{b})=-x_2(\mathbf{a}\times\mathbf{b})-x_3(\mathbf{a}\cdot\mathbf{b})\mathbf{a}+x_3|\mathbf{a}|^2\mathbf{b}$$

$$\Rightarrow x_3\{(\mathbf{a}\cdot\mathbf{b})-x_1\}\mathbf{a}+(1-x_2-x_3|\mathbf{a}|^2)\mathbf{b}+(x_2-x_3)(\mathbf{a}\times\mathbf{b})=0$$

Now, **a**, **b** and $\mathbf{a}\times\mathbf{b}$ are linearly independent.

Hence, $\quad x_3(\mathbf{a}\cdot\mathbf{b})=x_1,\ 1=x_2+x_3|\mathbf{a}|^2,\ x_2=x_3$

$$x_2=x_3=\frac{1}{1+|\mathbf{a}|^2} \text{ and } x_1=\frac{\mathbf{a}\cdot\mathbf{b}}{1+|\mathbf{a}|^2}$$

$$\mathbf{c}=\frac{\mathbf{a}\cdot\mathbf{b}}{(1+|\mathbf{a}|^2)}\mathbf{a}+\frac{1}{(1+|\mathbf{a}|^2)}[\mathbf{b}+(\mathbf{a}\times\mathbf{b})]$$

Product of Vectors Exercise 1 :
Single Correct Type Questions

1. If **a** has magnitude 5 and points North-East and vector **b** has magnitude 5 and points North-West, then $|\mathbf{a} - \mathbf{b}|$ is equal to
(a) 25 (b) 5
(c) $7\sqrt{3}$ (d) $5\sqrt{2}$

2. If $|\mathbf{a} + \mathbf{b}| > |\mathbf{a} - \mathbf{b}|$, then the angle between **a** and **b** is
(a) acute (b) obtuse
(c) $\pi/2$ (d) π

3. If **a** , **b** and **c** are three vectors such that $\mathbf{a} = \mathbf{b} + \mathbf{c}$ and the angle between **b** and **c** is $\frac{\pi}{2}$, then
(a) $a^2 = b^2 + c^2$ (b) $b^2 = c^2 + a^2$
(c) $c^2 = a^2 + b^2$ (d) $2a^2 - b^2 = c^2$
Note Here, $a = |\mathbf{a}|$, $b = |\mathbf{b}|$ and $c = |\mathbf{c}|$

4. If the angle between the vectors **a** and **b** be θ and $\mathbf{a} \cdot \mathbf{b} = \cos\theta$, then the true statement is
(a) **a** and **b** are equal vectors
(b) **a** and **b** are like vectors
(c) **a** and **b** are unlike vectors
(d) **a** and **b** are unit vectors

5. If the vector $\hat{\mathbf{i}} + \hat{\mathbf{j}} + \hat{\mathbf{k}}$ makes angles α, β and γ with vectors $\hat{\mathbf{i}}$, $\hat{\mathbf{j}}$ and $\hat{\mathbf{k}}$ respectively, then
(a) $\alpha = \beta \neq \gamma$ (b) $\alpha = \gamma \neq \beta$
(c) $\beta = \gamma \neq \alpha$ (d) $\alpha = \beta = \gamma$

6. $(\mathbf{r} \cdot \hat{\mathbf{i}})^2 + (\mathbf{r} \cdot \hat{\mathbf{j}})^2 + (\mathbf{r} \cdot \hat{\mathbf{k}})^2$ is equal to
(a) $3r^2$ (b) r^2
(c) 0 (d) None of these

7. Let **a** and **b** be two unit vectors inclined at an angle θ, then $\sin(\theta/2)$ is equal to
(a) $\frac{1}{2}|\mathbf{a} - \mathbf{b}|$ (b) $\frac{1}{2}|\mathbf{a} + \mathbf{b}|$
(c) $|\mathbf{a} - \mathbf{b}|$ (d) $|\mathbf{a} + \mathbf{b}|$

8. If $\mathbf{a} = 4\hat{\mathbf{i}} + 6\hat{\mathbf{j}}$ and $\mathbf{b} = 3\hat{\mathbf{j}} + 4\hat{\mathbf{k}}$, then the component of **a** along **b** is
(a) $\frac{18}{10\sqrt{3}}(3\hat{\mathbf{j}} + 4\hat{\mathbf{k}})$ (b) $\frac{18}{25}(3\hat{\mathbf{j}} + 4\hat{\mathbf{k}})$
(c) $\frac{18}{\sqrt{3}}(3\hat{\mathbf{j}} + 4\hat{\mathbf{k}})$ (d) $(3\hat{\mathbf{j}} + 4\hat{\mathbf{k}})$

9. If vector $\mathbf{a} = 2\hat{\mathbf{i}} - 3\hat{\mathbf{j}} + 6\hat{\mathbf{k}}$ and vector $\mathbf{b} = -2\hat{\mathbf{i}} + 2\hat{\mathbf{j}} - \hat{\mathbf{k}}$, then $\frac{\text{Projection of vector } \mathbf{a} \text{ on vector } \mathbf{b}}{\text{Projection of vector } \mathbf{b} \text{ on vector } \mathbf{a}}$ is equal to
(a) $\frac{3}{7}$ (b) $\frac{7}{3}$
(c) 3 (d) 7

10. If **a** and **b** are two vectors, then $(\mathbf{a} \times \mathbf{b})^2$ is equal to
(a) $\begin{vmatrix} \mathbf{a}\cdot\mathbf{b} & \mathbf{a}\cdot\mathbf{a} \\ \mathbf{b}\cdot\mathbf{b} & \mathbf{b}\cdot\mathbf{a} \end{vmatrix}$ (b) $\begin{vmatrix} \mathbf{a}\cdot\mathbf{a} & \mathbf{a}\cdot\mathbf{b} \\ \mathbf{b}\cdot\mathbf{a} & \mathbf{b}\cdot\mathbf{b} \end{vmatrix}$
(c) $\begin{vmatrix} \mathbf{a}\cdot\mathbf{b} \\ \mathbf{b}\cdot\mathbf{a} \end{vmatrix}$ (d) None of these

11. The moment of the force **F** acting at a point P, about the point C is
(a) $\mathbf{F} \times \mathbf{CP}$
(b) $\mathbf{CP} \cdot \mathbf{F}$
(c) a vector having the same direction as **F**
(d) $\mathbf{CP} \times \mathbf{F}$

12. The moment of a force represented by $\mathbf{F} = \hat{\mathbf{i}} + 2\hat{\mathbf{j}} + 3\hat{\mathbf{k}}$ about the point $2\hat{\mathbf{i}} - \hat{\mathbf{j}} + \hat{\mathbf{k}}$ is equal to
(a) $5\hat{\mathbf{i}} - 5\hat{\mathbf{j}} + 5\hat{\mathbf{k}}$ (b) $5\hat{\mathbf{i}} + 5\hat{\mathbf{j}} - 6\hat{\mathbf{k}}$
(c) $-5\hat{\mathbf{i}} - 5\hat{\mathbf{j}} + 5\hat{\mathbf{k}}$ (d) $-5\hat{\mathbf{i}} - 5\hat{\mathbf{j}} + 2\hat{\mathbf{k}}$

13. A force of magnitude 6 acts along the vector $(9, 6, -2)$ and passes through a point $A(4, -1, -7)$. Then moment of force about the point $O(1, -3, 2)$ is
(a) $\frac{150}{11}(2\hat{\mathbf{i}} - 3\hat{\mathbf{j}})$ (b) $\frac{6}{11}(50\hat{\mathbf{i}} - 75\hat{\mathbf{j}} + 36\hat{\mathbf{k}})$
(c) $150(2\hat{\mathbf{i}} - 3\hat{\mathbf{j}})$ (d) $6(50\hat{\mathbf{i}} - 75\hat{\mathbf{j}} + 36\hat{\mathbf{k}})$

14. A force $\mathbf{F} = 2\hat{\mathbf{i}} + \hat{\mathbf{j}} - \hat{\mathbf{k}}$ acts at a point A, whose position vector is $2\hat{\mathbf{i}} - \hat{\mathbf{j}}$. The moment of **F** about the origin is
(a) $\hat{\mathbf{i}} + 2\hat{\mathbf{j}} - 4\hat{\mathbf{k}}$ (b) $\hat{\mathbf{i}} - 2\hat{\mathbf{j}} - 4\hat{\mathbf{k}}$
(c) $\hat{\mathbf{i}} + 2\hat{\mathbf{j}} + 4\hat{\mathbf{k}}$ (d) $\hat{\mathbf{i}} - 2\hat{\mathbf{j}} + 4\hat{\mathbf{k}}$

15. If **a**, **b** and **c** are any three vectors and their inverse are $\mathbf{a}^{-1}, \mathbf{b}^{-1}$ and $\mathbf{c}^{-1}$ and $[\mathbf{abc}] \neq 0$, then $[\mathbf{a}^{-1}\mathbf{b}^{-1}\mathbf{c}^{-1}]$ will be
(a) zero (b) one
(c) non-zero (d) $[\mathbf{abc}]$

16. If **a**, **b** and **c** are three non-coplanar vectors, that $\frac{\mathbf{a}\cdot\mathbf{b}\times\mathbf{c}}{\mathbf{c}\times\mathbf{a}\cdot\mathbf{b}} + \frac{\mathbf{b}\cdot\mathbf{a}\times\mathbf{c}}{\mathbf{c}\cdot\mathbf{a}\times\mathbf{b}}$ is equal to
(a) 0 (b) 2
(c) -2 (d) None of these

17. $\mathbf{a} \times (\mathbf{b} \times \mathbf{c})$ is coplanar with
(a) **b** and **c** (b) **c** and **a**
(c) **a** and **b** (d) **a**, **b** and **c**

18. If $\mathbf{u} = \hat{\mathbf{i}} \times (\mathbf{a} \times \hat{\mathbf{i}}) + \hat{\mathbf{j}} \times (\mathbf{a} \times \hat{\mathbf{j}}) + \hat{\mathbf{k}} \times (\mathbf{a} \times \hat{\mathbf{k}})$, then
(a) $\mathbf{u} = 0$
(b) $\mathbf{u} = \hat{\mathbf{i}} + \hat{\mathbf{j}} + \hat{\mathbf{k}}$
(c) $\mathbf{u} = 2\mathbf{a}$
(d) $\mathbf{u} = \mathbf{a}$

19. If $\mathbf{a}=\hat{\mathbf{i}}+2\hat{\mathbf{j}}-2\hat{\mathbf{k}}$, $\mathbf{b}=2\hat{\mathbf{i}}-\hat{\mathbf{j}}+\hat{\mathbf{k}}$ and $\mathbf{c}=\hat{\mathbf{i}}+3\hat{\mathbf{j}}-\hat{\mathbf{k}}$, then $\mathbf{a}\times(\mathbf{b}\times\mathbf{c})$ is equal to

(a) $20\hat{\mathbf{i}}-3\hat{\mathbf{j}}+7\hat{\mathbf{k}}$
(b) $20\hat{\mathbf{i}}-3\hat{\mathbf{j}}-7\hat{\mathbf{k}}$
(c) $20\hat{\mathbf{i}}+3\hat{\mathbf{j}}-7\hat{\mathbf{k}}$
(d) None of the above

20. If $\mathbf{a}\times(\mathbf{b}\times\mathbf{c})=0$, then

(a) $|\mathbf{a}|=|\mathbf{b}|\cdot|\mathbf{c}|=1$ (b) $\mathbf{b}\parallel\mathbf{c}$
(c) $\mathbf{a}\parallel\mathbf{b}$ (d) $\mathbf{b}\perp\mathbf{c}$

21. A vector whose modulus is $\sqrt{51}$ and makes the same angle with $\mathbf{a}=\dfrac{\hat{\mathbf{i}}-2\hat{\mathbf{j}}+2\hat{\mathbf{k}}}{3}$, $\mathbf{b}=\dfrac{-4\hat{\mathbf{i}}-3\hat{\mathbf{k}}}{5}$ and $\mathbf{c}=\hat{\mathbf{j}}$, will be

(a) $5\hat{\mathbf{i}}+5\hat{\mathbf{j}}+\hat{\mathbf{k}}$
(b) $5\hat{\mathbf{i}}+\hat{\mathbf{j}}-5\hat{\mathbf{k}}$
(c) $5\hat{\mathbf{i}}+\hat{\mathbf{j}}+5\hat{\mathbf{k}}$
(d) $\pm(5\hat{\mathbf{i}}-\hat{\mathbf{j}}-5\hat{\mathbf{k}})$

22. The horizontal force and the force inclined at an angle 60° with the vertical, whose resultant is in vertical direction of P kg, are

(a) P, $2P$ (b) P, $P\sqrt{3}$
(c) $2P$, $P\sqrt{3}$ (d) None of these

23. If $x+y+z=0$, $|\mathbf{x}|=|\mathbf{y}|=|\mathbf{z}|=2$ and θ is angle between $\mathbf{y}$ and $\mathbf{z}$, then the value of $\operatorname{cosec}^2\theta+\cot^2\theta$ is equal to

(a) 4 / 3 (b) 5 / 3
(c) 1 / 3 (d) 1

24. The value of x for which the angle between the vectors $\mathbf{a}=-3\hat{\mathbf{i}}+x\hat{\mathbf{j}}+\hat{\mathbf{k}}$ and $\mathbf{b}=x\hat{\mathbf{i}}+2x\hat{\mathbf{j}}+\hat{\mathbf{k}}$ is acute and the angle between $\mathbf{b}$ and X-axis lies between $\pi/2$ and π satisfy

(a) $x>0$ (b) $x<0$
(c) $x>1$ only (d) $x<-1$ only

25. If $\mathbf{a}$, $\mathbf{b}$ and $\mathbf{c}$ are non-coplanar vectors and $\mathbf{d}=\lambda\mathbf{a}+\mu\mathbf{b}+\nu\mathbf{c}$, then λ is equal to

(a) $\dfrac{[\mathbf{dbc}]}{[\mathbf{bac}]}$ (b) $\dfrac{[\mathbf{bcd}]}{[\mathbf{bca}]}$
(c) $\dfrac{[\mathbf{bdc}]}{[\mathbf{abc}]}$ (d) $\dfrac{[\mathbf{cbd}]}{[\mathbf{abc}]}$

26. If the vectors $3\mathbf{p}+\mathbf{q}$, $5\mathbf{p}-3\mathbf{q}$ and $2\mathbf{p}+\mathbf{q}$, $4\mathbf{p}-2\mathbf{q}$ are pairs of mutually perpendicular vectors, then $\sin(\mathbf{pq})$ is

(a) $\sqrt{55}/4$ (b) $\sqrt{55}/8$
(c) 3/16 (d) $\sqrt{247}/16$

27. Let $\mathbf{u}=\hat{\mathbf{i}}+\hat{\mathbf{j}}$, $\mathbf{v}=\hat{\mathbf{i}}-\hat{\mathbf{j}}$ and $\mathbf{w}=\hat{\mathbf{i}}+2\hat{\mathbf{j}}+3\hat{\mathbf{k}}$. If $\hat{\mathbf{n}}$ is a unit vector such that $\mathbf{u}\cdot\hat{\mathbf{n}}=0$ and $\mathbf{v}\cdot\hat{\mathbf{n}}=0$, then $|\mathbf{w}\cdot\hat{\mathbf{n}}|$ is equal to

(a) 1 (b) 2 (c) 3 (d) 0

28. Given a parallelogram $ABCD$. If $|\mathbf{AB}|=a$, $|\mathbf{AD}|=b$ and $|\mathbf{AC}|=c$, then $\mathbf{DB}\cdot\mathbf{AB}$ has the value

(a) $\dfrac{3a^2+b^2-c^2}{2}$
(b) $\dfrac{a^2+3b^2-c^2}{2}$
(c) $\dfrac{a^2-b^2+3c^2}{2}$
(d) None of the above

29. For two particular vectors $\mathbf{A}$ and $\mathbf{B}$, it is known that $\mathbf{A}\times\mathbf{B}=\mathbf{B}\times\mathbf{A}$. What must be true about the two vectors?

(a) Atleast one of the two vectors must be the zero vector
(b) $\mathbf{A}\times\mathbf{B}=\mathbf{B}\times\mathbf{A}$ is true for any two vectors
(c) One of the two vectors is a scalar multiple of the other vector
(d) The two vectors must be perpendicular to each other

30. For some non-zero vector $\mathbf{V}$, if the sum of $\mathbf{V}$ and the vector obtained from $\mathbf{V}$ by rotating it by $\angle 2\alpha$ equals to the vector obtained from $\mathbf{V}$ by rotating it by $\angle\alpha$, then the value of α, is

(a) $2n\pi\pm\dfrac{\pi}{3}$ (b) $n\pi\pm\dfrac{\pi}{3}$
(c) $2n\pi\pm\dfrac{2\pi}{3}$ (d) $n\pi\pm\dfrac{2\pi}{3}$

31. In the isosceles ΔABC, $|\mathbf{AB}|=|\mathbf{BC}|=8$, a point E divides AB internally in the ratio 1 : 3, then the cosine of the angle between $\mathbf{CE}$ and $\mathbf{CA}$ is (where, $|\mathbf{CA}|=12$)

(a) $-\dfrac{3\sqrt{7}}{8}$ (b) $\dfrac{3\sqrt{8}}{17}$
(c) $\dfrac{3\sqrt{7}}{8}$ (d) $\dfrac{-3\sqrt{8}}{17}$

32. Given an equilateral ΔABC with side length equal to a. Let M and N be two points respectively, on the side AB and AC such that $\mathbf{AN}=k\mathbf{AC}$ and $\mathbf{AM}=\dfrac{\mathbf{AB}}{3}$. If $\mathbf{BN}$ and $\mathbf{CM}$ are orthogonal, then the value of k is

(a) $\dfrac{1}{5}$ (b) $\dfrac{1}{4}$
(c) $\dfrac{1}{3}$ (d) $\dfrac{1}{2}$

33. In a quadrilateral $ABCD$, $\mathbf{AC}$ is the bisector of the $(\mathbf{AB},\mathbf{AD})$ which is $\dfrac{2\pi}{3}$, $15|\mathbf{AC}|=3|\mathbf{AB}|=5|\mathbf{AD}|$, then $\cos(\mathbf{BA},\mathbf{CD})$ is equal to

(a) $-\dfrac{\sqrt{14}}{7\sqrt{2}}$ (b) $-\dfrac{\sqrt{21}}{7\sqrt{3}}$
(c) $\dfrac{2}{\sqrt{7}}$ (d) $\dfrac{2\sqrt{7}}{14}$

34. If the distance from the point $P(1, 1, 1)$ to the line passing through the points $Q(0, 6, 8)$ and $R(-1, 4, 7)$ is expressed in the form $\sqrt{p/q}$, where p and q are co-prime, then the value of $\frac{(p+q)(p+q-1)}{2}$ is equal to

(a) 4950 (b) 5050
(c) 5150 (d) None of these

35. Given the vectors

$$\mathbf{u} = 2\hat{\mathbf{i}} - \hat{\mathbf{j}} - \hat{\mathbf{k}}$$
$$\mathbf{v} = \hat{\mathbf{i}} - \hat{\mathbf{j}} + 2\hat{\mathbf{k}}$$
$$\mathbf{w} = \hat{\mathbf{i}} - \hat{\mathbf{k}}$$

If the volume of the parallelopiped having $-c\mathbf{u}, \mathbf{v}$ and $c\mathbf{w}$ as concurrent edges, is 8, then c is equal to

(a) ± 2 (b) 4
(c) 8 (d) Cannot be determined

36. The vector $\mathbf{c}$ is perpendicular to the vectors $\mathbf{a} = (2, -3, 1)$, $\mathbf{b} = (1, -2, 3)$ and satisfies the condition $\mathbf{c}\cdot(\hat{\mathbf{i}} + 2\hat{\mathbf{j}} - 7\hat{\mathbf{k}}) = 10$. Then, the vector $\mathbf{c}$ is equal to

(a) (7, 5, 1) (b) (− 7, − 5, − 1)
(c) (1, 1, − 1) (d) None of these

37. Let $\mathbf{a} = \hat{\mathbf{i}} + \hat{\mathbf{j}}$, $\mathbf{b} = \hat{\mathbf{j}} + \hat{\mathbf{k}}$ and $\mathbf{c} = \alpha\mathbf{a} + \beta\mathbf{b}$. If the vectors, $\hat{\mathbf{i}} - 2\hat{\mathbf{j}} + \hat{\mathbf{k}}$, $3\hat{\mathbf{i}} + 2\hat{\mathbf{j}} - \hat{\mathbf{k}}$ and $\mathbf{c}$ are coplanar, then $\frac{\alpha}{\beta}$ is equal to

(a) 1 (b) 2
(c) 3 (d) − 3

38. A rigid body rotates about an axis through the origin with an angular velocity $10\sqrt{3}$ rad/s. If ω points in the direction of $\hat{\mathbf{i}} + \hat{\mathbf{j}} + \hat{\mathbf{k}}$, then the equation to the locus of the points having tangential speed 20 m/s is

(a) $x^2 + y^2 + z^2 - xy - yz - zx - 1 = 0$
(b) $x^2 + y^2 + z^2 - 2xy - 2yz - 2zx - 1 = 0$
(c) $x^2 + y^2 + z^2 - xy - yz - zx - 2 = 0$
(d) $x^2 + y^2 + z^2 - 2xy - 2yz - 2zx - 2 = 0$

39. A rigid body rotates with constant angular velocity ω about the line whose vector equation is, $\mathbf{r} = \lambda(\hat{\mathbf{i}} + 2\hat{\mathbf{j}} + 2\hat{\mathbf{k}})$. The speed of the particle at the instant it passes through the point with position vector $(2\hat{\mathbf{i}} + 3\hat{\mathbf{j}} + 5\hat{\mathbf{k}})$ is equal to

(a) $\omega\sqrt{2}$ (b) 2ω
(c) $\omega/\sqrt{2}$ (d) None of these

40. Consider ΔABC with $A = (\mathbf{a})$, $B = (\mathbf{b})$ and $C = (\mathbf{c})$. If $\mathbf{b}\cdot(\mathbf{a} + \mathbf{c}) = \mathbf{b}\cdot\mathbf{b} + \mathbf{a}\cdot\mathbf{c}$, $|\mathbf{b} - \mathbf{a}| = 3$ and $|\mathbf{c} - \mathbf{b}| = 4$, then the angle between the medians **AM** and **BD** is

(a) $\pi - \cos^{-1}\left(\frac{1}{5\sqrt{13}}\right)$ (b) $\pi - \cos^{-1}\left(\frac{1}{13\sqrt{5}}\right)$
(c) $\cos^{-1}\left(\frac{1}{5\sqrt{13}}\right)$ (d) $\cos^{-1}\left(\frac{1}{13\sqrt{5}}\right)$

41. Given unit vectors **m**, **n** and **p** such that angle between **m** and **n** = Angle between **p** and $(\mathbf{m} \times \mathbf{n}) = \frac{\pi}{6}$, then $[\mathbf{npm}]$ is equal to

(a) $\sqrt{3}/4$ (b) 3/4
(c) 1/4 (d) None of these

42. If **a** and **b** are unit vectors, then the vector defined as $\mathbf{V} = (\mathbf{a} + \mathbf{b}) \times (\mathbf{a} + \mathbf{b})$ is collinear to the vector

(a) $\mathbf{a} + \mathbf{b}$ (b) $\mathbf{b} - \mathbf{a}$
(c) $2\mathbf{a} - \mathbf{b}$ (d) $\mathbf{a} + 2\mathbf{b}$

43. If **a** and **b** are orthogonal unit vectors, then for any non-zero vector **r**, the vector $(\mathbf{r} \times \mathbf{a})$ is equal to

(a) $[\mathbf{r}\ \hat{\mathbf{a}}\ \hat{\mathbf{b}}](\hat{\mathbf{a}} + \hat{\mathbf{b}})$
(b) $[\mathbf{r}\ \hat{\mathbf{a}}\ \hat{\mathbf{b}}]\hat{\mathbf{a}} + (\mathbf{r}\cdot\hat{\mathbf{a}})(\hat{\mathbf{a}} \times \hat{\mathbf{b}})$
(c) $[\mathbf{r}\ \hat{\mathbf{a}}\ \hat{\mathbf{b}}]\hat{\mathbf{b}} + (\mathbf{r}\cdot\hat{\mathbf{b}})(\hat{\mathbf{b}} \times \hat{\mathbf{a}})$
(d) $[\mathbf{r}\ \hat{\mathbf{a}}\ \hat{\mathbf{b}}]\hat{\mathbf{b}} + (\mathbf{r}\cdot\hat{\mathbf{a}})(\hat{\mathbf{a}} \times \hat{\mathbf{b}})$

44. If vector $\hat{\mathbf{i}} + 2\hat{\mathbf{j}} + 2\hat{\mathbf{k}}$ is rotated through an angle of 90°, so as to cross the positive direction of Y-axis, then the vector in the new position is

(a) $-\frac{2}{\sqrt{5}}\hat{\mathbf{i}} + \sqrt{5}\hat{\mathbf{j}} - \frac{4}{\sqrt{5}}\hat{\mathbf{k}}$ (b) $\frac{2}{\sqrt{5}}\hat{\mathbf{i}} - \sqrt{5}\hat{\mathbf{j}} + \frac{4}{\sqrt{5}}\hat{\mathbf{k}}$
(c) $4\hat{\mathbf{i}} - \hat{\mathbf{j}} - \hat{\mathbf{k}}$ (d) None of these

45. 10 different vectors are lying on a plane out of which four are parallel with respect to each other. Probability that three vectors chosen from them will satisfy the equation $\lambda_1\mathbf{a} + \lambda_2\mathbf{b} + \lambda_3\mathbf{c} = 0$, where λ_1, λ_2 and $\lambda_3 \neq 0$ is

(a) $\frac{{}^6C_2 \times {}^4C_1}{{}^{10}C_3}$ (b) $\frac{({}^6C_3 \times {}^4C_1) + {}^6C_3}{{}^{10}C_3}$
(c) $\frac{({}^6C_2 \times {}^4C_1) + ({}^4C_3)}{{}^{10}C_3}$ (d) $\frac{({}^6C_2 + {}^4C_1) + ({}^6C_2 \times {}^4C_1)}{{}^{10}C_3}$

46. If $\hat{\mathbf{a}}$ is a unit vector and projection of **x** along $\hat{\mathbf{a}}$ is 2 units and $(\hat{\mathbf{a}} \times \mathbf{x}) + \mathbf{b} = \mathbf{x}$, then **x** is equal to

(a) $\frac{1}{2}[\hat{\mathbf{a}} - \mathbf{b} + (\hat{\mathbf{a}} \times \mathbf{b})]$
(b) $\frac{1}{2}[2\hat{\mathbf{a}} - \mathbf{b} + (\hat{\mathbf{a}} \times \mathbf{b})]$
(c) $[\hat{\mathbf{a}} + (\hat{\mathbf{a}} \times \mathbf{b})]$
(d) None of the above

47. If $\mathbf{a}\cdot\mathbf{b}$ and **c** are any three non-zero vectors, then the component of $\mathbf{a} \times (\mathbf{b} \times \mathbf{c})$ perpendicular to **b** is

(a) $\mathbf{a} \times (\mathbf{b} \times \mathbf{c}) + \frac{(\mathbf{a} \times \mathbf{b})\cdot(\mathbf{c} \times \mathbf{a})}{|b|^2}\mathbf{b}$
(b) $\mathbf{a} \times (\mathbf{b} \times \mathbf{c}) + \frac{(\mathbf{a} \times \mathbf{c})\cdot(\mathbf{a} \times \mathbf{b})}{|b|^2}\mathbf{b}$
(c) $\mathbf{a} \times (\mathbf{b} \times \mathbf{c}) + \frac{(\mathbf{b} \times \mathbf{c})\cdot(\mathbf{b} \times \mathbf{a})}{|b|^2}\mathbf{b}$
(d) $\mathbf{a} \times (\mathbf{b} \times \mathbf{c}) + \frac{(\mathbf{a} \times \mathbf{b})\cdot(\mathbf{b} \times \mathbf{c})}{|b|^2}\mathbf{b}$

48. The position vector of a point P is $\mathbf{r} = x\hat{\mathbf{i}} + y\hat{\mathbf{j}} + z\hat{\mathbf{k}}$, where $x, y, z \in N$ and $\mathbf{a} = \hat{\mathbf{i}} + 2\hat{\mathbf{j}} + \hat{\mathbf{k}}$. If $\mathbf{r} \cdot \mathbf{a} = 20$, then the number of possible position of P is

(a) 81 (b) 49
(c) 100 (d) 36

49. Let $a, b > 0$ and $\alpha = \frac{\hat{\mathbf{i}}}{a} + \frac{4\hat{\mathbf{j}}}{b} + b\hat{\mathbf{k}}$ and $\beta = b\hat{\mathbf{i}} + a\hat{\mathbf{j}} + \frac{1}{b}\hat{\mathbf{k}}$, then the maximum value of $\frac{10}{5 + \alpha \cdot \beta}$ is

(a) 1 (b) 2
(c) 4 (d) 8

50. If **a**, **b** and **c** are any three vectors forming a linearly independent system, then $\forall\ \theta \in R$

$$\left[\mathbf{a}\cos\theta + \mathbf{b}\sin\theta + \mathbf{c}\cos 2\theta, \mathbf{a}\cos\left(\frac{2\pi}{3} + \theta\right) + \mathbf{b}\sin\left(\frac{2\pi}{3} + \theta\right) + \mathbf{c}\cos 2\left(\frac{2\pi}{3} + \theta\right), \mathbf{a}\cos\left(\theta - \frac{2\pi}{3}\right) + \mathbf{b}\sin\left(\theta - \frac{2\pi}{3}\right) + \mathbf{c}\cos 2\left(\theta - \frac{2\pi}{3}\right)\right]$$ equals

(a) $[\mathbf{a}\ \mathbf{b}\ \mathbf{c}]\cos\theta$
(b) $[\mathbf{a}\ \mathbf{b}\ \mathbf{c}]\cos 2\theta$
(c) $[\mathbf{a}\ \mathbf{b}\ \mathbf{c}]\cos 3\theta$
(d) None of the above

51. Two adjacent sides of a parallelogram $ABCD$ are given by $\mathbf{AB} = 2\hat{\mathbf{i}} + 10\hat{\mathbf{j}} + 11\hat{\mathbf{k}}$ and $\mathbf{AD} = -\hat{\mathbf{i}} + 2\hat{\mathbf{j}} + 2\hat{\mathbf{k}}$. The side AD is rotated by an acute angle α in the plane of parallelogram so that AD becomes AD'. If AD' makes a right angle with the side AB, then the cosine of angle α is given by

(a) $\frac{8}{9}$ (b) $\frac{\sqrt{17}}{9}$
(c) $\frac{1}{9}$ (d) $\frac{4\sqrt{5}}{9}$

52. If in a ΔABC, $\mathbf{BC} = \frac{\mathbf{e}}{|\mathbf{e}|} - \frac{\mathbf{f}}{|\mathbf{f}|}$ and $\mathbf{AC} = \frac{2\mathbf{e}}{|\mathbf{e}|}; |\mathbf{e}| \neq |\mathbf{f}|$, then the value of $\cos 2A + \cos 2B + \cos 2C$ must be

(a) -1 (b) 0
(c) 2 (d) $\frac{-3}{2}$

53. **a**, **b**, **c** are three unit of vectors, **a** and **b** are perpendicular to each other and vector **c** is equally inclined to both **a** and **b** at an angle θ. If $\mathbf{c} = \alpha\,\mathbf{a} + \beta\,\mathbf{b} + \gamma(\mathbf{a} \times \mathbf{b})$, where α, β, γ are constants , then

(a) $\alpha = \beta = -\cos\theta, \gamma^2 = \cos 2\theta$
(b) $\alpha = \beta = \cos\theta, \gamma^2 = \cos 2\theta$
(c) $\alpha = \beta = \cos\theta, \gamma^2 = -\cos 2\theta$
(d) $\alpha = \beta = -\cos\theta, \gamma^2 = -\cos 2\theta$

54. The ΔABC is such that the mid-points of the sides BC, CA, AB are $(l, 0, 0), (0, m, 0), (0, 0, n)$ respectively. Then, $\frac{AB^2 + BC^2 + CA^2}{l^2 + m^2 + n^2}$ is equal to

(a) 2 (b) 4
(c) 8 (d) 16

55. The angle between the lines whose direction cosines are given by $2l - m + 2n = 0$, $lm + mn + nl = 0$ is

(a) $\frac{\pi}{6}$ (b) $\frac{\pi}{4}$
(c) $\frac{\pi}{3}$ (d) $\frac{\pi}{2}$

56. A line makes an angle θ both with X and Y-axes. A possible range of θ is

(a) $\left[0, \frac{\pi}{4}\right]$ (b) $\left[0, \frac{\pi}{2}\right]$
(c) $\left[\frac{\pi}{4}, \frac{\pi}{2}\right]$ (d) $\left[\frac{\pi}{6}, \frac{\pi}{3}\right]$

57. Let **a**, **b** and **c** be the three vectors having magnitudes 1, 5 and 3 respectively such that the angle between **a** and **b** is θ and $\mathbf{a} \times (\mathbf{a} \times \mathbf{b}) = \mathbf{c}$, then $\tan\theta$ is equal to

(a) 0 (b) $\frac{2}{3}$
(c) $\frac{3}{5}$ (d) $\frac{3}{4}$

58. The perpendicular distance of a corner of a unit cube from a diagonal not passing through it is

(a) $\sqrt{\frac{3}{2}}$ (b) $\sqrt{\frac{2}{3}}$
(c) $\sqrt{\frac{3}{4}}$ (d) $\sqrt{\frac{4}{3}}$

59. If **p**, **q** are two non-collinear vectors such that $(b - c)\,\mathbf{p} \times \mathbf{q} + (c - a)\mathbf{p} + (a - b)\mathbf{q} = \mathbf{0}$ where a, b, c are lengths of sides of a triangle, then the triangle is

(a) right angled (b) obtuse angled
(c) equilateral (d) right angled isosceles triangle

60. Let $\mathbf{a} = \hat{\mathbf{i}} + \hat{\mathbf{j}} + \hat{\mathbf{k}}, \mathbf{b} = -\hat{\mathbf{i}} + \hat{\mathbf{j}} + \hat{\mathbf{k}}, \mathbf{c} = \hat{\mathbf{i}} - \hat{\mathbf{j}} + \hat{\mathbf{k}}$ and $\mathbf{d} = \hat{\mathbf{i}} + \hat{\mathbf{j}} - \hat{\mathbf{k}}$. Then, the line of intersection of planes one determined by **a**, **b** and other determined by **c**, **d** is perpendicular to

(a) X-axis (b) Y-axis
(c) both X-axis and Y-axis (d) both Y-axis and Z-axis

61. A parallelopiped is formed by planes drawn parallel to coordinate axes through the points $A = (1, 2, 3)$ and $B = (9, 8, 5)$. The volume of that parallelopiped is equal to (in cubic units)

(a) 192 (b) 48
(c) 32 (d) 96

62. Let **a**, **b** and **c** be three non-coplanar vectors and **d** be a non-zero vector, which is perpendicular to **a** + **b** + **c**. Now, if $\mathbf{d} = (\sin x)(\mathbf{a} \times \mathbf{b}) + (\cos y)(\mathbf{b} \times \mathbf{c}) + 2(\mathbf{c} \times \mathbf{a})$, then minimum value of $x^2 + y^2$ is equal to

(a) π^2 (b) $\frac{\pi^2}{2}$

(c) $\frac{\pi^2}{4}$ (d) $\frac{5\pi^2}{4}$

63. If $\alpha(\mathbf{a} \times \mathbf{b}) + \beta(\mathbf{b} \times \mathbf{c}) + \gamma(\mathbf{c} \times \mathbf{a}) = 0$, then

(a) *a*, *b*, *c* are coplanar if all of $\alpha, \beta, \gamma \neq 0$

(b) *a*, *b*, *c* are coplanar if any one $\alpha, \beta, \gamma = 0$

(c) *a*, *b* *c* are non-coplanar for any α, β, γ

(d) None of the above

64. Let area of faces,

$\Delta OAB = \lambda_1, \Delta OAC = \lambda_2, \Delta OBC = \lambda_3, \Delta ABC = \lambda_4$

and h_1, h_2, h_3, h_4 be perpendicular height from 0 to face ΔABC, *A* to the face ΔOBC, *B* to the face ΔOAC, *C* to the face ΔOAB, then the face

$$\frac{1}{3}\lambda_1 h_4 \cdot \frac{1}{3}\lambda_2 h_3 + \frac{1}{3}\lambda_3 h_2 + \frac{1}{3}\lambda_4 h_1$$

(a) $\frac{2}{3}|[\mathbf{AB\ AC\ OA}]|$ (b) $\frac{1}{3}|[\mathbf{AB\ AC\ OA}]|$

(c) $\frac{1}{3}|[\mathbf{OA\ OB\ OC}]|$ (d) None of these

65. Given four vectors **a**, **b**, **c** and **d**. The vectors **a**, **b**, **c** are coplanar but not collinear pair by pair and the vector **d** is not coplanar with the vecotrs **a**, **b** and **c**. If it is known that the angle between **a** and **b** is equal to that between **b** and **c**, each being equal to 60°. The angle between **d** and **a** is α and between **d** and **b** is β. Then, the angle between the vectors **d** and **c**.

(a) $\cos^{-1}(\cos\beta - \cos\alpha)$

(b) $\sin^{-1}(\cos\beta - \cos\alpha)$

(c) $\sin^{-1}(\sin\beta - \sin\alpha)$

(d) $\cos^{-1}(\tan\beta - \tan\alpha)$

66. The shortest distance between a diagonal of a unit cube and a diagonal of a face skew to it is

(a) $\frac{1}{2}$ (b) $\frac{1}{\sqrt{2}}$

(c) $\frac{1}{\sqrt{3}}$ (d) $\frac{1}{\sqrt{6}}$

67. Let $\mathbf{V} = 2\hat{\mathbf{i}} + \hat{\mathbf{j}} + \hat{\mathbf{k}}$ and $\mathbf{W} = \hat{\mathbf{i}} + 3\hat{\mathbf{k}}$. If *U* is a unit vector, then the maximum value of the scalar triple product $[\mathbf{U\ V\ W}]$ is

(a) −1 (b) $\sqrt{35}$

(c) $\sqrt{59}$ (d) $\sqrt{60}$

68. The length of the edge of the regular tetrahedron *ABCD* is 'a'. Points *E* and *F* are taken on the edges *AD* and *BD* respectively such that '*E*' divides **DA** and '*F*' divides **BD** in the ratio 2 : 1 each. Then, area of ΔCEF is

(a) $\frac{5a}{12\sqrt{3}}$ sq units (b) $\frac{a}{12\sqrt{3}}$ sq units

(c) $\frac{a^2}{12\sqrt{3}}$ sq units (d) $\frac{5a^2}{12\sqrt{3}}$ sq units

69. If two adjacent sides of two rectangles are represented by the vectors $\mathbf{p} = 5\mathbf{a} - 3\mathbf{b}$, $\mathbf{q} = -\mathbf{a} - 2\mathbf{b}$ and $\mathbf{r} = -4\mathbf{a} - \mathbf{b}; \mathbf{s} = -\mathbf{a} + \mathbf{b}$ respectively, then the angle between the vectors $\mathbf{x} = \frac{1}{3}(\mathbf{p} + \mathbf{r} + \mathbf{s})$ and $\mathbf{y} = \frac{1}{5}(\mathbf{r} + \mathbf{s})$

(a) $\pi - \cos^{-1}\left(\frac{19}{5\sqrt{43}}\right)$ (b) $\cos^{-1}\left(\frac{19}{5\sqrt{43}}\right)$

(c) $-\cos^{-1}\left(\frac{19}{5\sqrt{43}}\right)$ (d) $\pi - \cos^{-1}\left(\frac{19}{\sqrt{43}}\right)$

70. Let **a**, **b**, **c** are three vectors along the adjacent edges of a tetrahedron, if $|\mathbf{a}| = |\mathbf{b}| = |\mathbf{c}| = 2$ and $\mathbf{a} \cdot \mathbf{b} = \mathbf{b} \cdot \mathbf{c} = \mathbf{c} \cdot \mathbf{a} = 2$, then volume of tetrahedron is

(a) $\frac{1}{\sqrt{2}}$ (b) $\frac{2}{\sqrt{3}}$

(c) $\frac{\sqrt{3}}{2}$ (d) $\frac{2\sqrt{2}}{3}$

71. The angle θ between two non-zero vectors **a** and **b** satisfies the relation

$\cos\theta = (\mathbf{a} \times \hat{\mathbf{i}}) \cdot (\mathbf{b} \times \hat{\mathbf{i}}) + (\mathbf{a} \times \hat{\mathbf{j}}) \cdot (\mathbf{b} \times \hat{\mathbf{j}}) + (\mathbf{a} \times \hat{\mathbf{k}}) \cdot (\mathbf{b} \times \hat{\mathbf{k}})$,

then the least value of $|\mathbf{a}| + |\mathbf{b}|$ is equal to (where $\theta \neq 90°$)

(a) $\frac{1}{2}$ (b) 2

(c) $\sqrt{2}$ (d) 4

72. If the angle between the vectors $\mathbf{a} = \hat{\mathbf{i}} + (\cos x)\hat{\mathbf{j}} + \hat{\mathbf{k}}$ and $\mathbf{b} = (\sin^2 x - \sin x)\hat{\mathbf{i}} - (\cos x)\hat{\mathbf{j}} + (3 - 4\sin x)\hat{\mathbf{k}}$ is obtuse and $x \in \left(0, \frac{\pi}{2}\right)$, then the exhaustive set of values of '*x*' is equal to

(a) $x \in \left(0, \frac{\pi}{6}\right)$ (b) $x \in \left(\frac{\pi}{6}, \frac{\pi}{2}\right)$

(c) $x \in \left(\frac{\pi}{6}, \frac{\pi}{3}\right)$ (d) $x \in \left(\frac{\pi}{3}, \frac{\pi}{2}\right)$

73. If position vectors of the points *A*, *B* and *C* are **a**, **b**, **c** respectively and the points *D* and *E* divides line segments *AC* and *AB* in the ratio 2 : 1 and 1 : 3, respectively. Then, point of intersection of *BD* and *EC* divides *EC* in the ratio

(a) 2 : 1 (b) 1 : 3

(c) 1 : 2 (d) 3 : 2

Product of Vectors Exercise 2 :
More than One Option Correct Type Questions

74. If vectors $\mathbf{a}$ and $\mathbf{b}$ are non-collinear, than $\frac{\mathbf{a}}{|\mathbf{a}|}+\frac{\mathbf{b}}{|\mathbf{b}|}$ is
(a) a unit vector
(b) in the plane of $\mathbf{a}$ and $\mathbf{b}$
(c) equally inclined to $\mathbf{a}$ and $\mathbf{b}$
(d) perpendicular to $\mathbf{a}\times\mathbf{b}$

75. If $\mathbf{a}\times\mathbf{b}(\mathbf{b}\times\mathbf{c})=(\mathbf{a}\times\mathbf{b})\times\mathbf{c}$, then
(a) $(\mathbf{c}\times\mathbf{a})\times\mathbf{b}=0$ (b) $\mathbf{c}\times(\mathbf{a}\times\mathbf{b})=0$
(c) $\mathbf{b}\times(\mathbf{c}\times\mathbf{a})=0$ (d) $(\mathbf{c}\times\mathbf{a})\times\mathbf{b}=\mathbf{b}\times(\mathbf{c}\times\mathbf{a})=0$

76. Let $\mathbf{a}$ and $\mathbf{b}$ be two non-collinear unit vectors. If $\mathbf{u}=\mathbf{a}-(\mathbf{a}\,.\,\mathbf{b})\,\mathbf{b}$ and $\mathbf{v}=\mathbf{a}\times\mathbf{b}$, then $|\mathbf{v}|$ is
(a) $|\mathbf{u}|$ (b) $|\mathbf{u}|+|\mathbf{u}.\mathbf{a}|$
(c) $|\mathbf{u}|+|\mathbf{u}.\mathbf{b}|$ (d) $|\mathbf{u}|+\mathbf{u}.(\mathbf{a}+\mathbf{b})$

77. The scalars l and m such $l\mathbf{a}+m\mathbf{b}=\mathbf{c}$, where $\mathbf{a}, \mathbf{b}$ and $\mathbf{c}$ are give vectors, are equal to
(a) $l=\frac{(\mathbf{c}\times\mathbf{b}).(\mathbf{a}\times\mathbf{b})}{(\mathbf{a}\times\mathbf{b})^2}$ (b) $l=\frac{(\mathbf{c}\times\mathbf{a}).(\mathbf{b}\times\mathbf{a})}{(\mathbf{b}\times\mathbf{a})^2}$
(c) $m=\frac{(\mathbf{c}\times\mathbf{a}).(\mathbf{b}\times\mathbf{a})}{(\mathbf{b}\times\mathbf{a})^2}$ (d) $m=\frac{(\mathbf{c}\times\mathbf{a}).(\mathbf{b}\times\mathbf{a})}{(\mathbf{b}\times\mathbf{a})^2}$

78. Let $\hat{\mathbf{r}}$ be a unit vector satisfying $\hat{\mathbf{r}}\times\mathbf{a}=\mathbf{b}$, where $|\mathbf{a}|=\sqrt{3}$ and $|\mathbf{b}|=\sqrt{2}$. Then,
(a) $\hat{\mathbf{r}}=\frac{2}{3}(\mathbf{a}+\mathbf{a}\times\mathbf{b})$ (b) $\hat{\mathbf{r}}=\frac{1}{3}(\mathbf{a}+\mathbf{a}\times\mathbf{b})$
(c) $\hat{\mathbf{r}}=\frac{2}{3}(\mathbf{a}-\mathbf{a}\times\mathbf{b})$ (d) $\hat{\mathbf{r}}=\frac{1}{3}(-\mathbf{a}+\mathbf{a}\times\mathbf{b})$

79. $a_1, a_2, a_3 \in R-\{0\}$ and $a_1+a_2\cos 2x+a_3\sin^2 x=0$ for all $x\in R$, then
(a) vectors $\mathbf{a}=a_1\hat{\mathbf{i}}+a_2\hat{\mathbf{j}}+a_3\hat{\mathbf{k}}$ and $\mathbf{b}=4\hat{\mathbf{i}}+2\hat{\mathbf{j}}+\hat{\mathbf{k}}$ are perpendicular to each other
(b) vectors $\mathbf{a}=a_1\hat{\mathbf{i}}+a_2\hat{\mathbf{j}}+a_3\hat{\mathbf{k}}$ and $\mathbf{b}=-\hat{\mathbf{i}}+\hat{\mathbf{j}}+2\hat{\mathbf{k}}$ are perpendicular to each other
(c) if vectors $\mathbf{a}=a_1\hat{\mathbf{i}}+a_2\hat{\mathbf{j}}+a_3\hat{\mathbf{k}}$ is of length $\sqrt{6}$ units, then one of the ordered triplet $(a_1, a_2, a_3)=(1,-1,-2)$
(d) if vectors $2a_1+3a_2+6a_3$, then $|a_1\hat{\mathbf{i}}+a_2\hat{\mathbf{j}}+a_3\hat{\mathbf{k}}|$ is $2\sqrt{6}$

80. If $\mathbf{a}$ and $\mathbf{b}$ are two vectors and angles between them is θ, then
(a) $|\mathbf{a}\times\mathbf{b}|^2+(\mathbf{a}\cdot\mathbf{b})^2=|\mathbf{a}|^2|\mathbf{b}|^2$
(b) $|\mathbf{a}\times\mathbf{b}|=(\mathbf{a}\cdot\mathbf{b})$, if $\theta=\pi/4$
(c) $\mathbf{a}\times\mathbf{b}=(\mathbf{a}\cdot\mathbf{b})\hat{\mathbf{n}}$ (where $\hat{\mathbf{n}}$ is a normal unit vector), if $\theta=\pi/4$
(d) $|\mathbf{a}\times\mathbf{b}|\cdot(\mathbf{a}+\mathbf{b})=0$

81. If unit vectors $\mathbf{a}$ and $\mathbf{b}$ are inclined at an angle 2θ such that $|\mathbf{a}-\mathbf{b}|<1$ and $0\le\theta<=\pi$, then θ lies in the interval
(a) $[0,\pi/6)$ (b) $(5\pi/6\ \pi]$
(c) $[\pi/6,\pi/2)$ (d) $(\pi/2,5\pi/6]$

82. $\mathbf{b}$ and $\mathbf{c}$ are non-collinear if $\mathbf{a}\times(\mathbf{b}\times\mathbf{c})+(\mathbf{a}\cdot\mathbf{b})\mathbf{b}=(4-2x-\sin y)\mathbf{b}+(x^2-1)\mathbf{c}$ and $(\mathbf{c}\cdot\mathbf{c})\mathbf{a}=\mathbf{c}$. Then,
(a) $x=1$ (b) $x=-1$
(c) $y=(4n+1)\frac{\pi}{2}, n\in I$ (d) $y=(2n+1)\frac{\pi}{2}, n\in I$

83. If in triangle ABC, $\mathbf{AB}=\frac{\mathbf{u}}{|\mathbf{u}|}-\frac{\mathbf{v}}{|\mathbf{v}|}$ and $\mathbf{AC}=\frac{2\mathbf{u}}{|\mathbf{u}|}$, where $|\mathbf{u}|\ne|\mathbf{v}|$, then
(a) $1+\cos 2A+\cos 2B+\cos 2C=0$
(b) $\sin A=\cos C$
(c) projection of AC on BC is equal to BC
(d) projection of AB on BC is equal to AB

84. If $\mathbf{a}$, $\mathbf{b}$ and $\mathbf{c}$ be three non-zero vectors satisfying the condition $\mathbf{a}\times\mathbf{b}=\mathbf{c}$ and $\mathbf{b}\times\mathbf{c}=\mathbf{a}$, then which of the following always hold(s) good?
(a) $\mathbf{a}$, $\mathbf{b}$ and $\mathbf{c}$ are orthogonal in pairs.
(b) $[\mathbf{a}\ \mathbf{b}\ \mathbf{c}]=|\mathbf{b}|$
(c) $[\mathbf{a}\ \mathbf{b}\ \mathbf{c}]=|\mathbf{c}|^2$
(d) $|\mathbf{b}|=|\mathbf{c}|$

85. Given the following information about the non-zero vectors $\mathbf{A}$, $\mathbf{B}$ and $\mathbf{C}$
(i) $(\mathbf{A}\times\mathbf{B})\times\mathbf{A}=0$ (ii) $\mathbf{B}\cdot\mathbf{B}=4$
(iii) $\mathbf{A}\cdot\mathbf{B}=-6$ (iv) $\mathbf{B}\cdot\mathbf{C}=6$
Which one of the following holds good?
(a) $\mathbf{A}\times\mathbf{B}=0$ (b) $\mathbf{A}\cdot(\mathbf{B}\times\mathbf{C})=0$ (c) $\mathbf{A}\cdot\mathbf{A}=8$ (d) $\mathbf{A}\cdot\mathbf{C}=-9$

86. Let $\mathbf{a}$, $\mathbf{b}$ and $\mathbf{c}$ are non-zero vectors such that they are not orthogonal pairwise and such that $\mathbf{V}_1=\mathbf{a}\times(\mathbf{b}\times\mathbf{c})$ and $\mathbf{V}_2=(\mathbf{a}\times\mathbf{b})\times\mathbf{c}$, then which of the following hold(s) good?
(a) $\mathbf{a}$ and $\mathbf{b}$ are orthogonal (b) $\mathbf{a}$ and $\mathbf{c}$ are collinear
(c) $\mathbf{b}$ and $\mathbf{c}$ are orthogonal (d) $\mathbf{b}=\lambda\,(\mathbf{a}\times\mathbf{c})$ when λ is a scalar

87. Given three vectors
$\mathbf{U}=2\hat{\mathbf{i}}+3\hat{\mathbf{j}}-6\hat{\mathbf{k}}$, $\mathbf{V}=6\hat{\mathbf{i}}+2\hat{\mathbf{j}}+3\hat{\mathbf{k}}$ and $\mathbf{W}=3\hat{\mathbf{i}}-6\hat{\mathbf{j}}-2\hat{\mathbf{k}}$
which of the following hold good for the vectors $\mathbf{U}$, $\mathbf{V}$ and $\mathbf{W}$?
(a) $\mathbf{U}$, $\mathbf{V}$ and $\mathbf{W}$ are linearly dependent
(b) $(\mathbf{U}\times\mathbf{V})\times\mathbf{W}=\mathbf{0}$
(c) $\mathbf{U}$, $\mathbf{V}$ and $\mathbf{W}$ form a triplet of mutually perpendicular vectors
(d) $(\mathbf{U}\times(\mathbf{V}\times\mathbf{W})=0$

88. Let $\mathbf{a}=2\hat{\mathbf{i}}-\hat{\mathbf{j}}+\hat{\mathbf{k}}$, $\mathbf{b}=\hat{\mathbf{i}}+2\hat{\mathbf{j}}-\hat{\mathbf{k}}$ and $\mathbf{c}=\hat{\mathbf{i}}+\hat{\mathbf{j}}-2\hat{\mathbf{k}}$ be three vectors. A vector in the plane of $\mathbf{b}$ and $\mathbf{c}$ whose projection on $\mathbf{a}$ is of magnitude $\sqrt{\frac{2}{3}}$ is
(a) $2\hat{\mathbf{i}}+3\hat{\mathbf{j}}-3\hat{\mathbf{k}}$ (b) $2\hat{\mathbf{i}}+3\hat{\mathbf{j}}+3\hat{\mathbf{k}}$
(c) $-2\hat{\mathbf{i}}-\hat{\mathbf{j}}+5\hat{\mathbf{k}}$ (d) $2\hat{\mathbf{i}}+\hat{\mathbf{j}}+5\hat{\mathbf{k}}$

89. Three vectors $\mathbf{a}$ ($|\mathbf{a}| \neq 0$), $\mathbf{b}$ and $\mathbf{c}$ are such that $\mathbf{a} \times \mathbf{b} = 3\mathbf{a} \times \mathbf{c}$. Also, $|\mathbf{a}| = |\mathbf{b}| = 1$ and $|\mathbf{c}| = \frac{1}{3}$. If the angle between $\mathbf{b}$ and $\mathbf{c}$ is 60°, then

(a) $\mathbf{b} = 3\mathbf{c} + \mathbf{a}$ (b) $\mathbf{b} = 3\mathbf{c} - \mathbf{a}$
(c) $\mathbf{a} = 6\mathbf{c} + 2\mathbf{b}$ (d) $\mathbf{a} = 6\mathbf{c} - 2\mathbf{b}$

90. Let $\mathbf{a}, \mathbf{b}, \mathbf{c}$ be non-zero vectors and $|\mathbf{a}| = 1$ and $\mathbf{r}$ is a non-zero vector such that $\mathbf{r} \times \mathbf{a} = \mathbf{b}$ and $\mathbf{r} \cdot \mathbf{c} = 1$, then

(a) $\mathbf{a} \perp \mathbf{b}$ (b) $\mathbf{r} \perp \mathbf{b}$
(c) $\mathbf{r} \cdot \mathbf{a} = \dfrac{1 - [\mathbf{a}\ \mathbf{b}\ \mathbf{c}]}{\mathbf{a} \cdot \mathbf{b}}$ (d) $[\mathbf{r}\,\mathbf{a}\,\mathbf{b}] = 0$

91. If $\mathbf{a}$ and $\mathbf{b}$ are two unit vectors perpendicular to each other and $\mathbf{c} = \lambda_1 \mathbf{a} + \lambda_2 \mathbf{b} + \lambda_3 (\mathbf{a} \times \mathbf{b})$, then the following is (are) true

(a) $\lambda_1 = \mathbf{a} \cdot \mathbf{c}$
(b) $\lambda_2 = |\vec{\mathbf{b}} \times \vec{\mathbf{a}}|$
(c) $\lambda_3 = |(\mathbf{a} \times \mathbf{b}) \times \mathbf{c}|$
(d) $\lambda_1 + \lambda_2 + \lambda_3 = (\mathbf{a} + \mathbf{b} + \mathbf{a} \times \mathbf{b}) \cdot \mathbf{c}$

92. Given three non-coplanar vectors $\mathbf{OA} = \mathbf{a}, \mathbf{OB} = \mathbf{b}, \mathbf{OC} = \mathbf{c}$.
Let S be the centre of the sphere passing through the points, O, A, B, C if $\mathbf{OS} = \mathbf{x}$, then

(a) $\mathbf{x}$ must be linear combination of $\mathbf{a}, \mathbf{b}, \mathbf{c}$
(b) $\mathbf{x}$ must be linear combination of $\mathbf{b} \times \mathbf{c}$, $\mathbf{c} \times \mathbf{a}$ and $\mathbf{a} \times \mathbf{b}$
(c) $\mathbf{x} = \dfrac{a^2(\mathbf{b} \times \mathbf{c}) + b^2(\mathbf{c} \times \mathbf{a}) + c^2(\mathbf{a} \times \mathbf{b})}{2[\mathbf{a}\,\mathbf{b}\,\mathbf{c}]}$, $a = |\mathbf{a}|, b = |\mathbf{b}|, c = |\mathbf{c}|$
(d) $\mathbf{x} = \mathbf{a} + \mathbf{b} + \mathbf{c}$

93. If $\mathbf{a} = \hat{\mathbf{i}} + \hat{\mathbf{j}} + \hat{\mathbf{k}}$ and $\mathbf{b} = \hat{\mathbf{i}} - \hat{\mathbf{j}}$, then the vectors $(\mathbf{a} \cdot \hat{\mathbf{i}})\hat{\mathbf{i}} + (\mathbf{a} \cdot \hat{\mathbf{j}})\hat{\mathbf{j}} + (\mathbf{a} \cdot \hat{\mathbf{k}})\hat{\mathbf{k}}, (\mathbf{b} \cdot \hat{\mathbf{i}})\hat{\mathbf{i}} + (\mathbf{b} \cdot \hat{\mathbf{j}})\hat{\mathbf{j}} + (\mathbf{b} \cdot \hat{\mathbf{j}})\hat{\mathbf{k}}$ and $\hat{\mathbf{i}} + \hat{\mathbf{j}} - 2\hat{\mathbf{k}}$

(a) are mutually perpendicular
(b) are coplanar
(c) form a parallelopiped of volume 6 units
(d) form a parallelopiped of volume 3 units

94. If $\mathbf{a} = x\hat{\mathbf{i}} + y\hat{\mathbf{j}} + z\hat{\mathbf{k}}, \mathbf{b} = y\hat{\mathbf{i}} + z\hat{\mathbf{j}} + x\hat{\mathbf{k}}, \mathbf{c} = z\hat{\mathbf{i}} + x\hat{\mathbf{j}} + y\hat{\mathbf{k}}$, then $\mathbf{a} \times (\mathbf{b} \times \mathbf{c})$ is

(a) parallel to $(y - z)\hat{\mathbf{i}} + (z - x)\hat{\mathbf{j}} + (x - y)\hat{\mathbf{k}}$
(b) orthogonal to $\hat{\mathbf{i}} + \hat{\mathbf{j}} + \hat{\mathbf{k}}$
(c) orthogonal to $(y + z)\hat{\mathbf{i}} + (z + x)\hat{\mathbf{j}} + (x + y)\hat{\mathbf{k}}$
(d) parallel to $\hat{\mathbf{i}} + \hat{\mathbf{j}} + \hat{\mathbf{k}}$

95. If $\mathbf{a}, \mathbf{b}, \mathbf{c}$ are three non-zero vectors, then which of the following statement (s) is/are true?

(a) $\mathbf{a} \times (\mathbf{b} \times \mathbf{c}), \mathbf{b} \times (\mathbf{c} \times \mathbf{a}), \mathbf{c} \times (\mathbf{a} \times \mathbf{b})$ form a right handed system
(b) $\mathbf{c}, (\mathbf{a} \times \mathbf{b}) \times \mathbf{c}, \mathbf{a} \times \mathbf{b}$ form a right handed system
(c) $\mathbf{a} \cdot \mathbf{b} + \mathbf{b} \cdot \mathbf{c} + \mathbf{c} \cdot \mathbf{a} < 0$, if $\mathbf{a} + \mathbf{b} + \mathbf{c} = \mathbf{0}$
(d) $\dfrac{(\mathbf{a} \times \mathbf{b}) \cdot (\mathbf{b} \times \mathbf{c})}{(\mathbf{b} \times \mathbf{c}) \cdot (\mathbf{a} \times \mathbf{c})} = -1$, if $\mathbf{a} + \mathbf{b} + \mathbf{c} = \mathbf{0}$

96. Let the unit vectors $\mathbf{a}$ and $\mathbf{b}$ be perpendicular and unit vector $\mathbf{c}$ is inclined at angle α to $\mathbf{a}$ and $\mathbf{b}$. If $\mathbf{c} = l\mathbf{a} + m\mathbf{b} + n(\mathbf{a} \times \mathbf{b})$, then

(a) $l = m$ (b) $n^2 = 1 - 2l^2$
(c) $n^2 = -\cos 2\alpha$ (d) $m^2 = \dfrac{1 + \cos 2\alpha}{2}$

97. If $\mathbf{a}, \mathbf{b}, \mathbf{c}$ are three non-zero vectors, then which of the following statement(s) is/are true?

(a) $\mathbf{a} \times (\mathbf{b} \times \mathbf{c}), \mathbf{b} \times (\mathbf{c} \times \mathbf{a}), \mathbf{c} \times (\mathbf{a} \times \mathbf{b})$ form a right handed system
(b) $\mathbf{c}, (\mathbf{a} \times \mathbf{b}) \times \mathbf{c}, \mathbf{a} \times \mathbf{b}$ form a right handed system
(c) $\mathbf{a} \cdot \mathbf{b} + \mathbf{b} \cdot \mathbf{c} + \mathbf{c} \cdot \mathbf{a} < 0$ if $\mathbf{a} + \mathbf{b} + \mathbf{c} = \mathbf{0}$
(d) $\dfrac{(\mathbf{a} \times \mathbf{b}) \cdot (\mathbf{b} \times \mathbf{c})}{(\mathbf{b} \times \mathbf{c}) \cdot (\mathbf{a} \times \mathbf{c})} = -1$ if $\mathbf{a} + \mathbf{b} + \mathbf{c} = \mathbf{0}$

98. Let $\mathbf{a}$ and $\mathbf{b}$ be two given perpendicular vectors, which are non-zero. A vector $\mathbf{r}$ satisfying the equation $\mathbf{r} \times \mathbf{b} = \mathbf{a}$, can be ....

(a) $\mathbf{b} - \dfrac{\mathbf{a} \times \mathbf{b}}{|\mathbf{b}|^2}$ (b) $2\mathbf{b} - \dfrac{(\mathbf{a} \times \mathbf{b})}{|\mathbf{b}|^2}$
(c) $|\mathbf{a}|\mathbf{b} - \dfrac{\mathbf{a} \times \mathbf{b}}{|\mathbf{b}|^2}$ (d) $|\mathbf{b}|\mathbf{b} - \dfrac{(\mathbf{a} \times \mathbf{b})}{|\mathbf{b}|^2}$

99. If $\mathbf{a}$ and $\mathbf{b}$ are any two vectors, then possible integers(s) in the range of $\dfrac{3|\mathbf{a} + \mathbf{b}|}{2} + 2|\mathbf{a} - \mathbf{b}|$ is

(a) 2 (b) 3
(c) 4 (d) 5

100. If $\mathbf{a}$ is perpendicular to $\mathbf{b}$ and p is non-zero scalar such that $p\,\mathbf{r} + (\mathbf{r} \cdot \mathbf{b})\,\mathbf{a} = \mathbf{c}$, then $\mathbf{r}$

(a) $[\mathbf{r}\ \mathbf{a}\ \mathbf{c}] = 0$
(b) $p^2\mathbf{r} = p\mathbf{a} - (\mathbf{c} \cdot \mathbf{a})\mathbf{b}$
(c) $p^2\mathbf{r} = p\mathbf{b} - (\mathbf{a} \cdot \mathbf{b})\mathbf{c}$
(d) $p^2\mathbf{r} = p\mathbf{c} - (\mathbf{b} \cdot \mathbf{c})\mathbf{a}$

101. In a four-dimensional space where unit vectors along axes are $\hat{\mathbf{i}}, \hat{\mathbf{j}}, \hat{\mathbf{k}}$ and $\hat{\mathbf{l}}$ and $\mathbf{a}_1, \mathbf{a}_2, \mathbf{a}_3, \mathbf{a}_4$ are four non-zero vectors such that no vector can be expressed as linear combination of others and $(\lambda - 1)(\mathbf{a}_1 - \mathbf{a}_2) + \mu(\mathbf{a}_2 + \mathbf{a}_3) + \gamma(\mathbf{a}_3 + \mathbf{a}_4 - 2\mathbf{a}_2) + \mathbf{a}_3 + \delta\mathbf{a}_4 = \mathbf{0}$, then

(a) $\lambda = 1$ (b) $\mu = -\dfrac{2}{3}$
(c) $\lambda = \dfrac{2}{3}$ (d) $\delta = \dfrac{1}{3}$

102. A vector $(\mathbf{d})$ is equally inclined to three vectors $\mathbf{a} = \hat{\mathbf{i}} - \hat{\mathbf{j}} + \hat{\mathbf{k}}, \mathbf{b} = 2\hat{\mathbf{i}} + \hat{\mathbf{j}}$ and $\mathbf{c} = 3\hat{\mathbf{j}} - 2\hat{\mathbf{k}}$. Let $\mathbf{x}, \mathbf{y}, \mathbf{z}$ be three vectors in the plane of $\mathbf{a}, \mathbf{b}$; $\mathbf{b}, \mathbf{c}$; $\mathbf{c}, \mathbf{a}$ respectively, then

(a) $\mathbf{x} \cdot \mathbf{d} = 14$ (b) $\mathbf{y} \cdot \mathbf{d} = 3$
(c) $\mathbf{z} \cdot \mathbf{d} = 0$
(d) $\mathbf{r} \cdot \mathbf{d} = 0$, where $\mathbf{r} = \lambda\mathbf{x} + \mu\mathbf{y} + \delta\mathbf{z}$

103. If **a, b, c** are non-zero, non-collinear vectors such that a vctors such that a vector $\mathbf{p} = ab\cos(2\pi - (\mathbf{a}, \mathbf{c}))\,\mathbf{c}$ and a $\mathbf{q} = \mathbf{ac}\ \cos(\pi - (\mathbf{a}\ \mathbf{c}))$ then $\mathbf{b} + \mathbf{q}$ is

(a) parallel to **a** (b) perpendicular to **a**
(c) coplanar with **b** and **c** (d) coplanar with **a** and **c**

104. Given three vevtors **a, b, c** such that they are non-zero , non-coplanar vectors, then which of the following are coplanar.

(a) $\mathbf{a} + \mathbf{b}, \mathbf{b} + \mathbf{c}, \mathbf{c} + \mathbf{a}$ (b) $\mathbf{a} - \mathbf{b}, \mathbf{b} + \mathbf{c}, \mathbf{c} + \mathbf{a}$
(c) $\mathbf{a} + \mathbf{b}, \mathbf{b} - \mathbf{c}, \mathbf{c} + \mathbf{a}$ (d) $\mathbf{a} + \mathbf{b}, \mathbf{b} + \mathbf{c}, \mathbf{c} - \mathbf{a}$

105. If $\mathbf{r} = \hat{\mathbf{i}} + \hat{\mathbf{j}} + \lambda(2\hat{\mathbf{i}} + \hat{\mathbf{j}} + 4\hat{\mathbf{k}})$ and $\mathbf{r} \cdot (\hat{\mathbf{i}} + 2\hat{\mathbf{j}} - \hat{\mathbf{k}}) = 3$ are the equations of a line and a plane respectively, then which of the following is incorrect?

(a) line is perpendicular to the plane
(b) line lies in the plane
(c) line is parallel to the plane but does not lie in the plane
(d) line cuts the plane obliquely

106. If vectors **a** and **b** are two adjacent sides of a parallelogram, then the vector representing the altitude of the parallelogram which is perpendicular to **a** is

(a) $\mathbf{b} + \dfrac{\mathbf{b} \times \mathbf{a}}{|\mathbf{a}|^2}$ (b) $\dfrac{\mathbf{a} \cdot \mathbf{b}}{|\mathbf{b}|^2}\mathbf{b}$
(c) $\mathbf{b} - \dfrac{\mathbf{b} \cdot \mathbf{a}}{|\mathbf{a}|^2}\mathbf{a}$ (d) $\dfrac{\mathbf{a} \times (\mathbf{b} \times \mathbf{a})}{|\mathbf{a}|^2}$

107. Let **a, b, c** be three vectors such that each of them are non-collinear, $\mathbf{a} + \mathbf{b}$ and $\mathbf{b} + \mathbf{c}$ are collinear with **c** and **a** respectively and $\mathbf{a} + \mathbf{b} + \mathbf{c} = \mathbf{k}$. Then, $(|\mathbf{k}|, |\mathbf{k}|)$ lies on

(a) $y^2 = 4ax$ (b) $x^2 + y^2 - ax - by = 0$
(c) $x^2 - y^2 = 1$ (d) $|x| + |y| = 1$

108. If **a, b, c** are non-coplanar unit vectors also **b, c** are non-collinear and $2\mathbf{a} \times (\mathbf{b} \times \mathbf{c}) = \mathbf{b} + \mathbf{c}$, then

(a) angle between **a** and **c** is 60°
(b) angle between **b** and **c** is 30°
(c) angle between **a** and **b** is 120°
(d) **b** is perpendicular to **c**

109. If $\mathbf{a} = \dfrac{1}{7}(2\hat{\mathbf{i}} + 3\hat{\mathbf{j}} + 6\hat{\mathbf{k}})$; $\mathbf{b} = \dfrac{1}{7}(6\hat{\mathbf{i}} + 2\hat{\mathbf{j}} - 3\hat{\mathbf{k}})$;

$\mathbf{c} = c_1\hat{\mathbf{i}} + c_2\hat{\mathbf{j}} + c_3\hat{\mathbf{k}}$ and matrix $A = \begin{bmatrix} \frac{2}{7} & \frac{3}{7} & \frac{6}{7} \\ \frac{6}{7} & \frac{2}{7} & \frac{-3}{7} \\ c_1 & c_2 & c_3 \end{bmatrix}$

and $AA^T = I$, then **c**

(a) $\dfrac{3\hat{\mathbf{i}} + 6\hat{\mathbf{j}} + 2\hat{\mathbf{k}}}{7}$ (b) $\dfrac{1}{7}(3\hat{\mathbf{i}} - 6\hat{\mathbf{j}} + 2\hat{\mathbf{k}})$
(c) $\dfrac{1}{7}(-3\hat{\mathbf{i}} + 6\hat{\mathbf{j}} - 2\hat{\mathbf{k}})$ (d) $-\dfrac{1}{7}(3\hat{\mathbf{i}} + 6\hat{\mathbf{j}} + 2\hat{\mathbf{k}})$

Product of Vectors Exercise 3 : Statement I & II Type Questions

- **Directions** (Q. Nos. 110 to 121) Each of these questions contains two statements.

Statement I (Assertion) and **Statement II** (Reason) Each of these questions also has four alternatives choices, only one of which is the correct answer. You have to select the correct choice, as given below.

(a) Statement I is true, Statement II is true and Statement II is a correct explanation for Statement I.
(b) Statement I is true, Statement II is true but Statement II is not a correct explanation for Statement I.
(c) Statement I is true, Statement II is false.
(d) Statement I is false, Statement II is true.

110. Statement I A component of vector $\mathbf{b} = 4\hat{\mathbf{i}} + 2\hat{\mathbf{j}} + 3\hat{\mathbf{k}}$ in the direction perpendicular to the direction of vector $\mathbf{a} = \hat{\mathbf{i}} + \hat{\mathbf{j}} + \hat{\mathbf{k}}$ is $\hat{\mathbf{i}} - \hat{\mathbf{j}}$.

Statement II A component of vector in the direction $\mathbf{a} = \hat{\mathbf{i}} + \hat{\mathbf{j}} + \hat{\mathbf{k}}$ is $2\hat{\mathbf{i}} + 2\hat{\mathbf{j}} + 2\hat{\mathbf{k}}$.

111. Statement I $a_1\hat{\mathbf{i}} + a_2\hat{\mathbf{j}} + a_3\hat{\mathbf{k}}, b_1\hat{\mathbf{i}} + b_2\hat{\mathbf{j}} + b_3\hat{\mathbf{k}}$, and $c_1\hat{\mathbf{i}} + c_2\hat{\mathbf{j}} + c_3\hat{\mathbf{k}}$ are three mutually perpendicular unit vector, then $a_1\hat{\mathbf{i}} + b_1\hat{\mathbf{j}} + c_1\hat{\mathbf{k}}, a_2\hat{\mathbf{i}} + b_2\hat{\mathbf{j}} + c_2\hat{\mathbf{k}}$, and $a_3\hat{\mathbf{i}} + b_3\hat{\mathbf{j}} + c_3\hat{\mathbf{k}}$, may be mutually perpendicular unit vectors.

Statement II Value of determinant and its transpose are the same.

112. Consider the vector **a, b** and **c**.

Statement I $\mathbf{a} \times \mathbf{b} = (\hat{\mathbf{i}} \times \mathbf{b}).(\mathbf{b})\hat{\mathbf{i}} + (\hat{\mathbf{j}} \times \mathbf{a}).(\mathbf{b}\hat{\mathbf{j}} \times (\hat{\mathbf{k}} \times \mathbf{a}).\mathbf{b})\hat{\mathbf{k}}$

Statement II $\mathbf{c} = (\hat{\mathbf{i}}.\mathbf{c})\hat{\mathbf{i}} + (\hat{\mathbf{j}}.\mathbf{c})\hat{\mathbf{j}} + (\hat{\mathbf{k}}.\mathbf{c})\hat{\mathbf{k}}$

113. Statement I Distance of point $D(1, 0, -1)$ from the plane of points $A(1, -2, 0)$, $B(3, 1, 2)$ and $C(-1, 1, -1)$ is $\dfrac{8}{\sqrt{229}}$.

Statement II Volume of tetrahedron formed by the points A, B, C and D is $\dfrac{\sqrt{229}}{2}$.

114. Statement I $\mathbf{A} = 2\hat{\mathbf{i}} + 3\hat{\mathbf{j}} + 6\hat{\mathbf{k}}, \mathbf{B} = \hat{\mathbf{i}} + \hat{\mathbf{j}} - 2\hat{\mathbf{k}}$ and $\mathbf{C} = \hat{\mathbf{i}} + 2\hat{\mathbf{j}} + \hat{\mathbf{k}}$, then $|\mathbf{A} \times (\mathbf{A} \times (\mathbf{A} \times \mathbf{B})).\mathbf{C}| = 243$

Statement II $|\mathbf{A} \times (\mathbf{A} \times (\mathbf{A} \times \mathbf{B})).\mathbf{C}| = |\mathbf{A}|^2\,|[\mathbf{ABC}]|$

115. **Statement I** The number of vectors of unit length and perpendicular to both the vectors $\hat{\mathbf{i}}+\hat{\mathbf{j}}$ and $\hat{\mathbf{j}}+\hat{\mathbf{k}}$ is zero.

Statement II **a** and **b** are two non-zero and non-parallel vectors it is true that $\mathbf{a}\times\mathbf{b}$ is perpendicular to the plane containing **a** and **b**.

116. **Statement I** (S_1) : If $A(x_1, y_1)$, $B(x_2, y_2)$, $C(x_3, y_3)$ are non-collinear points. Then, every point (x, y) in the plane of ΔABC, can be expressed in the form

$$\left(\frac{kx_1+lx_2+mx_3}{k+l+m}, \frac{ky_1+ly_2+my_3}{k+l+m}\right)$$

Statement II (S_2) The condition for coplanarity of four points $A(\mathbf{a})$, $B(\mathbf{b})$, $C(\mathbf{c})$, $D(\mathbf{d})$ is that there exists scalars, l, m, n, p not all zeros such that

$$l\mathbf{a}+m\mathbf{b}+n\mathbf{c}+p\mathbf{d}=\mathbf{0}$$

where $l+m+n+p=0$

117. If **a**, **b** are non-zero vectors such that $|\mathbf{a}+\mathbf{b}|=|\mathbf{a}-2\mathbf{b}|$, then

Statement I Least value of $\mathbf{a}\cdot\mathbf{b}+\dfrac{4}{|\mathbf{b}|^2+2}$ is $2\sqrt{2}-1$

Statement II The expression $\mathbf{a}\cdot\mathbf{b}+\dfrac{4}{|\mathbf{b}|^2+2}$ is least when magnitude of **b** is $\sqrt{2\tan\left(\dfrac{\pi}{8}\right)}$

118. **Statement I** If $\mathbf{a}=3\hat{\mathbf{i}}-3\hat{\mathbf{j}}+\hat{\mathbf{k}}$, $\mathbf{b}=-\hat{\mathbf{i}}+2\hat{\mathbf{j}}+\hat{\mathbf{k}}$ and $\mathbf{c}=\hat{\mathbf{i}}+\hat{\mathbf{j}}+\hat{\mathbf{k}}$ and $\mathbf{d}=2\hat{\mathbf{i}}-\hat{\mathbf{j}}$, then there exist real numbers α, β, γ such that $\mathbf{a}=\alpha\mathbf{b}+\beta\mathbf{c}+\gamma\mathbf{d}$

Statement II **a**, **b**, **c**, **d** are four vectors in a 3-dimensional space. If **b**, **c**, **d** are non-coplanar, then there exist real numbers α, β, γ such that $\mathbf{a}=\alpha\mathbf{b}+\beta\mathbf{c}+\gamma\mathbf{d}$.

119. **Statement I** Let **a**, **b**, **c** and **d** are position vector four points A, B, C and D and $3\mathbf{a}-2\mathbf{b}+5\mathbf{c}-6\mathbf{d}=\mathbf{0}$, then points A, B, C and D are coplanar.

Statement II Three non-zero, linearly dependent coinitial vectors (**PQ**, **PR** and **PS**) are coplanar.

120. If $\mathbf{a}=\hat{\mathbf{i}}+\hat{\mathbf{j}}-\hat{\mathbf{k}}$, $\mathbf{b}=2\hat{\mathbf{i}}+\hat{\mathbf{j}}-3\hat{\mathbf{k}}$ and **r** is a vector satisfying $2\mathbf{r}+\mathbf{r}\times\mathbf{a}=\mathbf{b}$.

Statement I **r** can be expressed in terms of **a**, **b** and $\mathbf{a}\times\mathbf{b}$.

Statement II $\mathbf{r}=\dfrac{1}{7}(7\hat{\mathbf{i}}+5\hat{\mathbf{j}}-9\hat{\mathbf{k}}+\mathbf{a}\times\mathbf{b})$

121. Let $\hat{\mathbf{a}}$ and $\hat{\mathbf{c}}$ be units vectors at an angle $\dfrac{\pi}{3}$ with each other. If $(\hat{\mathbf{a}}\times(\hat{\mathbf{b}}\times\hat{\mathbf{c}}))\cdot(\hat{\mathbf{a}}\times\hat{\mathbf{c}})=5$ then

Statement I $[\hat{\mathbf{a}}\ \hat{\mathbf{b}}\ \hat{\mathbf{c}}]=10$

because

Statement II $[\mathbf{x}\ \mathbf{y}\ \mathbf{z}]=0$, if $\mathbf{x}=\mathbf{y}$ or $\mathbf{y}=\mathbf{z}$ or $\mathbf{x}=\mathbf{z}$

Product of Vectors Exercise 4 : Passage Based Questions

Passage I

(Q. Nos. 122-124)

Consider three vectors $\mathbf{p}=\hat{\mathbf{i}}+\hat{\mathbf{j}}+\hat{\mathbf{k}}$, $\mathbf{q}=2\hat{\mathbf{i}}+4\hat{\mathbf{j}}-\hat{\mathbf{k}}$ and $\mathbf{r}=\hat{\mathbf{i}}+\hat{\mathbf{j}}+3\hat{\mathbf{k}}$ and let **s** be a unit vector, then

122. **p**, **q** and **r** are

(a) linearly dependent
(b) can form the sides of a possible triangle
(c) such that the vectors $(\mathbf{q}-\mathbf{r})$ is orthogonal to **p**
(d) such that each one of these can be expressed as a linear combination of the other two

123. If $(\mathbf{p}\times\mathbf{q})\times\mathbf{r}=u\,\mathbf{p}+v\,\mathbf{q}+w\,\mathbf{r}$, then $(u+v+w)$ is equal to

(a) 8 (b) 2
(c) -2 (d) 4

124. The magnitude of the vector $(\mathbf{p}\cdot\mathbf{s})(\mathbf{q}\times\mathbf{r})+(\mathbf{q}\cdot\mathbf{s})(\mathbf{r}\times\mathbf{p})+(\mathbf{r}\cdot\mathbf{s})(\mathbf{p}\times\mathbf{q})$ is

(a) 4 (b) 8
(c) 18 (d) 2

Passage II

(Q. Nos. 125-127)

Consider the three vectors **p**, **q** and **r** such that $\mathbf{p}=\hat{\mathbf{i}}+\hat{\mathbf{j}}+\hat{\mathbf{k}}$ and $\mathbf{q}=\hat{\mathbf{i}}-\hat{\mathbf{j}}+\hat{\mathbf{k}}$; $\mathbf{p}\times\mathbf{r}=\mathbf{q}+c\mathbf{p}$ and $\mathbf{p}\cdot\mathbf{r}=2$

125. The value of $[\mathbf{p}\ \mathbf{q}\ \mathbf{r}]$ is

(a) $-\dfrac{5\sqrt{2}c}{|\mathbf{r}|}$ (b) $-\dfrac{8}{3}$
(c) 0 (d) greater than 0

126. If **x** is a vector such that $[\mathbf{p}\ \mathbf{q}\ \mathbf{r}]\,\mathbf{x}=(\mathbf{p}\times\mathbf{q})\times\mathbf{r}$, then **x** is

(a) $c\,(\hat{\mathbf{i}}-2\hat{\mathbf{j}}+\hat{\mathbf{k}})$ (b) a unit vector
(c) indeterminate, as $[\mathbf{p}\ \mathbf{q}\ \mathbf{r}]$ (d) $-(\hat{\mathbf{i}}-2\hat{\mathbf{j}}+\hat{\mathbf{k}})/2$

127. If **y** is a vector satisfying $(1+c)\,\mathbf{y}=\mathbf{p}\times(\mathbf{q}\times\mathbf{r})$, then the vectors **x**, **y** and **r**

(a) are collinear (b) are coplanar
(c) represent the coterminus edges of a tetrahedron whose volume is c cu units
(d) represent the coterminus edge of a paralloepiped whose volume is c cu units

Passage III

(Q. Nos. 128-130)

Let P and Q are two points on the curve $y = \log_{1/2}(x - 0.5) + \log_2 \sqrt{4x^2 - 4x + 1}$ and P is also on the circle $x^2 + y^2 = 10$, Q lies inside the given circle such that its abscissa is an integer.

128. The coordinates of P are given by

(a) (1, 2) (b) (2, 4) (c) (3, 1) (d) (3, 5)

129. $\mathbf{OP} \cdot \mathbf{OQ}$, O being the origin is

(a) 4 or 7 (b) 4 or 2 (c) 2 or 3 (d) 7 or 8

130. Max {| **PQ** |} is

(a) 1 (b) 4 (c) 0 (d) 2

Passage IV

(Q. Nos. 131 to 134)

If **a**, **b**, **c** are three given non-coplanar vectors and any arbitrary vector **r** is in space, where

$$\Delta_1 = \begin{vmatrix} \mathbf{r}\cdot\mathbf{a} & \mathbf{b}\cdot\mathbf{a} & \mathbf{c}\cdot\mathbf{a} \\ \mathbf{r}\cdot\mathbf{b} & \mathbf{b}\cdot\mathbf{b} & \mathbf{c}\cdot\mathbf{b} \\ \mathbf{r}\cdot\mathbf{c} & \mathbf{b}\cdot\mathbf{c} & \mathbf{c}\cdot\mathbf{c} \end{vmatrix};\ \Delta_2 = \begin{vmatrix} \mathbf{a}\cdot\mathbf{a} & \mathbf{r}\cdot\mathbf{a} & \mathbf{c}\cdot\mathbf{a} \\ \mathbf{a}\cdot\mathbf{b} & \mathbf{r}\cdot\mathbf{b} & \mathbf{c}\cdot\mathbf{b} \\ \mathbf{a}\cdot\mathbf{c} & \mathbf{r}\cdot\mathbf{c} & \mathbf{c}\cdot\mathbf{c} \end{vmatrix}$$

$$\Delta_3 = \begin{vmatrix} \mathbf{a}\cdot\mathbf{a} & \mathbf{b}\cdot\mathbf{a} & \mathbf{r}\cdot\mathbf{a} \\ \mathbf{a}\cdot\mathbf{b} & \mathbf{b}\cdot\mathbf{b} & \mathbf{r}\cdot\mathbf{b} \\ \mathbf{a}\cdot\mathbf{c} & \mathbf{b}\cdot\mathbf{c} & \mathbf{r}\cdot\mathbf{c} \end{vmatrix};\quad \Delta = \begin{vmatrix} \mathbf{a}\cdot\mathbf{a} & \mathbf{b}\cdot\mathbf{a} & \mathbf{c}\cdot\mathbf{a} \\ \mathbf{a}\cdot\mathbf{b} & \mathbf{b}\cdot\mathbf{b} & \mathbf{c}\cdot\mathbf{b} \\ \mathbf{a}\cdot\mathbf{c} & \mathbf{c}\cdot\mathbf{c} & \mathbf{c}\cdot\mathbf{c} \end{vmatrix}$$

131. The vector **r** is expressible in the form

(a) $\mathbf{r} = \frac{\Delta_1}{2\Delta}\mathbf{a} + \frac{\Delta_2}{2\Delta}\mathbf{b} + \frac{\Delta_3}{2\Delta}\mathbf{c}$

(b) $\mathbf{r} = \frac{2\Delta_1}{\Delta}\mathbf{a} + \frac{2\Delta_2}{\Delta}\mathbf{b} + \frac{2\Delta_3}{\Delta}\mathbf{c}$

(c) $\mathbf{r} = \frac{\Delta}{\Delta_1}\mathbf{a} + \frac{\Delta}{\Delta_2}\mathbf{b} + \frac{\Delta}{\Delta_3}\mathbf{c}$

(d) $\mathbf{r} = \frac{\Delta_1}{\Delta}\mathbf{a} + \frac{\Delta_2}{\Delta}\mathbf{b} + \frac{\Delta_3}{\Delta}\mathbf{c}$

132. The vector **r** is expressible as

(a) $\mathbf{r} = \frac{[\mathbf{r\,b\,c}]}{2[\mathbf{a\,b\,c}]}\mathbf{a} + \frac{[\mathbf{r\,c\,a}]}{2[\mathbf{a\,b\,c}]}\mathbf{b} + \frac{[\mathbf{r\,a\,b}]}{2[\mathbf{a\,b\,c}]}\mathbf{c}$

(b) $\mathbf{r} = \frac{2[\mathbf{r\,b\,c}]}{[\mathbf{a\,b\,c}]}\mathbf{a} + \frac{2[\mathbf{r\,c\,b}]}{[\mathbf{a\,b\,c}]}\mathbf{b} + \frac{2[\mathbf{r\,a\,b}]}{[\mathbf{a\,b\,c}]}\mathbf{c}$

(c) $\mathbf{r} = \frac{1}{[\mathbf{a\,b\,c}]}([\mathbf{r\,b\,c}]\mathbf{a} + [\mathbf{r\,c\,a}]\mathbf{b} + [\mathbf{r\,a\,b}]\,\mathbf{c})$

(d) None of the above

133. If vector is expressible as $\mathbf{r} = x\,\mathbf{a} + y\,\mathbf{b} + g\,\mathbf{c}$, then

(a) $\mathbf{a} = \frac{1}{[\mathbf{a\,b\,c}]}[(\mathbf{a}\,.\,\mathbf{a})(\mathbf{b}\times\mathbf{c}) + (\mathbf{b}\,.\,\mathbf{b})(\mathbf{c}\times\mathbf{a}) + (\mathbf{c}\,.\,\mathbf{c})\,(\mathbf{a}\times\mathbf{b})]$

(b) $\mathbf{a} = \frac{1}{[\mathbf{a\,b\,c}]}[(\mathbf{a}\,.\,\mathbf{a})(\mathbf{b}\times\mathbf{c}) + (\mathbf{b}\,.\,\mathbf{a})(\mathbf{c}\times\mathbf{a}) + (\mathbf{a}\,.\,\mathbf{a})\,(\mathbf{a}\times\mathbf{b})]$

(c) $\mathbf{a} = (\mathbf{a}\,.\,\mathbf{b})(\mathbf{b}\times\mathbf{c}) + (\mathbf{a}\,.\,\mathbf{b})(\mathbf{c}\times\mathbf{a}) + (\mathbf{c}\,.\,\mathbf{a})\,(\mathbf{a}\times\mathbf{b})$

(d) None of the above

134. The value of $\begin{vmatrix} \mathbf{a} & \mathbf{b} & \mathbf{c} \\ \mathbf{a}\,.\,\mathbf{p} & \mathbf{b}\,.\,\mathbf{p} & \mathbf{c}\,.\,\mathbf{p} \\ \mathbf{a}\,.\,\mathbf{q} & \mathbf{b}\,.\,\mathbf{q} & \mathbf{c}\,.\,\mathbf{q} \end{vmatrix}$ is

(a) $(\mathbf{p}\times\mathbf{q})\,[\mathbf{a}\times\mathbf{b}\ \mathbf{b}\times\mathbf{c}\ \mathbf{c}\times\mathbf{a}]$

(b) $2(\mathbf{p}\times\mathbf{q})\,[\mathbf{a}\times\mathbf{b}\ \mathbf{b}\times\mathbf{c}\ \mathbf{c}\times\mathbf{a}]$

(c) $4(\mathbf{p}\times\mathbf{q})[\mathbf{a}\times\mathbf{b}\ \mathbf{b}\times\mathbf{c}\ \mathbf{c}\times\mathbf{a}]$

(d) $(\mathbf{p}\times\mathbf{q})\,\sqrt{|[\mathbf{a}\times\mathbf{b}\ \mathbf{b}\times\mathbf{c}\ \mathbf{c}\times\mathbf{a}]|}$

Passage V

(Q. Nos. 135-136)

Let $g(x) = \int_0^x (3t^2 + 2t + 9)\,dt$ and $f(x)$ be a decreasing function, $\forall\, x \geq 0$ such that $\mathbf{AB} = f(x)\hat{\mathbf{i}} + g(x)\hat{\mathbf{j}}$ and $\mathbf{AC} = g(x)\hat{\mathbf{i}} + f(x)\hat{\mathbf{j}}$ are the two smallest sides of a ΔABC whose circumcentre lies outside the triangle, $\forall\, x > 0$.

135. Which of the following is true (for $x > 0$)?

(a) $f(x) > 0,\ g(x) < 0$

(b) $f(x) < 0,\ g(x) < 0$

(c) $f(x) > 0,\ g(x) > 0$

(d) $f(x) < 0, g(x) > 0$

136. $\lim_{t\to 0}\lim_{x\to\infty}\left(\cot\left(\frac{\pi}{4}(1-t^2)\right)\right)^{f(x)\cdot g(x)}$ is equal to

(a) 0 (b) 1

(c) e (d) does not exist

Passage VI

(Q. Nos 137-139)

Let **x**, **y**, **z** be the vector, such that $|\mathbf{x}| = |\mathbf{y}| = |\mathbf{z}| = \sqrt{2}$ and **x**, **y**, **z** make angles of 60° with each other also,

$$\mathbf{x}\times(\mathbf{y}\times\mathbf{z}) = \mathbf{a}$$

$$\mathbf{y}\times(\mathbf{z}\times\mathbf{x}) = \mathbf{b} \text{ and } \mathbf{x}\times\mathbf{y} = \mathbf{c}, \text{ then}$$

137. The value of **x** is

(a) $(\mathbf{a}+\mathbf{b})\times\mathbf{c} - (\mathbf{a}+\mathbf{b})$

(b) $(\mathbf{a}+\mathbf{b}) - (\mathbf{a}+\mathbf{b})\times\mathbf{c}$

(c) $\frac{1}{2}\{(\mathbf{a}+\mathbf{b})\times\mathbf{c} - (\mathbf{a}+\mathbf{b})\}$

(d) None of the above

138. The value of **y** is

(a) $\frac{1}{2}[(\mathbf{a}+\mathbf{b}) + (\mathbf{a}+\mathbf{b})\times\mathbf{c}]$ (b) $2[(\mathbf{a}+\mathbf{b}) + (\mathbf{a}+\mathbf{b})\times\mathbf{c}]$

(c) $4[(\mathbf{a}+\mathbf{b}) + (\mathbf{a}+\mathbf{b})\times\mathbf{c}]$ (d) None of these

139. The value of **z** is

(a) $\frac{1}{2}[(\mathbf{b}-\mathbf{a})\times\mathbf{c} + (\mathbf{a}+\mathbf{b})]$

(b) $\frac{1}{2}[(\mathbf{b}-\mathbf{a}) + (\mathbf{a}+\mathbf{b})\times\mathbf{c}]$

(c) $(\mathbf{b}-\mathbf{a})\times\mathbf{c} + (\mathbf{a}+\mathbf{b})$

(d) None of the above

Passage VII
(Q. Nos. 140-142)

a, **b**, **c** are non-zero unit vectors inclined pairwise with the same angle θ. p, q, r are non-zero scalars satisfying $\mathbf{a} \times \mathbf{b} + \mathbf{b} \times \mathbf{c} = p\mathbf{a} + q\mathbf{b} + r\mathbf{c}$. Now, answer the following questions.

140. Volume of parallelopiped with edges **a**, **b** and **c** is equal to

(a) $p + (q + r)\cos\theta$ (b) $(p + q + r)\cos\theta$
(c) $2p - (r + q)\cos\theta$ (d) None of these

141. $\frac{q}{p} + 2\cos\theta$ is equal to

(a) 1 (b) $2\,[\mathbf{a}\ \mathbf{b}\ \mathbf{c}]$
(c) 0 (d) None of these

142. $|(p + q)\cos\theta + r|$ is equal to

(a) $(1 + \cos\theta)\sqrt{1 - 2\cos\theta}$
(b) $2\sin^2\frac{\theta}{2}|\sqrt{1 + 2\cos\theta}|$
(c) $(1 - \sin\theta)|\sqrt{1 + 2\cos\theta}|$
(d) None of the above

Product of Vectors Exercise 5 : Matching Type Questions

143. Given two vectors $\mathbf{a} = \hat{\mathbf{i}} + 2\hat{\mathbf{j}} + 2\hat{\mathbf{k}}$ and $\mathbf{b} = \hat{\mathbf{i}} + \hat{\mathbf{j}} + 2\hat{\mathbf{k}}$. Match the Column I with Column II and mark the correct option from the codes given below.

	Column I		Column II
A.	A vector coplanar with **a** and **b**	p.	$-3\hat{\mathbf{i}} + 3\hat{\mathbf{j}} + 4\hat{\mathbf{k}}$
B.	A vector which is perpendicular to both **a** and **b**	q.	$2\hat{\mathbf{i}} - 2\hat{\mathbf{j}} + 3\hat{\mathbf{k}}$
C.	A vector which is equally inclined to **a** and **b**	r.	$\hat{\mathbf{i}} + \hat{\mathbf{j}}$
D.	A vector which forms a triangle with **a** and **b**	s.	$\hat{\mathbf{i}} - \hat{\mathbf{j}} + 5\hat{\mathbf{k}}$

144. Volume of parallelepiped formed by vectors $\mathbf{a} \times \mathbf{b}$, $\mathbf{b} \times \mathbf{c}$ and $\mathbf{c} \times \mathbf{a}$ is 36 sq units.

	Column I		Column II
A.	Volume of parallelopiped formed by vectors **a**, **b** and **c** is	p.	0 sq units
B.	Volume of tetrahedron formed by vectors **a**, **b** and **c** is	q.	12 sq units
C.	Volume of parallelopiped formed by vectors **a** + **b**, **b** + **c** and **c** + **a** is	r.	6 sq units
D.	Volume of parallelopiped formed by vectors **a** − **b**, **b** − **c** and **c** − **a** is	s.	1 sq units

145. Match the statement of Column I with values of Column II.

	Column I		Column II
A.	Let O be an interior point of ΔABC such that $\mathbf{OA} + 2\mathbf{OB} + 3\mathbf{OC} = 0$, then the ratio of the area of ΔABC to the area of ΔAOC, with O as the origin	p.	0
B.	$\mathbf{A}\cdot\mathbf{B} = \mathbf{A}\cdot\mathbf{C} = 0$, $\mathbf{B}\cdot\mathbf{C} = 3/2$ $\mathbf{A}\cdot\mathbf{A} = \mathbf{B}\cdot\mathbf{B} = \mathbf{C}\cdot\mathbf{C} = 1$, $[\mathbf{A}\ \mathbf{B}\ \mathbf{C}] =$	q.	$\frac{1}{2}$
C.	If **a**, **b**, **c** and **d** are non-zero vectors such that no three of them are in the same plane and no two are orthogonal, then the value of the scalar $\frac{(\mathbf{b}\times\mathbf{c})\cdot(\mathbf{a}\times\mathbf{d}) + (\mathbf{c}\times\mathbf{a})\cdot(\mathbf{b}\times\mathbf{d})}{(\mathbf{a}\times\mathbf{b})\cdot(\mathbf{d}\times\mathbf{c})}$ is	r.	2

146. Match the statement of Column I with values of Column II.

	Column I		Column II
A.	$|\mathbf{a}| = |\mathbf{b}| = 2$, $\mathbf{x} = \mathbf{a} + \mathbf{b}$, $\mathbf{y} = \mathbf{a} - \mathbf{b}$ If $|\mathbf{x} \times \mathbf{y}| = 2\{\lambda - (\mathbf{a}\cdot\mathbf{b})^2\}^{1/2}$, then value of λ is	p.	4
B.	The non-zero value of λ for which angle between $\lambda\hat{\mathbf{i}} + \hat{\mathbf{j}} + \hat{\mathbf{k}}$ and $\hat{\mathbf{i}} + \lambda\hat{\mathbf{j}} + \hat{\mathbf{k}}$ is $\frac{\pi}{3}$	q.	42

	Column I		Column II
C.	The non-zero value of k for which the lines $kx - 4y + 7z + 16 = 0 = 4x + 3y - 2z + 3$ and $x - 3y + 4z + 6 = 0 = x - y + z + 1$ are coplanar is	r.	16
D.	If $\|\mathbf{a}\| = \|\mathbf{b}\| = 1$ and $\|\mathbf{c}\| = 2$, then maximum value of $\|\mathbf{a} - 2\mathbf{b}\|^2 + \|\mathbf{b} - 2\mathbf{c}\|^2 + \|\mathbf{c} - 2\mathbf{a}\|^2$ is	s.	7
		t.	5

147. Match the statement of Column I with values of Column II.

	Column I		Column II
A.	Let $\mathbf{a}$, $\mathbf{b}$, $\mathbf{c}$ be the three vectors such that $\mathbf{a} \cdot (\mathbf{b} + \mathbf{c}) = \mathbf{b} \cdot (\mathbf{c} + \mathbf{a}) = \mathbf{c} \cdot (\mathbf{a} + \mathbf{b}) = 0$ and $\|\mathbf{a}\| = 1$, $\|\mathbf{b}\| = 4$, $\|\mathbf{c}\| = 8$, then $\|\mathbf{a} + \mathbf{b} + \mathbf{c}\|$ is equal to	p.	$\frac{5}{2}$
B.	If $\mathbf{a}_1, \mathbf{a}_2, \mathbf{a}_3$ are vectors reciprocal to the non-coplanar vectors $\mathbf{b}_1, \mathbf{b}_2, \mathbf{b}_3$ then $[\mathbf{a}_1\ \mathbf{a}_2\ \mathbf{a}_3][\mathbf{b}_1\ \mathbf{b}_2\ \mathbf{b}_3]$ is equal to	q.	9
C.	$ABCD$ is a quadrilateral with $AB = \mathbf{a}$, $AD = \mathbf{b}$ and $AC = 2\mathbf{a} + 3\mathbf{b}$. If its area is α times the area of the parallelogram with AB, AD as its adjacent sides, then α is equal to	r.	8
D.	If $\mathbf{d} = x(\mathbf{a} \times \mathbf{b}) + y(\mathbf{b} \times \mathbf{c}) + z(\mathbf{c} \times \mathbf{a})$ and $[\mathbf{a}\ \mathbf{b}\ \mathbf{c}] = \frac{1}{8}$, then $x + y + z = R(\mathbf{a} + \mathbf{b} + \mathbf{c}) \cdot \mathbf{d}$, where R = adjacent sides, then α is equal to	s.	1

Product of Vectors Exercise 6 : Integer Type Questions

148. Let $\hat{\mathbf{u}}$, $\hat{\mathbf{v}}$ and $\hat{\mathbf{w}}$ are three unit vectors, the angle between $\hat{\mathbf{u}}$ and $\hat{\mathbf{v}}$ is twice that of the angle between $\hat{\mathbf{u}}$ and $\hat{\mathbf{w}}$ and $\hat{\mathbf{v}}$ and $\hat{\mathbf{w}}$, then $[\hat{\mathbf{u}}\ \hat{\mathbf{v}}\ \hat{\mathbf{w}}]$ is equal. to

149. If $\mathbf{a}$, $\mathbf{b}$ and $\mathbf{c}$ are three vectors such that $[\mathbf{a}\ \mathbf{b}\ \mathbf{c}] = 1$, then find the value of $[\mathbf{a} + \mathbf{b}\ \mathbf{b} + \mathbf{c}\ \mathbf{c} + \mathbf{a}] + [\mathbf{a} \times \mathbf{b}\ \mathbf{b} \times \mathbf{c}\ \mathbf{c} \times \mathbf{a}] + [\mathbf{a} + (\mathbf{b} \times \mathbf{c})\ \mathbf{b} \times (\mathbf{c} \times \mathbf{a})\ \mathbf{c} \times (\mathbf{a} \times \mathbf{b})]$.

150. If $\hat{\mathbf{a}}$, $\hat{\mathbf{b}}$ and $\hat{\mathbf{c}}$ are the three unit vectors and α, β and γ are scalars such that $\hat{\mathbf{c}} = \alpha\hat{\mathbf{a}} + \beta\hat{\mathbf{b}} + \gamma(\hat{\mathbf{a}} \times \hat{\mathbf{b}})$. If is given that $\hat{\mathbf{a}} \cdot \hat{\mathbf{b}} = 0$ and $\hat{\mathbf{c}}$ makes equal angle with both $\hat{\mathbf{a}}$ and $\hat{\mathbf{b}}$, then evaluate $\alpha^2 + \beta^2 + \gamma^2$.

151. The three vectors $\hat{\mathbf{i}} + \hat{\mathbf{j}}$, $\hat{\mathbf{j}} + \hat{\mathbf{k}}$ and $\hat{\mathbf{k}} + \hat{\mathbf{i}}$ taken two at a time form three planes. If V be the volume of the tetrahedron having adjacent sides as the three unit vectors drawn perpendicular to those three planes, then find the value of $9\sqrt{3}V$.

152. Let $\hat{\mathbf{c}}$ be a unit vector coplanar with $\mathbf{a} = \hat{\mathbf{i}} - \hat{\mathbf{j}} + 2\hat{\mathbf{k}}$ and $\mathbf{b} = 2\hat{\mathbf{i}} - \hat{\mathbf{j}} + \hat{\mathbf{k}}$ such that $\hat{\mathbf{c}}$ is perpendicular to $\mathbf{a}$. If P be the projection of $\hat{\mathbf{c}}$ along $\mathbf{b}$, where $P = \frac{\sqrt{11}}{k}$ then find k.

153. Let $\mathbf{a}$, $\mathbf{b}$ and $\mathbf{c}$ are three vectors having magnitudes 1, 2 and 3, respectively satisfy the relation $[\mathbf{a}\ \mathbf{b}\ \mathbf{c}] = 6$. If $\hat{\mathbf{d}}$ is a unit vector coplanar with $\mathbf{b}$ and $\mathbf{c}$ such that $\mathbf{b} \cdot \hat{\mathbf{d}} = 1$, then evaluate $|(\mathbf{a} \times \mathbf{c}) \cdot \hat{\mathbf{d}}|^2 + |(\mathbf{a} \times \mathbf{c}) \times \hat{\mathbf{d}}|^2$.

154. Let $A(2\hat{\mathbf{i}} + 3\hat{\mathbf{j}} + 5\hat{\mathbf{k}})$, $B(-\hat{\mathbf{i}} + 3\hat{\mathbf{j}} + 2\hat{\mathbf{k}})$ and $C(\lambda\hat{\mathbf{i}} + 5\hat{\mathbf{j}} + \mu\hat{\mathbf{k}})$ are vertices of a triangle and its median through A is equally inclined to the positive directions of the axes. The value of $2\lambda - \mu$ is equal to

155. If V is the volume of the parallelopiped having three coterminous edges as $\mathbf{a}$, $\mathbf{b}$ and $\mathbf{c}$ are the volume of the parallelopiped having three coterminous edges as
$\alpha = (\mathbf{a} \cdot \mathbf{a})\mathbf{a} + (\mathbf{a} \cdot \mathbf{b})\mathbf{b} + (\mathbf{a} \cdot \mathbf{c})\mathbf{c}$,
$\beta = (\mathbf{b} \cdot \mathbf{a})\mathbf{a} + (\mathbf{b} \cdot \mathbf{b})\mathbf{b} + (\mathbf{b} \cdot \mathbf{c})\mathbf{c}$ and
$\gamma = (\mathbf{c} \cdot \mathbf{a})\mathbf{a} + (\mathbf{c} \cdot \mathbf{b})\mathbf{b} + (\mathbf{c} \cdot \mathbf{c})\mathbf{c}$ is V^λ, then $\lambda =$

156. If $\mathbf{a}$, $\mathbf{b}$ are vectors perpendicular to each other and $|\mathbf{a}| = 2$, $|\mathbf{b}| = 3$, $\mathbf{c} \times \mathbf{a} = \mathbf{b}$, then the least value of $2|\mathbf{c} - \mathbf{a}|$ is

157. If M and N are the mid-point of the diagonals AC and BD, respectively of a quadrilateral $ABCD$, then $AB + AD + CB + CD = kMN$, where $k =$ ……

158. If $\mathbf{a} \times \mathbf{b} = \mathbf{c}$, $\mathbf{b} \times \mathbf{c} = \mathbf{a}$, $\mathbf{c} \times \mathbf{a} = \mathbf{b}$. If vectors $\mathbf{a}$, $\mathbf{b}$ and $\mathbf{c}$ are forming a right handed system, then the volume of tetrahedron formed by vectors $3\mathbf{a} - 2\mathbf{b} + 2\mathbf{c}$, $-a - 2\mathbf{c}$ and $2\mathbf{a} - 3\mathbf{b} + 4\mathbf{c}$ is

159. Let $\mathbf{a}$ and $\mathbf{c}$ be unit vectors inclined at $\frac{\pi}{3}$ with each other. If $(\mathbf{a} \times (\mathbf{b} \times \mathbf{c})) \cdot (\mathbf{a} \times \mathbf{c}) = 5$, then $-[\mathbf{abc}] - 1 =$

160. Volume of parallelopiped formed by vectors $\mathbf{a} \times \mathbf{b}$, $\mathbf{b} \times \mathbf{c}$ and $\mathbf{c} \times \mathbf{a}$ is 36 sq units, then the volume of the parallelopiped formed by the vectors.

161. If α and β are two perpendicular unit vectors such that $\mathbf{x} = \hat{\beta} - (\alpha \times \mathbf{x})$; then the value of $4|\mathbf{x}|^2$ is.

162. The volume of the tetrahedron whose vertices are the points with position vectors $\hat{\mathbf{i}} + \hat{\mathbf{j}} + \hat{\mathbf{k}}$, $-\hat{\mathbf{i}} - 3\hat{\mathbf{j}} + 7\hat{\mathbf{k}}$, $\hat{\mathbf{i}} + 2\hat{\mathbf{j}} - 7\hat{\mathbf{k}}$ and $3\hat{\mathbf{i}} - 4\hat{\mathbf{j}} + \lambda\hat{\mathbf{k}}$ is 22, then the digit at unit place of λ is.

163. Volume of a tetrahedron whose coterminous edges are **a**, **b** and **c** is 3 and volume of a parallelopiped whose coterminous edges are $\mathbf{a} + \mathbf{b} - \mathbf{c}$, $\mathbf{a} - \mathbf{b}$, $\mathbf{b} - \mathbf{c}$ is V. Then, units digit of V is.

Product of Vectors Exercise 7 :
Subjective Type Questions

164. Prove Cauchy-Schwartz inequality $(\mathbf{a} \cdot \mathbf{b})^2 \le |\mathbf{a}|^2 \cdot |\mathbf{b}|^2$

165. Two points P and Q are given in the rectangular cartesian coordinates in the curve $y = 2^{x+2}$, such that $\mathbf{OP} \cdot \hat{\mathbf{i}} = -1$ and $\mathbf{OQ} \cdot \hat{\mathbf{i}} = 2$, where $\hat{\mathbf{i}}$ is a unit vector along the X-axis. Find the magnitude of $\mathbf{OQ} - 4\mathbf{OP}$.

166. O is the origin and A is a fixed point on the circle of radius **a** with centre O. The vector **OA** is denoted by **a**. A variable point P lie on the tangent at A and $\mathbf{OP} = \mathbf{r}$. Show that $\mathbf{a} \cdot \mathbf{r} = a^2$. Hence, if $P(x, y)$ and $A(x_1, y_1)$ deduce the equation of tangent at A to this circle.

167. If a is real constant and A, B and C are variable angles and $\sqrt{a^2 - 4}\tan A + a\tan B + \sqrt{a^2 + 4}\tan C = 6a$, then find the least value of $\tan^2 A + \tan^2 B + \tan^2 C$.

168. Given, the angles A, B and C of ΔABC. Find $\cos \angle BAM$, where M is mid-point of BC.

169. Find the perpendicular distance of $A(1, 4, -2)$ from the segment BC, where $B(2, 1, -2)$ and $C(0, -5, 1)$.

170. Given angles, A, B and C of ΔABC. Let M be the mid-point of segment **AB** and let D be the foot of the bisector of $\angle C$. Find the ratio of $\dfrac{\text{Area of } \Delta CDM}{\text{Area of } \Delta ABC}$ and also $\cos \phi = \cos \angle DCM$.

171. In the ΔABC a point P is taken on the side AB such that $AP : BP = 1 : 2$ and a point Q is taken on the side BC such that $CQ : BQ = 2 : 1$. If R be the point of intersection of lines AQ and CP, using vector find the area of ΔABC, if it is known that area of ΔABC is one unit.

172. If one diagonal of a quadrilateral bisects the other, then it also bisects the quadrilateral.

173. Two forces $F_1 = \{2, 3\}$ and $F_2 = \{4, 1\}$ are specified relative to a general cartesian form. Their points of application are respectively, $A = (1, 1)$ and $B = (2, 4)$. Find the coordinates of the resultant and the equation of the straight line l containing it.

174. A non-zero vector **a** is parallel to the line of intersection of the plane determined by the vectors, $\hat{\mathbf{i}}$, $\hat{\mathbf{i}} + \hat{\mathbf{j}}$ and the plane determined by the vectors $\hat{\mathbf{i}} - \hat{\mathbf{j}}$ and $\hat{\mathbf{i}} + \hat{\mathbf{k}}$. Find the angle between **a** and $\hat{\mathbf{i}} - 2\hat{\mathbf{j}} + 2\hat{\mathbf{k}}$

175. The vector $\hat{\mathbf{i}} + 2\hat{\mathbf{j}} + 2\hat{\mathbf{k}}$ turns through a right angle while passing through the positive X-axis on the way. Find the vector in its new position.

176. Let $\hat{\mathbf{u}}$ and $\hat{\mathbf{v}}$ are unit vectors and **w** is a vector such that $\hat{\mathbf{u}} \times \hat{\mathbf{v}} + \hat{\mathbf{u}} = \mathbf{w}$ and $\mathbf{w} \times \hat{\mathbf{u}} = \hat{\mathbf{v}}$, then find the value of $[\hat{\mathbf{u}}\ \hat{\mathbf{v}}\ \mathbf{w}]$.

177. **A**, **B** and **C** are three vectors given by $2\hat{\mathbf{i}} + \hat{\mathbf{k}}$, $\hat{\mathbf{i}} + \hat{\mathbf{j}} + \hat{\mathbf{k}}$ and $4\hat{\mathbf{i}} - 3\hat{\mathbf{j}} + 7\hat{\mathbf{k}}$ Then, find **R**, which satisfies the relation $\mathbf{R} \times \mathbf{B} = \mathbf{C} \times \mathbf{B}$ and $\mathbf{R} \cdot \mathbf{A} = 0$.

178. If $\mathbf{x} \cdot \mathbf{a} = 0$, $\mathbf{x} \cdot \mathbf{b} = 1$, $[\mathbf{x}\ \mathbf{a}\ \mathbf{b}] = 1$ and $\mathbf{a} \cdot \mathbf{b} \ne 0$, then find **x** in terms of **a** and **b**.

179. Let $\hat{\mathbf{x}}$, $\hat{\mathbf{y}}$ and z be unit vectors such that $\hat{\mathbf{x}} + \hat{\mathbf{y}} + \hat{\mathbf{z}} = \mathbf{a}$, $\hat{\mathbf{x}} \times (\hat{\mathbf{y}} \times \hat{\mathbf{z}}) = \mathbf{b}$, $(\hat{\mathbf{x}} \times \hat{\mathbf{y}}) \times \hat{\mathbf{z}} = \mathbf{c}$, $\mathbf{a} \cdot \hat{\mathbf{x}} = \dfrac{3}{2}$, $\mathbf{a} \cdot \hat{\mathbf{y}} = \dfrac{7}{4}$ and $|\mathbf{a}| = 2$. Find **x**, **y** and z in terms of **a**, **b** and **c**.

180. Let **a**, **b** and **c** be three mutually perpendicular vectors of equal magnitude. If the vector **x** satisfies the equation.
$$\mathbf{a} \times \{(\mathbf{x} - \mathbf{b}) \times \mathbf{a}\} + \mathbf{b} \times \{(\mathbf{x} - \mathbf{c}) \times \mathbf{b}\} + \mathbf{c} \times \{(\mathbf{x} - \mathbf{a}) \times \mathbf{c}\} = 0$$
then find **x**.

181. Given vectors $\mathbf{CB} = \mathbf{a}$, $\mathbf{CA} = \mathbf{b}$ and $\mathbf{CO} = \mathbf{x}$, where O is the centre of circle circumscribed about ΔABC, then find vector **x**.

Product of Vectors Exercise 8 : Questions Asked in Previous Years' Exam

(i) JEE Advanced & IIT-JEE

182. Let O be the origin and let PQR be an arbitrary triangle. The point S is such that

$$\mathbf{OP}\cdot\mathbf{OQ}+\mathbf{OR}\cdot\mathbf{OS}=\mathbf{OR}\cdot\mathbf{OP}+\mathbf{OQ}\cdot\mathbf{OS}=\mathbf{OQ}\cdot\mathbf{OR}+\mathbf{OP}\cdot\mathbf{OS}$$

Then the triangle PQR has S as its **[Single Correct Type, 2017 Adv.]**

(a) centroid (b) orthocentre
(c) incentre (d) circumcentre

Passage
(Q. Nos. 183-184)

Let O be the origin and OX, OY, OZ be three unit vectors in the directions of the sides, QR, RP, PQ respectively of a ΔPQR. **[Passage Type Question, 2017 Adv.]**

183. If the triangle PQR varies, then the minimum value of $\cos(P+Q)+\cos(Q+R)+\cos(R+P)$ is

(a) $-\frac{3}{2}$ (b) $\frac{3}{2}$ (c) $\frac{5}{3}$ (d) $-\frac{5}{3}$

184. $|\mathbf{OX}\times\mathbf{OY}|=$

(a) $\sin(P+Q)$ (b) $\sin(P+R)$
(c) $\sin(Q+R)$ (d) $\sin 2R$

185. Let $\mathbf{a}$, $\mathbf{b}$ and $\mathbf{c}$ be three unit vectors such that $\mathbf{a}\times(\mathbf{b}\times\mathbf{c})=\frac{\sqrt{3}}{2}(\mathbf{b}+\mathbf{c})$. If $\mathbf{b}$ is not parallel to $\mathbf{c}$, then the angle between $\mathbf{a}$ and $\mathbf{b}$ is **[Single Correct Type, 2016 Adv.]**

(a) $\frac{3\pi}{4}$ (b) $\frac{\pi}{2}$ (c) $\frac{2\pi}{3}$ (d) $\frac{5\pi}{6}$

186. Let $\mathbf{a}$, $\mathbf{b}$ and $\mathbf{c}$ be three non-zero vectors such that no two of them are collinear and $(\mathbf{a}\times\mathbf{b})\times\mathbf{c}=\frac{1}{3}|\mathbf{b}||\mathbf{c}|\mathbf{a}$. If θ is the angle between vectors $\mathbf{b}$ and $\mathbf{c}$, then a value of $\sin\theta$ is **[Single Correct Type, 2015 Adv.]**

(a) $\frac{2\sqrt{2}}{3}$ (b) $\frac{-\sqrt{2}}{3}$ (c) $\frac{2}{3}$ (d) $\frac{-2\sqrt{3}}{3}$

187. If $\mathbf{a}$, $\mathbf{b}$ and $\mathbf{c}$ are unit vectors satisfying $|\mathbf{a}-\mathbf{b}|^2+|\mathbf{b}-\mathbf{c}|^2+|\mathbf{c}-\mathbf{a}|^2=9$, then $|2\mathbf{a}+5\mathbf{b}+5\mathbf{c}|$ is equal to **[Subjective Type Question, 2012]**

188. The vector(s) which is/are coplanar with vectors $\hat{\mathbf{i}}+\hat{\mathbf{j}}+2\hat{\mathbf{k}}$ and $\hat{\mathbf{i}}+2\hat{\mathbf{j}}+\hat{\mathbf{k}}$, are perpendicular to the vector $\hat{\mathbf{i}}+\hat{\mathbf{j}}+\hat{\mathbf{k}}$ is/are **[More than One Option Correct Type, 2011]**

(a) $\hat{\mathbf{j}}-\hat{\mathbf{k}}$ (b) $-\hat{\mathbf{i}}+\hat{\mathbf{j}}$
(c) $\hat{\mathbf{i}}-\hat{\mathbf{j}}$ (d) $-\hat{\mathbf{j}}+\hat{\mathbf{k}}$

189. Let $\mathbf{a}=\hat{\mathbf{i}}+\hat{\mathbf{j}}+\hat{\mathbf{k}}$, $\mathbf{b}=\hat{\mathbf{i}}-\hat{\mathbf{j}}+\hat{\mathbf{k}}$ and $\mathbf{c}=\hat{\mathbf{i}}-\hat{\mathbf{j}}-\hat{\mathbf{k}}$ be three vectors. A vector $\mathbf{v}$ in the plane of $\mathbf{a}$ and $\mathbf{b}$, whose projection on $\mathbf{c}$ is $\frac{1}{\sqrt{3}}$, is given by **[Single Correct Type, 2011 Adv.]**

(a) $\hat{\mathbf{i}}-3\hat{\mathbf{j}}+3\hat{\mathbf{k}}$ (b) $-3\hat{\mathbf{i}}-3\hat{\mathbf{j}}-\hat{\mathbf{k}}$
(c) $3\hat{\mathbf{i}}-\hat{\mathbf{j}}+3\hat{\mathbf{k}}$ (d) $\hat{\mathbf{i}}+3\hat{\mathbf{j}}-3\hat{\mathbf{k}}$

190. Two adjacent sides of a parallelogram $ABCD$ are given by $\mathbf{AB}=2\hat{\mathbf{i}}+10\hat{\mathbf{j}}+11\hat{\mathbf{k}}$ and $\mathbf{AD}=-\hat{\mathbf{i}}+2\hat{\mathbf{j}}+2\hat{\mathbf{k}}$. The side AD is rotated by an acute angle α in the plane of the parallelogram so that AD becomes AD'. If AD' makes a right angle with the side AB, then the cosine of the angle α is given by **[Single Correct Type, 2010 Adv.]**

(a) $\frac{8}{9}$ (b) $\frac{\sqrt{17}}{9}$
(c) $\frac{1}{9}$ (d) $\frac{4\sqrt{5}}{9}$

191. Let P, Q, R and S be the points on the plane with position vectors $-2\hat{\mathbf{i}}-\hat{\mathbf{j}}$, $4\hat{\mathbf{i}}$, $3\hat{\mathbf{i}}+3\hat{\mathbf{j}}$ and $-3\hat{\mathbf{i}}+2\hat{\mathbf{j}}$, respectively. The quadrilateral $PQRS$ must be a **[Single Correct Type, IITJEE 2010]**

(a) parallelogram, which is neither a rhombus nor a rectangle
(b) square
(c) rectangle, but not a square
(d) rhombus, but not a square

192. If $\mathbf{a}$ and $\mathbf{b}$ are vectors in space given by $\mathbf{a}=\frac{\hat{\mathbf{i}}-2\hat{\mathbf{j}}}{\sqrt{5}}$ and $\mathbf{b}=\frac{2\hat{\mathbf{i}}+\hat{\mathbf{j}}+3\hat{\mathbf{k}}}{\sqrt{14}}$, then the value of $(2\mathbf{a}+\mathbf{b})\cdot[(\mathbf{a}\times\mathbf{b})\times(\mathbf{a}-2\mathbf{b})]$ is **[Integer Type Question, 2010]**

193. If $\mathbf{a}$, $\mathbf{b}$, $\mathbf{c}$ and $\mathbf{d}$ are the unit vectors such that $(\mathbf{a}\times\mathbf{b})\cdot(\mathbf{c}\times\mathbf{d})=1$ and $\mathbf{a}\cdot\mathbf{c}=\frac{1}{2}$, then **[More than One Option Correct Type, 2009]**

(a) $\mathbf{a}$, $\mathbf{b}$, $\mathbf{c}$ are non-coplanar
(b) $\mathbf{a}$, $\mathbf{b}$, $\mathbf{d}$ are non-coplanar
(c) $\mathbf{b}$, $\mathbf{d}$ are non-parallel
(d) $\mathbf{a}$, $\mathbf{d}$ are parallel and $\mathbf{b}$, $\mathbf{c}$ are parallel

194. The edges of a parallelopiped are of unit length and are parallel to non-coplanar unit vector $\hat{\mathbf{a}}$, $\hat{\mathbf{b}}$, $\hat{\mathbf{c}}$ such that $\hat{\mathbf{a}}\cdot\hat{\mathbf{b}}=\hat{\mathbf{b}}\cdot\hat{\mathbf{c}}=\hat{\mathbf{c}}\cdot\hat{\mathbf{a}}=\frac{1}{2}$. Then, the volume of the parallelopiped is **[Single Correct Type, IIT-JEE 2008]**

(a) $\frac{1}{\sqrt{2}}$ cu unit (b) $\frac{1}{2\sqrt{2}}$ cu unit
(c) $\frac{\sqrt{3}}{2}$ cu unit (d) $\frac{1}{\sqrt{3}}$ cu unit

195. Let two non-collinear unit vectors $\hat{\mathbf{a}}$ and $\hat{\mathbf{b}}$ form an acute angle. A point P moves, so that at any time t the position vector **OP** (where, O is the origin) is given by $\hat{\mathbf{a}}\cos t + \hat{\mathbf{b}}\sin t$. When P is farthest from origin O, let M be the length of **OP** and $\hat{\mathbf{u}}$ be the unit vector along **OP**. Then, **[Single Correct Type, IIT-JEE 2008]**

(a) $\hat{\mathbf{u}} = \frac{\hat{\mathbf{a}}+\hat{\mathbf{b}}}{|\hat{\mathbf{a}}+\hat{\mathbf{b}}|}$ and $M = (1+\hat{\mathbf{a}}\cdot\hat{\mathbf{b}})^{1/2}$
(b) $\hat{\mathbf{u}} = \frac{\hat{\mathbf{a}}-\hat{\mathbf{b}}}{|\hat{\mathbf{a}}-\hat{\mathbf{b}}|}$ and $M = (1+\hat{\mathbf{a}}\cdot\hat{\mathbf{b}})^{1/2}$
(c) $\hat{\mathbf{u}} = \frac{\hat{\mathbf{a}}+\hat{\mathbf{b}}}{|\hat{\mathbf{a}}+\hat{\mathbf{b}}|}$ and $M = (1+2\hat{\mathbf{a}}\cdot\hat{\mathbf{b}})^{1/2}$
(d) $\hat{\mathbf{u}} = \frac{\hat{\mathbf{a}}-\hat{\mathbf{b}}}{|\hat{\mathbf{a}}-\hat{\mathbf{b}}|}$ and $M = (1+2\hat{\mathbf{a}}\cdot\hat{\mathbf{b}})^{1/2}$

196. Let the vectors **PQ**, **QR**, **RS**, **ST**, **TU** and **UP** represent the sides of a regular hexagon.
Statement I $\mathbf{PQ}\times(\mathbf{RS}+\mathbf{ST}) \neq \mathbf{0}$. because
Statement II $\mathbf{PQ}\times\mathbf{RS} = \mathbf{0}$ and $\mathbf{PQ}\times\mathbf{ST} \neq \mathbf{0}$ **[Single Correct Type, 2007, 3M]**

(a) Statement I is true, Statement II is true and Statement II is a correct explanation for Statement I.
(b) Statement I is true, Statement II is true but Statement II is not a correct explanation for Statement I.
(c) Statement I is true, Statement II is false.
(d) Statement I is false, Statement II is true.

197. The number of distinct real values of λ, for which the vectors $-\lambda^2\hat{\mathbf{i}}+\hat{\mathbf{j}}+\hat{\mathbf{k}}$, $\hat{\mathbf{i}}-\lambda^2\hat{\mathbf{j}}+\hat{\mathbf{k}}$ and $\hat{\mathbf{i}}+\hat{\mathbf{j}}-\lambda^2\hat{\mathbf{k}}$ are coplanar, is **[Single Correct Type, IIT-JEE 2007]**

(a) 0 (b) 1
(c) $\pm\sqrt{2}$ (d) 3

198. Let **a**, **b**, **c** be unit vectors such that $\mathbf{a}+\mathbf{b}+\mathbf{c}=\mathbf{0}$. Which one of the following is correct? **[Single Correct Type, IIT-JEE 2007]**

(a) $\mathbf{a}\times\mathbf{b} = \mathbf{b}\times\mathbf{c} = \mathbf{c}\times\mathbf{a} = \mathbf{0}$
(b) $\mathbf{a}\times\mathbf{b} = \mathbf{b}\times\mathbf{c} = \mathbf{c}\times\mathbf{a} \neq \mathbf{0}$
(c) $\mathbf{b}\times\mathbf{b} = \mathbf{b}\times\mathbf{c} = \mathbf{a}\times\mathbf{c} = \mathbf{0}$
(d) $\mathbf{a}\times\mathbf{b}$, $\mathbf{b}\times\mathbf{c}$, $\mathbf{c}\times\mathbf{a}$ are mutually perpendicular

199. Let **A** be vector parallel to line of intersection of planes P_1 and P_2 through origin. P_1 is parallel to the vectors $2\hat{\mathbf{j}}+3\hat{\mathbf{k}}$ and $4\hat{\mathbf{j}}-3\hat{\mathbf{k}}$ and P_2 is parallel to $\hat{\mathbf{j}}-\hat{\mathbf{k}}$ and $3\hat{\mathbf{i}}+3\hat{\mathbf{j}}$, then the angle between vector **A** and $2\hat{\mathbf{i}}+\hat{\mathbf{j}}-2\hat{\mathbf{k}}$ is **[More than One Option Correct Type, 2006, 5M]**

(a) $\frac{\pi}{2}$ (b) $\frac{\pi}{4}$
(c) $\frac{\pi}{6}$ (d) $\frac{3\pi}{4}$

200. Let, $\mathbf{a}=\hat{\mathbf{i}}+2\hat{\mathbf{j}}+\hat{\mathbf{k}}$, $\mathbf{b}=\hat{\mathbf{i}}-\hat{\mathbf{j}}+\hat{\mathbf{k}}$, $\mathbf{c}=\hat{\mathbf{i}}+\hat{\mathbf{j}}-\hat{\mathbf{k}}$. A vector coplanar to **a** and **b** has a projection along **c** of magnitude $\frac{1}{\sqrt{3}}$, then the vector is **[Single Correct Type, IIT-JEE 2006]**

(a) $4\hat{\mathbf{i}}-\hat{\mathbf{j}}+4\hat{\mathbf{k}}$ (b) $4\hat{\mathbf{i}}+\hat{\mathbf{j}}-4\hat{\mathbf{k}}$
(c) $2\hat{\mathbf{i}}+\hat{\mathbf{j}}+\hat{\mathbf{k}}$ (d) None of these

201. If **a**, **b**, **c** are three non-zero, non-coplanar vectors and

$$\mathbf{b}_1 = \mathbf{b} - \frac{\mathbf{b}\cdot\mathbf{a}}{|\mathbf{a}|^2}\mathbf{a},\ \mathbf{b}_2 = \mathbf{b} + \frac{\mathbf{b}\cdot\mathbf{a}}{|\mathbf{a}|^2}\mathbf{a},$$

$$\mathbf{c}_1 = \mathbf{c} - \frac{\mathbf{c}\cdot\mathbf{a}}{|\mathbf{a}|^2}\mathbf{a} - \frac{\mathbf{c}\cdot\mathbf{b}}{|\mathbf{b}|^2}\mathbf{b},\ \mathbf{c}_2 = \mathbf{c} - \frac{\mathbf{c}\cdot\mathbf{a}}{|\mathbf{a}|^2}\mathbf{a} - \frac{\mathbf{c}\cdot\mathbf{b}_1}{|\mathbf{b}|^2}\mathbf{b}_1,$$

$$\mathbf{c}_3 = \mathbf{c} - \frac{\mathbf{c}\cdot\mathbf{a}}{|\mathbf{a}|^2}\mathbf{a} - \frac{\mathbf{a}\cdot\mathbf{b}_2}{|\mathbf{b}_2|^2}\mathbf{b}_2,\ \mathbf{c}_4 = \mathbf{a} - \frac{\mathbf{c}\cdot\mathbf{a}}{|\mathbf{a}|^2}\mathbf{a}.$$

Then, which of the following is a set of mutually orthogonal vectors? **[Single Correct Type, IIT-JEE 2005]**

(a) $\{\mathbf{a}, \mathbf{b}_1, \mathbf{c}_1\}$ (b) $\{\mathbf{a}, \mathbf{b}_1, \mathbf{c}_2\}$
(c) $\{\mathbf{a}, \mathbf{b}_2, \mathbf{a}_3\}$ (d) $\{\mathbf{a}, \mathbf{b}_2, \mathbf{c}_4\}$

202. The unit vector which is orthogonal to the vector $3\hat{\mathbf{i}}+2\hat{\mathbf{j}}+6\hat{\mathbf{k}}$ and is coplanar with the vectors $2\hat{\mathbf{i}}+\hat{\mathbf{j}}+\hat{\mathbf{k}}$ and $\hat{\mathbf{i}}-\hat{\mathbf{j}}+\hat{\mathbf{k}}$ is **[Single Correct Type, IIT-JEE 2004]**

(a) $\frac{2\hat{\mathbf{i}}-6\hat{\mathbf{j}}+\hat{\mathbf{k}}}{\sqrt{41}}$ (b) $\frac{2\hat{\mathbf{i}}-3\hat{\mathbf{j}}}{\sqrt{13}}$
(c) $\frac{3\hat{\mathbf{j}}-\hat{\mathbf{k}}}{\sqrt{10}}$ (d) $\frac{4\hat{\mathbf{i}}+3\hat{\mathbf{j}}-3\hat{\mathbf{k}}}{\sqrt{34}}$

203. The value of a, so that the volume of parallelopiped formed by $\hat{\mathbf{i}}+a\hat{\mathbf{j}}+\hat{\mathbf{k}}$, $\hat{\mathbf{j}}+a\hat{\mathbf{k}}$ and $a\hat{\mathbf{i}}+\hat{\mathbf{k}}$ become minimum, is **[Single Correct Type, IIT-JEE 2003]**

(a) -3 (b) 3
(c) $1/\sqrt{3}$ (d) $\sqrt{3}$

204. If $\mathbf{a} = (\hat{\mathbf{i}}+\hat{\mathbf{j}}+\hat{\mathbf{k}})$, $\mathbf{a}\cdot\mathbf{b}=1$ and $\mathbf{a}\times\mathbf{b} = \hat{\mathbf{j}}-\hat{\mathbf{k}}$, then **b** is equal to **[Single Correct Type, IIT-JEE 2003]**

(a) $\hat{\mathbf{i}}-\hat{\mathbf{j}}+\hat{\mathbf{k}}$ (b) $2\hat{\mathbf{j}}-\hat{\mathbf{k}}$
(c) $\hat{\mathbf{i}}$ (d) $2\hat{\mathbf{i}}$

205. If $\mathbf{V} = 2\hat{\mathbf{i}}+\hat{\mathbf{j}}-\hat{\mathbf{k}}$ and $\mathbf{W} = \hat{\mathbf{i}}+3\hat{\mathbf{k}}$. If **U** is a unit vector, then the maximum value of the scalar triple product [**U V W**] is **[Single Correct Type, IIT-JEE 2002]**

(a) -1 (b) $\sqrt{10}+\sqrt{6}$
(c) $\sqrt{59}$ (d) $\sqrt{60}$

206. If **a** and $\mathbf{b}_1$ are two unit vectors such that $\mathbf{a}+2\mathbf{b}$ and $5\mathbf{a}-4\mathbf{b}$, are perpendicular to each other, then the angle between **a** and **b** is **[Single Correct Type, 2002, 1M]**

(a) 45° (b) 60°
(c) $\cos^{-1}\left(\frac{1}{3}\right)$ (d) $\cos^{-1}\left(\frac{2}{7}\right)$

(ii) JEE Main & AIEEE

207. Let $\mathbf{a} = 2\hat{i} + \hat{j} - 2\hat{k}$, $\mathbf{b} = \hat{i} + \hat{j}$ and $\mathbf{c}$ be a vector such that $|\mathbf{c} - \mathbf{a}| = 3$, $|(\mathbf{a} \times \mathbf{b}) \times \mathbf{c}| = 3$ and the angle between $\mathbf{c}$ and $\mathbf{a} \times \mathbf{b}$ is 30°. Then, $\mathbf{a} \cdot \mathbf{c}$ is equal to **[JEE Main 2017]**

(a) $\frac{25}{8}$ (b) 2
(c) 5 (d) $\frac{1}{8}$

208. If $[\mathbf{a} \times \mathbf{b}\, \mathbf{b} \times \mathbf{c}\, \mathbf{c} \times \mathbf{a}] = \lambda\, [\mathbf{a}\ \mathbf{b}\ \mathbf{c}]^2$, then λ is equal to **[JEE Main 2014]**

(a) 0 (b) 1
(c) 2 (d) 3

209. Let $\mathbf{a}$ and $\mathbf{b}$ be two unit vectors. If the vectors $\mathbf{c} = \mathbf{a} + 2\mathbf{b}$ and $\mathbf{d} = 5\mathbf{a} - 4\mathbf{a}$ are perpendicular to each other, then the angle between $\mathbf{a}$ and $\mathbf{b}$ is **[AIEEE 2012]**

(a) $\frac{\pi}{6}$ (b) $\frac{\pi}{2}$
(c) $\frac{\pi}{3}$ (d) $\frac{\pi}{4}$

210. Let $ABCD$ be a parallelogram such that $\mathbf{AB} = \mathbf{q}$, $\mathbf{AD} = \mathbf{p}$ and $\angle BAD$ be an acute angle. If $\mathbf{r}$ is the vector that coincides with the altitude directed from the vertex B to the side AD, then $\mathbf{r}$ is given by **[AIEEE 2012]**

(a) $\mathbf{r} = 3\mathbf{q} + \frac{3(\mathbf{p} \cdot \mathbf{q})}{(\mathbf{p} \cdot \mathbf{p})}\mathbf{p}$ (b) $\mathbf{r} = -\mathbf{q} + \left(\frac{\mathbf{p} \cdot \mathbf{q}}{\mathbf{p} \cdot \mathbf{p}}\right)\mathbf{p}$
(c) $\mathbf{r} = \mathbf{q} - \left(\frac{\mathbf{p} \cdot \mathbf{q}}{\mathbf{p} \cdot \mathbf{p}}\right)\mathbf{p}$ (d) $\mathbf{r} = -3\mathbf{q} + \frac{3(\mathbf{p} \cdot \mathbf{q})}{(\mathbf{p} \cdot \mathbf{p})}\mathbf{p}$

211. If $\mathbf{a} = \frac{1}{\sqrt{10}}(3\hat{i} + \hat{k})$ and $\mathbf{b} = \frac{1}{7}(2\hat{i} + 3\hat{j} - 6\hat{k})$, then the value of $(2 - \mathbf{b}) \cdot [(\mathbf{a} \times \mathbf{b}) \times (\mathbf{a} + 2\mathbf{b})]$ is **[AIEEE 2011]**

(a) −3 (b) 5
(c) 3 (d) −5

212. The vectors $\mathbf{a}$ and $\mathbf{b}$ are not perpendicular and $\mathbf{c}$ and $\mathbf{d}$ are two vectors satisfying $\mathbf{b} \times \mathbf{c} = \mathbf{b} \times \mathbf{d}$ and $\mathbf{a} \cdot \mathbf{b} = 0$. Then, the vector $\mathbf{d}$ is equal to **[AIEEE 2011]**

(a) $\mathbf{c} + \left(\frac{\mathbf{a} \cdot \mathbf{c}}{\mathbf{a} \cdot \mathbf{b}}\right)\mathbf{b}$ (b) $\mathbf{b} + \left(\frac{\mathbf{b} \cdot \mathbf{c}}{\mathbf{a} \cdot \mathbf{b}}\right)\mathbf{c}$
(c) $\mathbf{c} - \left(\frac{\mathbf{a} \cdot \mathbf{c}}{\mathbf{a} \cdot \mathbf{b}}\right)\mathbf{b}$ (d) $\mathbf{b} - \left(\frac{\mathbf{b} \cdot \mathbf{c}}{\mathbf{a} \cdot \mathbf{b}}\right)\mathbf{c}$

213. If the vectors $p\hat{i} + \hat{j} + \hat{k}$, $\hat{i} + q\hat{j} + \hat{k}$ and $\hat{i} + \hat{j} + r\hat{k}$ (Where, $p \neq q \neq r \neq 1$) are coplanar, then the value of $pqr - (p + q + r)$ is **[AIEEE 2011]**

(a) − 2 (b) 2
(c) 0 (d) −1

214. Let $\mathbf{a} = \hat{j} - \hat{k}$ and $\mathbf{a} = \hat{i} - \hat{j} - \hat{k}$. Then, the vector $\mathbf{b}$ satisfying $\mathbf{a} \times \mathbf{b} + \mathbf{c} = 0$ and $\mathbf{a} \cdot \mathbf{b} = 3$, is **[AIEEE 2010]**

(a) $-\hat{i} + \hat{j} - 2\hat{k}$ (b) $2\hat{i} - \hat{j} + 2\hat{k}$
(c) $\hat{i} - \hat{j} - 2\hat{k}$ (d) $\hat{i} + \hat{j} - 2\hat{k}$

215. If the vectors $\mathbf{a} = \hat{i} - \hat{j} + 2\hat{k}$, $\mathbf{b} = 2\hat{i} + 4\hat{j} + \hat{k}$ and $\mathbf{c} = \lambda\hat{i} + \hat{j} + \mu\hat{k}$ are mutually orthogonal, then (λ, μ) is equal to **[AIEEE 2010]**

(a) (− 3, 2) (b) (2, − 3)
(c) (− 2, 3) (d) (3, − 2)

216. If $\mathbf{u}$, $\mathbf{v}$ and $\mathbf{w}$ are non-coplanar vectors and p, q are real numbers, then the equality $[3\mathbf{u}\ p\mathbf{v}\ p\mathbf{w}] - [p\mathbf{v}\ \mathbf{w}\ q\mathbf{u}] - [2\mathbf{w}\ q\mathbf{v}\ q\mathbf{u}] = 0$ holds for **[AIEEE 2009]**

(a) exactly two values of (p, q)
(b) more than two but not all values of (p, q)
(c) all values of (p, q)
(d) exactly one value of (p, q)

217. The vector $\mathbf{a} = \alpha\hat{i} + 2\hat{j} + \beta\hat{k}$ lies in the plane of the vectors $\mathbf{b} = \hat{i} + \hat{j}$ and $\mathbf{c} = \hat{j} + \hat{k}$ and bisects the angle between $\mathbf{b}$ and $\mathbf{c}$. Then, which one of the following gives possible values of α and β? **[AIEEE 2008]**

(a) $\alpha = 1, \beta = 1$ (b) $\alpha = 2, \beta = 2$
(c) $\alpha = 1, \beta = 2$ (d) $\alpha = 2, \beta = 1$

218. If $\mathbf{u}$ and $\mathbf{v}$ are unit vectors and θ is the acute angle between them, then $2\mathbf{u} \times 3\mathbf{v}$ is a unit vector for **[AIEEE 2007]**

(a) exactly two values of θ
(b) more than two values of θ
(c) no value of θ
(d) exactly one value of θ

219. Let $\mathbf{a} = \hat{i} + \hat{j} + \hat{k}$, $\mathbf{b} = \hat{i} - \hat{j} + 2\hat{k}$ and $\mathbf{c} = x\hat{i} + (x-2)\hat{j} - \hat{k}$. If the vector $\mathbf{c}$ lies in the plane of $\mathbf{a}$ and $\mathbf{b}$, then x equal to **[AIEEE 2007]**

(a) 0 (b) 1
(c) − 4 (d) −2

220. If $(\mathbf{a} \times \mathbf{b}) \times \mathbf{c} = \mathbf{a} \times (\mathbf{b} \times \mathbf{c})$, where $\mathbf{a}$, $\mathbf{b}$ and $\mathbf{c}$ are any three vectors such that $\mathbf{a} \cdot \mathbf{b} \neq 0$, $\mathbf{b} \cdot \mathbf{c} \neq 0$, then $\mathbf{a}$ and $\mathbf{c}$ are **[AIEEE 2006]**

(a) inclined at an angle of $\frac{\pi}{6}$ between them
(b) perpendicular
(c) parallel
(d) inclined at an angle of $\frac{\pi}{3}$ between them

221. The value of a, for which the points, A, B, C with position vectors $2\hat{i} - \hat{j} + \hat{k}$, $\hat{i} - 3\hat{j} - 5\hat{k}$ and $a\hat{i} - 3\hat{j} + \hat{k}$ respectively are the vertices of a right angled triangle with $C = \frac{\pi}{2}$ are **[AIEEE 2006]**

(a) −2 and −1 (b) −2 and 1
(c) 2 and −1 (d) 2 and 1

222. The distance between the line $\mathbf{r} = 2\hat{i} - 2\hat{j} + 3\hat{k} + \lambda(\hat{i} - \hat{j} + 4\hat{k})$ and the plane $\mathbf{r} \cdot (\hat{i} + 5\hat{j} + \hat{k}) = 5$ is **[AIEEE 2005]**

(a) $\frac{10}{3}$ (b) $\frac{3}{10}$ (c) $\frac{10}{3\sqrt{3}}$ (d) $\frac{10}{9}$

223. For any vector $\mathbf{a}$, the value of $(\mathbf{a} \times \hat{\mathbf{i}})^2 + (\mathbf{a} \times \hat{\mathbf{j}})^2 + (\mathbf{a} \times \hat{\mathbf{k}})^2$ is equal to **[AIEEE 2005]**

(a) $4\mathbf{a}^2$ (b) $2\mathbf{a}^2$
(c) $\mathbf{a}^2$ (d) $3\mathbf{a}^2$

224. If $\mathbf{a}, \mathbf{b}, \mathbf{c}$ are non-coplanar vectors and λ is a real number, then $[\lambda(\mathbf{a}+\mathbf{b}) \;\; \lambda^2\mathbf{b} \;\; \lambda\mathbf{c}] = [\mathbf{a} \;\; \mathbf{b}+\mathbf{c} \;\; \mathbf{b}]$ for **[AIEEE 2005]**

(a) exactly two values of λ
(b) exactly three values of λ
(c) no value of λ
(d) exactly one value of λ

225. Let $\mathbf{a} = \hat{\mathbf{i}} - \hat{\mathbf{k}}, \mathbf{b} = x\hat{\mathbf{i}} + \hat{\mathbf{j}} + (1-x)\hat{\mathbf{k}}$ and $\mathbf{c} = y\hat{\mathbf{i}} + x\hat{\mathbf{j}} + (1+x-y)\hat{\mathbf{k}}$. Then, $[\mathbf{a} \; \mathbf{b} \; \mathbf{c}]$ depends on **[AIEEE 2005]**

(a) Neither x nor y
(b) Both x and y
(c) Only x
(d) Only y

226. Let $\mathbf{u}, \mathbf{v}, \mathbf{w}$ be such that $|\mathbf{u}| = 1, |\mathbf{v}| = 2, |\mathbf{w}| = 3$. If the projection $\mathbf{v}$ along $\mathbf{u}$ is equal to that of $\mathbf{w}$ along $\mathbf{u}$ and $\mathbf{v}, \mathbf{w}$ are perpendicular to each other, then $|\mathbf{u} - \mathbf{v} + \mathbf{w}|$ equal to **[AIEEE 2004]**

(a) 2 (b) $\sqrt{7}$
(c) $\sqrt{14}$ (d) 14

227. Let $\mathbf{a}, \mathbf{b}$ and $\mathbf{c}$ be non-zero vectors such that $(\mathbf{a} \times \mathbf{b}) \times \mathbf{c} = \frac{1}{3}|\mathbf{b}||\mathbf{c}|\mathbf{a}$. If θ is an acute angle between the vectors $\mathbf{b}$ and $\mathbf{c}$, then $\sin\theta$ is equal to **[AIEEE 2004]**

(a) $\frac{1}{3}$ (b) $\frac{\sqrt{2}}{3}$
(c) $\frac{2}{3}$ (d) $\frac{2\sqrt{2}}{3}$

228. A particle is acted upon by constant forces $4\hat{\mathbf{i}} + \hat{\mathbf{j}} - 3\hat{\mathbf{k}}$ and $3\hat{\mathbf{i}} + \hat{\mathbf{j}} - \hat{\mathbf{k}}$ which displace it from a point $\hat{\mathbf{i}} + 2\hat{\mathbf{j}} + 3\hat{\mathbf{k}}$ to the point $5\hat{\mathbf{i}} + 4\hat{\mathbf{j}} + \hat{\mathbf{k}}$. The work done in standard units by the forces is given by **[AIEEE 2004]**

(a) 40 units (b) 30 units
(c) 25 units (d) 15 units

229. If $\mathbf{u}, \mathbf{v}$ and $\mathbf{w}$ are three non-coplanar vectors, then $(\mathbf{u}+\mathbf{v}-\mathbf{w}) \cdot [(\mathbf{u}-\mathbf{v}) \times (\mathbf{v}-\mathbf{w})]$ equal to **[AIEEE 2003]**

(a) 0 (b) $\mathbf{u} \cdot \mathbf{v} \times \mathbf{w}$
(c) $\mathbf{u} \cdot \mathbf{w} \times \mathbf{v}$ (d) $3\mathbf{u} \cdot \mathbf{v} \times \mathbf{w}$

230. $\mathbf{a}, \mathbf{b}, \mathbf{c}$ are three vectors, such that $\mathbf{a}+\mathbf{b}+\mathbf{c} = 0, |\mathbf{a}| = 1, |\mathbf{b}| = 2, |\mathbf{c}| = 3$, then $\mathbf{a} \cdot \mathbf{b} + \mathbf{b} \cdot \mathbf{c} + \mathbf{c} \cdot \mathbf{a}$ is equal to **[AIEEE 2003]**

(a) 0 (b) -7
(c) 7 (d) 1

231. A tetrahedron has vertices at $O(0, 0, 0), A(1, 2, 1), B(2, 1, 3)$ and $C(-1, 1, 2)$. Then, the angle between the faces OAB and ABC will be **[AIEEE 2003]**

(a) $\cos^{-1}\left(\frac{19}{35}\right)$ (b) $\cos^{-1}\left(\frac{17}{31}\right)$
(c) $30°$ (d) $90°$

232. Let $\mathbf{u} = \hat{\mathbf{i}} + \hat{\mathbf{j}}, \mathbf{v} = \hat{\mathbf{i}} - \hat{\mathbf{j}}$ and $\mathbf{w} = \hat{\mathbf{i}} + 2\hat{\mathbf{j}} + 3\hat{\mathbf{k}}$. If $\mathbf{n}$ is a unit vector such that $\mathbf{u} \cdot \mathbf{n} = 0$ and $\mathbf{v} \cdot \mathbf{n} = 0$, then $|\mathbf{w} \cdot \mathbf{n}|$ is equal to **[AIEEE 2003]**

(a) 0 (b) 1
(c) 2 (d) 3

233. Given, two vectors are $\hat{\mathbf{i}} - \hat{\mathbf{j}}$ and $\hat{\mathbf{i}} + 2\hat{\mathbf{j}}$, the unit vector coplanar with the two vectors and perpendicular to first is **[AIEEE 2002]**

(a) $\frac{1}{\sqrt{2}}(\hat{\mathbf{i}} + \hat{\mathbf{j}})$ (b) $\frac{1}{\sqrt{5}}(2\hat{\mathbf{i}} + \hat{\mathbf{j}})$
(c) $\pm\frac{1}{\sqrt{2}}(\hat{\mathbf{i}} + \hat{\mathbf{j}})$ (d) None of these

Answers

Exercise for Session 1

1. $\cos^{-1}\left(\frac{5}{7}\right)$ 2. $\frac{\pi}{4}$ 4. $\mathbf{r} = \pm\sqrt{3}(\hat{\mathbf{i}} + \hat{\mathbf{j}} + \hat{\mathbf{k}})$

5. $\frac{\pi}{2}$ 6. $\frac{60}{\sqrt{114}}$ 7. $-\frac{5}{2}$

8. $0°$ 9. $\frac{2\pi}{3}$ 10. $a > 2$

11. $\frac{1}{6}(\hat{\mathbf{i}} + \hat{\mathbf{j}} + \hat{\mathbf{k}})$ 12. $\frac{19}{9}(2\hat{\mathbf{i}} + \hat{\mathbf{j}} + 2\hat{\mathbf{k}}); \frac{1}{9}(-20\hat{\mathbf{i}} + 8\hat{\mathbf{j}} + 16\hat{\mathbf{k}})$

13. 40

Exercise for Session 2

1. $19\sqrt{2}$ 2. $\lambda = 3$ and $\mu = \frac{27}{2}$

3. -74 6. 3 7. $\frac{\pi}{6}$

8. $\frac{\pi}{4}$ 9. ± 7 10. $\frac{-\hat{\mathbf{i}}}{\sqrt{3}} + \frac{\hat{\mathbf{j}}}{\sqrt{3}} + \frac{\mathbf{k}}{\sqrt{3}}$

11. $\frac{5}{3}(7\hat{\mathbf{i}} - 4\hat{\mathbf{j}} - 4\hat{\mathbf{k}})$ 12. $\frac{1}{3}(160\,\hat{\mathbf{i}} - 4\hat{\mathbf{i}} - 70\hat{\mathbf{k}})$

13. $\pm 2\,(\mathbf{b} \times \mathbf{c})$ 14. $\frac{1}{2}\sqrt{65}$ sq. units

15. $\frac{\sqrt{61}}{2}$ sq. units. 17. $\hat{\mathbf{i}} + 2\hat{\mathbf{j}} + 4\hat{\mathbf{k}}$

18. $2\hat{\mathbf{i}} - 7\hat{\mathbf{j}} - 2\hat{\mathbf{k}}$ 19. $-20\hat{\mathbf{i}} + 16\hat{\mathbf{j}} + 12\hat{\mathbf{k}}$

Exercise for Session 3

1. 4 2. 4 cubic unit 3. $\frac{9}{2}$

4. 4 cubic unit 5. $\frac{2\sqrt{38}}{19}$

6. **a**, **b**, **c** form a right handed system. 9. 6 10. 1

Exercise for Session 4

1. 0 2. $3\frac{(-7\hat{\mathbf{i}} + 8\hat{\mathbf{j}} - \mathbf{k})}{\sqrt{114}}$

7. $\mathbf{a}' = \frac{1}{2}(-\mathbf{i} + \mathbf{k}), \mathbf{b}' = \frac{1}{2}(-\hat{\mathbf{j}} + \hat{\mathbf{k}})$ and $\mathbf{c}' = \frac{1}{2}(\mathbf{i} + \mathbf{j})$

9. $\mathbf{r} = y\,\mathbf{b} = \frac{1}{\mathbf{b}^2}(\mathbf{a} \times \mathbf{b})$ 10. $\mathbf{r} = \frac{1}{\mathbf{a} \cdot \mathbf{b}}(\mathbf{a} \times \mathbf{c} + m\mathbf{b})$

Chapter Exercises

1. (d) 2. (a) 3. (a) 4. (d) 5. (d) 6. (b)
7. (a) 8. (b) 9. (b) 10. (d) 11. (d) 12. (d)
13. (a) 14. (c) 15. (c) 16. (a) 17. (a) 18. (c)
19. (a) 20. (b) 21. (d) 22. (c) 23. (b) 24. (b)
25. (b) 26. (b) 27. (c) 28. (a) 29. (c) 30. (a)
31. (c) 32. (a) 33. (c) 34. (a) 35. (a) 36. (b)
37. (d) 38. (c) 39. (a) 40. (a) 41. (a) 42. (b)
43. (c) 44. (a) 45. (d) 46. (b) 47. (d) 48. (a)
49. (a) 50. (d) 51. (b) 52. (a) 53. (c) 54. (c)
55. (d) 56. (c) 57. (d) 58. (b) 59. (c) 60. (d)
61. (d) 62. (d) 63. (a) 64. (a) 65. (a) 66. (a)
67. (b) 68. (d) 69. (b) 70. (d) 71. (c) 72. (b)
73. (d) 74. (b,c,d) 75. (a,c,d) 76. (a,c) 77. (a,c) 78. (b,d)
79. (a,b,c,d) 80. (a,b,c,d) 81. (a,b) 82. (a,c)
83. (a,b,c) 84. (a,c) 85. (a,b,c) 86. (b,d)
87. (b,c,d) 88. (a,c) 89. (a,b)
90. (a,b,c) 91. (a,d) 92. (a,b,c) 93. (a,c) 94. (a,b,c)
95. (b,c,d) 96. (a,b,c,d) 97. (c,d) 98. (a,b,c,d)
99. (b,c,d) 100. (a,d) 101. (a,b,d)
102. (c,d) 103. (b,c) 104. (b,c,d) 105. (a,c,d)
106. (c,d) 107. (a,b)
108. (a,c) 109. (b,c) 110. (c) 111. (a) 112. (a) 113. (d)
114. (d) 115. (d) 116. (a) 117. (a) 118. (b) 119. (a)
120. (a) 121. (b) 122. (c) 123. (b) 124. (a) 125. (b)
126. (d) 127. (c) 128. (c) 129. (a) 130. (d) 131. (d)
132. (d) 133. (c) 134. (b) 135. (d) 136. (a) 137. (a)
138. (a) 139. (b) 140. (a) 141. (c) 142. (b)
143. (A) → (p,r), (B) → (q), (C) → (s), (D) → (p)
144. (A) → (r), (B) → (s), (C) → (q), (D) → (p)
145. (A) → (s), (B) → (r), (C) → (q)
146. (A) → (r), (B) → (p), (C) → (s), (D) → (q)
147. (A) → (q), (B) → (s), (C) → (p), (D) → (r)
148. (0) 149. (3) 150. (1) 151. (2) 152. (6) 153. (9)
154. (2) 155. (3) 156. (3) 157. (4) 158. (2) 159. (9)
160. (5) 161. (2) 162. (3) 163. (8) 165. (10)
166. $xx_1 + 2yy_1 = a$ 167. (12)

168. $\frac{\sin C + \sin B \cdot \sin A}{\sqrt{\sin^2 B + \sin^2 C + 2\sin B \cdot \sin C \cdot \cos A}}$

169. $\frac{3\sqrt{26}}{7}$ 170. $\frac{\sin A - \sin B}{2(\sin A + \sin B)}$ 171. $\frac{49}{28}$ sq units

173. $\{6, 4\}$ and $4x - 6y + 13 = 0$ 174. $\theta = \frac{\pi}{4}$

175. $\left(2\sqrt{2}\,\hat{\mathbf{i}} - \frac{1}{\sqrt{2}}\hat{\mathbf{j}} - \frac{1}{\sqrt{2}}\hat{\mathbf{k}}\right)$ 176. 1

177. $-\hat{\mathbf{i}} - 8\hat{\mathbf{j}} + 2\hat{\mathbf{k}}$ 178. $\frac{\mathbf{a} \cdot \mathbf{b}}{(\mathbf{a} \cdot \mathbf{b})^2 - \mathbf{a}^2\,\mathbf{b}^2}$

179. $\hat{\mathbf{x}} = \frac{1}{3}(3\mathbf{a} + 4\mathbf{b} + 8\mathbf{c}), \hat{\mathbf{y}} = -4\mathbf{c}, \hat{\mathbf{z}} = \frac{4}{3}(c - b)$

180. $x = \frac{a + b + c}{2}$

181. $x = \frac{1}{2}\frac{\mathbf{a}^2\mathbf{b}^2 - \mathbf{b}^2(\mathbf{a} \cdot \mathbf{b})}{2(\mathbf{a}^2\,\mathbf{b}^2) - (\mathbf{a} \cdot \mathbf{b})^2}\mathbf{a} + \frac{1}{2}\frac{\mathbf{a}^2\mathbf{b}^2 - \mathbf{a}^2(\mathbf{a} \cdot \mathbf{b})}{(\mathbf{a}^2\,\mathbf{b}^2) - (\mathbf{a} \cdot \mathbf{b})^2} \cdot \mathbf{b}$

182. (b) 183. (a) 184. (a) 185. (d) 186. (a) 187. (3)
188. (a) 189. (c) 190. (b) 191. (a) 192. (5) 193. (c)
194. (a) 195. (a) 196. (c) 197. (c) 198. (b) 199. (b,d)
200. (a) 201. (b) 202. (c) 203. (c) 204. (c) 205. (c)
206. (b) 207. (b) 208. (b) 209. (c) 210. (b) 211. (d)
212. (c) 213. (a) 214. (d) 215. (a) 216. (d) 217. (d)
218. (d) 219. (d) 220. (c) 221. (d) 222. (c) 223. (b)
224. (c) 225. (a) 226. (c) 227. (d) 228. (a) 229. (b)
230. (b) 231. (a) 232. (d) 233. (a)

Solutions

1. Since, $\mathbf{a} \perp \mathbf{b} \Rightarrow \mathbf{a} \cdot \mathbf{b} = 0$

$$|\mathbf{a} - \mathbf{b}|^2 = (\mathbf{a} - \mathbf{b})^2 = \mathbf{a}^2 + \mathbf{b}^2 - 2\,\mathbf{a} \cdot \mathbf{b} = 25 + 25$$

$$\Rightarrow \quad |\mathbf{a} - \mathbf{b}| = 5\sqrt{2}$$

2. $|\mathbf{a} + \mathbf{b}| > |\mathbf{a} - \mathbf{b}|$

On squaring both sides, we get

$$\mathbf{a}^2 + \mathbf{b}^2 + 2\mathbf{a} \cdot \mathbf{b} > \mathbf{a}^2 + \mathbf{b}^2 - 2\mathbf{a} \cdot \mathbf{b}$$

$$\Rightarrow \quad 4\mathbf{a} \cdot \mathbf{b} > 0 \Rightarrow \cos\theta > 0$$

Hence, $\theta < 90°$ (acute)

3. Given that, $\mathbf{a} = \mathbf{b} + \mathbf{c}$ and angle between $\mathbf{b}$ and $\mathbf{c}$ is $\frac{\pi}{2}$.

So, $\quad \mathbf{a}^2 = \mathbf{b}^2 + \mathbf{c}^2 + 2\mathbf{b} \cdot \mathbf{c}$

$\Rightarrow \quad \mathbf{a}^2 = \mathbf{b}^2 + \mathbf{c}^2 + 2|\mathbf{b}||\mathbf{c}|\cos\frac{\pi}{2}$

$\Rightarrow \quad \mathbf{a}^2 = \mathbf{b}^2 + \mathbf{c}^2 + 0$

$\therefore \quad \mathbf{a}^2 = \mathbf{b}^2 + \mathbf{c}^2$

i.e., $\quad \mathbf{a}^2 = \mathbf{b}^2 + \mathbf{c}^2$

4. Obviously, $\mathbf{a}$ and $\mathbf{b}$ are unit vectors.

5. Angle between $\hat{\mathbf{i}} + \hat{\mathbf{j}} + \hat{\mathbf{k}}$ and $\hat{\mathbf{i}}$ is equal to

$$\cos^{-1}\left\{\frac{(\hat{\mathbf{i}} + \hat{\mathbf{j}} + \hat{\mathbf{k}}) \cdot \hat{\mathbf{i}}}{|\hat{\mathbf{i}} + \hat{\mathbf{j}} + \hat{\mathbf{k}}||\hat{\mathbf{i}}|}\right\} \Rightarrow a = \cos^{-1}\left(\frac{1}{\sqrt{3}}\right)$$

Similarly, angle between $\hat{\mathbf{i}} + \hat{\mathbf{j}} + \hat{\mathbf{k}}$ and $\hat{\mathbf{j}}$ is

$$\beta = \cos^{-1}\left(\frac{1}{\sqrt{3}}\right)$$ and between $\hat{\mathbf{i}} + \hat{\mathbf{j}} + \hat{\mathbf{k}}$ and $\hat{\mathbf{k}}$ is

$$\gamma = \cos^{-1}\left(\frac{1}{\sqrt{3}}\right)$$

Hence, $\quad a = \beta = \gamma$

6. Let $\mathbf{r} = x\hat{\mathbf{i}} + y\hat{\mathbf{j}} + z\hat{\mathbf{i}} \Rightarrow \mathbf{r} \cdot \hat{\mathbf{i}} = x, \mathbf{r} \cdot \hat{\mathbf{j}} = y, \mathbf{r} \cdot \hat{\mathbf{k}} = z$

$$\Rightarrow \quad (\mathbf{r} \cdot \hat{\mathbf{i}})^2 + (\mathbf{r} \cdot \hat{\mathbf{j}})^2 + (\mathbf{r} \cdot \hat{\mathbf{k}})^2 = x^2 + y^2 + z^2 = r^2$$

7. $|\mathbf{a} - \mathbf{b}| = \sqrt{1^2 + 1^2 - 2 \cdot 1^2 \cos\theta} = \sqrt{2(1 - \cos\theta)}$

$$= \sqrt{2} \times \sqrt{2} \sin\frac{\theta}{2} = 2\sin\frac{\theta}{2} \Rightarrow \sin\frac{\theta}{2} = \frac{|\mathbf{a} - \mathbf{b}|}{2}$$

8. The component of vector $\mathbf{a}$ along $\mathbf{b}$ is

$$\frac{(\mathbf{a} \cdot \mathbf{b})\mathbf{b}}{|\mathbf{b}|^2} = \frac{18}{25}(3\hat{\mathbf{j}} + 4\hat{\mathbf{k}})$$

9. Required value $= \dfrac{\mathbf{b} \cdot \mathbf{a}}{|\mathbf{b}|} \Big/ \dfrac{\mathbf{a} \cdot \mathbf{b}}{|\mathbf{a}|} = \dfrac{|\mathbf{a}|}{|\mathbf{b}|} = \dfrac{7}{3}$

10. $(\mathbf{a} \times \mathbf{b})^2 = a^2b^2 - (\mathbf{a} \,.\, \mathbf{b})^2 = \begin{vmatrix} \mathbf{a} \cdot \mathbf{a} & \mathbf{a} \cdot \mathbf{b} \\ \mathbf{b} \cdot \mathbf{a} & \mathbf{b} \cdot \mathbf{b} \end{vmatrix}$

11. Torque $= \mathbf{r} \times \mathbf{F}$ or $\mathbf{CP} \times \mathbf{F}$

12. $\mathbf{r} \times \mathbf{F} = \begin{vmatrix} \hat{\mathbf{i}} & \hat{\mathbf{j}} & \hat{\mathbf{k}} \\ 2 & -1 & 1 \\ 1 & 2 & 3 \end{vmatrix} = -5\hat{\mathbf{i}} - 5\hat{\mathbf{j}} + 5\hat{\mathbf{k}}$

13. $\mathbf{OA} = 3\hat{\mathbf{i}} + 2\hat{\mathbf{j}} - 9\hat{\mathbf{k}}, \mathbf{F} = (9\hat{\mathbf{i}} + 6\hat{\mathbf{j}} - 2\hat{\mathbf{k}}) \times \frac{6}{11}$

$$\therefore \quad \text{Moment} = \mathbf{OA} \times \mathbf{F} = \frac{6}{11}\begin{vmatrix} \hat{\mathbf{i}} & \hat{\mathbf{j}} & \hat{\mathbf{k}} \\ 3 & 2 & -9 \\ 9 & 6 & -2 \end{vmatrix}$$

$$= \frac{6}{11}(50\hat{\mathbf{i}} - 75\hat{\mathbf{j}}) = \frac{150}{11}(2\hat{\mathbf{i}} - 3\hat{\mathbf{j}})$$

14. Force $(\mathbf{F}) = 2\hat{\mathbf{i}} + \hat{\mathbf{j}} - \hat{\mathbf{k}}$ and its position vector $= 2\hat{\mathbf{i}} - \hat{\mathbf{j}}$. We know that the position vector of a force about origin $(\mathbf{r}) = (2\hat{\mathbf{i}} - \hat{\mathbf{j}}) - (0\hat{\mathbf{i}} + 0\hat{\mathbf{j}} + 0\hat{\mathbf{k}})$ or $\mathbf{r} = 2\hat{\mathbf{i}} - \hat{\mathbf{j}}$ Therefore, moment of the force about origin

$$= \mathbf{r} \times \mathbf{F} = \begin{vmatrix} \hat{\mathbf{i}} & \hat{\mathbf{j}} & \hat{\mathbf{k}} \\ 2 & -1 & 0 \\ 2 & 1 & -1 \end{vmatrix} = \hat{\mathbf{i}} + 2\hat{\mathbf{j}} + 4\hat{\mathbf{k}}.$$

15. $\mathbf{a}^{-1} = \dfrac{\mathbf{b} \times \mathbf{c}}{[\mathbf{abc}]}, \mathbf{c}^{-1} = \dfrac{\mathbf{a} \times \mathbf{b}}{[\mathbf{abc}]}, \mathbf{b}^{-1} = \dfrac{\mathbf{c} \times \mathbf{a}}{[\mathbf{abc}]}$

$$\Rightarrow \quad [\mathbf{a}^{-1}\, \mathbf{b}^{-1}\, \mathbf{c}^{-1}] = \frac{(\mathbf{b} \times \mathbf{c})}{[\mathbf{abc}]} \cdot \left(\frac{(\mathbf{c} \times \mathbf{a})}{[\mathbf{abc}]} \times \frac{(\mathbf{a} \times \mathbf{b})}{[\mathbf{abc}]}\right)$$

$$= \frac{\mathbf{b} \times \mathbf{c}}{[\mathbf{abc}]} \cdot \frac{\mathbf{a}}{[\mathbf{abc}]} = \frac{1}{[\mathbf{abc}]} \neq 0$$

16. $\dfrac{\mathbf{a} \cdot \mathbf{b} \times \mathbf{c}}{\mathbf{c} \times \mathbf{a} \cdot \mathbf{b}} + \dfrac{\mathbf{b} \cdot \mathbf{a} \times \mathbf{c}}{\mathbf{c} \cdot \mathbf{a} \times \mathbf{b}} = \dfrac{[\mathbf{abc}]}{[\mathbf{cab}]} + \dfrac{[\mathbf{bac}]}{[\mathbf{cab}]}$

$$= \frac{[\mathbf{abc}]}{[\mathbf{cab}]} - \frac{[\mathbf{abc}]}{[\mathbf{cab}]} = 0$$

17. $\mathbf{b} \times \mathbf{c}$ is a vector perpendicular to $\mathbf{b}$, $\mathbf{c}$. Therefore, $\mathbf{a} \times (\mathbf{b} \times \mathbf{c})$ is a vector again in plane of $\mathbf{b}$, $\mathbf{c}$.

18. Let $\mathbf{a} = x\hat{\mathbf{i}} + y\hat{\mathbf{j}} + z\hat{\mathbf{k}}$

$$\mathbf{u} = \hat{\mathbf{i}} \times (\mathbf{a} \times \hat{\mathbf{i}}) + \hat{\mathbf{j}} \times (\mathbf{a} \times \hat{\mathbf{i}}) + \hat{\mathbf{k}} \times (\mathbf{a} \times \hat{\mathbf{k}})$$

$$= (\hat{\mathbf{i}} \cdot \hat{\mathbf{i}})\mathbf{a} - \hat{\mathbf{i}}(\mathbf{a} \cdot \hat{\mathbf{i}}) + (\hat{\mathbf{j}} \cdot \hat{\mathbf{j}})\mathbf{a} - \hat{\mathbf{j}}(\mathbf{a} \cdot \hat{\mathbf{j}})$$

$$+ (\hat{\mathbf{k}} \cdot \hat{\mathbf{k}})\mathbf{a} - \hat{\mathbf{k}}(\mathbf{a} \cdot \hat{\mathbf{k}}) = 3\mathbf{a} - \mathbf{a} = 2\mathbf{a}$$

19. $\mathbf{a} \times (\mathbf{b} \times \mathbf{c}) = \mathbf{a} \times \begin{vmatrix} \hat{\mathbf{i}} & \hat{\mathbf{j}} & \hat{\mathbf{k}} \\ 2 & -1 & 1 \\ 1 & 3 & -1 \end{vmatrix} = \mathbf{a} \times (-2\hat{\mathbf{i}} + 3\hat{\mathbf{j}} + 7\hat{\mathbf{k}})$

$$= \begin{vmatrix} \hat{\mathbf{i}} & \hat{\mathbf{j}} & \hat{\mathbf{k}} \\ 1 & 2 & -2 \\ -2 & 3 & 7 \end{vmatrix} = 20\hat{\mathbf{i}} - 3\hat{\mathbf{j}} + 7\hat{\mathbf{k}}$$

20. $\mathbf{a} \times (\mathbf{b} \times \mathbf{c}) = 0$

$\Rightarrow \quad \mathbf{a} \,||\, (\mathbf{b} \times \mathbf{c})$ or $\mathbf{b} \times \mathbf{c} = 0$

i.e., $\quad \mathbf{b} \,||\, \mathbf{c}$ or $\mathbf{a} = 0$

21. Let the required vector be $\alpha = d_1\hat{\mathbf{i}} + d_2\hat{\mathbf{j}} + d_3\hat{\mathbf{k}}$

where, $d_1^2 + d_2^2 + d_3^2 = 51$ (given) ...(i)

Now, each of the given vectors $\mathbf{a}$, $\mathbf{b}$ and $\mathbf{c}$ is a unit vectors.

$$\cos\theta = \frac{\mathbf{d} \cdot \mathbf{a}}{|\mathbf{d}||\mathbf{a}|} = \frac{\mathbf{d} \cdot \mathbf{b}}{|\mathbf{d}||\mathbf{b}|} = \frac{\mathbf{d} \cdot \mathbf{c}}{|\mathbf{d}||\mathbf{c}|}$$

or $\quad \mathbf{d} \cdot \mathbf{a} = \mathbf{d} \cdot \mathbf{b} = \mathbf{d} \cdot \mathbf{c}$

$|\mathbf{d}| = \sqrt{51}$ cancels out and $|\mathbf{a}| = |\mathbf{b}| = |\mathbf{c}| = 1$

Hence, $\frac{1}{3}(d_1 - 2d_2 + 2d_3) = \frac{1}{5}(-4d_1 + 0d_2 - 3d_3) = d_2$

$\Rightarrow \quad d_1 - 5d_2 + 2d_3 = 0$

and $\quad 4d_1 + 5d_2 + 3d_3 = 0$

On solving, we get $\quad \dfrac{d_1}{5} = \dfrac{d_2}{-1} = \dfrac{d_3}{-5} = \lambda \qquad$ (say)

Putting d_1, d_2 and d_3 in Eq. (i), we get $\lambda = \pm 1$

Hence, the required vectors are $\pm (5\hat{\mathbf{i}} - \hat{\mathbf{j}} - 5\hat{\mathbf{k}})$

22. Let $\mathbf{OA} = P_1\hat{\mathbf{i}}$, $\mathbf{CB} = -P_1\hat{\mathbf{i}}$ and $\mathbf{OB} = -P_1\hat{\mathbf{i}} + P\hat{\mathbf{j}}$

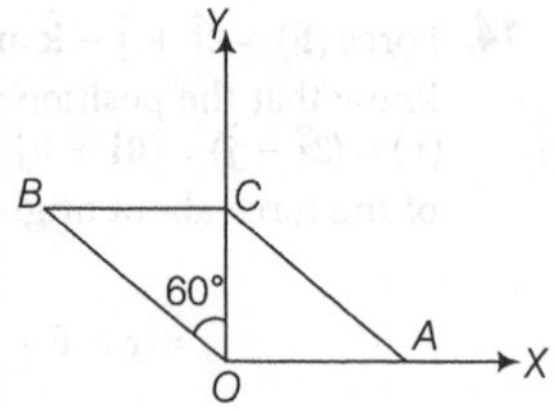

$$\frac{\mathbf{OB}\cdot\hat{\mathbf{j}}}{OB} = \cos 60^\circ$$

$$\Rightarrow \frac{(-P_1\hat{\mathbf{i}} + P\hat{\mathbf{j}})\cdot\hat{\mathbf{j}}}{\sqrt{P_1^2 + P^2}} = \frac{1}{2}$$

$$\Rightarrow \quad 2P = \sqrt{P^2 + P_1^2}$$

$$\Rightarrow \quad P_1 = P\sqrt{3}$$

$$|\mathbf{OB}| = \sqrt{P^2 + P_1^2} = \sqrt{P^2 + 3P^2} = 2P$$

23. $\mathbf{x} + \mathbf{y} + \mathbf{z} = 0 \Rightarrow \mathbf{x} = -(\mathbf{y} + \mathbf{z})$

$$|\mathbf{x}|^2 = (\mathbf{y} + \mathbf{z})\cdot(\mathbf{y} + \mathbf{z})$$

$$\Rightarrow \quad |\mathbf{x}|^2 = |\mathbf{y}|^2 + |\mathbf{z}|^2 + 2\mathbf{y}\cdot\mathbf{z}$$

$$\Rightarrow \quad |\mathbf{x}|^2 = |\mathbf{y}|^2 + |\mathbf{z}|^2 + 2|\mathbf{y}||\mathbf{z}|\cos\theta$$

$$\Rightarrow \quad 4 = 4 + 4 + 2\times 2\times 2\cos\theta$$

$$\Rightarrow \quad \cos\theta = \frac{-1}{2} \Rightarrow \theta = 120^\circ$$

$\therefore \operatorname{cosec}^2 120^\circ + \cot^2 120^\circ$

$$= \left(\frac{2}{\sqrt{3}}\right)^2 + \left(-\frac{1}{\sqrt{3}}\right)^2 = \frac{4}{3} + \frac{1}{3} = \frac{5}{3}$$

24. For acute angle $\mathbf{a}\cdot\mathbf{b} > 0$

i.e., $\quad -3x + 2x^2 + 1 > 0$

$\Rightarrow \quad (x-1)(2x-1) > 0$

For obtuse angle between $\mathbf{b}$ and X-axis $\mathbf{b}\,.\,\hat{\mathbf{i}} < 0 \Rightarrow x < 0$

25. Since, $\mathbf{d} = \lambda\mathbf{a} + \mu\mathbf{b} + \nu\mathbf{c}$

$$\therefore \quad \mathbf{d}\cdot(\mathbf{b}\times\mathbf{c}) = \lambda\mathbf{a}\cdot(\mathbf{a}\times\mathbf{c}) + \mu\mathbf{b}\cdot(\mathbf{b}\times\mathbf{c}) + \nu\mathbf{c}\cdot(\mathbf{b}\times\mathbf{c})$$

$$= \lambda\,[\mathbf{a}\;\mathbf{b}\;\mathbf{c}]$$

$$\Rightarrow \quad \lambda = \frac{[\mathbf{dbc}]}{[\mathbf{abc}]} = \frac{[\mathbf{bcd}]}{[\mathbf{bca}]}$$

26. $(3\mathbf{p} + \mathbf{q})\cdot(5\mathbf{p} - 3\mathbf{q}) = 0$

or $\quad 15\mathbf{p}^2 - 3\mathbf{q}^2 = 4\mathbf{p}\cdot\mathbf{q} \qquad$ ...(i)

$(2\mathbf{p} + \mathbf{q})\cdot(4\mathbf{p} - 2\mathbf{q}) = 0$ or $8\mathbf{p}^2 = 2\mathbf{q}^2$

$\Rightarrow \quad \mathbf{q}^2 = 4\mathbf{p}^2 \qquad$ ...(ii)

Now, $\quad \cos\theta = \dfrac{\mathbf{p}\cdot\mathbf{q}}{|\mathbf{p}||\mathbf{q}|}$

On substituting $\mathbf{q}^2 = 4\mathbf{p}^2$ in Eq. (i), we get

$$\Rightarrow \quad 3\mathbf{p}^2 = 4\mathbf{p}\cdot\mathbf{q}$$

$$\cos\theta = \frac{3}{4}\cdot\frac{\mathbf{p}^2}{|\mathbf{p}|2|\mathbf{p}|} = \frac{3}{8} \Rightarrow \sin\theta = \frac{\sqrt{55}}{8}$$

27. $\hat{\mathbf{n}} = a_1\hat{\mathbf{i}} + a_2\hat{\mathbf{j}} + a_3\hat{\mathbf{k}}$,

where, $a_1^2 + a_2^2 + a_3^2 = 1$

Also, $\quad \mathbf{u}\cdot\hat{\mathbf{n}} = 0 \Rightarrow a_1 + a_2 = 0$

$\quad \mathbf{v}\cdot\hat{\mathbf{n}} = 0 \Rightarrow a_1 - a_2 = 0$

Hence, $\quad a_1 = a_2 = 0$

$\therefore \quad a_3 = 1$ or -1

$\therefore \quad \hat{\mathbf{n}} = \hat{\mathbf{k}}$ or $-\hat{\mathbf{k}}$

$\quad |\mathbf{w}\cdot\hat{\mathbf{n}}| = 3$

28. To find $(\mathbf{a} - \mathbf{b})\cdot\mathbf{a}$

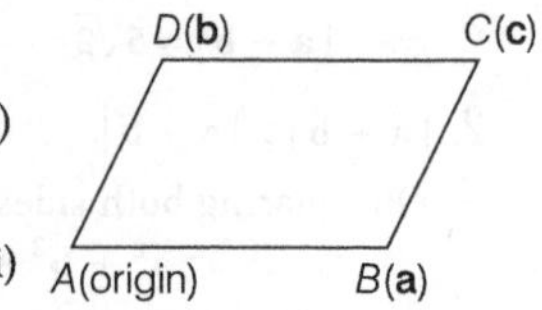

i.e., $\quad |\mathbf{a}|^2 - \mathbf{a}\cdot\mathbf{b} \qquad$ ...(i)

Now, $\quad \mathbf{a} + \mathbf{b} = \mathbf{c}$

$\Rightarrow |\mathbf{a}|^2 + |\mathbf{b}|^2 + 2\mathbf{a}\cdot\mathbf{b} = |\mathbf{c}|^2 \quad$ ...(ii)

On substituting the value of $\mathbf{a}\cdot\mathbf{b}$ from Eq. (ii) in Eq. (i), we get

$$a^2 - \frac{1}{2}(c^2 - a^2 - b^2)$$

$$\Rightarrow \quad \frac{3a^2 + b^2 - c^2}{2}$$

29. $\mathbf{A}\times\mathbf{B} = -\mathbf{A}\times\mathbf{B}$

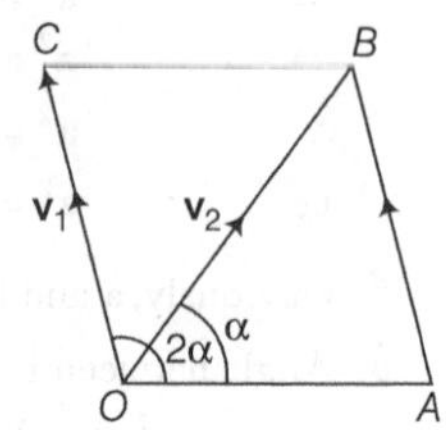

$\quad \mathbf{A}\times\mathbf{B} = 0$

either $\quad \mathbf{A} = 0$

or $\quad \mathbf{B} = 0$

or $\mathbf{A}$ and $\mathbf{B}$ are collinear

30. Given, $\quad \mathbf{V} + \mathbf{V}_1 = \mathbf{V}_2$

Also, $\quad \mathbf{V}\cdot\mathbf{V}_1 = 2\alpha$

$\quad (\mathbf{V}_1)^2 = (\mathbf{V}_2 - \mathbf{V})^2$ and $\mathbf{V}\cdot\mathbf{V}_2 = \alpha$

Also, $\quad |\mathbf{V}| = |\mathbf{V}_1| = |\mathbf{V}_2| = \lambda \qquad$ (say)

Hence, $\quad \lambda^2 = 2\lambda^2 - 2\lambda^2\cos\alpha$

$\Rightarrow \quad \cos\alpha = \dfrac{1}{2}$

31. Given, $\quad |\mathbf{b}| = |\mathbf{b} - \mathbf{c}| = 8$ and $|\mathbf{c}| = 12 \qquad$ ...(i)

$$AE = \frac{\mathbf{b}}{4} \text{ and } EC = \mathbf{c} - \frac{\mathbf{b}}{4}$$

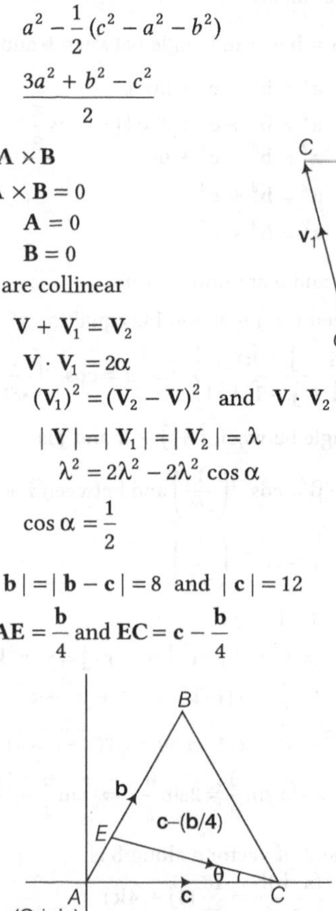

$$\cos\theta = \frac{\mathbf{c}\cdot\left(\mathbf{c} - \dfrac{\mathbf{b}}{4}\right)}{|\mathbf{c}|\left|\mathbf{c} - \dfrac{\mathbf{b}}{4}\right|} = \frac{\mathbf{c}^2 - \dfrac{\mathbf{c}\cdot\mathbf{b}}{4}}{12\left|\mathbf{c} - \dfrac{\mathbf{b}}{4}\right|} \qquad \text{...(ii)}$$

From Eq. (i), $|\mathbf{b}| = 8, |\mathbf{c}| = 12$

$$\therefore \quad |\mathbf{b} - \mathbf{c}|^2 = |\mathbf{b}|^2$$

$$\Rightarrow \quad |\mathbf{b}|^2 + |\mathbf{c}|^2 - 2\mathbf{b}\cdot\mathbf{c} = |\mathbf{b}|^2$$

$$\Rightarrow \qquad \mathbf{b}\cdot\mathbf{c} = 72 \qquad \text{...(iii)}$$

and $$\left|\mathbf{c} - \frac{\mathbf{b}}{4}\right|^2 = |\mathbf{c}|^2 + \frac{|\mathbf{b}|^2}{|16|} - \frac{\mathbf{b}\cdot\mathbf{c}}{2}$$

$$= 144 + 4 - 36 = 112$$

$$\therefore \qquad \left|\mathbf{c} - \frac{\mathbf{b}}{4}\right| = 4\sqrt{7} \qquad \text{...(iv)}$$

From Eqs. (ii), (iii) and (iv)

$$\cos\theta = \frac{144 - 10}{12 \times 4\sqrt{7}} = \frac{3\sqrt{7}}{8}$$

32. $\mathbf{BN}\cdot\mathbf{CM} = 0$

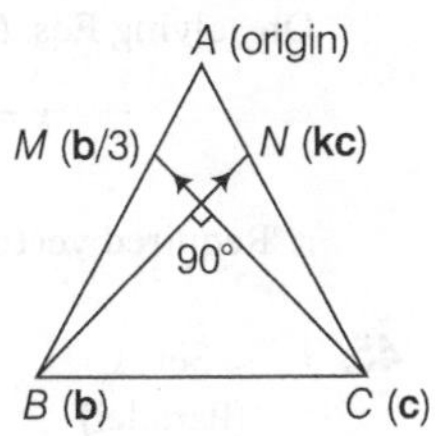

$$(k\mathbf{c} - \mathbf{b})\cdot\left(\frac{\mathbf{b}}{3} - \mathbf{c}\right) = 0$$

$$\frac{k}{3}\cdot\frac{a^2}{2} - ka^2 - \frac{a^2}{3} + \frac{a^2}{2} = 0$$

$$\frac{k}{6} - k + \frac{1}{6} = 0$$

$$\frac{5k}{6} = \frac{1}{6} \Rightarrow k = \frac{1}{5}$$

33. Given, $15\,|\mathbf{AC}| = 3\,|\mathbf{AB}| = 5\,|\mathbf{AD}|$

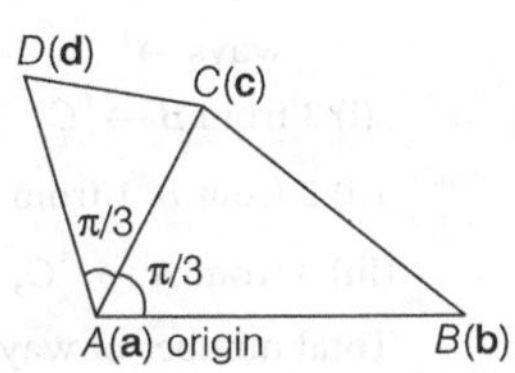

Let $|\mathbf{AC}| = \lambda > 0$

$\therefore$ $|\mathbf{AB}| = 5\lambda$

$|\mathbf{AD}| = 3\lambda$

Now, $$\cos(\mathbf{BA}\cdot\mathbf{CD}) = \frac{\mathbf{BA}\cdot\mathbf{CD}}{|\mathbf{BA}|\,|\mathbf{CD}|}$$

$$= \frac{\mathbf{b}\cdot(\mathbf{d} - \mathbf{c})}{|\mathbf{b}|\,|\mathbf{d} - \mathbf{c}|} \qquad \text{...(i)}$$

Now, numerator of Eq. (i), we get

$$\mathbf{b}\cdot\mathbf{c} - \mathbf{b}\cdot\mathbf{d} = |\mathbf{b}|\,|\mathbf{c}|\cos\frac{\pi}{3} - |\mathbf{b}|\,|\mathbf{d}|\cos\frac{2\pi}{3}$$

$$= (5\lambda)(\lambda)\frac{1}{2} + 5\lambda(3\lambda)\frac{1}{2}$$

$$= \frac{5\lambda^2 + 15\lambda^2}{2} = 10\lambda^2$$

Denominator of Eq. (i)

$$= |\mathbf{b}|\,|\mathbf{d} - \mathbf{c}|$$

Now, $$|\mathbf{d} - \mathbf{c}|^2 = \mathbf{d}^2 + \mathbf{c}^2 - 2\mathbf{c}\cdot\mathbf{d}$$

$$= 9\lambda^2 + \lambda^2 - 2(\lambda)(3\lambda)\frac{1}{2}$$

$$= 10\lambda^2 - 3\lambda^2 = 7\lambda^2$$

$$\therefore \qquad |\mathbf{d} - \mathbf{c}| = \sqrt{7}\lambda$$

Denominator of Eq. (i)

$$= (5\lambda)(7\lambda) = 5\sqrt{7}\lambda^2$$

$$\therefore \qquad \cos(\mathbf{BA}\cdot\mathbf{CD}) = \frac{10\lambda^2}{5\sqrt{7}\lambda^2} = \frac{2}{\sqrt{7}}$$

34. $\mathbf{a} = -\hat{\mathbf{i}} - 2\hat{\mathbf{j}} - \hat{\mathbf{k}}$

$$\mathbf{b} = \hat{\mathbf{i}} - 5\hat{\mathbf{j}} - 7\hat{\mathbf{k}}$$

$$|\mathbf{d}| = \frac{|\mathbf{a}\times\mathbf{b}|}{|\mathbf{a}|};\ |\mathbf{a}| = \sqrt{6}$$

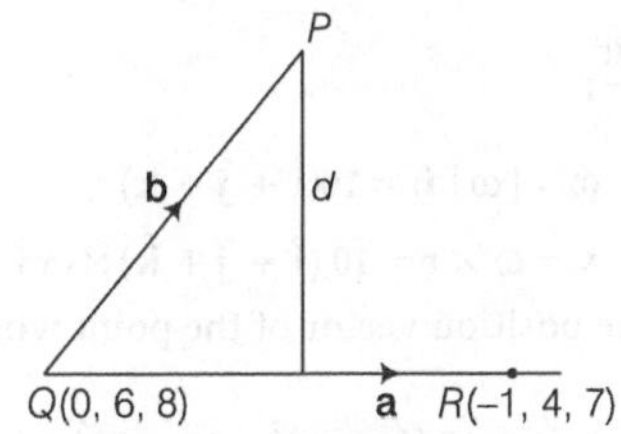

$$|\mathbf{a}\times\mathbf{b}|^2 = \mathbf{a}^2\,\mathbf{b}^2 - (\mathbf{a}\cdot\mathbf{b})^2 = (6)(75)$$

$$-(-1 + 10 + 7)^2 = 450 - 256 = 194$$

$$|\mathbf{a}\times\mathbf{b}| = \sqrt{194}$$

$$\therefore \qquad d = \sqrt{\frac{194}{6}} = \sqrt{\frac{97}{3}}$$

$$\therefore \qquad p + q = 100$$

$$\Rightarrow \qquad \frac{(p+q)(p+q-1)}{2} = \frac{100 \times 99}{2} = 4950$$

35. $$V = -c^2\,[\mathbf{u}\ \mathbf{v}\ \mathbf{w}] = -c^2\begin{vmatrix} 2 & -1 & -1 \\ 1 & -1 & 2 \\ 1 & 0 & -1 \end{vmatrix}$$

$$= -c^2[2(1-0) - 1(1) + (-2-1)]$$

$$= -c^2[2 - 1 - 3] = 8$$

$$\therefore \qquad 2c^2 = 8 \Rightarrow c = 2 \text{ or } -2$$

36. Let $\mathbf{c} = \lambda(\mathbf{a}\times\mathbf{b})$

Hence, $\lambda(\mathbf{a}\times\mathbf{b})\cdot(\hat{\mathbf{i}} + 2\hat{\mathbf{j}} - 7\hat{\mathbf{k}}) = 10$

$$\lambda\begin{vmatrix} 2 & -3 & 1 \\ 1 & -2 & 3 \\ 1 & 2 & -7 \end{vmatrix} = 10$$

$$\Rightarrow \qquad \lambda = -1 \Rightarrow \mathbf{c} = -(\mathbf{a}\times\mathbf{b})$$

$$\mathbf{a} = 2\hat{\mathbf{i}} - 3\hat{\mathbf{j}} + \hat{\mathbf{k}} \text{ and } \mathbf{b} = \hat{\mathbf{i}} - 2\hat{\mathbf{j}} + 3\hat{\mathbf{k}}$$

$$\begin{vmatrix} \hat{\mathbf{i}} & \hat{\mathbf{j}} & \hat{\mathbf{k}} \\ 2 & -3 & 1 \\ 1 & -2 & 3 \end{vmatrix} = (-9+2)\hat{\mathbf{i}} - (5)\hat{\mathbf{j}} + (-4+3)\hat{\mathbf{k}}$$

$$\Rightarrow \qquad (-7, -5, -1)$$

37. $\mathbf{V}_1 = \hat{\mathbf{i}} - 2\hat{\mathbf{j}} + \hat{\mathbf{k}}$

$$\mathbf{V}_2 = 3\hat{\mathbf{i}} + 2\hat{\mathbf{j}} - \hat{\mathbf{k}}$$

$$\mathbf{V}_3 = \mathbf{c} = \alpha\mathbf{a} + \beta\mathbf{b} = \alpha(\hat{\mathbf{i}} + \hat{\mathbf{j}}) + \beta(\hat{\mathbf{j}} + \hat{\mathbf{k}})$$

$$= \alpha\,\hat{\mathbf{i}} + (\alpha + \beta)\hat{\mathbf{j}} + \beta\hat{\mathbf{k}} = \mathbf{c}$$

Since, $\mathbf{V}_1, \mathbf{V}_2, \mathbf{V}_3$ are coplanar.

Now, $$\begin{vmatrix} 1 & -2 & 1 \\ 3 & 2 & -1 \\ \alpha & \alpha+\beta & \beta \end{vmatrix} = 0,$$ using $C_2 \to C_2 - (C_1 + C_3)$, we get

$$\begin{vmatrix} 1 & -4 & 1 \\ 3 & 0 & -1 \\ \alpha & 0 & \beta \end{vmatrix} = 0, \text{ hence } 4(3\beta + \alpha) = 0$$

$$\Rightarrow \qquad 3\beta + \alpha = 0$$

$$\Rightarrow \qquad \frac{\alpha}{\beta} = -3$$

38. $\hat{\mathbf{n}} = \dfrac{\hat{\mathbf{i}} + \hat{\mathbf{j}} + \hat{\mathbf{k}}}{\sqrt{3}}$;

$$\omega = |\omega|\,\hat{\mathbf{n}} = 10(\hat{\mathbf{i}} + \hat{\mathbf{j}} + \hat{\mathbf{k}})$$

Now, $\mathbf{v} = \omega \times \mathbf{r} = 10\,(\hat{\mathbf{i}} + \hat{\mathbf{j}} + \hat{\mathbf{k}}) \times (x\hat{\mathbf{i}} + y\hat{\mathbf{j}} + z\hat{\mathbf{k}})$,

where $\mathbf{r}$ is the position vector of the point whose locus is to be determined.

Hence, $\mathbf{v} = 10\,[(z-y)\hat{\mathbf{i}} - (z-x)\hat{\mathbf{j}} + (y-x)\hat{\mathbf{k}}]$

$\therefore$ $|\mathbf{v}| = 10\sqrt{(x-y)^2 + (y-z)^2 + (z-x)^2}$

Hence, $2(x^2 + y^2 + z^2 - xy - yz - zx) = 4$

$\Rightarrow$ $x^2 + y^2 + z^2 - xy - yz - zx - 2 = 0$

which is the equation of a cylinder.

39. $\hat{\mathbf{n}} = \dfrac{\hat{\mathbf{i}} + \hat{\mathbf{j}} + \hat{\mathbf{k}}}{3}; \omega = \dfrac{\omega}{3}(\hat{\mathbf{i}} + 2\hat{\mathbf{j}} + 2\hat{\mathbf{k}})$

$$\mathbf{v} = \omega \times \mathbf{r} = \frac{\omega}{3}\begin{vmatrix} \hat{\mathbf{i}} & \hat{\mathbf{j}} & \hat{\mathbf{k}} \\ 1 & 2 & 2 \\ 2 & 3 & 5 \end{vmatrix} = \frac{\omega}{3}(4\hat{\mathbf{i}} - \hat{\mathbf{j}} - \hat{\mathbf{k}})$$

$\therefore$ $|\mathbf{v}| = \dfrac{\omega}{3}\sqrt{18} = \omega\sqrt{2}$

40. $\mathbf{b}\cdot\mathbf{a} + \mathbf{b}\cdot\mathbf{c} = \mathbf{b}\cdot\mathbf{b} + \mathbf{a}\cdot\mathbf{c}$

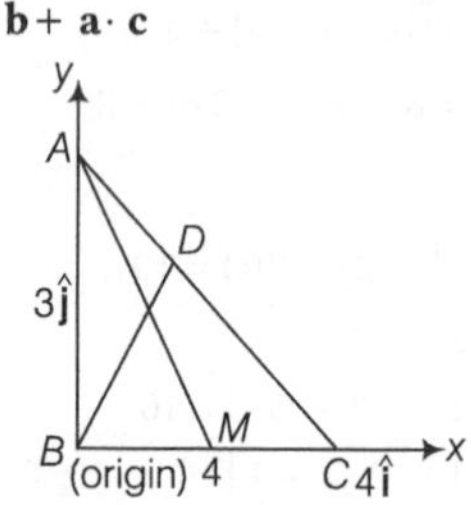

or $\mathbf{b}\cdot(\mathbf{a} - \mathbf{b}) - \mathbf{c}\cdot(\mathbf{a} - \mathbf{b}) = 0$

or $(\mathbf{b} - \mathbf{c})\cdot(\mathbf{a} - \mathbf{b}) = 0$

$\Rightarrow$ **BC** and **AB** are perpendicular.

Now, find angle between **AM** and **BD**.

where, $\mathbf{AM} = 2\hat{\mathbf{i}} - 3\hat{\mathbf{j}}$,

$\mathbf{BD} = \dfrac{4\hat{\mathbf{i}} + 3\hat{\mathbf{j}}}{2}$

$\therefore$ $\cos\theta = \dfrac{\mathbf{AM}\cdot\mathbf{BD}}{|\mathbf{AM}||\mathbf{BD}|} = \dfrac{-1}{5\sqrt{13}}$

$\Rightarrow$ $\theta = \pi - \cos^{-1}\left(\dfrac{1}{5\sqrt{13}}\right)$

41. $[\mathbf{npm}] = \sin\theta\cos\phi = \sin\dfrac{\pi}{6}\cdot\cos\dfrac{\pi}{6} = \dfrac{1}{2}\cdot\dfrac{\sqrt{3}}{2} = \dfrac{\sqrt{3}}{4}$

42. $\mathbf{V} = (\mathbf{a} \times \mathbf{b}) \times \mathbf{a} + (\mathbf{a} \times \mathbf{b}) \times \mathbf{b}$

$= \mathbf{b} - (\mathbf{a}\cdot\mathbf{b})\mathbf{a} + (\mathbf{a}\cdot\mathbf{b})\mathbf{b} - \mathbf{a} = (\mathbf{b} - \mathbf{a}) + (\mathbf{b} - \mathbf{a})(\mathbf{a}\cdot\mathbf{b})$

$\mathbf{V} = (\mathbf{b} - \mathbf{a})(1 + \mathbf{a}\cdot\mathbf{b}) = \lambda(\mathbf{b} - \mathbf{a})$

43. Since, **a** and **b** are perpendicular, hence **a**, **b** and $\mathbf{a} \times \mathbf{b}$ are non-coplanar. Hence, any vector say $(\mathbf{r} \times \mathbf{a})$ can be expressed as

$\therefore$ $\mathbf{r} \times \mathbf{a} = x\mathbf{a} + y\mathbf{b} + z(\mathbf{a} \times \mathbf{b})$ ...(i)

dot with **a** $\quad 0 = x + 0 + 0 \Rightarrow x = 0$

dot with **b** $\quad [\mathbf{r\,a\,b}] = 0 + y + 0 \Rightarrow y = [\mathbf{r\,a\,b}]$

dot with **r** $\quad 0 = x\mathbf{a}\cdot\mathbf{r} + y\,\mathbf{r}\cdot\mathbf{b} + z[\mathbf{r\,a\,b}]$

$0 = [\mathbf{r\,a\,b}](\mathbf{r}\cdot\mathbf{b}) + z[\mathbf{r\,a\,b}] \Rightarrow z = -(\mathbf{r}\cdot\mathbf{b})$

Hence, $\mathbf{r} \times \mathbf{a} = [\mathbf{r\,a\,b}]\,\mathbf{b} - (\mathbf{r}\cdot\mathbf{b})(\mathbf{a} \times \mathbf{b})$

$\mathbf{r} \times \mathbf{a} = [\mathbf{r\,a\,b}]\,\mathbf{b} + (\mathbf{r}\cdot\mathbf{b})(\mathbf{b} \times \mathbf{a})$

44. Since, $\hat{\mathbf{i}} + 2\hat{\mathbf{j}} + 2\hat{\mathbf{k}}$ is rotated so as to cross Y-axis, the vector in new position. Let the required vector be $\mathbf{x}\hat{\mathbf{i}} + y\hat{\mathbf{j}} + z\hat{\mathbf{k}}$

where, $x^2 + y^2 + z^2 = 9$ ...(i)

$x + 2y + 2z = 0$ ...(ii)

and $\begin{vmatrix} x & y & z \\ 1 & 2 & 2 \\ 0 & 1 & 0 \end{vmatrix} \Rightarrow 2x - z = 0$ ...(iii)

On solving Eqs. (i), (ii) and (iii), we get

$$x = -\frac{2}{\sqrt{5}}, y = \sqrt{5}, z = \frac{-4}{\sqrt{5}}$$

$\therefore$ Required vector, is $\dfrac{-2}{\sqrt{5}}\hat{\mathbf{i}} + \sqrt{5}\hat{\mathbf{j}} - \dfrac{4}{\sqrt{5}}\hat{\mathbf{k}}$

45.

Set A	$\rightarrow$	Set B
(Parallel)		(Non-parallel)
4		6

ways $\rightarrow$

(i) 3 from $B \rightarrow {}^6C_3$

(ii) 2 from B, 1 from $A \rightarrow {}^4C_1 \times {}^6C_2$

(iii) 3 from $A \rightarrow {}^4C_3$

Total number of ways

$$= {}^6C_3 + ({}^4C_1 \times {}^6C_2) + {}^4C_1$$

$\therefore$ $n(E) = {}^6C_{13} + {}^4C_1 + ({}^4C_1 \times {}^6C_{12})$

and $n(S) = {}^{10}C_3$

$\Rightarrow$ $P(E) = \dfrac{{}^6C_3 + {}^4C_1 + ({}^6C_2 \times {}^4C_1)}{{}^{10}C_3}$

46. $(\hat{\mathbf{a}} \times \mathbf{x}) + \mathbf{b} = \mathbf{x}$

$\Rightarrow$ $\hat{\mathbf{a}} \times (\hat{\mathbf{a}} \times \mathbf{x}) + (\hat{\mathbf{a}} \times \mathbf{b}) = \hat{\mathbf{a}} \times \mathbf{x}$

$(\hat{\mathbf{a}}\cdot\mathbf{x})\,\hat{\mathbf{a}} - (\hat{\mathbf{a}}\cdot\hat{\mathbf{a}})\,\mathbf{x} + (\hat{\mathbf{a}} \times \mathbf{b}) = \mathbf{x} - \mathbf{b}$

Projection of **x** along $\hat{\mathbf{a}}$ is 2 units

$\Rightarrow$ $\dfrac{(\hat{\mathbf{a}}\cdot\mathbf{x})}{|\hat{\mathbf{a}}|} = 2 \Rightarrow \hat{\mathbf{a}}\cdot\mathbf{x} = 2$

So, $\mathbf{x} = \dfrac{1}{2}[2\hat{\mathbf{a}} - \mathbf{b} + (\hat{\mathbf{a}} \times \mathbf{b})]$

47. We know, $\mathbf{a} \times (\mathbf{b} \times \mathbf{c}) = (\mathbf{a}\cdot\mathbf{c})\,\mathbf{b} - (\mathbf{a}\cdot\mathbf{b})\,\mathbf{c}$

Component of $\mathbf{a} \times (\mathbf{b} \times \mathbf{c})$ along **b** is

$$\left[\left\{\frac{(\mathbf{a}\cdot\mathbf{c})\mathbf{b} - (\mathbf{a}\cdot\mathbf{b})\mathbf{c}}{|\mathbf{b}|^2}\right\}\cdot\mathbf{b}\right]\mathbf{b} = \left(\frac{(\mathbf{a}\cdot\mathbf{c})(\mathbf{b}\cdot\mathbf{b}) - (\mathbf{a}\cdot\mathbf{b})(\mathbf{b}\cdot\mathbf{c})}{|\mathbf{b}|^2}\right)\mathbf{b}$$

So, component of $\mathbf{a} \times (\mathbf{b} \times \mathbf{c})$ perpendicular to **b** is

$$\mathbf{a} \times (\mathbf{b} \times \mathbf{c}) - \left(\frac{(\mathbf{a}\cdot\mathbf{c}\,(\mathbf{b}\cdot\mathbf{b}) - (\mathbf{a}\cdot\mathbf{b})(\mathbf{b}\cdot\mathbf{c})}{|\mathbf{b}|^2}\right)\mathbf{b}$$

$$= \mathbf{a} \times (\mathbf{b} \times \mathbf{c}) + \left(\frac{(\mathbf{a}\cdot\mathbf{b})(\mathbf{b}\cdot\mathbf{c}) - (\mathbf{a}\cdot\mathbf{c})(\mathbf{b}\cdot\mathbf{b})}{|\mathbf{b}|^2}\right)\mathbf{b}$$

$$= \mathbf{a} \times (\mathbf{b} \times \mathbf{c}) + \frac{(\mathbf{a} \times \mathbf{b})\cdot(\mathbf{b} \times \mathbf{c})}{|\mathbf{b}|^2}\,\mathbf{b}$$

48. $\mathbf{r}\cdot\mathbf{a} = 20, \Rightarrow x + 2y + z = 20, x, y, z \in N$

The number of non-negative integral solution are

$$^{17}C_1 + {}^{15}C_1 + \dots {}^{1}C_1 = 81$$

49. $\alpha\cdot\beta = \frac{b}{a} + \frac{4a}{b} + 1$

as $$\frac{b}{a} + \frac{4a}{b} + 1 \ge 5$$

So, $$\left(\frac{10}{5+\alpha\cdot\beta}\right)_{max} = 1$$

50. The system of vectors is coplanar.

$\because$ Their sum is zero.

51.

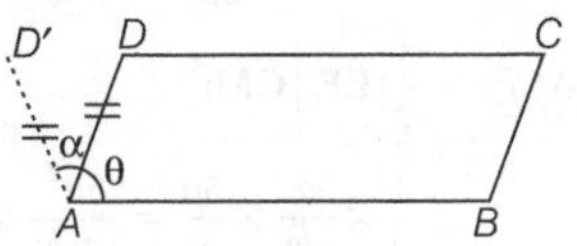

$$\cos\theta = \frac{\mathbf{AB}\cdot\mathbf{AD}}{|\mathbf{AB}||\mathbf{AD}|} = \frac{8}{9}$$

$\therefore$ $\cos^{-1}\left(\frac{8}{9}\right) + \alpha = \frac{\alpha}{2}$ by hypothesis

$\therefore$ $\sin\alpha = \frac{8}{9}$

$\therefore$ $\cos\alpha = \sqrt{1 - \frac{64}{81}} = \frac{\sqrt{17}}{9}$

52. $\mathbf{BA} + \mathbf{AC} = \mathbf{BC}$

$\Rightarrow$ $\mathbf{BA} = \mathbf{BC} - \mathbf{AC}$

$\Rightarrow$ $\frac{\mathbf{e}}{|\mathbf{e}|} - \frac{\mathbf{f}}{|\mathbf{f}|} - \frac{2\mathbf{e}}{|\mathbf{e}|} = -\left(\frac{\mathbf{e}}{|\mathbf{e}|} + \frac{\mathbf{f}}{|\mathbf{f}|}\right)$

Now, $\mathbf{BA}\cdot\mathbf{BC} = \left(\frac{\mathbf{e}}{|\mathbf{e}|} + \frac{\mathbf{f}}{|\mathbf{f}|}\right)\left(\frac{\mathbf{e}}{|\mathbf{e}|} - \frac{\mathbf{f}}{|\mathbf{e}|}\right) = 0$

$\Rightarrow$ $\angle B = 90^\circ$

$\Rightarrow$ $\cos 2B = -1$

and $\cos 2A + \cos 2C = 2\cos(A + C)\cos(A - C) = 0$ $(\because A + C = 90^\circ)$

$\Rightarrow \cos 2A + \cos 2B + \cos 2C = -1$

53. We have, $\mathbf{c}\cdot\mathbf{a} = \mathbf{c}\cdot\mathbf{b} = \cos\theta, \mathbf{a}\cdot\mathbf{b} = 0$

Now, $\mathbf{c} = \alpha\mathbf{a} + \beta\mathbf{b} + \gamma(\mathbf{a}\times\mathbf{b})$

Taking the dot product of both sides with $\mathbf{a}$, we get

$$\mathbf{c}\cdot\mathbf{a} = \alpha = \cos\theta \quad (\because |\mathbf{a}|^2 = 1, \mathbf{a}\cdot\mathbf{b} = 0)$$

Similarly, $\beta = \cos\theta$

Now, taking the dot product with $\mathbf{a}\times\mathbf{b}$, we get

$$[\mathbf{c}\ \mathbf{a}\ \mathbf{b}] = \gamma|\mathbf{a}\times\mathbf{b}|^2 = \gamma$$

Now, $[\mathbf{c}\ \mathbf{a}\ \mathbf{b}]^2 = [\mathbf{a}\ \mathbf{b}\ \mathbf{c}]^2$

$$= \begin{vmatrix} \mathbf{a}\cdot\mathbf{a} & \mathbf{a}\cdot\mathbf{b} & \mathbf{a}\cdot\mathbf{c} \\ \mathbf{b}\cdot\mathbf{a} & \mathbf{b}\cdot\mathbf{b} & \mathbf{b}\cdot\mathbf{c} \\ \mathbf{c}\cdot\mathbf{a} & \mathbf{a}\cdot\mathbf{b} & \mathbf{c}\cdot\mathbf{c} \end{vmatrix} = \begin{vmatrix} 1 & 0 & \cos\theta \\ 0 & 1 & \cos\theta \\ \cos\theta & \cos\theta & 1 \end{vmatrix}$$

$$= 1 - \cos^2\theta + \cos\theta(-\cos\theta)$$

Thus, $\alpha = \beta = \cos\theta, \gamma^2 = -\cos 2\theta.$

54. The mid-points of sides are $D(1, 0, 0)$, $F(0, 0, n)$.

$$EF^2 = \frac{BC^2}{4} \Rightarrow BC^2 = 4(m^2 + n^2)$$

$$\frac{AB^2 + BC^2 + CA^2}{l^2 + m^2 + n^2} = 8$$

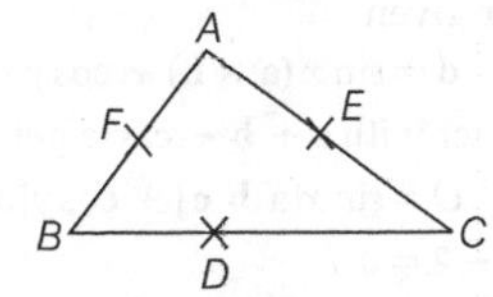

55. Eliminating m,

$$2(l + n)^2 + nl = 0$$

or $(2l + n)(l + 2n) = 0$

$n = -2l \Rightarrow m = -2l$

or $l = -2n \Rightarrow m = -2n$

The d.r's $1, -2, -2$ and $-2, -2, 1$. The lines are perpendicular.

56. $\cos^2\theta + \cos^2\theta + \cos^2\gamma = 1$

$\Rightarrow$ $\cos^2\gamma = -\cos 2\theta$

$\Rightarrow$ $\cos 2\theta \le 0 \Rightarrow \theta \in \left[\frac{\pi}{4}, \frac{\pi}{2}\right]$

57. $\mathbf{a}\times(\mathbf{a}\times\mathbf{b}) = \mathbf{c} \Rightarrow |\mathbf{a}||\mathbf{a}\times\mathbf{b}| = |\mathbf{c}|$

$$1(1\times 5)\sin\theta = 3$$

$$\sin\theta = \frac{3}{5} \text{ gives } \tan\theta = \frac{3}{4}.$$

58. From the figure the vector equation of OP is $r = \lambda(\hat{\mathbf{i}} + \hat{\mathbf{j}} + \hat{\mathbf{k}})$

$\therefore$ OM = projection of $\mathbf{OC}$ on $\mathbf{OP}$

$$= \mathbf{OC}\cdot\mathbf{OP} = \frac{1}{\sqrt{3}}$$

Now, $CM^2 = OC^2 - OM^2 = 1 - \frac{1}{3} = \frac{2}{3}$

$\therefore$ $CM = \sqrt{\frac{2}{3}}$

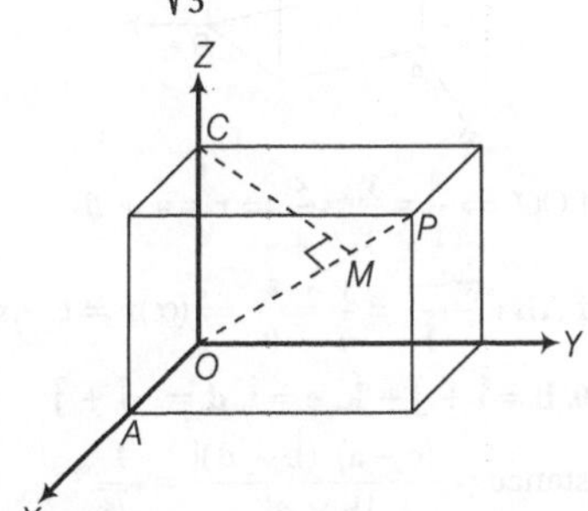

59. $\mathbf{p}\times\mathbf{q}, \mathbf{p}, \mathbf{q}$ are non-coplanar vectors

$\Rightarrow b - c = 0, c - a = 0, a - b = 0$

$\Rightarrow a = b = c$

$\Rightarrow \Delta$ is equilateral.

60. $(\mathbf{a}\times\mathbf{b})\times(\vec{\mathbf{c}}\times\mathbf{d}) = [\mathbf{abd}]\mathbf{c} - [\mathbf{a}\ \mathbf{b}\ \mathbf{c}]\mathbf{d}$

$$= 4\mathbf{c} - 4\mathbf{d}$$

$= -8\hat{\mathbf{i}}$ perpendicular to Y-axis, Z-axis

61. Translating the axes through $A(1, 2, 3)$.
A changes to $(0, 0, 0)$ B changes to $(8, 6, 2)$.
∴Coterminous edges are of lengths 8, 6, 2.
Volume of parallelopiped $= 8 \cdot 6 \cdot 2 = 96$ cu units

62. **a**, **b**, **c** are non-coplanar $\Rightarrow [\mathbf{a}, \mathbf{b}, \mathbf{c}] \neq 1$ Also, $\mathbf{a} \times \mathbf{b}$, $\mathbf{b} \times \mathbf{c}$, $\mathbf{c} \times \mathbf{a}$ are non-coplanar given

$$\mathbf{d} = \sin x\,(\mathbf{a} \times \mathbf{b}) + \cos y\,(\mathbf{b} \times \mathbf{c}) + 2(\mathbf{c} \times \mathbf{a}).$$

Taking dot product with $\mathbf{a} + \mathbf{b} + \mathbf{c}$, we get

$$O = \sin x[\mathbf{a}\ \mathbf{b}\ \mathbf{c}] + \cos y[\mathbf{a}\ \mathbf{b}\ \mathbf{c}] + 2[\mathbf{a}\ \mathbf{b}\ \mathbf{c}]$$

$\Rightarrow \sin x + \cos y + 2 = 0$

$\Rightarrow \quad \sin x + \cos y = -2$

$\Rightarrow \quad x = (4n-1)\frac{\pi}{2}, y = (2n-1)\pi, n \in z$

for least value of $x^2 + y^2$, $x = \frac{-\pi}{2}$, $y = \pi$ and least value is $\frac{5\pi^2}{4}$.

63. We have, $\alpha(\mathbf{a} \times \mathbf{b}) + \beta(\mathbf{b} \times \mathbf{c}) + \gamma(\mathbf{c} \times \mathbf{a}) = 0$
Taking dot product with c, we have

$$\alpha[\mathbf{a}\ \mathbf{b}\ \mathbf{c}] + \beta[\mathbf{b}\ \mathbf{c}\ \mathbf{c}] + \gamma[\mathbf{c}\ \mathbf{a}\ \mathbf{c}] = 0$$

i.e. $\quad \alpha[\mathbf{abc}] + 0 + 0 = 0$

$$\alpha[\mathbf{a}\ \mathbf{b}\ \mathbf{c}] = 0$$

Similarly, taking dot product with b and c, we have

$$\gamma\,[\mathbf{abc}] = 0, \beta\,[\mathbf{abc}] = 0$$

Now, even if one of $\alpha, \beta, \gamma \neq 0$, then we have $[\mathbf{abc}] = 0$
$\Rightarrow$ **a**, **b**, **c** are coplanar.

64. $\frac{1}{3}\lambda_1 h_4 + \frac{1}{3}\lambda_2 h_3 + \frac{1}{3}\lambda_3 h_2 + \frac{1}{3}\lambda_4 h_1 = 4$ area the tetrahedron $OABC$.

65. $\theta = \cos^{-1}(\cos\beta - \cos\alpha)$

66.

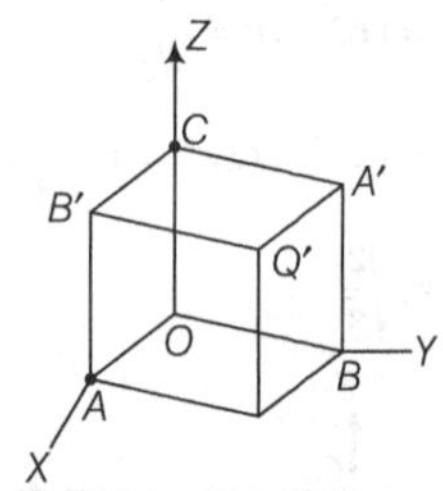

Equation of $OO' \Rightarrow \frac{x}{1} = \frac{y}{1} = \frac{z}{1} \Rightarrow \mathbf{r} = \mathbf{a} + t\mathbf{b}$

Equation of **AB** : $\frac{x-1}{-1} = \frac{y}{1} = \frac{z}{0} \Rightarrow (\alpha)\mathbf{r} = \mathbf{c} + s\mathbf{d}$,

where $\mathbf{a} = 0$, $\mathbf{b} = \hat{\mathbf{i}} + \hat{\mathbf{j}} + \hat{\mathbf{k}}$, $\mathbf{c} = \hat{\mathbf{i}}$, $\mathbf{d} = -\hat{\mathbf{i}} + \hat{\mathbf{j}}$

Shortest distance $= \frac{|(\mathbf{c} - \mathbf{a}) \cdot (\mathbf{b} \times \mathbf{d})|}{|\mathbf{b} \times \mathbf{d}|} = \frac{1}{\sqrt{6}}$

67. $v \times w = 3\hat{\mathbf{i}} - 5\hat{\mathbf{j}} - \hat{\mathbf{k}}$
Maximum value of $[uvw] = |u|\,|v \times w| = 1 \cdot \sqrt{35} = \sqrt{35}$

68. $|\mathbf{a}| = |\mathbf{b}| = |\mathbf{c}|$
$|\mathbf{b} - \mathbf{a}| = |\mathbf{b} - \mathbf{c}| = |\mathbf{c} - \mathbf{a}| = a$

$$\mathbf{a} \cdot \mathbf{b} = \mathbf{b} \cdot \mathbf{c} = \mathbf{c} \cdot \mathbf{a} = \mathbf{AB} \cdot \mathbf{AC} = \mathbf{CA} \cdot \mathbf{CB} = \mathbf{BA} \cdot \mathbf{BC}$$

$$= a^2 \cos\frac{\pi}{3}$$

$\Rightarrow \quad (\mathbf{EF})^2 = \frac{a^2}{\sqrt{3}} \Rightarrow \mathbf{EF} = \frac{a}{\sqrt{3}}$

$$|\mathbf{CF}| = |\mathbf{CE}| = \frac{\sqrt{7}\mathbf{a}}{3} \text{ and } |\mathbf{CM}| = \frac{5a}{6}$$

where 'm' is middle point of **EF**.

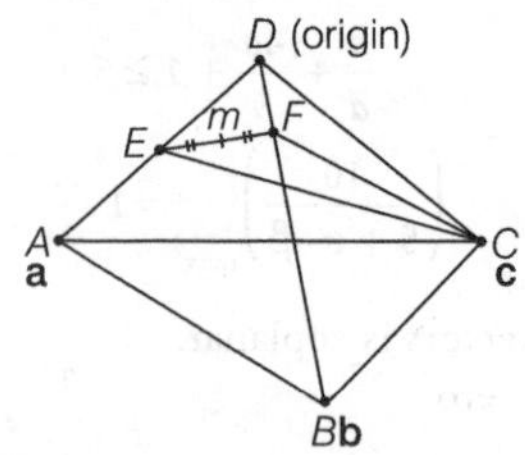

Area of $\Delta CEF = \frac{1}{2}|\mathbf{EF}|\,|\mathbf{CM}|$

$$= \frac{1}{2} \times \frac{a}{\sqrt{3}} \times \frac{5a}{6} = \frac{5a^2}{12\sqrt{3}} \text{ sq units}$$

69. $\mathbf{p} \cdot \mathbf{q} = 0$ and $\mathbf{r} \cdot \mathbf{s} = 0$

$\Rightarrow \quad (5\mathbf{a} - 3\mathbf{b}) \cdot (-\mathbf{a} - 2\mathbf{b}) = 0$

$6\mathbf{b}^2 - 7\mathbf{a} \cdot \mathbf{b} - 5\mathbf{a}^2 = 0 \quad$ ...(i)

$\Rightarrow \quad (-4\mathbf{a} - \mathbf{b}) \cdot (-\mathbf{a} + \mathbf{b}) = 0$

$4\mathbf{a}^2 - \mathbf{b}^2 - 3\mathbf{a} \cdot \mathbf{b} = 0 \quad$ ...(ii)

Now, $\quad \mathbf{x} = \frac{1}{3}(\mathbf{p} + \mathbf{r} + \mathbf{s})$

$$\mathbf{x} = \frac{1}{3}(5\mathbf{a} - 3\mathbf{b} - 4\mathbf{a} - \mathbf{b} - \mathbf{a} + \mathbf{b})$$

∴ $\quad \mathbf{x} = -\mathbf{b}, \mathbf{y} = \frac{1}{5}(\mathbf{r} + \mathbf{s}) = \frac{1}{5}(-5\mathbf{a}) = -\mathbf{a}$

Angle between **x** and **y** i.e $\cos\theta = \frac{\mathbf{x} \cdot \mathbf{y}}{|\mathbf{x}||\mathbf{y}|} = \frac{\mathbf{a} \cdot \mathbf{b}}{|\mathbf{a}|\,|\mathbf{b}|}$

From Eqs. (i) and (ii), we get

$$|\mathbf{a}| = \sqrt{\frac{25}{19}} \cdot \sqrt{\mathbf{a}\ \mathbf{b}} \text{ and } |\mathbf{b}| = \sqrt{\frac{43}{19}}\sqrt{\mathbf{a}\ \mathbf{b}}$$

$$|\mathbf{a}|\,|\mathbf{b}| = \frac{\sqrt{25 \times 43}}{19}(\vec{\mathbf{a}} \cdot \vec{\mathbf{b}}) = \frac{19}{5\sqrt{43}} = \frac{\mathbf{a} \cdot \mathbf{b}}{|\mathbf{a}|\,|\mathbf{b}|}$$

$\Rightarrow \quad \cos\theta = \frac{19}{5\sqrt{43}}$

∴ $\quad \theta = \cos^{-1}\left(\frac{19}{5\sqrt{43}}\right)$

70. Volume of tetrahedron $= \frac{1}{6}[\mathbf{a}\ \mathbf{b}\ \mathbf{c}]$

$$\text{Now, } [\mathbf{a}\ \mathbf{b}\ \mathbf{c}]^2 = \begin{vmatrix} \mathbf{a} \cdot \mathbf{a} & \mathbf{a} \cdot \mathbf{b} & \mathbf{a} \cdot \mathbf{c} \\ \mathbf{b} \cdot \mathbf{a} & \mathbf{b} \cdot \mathbf{b} & \mathbf{b} \cdot \mathbf{c} \\ \mathbf{c} \cdot \mathbf{a} & \mathbf{c} \cdot \mathbf{b} & \mathbf{c} \cdot \mathbf{c} \end{vmatrix} = \begin{vmatrix} 4 & 2 & 2 \\ 2 & 4 & 2 \\ 2 & 2 & 4 \end{vmatrix}$$

$$= 4(12) + 2(-4) + 2\,(-4)$$

$$\text{Volume} = \frac{1}{6} \times 4\sqrt{2} = \frac{2\sqrt{2}}{3}$$

71. Given, $\cos\theta = (\mathbf{a} \times \hat{\mathbf{i}}) \cdot (\mathbf{b} \times \hat{\mathbf{i}}) + (\mathbf{a} + \hat{\mathbf{j}}) \cdot (\mathbf{b} \times \hat{\mathbf{j}})$
$+ (\mathbf{a} \times \hat{\mathbf{k}}) \cdot (\mathbf{b} \times \hat{\mathbf{k}}) \quad$ ...(i)

Consider, $(\mathbf{a} \times \hat{\mathbf{i}}) \cdot (\mathbf{b} \times \hat{\mathbf{i}}) = [(\mathbf{a} \times \hat{\mathbf{i}})\mathbf{b}\ \hat{\mathbf{i}}] = ((\mathbf{a} \times \hat{\mathbf{i}}) \times \mathbf{b}) \cdot \hat{\mathbf{i}}$

$$((\mathbf{a}\cdot\mathbf{b})\hat{\mathbf{i}}) - (\hat{\mathbf{i}}\cdot\mathbf{b})\mathbf{a})\hat{\mathbf{i}} = (\mathbf{a}\cdot\mathbf{b})(\hat{\mathbf{i}}\cdot\hat{\mathbf{i}}) - (\hat{\mathbf{i}}\cdot\hat{\mathbf{b}})(\mathbf{a}\cdot\hat{\mathbf{i}})$$
$$= \mathbf{a}\cdot\mathbf{b} - \mathbf{a}_1\mathbf{b}_1$$

Similarly, $(\mathbf{a}\times\hat{\mathbf{j}})\cdot(\mathbf{b}\times\hat{\mathbf{j}}) = \mathbf{a}\cdot\mathbf{b} - \mathbf{a}_2\mathbf{b}_2$

and $(\mathbf{a}\times\hat{\mathbf{k}})(\mathbf{b}\times\hat{\mathbf{k}}) = \mathbf{a}\cdot\mathbf{b} - a_3b_3$

∴ From Eq. (i), we get

$$\cos\theta = 3\mathbf{a}\cdot\mathbf{b} - (\mathbf{a}_1\mathbf{b}_1 + \mathbf{a}_2\mathbf{b}_2 + \mathbf{a}_3\mathbf{b}_3)$$
$$= 3\mathbf{a}\cdot\mathbf{b} - \mathbf{a}\cdot\mathbf{b}$$

$$\Rightarrow \quad \frac{\mathbf{a}\cdot\mathbf{b}}{|\mathbf{a}||\mathbf{b}|} = 2\mathbf{a}\cdot\mathbf{b} \Rightarrow |\mathbf{a}||\mathbf{b}| = \frac{1}{2}$$

Now, use AM ≥ GM on $|\mathbf{a}|, |\mathbf{b}|$

$$\therefore \quad \frac{|\mathbf{a}|+|\mathbf{b}|}{2} \ge (|\mathbf{a}|\cdot|\mathbf{b}|)^{\frac{1}{2}} \Rightarrow |\mathbf{a}|+|\mathbf{b}| \ge \frac{2}{\sqrt{2}}$$

$$\Rightarrow \quad |\mathbf{a}|+|\mathbf{b}| \ge \sqrt{2}$$

72. $\because \mathbf{a}\cdot\mathbf{b} < 0$

$$\Rightarrow (\sin^2 x - \sin x) - \cos^2 x + 3 - 4\sin x < 0$$
$$\Rightarrow \quad 2\sin^2 x - 5\sin x + 2 < 0$$
$$\Rightarrow \quad \underbrace{(\sin x - 2)}_{(-)\text{ve}}(2\sin x - 1) < 0$$
$$\Rightarrow \quad \sin x > \frac{1}{2} \Rightarrow x \in \left(\frac{\pi}{6}, \frac{\pi}{2}\right)$$

73.

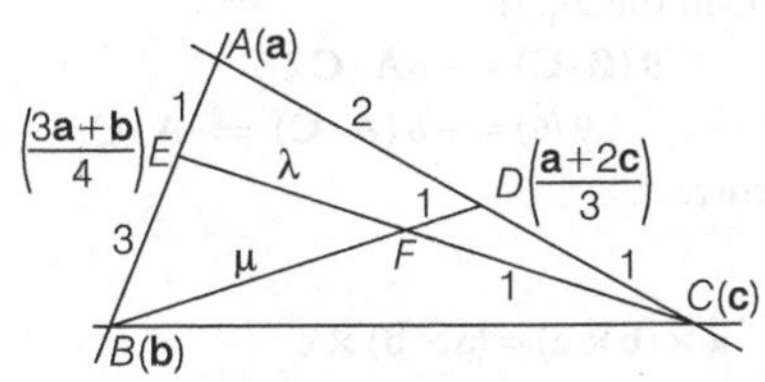

Positive vector of point $E = \frac{3\mathbf{a}+\mathbf{b}}{4}$

Position vector of point $D = \frac{\mathbf{a}+2\mathbf{c}}{3}$

Let point F divides EC in λ : 1 and BD in μ : 1,

then
$$\frac{\mathbf{b} + \frac{\mu}{3}(\mathbf{a}+2\mathbf{c})}{\mu+1} = \frac{\lambda\mathbf{c} + \frac{3\mathbf{a}+\mathbf{b}}{4}}{\lambda+1}$$

$$\left[\mathbf{b} + \frac{\mu}{3}(\mathbf{a}+2\mathbf{c})\right](\lambda+1) = \left(\lambda\mathbf{c} + \frac{3\mathbf{a}+\mathbf{b}}{4}\right)(\mu+1)$$

Comparing the coefficient of **a b** and **c**.

$$\Rightarrow \quad \frac{\mu(\lambda+1)}{3} = \frac{3(\mu+1)}{4} \quad \text{...(i)}$$

$$\Rightarrow \quad \lambda + 1 = \frac{\mu+1}{4} \quad \text{...(ii)}$$

$$\frac{2\mu}{3}(\lambda+1) = \lambda(\mu+1) \quad \text{...(iii)}$$

On solving, we get $\lambda = \frac{3}{2}$

74. Obviously, $\frac{\mathbf{a}}{|\mathbf{a}|} + \frac{\mathbf{b}}{|\mathbf{b}|}$ is a vector in the plane of **a** and **b** and hence perpendicular to $\mathbf{a}\times\mathbf{b}$. It is also equally inclined to **a** and **b** as it is along the angle bisector.

75.
$$\mathbf{a}\times(\mathbf{b}\times\mathbf{c}) = (\mathbf{a}\times\mathbf{b})\times\mathbf{c}$$
or $(\mathbf{a}\cdot\mathbf{c})\,\mathbf{b} - (\mathbf{a}\cdot\mathbf{b})\,\mathbf{c} = (\mathbf{a}\cdot\mathbf{c})\,\mathbf{b} - (\mathbf{c}\cdot\mathbf{b})\,\mathbf{a}$

or $(\mathbf{a}\cdot\mathbf{b})\,\mathbf{c} - (\mathbf{c}\cdot\mathbf{b})\,\mathbf{a} = \mathbf{0}$

or $\mathbf{b}\times(\mathbf{c}\times\mathbf{a}) = \mathbf{0}$

or $(\mathbf{c}\times\mathbf{a})\times\mathbf{b} = \mathbf{0}$

or $\mathbf{b}\times(\mathbf{c}\times\mathbf{a}) = (\mathbf{c}\times\mathbf{a})\times\mathbf{b} = \mathbf{0}$

76. Let angle between **a** and **b** be θ

$$\mathbf{v} = \mathbf{a}\times\mathbf{b} = |\mathbf{a}||\mathbf{b}|\sin\theta\,\hat{\mathbf{n}}$$

$$\therefore \quad |\mathbf{v}| = \sin\theta, \quad \left(\because |\mathbf{a}| = 1, |\mathbf{b}| = 1, \mathbf{n} = \frac{(\mathbf{a}\times\mathbf{b})}{|\mathbf{a}\times\mathbf{b}|} = \frac{\mathbf{v}}{|\mathbf{v}|}\right)$$

$$\mathbf{u} = \mathbf{a} - (\mathbf{a}\cdot\mathbf{b})\,\mathbf{b} = \mathbf{a} - \cos\theta\,\mathbf{b}$$
$$(\because \mathbf{a}\cdot\mathbf{b} = |\mathbf{a}||\mathbf{b}|\cos\theta = \cos\theta)$$

$$\mathbf{u}\cdot\mathbf{u} = |\mathbf{u}|^2 = 1 + \cos^2\theta - 2\cos\theta = \sin^2\theta$$

$$\therefore \quad |\mathbf{u}| = \sin\theta$$

$$\mathbf{u}\cdot\mathbf{a} = \mathbf{a}\cdot\mathbf{a} = -\cos\theta\,\mathbf{a}\cdot\mathbf{b} = 1 - \cos^2\theta = \sin^2\theta$$
$$\mathbf{u}\cdot\mathbf{b} = \mathbf{a}\cdot\mathbf{b} - \cos\theta\,\mathbf{b}\cdot\mathbf{b} = \cos\theta - \cos\theta = 0$$
$$\mathbf{u}\cdot(\mathbf{a}+\mathbf{b}) = (\mathbf{a} - \cos\theta\,\mathbf{b})\cdot(\mathbf{a}+\mathbf{b})$$
$$= 1 + \cos\theta - \cos^2\theta - \cos\theta$$
$$= 1 - \cos^2\theta = \sin^2\theta$$

77. Here, $(l\mathbf{a} + m\mathbf{b})\times\mathbf{b} = \mathbf{c}\times\mathbf{b}$

$$\Rightarrow \quad l\mathbf{a}\times\mathbf{b} = \mathbf{c}\times\mathbf{b}$$
$$\Rightarrow \quad l(\mathbf{a}\times\mathbf{b})^2 = (\mathbf{c}\times\mathbf{b})\cdot(\mathbf{a}\times\mathbf{b})$$
$$\Rightarrow \quad l = \frac{(\mathbf{c}\times\mathbf{b})\cdot(\mathbf{a}\times\mathbf{b})}{(\mathbf{a}\times\mathbf{b})^2}$$

Similarly, $m = \frac{(\mathbf{c}\times\mathbf{a})\cdot(\mathbf{b}\times\mathbf{a})}{(\mathbf{b}\times\mathbf{a})^2}$

78. $\mathbf{a}\times(\mathbf{r}\times\mathbf{a}) = \mathbf{a}\times\mathbf{b}$

$$3\,\mathbf{r} - (\mathbf{a}\cdot\mathbf{r})\,\mathbf{a} = \mathbf{a}\times\mathbf{b}$$

Also, $|\mathbf{r}\times\mathbf{a}| = |\mathbf{b}|$

$$\Rightarrow \quad \sin^2\theta = \frac{2}{3}$$

or $(1 - \cos^2\theta) = \frac{2}{3}$

or $\frac{1}{3} = \cos^2\theta \Rightarrow \mathbf{a}\cdot\mathbf{r} = \pm 1$

$$\Rightarrow \quad 3\mathbf{r} \pm \mathbf{a} = \mathbf{a}\times\mathbf{b}$$

or $\mathbf{r} = \frac{1}{3}(\mathbf{a}\times\mathbf{b} \pm \mathbf{a})$

79. $a_1 + a_2\cos 2x + a_3\sin^2 x = 0, \forall\, x \in R$

or $(a_1 + a_2) + \sin^2 x(a_3 - 2a_2) = 0$

$$\Rightarrow \quad a_1 + a_2 = 0 \text{ and } a_3 - 2a_2 = 0$$
$$\frac{a_1}{-1} = \frac{a^2}{1} = \frac{a_3}{2} = \lambda(\ne 0)$$
$$\Rightarrow \quad a^1 = -\lambda, a_2 = \lambda, a_3 = 2\lambda$$

80. $\mathbf{a}\times\mathbf{b} = |\mathbf{a}||\mathbf{b}|\sin\theta\,\hat{\mathbf{n}}$

or $|\mathbf{a}\times\mathbf{b}| = |\mathbf{a}||\mathbf{b}|\sin\theta$

or $\sin\theta = \frac{|\mathbf{a}\times\mathbf{b}|}{|\mathbf{a}|\times|\mathbf{b}|}$ ...(i)

$$\mathbf{a}\cdot\mathbf{b}=|\mathbf{a}||\mathbf{b}|\cos\theta$$

$$\Rightarrow \quad \cos\theta=\frac{|\mathbf{a}\cdot\mathbf{b}|}{|\mathbf{a}||\mathbf{b}|} \quad \text{...(ii)}$$

From Eqs. (i) and (ii),

$$\sin^2\theta+\cos^2\theta=1$$

$$\Rightarrow \quad |\mathbf{a}\times\mathbf{b}|^2+(\mathbf{a}\cdot\mathbf{b})^2=|\mathbf{a}|^2|\mathbf{b}|^2$$

If $\theta=\pi/4$, then $\sin\theta=\cos\theta=1/\sqrt{2}$. Therefore,

$$|\mathbf{a}\times\mathbf{b}|=\frac{|\mathbf{a}||\mathbf{b}|}{\sqrt{2}} \text{ and } \mathbf{a}\cdot\mathbf{b}=\frac{|\mathbf{a}||\mathbf{b}|}{\sqrt{2}}$$

$$|\mathbf{a}\times\mathbf{b}|=\mathbf{a}\cdot\mathbf{b}$$

$$\mathbf{a}\times\mathbf{b}=|\mathbf{a}||\mathbf{b}|\sin\theta\,\hat{\mathbf{n}}=\frac{|\mathbf{a}||\mathbf{b}|}{\sqrt{2}}\hat{\mathbf{n}}$$

$$=(\mathbf{a}\cdot\mathbf{b})\,\hat{\mathbf{n}}$$

81. We have, $|\mathbf{a}-\mathbf{b}|^2=|\mathbf{a}|^2+|\mathbf{b}|^2-2(\mathbf{a}\cdot\mathbf{b})$

or $|\mathbf{a}-\mathbf{b}|^2=|\mathbf{a}|^2+|\mathbf{b}|^2-2|\mathbf{a}||\mathbf{b}|\cos2\theta$

or $|\mathbf{a}-\mathbf{b}|^2=2-\cos2\theta \quad (\because |\mathbf{a}|=|\mathbf{b}|=1)$

$=4\sin^2\theta$ or $|\mathbf{a}-\mathbf{b}|=2|\sin\theta|$

Now, $|\mathbf{a}-\mathbf{b}|<1$

$\Rightarrow \quad 2|\sin\theta|<1$ or $|\sin\theta|<\frac{1}{2}$

$\Rightarrow \quad \theta\in[0,\pi/6)$ or $\theta\in(5\pi/6,\pi)$

82. $\mathbf{a}\times(\mathbf{b}\times\mathbf{c})+(\mathbf{a}\cdot\mathbf{b})\,\mathbf{b}$

$=(4-2x-\sin y)\,\mathbf{b}+(x^2-1)\,\mathbf{c}$

or $(\mathbf{a}\cdot\mathbf{c})\,\mathbf{b}-(\mathbf{a}\cdot\mathbf{b})\mathbf{c}+(\mathbf{a}\cdot\mathbf{b})\,\mathbf{b}$

$=(4-2x-\sin y)\,\mathbf{b}+(x^2-1)\,\mathbf{c}$

Now, $(\mathbf{c}\cdot\mathbf{c})\,\mathbf{a}=\mathbf{c}$.

Therefore, $(\mathbf{c}\cdot\mathbf{c})\,(\mathbf{a}\cdot\mathbf{c})=(\mathbf{c}\cdot\mathbf{c})$ or $\mathbf{a}\,.\,\mathbf{c}=1$

$\Rightarrow \quad 1+\mathbf{a}\cdot\mathbf{b}=4-2x-\sin y,\ x^2-1=-(\mathbf{a}\cdot\mathbf{b})$

or $\quad 1=4-2x-\sin y+x^2-1$

or $\quad \sin y=x^2-2x+2=(x-1)^2+1$

But, $\sin y\le1 \Rightarrow x=1, \sin y=1 \Rightarrow y=(4n+1)\frac{\pi}{2}, n\in I$

83. $\mathbf{AB}+\mathbf{BC}=\mathbf{AC}$

$$\mathbf{BC}=\frac{2\mathbf{u}}{|\mathbf{u}|}-\frac{\mathbf{u}}{|\mathbf{u}|}+\frac{\mathbf{v}}{|\mathbf{v}|}=\frac{\mathbf{u}}{|\mathbf{u}|}+\frac{\mathbf{v}}{|\mathbf{v}|}$$

$$\mathbf{AB}\cdot\mathbf{BC}=\left(\frac{\mathbf{u}}{|\mathbf{u}|}-\frac{\mathbf{v}}{|\mathbf{v}|}\right)\left(\frac{\mathbf{u}}{|\mathbf{u}|}+\frac{\mathbf{v}}{|\mathbf{v}|}\right)$$

$$=(\hat{\mathbf{u}}-\hat{\mathbf{v}})\cdot(\hat{\mathbf{u}}+\hat{\mathbf{v}})=1-1=0$$

$\Rightarrow \quad \angle B=90°$

$\Rightarrow \quad 1+\cos2A+\cos2B+\cos2C=0$

84. Clearly, $\mathbf{a}\cdot\mathbf{c}=0$ and $\mathbf{b}\cdot\mathbf{c}=0$. Also, $\mathbf{a}\cdot\mathbf{b}=0$

$\mathbf{a}\times\mathbf{b}=\mathbf{c}$

dot with $\mathbf{b}\Rightarrow\mathbf{b}\cdot\mathbf{c}=0$

Similarly, $\mathbf{b}\times\mathbf{c}=\mathbf{a}$

dot with $\mathbf{b}\Rightarrow\mathbf{a}\cdot\mathbf{b}=0$

dot with $\mathbf{c}\Rightarrow\mathbf{a}\cdot\mathbf{c}=0$

$\Rightarrow \quad \mathbf{a}\cdot\mathbf{b}=\mathbf{b}\cdot\mathbf{c}=\mathbf{c}\cdot\mathbf{a}=0$

Again, $\quad |\mathbf{a}||\mathbf{b}|=|\mathbf{c}|$

$|\mathbf{b}||\mathbf{c}|=|\mathbf{a}|$

$$\Rightarrow \quad \frac{|\mathbf{a}|}{|\mathbf{c}|}=\frac{|\mathbf{c}|}{|\mathbf{a}|}$$

$\Rightarrow \quad |\mathbf{a}|=|\mathbf{c}|$ and $|\mathbf{b}|=1$

$\Rightarrow \quad \mathbf{a}\times\mathbf{b}\cdot\mathbf{c}=|\mathbf{a}||\mathbf{b}||\mathbf{c}|=|\mathbf{a}|^2=|\mathbf{c}|^2$

(children will assume $\mathbf{a}=\hat{\mathbf{i}}$; $\mathbf{b}=\hat{\mathbf{j}}$ and $\mathbf{c}=\hat{\mathbf{k}}$ but in this case all the four will be correct which will be wrong).

85. Given, $|\mathbf{A}||\mathbf{B}|\cos\theta=-6;\ |\mathbf{B}|=2$ (given)

$\mathbf{B}\cdot\mathbf{C}=|\mathbf{B}||\mathbf{C}|\cos\phi=6$

and $\quad (\mathbf{A}\times\mathbf{B})\times\mathbf{A}=0$

$(\mathbf{A}\cdot\mathbf{A})\mathbf{B}-(\mathbf{B}\cdot\mathbf{A})\mathbf{A}=0$

$(\mathbf{A}\cdot\mathbf{A})=-6\mathbf{A} \quad$ ...(i)

$\therefore$ **A** and **B** are collinear and θ between **A** and **B** is π.

$\Rightarrow \quad \mathbf{A}\times\mathbf{B}=0$

$\Rightarrow$ (a) is correct.

$\Rightarrow \quad \mathbf{A}\cdot(\mathbf{B}\times\mathbf{C})=(\mathbf{A}\times\mathbf{B})\cdot\mathbf{C}=0$

$\Rightarrow$ (b) is correct.

Also, $\quad \mathbf{A}\cdot\mathbf{B}=-6$ and $|\mathbf{B}|=2$

$\therefore \quad |\mathbf{A}||\mathbf{B}|\cos\pi=-6|\mathbf{A}|\cdot(2)=6$

$\Rightarrow \quad |\mathbf{A}|=3\Rightarrow\mathbf{A}\cdot\mathbf{A}=9$

$\Rightarrow$ (c) is not correct.

Again, $\mathbf{A}\cdot\mathbf{C}=?$

dot with **C** in the Eq. (i)

$9\,(\mathbf{B}\cdot\mathbf{C})=-6\mathbf{A}\cdot\mathbf{C}$

$9\,(6)=-6\,(\mathbf{A}\cdot\mathbf{C})\Rightarrow\mathbf{A}\cdot\mathbf{C}=-9$

$\Rightarrow$ (d) is correct.

86. $\mathbf{V}_1=\mathbf{V}_2$

$\mathbf{a}\times(\mathbf{b}\times\mathbf{c})=(\mathbf{a}\times\mathbf{b})\times\mathbf{c}$

$(\mathbf{a}\cdot\mathbf{c})\,\mathbf{b}-(\mathbf{a}\cdot\mathbf{b})\,\mathbf{c}=(\mathbf{a}\cdot\mathbf{c})\,\mathbf{b}-(\mathbf{b}\cdot\mathbf{c})\,\mathbf{a}$

$\therefore \quad (\mathbf{a}\cdot\mathbf{b})\mathbf{c}=(\mathbf{b}\cdot\mathbf{c})\,\mathbf{a}$

$\Rightarrow$ Either **c** and **a** are collinear or **b** is perpendicular to both **a** and **c**

$\Rightarrow \quad \mathbf{b}=\lambda\,(\mathbf{a}\times\mathbf{c})$

87. It may be observed that

$$[\mathbf{U}\ \mathbf{V}\ \mathbf{W}]=\begin{vmatrix}2&3&-6\\6&2&3\\3&-6&-2\end{vmatrix}=343\ne0$$

$\Rightarrow$ **U**, **V** and **W** are non-coplanar, hence linearly independent

Further $\mathbf{U}\times\mathbf{V}=\mathbf{W}$ and $\mathbf{V}\times\mathbf{W}=\mathbf{U}$

They form a right handed triplet of mutually perpendicular vectors and of course!

$\Rightarrow \quad (\mathbf{U}\times\mathbf{V})\times\mathbf{W}=0$ and $\mathbf{U}\times(\mathbf{V}\times\mathbf{W})$

88. Let the required vector be $\mathbf{d}=x\hat{\mathbf{i}}+y\hat{\mathbf{j}}+z\hat{\mathbf{k}}$. For this to be coplanar wih **b** and **c**, we must have

$$\begin{vmatrix}x&y&z\\1&2&-1\\1&1&-2\end{vmatrix}=0$$

$\Rightarrow \quad x(-4+1)+y(-1+2)+z(1-2)=0$

$-3x+y-z=0$

The projection of **d** on **a** is $\frac{|\mathbf{a}\cdot\mathbf{d}|}{|\mathbf{a}|}$.

So, $\sqrt{\frac{2}{3}} = \frac{1}{\sqrt{6}}|2x - y + z|$

$\Rightarrow \quad 2x - y + z = \pm 2$

The choices (a) and (c) satisfy the Eqs. (i) and (ii).

89. $\mathbf{a} \times (\mathbf{b} - 3\mathbf{c}) = \mathbf{0}$

$\Rightarrow \quad \mathbf{b} - 3\mathbf{c} = \lambda \mathbf{a}$

$\Rightarrow \quad |\mathbf{b} - 3\mathbf{c}| = |\lambda \mathbf{a}|$

$\Rightarrow \quad 1 + 1 - 6.1 \cdot \frac{1}{3} \cdot \frac{1}{2} = |\lambda| \quad \Rightarrow \lambda \pm 1.$

$\therefore \quad \mathbf{b} - 3\mathbf{c} = \pm \mathbf{a}$

90. $(\mathbf{a} \times \mathbf{c}) \cdot (\mathbf{r} \times \mathbf{a}) = (\mathbf{a} \times \mathbf{c}) \cdot \mathbf{b}$

91. (a) is proved if we take dot product of both sides with **a**.

(b) If we take dot product with **b**, we get

$$\lambda_2 = \mathbf{b} \cdot \mathbf{c}$$

$\Rightarrow$ Option (b) is not true.

(c) If we take dot product of both sides with $\mathbf{a} \times \mathbf{b}$, we get

$$[\mathbf{c}\ \mathbf{b}\ \mathbf{a}] = \lambda_3 [\mathbf{a} \times \mathbf{b}]^2$$

$\Rightarrow \quad \lambda_3 = [\mathbf{a}\ \mathbf{b}\ \mathbf{c}]$ or $\mathbf{c} \cdot (\mathbf{a} \times \mathbf{b})$

$\Rightarrow$ Option (c) is wrong.

(d) is correct since $\lambda_1 + \lambda_2 + \lambda_3 = \mathbf{c} \cdot \mathbf{a} + \mathbf{b} \cdot \mathbf{c} + [\mathbf{a}\ \mathbf{b}\ \mathbf{c}]$.

92. (a) Since, **a**, **b**, **c**, are non-coplanar, option (a) is true.

Since, $\mathbf{b} \times \mathbf{c}$, $\mathbf{c} \times \mathbf{a}$, $\mathbf{a} \times \mathbf{b}$ are also non-coplanar.

(b) is also correct.

Since, $\mathbf{x} = \lambda(\mathbf{b} \times \mathbf{c}) + \mu(\mathbf{c} \times \mathbf{a}) + v(\mathbf{a} \times \mathbf{b})$

We have, $\lambda = \frac{\mathbf{a} \cdot \mathbf{x}}{[\mathbf{a}\ \mathbf{b}\ \mathbf{c}]}$, (on taking dot product with **a**)

μ and v have similar values.

Also, $\quad |\mathbf{x}| = |\mathbf{a} - \mathbf{x}|$

$\Rightarrow \quad \mathbf{a} \cdot \mathbf{x} = \frac{a^2}{2}$, etc.

$\Rightarrow$ Option (c) is correct.

If (c) is correct (d) is ruled out.

93. $\alpha = (\mathbf{a} \cdot \hat{\mathbf{i}})\hat{\mathbf{i}} + (\mathbf{a} \cdot \hat{\mathbf{j}})\hat{\mathbf{j}} + (\mathbf{a} \cdot \hat{\mathbf{k}})\hat{\mathbf{k}} = \mathbf{a} = (1, 1, 1)$

$\beta = (\mathbf{b} \cdot \hat{\mathbf{i}})\hat{\mathbf{i}} + (\mathbf{b} \cdot \hat{\mathbf{j}})\hat{\mathbf{j}} + (\mathbf{b} \cdot \hat{\mathbf{k}})\hat{\mathbf{k}} = \mathbf{b} = (1, -1, 0)$

$\gamma = (1, 1, -2)$

$\therefore \quad \alpha \cdot \beta = \beta \cdot \gamma = \gamma \cdot \alpha = 0$

$\Rightarrow \alpha, \beta, \gamma$ are mutually perpendicular $\alpha, \beta, \gamma = 6$

$\Rightarrow \alpha, \beta, \gamma$ form a parallelopiped of volume 6 units.

94. $\mathbf{a} \times (\mathbf{b} \times \mathbf{c}) = (\mathbf{a} \cdot \mathbf{c})\mathbf{b} - (\mathbf{a} \cdot \mathbf{b})\mathbf{c}$

$$= (xz + yx + yz)(y\hat{\mathbf{i}} + z\hat{\mathbf{j}} + x\hat{\mathbf{k}}) - (xy + yz + zx)(z\hat{\mathbf{i}} + x\hat{\mathbf{j}} + y\hat{\mathbf{k}})$$

$$= (xz + yx + yz)((y - z)\hat{\mathbf{i}} + (z - x)\hat{\mathbf{j}} + (x - y)\hat{\mathbf{k}})$$

Clearly, perpendicular to $\hat{\mathbf{i}} + \hat{\mathbf{j}} + \hat{\mathbf{k}}$ and also to $(y + z)\hat{\mathbf{i}} + (z + x)\hat{\mathbf{j}} + (x + y)\hat{\mathbf{k}}$ as dot products are zeros.

Clearly, parallel to $(y - z)\hat{\mathbf{i}} + (z - x)\hat{\mathbf{j}} + (x - y)\hat{\mathbf{k}}$

95. $A \rightarrow \mathbf{a} \times (\mathbf{b} \times \mathbf{c}) + \mathbf{b} \times (\mathbf{c} \times \mathbf{a}) + \mathbf{c} \times (\mathbf{a} \times \mathbf{b}) = \mathbf{O}$

$\Rightarrow$ Vectors are coplanar, so do not form RHS

$B \rightarrow (\mathbf{a} \times \mathbf{b}) \times \mathbf{c}, \mathbf{a} \times \mathbf{b}, \bar{\mathbf{c}} = \mathbf{0}$ in that order form RHS

$\Rightarrow \mathbf{c}, (\mathbf{a} \times \mathbf{b}) \times \mathbf{c}, \mathbf{a} \times \mathbf{b}$ also form RHS as they are in same cyclic order.

$C \rightarrow \mathbf{a} + \mathbf{b} + \mathbf{c} = \mathbf{0} \Rightarrow (\mathbf{a} + \mathbf{b} + \mathbf{c})^2 = \mathbf{0}$

$\Rightarrow \quad \mathbf{a}^2 + \mathbf{b}^2 + \mathbf{c}^2 = -2(\mathbf{a} \cdot \mathbf{b} + \mathbf{b} \cdot \mathbf{c} + \mathbf{c} \cdot \mathbf{a})$

Hence, $\quad \mathbf{a} \cdot \mathbf{b} + \mathbf{b} \cdot \mathbf{c} + \mathbf{c} \cdot \mathbf{a} < 0$

$D \rightarrow \mathbf{a} + \mathbf{b} + \mathbf{c} = \mathbf{0}$

$\Rightarrow \quad \mathbf{a} \times \mathbf{b} = \mathbf{b} \times \mathbf{c} = \mathbf{c} \times \mathbf{a}$

Using this we get result.

96. $\mathbf{a} \cdot \mathbf{b} = 0, \mathbf{c} \cdot \mathbf{a} = \mathbf{c} \cdot \mathbf{b} = \cos\alpha$

Take dot products with **a**, **b** and **c**, respectively.

$$l = m, n^2 + l^2 + m^2 = 1$$

$$n^2 = -\cos 2\alpha, m^2 = \frac{1 + \cos 2\alpha}{2}$$

97. $A \rightarrow \mathbf{a} \times (\mathbf{b} \times \mathbf{c}) + \mathbf{b}(\mathbf{c} \times \mathbf{a}) + \mathbf{c}(\mathbf{a} \times \mathbf{b}) = \mathbf{0}$

$\Rightarrow$ Vectors are coplanar, so do not form RHS

$B \rightarrow (\mathbf{a} \times \mathbf{b}) \times \mathbf{c}, \mathbf{a} \times \mathbf{b}, \mathbf{c}$ in that order form RHS

$\Rightarrow \mathbf{c}, (\mathbf{a} \times \mathbf{b}) \times \mathbf{c}, \mathbf{a} \times \mathbf{b}$ also form RHS as they are in same cyclic order

$C \rightarrow \mathbf{a} + \mathbf{b} + \mathbf{c} = \mathbf{0}$

$\Rightarrow \quad (\mathbf{a} + \mathbf{b} + \mathbf{c})^2 = 0$

$\Rightarrow \quad \mathbf{a}^2 + \mathbf{b}^2 + \mathbf{c}^2 = -2(\mathbf{a} \cdot \mathbf{b} + \mathbf{b} \cdot \mathbf{c} + \mathbf{c} \cdot \mathbf{a})$

Hence, $\mathbf{a} \cdot \mathbf{b} + \mathbf{b} \cdot \mathbf{c} + \mathbf{c} \cdot \mathbf{a} < 0$

$D \rightarrow \mathbf{a} + \mathbf{b} + \mathbf{c} = \mathbf{0}$

$\Rightarrow \quad \mathbf{a} \times \mathbf{b} = \mathbf{b} \times \mathbf{c} = \mathbf{c} \times \mathbf{a}$

Using this we get result.

98. Since **a**, **b**, $\mathbf{a} \times \mathbf{b}$ are non-coplanar,

$$\mathbf{r} = x\mathbf{a} + y\mathbf{b} + z(\mathbf{a} \times \mathbf{b}), \mathbf{r} \times \mathbf{b} = \mathbf{a}$$

$\Rightarrow \quad x\mathbf{a} \times \mathbf{b} + z(\mathbf{a} \times \mathbf{b}) \times \mathbf{b} = \mathbf{a}$

$\Rightarrow \quad x(\mathbf{a} \times \mathbf{b}) + z(\mathbf{b} \cdot \mathbf{a})\mathbf{b} - (\mathbf{b} \cdot \mathbf{b})\mathbf{a} = \mathbf{a}$

$\Rightarrow \quad x(\mathbf{a} \times \mathbf{b}) - \mathbf{a}(1 + |\mathbf{b}|^2 z) = 0$

$\Rightarrow \quad x = 0, z = -\frac{1}{|\mathbf{b}|^2}$

$\therefore \mathbf{r} = y\mathbf{b} - \frac{1}{|\mathbf{b}|^2}(\mathbf{a} \times \mathbf{b})$, where y is any scalar.

99. Let angle between **a** and **b** be θ.

We have, $\quad |\mathbf{a}| = |\mathbf{b}| = 1$

Now, $|\mathbf{a} + \mathbf{b}| = 2\cos\frac{\theta}{2}$ and $|\mathbf{a} - \mathbf{b}| = 2\sin\frac{\theta}{2}$

Consider, $F\theta = \frac{3}{2}\left(2\cos\frac{\theta}{2}\right) + 2\left(2\sin\frac{\theta}{2}\right)$

$\therefore \quad F(\theta) = 3\cos\frac{\theta}{2} + 4\sin\frac{\theta}{2}, \theta \in [0, \pi]$

$$F(\theta) = \frac{-3}{2}\sin\frac{\theta}{2} + 2\cos\frac{\theta}{2}$$

Now, $\quad F(\theta) = 0$

$\Rightarrow \quad \tan\frac{\theta}{2} = \frac{4}{3}$

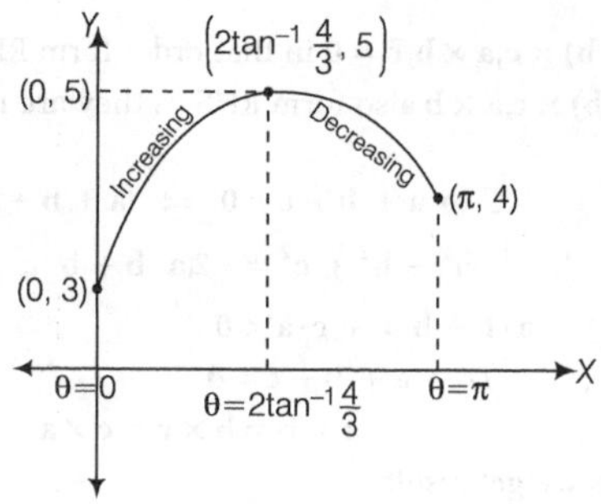

Clearly, $\quad F(\theta) = 3$

$$F\left(\theta = 2\tan^{-1}\frac{4}{3}\right) = 3\left(\frac{3}{5}\right) + 4\left(\frac{4}{5}\right) = \frac{9}{5} + \frac{16}{5} = \frac{25}{5} = 5$$

$$F(\theta = \pi) = 4$$

$\therefore \quad$ Range $= [3, 5]$

Hence, possible integer(s) in the range of $F(\theta)$ in $[0, \pi]$ are 3 *viz*, 3, 4 and 5.

100. Let $\mathbf{r} = x\mathbf{a} + y\mathbf{b} + z(\mathbf{a} \times \mathbf{b})$ with $\mathbf{b}$

$$\mathbf{r} \cdot \mathbf{b} = 0 + y|\mathbf{b}|^2$$

$$\mathbf{a}(\mathbf{r} \cdot \mathbf{b}) = y(\mathbf{b})^2\mathbf{a}$$

$\Rightarrow \quad \mathbf{c} - p\mathbf{r} = y|\mathbf{b}|^2\mathbf{a}$

$$\mathbf{r} = \frac{1}{p}\mathbf{c} - \frac{y|b|^2}{p}\mathbf{a}$$

$\therefore \quad [\mathbf{r}\ \mathbf{a}\mathbf{c}] = 0$

Now, $\quad \mathbf{r} \cdot \mathbf{b} = \frac{1}{p}\mathbf{c} \cdot \mathbf{b}$

$\therefore \quad y|b|^2 = \frac{\mathbf{b} \cdot \mathbf{c}}{p}$

$\therefore \quad \mathbf{r} = \frac{1}{p}\mathbf{c} - \frac{1}{p^2}(\mathbf{b} \cdot \mathbf{c})\mathbf{a}$

101. $(\lambda - 1)(\mathbf{a}_1 - \mathbf{a}_2) + \mu(\mathbf{a}_2 + \mathbf{a}_3) + \gamma(\mathbf{a}_3 + \mathbf{a}_4 - 2\mathbf{a}_2) + \mathbf{a}_3 + \delta\mathbf{a}_4 = 0$

i.e. $(\lambda - 1)\mathbf{a}_1 + (1 - \lambda + \mu - 2\gamma)\mathbf{a}_2 + (\mu + \gamma + 1)\mathbf{a}_3 + (\gamma + \delta)\mathbf{a}_4 = 0$

Since, $\mathbf{a}_1, \mathbf{a}_2, \mathbf{a}_3, \mathbf{a}_4$ are linearly independent

$\therefore \quad \lambda - 1 = 0, 1 - \lambda + \mu - 2\gamma = 0, \mu + \lambda + 1 = 0$

$\gamma + \delta = 0$

i.e. $\quad \lambda = 1, \mu = 2\gamma, \mu + \gamma + 1 = 0, \gamma + \delta = 0$

i.e. $\quad \lambda = 1, \mu = -\frac{2}{3}, \gamma = -\frac{1}{3}, \delta = \frac{1}{3}$

102. Since $[\mathbf{a}\ \mathbf{b}\ \mathbf{c}] = 0$

$\therefore \mathbf{a}, \mathbf{b}$ and $\mathbf{c}$ are complanar vectors

Further since $\mathbf{d}$ is equally inclined to $\mathbf{a}, \mathbf{b}$ and $\mathbf{c}$

$\therefore \mathbf{d} \cdot \mathbf{a} = \mathbf{d} \cdot \mathbf{b} = \mathbf{d} \cdot \mathbf{c} = 0$

$\therefore \mathbf{d} \cdot \mathbf{r} = 0$

103. $\mathbf{p} = ab\cos(2\pi - \theta)\mathbf{c}$, where θ is the angle between $\mathbf{a}$ and $\mathbf{b}$ and $\mathbf{q} = ac\cos(\pi - \phi)\mathbf{b}$ where ϕ is the angle between $\mathbf{a}$ and $\mathbf{b}$ now $\mathbf{p} + \mathbf{q} = (ab\cos\theta)\mathbf{c} - ac\cos\phi\mathbf{b} = (\mathbf{a} \cdot \mathbf{b})\mathbf{c} - (\mathbf{a} \cdot \mathbf{c})\mathbf{b} = \mathbf{a} \times (\mathbf{c} \times \mathbf{b})$

$\Rightarrow B$ and C

104. Verify $\mathbf{v}_1 + \mathbf{v}_2 = \mathbf{v}_3$ in order to quickly answer

105. Since, $(2\hat{\mathbf{i}} + \hat{\mathbf{j}} + 4\hat{\mathbf{k}}) \cdot (\hat{\mathbf{i}} + 2\hat{\mathbf{j}} - \hat{\mathbf{k}}) = 0$

and $\quad (\hat{\mathbf{i}} + \hat{\mathbf{j}}) \cdot (\hat{\mathbf{i}} + 2\hat{\mathbf{j}} - \hat{\mathbf{k}}), 1 + 2 = 3$

$\Rightarrow$ Line lies in the plane

106. $\quad \mathrm{OD} = \frac{(\mathbf{a} \cdot \mathbf{b})\mathbf{a}}{|\mathbf{a}|^2}$

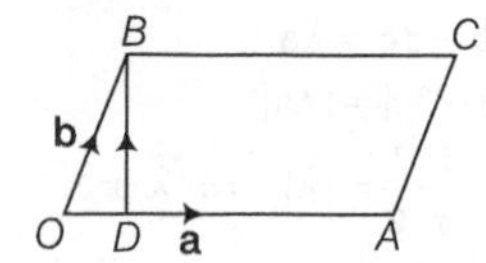

$\Rightarrow \quad \mathrm{DB} = \mathbf{b} - \mathrm{OD} = \mathbf{b} - \frac{(\mathbf{a} \cdot \mathbf{b})\mathbf{a}}{|\mathbf{a}|^2}$

$$= \frac{(\mathbf{a} \cdot \mathbf{a})\mathbf{b} - (\mathbf{a} \cdot \mathbf{b})\mathbf{a}}{|\mathbf{a}|^2} = \frac{\mathbf{a} \times (\mathbf{b} \times \mathbf{a})}{|\mathbf{a}|^2}$$

107. $\mathbf{a} + \mathbf{b} = \lambda\mathbf{c}$; $\mathbf{b} + \mathbf{c} = \mu\mathbf{a}$

$$\mathbf{a} - \mathbf{c} = \lambda\mathbf{c} - \mu\mathbf{a}$$

$$\mathbf{a}(1 + \mu) = \mathbf{c}(1 + \lambda)$$

but $\mathbf{a}$ and $\mathbf{c}$ are non-collinear $\Rightarrow \mu = -1, \lambda = -1$

$\therefore \quad \mathbf{a} + \mathbf{b} + \mathbf{c} = 0 = \hat{\mathbf{k}} \quad \Rightarrow |\mathbf{k}| = 0$

$\Rightarrow (k, k) \equiv (0, 0)$ all the given curves pass through $(0, 0)$

108. $(\mathbf{a} \cdot \mathbf{c})\mathbf{b} - (\mathbf{a} \cdot \mathbf{b})\mathbf{c} = \frac{\mathbf{b} + \mathbf{c}}{2}$; $\vec{\mathbf{b}}\left(\vec{\mathbf{a}} \cdot \vec{\mathbf{c}} - \frac{1}{2}\right) - \mathbf{c}\left(\mathbf{a} \cdot \mathbf{b} + \frac{1}{2}\right) = 0$

$\Rightarrow \quad \mathbf{a} \cdot \mathbf{c} = \frac{1}{2}$ and $\mathbf{a} \cdot \mathbf{c} = -\frac{1}{2}$

109. $AA^T = I \Rightarrow \mathbf{a}, \mathbf{b}, \mathbf{c}$ are orthogonal unit vectors

$$\mathbf{c} = \mathbf{a} \times \mathbf{b} = \frac{1}{49}\begin{vmatrix} \hat{\mathbf{i}} & \hat{\mathbf{j}} & \hat{\mathbf{k}} \\ 2 & 3 & 6 \\ 6 & 2 & -3 \end{vmatrix} = \frac{1}{7}(-3\hat{\mathbf{i}} + 6\hat{\mathbf{j}} - 2\hat{\mathbf{k}})$$

$\Rightarrow \quad \mathbf{c} = \pm\frac{1}{7}(3\hat{\mathbf{i}} - 6\hat{\mathbf{j}} + 2\hat{\mathbf{k}})$

110. Component of vector $\mathbf{b} = 4\hat{\mathbf{i}} + 2\hat{\mathbf{j}} + 3\hat{\mathbf{k}}$ in the direction of $\mathbf{a} = \hat{\mathbf{i}} + \hat{\mathbf{j}} + \hat{\mathbf{k}}$ is $\frac{\mathbf{a} \cdot \mathbf{b} \cdot \mathbf{a}}{|\mathbf{a}||\mathbf{a}|}$ or $3\hat{\mathbf{i}} + 3\hat{\mathbf{j}} + 3\hat{\mathbf{k}}$. Then, component in the direction perpendicular to the direction of $\mathbf{a} = \hat{\mathbf{i}} + \hat{\mathbf{j}} + \hat{\mathbf{k}}$ is $\mathbf{b} - 3\hat{\mathbf{i}} + 3\hat{\mathbf{j}} + 3\hat{\mathbf{k}} = \hat{\mathbf{i}} - \hat{\mathbf{j}}$

111. Le the three given unit vector be $\hat{\mathbf{a}}$, $\hat{\mathbf{b}}$ and $\hat{\mathbf{c}}$. Since, they are mutually perpendicular, $\mathbf{a} \cdot (\mathbf{b} \times \mathbf{c}) = 1$.

Therefore, $\begin{vmatrix} a_1 & a_2 & a_3 \\ b_1 & b_2 & b_3 \\ c_1 & c_2 & c_3 \end{vmatrix} = 1$

$$\begin{vmatrix} a_1 & b_1 & c_1 \\ a_2 & b_2 & c_2 \\ a_3 & b_3 & c_3 \end{vmatrix} = 1$$

Hence, $a_1\hat{\mathbf{i}} + b_1\hat{\mathbf{j}} + c_1\hat{\mathbf{k}}, a_2\hat{\mathbf{i}} + b_2\hat{\mathbf{j}} + c_2\hat{\mathbf{k}}$ and $a_3\hat{\mathbf{i}} + b_3\hat{\mathbf{j}} + c_3\hat{\mathbf{k}}$ may be mutually perpendicular.

112. Statement II is true (see properties of dot product)

Also, $(\hat{\mathbf{i}} \times \mathbf{b}) \cdot \mathbf{b} = \hat{\mathbf{i}} \cdot (\mathbf{a} \times \mathbf{b})$

$\Rightarrow \quad \mathbf{a} \times \mathbf{b} = (\hat{\mathbf{i}} \cdot (\mathbf{a} \times \mathbf{b}))\,\hat{\mathbf{i}} + (\hat{\mathbf{j}} \cdot (\mathbf{a} \times \mathbf{b}))\,\hat{\mathbf{j}} + (\hat{\mathbf{k}} \cdot (\mathbf{a} \times \mathbf{b}))\,\hat{\mathbf{k}}$

113. $\mathbf{AD} = 2\hat{j} - \hat{k}$, $\mathbf{BD} = -2\hat{i} - \hat{j} - 3\hat{k}$ and $\mathbf{CD} = 2\hat{i} - \hat{j}$

Volume of tetrahedron $= \frac{1}{6}[\mathbf{AD}\ \mathbf{BD}\ \mathbf{CD}]$

$$= \frac{1}{6}\left\|\begin{matrix} 0 & 2 & -1 \\ -2 & -1 & -3 \\ 2 & -1 & 0 \end{matrix}\right\| = \frac{8}{3}$$

Also, area of the triangle $ABC = \frac{1}{2}|\mathbf{AB} \times \mathbf{AC}|$

$$= \frac{1}{2}\left\|\begin{matrix} \hat{i} & \hat{j} & \hat{k} \\ 2 & 3 & 2 \\ -2 & 3 & -1 \end{matrix}\right\| = \frac{1}{2}|-9\hat{i} - 2\hat{j} + 12\hat{k}| = \frac{\sqrt{299}}{2}$$

Then, $\frac{8}{3} = \frac{1}{3} \times$(Distance of D from base ABC) $\times$(Area of triangle ABC)

Distance of D from base $ABC = 16/\sqrt{229}$

114. $\mathbf{A}\times(\mathbf{A}\cdot\mathbf{B})\,\mathbf{A} - (\mathbf{A}\cdot\mathbf{A})\,\mathbf{B})\cdot\mathbf{C}$

$$= \left(\underbrace{\mathbf{A}\times(\mathbf{A}\cdot\mathbf{B})\,\mathbf{A}}_{\text{zero}} - (\mathbf{A}\cdot\mathbf{A})\,\mathbf{A}\times\mathbf{B}\right)\cdot\mathbf{C} = -|\mathbf{A}|^2\,[\mathbf{A\ B\ C}]$$

Now, $|\mathbf{A}|^2 = 4 + 9 + 36 = 49$

$$[\mathbf{A\ B\ C}] = \begin{vmatrix} 2 & 3 & 6 \\ 1 & 1 & -2 \\ 1 & 2 & 1 \end{vmatrix} = 2(1+4) - 1(3-12) + 1(-6-6)$$

$$= 10 + 6 - 12 = 7 = 10 + 9 - 12 = 7$$

$\therefore\quad |-|\mathbf{A}|^2\,[\mathbf{A\ B\ C}]| = 49 \times 7 = 343$

115. $\pm\frac{\mathbf{a}\times\mathbf{b}}{\mathbf{a}\times\mathbf{b}} \perp \mathbf{a}$ and $\mathbf{b}$

$\therefore$ There are two such vectors

116. $\frac{k}{k+l+m} + \frac{l}{k+l+m} + \frac{m}{k+l+m} = 1$

$\Rightarrow$ Point lies in the plane of $\Delta\, ABC$.

117. $a^2 + b^2 + 2\mathbf{a}\cdot\mathbf{b} = a^2 + 4b^2 - 4\mathbf{a}\cdot\mathbf{b} \Rightarrow 6ab\cos\theta = 3b^2$

$\Rightarrow\ \mathbf{a}\cdot\mathbf{b} = \frac{1}{2}b^2$

$$GE = \frac{b^2}{2} + \frac{4}{b^2+2} = \frac{b^2+2}{2} + \frac{4}{b^2+2} - 1$$

$$= \sqrt{2}\left[\frac{b^2+2}{2\sqrt{2}} + \frac{2\sqrt{2}}{b^2+2}\right] - 1 = \sqrt{2}(\geq 2) - 1 \geq \sqrt{2} - 1$$

It is least when $\frac{b^2+2}{2\sqrt{2}} = 1$

$\Rightarrow\quad \mathbf{b} = \sqrt{2(\sqrt{2}-1)} = \sqrt{2\tan\pi/8}$

118. Both the statements are true and statement II is the not correct explanation of statement I. Because $\mathbf{b}$, $\mathbf{c}$, $\mathbf{d}$ in statement I are coplanar.

119. $3\mathbf{a} - 2\mathbf{b} + 5\mathbf{c} - 6\mathbf{d} = (2\mathbf{a} - 2\mathbf{b}) + (-5\mathbf{a} + 5\mathbf{c}) + (6\mathbf{a} - 6\mathbf{d}) = -2\mathbf{AB} + 5\mathbf{AC} - 6\mathbf{AD} = 0$

$\therefore$ **AB**, **AC** and **AD** are linearly dependent, hence by statement II, the statement I is true.

120. $\mathbf{r} = x\mathbf{a} + y\mathbf{b} + z(\mathbf{a}\times\mathbf{b})$

121. $(\hat{a}\times(\mathbf{b}\times\hat{c}))\cdot(\hat{a}\times\hat{c}) = (\hat{a}\cdot\hat{c})[\mathbf{b}\hat{a}\hat{c}] = 5$

$\Rightarrow\quad [\mathbf{b}\hat{a}\hat{c}] = 10 \Rightarrow [\hat{a}\mathbf{b}\hat{c}] = -10.$

Solutions (Q.Nos. 122-124)

122. $\underbrace{(\hat{i} + 3\hat{j} - 4\hat{k})}_{(\mathbf{q}-\mathbf{r})}\cdot(\hat{i} + \hat{j} + \hat{k}) = 1 + 3 - 4 = 0$

Since, $[\mathbf{q\ p\ r}] = \begin{vmatrix} 1 & 1 & 1 \\ 2 & 4 & -1 \\ 1 & 1 & 3 \end{vmatrix}$

$$= (12+1) - 1(6+1) + 1(2-4) = 13 - 7 - 2 = 4$$

$\Rightarrow$ (a) (b) (d) are wrong.

123. $(\mathbf{p}\times\mathbf{r})\times\mathbf{r} = (\mathbf{p}\cdot\mathbf{r})\,\mathbf{q} - (\mathbf{q}\cdot\mathbf{r})\,\mathbf{p}$

$\Rightarrow\quad u = -(\mathbf{q}\cdot\mathbf{r}) = -(2+4-3) = -3$

$v = \mathbf{p}\cdot\mathbf{r} = 1 + 1 + 3 = 5$ and $w = 0$

Hence, $u = -3, v = 5, w = 0 \Rightarrow u + v + w = 2$

124. $\because$ $\mathbf{p}$, $\mathbf{q}$ and $\mathbf{r}$ are non-coplanar, therefore $\mathbf{q}\times\mathbf{r}$, $\mathbf{r}\times\mathbf{p}$ and $\mathbf{p}\times\mathbf{q}$ are also non-coplanar

Hence, $\mathbf{s} = l(\mathbf{q}\times\mathbf{r}) + w(\mathbf{r}\times\mathbf{p}) + n(\mathbf{p}\times\mathbf{q})$

$\therefore\quad l = \frac{\mathbf{s}\cdot\mathbf{p}}{[\mathbf{p\ q\ r}]}, w = \frac{\mathbf{s}\cdot\mathbf{q}}{[\mathbf{p\ q\ r}]}, n = \frac{\mathbf{s}\cdot\mathbf{r}}{[\mathbf{p\ q\ r}]}$

Hence, $\mathbf{s}[\mathbf{p\ q\ r}] = (\mathbf{s}\cdot\mathbf{p})\,(\mathbf{q}\times\mathbf{r}) + (\mathbf{s}\cdot\mathbf{q})\,(\mathbf{r}\times\mathbf{p}) + (\mathbf{s}\cdot\mathbf{r})\,(\mathbf{p}\times\mathbf{q})$

$|(\mathbf{s}\cdot\mathbf{p})(\mathbf{q}\times\mathbf{r}) + (\mathbf{s}\cdot\mathbf{q})(\mathbf{r}\times\mathbf{p}) + (\mathbf{s}\cdot\mathbf{r})(\mathbf{p}\times\mathbf{q})|$
$= |\mathbf{s}\,[\mathbf{p\ q\ r}]| = [\mathbf{p\ q\ r}]$ (as $|\mathbf{s}| = 1$)

$$\begin{vmatrix} 1 & 1 & 1 \\ 2 & 4 & -1 \\ 1 & 1 & 3 \end{vmatrix} = (12+1) - 1(6+1) + 1(2-4) = 13 - 7 - 2 = 4$$

Solutions (Q.Nos. 125-127)

125. Given, $\mathbf{p} = \hat{i} + \hat{j} + \hat{k}; \mathbf{q} = \hat{i} - \hat{j} + \hat{k}$

$\mathbf{p}\times\mathbf{r} = \mathbf{q} + c\mathbf{p}$ and $\mathbf{p}\cdot\mathbf{r} = 2$

$\therefore\ \mathbf{p}\times(\mathbf{p}\times\mathbf{r}) = \mathbf{p}\times\mathbf{q}\,(\mathbf{q} + c\mathbf{p})$ and $\mathbf{p}\cdot(\mathbf{p}\times\mathbf{r}) = \mathbf{p}\cdot(\mathbf{q} + c\mathbf{p})$

$\therefore\ (\mathbf{p}\cdot\mathbf{r})\mathbf{p} - (\mathbf{p}\cdot\mathbf{p})\mathbf{r} = \mathbf{p}\times\mathbf{q} + c0$

$0 = \mathbf{p}\cdot\mathbf{q} + \mathbf{c}\,(\mathbf{p}\cdot\mathbf{q})$

$\therefore\quad (\mathbf{p}\cdot\mathbf{p})\mathbf{r} = (\mathbf{p}\cdot\mathbf{r})\mathbf{p} - \mathbf{p}\times\mathbf{q}$ ...(i)

$c = -\frac{\mathbf{p}\cdot\mathbf{q}}{\mathbf{p}\cdot\mathbf{p}}$ ...(ii)

But $\mathbf{p}\cdot\mathbf{p} = |\mathbf{p}| = \mathbf{p}^2 = 3$ ...(iii)

$\mathbf{p}\cdot\mathbf{p} = 1 - 1 + 1 = 1$ ...(iv)

$$\mathbf{p}\times\mathbf{q} = \begin{vmatrix} \hat{i} & \hat{j} & \hat{k} \\ 1 & 1 & 1 \\ 1 & -1 & 1 \end{vmatrix} = 2\hat{i} - 2\hat{k} \quad \text{...(v)}$$

Using Eqs. (iii), (iv) in Eq. (i) and Eq. (ii), we get

$3\mathbf{r} = 2\mathbf{p} - 2\hat{i} + 2\hat{k}$ and $c = -\frac{1}{3}$ ...(vi)

$\therefore\quad \mathbf{r} = \frac{1}{3}[2\hat{i} + 2\hat{j} + 2\hat{k} - 2\hat{i} + 2\hat{k}]$

$\therefore \quad \mathbf{r} = \frac{2}{3}(\hat{\mathbf{j}} + 2\hat{\mathbf{k}}) \quad \text{...(vii)}$

Now, $[\mathbf{p}\ \mathbf{q}\ \mathbf{r}] = \begin{vmatrix} 1 & 1 & 1 \\ 1 & -1 & 1 \\ 0 & \frac{1}{2} & \frac{4}{3} \end{vmatrix} = \left(\frac{-4}{3} - \frac{2}{3}\right) - 1\left(\frac{4}{3}\right) + 1\left(\frac{2}{3}\right)$

$\therefore \quad [\mathbf{p}\ \mathbf{q}\ \mathbf{r}] = -2 - \frac{2}{5} = -\frac{8}{3} \quad \text{...(viii)}$

126. $[\mathbf{p}\ \mathbf{q}\ \mathbf{r}]\ \mathbf{x} = (\mathbf{p} \times \mathbf{q}) \times \mathbf{r} \Rightarrow \frac{8}{3}\mathbf{x} = (\mathbf{p}\cdot\mathbf{r})\ \mathbf{q} - (\mathbf{q}\cdot\mathbf{r})\ \mathbf{p}$

$\Rightarrow \left(-\frac{8}{3}\right)\mathbf{x} = 2\mathbf{q} - \frac{2\mathbf{p}}{3} \quad \left(\mathbf{q}\cdot\mathbf{r} = \frac{2}{3}, \text{ verify yourself}\right)$

$\therefore \quad \mathbf{x} = -\frac{3}{8}\cdot\frac{2}{3}(3\mathbf{q} - \mathbf{p}) = \frac{1}{4}(3\hat{\mathbf{i}} - 3\hat{\mathbf{j}} + 3\hat{\mathbf{k}} - \hat{\mathbf{i}} - \hat{\mathbf{j}} - \hat{\mathbf{k}})$

$= -\frac{1}{4}(2\hat{\mathbf{i}} - 4\hat{\mathbf{j}} + 2\hat{\mathbf{k}})$

$\therefore \quad \mathbf{x} = -\frac{1}{2}(\hat{\mathbf{i}} - 2\hat{\mathbf{j}} + \hat{\mathbf{k}})$

127. As $c = -\frac{1}{3}$ from Eq. (vi)

$\therefore \quad \left(1 - \frac{1}{3}\right)\mathbf{y} = (\mathbf{p}\cdot\mathbf{r})\ \mathbf{q} - (\mathbf{p}\cdot\mathbf{q})\ \mathbf{r}$

$\frac{2}{3}\mathbf{y} = 2\mathbf{q} - (1)\mathbf{r} \quad$ [As $\mathbf{p}\cdot\mathbf{q} = 1$ from Eq. (iv)]

$\therefore \quad \mathbf{y} = \frac{3}{2}\left(2\hat{\mathbf{i}} - 2\hat{\mathbf{j}} + 2\hat{\mathbf{k}} - \frac{2}{3}\hat{\mathbf{j}} - \frac{4}{3}\hat{\mathbf{k}}\right) = \frac{3}{2}\,\frac{(6\hat{\mathbf{i}} - 8\hat{\mathbf{j}} + \hat{\mathbf{k}})}{3}$

$\mathbf{y} = 3\hat{\mathbf{i}} - 4\hat{\mathbf{j}} + \hat{\mathbf{k}}$

$\therefore \quad [\mathbf{x}\ \mathbf{y}\ \mathbf{r}] = \begin{vmatrix} -\frac{1}{2} & 1 & -\frac{1}{2} \\ 3 & -4 & 1 \\ 0 & \frac{2}{3} & \frac{4}{3} \end{vmatrix} = \frac{1}{2}\left(-\frac{16}{3} - \frac{2}{3}\right) - 1(4) - \frac{1}{2}(2)$

$\therefore \quad [\mathbf{x}\ \mathbf{y}\ \mathbf{r}] = 3 - 4 - 1 = -2$

$\therefore \quad \left|\frac{1}{6}[\mathbf{x}\ \mathbf{y}\ \mathbf{r}]\right| = \left|-\frac{1}{3}\right| = |\mathbf{c}|$

$\Rightarrow$ **x, y, r** are the coterminous edges of a tetrahedron whose volume is $|\mathbf{c}|$.

Solutions (Q.Nos. 128-130)

128. $y = \log_{1/2}\left(x - \frac{1}{2}\right) + \log_2 \sqrt{(2x-1)^2}$

But, $\quad x > \frac{1}{2} = \log_{1/2}\left(x - \frac{1}{2}\right) + \log_2(2x - 1)$

$y = 1$

$P \equiv (3, 1)$

129. $\mathbf{OP} = 3\hat{\mathbf{i}} + \hat{\mathbf{j}}$

$\mathbf{Q} \equiv (1, 1) \text{ or } (2, 1)$

$\mathbf{OQ} = \hat{\mathbf{i}} + \hat{\mathbf{j}} \text{ and } 2\hat{\mathbf{i}} + \hat{\mathbf{j}}$

$\mathbf{OP}\cdot\mathbf{OQ} = 3 + 1 = 4 \text{ and } 6 + 1 = 7$

130. $\mathbf{PQ} = \mathbf{OQ} - \mathbf{OP} = -2\hat{\mathbf{i}} \text{ or } \hat{\mathbf{i}}$

$|\mathbf{PQ}| = 2 \text{ or } 1$

Solutions (Q.Nos. 131-134)

131. Since, **a, b, c** are non-coplanar vectors, then

$[\mathbf{a}\ \mathbf{b}\ \mathbf{c}] \neq 0 \Rightarrow [\mathbf{a}\ \mathbf{b}\ \mathbf{c}]^2 \neq 0$

$\Rightarrow \quad [\mathbf{a}\ \mathbf{b}\ \mathbf{c}]^2 = \begin{vmatrix} \mathbf{a}\cdot\mathbf{a} & \mathbf{b}\cdot\mathbf{a} & \mathbf{c}\cdot\mathbf{a} \\ \mathbf{a}\cdot\mathbf{b} & \mathbf{b}\cdot\mathbf{b} & \mathbf{c}\cdot\mathbf{b} \\ \mathbf{a}\cdot\mathbf{c} & \mathbf{b}\cdot\mathbf{c} & \mathbf{c}\cdot\mathbf{c} \end{vmatrix} \neq 0$

Since, any vector **r** in space can be expressed as a linear combination of three non-coplanar vectors.

So, let $\quad \mathbf{r} = l\mathbf{a} + m\mathbf{b} + n\mathbf{c} \quad \text{...(i)}$

taking dot product by **a, b, c** successively, we get

$\mathbf{r}\cdot\mathbf{a} = l\mathbf{a}\cdot\mathbf{a} + m\mathbf{b}\cdot\mathbf{a} + n\mathbf{c}\cdot\mathbf{a} \quad \text{...(ii)}$

$\mathbf{r}\cdot\mathbf{b} = l\mathbf{a}\cdot\mathbf{b} + m\mathbf{b}\cdot\mathbf{b} + n\mathbf{c}\cdot\mathbf{b} \quad \text{...(iii)}$

$\mathbf{r}\cdot\mathbf{c} = l\mathbf{a}\cdot\mathbf{c} + m\mathbf{b}\cdot\mathbf{c} + n\mathbf{c}\cdot\mathbf{c} \quad \text{...(iv)}$

Now, eliminating l, m and n from above 4 relations, we get

$$\begin{vmatrix} \mathbf{r} & \mathbf{a} & \mathbf{b} & \mathbf{c} \\ \mathbf{r}\cdot\mathbf{a} & \mathbf{a}\cdot\mathbf{a} & \mathbf{b}\cdot\mathbf{a} & \mathbf{c}\cdot\mathbf{a} \\ \mathbf{r}\cdot\mathbf{b} & \mathbf{a}\cdot\mathbf{b} & \mathbf{b}\cdot\mathbf{b} & \mathbf{c}\cdot\mathbf{b} \\ \mathbf{r}\cdot\mathbf{c} & \mathbf{a}\cdot\mathbf{c} & \mathbf{b}\cdot\mathbf{c} & \mathbf{c}\cdot\mathbf{c} \end{vmatrix} = 0$$

Now, expanding along first row, we get

$$r = \left(\frac{\Delta_1}{\Delta}\right)\mathbf{a} + \left(\frac{\Delta_2}{\Delta}\right)\mathbf{b} + \left(\frac{\Delta_3}{\Delta}\right)\mathbf{c}$$

132. Since, **a, b, c** are three non-coplanar vectors, then On three exists scalars x, y, z, such that

$\mathbf{r} = x\mathbf{a} + y\mathbf{b} + z\mathbf{c} \quad \text{...(i)}$

Taking dot product by $\mathbf{b} \times \mathbf{c}$, $\mathbf{c} \times \mathbf{a}$ and $\mathbf{a} \times \mathbf{b}$ successively, we get

$\mathbf{r}\cdot(\mathbf{b} \times \mathbf{c}) = (x\mathbf{a} + y\mathbf{b} + z\mathbf{c})\cdot(\mathbf{b} \times \mathbf{c}) = x[\mathbf{a}\ \mathbf{b}\ \mathbf{c}]$

$\mathbf{r}\cdot(\mathbf{c} \times \mathbf{a}) = y[\mathbf{b}\ \mathbf{c}\ \mathbf{a}]$

$\mathbf{r}\cdot(\mathbf{a} \times \mathbf{b}) = z[\mathbf{c}\ \mathbf{a}\ \mathbf{b}]$

$\because \quad x = \frac{[\mathbf{r}\ \mathbf{b}\ \mathbf{c}]}{[\mathbf{a}\ \mathbf{b}\ \mathbf{c}]}, \ y = \frac{[\mathbf{r}\ \mathbf{c}\ \mathbf{a}]}{[\mathbf{a}\ \mathbf{b}\ \mathbf{c}]} \text{ and } z = \frac{[\mathbf{r}\ \mathbf{a}\ \mathbf{b}]}{[\mathbf{a}\ \mathbf{b}\ \mathbf{c}]}$

On substituting the values of x, y, z in Eq. (i), we get

$$r = \left\{\frac{[\mathbf{r}\ \mathbf{b}\ \mathbf{c}]}{[\mathbf{a}\ \mathbf{b}\ \mathbf{c}]}\right\}\mathbf{a} + \left\{\frac{[\mathbf{r}\ \mathbf{c}\ \mathbf{a}]}{[\mathbf{a}\ \mathbf{b}\ \mathbf{c}]}\right\}\mathbf{b} + \left\{\frac{[\mathbf{r}\ \mathbf{a}\ \mathbf{b}]}{[\mathbf{a}\ \mathbf{b}\ \mathbf{c}]}\right\}\mathbf{c}$$

or $\quad \mathbf{r} = \frac{1}{[\mathbf{a}\ \mathbf{b}\ \mathbf{c}]}\{[\mathbf{r}\ \mathbf{b}\ \mathbf{c}]\mathbf{a} + [\mathbf{r}\ \mathbf{c}\ \mathbf{a}]\ \mathbf{b} + [\mathbf{r}\ \mathbf{a}\ \mathbf{b}]\ \mathbf{c}\}$

133. We know that,

$[\mathbf{a} \times \mathbf{b}\ \ \mathbf{b} \times \mathbf{c}\ \ \mathbf{c} \times \mathbf{a}] = [\mathbf{a}\ \mathbf{b}\ \mathbf{c}]^2$

Clearly, $\quad [\mathbf{a} \times \mathbf{b}\ \ \mathbf{b} \times \mathbf{c}\ \ \mathbf{c} \times \mathbf{a}] \neq 0\{[\mathbf{a}\ \mathbf{b}\ \mathbf{c}] \neq 0\}$

$\Rightarrow \mathbf{a} \times \mathbf{b}\ \ \mathbf{b} \times \mathbf{c}\ \ \mathbf{c} \times \mathbf{a}$ are non-coplanar.

We also know that any vector in space can be expressed as a linear combination of any three non-coplanar vectors, so let.

$\mathbf{a} = l(\mathbf{b} \times \mathbf{c}) + m(\mathbf{c} \times \mathbf{a}) + n(\mathbf{a} \times \mathbf{b}) \quad \text{...(i)}$

On taking dot product on both sides by **a, b, c** successively, we get

$\mathbf{a}\cdot\mathbf{a} = l\ [\mathbf{a}\ \mathbf{b}\ \mathbf{c}]$

$\mathbf{a}\cdot\mathbf{b} = m\ [\mathbf{c}\ \mathbf{a}\ \mathbf{b}]$

$\mathbf{c}\cdot\mathbf{a} = n\ [\mathbf{c}\ \mathbf{a}\ \mathbf{b}]$

$\because \quad l = \frac{\mathbf{a}\cdot\mathbf{a}}{[\mathbf{a}\ \mathbf{b}\ \mathbf{c}]}, \ m = \frac{\mathbf{a}\cdot\mathbf{b}}{[\mathbf{a}\ \mathbf{b}\ \mathbf{c}]} \text{ and } n = \frac{\mathbf{a}\cdot\mathbf{c}}{[\mathbf{a}\ \mathbf{b}\ \mathbf{c}]}$

On substituting these values in Eq. (i), we get

$$\mathbf{a} = \frac{\mathbf{a}\cdot\mathbf{a}}{[\mathbf{a\ b\ c}]}(\mathbf{b}\times\mathbf{c}) + \frac{\mathbf{a}\cdot\mathbf{b}}{[\mathbf{a\ b\ c}]}(\mathbf{c}\times\mathbf{a}) + \frac{\mathbf{a}\cdot\mathbf{c}}{[\mathbf{a\ b\ c}}(\mathbf{a}\times\mathbf{b})$$

or $\mathbf{a} = \frac{1}{[\mathbf{a\ b\ c}]}\{\mathbf{a}\cdot\mathbf{a}\,(\mathbf{b}\times\mathbf{c}) + \mathbf{a}\cdot\mathbf{b}\,(\mathbf{c}\times\mathbf{a}) + \mathbf{a}\cdot\mathbf{c}\,(\mathbf{a}\times\mathbf{b})\}$

134. Let $\mathbf{a} = a_1\hat{\mathbf{i}} + a_2\hat{\mathbf{j}} + a_3\hat{\mathbf{k}}$, $\mathbf{b} = b_1\hat{\mathbf{i}} + b_2\hat{\mathbf{j}} + b_3\hat{\mathbf{k}}$,
$\mathbf{c} = c_1\hat{\mathbf{i}} + c_2\hat{\mathbf{j}} + c_3\hat{\mathbf{k}}$, $\mathbf{p} = p_1\hat{\mathbf{i}} + p_2\hat{\mathbf{j}} + p_3\hat{\mathbf{k}}$,

and $\mathbf{q} = q_1\hat{\mathbf{i}} + q_2\hat{\mathbf{j}} + q_3\hat{\mathbf{k}}$,

Then, $\begin{vmatrix} \mathbf{a} & \mathbf{b} & \mathbf{c} \\ \mathbf{a}\cdot\mathbf{p} & \mathbf{b}\cdot\mathbf{p} & \mathbf{c}\cdot\mathbf{p} \\ \mathbf{a}\cdot\mathbf{q} & \mathbf{b}\cdot\mathbf{q} & \mathbf{c}\cdot\mathbf{q} \end{vmatrix}$

$$\begin{vmatrix} a_1\hat{\mathbf{i}} + a_2\hat{\mathbf{j}} + a_3\hat{\mathbf{k}} & b_1\hat{\mathbf{i}} + b_2\hat{\mathbf{j}} + b_3\hat{\mathbf{k}} & c_1\hat{\mathbf{i}} + c_2\hat{\mathbf{j}} + c_3\hat{\mathbf{k}} \\ a_1p_1 + a_2p_2 + a_3p_3 & b_1p_1 + b_2p_2 + b_3p_3 & c_1p_1 + c_2p_2 + c_3p_3 \\ a_1q_1 + a_2q_2 + a_3q_3 & b_1q_1 + b_2q_2 + b_3q_3 & c_1q_1 + c_2q_2 + c_3q_3 \end{vmatrix}$$

$$\begin{vmatrix} \hat{\mathbf{i}} & \hat{\mathbf{j}} & \hat{\mathbf{k}} \\ p_1 & p_2 & p_3 \\ q_1 & q_2 & q_3 \end{vmatrix} \cdot \begin{vmatrix} a_1 & a_2 & a_3 \\ b_1 & b_2 & c_3 \\ c_1 & c_2 & c_3 \end{vmatrix}$$

$$= (\mathbf{p}\times\mathbf{q})\,[\mathbf{abc}]$$
$$= \sqrt{[\mathbf{abc}]^2}(\mathbf{p}\times\mathbf{q})$$
$$= \sqrt{[\mathbf{a}\times\mathbf{b}\mathbf{b}\times\mathbf{c}\mathbf{c}\times\mathbf{a}]}\,(\mathbf{p}\times\mathbf{q})$$

Solutions (Q.Nos. 135-136)

135. $\because g'(x) = 3x^2 + 2x + 0 > 0, \forall\, x \ge 0$

$\Rightarrow g(x)$ is an $\uparrow$ ing function.

If circumcentre lies outside, then triangle is obtuse angle triangle and angle containing the given sides is obtuse angle. Therefore,

$$(f(x)\hat{\mathbf{i}} + g(x)\hat{\mathbf{j}})\cdot(g(x)\hat{\mathbf{i}} + f(x)\hat{\mathbf{j}}) < 0$$

$\Rightarrow \quad f(x)\cdot g(x) < 0 \quad$ ...(i)

$\Rightarrow \quad g(x) \uparrow$ for $x \ge 0$

$\Rightarrow \quad g(x) > g(0) \forall x > 0$, Also, $g(0) = 0$

$\Rightarrow \quad g(x) > 0 \forall x > 0$, (i) $\Rightarrow f(x) < 0$

$\therefore \quad f(x) < 0$ and $g(x) > 0\ \forall\ x > 0$

136. If $x \to \infty$ then $g(x) \to \infty$ and $f(x)$ is some negative number, then

$$\lim_{t\to 0}\lim_{x\to\infty}\left(\cot\left(\underbrace{\frac{\pi}{4}\underbrace{(1-t^2)}_{1^-}}_{}\right)\right)_{1^+}^{\overbrace{f(x)\cdot g(x)}^{\to\infty}} = 0$$

Solutions (Q.Nos. 137 to 139)

We have $|\mathbf{x}| = |\mathbf{y}| = |\mathbf{z}| = \sqrt{2}$ and **x**, **y**, **z** make angle of 60° with each other.

$\therefore \quad \mathbf{x}\cdot\mathbf{y} = |\mathbf{x}|\,|\mathbf{y}|\cos 60° = \sqrt{2}(\sqrt{2})\cdot\frac{1}{2} = 1$

$\mathbf{y}\cdot\mathbf{z} = |\mathbf{y}|\,|\mathbf{z}|\cos 60° = \sqrt{2}(\sqrt{2})\left(\frac{1}{2}\right) = 1$

and $\quad \mathbf{x}\cdot\mathbf{z} = |\mathbf{x}|\,|\mathbf{z}|\cos 60° = \sqrt{2}(\sqrt{2})\left(\frac{1}{2}\right) = 1$

$\mathbf{x}\cdot\mathbf{x} = |\mathbf{x}|^2 = 2$

$\mathbf{y}\cdot\mathbf{y} = |\mathbf{y}|^2 = 2$ and $\mathbf{z}\cdot\mathbf{z} = |z|^2 = 2$

Now, $\mathbf{x}\times(\mathbf{y}\times\mathbf{z}) = \mathbf{a}$ and $\mathbf{y}\times(\mathbf{z}\times\mathbf{x}) = \mathbf{b}$ (given)

$\Rightarrow (\mathbf{x}\cdot\mathbf{z})\,\mathbf{y} - (\mathbf{x}\cdot\mathbf{y})\mathbf{z} = \mathbf{a}$ and $(\mathbf{y}\cdot\mathbf{x})\mathbf{z} - (\mathbf{y}\cdot\mathbf{z})\mathbf{x} = \mathbf{b}$

$\Rightarrow \quad \mathbf{y} - \mathbf{z} = \mathbf{a}$ and $\mathbf{z} - \mathbf{x} = \mathbf{b}$

$\Rightarrow \quad \mathbf{y} - \mathbf{x} = \mathbf{a} + \mathbf{b}$

Thus, we have

$\mathbf{y} - \mathbf{z} = \mathbf{a} \quad$ ...(i)

$\mathbf{z} - \mathbf{x} = \mathbf{b} \quad$ ...(ii)

$\mathbf{y} - \mathbf{x} = \mathbf{a} + \mathbf{b} \quad$ ...(iii)

Now, $\quad \mathbf{x}\times\mathbf{y} = \mathbf{c} \quad$ (given)

$\Rightarrow \quad \mathbf{x}\times(\mathbf{x}\times\mathbf{y}) = \mathbf{x}\times\mathbf{c}$ (taking cross-product with **x**)

$\Rightarrow \quad (\mathbf{x}\cdot\mathbf{y})\times\mathbf{x} - (\mathbf{x}\cdot\mathbf{x})\mathbf{y} = \mathbf{x}\times\mathbf{c}$

$\Rightarrow \quad \mathbf{x} - 2\mathbf{y} = \mathbf{x}\times\mathbf{c} \quad$ ...(iv)

Again, $\quad \mathbf{x}\times\mathbf{y} = \mathbf{c}$

$\Rightarrow \quad \mathbf{y}\times(\mathbf{x}\times\mathbf{y}) = \mathbf{y}\times\mathbf{c}$ (taking cross product with **y**)

$\Rightarrow \quad (\mathbf{y}\cdot\mathbf{y})\mathbf{x} - (\mathbf{y}\cdot\mathbf{x})\mathbf{y} = \mathbf{y}\times\mathbf{c}$

$\Rightarrow \quad 2\mathbf{x} - \mathbf{y} = \mathbf{y}\times\mathbf{c} \quad$ ...(iv)

On subtracting Eqs. (iv) and (v), we get

$\mathbf{x} - \mathbf{y} = (\mathbf{y}\times\mathbf{c}) - (\mathbf{x}\times\mathbf{c})$

$\Rightarrow \quad \mathbf{x}\times\mathbf{y} = (\mathbf{y} - \mathbf{x})\times\mathbf{c}$

$\Rightarrow \quad \mathbf{x} + \mathbf{y} = (\mathbf{a} + \mathbf{b})\times\mathbf{c} \quad$ ...(vi)

Adding Eqs. (iii) and (vi), we get

$$2\mathbf{y} = (\mathbf{a} + \mathbf{b}) + (\mathbf{a} + \mathbf{b})\times\mathbf{c},\ \mathbf{y} = \frac{1}{2}[(\mathbf{a} + \mathbf{b})\,(\mathbf{a} + \mathbf{b})\times\mathbf{c}$$

Substituting the value of **y** in Eq. (iii) in Eq. (i), we get

$$\mathbf{x} = \frac{1}{2}[(\mathbf{a} + \mathbf{b}) + (\mathbf{a} + \mathbf{b})\times\mathbf{c}] - (\mathbf{a} + \mathbf{b})$$

$$\Rightarrow \quad \mathbf{x} = \frac{1}{2}[(\mathbf{a} + \mathbf{b})\times\mathbf{c} - (\mathbf{a} + \mathbf{b})]$$

$$\mathbf{z} = \frac{1}{2}[(\mathbf{a} + \mathbf{b}) + (\mathbf{a} + \mathbf{b})\times\mathbf{c}] - \mathbf{a}$$

$$\mathbf{z} = \frac{1}{2}[(\mathbf{b} - \mathbf{a}) + (\mathbf{a} + \mathbf{b})\times\mathbf{c}].$$

137. (a) **138.** (a) **139.** (b)

Solutions (Q.Nos. 140 to 142)

Taking dot products with **a**, **b**, **c** respectively with given equation

$[\mathbf{abc}] = p + (q + r)\cos\theta \quad$ ...(i)

$0 = (p + r)\cos\theta + q \quad$ ...(ii)

$[\mathbf{abc}] = (p + q)\cos\theta + r \quad$ ...(iii)

Also, $\quad [\mathbf{abc}]^2 = \begin{vmatrix} 1 & \cos\theta & \cos\theta \\ \cos\theta & 1 & \cos\theta \\ \cos\theta & \cos\theta & 1 \end{vmatrix}$

$$= 2\cos^2\theta - 3\cos^3\theta + 1 = (1 - \cos\theta)^2\,(1 + 2\cos\theta)$$

$\therefore \quad v = |[\mathbf{abc}]| = |1 - \cos\theta|\,|\sqrt{1 + 2\cos\theta}|$

$$= 2\sin^2\frac{\theta}{2}|\sqrt{1 + 2\cos\theta}|$$

From Eqs. (i) and (iii) $p = r$; substituting in Eq. (ii), we get

$\therefore \quad 2p\cos\theta + q = 0 \Rightarrow \dfrac{q}{p} + 2\cos\theta = 0$

140. (a) **141.** (c) **142.** (b)

143. (A) We know that 3 vectors are coplanar, if $x\mathbf{a} + y\mathbf{b} + z\mathbf{c} = 0$

Clearly, $-3\hat{\mathbf{i}} + 3\hat{\mathbf{j}} + 4\hat{\mathbf{k}}$ and $\hat{\mathbf{i}} + \hat{\mathbf{j}}$ are two vectors lie in the plane ($\mathbf{a} + \mathbf{b}$ and $\mathbf{a} - \mathbf{b}$).

(B) $\mathbf{a} \times \mathbf{b}$ is a vector which perpendicular to both $\mathbf{a}$ and $\mathbf{b}$.

$$\therefore \quad \mathbf{a} \times \mathbf{b} = \begin{vmatrix} \hat{\mathbf{i}} & \hat{\mathbf{j}} & \hat{\mathbf{k}} \\ -1 & 2 & 2 \\ -2 & 1 & 2 \end{vmatrix} = \hat{\mathbf{i}}(4-2) - \hat{\mathbf{j}}(-2+4) + \hat{\mathbf{k}}(-1+4)$$

$$= 2\hat{\mathbf{i}} - 2\hat{\mathbf{j}} + 3\hat{\mathbf{k}}$$

(C) $\mathbf{c}$ is a vector which is equally inclined to $\mathbf{a}$ and $\mathbf{b}$

$\therefore \quad \mathbf{c} \cdot \mathbf{a} = \mathbf{c} \cdot \mathbf{b}$

Clearly $\hat{\mathbf{i}} - \hat{\mathbf{j}} + 5\hat{\mathbf{k}}$ satisfies the condition.

(D) $\mathbf{a}$, $\mathbf{b}$ and $\mathbf{c}$ are from the triangle

$\Rightarrow \quad \mathbf{AB} + \mathbf{BC} = \mathbf{AC}$

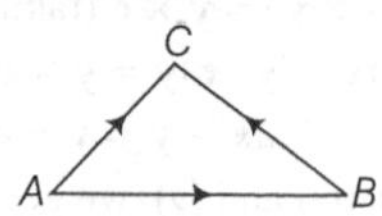

144. Given, $[\mathbf{a}\times\mathbf{b} \;\; \mathbf{b}\times\mathbf{c} \;\; \mathbf{c}\times\mathbf{a}] = 36$

$[\mathbf{a}\;\mathbf{b}\;\mathbf{c}] = 6$

$\Rightarrow$ Volume of tetrahedron from by vectors

$\mathbf{a}$, $\mathbf{b}$ and $\mathbf{c}$ is $\dfrac{1}{6}[\mathbf{a}\;\mathbf{b}\;\mathbf{c}]] = 1$

$[\mathbf{a}+\mathbf{b} \;\; \mathbf{b}+\mathbf{c} \;\; \mathbf{c}+\mathbf{a}] = 2[\mathbf{a}\;\mathbf{b}\;\mathbf{c}] = 12$

$\mathbf{a}-\mathbf{b}$, $\mathbf{b}-\mathbf{c}$ and $\mathbf{c}-\mathbf{a}$ are coplanar

$\Rightarrow \quad [\mathbf{a}-\mathbf{b} \;\; \mathbf{b}-\mathbf{c} \;\; \mathbf{c}-\mathbf{a}] = 0$

145. (A) $\dfrac{\text{Area of }\Delta ABC}{\text{Area of }\Delta AOC} = \dfrac{\frac{1}{2}|\mathbf{a}\times\mathbf{b} + \mathbf{b}\times\mathbf{c} + \mathbf{c}\times\mathbf{a}|}{\frac{1}{2}|\mathbf{a}\times\mathbf{c}|}$

Now, $\quad \mathbf{a} + 2\mathbf{b} + 3\mathbf{c} = 0$

Cross with $\mathbf{b}$, $\mathbf{a}\times\mathbf{b} + 3\mathbf{c}\times\mathbf{b} = 0$

$\Rightarrow \quad \mathbf{a}\times\mathbf{b} = 3(\mathbf{b}\times\mathbf{c})$

Cross with $\mathbf{a}$, $2\mathbf{a}\times\mathbf{b} + 3\mathbf{a}\times\mathbf{c} = 0$

$\Rightarrow \quad \mathbf{a}\times\mathbf{b} = \dfrac{3}{2}(\mathbf{c}\times\mathbf{a})$

$\therefore \quad \mathbf{a}\times\mathbf{b} = \dfrac{3}{2}(\mathbf{c}\times\mathbf{a}) = 3(\mathbf{b}\times\mathbf{c})$

Let $\quad (\mathbf{c}\times\mathbf{a}) = \mathbf{p}$

$\mathbf{a}\times\mathbf{b} = \dfrac{3\mathbf{p}}{2}$; $\mathbf{b}\times\mathbf{c} = \dfrac{\mathbf{p}}{2}$

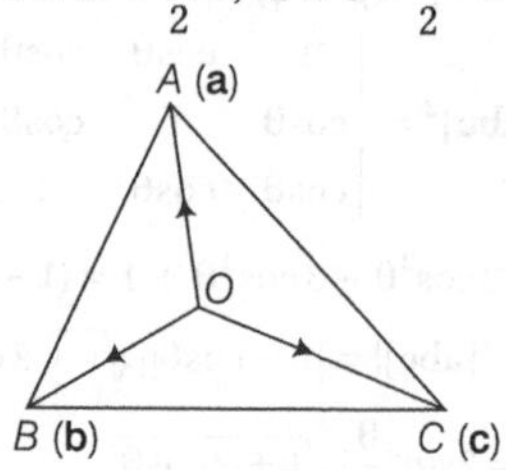

$$\therefore \quad \text{Ratio} = \frac{|\mathbf{a}\times\mathbf{b} + \mathbf{b}\times\mathbf{c} + \mathbf{c}\times\mathbf{a}|}{|\mathbf{c}\times\mathbf{a}|} = \frac{\left|\frac{3\mathbf{p}}{2} + \frac{\mathbf{p}}{2} + \mathbf{p}\right|}{|\mathbf{p}|} = \frac{3|\mathbf{p}|}{|\mathbf{p}|} = 3$$

$$\text{(B)}\ [\mathbf{ABC}]^2 = \begin{vmatrix} \mathbf{A}\cdot\mathbf{A} & \mathbf{A}\cdot\mathbf{B} & \mathbf{A}\cdot\mathbf{C} \\ \mathbf{B}\cdot\mathbf{A} & \mathbf{B}\cdot\mathbf{B} & \mathbf{B}\cdot\mathbf{C} \\ \mathbf{C}\cdot\mathbf{A} & \mathbf{C}\cdot\mathbf{B} & \mathbf{C}\cdot\mathbf{C} \end{vmatrix} = \begin{vmatrix} 1 & 0 & 0 \\ 0 & 1 & \frac{\sqrt{3}}{2} \\ 0 & \frac{\sqrt{3}}{2} & 1 \end{vmatrix}$$

$$1 = \left(1 - \frac{3}{4}\right) = \frac{1}{4}$$

$$[\mathbf{ABC}] = \frac{1}{2}$$

$$\text{(C)}\ (\mathbf{b}\times\mathbf{c})\cdot(\mathbf{a}\times\mathbf{d}) = \begin{vmatrix} \mathbf{b}\cdot\mathbf{a} & \mathbf{b}\cdot\mathbf{a} \\ \mathbf{c}\cdot\mathbf{a} & \mathbf{c}\cdot\mathbf{d} \end{vmatrix}$$

Similarly, compute others which gives (i).

146. (A) $\mathbf{x}\cdot\mathbf{y} = |\mathbf{a}|^2 - |\mathbf{b}|^2 = 0$

$\mathbf{x}$ is perpendicular to $\mathbf{y}$.

$|\mathbf{x}\times\mathbf{y}|^2 = |\mathbf{x}|^2|\mathbf{y}|^2$

$\{|\mathbf{a}|^2 + |\mathbf{b}|^2 + 2\mathbf{a}\cdot\mathbf{b}|\}\{|\mathbf{a}|^2 + |\mathbf{b}|^2 - 2\mathbf{a}\cdot\mathbf{b}\}$

$$= 64 - 4(\mathbf{a}\cdot\mathbf{b})^2 = 4\left\{16 - (\mathbf{a}\cdot\mathbf{b})^{\frac{1}{2}}\right\}$$

$$\lambda = 16$$

(B) $\dfrac{2\lambda + 1}{\sqrt{\lambda^2 + 2}\sqrt{\lambda^2 + 2}} = \dfrac{1}{2}$

$2(2\lambda + 1) = \lambda^2 + 2$

$\lambda^2 - 4\lambda = 0,\ \lambda = 0$ or 4

$\lambda = 4$ is non-zero value.

(C) If the lines are coplanar, all the 4 planes will have a common point.

Solving $4x + 3y - 2z + 3 = 0$

$x - 3y + 4z + 6 = 0$

$x - y + z + 1 = 0$

We get $x = -\dfrac{1}{3},\ y = -3,\ z = \dfrac{-11}{3}$

Substituting in $kx - 4y + 7z + 16 = 0$

We get $k = 7$

(D) $E = |\mathbf{a} - 2\mathbf{b}|^2 + |\mathbf{b} - 2\mathbf{c}|^2 + |\mathbf{c} - 2\mathbf{a}|^2$

$= 5(|\mathbf{a}|^2 + |\mathbf{b}|^2 + |\mathbf{c}|^2) - 4[\mathbf{a}\cdot\mathbf{b} + \mathbf{b}\cdot\mathbf{c} + \mathbf{c}\cdot\mathbf{a}]$

$= 5\cdot 6 - 4[\mathbf{a}\cdot\mathbf{b} + \mathbf{b}\cdot\mathbf{c} + \mathbf{c}\cdot\mathbf{a}]$

$\Rightarrow \mathbf{a}\cdot\mathbf{b} + \mathbf{b}\cdot\mathbf{c} + \mathbf{c}\cdot\mathbf{a} = \dfrac{30 - E}{4}$

Also, $\quad |\mathbf{a} + \mathbf{b} + \mathbf{c}|^2 \geq 0$

$6 + 2[\mathbf{a}\cdot\mathbf{b} + \mathbf{b}\cdot\mathbf{c} + \mathbf{c}\cdot\mathbf{a}] \geq 0$

$6 + 2\left[\dfrac{30 - E}{4}\right] \geq 0$

$12 + 30 - E \geq 0$

$42 \geq E$

$E \geq 42$

147. (A) $\mathbf{a}\cdot\mathbf{b}+\mathbf{a}\cdot\mathbf{c}=\mathbf{b}\cdot\mathbf{c}+\mathbf{b}\cdot\mathbf{a}=0$

$$\Rightarrow \mathbf{b}\cdot\mathbf{c}=\mathbf{a}\cdot\mathbf{c}=-\mathbf{a}\cdot\mathbf{b}=-\frac{17}{2}$$

$$\Rightarrow |\mathbf{a}+\mathbf{b}+\mathbf{c}|=\sqrt{a^2+b^2+c^2-2(\mathbf{a}\cdot\mathbf{b})}=9$$

(B) $[\mathbf{a}\ \mathbf{b}\ \mathbf{c}]$, write in terms of $(\mathbf{a}_1\ \mathbf{a}_2\ \mathbf{a}_3\ \mathbf{b}_1\mathbf{b}_2\mathbf{b}_3)$

$$[\mathbf{def}]=\left[\frac{\mathbf{b}\times\mathbf{c}}{[\mathbf{abc}]}\ \frac{\mathbf{c}\times\mathbf{a}}{[\mathbf{abc}]}\ \frac{\mathbf{a}\times\mathbf{b}}{[\mathbf{abc}]}\right]=\frac{1}{[\mathbf{abc}]}$$

$$\Rightarrow [\mathbf{abc}]\,[\mathbf{def}]=1$$

(C) $\alpha=\dfrac{\text{ar}(ABCD)}{\text{ar(parallelogram)}}=$

$$\left(\frac{\frac{1}{2}|\mathbf{a}\times(\mathbf{a}+3\mathbf{b})|+\frac{1}{2}|(2\mathbf{a}+3\mathbf{b})\times\mathbf{b}|}{|\mathbf{a}\times\mathbf{b}|}\right)=\frac{5}{2}$$

(D) $x=\dfrac{\mathbf{d}\cdot\mathbf{c}}{[\mathbf{abc}]},\ y=\dfrac{\mathbf{d}\cdot\mathbf{a}}{(\mathbf{a}\ \mathbf{b}\ \mathbf{c})},\ z=\dfrac{\mathbf{d}\cdot\mathbf{b}}{[\mathbf{abc}]}$

$$x+y+z=\frac{\mathbf{d}\cdot(\mathbf{a}+\mathbf{b}+\mathbf{c})}{[\mathbf{abc}]}$$

$$\Rightarrow \quad R=\frac{4}{[\mathbf{abc}]}=8$$

148. Let the angle between $\mathbf{u}$ and $\mathbf{v}$ is θ and $\mathbf{w}$ and $\mathbf{u}$ is θ

$$[\hat{\mathbf{u}}\ \hat{\mathbf{v}}\ \hat{\mathbf{w}}]^2=\begin{bmatrix}\hat{\mathbf{u}}\cdot\hat{\mathbf{u}} & \hat{\mathbf{u}}\cdot\hat{\mathbf{v}} & \hat{\mathbf{u}}\cdot\hat{\mathbf{w}}\\ \hat{\mathbf{v}}\cdot\hat{\mathbf{u}} & \hat{\mathbf{v}}\cdot\hat{\mathbf{v}} & \hat{\mathbf{v}}\cdot\hat{\mathbf{w}}\\ \hat{\mathbf{w}}\cdot\hat{\mathbf{u}} & \hat{\mathbf{w}}\cdot\hat{\mathbf{v}} & \hat{\mathbf{w}}\cdot\hat{\mathbf{w}}\end{bmatrix}$$

$$=\begin{bmatrix}1 & \cos2\theta & \cos\theta\\ \cos2\theta & 1 & \cos\theta\\ \cos\theta & \cos\theta & 1\end{bmatrix}=0$$

149. $2[\mathbf{a}\ \mathbf{b}\ \mathbf{c}]+[\mathbf{a}\ \mathbf{b}\ \mathbf{c}]^2+0=2\times1+1^2=3$

150. $\hat{\mathbf{a}}\cdot\hat{\mathbf{c}}=\hat{\mathbf{b}}\cdot\hat{\mathbf{c}}=\cos\theta$

$$\hat{\mathbf{c}}=\alpha\hat{\mathbf{a}}+\beta\hat{\mathbf{b}}+\gamma(\hat{\mathbf{a}}+\hat{\mathbf{b}})$$

Taking dot product with $\hat{\mathbf{a}}$ both sides $\cos\theta=\alpha$

Taking dot product with $\hat{\mathbf{b}}$ both sides $\cos\theta=\beta$

Taking dot product with $\mathbf{c}$ both sides

$$1=\alpha\cos\alpha+\beta\cos\theta+\gamma\,[\hat{\mathbf{a}}\ \hat{\mathbf{b}}\ \hat{\mathbf{c}}]$$

But $\quad [\hat{\mathbf{a}}\ \hat{\mathbf{b}}\ \hat{\mathbf{c}}]^2=\begin{bmatrix}1 & 0 & \cos\theta\\ 0 & 1 & \cos\theta\\ \cos\theta & \cos\theta & 1\end{bmatrix}$

$$=1-2\cos^2\theta$$

So, $\quad 1=\cos^2\theta+\cos^2\theta+\gamma\sqrt{1-2\cos^2\theta}$

$$\Rightarrow \quad \gamma=\sqrt{1-2\cos^2\theta}$$

So, $\alpha^2+\beta^2+\gamma^2=1$

151. The three adjacent sides of tetrahedron is given by

$$\frac{(\hat{\mathbf{i}}+\hat{\mathbf{j}})\times(\hat{\mathbf{j}}+\hat{\mathbf{k}})}{|(\hat{\mathbf{i}}+\hat{\mathbf{j}})\times(\hat{\mathbf{j}}+\hat{\mathbf{k}})|},\frac{(\hat{\mathbf{j}}+\hat{\mathbf{k}})(\hat{\mathbf{k}}+\hat{\mathbf{i}})}{|(\hat{\mathbf{j}}+\hat{\mathbf{k}})\times(\hat{\mathbf{k}}+\hat{\mathbf{i}})|},\frac{(\hat{\mathbf{k}}+\hat{\mathbf{i}})(\hat{\mathbf{i}}+\hat{\mathbf{j}})}{|(\hat{\mathbf{k}}+\hat{\mathbf{i}})\times(\hat{\mathbf{i}}+\hat{\mathbf{j}})|}$$

i.e., $\quad \dfrac{\hat{\mathbf{i}}+\hat{\mathbf{j}}+\hat{\mathbf{k}}}{\sqrt{3}},\dfrac{\hat{\mathbf{i}}+\hat{\mathbf{j}}-\hat{\mathbf{k}}}{\sqrt{3}},\dfrac{-\hat{\mathbf{i}}+\hat{\mathbf{j}}+\hat{\mathbf{k}}}{\sqrt{3}}$

$$V=\frac{1}{6}\times\frac{1}{3\sqrt{3}}\left\|\begin{matrix}1 & -1 & 1\\ 1 & 1 & -1\\ -1 & 1 & 1\end{matrix}\right\|=\frac{2}{9\sqrt{3}}$$

So, $\quad 9\sqrt{3}V=2$

152. Let $\quad \hat{\mathbf{c}}=x\mathbf{a}+y\,\mathbf{b}$, where x and y are scalars.

$$\Rightarrow \quad \hat{\mathbf{c}}=x(\hat{\mathbf{i}}-\hat{\mathbf{j}}+2\hat{\mathbf{k}})+y(2\hat{\mathbf{i}}-\hat{\mathbf{j}}+\hat{\mathbf{k}})$$

$$\Rightarrow \quad \hat{\mathbf{c}}=\hat{\mathbf{i}}(x+2y)+\hat{\mathbf{j}}(-x-y)+\hat{\mathbf{k}}(2x+y)$$

But, $\quad \hat{\mathbf{c}}\cdot\mathbf{a}=0$

$$6x+5y=0\ \Rightarrow\ y=-\frac{6x}{5}$$

So, $\quad \hat{\mathbf{c}}=\dfrac{-7x}{5}\hat{\mathbf{i}}+\dfrac{x}{5}\hat{\mathbf{j}}+\dfrac{4x}{5}\hat{\mathbf{k}}$

We have, $\dfrac{49x^2+x^2+16x^2}{25}=1\ \Rightarrow\ x^2=\dfrac{25}{66}$

$$\therefore \quad \hat{\mathbf{c}}=\pm\frac{5}{\sqrt{66}}\left(\frac{-7}{5}\hat{\mathbf{i}}+\frac{1}{5}\hat{\mathbf{j}}+\frac{4}{5}\hat{\mathbf{k}}\right)$$

$$p=|\hat{\mathbf{c}}\cdot\mathbf{b}|=\frac{\sqrt{11}}{6}$$

So, $\quad \dfrac{\sqrt{11}}{p}=6\ \Rightarrow\ k=6$

143. Let the angle between $\mathbf{a}$ and $\mathbf{b}$ is α and $\mathbf{a}\times\mathbf{b}$ and $\mathbf{c}$ is β.

$$\therefore \quad |[\mathbf{a}\ \mathbf{b}\ \mathbf{c}]|=6$$

$$\Rightarrow \quad \sin\alpha\cos\beta=1\ \Rightarrow\ \sin\alpha=1,\cos\beta=1$$

$$\Rightarrow \quad \alpha=90^\circ,\beta=0^\circ$$

$\Rightarrow$ **a, b and c are mutually perpendicular.**

Again, $[\mathbf{b}\ \mathbf{c}\ \hat{\mathbf{d}}]=0$

$$\Rightarrow \quad \begin{vmatrix}4 & 0 & 1\\ 0 & 9 & \mathbf{c}\cdot\hat{\mathbf{d}}\\ 1 & \mathbf{c}\cdot\hat{\mathbf{d}} & 1\end{vmatrix}=0$$

$$\Rightarrow \quad \mathbf{c}\cdot\hat{\mathbf{d}}=\pm\frac{3\sqrt{3}}{2}$$

We have, $\quad \mathbf{a}\cdot\mathbf{b}=0$

$$|\mathbf{a}\times\mathbf{c}\cdot\hat{\mathbf{d}}|^2=\begin{vmatrix}1 & 0 & 0\\ 0 & 9 & \frac{3\sqrt{3}}{2}\\ 0 & \frac{3\sqrt{3}}{2} & 1\end{vmatrix}$$

$$=9-\frac{27}{4}=\frac{9}{4}$$

$$|(\mathbf{a}\times\mathbf{c})\times\hat{\mathbf{d}}|^2=|(\mathbf{a}\cdot\hat{\mathbf{d}})\,\mathbf{c}-(\mathbf{c}\cdot\hat{\mathbf{d}})\mathbf{a}|^2$$

$$=\left|\frac{3\sqrt{3}}{2}\mathbf{a}\right|^2=\frac{27}{4}$$

So, $\quad |\mathbf{a}\times\mathbf{c}\cdot\hat{\mathbf{d}}|^2+|(\mathbf{a}\times\mathbf{c})\times\hat{\mathbf{d}}|^2=\dfrac{36}{4}=9$

154. P.V. of $D = \frac{\lambda - 1}{2}\hat{\mathbf{i}} + 4\hat{\mathbf{j}} + \frac{\mu + 2}{2}\hat{\mathbf{k}}$

D.R. of AD $= \frac{\lambda - 4}{2}, 1, \frac{\mu - 8}{2}$

But directions of AD should be $\frac{1}{\sqrt{3}}, \frac{1}{\sqrt{3}}, \frac{1}{\sqrt{3}}$.

$$\Rightarrow \quad \frac{\lambda - 4}{2} = 1 = \frac{\mu - 8}{2}$$

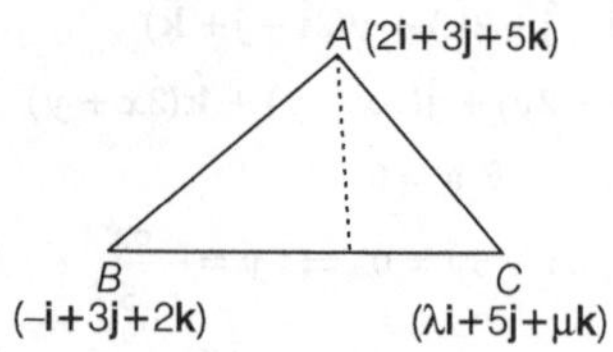

$$\lambda = 6, \mu = 10$$
$$2\lambda - \mu = 2$$

155. $v = [\mathbf{abc}]$

$$[\alpha\ \beta\ \gamma] = \begin{vmatrix} \mathbf{a}\cdot\mathbf{a} & \mathbf{a}\cdot\mathbf{b} & \mathbf{a}\cdot\mathbf{c} \\ \mathbf{b}\cdot\mathbf{a} & \mathbf{b}\cdot\mathbf{b} & \mathbf{b}\cdot\mathbf{c} \\ \mathbf{c}\cdot\mathbf{a} & \mathbf{c}\cdot\mathbf{b} & \mathbf{c}\cdot\mathbf{c} \end{vmatrix} [\mathbf{a\ b\ c}]$$

$$= [\mathbf{abc}]\,[\mathbf{abc}]\,[\mathbf{abc}] = v^3$$

$\therefore \quad \lambda = 3$

156. $\mathbf{c} \times \mathbf{a} = \mathbf{b} \Rightarrow |\mathbf{c} \times \mathbf{a}| = |\mathbf{b}|$

$\Rightarrow \quad |\mathbf{c}|\,|\mathbf{a}| \sin\theta = 3,$

$$|\mathbf{c}| = \frac{3}{2\sin\theta}\ |\mathbf{c} - \mathbf{a}|^2 = |\mathbf{c}|^2 + |\mathbf{a}|^2 - 2\mathbf{c}\cdot\mathbf{a}$$

$$= |\mathbf{c}|^2 + 4 - 2|\mathbf{c}|\cdot|\mathbf{a}|\cos\theta$$

$$= \frac{9}{4\sin^2\theta} + 4 - 2\cdot\frac{3}{2\sin\theta}\cdot 2\cdot\cos\theta$$

$$= 4 + \frac{9}{4}\operatorname{cosec}^2\theta - 6\cot\theta$$

$$= \frac{9}{4} + \left(\frac{3}{2}\cot\theta - 2\right)^2$$

$$|\mathbf{c} - \mathbf{a}|^2 \geq \frac{9}{4}$$

$$\Rightarrow \quad |\mathbf{c} - \mathbf{a}| \geq \frac{3}{2}$$

$$\Rightarrow \quad 2|\mathbf{c} - \mathbf{a}| \geq 3$$

$\therefore$ Min. of $2|\mathbf{c} - \mathbf{a}| = 3$

157. In ΔABD, N is the mid-point of BD.

$\therefore \quad AB + AD = 2AN \quad$ ...(i)

In ΔCBD, N is the mid-point of BD.

$\therefore \quad \mathbf{CB} + \mathbf{CD} = 2\mathbf{CN} \quad$ ...(ii)

Adding Eqs. (i) and (ii), we have

$AB + AD + CB + CD = 2(AN + CN) \quad$ ...(iii)

In $\Delta\ ANC$, M is the mid-point of AC

$\therefore \quad AN + CN = 2MN$

From Eq. (iii), we get

$AB + AD + CB + CD = 2\,(2MN) = 4MN$

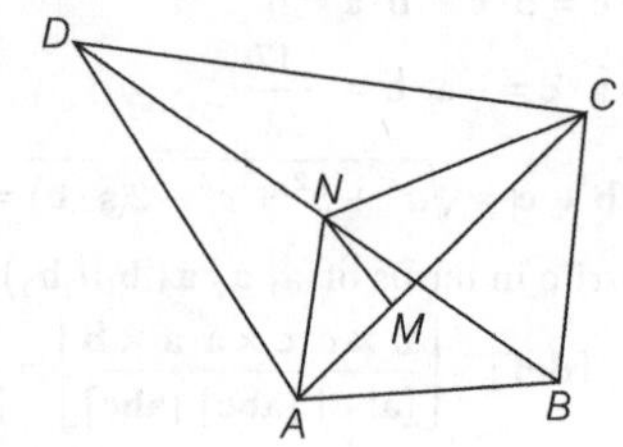

158. $|\mathbf{a}| = |\mathbf{b}| = |\mathbf{c}| = 1; [\mathbf{abc}] = 1$

Volume of the tetrahedron $= \frac{1}{6}$

$$= \frac{1}{6}\left\| \begin{vmatrix} 3 & -2 & 2 \\ -1 & 0 & -2 \\ 2 & -3 & 4 \end{vmatrix} [\mathbf{a\ b\ c}] \right\| = 2$$

159. $[\mathbf{a} \times (\mathbf{b} \times \mathbf{c})]\cdot(\mathbf{a} \times \mathbf{c}) = 5$

$\Rightarrow \quad [(\mathbf{a}\cdot\mathbf{c})\mathbf{b} - (\mathbf{a}\cdot\mathbf{b})\mathbf{c}]\cdot(\mathbf{a} \times \mathbf{c}) = 5 \Rightarrow (\mathbf{a}\cdot\mathbf{c})\,[\mathbf{bac}] = 5$

$\Rightarrow \quad [\mathbf{abc}] = -10$

$\Rightarrow \quad -[\mathbf{abc}] - 1 = +10 - 1 = 9$

160. $\begin{vmatrix} \mathbf{a}\cdot\mathbf{a} & \mathbf{a}\cdot\mathbf{b} & \mathbf{a}\cdot\mathbf{c} \\ \mathbf{b}\cdot\mathbf{a} & \mathbf{b}\cdot\mathbf{b} & \mathbf{b}\cdot\mathbf{c} \\ \mathbf{c}\cdot\mathbf{a} & \mathbf{c}\cdot\mathbf{b} & \mathbf{c}\cdot\mathbf{c} \end{vmatrix} = [\mathbf{abc}]^2 = [\mathbf{a \times b\ b \times c\ c \times a}]^2 = 36$

$$= 4 \times 9 = 2^2 \times 3^2$$

$\therefore \quad P + q = 5$

161. $\hat{\alpha}\cdot\hat{x} = \hat{\alpha}\cdot\hat{\beta} - \hat{\alpha}\cdot(\hat{\alpha} \times \mathbf{x}) = 0$

Also, $\hat{\alpha} \times \hat{x} = \hat{\alpha} \times \hat{\beta} - \hat{\alpha} \times (\hat{\alpha} \times \mathbf{x})$

$$= \alpha \times \hat{\beta} - (\hat{\alpha}\cdot\mathbf{x})\hat{\alpha} + |\alpha|^2\mathbf{x}$$

$\Rightarrow \quad \hat{\beta} - \mathbf{x} = \alpha \times \hat{\beta} + x \ \text{ or } \ 2\mathbf{x} = \hat{\beta} - \alpha \times \hat{\beta}$

$\Rightarrow \quad 4|\mathbf{x}|^2 = |\hat{\beta}|^2 + |\alpha \times \hat{\beta}|^2 - 2\hat{\beta}\cdot(\hat{\alpha} \times \hat{\beta}) = 2$

$\Rightarrow \quad |\mathbf{x}|^2 = \frac{1}{2} \qquad \Rightarrow 4|\mathbf{x}|^2 = 2$

162. $\frac{1}{6}\begin{vmatrix} -2 & -4 & 6 \\ 0 & 1 & -8 \\ 2 & -5 & \lambda - 1 \end{vmatrix} = 22 \Rightarrow \lambda = 133$

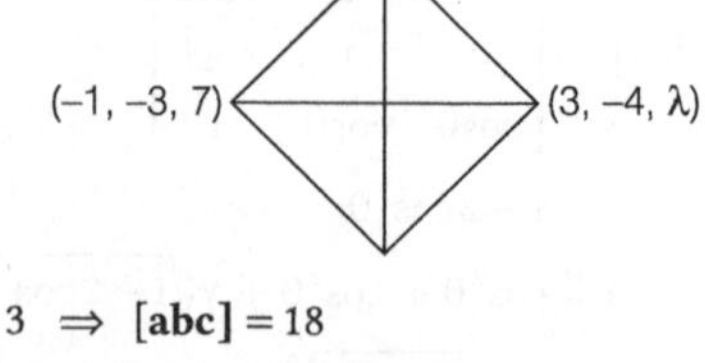

163. $\frac{1}{6}[\mathbf{abc}] = 3 \Rightarrow [\mathbf{abc}] = 18$

$$V = [\mathbf{a + b - c\ a - b\ b - c}]$$
$$= (\mathbf{a} + \mathbf{b} - \mathbf{c})\cdot(\mathbf{a} - \mathbf{b}) \times (\mathbf{b} - \mathbf{c}) = \mathbf{a}\cdot(\mathbf{b} \times \mathbf{c}) = [\mathbf{abc}] = 18$$

164. Let θ be the angle between vectors **a** and **b**. Then,

$\mathbf{a}\cdot\mathbf{b} = |\mathbf{a}||\mathbf{b}|\cos\theta$

$\Rightarrow \quad (\mathbf{a}\cdot\mathbf{b})^2 = |\mathbf{a}|^2|\mathbf{b}|^2\cos^2\theta$

Now, $\cos^2\theta \leq 1 \Rightarrow |\mathbf{a}|^2|\mathbf{b}|^2\cos^2\theta \leq |\mathbf{a}|^2|\mathbf{b}|^2$

$\therefore \quad (\mathbf{a}\cdot\mathbf{b})^2 \leq |\mathbf{a}|^2|\mathbf{b}|^2$

165. Let $P(x_1, y_1)$ and $Q(x_2, y_2)$ be the two points on $y = 2^{x+2}$

$\mathbf{OP} \cdot \hat{\mathbf{i}}$ = Projection of **OP** on the X-axis

$\Rightarrow \quad x_1 = -1 \qquad (\because \mathbf{OP} \cdot \hat{\mathbf{i}} = -1)$

Also, (x_1, y_1) lies on $y = 2^{x+2}$

$\therefore \quad y_1 = 2^{x_1 + 2} \Rightarrow y_1 = 2$

Also, $\mathbf{OQ} \cdot \hat{\mathbf{i}} =$ Projection of **OQ** on X-axis.

$\Rightarrow \quad x_2 = 2 \qquad (\text{given } \mathbf{OQ} \cdot \hat{\mathbf{i}} = 2)$

As (x_2, y_2) lies on $y = 2^{x+2}$

$$y_2 = 2^{x_2 + 2};\ y_2 = 16$$

Thus, $\quad \mathbf{OP} = x_1\hat{\mathbf{i}} + y_1\hat{\mathbf{j}} = -\hat{\mathbf{i}} + 2\hat{\mathbf{j}}$

and $\quad \mathbf{OQ} = x_2\hat{\mathbf{i}} + y_2\hat{\mathbf{j}} = 2\hat{\mathbf{i}} + 16\hat{\mathbf{j}}$

$\Rightarrow \quad \mathbf{OQ} - 4\mathbf{OP} = 6\hat{\mathbf{i}} + 8\hat{\mathbf{j}}$

$\Rightarrow \quad |\mathbf{OQ} - 4\mathbf{OP}| = \sqrt{36 + 64} = 10$

166. Let $A(x_1, y_1)$ in XY-plane.

$\therefore \quad \mathbf{OA} = \mathbf{a} = x_1\hat{\mathbf{i}} + y_1\hat{\mathbf{j}}$

$\quad \mathbf{OP} = \mathbf{r} = x\hat{\mathbf{i}} + y\hat{\mathbf{j}}$

$\because$ Point P lies on the tangent to the circle.

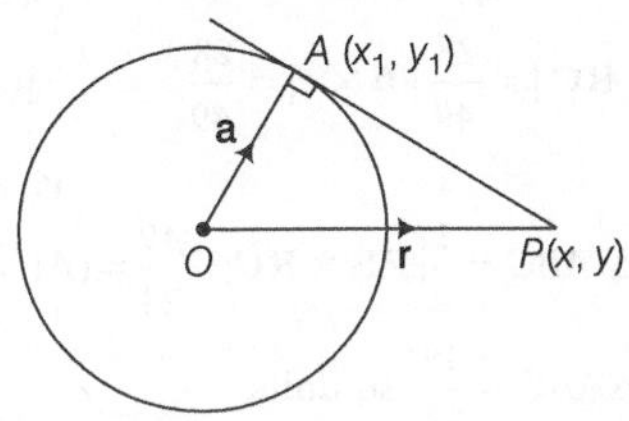

$\therefore$ **OA** is perpendicular to **AP**.

$\Rightarrow \quad \mathbf{OA} \cdot \mathbf{AP} = 0 \Rightarrow \mathbf{a} \cdot (\mathbf{r} - \mathbf{a}) = 0$

i.e., $\quad \mathbf{a} \cdot \mathbf{r} - \mathbf{a} \cdot \mathbf{a} = 0$

or $\quad \mathbf{a} \cdot \mathbf{r} = \mathbf{a} \cdot \mathbf{a}$

$\Rightarrow \quad \mathbf{a} \cdot \mathbf{r} = a^2$

$\Rightarrow \quad (x_1\hat{\mathbf{i}} + y_1\hat{\mathbf{j}}) \cdot (x\hat{\mathbf{i}} + y\hat{\mathbf{j}}) = a^2$

$\Rightarrow \quad xx_1 + yy_1 = a^2$

which is the equation of the tangent to the circle at the point A.

167. The given relation can be rewritten as,

$(\sqrt{a^2 - 4}\hat{\mathbf{i}} + a\hat{\mathbf{j}} + \sqrt{a^2 + 4}\hat{\mathbf{k}}) \cdot (\tan A\hat{\mathbf{i}} + \tan B\hat{\mathbf{j}} + \tan C\hat{\mathbf{k}}) = 6a$

$\Rightarrow \quad \sqrt{(a^2 - 4) + a^2 + (a^2 + 4)}$

$\sqrt{\tan^2 A + \tan^2 B + \tan^2 C} \cdot \cos\theta = 6a \quad (\because a \cdot b = |a||b|\cos\theta)$

$\sqrt{3}a \cdot \sqrt{\tan^2 A + \tan^2 B + \tan^2 C} \cdot \cos\theta = 6a$

$\Rightarrow \quad \tan^2 A + \tan^2 B + \tan^2 C = 12\sec^2\theta \qquad \text{...(i)}$

Also, $\quad 12\sec^2\theta \ge 12 \qquad (\because \sec^2\theta \ge 1) \text{ ...(ii)}$

From Eqs. (i) and (ii), $\tan^2 A + \tan^2 B + \tan^2 C \ge 12$

$\therefore$ Least value of $\tan^2 A + \tan^2 B + \tan^2 C = 12$

168. Here, M is the mid-point of BC.

$\therefore \quad \mathbf{AM} = \mathbf{AB} + \frac{1}{2}(\mathbf{AB} + \mathbf{AC}) \quad (\text{using } \mathbf{AB} + \mathbf{BC} = \mathbf{AC})$

$\mathbf{AB} \,||\, (\mathbf{AB} + \mathbf{AC})$

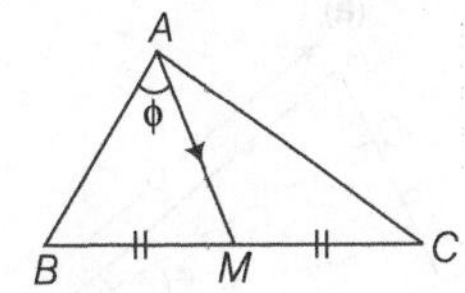

Let angle $BAM = \phi$

$$\therefore \quad \cos\phi = \frac{\mathbf{AB} \cdot (\mathbf{AB} + \mathbf{AC})}{|\mathbf{AB}||\mathbf{AB} + \mathbf{AC}|} = \frac{|\mathbf{AB}|^2 + \mathbf{AB} \cdot \mathbf{AC}}{c\sqrt{b^2 + c^2 + 2bc\cos A}}$$

$$= \frac{c^2 + c \cdot b\cos A}{c\sqrt{c^2 + b^2 + 2bc\cos A}} = \frac{c + b\cos A}{\sqrt{b^2 + c^2 + 2b\cos A}}$$

and $\quad \dfrac{b}{c} = \dfrac{\sin B}{\sin C} \Rightarrow \cos\phi$

$$= \frac{\sin C + \sin B \sin A}{\sqrt{\sin^2 B + \sin^2 C + 2\sin B \sin C \cos A}}$$

169. Let $\mathbf{p} = \mathbf{BA}$ and $\mathbf{q} = \mathbf{BC}$

Now, required perpendicular distance

$= AM = (BA)\sin\theta$

$= |\mathbf{p}|\sin\theta \qquad \text{...(i)}$

Consider, $|\mathbf{q} \times \mathbf{p}| = |\mathbf{q}||\mathbf{p}|\sin\theta$

On dividing by $|\mathbf{q}|$

$$\frac{|\mathbf{q} \times \mathbf{p}|}{|\mathbf{q}|} = |\mathbf{p}|\sin\theta \qquad \text{...(ii)}$$

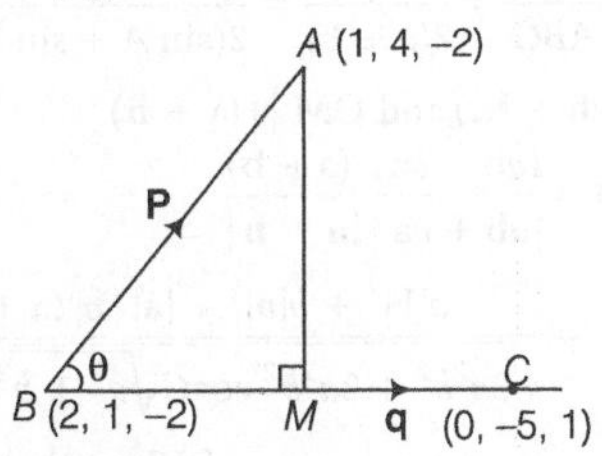

From Eqs. (i) and (ii), required perpendicular distance

$$= \frac{|\mathbf{q} \times \mathbf{p}|}{|\mathbf{q}|} \qquad \text{...(iii)}$$

where, $\quad \mathbf{q} = \mathbf{BC} = \mathbf{OC} - \mathbf{OB} = -2\hat{\mathbf{i}} - 6\hat{\mathbf{j}} + 3\hat{\mathbf{k}}$

$\quad \mathbf{P} = \mathbf{AB} = \mathbf{OA} - \mathbf{OB} = -\hat{\mathbf{i}} + 3\hat{\mathbf{j}}$

$\therefore \quad |\mathbf{q}| = \sqrt{4 + 36 + 9} = 7 \qquad \text{...(iv)}$

and $\quad \mathbf{q} \times \mathbf{p} = \begin{vmatrix} \hat{\mathbf{i}} & \hat{\mathbf{j}} & \hat{\mathbf{k}} \\ -2 & -6 & 3 \\ -1 & 3 & 0 \end{vmatrix} = -9\hat{\mathbf{i}} - 3\hat{\mathbf{j}} - 12\hat{\mathbf{k}}$

$$|\mathbf{q} \times \mathbf{p}| = \sqrt{81 + 9 + 144} = 3\sqrt{26} \qquad \text{...(v)}$$

From Eqs. (iii), (iv) and (v), we get

Perpendicular distance $= \dfrac{3\sqrt{26}}{7}$

170. Here, $AM = MD$ and CD is angle bisector of $\angle C$.

$$\mathbf{CD} = \frac{\mathbf{a}b + b\mathbf{a}}{a + b}$$

and $\quad \mathbf{CM} = \dfrac{\mathbf{a} + \mathbf{b}}{2}$

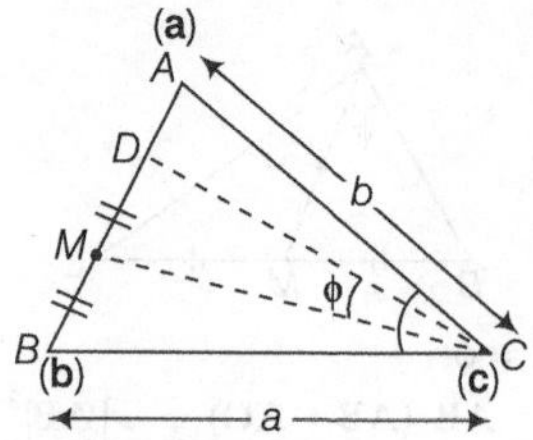

Where, $\mathbf{a} = \mathbf{CB}$ and $\mathbf{b} = \mathbf{CA}$

Consequently,

$$\text{Area of } \Delta\, CDM = \frac{1}{2}(\mathbf{CD} \times \mathbf{CM})$$

$$= \frac{1}{2}\frac{(a\mathbf{b} + b\mathbf{a}) \times (\mathbf{a} + \mathbf{b})}{1(a+b)}$$

$$= \frac{a(\mathbf{b} \times \mathbf{a}) + a(\mathbf{b} \times \mathbf{b}) + b\,(\mathbf{a} \times \mathbf{a}) + b(\mathbf{a} \times \mathbf{b})}{4(a+b)}$$

$$= \frac{(b-a)\,(\mathbf{a} \times \mathbf{b})}{4(a+b)}$$

(using $\mathbf{a} \times \mathbf{a} = \mathbf{b} \times \mathbf{b} = 0$ and $\mathbf{b} \times \mathbf{a} = -\mathbf{a} \times \mathbf{b}$)

$$= \frac{b-a}{2\,(a+b)}\left(\frac{\mathbf{a} \times \mathbf{b}}{2}\right)$$

$$= \frac{b-a}{2(a+b)}(\text{Area of } \Delta\, ABC)$$

$$\therefore \frac{\text{Area of } \Delta\, CDM}{\text{Area of } \Delta ABC} = \frac{(a-b)}{2(a+b)} = \frac{\sin A - \sin B}{2(\sin A + \sin B)}$$

Also, $\mathbf{CD} \,||\, (a\mathbf{b} + b\mathbf{a})$ and $\mathbf{CM} \,||\, (\mathbf{a} + \mathbf{b})$

$$\cos\phi = \frac{(a\mathbf{b} + b\mathbf{a})\cdot(\mathbf{a}+\mathbf{b})}{|a\mathbf{b} + b\mathbf{a}\,|\,|\mathbf{a} + \mathbf{b}|}$$

$$= \frac{a|\mathbf{b}|^2 + b|\mathbf{a}|^2 + |\mathbf{a}|\,|\mathbf{b}|\,(a+b)\cos C}{\sqrt{2a^2b^2 + 2a^2b^2\cos C}\sqrt{a^2 + b^2 + 2ab\cos CA}}$$

[Where, $|a| = a$ and $|b| = b$]

$$= \frac{(a+b)\cos(C/2)}{\sqrt{a^2 + b^2 + 2ab\cos C}}$$

$$= \frac{(\sin A + \sin B)\cos(C/2)}{\sqrt{\sin^2 A + \sin^2 B + 2\sin A \sin B \cos C}}$$

171. Let $A\,(\mathbf{O})$, $B(\mathbf{b})$, $C\,(\mathbf{c})$, $P(\mathbf{p})$, $Q(\mathbf{q})$, $R(\mathbf{r})$

We have, $p = \dfrac{b}{3}$

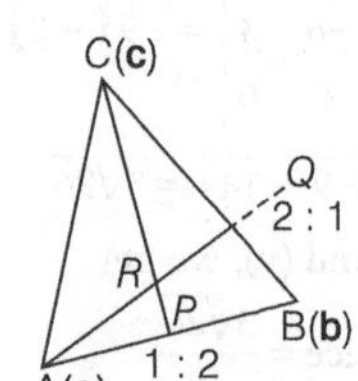

and $$\mathbf{Q} = \frac{2\mathbf{b} + \mathbf{c}}{3}$$

Equation of the line AQ, $\mathbf{r} = \lambda_1\left(\dfrac{2\mathbf{b} + \mathbf{c}}{3}\right)$

Equation of the line CP, $\mathbf{r} = \mathbf{c} + \lambda_2\left(\dfrac{\mathbf{b}}{3} - \mathbf{c}\right)$

R is the point of intersection of AQ and CP.

$\Rightarrow$ For point R , we have

$$\lambda_1\left(\frac{2\mathbf{b} + \mathbf{c}}{3}\right) = \mathbf{c} + \lambda_2\left(\frac{\mathbf{b}}{3} - \mathbf{c}\right)$$

$\Rightarrow \dfrac{2\lambda_2}{3} = \dfrac{\lambda_2}{3}$ (comparing coefficients of **b** and **c**)

and $$\frac{\lambda_1}{3} = 1 - \lambda_2$$

On solving, we get

$$\lambda_1 = 3/7, \lambda_2 = 6/7$$

$$\Rightarrow \quad \mathbf{R} = \frac{1}{7}(2\mathbf{b} + \mathbf{c})$$

Now, $$\mathbf{RB} = \mathbf{b} - \frac{1}{7}\cdot(2\mathbf{b} + \mathbf{c}) = \frac{5\mathbf{b} - \mathbf{c}}{7}$$

$$\mathbf{RC} = \mathbf{c} - \frac{1}{7}(2\mathbf{b} + \mathbf{c}) = \frac{(6\mathbf{c} - 2\mathbf{b})}{7}$$

$$\therefore \quad |\mathbf{RB} \times \mathbf{RC}| = \frac{1}{49}(5\mathbf{b} - \mathbf{c}) \times (6\mathbf{c} - 2\mathbf{b})$$

$$= \frac{1}{49}(30\mathbf{b} \times \mathbf{c} + 2\mathbf{c} \times \mathbf{b}) = \frac{28}{49}(\mathbf{b} \times \mathbf{c})$$

$$\Rightarrow \quad |\mathbf{RB} \times \mathbf{RC}| = \frac{28}{49}|\,\mathbf{b} \times \mathbf{c}| = \frac{28}{49} \qquad [(\text{area of } \Delta\, ABC)\cdot 2]$$

($\because$ area of $\Delta\, BRC = 1$)

$$\Rightarrow \text{Area of } \Delta\, ABC = \frac{1}{2}|\mathbf{RB} \times \mathbf{RC}| \cdot \frac{49}{28} = (\text{Area of } \Delta\, BRC)\cdot\frac{49}{28}$$

$$\therefore \quad \text{Area of } \Delta ABC = \frac{49}{28} \text{ sq units.}$$

172. Let $OABC$ be a given quadrilateral such that its diagonal **OB** bisects the diagonal **AC** let $\mathbf{OA} = \mathbf{a}$, $\mathbf{OB} = \mathbf{b}$, $\mathbf{OC} = \mathbf{c}$.

Since, the mid-point $\dfrac{\mathbf{a} + \mathbf{c}}{2}$ of **AC** lies on **OB**, there exits a scalar t such that,

$$\text{Area of } \Delta ABC = \frac{\mathbf{a} + \mathbf{c}}{2} = t\mathbf{b} \;\Rightarrow\; \mathbf{a} + \mathbf{c} = 2t\,\mathbf{b}$$

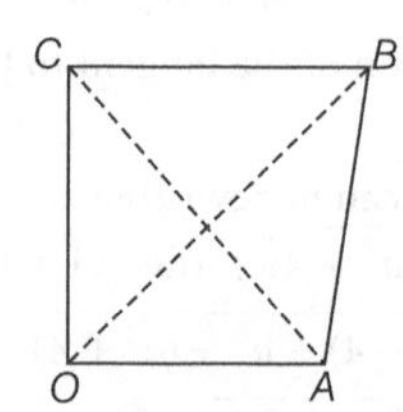

On multiplying both sides with **b**, we have

$$(\mathbf{a} + \mathbf{c}) \times \mathbf{b} = 2t\,\mathbf{b} \times \mathbf{b}$$

$$\Rightarrow \quad \mathbf{a} \times \mathbf{b} + \mathbf{c} \times \mathbf{b} = 0$$

$$\Rightarrow \quad \mathbf{a} \times \mathbf{b} = \mathbf{b} \times \mathbf{c}$$

$$\Rightarrow \quad \frac{1}{2}|\mathbf{a} \times \mathbf{b}| = \frac{1}{2}|\,\mathbf{b} \times \mathbf{c}|$$

$\Rightarrow$ Area of $\Delta\, OAB$ = Area of ΔOBC

Hence, the diagonal OB bisects the quadrilateral.

173. The coordinates of the resulting force $F = F_1 + F_2 = \{6, 4\}$ i.e., resultant F are 6 and 4. Now, let $M\,(a, y)$ be a arbitrary point of l. Then, the moment of the resultant about point M is equal to zero.

This moment is equal to sum of the moments $\mathbf{MA} \times \mathbf{F}_1$ and $\mathbf{MB} \times \mathbf{F}_2$ of component forces (the cross product of vectors is distributive.)

Since, $\mathbf{MA} = (1 - x, 1 - y)$, $\mathbf{MB} = \{2 - x, 4 - y\}$, if follows that

$$(\mathbf{MA} \times \mathbf{F}_1)(\hat{\mathbf{i}} + \hat{\mathbf{j}}) = 3(1 - x) - 2(1 - y) = 1 - 3x + 2y$$

$$(\mathbf{MB} \times \mathbf{F}_2) \cdot (\hat{\mathbf{i}} + \hat{\mathbf{j}}) = (2 - x) - 4(4 - y) = -14 - x + 4y$$

Hence, the equation of straight line l is

$(1 - 3x + 2y) + (-14 - x + 4y) = 0$

$\Rightarrow \quad -4x + 6y - 13 = 0$

$\Rightarrow \quad 4x - 6y + 13 = 0$

174. Let $\mathbf{a} = x\hat{\mathbf{i}} + y\hat{\mathbf{j}} + z\hat{\mathbf{k}}$

Now, $\mathbf{a}$, $\hat{\mathbf{i}}$ and $\hat{\mathbf{i}} + \hat{\mathbf{j}}$ are coplanar and $\mathbf{a}$, $\hat{\mathbf{i}} - \hat{\mathbf{j}}$ and $\hat{\mathbf{i}} + \hat{\mathbf{k}}$ are coplanar.

$\Rightarrow \quad [\mathbf{a}\ \hat{\mathbf{i}}\ \hat{\mathbf{i}} + \hat{\mathbf{j}}] = 0$ and $[\mathbf{a}\ \hat{\mathbf{i}} - \hat{\mathbf{j}}\ \hat{\mathbf{i}} + \hat{\mathbf{k}}] = 0$

$$\Rightarrow \begin{vmatrix} x & y & z \\ 1 & 0 & 0 \\ 1 & 1 & 0 \end{vmatrix} = 0 \text{ and } \begin{vmatrix} x & y & z \\ 1 & -1 & 0 \\ 1 & 0 & 1 \end{vmatrix} = 0$$

$\therefore \quad z = 0$ and $-x - y + z = 0$

$\Rightarrow \quad z = 0$ and $x + y = 0$

$\Rightarrow \quad y = -x$

$\therefore \quad \mathbf{a} = x\hat{\mathbf{i}} - x\hat{\mathbf{j}}$

$\Rightarrow \quad \mathbf{a} = \dfrac{\hat{\mathbf{i}} - \hat{\mathbf{j}}}{\sqrt{2}}$

Let the angle between $\mathbf{a}$ and $\hat{\mathbf{i}} - 2\hat{\mathbf{j}} + 2\hat{\mathbf{k}}$ be θ.

$$\cos\theta = \mathbf{a} \cdot \frac{\hat{\mathbf{i}} - 2\hat{\mathbf{j}} + 2\hat{\mathbf{k}}}{|\hat{\mathbf{i}} + 2\hat{\mathbf{j}} + 2\hat{\mathbf{k}}|} = \frac{(\hat{\mathbf{i}} - \hat{\mathbf{j}})}{\sqrt{2}} \cdot \frac{(\hat{\mathbf{i}} - 2\hat{\mathbf{j}} + 2\hat{\mathbf{k}})}{\sqrt{1 + 4 + 4}} = \frac{(1 - \hat{\mathbf{j}})(\hat{\mathbf{i}} - 2\hat{\mathbf{j}} + 2\hat{\mathbf{k}})}{\sqrt{2} \cdot 3} = \frac{1 + 2}{3\sqrt{2}} = \frac{1}{\sqrt{2}}$$

$\therefore \quad \theta = \dfrac{\pi}{4}$

175. In the new position, let the vector be $x\hat{\mathbf{i}} + y\hat{\mathbf{j}} + z\hat{\mathbf{k}}$. Since, it is perpendicular to the given vector.

$(x\hat{\mathbf{i}} + y\hat{\mathbf{j}} + z\hat{\mathbf{k}}) \cdot (\hat{\mathbf{i}} + 2\hat{\mathbf{j}} + 2\hat{\mathbf{k}}) = 0$

$\Rightarrow \quad x + 2y + 2z = 0 \quad$ ...(i)

The magnitude is the new position which also remains the same.

$\Rightarrow \quad x^2 + y^2 + z^2 = 1 + 4 + 4 = 9 \quad$ ...(ii)

The given vector, the vector in new position and the X-axis are coplanar.

$$\Rightarrow \begin{vmatrix} 1 & 0 & 0 \\ 1 & 2 & 2 \\ x & y & z \end{vmatrix} = 0$$

$\Rightarrow \quad y = z$ and $x = -4y$ (using $x + 2y + 22 = 0$)

Hence, $\quad x^2 + y^2 + z^2 = 9$

$\Rightarrow \quad 16y^2 + y^2 + y^2 = 9$

$\Rightarrow \quad y^2 = \dfrac{1}{2}, y = \pm \dfrac{1}{\sqrt{2}}$

$\Rightarrow \quad x = \pm 2\sqrt{2}$

It is given that the vector passes through the positive X-axis.

$\Rightarrow \quad x = 2\sqrt{2}$ and $y = 0 - \dfrac{1}{\sqrt{2}} = z$

Hence, required vector is $\left(2\sqrt{2}\hat{\mathbf{i}} - \dfrac{1}{\sqrt{2}}\hat{\mathbf{j}} - \dfrac{1}{\sqrt{2}}\hat{\mathbf{k}}\right)$.

176. $\hat{\mathbf{u}} + \hat{\mathbf{v}} + \hat{\mathbf{u}} + \mathbf{w}$ and $\mathbf{w} \times \hat{\mathbf{u}} = \hat{\mathbf{v}}$

$\Rightarrow \quad (\hat{\mathbf{u}} + \hat{\mathbf{v}} + \hat{\mathbf{u}}) \times \hat{\mathbf{u}} = \mathbf{w} \times \hat{\mathbf{u}}$

$\Rightarrow \quad (\hat{\mathbf{u}} + \hat{\mathbf{v}}) \times \hat{\mathbf{u}} + \hat{\mathbf{u}} \times \hat{\mathbf{u}} = \hat{\mathbf{v}}$

$\Rightarrow \quad (\hat{\mathbf{u}} \cdot \hat{\mathbf{u}})\hat{\mathbf{v}} - (\hat{\mathbf{v}} \cdot \hat{\mathbf{u}})\hat{\mathbf{u}} + \hat{\mathbf{u}} \times \hat{\mathbf{u}} = \hat{\mathbf{v}}$

(using $\hat{\mathbf{u}} \cdot \hat{\mathbf{u}} = 1$ and $\hat{\mathbf{u}} \times \hat{\mathbf{u}} = 0$, since unit vectors)

$\Rightarrow \quad \hat{\mathbf{v}} - (\hat{\mathbf{v}} \cdot \hat{\mathbf{u}})\hat{\mathbf{u}} = \hat{\mathbf{v}}$

$\Rightarrow \quad (\hat{\mathbf{u}} \cdot \hat{\mathbf{v}})\hat{\mathbf{u}} = 0 \Rightarrow \hat{\mathbf{u}} \cdot \hat{\mathbf{v}} = 0$

Now, $[\hat{\mathbf{u}}\ \hat{\mathbf{v}}\ \mathbf{w}]$

$\Rightarrow \quad \hat{\mathbf{u}} \cdot (\hat{\mathbf{v}} \times \mathbf{w}) \Rightarrow \hat{\mathbf{u}} \cdot [\hat{\mathbf{v}} \times (\hat{\mathbf{u}} \times \hat{\mathbf{v}} + \hat{\mathbf{u}})]$

(given $\mathbf{w} = \hat{\mathbf{u}} \times \hat{\mathbf{v}} + \hat{\mathbf{u}}$)

$\Rightarrow \quad \hat{\mathbf{u}} \cdot [\hat{\mathbf{v}} \times (\hat{\mathbf{u}} \times \hat{\mathbf{v}}) + \hat{\mathbf{v}} \times \hat{\mathbf{u}}]$

$\Rightarrow \quad \hat{\mathbf{u}} \cdot [(\hat{\mathbf{v}} \cdot \hat{\mathbf{v}})\hat{\mathbf{u}} - (\hat{\mathbf{v}} \cdot \hat{\mathbf{u}})\hat{\mathbf{v}} + \hat{\mathbf{v}} \times \hat{\mathbf{u}}]$

$[\because \hat{\mathbf{u}} \cdot \hat{\mathbf{v}} = 0$ from Eq. (i)$]$

$\Rightarrow \quad |\hat{\mathbf{v}}|^2 (\hat{\mathbf{u}} \cdot \hat{\mathbf{u}}) - \hat{\mathbf{u}} \cdot (\hat{\mathbf{v}} \times \hat{\mathbf{u}})$

$\Rightarrow \quad |\hat{\mathbf{v}}|^2 |\hat{\mathbf{u}}|^2 - 0 \quad (\because [\hat{\mathbf{u}}\ \hat{\mathbf{v}}\ \mathbf{w}] = 0)$

$(\because |\hat{\mathbf{u}}| = |\hat{\mathbf{v}}| = 1)$

$\Rightarrow \quad [\hat{\mathbf{u}}\ \hat{\mathbf{v}}\ \mathbf{w}] = 1$

177. We have,

$\mathbf{R} \times \mathbf{B} = \mathbf{C} \times \mathbf{B}$ and $\mathbf{R} \cdot \mathbf{A} = 0$

$\Rightarrow \quad \mathbf{A} \times (\mathbf{R} \times \mathbf{B}) = \mathbf{A} \times (\mathbf{C} \times \mathbf{B})$

$\Rightarrow \quad (\mathbf{A} \cdot \mathbf{B})\mathbf{R} - (\mathbf{A} \cdot \mathbf{R})\mathbf{B} = (\mathbf{A} \cdot \mathbf{B})\mathbf{C} - (\mathbf{A} \cdot \mathbf{C})\mathbf{B} \quad$ ...(i)

where, $\quad \mathbf{A} = 2\hat{\mathbf{i}} + \hat{\mathbf{k}}, \mathbf{B} = \hat{\mathbf{i}} + \hat{\mathbf{j}} + \hat{\mathbf{k}}$

and $\quad \mathbf{C} = 4\hat{\mathbf{i}} - 3\hat{\mathbf{j}} + 7\hat{\mathbf{k}}$

$\therefore \quad \mathbf{A} \cdot \mathbf{B} = 2 + 1 = 3, \mathbf{A} \cdot \mathbf{C} = 8 + 7 = 15$

Hence, Eq. (i) reduces to

$3\mathbf{R} - 0 \cdot \mathbf{B} = 3\mathbf{C} - 15\mathbf{B}$

or $\quad \mathbf{R} = \mathbf{C} - 5\mathbf{B} = (4\hat{\mathbf{i}} - 3\hat{\mathbf{j}} + 7\hat{\mathbf{k}}) - 5(\hat{\mathbf{i}} + \hat{\mathbf{j}} + \hat{\mathbf{k}})$

$\therefore \quad \mathbf{R} = -\hat{\mathbf{i}} - 8\hat{\mathbf{j}} + 2\hat{\mathbf{k}}$

178. Since, $\mathbf{a}$, $\mathbf{b}$ and $\mathbf{a} \times \mathbf{b}$ are non-coplanar vectors.

Let $\quad \mathbf{x} = \lambda + \mu\mathbf{b} + \gamma(\mathbf{a} \times \mathbf{b}) \quad$ ...(i)

$\mathbf{x} \cdot \mathbf{a} = \lambda\mathbf{a} \cdot \mathbf{a} + \mu\mathbf{b} \cdot \mathbf{a} + \gamma(\mathbf{a} \times \mathbf{b}) \cdot \mathbf{b}$

$\Rightarrow \quad 0 \times \lambda|\mathbf{a}|^2 + \mu\mathbf{a} \cdot \mathbf{b} \quad$ ...(ii)

Again from Eq. (i),

$\mathbf{x} \cdot \mathbf{b} = \lambda\mathbf{a} \cdot \mathbf{b} + \mu\mathbf{b} \cdot \mathbf{b} + \gamma(\mathbf{a} \times \mathbf{b}) \cdot \mathbf{b}$

$1 = \lambda\mathbf{a} \cdot \mathbf{b} + \mu|\mathbf{b}|^2 \quad$ ...(iii)

From Eq. (i)

$\mathbf{x} \cdot (\mathbf{a} \times \mathbf{b}) = \lambda\mathbf{a} \cdot (\mathbf{a} \times \mathbf{b}) + \mu\mathbf{b} \cdot (\mathbf{a} \times \mathbf{b}) + \gamma(\mathbf{a} \times \mathbf{b})^2$

$[\mathbf{x}\ \mathbf{a}\ \mathbf{b}] = \lambda[\mathbf{a}\ \mathbf{a}\ \mathbf{b}] + \mu[\mathbf{b}\ \mathbf{a}\ \mathbf{b}] + \gamma(\mathbf{a} \times \mathbf{b})^2$

$1 = \gamma(\mathbf{a} \times \mathbf{b})^2 \quad$ ...(iv)

From Eq. (ii), $\mu(\mathbf{a} \cdot \mathbf{b}) = -\lambda|\mathbf{a}|^2 \Rightarrow \mu = -\lambda\dfrac{|\mathbf{a}|^2}{\mathbf{a} \cdot \mathbf{b}}$

From Eq. (iii),

$$1 = \lambda \mathbf{a}\cdot\mathbf{b} - \lambda \frac{|\mathbf{a}|^2|\mathbf{b}|^2}{\mathbf{a}\cdot\mathbf{b}}$$

$$\Rightarrow \quad 1 = \lambda\left(\frac{(\mathbf{a}\cdot\mathbf{b})^2 - |\mathbf{a}|^2|\mathbf{b}|^2}{\mathbf{a}\cdot\mathbf{b}}\right)$$

$$\Rightarrow \quad \lambda = \frac{\mathbf{a}\cdot\mathbf{b}}{(\mathbf{a}\cdot\mathbf{b})^2 - a^2b^2}$$

179. $$\hat{\mathbf{x}} + \hat{\mathbf{y}} + \hat{\mathbf{z}} = \mathbf{a} \quad \text{...(i)}$$

$$\mathbf{a}\cdot\hat{\mathbf{x}} + \mathbf{a}\cdot\hat{\mathbf{y}} + \mathbf{a}\cdot\hat{\mathbf{z}} = \mathbf{a}\cdot\mathbf{a} = |\mathbf{a}|^2 = 4$$

$$\frac{3}{2} + \frac{7}{4} + \mathbf{a}\cdot\hat{\mathbf{z}} = 4 \Rightarrow \mathbf{a}\cdot\hat{\mathbf{z}} = \frac{3}{4} \quad \text{...(ii)}$$

From Eq. (i), $\hat{\mathbf{x}}\cdot(\hat{\mathbf{x}} + \hat{\mathbf{y}} + \hat{\mathbf{z}}) = \hat{\mathbf{x}}\cdot\mathbf{a} = \mathbf{a}\cdot\hat{\mathbf{x}} = \frac{3}{2}$

$$\Rightarrow \quad \hat{\mathbf{x}}\cdot\hat{\mathbf{x}} + \hat{\mathbf{x}}\cdot\hat{\mathbf{y}} + \hat{\mathbf{x}}\cdot\hat{\mathbf{z}} = \frac{3}{2}$$

$$\Rightarrow \quad 1 + \hat{\mathbf{x}}\cdot\hat{\mathbf{y}} + \hat{\mathbf{x}}\cdot\hat{\mathbf{z}} = \frac{3}{2}$$

$$\Rightarrow \quad \hat{\mathbf{x}}\cdot\hat{\mathbf{y}} + \hat{\mathbf{x}}\cdot\hat{\mathbf{z}} = \frac{1}{2} \quad \text{...(iii)}$$

From Eq. (i), $\hat{\mathbf{y}}\cdot(\hat{\mathbf{x}} + \hat{\mathbf{y}} + \hat{\mathbf{z}}) = \hat{\mathbf{y}}\cdot\mathbf{a} = \frac{7}{4}$

$$\hat{\mathbf{x}}\cdot\hat{\mathbf{y}} + 1 + \hat{\mathbf{y}}\cdot\hat{\mathbf{z}} = \frac{7}{4}$$

$$\hat{\mathbf{x}}.\hat{\mathbf{y}} + \hat{\mathbf{y}}\cdot\hat{\mathbf{z}} = \frac{3}{4} \quad \text{...(iv)}$$

From Eq. (i), $(\hat{\mathbf{x}} + \hat{\mathbf{y}} + \hat{\mathbf{z}})^2 = (\mathbf{a})^2$

$$\Rightarrow \hat{\mathbf{x}}\cdot\hat{\mathbf{x}} + \hat{\mathbf{y}}\cdot\hat{\mathbf{y}} + \hat{\mathbf{z}}\cdot\hat{\mathbf{z}} + 2(\hat{\mathbf{x}}\cdot\hat{\mathbf{y}} + \hat{\mathbf{y}}\cdot\hat{\mathbf{z}} + \hat{\mathbf{z}}\cdot\hat{\mathbf{x}}) = |\mathbf{a}|^2$$

$$3 + 2(\hat{\mathbf{x}}\cdot\hat{\mathbf{y}} + \hat{\mathbf{y}}\cdot\hat{\mathbf{z}} + \hat{\mathbf{z}}\cdot\hat{\mathbf{x}}) = 4 \quad \text{...(v)}$$

From Eqs. (iii), (iv) and (v), we get

$$\hat{\mathbf{y}}\cdot\hat{\mathbf{z}} = 0,\ \hat{\mathbf{x}}\cdot\hat{\mathbf{z}} = -\frac{1}{4}, \hat{\mathbf{x}}\cdot\hat{\mathbf{y}} = \frac{3}{4}$$

Now, $$\hat{\mathbf{x}}\times(\hat{\mathbf{y}}\times\hat{\mathbf{z}}) = \mathbf{b}$$

$$(\hat{\mathbf{x}}\cdot\hat{\mathbf{z}})\hat{\mathbf{y}} - (\hat{\mathbf{x}}\cdot\hat{\mathbf{y}})\hat{\mathbf{z}} = \mathbf{b}$$

$$-\frac{1}{4}\hat{\mathbf{y}} - \frac{3}{4}\hat{\mathbf{z}} = \mathbf{b} \quad \text{...(vi)}$$

Again, $$(\hat{\mathbf{x}}\times\hat{\mathbf{y}})\times\hat{\mathbf{z}} = \mathbf{c}$$

$$(\hat{\mathbf{x}}\cdot\hat{\mathbf{z}})\hat{\mathbf{y}} - (\hat{\mathbf{y}}\cdot\hat{\mathbf{z}})\hat{\mathbf{x}} = \hat{\mathbf{c}} \Rightarrow -\frac{1}{4}\hat{\mathbf{y}} = \mathbf{c}$$

$$\hat{\mathbf{y}} = -4\mathbf{c} \quad \text{[from Eq. (vi)]}$$

$$\hat{\mathbf{z}} = \frac{4}{3}(\mathbf{c} - \mathbf{b})$$

From Eq. (i), $\hat{\mathbf{x}} = \mathbf{a} - \hat{\mathbf{y}} - \hat{\mathbf{z}}$

$$\hat{\mathbf{x}} = \frac{1}{3}(3\mathbf{a} + 4\mathbf{b} + 8\mathbf{c})$$

$$\hat{\mathbf{x}} = \frac{1}{3}(3\mathbf{a} + 4\mathbf{b} + 8\mathbf{c})$$

$$\hat{\mathbf{y}} = -4\mathbf{c};\ \hat{\mathbf{z}} = \frac{4}{3}(\mathbf{c} - \mathbf{b})$$

180. Here,

$$\mathbf{a}\times\{(\mathbf{x} - \mathbf{b})\times a\} + \mathbf{b}\times\{(\mathbf{x} - \mathbf{c})\times\mathbf{b}\} + \mathbf{c}\times\{(\mathbf{x} - \mathbf{a})\times\mathbf{c}\} = 0$$

$$(\mathbf{a}\cdot\mathbf{a})(\mathbf{x} - \mathbf{b}) - \{\mathbf{a}\cdot(\mathbf{x} - \mathbf{b})\}\mathbf{a} + (\mathbf{b}\cdot\mathbf{b})(\mathbf{x} - \mathbf{c})$$

$$-\{\mathbf{b}\cdot(\mathbf{x} - \mathbf{c})\}\mathbf{b} + (\mathbf{c}\cdot\mathbf{c})(\mathbf{x} - \mathbf{a}) - \{\mathbf{c}\cdot(\mathbf{x} - a)\}\mathbf{c} = 0$$

$$\Rightarrow \lambda^2(\mathbf{x} - \mathbf{b}) - \{\mathbf{a}\cdot\mathbf{x} - 0\}\mathbf{a} + \lambda^2(\mathbf{x} - \mathbf{c}) - \{\mathbf{b}\cdot\mathbf{x} - 0\}\mathbf{b}$$

$$+ \lambda^2(\mathbf{x} - \mathbf{a}) - (\mathbf{c}\cdot\mathbf{x} - 0)\} = \mathbf{c} = 0$$

(using $\mathbf{a}\cdot\mathbf{b} = \mathbf{b}\cdot\mathbf{c} = \mathbf{c}\cdot\mathbf{a} = 0$ and $|\mathbf{a}| = |\mathbf{b}| = |\mathbf{c}| = \lambda$)

$$\Rightarrow \lambda^2\{\mathbf{x} - \mathbf{b} + \mathbf{x} - \mathbf{c} + \mathbf{x} - \mathbf{a}\}$$

$$= \{(\mathbf{a}\cdot\mathbf{x})\mathbf{a} + (\mathbf{b}\cdot\mathbf{x})\mathbf{b} + (\mathbf{c}\cdot\mathbf{x})\mathbf{c}\}$$

Let $\mathbf{x} = \alpha\mathbf{a} + \beta\mathbf{b} + \gamma\mathbf{c}$

($\mathbf{x}$ is linear combination of $\mathbf{a}$, $\mathbf{b}$ and $\mathbf{c}$)

$$\Rightarrow \quad \mathbf{a}\cdot\mathbf{x} = \alpha\mathbf{a}\cdot\mathbf{a} \Rightarrow \mathbf{a}\cdot\mathbf{x} = \alpha\lambda^2$$

$$\mathbf{b}\cdot\mathbf{x} = \beta\lambda^2$$

$$\mathbf{c}\cdot\mathbf{x} = \gamma\lambda^2$$

From Eq. (i), $\lambda^2\{3\mathbf{x} - (\mathbf{a} + \mathbf{b} + \mathbf{c})\}(\mathbf{a}\cdot\mathbf{x})$

$$\mathbf{a} + (\mathbf{b}\cdot\mathbf{x})\mathbf{b} + (\mathbf{c}\cdot\mathbf{x})\mathbf{c}$$

and from Eq. (iii)

$$\mathbf{a}\cdot\mathbf{x} = \lambda^2\alpha,\ \mathbf{b}\cdot\mathbf{x} = \lambda^2\beta,\ \mathbf{c}\cdot\mathbf{x} = \lambda^2\gamma$$

Above equation reduces to

$$\lambda^2(3\mathbf{x} - (\mathbf{a} + \mathbf{b} + \mathbf{c})\} = \lambda^2(\mathbf{a}\alpha + \mathbf{b}\beta + \mathbf{c}\gamma)$$

$$\Rightarrow \quad 3\mathbf{x} - (\mathbf{a} - \mathbf{b} + \mathbf{c})$$

$$\Rightarrow \quad 2\mathbf{x} = \mathbf{a} + \mathbf{b} + \mathbf{c}$$

$$\Rightarrow \quad \mathbf{x} = \frac{\mathbf{a} + \mathbf{b} + \mathbf{c}}{2}$$

181. Here, $\mathbf{OC} = \mathbf{x}$, $\mathbf{CA} = \mathbf{b}$, $\mathbf{CB} = \mathbf{a}$

$OA = (\mathbf{b} - x)$ and $OB = \mathbf{a} - x$

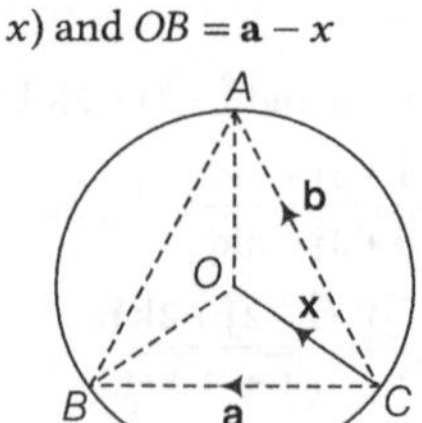

Now, $$\mathbf{OA}^2 = \mathbf{OB}^2 = \mathbf{OC}^2$$

$$\mathbf{x}^2 = (\mathbf{a} - \mathbf{x})^2 = (\mathbf{b} - \mathbf{x})^2$$

$$\Rightarrow \quad \mathbf{x}\cdot\mathbf{x} = (\mathbf{a} - \mathbf{x})\cdot(\mathbf{a} - \mathbf{x}) = (\mathbf{b} - \mathbf{x})(\mathbf{b} - \mathbf{x})$$

$$\Rightarrow \quad \mathbf{x}\cdot\mathbf{x} = \mathbf{a}\cdot\mathbf{a} - 2\mathbf{a}\cdot\mathbf{x} + \mathbf{x}\cdot\mathbf{x} = \mathbf{b}\cdot\mathbf{b} - 2\mathbf{b}\cdot\mathbf{x} + \mathbf{x}\cdot\mathbf{x}$$

$$\mathbf{a}\cdot\mathbf{x} = \frac{\mathbf{a}^2}{2} \text{ and } \mathbf{b}\cdot\mathbf{x} = \frac{\mathbf{b}^2}{2}$$

Now, if we take $\mathbf{x} = \lambda\mathbf{a} + \mu\mathbf{b}$, then from Eq. (i)

$$\lambda\mathbf{a}^2 + \mu\cdot\mathbf{a}\cdot\mathbf{b} = \frac{\mathbf{a}^2}{2}$$

and $$\lambda\mathbf{a}\cdot\mathbf{b} + \mu\mathbf{b}^2 = \frac{b^2}{2}$$

$\therefore$ On solving Eqs. (ii) and (iii),

$$\lambda = \frac{\mathbf{a}^2\mathbf{b}^2 - \mathbf{b}^2(\mathbf{a}\cdot\mathbf{b})}{2(\mathbf{a}^2\mathbf{b}^2) - (\mathbf{a}\cdot\mathbf{b})^2}$$

and $$\mu = \frac{\mathbf{a}^2\mathbf{b}^2 - a^2(\mathbf{a}\cdot\mathbf{b})}{2(\mathbf{a}^2\mathbf{b}^2) - (\mathbf{a}\cdot\mathbf{b})^2}$$

$$\Rightarrow \quad \mathbf{x} = \frac{1}{2}\frac{\mathbf{a}^2\mathbf{b}^2 - \mathbf{b}^2(\mathbf{a}\cdot\mathbf{b})}{(\mathbf{a}^2\mathbf{b}^2) - (\mathbf{a}\cdot\mathbf{b})^2}\mathbf{a} + \frac{1}{2}\frac{\mathbf{a}^2\mathbf{b}^2 - \mathbf{a}^2(\mathbf{a}\cdot\mathbf{b})}{(\mathbf{a}^2\mathbf{b}^2) - (\mathbf{a}\cdot\mathbf{b})^2}\cdot\mathbf{b}$$

182. $\mathbf{OP}\cdot\mathbf{OQ}+\mathbf{OR}\cdot\mathbf{OS}=\mathbf{OR}\cdot\mathbf{OP}+\mathbf{OQ}\cdot\mathbf{OS}$

$\Rightarrow \quad \mathbf{OP}(\mathbf{OQ}-\mathbf{OR})+\mathbf{OS}(\mathbf{OR}-\mathbf{OQ})=0$

$\Rightarrow \quad (\mathbf{OP}-\mathbf{OS})(\mathbf{OQ}-\mathbf{OR})=0$

$\Rightarrow \quad \mathbf{SP}\cdot\mathbf{RQ}=0$

Similarly $\mathbf{SR}\cdot\mathbf{PQ}=0$ and $\mathbf{SQ}\cdot\mathbf{PR}=0$

$\therefore$ S is orthocentre.

183. $\cos(P+Q)+\cos(Q+R)+\cos(R+P)$

$=-(\cos R+\cos P+\cos Q)$

Max. of $\cos P+\cos Q+\cos R=\frac{3}{2}$

Min. of $\cos(P+Q)+\cos(Q+R)+\cos(R+P)$ is

$=-\frac{3}{2}$

184. $\sin R=\sin(P+Q)$

185. Given, $|\hat{\mathbf{a}}|=|\hat{\mathbf{b}}|=|\hat{\mathbf{c}}|=1$

and $\quad \hat{\mathbf{a}}\times(\hat{\mathbf{b}}\times\hat{\mathbf{c}})=\frac{\sqrt{3}}{2}(\hat{\mathbf{b}}+\hat{\mathbf{c}})$

Now, consider $\hat{\mathbf{a}}\times(\hat{\mathbf{b}}\times\hat{\mathbf{c}})=\frac{\sqrt{3}}{2}(\hat{\mathbf{b}}+\hat{\mathbf{c}})$

$\Rightarrow \quad (\hat{\mathbf{a}}\cdot\hat{\mathbf{c}})\,\hat{\mathbf{b}}-(\hat{\mathbf{a}}\cdot\hat{\mathbf{b}})\,\hat{\mathbf{c}}=\frac{\sqrt{3}}{2}\hat{\mathbf{b}}+\frac{\sqrt{3}}{2}\hat{\mathbf{c}}$

On comparing, we get

$\hat{\mathbf{a}}\cdot\hat{\mathbf{b}}=-\frac{\sqrt{3}}{2}\Rightarrow|\hat{\mathbf{a}}||\hat{\mathbf{b}}|\cos\theta=-\frac{\sqrt{3}}{2}$

$\Rightarrow \quad \cos\theta=-\frac{\sqrt{3}}{2} \qquad [\because |\hat{\mathbf{a}}|=|\hat{\mathbf{b}}|=1]$

$\Rightarrow \quad \cos\theta=\cos\left(\pi-\frac{\pi}{6}\right)\Rightarrow\theta=\frac{5\pi}{6}$

186. Given, $(\mathbf{a}\times\mathbf{b})\times\mathbf{c}=\frac{1}{3}|\mathbf{b}|\;|\mathbf{c}|\,\mathbf{a}$

$\Rightarrow \quad -\mathbf{c}\times(\mathbf{a}\times\mathbf{b})=\frac{1}{3}|\mathbf{b}||\mathbf{c}|\,\mathbf{a}$

$\Rightarrow \quad -(\mathbf{c}\cdot\mathbf{b})\cdot\mathbf{a}+(\mathbf{c}\cdot\mathbf{a})\,\mathbf{b}=\frac{1}{3}|\mathbf{b}|\,|\mathbf{c}|\,\mathbf{a}$

$\left[\frac{1}{3}|\mathbf{b}||\mathbf{c}|+(\mathbf{c}\cdot\mathbf{b})\right]\mathbf{a}=(\mathbf{c}\cdot\mathbf{a})\mathbf{b}$

Since, **a** and **b** are not collinear.

$\mathbf{c}\cdot\mathbf{b}+\frac{1}{3}|\mathbf{b}||\mathbf{c}|=0$ and $\mathbf{c}\cdot\mathbf{a}=0$

$\Rightarrow \quad |\mathbf{c}||\mathbf{b}|\cos\theta+\frac{1}{3}|\mathbf{b}||\mathbf{c}|=0$

$\Rightarrow \quad |\mathbf{b}||\mathbf{c}|\left(\cos\theta+\frac{1}{3}\right)=0$

$\Rightarrow \quad \cos\theta+\frac{1}{3}=0 \qquad (\because |\mathbf{b}|\neq 0, |\mathbf{c}|\neq 0)$

$\Rightarrow \quad \cos\theta=-\frac{1}{3}\;;\;\sin\theta=\frac{\sqrt{8}}{3}=\frac{2\sqrt{2}}{3}$

187. If *a, b,c* are any three vectors

Then, $\quad |\mathbf{a}+\mathbf{b}+\mathbf{c}|^2\geq 0$

$\Rightarrow \quad |\mathbf{a}|^2+|\mathbf{b}|^2+|\mathbf{c}|^2+2(\mathbf{a}\cdot\mathbf{b}+\mathbf{b}\cdot\mathbf{c}+\mathbf{c}\cdot\mathbf{a})\geq 0$

$\therefore \quad \mathbf{a}\cdot\mathbf{b}+\mathbf{b}\cdot\mathbf{c}+\mathbf{c}\cdot\mathbf{a}\geq\frac{-1}{2}(|\mathbf{a}|^2+|\mathbf{b}|^2+|\mathbf{c}|^2)$

Given, $|\mathbf{a}-\mathbf{b}|^2+|\mathbf{b}-\mathbf{c}|^2+|\mathbf{c}-\mathbf{a}|^2=9$

$\Rightarrow |\mathbf{a}|^2+|\mathbf{b}|^2-2\mathbf{a}\cdot\mathbf{b}+|\mathbf{b}|^2+|\mathbf{c}|^2-2\mathbf{b}\cdot\mathbf{c}+|\mathbf{c}|^2+|\mathbf{a}|^2-2\mathbf{c}\cdot\mathbf{a}=9$

$\Rightarrow \; 6-2(\mathbf{a}\cdot\mathbf{b}+\mathbf{b}\cdot\mathbf{c}+\mathbf{c}\cdot\mathbf{a})=9 \qquad [\because |\mathbf{a}|=|\mathbf{b}|=|\mathbf{c}|=1]$

$\Rightarrow \; \mathbf{a}\cdot\mathbf{b}+\mathbf{b}\cdot\mathbf{c}+\mathbf{c}\cdot\mathbf{a}=-\frac{3}{2} \qquad \ldots(\text{i})$

Also, $\quad \mathbf{a}\cdot\mathbf{b}+\mathbf{b}\cdot\mathbf{c}+\mathbf{c}\cdot\mathbf{a}\geq\frac{-1}{2}(|\mathbf{a}|^2+|\mathbf{b}|^2+|\mathbf{c}|^2)$

$\geq-\frac{3}{2} \qquad \ldots(\text{ii})$

From Eqs. (i) and (ii), $|\mathbf{a}+\mathbf{b}+\mathbf{c}|=0$

as $\mathbf{a}\cdot\mathbf{b}+\mathbf{b}\cdot\mathbf{c}+\mathbf{c}\cdot\mathbf{a}$ is minimum when $|\mathbf{a}+\mathbf{b}+\mathbf{c}|=0$

$\Rightarrow \; \mathbf{a}+\mathbf{b}+\mathbf{c}=0$

$\therefore \quad |2\mathbf{a}+5\mathbf{b}+5\mathbf{c}|=|2\mathbf{a}+5(\mathbf{b}+\mathbf{c})|=|2\mathbf{a}-5\mathbf{a}|=3$

188. Let $\mathbf{a}=\hat{\mathbf{i}}+\hat{\mathbf{j}}+2\hat{\mathbf{k}},\;\mathbf{b}=\hat{\mathbf{i}}+2\hat{\mathbf{j}}+\hat{\mathbf{k}}$

and $\mathbf{c}=\hat{\mathbf{i}}+\hat{\mathbf{j}}+\hat{\mathbf{k}}$

$\therefore$ A vector coplanar to **a** and **b** and perpendicular to **c**

$=\lambda(\mathbf{a}\times\mathbf{b})\times\mathbf{c}=\lambda\{(\mathbf{a}\cdot\mathbf{c})\,\mathbf{v}-(\mathbf{b}\cdot\mathbf{c})\,\mathbf{a}\}$

$=\lambda\{(1+1+4)(\hat{\mathbf{i}}+2\hat{\mathbf{j}}+\hat{\mathbf{k}})-(1+2+1)(\hat{\mathbf{i}}+\hat{\mathbf{j}}+2\hat{\mathbf{k}})\}$

$=\lambda\{6\,\hat{\mathbf{i}}+12\,\hat{\mathbf{j}}+6\,\hat{\mathbf{k}}-6\hat{\mathbf{i}}-6\hat{\mathbf{j}}-12\hat{\mathbf{k}}\}$

$=\lambda\{6\,\hat{\mathbf{j}}-6\,\hat{\mathbf{k}}\}=6\lambda\{\hat{\mathbf{j}}-\hat{\mathbf{k}}\}$

For, $\lambda=\frac{1}{6}\Rightarrow$ Option (a) is correct.

and for $\lambda=-\frac{1}{6}\Rightarrow$ Option (d) is correct.

189. Let $\mathbf{v}=\mathbf{a}+\lambda\mathbf{b}$

$\mathbf{v}=(1+\lambda)\,\hat{\mathbf{i}}+(1-\lambda)\,\hat{\mathbf{j}}\,(1+\lambda)\,\hat{\mathbf{k}}$

Projection of **v** on $\mathbf{c}=\frac{1}{\sqrt{3}}$

$\Rightarrow \quad \frac{\mathbf{v}\cdot\mathbf{c}}{|\mathbf{c}|}=\frac{1}{\sqrt{3}}$

$\Rightarrow \quad \frac{(1+\lambda)-(1-\lambda)-(1+\lambda)}{\sqrt{3}}=\frac{1}{\sqrt{3}}$

$\Rightarrow \quad 1+\lambda-1+\lambda-1-\lambda=1$

$\Rightarrow \quad \lambda-1=1$

$\Rightarrow \quad \lambda=2$

$\therefore \quad \mathbf{v}=3\,\hat{\mathbf{i}}-\hat{\mathbf{j}}+3\hat{\mathbf{k}}$

190. $\mathbf{AB}=2\hat{\mathbf{i}}+10\hat{\mathbf{j}}+11\hat{\mathbf{k}}$

$\mathbf{AD}=-\hat{\mathbf{i}}+2\hat{\mathbf{j}}+2\hat{\mathbf{k}}$

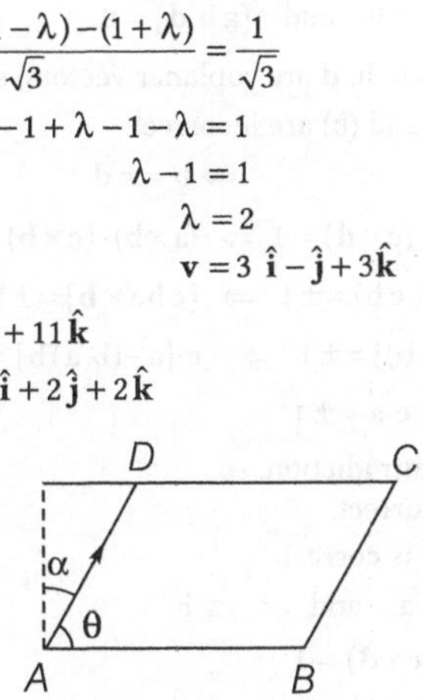

Angle 'θ' between **AB** and **AD** is

$$\cos(\theta)=\left|\frac{\mathbf{AB}\cdot\mathbf{AD}}{|\mathbf{AB}||\mathbf{AD}|}\right|$$

$$= \left|\frac{-2+20+22}{(15)(3)}\right| = \frac{8}{9} \Rightarrow \sin(\theta) = \frac{\sqrt{17}}{9}$$

Since, $\alpha + \theta = 90°$

$\therefore \quad \cos(\alpha) = \cos(90° - \theta) = \sin(\theta) = \frac{\sqrt{17}}{9}$

191. $m_{PQ} = \frac{1}{6}, m_{SR} = \frac{1}{6}, m_{RQ} = -3, m_{SP} = -3$

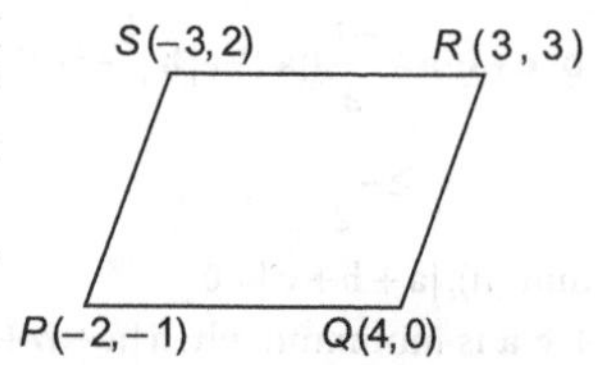

$\Rightarrow$ Parallelogram, but neither

$PR = SQ$ nor $PR \perp SQ$.

$\therefore$ So, it is a parallelogram, which is neither a rhombus nor a rectangle.

192. From the given information, it is clear that $\mathbf{a} = \frac{\hat{\mathbf{i}} - 2\hat{\mathbf{j}}}{\sqrt{5}}$

$\Rightarrow \quad |\mathbf{a}| = 1, |\mathbf{b}| = 1, \mathbf{a}\cdot\mathbf{b} = 0$

Now, $(2\mathbf{a} + \mathbf{b})\cdot[(\mathbf{a}\times\mathbf{b})\times(\mathbf{a} - 2\mathbf{b})]$

$= (2\mathbf{a} + \mathbf{b})\cdot[a^2\mathbf{b} - (\mathbf{a}\cdot\mathbf{b})\cdot\mathbf{a} + 2\,\mathbf{b}^2\cdot\mathbf{a} - 2(\mathbf{b}\cdot\mathbf{a})\cdot b]$

$= [2\mathbf{a} + \mathbf{b}]\cdot[\mathbf{b} + 2\mathbf{a}] = 4\mathbf{a}^2 + \mathbf{b}^2$

$= 4\cdot 1 + 1 = 5 \qquad$ [as $\mathbf{a}\cdot\mathbf{b} = 0$]

193. Let angle between $\mathbf{a}$ and $\mathbf{b}$ be θ_1, $\mathbf{c}$ and $\mathbf{d}$ be θ_2 and $\mathbf{a}\times\mathbf{b}$ and $\mathbf{b}\times\mathbf{d}$ be θ.

Since, $(\mathbf{a}\times\mathbf{b})\cdot(\mathbf{c}\times\mathbf{d}) = 1 \Rightarrow \sin\theta_1\cdot\sin\theta_2\cdot\cos\theta = 1$

$\Rightarrow \quad \theta_1 = 90°, \theta_2 = 90°, \theta = 0°$

$\Rightarrow \mathbf{a}\perp\mathbf{b}, \mathbf{c}\perp\mathbf{d}, (\mathbf{a}\times\mathbf{b})||(\mathbf{c}\times\mathbf{d})$

So, $\quad \mathbf{a}\times\mathbf{b} = k(\mathbf{c}\times\mathbf{d})$ and $\mathbf{a}\times\mathbf{b} = k(\mathbf{c}\times\mathbf{d})$

$\Rightarrow \quad (\mathbf{a}\times\mathbf{b})\cdot\mathbf{c} = k(\mathbf{c}\times\mathbf{d})\cdot\mathbf{c}$ and $(\mathbf{a}\times\mathbf{b})\cdot\mathbf{d} = k(\mathbf{c}\times\mathbf{d})\cdot\mathbf{d}$

$\Rightarrow \quad [\mathbf{a}\,\mathbf{b}\,\mathbf{c}] = 0$ and $[\mathbf{a}\,\mathbf{b}\,\mathbf{d}] = 0$

$\Rightarrow$ **a**, **b**, **c** and **a**, **b**, **d** are coplanar vectors, so options (a) and (b) are incorrect.

Let $\mathbf{b}\,||\,\mathbf{d} \qquad \Rightarrow \mathbf{b} = \pm\,\mathbf{d}$

As $\quad (\mathbf{a}\times\mathbf{b})\cdot(\mathbf{c}\times\mathbf{d}) = 1 \Rightarrow (\mathbf{a}\times\mathbf{b})\cdot(\mathbf{c}\times\mathbf{b}) = \pm 1$

$\Rightarrow \quad [\mathbf{a}\times\mathbf{b}\;\mathbf{c}\,\mathbf{b}] = \pm 1 \Rightarrow [\mathbf{c}\,\mathbf{b}\,\mathbf{a}\times\mathbf{b}] = \pm 1$

$\Rightarrow \mathbf{c}\cdot[\mathbf{b}\times(\mathbf{a}\times\mathbf{b})] = \pm 1 \Rightarrow \mathbf{c}\cdot[\mathbf{a} - (\mathbf{b}\cdot\mathbf{a})\mathbf{b}] = \pm 1$

$\Rightarrow \quad \mathbf{c}\cdot\mathbf{a} = \pm 1 \qquad [\because \mathbf{a}\cdot\mathbf{b} = 0]$

Which is a contradiction, so option (c) is correct.

Let option (d) is correct.

$\Rightarrow \quad \mathbf{d} = \pm\mathbf{a}$ and $\mathbf{c} = \pm\,\mathbf{b}$

As $\quad (\mathbf{a}\times\mathbf{b})\cdot(\mathbf{c}\times\mathbf{d}) = 1$

$\Rightarrow \quad (\mathbf{a}\times\mathbf{b})\cdot(\mathbf{b}\times\mathbf{a}) = \pm 1$

Which is a contradiction, so option (d) is incorrect.

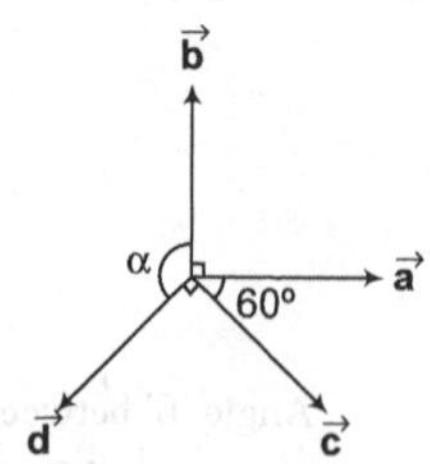

Alternatively options (c) and (d) may be observed from the above figure.

194. The volume of the parallelopiped with coterminous edges as $\hat{\mathbf{a}}, \hat{\mathbf{b}}, \hat{\mathbf{c}}$ is given by $[\hat{\mathbf{a}}\,\hat{\mathbf{b}}\,\hat{\mathbf{c}}] = \hat{\mathbf{a}}\cdot(\hat{\mathbf{b}}\times\hat{\mathbf{c}})$

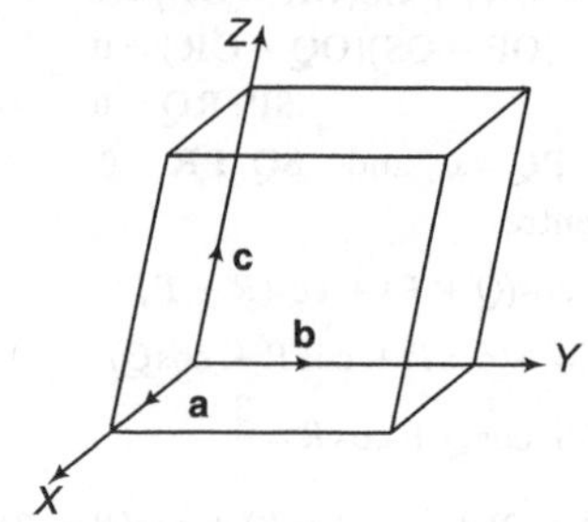

$$\text{Now, } [\hat{\mathbf{a}}\,\hat{\mathbf{b}}\,\hat{\mathbf{c}}]^2 = \begin{vmatrix} \hat{\mathbf{a}}\cdot\hat{\mathbf{a}} & \hat{\mathbf{a}}\cdot\hat{\mathbf{b}} & \hat{\mathbf{a}}\cdot\hat{\mathbf{c}} \\ \hat{\mathbf{b}}\cdot\hat{\mathbf{a}} & \hat{\mathbf{b}}\cdot\hat{\mathbf{b}} & \hat{\mathbf{b}}\cdot\hat{\mathbf{c}} \\ \hat{\mathbf{c}}\cdot\hat{\mathbf{a}} & \hat{\mathbf{c}}\cdot\hat{\mathbf{b}} & \hat{\mathbf{c}}\cdot\hat{\mathbf{c}} \end{vmatrix} = \begin{vmatrix} 1 & 1/2 & 1/2 \\ 1/2 & 1 & 1/2 \\ 1/2 & 1/2 & 1 \end{vmatrix}$$

$$\Rightarrow \quad [\hat{\mathbf{a}}\,\hat{\mathbf{b}}\,\hat{\mathbf{c}}]^2 = 1\left(1 - \frac{1}{4}\right) - \frac{1}{2}\left(\frac{1}{2} - \frac{1}{4}\right) + \frac{1}{2}\left(\frac{1}{4} - \frac{1}{2}\right) = \frac{1}{2}$$

Thus, the required volume of the parallelopiped

$= \frac{1}{\sqrt{2}}$ cu unit

195. Given, $\mathbf{OP} = \hat{\mathbf{a}}\cos t + \hat{\mathbf{b}}\sin t$

$\Rightarrow |\mathbf{OP}| = \sqrt{(\hat{\mathbf{a}}\cdot\hat{\mathbf{a}})\cos^2 t + (\hat{\mathbf{b}}\cdot\hat{\mathbf{b}})\sin^2 t + 2\hat{\mathbf{a}}\cdot\hat{\mathbf{b}}\sin t\cos t}$

$\Rightarrow |\mathbf{OP}| = \sqrt{1 + \hat{\mathbf{a}}\cdot\hat{\mathbf{b}}\sin 2t}$

$\Rightarrow \quad |\mathbf{OP}|_{\max} = M = \sqrt{1 + \hat{\mathbf{a}}\cdot\hat{\mathbf{b}}}$ at $\sin 2t = 1 \Rightarrow t = \frac{\pi}{4}$

At $\qquad t = \frac{\pi}{4}, \mathbf{OP} = \frac{1}{\sqrt{2}}(\hat{\mathbf{a}} + \hat{\mathbf{b}})$

Unit vector along **OP** at $\left(t = \frac{\pi}{4}\right) = \frac{\hat{\mathbf{a}} + \hat{\mathbf{b}}}{|\hat{\mathbf{a}} + \hat{\mathbf{b}}|}$

196. Since, **PQ** is not parallel to **TR**.

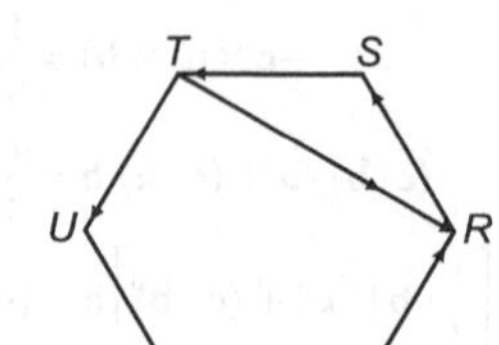

$\because$ **TR** is resultant of **RS** and **ST** vectors.

$\Rightarrow \qquad \mathbf{PQ}\times(\mathbf{RS} + \mathbf{ST}) \neq \mathbf{0}$.

But for Statement II, we have $\mathbf{PQ}\times\mathbf{RS} = \mathbf{0}$ which is not possible as **PQ** not parallel to **RS**.

Hence, Statement I is true and Statement II is false.

197. Since, given vectors are coplanar

$$\therefore \quad \begin{vmatrix} -\lambda^2 & 1 & 1 \\ 1 & -\lambda^2 & 1 \\ 1 & 1 & -\lambda^2 \end{vmatrix} = 0$$

$\Rightarrow \qquad \lambda^6 - 3\lambda^2 - 2 = 0$

$\Rightarrow (1 + \lambda^2)^2(\lambda^2 - 2) = 0 \Rightarrow \lambda = \pm\sqrt{2}$

198. Since, **a**, **b**, **c** are unit vectors and $\mathbf{a}+\mathbf{b}+\mathbf{c}=\mathbf{0}$, then **a**, **b**, **c** represent an equilateral triangle.

$\therefore \quad \mathbf{a}\times\mathbf{b}=\mathbf{b}\times\mathbf{c}=\mathbf{c}\times\mathbf{a}\neq\mathbf{0}.$

199. Let vector **AO** be parallel to line of intersection of planes P_1 and P_2 through origin.

Normal to plane p_1 is

$$\mathbf{n}_1=[(2\hat{\mathbf{j}}+3\hat{\mathbf{k}})\times(4\hat{\mathbf{j}}-3\hat{\mathbf{k}})]=-18\hat{\mathbf{i}}$$

Normal to plane p_2 is

$$\mathbf{n}_2=(\hat{\mathbf{j}}-\hat{\mathbf{k}})\times(3\hat{\mathbf{i}}+3\hat{\mathbf{j}})=3\hat{\mathbf{i}}-3\hat{\mathbf{j}}-3\hat{\mathbf{k}}$$

So, **OA** is parallel to $\pm(\mathbf{n}_1\times\mathbf{n}_2)=54\,\hat{\mathbf{j}}-54\hat{\mathbf{k}}$.

$\therefore$ Angle between $54\,(\hat{\mathbf{j}}-\hat{\mathbf{k}})$ and $(2\hat{\mathbf{i}}+\hat{\mathbf{j}}-2\hat{\mathbf{k}})$ is

$$\cos\theta=\pm\left(\frac{54+108}{3\cdot 54\cdot\sqrt{2}}\right)=\pm\frac{1}{\sqrt{2}}$$

$$\therefore \quad \theta=\frac{\pi}{4},\frac{3\pi}{4}$$

Hence, (b) and (d) are correct answers.

200. Let vector **r** be coplanar to **a** and **b**.

$$\therefore \quad \mathbf{r}=\mathbf{a}+t\mathbf{b}$$

$$\Rightarrow \quad \mathbf{r}=(\hat{\mathbf{i}}+2\hat{\mathbf{j}}+\hat{\mathbf{k}})+t\,(\hat{\mathbf{i}}-\hat{\mathbf{j}}+\hat{\mathbf{k}})$$

$$=(1+t)\hat{\mathbf{i}}+(2-t)\hat{\mathbf{j}}+(1+t)\hat{\mathbf{k}}$$

The projection of **r** on $\mathbf{c}=\frac{1}{\sqrt{3}}\cdot$ [given]

$$\Rightarrow \quad \frac{\mathbf{r}\cdot\mathbf{c}}{|\mathbf{c}|}=\frac{1}{\sqrt{3}}$$

$$\Rightarrow \quad \frac{|1\cdot(1+t)+1\cdot(2-t)-1\cdot(1+t)|}{\sqrt{3}}=\frac{1}{\sqrt{3}}$$

$\Rightarrow \quad (2-t)=\pm 1 \Rightarrow t=1$ or 3

When, $t=1$, we have $\mathbf{r}=2\hat{\mathbf{i}}+\hat{\mathbf{j}}+2\hat{\mathbf{k}}$

When, $t=3$, we have $\mathbf{r}=4\hat{\mathbf{i}}-\hat{\mathbf{j}}+4\hat{\mathbf{k}}$

201. Since, $\mathbf{b}_1=\mathbf{b}-\frac{\mathbf{b}\cdot\mathbf{a}}{|\mathbf{a}|^2}\mathbf{a},\ \mathbf{b}_1=\mathbf{b}+\frac{\mathbf{b}\cdot\mathbf{a}}{|\mathbf{a}|^2}\mathbf{a}$

and $\mathbf{c}=\mathbf{c}-\frac{\mathbf{c}\cdot\mathbf{a}}{|\mathbf{a}|^2}\mathbf{a}-\frac{\mathbf{c}\cdot\mathbf{b}}{|\mathbf{b}|^2}\mathbf{b},\ \mathbf{c}_2=\mathbf{c}-\frac{\mathbf{c}\cdot\mathbf{a}}{|\mathbf{a}|^2}\mathbf{a}-\frac{\mathbf{c}\cdot\mathbf{b}_1}{|\mathbf{b}|^2}\mathbf{b}_1$

$$\mathbf{c}_3=\mathbf{c}-\frac{\mathbf{c}\cdot\mathbf{a}}{|\mathbf{a}|^2}\mathbf{a}-\frac{\mathbf{c}\cdot\mathbf{b}_2}{|\mathbf{b}|^2}\mathbf{b}_2,\ \mathbf{c}_4=\mathbf{a}-\frac{\mathbf{c}\cdot\mathbf{a}}{|\mathbf{a}|^2}\mathbf{a}.$$

which shows $\mathbf{a}\cdot\mathbf{b}_1=0=\mathbf{a}\cdot\mathbf{c}=\mathbf{b}_1\cdot\mathbf{c}_2$

So, $\{\mathbf{a},\mathbf{b}_1,\mathbf{c}_2\}$ are mutually orthogonal vectors.

202. As we know that, a vector coplanar to **a**, **b** and orthogonal to **c** is $\lambda\{(\mathbf{a}\times\mathbf{b})\times\mathbf{c}\}$.

$\therefore$ A vector coplanar to $(2\hat{\mathbf{i}}+\hat{\mathbf{j}}+\hat{\mathbf{k}}),(\hat{\mathbf{i}}-\hat{\mathbf{j}}+\hat{\mathbf{k}})$ and orthogonal to $3\hat{\mathbf{i}}+2\hat{\mathbf{j}}+6\hat{\mathbf{k}}$

$$=\lambda\,[\{(2\hat{\mathbf{i}}+\hat{\mathbf{j}}+\hat{\mathbf{k}})\times(\hat{\mathbf{i}}-\hat{\mathbf{j}}+\hat{\mathbf{k}})\}\times(3\hat{\mathbf{i}}+2\hat{\mathbf{j}}+6\hat{\mathbf{k}})]$$

$$=\lambda\,[(2\hat{\mathbf{i}}-\hat{\mathbf{j}}-3\hat{\mathbf{k}})\times(3\hat{\mathbf{i}}+2\hat{\mathbf{j}}+6\hat{\mathbf{k}})]=\lambda\,(21\hat{\mathbf{j}}-7\hat{\mathbf{k}})$$

$$\therefore \text{ Unit vector }=+\frac{(21\hat{\mathbf{j}}-7\hat{\mathbf{k}})}{\sqrt{(21)^2+(7)^2}}$$

$$=+\frac{(3\hat{\mathbf{j}}-\hat{k})}{\sqrt{10}}$$

203. We know that, volume of parallelopiped whose edges are **a**, **b**, **c** $=[\mathbf{a}\,\mathbf{b}\,\mathbf{c}]$.

$$\therefore \quad [\mathbf{a}\,\mathbf{b}\,\mathbf{c}]=\begin{vmatrix}1 & a & 1\\ 0 & 1 & a\\ a & 0 & 1\end{vmatrix}=1+a^3-a$$

Let $f(a)=a^3-a+1$

$\Rightarrow \quad f'(a)=3a^2-1 \Rightarrow f''(a)=6a$

For maximum or minimum, put $f'(a)=0$

$\Rightarrow a=\pm\frac{1}{\sqrt{3}}$, which shows $f(a)$ is minimum at $a=\frac{1}{\sqrt{3}}$ and maximum at $a=-\frac{1}{\sqrt{3}}$.

204. We know that, $\mathbf{a}\times(\mathbf{a}\times\mathbf{b})=(\mathbf{a}\cdot\mathbf{b})\mathbf{a}-(\mathbf{a}\cdot\mathbf{a})\mathbf{b}$

$$\therefore \quad (\hat{\mathbf{i}}+\hat{\mathbf{j}}+\hat{\mathbf{k}})\times(\hat{\mathbf{j}}-\hat{\mathbf{k}})=(\hat{\mathbf{i}}+\hat{\mathbf{j}}+\hat{\mathbf{k}})-(\sqrt{3})^2\mathbf{b}$$

$$\Rightarrow \quad -2\hat{\mathbf{i}}+\hat{\mathbf{j}}+\hat{\mathbf{k}}=\hat{\mathbf{i}}+\hat{\mathbf{j}}+\hat{\mathbf{k}}-3\mathbf{b} \Rightarrow 3\mathbf{b}=+3\hat{\mathbf{i}}$$

$$\therefore \quad \mathbf{b}=\hat{\mathbf{i}}$$

205. Given, $\mathbf{V}=2\hat{\mathbf{i}}+\hat{\mathbf{j}}-\hat{\mathbf{k}}$ and $\mathbf{W}=\hat{\mathbf{i}}+3\hat{\mathbf{k}}$

$$[\mathbf{U}\,\mathbf{V}\,\mathbf{W}]=\mathbf{U}\cdot[(2\hat{\mathbf{i}}+\hat{\mathbf{j}}-\hat{\mathbf{k}})\times(\hat{\mathbf{i}}+3\hat{\mathbf{k}})]$$

$$=\mathbf{U}\cdot(3\hat{\mathbf{i}}-7\hat{\mathbf{j}}-\hat{\mathbf{k}})=|\mathbf{U}||3\hat{\mathbf{i}}-7\hat{\mathbf{j}}-\hat{\mathbf{k}}|\cos\theta$$

Which is maximum, if angle between **U** and $3\hat{\mathbf{i}}-7\hat{\mathbf{j}}-\hat{\mathbf{k}}$ is 0 and maximum value

$$=|3\hat{\mathbf{i}}-7\hat{\mathbf{j}}-\hat{\mathbf{k}}|=\sqrt{59}$$

206. Since, $(\mathbf{a}+2\mathbf{b})\cdot(5\mathbf{a}-4\mathbf{a})=0$

$\Rightarrow \quad 5|\mathbf{a}|^2+6\mathbf{a}\cdot\mathbf{b}-8|\mathbf{b}|^2=0$

$\Rightarrow \quad 6\mathbf{a}\cdot\mathbf{b}=3 \quad [\because |\mathbf{a}|=|\mathbf{b}|=1]$

$\Rightarrow \quad \cos\theta=\frac{1}{2} \Rightarrow \theta=60^\circ$

207. We have, $\mathbf{a}=2\hat{\mathbf{i}}+\hat{\mathbf{j}}-2\hat{\mathbf{k}}$

$\Rightarrow |\mathbf{a}|=\sqrt{4+1+4}=3$

and $\quad \mathbf{b}=\hat{\mathbf{i}}+\hat{\mathbf{j}}\Rightarrow|\mathbf{b}|=\sqrt{1+1}=\sqrt{2}$

Now, $\quad |\mathbf{c}-\mathbf{a}|=3\Rightarrow|\mathbf{c}-\mathbf{a}|^2=9$

$\Rightarrow \quad (\mathbf{c}-\mathbf{a})\cdot(\mathbf{c}-\mathbf{a})=9$

$\Rightarrow \quad |\mathbf{c}|^2+|\mathbf{a}|^2-2\mathbf{c}\cdot\mathbf{a}=9 \quad$ ...(i)

Again, $\quad |\mathbf{a}\times\mathbf{b}|\times\mathbf{c}|=3$

$\Rightarrow \quad |\mathbf{a}\times\mathbf{b}||\mathbf{c}|\sin 30^\circ=3\Rightarrow|\mathbf{c}|=\frac{6}{|\mathbf{a}\times\mathbf{b}|}$

But $\quad \mathbf{a}\times\mathbf{b}=\begin{vmatrix}\hat{\mathbf{i}} & \hat{\mathbf{j}} & \hat{\mathbf{k}}\\ 2 & 1 & -2\\ 1 & 1 & 0\end{vmatrix}=2\hat{\mathbf{i}}-2\hat{\mathbf{j}}+\hat{\mathbf{k}}$

$\therefore \quad |\mathbf{c}|\frac{6}{\sqrt{4+4+1}}=2 \quad$ ...(ii)

From Eqs. (i) and (ii), we get

$$(2)^2+(3)^2-2\mathbf{c}\cdot\mathbf{a}=9 \Rightarrow 4+9-2\mathbf{c}\cdot\mathbf{a}=9\Rightarrow\mathbf{c}\cdot\mathbf{a}=2$$

208. Use the formulae, $\mathbf{a}\times(\mathbf{b}\times\mathbf{c})=(\mathbf{a}\cdot\mathbf{c})\mathbf{b}-(\mathbf{a}\cdot\mathbf{b})\,\mathbf{c}$,

$$[\mathbf{a}\ \mathbf{b}\ \mathbf{c}]=[\mathbf{b}\ \mathbf{c}\ \mathbf{a}]=[\mathbf{c}\ \mathbf{a}\ \mathbf{b}]$$

and $[\mathbf{a}\ \mathbf{a}\ \mathbf{b}]=[\mathbf{a}\ \mathbf{b}\ \mathbf{b}]=[\mathbf{a}\ \mathbf{c}\ \mathbf{c}]=0$

Further, simplify it and get the result.

Now, $[\mathbf{a}\times\mathbf{b}\ \mathbf{b}\times\mathbf{c}\ \mathbf{c}\times\mathbf{a}]$

$$=\mathbf{a}\times\mathbf{b}\cdot((\mathbf{b}\times\mathbf{c})\times(\mathbf{c}\times\mathbf{a}))$$
$$=\mathbf{a}\times\mathbf{b}\cdot((\mathbf{k}\times\mathbf{c}\times\mathbf{a}))\qquad[\text{Here, }\mathbf{k}=\mathbf{b}\times\mathbf{c}]$$
$$=\mathbf{a}\times\mathbf{b}\cdot[(\mathbf{k}\cdot\mathbf{a})\,\mathbf{c}-(\mathbf{k}\cdot\mathbf{c})\,\mathbf{a}]$$
$$=(\mathbf{a}\times\mathbf{b})\cdot((\mathbf{b}\times\mathbf{c}\cdot\mathbf{a})\,\mathbf{c}-(\mathbf{b}\times\mathbf{c}\cdot\mathbf{c})\mathbf{a})$$
$$=(\mathbf{a}\times\mathbf{b})\cdot([\mathbf{b}\ \mathbf{c}\ \mathbf{a}]\,\mathbf{c})-0\qquad[\because[\mathbf{b}\times\mathbf{c}\cdot\mathbf{c}]=0]$$
$$=\mathbf{a}\times\mathbf{b}\cdot\mathbf{c}\,[\mathbf{b}\ \mathbf{c}\ \mathbf{a}]=[\mathbf{a}\ \mathbf{b}\ \mathbf{c}]\,[\mathbf{b}\ \mathbf{c}\ \mathbf{a}]$$
$$=[\mathbf{a}\ \mathbf{b}\ \mathbf{c}]^2\qquad\{\because[\mathbf{a}\ \mathbf{b}\ \mathbf{c}]=[\mathbf{b}\ \mathbf{c}\ \mathbf{a}]\}$$

Hence, $[\mathbf{a}\times\mathbf{b}\ \mathbf{b}\times\mathbf{c}\ \mathbf{c}\times\mathbf{a}]=\lambda\,[\mathbf{a}\ \mathbf{b}\ \mathbf{c}]^2$

$$\Rightarrow\quad[\mathbf{a}\ \mathbf{b}\ \mathbf{c}]^2=\lambda\,[\mathbf{a}\ \mathbf{b}\ \mathbf{c}]^2$$
$$\Rightarrow\quad\lambda=1$$

209. Given that,

(i) **a** and **b** are unit vectors, i.e. $|\mathbf{a}|=|\mathbf{b}|=1$

(ii) $\mathbf{c}=\mathbf{a}+2\mathbf{b}$ and $\mathbf{d}=5\mathbf{a}-4\mathbf{b}$

(iii) **c** and **d** are perpendicular to each other. i.e. $\mathbf{c}\cdot\mathbf{d}=0$

To find Angle between **a** and **b**.

Now, $\mathbf{c}\cdot\mathbf{d}=0\Rightarrow(\mathbf{a}+2\,\mathbf{b})\cdot(5\,\mathbf{a}-4\,\mathbf{b})=0$

$$\Rightarrow 5\,\mathbf{a}\cdot\mathbf{a}-4\,\mathbf{a}\cdot\mathbf{b}+10\,\mathbf{b}\cdot\mathbf{a}-8\,\mathbf{b}\cdot\mathbf{b}=0$$
$$\Rightarrow\quad 6\,\mathbf{a}\cdot\mathbf{b}=3$$
$$\Rightarrow\quad \mathbf{a}\cdot\mathbf{b}=\frac{1}{2}$$

So, the angle between **a** and **b** is $\frac{\pi}{3}$.

210. Given,

(i) A parallelogram $ABCD$ such that $\mathbf{AB}=\mathbf{q}$ and $\mathbf{AD}=\mathbf{p}$.

(ii) The altitude from vertex B to side AD coincides with a vector **r**.

To find The vector **r** in terms of **p** and **q**.

Let E be the foot of perpendicular from B to side AD.

AE = Projection of vector **q** on $\mathbf{p}=\mathbf{q}\cdot\hat{\mathbf{p}}=\dfrac{\mathbf{q}\cdot\mathbf{p}}{|\mathbf{p}|}$

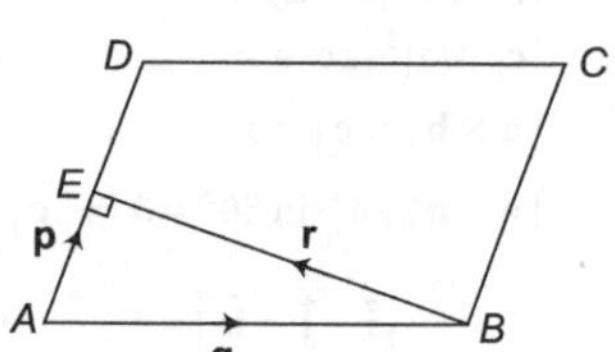

AE = Vector along AE of length AE

$$=|\mathbf{AE}|\,\hat{\mathbf{AE}}=\left(\frac{\mathbf{q}\cdot\mathbf{p}}{|\mathbf{p}|}\right)\hat{\mathbf{p}}=\frac{(\mathbf{q}\cdot\mathbf{p})\mathbf{p}}{|\mathbf{p}|^2}$$

Now, applying triangles law in ΔABE, we get

$$\mathbf{AB}+\mathbf{BE}=\mathbf{AE}$$
$$\Rightarrow\quad\mathbf{q}+\mathbf{r}=\frac{(\mathbf{q}\cdot\mathbf{p})\mathbf{p}}{|\mathbf{p}|^2}\Rightarrow\mathbf{r}=\frac{(\mathbf{q}\cdot\mathbf{p})\mathbf{p}}{|\mathbf{p}|^2}-\mathbf{q}$$
$$\Rightarrow\quad\mathbf{r}=-\mathbf{q}+\left(\frac{\mathbf{q}\cdot\mathbf{p}}{\mathbf{p}\cdot\mathbf{p}}\right)\mathbf{p}$$

211. $\mathbf{a}=\dfrac{1}{\sqrt{10}}(3\hat{\mathbf{i}}+\hat{\mathbf{k}})$ and $\mathbf{b}=\dfrac{1}{7}(2\,\hat{\mathbf{i}}+3\,\hat{\mathbf{j}}-6\,\hat{\mathbf{k}})$

$$\therefore\quad(2\mathbf{a}-\mathbf{b})\cdot\{(\mathbf{a}\times\mathbf{b})\times(\mathbf{a}+2\mathbf{b})\}$$
$$=(2\mathbf{a}-\mathbf{b})\cdot\{(\mathbf{a}\times\mathbf{b})\times\mathbf{a}+(\mathbf{a}\times\mathbf{b})\times2\mathbf{b}\}$$
$$=(2\mathbf{a}-\mathbf{b})\cdot\{(\mathbf{a}\cdot\mathbf{a})\,\mathbf{b}-(\mathbf{b}\cdot\mathbf{a})\,\mathbf{a}+2\,(\mathbf{a}\cdot\mathbf{b})\,\mathbf{b}-2\,(\mathbf{b}\cdot\mathbf{b})\,\mathbf{a}\}$$
$$=(2\mathbf{a}-\mathbf{b})\cdot\{1\,(\mathbf{b})-(0)\,\mathbf{a}+2\,(0)\,\mathbf{b}-2\,(1)\,\mathbf{a}\}$$
$$[\text{as }\mathbf{a}\cdot\mathbf{b}=0\text{ and }\mathbf{a}\cdot\mathbf{a}=\mathbf{b}\cdot\mathbf{b}=1]$$
$$=(2\mathbf{a}-\mathbf{b})\,(\mathbf{b}-2\mathbf{a})$$
$$=-(4\,|\mathbf{a}|^2-4\,\mathbf{a}\cdot\mathbf{b}+|\mathbf{b}|^2)=-\{4-0+1\}=-5$$

212. Given, $\mathbf{a}\cdot\mathbf{b}\neq0$, $\mathbf{a}\cdot\mathbf{d}=0$ ...(i)

and $\mathbf{b}\times\mathbf{c}=\mathbf{b}\times\mathbf{d}$

$$\Rightarrow\quad\mathbf{b}\times(\mathbf{c}-\mathbf{d})=0$$
$$\therefore\quad\mathbf{b}\,||\,(\mathbf{c}-\mathbf{d})$$
$$\Rightarrow\quad\mathbf{c}-\mathbf{d}=\lambda\mathbf{b}$$
$$\Rightarrow\quad\mathbf{d}=\mathbf{c}-\lambda\mathbf{b}\qquad\text{...(ii)}$$

Taking dot product with **a**, we get

$$\mathbf{a}\cdot\mathbf{d}=\mathbf{a}\cdot\mathbf{c}-\lambda\mathbf{a}\cdot\mathbf{b}$$
$$\Rightarrow\quad0=\mathbf{a}\cdot\mathbf{c}-\lambda\,(\mathbf{a}\cdot\mathbf{b})$$
$$\therefore\quad\lambda=\frac{\mathbf{a}\cdot\mathbf{c}}{\mathbf{a}\cdot\mathbf{b}}\qquad\text{...(iii)}$$
$$\therefore\quad\mathbf{d}=\mathbf{c}-\frac{(\mathbf{a}\cdot\mathbf{c})}{(\mathbf{a}\cdot\mathbf{b})}\,\mathbf{b}$$

213. Given, $\mathbf{a}=p\,\hat{\mathbf{i}}+\hat{\mathbf{j}}+\hat{\mathbf{k}}$, $\mathbf{b}=\hat{\mathbf{i}}+q\hat{\mathbf{j}}+\hat{\mathbf{k}}$ and $\mathbf{c}=\hat{\mathbf{i}}+\hat{\mathbf{j}}+r\,\hat{\mathbf{k}}$ are coplanar and $p\neq q\neq r\neq1$.

Since, **a**, **b** and **c** are coplanar.

$$\Rightarrow\quad[\mathbf{a}\ \mathbf{b}\ \mathbf{c}]=0$$
$$\Rightarrow\quad\begin{vmatrix}p&1&1\\1&q&1\\1&1&r\end{vmatrix}=0$$
$$\Rightarrow\quad p\,(qr-1)-1\,(r-1)+1\,(1-q)=0$$
$$\Rightarrow\quad pqr-p-r+1+1-q=0$$
$$\therefore\quad pqr-(p+q+r)=-2$$

214. We have, $\mathbf{a}\times\mathbf{b}+\mathbf{c}=0$

$$\Rightarrow\quad\mathbf{a}\times(\mathbf{a}\times\mathbf{b})+\mathbf{a}\times\mathbf{c}=0$$
$$\Rightarrow\quad(\mathbf{a}\cdot\mathbf{b})\mathbf{a}-(\mathbf{a}\cdot\mathbf{a})\mathbf{b}+\mathbf{a}\times\mathbf{c}=0$$
$$\Rightarrow\quad3\mathbf{a}-2\mathbf{b}+\mathbf{a}\times\mathbf{c}=0$$
$$\Rightarrow\quad2\mathbf{b}=3\mathbf{a}+\mathbf{a}\times\mathbf{c}$$
$$\Rightarrow\quad2\mathbf{b}=3\hat{\mathbf{j}}-3\hat{\mathbf{k}}-2\hat{\mathbf{i}}-\hat{\mathbf{j}}-\hat{\mathbf{k}}=-2\hat{\mathbf{i}}+2\hat{\mathbf{j}}-4\hat{\mathbf{k}}$$
$$\therefore\quad\mathbf{b}=-\hat{\mathbf{i}}+\hat{\mathbf{j}}-2\,\hat{\mathbf{k}}$$

215. Since, the given vectors are mutually orthogonal, therefore

$$\mathbf{a}\cdot\mathbf{b}=2-4+2=0$$
$$\mathbf{a}\cdot\mathbf{c}=\lambda-1+2\mu=0\qquad\text{...(i)}$$
$$\text{and}\quad\mathbf{b}\cdot\mathbf{c}=2\lambda+4+\mu=0\qquad\text{...(ii)}$$

On solving Eqs. (i) and (ii), we get

$$\mu=2\text{ and }\lambda=-3$$

Hence, $(\lambda,\mu)=(-3,2)$

216. Since, $[3\mathbf{u}\ p\mathbf{v}\ p\mathbf{w}]-[p\mathbf{v}\ \mathbf{w}\ q\mathbf{u}]-[2\mathbf{w}\ q\mathbf{v}\ q\mathbf{u}]=0$

$$\therefore 3p^2[\mathbf{u}\cdot(\mathbf{v}\times\mathbf{w})]-pq\,[\mathbf{v}\cdot(\mathbf{w}\times\mathbf{u})]-2q^2\,[\mathbf{w}\cdot(\mathbf{v}\times\mathbf{u})]=0$$

$\Rightarrow \quad (3p^2 - pq + 2q^2)[\mathbf{u} \cdot (\mathbf{v} \times \mathbf{w})] = 0$

But $\quad [\mathbf{u\,v\,w}] \neq 0$

$\Rightarrow \quad 3p^2 - pq + 2q^2 = 0$

$\therefore \quad p = q = 0$

217. Given that, $\mathbf{b} = \hat{\mathbf{i}} + \hat{\mathbf{j}}$ and $\mathbf{c} = \hat{\mathbf{j}} + \hat{\mathbf{k}}$.
The equation of bisector of **b** and **c** is

$$\mathbf{r} = \lambda(\mathbf{b} + \mathbf{c}) = \lambda\left(\frac{\hat{\mathbf{i}} + \hat{\mathbf{j}}}{\sqrt{2}} + \frac{\hat{\mathbf{j}} + \hat{\mathbf{k}}}{\sqrt{2}}\right)$$

$$= \frac{\lambda}{\sqrt{2}}(\hat{\mathbf{i}} + 2\hat{\mathbf{j}} + \hat{\mathbf{k}}) \qquad \ldots \text{(i)}$$

Since, vector **a** lies in plane of **b** and **c**.

$\therefore \quad \mathbf{a} = \mathbf{b} + \mu\mathbf{c}$

$\Rightarrow \quad \frac{\lambda}{\sqrt{2}}(\hat{\mathbf{i}} + 2\hat{\mathbf{j}} + \hat{\mathbf{k}}) = (\hat{\mathbf{i}} + \hat{\mathbf{j}}) + \mu(\hat{\mathbf{j}} + \hat{\mathbf{k}})$

On equating the coefficient of $\hat{\mathbf{i}}$ both sides, we get

$$\frac{\lambda}{\sqrt{2}} = 1 \Rightarrow \lambda = \sqrt{2}$$

On putting $\lambda = \sqrt{2}$ in Eq. (i), we get

$$\mathbf{r} = \hat{\mathbf{i}} + 2\hat{\mathbf{j}} + \hat{\mathbf{k}}$$

Since, the given vector a represents the same bisector equation $\hat{\mathbf{r}}$.

$\therefore \quad \alpha = 1$ and $\beta = 1$

218. Since, $(2\mathbf{u} \times 3\mathbf{v})$ is a unit vector.

$\Rightarrow \quad |2\mathbf{u} \times 3\mathbf{v}| = 1$

$\Rightarrow \quad 6|\mathbf{u}||\mathbf{v}|\sin\theta| = 1$

$\Rightarrow \quad \sin\theta = \frac{1}{6} \qquad [\because |\mathbf{u}| = |\mathbf{v}| = 1]$

Since, θ is an acute angle, then there is exactly one value of θ for which $(2\mathbf{u} \times 3\mathbf{v})$ is a unit vector.

219. Since, given vectors **v**, **b** and **c** are coplanar.

$$\therefore \quad \begin{vmatrix} 1 & 1 & 1 \\ 1 & -1 & 2 \\ x & x-2 & -1 \end{vmatrix} = 0$$

$\Rightarrow \quad 1\{1 - 2(x-2)\} - 1(-1 - 2x) + 1(x - 2 + x) = 0$

$\Rightarrow \quad 1 - 2x + 4 + 1 + 2x + 2x - 2 = 0$

$\Rightarrow \quad 2x = -4 \Rightarrow x = -2$

220. Since, $\quad (\mathbf{a} \times \mathbf{b}) \times \mathbf{c} = \mathbf{a} \times (\mathbf{b} \times \mathbf{c})$

$\therefore \quad (\mathbf{a} \cdot \mathbf{c})\,\mathbf{b} - (\mathbf{b} \cdot \mathbf{c})\,\mathbf{a} = (\mathbf{a} \cdot \mathbf{c})\,\mathbf{b} - (\mathbf{a} \cdot \mathbf{b})\,\mathbf{c}$

$\Rightarrow \quad (\mathbf{b} \cdot \mathbf{c})\mathbf{a} = (\mathbf{a} \cdot \mathbf{b})\mathbf{c}$

$\Rightarrow \quad \mathbf{a} = \frac{(\mathbf{a} \cdot \mathbf{b})}{(\mathbf{b} \cdot \mathbf{c})} \cdot \mathbf{c}$

Hence, **a** is parallel to **c**.

221. Since, position vectors of A, B, C are $2\hat{\mathbf{i}} - \hat{\mathbf{j}} + \hat{\mathbf{k}}, \hat{\mathbf{i}} - 3\hat{\mathbf{j}} - 5\hat{\mathbf{k}}$ and $a\hat{\mathbf{i}} - 3\hat{\mathbf{j}} + \hat{\mathbf{k}}$, respectively.

Now, $\quad \mathbf{AC} = (a\hat{\mathbf{i}} - 3\hat{\mathbf{j}} + \hat{\mathbf{k}}) - (2\hat{\mathbf{i}} - \hat{\mathbf{j}} + \hat{\mathbf{k}})$

$= (a-2)\hat{\mathbf{i}} - 2\hat{\mathbf{j}}$

and $\quad \mathbf{BC} = (a\hat{\mathbf{i}} - 3\hat{\mathbf{j}} + \hat{\mathbf{k}}) - (\hat{\mathbf{i}} - 3\hat{\mathbf{j}} - 5\hat{\mathbf{k}})$

$= (a-1)\hat{\mathbf{i}} + 6\hat{\mathbf{k}}$

Since, the ΔABC is right angled at C, then

$$\mathbf{AC} \cdot \mathbf{BC} = 0$$

$\Rightarrow \quad \{(a-2)\hat{\mathbf{i}} - 2\hat{\mathbf{j}}\} \cdot \{(a-1)\hat{\mathbf{i}} + 6\hat{\mathbf{k}}\} = 0$

$\Rightarrow \quad (a-2)(a-1) = 0$

$\therefore \quad a = 1$ and $a = 2$

222. Line is parallel to plane as

$$(\hat{\mathbf{i}} - \hat{\mathbf{j}} + 4\hat{\mathbf{k}}) \cdot (\hat{\mathbf{i}} + 5\hat{\mathbf{j}} + \hat{\mathbf{k}}) = 1 - 5 + 4 = 0$$

General point on the line is

$$(\lambda + 2, -\lambda - 2, 4\lambda + 3).$$

For $\lambda = 0$, a point on this line is $(2, -2, 3)$ and distance from $\mathbf{r} \cdot (\hat{\mathbf{i}} + 5\hat{\mathbf{j}} + \hat{\mathbf{k}}) = 5$ or $x + 5y + z = 5$ is

$$d = \left|\frac{2 + 5(-2) + 3 - 5}{\sqrt{1 + 25 + 1}}\right| \Rightarrow d = \left|\frac{-10}{3\sqrt{3}}\right| = \frac{10}{3\sqrt{3}}$$

223. Let $\mathbf{a} = a_1\hat{\mathbf{i}} + a_2\hat{\mathbf{j}} + a_3\hat{\mathbf{k}}$

Then, $\quad \mathbf{a} \times \hat{\mathbf{i}} = -a_2\hat{\mathbf{k}} + a_3\hat{\mathbf{j}}$

$\mathbf{a} \times \hat{\mathbf{j}} = a_1\hat{\mathbf{k}} - a_3\hat{\mathbf{i}}$

$\mathbf{a} \times \hat{\mathbf{k}} = -a_1\hat{\mathbf{j}} + a_2\hat{\mathbf{i}}$

$\therefore \quad (\mathbf{a} \times \hat{\mathbf{i}})^2 + (\mathbf{a} \times \hat{\mathbf{j}})^2 + (\mathbf{a} \times \hat{\mathbf{k}})^2$

$= a_2^2 + a_3^2 + a_1^2 + a_3^2 + a_1^2 + a_2^2$

$= 2(a_1^2 + a_2^2 + a_3^2) = 2a^2$

224. Given that, $[\lambda(\mathbf{a} + \mathbf{b}) \;\; \lambda^2\mathbf{b} \;\; \lambda\mathbf{c}] = [\mathbf{a} \;\; \mathbf{b} + \mathbf{c} \;\; \mathbf{b}]$

$$\therefore \quad \begin{vmatrix} \lambda(a_1 + b_1) & \lambda(a_2 + b_2) & \lambda(a_3 + b_3) \\ \lambda^2 b_1 & \lambda^2 b_2 & \lambda^2 b_3 \\ \lambda c_1 & \lambda c_2 & \lambda c_3 \end{vmatrix} = \begin{vmatrix} a_1 & a_2 & a_3 \\ b_1 + c_1 & b_2 + c_2 & b_3 + c_3 \\ b_1 & b_2 & b_3 \end{vmatrix}$$

$$\Rightarrow \quad \lambda^4 \begin{vmatrix} a_1 & a_2 & a_3 \\ b_1 & b_2 & b_3 \\ c_1 & c_2 & c_3 \end{vmatrix} = -\begin{vmatrix} a_1 & a_2 & a_3 \\ b_1 & b_2 & b_3 \\ c_1 & c_2 & c_3 \end{vmatrix}$$

$\Rightarrow \quad \lambda^4 = -1$

So, no real value of λ exists.

225. Given, vectors are

$$\mathbf{a} = \hat{\mathbf{i}} - \hat{\mathbf{k}}, \;\; \mathbf{b} = x\hat{\mathbf{i}} + \hat{\mathbf{j}} + (1 - x)\hat{\mathbf{k}}$$

and $\quad \mathbf{c} = y\hat{\mathbf{i}} + x\hat{\mathbf{j}} + (1 + x - y)\hat{\mathbf{k}}$

$$\therefore \quad [\mathbf{a\,b\,c}] = \begin{vmatrix} 1 & 0 & -1 \\ x & 1 & 1-x \\ y & x & 1+x-y \end{vmatrix}$$

Applying $C_3 \to C_3 + C_1$, we get

$$= \begin{vmatrix} 1 & 0 & 0 \\ x & 1 & 1 \\ y & x & 1+x \end{vmatrix} = 1(1 + x) - x = 1$$

Thus, $[\mathbf{a\,b\,c}]$ depends upon neither x nor y.

226. Since, $\quad |\mathbf{u}| = 1, |\mathbf{v}| = 2, |\mathbf{w}| = 3$

The projection of **v** along $\mathbf{u} = \frac{\mathbf{v} \cdot \mathbf{u}}{|\mathbf{u}|}$

and the projection of **w** along $\mathbf{u} = \frac{\mathbf{w} \cdot \mathbf{u}}{|\mathbf{u}|}$

According to given condition,

$$\frac{\mathbf{v}\cdot\mathbf{u}}{|\mathbf{u}|}=\frac{\mathbf{w}\cdot\mathbf{u}}{|\mathbf{u}|}$$

$$\Rightarrow \quad \mathbf{v}\cdot\mathbf{u}=\mathbf{w}\cdot\mathbf{u} \qquad \text{...(i)}$$

Since, $\mathbf{v},\mathbf{w}$ are perpendicular to each other.

$$\therefore \quad \mathbf{v}\cdot\mathbf{w}=0 \qquad \text{...(ii)}$$

Now, $|\mathbf{u}-\mathbf{v}+\mathbf{w}|^2=|\mathbf{u}|^2+|\mathbf{v}|^2+|\mathbf{w}|^2 -2\mathbf{u}\cdot\mathbf{v}-2\,\mathbf{v}\cdot\mathbf{w}+2\,\mathbf{u}\cdot\mathbf{w}$

$$\Rightarrow \quad |\mathbf{u}-\mathbf{v}+\mathbf{w}|^2=1+4+9-2\mathbf{u}\cdot\mathbf{v}+2\mathbf{v}\cdot\mathbf{u}$$

[from Eqs. (i) and (ii)]

$$\Rightarrow \quad |\mathbf{u}-\mathbf{v}+\mathbf{w}|^2=1+4+9$$

$$\Rightarrow \quad |\mathbf{u}-\mathbf{v}+\mathbf{w}|=\sqrt{14}$$

227. Given that, $\frac{1}{3}|\mathbf{b}||\mathbf{c}|\mathbf{a}=(\mathbf{a}\times\mathbf{b})\times\mathbf{c}$

We know that,

$$(\mathbf{a}\times\mathbf{b})\times\mathbf{c}=(\mathbf{a}\cdot\mathbf{c})\,\mathbf{b}-(\mathbf{b}\cdot\mathbf{c})\mathbf{a}$$

$$\therefore \quad \frac{1}{3}|\mathbf{b}||\,\mathbf{c}|\mathbf{a}=(\mathbf{a}\cdot\mathbf{c})\,\mathbf{b}-(\mathbf{b}\cdot\mathbf{c})\mathbf{a}$$

On comparing the coefficients of **a** and **b**, we get

$$\frac{1}{3}|\mathbf{b}||\,\mathbf{c}|=-\mathbf{b}\cdot\mathbf{c} \text{ and } \mathbf{a}\cdot\mathbf{c}=0$$

$$\Rightarrow \quad \frac{1}{3}|\mathbf{b}|\,|\mathbf{c}|=-|\mathbf{b}||\mathbf{c}|\cos\theta$$

$$\Rightarrow \quad \cos\theta=-\frac{1}{3} \quad \Rightarrow 1-\sin^2\theta=\frac{1}{9}$$

$$\Rightarrow \quad \sin^2\theta=\frac{8}{9}$$

$$\therefore \quad \sin\theta=\frac{2\sqrt{2}}{3} \qquad \left[\because\ 0\le\theta\le\frac{\pi}{2}\right]$$

228. Total force, $\mathbf{F}=(4\hat{\mathbf{i}}+\hat{\mathbf{j}}-3\hat{\mathbf{k}})+(3\hat{\mathbf{i}}+\hat{\mathbf{j}}-\hat{\mathbf{k}})$

$$\therefore \quad \mathbf{F}=7\hat{\mathbf{i}}+2\hat{\mathbf{j}}-4\hat{\mathbf{k}}$$

The particle is displaced from $A(\hat{\mathbf{i}}+2\hat{\mathbf{j}}+3\hat{\mathbf{k}})$ to $B(5\hat{\mathbf{i}}+4\hat{\mathbf{j}}+\hat{\mathbf{k}})$.

Now, displacement,

$$\mathbf{AB}=(5\,\hat{\mathbf{i}}+4\,\hat{\mathbf{j}}+\hat{\mathbf{k}})-(\hat{\mathbf{i}}+2\,\hat{\mathbf{j}}+3\,\hat{\mathbf{k}})=4\,\hat{\mathbf{i}}+2\,\hat{\mathbf{j}}-2\,\hat{\mathbf{k}}$$

$\therefore$ Work done $=\mathbf{F}\cdot\mathbf{AB}$

$$=(7\hat{\mathbf{i}}+2\hat{\mathbf{j}}-4\hat{\mathbf{k}})\cdot(4\hat{\mathbf{i}}+2\hat{\mathbf{j}}-2\hat{\mathbf{k}})$$

$$=28+4+8=40 \text{ units}$$

229. $(\mathbf{u}+\mathbf{v}-\mathbf{w})\cdot[(\mathbf{u}-\mathbf{v})\times(\mathbf{v}-\mathbf{w})]$

$$=(\mathbf{u}+\mathbf{v}-\mathbf{w})\cdot[\mathbf{u}\times\mathbf{v}-\mathbf{u}\times\mathbf{w}-\mathbf{v}\times\mathbf{v}+\mathbf{v}\times\mathbf{w}]$$

$$=\mathbf{u}\cdot(\mathbf{u}\times\mathbf{v})-\mathbf{u}\cdot(\mathbf{u}\times\mathbf{w})+\mathbf{u}\cdot(\mathbf{v}\times\mathbf{w})+\mathbf{v}\cdot(\mathbf{u}\times\mathbf{v})$$

$$-\mathbf{v}\cdot(\mathbf{u}\times\mathbf{w})+\mathbf{v}\cdot(\mathbf{v}\times\mathbf{w})-\mathbf{w}\cdot(\mathbf{u}\times\mathbf{v})+\mathbf{w}\cdot(\mathbf{u}\times\mathbf{w})-\mathbf{w}\cdot(\mathbf{v}\times\mathbf{w})$$

$$=\mathbf{u}\cdot\mathbf{v}\times\mathbf{w}-\mathbf{v}\cdot\mathbf{u}\times\mathbf{w}-\mathbf{w}\cdot\mathbf{u}\times\mathbf{v} \qquad \{[\mathbf{a},\mathbf{a}\ \mathbf{b}]=0\}$$

$$=\mathbf{u}\cdot\mathbf{v}\times\mathbf{w}+\mathbf{w}\cdot\mathbf{u}\times\mathbf{v}-\mathbf{w}\cdot\mathbf{u}\times\mathbf{v}=\mathbf{u}\cdot\mathbf{v}\times\mathbf{w}$$

230. Given that, $|\mathbf{a}|=1, |\mathbf{b}|=2, |\mathbf{c}|=3$

and $\quad \mathbf{a}+\mathbf{b}+\mathbf{c}=\mathbf{0}$

Now, $\quad (\mathbf{a}+\mathbf{b}+\mathbf{c})^2=|\mathbf{a}|^2+|\mathbf{b}|^2+|\mathbf{c}|^2+2(\mathbf{a}\cdot\mathbf{b}+\mathbf{b}\cdot\mathbf{c}+\mathbf{c}\cdot\mathbf{a})$

$$\Rightarrow \quad 0=1^2+2^2+3^2+2\,(\mathbf{a}\cdot\mathbf{b}+\mathbf{b}\cdot\mathbf{c}+\mathbf{c}\cdot\mathbf{a})$$

$$\Rightarrow \quad 2(\mathbf{a}\cdot\mathbf{b}+\mathbf{b}\cdot\mathbf{c}+\mathbf{c}\cdot\mathbf{a})=-14$$

$$\Rightarrow \quad \mathbf{a}\cdot\mathbf{b}+\mathbf{b}\cdot\mathbf{c}+\mathbf{c}\cdot\mathbf{a}=-7$$

231. Vector perpendicular to face OAB is $\mathbf{n}_1$.

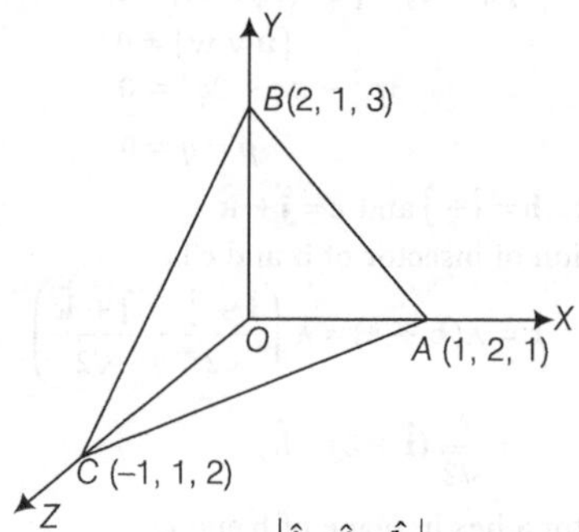

$$=\mathbf{OA}\times\mathbf{OB}=\begin{vmatrix}\hat{\mathbf{i}}&\hat{\mathbf{j}}&\hat{\mathbf{k}}\\1&2&1\\2&1&3\end{vmatrix}=5\hat{\mathbf{i}}-\hat{\mathbf{j}}-3\hat{\mathbf{k}}$$

Vector perpendicular to face ABC is $\mathbf{n}_2$

$$=\mathbf{AB}\times\mathbf{AC}=\begin{vmatrix}\hat{\mathbf{i}}&\hat{\mathbf{j}}&\hat{\mathbf{k}}\\1&-1&2\\-2&-1&1\end{vmatrix}=\hat{\mathbf{i}}-5\hat{\mathbf{j}}-3\hat{\mathbf{k}}$$

Since, angle between faces is equal to the angle between their normals.

$$\therefore \quad \cos\theta=\frac{\mathbf{n}_1\cdot\mathbf{n}_2}{|\mathbf{n}_1||\mathbf{n}_2|}=\frac{5\times1+(-1)\times(-5)+(-3)\times(-3)}{\sqrt{5^2+(-1)^2+(-3)^2}\ \sqrt{1^2+(-5)^2+(-3)^2}}$$

$$=\frac{5+5+9}{\sqrt{35}\ \sqrt{35}}=\frac{19}{35} \Rightarrow \theta=\cos^{-1}\left(\frac{19}{35}\right)$$

232. Given that, $\mathbf{u}=\hat{\mathbf{i}}+\hat{\mathbf{j}},\ \mathbf{v}=\hat{\mathbf{i}}-\hat{\mathbf{j}}, \mathbf{w}=\hat{\mathbf{i}}+2\,\hat{\mathbf{j}}+3\,\hat{\mathbf{k}},$

$\mathbf{u}\cdot\mathbf{n}=0$ and $\mathbf{v}\cdot\mathbf{n}=0$

i.e. $\quad \mathbf{n}=\dfrac{\mathbf{u}\times\mathbf{v}}{|\mathbf{u}\times\mathbf{v}|}$

Now, $\quad \mathbf{u}\times\mathbf{v}=\begin{vmatrix}\hat{\mathbf{i}}&\hat{\mathbf{j}}&\hat{\mathbf{k}}\\1&1&0\\1&-1&0\end{vmatrix}=0\hat{\mathbf{i}}-0\hat{\mathbf{j}}-2\hat{\mathbf{k}}=-2\hat{\mathbf{k}}$

$$\therefore \quad |\mathbf{w}\cdot\mathbf{n}|=\frac{|\mathbf{w}\cdot\mathbf{u}\times\mathbf{v}|}{|\mathbf{u}\times\mathbf{v}|}=\frac{|-6\hat{\mathbf{k}}|}{|-2\hat{\mathbf{k}}|}=3$$

$$[\because\ \mathbf{w}\cdot(\mathbf{u}\times\mathbf{w})=(\hat{\mathbf{i}}+2\hat{\mathbf{j}}+3\hat{\mathbf{k}})\cdot(-2\hat{\mathbf{k}})=-6\hat{\mathbf{k}}]$$

Hence, $\quad |\mathbf{w}\cdot\mathbf{n}|=3$

233. Given two vectors lie in XY-plane. So, a vector coplanar with them is $\quad \mathbf{a}=x\hat{\mathbf{i}}+y\hat{\mathbf{j}}$

Since, $\quad \mathbf{a}\perp(\hat{\mathbf{i}}-\hat{\mathbf{j}})$

$$\Rightarrow \quad (x\hat{\mathbf{i}}+y\hat{\mathbf{j}})\cdot(\hat{\mathbf{i}}-\hat{\mathbf{j}})=0$$

$$\Rightarrow \quad x-y=0$$

$$\Rightarrow \quad x=y$$

$$\therefore \quad \mathbf{a}=x\hat{\mathbf{i}}+x\hat{\mathbf{j}}$$

and $\quad 4|\mathbf{a}|=\sqrt{x^2+x^2}=x\sqrt{2}$

$\therefore$ Required unit vector

$$=\frac{\mathbf{a}}{|\mathbf{a}|}=\frac{x(\hat{\mathbf{i}}+\hat{\mathbf{j}})}{x\sqrt{2}}=\frac{1}{\sqrt{2}}(\hat{\mathbf{i}}+\hat{\mathbf{j}})$$

CHAPTER

03

Three Dimensional Coordinate System

Learning Part

Session 1

Introduction, Position Vector of a Point in Space, Shifting of Origin, Distance Formula, Section Formula,Direction Cosines and Direction Ratios of a Vector, Projection of the Line Segment Joining Two Points on a Given Line

Introduction

Let OY and OZ be two perpendicular lines which intersect at O and let a third straight line OX be perpendicular to the plane in which they lie. The three mutually perpendicular lines form a set of coordinate axis. They determine three mutually perpendicular planes called coordinate planes.

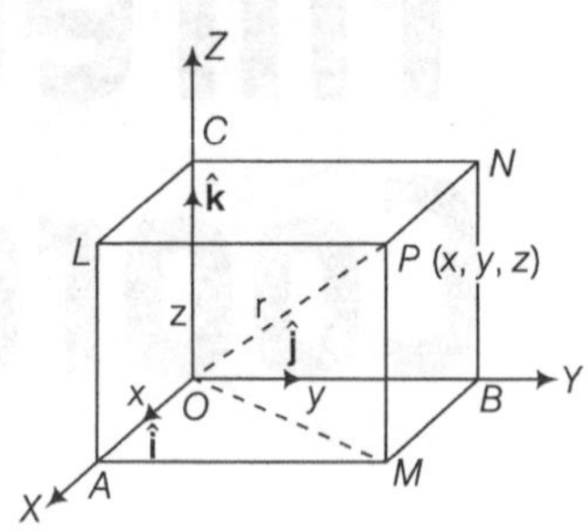

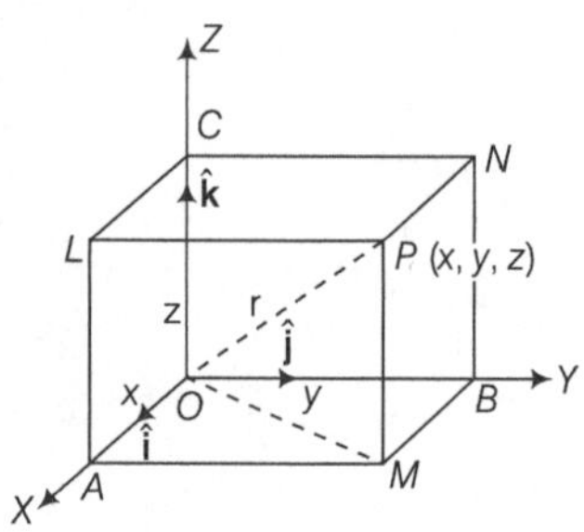

Remarks

1. The axes to coordinate form a right handed set (in the figure) i.e. a right handed screw, driven from O to X would rotate in the sense from OY to OZ.
2. The points A, B and C are the orthogonal projections of P on the X, Y and Z-axes.
3. Points L, M and N are $(x, 0, z)$, $(x, y, 0)$, $(0, y, z)$ and A, B and C are $(x, 0, 0)$, $(0, y, 0)$, $(0, 0, z)$, respectively.

Signs of Coordinates of a Point in Various Octants

Octant/ Coordinates	$OXYZ$	$OX'YZ$	$OXY'Z$	$OXYZ$	$OX'YZ$	$OX'YZ'$	$OXY'Z'$	$OX'Y'Z'$
x	+	−	+	+	−	−	+	−
y	+	+	−	+	−	+	−	−
z	+	+	+	−	+	−	−	−

Note

Any point on X-axis $= (x, 0, 0)$

Y-axis $= (0, y, 0)$

Z-axis $= (0, 0, z)$

XY-plane $= (x, y, 0)$

YZ-plane $= (0, y, z)$

ZX-plane $= (x, 0, z)$

$$\therefore \quad OP = \sqrt{x^2 + y^2 + z^2}$$

Position Vector of a Point in Space

Let $\hat{\mathbf{i}}, \hat{\mathbf{j}}, \hat{\mathbf{k}}$ be unit vector (called base vector) along OX, OY and OZ, respectively.

Let $P(x, y, z)$ be a point in space, let the position of P be $\mathbf{r}$.

Then,
$$\mathbf{r} = \mathbf{OP} = \mathbf{OM} + \mathbf{MP}$$
$$= (\mathbf{OA} + \mathbf{AM}) + \mathbf{MP} = \mathbf{OA} + \mathbf{OB} + \mathbf{OC}$$
$$\mathbf{r} = x\hat{\mathbf{i}} + y\hat{\mathbf{j}} + z\hat{\mathbf{k}}$$

Thus, the position vector of a point P is, $x\hat{\mathbf{i}} + y\hat{\mathbf{j}} + z\hat{\mathbf{k}}$.

Example 1. Planes are drawn parallel to the coordinate planes through the points $(1, 2, 3)$ and $(3, -4, -5)$. Find the lengths of the edges of the parallelopiped so formed.

Sol. Let $P = (1, 2, 3)$, $Q = (3, -4, -5)$ through which planes are drawn parallel to the coordinate planes shown as,

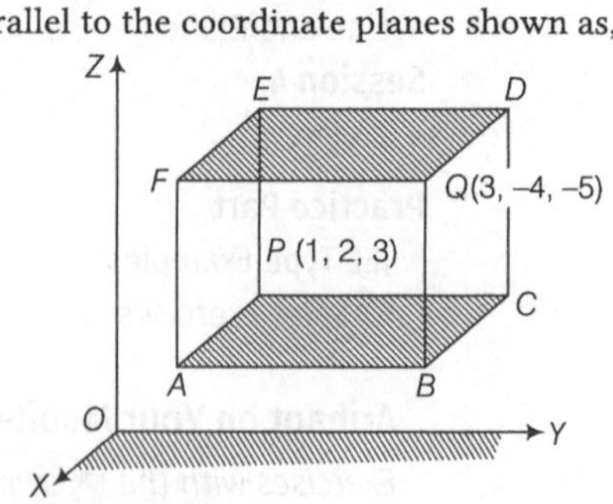

$\therefore PE$ = Distance between parallel planes $ABCP$ and $FQDE$, i.e. (along Z-axis)

$$= |-5-3| = 8$$

PA = Distance between parallel planes $ABQF$ and $PCDE$, i.e. (along X-axis)

$$= |3-1| = 2$$

PC = Distance between parallel planes $BCDQ$ and $APEF$, i.e. (along Y-axis)

$$= |-4-2| = 6$$

$\therefore$ Lengths of edges of the parallelopiped are 2, 6, 8.

Shifting of Origin

Shifting the origin to another point without changing the directions of the axes is called the translation of axes.

Let the origin $O(0, 0, 0)$ be shifted to another point $O'(x', y', z')$ without changing the direction of axes. Let the new coordinate frame be $O'X'Y'Z'$. Let $P(x, y, z)$ be a point with respect to the coordinate frame $OXYZ$. Then, coordinate of point P with respect to new coordinate frame $O'X'Y'Z'$ is (x_1, y_1, z_1), where $x_1 = x - x'$,

$$y_1 = y - y', z_1 = z - z'$$

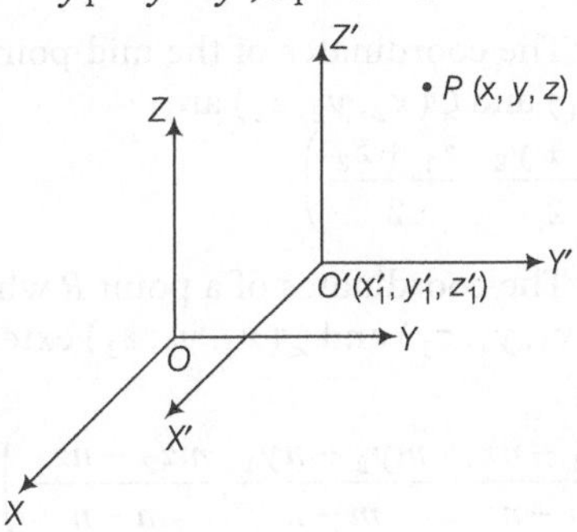

Example 2. If the origin is shifted $(1, 2, -3)$ without changing the directions of the axis, then find the new coordinates of the point $(0, 4, 5)$ with respect to new frame.

Sol. In the new frame $x' = x - x_1, y' = y - y_1, z' = z - z_1$, where (x_1, y_1, z_1) is shifted origin.

$$\Rightarrow \quad x' = 0 - 1 = -1,$$

$$y' = 4 - 2 = 2, z' = 5 + 3 = 8$$

Hence, the coordinates of the point with respect to the new coordinates frame are $(-1, 2, 8)$.

Distance Formula

The distance between the points $P(x_1, y_1, z_1)$ and $Q(x_2, y_2, z_2)$ is given by

$$PQ = \sqrt{(x_2 - x_1)^2 + (y_2 - y_1)^2 + (z_2 - z_1)^2}$$

Proof. Let O be the origin and let $P(x_1, y_1, z_1)$ and $Q(x_2, y_2, z_2)$ be two given points.

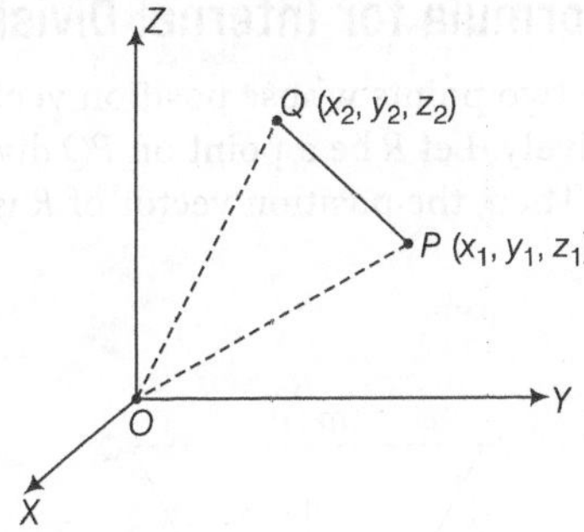

The, $\mathbf{OP} = x_1\hat{\mathbf{i}} + y_1\hat{\mathbf{j}} + z_1\hat{\mathbf{k}}$

$$\mathbf{OQ} = x_2\hat{\mathbf{i}} + y_2\hat{\mathbf{j}} + z_2\hat{\mathbf{k}}$$

Now, $\mathbf{PQ}$ = Position vector of Q − Position vector of P

$$= \mathbf{OQ} - \mathbf{OP}$$

$$= (x_2\hat{\mathbf{i}} + y_2\hat{\mathbf{j}} + z_2\hat{\mathbf{k}}) - (x_1\hat{\mathbf{i}} + y_2\hat{\mathbf{j}} + z_1\hat{\mathbf{k}})$$

$$= (x_2 - x_1)\hat{\mathbf{i}} + (y_2 - y_1)\hat{\mathbf{j}} + (z_2 - z_1)\hat{\mathbf{k}}$$

$$\therefore \quad PQ = |\mathbf{PQ}|$$

$$= \sqrt{(x_2 - x_1)^2 + (y_2 - y_1)^2 + (z_2 - z_1)^2}$$

Hence, $PQ = \sqrt{(x_2 - x_1)^2 + (y_2 - y_1)^2 + (z_2 - z_1)^2}$

Example 3. Find the distance between the points $P(-2, 4, 1)$ and $Q(1, 2, -5)$.

Sol. We have, $PQ = \sqrt{(1+2)^2 + (2-4)^2 + (-5-1)^2}$

$$PQ = \sqrt{3^2 + (-2)^2 + (-6)^2}$$

$$= \sqrt{9 + 4 + 36}$$

$$= \sqrt{49} = 7$$

Example 4. Prove by using distance formula that the points $P(1, 2, 3)$, $Q(-1, -1, -1)$ and $R(3, 5, 7)$ are collinear.

Sol. We have,

$$PQ = \sqrt{(-1-1)^2 + (-1-2)^2 + (-1-3)^2}$$

$$= \sqrt{4 + 9 + 16} = \sqrt{29}$$

$$QR = \sqrt{(3+1)^2 + (5+1)^2 + (7+1)^2}$$

$$= \sqrt{16 + 36 + 64}$$

$$= \sqrt{116} = 2\sqrt{29}$$

and $\quad PR = \sqrt{(3-1)^2 + (5-2)^2 + (7-3)^2}$

$$= \sqrt{4 + 9 + 16} = \sqrt{29}$$

Since, $QR = QP + PR$

Therefore, the given points are collinear.

Section Formula

1. Section Formula for Internal Division

Let P and Q be two points whose position vectors are $\mathbf{r}_1$ and $\mathbf{r}_2$ respectively. Let R be a point on PQ dividing it in the ratio $m:n$. Then, the position vector of R is given by $\mathbf{r}=\dfrac{m\mathbf{r}_2+n\mathbf{r}_1}{m+n}$.

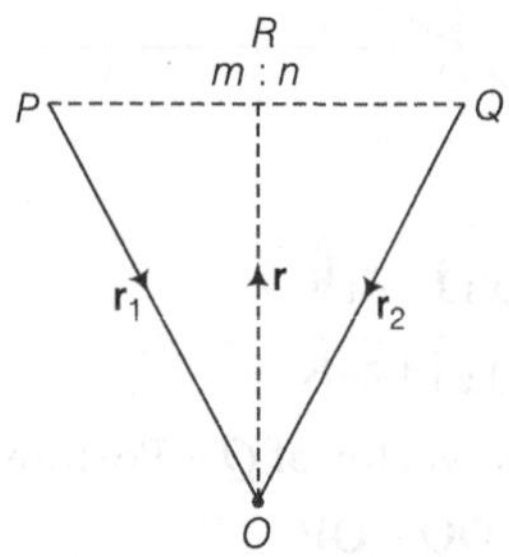

Proof. Let O be the origin. Then, $\mathbf{OP}=\mathbf{r}_1$, $\mathbf{OQ}=\mathbf{r}_2$ and $\mathbf{OR}=\mathbf{r}$

Now, $$\frac{PR}{RQ}=\frac{m}{n}$$

$$\Rightarrow \quad n\mathbf{PR}=m\mathbf{RQ}$$
$$\Rightarrow \quad n(\mathbf{OR}-\mathbf{OP})=m(\mathbf{OQ}-\mathbf{OR})$$
$$\Rightarrow \quad n(\mathbf{r}-\mathbf{r}_1)=m(\mathbf{r}_2-\mathbf{r})$$
$$\Rightarrow \quad (m+n)\mathbf{r}=m\mathbf{r}_2+n\mathbf{r}_1$$
$$\Rightarrow \quad \mathbf{r}=\frac{m\mathbf{r}_2+n\mathbf{r}_1}{m+n}$$

Corollary

Mid-point formula Let P and Q be two points whose position vectors are given by $\mathbf{r}_1$ and $\mathbf{r}_2$ respectively. Then, the position vector of the mid-point R of PQ is given by, $\mathbf{r}=\dfrac{\mathbf{r}_1+\mathbf{r}_2}{2}$.

2. Section Formula for External Division

Let P and Q be the points whose position vectors are $\mathbf{r}_1$ and $\mathbf{r}_2$ respectively. Let R be a point on PQ dividing it externally in the ratio $m:n$. Then, the position vector of R is given by,

$$\mathbf{r}=\frac{m\mathbf{r}_2-n\mathbf{r}_1}{m-n}$$

Proof. Let O be the origin. Then, $\mathbf{OP}=\mathbf{r}_1$, $\mathbf{OQ}=\mathbf{r}_2$ and $\mathbf{OR}=\mathbf{r}$

Now, $$\frac{PR}{QR}=\frac{m}{n}$$

$$\Rightarrow \quad n\mathbf{PR}=m\mathbf{QR}$$
$$\Rightarrow \quad n(\mathbf{OR}-\mathbf{OP})=m(\mathbf{OR}-\mathbf{OQ})$$
$$\Rightarrow \quad n(\mathbf{r}-\mathbf{r}_1)=m(\mathbf{r}-\mathbf{r}_2)$$
$$\Rightarrow \quad (m-n)\mathbf{r}=m\mathbf{r}_2-n\mathbf{r}_1$$
$$\Rightarrow \quad \mathbf{r}=\left(\frac{m\mathbf{r}_2-n\mathbf{r}_1}{m-n}\right)$$

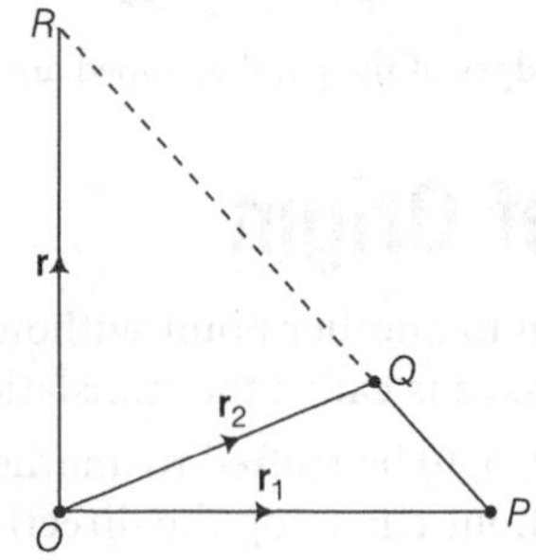

Corollary 1. If $R(x,y,z)$ is a point dividing the join of $P(x_1,y_1,z_1)$ and $Q(x_2,y_2,z_2)$ in the ratio $m:n$.

Then, $x=\dfrac{mx_2+nx_1}{m+n}$, $y=\dfrac{my_2+ny_1}{m+n}$, $z=\dfrac{mz_2+nz_1}{m+n}$

Corollary 2. The coordinates of the mid-point of the joint of $P(x_1,y_1,z_1)$ and $Q(x_2,y_2,z_2)$ are $\left(\dfrac{x_1+x_2}{2},\dfrac{y_1+y_2}{2},\dfrac{z_1+z_2}{2}\right)$

Corollary 3. The coordinates of a point R which divides the join of $P(x_1,y_1,z_1)$ and $Q(x_2,y_2,z_2)$ externally in the ratio $m:n$ are

$$\left(\frac{mx_2-nx_1}{m-n},\frac{my_2-ny_1}{m-n},\frac{mz_2-nz_1}{m-n}\right)$$

Corollary 4. The coordinate of centroid of triangle with vertices $(x_1,y_1,z_1),(x_2,y_2,z_2),(x_3,y_3,z_3)$ is

$$\left(\frac{x_1+x_2+x_3}{3},\frac{y_1+y_2+y_3}{3},\frac{z_1+z_2+z_3}{3}\right)$$

Corollary 5. Centroid of tetrahedron with vertices $(x_1,y_1,z_1),(x_2,y_2,z_2),(x_3,y_3,z_3),(x_4,y_4,z_4)$ is

$$\left(\frac{x_1+x_2+x_3+x_4}{4},\frac{y_1+y_2+y_3+y_4}{4},\frac{z_1+z_2+z_3+z_4}{4}\right)$$

| Example 5. Find the ratio in which $2x+3y+5z=1$ divides the line joining the points $(1, 0, -3)$ and $(1, -5, 7)$.

Sol. Let $2x+3y+5z=1$ divides $(1, 0, -3)$ and $(1, -5, 7)$ in the ratio of $k:1$ at point P.

Then, $P = \left(\frac{k+1}{k+1}, \frac{-5k}{k+1}, \frac{7k-3}{k+1}\right)$ which must satisfy

$2x + 3y + 5z = 1$

$$\Rightarrow 2\left(\frac{k+1}{k+1}\right) + 3\left(\frac{-5k}{k+1}\right) + 5\left(\frac{7k-3}{k+1}\right) = 1$$

$$\Rightarrow \quad 2k + 2 - 15k + 35k - 15 = k + 1$$

$$\Rightarrow \quad 21k = 14 \Rightarrow k = \frac{2}{3}$$

$\therefore$ $2x + 3y + 5z = 1$, divides $(1, 0, -3)$ and $(1, -5, 7)$ in the ratio of $2:3$.

Example 6. If $A(3, 2, -4)$, $B(5, 4, -6)$ and $C(9, 8, -10)$ are three collinear points, then find the ratio in which point C divides AB.

Sol. Let C divide AB in the ratio $\lambda : 1$. Then,

$$C \equiv \left(\frac{5\lambda + 3}{\lambda + 1}, \frac{4\lambda + 2}{\lambda + 1}, \frac{-6\lambda - 4}{\lambda + 1}\right) = (9, 8, -10)$$

Comparing, $5\lambda + 3 = 9\lambda + 9$ or $4\lambda = -6$

$$\therefore \quad \lambda = -\frac{3}{2}$$

Also, from $4\lambda + 2 = 8\lambda + 8$ and $-6\lambda - 4 = -10\lambda - 10$, we get the same value of λ.

$\therefore$ C divides AB in the ratio $3:2$ externally.

Example 7. Show that the plane $ax + by + cz + d = 0$ divides the line joining (x_1, y_1, z_1) and (x_2, y_2, z_2) in the ratio of $\left(-\frac{ax_1 + ay_1 + cz_1 + d}{ax_2 + by_2 + cz_2 + d}\right)$.

Sol. Let the plane $ax + by + cz + d = 0$ divides the line joining (x_1, y_1, z_1) and (x_2, y_2, z_2) in the ratio $k : 1$ as shown in figure.

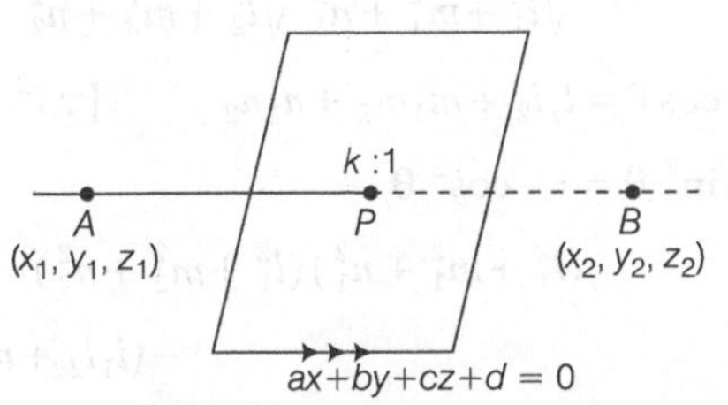

$\therefore$ Coordinates of $P\left(\frac{kx_2 + x_1}{k+1}, \frac{ky_2 + y_1}{k+1}, \frac{kz_2 + z_1}{k+1}\right)$

must satisfy $ax + by + cz + d = 0$

i.e., $a\left(\frac{kx_2 + x_1}{k+1}\right) + b\left(\frac{ky_2 + y_1}{k+1}\right) + c\left(\frac{kz_2 + z_1}{k+1}\right) + d = 0$

$$\Rightarrow a(kx_2 + x_1) + b(ky_2 + y_1) + c(kz_2 + z_1) + d(k+1) = 0$$

$$\Rightarrow \quad k(ax_2 + by_2 + cz_2 + d) + (ax_1 + by_1 + cz_1 + d) = 0$$

$$\Rightarrow \quad k = -\frac{(ax_1 + by_1 + cz_1 + d)}{(ax_2 + by_2 + cz_2 + d)}$$

Remark

Students are advised to learn above result as a formula i.e. $ax + by + cz + d = 0$ divides join of (x_1, y_1, z_1) and (x_2, y_2, z_2) in ratio of $-\frac{(ax_1 + by_1 + cz_1 + d)}{(ax_2 + by_2 + cz_2 + d)}$.

Example 8. Find the ratio in which the join of $(2, 1, 5)$, $(3, 4, 3)$ is divided by the plane $2x + 2y - 2z - 1 = 0$.

Sol. Using above result,

$$\text{Required ratio} = \frac{\{(2(2) + 2(1) - 2(5) - 1\}}{\{2(3) + 2(4) - 2(3) - 1\}}$$

$$= \frac{\{6 - 11\}}{\{14 - 7\}} = \frac{-5}{7}$$

$\Rightarrow 2x + 2y - 2z - 1 = 0$ divides $(2, 1, 5)$ and $(3, 4, 5)$ externally in ratio of $5:7$.

Direction Cosines and Direction Ratios of a Vector

1. Direction Cosines (DC's)

If α, β and γ are the angles which a vector **OP** makes with the positive directions of the coordinate axes OX, OY and OZ respectively. Then $\cos\alpha$, $\cos\beta$ and $\cos\gamma$ are known as direction cosines of **OP** and are generally denoted by letters l, m and n, respectively.

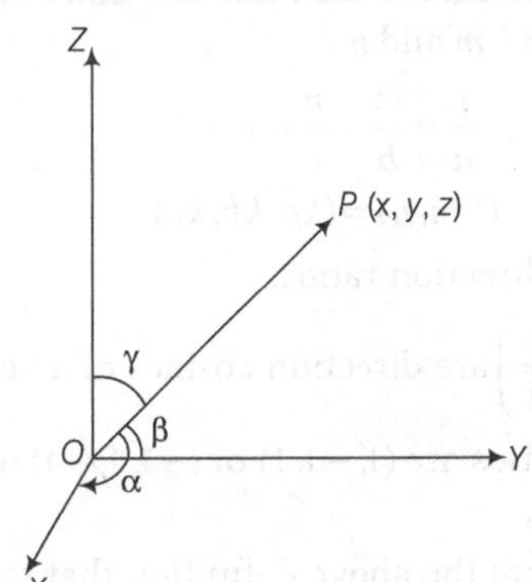

Thus, $l = \cos\alpha$; $m = \cos\beta$; $n = \cos\gamma$. The angles α, β and γ are known as direction angles and they satisfy the condition $0 \le \alpha, \beta, \gamma \le \pi$.

It can be seen form the figure

$$\cos\alpha = \frac{x}{OP}$$

Similarly, $$\cos\beta = \frac{y}{OP}$$

$$\cos\gamma = \frac{z}{OP}$$

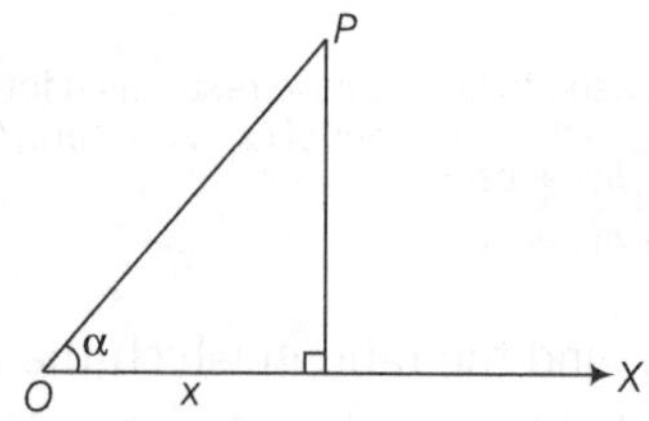

Where, OP is the modulus of positive vector of P.

Clearly, $OP = \sqrt{x^2 + y^2 + z^2}$

So, $l^2 + m^2 + n^2 = \cos^2 \alpha + \cos^2 \beta + \cos^2 \gamma$

$$= \frac{x^2 + y^2 + z^2}{OP^2} = \frac{x^2 + y^2 + z^2}{x^2 + y^2 + z^2} = 1$$

$\therefore \quad l^2 + m^2 + n^2 = \cos^2 \alpha + \cos^2 \beta + \cos^2 \gamma = 1$

$\therefore$ If $\quad \mathbf{OP} = \mathbf{r} = x\hat{\mathbf{i}} + y\hat{\mathbf{j}} + z\hat{\mathbf{k}}$

Then, $\quad \hat{\mathbf{r}} = l\,\hat{\mathbf{i}} + m\,\hat{\mathbf{j}} + n\,\hat{\mathbf{k}}$

By definition it follows that the direction cosine of the axis x are $\cos 0°, \cos 90°, \cos 90°$, i.e. $(1, 0, 0)$.

Similarly, direction cosine of the axes Y and Z are $(0, 1, 0)$ and $(0, 0, 1)$, respectively.

2. Direction Ratios (DR's)

Let l, m and n be the direction cosines of a vector $\mathbf{r}$ and a, b and c be three numbers such that a, b and c are proportional to l, m and n

i.e. $$\frac{l}{a} = \frac{m}{b} = \frac{n}{c} = k$$

or $$(l, m, n) = (ka, kb, kc)$$

$\Rightarrow (a, b, c)$ are direction ratios.

If $\left(\frac{1}{\sqrt{3}}, \frac{-1}{\sqrt{3}}, \frac{1}{\sqrt{3}}\right)$ are direction cosines of a vector $\mathbf{r}$, then its direction ratios are $(1, -1, 1)$ or $(-1, 1, -1)$ or $(2, -2, 2)$ or $(\lambda, -\lambda, \lambda)$... etc.

It is evident from the above definition that to obtain direction ratios of a vector from its direction cosines, we just multiply them by a common number.

"That shows there can be infinitely many direction ratios for a given vector but the direction cosines are unique".

To obtain direction cosines from direction ratios.

Let a, b and c be direction ratios of a vector $\mathbf{r}$ having direction cosines l, m and n. Then,

$$l = \lambda a, m = \lambda b, n = \lambda c \qquad \text{(by definition)}$$

$$\therefore \quad l^2 + m^2 + n^2 = 1$$

$$\Rightarrow \quad a^2\lambda^2 + b^2\lambda^2 + c^2\lambda^2 = 1$$

$$\Rightarrow \quad \lambda = \pm \frac{1}{\sqrt{a^2 + b^2 + c^2}}$$

So, $$l = \pm \frac{a}{\sqrt{a^2 + b^2 + c^2}},$$

$$m = \pm \frac{b}{\sqrt{a^2 + b^2 + c^2}},$$

$$n = \pm \frac{c}{\sqrt{a^2 + b^2 + c^2}}$$

For example, let the direction ratios of a point be $(3, 1, -2)$.

$\Rightarrow$ Direction cosines are

$$\left(\frac{3}{\sqrt{3^2 + 1^2 + (-2)^2}}, \frac{1}{\sqrt{3^2 + 1^2 + (-2)^2}}, \frac{-2}{\sqrt{3^2 + 1^2 + (-2)^2}}\right)$$

$$\Rightarrow \left(\frac{3}{\sqrt{14}}, \frac{1}{\sqrt{14}}, -\frac{2}{\sqrt{14}}\right)$$

3. Angle between Two Vectors in Terms of Direction Cosines and Direction Ratios

Let $\mathbf{a}$ and $\mathbf{b}$ be two given vectors with direction cosines l_1, m_1, n_1 and l_2, m_2, n_2 respectively. Then,

$\mathbf{a} = l_1\hat{\mathbf{i}} + m_1\hat{\mathbf{j}} + n_1\hat{\mathbf{k}}$ and $\mathbf{b} = l_2\hat{\mathbf{i}} + m_2\hat{\mathbf{j}} + n_2\hat{\mathbf{k}}$

$\therefore \cos\theta = \dfrac{\mathbf{a}\cdot\mathbf{b}}{|\mathbf{a}|\cdot|\mathbf{b}|}$, where θ be the angle between $\mathbf{a}$ and $\mathbf{b}$.

$$\Rightarrow \quad \cos\theta = \frac{|l_1\hat{\mathbf{i}} + m_1\hat{\mathbf{j}} + n_1\hat{\mathbf{k}}|\cdot|l_2\hat{\mathbf{i}} + m_2\hat{\mathbf{j}} + n_2\hat{\mathbf{k}}|}{|l_1\hat{\mathbf{i}} + m_1\hat{\mathbf{j}} + n_1\hat{\mathbf{k}}||l_2\hat{\mathbf{i}} + m_2\hat{\mathbf{j}} + n_2\hat{\mathbf{k}}|}$$

$$\Rightarrow \quad \cos\theta = \frac{l_1 l_2 + m_1 m_2 + n_1 n_2}{\sqrt{l_1^2 + m_1^2 + n_1^2}\sqrt{l_2^2 + m_2^2 + n_2^2}}$$

$$\Rightarrow \quad \cos\theta = l_1 l_2 + m_1 m_2 + n_1 n_2 \qquad [\because l^2 + m^2 + n^2 = 1]$$

Also, $$\sin^2\theta = 1 - \cos^2\theta$$

$$= (l_1^2 + m_1^2 + n_1^2)(l_2^2 + m_2^2 + n_2^2) - (l_1 l_2 + m_1 m_2 + n_1 n_2)^2$$

$$\Rightarrow \sin^2\theta = (m_1 n_2 - m_2 n_1)^2 + (n_1 l_2 - n_2 l_1)^2 + (l_1 m_2 - l_2 m_1)^2$$

$$\Rightarrow \quad \sin\theta = \pm\sqrt{(m_1 n_2 - m_2 n_1)^2 + (n_1 l_2 - n_2 l_1)^2 + (l_1 m_2 - l_2 m_1)^2}$$

Remarks

1. Acute angle θ between the two lines having direction cosines l_1, m_1, n_1 and l_2, m_2, n_2 is given by

$$\cos\theta = |l_1 l_2 + m_1 m_2 + n_1 n_2|$$

$$\sin\theta = \sqrt{(l_1 m_2 - l_2 m_1)^2 + (m_1 n_2 - m_2 n_1)^2 + (n_1 l_2 - n_2 l_1)^2}$$

2. If a_1, b_1, c_1 and a_2, b_2, c_2 are the direction ratios of two lines, then the acute angle θ between them is given by

$$\cos\theta = \frac{|a_1a_2 + b_1b_2 + c_1c_2|}{\sqrt{a_1^2 + b_1^2 + c_1^2}\sqrt{a_2^2 + b_2^2 + c_2^2}}$$

$$\sin\theta = \frac{\sqrt{(a_1b_2 - a_2b_1)^2 + (b_1c_2 - b_2c_1)^2 + (c_1a_2 - c_2a_1)^2}}{\sqrt{a_1^2 + b_1^2 + c_1^2}\sqrt{a_2^2 + b_2^2 + c_2^2}}$$

3. The two lines with direction cosines l_1, m_1, n_1 and l_2, m_2, n_2 are perpendicular to each other if $\theta = \frac{\pi}{2}$

$$\Rightarrow \quad \cos\theta = l_1l_2 + m_1m_2 + n_1n_2 = 0$$

4. The two lines with direction cosines l_1, m_1, n_1 and l_2, m_2, n_2 are parallel to each other if $\theta = 0$

or $\quad \pi \Rightarrow \sin\theta = 0$

$$\Rightarrow (m_1n_2 - m_2n_1)^2 + (n_1l_2 - n_2l_1)^2 + (l_1m_2 - l_2m_1)^2 = 0$$

$$\Rightarrow \quad \frac{l_1}{l_2} = \frac{m_1}{m_2} = \frac{n_1}{n_2}$$

5. The angle between two lines having direction ratios a_1, b_1, c_1 and a_2, b_2, c_2 is given by

$$\cos\theta = \frac{a_1a_2 + b_1b_2 + c_1c_2}{\sqrt{(a_1^2 + b_1^2 + c_1^2}\sqrt{a_2^2 + b_2^2 + c_2^2)}}$$

Thus, the two straight lines are perpendicular, if

$$a_1a_2 + b_1b_2 + c_1c_2 = 0$$

The two straight lines are parallel if $\frac{a_1}{a_2} = \frac{b_1}{b_2} = \frac{c_1}{c_2}$

Projection of the Line Segment Joining Two Points on a Given Line

The projection of the line segment joining two given points (x_1, y_1, z_1) and (x_2, y_2, z_2) on the line having direction cosines l, m, n is given by $l(x_2 - x_1) + m(y_2 - y_1) + n(z_2 - z_1)$, which is clear from the vector.

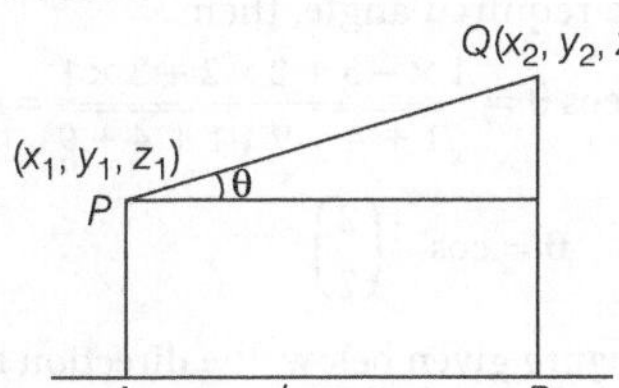

Clearly, $\quad \mathbf{PQ} = (x_2 - x_1)\hat{\mathbf{i}} + (y_2 - y_1)\hat{\mathbf{j}} + (z_2 - z_1)\hat{\mathbf{k}}$

and the line $\mathbf{AB} = l\hat{\mathbf{i}} + m\hat{\mathbf{j}} + n\hat{\mathbf{k}}$

The projection of **PQ** on **AB**

$$= \frac{\mathbf{PQ} \cdot \mathbf{AB}}{|\mathbf{AB}|} = \frac{l(x_2 - x_1) + m(y_2 - y_1) + n(z_2 - z_1)}{\sqrt{l^2 + m^2 + n^2}}$$

$$= l(x_2 - x_1) + m(y_2 - y_1) + n(z_2 - z_1)$$

Example 9. What are the direction cosines of a line which is equally inclined to the coordinate axes?

Sol. If α, β and γ are the angles that a line makes with the coordinate axes, then if they are equally inclined.

$$\Rightarrow \quad \alpha = \beta = \gamma$$

Also, $\quad l^2 + m^2 + n^2 = 1$

$$\Rightarrow \quad \cos^2\alpha + \cos^2\beta + \cos^2\gamma = 1$$

$$\Rightarrow \quad \cos^2\alpha + \cos^2\alpha + \cos^2\alpha = 1$$

$$\Rightarrow \quad 3\cos^2\alpha = 1$$

$$\Rightarrow \quad \cos\alpha = \pm\frac{1}{\sqrt{3}} = \cos\beta = \cos\gamma$$

$\therefore$ Direction cosines are $\left(\frac{1}{\sqrt{3}}, \frac{1}{\sqrt{3}}, \frac{1}{\sqrt{3}}\right)$

or $\quad \left(-\frac{1}{\sqrt{3}}, -\frac{1}{\sqrt{3}}, -\frac{1}{\sqrt{3}}\right)$.

Example 10. If a line makes angles α, β and γ with the coordinates axes, prove that $\sin^2\alpha + \sin^2\beta + \sin^2\gamma = 2$.

Sol. Let l, m and n be the direction cosines of the given vector.

Then, $l = \cos\alpha, m = \cos\beta, n = \cos\gamma$

Now, $\quad l^2 + m^2 + n^2 = 1$

$$\Rightarrow \quad \cos^2\alpha + \cos^2\beta + \cos^2\gamma = 1$$

$$\Rightarrow \quad 1 - \sin^2\alpha + 1 - \sin^2\beta + 1 - \sin^2\gamma = 1$$

$$\Rightarrow \quad \sin^2\alpha + \sin^2\beta + \sin^2\gamma = 2$$

Example 11. A line OP through origin O is inclined at $30°$ and $45°$ to OX and OY, respectively. Find the angle at which it is inclined to OZ.

Sol. Let l, m and n be the direction cosines of the given vector.

$$l^2 + m^2 + n^2 = 1$$

where, $\quad \alpha = 30°, \beta = 45°$

$$\therefore \quad \cos^2\alpha + \cos^2\beta + \cos^2\gamma = 1$$

$$\Rightarrow \quad \cos^2 30° + \cos^2 45° + \cos^2\gamma = 1$$

$$\Rightarrow \quad \left(\frac{\sqrt{3}}{2}\right)^2 + \left(\frac{1}{\sqrt{2}}\right)^2 + \cos^2\gamma = 1$$

$$\Rightarrow \quad \cos^2\gamma = 1 - \frac{3}{4} - \frac{1}{2}$$

$$\Rightarrow \quad \cos^2\gamma = \frac{4 - 3 - 2}{4}$$

$$\Rightarrow \quad \cos^2\gamma = -\frac{1}{4} \text{ which is not possible.}$$

$\therefore$ There exists no point which is inclined to $30°$ to X-axis and $45°$ to Y-axis.

Example 12. Find the direction cosines of a vector **r** which is equally inclined to *OX*, *OY* and *OZ*. If $|\mathbf{r}|$ is given, find the total number of such vectors.

Sol. Let l, m and n be the direction cosines of **r**.

Since, **r** is equally inclined with X, Y and Z-axes.

$$\therefore \quad l^2 + m^2 + n^2 = 1$$

$$\Rightarrow \quad 3l^2 = 1 \quad (\because l = m = n)$$

$$\Rightarrow \quad l = \pm \frac{1}{\sqrt{3}}$$

$\therefore$ Direction cosines of **r** are $\pm \frac{1}{\sqrt{3}}, \pm \frac{1}{\sqrt{3}}, \pm \frac{1}{\sqrt{3}}$.

Now, $\mathbf{r} = |\mathbf{r}|(l\hat{\mathbf{i}} + m\hat{\mathbf{j}} + n\hat{\mathbf{k}})$

$$\Rightarrow \quad \mathbf{r} = |\mathbf{r}|\left(\pm \frac{1}{\sqrt{3}}\hat{\mathbf{i}} \pm \frac{1}{\sqrt{3}}\hat{\mathbf{j}} \pm \frac{1}{\sqrt{3}}\hat{\mathbf{k}}\right)$$

Since, '+' and '–' signs can be arranged at three planes.

There are eight vectors (i.e. $2 \times 2 \times 2$) which are equally inclined to axes.

Example 13. If the points $(0, 1, -2)$, $(3, \lambda, -1)$ and $(\mu, -3, -4)$ are collinear, verify whether the point $(12, 9, 2)$ is also on the same line.

Sol. Let the points be A, B and C, whose coordinates are $(0, 1, -2)$, $(3, \lambda, -1)$ and $(\mu, -3, -4)$ respectively.

Let $D = (12, 9, 2)$

$\Rightarrow$ DR's of $AB = (3 - 0, \lambda - 1, -1 + 2)$

$= (3, \lambda - 1, 1)$

DR's of $AC = (\mu - 0, -3 - 1, -4 - (-2))$

$= (\mu, -4, -2)$

Since, A, B and C are collinear.

$$\Rightarrow \quad \frac{3}{\mu} = \frac{\lambda - 1}{-4} = \frac{1}{-2}$$

$$\Rightarrow \quad \mu = -6, \lambda = 3$$

$\therefore$ Direction ratios of AB are $(3, 2, 1)$.

Now, direction ratios of AD are $(12 - 0, 9 - 1, 2 - (2))$ or $(12, 8, 4)$

Here, $\frac{3}{12} = \frac{2}{8} = \frac{1}{4}$

$\therefore \quad AB \parallel AD$

Since, AB and AD lie on same straight line.

Hence, the point $(12, 9, 2)$ is on the same line.

Example 14. A vector **r** has length 21 and direction ratios $2, -3, 6$. Find the direction cosines and components of **r**, given that **r** makes an obtuse angle with X-axis.

Sol. Here, direction ratio's are $2, -3, 6$.

$\therefore$ Direction cosines can be written as $(2\lambda, -3\lambda, 6\lambda)$.

where, $(2\lambda)^2 + (-3\lambda)^2 + (6\lambda)^2 = 1 \quad (\because l^2 + m^2 + n^2 = 1)$

as $49\lambda^2 = 1$

$$\Rightarrow \quad \lambda = \mp \frac{1}{7}$$

$\therefore$ Direction cosines are $\left(\pm \frac{2}{7}, \mp \frac{3}{7}, \pm \frac{6}{7}\right)$.

But it makes obtuse angle with X-axis $\Rightarrow l < 0$.

$\therefore$ Direction cosines are $\left(-\frac{2}{7}, \frac{3}{7}, -\frac{6}{7}\right)$

Also, $\mathbf{r} = |\mathbf{r}|(l\hat{\mathbf{i}} + m\hat{\mathbf{j}} + n\hat{\mathbf{k}})$

$$\Rightarrow \quad \mathbf{r} = 21\left(-\frac{2}{7}\hat{\mathbf{i}} + \frac{3}{7}\hat{\mathbf{j}} - \frac{6}{7}\hat{\mathbf{k}}\right) \quad (\text{given, } |\mathbf{r}| = 21)$$

$$\mathbf{r} = 3(-2\hat{\mathbf{i}} + 3\hat{\mathbf{j}} - 6\hat{\mathbf{k}})$$

So, the component of **r** along X, Y and Z-axes are $-6\hat{\mathbf{i}}$, $9\hat{\mathbf{j}}$ and $-18\hat{\mathbf{k}}$, respectively.

Example 15. Find the angle between the lines whose direction cosines are $\left(-\frac{\sqrt{3}}{4}, \frac{1}{4}, -\frac{\sqrt{3}}{2}\right)$ and $\left(-\frac{\sqrt{3}}{4}, \frac{1}{4}, \frac{\sqrt{3}}{2}\right)$.

Sol. Let θ be the required angle, then

$$\cos\theta = l_1l_2 + m_1m_2 + n_1n_2$$

$$= \frac{3}{16} + \frac{1}{16} - \frac{3}{4} = -\frac{1}{2}$$

$$\therefore \quad \cos\theta = -\frac{1}{2} \Rightarrow \theta = 120°$$

Example 16.

(i) Find the angle between the lines whose direction ratios are $1, 2, 3$ and $-3, 2, 1$.

(ii) Find the acute angle between two diagonal of a cube.

Sol. (i) Let θ be the required angle, then

$$\cos\theta = \frac{1 \times -3 + 2 \times 2 + 3 \times 1}{\sqrt{1 + 4 + 9}\sqrt{1 + 4 + 9}} = \frac{4}{14} = \frac{2}{7}$$

$$\Rightarrow \quad \theta = \cos^{-1}\left(\frac{2}{7}\right)$$

(ii) From the figure given below, the direction ratios of the diagonals OP and CD of a given cube are given by

$$a - 0, a - 0, a - 0$$

and

$$a - 0, a - 0, 0 - a$$

and hence their respective direction cosines are

$$\frac{a}{\sqrt{a^2 + a^2 + a^2}}, \frac{a}{\sqrt{a^2 + a^2 + a^2}}, \frac{-a}{\sqrt{a^2 + a^2 + a^2}}$$

i.e. $\frac{1}{\sqrt{3}}, \frac{1}{\sqrt{3}}, \frac{-1}{\sqrt{3}}$

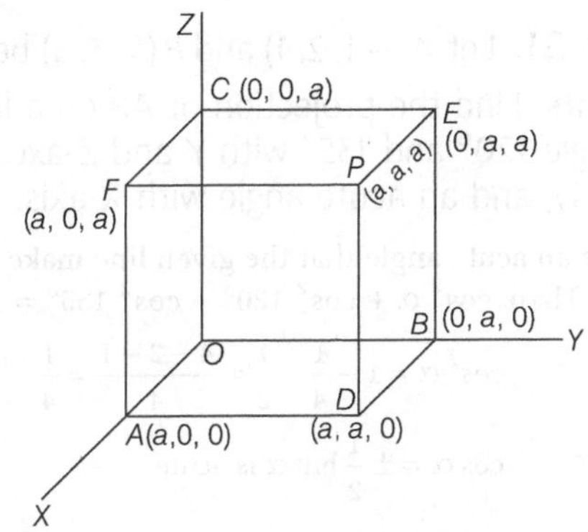

and $\dfrac{a}{\sqrt{a^2+a^2+a^2}}, \dfrac{a}{\sqrt{a^2+a^2+a^2}}, \dfrac{-a}{\sqrt{a^2+a^2+a^2}}$

i.e. $\dfrac{1}{\sqrt{3}}, \dfrac{1}{\sqrt{3}}, \dfrac{-1}{\sqrt{3}}$

Let θ be the angle between these diagonals, then

$$\cos\theta = \frac{1}{\sqrt{3}} \times \frac{1}{\sqrt{3}} + \frac{1}{\sqrt{3}} \times \frac{1}{\sqrt{3}} + \frac{1}{\sqrt{3}} \times \frac{-1}{\sqrt{3}} = \frac{1}{3} + \frac{1}{3} - \frac{1}{3} = \frac{1}{3}$$

$$\Rightarrow \theta = \cos^{-1}\left(\frac{1}{3}\right)$$

Example 17. Find the angle between the lines whose direction cosines are given by $l + m + n = 0$ and $2l^2 + 2m^2 - n^2 = 0$.

Sol. $l^2 + m^2 + n^2 = 1$

$$l + m + n = 0 \quad \text{...(i)}$$

$$2l^2 + 2m^2 - n^2 = 0 \quad \text{...(ii)}$$

$$2(l^2 + m^2) - n^2 = 0$$

$$2(1 - n^2) = n^2 \Rightarrow 3n^2 = 2 \Rightarrow n = \pm\sqrt{\frac{2}{3}} \quad \text{...(iii)}$$

$$2(l^2 + m^2) = n^2 = (-(l + m))^2 \quad \text{...(iv)}$$

$$\Rightarrow 2l^2 + 2m^2 = l^2 + m^2 + 2lm$$

$$\Rightarrow l^2 + m^2 - 2lm = 0$$

$$\Rightarrow (l - m)^2 = 0 \Rightarrow l = m$$

$$\Rightarrow l + m = \mp\sqrt{\frac{2}{3}}$$

$$\Rightarrow 2l = \mp\sqrt{\frac{2}{3}}$$

$$\therefore l = \pm\frac{1}{\sqrt{6}}, m = \pm\frac{1}{\sqrt{6}}$$

Direction cosines are $\left(\dfrac{1}{\sqrt{6}}, \dfrac{1}{\sqrt{6}}, \sqrt{\dfrac{2}{3}}\right)$ and $\left(\dfrac{1}{\sqrt{6}}, \dfrac{1}{\sqrt{6}}, -\sqrt{\dfrac{2}{3}}\right)$

or $\left(-\dfrac{1}{\sqrt{6}}, -\dfrac{1}{\sqrt{6}}, \sqrt{\dfrac{2}{3}}\right)$ and $\left(-\dfrac{1}{\sqrt{6}}, -\dfrac{1}{\sqrt{6}}, -\sqrt{\dfrac{2}{3}}\right)$

The angle between these lines in both the cases is $\cos^{-1}\left(-\dfrac{1}{3}\right)$.

Example 18. If the direction cosines of a variable line in two adjacent points be l, m, n and $l + \delta l, m + \delta m, n + \delta n$, show that the small angle $\delta\theta$ between the two positions, is given by $\delta\theta^2 = \delta l^2 + \delta m^2 + \delta n^2$.

Sol. We have, $l^2 + m^2 + n^2 = 1$

and $(l + \delta l)^2 + (m + \delta m)^2 + (n + \delta n)^2 = 1$

$$\Rightarrow l^2 + m^2 + n^2 + (\delta l)^2 + (\delta m)^2 + (\delta n)^2 + 2(l\delta l + m\delta m + n\delta n) = 1$$

$$\Rightarrow 1 + (\delta l)^2 + (\delta m)^2 + (\delta n)^2 + 2(l\delta l + m\delta m + n\delta n) = 1$$

$$\Rightarrow (\delta l)^2 + (\delta m)^2 + (\delta n)^2 = -2(l\delta l + m\delta m + n\delta n) \quad \text{...(i)}$$

Let δθ be angle between the two positions.

$$\cos\delta\theta = l(l + \delta l) + m(m + \delta m) + n(n + \delta n)$$

$$\Rightarrow 1 - 2\sin^2\frac{\delta\theta}{2} = 1 + l\delta l + m\delta m + n\delta n \quad \text{...(ii)}$$

From Eqs. (i) and (ii), we get

$$(\delta l)^2 + (\delta m)^2 + (\delta n)^2 = 4\sin^2\frac{\delta\theta}{2}$$

$$\Rightarrow 4\left(\frac{\delta\theta}{2}\right)^2 = l\delta l + m\delta m + n\delta n$$

$$\Rightarrow l\delta l + m\delta m + n\delta n = (\delta\theta)^2,$$

$$\left(\text{since, } \sin\frac{\delta\theta}{2} \to \frac{\delta\theta}{2} \text{ as } \delta\theta \text{ is very small}\right)$$

Example 19. If l_1, m_1, n_1 and l_2, m_2, n_2 are the direction cosines of two mutually perpendicular lines, shows that the direction cosines of the line perpendicular to both of them are $m_1n_2 - m_2n_1$; $n_1l_2 - n_2l_1$; $l_1m_2 - l_2m_1$.

Sol. Let l, m and n be the direction cosines of the line perpendicular to both the given lines.

$$\therefore ll_1 + mm_1 + nn_1 = 0 \text{ and } ll_2 + mm_2 + nn_2 = 0$$

Solving them, we get $\dfrac{l}{\begin{vmatrix} m_1 & n_1 \\ m_2 & n_2 \end{vmatrix}} = \dfrac{m}{\begin{vmatrix} n_1 & l_1 \\ n_2 & l_2 \end{vmatrix}} = \dfrac{n}{\begin{vmatrix} l_1 & m_1 \\ l_2 & m_2 \end{vmatrix}}$

$$\Rightarrow \frac{l}{m_1n_2 - m_2n_1} = \frac{m}{n_1l_2 - n_2l_1} = \frac{n}{l_1m_2 - l_2m_1} = k$$

$$\therefore l = k(m_1n_2 - m_2n_1), m = k(n_1l_2 - n_2l_1), n = k(l_1m_2 - l_2m_1)$$

On squaring and adding, we get

$$l^2 + m^2 + n^2 = k^2\{(m_1n_2 - m_2n_1)^2 + (n_1l_2 - n_2l_1)^2\} + (l_1m_2 - l_2m_1)^2$$

$$\Rightarrow 1 = k^2\{\sin^2\theta\}$$

where, θ is the angle between the given lines as we know,

$$\sin\theta = \sqrt{(a_1b_2 - a_2b_1)^2 + (b_1c_2 - b_2c_1)^2 + (c_1a_2 - c_2a_1)^2}$$

where, a_1, b_1, c_1 and a_2, b_2, c_2 are direction cosines.

$$\Rightarrow 1 = k^2 \cdot 1 \quad (\because \theta = 90°, \text{ given})$$

$$\Rightarrow k = 1$$

Hence, direction cosines of a line perpendicular to both of them are $m_1n_2 - m_2n_1, n_1l_2 - n_1l_2, l_1m_2 - l_2m_1$.

Example 20. Find the direction cosines of the line which is perpendicular to the lines with direction cosines proportional to $(1, -2, -2)$ and $(0, 2, 1)$.

Sol. If l, m and n are the direction cosines of the line perpendicular to the given line, then

$$l \cdot (1) + m \cdot (-2) + n \cdot (-2) = 0$$

$$\Rightarrow \quad l - 2m - 2n = 0 \qquad \text{...(i)}$$

$$\text{and} \quad l \cdot 0 + m \cdot 2 + n \cdot 1 = 0$$

$$0 + 2m + n = 0 \qquad \text{...(ii)}$$

Then, from Eqs. (i) and (ii) by cross multiplication, we get

$$\frac{l}{2} = \frac{m}{-1} = \frac{n}{2}$$

$$\Rightarrow \quad \frac{l}{2} = \frac{m}{-1} = \frac{n}{2}$$

$$= \frac{\sqrt{l^2 + m^2 + n^2}}{\sqrt{4 + 1 + 4}} = \frac{1}{3} \qquad (\because l^2 + m^2 + n^2 = 1)$$

$$\Rightarrow \quad l = \frac{2}{3}, m = -\frac{1}{3},$$

$$n = \frac{2}{3}$$

Example 21. Let $A(-1, 2, 1)$ and $B(4, 3, 5)$ be two given points. Find the projection of AB on a line which makes angle $120°$ and $135°$ with Y and Z-axes respectively, and an acute angle with X-axis.

Sol. Let α be an acute angle that the given line make with X-axis. Then, $\cos^2 \alpha + \cos^2 120° + \cos^2 135° = 1$

$$\Rightarrow \quad \cos^2 \alpha = 1 - \frac{1}{4} - \frac{1}{2} = \frac{4 - 2 - 1}{4} = \frac{1}{4}$$

$$\Rightarrow \quad \cos \alpha = \pm \frac{1}{2} \text{ but } \alpha \text{ is acute}$$

$$\therefore \quad \cos \alpha = +\text{ve}$$

$$\Rightarrow \quad \cos \alpha = \frac{1}{2} = \cos 60° \Rightarrow \alpha = 60°$$

Thus, the direction cosines of the given straight line are $\cos 60°, \cos 120°, \cos 135°$, i.e. $\frac{1}{2}, -\frac{1}{2}, -\frac{1}{\sqrt{2}}$

Hence the projection of AB on the line

$$= \frac{1}{2}(4 + 1) - \frac{1}{2}(3 - 2) - \frac{1}{\sqrt{2}}(5 - 1) = \frac{5}{2} - \frac{1}{2} - 2\sqrt{2}$$

$$= (2 - 2\sqrt{2}) \text{ units}$$

Exercise for Session 1

1. In how many disjoint parts does the three dimensional rectangular cartesian coordinate system divide the space.
2. Find the distance between the points $(k, k + 1, k + 2)$ and $(0, 1, 2)$.
3. Show that the points $(1, 2, 3)$, $(-1, -2, -1)$, $(2, 3, 2)$ and $(4, 7, 6)$ are the vertices of a parallelogram.
4. If the mid-points of the sides of a triangle are $(1, 5, -1)$, $(0, 4, -2)$ and $(2, 3, 4)$. Find its vertices.
5. Find the maximum distance between the points $(3 \sin \theta, 0, 0)$ and $(4 \cos \theta, 0, 0)$.
6. If $A = (1, 2, 3)$, $B = (4, 5, 6)$, $C = (7, 8, 9)$ and D, E, F are the mid-points of the triangle ABC, then find the centroid of the triangle DEF.
7. A line makes angles α, β and γ with the coordinate axes. If $\alpha + \beta = 90°$, then find γ.
8. If α, β and γ are angles made by a line with positive direction of X-axis, Y-axis and Z-axis respectively, then find the value of $\cos 2\alpha + \cos 2\beta + \cos 2\gamma$.
9. If $\cos \alpha$, $\cos \beta$, $\cos \gamma$ are the direction cosine of a line, then find the value of $\cos^2 \alpha + (\cos \beta + \sin \gamma)(\cos \beta - \sin^2 \gamma)$.
10. A line makes angles $\alpha, \beta, \gamma, \delta$ with the four diagonals of a cube, then prove that $\cos^2 \alpha + \cos^2 \beta + \cos^2 \gamma + \cos^2 \delta = \frac{4}{3}$.
11. Find the direction cosine of line which is perpendicular to the lines with direction ratio $[1, -2, -2]$ and $[0, 2, 1]$.
12. The projection of a line segment on the axis 1, 2, 3 respectively. Then find the length of line segment.

Session 2

Equation of a Straight Line in Space, Angle between Two Lines, Perpendicular Distance of a Point from a Line, Shortest Distance between Two Lines

Equation of a Straight Line in Space

A straight line in space is specified basically in two ways

(i) A line passing through a given point and parallel to a given vector.

(ii) A line passing through any two given points.

1. Vector Equation of a Line Passing Through a Given Point and Parallel to a Given Vector

To find the vector equation of a straight line which passes through a given point and is parallel to a given vector.

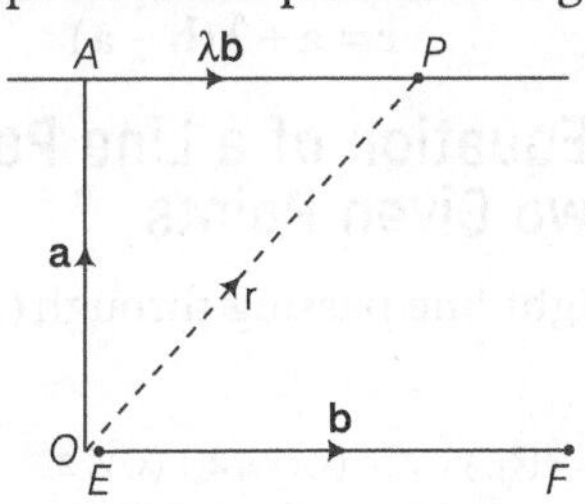

Let A be the given point and let EF be the given line, then through A draw AP parallel to given line EF.

Let **b** any vector parallel to the given line. Take any point O as the origin of reference. Let a the position vector of the given point A.

Let P be any point on the AP and let its position vector be **r**. Then, we have

$$\mathbf{r} = \mathbf{OP} = \mathbf{OA} + \mathbf{AP} = \mathbf{a} + \lambda\mathbf{b} \qquad \text{(where, } \mathbf{AP} = \lambda\mathbf{b})$$

Hence, the vector equation of straight line

$$\mathbf{r} = \mathbf{a} + \lambda\mathbf{b} \qquad \text{...(i)}$$

Remarks

1. Here, **r** is the position vector of any point $P(x, y, z)$ on the line
$\therefore \quad \mathbf{r} = x\hat{\mathbf{i}} + y\hat{\mathbf{j}} + z\hat{\mathbf{k}}$.
2. In particular, the equation of the straight line through origin and parallel to **b** is $\mathbf{r} = \lambda\mathbf{b}$.

2. Cartesian Equation of a Line Passing Through a Given Point and Given Direction Ratios

Let the coordinates of the given point A be (x_1, y_1, z_1) and the direction ratios of the line be a, b and c. Consider the coordinate of any point P be (x, y, z). Then,

$$\mathbf{r} = x\hat{\mathbf{i}} + y\hat{\mathbf{j}} + z\hat{\mathbf{k}};$$

$$\mathbf{a} = x_1\hat{\mathbf{i}} + y_1\hat{\mathbf{j}} + z_1\hat{\mathbf{k}}$$

and

$$\mathbf{b} = a\hat{\mathbf{i}} + b\hat{\mathbf{j}} + c\hat{\mathbf{k}}$$

Substituting these values in (i) and equating the coefficients of $\hat{\mathbf{i}}$, $\hat{\mathbf{j}}$ and $\hat{\mathbf{k}}$, we get

$$x = x_1 + \lambda a$$
$$y = y_1 + \lambda b$$
$$z = z_1 + \lambda c$$

These are parametric equations of the line.

Eliminating the parameter λ, we get

$$\frac{x - x_1}{a} = \frac{y - y_1}{b} = \frac{z - z_1}{c}.$$

Remarks

1. Parametric equation of straight line
$$\frac{x - x_1}{a} = \frac{y - y_1}{b} = \frac{z - z_1}{c} = \lambda$$
$\Rightarrow \quad x = x_1 + \lambda a,\ y = y_1 + \lambda b,\ z = z_1 + \lambda c$
(where, λ being parameter)
2. Since, X, Y and Z-axes pass through origin and have direction cosines (1, 0, 0), (0, 1, 0) and (0, 0, 1).
$\therefore$ Their equations are
Equation of X-axis, $\dfrac{x - 0}{1} = \dfrac{y - 0}{0} = \dfrac{z - 0}{0}$
$\Rightarrow \quad y = 0$ and $z = 0$
Equation of Y-axis, $\dfrac{x - 0}{0} = \dfrac{y - 0}{1} = \dfrac{z - 0}{0}$
$\Rightarrow \quad x = 0$ and $z = 0$
Equation of Z-axis, $\dfrac{x - 0}{0} = \dfrac{y - 0}{0} = \dfrac{z - 0}{1}$
$\Rightarrow \quad x = 0$ and $y = 0$

Example 22. Find the equation of straight line parallel to $2\hat{i} - \hat{j} + 3\hat{k}$ and passing through the point $(5, -2, 4)$.

Sol. **Vector form** Let $P = (5, -2, 4)$, then $\mathbf{OP} = 5\hat{\mathbf{i}} - 2\hat{\mathbf{j}} + 4\hat{\mathbf{k}} = \mathbf{a}$

Also, $\quad \mathbf{b} = 2\hat{\mathbf{i}} - \hat{\mathbf{j}} + 3\hat{\mathbf{k}}$

So, equation of straight line passing through **a** and parallel to straight line whose direction ratios are **b** is given as

$$\mathbf{r} = \mathbf{a} + \lambda\mathbf{b}$$

$$\Rightarrow \quad \mathbf{r} = (5\hat{\mathbf{i}} - 2\hat{\mathbf{j}} + 4\hat{\mathbf{k}}) + \lambda(2\hat{\mathbf{i}} - \hat{\mathbf{j}} + 3\hat{\mathbf{k}})$$

Cartesian form Here, $(x_1, y_1, z_1) = (5, -2, 4)$ and parallel to straight line whose DR's are $(2, -1, 3)$, so equation of the straight line is $\dfrac{x-5}{2} = \dfrac{y+2}{-1} = \dfrac{z-4}{3}$.

Example 23. Find the vector equation of a line passing through $(2, -1, 1)$ and parallel to the line whose equation is $\dfrac{x-3}{2} = \dfrac{y+1}{7} = \dfrac{z-2}{-3}$

Sol. Since, the required line is parallel to

$$\frac{x-3}{2} = \frac{y+1}{7} = \frac{z-2}{-3}$$

it follows that the required line passing through $A(2\hat{\mathbf{i}} - \hat{\mathbf{j}} + \hat{\mathbf{k}})$ has the direction of $2\hat{\mathbf{i}} + 7\hat{\mathbf{j}} - 3\hat{\mathbf{k}}$. Hence, the vector equation of the required line is

$\mathbf{r} = 2\hat{\mathbf{i}} - \hat{\mathbf{j}} + \hat{\mathbf{k}} + \lambda(2\hat{\mathbf{i}} + 7\hat{\mathbf{j}} - 3\hat{\mathbf{k}})$ where λ is a parameter.

Example 24. The cartesian equation of a line are $6x - 2 = 3y + 1 = 2z - 2$. Find its direction ratios and also find the vector equation of the line.

Sol. We know that,

$\dfrac{x - x_1}{a} = \dfrac{y - y_1}{b} = \dfrac{z - z_1}{c}$ is cartesian equation of straight line.

$$\therefore \quad 6x - 2 = 3y + 1 = 2z - 2$$

$$\Rightarrow \quad 6\left(x - \frac{1}{3}\right) = 3\left(y + \frac{1}{3}\right) = 2(z - 1)$$

$$\Rightarrow \quad \frac{x - \frac{1}{3}}{\frac{1}{6}} = \frac{y + \frac{1}{3}}{\frac{1}{3}} = \frac{z - 1}{\frac{1}{2}}$$

$$\Rightarrow \quad \frac{x - \frac{1}{3}}{1} = \frac{y + \frac{1}{3}}{2} = \frac{z - 1}{3}$$

which shows given line passes through $\left(\dfrac{1}{3}, -\dfrac{1}{3}, 1\right)$ and has direction ratios $(1, 2, 3)$.

$\therefore$ Its vector equation is

$$\mathbf{r} = \left(\frac{1}{3}\hat{\mathbf{i}} - \frac{1}{3}\hat{\mathbf{j}} + \hat{\mathbf{k}}\right) + \lambda(\hat{\mathbf{i}} + 2\hat{\mathbf{j}} + 3\hat{\mathbf{k}})$$

3. Vector Equation of a Line Passing Through Two Given Points

The vector equation of a line passing through two points whose position vectors **a** and **b** is

$$\mathbf{r} = a + \lambda(\mathbf{b} - \mathbf{a})$$

Let O be the origin and A and B be the given points with position vectors **a** and **b**, respectively.

Then, $\quad \mathbf{OP} = \mathbf{r}, \mathbf{OA} = \mathbf{a}$ and $\mathbf{OB} = \mathbf{b}$

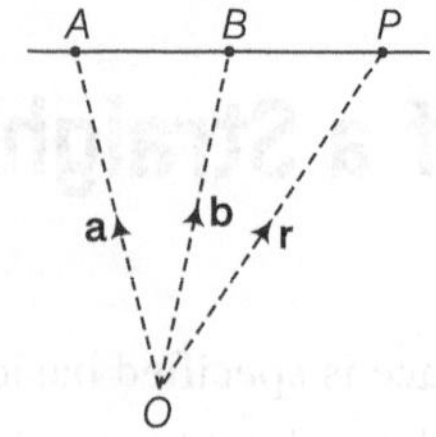

Since, **AP** is collinear with **AB**.

$$\therefore \quad \mathbf{AP} = \lambda\,\mathbf{AB} \text{ for some scalar } \lambda$$

$$\Rightarrow \quad \mathbf{OP} - \mathbf{OA} = \lambda(\mathbf{OB} - \mathbf{OA})$$

$$\Rightarrow \quad \mathbf{r} - \mathbf{a} = \lambda(\mathbf{b} - \mathbf{a})$$

$$\Rightarrow \quad \mathbf{r} = \mathbf{a} + \lambda(\mathbf{b} - \mathbf{a})$$

$\therefore$ Equation of straight line passing through **a** and **b**.

$$\Rightarrow \quad \mathbf{r} = \mathbf{a} + \lambda(\mathbf{b} - \mathbf{a})$$

4. Cartesian Equation of a Line Passing Through Two Given Points

Equation of straight line passing through (x_1, y_1, z_1), (x_2, y_2, z_2).

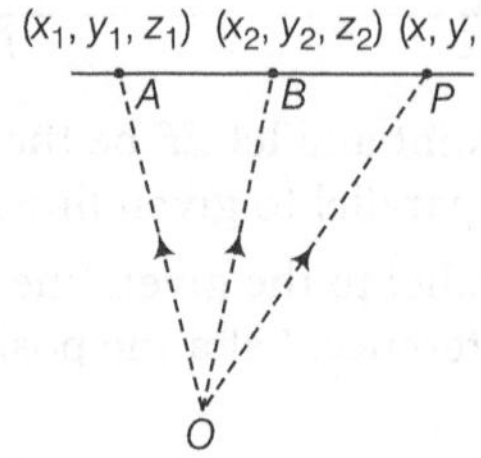

The direction ratios of

$$AB = (x_2 - x_1, y_2 - y_1, z_2 - z_1)$$

The direction ratios of

$$AP = (x - x_1, y - y_1, z - z_1)$$

Since, they are proportional

$$\frac{x - x_1}{x_2 - x_1} = \frac{y - y_1}{y_2 - y_1} = \frac{z - z_1}{z_2 - z_1}$$

Example 25. Find the vector equation of line passing through $A(3,4,-7)$ and $B(1,-1,6)$. Also, find its cartesian equations.

Sol. Since, the line passes through $A(3\hat{\mathbf{i}}+4\hat{\mathbf{j}}-7\hat{\mathbf{k}})$ and $B(\hat{\mathbf{i}}-\hat{\mathbf{j}}+6\hat{\mathbf{k}})$, its vector equation is

$$\mathbf{r}=3\hat{\mathbf{i}}+4\hat{\mathbf{j}}-7\hat{\mathbf{k}}+\lambda[(\hat{\mathbf{i}}-\hat{\mathbf{j}}+6\hat{\mathbf{k}})-(3\hat{\mathbf{i}}+4\hat{\mathbf{j}}-7\hat{\mathbf{k}})]$$

or $\mathbf{r}=3\hat{\mathbf{i}}+4\hat{\mathbf{j}}-7\hat{\mathbf{k}}-\lambda(2\hat{\mathbf{i}}+5\hat{\mathbf{j}}-13\hat{\mathbf{k}})$ ...(i)

where λ is a parameter.

The cartesian equivalent of (i) is $\frac{x-3}{2}=\frac{y-4}{5}=\frac{z+7}{-13}$.

Example 26. Find the equation of a line which passes through the point $(2,3,4)$ and which has equal intercepts on the axes.

Sol. Since, lines has equal intercepts on axes, it is equally inclined to axes.

$\Rightarrow$ line is along the vector $a(\hat{\mathbf{i}}+\hat{\mathbf{j}}+\hat{\mathbf{k}})$

$\Rightarrow$ Equation of line is $\frac{x-2}{1}=\frac{y-3}{1}=\frac{z-4}{1}$

Angle between Two Lines

Vector Form

Let $\mathbf{r}=\mathbf{a}+\lambda\mathbf{b}$ ...(i)

and $\mathbf{r}=\mathbf{a}'+\mu\mathbf{b}'$ ...(ii)

be two straight line in space.

Clearly, Eqs. (i) and (ii) are straight line in the directions of $\mathbf{b}$ and $\mathbf{b}'$, respectively.

Let θ be the between the straight lines (i) and (ii).

Then, θ is the angle between the vectors $\mathbf{b}$ and $\mathbf{b}'$ also

$$\mathbf{b}\cdot\mathbf{b}'=|\mathbf{b}||\mathbf{b}'|\cos\theta$$

$$\Rightarrow \cos\theta=\frac{\mathbf{b}\cdot\mathbf{b}'}{|\mathbf{b}||\mathbf{b}'|}$$

Cartesian Form

Let $\frac{x-x_1}{a_1}=\frac{y-y_1}{b_1}=\frac{z-z_1}{c_1}$ ...(i)

and $\frac{x-x_2}{a_2}=\frac{y-y_2}{b_1}=\frac{z-z_2}{c_2}$ ...(ii)

be two straight lines. Then, $\mathbf{b}=a_1\hat{\mathbf{i}}+b_1\hat{\mathbf{j}}+c_1\hat{\mathbf{k}}$

$$\mathbf{b}'=a_2\hat{\mathbf{i}}+b_2\hat{\mathbf{j}}+c_2\hat{\mathbf{k}}$$

So, that $\mathbf{b}\cdot\mathbf{b}'=a_1a_2+b_1b_2+c_1c_2$

and $|\mathbf{b}|=\sqrt{a_1^2+b_1^2+c_1^2}\,;\ |\mathbf{b}'|=\sqrt{a_2^2+b_2^2+c_2^2}$

$$\therefore \cos\theta=\frac{a_1a_2+b_1b_2+c_1c_2}{\sqrt{a_1^2+b_1^2+c_1^2}\sqrt{a_2^2+b_2^2+c_2^2}}$$

Condition for Perpendicularity

The lines are perpendicular, then

$$\mathbf{b}\cdot\mathbf{b}'=0$$

$$\Rightarrow a_1a_2+b_1b_2+c_1c_2=0$$

Condition for Parallelism

The lines are parallel, then $\mathbf{b}=(\mathbf{b}')\lambda$, for some scalar λ

$$\Rightarrow \frac{a_1}{a_2}=\frac{b_1}{b_2}=\frac{c_1}{c_2}$$

Example 27. Find the angle between the pair of lines

$$\mathbf{r}=3\hat{\mathbf{i}}+2\hat{\mathbf{j}}-4\hat{\mathbf{k}}+\lambda(\hat{\mathbf{i}}+2\hat{\mathbf{j}}+2\hat{\mathbf{k}})$$

and $$\mathbf{r}=5\hat{\mathbf{i}}-2\hat{\mathbf{k}}+\mu(3\hat{\mathbf{i}}+2\hat{\mathbf{j}}+6\hat{\mathbf{k}})$$

Sol. Given line are

$$\mathbf{r}=(3\hat{\mathbf{i}}+2\hat{\mathbf{j}}-4\hat{\mathbf{k}})+\lambda(\hat{\mathbf{i}}+2\hat{\mathbf{j}}+2\hat{\mathbf{k}})$$

and $$\mathbf{r}=(5\hat{\mathbf{i}}-2\hat{\mathbf{k}})+\mu(3\hat{\mathbf{i}}+2\hat{\mathbf{j}}+6\hat{\mathbf{k}})$$

We know that, angle between $\mathbf{r}=\mathbf{a}_1+\lambda\mathbf{b}_1$ and $\mathbf{r}=\mathbf{a}_2+\mu\mathbf{b}_2$,

$$\cos\theta=\frac{\mathbf{b}_1\cdot\mathbf{b}_2}{|\mathbf{b}_1||\mathbf{b}_2|}$$

$$\therefore \cos\theta=\frac{(\hat{\mathbf{i}}+2\hat{\mathbf{j}}+2\hat{\mathbf{k}})\cdot(3\hat{\mathbf{i}}+2\hat{\mathbf{j}}+6\hat{\mathbf{k}})}{\sqrt{1^2+2^2+2^2}\sqrt{3^2+2^2+6^2}}=\frac{3+4+12}{\sqrt{9}\cdot\sqrt{49}}=\frac{19}{21}$$

$$\therefore \theta=\cos^{-1}\left(\frac{19}{21}\right)$$

Example 28. Prove that the line $x=ay+b$, $z=cy+d$ and $x=a'y+b'$, $z=dy+d'$ are perpendicular, if $aa'+cc'+1=0$.

Sol. We can write the equations of straight line as

$$\frac{x-b'}{a'}=y,\ y=\frac{z-d'}{c'}$$

$$\Rightarrow \frac{x-b'}{a'}=\frac{y-0}{1}=\frac{z-d'}{c'} \quad ...(i)$$

and $$\frac{x-b}{a}=y,\ y=\frac{z-d}{c}$$

$$\Rightarrow \frac{x-b}{a}=\frac{y-0}{1}=\frac{z-d_2}{c} \quad ...(ii)$$

We know that, $\frac{x-x_1}{a_1}=\frac{y-y_1}{b_1}=\frac{z-z_1}{c_1}$

and $\frac{x-x_2}{a_2}=\frac{y-y_2}{b_2}=\frac{z-z_2}{c_2}$

are perpendicular, if $a_1a_2+b_1b_2+c_1c_2=0$

$\therefore$ For the straight lines given by Eqs. (i) and (ii), to be perpendicular.

$$a'a+1\cdot1+c'c=0$$

$$\Rightarrow aa'+cc'+1=0$$

Perpendicular Distance of a Point from a Line

1. Foot of Perpendicular from a Point on the Given Line

(i) **Cartesian Form** Here, the equation of line AB is

$$\frac{x-x_1}{a}=\frac{y-y_1}{b}=\frac{z-z_1}{c}$$

Let L be the foot of the perpendicular drawn from $P(\alpha, \beta, \gamma)$ on the line $\frac{x-x_1}{a}=\frac{y-y_1}{b}=\frac{z-z_1}{c}$.

Let the coordinates of L be $(x_1+a\lambda, y_1+b\lambda, z_1+c\lambda)$.

Then, the direction ratios of PL are $(x_1+a\lambda-\alpha, y_1+b\lambda-\beta, z_1+c\lambda-\gamma)$.

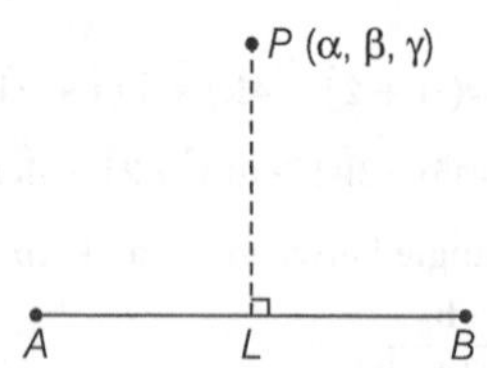

Direction ratios of AB are (a, b, c).

Since PL is perpendicular to AB.

$$a(x_1+a\lambda-\alpha)+b(y_1+b\lambda-\beta)+c(z_1+c\lambda-\gamma)=0$$

$$\lambda=\frac{a(\alpha-x_1)+b(\beta-y_1)+c(\gamma+z_1)}{a^2+b^2+c^2}$$

Putting the value of λ in $(x_1+a\lambda, y_1+b\lambda, z_1+c\lambda)$, we get the foot of the perpendicular. Now, we can get distance PL using distance formula.

(ii) **Vector Form** Let L be the foot of the perpendicular drawn from $P(\alpha)$ on the line $\mathbf{r}=\mathbf{a}+\lambda\mathbf{b}$.

Since, $\mathbf{r}$ denotes the position vector of any point on the line $\mathbf{r}=\mathbf{a}+\lambda\mathbf{b}$, the position vector of L will be $(\mathbf{a}+\lambda\mathbf{b})$

Directions ratios of $PL=\mathbf{a}-\alpha+\lambda\mathbf{b}$

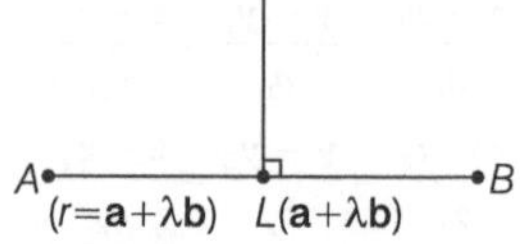

Since, **PL** is perpendicular to **b**,

$$(\mathbf{a}-\alpha+\lambda\mathbf{b})\cdot\mathbf{b}=0$$

$$\Rightarrow \quad (\mathbf{a}-\alpha)\cdot\mathbf{b}+\lambda\mathbf{b}\cdot\mathbf{b}=0$$

$$\Rightarrow \quad \lambda=\frac{-(\mathbf{a}-\alpha)\cdot\mathbf{b}}{|\mathbf{b}|^2}$$

$\Rightarrow$ Position vector of L is $\mathbf{a}-\left(\frac{(\mathbf{a}-\alpha)\cdot\mathbf{b}}{|\mathbf{b}|^2}\right)\mathbf{b}$, which is the foot of the perpendicular.

(iii) The distance of the point (x_2, y_2, z_2) from the line $\frac{x-x_1}{l}=\frac{y-y_1}{m}=\frac{z-z_1}{n}$, (where l, m and n are direction cosines of the line), is

$$[(x_2-x_1)^2+(y_2-y_1)^2+(z_2-z_1)^2 - \{l(x_2-x_1)+m(y_2-y_1)+n(z_2-z_1)\}^2]^{1/2}$$

Let $\quad \mathbf{r}_1=(x_2-x_1)\hat{\mathbf{i}}+(y_2-y_1)\hat{\mathbf{j}}+(z_2-z_1)\hat{\mathbf{k}}$

$\quad \mathbf{r}_2=l\hat{\mathbf{i}}+m\hat{\mathbf{j}}+n\hat{\mathbf{k}}$

$$\therefore \quad \cos\theta=\frac{\mathbf{r}_2\cdot\mathbf{r}_1}{|\mathbf{r}_2|\cdot|\mathbf{r}_1|}$$

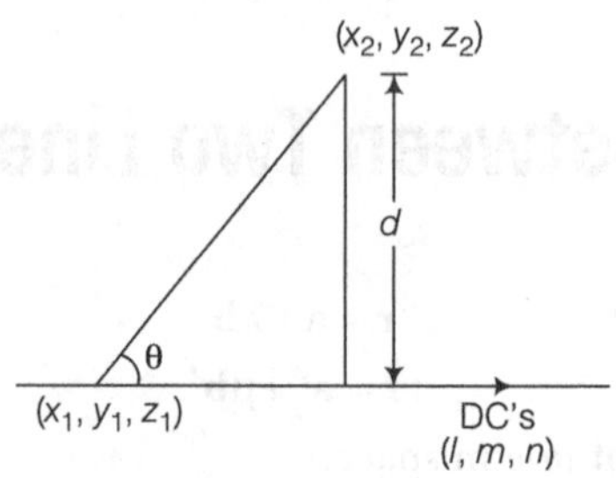

Also, $\quad d=|\mathbf{r}_1|\sin\theta$

$$d^2=|\mathbf{r}_1|^2\sin^2\theta$$

$$=|\mathbf{r}_1|^2(1-\cos^2\theta)$$

$$=|\mathbf{r}_1|^2\left(1-\frac{(\mathbf{r}_1\cdot\mathbf{r}_2)^2}{|\mathbf{r}_1|^2\cdot|\mathbf{r}_2|^2}\right)$$

$$d^2=|\mathbf{r}_1|^2-(\mathbf{r}_1\cdot\mathbf{r}_2)^2 \quad \text{(where, } |\mathbf{r}_2|=1\text{)}$$

$$\Rightarrow \quad d=\sqrt{|\mathbf{r}_1|^2-(\mathbf{r}_1\cdot\mathbf{r}_2)^2}$$

$$d=\sqrt{\{(x_2-x_1)^2+(y_2-y_1)^2+(z_2-z_1)^2 - \{l(x_2-x_1)+m(y_2-y_1)+n(z_2-z_1)\}^2}$$

2. Reflection or Image of a Point in a Straight Line

(i) **Cartesian Form** To find the reflection or image of a point in a straight line in cartesian form.

Let $P(\alpha, \beta, \gamma)$ be the point and $\frac{x-x_1}{a}=\frac{y-y_1}{b}=\frac{z-z_1}{c}$ be the given line.

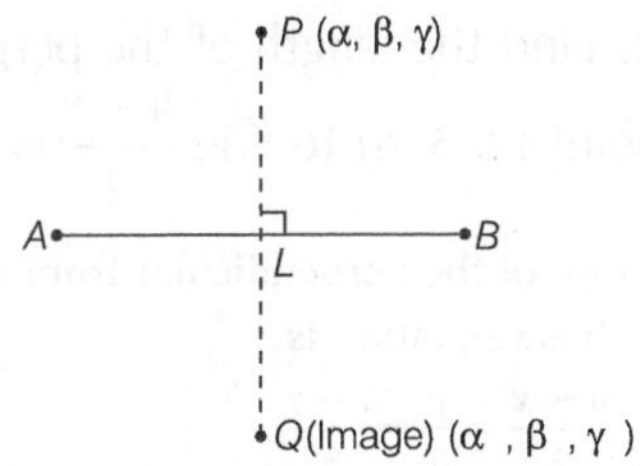

Let L be the foot of perpendicular from P to AB and let Q be the image of the point in the given line, where $PL = LQ$.

Let the coordinates of L be

$$(x_1 + a\lambda, y_1 + b\lambda, z_1 + c\lambda)$$

Then, direction ratios of PL are

$$(x_1 + a\lambda - \alpha, y_1 + b\lambda - \beta, z_1 + c\lambda - \gamma)$$

Since, PL is perpendicular to the given line, whose direction ratios are a, b and c.

$$\therefore (x_1 + a\lambda - \alpha) \cdot a + (y_1 + b\lambda - \beta) \cdot b + (z_1 + c\lambda - \gamma) \cdot c = 0$$

$$\Rightarrow \quad \lambda = \frac{\{a(\alpha - x_1) + b(\beta - y_1) + c(\gamma - z_1)\}}{a^2 + b^2 + c^2}$$

Substituting λ, we get L, (foot of perpendicular)

Let coordinates of $Q(\alpha', \beta', \gamma')$ be image.

$\therefore$ Mid-point of PQ is L.

$$\therefore \quad \frac{\alpha + \alpha'}{2} = x_1 + a\lambda, \frac{\beta + \beta'}{2} = y_1 + b\lambda, \frac{\gamma + \gamma'}{2} = z_1 + c\lambda$$

$$\therefore \quad \alpha' = 2(x_1 + a\lambda) - \alpha, \beta' = 2(y_1 + b\lambda) - \beta, \gamma' = 2(z_1 + c\lambda) - \gamma$$

(ii) **Vector Form** To find the reflection or image of a point in a straight line in vector form. Let $P(\alpha)$ be the given point and $\mathbf{r} = \mathbf{a} + \lambda \mathbf{b}$ be the given line.

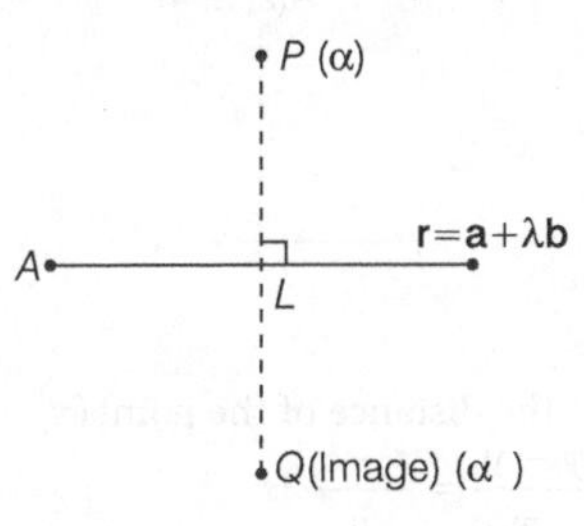

Let Q be the image of P in $\mathbf{r} = \mathbf{a} + \lambda \mathbf{b}$

$$\therefore \quad \mathbf{PL} = \mathbf{a} + \lambda \mathbf{b} - \alpha$$

Since, **PL** is perpendicular to the given line,

$$\therefore \quad \mathbf{PL} \perp \mathbf{b}$$

$$\Rightarrow \quad \mathbf{PL} \cdot \mathbf{b} = 0$$

$$\Rightarrow \quad (\mathbf{a} + \lambda \mathbf{b} - \alpha) \cdot \mathbf{b} = 0$$

$$\Rightarrow \quad \lambda = -\frac{[(\mathbf{a} - \alpha) \cdot \mathbf{b}]}{|\mathbf{b}|^2}$$

$\therefore$ Position vector of L,

$$\mathbf{a} + \lambda \mathbf{b} = \mathbf{a} - \left(\frac{(\mathbf{a} - \alpha) \cdot \mathbf{b}}{|\mathbf{b}|^2}\right) \mathbf{b}$$

Let Q be the image of point P and α' be the position vector.

Since, L is mid-point of PQ.

$$\Rightarrow \quad \frac{\alpha + \alpha'}{2} = \mathbf{a} - \left(\frac{(\mathbf{a} - \alpha) \cdot \mathbf{b}}{|\mathbf{b}|^2}\right) \mathbf{b}$$

$$\Rightarrow \quad \alpha' = 2\mathbf{a} - \left(\frac{2(\mathbf{a} - \alpha) \cdot \mathbf{b}}{|\mathbf{b}|^2}\right) \mathbf{b} - \alpha$$

which is image of P on $\mathbf{r}$.

Example 29. Find the foot of perpendicular drawn from the point $2\hat{\mathbf{i}} - \hat{\mathbf{j}} + 5\hat{\mathbf{k}}$ to the line $\mathbf{r} = (11\hat{\mathbf{i}} - 2\hat{\mathbf{j}} - 8\hat{\mathbf{k}}) + \lambda(10\hat{\mathbf{i}} - 4\hat{\mathbf{j}} - 11\hat{\mathbf{k}})$. Also, find the length of the perpendicular.

Sol. Let L be the foot of the perpendicular drawn from $P(2\hat{\mathbf{i}} - \hat{\mathbf{j}} + 5\hat{\mathbf{k}})$ on the line

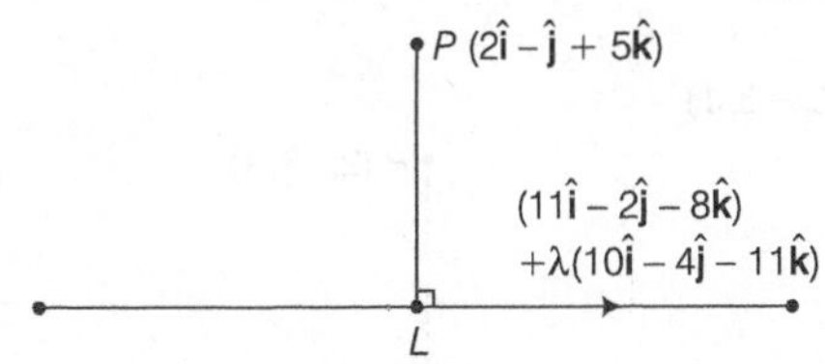

$$\mathbf{r} = (11\hat{\mathbf{i}} - 2\hat{\mathbf{j}} - 8\hat{\mathbf{k}}) + \lambda(10\hat{\mathbf{i}} - 4\hat{\mathbf{j}} - 11\hat{\mathbf{k}})$$

Let the position vector of L is

$$(11\hat{\mathbf{i}} - 2\hat{\mathbf{j}} - 8\hat{\mathbf{k}}) + \lambda(10\hat{\mathbf{i}} - 4\hat{\mathbf{j}} - 11\hat{\mathbf{k}})$$
$$= (11 + 10\lambda)\hat{\mathbf{i}} + (-2 - 4\lambda)\hat{\mathbf{j}} + (-8 - 11\lambda)\hat{\mathbf{k}}$$

$\therefore \mathbf{PL}$ = Position vector of L − Position vector of P

$$= (9 + 10\lambda)\hat{\mathbf{i}} + (-1 - 4\lambda)\hat{\mathbf{j}} + (-13 - 11\lambda)\hat{\mathbf{k}}$$

Since, PL is perpendicular to the given line and parallel to

$$\mathbf{b} = 10\hat{\mathbf{i}} - 4\hat{\mathbf{j}} - 11\hat{\mathbf{k}}. \quad \Rightarrow \mathbf{PL} \cdot \mathbf{b} = 0$$

$$\Rightarrow \{(9 + 10\lambda)\hat{\mathbf{i}} + (-1 - 4\lambda)\hat{\mathbf{j}} + (-13 - 11\lambda)\hat{\mathbf{k}}\} \cdot (10\hat{\mathbf{i}} - 4\hat{\mathbf{j}} - 11\hat{\mathbf{k}}) = 0$$

$$\Rightarrow 10(9 + 10\lambda) - 4(-1 - 4\lambda) - 11(-13 - 11\lambda) = 0$$

$$\Rightarrow \quad \lambda = -1$$

On putting $\lambda = -1$, we get L as $(\hat{\mathbf{i}} + 2\hat{\mathbf{j}} + 3\hat{\mathbf{k}})$

Now, $\quad \mathbf{PL} = (\hat{\mathbf{i}} + 2\hat{\mathbf{j}} + 3\hat{\mathbf{k}}) - (2\hat{\mathbf{i}} - \hat{\mathbf{j}} + 5\hat{\mathbf{k}})$

$$= (-\hat{\mathbf{i}} + 3\hat{\mathbf{j}} - 2\hat{\mathbf{k}})$$

Hence, the length of perpendicular from P on the given line

$$= |\mathbf{PL}| = \sqrt{1 + 9 + 4} = \sqrt{14}$$

Example 30. Find the coordinates of the foot of the perpendicular drawn from point $A(1, 0, 3)$ to the join of points $B(4, 7, 1)$ and $C(3, 5, 3)$.

Sol. Let D be the foot of the perpendicular and let it divide BC in the ratio $\lambda : 1$. Then, the coordinates of D are $\frac{3\lambda + 4}{\lambda + 1}$, $\frac{5\lambda + 7}{\lambda + 1}$ and $\frac{3\lambda + 1}{\lambda + 1}$.

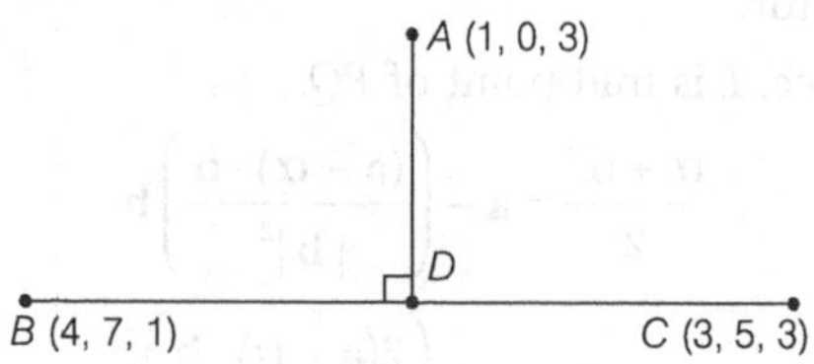

Now, $\quad \mathbf{AD} \perp \mathbf{BC} \Rightarrow \mathbf{AD} \cdot \mathbf{BC} = 0$

$$\Rightarrow \quad (2\lambda + 3) + 2(5\lambda + 7) + 4 = 0 \Rightarrow \lambda = -\frac{7}{4}$$

$\Rightarrow$ Coordinates of D are $\frac{5}{3}, \frac{7}{3}$ and $\frac{17}{3}$.

Example 31. Find the length of perpendicular from $P(2, -3, 1)$ to the line $\frac{x+1}{2} = \frac{y-3}{3} = \frac{z+2}{-1}$.

Sol. Given line is $\frac{x+1}{2} = \frac{y-3}{3} = \frac{z+2}{-1} = r$ ...(i)

and $P(2, -3, 1)$

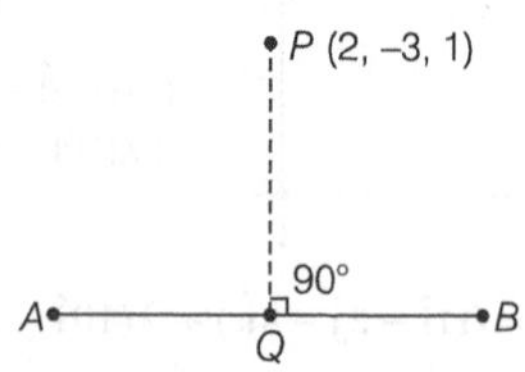

Coordinates of any point on line (i) may be taken as

$$(2r - 1, 3r + 3, -r - 2)$$

Let $\quad Q = (2r - 1, 3r + 3, -r - 2)$

Direction ratio's of PQ are $(2r - 3, 3r + 6, -r - 3)$.

Direction ratio's of AB are $(2, 3, -1)$.

Since, $\quad PQ \perp AB$

$$2(2r - 3) + 3(3r + 6) - 1(-r - 3) = 0$$

$$\Rightarrow \quad r = -\frac{15}{14}$$

$$\therefore \quad Q = \left(-\frac{22}{7}, -\frac{3}{14}, -\frac{13}{14}\right)$$

$$PQ^2 = \left(2 + \frac{22}{7}\right)^2 + \left(-3 + \frac{3}{14}\right)^2 + \left(1 + \frac{13}{14}\right)^2 = \frac{531}{14}$$

$$PQ = \sqrt{\frac{531}{14}} \text{ units}$$

Example 32. Find the length of the perpendicular drawn from point $(2, 3, 4)$ to line $\frac{4-x}{2} = \frac{y}{6} = \frac{1-z}{3}$.

Sol. Let P be the foot of the perpendicular from $A(2, 3, 4)$ to the given line l whose equation is

$$\frac{4-x}{2} = \frac{y}{6} = \frac{1-z}{3}$$

or $\quad \frac{x-4}{-2} = \frac{y}{6} = \frac{z-1}{-3} = k$ (say). ...(i)

Therefore, $x = 4 - 2k$, $y = 6k$, $z = 1 - 3k$

As P lies on (i), coordinates of P are $(4 - 2k, 6k, 1 - 3k)$ for some value of k.

The direction ratios of AP are

$$(4 - 2k - 2, 6k - 3, 1 - 3k - 4)$$

or $\quad (2 - 2k, 6k - 3, -3 - 3k)$.

Also, the direction ratios of l are -2, 6 and -3.

Since, $AP \perp l$

$$\Rightarrow \quad -2(2 - 2k) + 6(6k - 3) - 3(-3 - 3k) = 0$$

$$\Rightarrow \quad -4 + 4k + 36k - 18 + 9 + 9k = 0$$

or $\quad 49k - 13 = 0$ or $k = \frac{13}{49}$

We have,
$$AP^2 = (4 - 2k - 2)^2 + (6k - 3)^2 + (1 - 3k - 4)^2$$
$$= (2 - 2k)^2 + (6k - 3)^2 + (-3 - 3k)^2$$
$$= 4 - 8k + 4k^2 + 36k^2 - 36k + 9 + 9 + 18k + 9k^2$$
$$= 22 - 26k + 49k^2$$
$$= 22 - 26\left(\frac{13}{49}\right) + 49\left(\frac{13}{49}\right)^2$$
$$= \frac{22 \times 49 - 26 + 13 + 13^2}{49} = \frac{909}{49}$$

$$AP = \frac{3}{7}\sqrt{101}$$

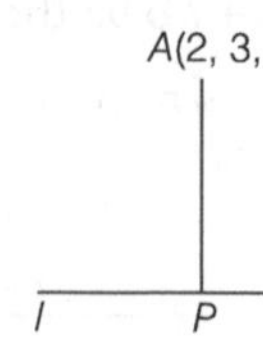

Aliter

We know that the distance of the point (x_2, y_2, z_2) from the line $\frac{x - x_1}{l} = \frac{y - y_1}{m} = \frac{z - z_1}{n}$ is

$$\sqrt{(x_2 - x_1)^2 + (y_2 - y_1)^2 + (z_2 - z_1)^2 - (l(x_2 - x_1) + m(y_2 - y_1) + n(z_2 - z_1))^2}$$

Here, (x_2, y_2, z_2) are $(2, 3, 4)$ and (x_1, y_1, z_1) are $(4, 0, 1)$ and $(l, m, n) = \left(\frac{-2}{7}, \frac{6}{7}, \frac{-3}{7}\right)$.

$$\therefore \quad d = \sqrt{(2-4)^2 + (3-0)^2 + (4-1)^2 - \left[\frac{-2}{7}(2-4) + \frac{6}{7}(3-0) - \frac{3}{7}(4-1)\right]^2}$$

$$= \sqrt{4+9+9-\left(\frac{4+18-9}{7}\right)^2}$$

$$= \sqrt{22 - \frac{169}{49}} = \sqrt{\frac{1078-169}{49}}$$

$$= \frac{\sqrt{909}}{7} = \frac{3}{7}\sqrt{101}$$

Example 33. Find the image of the point (1, 6, 3) in the line $\frac{x}{1} = \frac{y-1}{2} = \frac{z-2}{3}$.

Sol. Let P be the given point and let L be the foot of perpendicular from P to the given line.

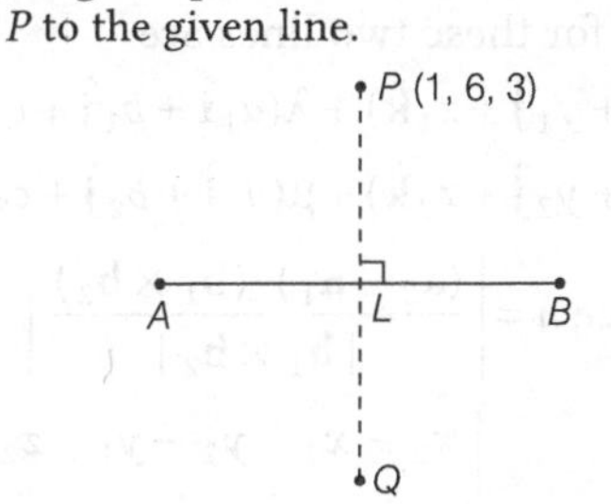

The coordinates of a general point on the given line are given by

$$\frac{x-0}{1} = \frac{y-1}{2} = \frac{z-2}{3} = \lambda$$

i.e. $\quad x = \lambda, y = 2\lambda + 1, z = 3\lambda + 2$

Let the coordinates of L be

$$(\lambda, 2\lambda + 1, 3\lambda + 2) \quad \ldots\text{(i)}$$

So, direction ratios of PL are

$$(\lambda - 1, 2\lambda - 5, 3\lambda - 1)$$

Direction ratios of the given line are (1, 2, 3) which is perpendicular to PL

$$(\lambda - 1)\cdot 1 + (2\lambda - 5)\cdot 2 + (3\lambda - 1)\cdot 3 = 0$$

$$\Rightarrow \quad \lambda = 1$$

So, coordinates of L are (1, 3, 5).

Let $Q(x_1, y_1, z_1)$ be the image of $P(1, 6, 3)$ on given line.

Since, L is mid-point of PQ.

$$\therefore \quad 1 = \frac{x_1+1}{2}, 3 = \frac{y_1+6}{2}, 5 = \frac{z_1+3}{2}$$

$$\Rightarrow \quad x_1 = 1, y_1 = 0, z_1 = 7$$

$\therefore$ Image of $P(1, 6, 3)$ in the given line is (1, 0, 7).

Example 34. Find the coordinates of those points on the line $\frac{x-1}{2} = \frac{y+2}{3} = \frac{z-3}{6}$ which are at a distance of 3 units from points (1, – 2, 3).

Sol. Here, $\frac{x-1}{2} = \frac{y+2}{3} = \frac{z-3}{6}$,

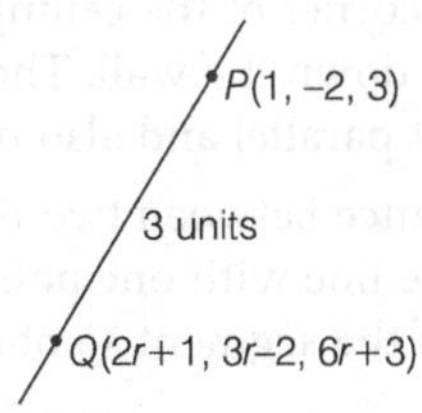

is the given straight line ...(i)

Let $P = (1, -2, 3)$ on the straight line.

Here, direction ratios of line (i) are (2, 3, 6).

$\therefore$ Direction cosines of line (i) are $\frac{2}{7}, \frac{3}{7}, \frac{6}{7}$

Equation of line (i) may be written as

$$\frac{x-1}{2/7} = \frac{y+2}{3/7} = \frac{z-3}{6/7} \quad \ldots\text{(ii)}$$

Coordinates of any point on the line (ii) may be taken as

$$\left(\frac{2}{7}r + 1, \frac{3}{7}r - 2, \frac{6}{7}r + 3\right)$$

Let $Q\left(\frac{2}{7}r + 1, \frac{3}{7}r - 2, \frac{6}{7}r + 3\right)$

Given, $\quad |\mathbf{r}| = 3$

$\therefore \quad r = \pm 3$

Putting the value of r, we have $Q = \left(\frac{13}{7}, \frac{-5}{7}, \frac{39}{7}\right)$ or $\left(\frac{1}{7}, \frac{-23}{7}, \frac{3}{7}\right)$

Shortest Distance between Two Lines

If two lines in space intersect at a point, then the shortest distance between them is zero. Also, if two lines in space are parallel, then the shortest distance between them will be the perpendicular distance, i.e. the length of the perpendicular drawn from any point on one line onto the other line. Further, in a space, there are lines which are neither intersecting nor parallel. In fact, such pair of lines are non-coplanar and are called skew lines.

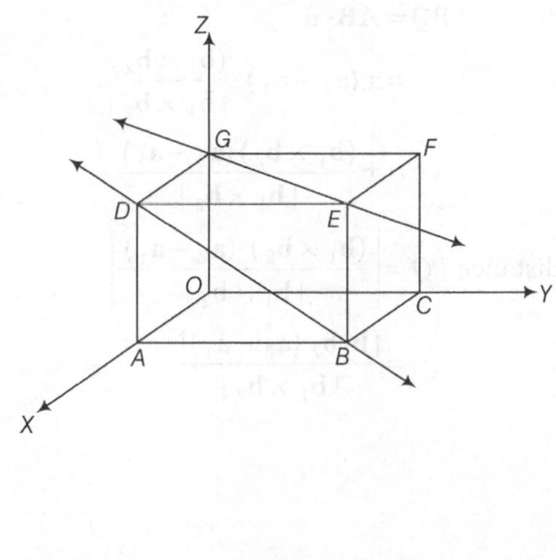

Line GE goes diagonally across the ceiling and line DB passes through one corner of the ceiling directly above A and goes diagonally down the wall. These lines are skew because they are not parallel and also never meet.

By the shortest distance between two lines, we mean the join of a point in one line with one point on the other line so that the length of the segment so obtained is the smallest.

1. Shortest Distance between Two Skew Straight Lines

Line of Shortest Distance

If l_1 and l_2 are two skew lines, then there is one and only one line perpendicular to each of lines l_1 and l_2 which is known as the line of shortest distance.

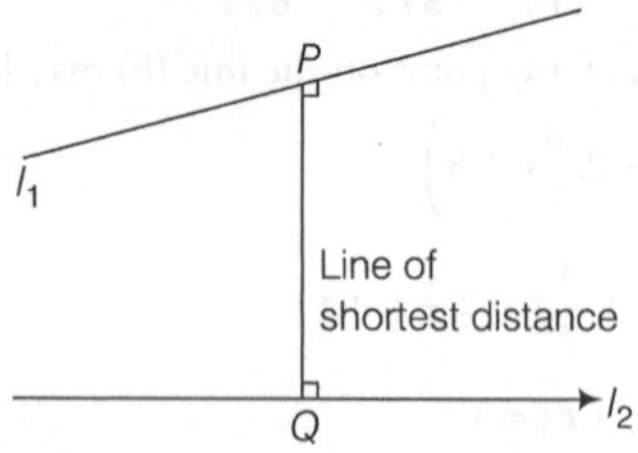

Here, distance PQ is called to be shortest distance.

Vector Form

Let l_1 and l_2 be two lines whose equations are $\mathbf{r}=\mathbf{a}_1+\lambda\mathbf{b}_1$ and $\mathbf{r}=\mathbf{a}_2+\mu\mathbf{b}_2$, respectively.

Clearly, l_1 and l_2 pass through the points A and B with $\mathbf{a}_1$ and $\mathbf{a}_2$, respectively and are parallel to the vectors $\mathbf{b}_1$ and $\mathbf{b}_2$, respectively.

Since, **PQ** is perpendicular to both l_1 and l_2 which are parallel to $\mathbf{b}_1$ and $\mathbf{b}_2$.

$\therefore$ **PQ** is parallel to $\mathbf{b}_1\times\mathbf{b}_2$.

Let $\hat{\mathbf{n}}$ be a unit vector along **PQ**, then $\hat{\mathbf{n}}=\pm\dfrac{\mathbf{b}_1\times\mathbf{b}_2}{|\mathbf{b}_1\times\mathbf{b}_2|}$

$$\therefore \quad PQ = \text{Projection of } \mathbf{AB} \text{ on } \mathbf{PQ}$$

$$\Rightarrow \quad \mathbf{PQ}=\mathbf{AB}\cdot\hat{\mathbf{n}}$$

$$=\pm(\mathbf{a}_2-\mathbf{a}_1)\cdot\frac{(\mathbf{b}_1\times\mathbf{b}_2)}{|\mathbf{b}_1\times\mathbf{b}_2|}$$

$$=\pm\frac{(\mathbf{b}_1\times\mathbf{b}_2)(\mathbf{a}_2-\mathbf{a}_1)}{|\mathbf{b}_1\times\mathbf{b}_2|}$$

Hence, distance $PQ=\left|\dfrac{(\mathbf{b}_1\times\mathbf{b}_2)\cdot(\mathbf{a}_2-\mathbf{a}_1)}{|\mathbf{b}_1\times\mathbf{b}_2|}\right|$

$$=\frac{[\mathbf{b}_1\mathbf{b}_2\,(\mathbf{a}_2-\mathbf{a}_1)]}{|\mathbf{b}_1\times\mathbf{b}_2|}$$

Condition for Lines to Intersecting

The two lines are intersecting, if

$$\left|\frac{(\mathbf{b}_1\times\mathbf{b}_2)\cdot(\mathbf{a}_2-\mathbf{a}_1)}{|\mathbf{b}_1\times\mathbf{b}_2|}\right|=0$$

$$\Rightarrow \quad (\mathbf{b}_1\times\mathbf{b}_2)\cdot(\mathbf{a}_2-\mathbf{a}_1)=0$$

$$\Rightarrow \quad [\mathbf{b}_1\mathbf{b}_2(\mathbf{a}_2-\mathbf{a}_1)]=0$$

Cartesian Form

Let the two skew lines be

$$\frac{x-x_1}{a_1}=\frac{y-y_1}{b_1}=\frac{z-z_1}{c_1}$$

and $$\frac{x-x_2}{a_2}=\frac{y-y_2}{b_2}=\frac{z-z_2}{c_2}$$

Vector equations for these two lines are

$$\mathbf{r}=(x_1\hat{\mathbf{i}}+y_1\hat{\mathbf{j}}+z_1\hat{\mathbf{k}})+\lambda(a_1\hat{\mathbf{i}}+b_1\hat{\mathbf{j}}+c_1\hat{\mathbf{k}})$$

and $$\mathbf{r}=(x_2\hat{\mathbf{i}}+y_2\hat{\mathbf{j}}+z_2\hat{\mathbf{k}})+\mu(a_2\hat{\mathbf{i}}+b_2\hat{\mathbf{j}}+c_2\hat{\mathbf{k}})$$

$\because$ Shortest distance $n=\left|\dfrac{(\mathbf{a}_2-\mathbf{a}_1)\cdot(\mathbf{b}_1\times\mathbf{b}_2)}{|\mathbf{b}_1\times\mathbf{b}_2|}\right|$

$$\Rightarrow \quad d=\frac{\begin{vmatrix} x_2-x_1 & y_2-y_1 & z_2-z_1\\ a_1 & b_1 & c_1\\ a_2 & b_2 & c_2\end{vmatrix}}{\sqrt{(b_1c_2-b_2c_1)^2+(c_1a_2-a_1c_2)^2+(a_1b_2-a_2b_1)^2}}$$

Conditions for Lines to Intersect

The lines are intersecting, if shortest distance $=0$

$$\Rightarrow \quad \begin{vmatrix} x_2-x_1 & y_2-y_1 & z_2-z_1\\ a_1 & b_1 & c_1\\ a_2 & b_2 & c_2\end{vmatrix}=0$$

2. Shortest Distance between Parallel Lines

Let l_1 and l_2 be two parallel lines whose equations are

$$\mathbf{r}=\mathbf{a}_1+\lambda\mathbf{b} \quad \text{or} \quad \mathbf{r}=\mathbf{a}_2+\mu\mathbf{b}, \text{ respectively.}$$

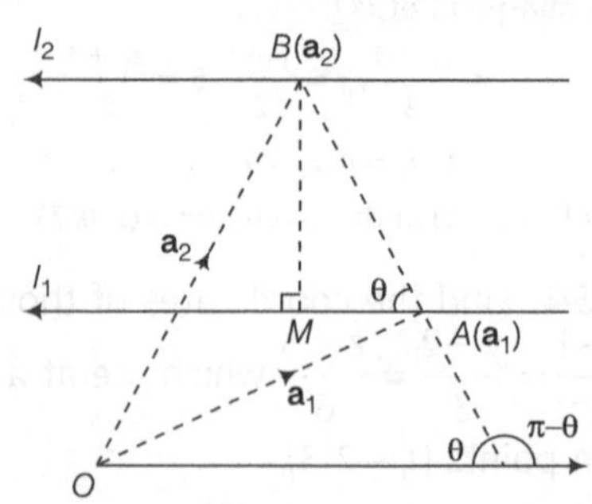

Clearly, l_1 and l_2 pass through the points A and B with position vectors $\mathbf{a}_1$ and $\mathbf{a}_2$, respectively and both are parallel to the vector $\mathbf{b}$, where BM is the shortest distance between l_1 and l_2.

Let θ be the angle between AB and l_1.

$$\therefore \quad \sin\theta = \frac{BM}{AB}$$

$$\Rightarrow \quad BM = AB\sin\theta = |\mathbf{AB}|\sin\theta$$

Now, $\quad |\mathbf{AB}\times\mathbf{b}| = |\mathbf{AB}||\mathbf{b}|\sin(\pi-\theta)$

$$|\mathbf{AB}||\mathbf{b}|\sin\theta$$

$$= (|\mathbf{AB}|\sin\theta)|\mathbf{b}| = BM|\mathbf{b}|$$

$$\therefore \quad BM = \frac{|\mathbf{AB}\times\mathbf{b}|}{|\mathbf{b}|} = \frac{|(\mathbf{a}_2-\mathbf{a}_1)\times\mathbf{b}|}{|\mathbf{b}|}$$

$\therefore$ Shortest distance between parallel lines

$\mathbf{r} = \mathbf{a}_1 + \lambda\mathbf{b}_1$ and $\mathbf{r} = \mathbf{a}_2 + \mu\mathbf{b}$ is

$$d = \frac{|(\mathbf{a}_2-\mathbf{a}_1)\times\mathbf{b}|}{|\mathbf{b}|}$$

Example 35. Show that the two lines $\frac{x-1}{2} = \frac{y-2}{3} = \frac{z-3}{4}$ and $\frac{x-4}{5} = \frac{y-1}{2} = z$, intersect. Also, find the point of intersection of these lines.

Sol. Here, $\frac{x-1}{2} = \frac{y-2}{3} = \frac{z-3}{4}$ ...(i)

and $\frac{x-4}{5} = \frac{y-1}{2} = \frac{z-0}{1}$ ...(ii)

Any point on line (i) is $P(2r+1, 3r+2, 4r+3)$ and any point on the line (ii) is $Q(5\lambda+4, 2\lambda+1, \lambda)$.

They intersect if and only if

$$2r+1 = 5\lambda+4,\ 3r+2 = 2\lambda+1,\ 4r+3 = \lambda$$

On solving, $\quad r = -1, \lambda = -1$

Clearly, for these values of λ and r $P(-1, -1, -1)$

Hence, lines (i) and (ii) intersect at $(-1, -1, -1)$.

Example 36. Find the shortest distance between the lines $\mathbf{r} = (4\hat{\mathbf{i}}-\hat{\mathbf{j}}) + \lambda(\hat{\mathbf{i}}+2\hat{\mathbf{j}}-3\hat{\mathbf{k}})$ and $\mathbf{r} = (\hat{\mathbf{i}}-\hat{\mathbf{j}}+2\hat{\mathbf{k}}) + \mu(2\hat{\mathbf{i}}+4\hat{\mathbf{j}}-5\hat{\mathbf{k}})$.

Sol. We know, the shortest distance between the lines

$\mathbf{r} = \mathbf{a}_1 + \lambda\mathbf{b}_1$ and $\mathbf{r} = \mathbf{a}_1 + \lambda\mathbf{b}_2$

$$\Rightarrow \quad d = \left|\frac{(\mathbf{a}_2-\mathbf{a}_1)\cdot(\mathbf{b}_1\times\mathbf{b}_2)}{|\mathbf{b}_1\times\mathbf{b}_2|}\right|$$

On comparing the given equation with the equations

$\mathbf{r} = \mathbf{a}_1 + \lambda\mathbf{b}_1$ and $\mathbf{r} = \mathbf{a}_1 + \lambda\mathbf{b}_2$ respectively, we have

$\mathbf{a}_1 = 4\hat{\mathbf{i}}-\hat{\mathbf{j}}$, $\mathbf{a}_2 = \hat{\mathbf{i}}-\hat{\mathbf{j}}+2\hat{\mathbf{k}}$, $\mathbf{b}_1 = \hat{\mathbf{i}}+2\hat{\mathbf{j}}-3\hat{\mathbf{k}}$ and $\mathbf{b}_2 = 2\hat{\mathbf{i}}+4\hat{\mathbf{j}}-5\hat{\mathbf{k}}$

Now, $\quad \mathbf{a}_2-\mathbf{a}_1 = -3\hat{\mathbf{i}}+0\hat{\mathbf{j}}+2\hat{\mathbf{k}}$

and $\quad \mathbf{b}_1\times\mathbf{b}_2 = \begin{vmatrix} \hat{\mathbf{i}} & \hat{\mathbf{j}} & \hat{\mathbf{k}} \\ 1 & 2 & -3 \\ 2 & 4 & -5 \end{vmatrix} = 2\hat{\mathbf{i}}-\hat{\mathbf{j}}+0\hat{\mathbf{k}}$

$$\therefore \quad (\mathbf{a}_2-\mathbf{a}_1)\cdot(\mathbf{b}_1\times\mathbf{b}_2) = (-3\hat{\mathbf{i}}+0\hat{\mathbf{j}}+2\hat{\mathbf{k}})\cdot(2\hat{\mathbf{i}}-\hat{\mathbf{j}}+0\hat{\mathbf{k}}) = -6$$

and $\quad |\mathbf{b}_1\times\mathbf{b}_2| = \sqrt{4+1+0} = \sqrt{5}$

$$\therefore \text{ Shortest distance, } d = \left|\frac{(\mathbf{a}_2-\mathbf{a}_1)\cdot(\mathbf{b}_1\times\mathbf{b}_2)}{|\mathbf{b}_1\times\mathbf{b}_2|}\right| = \left|\frac{-6}{\sqrt{5}}\right| = \frac{6}{\sqrt{5}}$$

Example 37. Find the shortest distance between the lines

$$\frac{x-1}{2} = \frac{y-2}{3} = \frac{z-3}{4} \text{ and } \frac{x-2}{3} = \frac{y-4}{4} = \frac{z-5}{5}.$$

Sol. Given lines are

$$\frac{x-1}{2} = \frac{y-2}{3} = \frac{z-3}{4} \quad \text{...(i)}$$

and $$\frac{x-2}{3} = \frac{y-4}{4} = \frac{z-5}{5} \quad \text{...(ii)}$$

Here, $x_1 = 1, y_1 = 2, z_3 = 3; x_2 = 2, y_2 = 4, z_2 = 5$

$l_1 = 2, m_1 = 3, n_1 = 4; l_2 = 3, m_2 = 4, n_2 = 5$

Shortest distance between the lines (i) and (ii) are modulus of

$$= \frac{\begin{vmatrix} x_2-x_1 & y_2-y_1 & z_2-z_1 \\ l_1 & m_1 & n_1 \\ l_2 & m_2 & n_2 \end{vmatrix}}{\sqrt{(l_1m_2-l_2m_1)+(m_1n_2-m_2n_1)^2 + (l_1n_2-l_2n_1)^2}} \quad \text{...(iii)}$$

Now, $= \begin{vmatrix} x_2-x_1 & y_2-y_1 & z_2-z_1 \\ l_1 & m_1 & n_1 \\ l_2 & m_2 & n_2 \end{vmatrix} = \begin{vmatrix} 1 & 2 & 2 \\ 2 & 3 & 4 \\ 3 & 4 & 5 \end{vmatrix}$

$$= 1(15-16) - 2(10-12) + 2(8-9) = 1$$

Also, $(l_1m_2-l_2m_1)^2 + (m_1n_2-m_2n_1)^2 + (n_1l_2-n_2l_1)^2$

$$= (8-9)^2 + (15-16)^2 + (10-12)^2$$

$$= 6$$

From Eq. (iii) shortest distance between lines (i) (ii), we get

$$= \left|\frac{1}{\sqrt{6}}\right| = \frac{1}{\sqrt{6}}$$

Example 38. Find the shortest distance and the vector equation of the line of shortest distance between the lines given by

$$\mathbf{r} = (3\hat{\mathbf{i}}+8\hat{\mathbf{j}}+3\hat{\mathbf{k}}) + \lambda(3\hat{\mathbf{i}}-\hat{\mathbf{j}}+\hat{\mathbf{k}})$$

and $$\mathbf{r} = (-3\hat{\mathbf{i}}-7\hat{\mathbf{j}}+6\hat{\mathbf{k}}) + \mu(-3\hat{\mathbf{i}}+2\hat{\mathbf{j}}+4\hat{\mathbf{k}})$$

Sol. Given lines are

$$\mathbf{r} = (3\hat{\mathbf{i}}+8\hat{\mathbf{j}}+3\hat{\mathbf{k}}) + \lambda(3\hat{\mathbf{i}}-\hat{\mathbf{j}}+\hat{\mathbf{k}}) \quad \text{...(i)}$$

and $$\mathbf{r} = (-3\hat{\mathbf{i}}-7\hat{\mathbf{j}}+6\hat{\mathbf{k}}) + \mu(-3\hat{\mathbf{i}}+2\hat{\mathbf{j}}+4\hat{\mathbf{k}}) \quad \text{...(ii)}$$

Equation of lines (i) and (iii) in cartesian form,

$$AB: \frac{x-3}{3} = \frac{y-8}{-1} = \frac{z-3}{1} = \lambda \quad \text{...(iii)}$$

and $$CD: \frac{x+3}{-3} = \frac{y+7}{2} = \frac{z-6}{4} = \mu \quad \text{...(iv)}$$

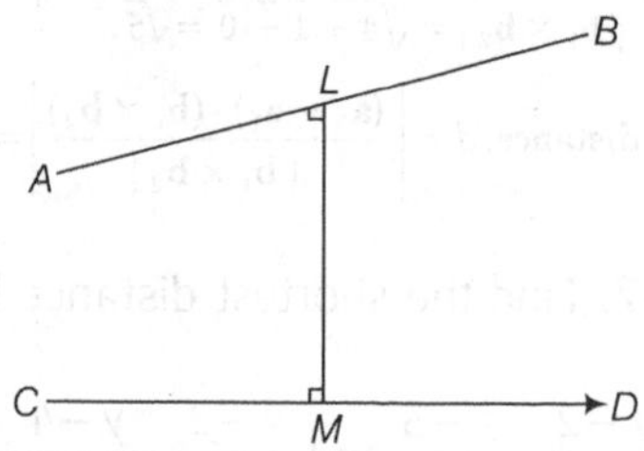

Let $L(3\lambda + 3), -\lambda + 8, \lambda + 3), M(-3\mu - 3, 2\mu - 7, 4\mu + 6)$

Direction ratios of LM are

$$(3\lambda + 3\mu + 6, -\lambda - 2\mu + 15, \lambda - 4\mu - 3)$$

Since, $LM \perp AB$

$$\therefore \quad 3(3\lambda + 3\mu + 6) - 1(-\lambda - 2\mu + 15) + 1(\lambda - 4\mu - 3) = 0$$

or $$11\lambda + 7\mu = 0 \quad \text{...(v)}$$

Again, $LM \perp CD$

$$-3(3\lambda + 3\mu + 6) - 2(-\lambda - 2\mu + 15) + 4(\lambda - 4\mu - 3) = 0$$

or $$-7\lambda - 29\mu = 0 \quad \text{...(vi)}$$

Solving Eqs. (v) and (vi), we get

$$\lambda = 0 = \mu$$

$$\therefore \quad L \equiv (3, 8, 3),$$

$$M \equiv (-3, -7, 6)$$

Hence, the shortest distance,

$$LM = \sqrt{(3+3)^2 + (8+7)^2 + (3-6)^2} = 3\sqrt{30} \text{ units}$$

$\therefore$ Vector equation of LM is

$$\mathbf{r} = 3\hat{\mathbf{i}} + 8\hat{\mathbf{j}} + 3\hat{\mathbf{k}} + t(6\hat{\mathbf{i}} + 15\hat{\mathbf{j}} - 3\hat{\mathbf{k}})$$

Also, the cartesian equation of LM is

$$\frac{x-3}{6} = \frac{y-8}{15} = \frac{z-3}{-3}$$

Example 39. Find the shortest distance between lines $\mathbf{r} = (\hat{\mathbf{i}} + 2\hat{\mathbf{j}} + \hat{\mathbf{k}}) + \lambda(2\hat{\mathbf{i}} + \hat{\mathbf{j}} + 2\hat{\mathbf{k}})$ and $\mathbf{r} = 2\hat{\mathbf{i}} - \hat{\mathbf{j}} - \hat{\mathbf{k}} + \mu(2\hat{\mathbf{i}} + \hat{\mathbf{j}} + 2\hat{\mathbf{k}})$.

Sol. Here lines (i) and (ii) are passing through the points $\mathbf{a}_1 = \hat{\mathbf{i}} + 2\hat{\mathbf{j}} + \hat{\mathbf{k}}$ and $\mathbf{a}_2 = 2\hat{\mathbf{i}} - \hat{\mathbf{j}} - \hat{\mathbf{k}}$, respectively, and are parallel to the vector $\mathbf{b} = 2\hat{\mathbf{i}} + \hat{\mathbf{j}} + 2\hat{\mathbf{k}}$.

Hence, the distance between the lines using the formula

$$\frac{|\mathbf{b} \times (\mathbf{a}_2 - \mathbf{a}_1)|}{|\mathbf{b}|} = \frac{\begin{vmatrix} \hat{\mathbf{i}} & \hat{\mathbf{j}} & \hat{\mathbf{k}} \\ 2 & 1 & 2 \\ 1 & -3 & -2 \end{vmatrix}}{3}$$

$$= \frac{|4\hat{\mathbf{i}} - 6\hat{\mathbf{j}} - 7\hat{\mathbf{k}}|}{3} = \frac{\sqrt{16 + 36 + 49}}{3} = \sqrt{\frac{101}{3}}$$

Example 40. Find the equation of a line which passes through the point (1, 1, 1) and intersects the lines $\frac{x-1}{2} = \frac{y-2}{3} = \frac{z-3}{4}$ and $\frac{x+2}{1} = \frac{y-3}{2} = \frac{z+1}{4}$.

Sol. Any line passing through the point (1, 1, 1) is

$$\frac{x-1}{a} = \frac{y-1}{b} = \frac{z-1}{c} \quad \text{...(i)}$$

This line intersects the line $\frac{x-1}{2} = \frac{y-2}{3} = \frac{z-3}{4}$.

If $a : b : c \neq 2 : 3 : 4$ and $\begin{vmatrix} 1-1 & 2-1 & 3-1 \\ a & b & c \\ 2 & 3 & 4 \end{vmatrix} = 0$

$$\Rightarrow \quad a - 2b + c = 0$$

Again, line (i) intersects line

$$\frac{x-(-2)}{1} = \frac{y-3}{2} = \frac{z-(-1)}{4} \quad \text{...(ii)}$$

If $a : b : c \neq 1 : 2 : 4$ and $\begin{vmatrix} -2-1 & 3-1 & -1-1 \\ a & b & c \\ 1 & 2 & 4 \end{vmatrix} = 0$

$$\Rightarrow \quad 6a + 5b - 4c = 0 \quad \text{...(iii)}$$

From (ii) and (iii) by cross multiplication, we have

$$\frac{a}{8-5} = \frac{b}{6+4} = \frac{c}{5+12}$$

$$\Rightarrow \quad \frac{a}{3} = \frac{b}{10} = \frac{c}{17}$$

So, the required lines is $\frac{x-1}{3} = \frac{y-1}{10} = \frac{z-1}{17}$

Example 41. If the straight lines $x = -1 + s$, $y = 3 - \lambda s$, $z = 1 + \lambda s$ and $x = \frac{t}{2}$, $y = 1 + t$, $z = 2 - t$, with parameters s and t, respectively, are coplanar, then find λ.

Sol. The given lines $\frac{x+1}{1} = \frac{y-3}{-\lambda} = \frac{z-1}{\lambda} = s$

$\frac{x-0}{y_2} = \frac{y-1}{1} = \frac{y-2}{-1} = t$ are coplanar if

$$\begin{vmatrix} 0+1 & 1-3 & 2-1 \\ 1 & -\lambda & \lambda \\ 1/2 & 1 & -1 \end{vmatrix} = 0$$

$$\begin{vmatrix} 1 & -2 & 1 \\ 1 & -\lambda & \lambda \\ 1/2 & 1 & -1 \end{vmatrix} = 0$$

$$\Rightarrow \quad 1(\lambda - \lambda) + 2\left(1 - \frac{\lambda}{2}\right) + 1\left(1 + \frac{\lambda}{2}\right) = 0$$

$$\Rightarrow \quad \lambda = -2$$

Exercise for Session 2

1. The cartesian equation of a line is $\frac{x-3}{2}=\frac{y+1}{-2}=\frac{z-3}{5}$. Find the vector equation of the line.

2. A line passes through the point with position vector $2\hat{\mathbf{i}}-3\hat{\mathbf{j}}+4\hat{\mathbf{k}}$ and is in the direction of $3\hat{\mathbf{i}}+4\hat{\mathbf{j}}-5\hat{\mathbf{k}}$. Find the equation of the line is vector and cartesian forms.

3. Find the coordinates of the point where the line through (3, 4, 1) and (5, 1, 6) crosses *XY*-plane.

4. Find the angle between the pairs of line $\mathbf{r}=3\hat{\mathbf{i}}+2\hat{\mathbf{j}}-4\hat{\mathbf{k}}+\lambda(\hat{\mathbf{i}}+2\hat{\mathbf{j}}+2\hat{\mathbf{k}})$, $\hat{\mathbf{r}}=5\hat{\mathbf{i}}-2\hat{\mathbf{j}}+\mu(3\hat{\mathbf{i}}+2\hat{\mathbf{j}}+6\hat{\mathbf{k}})$

5. Show that the two lines $\frac{x-1}{2}=\frac{y-2}{3}=\frac{z-3}{4}$ and $\frac{x-4}{5}=\frac{y-1}{2}=z$ intersect. Find also the point of intersection of these lines.

6. Find the magnitude of the shortest distance between the lines $\frac{x}{2}=\frac{y}{-3}=\frac{z}{1}$ and $\frac{x-2}{3}=\frac{y-1}{-5}=\frac{z+2}{2}$.

7. Find the perpendicular distance of the point (1, 1, 1) from the line $\frac{x-2}{2}=\frac{y+3}{2}=\frac{z}{-1}$.

8. Find the equation of the line drawn through the point (1, 0, 2) to meet at right angles the line $\frac{x+1}{3}=\frac{y-2}{-2}=\frac{z+1}{-1}$.

9. Find the equation of line through (1, 2, – 1) and perpendicular to each of the lines $\frac{x}{1}=\frac{y}{0}=\frac{z}{-1}$ and $\frac{x}{3}=\frac{y}{4}=\frac{z}{5}$.

10. Find the image of the point (1, 2, 3) in the line $\frac{x-6}{3}=\frac{y-7}{2}=\frac{z-7}{-2}$.

Session 3

Plane, Equation of Plane in Various Form, Angle between Two Planes, Family of Planes, Two Sides of a Plane, Distance of a Point from a Plane, Equation of Planes Bisecting the Angle between Two Planes, Line and Plane

Plane

A plane is a surface such that if any two points are taken on it, the line segment joining them lies completely on the surface.

General Form General equation of the first degree in x, y, z always represents a plane.

The general equation of plane is $ax + by + cz + d = 0$.

Proof. Let first degree equation in x, y and z be

$$ax + by + cz + d = 0 \quad \text{...(i)}$$

In order to prove that Eq. (i) is the equation of plane, it is sufficient to show that every point on the line joining two points lies on the surface represented by Eq. (i).

Let $P(x_1, y_1, z_1)$ and $Q(x_2, y_2, z_2)$ be two points on the surface represented by Eq. (i).

Then, $$ax_1 + by_1 + cz_1 + d = 0 \quad \text{...(ii)}$$

and $$ax_2 + by_2 + cz_2 + d = 0 \quad \text{...(iii)}$$

Let R be any arbitrary point on the line segment joining P and Q. Suppose R divides PQ in the ratio $\lambda : 1$.

$\therefore R$ is $\left(\dfrac{x_1 + \lambda x_2}{1+\lambda}, \dfrac{y_1 + \lambda y_2}{1+\lambda}, \dfrac{z_1 + \lambda z_2}{1+\lambda}\right)$

We are to show that R lies on the surface represented by Eq. (i) for all values of λ. For this, it is sufficient to prove that R satisfy Eq. (i)

On putting this value of R in LHS of Eq. (i), we obtain

$$a\left(\frac{x_1 + \lambda x_2}{\lambda+1}\right) + b\left(\frac{y_1 + \lambda y_2}{\lambda+1}\right) + c\left(\frac{z_1 + \lambda z_2}{\lambda+1}\right) + d$$

$$= \frac{1}{\lambda+1}\{(ax_1 + by_1 + cz_1) + \lambda(ax_2 + by_2 + cz_2)\}$$

$$= \frac{1}{\lambda+1}[0 + 0] \quad \text{[using Eqs. (ii) and (iii)]}$$

$$= 0$$

which shows that the point R lies on Eq. (i). Since, R is an arbitrary point on the line segment joining P and Q.

$\therefore$ Every point on PQ lies on the surface represented by Eq. (i).

Hence, $ax + by + cz + d = 0$ is equation of plane.

Equation of a Plane Passing Through a Given Point

The general equation of a plane passing through a given point (x_1, y_1, z_1) is $a(x - x_1) + b(y - y_1) + c(z - z_1) = 0$, where a, b and c are constants.

Proof. The general equation of plane is

$$ax + by + cz + d = 0 \quad \text{...(i)}$$

If it passes through (x_1, y_1, z_1)

$$\Rightarrow \quad ax_1 + by_1 + cz_1 + d = 0 \quad \text{...(ii)}$$

On subtracting Eq. (i) from Eq. (ii), we get

$$a(x - x_1) + b(y - y_1) + c(z - z_1) = 0$$

which is the equation of a plane passing through (x_1, y_1, z_1).

Example 42. Show that the four points $(0, -1, -1)$, $(-4, 4, 4)$, $(4, 5, 1)$ and $(3, 9, 4)$ are coplanar. Find the equation of the plane containing them.

Sol. We shall find the equation of a plane passing through any three out of the given four points and show that the fourth point satisfies the equation.

Now, any plane passing through $(0, -1, -1)$ is

$$a(x - 0) + b(y + 1) + c(z + 1) = 0 \quad \text{...(i)}$$

If it passes through $(-4, 4, 4)$, we have

$$a(-4) + b(5) + c(5) = 0 \quad \text{...(ii)}$$

Also, if plane passes through $(4, 5, 1)$, we have

$$a(4) + b(6) + c(2) = 0$$

$$\Rightarrow \quad 2a + 3b + c = 0 \quad \text{...(iii)}$$

On solving Eqs. (ii) and (iii) by cross multiplication method, we obtain

$$\frac{a}{-5} = \frac{b}{7} = \frac{c}{-11} = k$$

On putting in Eq. (i), we get

$$-5kx + 7k(y+1) - 11k(z+1) = 0$$

$$\Rightarrow \quad -5x + 7y - 11z - 4 = 0$$

(required equation of plane)

Clearly, the fourth point namely (3, 9, 4) satisfies this equation. Hence, the given points are coplanar and the equation of plane containing those points, is $5x - 7y + 11z + 4 = 0$

Equation of Plane in Various Form

A plane is determined uniquely if

(i) The normal to the plane and its distance from the origin is given, i.e. the equation of a plane in normal form.

(ii) It passes through a point and is perpendicular to a given direction.

(iii) It passes through three given non-collinear points.

Equation of Plane in Normal Form

Vector Form

The vector equation of a plane normal to unit vector $\hat{\mathbf{n}}$ and at a distance d from the origin is $\mathbf{r} \cdot \hat{\mathbf{n}} = d$.

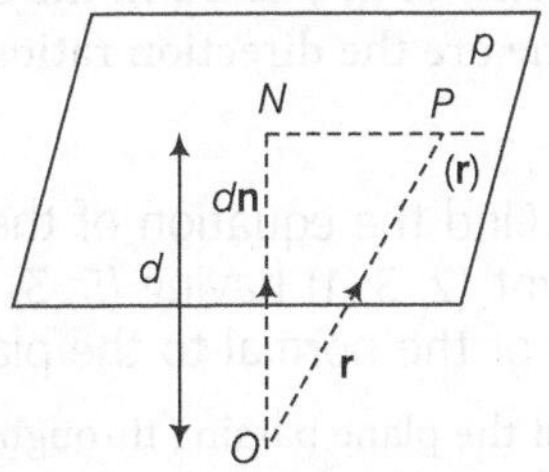

Proof. Let O be the origin and let ON be the perpendicular from O to the given plane π such that $\mathbf{ON} = d\hat{\mathbf{n}}$, where d is perpendicular distance of plane from origin.

Let P be a point on the plane, with position vector $\mathbf{r}$ so that $\mathbf{OP} = \mathbf{r}$

Now, $\quad \mathbf{NP} \perp \mathbf{ON}$

$$\Rightarrow \quad \mathbf{NP} \cdot \mathbf{ON} = 0 \qquad \text{...(i)}$$

$$\Rightarrow \quad (\mathbf{OP} - \mathbf{ON}) \cdot \mathbf{ON} = 0$$

$$\Rightarrow \quad (\mathbf{r} - d\hat{\mathbf{n}}) \cdot d\hat{\mathbf{n}} = 0$$

$$\Rightarrow \quad \mathbf{r} \cdot d\hat{\mathbf{n}} - d^2 \hat{\mathbf{n}} \cdot \hat{\mathbf{n}} = 0$$

$$\Rightarrow \quad d\mathbf{r} \cdot \hat{\mathbf{n}} - d^2 |\mathbf{n}|^2 = 0 \qquad (\because d \neq 0)$$

$$\Rightarrow \quad \mathbf{r} \cdot \hat{\mathbf{n}} - d = 0 \qquad (\because |\hat{\mathbf{n}}| = 1)$$

$$\Rightarrow \quad \mathbf{r} \cdot \hat{\mathbf{n}} = d \qquad \text{...(ii)}$$

Thus, the required equation of the plane is $\mathbf{r} \cdot \hat{\mathbf{n}} = d$.

Cartesian Form

Equation (ii) gives the vector equation of a plane, where $\hat{\mathbf{n}}$ is the unit vector normal to the plane. Let $P(x, y, z)$ be any point on the plane. Then

$$\mathbf{OP} = \mathbf{r} = x\hat{\mathbf{i}} + y\hat{\mathbf{j}} + z\hat{\mathbf{k}}$$

Let l, m and n be the direction cosines of $\hat{\mathbf{n}}$.

Then, $\quad \hat{\mathbf{n}} = (l\hat{\mathbf{i}} + m\hat{\mathbf{j}} + n\hat{\mathbf{k}})$

Therefore, (ii) gives

$$(x\hat{\mathbf{i}} + y\hat{\mathbf{j}} + z\hat{\mathbf{k}}) \cdot (l\hat{\mathbf{i}} + m\hat{\mathbf{j}} + n\hat{\mathbf{k}}) = d$$

or $$lx + my + nz = d \qquad \text{...(iii)}$$

This is the cartesian equation of the plane in the normal form.

Note

Equation (iii) shows that if $\mathbf{r} \cdot (a\hat{\mathbf{i}} + b\hat{\mathbf{j}} + c\hat{\mathbf{k}}) = d$ is the vector equation of a plane, then $ax + by + cz = d$ is the cartesian equation of the plane, where a, b and c are the direction ratios of the normal to the plane.

The equation $\mathbf{r} \cdot \mathbf{n} = d$ is in normal form, if $\mathbf{n}$ is a unit vector and d is the distance of the plane from the origin. If $\mathbf{n}$ is not a unit vector, then to reduce the equation $\mathbf{r} \cdot \mathbf{n} = d$ to normal form, we reduce the equation $\mathbf{r} \cdot \mathbf{n} = d$ to normal form by dividing both sides by $|\mathbf{n}|$, we get

$$\frac{\mathbf{r} \cdot \mathbf{n}}{|\mathbf{n}|} = \frac{d}{|\mathbf{n}|} \Rightarrow \mathbf{r} \cdot \mathbf{n} = \frac{d}{|\mathbf{n}|} = p \text{ (distance from the origin)}$$

Example 43. Find the vector equation of plane which is at a distance of 8 units from the origin and which is normal to the vector $2\hat{\mathbf{i}} + \hat{\mathbf{j}} + 2\hat{\mathbf{k}}$.

Sol. Here, $d = 8$ and $n = 2\hat{\mathbf{i}} + \hat{\mathbf{j}} + 2\hat{\mathbf{k}}$

$$\mathbf{n} = \frac{\mathbf{n}}{|\mathbf{n}|} = \frac{2\hat{\mathbf{i}} + \hat{\mathbf{j}} + 2\hat{\mathbf{k}}}{\sqrt{2^2 + 1^2 + 2^2}} = \frac{2\hat{\mathbf{i}} + \hat{\mathbf{j}} + 2\hat{\mathbf{k}}}{3}$$

Hence, the required equation of plane is, $\mathbf{r} \cdot \mathbf{n} = d$

$$\Rightarrow \quad \mathbf{r} \cdot \left(\frac{2\hat{\mathbf{i}} + \hat{\mathbf{j}} + 2\hat{\mathbf{k}}}{3} \right) = 8$$

$$\Rightarrow \quad \mathbf{r} \cdot (2\hat{\mathbf{i}} + \hat{\mathbf{j}} + 2\hat{\mathbf{k}}) = 24$$

Example 44. Reduce the equation $\mathbf{r} \cdot (3\hat{\mathbf{i}} - 4\hat{\mathbf{j}} + 12\hat{\mathbf{k}}) = 5$ to normal form and hence find the length of perpendicular from the origin to the plane.

Sol. The given equation of plane is

$$\mathbf{r} \cdot (3\hat{\mathbf{i}} - 4\hat{\mathbf{j}} + 12\hat{\mathbf{k}}) = 5 \text{ or } \mathbf{r} \cdot \mathbf{n} = 5$$

where, $\quad \mathbf{n} = 3\hat{\mathbf{i}} - 4\hat{\mathbf{j}} + 12\hat{\mathbf{k}}$

Since, $|\mathbf{n}| = \sqrt{9 + 16 + 144} = 13 \neq 1$, therefore the given equation is not the normal form. To reduce to normal form we divide both sides by $|\mathbf{n}|$ i.e. $\frac{\mathbf{r}\cdot\mathbf{n}}{|\mathbf{n}|}=\frac{5}{|\mathbf{n}|}$ or

$\mathbf{r}\cdot\left(\frac{3}{13}\hat{\mathbf{i}}-\frac{4}{13}\hat{\mathbf{j}}+\frac{12}{13}\hat{\mathbf{k}}\right)=\frac{5}{13}$. This is the normal form of the equation of given plane and length perpendicular $=\frac{5}{13}$.

Example 45. Find the distance of the plane $2x-y-2z-9=0$ from the origin.

Sol. The plane can be put in vector form as $\mathbf{r}\cdot(2\hat{\mathbf{i}}-\hat{\mathbf{j}}-2\hat{\mathbf{k}})=9$.

Here, $\mathbf{n}=2\hat{\mathbf{i}}-\hat{\mathbf{j}}-2\hat{\mathbf{k}}$

$\Rightarrow \quad \frac{\mathbf{n}}{|\mathbf{n}|}=\frac{2\hat{\mathbf{i}}-\hat{\mathbf{j}}-2\hat{\mathbf{k}}}{3}$

Dividing equation throughout by 3, we have equation of plane in normal form as $\mathbf{r}\cdot\frac{(2\hat{\mathbf{i}}-\hat{\mathbf{j}}-2\hat{\mathbf{k}})}{3}=3$, in which 3 is the distance of the plane from the origin.

Example 46. Find the vector equation of a line passing through $3\hat{\mathbf{i}}-5\hat{\mathbf{j}}+7\hat{\mathbf{k}}$ and perpendicular to the plane $3x-4y+5z=8$.

Sol. The given plane $3x-4y+5z=8$.

or $\quad (3\hat{\mathbf{i}}-4\hat{\mathbf{j}}+5\hat{\mathbf{k}})(x\hat{\mathbf{i}}+y\hat{\mathbf{j}}+z\hat{\mathbf{k}})=8$.

This shows that $\mathbf{d}=3\hat{\mathbf{i}}-4\hat{\mathbf{j}}+5\hat{\mathbf{k}}$ is normal to the given plane.

Therefore, the required line is parallel to $3\hat{\mathbf{i}}-4\hat{\mathbf{j}}+5\hat{\mathbf{k}}$.

Since, the required line passes through $3\hat{\mathbf{i}}-5\hat{\mathbf{j}}+7\hat{\mathbf{k}}$, its equation is given by $\mathbf{r}=3\hat{\mathbf{i}}-5\hat{\mathbf{j}}+7\hat{\mathbf{k}}+\lambda(3\hat{\mathbf{i}}-4\hat{\mathbf{j}}+5\hat{\mathbf{k}})$, where λ is a parameter.

Example 47. Find the unit vector perpendicular to the plane $\mathbf{r}\cdot(2\hat{\mathbf{i}}+\hat{\mathbf{j}}+2\hat{\mathbf{k}})=5$.

Sol. Vector normal to the plane is $\mathbf{n}=2\hat{\mathbf{i}}+\hat{\mathbf{j}}+2\hat{\mathbf{k}}$

Hence, unit vector perpendicular to the plane is

$$\frac{\mathbf{n}}{|\mathbf{n}|}=\frac{2\hat{\mathbf{i}}+\hat{\mathbf{j}}+2\hat{\mathbf{k}}}{\sqrt{2^2+1^2+2^2}}$$

$$=\frac{1}{3}(2\hat{\mathbf{i}}+\hat{\mathbf{j}}+2\hat{\mathbf{k}})$$

Vector Equation of a Plane Passing Through a Given Point and Normal to a Given Vector

The vector equation of a plane passing through a point having position vector **a** and normal to vector **n** is $(\mathbf{r}-\mathbf{a})\cdot\mathbf{n}=0$.

Proof Suppose the planer π passes through a point having position vector **a** and is normal to the vector **n**. Let O be the origin and **r** be the position vector of any point P on the plane π. Then, $\mathbf{OP}=\mathbf{r}$.

Since, **AP** lies in the plane and **n** is a normal to the plane π.

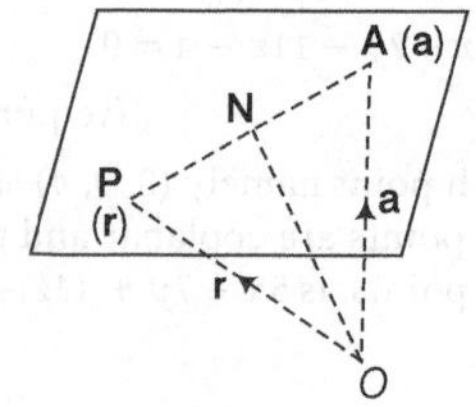

$\therefore \quad \mathbf{AP}\perp\mathbf{n}$

$\Rightarrow \quad \mathbf{AP}\cdot\mathbf{n}=0 \Rightarrow (\mathbf{r}\cdot\mathbf{a})\cdot\mathbf{n}=0 \quad (\because \mathbf{AP}=\mathbf{r}-\mathbf{a})$

Hence, the required equation of the plane is

$$(\mathbf{r}-\mathbf{a})\cdot\mathbf{n}=0$$

Note

The above equation can be written as $\mathbf{r}\cdot\mathbf{n}=d$, where $\mathbf{d}=\mathbf{a}\cdot\mathbf{n}$ (known as scalar product form of plane).

Cartesian Form

If $\mathbf{r}=x\hat{\mathbf{i}}+y\hat{\mathbf{j}}+z\hat{\mathbf{k}}$, $\mathbf{a}=x_1\hat{\mathbf{i}}+y_1\hat{\mathbf{j}}+z_1\hat{\mathbf{k}}$ and $\mathbf{n}=a\hat{\mathbf{i}}+b\hat{\mathbf{j}}+c\hat{\mathbf{k}}$, then $(\mathbf{r}-\mathbf{a})=(x-x_1)\hat{\mathbf{i}}+(y-y_1)\hat{\mathbf{j}}+(z-z_1)\hat{\mathbf{k}}$

Then equation of the plane can be written as

$(x-x_1)\hat{\mathbf{i}}+(y-y_1)\hat{\mathbf{j}}+(z-z_1)\hat{\mathbf{k}}\cdot(a\hat{\mathbf{i}}+b\hat{\mathbf{j}}+c\hat{\mathbf{k}})=0$

$\Rightarrow a(x-x_1)+b(y-y_1)+c(z-z_1)=0$

Thus, the coefficients of x, y and z in the cartesian equation of a plane are the direction ratios of the normal to the plane.

Example 48. Find the equation of the plane passing through the point (2, 3, 1) having (5, 3, 2) as the direction ratios of the normal to the plane.

Sol. The equation of the plane passing through (x_1, y_1, z_1) and perpendicular to the line with direction ratios a, b and c is given by $a(x-x_1)+b(y-y_1)+c(z-z_1)=0$.

Now, since the plane passes through (2, 3, 1) and is perpendicular to the line having direction ratios (5, 3, 2), the equation of the plane is given by $5(x-2)+3(y-3)+2(z-1)=0$ or $5x+3y+2z=21$.

Example 49. The foot of the perpendicular drawn from the origin to a plane is (12, –4, 3). Find the equation of the plane.

Sol. Since $P(12,-4,3)$ is the foot of the perpendicular from the origin to the plane OP is normal to the plane π. Thus, the direction ratios of normal to the plane are 12, –4 and 3. Now, since the plane passes through (12, – 4 , 3), its equation is given by

$$12(x-12)-4(y+4)+3(z-3)=0$$

or $\quad 12x-4y+3z-169=0$

Example 50. A vector **n** of magnitude 8 units is inclined to X-axis at 45°, Y-axis at 60° and an acute angle with Z-axis. If a plane passes through a point $(\sqrt{2}, -1, 1)$ and is normal to **n**, then find its equation in vector form.

Sol. Let γ be the angle made by **n** with Z-axis, then direction cosines of **n** are

$$l = \cos 45° = \frac{1}{\sqrt{2}}, m = \cos 60° = \frac{1}{2} \text{ and } n = \cos \gamma$$

$$\therefore \quad l^2 + m^2 + n^2 = 1 \Rightarrow \frac{1}{2} + \frac{1}{4} + n^2 = 1$$

$$\Rightarrow \quad n^2 = \frac{1}{4}$$

$$n = \frac{1}{2} \text{ (neglecting } n = -\frac{1}{2} \text{ as } \gamma \text{ is acute} : n > 0)$$

We have, $|\mathbf{n}| = 8$

$$\mathbf{n} = |\mathbf{n}|(l\hat{\mathbf{i}} + m\hat{\mathbf{j}} + n\hat{\mathbf{k}})$$

$$\mathbf{n} = 8\left(\frac{1}{\sqrt{2}}\hat{\mathbf{i}} + \frac{1}{2}\hat{\mathbf{j}} + \frac{1}{2}\hat{\mathbf{k}}\right) = 4\sqrt{2}\hat{\mathbf{i}} + 4\hat{\mathbf{j}} + 4\hat{\mathbf{k}}$$

The required plane passes through the point $(\sqrt{2}, -1, 1)$ having position vector

$$\mathbf{a} = \sqrt{2}\hat{\mathbf{i}} - \hat{\mathbf{j}} + \hat{\mathbf{k}}$$

So, its vector equation is $(\mathbf{r} - \mathbf{a}) \cdot \mathbf{n} = 0$

$$\Rightarrow \quad \hat{\mathbf{r}} \cdot \hat{\mathbf{n}} = \hat{\mathbf{a}} \cdot \hat{\mathbf{n}}$$

$$\Rightarrow \mathbf{r} \cdot (4\sqrt{2}\hat{\mathbf{i}} + 4\hat{\mathbf{j}} + 4\hat{\mathbf{k}}) = (\sqrt{2}\hat{\mathbf{i}} - \hat{\mathbf{j}} + \hat{\mathbf{k}}) \cdot (4\sqrt{2}\hat{\mathbf{i}} + 4\hat{\mathbf{j}} + 4\hat{\mathbf{k}})$$

$$\Rightarrow \mathbf{r} \cdot (4\sqrt{2}\hat{\mathbf{i}} + 4\hat{\mathbf{j}} + 4\hat{\mathbf{k}}) = 8$$

$$\Rightarrow \quad \mathbf{r} \cdot (\sqrt{2}\hat{\mathbf{i}} + \hat{\mathbf{j}} + \hat{\mathbf{k}}) = 2$$

Example 51. Find the equation of the plane such that image of point (1, 2, 3) in it is (– 1, 0, 1).

Sol. Since, the image of $A(1, 2, 3)$ in the plane is $B(-1, 0, 1)$, the plane passes through the mid-point (0, 1, 2) of AB and is normal to the vector $\mathbf{AB} = -2\hat{\mathbf{i}} - 2\hat{\mathbf{j}} - 2\hat{\mathbf{k}}$.

Hence, the equation of the plane is $-2(x - 0) - 2(y - 1) - 2(z - 2) = 0$

or $\quad x + y + z = 3.$

Equation of a Plane Passing through Three Given Points

Cartesian Form

Let the plane be passing through points $A(x_1, y_1, z_1)$, $B(x_2, y_2, z_2)$ and $C(x_3, y_3, z_3)$.

Let $P(x, y, z)$ be any point on the plane.

Then, vectors **PA**, **BA** and **CA** are coplanar.

$$[\mathbf{PA}\ \mathbf{BA}\ \mathbf{CA}] = 0$$

$$\Rightarrow \begin{vmatrix} x - x_1 & y - y_1 & z - z_1 \\ x_2 - x_1 & y_2 - y_1 & z_2 - z_1 \\ x_3 - x_1 & y_3 - y_1 & z_3 - z_1 \end{vmatrix} = 0,$$ which is the required equation of the plane

Vector Form

Vector form of the equation of the plane passing through three points A, B and C having position vectors **a**, **b** and **c**, respectively.

Let **r** be the position vector of any point P in the plane.

Hence, vector $\mathbf{AP} = \mathbf{r} - \mathbf{a}$ $\mathbf{AB} = \mathbf{b} - \mathbf{a}$ and $\mathbf{AC} = \mathbf{c} - \mathbf{a}$ are coplanar.

Hence, $(\mathbf{r} - \mathbf{a}) \cdot \{(\mathbf{b} - \mathbf{a}) \times (\mathbf{c} - \mathbf{a})\} = 0$

$$\Rightarrow \quad (\mathbf{r} - \mathbf{a}) \cdot (\mathbf{b} \times \mathbf{c} - \mathbf{b} \times \mathbf{a} - \mathbf{a} \times \mathbf{c} + \mathbf{a} \times \mathbf{a}) = 0$$

$$\Rightarrow \quad (\mathbf{r} - \mathbf{a}) \cdot (\mathbf{b} \times \mathbf{c} + \mathbf{a} \times \mathbf{b} + \mathbf{c} \times \mathbf{a}) = 0$$

$$\Rightarrow \quad \mathbf{r} \cdot (\mathbf{b} \times \mathbf{c} + \mathbf{a} \times \mathbf{b} + \mathbf{c} \times \mathbf{a}) = \mathbf{a} \cdot (\mathbf{b} \times \mathbf{c}) + \mathbf{a} \cdot (\mathbf{a} \times \mathbf{b}) + \mathbf{a} \cdot (\mathbf{c} \times \mathbf{a})$$

$$\Rightarrow \quad [\mathbf{r}\ \mathbf{b}\ \mathbf{c}] + [\mathbf{r}\ \mathbf{a}\ \mathbf{b}] + [\mathbf{r}\ \mathbf{c}\ \mathbf{a}] = [\mathbf{a}\ \mathbf{b}\ \mathbf{c}]$$

which is the required equation of the plane.

> ***Note***
> 1. If p is the length of perpendicular from the origin on this plane, then $p = [\mathbf{a}\ \mathbf{b}\ \mathbf{c}] / n$, where $n = |\mathbf{a} \times \mathbf{b} + \mathbf{b} \times \mathbf{c} + \mathbf{c} \times \mathbf{a}|$.
> 2. Four points **a, b, c** and **d** are coplanar if **d** lies on the plane containing **a, b** and **c**.
>
> or $\quad \mathbf{d} \cdot [\mathbf{a} \times \mathbf{b} + \mathbf{b} \times \mathbf{c} + \mathbf{c} \times \mathbf{a}] = [\mathbf{a}\ \mathbf{b}\ \mathbf{c}]$
>
> or $\quad [\mathbf{d}\ \mathbf{a}\ \mathbf{b}] + [\mathbf{d}\ \mathbf{b}\ \mathbf{c}] + [\mathbf{d}\ \mathbf{c}\ \mathbf{a}] = [\mathbf{a}\ \mathbf{b}\ \mathbf{c}]$

Example 52. Find the equation of the plane passing through $A(2, 2, -1)$, $B(3, 4, 2)$ and $C(7, 0, 6)$. Also find a unit vector perpendicular to this plane.

Sol. Here, $(x_1, y_1, z_1) \equiv (2, 2, -1)$, $(x_2, y_2, z_2) \equiv (3, 4, 2)$ and $(x_3, y_3, z_3) \equiv (7, 0, 6)$.

Then, the equation of the plane is

$$\begin{vmatrix} x - x_1 & y - y_1 & z - z_1 \\ x_2 - x_1 & y_2 - y_1 & z_2 - z_1 \\ x_3 - x_1 & y_3 - y_1 & z_3 - z_1 \end{vmatrix} = 0$$

or
$$\begin{vmatrix} x - 2 & y - 2 & z - (-1) \\ 3 - 2 & 4 - 2 & 2 - (-1) \\ 7 - 2 & 0 - 2 & 6 - (-1) \end{vmatrix} = 0$$

or $\quad 5x + 2y - 3z = 17$

A normal vector to this plane is $\mathbf{d} = 5\hat{\mathbf{i}} + 2\hat{\mathbf{j}} - 3\hat{\mathbf{k}}$ ...(i)

Therefore, a unit vector normal to (i) is given by

$$\hat{\mathbf{n}} = \frac{\mathbf{d}}{|\mathbf{d}|} = \frac{5\hat{\mathbf{i}} + 2\hat{\mathbf{j}} - 3\hat{\mathbf{k}}}{\sqrt{25 + 4 + 9}}$$

$$= \frac{1}{\sqrt{38}}(5\hat{\mathbf{i}} + 2\hat{\mathbf{j}} - 3\hat{\mathbf{k}})$$

Example 53. Find equation of plane passing through the points $P(1, 1, 1)$, $Q(3, -1, 2)$ and $R(-3, 5, -4)$.

Sol. Let the equation of plane passing through (1, 1, 1) be $a(x-1)+b(y-1)+c(z-1)=0$, as it passes through the points Q and R.

$$\therefore \quad 2a-2b+c=0$$

and $$-4a+4b-5c=0$$

Hence, solving by cross multiplication method, we get

$$\frac{a}{10-4}=\frac{b}{-4+10}=\frac{c}{8-8}=k$$

$$\therefore \quad a=6k,\ b=6k,\ c=0$$

On substituting in Eq. (i), we get

$$6(x-1)+6(y-1)+0=0$$

i.e. $x+y=2$; which is the required equation.

Aliter Equation of plane passing through (x_1, y_1, z_1), (x_2, y_2, z_2) and (x_3, y_3, z_3) is

$$\begin{vmatrix} x-x_1 & y-y_1 & z-z_1 \\ x_2-x_1 & y_2-y_1 & z_2-z_1 \\ x_3-x_1 & y_3-y_1 & z_3-z_1 \end{vmatrix}=0$$

i.e. $$\begin{vmatrix} x-1 & y-1 & z-1 \\ 3-1 & -1-1 & 2-1 \\ -3-1 & 5-1 & -4-1 \end{vmatrix}=0$$

On solving, we get $x+y=2$

Equation of a Plane Passing Through a Given Point and Parallel to Two Given Vectors

Let a plane pass through $A(\mathbf{a})$ and is parallel to the plane formed by two vectors $\mathbf{b}$ and $\mathbf{c}$. Since, $\mathbf{AP}$ lies in the plane and $\mathbf{b}$ and $\mathbf{c}$ are two non-collinear vectors,

$$\mathbf{AP}=\lambda\mathbf{b}+\mu\mathbf{c}$$

$$\Rightarrow \quad \mathbf{r}-\mathbf{a}=\lambda\mathbf{b}+\mu\mathbf{c}$$

$$\Rightarrow \quad \mathbf{r}=\mathbf{a}+\lambda\mathbf{b}+\mu\mathbf{c}$$

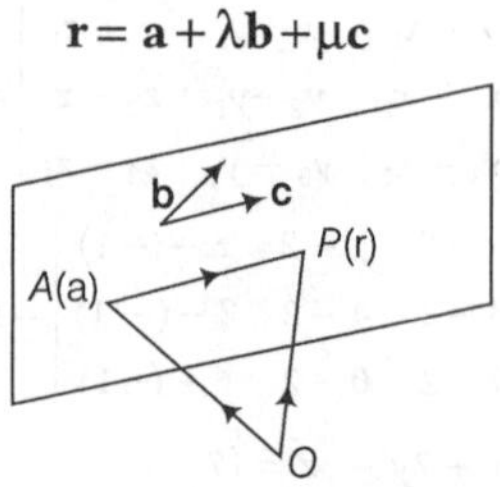

Here, λ and μ are arbitrary scalars.

This form is also called the parametric form of the plane. It can also be written in the non-parametric form as

$$(\mathbf{r}-\mathbf{a})\cdot(\mathbf{b}\times\mathbf{c})=0$$

or $$[\mathbf{r}\ \mathbf{b}\ \mathbf{c}]=[\mathbf{a}\ \mathbf{b}\ \mathbf{c}]$$

Cartesian Form

From $(\mathbf{r}-\mathbf{a})\cdot(\mathbf{b}\times\mathbf{c})=0$, we have $[\mathbf{r}-\mathbf{a}\ \mathbf{b}\ \mathbf{c}]$

$$\Rightarrow \begin{vmatrix} x-x_1 & y-y_1 & z-z_1 \\ x_2 & y_2 & z_2 \\ x_3 & y_3 & z_3 \end{vmatrix}=0,$$ which is the required

equation of the plane, where $\mathbf{b}=x_2\hat{\mathbf{i}}+y_2\hat{\mathbf{j}}+z_2\hat{\mathbf{k}}$ and $\mathbf{c}=x_3\hat{\mathbf{i}}+y_3\hat{\mathbf{j}}+z_3\hat{\mathbf{k}}$.

Example 54. Find the vector equation of the following planes in cartesian form $\mathbf{r}=\hat{\mathbf{i}}-\hat{\mathbf{j}}+\lambda(\hat{\mathbf{i}}+\hat{\mathbf{j}}+\hat{\mathbf{k}})+\mu(\hat{\mathbf{i}}-2\hat{\mathbf{j}}+3\hat{\mathbf{k}})$.

Sol. The equation of the plane is

$$\mathbf{r}=\hat{\mathbf{i}}-\hat{\mathbf{j}}+\lambda(\hat{\mathbf{i}}+\hat{\mathbf{j}}+\hat{\mathbf{k}})+\mu(\hat{\mathbf{i}}-2\hat{\mathbf{j}}+3\hat{\mathbf{k}}).$$

Let $\mathbf{r}=x\hat{\mathbf{i}}+y\hat{\mathbf{j}}+z\hat{\mathbf{k}}$

Hence, the equation is

$$(x\hat{\mathbf{i}}+y\hat{\mathbf{j}}+z\hat{\mathbf{k}})-(\hat{\mathbf{i}}-\hat{\mathbf{j}})=\lambda(\hat{\mathbf{i}}+\hat{\mathbf{j}}+\hat{\mathbf{k}})+\mu(\hat{\mathbf{i}}-2\hat{\mathbf{j}}+3\hat{\mathbf{k}})$$

Thus, vectors $(x\hat{\mathbf{i}}+y\hat{\mathbf{j}}+z\hat{\mathbf{k}})-(\hat{\mathbf{i}}-\hat{\mathbf{j}})$, $\hat{\mathbf{i}}+\hat{\mathbf{j}}+\hat{\mathbf{k}}$, $\hat{\mathbf{i}}-2\hat{\mathbf{j}}+3\hat{\mathbf{k}}$ are coplanar.

Therefore, the equation of the plane is

$$\begin{vmatrix} x-1 & y-(-1) & z-0 \\ 1 & 1 & 1 \\ 1 & -2 & 3 \end{vmatrix}=0$$

or $$5x-2y-3z-7=0$$

Intercept Form of a Plane

The equation of a plane having intercepting lengths a, b and c with X-axis, Y-axis and Z-axis, respectively is

$$\frac{x}{a}+\frac{y}{b}+\frac{z}{c}=1$$

Proof Let O be the origin and let OX, OY and OZ be the coordinate axes.

Let the plane meets the coordinate axes at the points P, Q and R, respectively such that

$OP=a$, $OQ=b$ and $P=c$. Then, the coordinates of the points are $P(a,0,0)$, $Q(0,b,0)$ and $R(0,0,c)$.

Let the equation of plane be

$$Ax+By+Cz+D=0 \quad \text{...(i)}$$

Since, Eq. (i) passes through $(a,0,0)$, $(0,b,0)$ and $(0,0,c)$, we have

$$Aa+D=0 \Rightarrow A=\frac{-D}{a}$$

$$Bb+D=0 \Rightarrow B=\frac{-D}{b}$$

$$Cc + D = 0 \Rightarrow C = \frac{-D}{b}$$

On putting these values in Eq. (i),we get required equation of plane as

$$\frac{-D}{a}x - \frac{D}{b}y - \frac{D}{c}z = -D$$

$$\Rightarrow \quad \frac{x}{a} + \frac{y}{b} + \frac{z}{c} = 1$$

Example 55. A plane meets the coordinates axes in *A, B* and *C* such that the centroid of the ΔABC is the point (p, q, r), show that the equation of the plane is $\frac{x}{p} + \frac{y}{q} + \frac{z}{r} = 3$

Sol. Let the required equation of plane be

$$\frac{x}{a} + \frac{y}{b} + \frac{z}{c} = 1 \quad \text{...(i)}$$

Then, the coordinates of *A, B* and *C* are $A(a, 0, 0)$, $B(0, b, 0)$ and $C(0, 0, c)$, respectively. So, the centroid of the ΔABC,

$$\left(\frac{a}{3}, \frac{b}{3}, \frac{c}{3}\right)$$

But the coordinate of the centroid are (p, q, r).

$$\therefore \quad \frac{a}{3} = p, \frac{b}{3} = q, \frac{c}{3} = r$$

On putting the values of *a, b* and *c* in Eq. (i), we get

The required plane as

$$\frac{x}{3p} + \frac{y}{3q} + \frac{z}{3r} = 1$$

$$\Rightarrow \quad \frac{x}{p} + \frac{y}{q} + \frac{z}{r} = 3$$

Example 56. A variable plane moves in such a way that the sum of the reciprocals of its intercepts on the three coordinate axes is constant. Show that the plane passes through the fixed point.

Sol. Let the equation of the plane be $\frac{x}{a} + \frac{y}{b} + \frac{z}{c} = 1$. Then, the intercepts made by the plane with axes are a, b and c.

$$\therefore \quad \frac{1}{a} + \frac{1}{b} + \frac{1}{c} = \text{constant } (k) \quad \text{...(i) (given)}$$

$$\Rightarrow \quad \frac{1}{ak} + \frac{1}{bk} + \frac{1}{ck} = 1 \text{ comparing with } \frac{x}{a} + \frac{y}{b} + \frac{z}{c} = 1$$

$$x = \frac{1}{k}, y = \frac{1}{k}$$

and $$z = \frac{1}{k}$$

This shows Eq. (i) passes through the fixed point $\left(\frac{1}{k}, \frac{1}{k}, \frac{1}{k}\right)$.

Angle between Two Planes

Vector Form

The angle between two planes is defined as the angle between their normals.

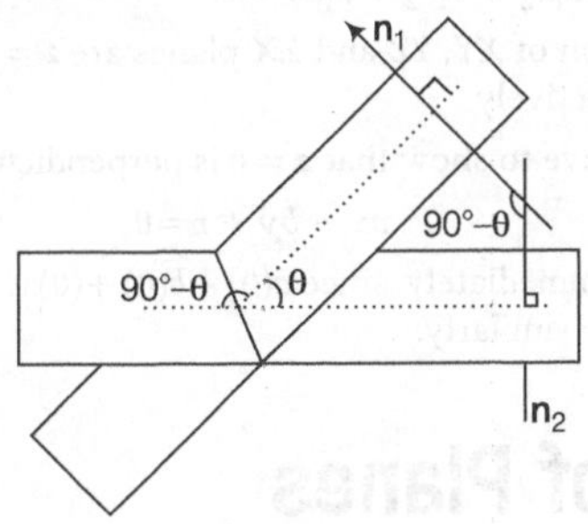

Let θ be the angle between planes $\mathbf{r} \cdot \mathbf{n}_1 = d_1$ and $\mathbf{r} \cdot \mathbf{n}_2 = d_2$ then $\cos\theta = \frac{\mathbf{n}_1 \cdot \mathbf{n}_2}{|\mathbf{n}_1||\mathbf{n}_2|}$

Condition for Perpendicularity

If the planes $\mathbf{r} \cdot \mathbf{n}_1 = d_1$ and $\mathbf{r} \cdot \mathbf{n}_2 = \mathbf{d}_2$ are perpendicular, then $\mathbf{n}_1$ and $\mathbf{n}_2$ are perpendicular. Therefore, $\mathbf{n}_1 \cdot \mathbf{n}_2 = 0$

Condition for Parallelism

If the planes $\mathbf{r} \cdot \mathbf{n}_1 = d_1$ and $\mathbf{r} \cdot \mathbf{n}_2 = d_2$ are parallel, there exists the scalar λ such that $\mathbf{n}_1 = \lambda \mathbf{n}_2$.

Cartesian Form

If the planes are $a_1 x + b_1 y + c_1 z + d = 0$

and $$a_2 x + b_2 y + c_2 z + d_2 = 0$$

$$\Rightarrow \quad \cos\theta = \frac{a_1 a_2 + b_1 b_2 + c_1 c_2}{\sqrt{a_1^2 + b_1^2 + c_1^2}\sqrt{a_2^2 + b_2^2 + c_2^2}}$$

Condition for parallelism

$$\frac{a_1}{a_2} = \frac{b_1}{b_2} = \frac{c_1}{c_2} = \lambda$$

Condition for perpendicularity

$$a_1 a_2 + b_1 b_2 + c_1 c_2 = 0$$

Example 57. Find the angle between the planes $2x + y - 2x + 3 = 0$ and $\mathbf{r} \cdot (6\hat{\mathbf{i}} + 3\hat{\mathbf{j}} + 2\hat{\mathbf{k}}) = 5$.

Sol. Normals along the given planes are $2\hat{\mathbf{i}} + \hat{\mathbf{j}} - 2\hat{\mathbf{k}}$ and $6\hat{\mathbf{i}} + 3\hat{\mathbf{j}} + 2\hat{\mathbf{k}}$

Then angle between planes,

$$\theta = \cos^{-1}\frac{(2\hat{\mathbf{i}} + \hat{\mathbf{j}} - 2\hat{\mathbf{k}}) \cdot (6\hat{\mathbf{i}} + 3\hat{\mathbf{j}} + 2\hat{\mathbf{k}})}{\sqrt{(2)^2 + (1)^2 + (-2)^2}\sqrt{(6)^2 + (3)^2 + (2)^2}} = \cos^{-1}\frac{11}{21}$$

Example 58. Show that $ax+by+r=0$, $by+cz+p=0$ and $cz+ax+q=0$ are perpendicular to *XY*, *YZ* and *ZX* planes, respectively.

Sol. The planes $a_1x+b_1y+c_1z+d_1=0$ and $a_2x+b_2y+c_2z+d_2=0$ are perpendicular to each other if and only if $a_1a_2+b_1b_2+c_1c_2=0$.

The equation of *XY*, *YZ* and *ZX* planes are $z=0$, $x=0$ and $y=0$, respectively.

Now, we have to show that $z=0$ is perpendicular to

$$ax+by+r=0.$$

It follows immediately, since $a(0)+b(0)+(0)(1)=0$, other parts can be done similarly.

Family of Planes

Plane Parallel to a Given Plane

Since parallel planes have the same normal vector, so equation of a plane parallel to $\mathbf{r}\cdot\mathbf{n}=d_1$ is of the form $\mathbf{r}\cdot\mathbf{n}=d_2$, where d_2 is determined by the given conditions.

In cartesian form, if $ax+by+cz+d=0$ be the given plane then the plane parallel to this plane is $ax+by+cz+k=0$.

Example 59. Find the equation of the plane through the point $(1,4,-2)$ and parallel to the plane $-2x+y-3z=7$.

Sol. Let the equation of a plane parallel to the plane $-2x+y-3z=7$ be

$$-2x+y-3z+k=0 \quad \text{...(i)}$$

This passes through $(1,4,-2)$, therefore

$$(-2)(1)+4-3(-2)+k=0$$

$$\Rightarrow \quad -2+4+6+k=0 \Rightarrow k=-8$$

Putting $k=-8$ in Eq. (i), we obtain

$$-2x+y-3z-8=0 \text{ or } -2x+y-3z=8$$

This is the equation of the required plane.

Example 60. Find the equation of the plane passing through $(3,4,-1)$, which is parallel to the plane $\mathbf{r}\cdot(2\hat{\mathbf{i}}-3\hat{\mathbf{j}}+5\hat{\mathbf{k}})+7=0$.

Sol. The equation of any plane which is parallel to $\mathbf{r}\cdot(2\hat{\mathbf{i}}-3\hat{\mathbf{j}}+5\hat{\mathbf{k}})+7=0$ is

$$\mathbf{r}\cdot(2\hat{\mathbf{i}}-3\hat{\mathbf{j}}+5\hat{\mathbf{k}})+\lambda=0$$

or $\quad 2x-3y+5z+\lambda=0$

Further (i) will pass through $(3,4,-1)$

if $(2)(3)+(-3)(4)+5(-1)+\lambda=0$

or $\quad -11+\lambda=0 \Rightarrow \lambda=11$

Thus, equation of the required plane is

$$\mathbf{r}\cdot(2\hat{\mathbf{i}}-3\hat{\mathbf{j}}+5\hat{\mathbf{k}})+11=0$$

Equation of any Plane Passing Through the Line of Intersection of Two Plane

The equation of the plane passing through the line of intersection of the planes

$a_1x+b_1y+c_1z+d_1=0$ and $a_2x+b_2y+c_2z+d_2=0$ is $(a_1x+b_1y+c_1z+d_1)+k(a_2x+b_2y+c_2z+d_2)=0$

Proof Let the given plane be

$$a_1x+b_1y+c_1z+d_1=0 \quad \text{...(i)}$$

and $\quad a_2x+b_2y+c_2z+d_2=0 \quad$ ...(ii)

$\therefore$ Required plane is $(a_1x+b_1y+c_1z+d_1)$

$$+k(a_2x+b_2y+c_2z+d_2)=0 \quad \text{...(iii)}$$

Clearly, plane (iii) represents the equation of plane.

Let (α,β,γ) be a point on the line of intersection of planes (i) and (ii), then *P* lies on planes (i) and (ii).

$\therefore \quad a_1\alpha+b_1\beta+c_1\gamma+d_1=0 \quad$ ...(iv)

and $\quad a_2\alpha+b_2\beta+c_2\gamma+d_2=0 \quad$ ...(v)

Now, multiply by *k* in plane (v) and then adding planes (iv) and (v), we get

$$(a_1\alpha+b_1\beta+c_1\gamma+d_1)+k(a_2\alpha+b_2\beta+c_2\gamma+d_2)=0$$

$\Rightarrow P(\alpha,\beta,\gamma)$ lies on plane (iii).

Hence, plane (iii) passes through each point on the line of intersection of planes (i) and (ii).

Thus, plane (iii) is the equation of plane passing through the line of intersection of planes (i) and (ii).

Vector Form

Equation of planes passing through the line of intersection of planes

$$\mathbf{r}\cdot\mathbf{n}_1=d_1 \text{ and } \mathbf{r}\cdot\mathbf{n}_2=d_2 \text{ is}$$

$$(\mathbf{r}\cdot\mathbf{n}_1-d_1)+k(\mathbf{r}\cdot\mathbf{n}_2-d_2)=0$$

or $\quad \mathbf{r}\cdot(\mathbf{n}_1+k\mathbf{n}_2)=d_1+kd_2$, k being any scalar.

Example 61. Find the equation of the plane containing the line of intersection of the plane $x+y+z-6=0$ and $2x+3y+4z+5=0$ and passing through the points $(1,1,1)$.

Sol. The equation of a plane through the line of intersection of the given plane is

$$(x+y+z-6)+\lambda(2x+3y+4z+5)=0 \quad \text{...(i)}$$

If line (i) passes through $(1,1,1)$, we have

$$-3+14\lambda=0$$

$$\Rightarrow \quad \lambda=\frac{3}{14}.$$

Putting $\lambda = \frac{3}{14}$ in line (i), we obtain the equation of the required plane as

$$(x + y + z - 6) + \frac{3}{14}(2x + 3y + 4z + 5) = 0$$

$$\Rightarrow \quad 20x + 23y + 26z - 69 = 0$$

Example 62. Find the planes passing through the intersection of planes $\mathbf{r}\cdot(2\hat{\mathbf{i}} - 3\hat{\mathbf{j}} + 4\hat{\mathbf{k}}) = 1$ and $\mathbf{r}\cdot(\hat{\mathbf{i}} - \hat{\mathbf{j}}) + 4 = 0$ and perpendicular to planes $\mathbf{r}\cdot(2\hat{\mathbf{i}} - \hat{\mathbf{j}} + \hat{\mathbf{k}}) = -8$.

Sol. The equation of any plane through the line of intersection of the given planes is

$$\{\mathbf{r}\cdot(2\hat{\mathbf{i}} - 3\hat{\mathbf{j}} + 4\hat{\mathbf{k}}) - 1\} + \lambda\{\mathbf{r}\cdot(\hat{\mathbf{i}} - \hat{\mathbf{j}}) + 4\} = 0$$

$$\mathbf{r}\cdot\{(2 + \lambda)\hat{\mathbf{i}} - (3 + \lambda)\hat{\mathbf{j}} + 4\hat{\mathbf{k}}\} = 1 - 4\lambda \quad \text{...(i)}$$

If it is perpendicular to $\mathbf{r}\cdot(2\hat{\mathbf{i}} - \hat{\mathbf{j}} + \hat{\mathbf{k}}) + 8 = 0$, then

$$\{(2 + \lambda)\hat{\mathbf{i}} - (3 + \lambda)\hat{\mathbf{j}} + 4\hat{\mathbf{k}}\}\cdot(2\hat{\mathbf{i}} - \hat{\mathbf{j}} + \hat{\mathbf{k}}) = 0$$

$$2(2 + \lambda) + (3 + \lambda) + 4 = 0$$

$$\lambda = -\frac{11}{3}$$

Putting $\lambda = -\frac{11}{3}$ in line (i), we obtain the equation of the required plane as $\mathbf{r}\cdot(-5\hat{\mathbf{i}} + 2\hat{\mathbf{j}} + 12\hat{\mathbf{k}}) = 47$

Two Sides of a Plane

Let $ax + by + cz + d = 0$ be the plane, then the points (x_1, y_1, z_1) and (x_2, y_2, z_2) lie on the same side or opposite side according as

$$\frac{ax_1 + by_1 + cz_1 + d}{ax_2 + by_2 + cz_2 + d} > 0 \text{ or } < 0$$

Proof Here equation of plane is,

$$ax + by + cz + d = 0 \quad \text{...(i)}$$

Let Eq. (i) divide the line segment joining P and Q at R internally in the ratio $m : n$.

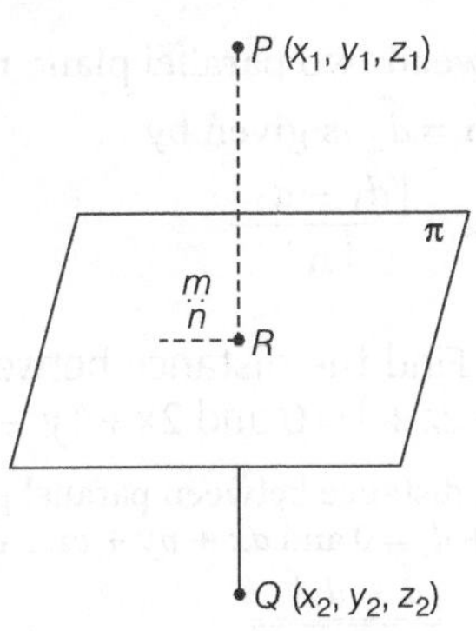

Then, $R\left(\frac{mx_2 + nx_1}{m + n}, \frac{my_2 + ny_1}{m + n}, \frac{mz_2 + nz_1}{m + n}\right)$

Since, R lies on the plane (i).

$$\therefore a\left(\frac{mx_2 + nx_1}{m + n}\right) + b\left(\frac{my_2 + ny_1}{m + n}\right) + c\left(\frac{mz_2 + nz_1}{m + n}\right) + d = 0$$

$$\Rightarrow \quad a(mx_2 + nx_1) + b(my_2 + ny_1) + c(mz_2 + nz_1) + d(m + n) = 0$$

$$\Rightarrow \quad m(ax_2 + by_2 + cz_2 + d) + n(ax_1 + by_1 + cz_1 + d) = 0$$

$$\Rightarrow \quad \frac{m}{n} = -\frac{(ax_1 + by_1 + cz_1 + d)}{(ax_2 + by_2 + cz_2 + d)} \quad \text{...(ii)}$$

Now, if $ax_1 + by_1 + cz_1 + d$ and $ax_2 + by_2 + cz_2 + d$

are of same sign $\quad \frac{m}{n} < 0 \quad$ (external division)

are of opposite sign $\quad \frac{m}{n} > 0 \quad$ (internal division)

$$\therefore \text{ If } \quad \frac{ax_1 + by_1 + cz_1 + d}{ax_2 + by_2 + cz_2 + d} > 0 \quad \text{(same side)}$$

$$\frac{ax_1 + by_1 + cz_1 + d}{ax_2 + by_2 + cz_2 + d} < 0 \quad \text{(opposite side)}$$

Example 63. Find the interval of α for which $(\alpha, \alpha^2, \alpha)$ and $(3, 2, 1)$ lies on same side of $x + y - 4z + 2 = 0$.

Sol. $(\alpha, \alpha^2, \alpha)$ and $(3, 2, 1)$ lies on same side of $x + y - 4z + 2 = 0$

$$\therefore (\alpha + \alpha^2 - 4\alpha + 2)(3 + 2 - 4 + 2) > 0$$

$$\Rightarrow \quad \alpha^2 - 3\alpha + 2 > 0$$

$$(\alpha - 1)(\alpha - 2) > 0 \Rightarrow \alpha \in (-\infty, 1] \cup (2, \infty]$$

Distance of a Point from a Plane

Vector Form

The length of the perpendicular from a point having position vector $\mathbf{a}$ to the plane $\mathbf{r}\cdot\mathbf{n} = d$ is given by

$$P = \frac{|\mathbf{a}\cdot\mathbf{n} - d|}{|\mathbf{n}|}$$

Proof. Let π be the given plane and $P(\mathbf{a})$ be the given point. Let PM be the length of perpendicular from P to the plane π.

Since, line PM passes through $P(\mathbf{a})$ and is parallel to the vector $\mathbf{n}$ which is normal to the plane π. So, vector equation of line PM is

$$\mathbf{r} = \mathbf{a} + \lambda\mathbf{n} \quad \text{...(i)}$$

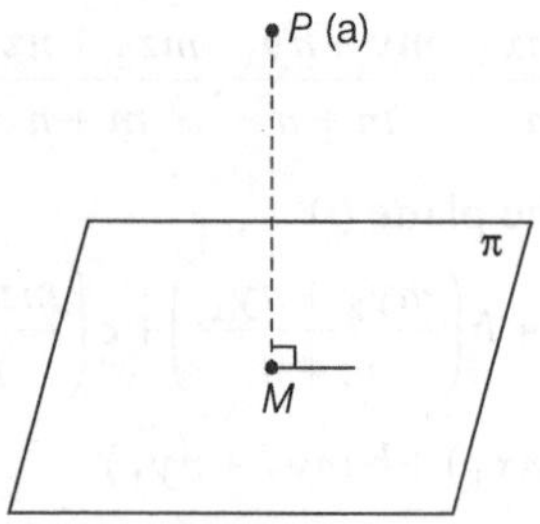

Point M is the intersection of Eq. (i) and the given plane π.

$$\therefore \quad (\mathbf{a}+\lambda\mathbf{n})\cdot\mathbf{n} = d$$

$$\Rightarrow \quad \mathbf{a}\cdot\mathbf{n}+\lambda\mathbf{n}\cdot\mathbf{n} = d \Rightarrow \lambda = \frac{d-(\mathbf{a}\cdot\mathbf{n})}{|\mathbf{n}|^2}$$

On putting the value of λ in Eq. (i), we obtain the position vector of M given by

$$\mathbf{r} = \mathbf{a} + \left(\frac{d-\mathbf{a}\cdot\mathbf{n}}{|\mathbf{n}|^2}\right)\mathbf{n}$$

$$\mathbf{PM} = \text{Position vector of } M - \text{Position vector of } P$$

$$= \mathbf{a} + \left(\frac{d-(\mathbf{a}\cdot\mathbf{n})}{|\mathbf{n}|^2}\right)\mathbf{n} - \mathbf{a}$$

$$\mathbf{PM} = \left(\frac{d-(\mathbf{a}\cdot\mathbf{n})}{|\mathbf{n}|^2}\right)\mathbf{n}$$

$$\Rightarrow \quad PM = |\mathbf{PM}| = \left|\frac{(d-\mathbf{a}\cdot\mathbf{n})\,\mathbf{n}}{|\mathbf{n}|^2}\right|$$

$$= \frac{|d-(\mathbf{a}\cdot\mathbf{n})|\,|\mathbf{n}|}{|\mathbf{n}|^2} = \frac{|d-(\mathbf{a}\cdot\mathbf{n})|}{|\mathbf{n}|}$$

Thus, the length of perpendicular from a point having position vector $\mathbf{a}$ on the plane $\mathbf{r}\cdot\mathbf{n} = d$ is $\dfrac{|\mathbf{a}\cdot\mathbf{n} - d|}{|\mathbf{n}|}$

Cartesian Form

The length of perpendicular from a point $P(x_1, y_1, z_1)$ to the plane $ax + by + cz + d = 0$. Then, the equation of PM is

$$\frac{x-x_1}{a} = \frac{y-y_1}{b} = \frac{z-z_1}{c} \qquad \text{...(i)}$$

The coordinates of any point on this line are

$$(x_1 + ar, y_1 + br, z_1 + cr)$$

Thus, the point coincides with M iff it lies on plane.

i.e. $a(x_1 + ar) + b(y_1 + br) + c(z_1 + cr) + d = 0$

$$\text{i.e.} \quad r = -\frac{(ax_1 + by_1 + cz_1 + d)}{a^2 + b^2 + c^2} \qquad \text{...(ii)}$$

$$\text{Now,} \quad PM = \sqrt{(x_1 + ar - x_1)^2 + (y_1 + br - y_1)^2 + (z_1 + cr - z_1)^2}$$

$$= \sqrt{(a^2 + b^2 + c^2)\,r^2} = \sqrt{a^2 + b^2 + c^2}\,|r|$$

$$= \sqrt{a^2 + b^2 + c^2}\left|\frac{-(ax_1 + by_1 + cz_1 + d)}{a^2 + b^2 + c^2}\right|$$

[from Eq. (ii)]

$$\therefore \quad PM = \frac{|(ax_1 + by_1 + cz_1 + d)|}{\sqrt{a^2 + b^2 + c^2}}$$

Example 64. Find the distance of the point (2, 1, 0) from the plane $2x + y + 2z + 5 = 0$.

Sol. We know that the distance of the point (x_1, y_1, z_1) from the plane $ax + by + cz + d = 0$ is

$$\frac{|ax_1 + by_1 + cz_1 + d|}{\sqrt{a^2 + b^2 + c^2}}$$

$$\text{So, required distance} = \frac{|2\times 2 + 1 + 2\times 0 + 5|}{\sqrt{2^2 + 1^2 + 2^2}} = \frac{10}{3}.$$

Distance between the Parallel Planes

The distance between two parallel planes

$$ax + by + cz + d_1 = 0$$

$$\text{and} \quad ax + by + cz + d_2 = 0$$

$$\text{is given by} \quad d = \left|\frac{(d_2 - d_1)}{\sqrt{a^2 + b^2 + c^2}}\right|$$

Proof. Let d = Difference of the length of perpendicular from origin to the two planes.

$$= \left|\frac{|d_1|}{\sqrt{a^2 + b^2 + c^2}} - \frac{|d_2|}{\sqrt{a^2 + b^2 + c^2}}\right|.$$

$$\text{if } d_1 \text{ and } d_2 \text{ are of same side} = \left|\frac{d_1 - d_2}{\sqrt{a^2 + b^2 + c^2}}\right|$$

Vector Form

The distance between two parallel plane $\mathbf{r}\cdot\mathbf{n} = d_1$ and $\mathbf{r}\cdot\mathbf{n} = d_2$ is given by

$$d = \frac{|d_1 - d_2|}{|\mathbf{n}|}$$

Example 65. Find the distance between the parallel planes $x + 2y - 2z + 1 = 0$ and $2x + 4y - 4z + 5 = 0$.

Sol. We know that, distance between parallel planes $ax + by + cz + d_1 = 0$ and $ax + by + cz + d_2 = 0$ is,

$$\frac{|d_1 - d_2|}{\sqrt{a^2 + b^2 + c^2}}$$

$\therefore$ Distance between $x + 2y - 2z + 1 = 0$

and $x + 2y - 2z + \frac{5}{2} = 0$ is

$$\frac{\left|\frac{5}{2} - 1\right|}{\sqrt{1 + 4 + 4}} = \frac{1}{2}$$

Equation of Planes Bisecting the Angle between Two Planes

Equation of the planes bisecting the angle between the planes.

$a_1x + b_1y + c_1z + d_1 = 0$ and $a_2x + b_2y + c_2z + d_2 = 0$ is

$$\frac{a_1x + b_1y + c_1z + d_1}{\sqrt{a_1^2 + b_1^2 + c_1^2}} = \pm \frac{a_2x + b_2y + c_2z + d_2}{\sqrt{a_2^2 + b_2^2 + c_2^2}}$$

Proof. Given planes are

$$a_1x + b_1y + c_1z + d_1 = 0 \quad \text{...(i)}$$

and $$a_2x + b_2y + c_2z + d_2 = 0 \quad \text{...(ii)}$$

Let $P(x, y, z)$ be a point on the plane bisecting the angle between planes (i) and (ii).

Let PL and PM be the length of perpendiculars from P to planes (i) and (ii).

$\therefore \quad PL = PM$

$$\Rightarrow \quad \left|\frac{a_1x + b_1y + c_1z + d_1}{\sqrt{a_1^2 + b_1^2 + c_1^2}}\right| = \left|\frac{a_2x + b_2y + c_2z + d_2}{\sqrt{a_2^2 + b_2^2 + c_2^2}}\right|$$

$$\frac{a_1x + b_1y + c_1z + d_1}{\sqrt{a_1^2 + b_1^2 + c_1^2}} = \pm \frac{a_2x + b_2y + c_2z + d_2}{\sqrt{a_2^2 + b_2^2 + c_2^2}}$$

This is equation of planes bisecting the angles between the planes (i) and (ii).

Vector Form

Equation of planes bisecting the angle between planes $\mathbf{r} \cdot \mathbf{n}_1 = d_1$ and $\mathbf{r} \cdot \mathbf{n}_2 = d_2$ are

$$\left|\frac{\mathbf{r} \cdot \mathbf{n}_1 - d_1}{\mathbf{n}_1}\right| = \left|\frac{\mathbf{r} \cdot \mathbf{n}_2 - d_2}{\mathbf{n}_2}\right|$$

$$\Rightarrow \quad \frac{\mathbf{r} \cdot \mathbf{n}_1 - d_1}{\mathbf{n}_1} = \pm \frac{\mathbf{r} \cdot \mathbf{n}_2 - d_2}{\mathbf{n}_2}$$

$$\Rightarrow \quad \mathbf{r} \cdot \frac{\mathbf{n}_1}{|\mathbf{n}_1|} \pm \mathbf{r} \cdot \frac{\mathbf{n}_2}{|\mathbf{n}_1|} = \frac{d_1}{|\mathbf{n}_1|} \pm \frac{d_2}{|\mathbf{n}_2|}$$

$$\Rightarrow \quad \mathbf{r} \cdot (\hat{\mathbf{n}}_1 \pm \hat{\mathbf{n}}_2) = \frac{d_1}{|\mathbf{n}_1|} \pm \frac{d_2}{|\mathbf{n}_2|}$$

Bisector of the Angle between the Two Planes Containing the Origin

Let the equation of the two planes be

$$a_1x + b_1y + c_1z + d_1 = 0 \quad \text{...(i)}$$

and $$a_2x + b_2y + c_2z + d_2 = 0 \quad \text{...(ii)}$$

where, d_1 and d_2 are positive, then equation of the bisector of the angle between the planes (i) and (ii) containing the origin is

$$\frac{a_1x + b_1y + c_1z + d_1}{\sqrt{a_1^2 + b_1^2 + c_1^2}} = \frac{a_2x + b_2y + c_2z + d_2}{\sqrt{a_2^2 + b_2^2 + c_2^2}}$$

Bisector of the Acute and Obtuse Angles between Two Planes

Let the two planes be

$$a_1x + b_1y + c_1z + d_1 = 0 \quad \text{...(i)}$$

and $$a_2x + b_2y + c_2z + d_2 = 0 \quad \text{...(ii)}$$

where, d_1 and $d_2 > 0$

(a) If $a_1a_2 + b_1b_2 + c_1c_2 > 0$, the origin lies in the obtuse angle between the two planes and the equation of bisector of the acute angle is,

$$\frac{a_1x + b_1y + c_1z + d_1}{\sqrt{a_1^2 + b_1^2 + c_1^2}} = -\frac{a_2x + b_2y + c_2z + d_2}{\sqrt{a_2^2 + b_2^2 + c_2^2}}$$

(b) If $a_1a_2 + b_1b_2 + c_1c_2 < 0$, then origin lies in the acute angle between the two planes and the equation of bisector of the acute angle between two planes is

$$\frac{a_1x + b_1y + c_1z + d_1}{\sqrt{a_1^2 + b_1^2 + c_1^2}} = +\frac{a_2x + b_2y + c_2z + d_2}{\sqrt{a_2^2 + b_2^2 + c_2^2}}$$

Example 66. Find the equation of the bisector planes of the angles between the planes $2x - y + 2z + 3 = 0$ and $3x - 2y + 6z + 8 = 0$ and specify the plane which bisects the acute angle and the plane which bisects the obtuse angle.

Sol. The two given planes are

$$2x - y + 2z + 3 = 0 \text{ and } 3x - 2y + 6z + 8 = 0$$

where, d_1 and $d_2 > 0$

and $$a_1a_2 + b_1b_2 + c_1c_2 = 6 + 2 + 12 > 0$$

$$\therefore \quad \frac{a_1x + b_1y + c_1z + d_1}{\sqrt{a_1^2 + b_1^2 + c_1^2}} = -\frac{a_2x + b_2y + c_2z + d_2}{\sqrt{a_2^2 + b_2^2 + c_2^2}}$$

(obtuse angle bisector)

and $$\frac{a_1x + b_1y + c_1z + d_1}{\sqrt{a_1^2 + b_1^2 + c_1^2}} = \frac{a_2x + b_2y + c_2z + d_2}{\sqrt{a_2^2 + b_2^2 + c_2^2}}$$

(acute angle bisector)

i.e. $\dfrac{2x-y+2z+3}{\sqrt{4+1+4}}=\pm\dfrac{3x-2y+6z+8}{\sqrt{9+4+36}}$

$\Rightarrow (14x-7y+14z+21)=\pm(9x-6y+18z+24)$

Taking positive sign on the right hand side, we get

$5x-y-4z-3=0$ (obtuse angle bisector)

and taking negative sign on the right hand side, we get

$23x-13y+32z+45=0$ (acute angle bisector)

Line and Plane

Line of Intersection of Two Planes

Let two non-parallel planes are $\mathbf{r}\cdot\mathbf{n}_1=d_1$ and $\mathbf{r}\cdot\mathbf{n}_2=d_2$

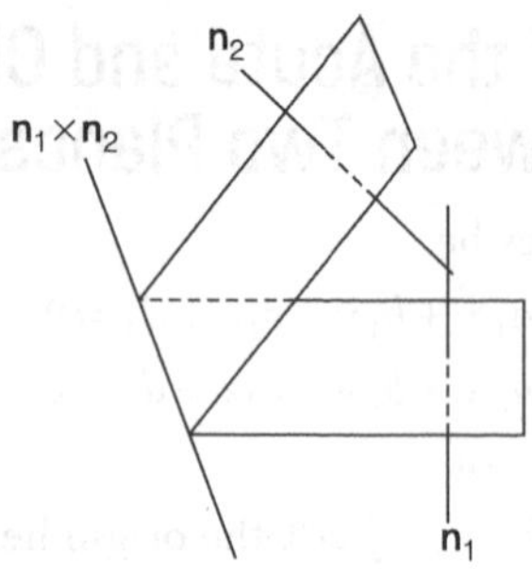

Now line of intersection of planes is perpendicular to vector $\mathbf{n}_1$ and $\mathbf{n}_2$.

$\therefore$ Line of intersection is parallel to vector $\mathbf{n}_1\times\mathbf{n}_2$.

If we wish to find the equation of line of intersection of planes $a_1x+b_1y+c_1z-d_1=0$ and $a_2x+b_2y+c_2z-d_2=0$, then we find any point on the line by putting $z=0$ (say), then we can find corresponding values of x and y be solving equations $a_1x+b_1y-d_1=0$ and $a_2x+b_2y-d_2=0$. Thus, by fixing the value of $z=\lambda$, we can find the corresponding value of x and y in terms of λ. After getting x, y and z in terms of λ, we can find the equation of line in symmetric form.

Example 67. Reduce the equation of line $x-y+2z=5$ and $3x+y+z=6$ in symmetrical form.
Or
Find the line of intersection of planes $x-y+2z=5$ and $3x+y+z=6$.

Sol. Given $x-y+2z=5$, $3x+y+z=6$.

Let $z=\lambda$

Then, $x-y=5-2\lambda$

and $3x+y=6-\lambda$.

Solving these two equations, $4x=11-3\lambda$

and $4y=4x-20+8\lambda=-9+5\lambda$

The equation of the line is $\dfrac{4x-11}{-3}=\dfrac{4y+9}{5}=\dfrac{z-0}{1}$.

Angle between a Line and a Plane

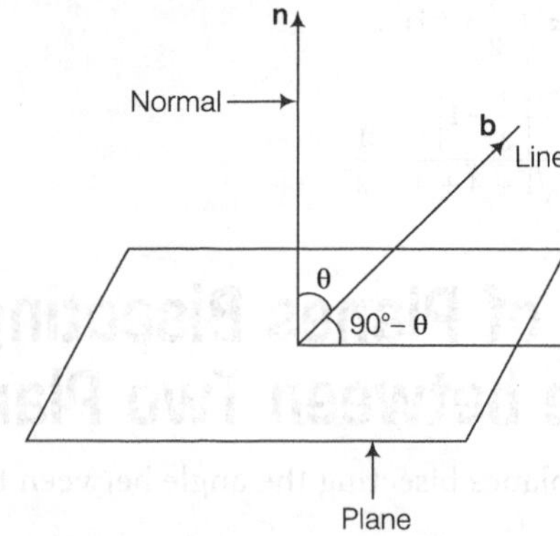

The angle between a line and a plane is the complement of the angle between the line and the normal to the plane.

If the equation of the line is $\mathbf{r}=\mathbf{a}+\lambda\mathbf{b}$ and that of the plane is $\mathbf{r}\cdot\mathbf{n}=d$, then angle θ between the line and the normal to the plane is $\cos\theta=\left|\dfrac{\mathbf{b}\cdot\mathbf{n}}{|\mathbf{b}||\mathbf{n}|}\right|$.

So, the angle ϕ between the line and the plane is given by $90°-\theta$.

$$\sin\phi=\left|\frac{\mathbf{b}\cdot\mathbf{n}}{|\mathbf{b}||\mathbf{n}|}\right| \text{ or } \phi=\sin^{-1}\left|\frac{\mathbf{b}\cdot\mathbf{n}}{|\mathbf{b}||\mathbf{n}|}\right|$$

Line $\mathbf{r}=\mathbf{a}+\lambda\mathbf{b}$ and plane $\mathbf{r}\cdot\mathbf{n}=d$ are perpendicular if $\mathbf{b}=\lambda\mathbf{n}$ or $\mathbf{b}\times\mathbf{n}=\mathbf{0}$ and parallel if $\mathbf{b}\perp\mathbf{n}$ or $\mathbf{b}\cdot\mathbf{n}=0$.

Example 68. Find the angle between the line $\mathbf{r}=\hat{\mathbf{i}}+2\hat{\mathbf{j}}-\hat{\mathbf{k}}+\lambda(\hat{\mathbf{i}}-\hat{\mathbf{j}}+\hat{\mathbf{k}})$ and the plane $\mathbf{r}\cdot(2\hat{\mathbf{i}}-\hat{\mathbf{j}}+\hat{\mathbf{k}})=4$.

Sol. We know that if θ is the angle between the lines $\mathbf{r}=\mathbf{a}+\lambda\mathbf{b}$ and $\mathbf{r}\cdot\mathbf{n}=p$, then $\sin\theta=\left|\dfrac{\mathbf{b}\cdot\mathbf{n}}{|\mathbf{b}||\mathbf{n}|}\right|$

Therefore, if θ is the angle between $\mathbf{r}=\hat{\mathbf{i}}+2\hat{\mathbf{j}}-\hat{\mathbf{k}}+\lambda(\hat{\mathbf{i}}-\hat{\mathbf{j}}+\hat{\mathbf{k}})$ and $\mathbf{r}\cdot(2\hat{\mathbf{i}}-\hat{\mathbf{j}}+\hat{\mathbf{k}})=4$, then

$$\sin\theta=\left|\frac{(\hat{\mathbf{i}}-\hat{\mathbf{j}}+\hat{\mathbf{k}})\cdot(2\hat{\mathbf{i}}-\hat{\mathbf{j}}+\hat{\mathbf{k}})}{|\hat{\mathbf{i}}-\hat{\mathbf{j}}+\hat{\mathbf{k}}||2\hat{\mathbf{i}}-\hat{\mathbf{j}}+\hat{\mathbf{k}}|}\right|$$

$$=\frac{2+1+1}{\sqrt{1+1+1}\sqrt{4+1+1}}=\frac{4}{\sqrt{3}\sqrt{6}}=\frac{4}{3\sqrt{2}}$$

$$\Rightarrow \theta=\sin^{-1}\left(\frac{4}{3\sqrt{2}}\right)$$

Intersection of a Line and a Plane

To find the point of intersection of the line $\dfrac{x-x_1}{l}=\dfrac{y-y_1}{m}=\dfrac{z-z_1}{n}$ and the plane

$$ax + by + cz + d = 0.$$

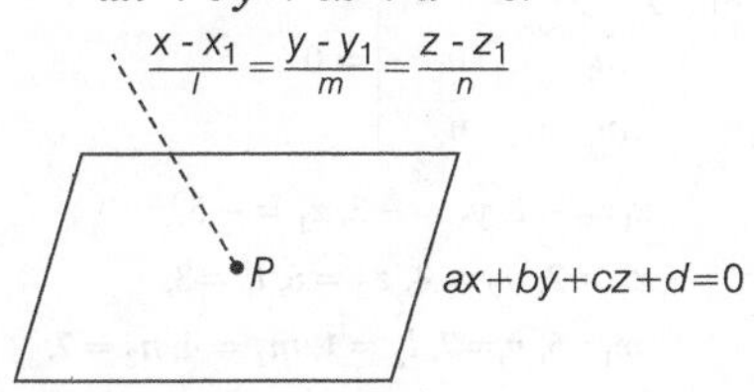

Let $\quad \dfrac{x - x_1}{l} = \dfrac{y - y_1}{m} = \dfrac{z - z_1}{n} = r$

$\therefore \quad (x = rl + x_1, y = mr + y_1, z = nr + z_1) \quad$ ...(i)

be a point in the plane say P.

It must satisfy the equation of plane.

$\therefore \quad a(x_1 + lr) + b(y_1 + mr) + c(z_1 + nr) + d = 0$

$\Rightarrow \quad (ax_1 + by_1 + cz_1 + d) + r(al + bm + cn) = 0$

$\Rightarrow \quad r = -\dfrac{(ax_1 + by_1 + cz_1 + d)}{al + bm + cn}$

On substituting the value of r Eq. (i), we get the coordinates of the required point of intersection.

(i) Condition for a Line to be Parallel to a Plane

Let line $\dfrac{x - x_1}{l} = \dfrac{y - y_1}{m} = \dfrac{z - z_1}{n}$ be parallel to plane $ax + by + cz + d = 0$ iff;

$$\theta = 0 \text{ or } \pi \text{ or } \sin\theta = 0 \Rightarrow al + bm + cm = 0$$

(ii) Condition for a Line to Lie in the Plane

Condition for $\dfrac{x - x_1}{l} = \dfrac{y - y_1}{m} = \dfrac{z - z_1}{n}$ to lie in the plane $\quad ax + by + cz + d = 0$ are

$$al + bm + cn = 0 \text{ and } ax_1 + by_1 + cz_1 + d = 0$$

Note

A line will be in a plane iff

(i) the normal to the plane is perpendicular to the line.

(ii) a point on the line in the plane.

Example 69. Find the distance between the point with position vector $-\hat{\mathbf{i}} - 5\hat{\mathbf{j}} - 10\hat{\mathbf{k}}$ and the point of intersection of the line $\dfrac{x-2}{3} = \dfrac{y+1}{4} = \dfrac{z-2}{12}$ with the plane $x - y + z = 5$.

Sol. The coordinates of any point on the line

$\dfrac{x-2}{3} = \dfrac{y+1}{4} = \dfrac{z-2}{2} = r$ (say) are $(3r + 2, 4r - 1, 12r + 2)$

If it lies on the plane $x - y + z = 5$, then

$3r + 2 - 4r + 1 + 12r + 2 = 5 \Rightarrow 11r = 0 \Rightarrow r = 0.$

Putting $r = 0$ in (i), we obtain $(2, -1, 2)$ as the coordinates of the point of intersection of the given line and plane.

Required distance = distance between points $(-1, -5, -10)$ and $(2, -1, 2) = \sqrt{(2+1)^2 + (-1+5)^2 + (2+10)^2}$

$$= \sqrt{9 + 16 + 144} = \sqrt{169} = 13.$$

Coplanarity of Two Lines

The straight line $\dfrac{x - x_1}{l} = \dfrac{y - y_1}{m} = \dfrac{z - z_1}{n}$ lies in a given plane $ax + by + cz + d = 0$ if $ax_1 + by_1 + cz_1 + d = 0$ and $al + bm + cn = 0$

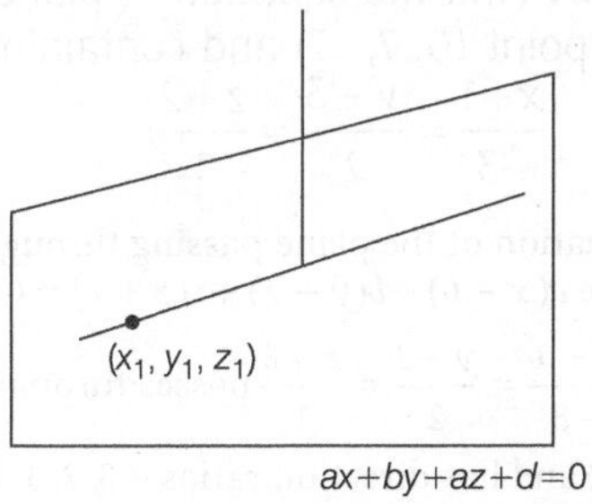

Thus, the general equation of the plane containing a straight line

$$\frac{x - x_1}{l} = \frac{y - y_1}{m} = \frac{z - z_1}{n} \text{ is}$$

$$a(x - x_1) + b(y - y_1) + c(z - z_1) = 0$$

where, $\quad al + bm + cn = 0$

The equation of the plane containing a straight line $\dfrac{x - x_1}{l} = \dfrac{y - y_1}{m} = \dfrac{z - z_1}{n}$ and parallel to the straight line $\dfrac{x - x_2}{l_1} = \dfrac{y - y_2}{m_1} = \dfrac{z - z_2}{n_1}$ is

$$\begin{vmatrix} x - x_1 & y - y_1 & z - z_1 \\ l & m & n \\ l_1 & m_1 & n_1 \end{vmatrix} = 0$$

Hence, the equation of the plane containing two given straight lines

$$\frac{x - x_1}{l} = \frac{y - y_1}{m} = \frac{z - z_1}{n}$$

and $\quad \dfrac{x - x_2}{l_1} = \dfrac{y - y_2}{m_1} = \dfrac{z - z_2}{n_1}$

$$\begin{vmatrix} x - x_1 & y - y_1 & z - z_1 \\ l & m & n \\ l_1 & m_1 & n_1 \end{vmatrix} = 0$$

$$\text{or} \quad \begin{vmatrix} x - x_2 & y - y_2 & z - z_2 \\ l & m & n \\ l_1 & m_1 & n_1 \end{vmatrix} = 0$$

If the lines $\mathbf{r} = \mathbf{a}_1 + \lambda \mathbf{b}_1$ and $\mathbf{r} = \mathbf{a}_2 + \lambda \mathbf{b}_2$ are coplanar, then

$$[\mathbf{a}_1 \ \mathbf{b}_1 \ \mathbf{b}_2] = [\mathbf{a}_2 \ \mathbf{b}_1 \ \mathbf{b}_2]$$

and the equation of plane containing them is

$$[\mathbf{r} \ \mathbf{b}_1 \ \mathbf{b}_2] = [\mathbf{a}_1 \ \mathbf{b}_1 \ \mathbf{b}_2]$$

$$\text{or} \quad [\mathbf{r} \ \mathbf{b}_1 \ \mathbf{b}_2] = [\mathbf{a}_2 \mathbf{b}_1 \ \mathbf{b}_2]$$

Example 70. Find the equation of plane passing through the point (0, 7, –7) and containing the line $\frac{x+1}{-3} = \frac{y-3}{2} = \frac{z+2}{1}$.

Sol. Let the equation of the plane passing through the point $(0, 7, -7)$ be $a(x-0) + b(y-7) + c(z+7) = 0$ ...(i)

The line $\frac{x+1}{-3} = \frac{y-3}{2} = \frac{z+2}{1}$ passes through the point $(-1, 3, -2)$ and has direction ratios $-3, 2, 1$. If (i) contains this line, it must pass through $(-1, 3, -2)$ and must be parallel to the line. Therefore,

$$a(-1) + b(3-7) + c(-2+7) = 0$$

i.e. $\quad a(-1) + b(-4) + c(5) = 0$ ...(ii)

and $\quad -3a + 2b + 1c = 0$ ...(iii)

On solving Eqs. (ii) and (iii) by cross multiplication, we get

$$\frac{a}{-14} = \frac{b}{-14} = \frac{c}{-14} \Rightarrow \frac{a}{1} = \frac{b}{1} = \frac{c}{1} = \lambda \quad \text{(say)}$$

$\Rightarrow \quad a = \lambda, b = \lambda, c = \lambda$

Putting the values of a, b, c in (i), we obtain

$$\lambda(x-0) + \lambda(y-7) + \lambda(z+7) = 0$$

$\Rightarrow \quad x + y + z = 0$

This is the equation of the required plane.

Example 71. Prove that the lines $\frac{x+1}{3} = \frac{y+3}{5} = \frac{z+5}{7}$ and $\frac{x-2}{1} = \frac{y-4}{4} = \frac{z-6}{7}$ are coplanar. Also, find the plane containing these two lines.

Sol. We know that, the line $\frac{x-x_1}{l_1} = \frac{y-y_1}{m_1} = \frac{z-z_1}{n_1}$

and $\frac{x-x_2}{l_2} = \frac{y-y_2}{m_2} = \frac{z-z_2}{n_2}$ are coplanar if

$$\begin{vmatrix} x_2 - x_1 & y_2 - y_1 & z_2 - z_1 \\ l_1 & m_1 & n_1 \\ l_2 & m_2 & n_2 \end{vmatrix} = 0$$

and the equation of the plane containing these two lines is

$$\begin{vmatrix} x - x_1 & y - y_1 & z - z_1 \\ l_1 & m_1 & n_1 \\ l_2 & m_2 & n_2 \end{vmatrix} = 0$$

Here, $\quad x_1 = -1, y_1 = -3, z_1 = -5,$
$x_2 = 2, y_2 = 4, z_2 = 6, l_1 = 3,$
$m_1 = 5, n_1 = 7, l_2 = 1, m_2 = 4, n_2 = 7.$

$$\therefore \begin{vmatrix} x_2 - x_1 & y_2 - y_1 & z_2 - z_1 \\ l_1 & m_1 & n_1 \\ l_2 & m_2 & n_2 \end{vmatrix} = \begin{vmatrix} 3 & 7 & 11 \\ 3 & 5 & 7 \\ 1 & 4 & 7 \end{vmatrix} = 0$$

so, the given lines are coplanar.

The equation of the plane containing the lines is

$$\begin{vmatrix} x+1 & y+3 & z+5 \\ 3 & 5 & 7 \\ 1 & 4 & 7 \end{vmatrix} = 0$$

ar $(x+1)(35-28) - (y+3)(21-7) + (z+5)(12-5) = 0$

or $\quad x - 2y + z = 0.$

Image of a Point in a Plane

To find the image of the point (α, β, γ) in the plane

$$ax + by + cz + d = 0 \quad \text{...(i)}$$

Let $Q(x_1, y_1, z_1)$ be the image of point P in the plane (i).

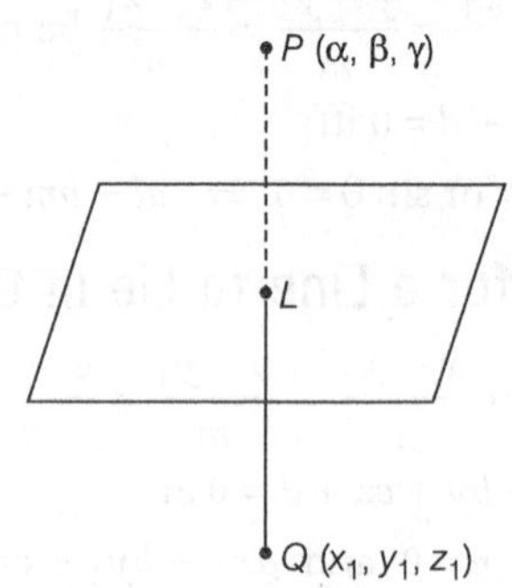

Let PQ meet plane (i) at L, direction ratios of normal to plane (i) are (a, b, c), since PQ perpendicular of plane (i). So, direction ratios of PQ are a, b, c.

$\Rightarrow$ Equation of line PQ is,

$$\frac{x-\alpha}{a} = \frac{y-\beta}{b} = \frac{z-\gamma}{c} = r \quad \text{(say)}$$

Coordinate of any point on line PQ may be taken as

$$(ar + \alpha, br + \beta, cr + \gamma)$$

Let $\quad Q(ar + \alpha, br + \beta, cr + \gamma)$

Since, L is the middle point of PQ

$$\therefore \quad L = \left(\alpha + \frac{ar}{2}, \beta + \frac{br}{2}, \gamma + \frac{cr}{2}\right)$$

Since, L lies on plane (i), we get

$$a\left(\frac{ar}{2}+\alpha\right)+b\left(\frac{br}{2}+\beta\right)+c\left(\frac{cr}{2}+\gamma\right)+d=0$$

$$\Rightarrow \quad (a^2+b^2+c^2)\frac{r}{2}=-(a\alpha+b\beta+c\gamma+d)$$

$$\Rightarrow \quad r=\frac{-2(a\alpha+b\beta+c\gamma+d)}{a^2+b^2+c^2}$$

Example 72. Find the image of the point $P(3, 5, 7)$ in the plane $2x+y+z=0$.

Sol. Given plane is $2x+y+z=0$ ...(i)

and the point $P(3, 5, 7)$

DR's of normal to the plane (i) are 2, 1, 1.

Let Q be the image of a point P in plane (i).

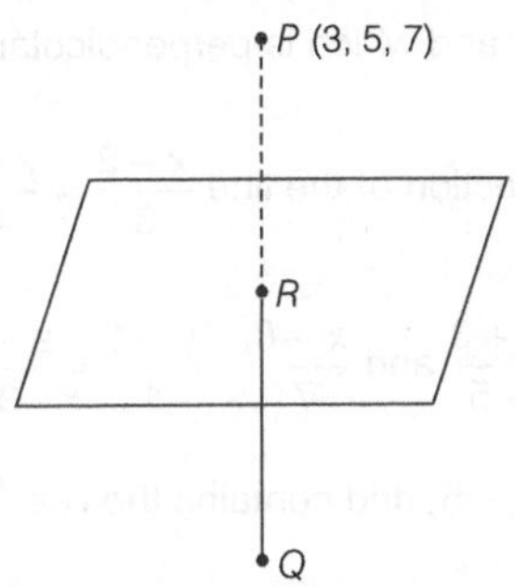

$\therefore$ Equation of line PR is $\dfrac{x-3}{2}=\dfrac{y-5}{1}=\dfrac{z-7}{1}=r$

Let $R(2r+3, r+5, r+7)$

Since, R lies on plane (i).

$\therefore \quad 2(2r+3)+(r+5)+(r+7)=0;\ 6r+18=0$

$\therefore \quad r=-3 \ \therefore \ R\equiv(-3, 2, 4)$

Let $\quad Q\equiv(\alpha, \beta, \gamma)$

Since, R is mid-point of PQ.

$$\therefore \quad -3=\frac{\alpha+3}{2} \Rightarrow \alpha=-9$$

$$2=\frac{\beta+5}{2} \Rightarrow \beta=-1$$

$$4=\frac{\gamma+7}{2} \Rightarrow \gamma=1$$

$\therefore \quad Q\equiv(-9, 1, 1)$

Example 73. Find the length and the foot of the perpendicular from the point $(7, 14, 5)$ to the plane $2x+4y-z=2$.

Sol. The required length $=\dfrac{2(7)+4(14)-(5)-2}{\sqrt{2^2+4^2+1^2}}$

$$=\frac{14+56-5-2}{\sqrt{4+16+1}}=\frac{63}{\sqrt{21}}$$

Let the coordinates of the foot of the perpendicular from the point $P(7, 14, 5)$ be $M(\alpha, \beta, \gamma)$.

Then, the direction ratios of PM are $\alpha-7, \beta-14$ and $\gamma-5$.

Therefore, the direction ratios of the normal to the plane are $\alpha-7, \beta-14$ and $\gamma-5$.

But the direction ratios of normal to the given plane $2x+4y-z=2$ are 2, 4 and -1.

Hence, $\quad \dfrac{\alpha-7}{2}=\dfrac{\beta-14}{4}=\dfrac{\gamma-5}{-1}=k$

$\therefore \quad \alpha=2k+7, \beta=4k+14$ and $\gamma=-k+5$ ...(i)

Since, α, β and γ lie on the plane $2x+4y-z=2$,

$2\alpha+4\beta-\gamma=2$

$\Rightarrow 2(7+2k)+4(14+4k)-(5-k)=2$

$\Rightarrow \quad 14+4k+56+16k-5+k=2$

$\Rightarrow \quad 21k=-63 \Rightarrow k=-3$

Now, putting $k=-3$ in (i), we get $\alpha=1, \beta=2, \gamma=8$

Hence, the foot of the perpendicular is (1, 2, 8).

Example 74. Find the image of the line $\dfrac{x-1}{9}=\dfrac{y-2}{-1}=\dfrac{z+3}{-3}$ in the plane $3x-3y+10z-26=0$.

Sol.

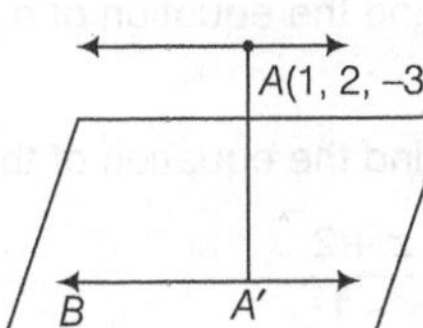

$$\frac{x-1}{9}=\frac{y-2}{-1}=\frac{z+3}{-3} \quad \text{...(i)}$$

$$3x-3y+10z-26=0 \quad \text{...(ii)}$$

The direction ratios of the line are 9, -1 and -3 and direction ratios of the normal to the given plane are 3, -3 and 10.

Since, $9\cdot3+(-1)(-3)+(-3)10=0$ and the point $(1, 2, -3)$ of line (i) does not lie in plane (ii) for $3\cdot1-3\cdot2+10\cdot(-3)-26\neq0$, line (i) is parallel to plane (ii). Let A' be the image of point $A(1, 2, -3)$ in plane (ii). Then the image of the line (i) in the plane (ii) is the line through A' and parallel to the line (i).

Let point A' be (p, q, r). Then

$$\frac{p-1}{3}=\frac{q-2}{-3}=\frac{r+3}{10}$$

$$=-\frac{(3(1)-3(2)+10(-3)-26)}{9+9+100}=\frac{1}{2}$$

The point $A'\left(\dfrac{5}{2}, \dfrac{1}{2}, 2\right)$

The equation of line BA' is $\dfrac{x-\left(\frac{5}{2}\right)}{9}=\dfrac{y-\left(\frac{1}{2}\right)}{-1}=\dfrac{z-2}{-3}$

Exercise for Session 3

1. Find the equation of plane passing through the point (1, 2, 3) and having the vector $\mathbf{r} = 2\hat{\mathbf{i}} - \hat{\mathbf{j}} + 3\hat{\mathbf{k}}$ normal to it.

2. Find a unit vector normal to the plane through the points (1, 1, 1), (– 1, 2, 3) and (2, – 1, 3).

3. Show that the four points (0, – 1, 0), (2, 1, – 1), (1, 1, 1) and (3, 3, 0) are coplaner. Also, find equation of plane through them.

4. Find the equation of plane passing through the line of intersection of planes $3x + 4y - 4 = 0$ and $x + 7y + 3z + z = 0$ and also through origin.

5. Find equation of angle bisector of plane $x + 2y + 3z - z = 0$ and $2x - 3y + z + 4 = 0$.

6. Find image of point (1, 3, 4) in the plane $2x - y + z + 3 = 0$.

7. Find the angle between the line $\frac{x+1}{2} = \frac{y}{3} = \frac{z-3}{6}$ and the plane $3x + y + z = 7$.

8. Find the equation of plane which passes through the point (1, 2, 0) and which is perpendicular to the plane $x - y + z = 3$ and $2x + y - z + 4 = 0$.

9. Find the distance of the point (– 1, – 5, – 10) from the point of intersection of the line $\frac{x-2}{3} = \frac{y+1}{4} = \frac{z-2}{12}$ and plane $x - y + z = 5$.

10. Find the equation of a plane containing the lines $\frac{x-5}{4} = \frac{y+7}{4} = \frac{z+3}{-5}$ and $\frac{x-8}{7} = \frac{y-4}{1} = \frac{z-5}{3}$.

11. Find the equation of the plane which passes through the point (3, 4, – 5) and contains the line $\frac{x+1}{2} = \frac{y-1}{3} = \frac{z+2}{-1}$.

12. Find the equation of the planes parallel to the plane $x - 2y + 2z - 3 = 0$. Which is at a unit distance from the point (1, 2, 3).

13. Find the equation of the bisector planes of the angles between the plane $x + 2y + 2z = 19$ and $4x - 3y + 12z + 3 = 0$ and specify the plane which bisects the acute angle and the plane which bisects the obtuse angle.

14. Find the equation of the image of the plane $x - 2y + 2z = 3$ in the plane $x + y + z = 1$.

15. Find the equation of a plane which passes through the point (1, 2, 3) and which is at the maximum distance from the point (– 1, 0, 2).

Session 4

Sphere

Sphere

A sphere is the locus of a point which moves in space in such a way that its distance from a fixed point always remains constant. The fixed point is called the centre of the sphere and the fixed distance is called the radius of sphere. Shown as in adjoining figure.

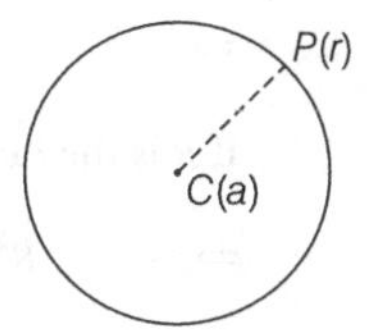

Equation of Sphere whose Centre *c* and Radius is *a*

Let O be the origin of reference and $\mathbf{C}$ be the centre of sphere whose position vector is $\mathbf{c}$. Let P be any point on the surface of the sphere whose position vector is $\mathbf{r}$.

Thus, $\quad \mathbf{OP} = \mathbf{r}$ and $\mathbf{OC} = \mathbf{c}$

$\therefore \quad \mathbf{CP} = \mathbf{OP} - \mathbf{OC} = \mathbf{r} - \mathbf{c}$

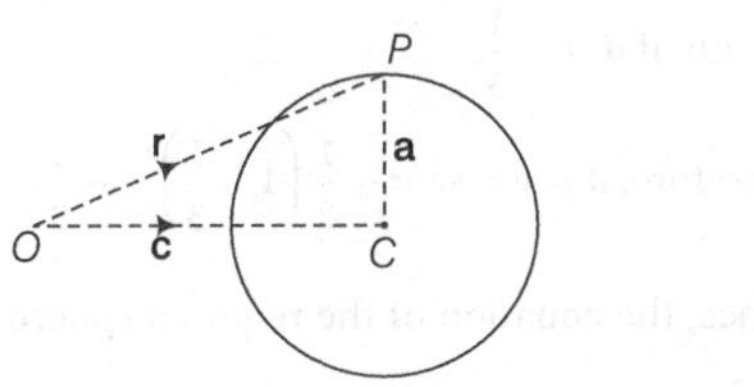

Now, $\quad |\mathbf{r} - \mathbf{c}| = \mathbf{a} \quad$ [radius of sphere]

$\Rightarrow \quad |\mathbf{r} - \mathbf{c}|^2 = \mathbf{a}^2$

$\Rightarrow \quad (\mathbf{r} - \mathbf{c}) \cdot (\mathbf{r} - \mathbf{c}) = \mathbf{a}^2$

$\Rightarrow \quad r^2 - 2\mathbf{r} \cdot \mathbf{c} + c^2 = \mathbf{a}^2$

$\Rightarrow \quad r^2 - 2r \cdot \mathbf{c} + (c^2 - a^2) = 0$

which is the required equation of sphere.

Cartesian Equation of a Sphere

The equation of sphere with centre (a, b, c) and radius R is

$$(x - a)^2 + (y - b)^2 + (z - c)^2 = R^2$$

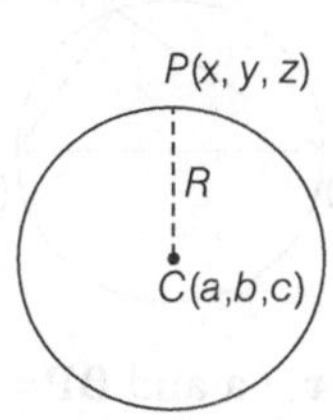

Proof. Let C be the centre of the sphere.

Then, coordinates of C are (a, b, c). Let $P(x, y, z)$ be any point on the sphere, then

$$CP = R$$

$\Rightarrow \quad CP^2 = R^2$

$\Rightarrow \quad (x - a)^2 + (y - b)^2 + (z - c)^2 = R^2$

Since, $P(x, y, z)$ is an arbitrary point on the sphere, therefore required equation of the sphere is

$(x - a)^2 + (y - b)^2 + (z - c)^2 = R^2$

Remarks

1. The above equation is called the central form of a sphere. If the centre is at the origin, then equation of sphere is,
   $x^2 + y^2 + z^2 = R^2$
   (known as the standard form of the sphere)
2. Above equation can also be written as
   $x^2 + y^2 + z^2 - 2ax - 2by - 2cz + (a^2 + b^2 + c^2 - R^2) = 0$
   which has the following characteristics of the equation of sphere
   (i) It is a second degree equation in x, y and z.
   (ii) The coefficient of x^2, y^2 and z^2 are all equal.
   (iii) The term containing the product of xy, yz and zx are absent.

Example 75. Find the vector equation of a sphere with centre having the position vector $\hat{\mathbf{i}} + \hat{\mathbf{j}} + \hat{\mathbf{k}}$ and radius $\sqrt{3}$.

Sol. We know that equation of sphere is

$|\mathbf{r} - \mathbf{c}| = a \quad$ (vector form)

$\Rightarrow \quad |\mathbf{r} - (\hat{\mathbf{i}} + \hat{\mathbf{j}} + \hat{\mathbf{k}})| = \sqrt{3}$

which is the required equation of sphere.

Example 76. Find the equation of sphere whose centre is (5, 2, 3) and radius is 2 in cartesian form.

Sol. The required equation of the sphere is

$(x - 5)^2 + (y - 2)^2 + (z - 3)^2 = 2^2$

$\Rightarrow \quad x^2 + y^2 + z^3 - 10x - 4y - 6z + 34 = 0$

Example 77. Find the equation of a sphere whose centre is (3, 1, 2) and radius is 5.

Sol. The equation of the sphere whose centre is (3, 1, 2) and radius is 5, is

$$(x-3)^2 + (y-1)^2 + (z-2)^2 = 5^2$$

$$x^2 + y^2 + z^2 - 6x - 2y - 4z - 11 = 0$$

General Equation of Sphere

The equation $x^2 + y^2 + z^2 + 2ux + 2vy + 2wz + d = 0$ represents a sphere with centre $(-u, -v, -w)$ i.e.

$$\left(-\frac{1}{2}\text{ coefficient of } x, -\frac{1}{2}\text{ coefficient of } y, -\frac{1}{2}\text{ coefficient of } z\right)$$

and radius $= \sqrt{u^2 + v^2 + w^2 - d}$.

Note

The equation $x^2 + y^2 + z^2 + 2ux + 2vy + 2wz + d = 0$ represents a real sphere, if $u^2 + v^2 + w^2 - d > 0$. If $u^2 + v^2 + w^2 - d = 0$, then it represents a point sphere. The sphere is imaginary, if $u^2 + v^2 + w^2 - d < 0$.

Example 78. Find the centre and radius of the sphere $2x^2 + 2y^2 + 2z^2 - 2x - 4y + 2z + 3 = 0$.

Sol. The given equation

$$x^2 + y^2 + z^2 - x - 2y + z + \frac{3}{2} = 0;$$

where cetnre is

$$\left(-\frac{1}{2}\text{ coefficient of } x, -\frac{1}{2}\text{ coefficient of } y, -\frac{1}{2}\text{ coefficient of } z\right)$$

$\therefore$ Centre $= \left(\frac{1}{2}, -1, -\frac{1}{2}\right)$

and Radius $= \sqrt{\left(\frac{1}{2}\right)^2 + (-1)^2 + \left(-\frac{1}{2}\right)^2 - \frac{3}{2}}$

$$= \sqrt{\frac{1}{4} + 1 + \frac{1}{4} - \frac{3}{2}} = 0$$

$\therefore$ Given sphere represents a point sphere $\left(\frac{1}{2}, -1, -\frac{1}{2}\right)$.

Example 79. Find the equation of the sphere passing through (0, 0, 0), (1, 0, 0), (0, 1, 0) and (0, 0, 1).

Sol. Let the equation of the sphere be

$$x^2 + y^2 + z^2 + 2ux + 2vy + 2wz + d = 0 \quad \text{...(i)}$$

As (i) passes through (0, 0, 0), (1, 0, 0), (0,1, 0) and (0, 0, 1), we must have $d = 0, 1 + 2u + d = 0$

$$1 + 2v + d = 0 \text{ and } 1 + 2w + d = 0$$

Since, $d = 0$, we get $2u = 2v = 2w = -1$

Thus, the equation of the required sphere is

$$x^2 + y^2 + z^2 - x - y - z = 0.$$

Example 80. Find the equation of a sphere which passes through (1, 0, 0), (0, 1, 0) and (0, 0, 1) and has radius as small as possible.

Sol. Let the equation of the required sphere be

$$x^2 + y^2 + z^2 + 2ux + 2vy + 2wz + d = 0 \quad \text{...(i)}$$

As the sphere passes through (1, 0, 0), (0,1, 0) and (0, 0, 1), we get

$$1 + 2u + d = 0, 1 + 2v + d = 0 \text{ and } 1 + 2w + d = 0$$

$$\Rightarrow \quad u = v = w = -\frac{1}{2}(d+1)$$

If R is the radius of the sphere, then $R^2 = u^2 + v^2 + w^2 - d$

$$\Rightarrow \quad R^2 = \frac{3}{4}(d+1)^2 - d$$

$$= \frac{3}{4}\left[d^2 + 2d + 1 - \frac{4}{3}d\right]$$

$$= \frac{3}{4}\left[d^2 + \frac{2}{3}d + 1\right]$$

$$= \frac{3}{4}\left[\left(d + \frac{1}{3}\right)^2 + 1 - \frac{1}{9}\right]$$

$$= \frac{3}{4}\left[\left(d + \frac{1}{3}\right)^2 + \frac{8}{9}\right]$$

The last equation shows that R^2 (and thus R) will be the least if an only if $d = -\frac{1}{3}$.

Therefore, $u = v = w = -\frac{1}{2}\left(1 - \frac{1}{3}\right) = -\frac{1}{3}$

Hence, the equation of the required sphere is $x^2 + y^2 + z^2 - \frac{2}{3}(x + y + z) - \frac{1}{3} = 0$ or $3(x^2 + y^2 + z^2) - 2(x + y + z) - 1 = 0$.

Diameter Form of the Equation of a Sphere

If the position vectors of the extremities of a diameter of a sphere are **a** and **b**, then its equation is

$$(\mathbf{r} - \mathbf{a}) \cdot (\mathbf{r} - \mathbf{b}) = 0$$

$$\Rightarrow \quad |\mathbf{r}|^2 - \mathbf{r} \cdot (\mathbf{a} + \mathbf{b}) + \mathbf{a} \cdot \mathbf{b} = 0$$

Proof. Let **a** and **b** be the position vectors of the extremities A and B of a diameter AB of sphere. Let **r** be the position vector of any point P on the sphere. Then,

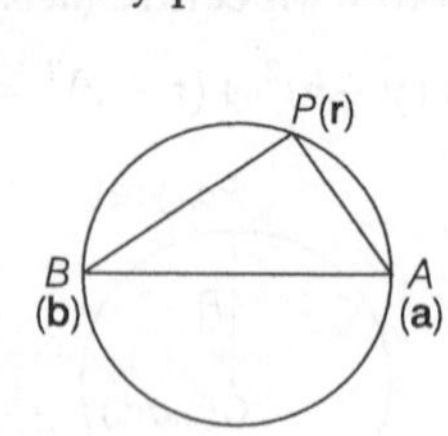

$$\mathbf{AP} = \mathbf{r} - \mathbf{a} \text{ and } \mathbf{BP} = \mathbf{r} - \mathbf{b}$$

Since, the diameter of a sphere subtends a right at any point on the sphere, therefore

$$\Rightarrow \quad \angle APB = \frac{\pi}{2}$$

$$\Rightarrow \quad \mathbf{AP} \cdot \mathbf{BP} = 0$$

$$\Rightarrow \quad (\mathbf{r} - \mathbf{a}) \cdot (\mathbf{r} - \mathbf{b}) = 0$$

$$\mathbf{r} \cdot \mathbf{r} - \mathbf{r} \cdot \mathbf{b} - \mathbf{r} \cdot \mathbf{a} + \mathbf{a} \cdot \mathbf{b} = 0$$

$$|\mathbf{r}|^2 - (\mathbf{a} + \mathbf{b}) \cdot \mathbf{r} + \mathbf{a} \cdot \mathbf{b} = 0$$

This is the required equation of sphere.

Vector Form

If the position vectors of the extremities of a diameter of a sphere are **a** and **b**, then its equation is

$$|\mathbf{r} - \mathbf{a}|^2 + |\mathbf{r} - \mathbf{b}|^2 = |\mathbf{a} - \mathbf{b}|^2$$

Proof. Let **a** and **b** be the position vectors of the extremities A and B of a diameter of a sphere. Let **r** be the position vector of any point P on the sphere, then

$$\mathbf{AP} = \mathbf{r} - \mathbf{a}$$

and $$\mathbf{BP} = \mathbf{r} - \mathbf{b}$$

Since, ΔAPB is a right angled triangle.

$$\therefore \quad AP^2 + BP^2 = AB^2$$

$$\Rightarrow \quad |\mathbf{AP}|^2 + |\mathbf{BP}|^2 = |\mathbf{AB}|^2$$

$$\Rightarrow \quad |\mathbf{r} - \mathbf{a}|^2 + |\mathbf{r} - \mathbf{b}|^2 = |\mathbf{a} - \mathbf{b}|^2$$

This is the required equation of the sphere.

Cartesian Form

If (x_1, y_1, z_1) and (x_2, y_2, z_2) are the coordinates of the extremities of a diameter of a sphere, then its equation is,

$$(x - x_1)(x - x_2) + (y - y_1)(y - y_2) + (z - z_1)(z - z_2) = 0$$

Example 81. Find the equation of the sphere described on the joint of points A and B having Position position vectors $2\hat{\mathbf{i}} + 6\hat{\mathbf{j}} - 7\hat{\mathbf{k}}$ and $-2\hat{\mathbf{i}} + 4\hat{\mathbf{j}} - 3\hat{\mathbf{k}}$, respectively, as the diameter. Find the centre and the radius of the sphere.

Sol. If point P with position vector $\mathbf{r} = x\hat{\mathbf{i}} + y\hat{\mathbf{j}} + z\hat{\mathbf{k}}$ is any point on the sphere, then $\mathbf{AP} \cdot \mathbf{BP} = 0$

$$(x - 2)(x + 2) + (y - 6)(y - 4) + (z + 7)(z + 3) = 0$$

$$\Rightarrow (x^2 - 4) + (y^2 - 10y + 24) + (z^2 + 10z + 21) = 0$$

$$\Rightarrow \quad x^2 + y^2 + z^2 - 10y + 10z + 41 = 0$$

The centre of this sphere is (0, 5, – 5) and its radius is

$$\sqrt{5^2 + (-5)^2 - 41} = \sqrt{9} = 3$$

Section of a Sphere by a Plane

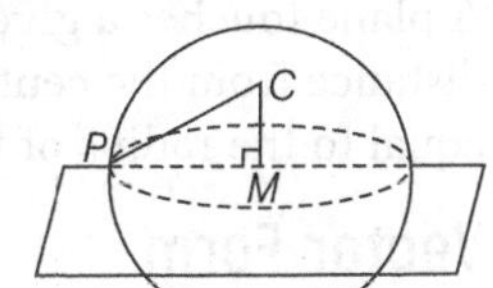

Consider a sphere intersected by a plane. The set of points common to both sphere and plane is called a plane section of a sphere.

It can be easily seen the plane section of sphere is a circle.

Let C be the centre of the sphere and M be the foot of the perpendicular from C on the plane. Then, M is the centre of the circle and radius of circle is given by PM.

i.e. $$PM = \sqrt{CP^2 - CM^2}$$

The centre M of the circle is the point of intersection of the plane and line CM, which passes through C and is perpendicular to given plane.

Example 82. Find the radius of the circular section in which the sphere $|\mathbf{r}| = 5$ is cut by the plane $\mathbf{r} \cdot (\hat{\mathbf{i}} + \hat{\mathbf{j}} + \hat{\mathbf{k}}) = 3\sqrt{3}$.

Sol. Let A be the foot of the perpendicular from the centre O to the plane $\mathbf{r} \cdot (\hat{\mathbf{i}} + \hat{\mathbf{j}} + \hat{\mathbf{k}}) - 3\sqrt{3} = 0$

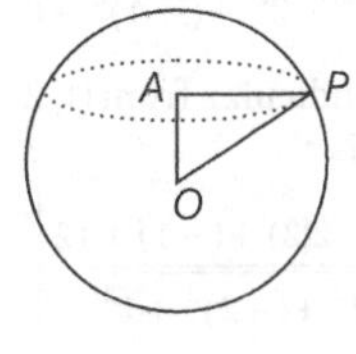

Then, $|OA| = \left| \dfrac{0 \cdot (\hat{\mathbf{i}} + \hat{\mathbf{j}} + \hat{\mathbf{k}}) - 3\sqrt{3}}{|\hat{\mathbf{i}} + \hat{\mathbf{j}} + \hat{\mathbf{k}}|} \right| = \dfrac{3\sqrt{3}}{\sqrt{3}} = 3$ (perpendicular distance of a point from the plane)

If P is any point on the circle, then P lies on the plane as well as on the sphere. Therefore, OP = radius of the sphere = 5

Now, $$AP^2 = OP^2 - OA^2 = 5^2 - 3^2 = 16$$

$$\Rightarrow \quad AP = 4$$

Example 83. Find the centre of the circle given by $\mathbf{r} \cdot (\hat{\mathbf{i}} + 2\hat{\mathbf{j}} + 2\hat{\mathbf{k}}) = 15$ and $|\mathbf{r} - (\hat{\mathbf{j}} + 2\hat{\mathbf{k}})| = 4$.

Sol. The equation of a line through the centre $\hat{\mathbf{j}} + 2\hat{\mathbf{k}}$ and normal to the given plane is

$$\mathbf{r} = (\hat{\mathbf{j}} + 2\hat{\mathbf{k}}) + \lambda(\hat{\mathbf{i}} + 2\hat{\mathbf{j}} + 2\hat{\mathbf{k}}) \quad \text{...(i)}$$

This meets the plane at a point for which we must have

$$[(\hat{\mathbf{j}} + 2\hat{\mathbf{k}}) + \lambda(\hat{\mathbf{i}} + 2\hat{\mathbf{j}} + 2\hat{\mathbf{k}})] \cdot (\hat{\mathbf{i}} + 2\hat{\mathbf{j}} + 2\hat{\mathbf{k}}) = 15$$

$$\Rightarrow \quad 6 + 9\lambda = 15$$

$$\Rightarrow \quad \lambda = 1$$

On putting $\lambda = 1$ in Eq. (i), we obtain the position vectors of the centre as $\hat{\mathbf{i}} + 3\hat{\mathbf{j}} + 4\hat{\mathbf{k}}$, Hence, the coordinates of the centre of the circle are (1, 3, 4).

Condition of Tengency of a Plane to a Sphere

A plane touches a given sphere, if the perpendicular distance from the centre of the sphere to the planes is equal to the radius of the sphere.

Vector Form

The plane $\mathbf{r}\cdot\mathbf{n}=d$ touches the sphere $|\mathbf{r}-\mathbf{a}|=R$,

if $\dfrac{|\mathbf{a}\cdot\mathbf{n}-d|}{|\mathbf{n}|}=R$

Cartesian Form

The plane $lx+my+nz=p$ touches the sphere $x^2+y^2+z^2+2ux+2vy+2wz+d=0$,

if $(ul+vm+wn+p)^2=(l^2+m^2+n^2)(u^2+v^2+w^2-d)$

Example 84. Show that the plane $2x-2y+z+12=0$ touches the sphere $x^2+y^2+z^2-2z-4y+2z-3=0$.

Sol. The given plane will touch the given sphere if the perpendicular distance from the centre of the sphere to the plane is equal to the radius of the sphere. The centre of the given sphere $x^2+y^2+z^2-2x-4y+2z-3=0$ is $(1, 2, -1)$ and its radius is $\sqrt{1^2+2^2+(-1)^2-(-3)}=3$.

Length of the perpendicular from $(1, 2, -1)$ to the plane $2x-2y+z+12=0$ is

$$\left|\frac{2(1)-2(2)+(-1)+12}{\sqrt{2^2+(-2)^2+1^2}}\right|=\frac{9}{3}=3$$

Thus, the given plane touches the given sphere.

Example 85. Find the equation of the sphere whose centre has the position vector $3\hat{\mathbf{i}}+6\hat{\mathbf{j}}-4\hat{\mathbf{k}}$ and which touches the plane $\mathbf{r}\cdot(2\hat{\mathbf{i}}-2\hat{\mathbf{j}}-\hat{\mathbf{k}})=10$.

Sol. Let the radius of the required sphere be R. Then, its equation is

$$|\mathbf{r}-(3\hat{\mathbf{i}}+6\hat{\mathbf{j}}-4\hat{\mathbf{k}})|=R \quad \text{...(i)}$$

Since, the plane $\mathbf{r}\cdot(2\hat{\mathbf{i}}-2\hat{\mathbf{j}}-\hat{\mathbf{k}})=10$ touches the sphere (i), therefore length of perpendicular from the centre to the plane $\mathbf{r}\cdot(2\hat{\mathbf{i}}-2\hat{\mathbf{j}}-\hat{\mathbf{k}})=10$ is equal to R.

i.e. $\dfrac{|(3\hat{\mathbf{i}}-6\hat{\mathbf{j}}-4\hat{\mathbf{k}})\cdot(2\hat{\mathbf{i}}-2\hat{\mathbf{j}}-\hat{\mathbf{k}})-10|}{|2\hat{\mathbf{i}}-2\hat{\mathbf{j}}-\hat{\mathbf{k}}|}=R \Rightarrow R=4$

On putting $R=4$ in Eq. (i), we obtain $|\mathbf{r}-(3\hat{\mathbf{i}}+6\hat{\mathbf{j}}-4\hat{\mathbf{k}})|=4$ as the equation of the required sphere.

Example 86. A variable plane passes through a fixed point (a, b, c) and cuts the coordinate axes at points A, B and C. Show that the locus of the centre of the sphere $OABC$ is $\dfrac{a}{x}+\dfrac{b}{y}+\dfrac{c}{z}=2$.

Sol. Let (α, β, γ) be any point on the locus. Then according to the given condition, (α, β, γ) is the centre of the sphere through the origin. Therefore, its equation is given by

$$(x-\alpha)^2+(y-\beta)^2+(z-\gamma)^2=(0-\alpha)^2+(0-\beta)^2+(0-\gamma)^2$$

$$x^2+y^2+z^2-2\alpha x-2\beta y-2\gamma z=0$$

To obtain its point of intersection with the X-axis, we put $y=0$ and $z=0$, so that

$$x^2-2\alpha x=0$$

$$\Rightarrow \quad x(x-2\alpha)=0$$

$$\Rightarrow \quad x=0 \text{ or } x=2\alpha$$

Thus, the plane meets X-axis at $O(0, 0, 0)$ and $A(2\alpha, 0, 0)$. Similarly, it meets Y-axis at $O(0, 0, 0)$ and $B(0, 2\beta, 0)$, and Z-axis at $O(0, 0, 0)$ and $C(0, 0, 2\gamma)$.

The equation of the plane through A, B and C is

$$\frac{x}{2\alpha}+\frac{y}{2\beta}+\frac{z}{2\gamma}=1 \quad \text{(intercept form)}$$

Since, it passes through (a, b, c), we get

$$\frac{a}{2\alpha}+\frac{b}{2\beta}+\frac{c}{2\gamma}=1$$

or
$$\frac{a}{\alpha}+\frac{b}{\beta}+\frac{c}{\gamma}=2$$

Hence, locus of (α, β, γ) is $\dfrac{a}{x}+\dfrac{b}{y}+\dfrac{c}{z}=2$.

Example 87. A sphere of constant radius k passes through the origin and meets the axis at A, B and C. Prove that the centroid of triangle ABC lies on the sphere $9(x^2+y^2+z^2)=4k^2$.

Sol. Let the equation of any sphere passing through the origin and having radius k be

$$x^2+y^2+z^2+2ux+2vy+2wz=0$$

As the radius of the sphere is k, we get

$$u^2+v^2+w^2=k^2$$

Note that (i) meets the X-axis at $O(0, 0, 0)$ and $A(-2u, 0, 0)$; Y-axis at $O(0, 0, 0)$ and $B(0, -2v, 0)$, and Z-axis at $O(0, 0, 0)$ and $C(0, 0, -2w)$.

Let the centroid of the triangle ABC be (α, β, γ), Then

$$\alpha=-\frac{2u}{3}, \beta=-\frac{2v}{3}, \gamma=-\frac{2w}{3}$$

$$\Rightarrow \quad u=-\frac{3\alpha}{2}, v=-\frac{3\beta}{2}, w=-\frac{3v}{2}$$

Putting this in (ii), we get

$$\left(\frac{-3}{2}\alpha\right)^2+\left(\frac{-3}{2}\beta\right)^2+\left(\frac{-3}{2}\gamma\right)^2=k^2$$

$$\Rightarrow \quad \alpha^2+\beta^2+\gamma^2=\frac{4}{9}k^2$$

This shows that the centroid of triangle ABC lies on

$$x^2+y^2+z^2=\frac{4}{9}k^2.$$

Exercise for Session 4

1. Find the centre and radius of sphere $2(x-5)(x+1)+2(y+5)(y-1)+2(z-2)(z+2)=7$

2. Obtain the equation of the sphere with the points $(1,-1,1)$ and $(3,-3,3)$ as the extremities of a diametre and find the coordinates of its centre.

3. Find the equation of sphere which passes through (1, 0, 0) and has its centre on the positive direction of Y-axis and has radius 2.

4. Find the equation of sphere if it touches the plane $\mathbf{r}\cdot(2\hat{\mathbf{i}}-2\hat{\mathbf{j}}-\hat{\mathbf{k}})=0$ and the position vector of its centre is $3\hat{\mathbf{i}}+6\hat{\mathbf{j}}-4\hat{\mathbf{k}}$.

5. Find the value of λ for which the plane $x+y+z=\sqrt{3}\lambda$ touches the sphere $x^2+y^2+z^2-2x-2y-2z=6$.

6. Find the equation of sphere concentric with sphere $2x^2+2y^2+2z^2-6x+2y-4z=1$ and double its radius.

7. A sphere has the equation $|\mathbf{r}-\mathbf{a}|^2+|\mathbf{r}-\mathbf{b}|^2=72$, where $\mathbf{a}=\hat{\mathbf{i}}+3\hat{\mathbf{j}}-6\hat{\mathbf{k}}$ and $\mathbf{b}=2\hat{\mathbf{i}}+4\hat{\mathbf{j}}+2\hat{\mathbf{k}}$

   Find

   (i) The centre of sphere

   (ii) The radius of sphere

   (iii) Perpendicular distance from the centre of the sphere to the plane $\mathbf{r}\cdot(2\hat{\mathbf{i}}+2\hat{\mathbf{j}}-\hat{\mathbf{k}})+3=0$.

JEE Type Solved Examples :
Single Option Correct Type Questions

Ex. 1 *If a line makes angle α, β and γ with the coordinates axes, then*

(a) $\cos 2\alpha + \cos 2\beta + \cos 2\gamma - 1 = 0$
(b) $\cos 2\alpha + \cos 2\beta + \cos 2\gamma - 2 = 0$
(c) $\cos 2\alpha + \cos 2\beta + \cos 2\gamma + 1 = 0$
(d) $\cos 2\alpha + \cos 2\beta + \cos 2\gamma + 2 = 0$

Sol. (c) If $\cos\alpha$, $\cos\beta$ and $\cos\gamma$ are the DC's of a line, then

$$2\cos^2\alpha + 2\cos^2\beta + 2\cos^2\gamma = 2$$

$$\Rightarrow \quad 1 + \cos 2\alpha + 1 + \cos 2\beta + 1 + \cos 2\gamma = 2$$

$$\Rightarrow \quad \cos 2\alpha + \cos 2\beta + \cos 2\gamma + 1 = 0$$

Ex. 2 *The points* $(5, -4, 2)$ $(4, -3, 1)$, $(7, -6, 4)$ *and* $(8, -7, 5)$ *are the vertices of*

(a) a rectangle (b) a square
(c) a parallelogram (d) None of these

Sol. (c) Let $A(5, -4, 2)$, $B(4, -3, 1)$, $C(7, -6, 4)$ and $D(8, -7, 5)$

$$\mathbf{AB} = -\hat{\mathbf{i}} + \hat{\mathbf{j}} - \hat{\mathbf{k}}$$

$$\mathbf{BC} = 3\hat{\mathbf{i}} - 3\hat{\mathbf{j}} + 3\hat{\mathbf{k}}$$

$$\mathbf{CD} = \hat{\mathbf{i}} - \hat{\mathbf{j}} + \hat{\mathbf{k}}$$

and $\mathbf{DA} = -3\hat{\mathbf{i}} + 3\hat{\mathbf{j}} - 3\hat{\mathbf{k}}$

Clearly $\mathbf{AB} \parallel \mathbf{CD}$ and $\mathbf{BC} \parallel \mathbf{DA}$

Also, $\mathbf{AB} \cdot \mathbf{BC} = -9 \neq 0$

$\therefore$ $ABCD$ is a parallelogram.

Ex. 3 *In* ΔABC *the mid-point of the sides AB, BC and CA are respectively* $(l, 0, 0)$, $(0, m, 0)$ *and* $(0, 0, n)$. *Then,* $\dfrac{AB^2 + BC^2 + CA^2}{l^2 + m^2 + n^2}$ *is equal to*

(a) 2 (b) 4
(c) 8 (d) 16

Sol. (c) From the figure,

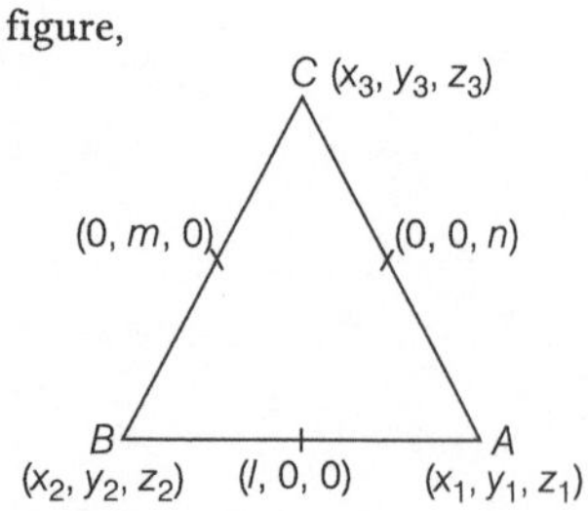

$$x_1 + x_2 = 2l, \; y_1 + y_2 = 0, \; z_1 + z_2 = 0$$

$$x_2 + x_3 = 0, \; y_2 + y_3 = 2m, \; z_2 + z_3 = 0$$

and $x_1 + x_3 = 0$, $y_1 + y_3 = 0$, $z_1 + z_3 = 2n$

On solving, we get

$$x_1 = l, \; x_2 = l, \; x_3 = -l$$

$$y_1 = -m, \; y_2 = m, \; y_3 = m$$

and $z_1 = n$, $z_2 = -n$, $z_3 = n$

$\therefore$ Coordinates are $A(l, -m, n)$, $B(l, m, -n)$ and $C(-l, m, n)$

$$\therefore \quad \frac{AB^2 + BC^2 + CA^2}{l^2 + m^2 + n^2} = \frac{(4m^2 + 4n^2) + (4l^2 + 4n^2) + (4l^2 + 4m^2)}{l^2 + m^2 + n^2} = 8$$

Ex. 4 *The angle between a line with direction ratios proportional to 2, 2, 1 and a line joining* $(3, 1, 4)$ *to* $(7, 2, 12)$, *is*

(a) $\cos^{-1}\left(\frac{2}{3}\right)$ (b) $\cos^{-1}\left(-\frac{2}{3}\right)$
(c) $\tan^{-1}\left(\frac{2}{3}\right)$ (d) None of these

Sol. (a) A line with direction ratios proportional to 2, 2, 1 is parallel to the vector $\mathbf{a} = 2\hat{\mathbf{i}} + 2\hat{\mathbf{j}} + \hat{\mathbf{k}}$.

Line joining $P(3, 1, 4)$ to $Q(7, 2, 12)$ is parallel to the vector

$$\mathbf{PQ} = 4\hat{\mathbf{i}} + \hat{\mathbf{j}} + 8\hat{\mathbf{k}}.$$

Let θ be the required angle. Then,

$$\cos\theta = \frac{\mathbf{a} \cdot \mathbf{PQ}}{|\mathbf{a}||\mathbf{PQ}|} = \frac{8 + 2 + 8}{\sqrt{4 + 4 + 1}\sqrt{16 + 1 + 64}}$$

$$\Rightarrow \quad \cos\theta = \frac{18}{3 \times 9} = \frac{2}{3} \Rightarrow \theta = \cos^{-1}\left(\frac{2}{3}\right)$$

Ex. 5 *The angle between the lines* $2x = 3y = -z$ *and* $6x = -y = -4z$ *is*

(a) $30°$ (b) $45°$
(c) $60°$ (d) $90°$

Sol. (d) Given, equation of lines can be rewritten as

$$\frac{x}{1/2} = \frac{y}{1/3} = \frac{z}{-1}$$

and $$\frac{x}{1/6} = \frac{y}{-1} = \frac{z}{-1/4}$$

$$\therefore \quad \cos\theta = \frac{a_1a_2 + b_1b_2 + c_1c_2}{\sqrt{a_1^2 + b_1^2 + c_1^2}\sqrt{a_2^2 + b_2^2 + c_2^2}}$$

$$= \frac{\frac{1}{2} \times \frac{1}{6} + \frac{1}{3} \times (-1) - 1 \times \left(-\frac{1}{4}\right)}{\sqrt{\frac{1}{4} + \frac{1}{9} + 1}\sqrt{\frac{1}{36} + 1 + \frac{1}{16}}}$$

$$= \frac{\frac{1}{12} - \frac{1}{3} + \frac{1}{4}}{\sqrt{\frac{1}{4} + \frac{1}{9} + 1}\sqrt{\frac{1}{36} + 1 + \frac{1}{16}}}$$

$$\Rightarrow \quad \cos\theta = 0 \Rightarrow \theta = 90°$$

Ex. 6 *A line makes the same angle θ with X-axis and Z-axis. If the angle β, which it makes with Y-axis, is such that $\sin^2\beta = 3\sin^2\theta$, then the value of $\cos^2\theta$ is*

(a) $\frac{1}{5}$ (b) $\frac{2}{5}$

(c) $\frac{3}{5}$ (d) $\frac{2}{3}$

Sol. (c) Since, $\cos^2\theta + \cos^2\beta + \cos^2\theta = 1$ $[\because l^2 + m^2 + n^2 = 1]$

$\Rightarrow 2\cos^2\theta + 1 - 3\sin^2\theta = 1$ $[\because \sin^2\beta = 3\sin^2\theta]$

$\Rightarrow 2\cos^2\theta - 3(1 - \cos^2\theta) = 0$

$\Rightarrow 5\cos^2\theta = 3 \Rightarrow \cos^2\theta = \frac{3}{5}$

Ex. 7 *The projection of a line segment on the coordinate axes are 2, 3, 6. Then, the length of the line segment is*

(a) 7 (b) 5

(c) 1 (d) 11

Sol. (a) Let the length of the line segment be r and its direction cosines be l, m, n. Then, its projections on the coordinate axes are lr, mr, nr.

$\therefore lr = 2, mr = 3$ and $nr = 6$

$\Rightarrow l^2r^2 + m^2r^2 + n^2r^2 = 4 + 9 + 36$

$\Rightarrow r^2(l^2 + m^2 + n^2) = 49$

$\Rightarrow r^2 = 49 \Rightarrow r = 7$ $[\because l^2 + m^2 + n^2 = 1]$

Ex. 8 *The equation of the straight line through the origin and parallel to the line $(b+c)x + (c+a)y + (a+b)z = k = (b-c)x + (c-a)y + (a-b)z$ are*

(a) $\frac{x}{b^2 - c^2} = \frac{y}{c^2 - a^2} = \frac{z}{a^2 - b^2}$

(b) $\frac{x}{b} = \frac{y}{b} = \frac{z}{a}$

(c) $\frac{x}{a^2 - bc} = \frac{y}{b^2 - ca} = \frac{z}{c^2 - ab}$

(d) None of the above

Sol. (c) Equations of straight line through the origin are

$$\frac{x-0}{l} = \frac{y-0}{m} = \frac{z-0}{n}$$

where, $l(b+c) + m(c+a) + n(a+b) = 0$

and $l(b-c) + m(c-a) + n(a-b) = 0$

On solving, $\frac{1}{2(a^2 - bc)} = \frac{m}{2(b^2 - ca)} = \frac{n}{2(c^2 - ab)}$

Equations of the straight line are

$$\frac{x}{a^2 - bc} = \frac{y}{b^2 - ca} = \frac{z}{c^2 - ab}.$$

Ex. 9 *The coordinates of the foot of the perpendicular drawn from the point $A(1, 0, 3)$ to the join of the points $B(4, 7, 1)$ and $C(3, 5, 3)$ are*

(a) $\left(\frac{5}{3}, \frac{7}{3}, \frac{17}{3}\right)$ (b) $(5, 7, 17)$

(c) $\left(\frac{5}{7}, -\frac{7}{3}, \frac{17}{3}\right)$ (d) $\left(-\frac{5}{3}, \frac{7}{3}, -\frac{17}{3}\right)$

Sol. (a) Let D be the foot of the perpendicular and let it divide BC in the ratio $\lambda : 1$. Then, the coordinates of D are

$$\left(\frac{3\lambda + 4}{\lambda + 1}, \frac{5\lambda + 7}{\lambda + 1}, \frac{3\lambda + 1}{\lambda + 1}\right)$$

Now, $\mathbf{AD} \perp \mathbf{BC} \Rightarrow \mathbf{AD} \cdot \mathbf{BC} = 0$

$\Rightarrow -(2\lambda + 3) - 2(5\lambda + 7) - 4 = 0 \Rightarrow \lambda = -\frac{7}{4}$

So, the coordinates of D are $\left(\frac{5}{3}, \frac{7}{3}, \frac{17}{3}\right)$

Ex. 10 *A mirror and a source of light are situated at the origin O and at a point on OX, respectively. A ray of light from the source strikes the mirror and is reflected. If the direction ratios of the normal to the plane are proportional to 1, −1, 1, then direction cosines of the reflected ray are*

(a) $\frac{1}{3}, \frac{2}{3}, \frac{2}{3}$ (b) $-\frac{1}{3}, \frac{2}{3}, \frac{2}{3}$

(c) $-\frac{1}{3}, -\frac{2}{3}, -\frac{2}{3}$ (d) $-\frac{1}{3}, -\frac{2}{3}, \frac{2}{3}$

Sol. (d) Let the source of light be situated at $A(a, 0, 0)$, where, $a \neq 0$.

Let OA be the incident ray, OB be the reflected ray and ON be the normal to the mirror at O.

$\therefore \angle AON = \angle NOB = \frac{\theta}{2}$ (say)

Direction ratios of **OA** are proportional to a, 0, 0 and so its direction cosines are 1, 0, 0.

Direction cosines of ON are $\frac{1}{\sqrt{3}}, \frac{-1}{\sqrt{3}}, \frac{1}{\sqrt{3}}$

$\therefore \cos\frac{\theta}{2} = \frac{1}{\sqrt{3}}$

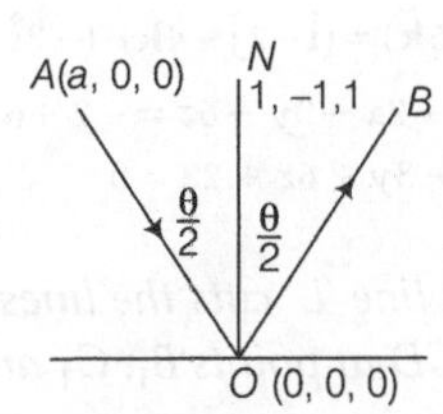

Let l, m, n be the direction cosines of the reflected ray OB.

Then, $\frac{l+1}{2\cos\frac{\theta}{2}} = \frac{1}{\sqrt{3}}, \frac{m+0}{2\cos\frac{\theta}{2}} = -\frac{1}{\sqrt{3}}$

and $\frac{n+0}{2\cos\frac{\theta}{2}} = \frac{1}{\sqrt{3}}$

$\Rightarrow \quad l = \frac{2}{3} - 1, m = -\frac{2}{3}, n = \frac{2}{3}$

$\Rightarrow \quad l = -\frac{1}{3}, m = -\frac{2}{3}, n = \frac{2}{3}$

Hence, direction cosines of the reflected ray are $-\frac{1}{3}, -\frac{2}{3}, \frac{2}{3}$.

Ex. 11 *Equation of plane passing through the points (2, 2, 1), (9, 3, 6) and perpendicular to the plane $2x + 6y + 6z - 1 = 0$, is*

(a) $3x + 4y + 5z = 9$

(b) $3x + 4y - 5z + 9 = 0$

(c) $3x + 4y - 5z - 9 = 0$

(d) None of the above

Sol. (c) Equation of a plane passing through (2, 2, 1) is

$$a(x-2) + b(y-2) + c(z-1) = 0 \quad \text{...(i)}$$

This passes through (9, 3, 6) and is perpendicular to

$$2x + 6y + 6z - 1 = 0$$

$\therefore \quad 7a + b + 5c = 0$ and $2a + 6b + 6c = 0$

Solving these two by cross-multiplication, we get

$$\frac{a}{-24} = \frac{b}{-32} = \frac{c}{40}$$

$$\Rightarrow \quad \frac{a}{-3} = \frac{b}{-4} = \frac{c}{5}$$

Substituting the values of a, b, c in Eq. (i), we get $3x + 4y - 5z - 9 = 0$ as the required plane.

Ex. 12 *If the position vectors of the points A and B are $3\hat{i} + \hat{j} + 2\hat{k}$ and $\hat{i} - 2\hat{j} - 4\hat{k}$ respectively, then the equation of the plane through B and perpendicular to AB is*

(a) $2x + 3y + 6z + 28 = 0$

(b) $3x + 2y + 6z = 28$

(c) $2x - 3y + 6z + 28 = 0$

(d) $3x - 2y + 6z = 28$

Sol. (a) We have, $\mathbf{AB} = -2\hat{i} - 3\hat{j} - 6\hat{k}$

So, vector equation of the plane is

$$|\mathbf{r} - (\hat{i} - 2\hat{j} - 4\hat{k})| \cdot \mathbf{AB} = 0$$

$$\Rightarrow \mathbf{r} \cdot (-2\hat{i} - 3\hat{j} - 6\hat{k}) = (\hat{i} - 2\hat{j} - 4\hat{k}) \cdot (-2\hat{i} - 3\hat{j} - 6\hat{k})$$

$$\Rightarrow \quad -2x - 3y - 6z = -2 + 6 + 24$$

$$\Rightarrow \quad 2x + 3y + 6z + 28 = 0$$

Ex. 13 *A straight line 'L' cuts the lines AB, AC and AD of a parallelogram ABCD at points B_1, C_1 and D_1, respectively. If $\mathbf{AB}_1 = \lambda_1 \mathbf{AB}$, $\mathbf{AD}_1 = \lambda_2 \mathbf{AD}$ and $\mathbf{AC}_1 = \lambda_3 \mathbf{AC}$, then $\frac{1}{\lambda_3}$ is equal to*

(a) $\frac{1}{\lambda_1} + \frac{1}{\lambda_2}$

(b) $\frac{1}{\lambda_1} - \frac{1}{\lambda_2}$

(c) $-\lambda_1 + \lambda_2$

(d) $\lambda_1 + \lambda_2$

Sol. (a) Let $\mathbf{AB} = \mathbf{a}$, $\mathbf{AD} = \mathbf{b}$, then $\mathbf{AC} = \mathbf{a} + \mathbf{b}$

Given, $\quad \mathbf{AB}_1 = \lambda_1 \mathbf{a}$, $\mathbf{AD}_1 = \lambda_2 \mathbf{b}$, $\mathbf{AC}_1 = \lambda_3(\mathbf{a} + \mathbf{b})$

$$\mathbf{B}_1\mathbf{D}_1 = \mathbf{AD}_1 - \mathbf{AB}_1 = \lambda_2 \mathbf{b} - \lambda_1 \mathbf{a}$$

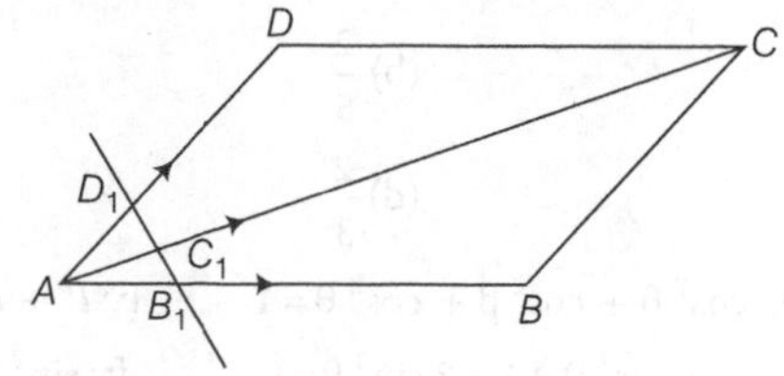

Since, vectors $\mathbf{D}_1\mathbf{C}_1$ and $\mathbf{B}_1\mathbf{D}_1$ are collinear, we have $\mathbf{D}_1\mathbf{C}_1 = k\, \mathbf{B}_1\mathbf{D}_1$ for some $k \in R$.

$$\Rightarrow \quad \mathbf{AC}_1 - \mathbf{AD}_1 = k \cdot \mathbf{B}_1\mathbf{D}_1$$

$$\Rightarrow \quad \lambda_3(\mathbf{a} + \mathbf{b}) - \lambda_2 \mathbf{b} = k \cdot (\lambda_2 \mathbf{b} - \lambda_1 \mathbf{a})$$

$$\Rightarrow \quad \lambda_3 \mathbf{a} + (\lambda_3 - \lambda_2)\mathbf{b} = k \cdot \lambda_2 \mathbf{b} - k \cdot \lambda_1 \mathbf{a}$$

Thus, $\lambda_3 = -k\lambda_1$ and $\lambda_3 - \lambda_2 = k\lambda_2$

$$\Rightarrow \quad k = \frac{-\lambda_3}{\lambda_1} = \frac{\lambda_3 - \lambda_2}{\lambda_2} \Rightarrow \lambda_1\lambda_2 = \lambda_1\lambda_3 + \lambda_2\lambda_3$$

$$\Rightarrow \quad \frac{1}{\lambda_3} = \frac{1}{\lambda_1} + \frac{1}{\lambda_2}$$

Ex. 14 *If the direction cosines of two lines are such that $l + m + n = 0$, $l^2 + m^2 - n^2 = 0$, then the angle between them is*

(a) π

(b) $\frac{\pi}{3}$

(c) $\frac{\pi}{4}$

(d) $\frac{\pi}{6}$

Sol. (b) If l, m, n are direction cosines of two lines are such that

$$l + m + n = 0 \quad \text{...(i)}$$

and $$l^2 + m^2 - n^2 = 0 \quad \text{...(ii)}$$

$$\Rightarrow \quad l^2 + m^2 - (-l - m)^2 = 0$$

$$\Rightarrow \quad 2lm = 0 \Rightarrow l = 0 \text{ or } m = 0$$

If $l = 0$, then $n = -m$

$$\Rightarrow \quad l : m : n = 0 : 1 : -1$$

and if $m = 0$, then $n = -l$

$$\Rightarrow \quad l : m : n = 1 : 0 : -1$$

$$\therefore \quad \cos\theta = \frac{0 + 0 + 1}{\sqrt{0 + 1 + 1}\sqrt{0 + 1 + 1}} = \frac{1}{2}$$

$$\Rightarrow \quad \theta = \frac{\pi}{3}$$

Ex. 15 *The equation of the plane passing through the mid-point of the line points (1, 2, 3) and (3, 4, 5) and perpendicular to it is*

(a) $x + y + z = 9$

(b) $x + y + z = -9$

(c) $2x + 3y + 4z = 9$

(d) $2x + 3y + 4z = -9$

Sol. (a) The DR's of the joining of the points (1,2, 3) and (3, 4, 5) and $(3 - 1, 4 - 2, 5 - 3)$, ie. (2, 2, 2)

Also, the mid-point of the join of the points (1, 2, 3) and (3, 4, 5) is (2, 3, 4).

$\therefore$ Equation of plane which passes through (2, 3, 4) and the DR's of its normal are (2, 2, 2) is

$$2(x-2)+2(y-3)+2(z-4)=0$$
$$\Rightarrow \quad x+y+z-9=0$$
$$\Rightarrow \quad x+y+z=9$$

Ex. 16 *Equation of the plane that contains the lines* $\mathbf{r}=(\hat{\mathbf{i}}+\hat{\mathbf{j}})+\lambda(\hat{\mathbf{i}}+2\hat{\mathbf{j}}-\hat{\mathbf{k}})$ *and,* $\mathbf{r}=(\hat{\mathbf{i}}+\hat{\mathbf{j}})+\mu(-\hat{\mathbf{i}}+\hat{\mathbf{j}}-2\hat{\mathbf{k}})$ *is*

(a) $\mathbf{r}\cdot(2\hat{\mathbf{i}}+\hat{\mathbf{j}}-3\hat{\mathbf{k}})=-4$

(b) $\mathbf{r}\times(-\hat{\mathbf{i}}+\hat{\mathbf{j}}+\hat{\mathbf{k}})=0$

(c) $\mathbf{r}\cdot(-\hat{\mathbf{i}}+\hat{\mathbf{j}}+\hat{\mathbf{k}})=0$

(d) None of the above

Sol. (c) The lines are parallel to the vectors $\mathbf{b}_1=\hat{\mathbf{i}}+2\hat{\mathbf{j}}-\hat{\mathbf{k}}$ and $\mathbf{b}_2=-\hat{\mathbf{i}}+\hat{\mathbf{j}}-2\hat{\mathbf{k}}$. Therefore, the plane is normal to the vector

$$\mathbf{n}=\mathbf{b}_1\times\mathbf{b}_2=\begin{vmatrix}\hat{\mathbf{i}} & \hat{\mathbf{j}} & \hat{\mathbf{k}}\\ 1 & 2 & -1\\ -1 & 1 & -2\end{vmatrix}=-3\hat{\mathbf{i}}+3\hat{\mathbf{j}}+3\hat{\mathbf{k}}$$

The required plane passes through $(\hat{\mathbf{i}}+\hat{\mathbf{j}})$ and is normal to the vector $\mathbf{n}$. Therefore, its equation is

$\mathbf{r}\cdot\mathbf{n}=\mathbf{a}\cdot\mathbf{n}$

$\Rightarrow \mathbf{r}\cdot(-3\hat{\mathbf{i}}+3\hat{\mathbf{j}}+3\hat{\mathbf{k}})=(\hat{\mathbf{i}}+\hat{\mathbf{j}})\cdot(-3\hat{\mathbf{i}}+3\hat{\mathbf{j}}+3\hat{\mathbf{k}})$

$\Rightarrow \mathbf{r}\cdot(-3\hat{\mathbf{i}}+3\hat{\mathbf{j}}+3\hat{\mathbf{k}})=-3+3$

$\Rightarrow \mathbf{r}\cdot(-\hat{\mathbf{i}}+\hat{\mathbf{j}}+\hat{\mathbf{k}})=0$

Ex. 17 *The line* $\dfrac{x-2}{3}=\dfrac{y+1}{2}=\dfrac{z-1}{-1}$ *intersects the curve* $xy=c^2, z=0$, *if c is equal to*

(a) ± 1 (b) $\pm \dfrac{1}{3}$

(c) $\pm\sqrt{5}$ (d) None of these

Sol. (c) At the point on the line where it intersects the given curve, we have $z=0$, so that

$$\frac{x-2}{3}=\frac{y+1}{2}=\frac{0-1}{-1}$$

$$\Rightarrow \quad \frac{x-2}{3}=1 \text{ and } \frac{y+1}{2}=1$$

$$\Rightarrow \quad x=5 \text{ and } y=1.$$

Putting these values of x and y in $xy=c^2$, we get $c^2=5 \Rightarrow c=\pm\sqrt{5}$.

Ex. 18 *The distance between the line* $\mathbf{r}=2\hat{\mathbf{i}}-2\hat{\mathbf{j}}+3\hat{\mathbf{k}}+\lambda(\hat{\mathbf{i}}-\hat{\mathbf{j}}+4\hat{\mathbf{k}})$ *and the plane* $\mathbf{r}\cdot(\hat{\mathbf{i}}+5\hat{\mathbf{j}}+\hat{\mathbf{k}})=5$, *is*

(a) $\dfrac{10}{9}$ (b) $\dfrac{10}{3\sqrt{3}}$

(c) $\dfrac{10}{3}$ (d) None of these

Sol. (b) Clearly, the given line passes through the point $\mathbf{a}=2\hat{\mathbf{i}}-2\hat{\mathbf{j}}+3\hat{\mathbf{k}}$ and is parallel to the vector $\mathbf{b}=\hat{\mathbf{i}}-\hat{\mathbf{j}}+4\hat{\mathbf{k}}$.

The plane is normal to the vector $\mathbf{n}=\hat{\mathbf{i}}+5\hat{\mathbf{j}}+\hat{\mathbf{k}}$.

We have, $\quad \mathbf{b}\cdot\mathbf{n}=1-5+4=0$

So, the line is parallel to the plane.

$\therefore$ Required distance

= Length of the perpendiculars from a point on the line to the given plane.

= Length of the perpendicular from $(2\hat{\mathbf{i}}-2\hat{\mathbf{j}}+3\hat{\mathbf{k}})$ to the given plane.

$$=\left|\frac{(2\hat{\mathbf{i}}-2\hat{\mathbf{j}}+3\hat{\mathbf{k}})\cdot(\hat{\mathbf{i}}+5\hat{\mathbf{j}}+\hat{\mathbf{k}})-5}{\sqrt{1+25+1}}\right|$$

$$=\left|\frac{2-10+3-5}{3\sqrt{3}}\right|=\frac{10}{3\sqrt{3}}$$

Ex. 19 *If the plane* $\dfrac{x}{2}+\dfrac{y}{3}+\dfrac{z}{4}=1$, *cuts the coordinate axes in A, B, C, then the area of* ΔABC *is*

(a) $\sqrt{29}$ sq. units (b) $\sqrt{41}$ sq. units

(c) $\sqrt{61}$ sq. units (d) None of these

Sol. (c) The given plane cuts the coordinate axes in $A(2,0,0)$, $B(0,3,0)$ and $C(0,0,4)$.

$\therefore$ Area of $\Delta ABC=\dfrac{1}{2}AB\times AC\times\sin\angle BAC$

Now, $AB=\sqrt{4+9+0}=\sqrt{13}$, $AC=\sqrt{4+0+16}=\sqrt{20}$.

$$\cos\angle BAC=\frac{\mathbf{AB}\cdot\mathbf{AC}}{|\mathbf{AB}||\mathbf{AC}|}=\frac{(-2\hat{\mathbf{i}}+3\hat{\mathbf{j}})\cdot(-2\hat{\mathbf{i}}+4\hat{\mathbf{k}})}{\sqrt{4+9}\sqrt{4+16}}$$

$$\Rightarrow \cos\angle BAC=\frac{4+0+0}{\sqrt{13}\sqrt{20}}=\frac{4}{\sqrt{13}\sqrt{20}}=\frac{2}{\sqrt{65}}$$

$$\Rightarrow \sin\angle BAC=\sqrt{1-\frac{4}{65}}=\sqrt{\frac{61}{65}}$$

Hence, Area of $\Delta ABC=\dfrac{1}{2}\times\sqrt{13}\times\sqrt{20}\times\sqrt{\dfrac{61}{65}}=\sqrt{61}$ sq. units.

Ex. 20 *The distance of the point* $(1,-2,3)$ *from the plane* $x-y+z=5$ *measured parallel to the line* $\dfrac{x}{2}=\dfrac{y}{3}=\dfrac{z-1}{-6}$ *is*

(a) 1 (b) 2

(c) 4 (d) None of these

Sol. (a) The equation of the line passing through $P(1,-2,3)$ and parallel to the given line is

$$\frac{x-1}{2}=\frac{y+2}{3}=\frac{z-3}{-6}$$

Suppose it meets the plane $x-y+z=5$ at the point Q given by

$$\frac{x-1}{2}=\frac{y+2}{3}=\frac{z-3}{-6}$$
$$=\lambda \text{ i.e. } (2\lambda+1, 3\lambda-2, -6\lambda+3)$$

This lies on $x - y + z = 5$. Therefore,

$$2\lambda + 1 - 3\lambda + 2 - 6\lambda + 3 = 5$$

$$\Rightarrow \quad -7\lambda = -1 \Rightarrow \lambda = \frac{1}{7}$$

So, the coordinates of Q are $\left(\frac{9}{7}, -\frac{11}{7}, \frac{15}{7}\right)$.

Hence, required distance $= PQ = \sqrt{\frac{4}{49} + \frac{9}{49} + \frac{36}{49}} = 1.$

Ex. 21 *The length of the perpendicular from the origin to the plane passing through the point* **a** *and containing the line* $\mathbf{r} = \mathbf{b} + \lambda\mathbf{c}$ *is*

(a) $\dfrac{[\mathbf{a}\ \mathbf{b}\ \mathbf{c}]}{|\mathbf{a}\times\mathbf{b} + \mathbf{b}\times\mathbf{c} + \mathbf{c}\times\mathbf{a}|}$

(b) $\dfrac{[\mathbf{a}\ \mathbf{b}\ \mathbf{c}]}{|\mathbf{a}\times\mathbf{b} + \mathbf{b}\times\mathbf{c}|}$

(c) $\dfrac{[\mathbf{a}\ \mathbf{b}\ \mathbf{c}]}{|\mathbf{b}\times\mathbf{c} + \mathbf{c}\times\mathbf{a}|}$

(d) $\dfrac{[\mathbf{a}\ \mathbf{b}\ \mathbf{c}]}{|\mathbf{a}\times\mathbf{b} + \mathbf{c}\times\mathbf{a}|}$

Sol. (c) The plane passing through **a** and containing the line $\mathbf{r} = \mathbf{b} + \lambda\mathbf{c}$ also passes through the point **b** and is parallel to the vector **c**. So, it is normal to the vector $(\mathbf{a} - \mathbf{b}) \times \mathbf{c}$.

Thus, the equation of the plane is

$$(\mathbf{r} - \mathbf{a}) \cdot |(\mathbf{a} - \mathbf{b}) \times \mathbf{c}| = 0$$

$$\Rightarrow \quad (\mathbf{r} - \mathbf{a}) \cdot (\mathbf{a}\times\mathbf{c} - \mathbf{b}\times\mathbf{c}) = 0$$

$$\Rightarrow \quad \mathbf{r} \cdot (\mathbf{a}\times\mathbf{c} - \mathbf{b}\times\mathbf{c}) = \mathbf{a} \cdot (\mathbf{a}\times\mathbf{c} - \mathbf{b}\times\mathbf{c})$$

$$\Rightarrow \quad \mathbf{r} \cdot (\mathbf{a}\times\mathbf{c} - \mathbf{b}\times\mathbf{c}) = -\mathbf{a} \cdot (\mathbf{b}\times\mathbf{c})$$

$$\Rightarrow \quad \mathbf{r} \cdot (\mathbf{b}\times\mathbf{c} + \mathbf{c}\times\mathbf{a}) - [\mathbf{a}\ \mathbf{b}\ \mathbf{c}] = 0$$

$\therefore$ Length of the perpendicular from the origin to this plane

$$= \left|\frac{\mathbf{0} \cdot (\mathbf{b}\times\mathbf{c} + \mathbf{c}\times\mathbf{a}) - [\mathbf{a}\ \mathbf{b}\ \mathbf{c}]}{|\mathbf{b}\times\mathbf{c} + \mathbf{c}\times\mathbf{a}|}\right|$$

$$= \frac{[\mathbf{a}\ \mathbf{b}\ \mathbf{c}]}{|\mathbf{b}\times\mathbf{c} + \mathbf{c}\times\mathbf{a}|}$$

Ex. 22 *If* $P(0, 1, 0)$ *and* $Q(0, 0, 1)$ *are two points, then the projection of* **PQ** *on the plane* $x + y + z = 3$ *is*

(a) 2 (b) 3

(c) $\sqrt{2}$ (d) $\sqrt{3}$

Sol. (c) The projection of PQ on the given plane is $PQ \cos\theta$, where θ is the angle between PQ and the plane.

Let **n** be a vector normal to the plane.

We have, $\mathbf{PQ} = -\hat{\mathbf{j}} + \hat{\mathbf{k}}$ and $\mathbf{n} = \hat{\mathbf{i}} + \hat{\mathbf{j}} + \hat{\mathbf{k}}$

$$\therefore \quad \sin\theta = \frac{\mathbf{PQ} \cdot \mathbf{na}}{|\mathbf{PQ}||\mathbf{n}|} = 0$$

$\Rightarrow$ **PQ** is parallel to the plane.

Hence, projection of **PQ** on the given plane

$$= |\mathbf{PQ}| \cos\theta$$

$$= |\mathbf{PQ}| = \sqrt{1 + 1} = \sqrt{2}$$

Ex. 23 *The equation of the plane through the intersection of the planes* $x + y + z = 1$ *and* $2x + 3y - z + 4 = 0$ *and parallel to X-axis, is*

(a) $y - 3z + 6 = 0$ (b) $3y - z + 6 = 0$

(c) $y + 3z + 6 = 0$ (d) $3y - 2z + 6 = 0$

Sol. (a) The equation of the plane through the intersection of the planes $x + y + z = 1$ and $2x + 3y - z + 4 = 0$ is

$$(x + y + z - 1) + \lambda(2x + 3y - z + 4) = 0$$

or, $\quad (2\lambda + 1)x + (3\lambda + 1)y + (1 - \lambda)z + 4\lambda - 1 = 0 \quad$ ...(i)

It is parallel to X-axis, i.e. $\frac{x}{1} = \frac{y}{0} = \frac{z}{0}$

$$\therefore \quad 1(2\lambda + 1) + 0 \times (3\lambda + 1) + 0(1 - \lambda) = 0$$

$$\Rightarrow \quad \lambda = -\frac{1}{2}$$

Substituting $\lambda = -\frac{1}{2}$ in Eq. (i), we get

$y - 3z + 6 = 0$ as the equation of the required plane.

Ex. 24 *A plane passes through the point* $(1, 1, 1)$. *If* b, c, a *are the direction ratios of a normal to the plane where* $a, b, c\ (a < b < c)$ *are the prime factors of* 2001, *then the equation of the plane* II *is*

(a) $29x + 31y + 3z = 63$

(b) $23x + 29y - 29z = 23$

(c) $23x + 29y + 3z = 55$

(d) $31x + 37y + 3z = 71$

Sol. (c) The equation of the plane is

$$b(x - 1) + c(y - 1) + a(z - 1) = 0 \quad \text{...(i)}$$

Now, $2001 = 3 \times 23 \times 29$

$$\therefore \quad a < b < c \Rightarrow a = 3, b = 23 \text{ and } c = 29.$$

Substituting the values of a, b, c in Eq. (i), we obtain $23x + 29y + 3z = 55$ as the equation of the required plane.

Ex. 25 *If the direction ratios of two lines are given by* $a + b + c = 0$ *and* $2ab + 2ac - bc = 0$, *then the angle between the lines is*

(a) π (b) $\frac{2\pi}{3}$

(c) $\frac{\pi}{2}$ (d) $\frac{\pi}{3}$

Sol. (b) We have,

$$a + b + c = 0 \text{ and } 2ab + 2ac - bc = 0$$

$$\Rightarrow \quad a = -(b + c) \text{ and } 2a(b + c) - bc = 0$$

$$\Rightarrow \quad -2(b + c)^2 - bc = 0$$

$$\Rightarrow \quad 2b^2 + 5bc + 2c^2 = 0$$

$$\Rightarrow \quad (2b + c)(b + 2c) = 0$$

$$\Rightarrow \quad 2b + c = 0 \text{ or, } b + 2c = 0$$

If $2b + c = 0$, then $a = -(b + c) \quad \Rightarrow a = b$

$$\therefore \quad a = b \text{ and } c = -2b \Rightarrow \frac{a}{1} = \frac{b}{1} = \frac{c}{-2}$$

If $b + 2c = 0$, then $a = -(b + c) \Rightarrow a = c$

$$\therefore \quad a = c \text{ and } b = -2c \Rightarrow \frac{a}{1} = \frac{b}{-2} = \frac{c}{1}$$

Thus, the direction ratios of two lines are proportional to 1, 1, − 2 and 1, − 2, 1, respectively. So, the angle θ between them is given by

$$\cos\theta = \frac{1-2-2}{\sqrt{1+1+4}\sqrt{1+4+1}} = \frac{-1}{2} \Rightarrow \theta = \frac{2\pi}{3}$$

Ex. 26 *A tetrahedron has vertices at $O(0, 0, 0)$, $A(1, 2, 1)$, $B(2, 1, 3)$ and $C(-1, 1, 2)$. Then, the angle between the faces OAB and ABC will be*

(a) 90° (b) $\cos^{-1}\left(\frac{19}{35}\right)$

(c) $\cos^{-1}\left(\frac{17}{31}\right)$ (d) 30°

Sol. (b) Let $\mathbf{n}_1$ and $\mathbf{n}_2$ be the vectors normal to the faces OAB and ABC. Then,

$$\mathbf{n}_1 = \mathbf{OA} \times \mathbf{OB} = \begin{vmatrix} \hat{\mathbf{i}} & \hat{\mathbf{j}} & \hat{\mathbf{k}} \\ 1 & 2 & 1 \\ 2 & 1 & 3 \end{vmatrix} = 5\hat{\mathbf{i}} - \hat{\mathbf{j}} - 3\hat{\mathbf{k}}$$

and $$\mathbf{n}_2 = \mathbf{AB} \times \mathbf{AC} = \begin{vmatrix} \hat{\mathbf{i}} & \hat{\mathbf{j}} & \hat{\mathbf{k}} \\ 1 & -1 & 2 \\ -2 & -1 & 1 \end{vmatrix} = \hat{\mathbf{i}} - 5\hat{\mathbf{j}} - 3\hat{\mathbf{k}}$$

If θ is the angle between the faces OAB and ABC, then

$$\cos\theta = \frac{\mathbf{n}_1 \cdot \mathbf{n}_2}{|\mathbf{n}_1||\mathbf{n}_2|}$$

$$\Rightarrow \quad \cos\theta = \frac{5+5+9}{\sqrt{25+1+9}\sqrt{1+25+9}} = \frac{19}{35}$$

$$\Rightarrow \quad \theta = \cos^{-1}\left(\frac{19}{35}\right)$$

Ex. 27 *The vector equation of the plane through the point $(2, 1, -1)$ and passing through the line of intersection of the plane $\mathbf{r} \cdot (\hat{\mathbf{i}} + 3\hat{\mathbf{j}} - \hat{\mathbf{k}}) = 0$ and $\mathbf{r} \cdot (\hat{\mathbf{j}} + 2\hat{\mathbf{k}}) = 0$, is*

(a) $\mathbf{r} \cdot (\hat{\mathbf{i}} + 9\hat{\mathbf{j}} + 11\hat{\mathbf{k}}) = 0$ (b) $\hat{\mathbf{r}} \cdot (\hat{\mathbf{i}} + 9\hat{\mathbf{j}} + 11\hat{\mathbf{k}}) = 6$

(c) $\hat{\mathbf{r}} \cdot (\hat{\mathbf{i}} - 3\hat{\mathbf{j}} - 13\hat{\mathbf{k}}) = 0$ (d) None of these

Sol. (a) The vector equation of a plane through the line of intersection of the plane $\mathbf{r} \cdot (\hat{\mathbf{i}} + 3\hat{\mathbf{j}} - \hat{\mathbf{k}}) = 0$ and $\mathbf{r} \cdot (\hat{\mathbf{j}} + 2\hat{\mathbf{k}}) = 0$ can be written as

$$\mathbf{r} \cdot (\hat{\mathbf{i}} + 3\hat{\mathbf{j}} + -\hat{\mathbf{k}}) + \lambda\{\mathbf{r} \cdot (\hat{\mathbf{j}} + 2\hat{\mathbf{k}})\} = 0 \quad \text{...(i)}$$

This passes through $2\hat{\mathbf{i}} + \hat{\mathbf{j}} - \hat{\mathbf{k}}$.

$$\therefore \quad (2\hat{\mathbf{i}} + \hat{\mathbf{j}} - \hat{\mathbf{k}}) \cdot (\hat{\mathbf{i}} + 3\hat{\mathbf{j}} - \hat{\mathbf{k}}) + \lambda(2\hat{\mathbf{i}} + \hat{\mathbf{j}} - \hat{\mathbf{k}}) \cdot (\hat{\mathbf{j}} + 2\hat{\mathbf{k}}) = 0$$

$$\Rightarrow \quad (2 + 3 + 1) + \lambda(0 + 1 - 2) = 0 \Rightarrow \lambda = 6.$$

Putting the value of λ in Eq. (i), we get the equation of the required plane as

$$\mathbf{r} \cdot (\hat{\mathbf{i}} + 9\hat{\mathbf{j}} + 11\hat{\mathbf{k}}) = 0$$

Ex. 28 *The vector equation of the plane through the point $\hat{\mathbf{i}} + 2\hat{\mathbf{j}} - \hat{\mathbf{k}}$ and perpendicular to the line of intersection of the plane $\mathbf{r} \cdot (3\hat{\mathbf{i}} - \hat{\mathbf{j}} + \hat{\mathbf{k}}) = 1$ and $\mathbf{r} \cdot (\hat{\mathbf{i}} + 4\hat{\mathbf{j}} - 2\hat{\mathbf{k}}) = 2$, is*

(a) $\mathbf{r} \cdot (2\hat{\mathbf{i}} + 7\hat{\mathbf{j}} - 13\hat{\mathbf{k}}) = 1$ (b) $\mathbf{r} \cdot (2\hat{\mathbf{i}} - 7\hat{\mathbf{j}} - 13\hat{\mathbf{k}}) = 1$

(c) $\mathbf{r} \cdot (2\hat{\mathbf{i}} + 7\hat{\mathbf{j}} + 13\hat{\mathbf{k}}) = 0$ (d) None of these

Sol. (b) The line of intersection of the planes

$$\mathbf{r} \cdot (3\hat{\mathbf{i}} - \hat{\mathbf{j}} + \hat{\mathbf{k}}) = 1$$

and $$\mathbf{r} \cdot (\hat{\mathbf{i}} + 4\hat{\mathbf{j}} - 2\hat{\mathbf{k}}) = 2$$

is common to both the planes. Therefore, it is perpendicular to normals to the two planes, i.e.

$$\mathbf{n}_1 = 3\hat{\mathbf{i}} - \hat{\mathbf{j}} + \hat{\mathbf{k}}$$

and $$\mathbf{n}_2 = \hat{\mathbf{i}} + 4\hat{\mathbf{j}} - 2\hat{\mathbf{k}}$$

Hence, it is parallel to the vector $\mathbf{n}_1 \times \mathbf{n}_2 = -2\hat{\mathbf{i}} + 7\hat{\mathbf{j}} + 13\hat{\mathbf{k}}$.

Thus, we have to find the equation of the plane passing through $\mathbf{a} = \hat{\mathbf{i}} + 2\hat{\mathbf{j}} - \hat{\mathbf{k}}$ and normal to the vector $\mathbf{n} = \mathbf{n}_1 \times \mathbf{n}_2$.

The equation of the required plane is

$$(\mathbf{r} - \mathbf{a}) \cdot \mathbf{n} = 0$$

$$\Rightarrow \quad \mathbf{r} \cdot \mathbf{n} = \mathbf{a} \cdot \mathbf{n}$$

$$\Rightarrow \quad \mathbf{r} \cdot (-2\hat{\mathbf{i}} + 7\hat{\mathbf{j}} + 13\hat{\mathbf{k}}) = (\hat{\mathbf{i}} + 2\hat{\mathbf{j}} - \hat{\mathbf{k}}) \cdot (-2\hat{\mathbf{i}} + 7\hat{\mathbf{j}} + 13\hat{\mathbf{k}})$$

$$\Rightarrow \quad \mathbf{r} \cdot (2\hat{\mathbf{i}} - 7\hat{\mathbf{j}} - 13\hat{\mathbf{k}}) = 1$$

Ex. 29 *The cartesian equation of the plane $\mathbf{r} = (1 + \lambda - \mu)\hat{\mathbf{i}} + (2 - \lambda)\hat{\mathbf{j}} + (3 - 2\lambda + 2\mu)\hat{\mathbf{k}}$, is*

(a) $2x + y = 5$ (b) $2x - y = 5$

(c) $2x + z = 5$ (d) $2x - z = 5$

Sol. (c) We have,

$$\mathbf{r} = (1 + \lambda - \mu)\hat{\mathbf{i}} + (2 - \lambda)\hat{\mathbf{j}} + (3 - 2\lambda + 2\mu)\hat{\mathbf{k}}$$

$$\Rightarrow \quad \mathbf{r} = (\hat{\mathbf{i}} + 2\hat{\mathbf{j}} + 3\hat{\mathbf{k}}) + \lambda(\hat{\mathbf{i}} - \hat{\mathbf{j}} - 2\hat{\mathbf{k}}) + \mu(-\hat{\mathbf{i}} + 2\hat{\mathbf{k}})$$

which is a plane passing through $\mathbf{a} = \hat{\mathbf{i}} + 2\hat{\mathbf{j}} + 3\hat{\mathbf{k}}$ and parallel to the vectors $\mathbf{b} = \hat{\mathbf{i}} - \hat{\mathbf{j}} - 2\hat{\mathbf{k}}$ and $\mathbf{c} = -\hat{\mathbf{i}} + 2\hat{\mathbf{k}}$.

Therefore, it is normal to the vector

$$\mathbf{n} = \mathbf{b} \times \mathbf{c} = -2\hat{\mathbf{i}} - \hat{\mathbf{k}}$$

Hence, its vector equation is

$$(\mathbf{r} - \mathbf{a}) \cdot \mathbf{n} = 0$$

$$\Rightarrow \quad \mathbf{r} \cdot \mathbf{n} = \mathbf{a} \cdot \mathbf{n}$$

$$\Rightarrow \quad \mathbf{r} \cdot (-2\hat{\mathbf{i}} - \hat{\mathbf{k}}) = -2 - 3$$

$$\Rightarrow \quad \mathbf{r} \cdot (2\hat{\mathbf{i}} + \hat{\mathbf{k}}) = 5$$

So, the cartesian equation of the plane is

$$(x\hat{\mathbf{i}} + y\hat{\mathbf{j}} + z\hat{\mathbf{k}}) \cdot (2\hat{\mathbf{i}} + \hat{\mathbf{k}}) = 5 \Rightarrow 2x + z = 5$$

Ex. 30 *A variable plane is at a distance, k from the origin and meets the coordinates axis in A, B, C. Then, the locus of the centroid of ΔABC is*

(a) $x^{-2} + y^{-2} + z^{-2} = k^{-2}$

(b) $x^{-2} + y^{-2} + z^{-2} = 4k^{-2}$

(c) $x^{-2} + y^{-2} + z^{-2} = 16k^{-2}$

(d) $x^{-2} + y^{-2} + z^{-2} = 9k^{-2}$

Sol. (d) Let the equation of the variable plane be $\frac{x}{a}+\frac{y}{b}+\frac{z}{c}=1$.

This meets the coordinates axes at $A(a, 0, 0)$, $B(0, b, 0)$ and $C(0, 0, c)$.

Let (α, β, γ) be the coordinates of the centroid of ΔABC. Then,

$$\alpha=\frac{a}{3}, \beta=\frac{b}{3}, \gamma=\frac{c}{3}$$

$$\Rightarrow \quad a=3\alpha, b=3\beta, c=3\gamma \quad \text{...(i)}$$

The plane is at a distance, k from the origin.

$$\therefore \quad \left|\frac{\frac{0}{a}+\frac{0}{b}+\frac{0}{c}-1}{\sqrt{\frac{1}{a^2}+\frac{1}{b^2}+\frac{1}{c^2}}}\right|=k$$

$$\Rightarrow \quad \frac{1}{a^2}+\frac{1}{b^2}+\frac{1}{c^2}=\frac{1}{k^2}$$

$$\Rightarrow \quad \alpha^{-2}+\beta^{-2}+\gamma^{-2}=9k^{-2}$$

Hence, the locus of (α, β, γ) is $x^{-2}+y^{-2}+z^{-2}=9k^{-2}$

JEE Type Solved Examples : More than One Correct Option Type Questions

- ***Ex. 31*** *The direction ratios of the line $x-y+z-5=0=x-3y-6$ are*

(a) 3, 1, – 2 (b) 2, – 4, 1

(c) $\frac{3}{\sqrt{14}}, \frac{1}{\sqrt{14}}, \frac{-2}{\sqrt{14}}$ (d) $\frac{2}{\sqrt{21}}, \frac{-4}{\sqrt{21}}, \frac{1}{\sqrt{21}}$

Sol. (a, c) Let the DR's of line are a, b and c.

As the line is perpendicular to both the planes

$$\Rightarrow \quad a-b+c=0$$
$$a-3b+0\cdot c=0$$
$$\frac{a}{3}=\frac{b}{1}=\frac{c}{-2}$$

Hence, (a) and (c) are correct answers.

- ***Ex. 32*** *The equation of the line $x+y+z-1=0$, $4x+y-2z+2=0$ written in the symmetrical form is*

(a) $\frac{x+1}{1}=\frac{y-2}{-2}=\frac{z-0}{1}$

(b) $\frac{x}{1}=\frac{y}{-2}=\frac{z-1}{1}$

(c) $\frac{\frac{x+1}{2}}{1}=\frac{y-1}{-2}=\frac{\frac{z-1}{2}}{1}$

(d) $\frac{x-1}{2}=\frac{y+2}{-1}+\frac{z-2}{2}$

Sol. (a, b, c, d) $x+y+z-1=0$

$$4x+y-2z+2=0$$

∴ Direction ratios of the line are $(-3, 6, -3)$.

i.e. $\quad <1, -2, 1>$

Let $z=k$, then $x=k-1$, $y=2-2k$

i.e. $(k-1, 2-2k, k)$ is any point on the line.

∴ $(-1, 2, 0)$, $(0, 0, 1)$, $\left(-\frac{1}{2}, 1, \frac{1}{2}\right)$ and $(1, -2, 2)$ are points on the line.

Hence, (a), (b), (c) and (d) are the correct answers.

- ***Ex. 33*** *The direction cosines of the lines bisecting the angle between the line whose direction cosines are l_1, m_1, n_1 and l_2, m_2, n_2 and the angle between these lines is θ, are*

(a) $\frac{l_1+l_2}{\cos\frac{\theta}{2}}, \frac{m_1+m_2}{\cos\frac{\theta}{2}}, \frac{n_1+n_2}{\cos\frac{\theta}{2}}$

(b) $\frac{l_1+l_2}{2\cos\frac{\theta}{2}}, \frac{m_1+m_2}{2\cos\frac{\theta}{2}}, \frac{n_1+n_2}{2\cos\frac{\theta}{2}}$

(c) $\frac{l_1+l_2}{\sin\frac{\theta}{2}}, \frac{m_1+m_2}{\sin\frac{\theta}{2}}, \frac{n_1+n_2}{\sin\frac{\theta}{2}}$

(d) $\frac{l_1-l_2}{2\sin\frac{\theta}{2}}, \frac{m_1-m_2}{2\sin\frac{\theta}{2}}, \frac{n_1-n_2}{2\sin\frac{\theta}{2}}$

Sol. (b, d) Distance ratio of the bisector are

$$<l_1+l_2, m_1+m_2, n_1+n_2>$$
$$=\sqrt{(l_1+l_2)^2+(m_1+m_2)^2+(n_1+n_2)^2}$$
$$=\sqrt{2+2(l_1l_2+m_1m_2+n_1n_2)}$$
$$=\sqrt{2+2\cos\theta}=2\cos\frac{\theta}{2}$$

∴ Direction cosines are $\left\langle\frac{l_1+l_2}{2\cos\frac{\theta}{2}}, \frac{m_1+m_2}{2\cos\frac{\theta}{2}}, \frac{n_1+n_2}{2\cos\frac{\theta}{2}}\right\rangle$

Distance ratio of the other bisector are

$$<l_1-l_2, m_1-m_2, n_1-n_2> \sqrt{(l_1-l_2)^2+(m_1-m_2)^2+(n_1-n_2)^2}$$
$$=2\sin\frac{\theta}{2}$$

∴ Direction cosines of the bisector are

$$\left\langle\frac{l_1-l_2}{2\sin\frac{\theta}{2}}, \frac{m_1-m_2}{2\sin\frac{\theta}{2}}, \frac{n_1-n_2}{2\sin\frac{\theta}{2}}\right\rangle$$

Hence, (b) and (d) are correct answers.

Ex. 34 *Consider the planes $3x-6y+2z+5=0$ and $4x-12y+3z=3$. The planes $67x+162y+47z+44=0$ bisects the angle between the planes which*

(a) contains origin (b) is acute
(c) is obtuse (d) None of these

Sol. (a,b) $3x-6y+2z+5=0$ ...(i)

$-4x+12y-3z+3=0$ ...(ii)

$$\frac{3x-6y+2z+5}{\sqrt{9+36+4}}=\frac{-4x+12y-3z+3}{\sqrt{16+144+9}}$$

Bisects the angle between the planes that contains the origin.

$$13(3x-6y+2z+5)=7(-4x+12y-3z+3)$$
$$39x-78y+26z+65=-28x+84y-21z+21$$
$$67x-162y+47z+44=0 \quad ...(iii)$$

Let θ be the angle between Eqs. (i) and (iii), then find $\cos\theta$ and then we obtain $|\tan\theta|<1$.

Hence, (a) and (b) are the correct answer.

Ex. 35 *Consider the equation of line AB is $\frac{x}{2}=\frac{y}{-3}=\frac{z}{6}$. Through a point $P(1,2,5)$ line PN is drawn perpendicular to AB and line PQ is drawn parallel to the plane $3x+4y+5z=0$ to meet AB is Q. Then,*

(a) coordinate of N are $\left(\frac{52}{49},-\frac{78}{49},\frac{156}{49}\right)$

(b) the coordinates of Q are $\left(3,-\frac{9}{2},9\right)$

(c) the equation of PN is $\frac{x-1}{3}=\frac{y-2}{-176}=\frac{z-5}{-89}$

(d) coordinates of N are $\left(\frac{156}{49},\frac{52}{49},-\frac{78}{49}\right)$

Sol. (a,b,c) Equation of line AB is $\frac{x}{2}=\frac{y}{-3}=\frac{z}{6}$

Its DR's are $<2,-3,6>$

Let the coordinates be $<2r,-3r,6r>$

DR's of PN are $<2r-1,-3r-2,6r-5>$

It is perpendicular to AB

$\therefore \quad 2(2r-1)-3(-3r-2)+6(6r-5)=0$

$$4r-2+9r+6+36r-30=0$$

$$49r=26 \text{ i.e. } r=\frac{26}{49}$$

(a) $\therefore$ Coordinates of N are $\left(\frac{52}{49},-\frac{78}{49},\frac{156}{49}\right)$

(b) Let the coordinates of Q be $(2r,-3r,6r)$, then DR's of PQ are $<2r-1,-3r-2,6r-5>$. Since, PQ is parallel to the plane.

$\therefore \quad 3(2r-1)+4(-3r-2)+5(6r-5)=0$

$$6r-3-12r-8+30r-25=0$$

$$24r=36, r=\frac{3}{2}$$

$\therefore$ Coordinates of Q are $\left(3,-\frac{9}{2},9\right)$.

Equation of PN is $\frac{x-1}{3}=\frac{y-2}{-176}=\frac{z-5}{-89}$.

Ex. 36 *The equation of a plane is $2x-y-3z=5$ and $A(1,1,1)$, $B(2,1,-3)$, $C(1,-2,-2)$ and $D(-3,1,2)$ are four points. Which of the following line segment are intersected by the plane?*

(a) AD (b) AB
(c) AC (d) BC

Sol. (b, c) For $A(1,1,1)$, $2x-y-3z-5=2-1-3-5<0$

For $B(2,1,-3)$, $2x-y-3z-5=0-1+9-5>0$

For $C(1,-2,-2)$, $2x-y-3z-5=2+2+6-5>0$. A, D

For $D(-3,1,2)$, $2x-y-3z-5=-6-1-6-5=-18<0$

are on one side of the plane and B, C are on the other side, the line segments AB, AC, BD, CD intercept the plane.

Ex. 37 *The coordinates of a point on the line $\frac{x-1}{2}=\frac{y+1}{-3}=z$ at a distance $4\sqrt{14}$ from the point $(1,-1,0)$ are*

(a) $(9,-13,4)$
(b) $(8\sqrt{14}+1,-12\sqrt{14}-1,4\sqrt{14})$
(c) $(-7,11,-4)$
(d) $(-8\sqrt{14}+1,12\sqrt{14}-1,-4\sqrt{14})$

Sol. (a, c) The coordinates of any point on the given line are $(2r+1,-3r-1,r)$

The distance of this point from the point $(1,-1,0)$ is given to be $4\sqrt{14}$.

$$\Rightarrow \quad (2r)^2+(-3r)^2+(r)^2=(4\sqrt{14})^2$$
$$\Rightarrow \quad 14r^2=16\times 14$$
$$\Rightarrow \quad r=\pm 4$$

So, the coordinate of the required point are

$(9,-13,4)$ or $(-7,11,-4)$.

Ex. 38 *The line whose vector equation are*

$$\mathbf{r}=2\hat{\mathbf{i}}-3\hat{\mathbf{j}}+7\hat{\mathbf{k}}+\lambda(2\hat{\mathbf{i}}+p\hat{\mathbf{j}}+5\hat{\mathbf{k}})$$

and

$$\mathbf{r}=\hat{\mathbf{i}}+2\hat{\mathbf{j}}+3\hat{\mathbf{k}}+\mu(3\hat{\mathbf{i}}-p\hat{\mathbf{j}}+p\hat{\mathbf{k}})$$

are perpendicular for all values of λ and μ if p equals to

(a) -1 (b) 2
(c) 5 (d) 6

Sol. (a, d) The given lines are perpendicular for all values of λ and μ if the vectors.

$2\hat{\mathbf{i}}+p\hat{\mathbf{j}}+5\hat{\mathbf{k}}$ and $3\hat{\mathbf{i}}-p\hat{\mathbf{j}}+p\hat{\mathbf{k}}$ are perpendicular

$$\Rightarrow \quad 2\times 3+p(-p)+5p=0$$
$$\Rightarrow \quad p^2-5p-6=0$$
$$\Rightarrow \quad p=-1 \text{ or } 6$$

Ex. 39 *Equation of a plane passing through the lines* $2x-y+z-3=0, 3x+y+z-5=0$ *and which is at a distance of* $\frac{1}{\sqrt{6}}$ *from the point* $(2,1,-1)$ *is*

(a) $2x-y+z-3=0$
(b) $3x+y+z-5=0$
(c) $62x+29y+19z-105=0$
(d) $x+2y-2=0$

Sol. (a,c) Equation of a plane through the given line is

$2x-y+z-3+\lambda(3x+y+z-5)=0$

$\Rightarrow \quad (2+3\lambda)x+(\lambda-1)y+(\lambda+1)z-(3+5\lambda)=0$

So, $$\frac{1}{\sqrt{6}}=\frac{2(2+3\lambda)+(\lambda-1)-(\lambda+1)-3-5\lambda}{\sqrt{(2+3\lambda)^2+(\lambda-1)^2+(\lambda+1)^2}}$$

$\Rightarrow \quad 11\lambda^2+12\lambda+6=6(\lambda-1)^2$

$=6(\lambda^2-2\lambda+1)$

$\Rightarrow \quad 5\lambda^2+24\lambda=0$

$\Rightarrow \quad \lambda=0$ or $\lambda=-\frac{24}{5}$.

Thus, the equation of the required planes are

$2x-y+z-3=0$ or $62x+29y+19z-105=0$.

Ex. 40 *The plane passing through the point* $(-2,-2,2)$ *and containing the line joining the points* $(1,1,1)$ *and* $(1,-1,2)$ *makes intercepts of lengths a, b, c respectively on the axes of x, y and z respectively, then*

(a) $a=3b$ (b) $b=2c$
(c) $a+b+c=12$ (d) $a+2b+2c=0$

Sol. (a,b,c) Equation of any plane passing through $(-2,-2,2)$ is

$$A(x+2)+B(y+2)+C(z-2)=0$$

Since it contains the line joining $(1,1,1)$ and $(1,-1,2)$ these points also lie on this plane.

$\Rightarrow \quad 3A+3B-C=0$ and $3A+B+0=0$

$\Rightarrow \quad \frac{A}{1}=\frac{B}{-3}=\frac{C}{-6}$.

So, the equation of the plane is

$$(x+2)-3(y+2)-6(z-2)=0$$

or $$x-3y-6z+8=0$$

or $$\frac{x}{-8}+\frac{y}{\frac{8}{3}}+\frac{z}{\frac{8}{6}}=1 \Rightarrow a=8, b=\frac{8}{3}, c=\frac{8}{6}$$

$\Rightarrow \quad a=3b, b=2c, a+b+c=12$

and $\quad a+2b+2c=16$

JEE Type Solved Examples : Statement I and II Type Questions

- **Directions** (Q.Nos. 41-45) For the following questions, choose the correct answers from the codes (a), (b), (c) and (d) defined as follows:
  (a) Statement I is true, Statement II is also true; Statement II is the correct explanation of Statement I.
  (b) Statement I is true, Statement II is also true; Statement II is not the correct explanation of Statement I.
  (c) Statement I is true, Statement II is false.
  (d) Statement I is false, Statement II is true.

Ex. 41 Statement I *A line L is perpendicular to the plane* $3x-4y+5z=10$.

Statement II *Direction cosines of L be* $<\frac{3}{5\sqrt{2}},-\frac{4}{5\sqrt{2}},\frac{1}{\sqrt{2}}>$.

Sol. (a) $lx+my+nz=P$ be the equation of a plane in the normal form.

$\therefore$DR of the plane

$3x-4y+5z=10$ be $<3,-4,5>$.

$\Rightarrow$ Direction cosines

$$<\frac{3}{5\sqrt{2}},\frac{-4}{5\sqrt{2}},\frac{1}{\sqrt{2}}>.$$

Ex. 42 *The equation of two straight line are* $\frac{x-1}{2}=\frac{y+3}{1}=\frac{z-2}{-3}$ *and* $\frac{x-2}{1}=\frac{y-1}{-3}=\frac{z+3}{2}$

Statement I *The given lines are coplanar.*

Statement II *The equation* $2x_1-y_1=1, x_1+3y_1=4, 3x_1+2y_1=5$ *are consistent.*

Sol. (a) Any point on the first line is $(2x_1+1, x_2-3, -3x_1-2)$.

Any point on the second line is $(y_1-2, 3y_1+1, 2y_1-3)$.

If two lines are coplanar, then $2x_1-y_1=1, x_1+3y_1=4, 3x_1+2y_1=5$ are consistent.

Ex. 43 Statement I *The distance between the planes* $4x-5y+3z=5$ *and* $4x-5y+3z+2=0$ *is* $\frac{3}{5\sqrt{2}}$.

Statement II *The distance between* $ax+by+cz+d_1=0$ *and* $ax+by+cz+d_2=0$ *is* $\left|\frac{d_1-d_2}{\sqrt{a^2+b^2+c^2}}\right|$.

Sol. (d) Distance $=\left|\frac{5+2}{\sqrt{50}}\right|=\frac{7}{5\sqrt{2}}$

Ex. 44 *Given the line* $L: \frac{x-1}{3} = \frac{y+1}{2} = \frac{z-3}{-1}$ *and the plane* $\pi : x - 2y - z = 0$

Statement I *L lies in* π.

Statement II *L is parallel to* π.

Sol. (c) $x = 1 + 3r, y = -1 + 2r, t = 3 - r$

$\therefore \quad 1 + 3r - 2(-1 + 2r) - 3 + r = 0$

$3 \times 1 - 2 \times 2 + 1 \times 1 = 0$

Hence, L is parallel to π.

Ex. 45 Statement I *Line* $\frac{x-1}{3} = \frac{y-2}{11} = \frac{z+1}{11}$ *lies in the plane* $11x - 3z - 14 = 0$.

Statement II *A straight line lies in a plane, if the line is parallel to plane and a point of the line in the plane.*

Sol. (a) Statement I $(1, 2, -1)$ is a point on the line and $11 + 3 - 14 = 0$.

$\therefore$ The point lies on the plane $11x - 3z - 14 = 0$.

Further $3 \times 11 + 11(-3) = 0$.

$\therefore$ The line lies in the plane.

Statement II is also true.

JEE Type Solved Examples : Passage Based Questions

Passage I

(Ex. Nos. 46 to 48)

Two line whose equation are $\frac{x-3}{2} = \frac{y-2}{3} = \frac{z-1}{\lambda}$ *and* $\frac{x-2}{3} = \frac{y-3}{2} = \frac{z-2}{3}$ *lie in the same plane, then.*

Ex. 46 *The value of* $\sin^{-1} \sin \lambda$ *is equal to*

(a) 3 (b) $\pi - 3$

(c) 4 (d) $\pi - 4$

Ex. 47 *Point of intersection of the lines lies on*

(a) $3x + y + z = 20$ (b) $2x + y + z = 25$

(c) $3x + 2y + z = 24$ (d) $x = y = z$

Ex. 48 *Angle between the plane containing both the lines and the plane* $4x + y + 2z = 0$ *is equal to*

(a) $\frac{\pi}{3}$ (b) $\frac{\pi}{2}$

(c) $\frac{\pi}{6}$ (d) $\cos^{-1} \frac{2}{\sqrt{186}}$

Sol. (Ex. 46-48)

46. (d) Both lines and coplanar

$$\begin{vmatrix} 2 & 3 & \lambda \\ 3 & 2 & 3 \\ 1 & -1 & -1 \end{vmatrix} = 0$$

$\Rightarrow \quad 2(-2 + 3) + 3(3 + 3) + \lambda(-3 - 2) = 0$

$\Rightarrow \quad \lambda = 4$

$\sin^{-1}\sin 4 = \sin^{-1}\sin(\pi - 4) = \pi - 4$

47. (d) Let $\frac{x-3}{2} = \frac{y-2}{3} = \frac{z-1}{4} = r_1$

$\Rightarrow \quad x = 3 + 2r_1$

$y = 2 + 3r_1$

$z = 1 + 4r_1$

It will lie on $\frac{x-2}{3} = \frac{y-3}{2} = \frac{z-2}{3} \Rightarrow r_1 = 1$

So, point of intersection is $(5, 5, 5)$.

48. (b) Equation of plane contains both lines

$$\begin{vmatrix} x-3 & y-2 & z-1 \\ 2 & 3 & 4 \\ 3 & 2 & 3 \end{vmatrix} = 0$$

$(x - 3)(1) + (y - 2)(12 - 6) + (z - 1)(4 - 9) = 0$

$x + 6y - 5z = 10$

Thus, the angle is $\frac{\pi}{2}$.

Passage II

(Ex. Nos. 49 to 51)

Let $a_1x + b_1y + c_1z + d_1 = 0$ *and* $a_2x + b_2y + c_2z = d_2 = 0$ *be two planes, where* $d_1, d_2 > 0$. *Then, origin lies in acute angle, if* $a_1a_2 + b_1b_2 + c_1c_2 < 0$ *and origin lies in obtuse angle, if* $a_1a_2 + b_1b_2 + c_1c_2 > 0$.

Further point (x_1, y_1, z_1) *and origin both lie either in acute angle or in obtuse angle. If* $(a_1x_1 + b_1y_1 + c_1z_1 + d_1)(a_2x_1 + b_2y_1 + c_2z_1 + d_2) > 0$.

One of (x_1, y_1, z_1) *and origin lie in acute and the other in obtuse angle; if* $(a_1x_1 + b_1y_1 + c_1z_1 + d_1)(a_2x_1 + b_2y_1 + c_2z_1 + d_2) < 0$.

Ex. 49 *Given that planes* $2x + 3y - 4z + 7 = 0$ *and* $x - 2y + 3z - 5 = 0$. *If a point P is* $(1, -2, 3)$, *then*

(a) O and P both lie in acute angle between the planes

(b) O and P both lie in obtuse angle

(c) O lies in acute angle, P lies in obtuse angle

(d) O lies in obtuse angle, P lies an acute angle

Ex. 50 *Given the planes $x+2y-3z+5=0$ and $2x+y+3z+1=0$. If a point P is $(2,-1,2)$, then*

(a) O and P both lie in acute angle between the planes
(b) O and P both lie in obtuse angle
(c) O lies in acute angle, P lies in obtuse angle
(d) O lies in obtuse angle, P lies an acute angle

Ex. 51 *Given the planes $x+2y-3z+2=0$ and $x-2y+3z+7=0$, if the point P is $(1, 2, 2)$ then*

(a) O and P both lie in acute angle between the planes
(b) O and P both lie in obtuse angle
(c) O lies in acute angle, P lie in obtuse angle
(d) O lies in obtuse angle, P lies in acute angle

Sol. (Ex. 49-51)

49. (b) Equation of the second plane is $-x+2y-3z+5=0$

$2(-1)+3\cdot 2+(-4)(-3)>0$

$\therefore$ Origin lies in obtuse angle.

$(2\times 1+3(-2)-4\times 3+7)(-1+2(-2)-3\times 3+5)$
$=(2-6-12+7)(-1-4-9+5)>0$

$\therefore P$ lies in obtuse angle.

50. (c) $1\times 2+2\times 1-3\times 3<0$

$\therefore$ Origin lies in acute angle.

Also, $(2+2(-1)-3(2)+5)(2\times 2-1+3\times 2+1)=(-1)(10)<0$

$\therefore$ P lies in obtuse angle.

51. (a) $1-4-9<0$

$\therefore$ Origin lies in acute angle.

Further $(1+4-6+2)(1-4+6+7)>0$

$\therefore$ The point P lies in acute angle.

Passage III

(Ex. Nos. 52 to 54)

In a parallelogram OABC with position vectors of A is $3\hat{\mathbf{i}}+4\hat{\mathbf{j}}$ and C is $4\hat{\mathbf{i}}+3\hat{\mathbf{j}}$ with reference to O as origin. A point E is taken on the side BC which divides it in the ratio of 2 : 1. Also, the line segment AE intersects the line bisecting the $\angle AOC$ internally at P. CP when extended meets AB at point F.

Ex. 52 *The position vector of P is*

(a) $\hat{\mathbf{i}}+\hat{\mathbf{j}}$ (b) $\frac{2}{3}(\hat{\mathbf{i}}+\hat{\mathbf{j}})$
(c) $\frac{13}{3}(\hat{\mathbf{i}}+\hat{\mathbf{j}})$ (d) $\frac{21}{5}(\hat{\mathbf{i}}+\hat{\mathbf{j}})$

Ex. 53 *The equation of line parallel to* **CP** *and passing through $(2, 3, 4)$ is*

(a) $\frac{x-2}{1}=\frac{y-3}{5}, z=4$ (b) $\frac{x-2}{1}=\frac{y-3}{6}, z=4$
(c) $\frac{x-2}{2}=\frac{y-3}{5}, z=3$ (d) $\frac{x-2}{3}=\frac{y-3}{5}, z=3$

Ex. 54 *The equation of plane containing line AC and at a maximum distance from B is*

(a) $\mathbf{r}\cdot(\hat{\mathbf{i}}+\hat{\mathbf{j}})=7$ (b) $\mathbf{r}\cdot(\hat{\mathbf{i}}-\hat{\mathbf{j}})=7$
(c) $\mathbf{r}\cdot(2\hat{\mathbf{i}}-\hat{\mathbf{j}})=7$ (d) $\mathbf{r}\cdot(3\hat{\mathbf{i}}+4\hat{\mathbf{j}})=7$

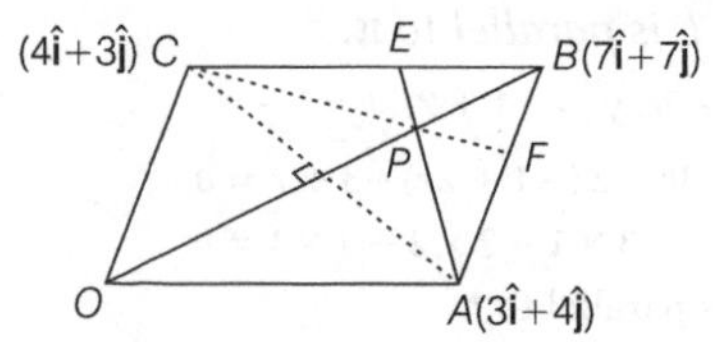

Sol. (Ex 52-54)

52. (d) $\mathbf{OB}=7\hat{\mathbf{i}}+7\hat{\mathbf{j}}$, $\mathbf{OE}=5\hat{\mathbf{i}}+\frac{13}{3}\hat{\mathbf{j}}$, $\mathbf{OP}=\frac{21}{5}(\hat{\mathbf{i}}+\hat{\mathbf{j}})$

53. (b) Direction ratio of **CP** is $(1, 6, 0)$, then equation of line passing through $(2, 3, 4)$ and parallel to **CP** is

$$\frac{x-2}{1}=\frac{y-3}{6}=\frac{z-4}{0}$$

54. (a) The plane containing line AC and at a maximum distance from B must be perpendicular to the plane $OABC$.

Since, $OABC$ is rhombus, so **OB** must normal to the plane. So, equation of required plane is

$$[\mathbf{r}-4\hat{\mathbf{i}}-3\hat{\mathbf{j}}]\cdot(\hat{\mathbf{i}}+\hat{\mathbf{j}})=0$$
$$\Rightarrow \quad (\hat{\mathbf{i}}+\hat{\mathbf{j}})=7$$

Passage IV

(Ex. Nos. 55 to 57)

A ray of light comes along the line $L=0$ and strikes the plane mirror kept along the plane $P=0$ at B. $A(2, 1, 6)$ is a point on the line $L=0$ whose image about $P=0$ is A'. It is given that $L=0$ is $\frac{x-2}{3}=\frac{y-1}{4}=\frac{z-6}{5}$ and $P=0$ is $x+y-2z=3$.

Ex. 55 *The coordinates of A' are*

(a) $(6, 5, 2)$ (b) $(6, 5, -2)$
(c) $(6, -5, 2)$ (d) None of these

Ex. 56 *The coordinates of B are*

(a) $(5, 10, 6)$ (b) $(10, 15, 11)$
(c) $(-10, -15, -14)$ (d) None of these

Ex. 57 *If $L_1=0$ is the reflected ray, then its equation is*

(a) $\frac{x+10}{4}=\frac{y-5}{4}=\frac{z+2}{3}$

(b) $\frac{x+10}{3}=\frac{y+15}{5}=\frac{z+14}{5}$

(c) $\frac{x+10}{4}=\frac{y+15}{5}=\frac{z+14}{3}$

(d) None of the above

Sol. (Ex 55-57)

55. (b) Let $Q(x_2, y_2, z_2)$ be the image of $A(2, 1, 6)$ about mirror $x + y - 2z = 3$. Then,

$$\frac{x_2 - 2}{1} = \frac{y_2 - 1}{1} = \frac{z_2 - 6}{-2}$$

$$= \frac{-2(2 + 1 - 12 - 3)}{1^2 + 1^2 + 2^2} = 4$$

$$\Rightarrow \quad (x_2, y_2, z_2) \equiv (6, 5, -2)$$

56. (c) Let $\frac{x - 2}{3} = \frac{y - 1}{4} = \frac{z - 6}{5} = \lambda$

$x = 2 + 3\lambda, y = 1 + 4\lambda, z = 6 + 5\lambda$ lies on plane $x + y - 2z = 3$

$$\Rightarrow \quad 2 + 3\lambda + 1 + 4\lambda - 2(6 + 5\lambda) = 3$$

$$\Rightarrow \quad 3 + 7\lambda - 12 - 10\lambda = 3$$

$$\Rightarrow \quad -3\lambda = 12$$

$$\Rightarrow \quad \lambda = -4$$

Point $\quad B \equiv (-10, -15, -14)$

57. (c) The equation of the reflected ray $L_1 = 0$ is the line joining $Q(x_2, y_2, z_2)$ and $B(-10, -15, -14)$.

$$\frac{x + 10}{16} = \frac{y + 15}{20} = \frac{z + 14}{12}$$

or

$$\frac{x + 10}{4} = \frac{y + 15}{5} = \frac{z + 14}{3}$$

Passage V

(Ex. Nos. 58 to 60)

The line of greatest slope on an inclined plane P_1 is the line in the plane P_1 which is perpendicular to the line of intersection of the plane P_1 and a horizontal plane P_2.

● **Ex. 58** *Assuming the plane $4x - 3y + 7z = 0$ to be horizontal, the direction cosines of the line of greatest slope in the plane $2x + y - 5z = 0$ are*

(a) $\frac{3}{\sqrt{11}}, \frac{-1}{\sqrt{11}}, \frac{1}{\sqrt{11}}$ (b) $\frac{3}{\sqrt{11}}, \frac{1}{\sqrt{11}}, \frac{-1}{\sqrt{11}}$

(c) $\frac{-3}{\sqrt{11}}, \frac{1}{\sqrt{11}}, \frac{1}{\sqrt{11}}$ (d) None of these

● **Ex. 59** *The equation of a line of greatest slope can be*

(a) $\frac{x}{3} = \frac{y}{1} = \frac{z}{-1}$ (b) $\frac{x}{3} = \frac{y}{-1} = \frac{z}{1}$

(c) $\frac{x}{-3} = \frac{y}{1} = \frac{z}{1}$ (d) $\frac{x}{1} = \frac{y}{3} = \frac{z}{-1}$

● **Ex. 60** *The coordinates of a point on the plane $2x + y - 5z = 0$, $2\sqrt{11}$ unit away from the line of intersection of $2x + y - 5z = 0$ and $4x - 3y + 7z = 0$ are*

(a) (6, 2, – 2) (b) (3, 1, – 1)

(c) (6, – 2, 2) (d) (1, 3, – 1)

Sol. (Ex. 58-60)

58. (a) Plane P_1 is of the form $\mathbf{r} \cdot \mathbf{n}_1 = 0$, where $\mathbf{n}_1 = (4, -3, 7)$

Plane P_2 is of the form $\mathbf{r} \cdot \mathbf{n}_2 = 0$,

where $\quad \mathbf{n}_2 = (2, 1, -5)$

The vector **b** along the line of intersection of planes is

$$\mathbf{n}_1 \times \mathbf{n}_1 = (4, 17, 5) = \mathbf{n}_3$$

Since the line of greatest slope is perpendicular to $\mathbf{n}_3$ and $\mathbf{n}_2$ the vector along the line of greatest slope

$$= \mathbf{n}_2 \times \mathbf{n}_3 = (3, -1, 1) = \mathbf{n}_4$$

and the unit vector

$$\mathbf{n}_4 = \hat{\mathbf{n}} = \left(\frac{3}{\sqrt{11}}, \frac{-1}{\sqrt{11}}, \frac{1}{\sqrt{11}}\right)$$

59. (b) Since, (0, 0, 0) is a point on both planes, it lies on the line of intersection.

Hence, the equation a line of greatest slope can be

$$\frac{x}{3} = \frac{y}{-1} = \frac{z}{1}$$

60. (c) The point on the line $\frac{x}{3} = \frac{y}{-1} = \frac{z}{1}$ at a distance $2\sqrt{11}$ unit from the origin is given by

$$\frac{x}{\frac{3}{\sqrt{11}}} = \frac{y}{\frac{-1}{\sqrt{11}}} = \frac{z}{1} = 2\sqrt{11}$$

The point is (6, – 2, 2).

JEE Type Solved Examples : Matching Type Questions

Ex. 61 *Match the entries between following two columns.*

Column I	Column II
A. If the line $\frac{x-1}{1}=\frac{y+1}{-2}=\frac{z+1}{\lambda}$ lies in the plane $3x-2y+5z=0$, then λ is equal to	p. $\sin^{-1}\sqrt{\frac{6}{25}}$
B. If $(3, \lambda, \mu)$ is a point on the line $2x+y+z-3=0=x-2y+z-1$, then $\lambda+\mu$ is equal to	q. $-\frac{7}{5}$
C. The angle between the line $x=y=z$ and the plane $4x-3y+5z=2$ is	r. -3
D. The angle between the planes $x+y+z=0$ and $3x-4y+5z=0$ is	s. $\cos^{-1}\sqrt{\frac{8}{75}}$

Sol. (A) → (q), (B) → (r), (C) → (p), (D) → (s)

(A) $3\cdot 1-2(-2)+5(\lambda)=0 \Rightarrow \lambda=-\frac{7}{5}$

(B) Point $(3, \lambda, \mu)$ lies on $2x+y+z-3=x-2y+z-1$

$\Rightarrow 3\cdot 2+\lambda+\mu-3=0$ and $3-2\lambda+\mu-1=0$

$\Rightarrow \lambda+\mu+3=0$ and $2\lambda-\mu-2=0$

So, $\lambda+\mu=-3$

(C) $\sin\theta=\frac{1\cdot 4+1(-3)+1\cdot 5}{\sqrt{1^2+1^2+1^2}\sqrt{16+9+25}}=\frac{6}{\sqrt{3}\sqrt{50}}$

$$\theta=\sin^{-1}\sqrt{\frac{6}{25}}$$

(D) $\cos\theta=\frac{1\cdot 3+1(-4)+1\cdot 5}{\sqrt{3}\sqrt{16+9+25}}=\frac{4}{\sqrt{3}\sqrt{50}}$

$$\theta=\cos^{-1}\sqrt{\frac{8}{75}}$$

Ex. 62 *Match the following*

Column I	Column II
A. $\frac{x-1}{2}=\frac{y-2}{3}=\frac{z-3}{4}$ and $\frac{x-1}{3}=\frac{y-3}{4}=\frac{z-5}{5}$ are	p. coincident
B. $\frac{x-1}{2}=\frac{y-2}{3}=\frac{z-3}{4}$ and $\frac{x-3}{2}=\frac{y-5}{3}=\frac{z-7}{4}$ are	q. parallel and different
C. $\frac{x-2}{5}=\frac{y+3}{4}=\frac{5-z}{2}$ and $\frac{x-7}{5}=\frac{y-1}{4}=\frac{z-2}{-2}$ are	r. skew
D. $\frac{x-3}{2}=\frac{y+2}{3}=\frac{z-4}{5}$ and $\frac{x-3}{3}=\frac{y-2}{2}=\frac{z-7}{5}$ are	s. intersecting in a point

Sol. (A) → (s), (B) → (p), (C) → (q), (D) → (r)

(A) Both the lines pass through the point (7, 11, 15).

(B) < 2, 3, 4 > are direction ratios of both the lines. Also, the point (1, 2, 3) is common to both.

∴The lines are coincident.

(c) < 5, 4, – 2 > are direction ratios of both the lines.

∴The lines are parallel.

Also, $x=2+5\lambda, y=-3+4\lambda, z=5-2\lambda$

$$\therefore \quad \frac{2+5\lambda-7}{5}=\frac{-3+4\lambda-1}{4}=\frac{5-2\lambda-2}{-2}$$

i.e. $\lambda-1=\lambda-1=\frac{3-2\lambda}{-2}$

∴No value of λ.

Thus, the lines are parallel and different.

(D) < 2, 3, 5 > and <3, 2, 5 > are direction ratios of first and second line, respectively.

Ex. 63 *Match the followings*

Column I	Column II
A. The coordinates of a point on the line $x=4y+5$, $z=3y-6$ at a distance 3 from the point $(5, 3, -6)$ is/are	p. $(-1, -2, 0)$
B. The plane containing the lines $\frac{x-2}{3}=\frac{y+3}{5}=\frac{z+5}{7}$ and parallel to $\hat{\mathbf{i}}+4\hat{\mathbf{j}}+7\hat{\mathbf{k}}$ has the point	q. $(5, 0, -6)$
C. A line passes through two points $A(2, -3, -1)$ and $B(8, -1, 2)$. The coordinates of a point on this line nearer to the origin and at a distance of 14 units from A is/are	r. $(2, 5, 7)$
D. The coordinates of the foot of the perpendicular from the point $(3, -1, 11)$ on the line $\frac{x}{2}=\frac{y-2}{3}=\frac{z-3}{4}$ is/are	s. $(14, 1, 5)$

Sol. (A) → (q), (B) → (p), (C) → (s), (D) → (r)

(A) The given line is $x=4y+5, z=3y-6$,

or $\frac{x-5}{4}=y, \quad \frac{z+6}{3}=y$

or $\frac{x-5}{4}=\frac{y}{1}=\frac{z+6}{3}=\lambda$ (say)

Any point on the line is of the form $(4\lambda+5, \lambda, 3\lambda-6)$.

The distance between $(4\lambda+5, \lambda, 3\lambda-6)$ and $(5, 3, -6)$ is 3 units (given).

Therefore, $(4\lambda+5-5)^2+(\lambda-3)^2+(3\lambda-6+6)^2=9$

$\Rightarrow 16\lambda^2+\lambda^2+9-6\lambda+9\lambda^2=9$

$\Rightarrow 26\lambda^2-6\lambda=0$

$\Rightarrow \lambda=0, \frac{3}{13}$

The point is (5, 0, – 6)

(B) The equation of the plane containing the lines $\frac{x-2}{3}=\frac{y+3}{5}=\frac{z+5}{7}$ and parallel to $\hat{\mathbf{i}}+4\hat{\mathbf{j}}+7\hat{\mathbf{k}}$.

$$\begin{vmatrix} x-2 & y+3 & z+5 \\ 1 & 4 & 7 \\ 3 & 5 & 7 \end{vmatrix} = 0 \Rightarrow x - 2y + z - 3 = 0$$

Point $(-1, -2, 0)$ lies on this plane.

(C) The line passing through points $A(2, -3, -1)$

and $B(8, -1, 2)$ is $\dfrac{x-2}{8-2} = \dfrac{y+3}{-1+3} = \dfrac{z+1}{2+1}$

or $\dfrac{x-2}{6} = \dfrac{y+3}{2} = \dfrac{z+1}{3} = \lambda$ (say)

Any point on this line is of the form $P(6\lambda + 2, 2\lambda - 3, 3\lambda - 1)$, whose distance from point $A(2, -3, -1)$ is 14 units. Therefore,

$\Rightarrow \quad PA = 14 \Rightarrow PA^2 = (14)^2$

$\Rightarrow \quad (6\lambda)^2 + (2\lambda)^2 + (3\lambda)^2 = 196$

$\Rightarrow \quad 49\lambda^2 = 196 \Rightarrow \lambda^2 = 4 \Rightarrow \lambda = \pm 2$

Therefore, the required points are $(14, 1, 5)$ and $(-10, -7, -7)$. The point nearer to the origin is $(14, 1, 5)$.

(D) Any point on line AB, $\dfrac{x}{2} = \dfrac{y-2}{3} = \dfrac{z-3}{4} = \lambda$ is

$M(2\lambda, 3\lambda + 2, 4\lambda + 3)$. Therefore, the direction ratios of PM are $2\lambda - 3, 3\lambda + 3$ and $4\lambda - 8$.

But $\quad PM \perp AB$

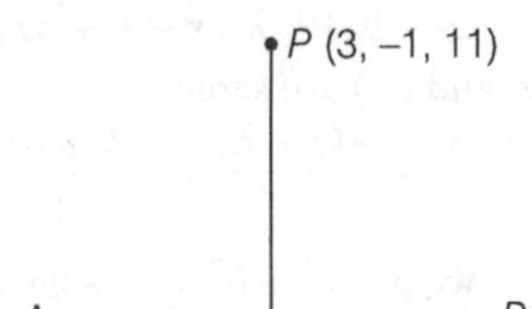

$\therefore \quad 2(2\lambda - 3) + 3(3\lambda + 3) + 4(4\lambda - 8) = 0$

$4\lambda - 6 + 9\lambda + 9 + 16\lambda - 32 = 0$

$29\lambda - 29 = 0; \ \lambda = 1$

Therefore, foot of the perpendicular is $M(2, 5, 7)$.

Ex. 64 *Match the followings*

Column I	Column II
A. Image of the point (3, 5, 7) in the plane $2x + y + z = -18$ is	p. $(-1, -1, -1)$
B. The point of intersection of the line $\dfrac{x-2}{-3} = \dfrac{y-1}{-2} = \dfrac{z-3}{2}$ and the plane $2x + y - z = 3$ is	q. $(-21, -7, -5)$
C. The foot of the perpendicular from the point (1, 1, 2) to the plane $2x - 2y + 4z + 5 = 0$ is	r. $\left(\dfrac{5}{2}, \dfrac{2}{3}, \dfrac{8}{3}\right)$
D. The intersection point of the lines $\dfrac{x-1}{2} = \dfrac{y-2}{3} = \dfrac{z-3}{4}$ and $\dfrac{x-4}{5} = \dfrac{y-1}{2} = z$ is	s. $\left(-\dfrac{1}{12}, \dfrac{25}{12}, \dfrac{2}{12}\right)$

Sol. (A) → (q), (B) → (r), (C) → (s), (D) → (p)

(A) If the required image is (x, y, z), then

$$\frac{x-3}{2} = \frac{y-5}{1} = \frac{z-7}{1} = -\frac{2(6+5+7+18)}{2^2+1^2+1^2}$$

$= -12$ or $(-21, -7, -5)$.

(B) Any point on the line $\dfrac{x-2}{-3} = \dfrac{y-1}{2} = \dfrac{z-3}{2} = \lambda$ is $(-3\lambda + 2, 2\lambda + 1, 2\lambda + 3)$, which lies on plane $2x + y - z = 3$. Therefore,

$$-6\lambda + 4 + 2\lambda + 1 - 2\lambda - 3 = 3$$

$$-6\lambda = 1$$

$$\lambda = -\frac{1}{6}$$

Therefore, the point is $\left(\dfrac{5}{2}, \dfrac{2}{3}, \dfrac{8}{3}\right)$

(C) If (x, y, z) is required foot of the perpendicular, then

$$\frac{x-1}{2} = \frac{y-1}{-2} = \frac{z-2}{4} = \frac{(2-2+8+5)}{2^2+(-2)^2+4^2}$$

or $\quad (x, y, z) \equiv \left(\dfrac{-1}{12}, \dfrac{25}{12}, \dfrac{-2}{12}\right)$

(D) Any point on the line $\dfrac{x-1}{2} = \dfrac{y-2}{3} = \dfrac{z-3}{4} = \lambda$ is

$P(2\lambda + 1, 3\lambda + 2, 4\lambda + 3)$, which satisfies the line

$$\frac{x-4}{5} = \frac{y-1}{2} = \frac{z}{1}$$

or $\quad \dfrac{2\lambda + 1 - 4}{5} = \dfrac{3\lambda + 2 - 1}{2} = \dfrac{4\lambda + 3}{1}$

$\Rightarrow \quad \lambda = -1$

The required point is $(-1, -1, -1)$.

Ex. 65 $\dfrac{x-2}{3} = \dfrac{y-3}{4} = \dfrac{z-4}{5}$

Column I	Column II
A. Point on the line at a distance $10\sqrt{2}$ from (2, 3, 4)	p. $(-1, -1, -1)$
B. Point on the line common to the plane $x + y + z + 3 = 0$	q. (2, 3, 4)
C. Point on the line at a distance $\sqrt{29}$ from the origin.	r. (8, 11, 14)
D. Point on the line common to the plane $x + y - z + 3 = 0$	s. $(-4, -5, -6)$

Sol. (A) → (r, s), (B) → (p), (C) → (q), (D) → (s)

Any point on the line is $(3r + 2, 4r + 3, 5r + 4$

(A) $(3r + 2 - 2)^2 + (4r + 3 - 3)^2 + (5r + 4 - 4)^2 = 200$

$\Rightarrow \quad (9 + 16 + 25)r^2 = 200 \Rightarrow r = \pm 2$

For $r = 2$, the point is $(8, 11, 14)$, For $r = -2$ it is $(-4, -5, -6)$

(B) $3r + 2 + 4r + 3 + 5r + 4 + 3 = 0$

$\Rightarrow \quad 12r + 12 = 0 \Rightarrow r = -1$

and the point on the line common to the plane is $(-1, -1, -1)$.

(C) $(3r + 2)^2 + (4r + 3)^2 + (5r + 4)^2 = 29$

$$50r^2 + 76r = 0 \Rightarrow r = 0, r = -\frac{76}{50}$$

For $r = 0$, the point is $(2, 3, 4)$.

(D) $3r + 2 + 4r + 3 - 5r - 4 + 3 = 0$

$\Rightarrow \quad 2r + 4 = 0 \Rightarrow r = -2$

$\therefore$ The point on the line common to the plane is $(-4, -5, -6)$

JEE Type Solved Examples : Single Integer Answer Type Questions

Ex. 66 *If the perpendicular distance of the point (6, 5, 8) from the Y-axis is 5λ unit, then λ is equal to*

Sol. (2) Foot of perpendicular from (6, 5, 8) on Y-axis is (0, 5, 0).

$$\text{Required distance} = \sqrt{(6-0)^2+(5-5)^2+(8-0)^2} = 10 \text{ unit}$$

$$\Rightarrow \quad 5\lambda = 10 \Rightarrow \lambda = \frac{10}{5} = 2$$

Ex. 67 *A parallelopiped is formed by planes drawn through the points (2, 4, 5) and (5, 9, 7) parallel to the coordinate planes. The length of the diagonal of the parallelopiped is*

Sol. (7) The length of the edges are given by $a = 5-2 = 3$, $b = 9-3 = 6$, $c = 7-5 = 2$, so length of the diagonal

$$= \sqrt{a^2+b^2+c^2} = \sqrt{9+36+4} = 7 \text{ units}$$

Ex. 68 *If the shortest distance between the lines $\frac{x-3}{3} = \frac{y-8}{-1} = \frac{z-3}{1}$ and $\frac{x+3}{-3} = \frac{y+7}{2} = \frac{z-6}{4}$ is $\lambda\sqrt{30}$ unit, then the value of λ is*

Sol. (3) Given, lines are

$$\mathbf{r} = 3\hat{\mathbf{i}}+8\hat{\mathbf{j}}+3\hat{\mathbf{k}}+\lambda(3\hat{\mathbf{i}}-\hat{\mathbf{j}}+\hat{\mathbf{k}})$$

$$\mathbf{r} = (-3\hat{\mathbf{i}}-7\hat{\mathbf{j}}+6\hat{\mathbf{k}})+\mu(-3\hat{\mathbf{i}}+2\hat{\mathbf{j}}+4\hat{\mathbf{k}})$$

where λ, μ are parameters.

Shortest distance

$$= \frac{[(-3-3)\hat{\mathbf{i}}+(7-8)\hat{\mathbf{j}}+(6-3)\hat{\mathbf{k}}]}{[(3\hat{\mathbf{i}}-\hat{\mathbf{j}}+\hat{\mathbf{k}})\times(-3\hat{\mathbf{i}}+2\hat{\mathbf{j}}+4\hat{\mathbf{k}})]}$$

$$= \frac{(-6\hat{\mathbf{i}}-15\hat{\mathbf{j}}+3\hat{\mathbf{k}})\times(-6\hat{\mathbf{j}}-15\hat{\mathbf{k}}+3\hat{\mathbf{k}})}{\sqrt{36+225+9}}$$

$$= \sqrt{270} = 3\sqrt{30} \text{ unit}$$

Ex. 69 *If the planes $x - cy - bz = 0$, $cx - y + az = 0$ and $bx + ay - z = 0$ pass through a line, then the value of $a^2+b^2+c^2+2abc$ is*

Sol. (1) Given, planes are

$$x - cy - bz = 0 \quad \text{...(i)}$$

$$cx - y + az = 0 \quad \text{...(ii)}$$

$$bx + ay - z = 0 \quad \text{(iii)}$$

Equation of planes passing through the line of intersection of planes (i) and (ii) may be taken as

$$(x - cy - bz) + \lambda(cx - y + az) = 0 \quad \text{...(iv)}$$

Now, planes (iii) and (iv) are same

$$\therefore \quad \frac{1+c\lambda}{b} = \frac{-(c+\lambda)}{a} = \frac{-b+a\lambda}{-1}$$

By eliminating λ, we get $a^2+b^2+c^2+2abc = 1$

Ex. 70 *If the line $\frac{x-4}{1} = \frac{y-2}{1} = \frac{z-k}{2}$ lies exactly on the plane $2x - 4y + z = 7$, the value of k is*

Sol. (7) The point (4, 2, k) must satisfy the plane.

So, $\quad 8 - 8 + k = 7 \Rightarrow k = 7$

Subjective Type Questions

Ex. 71 *The equation of motion of rockets are*

$$x = 2t,\ y = -4t,\ z = 4t,$$

where the time 't' is given in second and the coordinates of a moving point in kilometres.

What is the path of the rocket? At what distance will the rocket be from the starting point O(0, 0, 0) in 10s.

Sol. Eliminating 't' from the given equations, we get the equation of the path,

$$\frac{x}{2} = \frac{y}{-4} = \frac{z}{4}$$

or

$$\frac{x}{1} = \frac{y}{-2} = \frac{z}{2}$$

Thus, the path of the rocket represents a straight line passing through the origin.

For $\quad t = 10$ s

We have, $x = 20$, $y = -40$, $z = 40$

and $\quad |\mathbf{r}| = |\mathbf{OM}| = \sqrt{x^2+y^2+z^2} = \sqrt{400+1600+1600} = 60$ km

Ex. 72 *Write the equation of a tangent to the curve $x = t$, $y = t^2$, $z = t^3$ at its point M(1, 1, 1); ($t = 1$).*

Sol. Here, $\mathbf{r} = t\hat{\mathbf{i}} + t^2\hat{\mathbf{j}} + t^3\hat{\mathbf{k}}$

$$\frac{dr}{dt} = \hat{\mathbf{i}} + 2t\hat{\mathbf{j}} + 3t^2\hat{\mathbf{k}}$$

Hence, the direction of the tangent at the point M is determined by the vector.

$$\left(\frac{dr}{dt}\right)_M = \hat{\mathbf{i}} + 2\hat{\mathbf{j}} + 3\hat{\mathbf{k}}$$

Thus, the equation of the desired tangent is,

$$\frac{x-1}{1} = \frac{y-1}{2} = \frac{z-1}{3}$$

Ex. 73 *Find the locus of a point, the sum of squares of whose distances from the planes $x - z = 0$, $x - 2y + z = 0$ and $x + y + z = 0$ is 36.*

Sol. Given planes are $x - z = 0$, $x - 2y + z = 0$

and $x + y + z = 0$.

Let the point whose locus is required be $P(\alpha, \beta, \gamma)$. According to question,

$$\frac{|\alpha + \gamma|^2}{2} + \frac{|\alpha - 2\beta + \gamma|^2}{6} + \frac{|\alpha + \beta + \gamma|^2}{3} = 36$$

or $3(\alpha^2 + \gamma^2 - 2\alpha\gamma) + \alpha^2 + 4\beta^2 + \gamma^2 - 4\alpha\beta - 4\beta\gamma + 2\alpha\gamma$
$+ 2(\alpha^2 + \beta^2 + \gamma^2 + 2\alpha\beta + 2\beta\gamma + 2\alpha\gamma) = 36 \times 6$

or $6\alpha^2 + 6\beta^2 + 6\gamma^2 = 36 \times 6$

or $\alpha^2 + \beta^2 + \gamma^2 = 36$

Hence, the required equation of locus is

$$x^2 + y^2 + z^2 = 36$$

Ex. 74 *The plane $ax + by = 0$ is rotated through an angle α about its line of intersection with the plane $z = 0$. Show that the equation to the plane in new position is $ax + by \pm z\sqrt{a^2 + b^2}\tan\alpha = 0$*

Sol. Given planes are

$$ax + by = 0 \quad \text{...(i)}$$

and $$z = 0 \quad \text{...(ii)}$$

$\therefore$ Equation of any plane passing through the line of intersection of planes (i) and (ii) may be taken as,

$$ax + by + kz = 0 \quad \text{...(iii)}$$

The direction cosines of a normal to the plane (iii) are

$$\frac{a}{\sqrt{a^2 + b^2 + k^2}}, \frac{b}{\sqrt{a^2 + b^2 + k^2}}, \frac{k}{\sqrt{a^2 + b^2 + k^2}}$$

The direction cosines of a normal to the plane (i) are

$$\frac{a}{\sqrt{a^2 + b^2}}, \frac{b}{\sqrt{a^2 + b^2}}, 0$$

Since, the angle between the planes (i) and (iii) is α.

$$\therefore \quad \cos\alpha = \frac{a \cdot a + b \cdot b + k \cdot 0}{\sqrt{a^2 + b^2 + k^2}\sqrt{a^2 + b^2}}$$

$$= \sqrt{\frac{a^2 + b^2}{a^2 + b^2 + k^2}}$$

$$k^2 \cos^2\alpha = \alpha^2(1 - \cos^2\alpha) + b^2(1 - \cos^2\alpha)$$

$$k^2 = \frac{(a^2 + b^2)\sin^2\alpha}{\cos^2\alpha}$$

On putting in Eq. (iii) $k = \pm\sqrt{a^2 + b^2}\tan\alpha$, we get equation of plane as,

$$ax + by \pm z\sqrt{a^2 + b^2}\tan\alpha = 0.$$

Ex. 75 *Assuming the plane $4x - 3y + 7z = 0$ to be horizontal, find the equation of the line of greatest slope through the point $(2, 1, 1)$ in the plane $2x + y - 5z = 0$.*

Sol. The required line passing through the point (2, 1, 1) in the plane $2x + y - 5z = 0$ and is having greatest slope, so it must be perpendicular to the line of intersection of the planes

$$2x + y - 5z = 0 \quad \text{...(i)}$$

and $$4x - 3y + 7z = 0 \quad \text{...(ii)}$$

Let the DR's of the line of intersection of Eqs. (i) and (ii) are a, b, c.

$$2a + b - 5c = 0$$

and $$4a - 3b + 7c = 0$$

(as DR's of straight line (a, b, c) is perpendicular to DR's of normal to both the planes)

$$\frac{a}{4} = \frac{b}{17} = \frac{c}{5}$$

Now, let the direction ratio of required line be proportional to l, m and n, then its equation be

$$\frac{x-2}{l} = \frac{y-1}{m} = \frac{z-1}{n}$$

where, $2l + m - 5n = 0$ and $4l + 17m + 5n = 0$

So, $$\frac{l}{3} = \frac{m}{-1} = \frac{n}{1}$$

Thus, the required line is $\dfrac{x-2}{3} = \dfrac{y-1}{-1} = \dfrac{z-1}{1}$

Ex. 76 *Does $\dfrac{a}{x-y} + \dfrac{b}{y-z} + \dfrac{c}{z-x} = 0$ represents a pair of planes?*

Sol. Here, given equation is $\dfrac{a}{x-y} + \dfrac{b}{y-z} + \dfrac{c}{z-x} = 0$

$\Rightarrow a(y-z)(z-x) + b(x-y)(z-x) + c(x-y)(y-z) = 0$

$\Rightarrow -axy + ayz - az^2 + axz + bxz - bx^2 - byz$
$+ byx + cxy - cxz - cy^2 + cyz = 0$

$\Rightarrow bx^2 + cy^2 + az^2 - (b + c - a)xy -$
$(c + a - b)yz - (a + b - c)zx = 0$

$\therefore$ Value of determinant;

$$\begin{vmatrix} b & -\frac{1}{2}(b+c-a) & -\frac{1}{2}(a+b-c) \\ -\frac{1}{2}(b+c-a) & c & -\frac{1}{2}(c+a-b) \\ -\frac{1}{2}(a+b-c) & \frac{1}{2}(c+a-b) & a \end{vmatrix}$$

$$= \begin{vmatrix} 0 & 0 & 0 \\ -\frac{1}{2}(b+c-a) & c & -\frac{1}{2}(c+a-b) \\ -\frac{1}{2}(a+b-c) & -\frac{1}{2}(c+a-b) & a \end{vmatrix}$$

$[R_1 \to R_1 + R_2 + R_3]$

$= 0$

Hence, the given equation represents a pair of planes.

Ex. 77 *If the straight line* $\frac{x-\alpha}{l} = \frac{y-\beta}{m} = \frac{z-\gamma}{n}$ *intersect the curve* $ax^2 + by^2 = 1$, $z = 0$, *then prove that* $a(\alpha n - \gamma l)^2 + b(\beta n - \gamma m)^2 = n^2$.

Sol. Here, $\frac{x-\alpha}{l} = \frac{y-\beta}{m} = \frac{z-\gamma}{n} = \lambda$

$\therefore$ Any point on the given line

$(\lambda l + \alpha, \lambda m + \beta + \lambda n + y)$

If it lies on the curve $ax^2 + by^2 = 1, z = 0$

(as the point of intersection)

$a(\alpha + l\lambda)^2 + b(\beta + m\lambda)^2 = 1$ ...(i)

and $\lambda n + \gamma = 0$ ...(ii)

From Eq. (ii), $\lambda = \frac{-\gamma}{n}$ must satisfy Eq. (i), we get

$$a\left(\alpha - \frac{l\gamma}{n}\right)^2 + b\left(\beta - \frac{m\gamma}{n}\right)^2 = 1$$

$$a(n\alpha - l\gamma)^2 + b(n\beta - m\gamma)^2 = n^2$$

Ex. 78 *Prove that the three lines from O with direction cosines* $l_1, m_1, n_1; l_2, m_2, n_2$ *and* l_3, m_3, n_3 *are coplanar, if*

$$l_1(m_2n_3 - n_2m_3) + m_1(n_2l_3 - l_2n_3) + n_1(l_2m_3 - l_3m_2) = 0.$$

Sol. Here, three given lines are coplanar, if they have common perpendicular.

Let DC's of common perpendicular be l, m and n.

$ll_1 + mm_1 + nn_1 = 0$ ...(i)

$ll_2 + mm_2 + nn_2 = 0$ ...(ii)

and $ll_3 + mm_3 + nn_3 = 0$ ...(iii)

Solving Eqs. (ii) and (iii) by cross multiplication method, we get

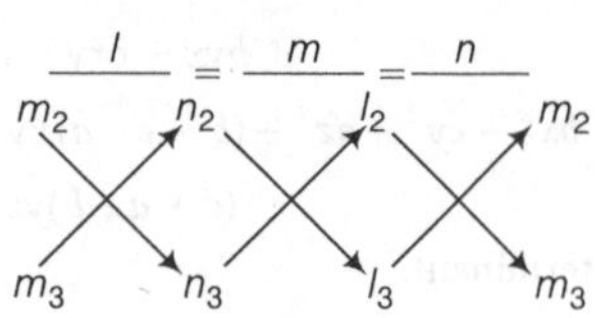

$$\Rightarrow \frac{l}{m_2n_3 - n_2m_3} = \frac{m}{n_2l_3 - n_3l_2} = \frac{n}{l_2m_3 - l_3m_2} = k$$

$\Rightarrow l = k(m_2n_3 - n_2m_3), m = k(n_2l_3 - n_3l_2), n = k(l_2m_3 - l_3m_2)$

Substituting in Eq. (i), we get

$k(m_2n_3 - n_2m_3)l_1 + k(n_2l_3 - n_3l_2)m_1 + k(l_2m_3 - l_3m_2)n_2 = 0$

$\Rightarrow l_1(m_2n_3 - n_2m_2) + m_1(n_2l_3 - n_3l_2) + n_1(l_2m_3 - l_3m_2)$

Ex. 79 *A line makes angle,* α, β, γ *and* δ *with the four diagonals of cube, prove that*

$$\cos^2\alpha + \cos^2\beta + \cos^2\gamma + \cos^2\delta = \frac{4}{3}.$$

Sol. Let the cube be shown in the figure, where four diagonals are OP, AL, BM and CN and $A(a, 0, 0)$, $B(0, a, 0)$, $C(0, 0, a)$, $L(0, a, a)$, $M(a, 0, a)$, $N(a, a, 0)$ and $P(a, a, a)$, hence direction cosines of OP are

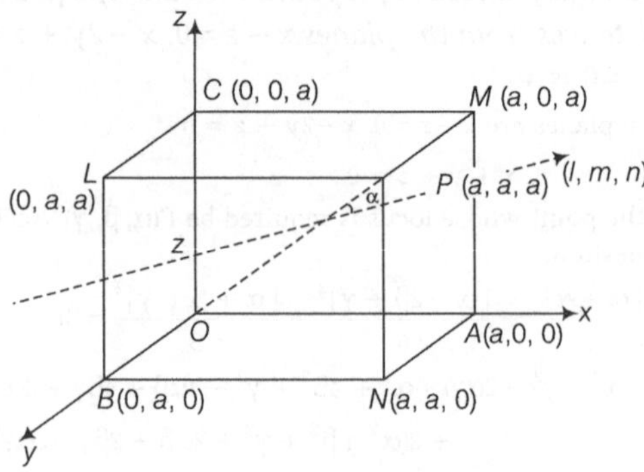

$$\frac{a}{\sqrt{a^2+a^2+a^2}}, \frac{a}{\sqrt{a^2+a^2+a^2}}, \frac{a}{\sqrt{a^2+a^2+a^2}}$$

$$= \left(\frac{1}{\sqrt{3}}, \frac{1}{\sqrt{3}}, \frac{1}{\sqrt{3}}\right)$$

The DC's of AL are $\left(-\frac{1}{\sqrt{3}}, \frac{1}{\sqrt{3}}, \frac{1}{\sqrt{3}}\right)$.

The DC's of BM are $\left(\frac{1}{\sqrt{3}}, -\frac{1}{\sqrt{3}}, \frac{1}{\sqrt{3}}\right)$.

The DC's of CN are $\left(\frac{1}{\sqrt{3}}, \frac{1}{\sqrt{3}}, -\frac{1}{\sqrt{3}}\right)$.

Let the DC's of required line be (l, m, n).

$\therefore \quad \cos\alpha = \frac{l+m+n}{\sqrt{3}}, \cos\beta = \frac{-l+m+n}{\sqrt{3}},$

$\cos\gamma = \frac{l-m+n}{\sqrt{3}}$ and $\cos\delta = \frac{l+m-n}{\sqrt{3}}$

$\cos^2\alpha + \cos^2\beta + \cos^2\gamma + \cos^2\delta$

$$= \frac{1}{3}\{(l+m+n)^2 + (-l+m+n)^2 + (l-m+n)^2 + (l+m-n)^2\}$$

$$= \frac{4}{3}(l^2+m^2+n^2) = \frac{4}{3}$$

Ex. 80 *Let PM be the perpendicular from the point P(1, 2, 3) to XY-plane. If OP makes an angle* θ *with the positive direction of the Z-axis and OM makes an angle* ϕ *with the positive direction of X-axis, where O is the origin, then find* θ *and* ϕ.

Sol. Here, P be (x, y, z) shown as,

then, $x = r\sin\theta \cdot \cos\phi$, $y = r\sin\theta\sin\phi$, $z = r\cos\theta$ ...(i)

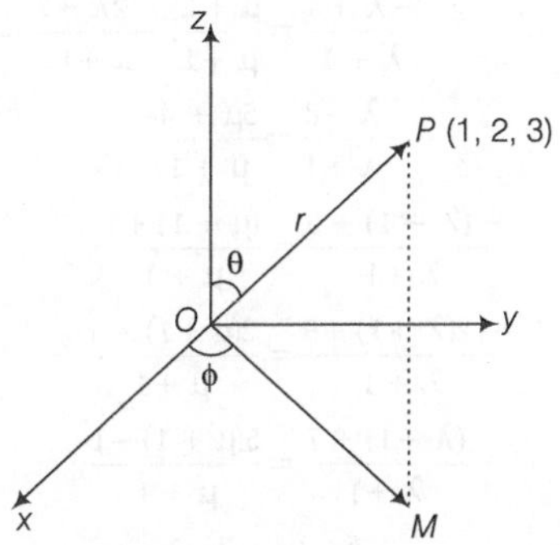

$\Rightarrow \; 1 = r\sin\theta\cdot\cos\phi, 2 = r\sin\theta\sin\phi, 3 = r\cos\theta$

$\Rightarrow \; 1^2+2^2+3^2 = r^2\sin^2\theta\cos^2\phi + r^2\sin^2\theta\sin^2\phi + r^2\cos^2\theta$

$= r^2\sin^2\theta(\cos^2\phi+\sin^2\phi)+r^2\cos^2\theta$

$= r^2\sin^2\theta + r^2\cos^2\theta = r^2$

$\Rightarrow \quad r = \pm\sqrt{14}$

$\therefore$ From Eq. (i), we have

$$\sin\theta\cos\phi = +\frac{1}{\sqrt{14}},$$

$$\sin\theta\sin\phi = \frac{2}{\sqrt{14}}, \cos\theta = \frac{3}{\sqrt{14}}$$

(neglecting – ve sign as acute angles)

$$\therefore \quad \frac{\sin\theta\sin\phi}{\sin\theta\cos\phi} = \frac{2}{1}$$

and $$\tan\theta = \frac{\sin\theta}{\cos\theta} = \frac{\sqrt5}{3}$$

$$\Rightarrow \quad \tan\phi = 2 \text{ and } \tan\theta = \frac{\sqrt5}{3}$$

$$\Rightarrow \quad \phi = \tan^{-1}2 \text{ and } \theta = \tan^{-1}\left(\frac{\sqrt5}{3}\right)$$

Ex. 81 *Find the distance of the point $(1, 0, -3)$ from the plane $x-y-z=9$ measured parallel to the line, $\frac{x-2}{2}=\frac{y+2}{3}=\frac{2-6}{-6}$.*

Sol. Given plane is

$$x-y-z=9 \qquad \text{...(i)}$$

Given line AB is $\frac{x-2}{2}=\frac{y+2}{3}=\frac{z-6}{-6}$ ...(ii)

Equation of line passing through $(1, 0, -3)$ and parallel to

$$\frac{x-2}{2}=\frac{y+2}{3}=\frac{z-6}{-6}$$

is $$\frac{x-1}{2}=\frac{y-0}{3}=\frac{z+3}{-6}=r \qquad \text{...(iii)}$$

Coordinate of any point on Eq. (iii) may be given as

$$P(2r+1, 3r, -6r-3)$$

If P is intersection of Eqs. (i) and (iii), then it must lie on Eq. (i).

$$(2r+1)-(3r)-(-6r-3)=9$$

$$2r+1-3r+6r+3=9$$

$$5r=5$$

$$\Rightarrow \quad r=1$$

$\therefore$ Coordinate of $P(3, 3, -9)$.

$\Rightarrow$ Distance between $(1, 0, -3)$ and $(3, 3, -9)$

$$= \sqrt{(3-1)^2+(3-0)^2+(-9+3)^2}$$

$$= \sqrt{4+9+36} = 7$$

Ex. 82 *Find the equation of the plane which passes through $a_1x+b_1y+c_1z+d_1=0, a_2x+b_2y+c_2z+d_2=0$ and which is parallel to the line $\frac{x-a}{l}=\frac{y-\beta}{m}=\frac{z-\gamma}{n}$.*

Sol. Given, $a_1x+b_1y+c_1z+d_1=0$

$$a_2x+b_2y+c_2z+d_2=0 \qquad \text{...(i)}$$

and $$\frac{x-\alpha}{l}=\frac{y-\beta}{m}=\frac{z-\gamma}{n} \qquad \text{...(ii)}$$

Equation of plane through the intersection of plane (i) is given by

$$(a_1x+b_1y+c_1z+d_1)+\lambda(a_2x+b_2y+c_2z+d_2)=0$$

or $(a_1+\lambda a_2)x+(b_1+\lambda b_2)y+(c_1+\lambda c_2)z + (d_1+\lambda d_2)=0$ ...(iii)

DR's of normal to Eq. (iii) are

$$(a_1+\lambda a_2), (b_1+\lambda b_2), (c_1+\lambda c_2)$$

$\therefore$ Eq. (iii) is parallel to Eq. (ii).

$\Rightarrow$ Normal to plane (iii) should be perpendicular to line (ii).

$\therefore \quad (a_1+\lambda a_2)l+(b_1+\lambda b_2)m+(c_1+\lambda c_2)n=0$

$\Rightarrow \quad \lambda = -\frac{(a_1l+b_1m+c_1n)}{(a_2l+b_2m+c_2n)}$, putting in (iii), we get

$$(a_1x+b_1y+c_1z+d_1)(a_2l+b_2m+c_2n)-(a_1l+b_1+c_1n)(a_2x+b_2y+c_2z+d_2)=0$$

Hence, the equation of required plane.

Ex. 83 *Find the perpendicular distance of a corner of a unit cube from a diagonal not passing through it.*

Sol. Let the edges OA, OB, OC of the unit cube be along OX, OY and OZ, respectively. Since, $OA=OB=OC=1$ unit

$$\therefore \quad \mathbf{OA}=\hat{\mathbf{i}}, \mathbf{OB}=\hat{\mathbf{j}}, \mathbf{OC}=\hat{\mathbf{k}}$$

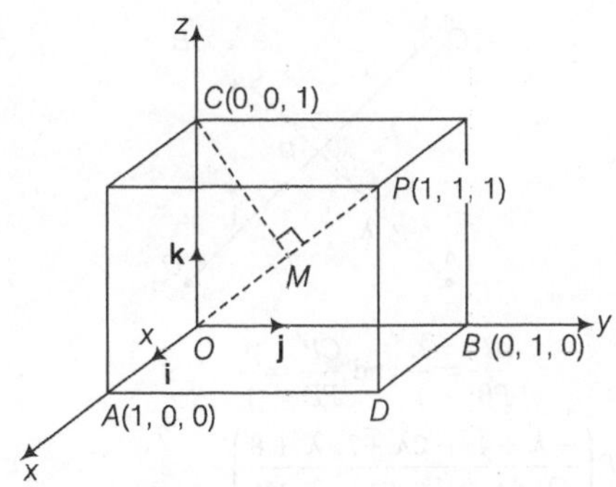

Let CM be perpendicular from the corner C on the diagonal OP. The vector equation of OP is

$$\mathbf{r}=\lambda(\hat{\mathbf{i}}+\hat{\mathbf{j}}+\hat{\mathbf{k}})$$

$\therefore \quad OM = $ Projection of **OC** on **OP**

$$= \mathbf{OC} \cdot \mathbf{OP}$$

$$= \hat{\mathbf{k}} \cdot \frac{(\hat{\mathbf{i}} + \hat{\mathbf{j}} + \hat{\mathbf{k}})}{\sqrt{3}} = \frac{1}{\sqrt{3}}$$

Now, $\quad OC^2 = OM^2 + CM^2$

$$\Rightarrow \quad CM^2 = |OC|^2 - OM^2 = 1 - \frac{1}{3} = \frac{2}{3}$$

$$\Rightarrow \quad CM = \sqrt{\frac{2}{3}}$$

Ex. 84 *If a variable plane forms a tetrahedron of constant volume $64k^3$ with the coordinate planes, then find the locus of the centroid of the tetrahedron.*

Sol. Let the variable plane intersects the coordinate axes at $A(a, 0, 0)$, $B(0, b, 0)$ and $C(0, 0, c)$. Then, the equation of the plane will be

$$\frac{x}{a} + \frac{y}{b} + \frac{z}{c} = 1 \quad \text{...(i)}$$

Let $P(\alpha, \beta, \gamma)$ be the centroid of tetrahedron $OABC$, then

$$\alpha = \frac{a}{4}, \beta = \frac{b}{4}, \gamma = \frac{c}{4}$$

or $\quad a = 4\alpha, b = 4\beta, c = 4\gamma$

$\Rightarrow$ Volume of tetrahedron $= \frac{1}{3}$ (Area of ΔAOB) $\cdot OC$

$$\Rightarrow \quad 64k^3 = \frac{1}{3}\left(\frac{1}{2}ab\right)c = \frac{abc}{6}$$

$$\Rightarrow \quad 64k^3 = \frac{(4\alpha)(4\beta)(4\gamma)}{6}$$

$$\Rightarrow \quad k^3 = \frac{\alpha\beta\gamma}{6}$$

$\therefore$ Required locus of $P(\alpha, \beta, \gamma)$ is $xyz = 6k^3$.

Ex. 85 *Show that the line segments joining the points $(4, 7, 8), (-1, -2, 1)$ and $(2, 3, 4)$ $(1, 2, 5)$ intersect. Verify whether the four points concyclic.*

Sol. Here, $A(4, 7, 8)$, $B(-1, -2, 1)$, $C(2, 3, 4)$ and $D(1, 2, 5)$. If the lines AB and CD intersect at P, then let

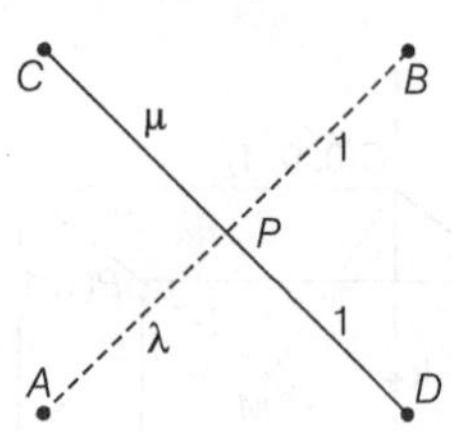

$$\frac{AP}{PB} = \frac{\lambda}{1} \text{ and } \frac{CP}{PD} = \frac{\mu}{1}$$

$$\text{Then, } P\left(\frac{-\lambda + 4}{\lambda + 1}, \frac{-2\lambda + 7}{\lambda + 1}, \frac{\lambda + 8}{\lambda + 1}\right)$$

$$= \left(\frac{\mu + 2}{\mu + 1}, \frac{2\mu + 3}{\mu + 1}, \frac{5\mu + 4}{\mu + 1}\right)$$

$$\Rightarrow \quad \frac{-\lambda + 4}{\lambda + 1} = \frac{\mu + 2}{\mu + 1}, \frac{-2\lambda + 7}{\lambda + 1} = \frac{2\mu + 3}{\mu + 1}$$

$$\frac{\lambda + 8}{\lambda + 1} = \frac{5\mu + 4}{\mu + 1}$$

$$\Rightarrow \quad \frac{-(\lambda + 1) + 5}{\lambda + 1} = \frac{(\mu + 1) + 1}{\mu + 1}$$

$$\frac{-2(\lambda + 1) + 9}{\lambda + 1} = \frac{2(\mu + 1) + 1}{\mu + 1}$$

$$\frac{(\lambda + 1) + 7}{\lambda + 1} = \frac{5(\mu + 1) - 1}{\mu + 1}$$

$$\Rightarrow \quad -1 + \frac{5}{\lambda + 1} = 1 + \frac{1}{\mu + 1}$$

$$-2 + \frac{9}{\lambda + 1} = 2 + \frac{1}{\mu + 1};$$

$$1 + \frac{7}{\lambda + 1} = 5 - \frac{1}{(\mu + 1)}$$

Let $\frac{1}{\lambda + 1} = x$ and $\frac{1}{\mu + 1} = y$

$$\Rightarrow \quad 5x - y = 2; 9x - y = 4; 7x + y = 4$$

On solving, $x = \frac{1}{2}, y = \frac{1}{2}$

$$\Rightarrow \quad \lambda + 1 = 2, \mu + 1 = 2$$

$$\lambda = 1, \mu = 1$$

Clearly, if $\lambda = 1$ and $\mu = 1$,
AB and CD bisects each other.

$$\therefore \quad P = \left(\frac{3}{2}, \frac{5}{2}, \frac{9}{2}\right)$$

Now, $\quad AP = \sqrt{\left(4 - \frac{3}{2}\right)^2 + \left(7 - \frac{5}{2}\right)^2 + \left(8 - \frac{9}{2}\right)^2}$

$$= \frac{\sqrt{155}}{2} = PB$$

Also, $\quad CP = \sqrt{\left(2 - \frac{3}{2}\right)^2 + \left(3 - \frac{5}{2}\right)^2 + \left(4 - \frac{9}{2}\right)^2}$

$$= \frac{\sqrt{3}}{2} = PD$$

We know four points A, B, C and D are concylic, if

$$AP \cdot PB = PC \cdot PD$$

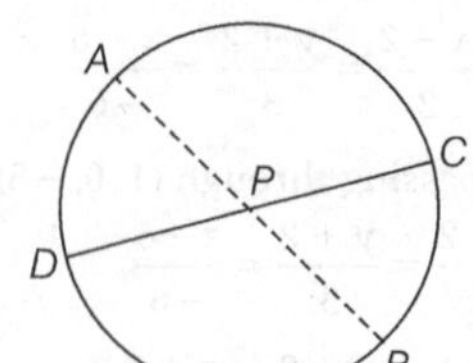

But here,

$$AP \cdot PB = \frac{155}{4} \text{ and } PC \cdot PD = \frac{3}{4}$$

$\therefore$ Points are non-concyclic.

Ex. 86 *If P be any point on the plane $lx + my + nz = p$ and Q be a point on the line OP such that $OP \cdot OQ = p^2$, show that the locus of the point Q is $p(lx + my + nz) = x^2 + y^2 + z^2$.*

Sol. Let $P(\alpha, \beta, \gamma)$ and $Q(x_1, y_1, z_1)$

DR's of OP are (α, β, γ) and DR's of OQ are (x_1, y_1, z_1).

$\because$ O, Q and P are collinear.

$$\therefore \quad \frac{\alpha}{x_1} = \frac{\beta}{y_1} = \frac{\gamma}{z_1} = k \quad \text{(say)} \quad \text{...(i)}$$

Since, $P(\alpha, \beta, \gamma)$ lie on the plane

$$lx + my + nz = p,$$
$$l\alpha + m\beta + n\gamma = p$$

Since, $P(\alpha, \beta, \gamma)$ lie on the plane $lx + my + nz = p$,

$l\alpha + m\beta + n\gamma \pm p$

$$\Rightarrow \quad klx_1 + kmy_1 + knz_1 = p \quad \text{[using Eq. (i)]} \quad \text{...(ii)}$$

Since, $OP \cdot OQ = p^2$

$$\therefore \quad \sqrt{\alpha^2 + \beta^2 + \gamma^2} \cdot \sqrt{x_1^2 + y_1^2 + z_1^2} = p^2$$

$$\sqrt{k^2x_1^2 + k^2y_1^2 + k^2z_1^2} \cdot \sqrt{x_1^2 + y_1^2 + z_1^2} = p^2$$

$$\Rightarrow \quad k(x_1^2 + y_1^2 + z_1^2) = p^2 \quad \text{...(iii)}$$

From Eqs. (ii) and (iii), we get

$$\frac{lx_1 + my_1 + nz_1}{x_1^2 + y_1^2 + z_1^2} = \frac{1}{p}$$

$$\Rightarrow \quad p(lx_1 + my_1 + nz_1) = (x_1^2 + y_1^2 + z_1^2)$$

Hence, locus of Q

$$\Rightarrow \quad p(lx + my + nz) = x^2 + y^2 + z^2$$

Ex. 87 *Find the reflection of the plane $ax + by + cz + d = 0$ in the plane $a'x + b'y + c'z + d' = 0$*

Sol. Given planes are

$$ax + by + cz + d = 0 \quad \text{...(i)}$$

$$\text{and} \quad a'x + b'y + c'z + d' = 0 \quad \text{...(ii)}$$

Let $P(\alpha, \beta, \gamma)$ be an arbitrary point in the plane (i) and $Q(p, q, r)$ be the reflection of the point P in plane (ii). Locus of Q will be the required reflection of plane (i) in plane (ii), let L be the mid-point of PQ.

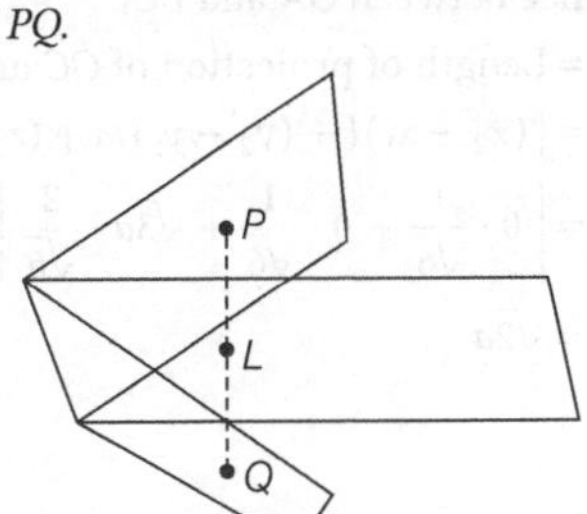

$$\text{Then,} \quad L\left(\frac{p+\alpha}{2}, \frac{q+\beta}{2}, \frac{r+\gamma}{2}\right)$$

L lies in plane (ii), we get

$$\therefore \quad a'\left(\frac{p+\alpha}{2}\right) + b'\left(\frac{q+\beta}{2}\right) + c'\left(\frac{r+\gamma}{2}\right) + d' = 0$$

$$\Rightarrow \quad a'(p + \alpha) + b'(q + \beta) + c'(r + \gamma) + 2d' = 0 \quad \text{...(iii)}$$

DR's of PQ are $\alpha - p, \beta - q, \gamma - r$.

Since, PQ perpendicular on plane Eq. (iii), we get

$$\frac{\alpha - p}{a'} = \frac{\beta - q}{b'} = \frac{\gamma - r}{c'} = k \quad \text{(say)}$$

$$\therefore \quad \alpha = p + a'k, \beta = q + b'k, \gamma = r + c'k$$

Putting the values of α, β and γ in Eq. (iii), we get

$$a'(2p + ka') + b'(2q + kb') + c(2r + kc') + 2d' = 0$$

$$\Rightarrow \quad 2(a'p + b'q + c'r + d') = -k(a'^2 + b'^2 + c'^2) \quad \text{...(iv)}$$

Since, $P(\alpha, \beta, \gamma)$ lies on plane (i), we get

$$a\alpha + b\beta + c\gamma + d = 0$$

$$\Rightarrow \quad a(p + ka') + n(q + kb') + c(r + kc') + d = 0$$

$$\therefore \quad k = -\frac{(ap + bq + cr + d)}{(aa' + bb' + cc')}$$

Putting the value of k in Eq. (iv), we get

$$2(a'p + b'q + c'r + d') = \frac{(a'^2 + b'^2 + c'^2)(ap + bq + cr + d)}{aa' + bb' + cc'}$$

$\therefore$ Locus of $Q(p, q, r)$.

i.e. equation reflection of plane (i) in plane (ii) is,

$$2(aa' + bb' + cc')(a'x + b'y + c'z + d') = (a'^2 + b'^2 + c'^2)(ax + by + cz + d)$$

Ex. 88 *A point P moves on a plane $\frac{x}{a} + \frac{y}{b} + \frac{z}{c} = 1$. A plane through P and perpendicular to OP meets the coordinate axes in A, B and C. If the planes through A, B and C parallel to the planes $x = 0$, $y = 0$ and $z = 0$ intersect in Q, then find the locus of Q.*

Sol. Given plane is $\frac{x}{a} + \frac{y}{b} + \frac{z}{c} = 1$ ...(i)

Let $P(h, k, l)$ be the point on plane.

$$\therefore \quad \frac{h}{a} + \frac{k}{b} + \frac{1}{c} = 1 \quad \text{...(ii)}$$

$$\Rightarrow \quad OP = \sqrt{h^2 + k^2 + l^2}$$

$$\text{DC's of } OP = \left(\frac{h}{\sqrt{h^2 + k^2 + l^2}}, \frac{k}{\sqrt{h^2 + k^2 + l^2}}, \frac{l}{\sqrt{h^2 + k^2 + l^2}}\right)$$

$\therefore$ Equation of the plane through P and normal to OP is,

$$\frac{hx}{\sqrt{h^2 + k^2 + l^2}} + \frac{ky}{\sqrt{h^2 + k^2 + l^2}} + \frac{lz}{\sqrt{h^2 + k^2 + l^2}} = \sqrt{h^2 + k^2 + l^2}$$

$$\Rightarrow \quad hx + ky + lz = h^2 + k^2 + l^2$$

$$\therefore \quad A \equiv \left(\frac{h^2 + k^2 + l^2}{h}, 0, 0\right)$$

$$B \equiv \left(0, \frac{h^2 + k^2 + l^2}{k}, 0\right)$$

$$C \equiv \left(0, 0, \frac{h^2+k^2+l^2}{l}\right)$$

Let $Q(\alpha, \beta, \gamma)$, then

$$\alpha = \frac{h^2+k^2+l^2}{h}, \quad \beta = \frac{h^2+k^2+l^2}{k}, \quad \gamma = \frac{h^2+k^2+l^2}{l} \quad \text{...(iii)}$$

Now, $$\frac{1}{\alpha^2}+\frac{1}{\beta^2}+\frac{1}{\gamma^2} = \frac{h^2+k^2+l^2}{(h^2+k^2+l^2)^2} = \frac{1}{h^2+k^2+l^2} \quad \text{...(iv)}$$

From Eq. (iii), we get

$$h = \frac{h^2+k^2+l^2}{\alpha}$$

$$\Rightarrow \quad \frac{h}{a} = \frac{h^2+k^2+l^2}{a\alpha}$$

Similarly, $$\frac{k}{b} = \frac{h^2+k^2+l^2}{b\beta}$$

and $$\frac{l}{c} = \frac{h^2+k^2+l^2}{c\gamma}$$

$$\frac{h^2+k^2+l^2}{a\alpha}+\frac{h^2+k^2+l^2}{b\beta}+\frac{h^2+k^2+l^2}{c\gamma} = \frac{h}{a}+\frac{k}{b}+\frac{1}{c}=1 \quad \text{[from Eq. (ii)]}$$

$$\Rightarrow \quad \frac{1}{a\alpha}+\frac{1}{b\beta}+\frac{1}{c\gamma} = \frac{1}{h^2+k^2+l^2} = \frac{1}{\alpha^2}+\frac{1}{\beta^2}+\frac{1}{\gamma^2} \quad \text{[from Eq. (iv)]}$$

∴Required equation of locus is

$$\frac{1}{ax}+\frac{1}{by}+\frac{1}{cz} = \frac{1}{x^2}+\frac{1}{y^2}+\frac{1}{z^2}.$$

Ex. 89 *Prove that the shortest distance between any two opposite edges of a tetrahedron formed by the planes $y+z=0, x+z=0, x+y=0, x+y+z=\sqrt{3}a$ is $\sqrt{2}a$.*

Sol. Here, planes

$$y+z=0, z+x=0, x+y=0 \text{ meet at } O(0, 0, 0).$$

Let the tetrahedron be $OABC$.

Let the equation of one of the pair of opposite edges OA and BC be

$$y+z=0, x+z=0 \quad \text{...(i)}$$

and $$x+y=0, x+y+z=\sqrt{3}a \quad \text{...(ii)}$$

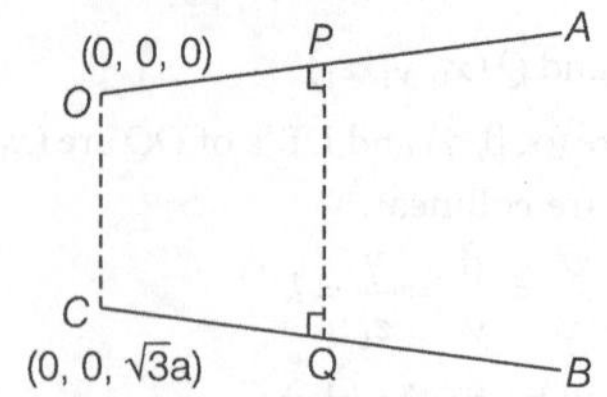

Eqs. (i) and (ii) can be expressed in symmetrical form as

$$\frac{x-0}{1} = \frac{y-0}{1} = \frac{z-0}{-1} \quad \text{...(iii)}$$

and $$\frac{x-0}{1} = \frac{y-0}{-1} = \frac{z-\sqrt{3}a}{0} \quad \text{...(iv)}$$

DR's of OA and BC are $(1, 1, -1)$ and $(-1, 1, 0)$.

Let PQ be the shortest distance between OA and BC having direction cosine (l, m, n).

∴ PQ is perpendicular to both OA and BC.

$$l+m+n=0 \quad \text{...(v)}$$

$$l-m=0 \quad \text{...(vi)}$$

On solving Eqs. (v) and (vi), we get

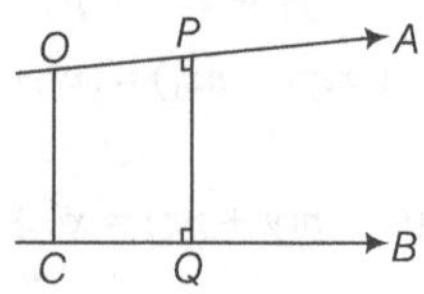

$$\frac{l}{1} = \frac{m}{1} = \frac{n}{2} = k$$

Also, $$l^2+m^2+n^2=1$$

$$\therefore \quad k^2+k^2+4k^2=1 \Rightarrow k = \frac{1}{\sqrt{6}}$$

$$\therefore \quad l = \frac{1}{\sqrt{6}} = m \text{ and } n = \frac{2}{\sqrt{6}}$$

Shortest distance between OA and BC,

i.e. $$PQ = \text{Length of projection of } OC \text{ and } PQ$$

$$= |(x_2-x_1)l+(y_2-y_1)m+(z_2-z_1)n|$$

$$= \left|0\cdot\frac{1}{\sqrt{6}}+0\cdot\frac{1}{\sqrt{6}}+\sqrt{3}a\cdot\frac{2}{\sqrt{6}}\right|$$

$$= \sqrt{2}a$$

Three Dimensional Coordinate System Exercise 1 :
Single Option Correct Type Questions

1. The xy-plane divides the line joining the points $(-1, 3, 4)$ $(2, -5, 6)$.
(a) Internally in the ratio 2 : 3
(b) externally in the ratio 2 : 3
(c) internally in the ratio 3 : 2
(d) externally in the ratio 3 : 2

2. Ratio in which the zx-plane divides the join of (1, 2, 3) and (4, 2, 1).
(a) 1 : 1 internally (b) 1 : 1 externally
(c) 2 : 1 internally (d) 2 : 1 externally

3. If $P(3, 2, -4)$, $Q(5, 4, -6)$ and $R(9, 8, -10)$ are collinear, then R divides PQ in the ratio
(a) 3 : 2 internally (b) 3 : 2 externally
(c) 2 : 1 internally (d) 2 : 1 externally

4. $A(3, 2, 0)$, $B(5, 3, 2)$ and $C(-9, 6, -3)$ are the vertices of a triangle ABC. If the bisector of $\angle ABC$ meets BC at D, then coordinates of D are
(a) $\left(\frac{19}{8}, \frac{57}{16}, \frac{17}{16}\right)$ (b) $\left(-\frac{19}{8}, \frac{57}{16}, \frac{17}{16}\right)$
(c) $\left(\frac{19}{8}, -\frac{57}{16}, \frac{17}{16}\right)$ (d) None of these

5. A line passes through the points $(6, -7, -1)$ and $(2, -3, 1)$. The direction cosines of the line so directed that the angle made by it with the positive direction of x-axis is acute, are
(a) $\frac{2}{3}, -\frac{2}{3}, -\frac{1}{3}$ (b) $-\frac{2}{3}, \frac{2}{3}, \frac{1}{3}$
(c) $\frac{2}{3}, -\frac{2}{3}, \frac{1}{3}$ (d) $\frac{2}{3}, \frac{2}{3}, \frac{1}{3}$

6. If P is a point in space such that OP is inclined to OX at 45° and OY to 60° then **OP** is inclined to OZ at
(a) 75°
(b) 60° and 120°
(c) 75° and 105°
(d) 255°

7. l_1, m_1, n_1 and l_2, m_2, n_2 are direction cosines of the two lines inclined to each other at an angle θ, then the direction cosines of the internal bisector of the angle between these lines are
(a) $\frac{l_1 + l_2}{2\sin\frac{\theta}{2}}, \frac{m_1 + m_2}{2\sin\frac{\theta}{2}}, \frac{n_1 + n_2}{2\sin\frac{\theta}{2}}$ (b) $\frac{l_1 + l_2}{2\cos\frac{\theta}{2}}, \frac{m_1 + m_2}{2\cos\frac{\theta}{2}}, \frac{n_1 + n_2}{2\cos\frac{\theta}{2}}$
(c) $\frac{l_1 - l_2}{2\sin\frac{\theta}{2}}, \frac{m_1 - m_2}{2\sin\frac{\theta}{2}}, \frac{n_1 - n_2}{2\sin\frac{\theta}{2}}$ (d) $\frac{l_1 - l_2}{2\cos\frac{\theta}{2}}, \frac{m_1 - m_2}{2\cos\frac{\theta}{2}}, \frac{n_1 - n_2}{2\cos\frac{\theta}{2}}$

8. The equation of the plane perpendicular to the line $\frac{x-1}{1}, \frac{y-2}{-1}, \frac{z+1}{2}$ and passing through the point (2, 3, 1), is
(a) $\mathbf{r}.(\hat{\mathbf{i}} + \hat{\mathbf{j}} + 2\hat{\mathbf{k}}) = 1$ (b) $\mathbf{r}.(\hat{\mathbf{i}} - \hat{\mathbf{j}} + 2\hat{\mathbf{k}}) = 1$
(c) $\mathbf{r}.(\hat{\mathbf{i}} - \hat{\mathbf{j}} + 2\hat{\mathbf{k}}) = 7$ (d) None of these

9. The locus of a point which moves so that the difference of the squares of its distances from two given points is constant, is a
(a) straight line (b) plane
(c) sphere (d) None of these

10. The position vectors of points **a** and **b** are $\hat{\mathbf{i}} - \hat{\mathbf{j}} + 3\hat{\mathbf{k}}$ and $3\hat{\mathbf{i}} + 3\hat{\mathbf{j}} + 3\hat{\mathbf{k}}$ respectively. The equation of a plane is $\mathbf{r}.(5\hat{\mathbf{i}} + 2\hat{\mathbf{j}} - 7\hat{\mathbf{k}}) + 9 = 0$. The points **a** and **b**
(a) lie on the plane
(b) are on the same side of the plane
(c) are on the opposite side of the plane
(d) None of the above

11. The vector equation of the plane through the point $2\hat{\mathbf{i}} - \hat{\mathbf{j}} - 4\hat{\mathbf{k}}$ and parallel to the plane $\mathbf{r}.(4\hat{\mathbf{i}} - 12\hat{\mathbf{j}} - 3\hat{\mathbf{k}}) - 7 = 0$, is
(a) $\mathbf{r}.(4\hat{\mathbf{i}} - 12\hat{\mathbf{j}} - 3\hat{\mathbf{k}}) = 0$ (b) $\mathbf{r}.(4\hat{\mathbf{i}} - 12\hat{\mathbf{j}} - 3\hat{\mathbf{k}}) = 32$
(c) $\mathbf{r}.(4\hat{\mathbf{i}} - 12\hat{\mathbf{j}} - 3\hat{\mathbf{k}}) = 12$ (d) None of these

12. Let L_1 be the line $\mathbf{r}_1 = 2\hat{\mathbf{i}} + \hat{\mathbf{j}} - \hat{\mathbf{k}} + \lambda(\hat{\mathbf{i}} + 2\hat{\mathbf{k}})$ and let L_2 be the another line $\mathbf{r}_2 = 3\hat{\mathbf{i}} + \hat{\mathbf{j}} + \mu(\hat{\mathbf{i}} + \hat{\mathbf{j}} - \hat{\mathbf{k}})$. Let π be the plane which contains the line L_1 and is parallel to L_2. The distance of the plane π from the origin is
(a) $\sqrt{\frac{2}{7}}$ (b) $\frac{1}{7}$
(c) $\sqrt{6}$ (d) None of these

13. For the line $\frac{x-1}{1} = \frac{y-2}{2} = \frac{z-3}{3}$, which one of the following is incorrect ?
(a) it lies in the plane $x - 2y + z = 0$
(b) it is same as line $\frac{x}{1} = \frac{y}{2} = \frac{z}{3}$
(c) it passes through (2, 3, 5)
(d) it is parallel to the plane $x - 2y + z - 6 = 0$

14. The value of m for which straight line $3x - 2y + z + 3 = 0 = 4x - 3y + 4z + 1$ is parallel to the plane $2x - y + mz - 2 = 0$ is
(a) -2 (b) 8
(c) -18 (d) 11

15. The length of projection of the line segment joining the points (1, 0, −1) and (−1, 2, 2) on the plane $x + 3y - 5z = 6$, is equal to

(a) 2 (b) $\sqrt{\frac{271}{53}}$

(c) $\sqrt{\frac{472}{31}}$ (d) $\sqrt{\frac{474}{35}}$

16. The number of planes that are equidistant from four non-coplanar points is

(a) 3 (b) 4

(c) 7 (d) 9

17. In a three dimensional co-ordinate system, P, Q and R are images of a point $A(a, b, c)$ in the xy, yz and zx planes, respectively. If G is the centroid of triangle PQR, then area of triangle AOG is (O is the origin)

(a) 0 (b) $a^2 + b^2 + c^2$

(c) $\frac{2}{3}(a^2 + b^2 + c^2)$ (d) None of these

18. A plane passing through (1, 1, 1) cuts positive direction of coordinate axes at A, B and C, then the volume of tetrahedron $OABC$ satisfies

(a) $V \le \frac{9}{2}$ (b) $V \ge \frac{9}{2}$

(c) $V = \frac{9}{2}$ (d) None of these

19. If lines $x = y = z$ and $x = \frac{y}{2} = \frac{z}{3}$ and third line passing through (1, 1, 1) form a triangle of area $\sqrt{6}$ units, then point of intersection of third line with second line will be

(a) (1, 2, 3) (b) (2, 4, 6)

(c) $\left(\frac{4}{3}, \frac{8}{3}, \frac{12}{3}\right)$ (d) None of these

20. The point of intersection of the line passing through (0, 0, 1) and intersecting the lines $x + 2y + z = 1$, $-x + y - 2z = 2$ and $x + y = 2$, $x + z = 2$ with xy plane is

(a) $\left(\frac{5}{3}, -\frac{1}{3}, 0\right)$ (b) (1, 1, 0)

(c) $\left(\frac{2}{3}, -\frac{1}{3}, 0\right)$ (d) $\left(-\frac{5}{3}, \frac{1}{3}, 0\right)$

21. Two systems of rectangular axes have the same origin. If a plane cuts them at distance a, b, c and a', b', c' from the origin, then :

(a) $\frac{1}{a^2} + \frac{1}{b^2} + \frac{1}{c^2} + \frac{1}{a'^2} + \frac{1}{b'^2} + \frac{1}{c'^2} = 0$

(b) $\frac{1}{a^2} - \frac{1}{b^2} - \frac{1}{c^2} + \frac{1}{a'^2} - \frac{1}{b'^2} - \frac{1}{c'^2} = 0$

(c) $\frac{1}{a^2} + \frac{1}{b^2} + \frac{1}{c^2} - \frac{1}{a'^2} - \frac{1}{b'^2} - \frac{1}{c'^2} = 0$

(d) $\frac{1}{a^2} - \frac{1}{b^2} + \frac{1}{c^2} - \frac{1}{a'^2} + \frac{1}{b'^2} - \frac{1}{c'^2} = 0$

22. The line $\frac{x+6}{5} = \frac{y+10}{3} = \frac{z+14}{8}$ is the hypotenuse of an isosceles right angled triangle whose opposite vertex is (7, 2, 4). Then which of the following is not the side of the triangle ?

(a) $\frac{x-7}{2} = \frac{y-2}{-3} = \frac{z-4}{6}$

(b) $\frac{x-7}{3} = \frac{y-2}{6} = \frac{z-4}{2}$

(c) $\frac{x-7}{3} = \frac{y-2}{5} = \frac{z-4}{-1}$

(d) None of these

23. Consider the following 3 lines in space

$L_1 : \mathbf{r} = 3\mathbf{i} - \mathbf{j} + 2\mathbf{k} + \lambda(2\mathbf{i} + 4\mathbf{j} - \mathbf{k})$

$L_2 : \mathbf{r} = \mathbf{i} + \mathbf{j} - 3\mathbf{k} + \mu(4\mathbf{i} + 2\mathbf{j} + 4\mathbf{k})$

$L_3 : \mathbf{r} = 3\mathbf{i} + 2\mathbf{j} - 2\mathbf{k} + t(2\mathbf{i} + \mathbf{j} + 2\mathbf{k})$

Then, which one of the following pair(s) is/are in the same plane ?

(a) Only L_1L_2 (b) Only L_2L_3

(c) Only L_3L_1 (d) L_1L_2 and L_2L_3

24. Let $\mathbf{r} = \mathbf{a} + \lambda\mathbf{l}$ and $\mathbf{r} = \mathbf{b} + \mu\mathbf{m}$ be two lines in space, where $\mathbf{a} = 5\mathbf{i} + \mathbf{j} + 2\mathbf{k}$, $\mathbf{b} = -\mathbf{i} + 7\mathbf{j} + 8\mathbf{k}$, $\mathbf{l} = -4\mathbf{i} + \mathbf{j} - \mathbf{k}$, and $\mathbf{m} = 2\mathbf{i} - 5\mathbf{j} - 7\mathbf{k}$, then the position vector of a point which lies on both of these lines, is

(a) $\mathbf{i} + 2\mathbf{j} + \mathbf{k}$

(b) $2\mathbf{i} + \mathbf{j} + \mathbf{k}$

(c) $\mathbf{i} + \mathbf{j} + 2\mathbf{k}$

(d) non-existent as the lines are skew

25. L_1 and L_2 are two lines whose vector equations are

$L_1 : \mathbf{r} = \lambda[(\cos\theta + \sqrt{3})\mathbf{i} + (\sqrt{2}\sin\theta)\mathbf{j} + (\cos\theta - \sqrt{3})\mathbf{k}]$

and $L_2 : \mathbf{r} = \mu(a\mathbf{i} + b\mathbf{j} + c\mathbf{k})$

where, λ and μ are scalars and α is the acute angle between L_1 and L_2. If the angle α is independent of θ, then the value of α is

(a) $\frac{\pi}{6}$ (b) $\frac{\pi}{4}$

(c) $\frac{\pi}{3}$ (d) $\frac{\pi}{2}$

26. The vector equations of two lines L_1 and L_2 are respectively,

$$\mathbf{r} = 17\mathbf{i} - 9\mathbf{j} + 9\mathbf{k} + \lambda(3\mathbf{i} + \mathbf{j} + 5\mathbf{k})$$

and $\mathbf{r} = 15\mathbf{i} - 8\mathbf{j} - \mathbf{k} + \mu(4\mathbf{i} + 3\mathbf{j})$

I. L_1 and L_2 are skew lines.

II. (11, −11, −1) is the point of intersection of L_1 and L_2.

III. (−11, 11, 1) is the point of intersection of L_1 and L_2.

IV. $\cos^{-1}\left(\frac{3}{\sqrt{35}}\right)$ is the acute angle between, L_1 and L_2.

Then, which of the following is true ?

(a) II and IV (b) I and IV

(c) Only IV (d) III and IV

27. Consider three vectors $\mathbf{p} = \mathbf{i} + \mathbf{j} + \mathbf{k}, \mathbf{q} = 2\mathbf{i} + 4\mathbf{j} - \mathbf{k}$ and $\mathbf{r} = \mathbf{i} + \mathbf{j} + 3\mathbf{k}$. If $\mathbf{p}$, $\mathbf{q}$ and $\mathbf{r}$ denotes the position vector of three non-collinear points, then the equation of the plane containing these points is

(a) $2x - 3y + 1 = 0$ (b) $x - 3y + 2z = 0$
(c) $3x - y + z - 3 = 0$ (d) $3x - y - 2 = 0$

28. The intercept made by the plane $\mathbf{r}.\mathbf{n} = q$ on the x-axis is

(a) $\frac{q}{\mathbf{i}.\mathbf{n}}$ (b) $\frac{\mathbf{i}.\mathbf{n}}{q}$
(c) $(\mathbf{i}.\mathbf{n})\, q$ (d) $\frac{q}{|\mathbf{n}|}$

29. If the distance between the planes

$$8x + 12y - 14z = 2 \text{ and } 4x + 6y - 7z = 2$$

can be expressed in the form $\frac{1}{\sqrt{N}}$, where N is natural, then the value of $\frac{N(N+1)}{2}$ is

(a) 4950 (b) 5050
(c) 5150 (d) 5151

30. A plane passes through the points $P(4, 0, 0)$ and $Q(0, 0, 4)$ and is parallel to the Y-axis. The distance of the plane from the origin is

(a) 2 (b) 4
(c) $\sqrt{2}$ (d) $2\sqrt{2}$

31. If from the point $P(f, g, h)$ perpendiculars PL and PM be drawn to yz and zx-planes, then the equation to the plane OLM is

(a) $\frac{x}{f} + \frac{y}{g} - \frac{z}{h} = 0$ (b) $\frac{x}{f} + \frac{y}{g} + \frac{z}{h} = 0$
(c) $\frac{x}{f} - \frac{y}{g} + \frac{z}{h} = 0$ (d) $-\frac{x}{f} + \frac{y}{g} + \frac{z}{h} = 0$

32. The plane XOZ divides the join of $(1, -1, 5)$ and $(2, 3, 4)$ in the ratio $\lambda : 1$, then λ is

(a) -3 (b) $-\frac{1}{3}$
(c) 3 (d) $\frac{1}{3}$

33. A variable plane forms a tetrahedron of constant volume $64K^3$ with the coordinate planes and the origin, then locus of the centroid of the tetrahedron is

(a) $x^3 + y^3 + z^3 = 6k^3$ (b) $xyz = 6k^3$
(c) $x^2 + y^2 + z^2 = 4k^2$ (d) $x^{-2} + y^{-2} + z^{-2} = 4k^{-2}$

34. Let $ABCD$ be a tetrahedron such that the edges AB, AC and AD are mutually perpendicular. Let the area of ΔABC, ΔACD and ΔADB be 3, 4 and 5 sq units, respectively. Then, the area of the ΔBCD, is

(a) $5\sqrt{2}$ (b) 5
(c) $5/\sqrt{2}$ (d) $\frac{5}{2}$

35. Equation of the line which passes through the point with position vector (2, 1, 0) and perpendicular to the plane containing the vectors $\mathbf{i} + \mathbf{j}$ and $\mathbf{j} + \mathbf{k}$ is

(a) $\mathbf{r} = (2, 1, 0) + t(1, -1, 1)$
(b) $\mathbf{r} = (2, 1, 0) + t(-1, 1, 1)$
(c) $\mathbf{r} = (2, 1, 0) + t(1, 1, -1)$
(d) $\mathbf{r} = (2, 1, 0) + t(1, 1, 1)$

Where, t is a parameter.

36. Which of the following planes are parallel but not identical ?

$$P_1 : 4x - 2y + 6z = 3$$
$$P_2 : 4x - 2y - 2z = 6$$
$$P_3 : -6x + 3y - 9z = 5$$
$$P_4 : 2x - y - z = 3$$

(a) P_2 and P_3 (b) P_2 and P_4
(c) P_1 and P_3 (d) P_1 and P_4

37. A parallelopiped is formed by planes drawn through the points (1, 2, 3) and (9, 8, 5) parallel to the coordinate planes, then which of the following is not the length of an edge of this rectangular parallelopiped ?

(a) 2 (b) 4
(c) 6 (d) 8

38. vector equation of the plane $\mathbf{r} = \mathbf{i} - \mathbf{j} + \lambda(\mathbf{i} + \mathbf{j} + \mathbf{k}) + \mu(\mathbf{i} - 2\mathbf{j} + 3\mathbf{k})$ in the scalar dot product form is

(a) $\mathbf{r}.(5\mathbf{i} - 2\mathbf{j} + 3\mathbf{k}) = 7$
(b) $\mathbf{r}.(5\mathbf{i} + 2\mathbf{j} - 3\mathbf{k}) = 7$
(c) $\mathbf{r}.(5\mathbf{i} - 2\mathbf{j} - 3\mathbf{k}) = 7$
(d) $\mathbf{r}.(5\mathbf{i} + 2\mathbf{j} + 3\mathbf{k}) = 7$

39. The vector equations of the two lines L_1 and L_2 are given by $L_1 : \mathbf{r} = (2\mathbf{i} + 9\mathbf{j} + 13\mathbf{k}) + \lambda(\mathbf{i} + 2\mathbf{j} + 3\mathbf{k})$ and $L_2 : \mathbf{r} = (-3\mathbf{i} + 7\mathbf{j} + p\mathbf{k}) + \mu(-\mathbf{i} + 2\mathbf{j} - 3\mathbf{k})$. Then, the lines L_1 and L_2 are

(a) skew lines for all $p \in R$
(b) intersecting for all $p \in R$ and the point of intersection is $(-1, 3, 4)$
(c) intersecting lines for $p = -2$
(d) intersecting for all real $p \in R$

40. Consider the plane $(x, y, z) = (0, 1, 1) + \lambda(1, -1, 1) + \mu(2, -1. 0)$. The distance of this plane from the origin is

(a) $\frac{1}{3}$ (b) $\frac{\sqrt{3}}{2}$
(c) $\sqrt{\frac{3}{2}}$ (d) $\frac{2}{\sqrt{3}}$

41. The value of a for which the lines $\frac{x-2}{1} = \frac{y-9}{2} = \frac{z-13}{3}$ and $\frac{x-a}{-1} = \frac{y-7}{2} = \frac{z+2}{-3}$ intersect, is

(a) -5 (b) -2
(c) 5 (d) -3

42. For the line $\frac{x-1}{1}=\frac{y-2}{2}=\frac{z-3}{3}$, which one of the following is incorrect ?

(a) It lies in the plane $x-2y+z=0$.

(b) It is same as line $\frac{x}{1}=\frac{y}{2}=\frac{z}{3}$.

(c) It passes through (2, 3, 5).

(d) It is parallel to the plane $x-2y+z-6=0$.

43. Given planes $P_1 : cy+bz=x$;

$$P_2 : az+cx=y$$
$$P_3 : bx+ay=z$$

P_1, P_2 and P_3 pass through one line, if

(a) $a^2+b^2+c^2=ab+bc+ca$

(b) $a^2+b^2+c^2+2abc=1$

(c) $a^2+b^2+c^2=1$

(d) $a^2+b^2+c^2+2ab+2bc+2ca+2abc=1$

44. The lines $\frac{x-2}{1}=\frac{y-3}{1}=\frac{z-4}{-k}$ and $\frac{x-1}{k}=\frac{y-4}{2}=\frac{z-5}{1}$ are coplanar, if

(a) $k=0$ or -1 (b) $k=1$ or -1

(c) $k=0$ or -3 (d) $k=3$ or -3

45. The line $\frac{x-2}{3}=\frac{y+1}{2}=\frac{z-1}{-1}$ intersects the curve $xy=c^2$, in xy-plane, if c is equal to

(a) ± 1 (b) $\pm\frac{1}{3}$

(c) $\pm\sqrt{5}$ (d) None of these

46. The line which contains all points (x, y, z) which are of the form $(x, y, z)=(2,-2,5)+\lambda(1,-3,2)$ intersects the plane $2x-3y+4z=163$ at P and intersects the YZ-plane at Q. If the distance PQ is $a\sqrt{b}$, where $a, b \in N$ and $a>3$, then $(a+b)$ is equal to

(a) 23 (b) 95

(c) 27 (d) None of these

47. If the three planes $\mathbf{r}.\mathbf{n}_1=p_1$, $\mathbf{r}.\mathbf{n}_2=p_2$ and $\mathbf{r}.\mathbf{n}_3=p_3$ have a common line of intersection, then $p_1(\mathbf{n}_2\times\mathbf{n}_3)+p_2(\mathbf{n}_3\times\mathbf{n}_1)+p_3(\mathbf{n}_1\times\mathbf{n}_2)$ is equal to

(a) 1 (b) 2

(c) 0 (d) -1

48. The equation of the plane which passes through the line of intersection of the planes $\mathbf{r}.\mathbf{n}_1=q_1$, $\mathbf{r}.\mathbf{n}_2=q_2$ and is parallel to the line of intersection of the planes $\mathbf{r}.\mathbf{n}_3=q_3$ and $\mathbf{r}.\mathbf{n}_4=q_4$, is

(a) $[\mathbf{n}_2\ \mathbf{n}_3\ \mathbf{n}_4](\mathbf{r}.\mathbf{n}_1-q_1)=[\mathbf{n}_1\ \mathbf{n}_3\ \mathbf{n}_4](\mathbf{r}.\mathbf{n}_2-q_2)$

(b) $[\mathbf{n}_1\ \mathbf{n}_2\ \mathbf{n}_3](\mathbf{r}.\mathbf{n}_4-q_4)=[\mathbf{n}_4\ \mathbf{n}_3\ \mathbf{n}_1](\mathbf{r}.\mathbf{n}_2-q_2)$

(c) $[\mathbf{n}_4\ \mathbf{n}_3\ \mathbf{n}_1](\mathbf{r}.\mathbf{n}_4-q_4)=[\mathbf{n}_1\ \mathbf{n}_2\ \mathbf{n}_3](\mathbf{r}.\mathbf{n}_2-q_2)$

(d) None of the above

49. A straight line is given by $\mathbf{r}=(1+t)\,\mathbf{i}+3t\,\mathbf{j}+(1-t)\,\mathbf{k}$, where $t \in R$. If this line lies in the plane $x+y+cz=d$, then the value of $(c+d)$ is

(a) -1 (b) 1

(c) 7 (d) 9

50. The distance of the point $(-1,-5,-10)$ from the point of intersection of the line $\frac{x-2}{2}=\frac{y+1}{4}=\frac{z-2}{12}$ and the plane $x-y+z=5$ is

(a) $2\sqrt{11}$ (b) $\sqrt{126}$

(c) 13 (d) 14

51. $P(\mathbf{p})$ and $Q(\mathbf{q})$ are the position vector of two fixed points and $R(\mathbf{r})$ is the position vector of a variable point. If R moves such that $(\mathbf{r}-\mathbf{p})\times(\mathbf{r}-\mathbf{q})=0$, then the locus of R is

(a) a plane containing the origin O and parallel to two non-collinear vector **OP** and **OQ**.

(b) the surface of a sphere described on PQ as its diameter.

(c) a line passing through the points P and Q.

(d) a set of lines parallel to the line PQ.

52. The three vectors $\mathbf{i}+\mathbf{j}$, $\mathbf{j}+\mathbf{k}$, $\mathbf{k}+\mathbf{i}$ taken two at a time form three planes. The three unit vectors drawn perpendicular to these three planes form a parallelopiped of volume

(a) $\frac{1}{3}$ (b) 4

(c) $3\frac{\sqrt{3}}{4}$ (d) $\frac{4}{3\sqrt{3}}$

53. The orthogonal projection A' of the point A with position vector (1, 2, 3) on the plane $3x-y+4z=0$ is

(a) $(-1, 3, -1)$ (b) $\left(-\frac{1}{2}, \frac{5}{2}, 1\right)$

(c) $\left(\frac{1}{2}, -\frac{5}{2}, -1\right)$ (d) (6., -7, -5)

54. The equation of the line passing through (1, 1, 1) and perpendicular to the line of intersection of the planes $x+2y-4z=0$ and $2x-y+2z=0$ is

(a) $\frac{x-1}{5}=\frac{1-y}{1}=\frac{z-1}{2}$ (b) $\frac{x-1}{-5}=\frac{1-y}{1}=\frac{z-1}{2}$

(c) $\frac{x-1}{0}=\frac{1-y}{-10}=\frac{z-1}{-5}$ (d) $\frac{x-1}{-10}=\frac{y-1}{0}=\frac{z-1}{-5}$

55. A variable plane at a distance of 1 unit from the origin cuts the axes at A, B and C. If the centroid $D(x, y, z)$ of ΔABC satisfies the relation $\frac{1}{x^2}+\frac{1}{y^2}+\frac{1}{z^2}=K$, then the value of K is

(a) 3 (b) 1

(c) $\frac{1}{3}$ (d) 9

56. The angle between the lines AB and CD, where $A=(0,0,0)$, $B=(1,1,1)$, $C=(-1,-1,-1)$ and $D=(0,1,0)$ is given by

(a) $\cos\theta=\frac{1}{\sqrt{3}}$ (b) $\cos\theta=\frac{4}{3\sqrt{2}}$

(c) $\cos\theta=\frac{1}{\sqrt{5}}$ (d) $\cos\theta=\frac{1}{2\sqrt{2}}$

57. The shortest distance of a point $(1, 2, -3)$ from a plane making intercepts 1, 2 and 3 units on position X, Y and Z-axes respectively, is

(a) 2 (b) 0

(c) $\frac{13}{12}$ (d) $\frac{12}{7}$

58. A tetrahedron has vertices $O(0,0,0)$, $A(1,2,1)$, $B(2,1,3)$ and $C(-1,1,2)$. Then the angle between the faces OAB and ABC will be

(a) $\cos^{-1}\left(\frac{19}{35}\right)$ (b) $\cos^{-1}\left(\frac{17}{31}\right)$

(c) 30° (d) 90°

59. The direction ratios of line I_1 passing through $P(1,3,4)$ and perpendicular to line $I_2 \frac{x-1}{2}=\frac{y-2}{3}=\frac{z-3}{4}$ (where, I_1 and I_2 are coplanar) is

(a) 14, 8, 1 (b) −14, 8, −1

(c) 14, −8, −1 (d) −14, −8, 1

60. Equation of the plane through three points A, B and C with position vectors $-6\mathbf{i}+3\mathbf{j}+2\mathbf{k}$, $3\mathbf{i}-2\mathbf{j}+4\mathbf{k}$ and $5\mathbf{i}+7\mathbf{j}+3\mathbf{k}$ is equal to

(a) $\mathbf{r}.(\mathbf{i}-\mathbf{j}+7\mathbf{k})+23=0$ (b) $\mathbf{r}.(\mathbf{i}+\mathbf{j}+7\mathbf{k})=23$

(c) $\mathbf{r}.(\mathbf{i}+\mathbf{j}-7\mathbf{k})+23=0$ (d) $\mathbf{r}.(\mathbf{i}-\mathbf{j}-7\mathbf{k})=23$

61. $OABC$ is a tetrahedron. The position vectors of A, B and C are $\mathbf{i}$, $\mathbf{i}+\mathbf{j}$ and $\mathbf{j}+\mathbf{k}$, respectively. O is origin. The height of the tetrahedron (taking plane ABC as base) is

(a) $\frac{1}{2}$ (b) $\frac{1}{\sqrt{2}}$

(c) $\frac{1}{2\sqrt{2}}$ (d) None of these

62. The plane $x-y-z=4$ is rotated through an angle 90° about its line of intersection with the plane $x+y+2z=4$. Then the equation of the plane in its new position is

(a) $x+y+4z=20$ (b) $x+5y+4z=20$

(c) $x+y-4z=20$ (d) $5x+y+4z=20$

63. Let A_{xy}, A_{yz}, A_{zx} be the area of the projection of a plane area A on the xy, yz, zx plane respectively Then $A^2=$

(a) $A_{xy}^2+A_{yz}^2+A_{zx}^2$ (b) $\sqrt{A_{xy}^2+A_{yz}^2+A_{zx}^2}$

(c) $A_{xy}+A_{yz}+A_{xz}$ (d) $\sqrt{A_{xy}+A_{yz}+A_{zx}}$

64. Through the point $P(h,k,l)$ a plane is drawn at right angles to OP to meet co-ordinate axes at A, B and C. If $OP=p$ then the area of the ΔABC is

(a) $\frac{p^5}{2hkl}$ (b) $\frac{p^5}{hkl}$

(c) $\frac{p^3}{2hkl}$ (d) $\frac{p^3}{hkl}$

65. The volume of the tetrahedron included between the plane $3x+4y-5z-60=0$ and the co-ordinate planes is

(a) 60 (b) 600

(c) 720 (d) 400

66. The angle between the lines whose direction cosines are given by the equations $l^2+m^2-n^2=0, l+m+n=0$ is

(a) $\cos^{-1}(2\sqrt{3})$ (b) $\cos^{-1}\sqrt{3}$

(c) $\frac{\pi}{3}$ (d) $\frac{\pi}{2}$

67. The distance between the line $\mathbf{r}=2\hat{\mathbf{i}}-2\hat{\mathbf{j}}+3\hat{\mathbf{k}}+\lambda(\hat{\mathbf{i}}-\hat{\mathbf{j}}+4\hat{\mathbf{k}})$ and the plane $\mathbf{r}\cdot(\hat{\mathbf{i}}+5\hat{\mathbf{j}}+\hat{\mathbf{k}})=5$ is

(a) $\frac{10}{3\sqrt{3}}$ (b) $\frac{10}{3}$

(c) $\frac{10}{9}$ (d) $\frac{10}{\sqrt{3}}$

68. The Cartesian equation of the plane perpendicular to the line $\frac{x-1}{2}=\frac{y-3}{-1}=\frac{z-4}{2}$ and passing through the origin is

(a) $2x-y+2z-7=0$ (b) $2x+y+2z=0$

(c) $2x-y+2z=0$ (d) $2x-y-z=0$

69. Let $P(3,2,6)$ be a point in space and Q be a point on the line $\mathbf{r}=(\hat{\mathbf{i}}-\hat{\mathbf{j}}+2\hat{\mathbf{k}})+\mu(-3\hat{\mathbf{i}}+\hat{\mathbf{j}}+5\hat{\mathbf{k}})$. Then the value of μ for which the vector $\mathbf{PQ}$ is parallel to the plane $x-4y+3z=1$ is

(a) $\frac{1}{4}$ (b) $-\frac{1}{4}$

(c) $\frac{1}{8}$ (d) $-\frac{1}{8}$

70. A plane makes intercepts OA, OB and OC whose measurements are a, b and c on the OX, OY and OZ axes. The area of triangle ABC is

(a) $\frac{1}{2}(ab+bc+ca)$ (b) $\frac{1}{2}abc(a+b+c)$

(c) $\frac{1}{2}(a^2b^2+b^2c^2+c^2a^2)^{1/2}$ (d) $\frac{1}{2}(a+b+c)^2$

71. The radius of the circle in which the sphere $x^2+y^2+z^2+2x-2y-4z-19=0$ is cut by the plane $x+2y+2z+7=0$ is

(a) 2 (b) 3

(c) 4 (d) 1

72. Let $\mathbf{a} = \hat{\mathbf{i}} + \hat{\mathbf{j}}$ and $\mathbf{b} = 2\hat{\mathbf{i}} - \hat{\mathbf{k}}$, then the point of intersection of the lines $\mathbf{r} \times \mathbf{a} = \mathbf{b} \times \mathbf{a}$ and $\mathbf{r} \times \mathbf{b} = \mathbf{a} \times \mathbf{b}$ is

(a) $(3, -1, 1)$ (b) $(3, 1, -1)$
(c) $(-3, 1, 1)$ (d) $(-3, -1, -1)$

73. The co-ordinates of the point P on the line $\mathbf{r} = (\hat{\mathbf{i}} + \hat{\mathbf{j}} + \hat{\mathbf{k}}) + \lambda(-\hat{\mathbf{i}} + \hat{\mathbf{j}} - \hat{\mathbf{k}})$ which is nearest to the origin is

(a) $\left(\frac{2}{3}, \frac{4}{3}, \frac{2}{3}\right)$ (b) $\left(-\frac{2}{3}, -\frac{4}{3}, \frac{2}{3}\right)$
(c) $\left(\frac{2}{3}, \frac{4}{3}, -\frac{2}{3}\right)$ (d) None of these

74. The 3-dimensional vectors $\mathbf{v}_1, \mathbf{v}_2, \mathbf{v}_3$ satisfying $\mathbf{v}_1 . \mathbf{v}_1 = 4, \mathbf{v}_1 . \mathbf{v}_2 = -2, \mathbf{v}_1 . \mathbf{v}_3 = 6, \mathbf{v}_2 . \mathbf{v}_2 = 2, \mathbf{v}_2 . \mathbf{v}_3 = -5$, $\mathbf{v}_3 . \mathbf{v}_3 = 29$, then $\mathbf{v}_3$ may be

(a) $-3\hat{\mathbf{i}} + 2\hat{\mathbf{j}} \pm 4\hat{\mathbf{k}}$ (b) $3\hat{\mathbf{i}} - 2\hat{\mathbf{j}} \pm 4\hat{\mathbf{k}}$
(c) $-2\hat{\mathbf{i}} + 3\hat{\mathbf{j}} \pm 4\hat{\mathbf{k}}$ (d) $2\hat{\mathbf{i}} + 3\hat{\mathbf{j}} \pm 4\hat{\mathbf{k}}$

75. The points $\hat{\mathbf{i}} - \hat{\mathbf{j}} + 3\hat{\mathbf{k}}$ and $3\hat{\mathbf{i}} + 3\hat{\mathbf{j}} + 3\hat{\mathbf{k}}$ are equidistant from the plane $\mathbf{r}.(5\hat{\mathbf{i}} + 2\hat{\mathbf{j}} - 7\hat{\mathbf{k}}) + 9 = 0$, then they are

(a) on the same sides of the plane
(b) parallel of the plane
(c) on the opposite sides of the plane
(d) None of the above

76. A, B, C, D are four points in space. Then, $AC^2 + BD^2 + AD^2 + BC^2 \geq$

(a) $AB^2 + CD^2$ (b) $\frac{1}{AB^2} - \frac{1}{CD^2}$
(c) $\frac{1}{CD^2} - \frac{1}{AB^2}$ (d) None of these

77. If $|x_1| > |y_1| + |z_1|, |y_2| > |x_2| + |z_2|, |z_3| > |x_3| + |y_3|$, then $x_1\hat{\mathbf{i}} + y_1\hat{\mathbf{j}} + z_1\hat{\mathbf{k}}$, $x_2\hat{\mathbf{i}} + y_2\hat{\mathbf{j}} + z_2\hat{\mathbf{k}}$ and $x_3\hat{\mathbf{i}} + y_3\hat{\mathbf{j}} + z_3\hat{\mathbf{k}}$ are

(a) perpendicular (b) collinear
(c) coplanar (d) non-coplanar

78. The position vector of the point of intersection of three planes $\mathbf{r}.\mathbf{n}_1 = q_1, \mathbf{r}.\mathbf{n}_2 = q_2, \mathbf{r}.\mathbf{n}_3 = q_3$, where $\mathbf{n}_1, \mathbf{n}_2$ and $\mathbf{n}_3$ are non-coplanar vectors, is

(a) $\frac{1}{[\mathbf{n}_1\ \mathbf{n}_2\ \mathbf{n}_3]}[q_3(\mathbf{n}_1 \times \mathbf{n}_2) + q_1(\mathbf{n}_2 \times \mathbf{n}_3) + q_2(\mathbf{n}_3 \times \mathbf{n}_1)]$

(b) $\frac{1}{[\mathbf{n}_1\ \mathbf{n}_2\ \mathbf{n}_3]}[q_1(\mathbf{n}_1 \times \mathbf{n}_2) + q_2(\mathbf{n}_2 \times \mathbf{n}_3) + q_3(\mathbf{n}_3 \times \mathbf{n}_1)]$

(c) $-\frac{1}{[\mathbf{n}_3\ \mathbf{n}_1\ \mathbf{n}_2]}[q_1(\mathbf{n}_1 \times \mathbf{n}_2) + q_2(\mathbf{n}_2 \times \mathbf{n}_3) + q_3(\mathbf{n}_3 \times \mathbf{n}_1)]$

(d) None of the above

79. A pentagon is formed by cutting a triangular corner from a rectangular piece of paper. The five sides of the pentagon have length 13, 19, 20, 25 and 31 not necessarily in that order. The area of the pentagon is

(a) 459 sq units (b) 600 sq units
(c) 680 sq units (d) 745 sq units

80. In a three dimensional coordinate system P, Q and R are images of a point $A(a, b, c)$ in the XY they YZ and the ZX planes respectively. If G is the centroid of triangle PQR, then area of triangle AOG is (O is the origin)

(a) 0 (b) $a^2 + b^2 + c^2$
(c) $\frac{2}{3}(a^2 + b^2 + c^2)$ (d) None of these

81. A plane $2x + 3y + 5z = 1$ has a point P which is at minimum distance from line joining $A(1, 0, -3)$, $B(1, -5, 7)$, then distance AP is equal to

(a) $3\sqrt{5}$ (b) $2\sqrt{5}$
(c) $4\sqrt{5}$ (d) None of these

82. The locus of a point which moves in such a way that its distance from the line $\frac{x}{1} = \frac{y}{1} = \frac{z}{-1}$ is twice the distance from the plane $x + y + z = 0$ is

(a) $x^2 + y^2 + z^2 - 5x - 3y - 3z = 0$
(b) $x^2 + y^2 + z^2 + 5x + 3y + 3z = 0$
(c) $x^2 + y^2 + z^2 - 5xy - 3yz - 3zx = 0$
(d) $x^2 + y^2 + z^2 + 5xy + 3yz + 3zx = 0$

83. A cube $C = \{(x, y, z) \mid 0 \leq x, y, z \leq 1\}$ is cut by a sharp knife along the plane $x = y$, $y = z$, $z = x$. If no piece is moved until all three cuts are made, the number of pieces is

(a) 6 (b) 7
(c) 8 (d) 27

84. A ray of light is sent through the point $P(1, 2, 3)$ and is reflected on the XY-plane. If the reflected ray passes through the point $Q(3, 2, 5)$, then the equation of the reflected ray is

(a) $\frac{x-3}{1} = \frac{y-2}{0} = \frac{z-5}{1}$ (b) $\frac{x-3}{1} = \frac{y-2}{0} = \frac{z-5}{-4}$
(c) $\frac{x-3}{1} = \frac{y-2}{0} = \frac{z-5}{4}$ (d) $\frac{x-1}{1} = \frac{y-2}{0} = \frac{z-3}{4}$

85. A plane cutting the axes in P, Q, R passes through $(\alpha - \beta, \beta - \gamma, \gamma - \alpha)$. If O is the origin, then locus of centre of sphere $OPQR$ is

(a) $\alpha x + \beta y + \gamma z = 4$
(b) $(\alpha - \beta)x + (\beta - \gamma)y + (\gamma - \alpha)z = 0$
(c) $(\alpha - \beta)yz + (\beta - \gamma)zx + (\gamma - \alpha)xy = 2xyz$
(d) $\left(\frac{1}{\alpha^2} + \frac{1}{\beta^2} + \frac{1}{\gamma^2}\right)(x^2 + y^2 + z^2) = xyz$

86. The shortest distance between any two opposite edges of the tetrahedron formed by planes $x + y = 0$, $y + z = 0$, $z + x = 0$, $x + y + z = a$ is constant, equal to

(a) $2a$ (b) $\frac{2a}{\sqrt{6}}$
(c) $\frac{a}{\sqrt{6}}$ (d) $\frac{2a}{\sqrt{3}}$

87. The angle between the pair of planes represented by equation $2x^2 - 2y^2 + 4z^2 + 6xz + 2yz + 3xy = 0$ is

(a) $\cos^{-1}\left(\frac{1}{3}\right)$ (b) $\cos^{-1}\left(\frac{4}{21}\right)$

(c) $\cos^{-1}\left(\frac{4}{9}\right)$ (d) $\cos^{-1}\left(\frac{7}{\sqrt{84}}\right)$

88. Let (p, q, r) be a point on the plane $2x + 2y + z = 6$, then the least value of $p^2 + q^2 + r^2$ is equal to

(a) 4 (b) 5 (c) 6 (d) 8

89. The four lines drawing from the vertices of any tetrahedron to the centroid of the opposite faces meet in a point whose distance from each vertex is 'k' times the distance from each vertex to the opposite face, where k is

(a) $\frac{1}{3}$ (b) $\frac{1}{2}$ (c) $\frac{3}{4}$ (d) $\frac{5}{4}$

90. The shortest distance from (1, 1, 1) to the line of intersection of the pair of planes $xy + yz + zx + y^2 = 0$ is

(a) $\sqrt{\frac{8}{3}}$ (b) $\frac{2}{\sqrt{3}}$ (c) $\frac{1}{\sqrt{3}}$ (d) $\frac{2}{3}$

91. The shortest distance between the two lines $L_1 : x = k_1$; $y = k_2$ and $L_2 : x = k_3; y = k_4$ is equal to

(a) $\left|\sqrt{k_1^2 + k_2^2} - \sqrt{k_3^2 + k_4^2}\right|$ (b) $\sqrt{k_1k_3 + k_2k_4}$

(c) $\sqrt{(k_1 + k_3)^2 + (k_2 + k_4)^2}$ (d) $\sqrt{(k_1 - k_3)^2 + (k_2 - k_4)^2}$

92. $A = \begin{bmatrix} l_1 & m_1 & n_1 \\ l_2 & m_2 & n_2 \\ l_3 & m_3 & n_3 \end{bmatrix}$ and $B = \begin{bmatrix} p_1 & q_1 & r_1 \\ p_2 & q_2 & r_2 \\ p_3 & q_3 & r_3 \end{bmatrix}$, where p_i, q_i, r_i are the cofactors of the elements l_i, m_i, n_i for $i = 1, 2, 3$. If $(l_1, m_1, n_1), (l_2, m_2, n_2)$ and (l_3, m_3, n_3) are the direction cosines of three mutually perpendicular lines, then $(p_1, q_1, r_1), (p_2, q_2, r_2)$ and (p_3, q_3, r_3) are

(a) the direction cosines of three mutually perpendicular lines

(b) the direction ratios of three mutually perpendicular lines which are not direction cosines

(c) the direction cosines of three lines which need not be perpendicular

(d) the direction ratios but not the direction cosines of three lines which need not be perpendicular

93. If $ABCD$ is a tetrahedron such that each ΔABC, ΔABD and ΔACD has a right angle at A. If ar$(\Delta ABC) = k_1$, ar$(\Delta ABD) = k_2$, ar$(\Delta BCD) = k_3$, then ar(ΔACD) is

(a) $\sqrt{k_1^2 + k_2^2 + k_3^2}$ (b) $\sqrt{\frac{k_1k_2k_3}{k_1 + k_2 + k_3}}$

(c) $\sqrt{|k_1^2 + k_2^2 - k_3^2|}$ (d) $\sqrt{|k_2^2 - k_1^2 - k_3^2|}$

94. In a regular tetrahedron, if the distance between the mid-points of opposite edges is unity, its volume is

(a) $\frac{1}{3}$ (b) $\frac{1}{2}$

(c) $\frac{1}{\sqrt{2}}$ (d) $\frac{1}{6\sqrt{2}}$

95. A variable plane makes intercepts on X, Y and Z-axes and it makes a tetrahedron of volume 64 cu. u. The locus of foot of perpendicular from origin on this plane is

(a) $(x^2 + y^2 + z^2)^3 = 384\,xyz$

(b) $xyz = 681$

(c) $(x + y + z)\left(\frac{1}{x} + \frac{1}{y} + \frac{1}{z}\right)^2 = 16$

(d) $xyz\,(x + y + z) = 81$

96. If P, Q, R, S are four coplanar points on the sides AB, BC, CD, DA of a skew quadrilateral, then $\frac{AP}{PB} \cdot \frac{BQ}{QC} \cdot \frac{CR}{RD} \cdot \frac{DS}{SA}$ equals

(a) 1 (b) -1 (c) 3 (d) -3

Three Dimensional Coordinate System Exercise 2 : More than One Correct Option Type Questions

97. Given the equations of the line $3x - y + z + 1 = 0$ and $5x + y + 3z = 0$. Then, which of the following is correct ?

(a) Symmetrical form of the equations of line is

$$\frac{x}{2} = \frac{y - \frac{1}{8}}{-1} = \frac{z + \frac{5}{8}}{1}.$$

(b) Symmetrical form of the equations of line is

$$\frac{x + \frac{1}{8}}{1} = \frac{y - \frac{5}{8}}{-1} = \frac{z}{-2}.$$

(c) Equation of the plane through (2, 1, 4) and perpendicular to the given lines is $2x - y + z - 7 = 0$.

(d) Equation of the plane through (2, 1, 4) and perpendicular to the given lines is $x + y - 2z + 5 = 0$.

98. Consider the family of planes $x + y + z = c$ where c is a parameter intersecting the coordinate axes at P, Q and R and α, β and γ are the angles made by each member of this family with positive x, y and z-axes. Which of the following interpretations hold good for this family?

(a) Each member of this family is equally inclined with coordinate axes.
(b) $\sin^2 \alpha + \sin^2 \beta + \sin^2 \gamma = 1$
(c) $\cos^2 \alpha + \cos^2 \beta + \cos^2 \gamma = 2$
(d) For $c = 3$ area of the ΔPQR is $3\sqrt{3}$ sq units.

99. Equation of the line through the point (1, 1, 1) and intersecting the lines $2x - y - z - 2 = 0 = x + y + z - 1$ and $x - y - z - 3 = 0 = 2x + 4y - z - 4$
(a) $x - 1 = 0, 7x + 17y - 3z - 134 = 0$
(b) $x - 1 = 0, 9x + 15y - 5z - 19 = 0$
(c) $x - 1 = 0, \dfrac{y-1}{1} = \dfrac{z-1}{3}$
(d) $x - 2y + 2z - 1 = 0, 9x + 15y - 5z - 19 = 0$

100. Through a point $P(h, k, l)$ a plane is drawn at right angles to OP to meet the coordinate axes in A, B and C. If $OP = p$, A_{xy} is area of projection of ΔABC on xy-plane, A_{yz} is area of projection of ΔABC on yz-plane, then
(a) $\Delta = \left|\dfrac{p^5}{hkl}\right|$ (b) $\Delta = \left|\dfrac{p^5}{2hkl}\right|$ (c) $\dfrac{A_{xy}}{A_{yz}} = \left|\dfrac{l}{h}\right|$ (d) $\dfrac{A_{xy}}{A_{yz}} = \left|\dfrac{h}{l}\right|$

101. Which of the following statements is/are correct?
(a) If $\mathbf{n}.\mathbf{a} = 0$, $\mathbf{n}.\mathbf{b} = 0$ and $\mathbf{n}.\mathbf{c} = 0$ for some non-zero vector $\mathbf{n}$, then $[\mathbf{a}\ \mathbf{b}\ \mathbf{c}] = 0$
(b) There exist a vector making angles 30° and 45° with x-axis and Y-axis.
(c) Locus of point for which $x = 3$ and $y = 4$ is a line parallel to the Z-axis whose distance from the Z-axis is 5
(d) The vertices of regular tetrahedron are O, A, B, C where 'O' is the origin. The vector $\mathbf{OA} + \mathbf{OB} + \mathbf{OC}$ is perpendicular to the plane ABC

102. Which of the following is/are correct about a tetrahedron ?
(a) Centroid of a tetrahedron lies on lines joining any vertex to the centroid opposite face
(b) Centroid of a tetrahedron lies on the lines joining the mid point of the opposite faces
(c) Distance of centroid from all the vertices are equal
(d) None of the above

103. A variable plane cutting coordinate axes in A, B, C is at a constant distance from the origin. Then the locus of centroid of the ΔABC is
(a) $x^{-2} + y^{-2} + z^{-2} = 16$ (b) $x^{-2} + y^{-2} + z^{-2} = 9$
(c) $\dfrac{1}{9}\left\{\dfrac{1}{x^2} + \dfrac{1}{y^2} + \dfrac{1}{z^2}\right\}^{-1} = 0$ (d) $X + Y = 0$

104. Equation of any plane containing the line $\dfrac{x - x_1}{a} = \dfrac{y - y_1}{b} = \dfrac{z - z_1}{c}$ is $A(x - x_1) + B(y - y_1) + C(z - z_1) = 0$, then pick correct alternatives
(a) $\dfrac{A}{a} = \dfrac{B}{b} = \dfrac{C}{c}$ is true for the line to be perpendicular to the plane
(b) $A(a + 3) + B(b - 1) + C(c - 2) = 0$
(c) $2aA + 3bB + 4cC = 0$
(d) $Aa + Bb + Cc = 0$

105. The line $\dfrac{x-2}{3} = \dfrac{y+1}{2} = \dfrac{z-1}{-1}$ intersects the curve $x^2 + y^2 = r^2, z = 0$ then
(a) Equation of the plane through (0, 0, 0) perpendicular to the given line is $3x + 2y - z = 0$
(b) $r = \sqrt{26}$ (c) $r = 6$
(d) $r = 7$

106. A vector equally inclined to the vectors $\hat{\mathbf{i}} - \hat{\mathbf{j}} + \hat{\mathbf{k}}$ and $\hat{\mathbf{i}} + \hat{\mathbf{j}} - \hat{\mathbf{k}}$ then the plane containing them is
(a) $\dfrac{\hat{\mathbf{i}} + \hat{\mathbf{j}} - \hat{\mathbf{k}}}{\sqrt{3}}$ (b) $\hat{\mathbf{j}} - \hat{\mathbf{k}}$ (c) $2\hat{\mathbf{i}}$ (d) $\hat{\mathbf{i}}$

107. Consider the plane through $(2, 3, -1)$ and at right angles to the vector $3\hat{\mathbf{i}} - 4\hat{\mathbf{j}} + 7\hat{\mathbf{k}}$ from the origin is
(a) The equation of the plane through the given point is $3x - 4y + 7z + 13 = 0$
(b) perpendicular distance of plane from origin $\dfrac{1}{\sqrt{74}}$
(c) perpendicular distance of plane from origin $\dfrac{13}{\sqrt{74}}$
(d) perpendicular distance of plane from origin $\dfrac{21}{\sqrt{74}}$

108. A plane passes through a fixed point (a, b, c) and cuts the axes in A, B, C. The locus of a point equidistant from origin, A, B and C must be
(a) $ayz + bzx + cxy = 2xyz$ (b) $\dfrac{a}{x} + \dfrac{b}{y} + \dfrac{c}{z} = 1$
(c) $\dfrac{a}{x} + \dfrac{b}{y} + \dfrac{c}{z} = 2$ (d) $\dfrac{a}{x} + \dfrac{b}{y} + \dfrac{c}{z} = 3$

109. Let $\mathbf{A}$ be vector parallel to line of intersection of planes P_1 and P_2. Plane P_1 is parallel to the vectors $2\hat{\mathbf{j}} + 3\hat{\mathbf{k}}$ and $4\hat{\mathbf{j}} - 3\hat{\mathbf{k}}$ and that P_2 is parallel to $\hat{\mathbf{j}} - \hat{\mathbf{k}}$ and $3\hat{\mathbf{i}} + 3\hat{\mathbf{j}}$, then the angle between vector $\mathbf{A}$ and a given vector $2\hat{\mathbf{i}} + \hat{\mathbf{j}} - 2\hat{\mathbf{k}}$ is
(a) $\dfrac{\pi}{2}$ (b) $\dfrac{\pi}{4}$ (c) $\dfrac{\pi}{6}$ (d) $\dfrac{3\pi}{4}$

110. Consider the lines $x = y = z$ and the line $2x + y + z - 1 = 0 = 3x + y + 2z - 2$, then
(a) the shortest distance between the two lines is $\dfrac{1}{\sqrt{2}}$
(b) the shortest distance between the two lines is $\sqrt{2}$
(c) plane containing 2nd line parallel to 1st line is $y - z + 1 = 0$
(d) the shortest distance between the two lines $\dfrac{\sqrt{3}}{2}$

111. If p_1, p_2, p_3 denote the perpendicular distances of the plane $2x - 3y + 4z + 2 = 0$ from the parallel planes.
(a) $p_1 + 8p_2 - p_3 = 0$
(b) $p_3 = 16p_2$
(c) $8p_2 = p_1$
(d) $p_1 + 2p_2 + 3p_3 = \sqrt{29}$

112. A line segment has length 63 and direction ratios are 3, −2,6. The components of the line vector are
(a) −27, 18, 54
(b) 27, −18, −54
(c) 27, −18, 54
(d) −27, 18, −54

113. The lines $\frac{x-2}{1} = \frac{y-3}{1} = \frac{z-4}{-k}$ and $\frac{x-1}{k} = \frac{y-4}{2} = \frac{z-5}{1}$ are coplanar if
(a) $k = 0$
(b) $k = -1$
(c) $k = 2$
(d) $k = -3$

114. The points $A(4, 5, 10)$, $B(2, 3, 4)$ and $C(1, 2, -1)$ are three vertices of a parallelogram $ABCD$, then
(a) Vector equation of AB is $2\mathbf{i} + 3\mathbf{j} + 4\mathbf{k} + \lambda(\mathbf{i} + \mathbf{j} + 3\mathbf{k})$
(b) Cartesian equation of BC is $\frac{x-2}{1} = \frac{y-3}{1} = \frac{z-4}{5}$
(c) Coordinates of D are (3, 4, 5)
(d) $ABCD$ is a rectangle

115. The line $x = y = z$ meets the plane $x + y + z = 1$ at the point P and the sphere $x^2 + y^2 + z^2 = 1$ at the points R and S, then
(a) $PR + PS = 2$
(b) $PR \times PS = \frac{2}{3}$
(c) $PR = PS$
(d) $PR + PS = RS$

116. A rod of length 2 units whose one end is $(1, 0, -1)$ and other end touches the plane $x - 2y + 2z + 4 = 0$, then
(a) The rod sweeps the figure whose volume is π cubic units.
(b) The area of the region which the rod traces on the plane is 2π.
(c) The length of projection of the rod on the plane is $\sqrt{3}$ units.
(d) The centre of the region which the rod traces on the plane is $\left(\frac{2}{3}, \frac{2}{3}, \frac{-5}{3}\right)$.

117. Consider the planes $P_1 : 2x + y + z + 4 = 0$
$P_2 : y - z + 4 = 0$ and $P_3 : 3x + 2y + z + 8 = 0$
Let L_1, L_2, L_3 be the lines of intersection of the planes P_2 and P_3, P_3 and P_1, and P_1 and P_2 respectively. Then,
(a) at least two of the lines L_1, L_2 and L_3 are non-parallel
(b) at least two of the lines L_1, L_2 and L_3 are parallel
(c) the three planes intersect in a line
(d) the three planes form a triangular prism

118. The volume of a right triangular prism $ABCA_1B_1C_1$ is equal to 3. Find the coordinates of the vertex A_1, if the coordinates of the base vertices of the prism are $A(1, 0, 1)$, $B(2, 0, 0)$ and $C(0, 1, 0)$.
(a) (−2, 0, 2)
(b) (0, −2, 0)
(c) (0, 2, 0)
(d) (2, 2, 2)

119. Let a plane pass through origin and is parallel to the line $\frac{x-1}{2} = \frac{y+3}{-1} = \frac{z+1}{-2}$ such that distance between plane and the line is $\frac{5}{3}$. Then, equation of the plane is ....
(a) $x - 2y + 2z = 0$
(b) $x - 2y - 2z = 0$
(c) $2x + 2y + z = 0$
(d) $x + y + z = 0$

120. $OABC$ is a regular tetrahedron of side unity, then
(a) the length of perpendicular from one vertex to opposite face is $\sqrt{2/3}$
(b) the perpendicular distance from mid-point of $\overline{OA}$ to the plane ABC is $1/\sqrt{6}$
(c) the angle between two skew edges is $\pi/2$
(d) the distance of centroid of the tetrahedron form any vertex is $\sqrt{3/8}$

121. If $OABC$ is a tetrahedron such that $OA^2 + BC^2 = OB^2 + CA^2 = OC^2 + AB^2$, then
(a) $OA \perp BC$
(b) $OB \perp AC$
(c) $OC \perp AB$
(d) $AB \perp AC$

122. If the line $\frac{x}{1} = \frac{y}{2} = \frac{z}{3}$ intersects the line $3\beta^2 x + 3(1-2\alpha)y + z = 3 = -\frac{1}{2}\{6\alpha^2 x + 3(1-2\beta)y + 2z\}$ then point $(\alpha, \beta, 1)$ lie on the plane
(a) $2x - y + z = 4$
(b) $x + y - z = 2$
(c) $x - 2y = 0$
(d) $2x - y = 0$

123. Let PM be the perpendicular from the point $P(1, 2, 3)$ to XY plane. If $\overline{OP}$ makes an angle θ with the positive direction of Z-axis and $\overline{OM}$ makes an angle ϕ with the positive direction of X-axis, where O is the origin and θ and ϕ are acute angles, then
(a) $\tan\theta = \frac{\sqrt{5}}{3}$
(b) $\sin\theta \sin\phi = \frac{2}{\sqrt{14}}$
(c) $\tan\phi = 2$
(d) $\cos\theta \cos\phi = \frac{1}{\sqrt{14}}$

124. A variable plane which remains at a constant distance P from the origin (0) cuts the coordinate axes in A, B, C
(a) Locus of centroid of tetrahedron $OABC$ is
$$x^2y^2 + y^2z^2 + z^2x^2 = \frac{16}{p^2}x^2y^2z^2$$
(b) Locus of centroid of tetrahedron $OABC$ is
$$x^2y^2 + y^2z^2 + z^2x^2 = \frac{4}{p^2}x^2y^2z^2$$
(c) Parametric equation of the centroid of the tetrahedron is of the form $\left(\frac{P}{4}\sec\alpha\sec\beta, \frac{P}{4}\sec\alpha \operatorname{cosec}\beta, \frac{P}{4}\operatorname{cosec}\alpha\right)$,
$\alpha, \beta \in (0, 2\pi) - \{\pi/2, \pi, 3\pi/2\}$
(d) None of the above

Three Dimensional Coordinate System Exercise 3 : Statement I and II Type Questions

- **Directions** (Q. Nos. 125 to 138) For the following questions, choose the correct answers from the codes (a), (b), (c) and (d) defined as follows :
  - (a) Statement I is true, Statement II is also true; Statement II is the correct explanation of Statement I
  - (b) Statement I is true, Statement II is also true; Statement II is not the correct explanation of Statement I
  - (c) Statement I is true, Statement II is false
  - (d) Statement I is false, Statement II is true

125. **Statement I** Let $A(\hat{i}+\hat{j}+\hat{k})$ and $B(\hat{i}-\hat{j}+\hat{k})$ be two points, $P(2\hat{i}+3\hat{j}+\hat{k})$ lies exterior to the sphere with AB as one of its diameters.

Statement II If A and B are any two points and P is a point in space such that $\mathbf{PA}.\mathbf{PB} > 0$, then the point P lies exterior to the sphere with AB as one of its diameters.

126. **Statement I** If $\mathbf{r}=x\hat{i}+y\hat{j}+z\hat{k}$, then equation $\mathbf{r}\times(2\hat{i}-\hat{j}+3\hat{k})=3\hat{i}+\hat{k}$ represents a straight line.

Statement II If $\mathbf{r}=x\hat{i}+y\hat{j}+z\hat{k}$, then equation $\mathbf{r}\times(\hat{i}+2\hat{j}-3\hat{k})=2\hat{i}-\hat{j}$ represents a straight line.

127. **Statement I** Let θ be the angle between the line $\frac{x-2}{2}=\frac{y-1}{-3}=\frac{z+2}{-2}$ and the plane $x+y-z=5$. Then, $\theta=\sin^{-1}\left(\frac{1}{\sqrt{51}}\right)$

Statement II Angle between a straight line and a plane is the complement of angle between the line and normal to the plane.

128. **Statement I** A point on the straight line $2x+3y-4z=5$ and $3x-2y+4z=7$ can be determined by taking $x=k$ and then solving the two equations for y and z, where k is any real number.

Statement II If $c'\neq kc$, then the straight line $ax+by+cz+d=0$, $Kax+Kby+c'z+d'=0$, does not intersect the plane $z=\alpha$, where α is any real number.

129. Let the line L having equation $\frac{x-1}{2}=\frac{y-3}{5}=\frac{z-1}{3}$, intersects the plane P, having equation $x-y+z=5$ at the point A.

Statement I Equation of the line L' through the point A, lying in the plane P and having minimum inclination with line L is $8x+y-72-4=0=x-y+z-5$

Statement II Line L' must be the projection of the line L in the plane P.

130. Given lines $\frac{x-4}{2}=\frac{y+5}{4}=\frac{z-1}{-3}$ and $\frac{x-2}{1}=\frac{y+1}{3}=\frac{z}{2}$

Statement I The lines intersect.

Statement II They are not parallel.

131. Consider the lines $L_1:\mathbf{r}=\mathbf{a}+\lambda\,\mathbf{b}$ and $L_2:\mathbf{r}=\mathbf{b}+\mu\,\mathbf{a}$, where $\mathbf{a}$ and $\mathbf{b}$ are non-zero and non-collinear vectors.

Statement I L_1 and L_2 are coplanar and the plane containing these lines passes through origin.

Statement II $(a-b).(b\times a)=0$ and the plane containing L_1 and L_2 is $[\mathbf{r}\ \mathbf{a}\ \mathbf{b}]=0$ which passes through origin.

132. **Statement I** P is a point (a,b,c). Let A, B, C be the images of P in yz, zx and xy planes respectively, then equation of the plane passing through the points A, B and C is $\frac{x}{a}+\frac{y}{b}+\frac{z}{c}=1$

Statement II The image of a point P in a plane is the foot of the perpendicular drawn from P on the plane.

133. **Statement I** The locus of a point which is equidistant from the points whose position vectors are $3\hat{i}-2\hat{j}+5\hat{k}$ and $\hat{i}+2\hat{j}-\hat{k}$ is $\mathbf{r}(\hat{i}-2\hat{j}+3\hat{k})=8$.

Statement II The locus of a point which is equidistant from the points whose position vectors are $\mathbf{a}$ and $\mathbf{b}$ is $\left|\mathbf{r}-\frac{\mathbf{a}+\mathbf{b}}{2}\right|.(\mathbf{a}-\mathbf{b})=0$

134. **Statement I** If the vectors $\mathbf{a}$ and $\mathbf{c}$ are non-collinear then the lines $\mathbf{r}=6\mathbf{a}-\mathbf{c}+\lambda(2\mathbf{c}-\mathbf{a})$ and $\mathbf{r}=\mathbf{a}-\mathbf{c}+\mu(\mathbf{a}+3\mathbf{c})$ are coplanar.

Statement II There exist λ and μ such that the two values of $\mathbf{r}$ in Statement I becomes same.

135. **Statement I** The lines $\frac{x-1}{1}=\frac{y}{-1}=\frac{z+1}{1}$ and $\frac{x-2}{1}=\frac{y+1}{2}=\frac{z}{3}$ are coplanar and equation of the plane containing them is $5x+2y-3z-8=0$

Statement II The line $\frac{x-2}{1}=\frac{y+1}{2}=\frac{z}{3}$ is perpendicular to the plane $3x+6y+9z-8=0$ and parallel to the plane $x+y-z=0$

136. The equation of two straight lines are $\frac{x-1}{2}=\frac{y+3}{1}=\frac{z-2}{-3}$ and $\frac{x-2}{1}=\frac{y-1}{-3}=\frac{z+3}{2}$.

Statement I The given lines are coplanar.

Statement II The equations $2x_1-y_1=1$, $x_1+3y_1=4$ and $3x_1+2y_1=5$ are consistent.

137. **Statement I** A plane passes through the point $A(2, 1, -3)$. If distance of this plane from origin is maximum, then its equation is $2x + y - 3z = 14$.

Statement II If the plane passing through the point $A(\mathbf{a})$ is at maximum distance from origin, then normal to the plane is vector $\mathbf{a}$.

138. **Statement I** At least two of the lines L_1, L_2 and L_3 are non-parallel.

Statement II The three planes do not have a common point.

Three Dimensional Coordinate System Exercise 4 : Passage Based Questions

Passage I

(Q. Nos. 139 to 142)

Let $A(1, 2, 3)$, $B(0, 0, 1)$ and $C(-1, 1, 1)$ are the vertices of ΔABC.

139. The equation of internal angle bisector through A to side BC is

(a) $\mathbf{r} = \hat{\mathbf{i}} + 2\hat{\mathbf{j}} + 3\hat{\mathbf{k}} + \mu(3\hat{\mathbf{i}} + 2\hat{\mathbf{j}} + 3\hat{\mathbf{k}})$

(b) $\mathbf{r} = \hat{\mathbf{i}} + 2\hat{\mathbf{j}} + 3\hat{\mathbf{k}} + \mu(3\hat{\mathbf{i}} + 4\hat{\mathbf{j}} + 3\hat{\mathbf{k}})$

(c) $\mathbf{r} = \hat{\mathbf{i}} + 2\hat{\mathbf{j}} + 3\hat{\mathbf{k}} + \mu(3\hat{\mathbf{i}} + 3\hat{\mathbf{j}} + 2\hat{\mathbf{k}})$

(d) $\mathbf{r} = \hat{\mathbf{i}} + 2\hat{\mathbf{j}} + 3\hat{\mathbf{k}} + \mu(3\hat{\mathbf{i}} + 3\hat{\mathbf{j}} + 4\hat{\mathbf{k}})$

140. The equation of altitude through B to side AC is

(a) $\mathbf{r} = \mathbf{k} + t(7\hat{\mathbf{i}} - 10\hat{\mathbf{j}} + 2\hat{\mathbf{k}})$

(b) $\mathbf{r} = \mathbf{k} + t(-7\hat{\mathbf{i}} + 10\hat{\mathbf{j}} + 2\hat{\mathbf{k}})$

(c) $\mathbf{r} = \mathbf{k} + t(7\hat{\mathbf{i}} - 10\hat{\mathbf{j}} - 2\hat{\mathbf{k}})$

(d) $\mathbf{r} = \mathbf{k} + t(7\hat{\mathbf{i}} + 10\hat{\mathbf{j}} + 2\hat{\mathbf{k}})$

141. The equation of median through C to side AB is

(a) $\mathbf{r} = -\hat{\mathbf{i}} + \hat{\mathbf{j}} + \hat{\mathbf{k}} + p(3\hat{\mathbf{i}} - 2\hat{\mathbf{k}})$

(b) $\mathbf{r} = -\hat{\mathbf{i}} + \hat{\mathbf{j}} + \hat{\mathbf{k}} + p(3\hat{\mathbf{i}} + 2\hat{\mathbf{k}})$

(c) $\mathbf{r} = -\hat{\mathbf{i}} + \hat{\mathbf{j}} + \hat{\mathbf{k}} + p(-3\hat{\mathbf{i}} + 2\hat{\mathbf{k}})$

(d) $\mathbf{r} = -\hat{\mathbf{i}} + \hat{\mathbf{j}} + \hat{\mathbf{k}} + p(3\hat{\mathbf{i}} + 2\hat{\mathbf{j}})$

142. The area of (ΔABC) is equal to

(a) $\frac{9}{2}$ (b) $\frac{\sqrt{17}}{2}$ (c) $\frac{17}{2}$ (d) $\frac{7}{2}$

Passage II

(Q. Nos. 143 to 144)

Consider a plane $x + y - z = 1$ and the point $A(1, 2, -3)$. A line L has the equation $x = 1 + 3r$, $y = 2 - r$, $z = 3 + 4r$

143. The coordinate of a point B of line L, such that AB is parallel to the plane, is

(a) $(10, -1, 15)$ (b) $(-5, 4, -5)$

(c) $(4, 1, 7)$ (d) $(-8, 5, -9)$

144. Equation of the plane containing the line L and the point A has the equation

(a) $x - 3y + 5 = 0$ (b) $x + 3y - 7 = 0$

(c) $3x - y - 1 = 0$ (d) $3x + y - 5 = 0$

Passage III

(Q. Nos. 145 to 148)

Consider a triangular pyramid $ABCD$ the position vector of whose angular points are $A(3, 0, 1)$, $B(-1, 4, 1)$, $C(5, 2, 3)$ and $D(0, -5, 4)$. Let G be the point of intersection of the medians of the ΔBCD.

145. The length of the vector $\mathbf{AG}$ is

(a) $\sqrt{17}$ (b) $\frac{\sqrt{51}}{3}$ (c) $\frac{\sqrt{51}}{9}$ (d) $\frac{\sqrt{59}}{4}$

146. Area of the ΔABC (in sq units) is

(a) 24 (b) $8\sqrt{6}$

(c) $4\sqrt{6}$ (d) None of these

147. The length of the perpendicular from the vertex D on the opposite face is

(a) $\frac{14}{\sqrt{6}}$ (b) $\frac{2}{\sqrt{6}}$

(c) $\frac{3}{\sqrt{6}}$ (d) None of these

148. Equation of the plane ABC is

(a) $x + y + 2z = 5$ (b) $x - y - 2z = 1$

(c) $2x + y - 2z = 4$ (d) $x + y - 2z = 1$

Passage IV

(Q. Nos. 149 to 151)

A line L_1 passing through a point with position vector $\mathbf{p} = \mathbf{i} + 2\mathbf{j} + 3\mathbf{k}$ and parallel $\mathbf{a} = \mathbf{i} + 2\mathbf{j} + 3\mathbf{k}$, Another line L_2 passing through a point with position vector $= 2\mathbf{i} + 3\mathbf{j} + 3\mathbf{k}$ and parallel to $\mathbf{b} = 3\mathbf{i} + \mathbf{j} + 2\mathbf{k}$.

149. Equation of plane equidistant from line L_1 and L_2 is

(a) $\hat{\mathbf{r}} \cdot (\mathbf{i} - 7\mathbf{j} - 5\mathbf{k}) = 3$ (b) $\hat{\mathbf{r}} \cdot (\mathbf{i} + 7\mathbf{j} + 5\mathbf{k}) = 3$

(c) $\hat{\mathbf{r}} \cdot (\mathbf{i} - 7\mathbf{j} - 5\mathbf{k}) = 9$ (d) $\hat{\mathbf{r}} \cdot (\mathbf{i} + 7\mathbf{j} - 5\mathbf{k}) = 9$

150. Equation of a line passing through the point $(2, -3, 2)$ and equally inclined to the line L_1 and L_2 may be equal to

(a) $\frac{x-2}{2} = \frac{y-3}{-1} = \frac{z-2}{1}$ (b) $\frac{x-2}{-2} = y + 3 = z - 2$

(c) $\frac{x-2}{-4} = \frac{y+3}{3} = \frac{z-5}{2}$ (d) $\frac{x+2}{4} = \frac{y+3}{3} = \frac{z-2}{-5}$

151. The minimum distance of origin from the plane passing through the point with position vector $\mathbf{p}$ and perpendicular to the line L_2 is

(a) $\sqrt{14}$ (b) $\frac{7}{\sqrt{14}}$

(c) $\frac{11}{\sqrt{14}}$ (d) None of these

Passage V

(Q. Nos. 152 to 154)

For positive l, m and n, if the planes $x = ny + mz$, $y = lz + nx$, $z = mz + ly$ intersect in a straight line, then

152. l, m and n satisfy the equation

(a) $l^2 + m^2 + n^2 = 2$ (b) $l^2 + m^2 + n^2 + 2lmn = 1$
(c) $l^2 + m^2 + n^2 = 1$ (d) None of these

153. $\cos^{-1} l + \cos^{-1} m + \cos^{-1} n$ is equal to

(a) 90° (b) 50°
(c) 180° (d) None of these

154. The equation of the straight line is $\frac{x}{a} = \frac{y}{b} = \frac{z}{c}$, where the ordered traid (a, b, c) is

(a) $\sqrt{1 - l^2}, \sqrt{1 - m^2}, \sqrt{1 - n^2}$
(b) l, m and n
(c) $\frac{1}{\sqrt{1 - l^2}}, \frac{m}{\sqrt{1 - m^2}}$ and $\frac{n}{\sqrt{1 - n^2}}$
(d) None of the above

Passage VI

(Q. Nos. 155 to 157)

If $\mathbf{a} = 6\hat{\mathbf{i}} + 7\hat{\mathbf{j}} + 7\hat{\mathbf{k}}$, $\mathbf{b} = 3\hat{\mathbf{i}} + 2\hat{\mathbf{j}} - 2\hat{\mathbf{k}}$, $P(1, 2, 3)$

155. The position vector of L, the foot of the perpendicular from P on the line $\mathbf{r} = \mathbf{a} + \lambda\mathbf{b}$ is

(a) $6\hat{\mathbf{i}} + 7\hat{\mathbf{j}} + 7\hat{\mathbf{k}}$ (b) $3\hat{\mathbf{i}} + 2\hat{\mathbf{j}} - 2\hat{\mathbf{k}}$
(c) $3\hat{\mathbf{i}} + 5\hat{\mathbf{j}} + 9\hat{\mathbf{k}}$ (d) $9\hat{\mathbf{i}} + 9\hat{\mathbf{j}} + 5\hat{\mathbf{k}}$

156. The image of the point P in the line $\mathbf{r} = \mathbf{a} + \lambda\mathbf{b}$ is

(a) (11, 12, 11) (b) (5, 2, −7)
(c) (5, 8, 15) (d) (17, 16, 7)

157. If A is the point with position vector $\mathbf{a}$ then area of the triangle ΔPLA is sq. units is equal to

(a) $3\sqrt{6}$ (b) $\frac{7\sqrt{17}}{2}$

(c) $\sqrt{17}$ (d) $\frac{7}{2}$

Passage VII

(Q. Nos. 158 to 160)

$A(-2, 2, 3)$ and $B(13, -3, 13)$ and L is a line through A.

158. A point P moves in the space such that $3PA = 2PB$, then the locus of P is

(a) $x^2 + y^2 + z^2 + 28x - 12y + 10z - 247 = 0$
(b) $x^2 + y^2 + z^2 - 28x + 12y + 10z - 247 = 0$
(c) $x^2 + y^2 + z^2 + 28x - 12y - 10z + 247 = 0$
(d) $x^2 + y^2 + z^2 - 28x + 12y - 10z + 247 = 0$

159. Coordinates of the point P which divides the join of A and B in the ratio 2 : 3 internally are

(a) $\left(\frac{33}{5}, -\frac{2}{5}, 9\right)$ (b) (4, 0, 7)

(c) $\left(\frac{32}{5}, -\frac{12}{5}, \frac{17}{5}\right)$ (d) (20, 0, 35)

160. Equation of a line L, perpendicular to the line AB is

(a) $\frac{x+2}{15} = \frac{y-2}{-5} = \frac{z-3}{10}$
(b) $\frac{x-2}{3} = \frac{y+2}{13} = \frac{z+3}{2}$
(c) $\frac{x+2}{3} = \frac{y-2}{13} = \frac{z-3}{2}$
(d) $\frac{x-2}{15} = \frac{y+2}{-5} = \frac{z+3}{10}$

Passage VIII

(Q. Nos. 161 to 163)

The vector equation of a plane is a relation satisfied by position vectors of all the points on the plane. If P is a plane and $\hat{\mathbf{n}}$ is a unit vector through origin which is perpendicular to the plane P then vector equation of the plane must be $\mathbf{r}.\hat{\mathbf{n}} = d$ where d represents perpendicular distance of plane p from origin.

161. If A is a point vector $\mathbf{a}$ then perpendicular distance of A from the plane $\mathbf{r}.\hat{\mathbf{n}} = d$ must be

(a) $|d + \mathbf{a}\hat{\mathbf{n}}|$ (b) $|d - \mathbf{a}\hat{\mathbf{n}}|$
(c) $|\mathbf{a}| - d|$ (d) $|d - \hat{\mathbf{a}}|$

162. If $\mathbf{b}$ be the foot of perpendicular from A to the plane $\mathbf{r}.\hat{\mathbf{n}} = d$ then $\mathbf{b}$ must be

(a) $\mathbf{a} + (d - \mathbf{a}.\hat{\mathbf{n}})\,\hat{\mathbf{n}}$ (b) $\mathbf{a} - (d - \mathbf{a}.\hat{\mathbf{n}})\,\hat{\mathbf{n}}$
(c) $\mathbf{a} + \mathbf{a}.\hat{\mathbf{n}}$ (d) $\mathbf{a} - \mathbf{a}.\hat{\mathbf{n}}$

163. The position vector of the image of the point $\mathbf{a}$ in the plane $\mathbf{r}.\hat{\mathbf{n}} = d$ must be $(d \neq 0)$

(a) $-\mathbf{a}.\hat{\mathbf{n}}$ (b) $\mathbf{a} - 2(d - \mathbf{a}.\hat{\mathbf{n}})\,\hat{\mathbf{n}}$
(c) $\mathbf{a} + 2(d - \mathbf{a}.\hat{\mathbf{n}})\,\hat{\mathbf{n}}$ (d) $\mathbf{a} + d(-\mathbf{a}.\hat{\mathbf{n}})$

Passage IX

(Q. Nos. 164 to 166)

A circle is the locus of a point in a plane such that its distance from a fixed point in the plane is constant. Anologously, a sphere is the locus of a point in space such that its distance from a fixed point in space is constant.

The fixed point is called the centre and the constant distance is called the radius of the circle/sphere.

In anology with the equation of the circle $|z - c| = a$, *the equation of a sphere of radius a is* $|\mathbf{r} - \mathbf{c}| = a$, *where* **c** *is the position vector of the centre and* **r** *is the position vector of any point on the surface of the sphere. In Cartesian system, the equation of the sphere, with centre at* $(-g, -f, -h)$ *is* $x^2 + y^2 + z^2 + 2gx + 2fy + 2hz + c = 0$ *and its radius is* $\sqrt{f^2 + g^2 + h^2 - c}$.

164. Radius of the sphere, with $(2, -3, 4)$ and $(-5, 6, -7)$ as extremities of a diameter, is

(a) $\sqrt{\frac{251}{2}}$ (b) $\sqrt{\frac{251}{3}}$

(c) $\sqrt{\frac{251}{4}}$ (d) $\sqrt{\frac{251}{5}}$

165. The centre of the sphere $(x-4)(x+4)+(y-3)(y+3)+z^2 = 0$ is

(a) (4, 3, 0) (b) (– 4, – 3, 0)

(c) (0, 0, 0) (d) None of these

166. Equation of the sphere having centre at (3, 6, – 4) and touching the plane $\mathbf{r}\cdot(2\hat{\mathbf{i}} - 2\hat{\mathbf{j}} - \hat{\mathbf{k}}) = 10$, is $(x-3)^2 + (y-6)^2 + (z+4)^2 = k^2$, where k is equal to

(a) 3 (b) 4

(c) 6 (d) $\sqrt{17}$

Passage X

(Q. Nos. 167 to 168)

Let $A(2, 3, 5)$, $B(-1, 3, 2)$, $C(\lambda, 5, \mu)$ *are the vertices of a triangle and its median through A (i.e.) AD is equally inclined to the coordinates axes.*

On the basis of the above information answer the following

167. The value of $2\lambda - \mu$ is equal to

(a) 13 (b) 4

(c) 3 (d) None of these

168. Projection of AB on BC is

(a) $\frac{8\sqrt{3}}{11}$ (b) $\frac{-8\sqrt{3}}{11}$

(c) – 48 (d) 48

Passage XI

(Q. Nos. 169 to 171)

The line of greatest slope on an inclined plane P_1 *is that line in the plane which is perpendicular to the line of intersection of plane* P_1 *and a horizontal plane* P_2.

169. Assuming the plane $4x - 3y + 7z = 0$ to be horizontal, the direction cosines of the line of greatest slope in the plane $2x + y - 5z = 0$ are

(a) $\left(\frac{3}{\sqrt{11}}, \frac{-1}{\sqrt{11}}, \frac{1}{\sqrt{11}}\right)$ (b) $\left(\frac{3}{\sqrt{11}}, \frac{1}{\sqrt{11}}, \frac{-1}{\sqrt{11}}\right)$

(c) $\left(\frac{-3}{\sqrt{11}}, \frac{1}{\sqrt{11}}, \frac{1}{\sqrt{11}}\right)$ (d) $\left(\frac{1}{\sqrt{11}}, \frac{3}{\sqrt{11}}, \frac{-1}{\sqrt{11}}\right)$

170. The equation of a line of greatest slope can be

(a) $\frac{x}{3} = \frac{y}{1} = \frac{z}{-1}$ (b) $\frac{x}{3} = \frac{y}{-1} = \frac{z}{1}$

(c) $\frac{x}{-3} = \frac{y}{1} = \frac{z}{1}$ (d) $\frac{x}{1} = \frac{y}{3} = \frac{z}{-1}$

171. The coordinates of a point on the plane $2x + y - 5z = 0$ is $\sqrt{11}$ units away from the line of intersection of the given two planes are

(a) (3, 1, – 1) (b) (– 3, 1, 1)

(c) (3, – 1, 1) (d) (1, 3, – 1)

Passage XII

(Q. Nos. 172 to 174)

Given four points $A(2, 1, 0)$, $B(1, 0, 1)$, $C(3, 0, 1)$ *and* $D(0, 0, 2)$. *Point D lies on a line L orthogonal to the plane determined by the points A, B and C.*

172. The equation of the plane ABC is

(a) $x + y + z - 3 = 0$ (b) $y + z - 1 = 0$

(c) $x + z - 1 = 0$ (d) $2y + z - 1 = 0$

173. The equation of the line L is

(a) $\mathbf{r} = 2\hat{\mathbf{k}} + \lambda(\hat{\mathbf{i}} + \hat{\mathbf{k}})$

(b) $\mathbf{r} = 2\hat{\mathbf{k}} + \lambda(2\hat{\mathbf{j}} + \hat{\mathbf{k}})$

(c) $\mathbf{r} = 2\hat{\mathbf{k}} + \lambda(\hat{\mathbf{j}} + \hat{\mathbf{k}})$

(d) None of the above

174. The perpendicular distance of D from the plane ABC is

(a) $\sqrt{2}$ (b) 1/2

(c) 2 (d) $1/\sqrt{2}$

Three Dimensional Coordinate System Exercise 5 : Matching Type Questions

175. Consider the following four pairs of line in Column I and match them with one or more entries in Column II.

Column I	Column II
(A) $L_1: x = 1 + t, y = t, z = 2 - 5t$ $L_2: r = (2, 1, -3) + \lambda(2, 2, -10)$	(p) non-coplanar lines
(B) $L_1: \frac{x-1}{2} = \frac{y-3}{2} = \frac{z-2}{-1}$ $L_2: \frac{x-2}{1} = \frac{y-6}{-1} = \frac{z+2}{-1}$	(q) lines lie in a unique plane
(C) $L_1: x = -6t, y = 1 + 9t, z = -3t$ $L_2: x = 1 + 2s, y = 4 - 3s, z = s$	(r) infinite planes containing both the lines
(D) $L_1: \frac{x}{1} = \frac{y-1}{2} = \frac{z-2}{3}$ $L_2: \frac{x-3}{-4} = \frac{y-2}{-3} = \frac{z-1}{2}$	(s) lines are not intersecting at a unique point

176. $P(0, 3, -2)$, $Q(3, 7, -1)$ and $R(1, -3, -1)$ are 3 given points. Let L_1 be the line passing through P and Q and L_2 be the line through R and parallel to the vector $\mathbf{V} = \hat{\mathbf{i}} + \hat{\mathbf{k}}$.

Column I	Column II
(A) Perpendicular distance of P from L_1	(p) $7\sqrt{3}$
(B) Shortest distance between L_1 and L_2	(q) 2
(C) Area of the ΔPQR	(r) 6
(D) Distance from (0, 0, 0) to the plane PQR	(s) $\frac{19}{\sqrt{147}}$

177. Match the statements of Column I with values of Column II.

Column I	Column II
(A) If the line $\frac{x-1}{1} = \frac{y+1}{-2} = \frac{z+1}{\lambda}$ lies in the plane $3x - 2y + 5z = 0$, then λ is equal to	(p) $\sin^{-1}\sqrt{\frac{6}{25}}$
(B) If $(3, \lambda, \mu)$ is a point on the line $2x + y + z - 3 = 0 = x - 2y + z - 1$, then $\lambda + \mu$ is equal to	(q) $-\frac{7}{5}$
(C) The angle between the line $x = y = z$ and the plane $4x - 3y + 5z = 2$ is	(r) -3
(D) The angle between the planes $x + y + z = 0$ and $3x - 4y + 5z = 0$	(s) $\cos^{-1}\sqrt{\frac{8}{75}}$

178. Consider the lines given by $L_1 : x + 3y - 5 = 0$, $L_2 : 3x - ky - 1 = 0$ and $L_3 : 5x + 2y - 12 = 0$.

Match the statement of Column I with values of Column II.

Column I	Column II
(A) L_1, L_2 and L_3 are concurrent, if	(p) $k = -9, -\frac{6}{5}, 5$
(B) One of L_1, L_2 and L_3 is parallel to atleast one of the other two, if	(q) $k = -\frac{6}{5}, -9$
(C) L_1, L_2 and L_3 form a triangle, if	(r) $k = \frac{5}{6}$
(D) L_1, L_2 and L_3 do not form a triangle, if	(s) $k = 5$
	(t) $k = 0$

179. A variable plane cuts the x, y and z-axes at the points, A, B and C, respectively such that the volume of the tetrahedron $OABC$ remain constant equal to 32 cu units and O is the origin of the coordinate system.

Column I	Column II
(A) The locus of the centroid of the tetrahedron is	(p) $xyz = 24$
(B) The locus of the point equidistant from O, A, B and C is	(q) $(x^2 + y^2 + z^2) = 192\,xyz$
(C) The locus of the foot of perpendicular from origin to the plane is	(r) $xyz = 3$
(D) If PA, PB and PC are mutually perpendicular, then the locus of P is	(s) $(x^2 + y^2 + z^2)^3 = 1536\,xyz$

180. Match the statements of Column I with values of Column II.

Column I	Column II
(A) The area of the triangle whose vertices are (0, 0, 0) (3, 4, 7) and (5, 2, 6) is	(p) 0
(B) The smallest radius of the sphere passing through (1, 0, 0), (0, 1, 0) and (0, 0, 1) is	(q) $\frac{70}{3}$
(C) The value(s) of λ for which the triangle with vertices $A(6, 10, 10)$ $B(1, 0, -5)$ and $C(6, -10, \lambda)$ will be a right angled triangle (right angled at A) is/are	(r) $\sqrt{\frac{2}{3}}$
(D) d is the perpendicular distance from (1, 3, 4) to $\frac{x-1}{-1} = \frac{y-1}{1} = \frac{z}{1}$, then value of $\frac{d}{2\sqrt{3}}$	(s) $\frac{3}{2}\sqrt{65}$

181. Match the statements of Column I with values of Column II. Consider the cube

Column I	Column II
(A) Angle between any two solid diagonal	(p) $\cos^{-1}\dfrac{2}{\sqrt{6}}$
(B) Angle between a solid diagonal and a plane	(q) $\cos^{-1}\left(+\dfrac{1}{2}\right)$
(C) Angle between plane diagonals of adjacent faces	(r) $\cos^{-1}\dfrac{1}{3}$
(D) The values of $\lvert\mathbf{a}\times\mathbf{b}\rvert$	(s) $\dfrac{1}{2}$

Three Dimensional Coordinate System Exercise 6 :
Single Integer Answer Type Questions

182. In a tetrahedron $OABC$, if $\mathbf{OA}=\hat{\mathbf{i}}$, $\mathbf{OB}=\hat{\mathbf{i}}+\hat{\mathbf{j}}$ and $\mathbf{OC}=\hat{\mathbf{i}}+2\hat{\mathbf{j}}+\hat{\mathbf{k}}$, if shortest distance between edges OA and BC is m, then $\sqrt{2}m$ is equal to ... (Where O is the origin)

183. A parallelopiped is formed by planes drawn through the points (2, 4, 5) and (5, 9, 7) parallel to the coordinate planes. The length of the diagonal of the parallelopiped is ....

184. If the perpendicular distance of the point (6, 5, 8) from the Y-axis is 5λ units, then λ is equal to ....

185. If the shortest distance between the lines $\dfrac{x-3}{3}=\dfrac{y-8}{-1}=\dfrac{z-3}{1}$ and $\dfrac{x+3}{-3}=\dfrac{y+7}{2}=\dfrac{z-6}{2}$ is $\lambda\sqrt{30}$ units, then the value of λ is ....

186. If the planes $x-cy-bz=0$, $cx-y+az=0$ and $bx+ay-z=0$ pass through a line then the values of $a^2+b^2+c^2+2abc$ is ....

187. If xz-plane divide the join of point (2, 3, 4) and (1, –1, 5) in the ratio $\lambda:1$, then the integer λ should be equal to

188. If the triangle ABC whose vertices are $A(-1,1,1)$, $B(1,-1,1)$ and $C(1,1,-1)$ is projected on xy-plane, then the area of the projected triangles is ....

189. The equation of a plane which bisects the line joining (1, 5, 7) and (–3, 1, –1) is $x+y+2z=\lambda$, then λ must be ....

190. The shortest distance between origin and a point on the space curve $x=2\sin t$, $y=2\cos t$, $z=3t$ is ....

191. The plane $2x-2y+z+12=0$ touches the surface $x^2+y^2+z^2-2x-4y+2z-3=0$ only at point $(-1,\lambda,-2)$. The value of λ must be

192. If the centroid of the tetrahedron $OABC$ where A, B, C are the points $(a, 2, 3)$, $(1, b, 2)$ and $(2, 1, c)$ be (1, 2, 3), then the point (a, b, c) is at distance $5\sqrt{\lambda}$ from origin, then λ must be equal to ....

193. If the circumcentre of the triangle whose vertices are $(3,2,-5)$, $(-3,8,-5)$ and $(-3,2,1)$ is $(-1,\lambda,-3)$ the integer λ must be equal to ....

194. If $\overline{P_1P_2}$ is perpendicular to $\overline{P_2P_3}$, then the value of k is, where $P_1(k,1,-1)$, $P_2(2k,0,2)$ and $P_3(2+2k,k,1)$ is ....

195. Let the equation of the plane containing line $x-y-z-4=0=x+y+2z-4$ and parallel to the line of intersection of the planes $2x+3y+z=1$ and $x+3y+2z=2$ be $x+Ay+Bz+C=0$. Then the values of $\lvert A+B+C-4\rvert$ is ....

196. Let $P(a,b,c)$ be any point on the plane $3x+2y+z=7$, then find the least value of $2(a^2+b^2+c^2)$.

197. The plane denoted by $P_1: 4x+7y+4z+81=0$ is rotated through a right angle about its line of intersection with the plane $P_2: 5x+3y+10z=25$. If the plane in its new position be denoted by P, and the distance of this plane from the origin is d, then the value of $\left[\dfrac{k}{2}\right]$ (where [.] represents greatest integer less than or equal to k) is ......

198. The distance of the point $P(-2,3,-4)$ from the line $\dfrac{x+2}{3}=\dfrac{2y+3}{4}=\dfrac{3z+4}{5}$ measured parallel to the plane $4x+12y-3z+1=0$ is d, then find the value of $(2d-8)$. is ....

199. The position vectors of the four angular points of a tetrahedron $OABC$ are (0, 0, 0), (0, 0, 2), (0, 4, 0) and (6, 0, 0), respectively. A point P inside the tetrahedron is at the same distance 'r' from the four plane faces of the tetrahedron. Then, the value of $9r$ is ....

200. Value of λ do the planes $x-y+z+1=0$, $\lambda x+3y+2z-3=0$, $3x+\lambda y+z-2=0$ form a triangular prism must be

201. If the lattice point $P(x, y, z)$; $x, y, z > 0$ and $x, y, z \in I$ with least value of z such that the 'P' lies on the planes $7x+6y+2z=272$ and $x-y+z=16$, then the value of $(x+y+z-42)$ is equal to

202. If the line $x=y=z$ intersect the line $x\sin A+y\sin B+z\sin C-2d^2=0$ $=x\sin 2A+y\sin 2B+z\sin C-d^2$, where A, B, C are the internal angles of a triangle and $\sin\frac{A}{2}\sin\frac{B}{2}\sin\frac{C}{2}=k$, then the value of $64k$ is equal to

203. The number of real values of k for which the lines $\frac{x}{1}=\frac{y-1}{k}=\frac{z}{-1}$ and $\frac{x-k}{2k}=\frac{y-k}{3k-1}=\frac{z-2}{k}$ are coplanar, is

204. Let G_1, G_2 and G_3 be the centroids of the triangular faces OBC, OCA and OAB of a tetrahedron $OABC$. If V_1 denotes the volume of tetrahedron $OABC$ and V_2 that of the parallelepiped with OG_1, OG_2 and OG_3 as three concurrent edges, then the value of $4V_1/V_2$ is (where O is the origin)

205. A variable plane which remains at a constant distance p from the origin cuts the coordinate axes in A, B, C. The locus of the centroid of the tetrahedron $OABC$ is $x^2y^2+y^2z^2+z^2x^2=\frac{k}{p^2}x^2y^2z^2$, then $\sqrt[5]{2k}$ is

206. If (l_1, m_1, n_1); (l_2, m_2, n_2) are D.C's of two lines, then $(l_1m_2-l_2m_1)^2+(m_1n_2-m_2n_1)^2$ $+(n_1l_2-n_2l_1)^2+(l_1l_2+m_1m_2+n_1n_2)^2=$

207. If the coordinates (x, y, z) of the point S which is equidistant from the points $O(0, 0, 0)$, $A(n^5, 0, 0)$ $B(0, n^4, 0)$, $C(0, 0, n)$ obey the relation $2(x+y+z)+1=0$. If $n \in Z$, then $|n|=$ ________ ($|\cdot|$ is the modulus function).

Three Dimensional Coordinate System Exercise 7 : Subjective Type Questions

208. Find the angle between the lines whose direction cosines has the relation $l+m+n=0$ and $2l^2+2m^2-n^2=0$.

209. Prove that the two lines whose direction cosines are connected by the two relations $al+bm+cn=0$ and $ul^2+vm^2+wn^2=0$ are perpendicular if

$$a^2(v+w)+b^2(w+u)+c^2(u+v)=0$$

and parallel if $\frac{a^2}{u}+\frac{b^2}{v}+\frac{c^2}{w}=0$.

210. Find the point on the line $\frac{x+2}{3}=\frac{y+1}{2}=\frac{z-3}{2}$ at a distance of $3\sqrt{2}$ from the point $(1, 2, 3)$.

211. A line passes through $(2, -1, 3)$ and is perpendicular to the lines $\mathbf{r}\cdot(\hat{\mathbf{i}}+\hat{\mathbf{j}}-\hat{\mathbf{k}})+\lambda(2\hat{\mathbf{i}}-2\hat{\mathbf{j}}+\hat{\mathbf{k}})$ and $\mathbf{r}=(2\hat{\mathbf{i}}-\hat{\mathbf{j}}-3\hat{\mathbf{k}})+\mu(\hat{\mathbf{i}}+2\hat{\mathbf{j}}+2\hat{\mathbf{k}})$ obtain its equation.

212. Find the equations of the two lines through the origin which intersect the line $\frac{x-3}{2}=\frac{y-3}{1}=\frac{z}{1}$ at angle of $\frac{\pi}{3}$ each.

213. Vertices B and C of ΔABC lie along the line $\frac{x+2}{2}=\frac{y-1}{1}=\frac{z-0}{4}$. Find the area of the triangle given that A has coordinates $(1, -1, 2)$ and line segment BC has length 5.

214. A point P moves on the plane $\frac{x}{a}+\frac{y}{b}+\frac{z}{c}=1$ which is fixed. The plane through P perpendicular to OP meets the axes in A, B and C. The planes through A, B, C parallel to yz, zx, xy planes intersect in Q. Prove that if the axis bc rectangular, then the locus of Q is

$$\frac{1}{x^2}+\frac{1}{y^2}+\frac{1}{z^2}=\frac{1}{ax}+\frac{1}{by}+\frac{1}{cz}$$

215. Prove that the distance of the point of intersection of the line $\frac{x-2}{3}=\frac{y+1}{4}=\frac{z-2}{12}$ and the plane $x-y+z=5$ from the point $(-1, -5, -10)$ is 13.

216. Find the equation of the plane through the intersection of the planes $x+3y+6=0$ and $3x-y-4z=0$, whose perpendicular distance from the origin is unity.

217. Find the equation of the image of the plane $x-2y+2z-3=0$ in the plane $x+y+z-1=0$.

Three Dimensional Coordinate System Exercise 8 : Questions Asked in Previous Years Exam

(i) JEE Advanced & IIT JEE

218. Consider a pyramid $OPQRS$ located in the first octant $(x \geq 0, y \geq 0, z \geq 0)$ with O as origin and OP and OR along the X-axis and the Y-axis, respectively. The base $OPQR$ of the pyramid is a square with $OP = 3$. The point S is directly above the mid-point T of diagonal OQ such that $TS = 3$. Then,

[More than One Correct Type Question, 2016 Adv.]

(a) the acute angle between OQ and OS is $\frac{\pi}{3}$

(b) the equation of the plane containing the ΔOQS is $x - y = 0$

(c) the length of the perpendicular from P to the plane containing the ΔOQS is $\frac{3}{\sqrt{2}}$

(d) the perpendicular distance from O to the straight line containing RS is $\sqrt{\frac{15}{2}}$

219. Let P be the image of the point (3, 1, 7) with respect to the plane $x - y + z = 3$. Then, the equation of the plane passing through P and containing the straight line $\frac{x}{1} = \frac{y}{2} = \frac{z}{1}$ is

[Single Option Correct Type Question, 2016 Adv.]

(a) $x + y - 3z = 0$ (b) $3x + z = 0$
(c) $x - 4y + 7z = 0$ (d) $2x - y = 0$

220. From a point $P(\lambda, \lambda, \lambda)$, perpendiculars PQ and PR are drawn respectively on the lines $y = x, z = 1$ and $y = -x, z = -1$. If P is such that $\angle QPR$ is a right angle, then the possible value(s) of λ is (are)

[Single Option Correct Type Question, 2014 Adv.]

(a) $\sqrt{2}$ (b) 1 (c) -1 (d) $-\sqrt{2}$

221. Two lines $L_1: x = 5, \frac{y}{3-\alpha} = \frac{z}{-2}$ and $L_2 : x = \alpha, \frac{y}{-1} = \frac{z}{2-\alpha}$ are coplanar. Then, α can take value(s)

[More than One Correct Type Question, 2013 Adv.]

(a) 1 (b) 2
(c) 3 (d) 4

222. A line l passing through the origin is perpendicular to the lines **[More than One Correct Type Question, 2013 Adv.]**

$l_1 : (3 + t)\hat{\mathbf{i}} + (-1 + 2t)\hat{\mathbf{j}} + (4 + 2t)\hat{\mathbf{k}}, -\infty < t < \infty$

$l_2 : (3 + 2s)\hat{\mathbf{i}} + (3 + 2s)\hat{\mathbf{j}} + (2 + s)\hat{\mathbf{k}}, -\infty < s < \infty$

Then, the coordinate(s) of the point(s) on l_2 at a distance of $\sqrt{17}$ from the point of intersection of l and l_1 is (are)

(a) $\left(\frac{7}{3}, \frac{7}{3}, \frac{5}{3}\right)$ (b) $(-1, -1, 0)$

(c) $(1, 1, 1)$ (d) $\left(\frac{7}{9}, \frac{7}{9}, \frac{8}{9}\right)$

223. Perpendicular are drawn from points on the line $\frac{x+2}{2} = \frac{y+1}{-1} = \frac{z}{3}$ to the plane $x + y + z = 3$. The feet of perpendiculars lie on the line

[Single Option Correct Type Question, 2013 Adv.]

(a) $\frac{x}{5} = \frac{y-1}{8} = \frac{z-2}{-13}$ (b) $\frac{x}{2} = \frac{y-1}{3} = \frac{z-2}{-5}$

(c) $\frac{x}{4} = \frac{y-1}{3} = \frac{z-2}{-7}$ (d) $\frac{x}{2} = \frac{y-1}{-7} = \frac{z-2}{5}$

224. Consider the lines $L_1 : \frac{x-1}{2} = \frac{y}{-1} = \frac{z+3}{1}$, $L_2 : \frac{x-4}{1} = \frac{y+3}{1} = \frac{z+3}{2}$ and the planes $P_1 : 7x + y + 2z = 3, P_2 : 3x + 5y - 6z = 4$. Let $ax + by + cz = d$ the equation of the plane passing through the point of intersection of lines L_1 and L_2 and perpendicular to planes P_1 and P_2.

Match List I with List II and select the correct answer using the code given below the lists.

[Single Option Correct Type Question, 2013 Adv.]

	List I		List II
P.	$a =$	1.	13
Q.	$b =$	2.	-3
R.	$c =$	3.	1
S.	$d =$	4.	-2

Codes

	P	Q	R	S		P	Q	R	S
(a)	3	2	4	1	(b)	1	3	4	2
(c)	3	2	1	4	(d)	2	4	1	3

225. If the straight lines $\frac{x-1}{2} = \frac{y+1}{k} = \frac{z}{2}$ and $\frac{x+1}{5} = \frac{y+1}{2} = \frac{z}{k}$ are coplanar, then the plane(s) containing these two lines is/are

[More than One Correct Type Question, IIT-JEE 2012]

(a) $y + 2z = -1$ (b) $y + z = -1$
(c) $y - z = -1$ (d) $y - 2z = -1$

226. If the distance between the plane $Ax - 2y + z = d$ and the plane containing the lines $\frac{x-1}{2} = \frac{y-2}{3} = \frac{z-3}{4}$ and $\frac{x-2}{3} = \frac{y-3}{4} = \frac{z-4}{5}$ is $\sqrt{6}$, then $|d|$ is equal to....

[Single Option Correct Type Question, IIT-JEE 2010]

Passage

(Q. Nos. 227-229)

Read the following passage and answer the questions.
Consider the lines

$$L_1 : \frac{x+1}{3} = \frac{y+2}{1} = \frac{z+1}{2}$$

$$L_2 : \frac{x-2}{1} = \frac{y+2}{2} = \frac{z-3}{3}$$

[Passage Type Question, IIT-JEE 2008]

227. The distance of the point (1, 1, 1) from the plane passing through the point (−1, −2, −1) and whose normal is perpendicular to both the lines L_1 and L_2, is

(a) $2/\sqrt{75}$ unit (b) $7/\sqrt{75}$ unit
(c) $13/\sqrt{75}$ units (d) $23/\sqrt{75}$ units

228. The shortest distance between L_1 and L_2 is

(a) 0 unit (b) $17/\sqrt{3}$ units
(c) $41/5\sqrt{3}$ units (d) $17/5\sqrt{3}$ units

229. The unit vector perpendicular to both L_1 and L_2 is

(a) $\frac{-\hat{i}+7\hat{j}+7\hat{k}}{\sqrt{99}}$ (b) $\frac{-\hat{i}-7\hat{j}+5\hat{k}}{5\sqrt{3}}$
(c) $\frac{-\hat{i}+7\hat{j}+5\hat{k}}{5\sqrt{3}}$ (d) $\frac{7\hat{i}-7\hat{j}-\hat{k}}{\sqrt{99}}$

■ **Directions** (Q. Nos. 230-231) For the following questions, choose the correct answer from the codes (a), (b), (c) and (d) defined as follows.

(a) Statement I is true, Statement II is also true; Statement II is the correct explanation of Statement I
(b) Statement I is true, Statement II is also true; Statement II is not the correct explanation of Statement I
(c) Statement I is true; Statement II is false
(d) Statement I is false; Statement II is true

230. Consider three planes

$$P_1 : x - y + z = 1,\ P_2 : x + y - z = -1$$

and $P_3 : x - 3y + 3z = 2$

Let L_1, L_2, L_3 be the lines of intersection of the planes P_2 and P_3, P_3 and P_1, P_1 and P_2, respectively.

Statement I Atleast two of the lines L_1, L_2 and L_3 are non-parallel.

Statement II The three planes do not have a common point. **[Assertion and Reason Type Question, IIT-JEE 2008]**

231. Consider the planes

$$3x - 6y - 2z = 15 \text{ and } 2x + y - 2z = 5.$$

Statement I The parametric equations of the line of intersection of the given planes are $x = 3 + 14t$, $y = 1 + 2t$, $z = 15t$.

Statement II The vectors $14\hat{i} + 2\hat{j} + 15\hat{k}$ is parallel to the line of intersection of the given planes.

[Assertion and Reason Type Question, IIT-JEE 2007]

232. Consider the following linear equations

$ax + by + cz = 0,\ bx + cy + az = 0,\ cx + ay + bz = 0$

[Matching Type Question, IIT-JEE 2007]

	Column I		Column II
A.	$a+b+c \neq 0$ and $a^2+b^2+c^2 = ab+bc+ca$	p.	The equations represent planes meeting only at a single point
B.	$a+b+c = 0$ and $a^2+b^2+c^2 \neq ab+bc+ca$	q.	The equations represent the line $x = y = z$
C.	$a+b+c \neq 0$ and $a^2+b^2+c^2 \neq ab+bc+ca$	r.	The equations represent identical planes
D.	$a+b+c = 0$ and $a^2+b^2+c^2 = ab+bc+ca$	r.	The equations represent the whole of the three-dimensional space

(ii) JEE Main and AIEEE

233. If the image of the point $P(1, -2, 3)$ in the plane $2x + 3y - 4z + 22 = 0$ measured parallel to the line $\frac{x}{1} = \frac{y}{4} = \frac{z}{5}$ is Q, then PQ is equal to **[2017 JEE Main]**

(a) $3\sqrt{5}$ (b) $2\sqrt{42}$
(c) $\sqrt{42}$ (d) $6\sqrt{5}$

234. The distance of the point (1, 3, −7) from the plane passing through the point (1, −1, −1) having normal perpendicular to both the lines $\frac{x-1}{1} = \frac{y+2}{-2} = \frac{z-4}{3}$ and $\frac{x-2}{2} = \frac{y+1}{-1} = \frac{z+7}{-1}$, is

(a) $\frac{20}{\sqrt{74}}$ units (b) $\frac{10}{\sqrt{83}}$ units **[2017 JEE Main]**
(c) $\frac{5}{\sqrt{83}}$ units (d) $\frac{10}{\sqrt{74}}$ units

235. The distance of the point (1, −5, 9) from the plane $x - y + z = 5$ measured along the line $x = y = z$ is

(a) $3\sqrt{10}$ (b) $10\sqrt{3}$ **[2016 JEE Main]**
(c) $\frac{10}{\sqrt{3}}$ (d) $\frac{20}{3}$

236. If the line, $\frac{x-3}{2} = \frac{y+2}{-1} = \frac{z+4}{3}$ lies in the plane, $lx + my - z = 9$, then $l^2 + m^2$ is equal to **[2016 JEE Main]**

(a) 26 (b) 18 (c) 5 (d) 2

237. The distance of the point (1, 0, 2) from the point of intersection of the line $\frac{x-2}{3} = \frac{y+1}{4} = \frac{z-2}{12}$ and the plane $x - y + z = 16$, is **[2015 JEE Main]**

(a) $2\sqrt{14}$ (b) 8
(c) $3\sqrt{21}$ (d) 13

238. The equation of the plane containing the line $2x - 5y + z = 3$, $x + y + 4z = 5$ and parallel to the plane $x + 3y + 6z = 1$, is **[2015 JEE Main]**

(a) $2x + 6y + 12z = 13$ (b) $x + 3y + 6z = -7$
(c) $x + 3y + 6z = 7$ (d) $2x + 6y + 12z = -13$

239. The angle between the lines whose direction cosines satisfy the equations $l + m + n = 0$ and $l^2 = m^2 + n^2$ is **[2014 JEE Main]**

(a) $\frac{\pi}{3}$ (b) $\frac{\pi}{4}$
(c) $\frac{\pi}{6}$ (d) $\frac{\pi}{2}$

240. The image of the line $\frac{x-1}{3} = \frac{y-3}{1} = \frac{z-4}{-5}$ in the plane $2x - y + z + 3 = 0$ is the line **[2014 JEE Main]**

(a) $\frac{x+3}{3} = \frac{y-5}{1} = \frac{z-2}{-5}$ (b) $\frac{x+3}{-3} = \frac{y-5}{-1} = \frac{z+2}{5}$
(c) $\frac{x-3}{3} = \frac{y+5}{1} = \frac{z-2}{-5}$ (d) $\frac{x-3}{-3} = \frac{y+5}{-1} = \frac{z-2}{5}$

241. Distance between two parallel planes $2x + y + 2z = 8$ and $4x + 2y + 4z + 5 = 0$ is **[2013 JEE Main]**

(a) $\frac{3}{2}$ (b) $\frac{5}{2}$
(c) $\frac{7}{2}$ (d) $\frac{9}{2}$

242. If the lines

$$\frac{x-2}{1} = \frac{y-3}{1} = \frac{z-4}{-k} \text{ and } \frac{x-1}{k} = \frac{y-4}{2} = \frac{z-5}{1}$$

are coplanar, then k can have **[2013 JEE Main]**

(a) any value (b) exactly one value
(c) exactly two values (d) exactly three values

243. An equation of a plane parallel to the plane $x - 2y + 2z - 5 = 0$ and at a unit distance from the origin is **[AIEEE 2012]**

(a) $x - 2y + 2z - 3 = 0$ (b) $x - 2y + 2z + 1 = 0$
(c) $x - 2y + 2z - 1 = 0$ (d) $x - 2y + 2z + 5 = 0$

244. If the line $\frac{x-1}{2} = \frac{y+1}{3} = \frac{z-1}{4}$ and $\frac{x-3}{1} = \frac{y-k}{2} = \frac{z}{1}$ intersect, then k is equal to **[AIEEE 2012]**

(a) -1 (b) $\frac{2}{9}$
(c) $\frac{9}{2}$ (d) 0

245. If the angle between the line $x = \frac{y-1}{2} = \frac{z-3}{\lambda}$ and the plane $x + 2y + 3z = 4$ is $\cos^{-1}\left(\sqrt{\frac{5}{14}}\right)$, then λ equals **[AIEEE 2011]**

(a) $\frac{3}{2}$ (b) $\frac{2}{5}$
(c) $\frac{5}{3}$ (d) $\frac{2}{3}$

246. Statement I The point $A(1, 0, 7)$ is the mirror image of the point $B(1, 6, 3)$ in the line $\frac{x}{1} = \frac{y-1}{2} = \frac{z-2}{3}$.

Statement II The line $\frac{x}{1} = \frac{y-1}{2} = \frac{z-2}{3}$ bisects the line segment joining $A(1, 0, 7)$ and $B(1, 6, 3)$. **[AIEEE 2011]**

(a) Statement I is true, Statement II is true; Statement II is not a correct explanation for Statement I
(b) Statement I is true, Statement II is false
(c) Statement I is false, Statement II is true
(d) Statement I is true, Statement II is true; Statement II is a correct explanation for Statement I

247. The length of the perpendicular drawn from the point $(3, -1, 11)$ to the line $\frac{x}{2} = \frac{y-2}{3} = \frac{z-3}{4}$ is **[AIEEE 2011]**

(a) $\sqrt{66}$ (b) $\sqrt{29}$
(c) $\sqrt{33}$ (d) $\sqrt{53}$

248. The distance of the point $(1, -5, 9)$ from the plane $x - y + z = 5$ measured along a straight line $x = y = z$, is **[AIEEE 2010]**

(a) $3\sqrt{5}$ (b) $10\sqrt{3}$
(c) $5\sqrt{3}$ (d) $3\sqrt{10}$

249. A line AB in three-dimensional space makes angles 45° and 120° with the positive X-axis and the positive Y-axis, respectively. If AB makes an acute angle θ with the positive Z-axis, then θ equals **[AIEEE 2010]**

(a) 30° (b) 45°
(c) 60° (d) 75°

250. Statement I The point $A(3, 1, 6)$ is the mirror image of the point $B(1, 3, 4)$ in the plane $x - y + z = 5$.

Statement II The plane $x - y + z = 5$ bisects the line segment joining $A(3, 1, 6)$ and $B(1, 3, 4)$. **[AIEEE 2010]**

(a) Statement I is true, Statement II is true; Statement II is the correct explanation of Statement I
(b) Statement I is true, Statement II is true; Statement II is not the correct explanation of Statement I
(c) Statement I is true, Statement II is false
(d) Statement I is false, Statement II is true

251. Let the line $\frac{x-2}{3} = \frac{y-1}{-5} = \frac{z+2}{2}$ lies in the plane $x + 3y - \alpha z + \beta = 0$. Then, (α, β) equals **[AIEEE 2009]**

(a) $(6, -17)$ (b) $(-6, 7)$ (c) $(5, -15)$ (d) $(-5, 15)$

252. The projections of a vector on the three coordinate axes are $6, -3, 2$, respectively. The direction cosines of the vector are **[AIEEE 2009]**

(a) $6, -3, 2$ (b) $\frac{6}{5}, -\frac{3}{5}, \frac{2}{5}$
(c) $\frac{6}{7}, -\frac{3}{7}, \frac{2}{7}$ (d) $-\frac{6}{7}, -\frac{3}{7}, \frac{2}{7}$

253. The line passing through the points (5, 1 a), and (3, b, 1) crosses the YZ-plane at the point $\left(0, \frac{17}{2}, \frac{-13}{2}\right)$. Then, **[AIEEE 2008]**

(a) $a=8, b=2$ (b) $a=2, b=8$
(c) $a=4, b=6$ (d) $a=6, b=4$

254. If the straight lines

$$\frac{x-1}{k}=\frac{y-2}{2}=\frac{z-3}{3} \text{ and } \frac{x-2}{3}=\frac{y-3}{k}=\frac{z-1}{2}$$

intersect at a point, then the integer k is equal to **[AIEEE 2008]**

(a) -2 (b) -5
(c) 5 (d) 2

255. Let L be the line of intersection of the planes $2x+3y+z=1$ and $x+3y+2z=2$. If L makes an angle α with the positive X-axis, then $\cos\alpha$ equals **[AIEEE 2007]**

(a) $1/\sqrt{3}$ (b) $1/2$
(c) 1 (d) $1/\sqrt{2}$

256. If a line makes an angle of $\frac{\pi}{4}$ with the positive directions of each of X-axis and Y-axis, then the angle that the line makes with the positive direction of the Z-axis is **[AIEEE 2007]**

(a) $\pi/6$ (b) $\pi/3$
(c) $\pi/4$ (d) $\pi/2$

257. If (2, 3, 5) is one end of a diameter of the sphere $x^2+y^2+z^2-6x-12y-2z+20=0$, then the coordinates of the other end of the diameter are **[AIEEE 2007]**

(a) (4, 9, −3) (b) (4, −3, 3)
(c) (4, 3, 5) (d) (4, 3, −3)

258. The two lines $x=ay+b, z=cy+d$ and $x=a'y+b', z=c'y+d'$ are perpendicular to each other, if **[AIEEE 2006,2003]**

(a) $aa'+cc'=1$ (b) $\frac{a}{a'}+\frac{c}{c'}=-1$
(c) $\frac{a}{a'}+\frac{c}{c'}=1$ (d) $aa'+cc'=-1$

259. The image of the point $(-1, 3, 4)$ in the plane $x-2y=0$ is **[AIEEE 2006]**

(a) (15, 11, 4) (b) $\left(-\frac{17}{3}, -\frac{19}{3}, 1\right)$
(c) (8, 4, 4) (d) $\left(\frac{9}{5}, -\frac{13}{5}, 4\right)$

260. If the plane $2ax-3ay+4az+6=0$ passes through the mid-point of the line joining the centres of the spheres $x^2+y^2+z^2+6x-8y-2z=13$ and $x^2+y^2+z^2-10x+4y-2z=8$, then a equals **[AIEEE 2005]**

(a) 2 (b) -2
(c) 1 (d) -1

261. If the angle θ between the line $\frac{x+1}{1}=\frac{y-1}{2}=\frac{z-2}{2}$ and the plane $2x-y+\sqrt{\lambda}z+4=0$ is such that $\sin\theta=\frac{1}{3}$. The value of λ is **[AIEEE 2005]**

(a) $-\frac{4}{3}$ (b) $\frac{3}{4}$
(c) $-\frac{3}{5}$ (d) $\frac{5}{3}$

262. The angle between the lines $2x=3y=-z$ and $6x=-y=-4z$ is **[AIEEE 2005]**

(a) 30° (b) 45°
(c) 90° (d) 0°

263. The plane $x+2y-z=4$ cuts the sphere $x^2+y^2+z^2-x+z-2=0$ in a circle of radius **[AIEEE 2005]**

(a) $\sqrt{2}$ (b) 2
(c) 1 (d) 3

Answers

Exercise for Session 1

1. 8 2. $\sqrt{3}\,|k|$ 4. (1,2, 3), (3, 4, 5), (– 1, 6, – 7)
5. 5 6. (4, 5, 6) 7. 90°
8. –1 9. 0 11. $\left(\frac{2}{3}, \frac{-1}{3}, \frac{2}{3}\right)$ or $\left(\frac{-2}{3}, \frac{1}{3}, \frac{-2}{3}\right)$
12. $\sqrt{14}$

Exercise for Session 2

1. $\mathbf{r} = 3\hat{i} - \hat{j} + 3\hat{k} + \lambda(2\hat{i} - 2\hat{j} + 5\hat{k})$
2. $\mathbf{r} = 2\hat{i} - 3\hat{j} + 4\hat{k} + \lambda(3\hat{i} + 4\hat{j} - 5\hat{k})$, $\frac{x-2}{3} = \frac{y+3}{4} = \frac{z-4}{-5}$
3. $\left(\frac{13}{5}, \frac{23}{5}, 0\right)$ 4. $\cos^{-1}\left(\frac{19}{21}\right)$ 5. (– 1, – 1, – 1)
6. $\frac{1}{\sqrt{3}}$ 7. $\frac{41}{10}$
8. $\frac{x-1}{-1} = \frac{y}{2} = \frac{z-2}{-7}$ or $\frac{x-1}{1} = \frac{y}{-2} = \frac{z-2}{7}$
9. $\frac{x-1}{1} = \frac{y-2}{-2} = \frac{z+1}{1}$ 10. (5, 8, 15)

Exercise for Session 3

1. $2x - y + 3z = 9$ 2. $\pm\frac{1}{3}(2\hat{i} + 2\hat{j} + \hat{k})$
3. $4x - 3y + 2z = 3$ 4. $5x + 18y + 6z = 0$
5. $x - 5y - 2z + 6 = 0$, $3x - y + 4z - 2 = 0$
6. (– 3, 5, 2) 7. $\sin^{-1}\left(\frac{15}{7\sqrt{11}}\right)$
8. $y + z = 2$ 9. 13
10. $17x - 47y - 24z + 172 = 0$
11. $3x - y + 3z + 10 = 0$
12. $x - 2y + 2z = 0$ and $x - 2y + 2z - 6 = 0$
13. $25x + 17y + 62z = 238$ (acute angle bisector)
$x + 35y - 10z = 256$ (obtuse angle bisector)
14. $x - 8y + 4z = 7$
15. $2x + 2y + z = 9$

Exercise for Session 4

1. Centre (2, – 2, 0), Radius = $\sqrt{\frac{51}{2}}$
2. $x^2 + y^2 + z^2 - 4x + 4y - 4z + 9 = 0$, Centre (2, – 2, 2)
3. $x^2 + y^2 + z^2 - 2\sqrt{3}y - 1 = 0$
4. $9x^2 + 9y^2 + 9z^2 - 54x - 108y + 72z + 545 = 0$
5. $\lambda = \sqrt{3} \pm 3$
6. $2x^2 + 2y^2 + 2z^2 - 6x + 2y - 4z = 25$
7. (i) $\left(\frac{3}{2}, \frac{7}{2}, -2\right)$ (ii) $\frac{\sqrt{78}}{2}$ (iii) 5

Chapter Exercises

1. (b) 2. (b) 3. (b) 4. (a) 5. (a) 6. (b)
7. (b) 8. (b) 9. (b) 10. (c) 11. (b) 12. (a)
13. (c) 14. (a) 15. (d) 16. (c) 17. (a) 18. (b)
19. (b) 20. (a) 21. (c) 22. (c) 23. (d) 24. (a)
25. (a) 26. (b) 27. (d) 28. (a) 29. (d) 30. (d)
31. (a) 32. (d) 33. (b) 34. (a) 35. (a) 36. (c)
37. (b) 38. (c) 39. (c) 40. (c) 41. (d) 42. (c)
43. (c) 44. (c) 45. (c) 46. (a) 47. (b) 48. (a)
49. (d) 50. (c) 51. (c) 52. (d) 53. (b) 54. (a)
55. (d) 56. (b) 57. (b) 58. (a) 59. (c) 60. (a)
61. (b) 62. (d) 63. (a) 64. (a) 65. (b) 66. (c)
67. (a) 68. (c) 69. (a) 70. (c) 71. (b) 72. (b)
73. (a) 74. (b) 75. (c) 76. (a) 77. (d) 78. (a)
79. (d) 80. (a) 81. (b) 82. (c) 83. (a) 84. (c)
85. (c) 86. (b) 87. (c) 88. (a) 89. (c) 90. (a)
91. (d) 92. (a) 93. (c) 94. (a) 95. (a) 96. (a)
97. (b,d) 98. (a,b,c) 99. (b,c)
100. (b,e) 101. (a,c,d) 102. (a,b)
103. (b,c) 104. (a,b) 105. (a,b)
106. (c,d) 107. (b,c) 108. (a,c)
109. (b,d) 110. (a,c) 111. (a,b,c,d)
112. (c,d) 113. (a,d) 114. (a,b,c)
115. (a,b,d) 116. (a,c,d) 117. (b,c)
118. (b,d) 119. (a,c) 120. (a,b,c,d)
121. (a,b,c) 122. (a,b,c) 123. (a,b,c)
124. (a,b) 125. (d) 126. (d)
127. (a) 128. (b) 129. (b) 130. (d) 131. (a) 132. (c)
133. (a) 134. (a) 135. (b) 136. (a) 137. (a) 138. (d)
139. (d) 140. (b) 141. (b) 142. (b) 143. (d) 144. (b)
145. (b) 146. (c) 147. (a) 148. (d) 149. (d) 150. (b)
151. (c) 152. (b) 153. (c) 154. (a) 155. (c) 156. (c)
157. (b) 158. (a) 159. (b) 160. (c) 161. (b) 162. (a)
163. (c) 164. (c) 165. (c) 166. (b) 167. (b) 168. (b)
169. (a) 170. (b) 171. (c) 172. (b) 173. (c) 174. (d)
175. (A) $\rightarrow r$ (B) $\rightarrow q$, (C) $\rightarrow (q,s)$ (D) $\rightarrow (p,s)$
176. (A) $\rightarrow r$ (B) $\rightarrow q$, (C) $\rightarrow p$, (D) $\rightarrow s$
177. (A) $\rightarrow q$ (B) $\rightarrow r$, (C) $\rightarrow p$ (D) $\rightarrow s$
178. (A) $\rightarrow s$ (B) $\rightarrow q$, (C) $\rightarrow (r,t)$ (D) $\rightarrow (p,s)$
179. (A) $\rightarrow r$ (B) $\rightarrow q$, (C) $\rightarrow q$, (D) $\rightarrow (s)$
180. (A) $\rightarrow s$, (B) $\rightarrow r$, (C) $\rightarrow q$, (D) $\rightarrow (s)$
181. (A) $\rightarrow r$ (B) $\rightarrow p$, (C) $\rightarrow q$
182. (1) 183. (7) 184. (2) 185. (3) 186. (1) 187. (3)
188. (2) 189. (8) 190. (2) 191. (4) 192. (3) 193. (4)
194. (3) 195. (7) 196. (7) 197. (7) 198. (9) 199. (6)
200. (4) 201. (4) 202. (4) 203. (1) 204. (9) 205. (2)
206. (1) 207. (1)
208. $\cos^{-1}\left(\frac{-1}{3}\right)$ 210. (–2, – 1, 3) and $\left(\frac{56}{17}, \frac{43}{17}, \frac{111}{17}\right)$
211. $\mathbf{r} = (2\hat{i} - \hat{j} + 3\hat{k}) + \mu(2\hat{j} + \hat{i} - 2\hat{k})$
212. $\frac{x}{1} = \frac{y}{2} = \frac{z}{-1}$ and $\frac{x}{-1} = \frac{y}{1} = \frac{z}{-2}$
213. $\sqrt{\frac{1775}{28}}$ sq units 216. $2x + y - 2z + 3 = 0$ and $x - 2y - 2z - 3 = 0$
217. $x - 8y + 4z - 7 = 0$
218. (b, c, d) 219. (c) 220. (c) 221. (a,d) 222. (b,d)
223. (d) 224. (a) 225. (b, c) 226. (b) 227. (c) 228. (d)
229. (b) 230. (d) 231. (d)
232. (A) $\rightarrow$ (r); (B) $\rightarrow$ (q); (C) $\rightarrow$ (b); (D) $\rightarrow$ (s)
233. (b) 234. (b) 235. (b) 236. (d) 237. (d) 238. (c)
239. (a) 240. (a) 241. (c) 242. (c) 243. (a) 244. (c)
245. (d) 246. (d) 247. (d) 248. (b) 249. (c) 250. (a)
251. (b) 252. (c) 253. (d) 254. (b) 255. (a) 256. (d)
257. (a) 258. (d) 259. (d) 260. (b) 261. (d) 262. (c)
263. (c)

Solutions

1. Suppose xy-plane divides the line joining the given points in the ratio $\lambda : 1$. The coordinate of the point of division are $\left(\frac{2\lambda-1}{\lambda+1}, \frac{-5\lambda+3}{\lambda+1}, \frac{6\lambda+4}{\lambda+1}\right)$.

This point lies on xy-plane.

$$\therefore \quad \frac{6\lambda+4}{\lambda+1} = 0 \Rightarrow \lambda = -\frac{2}{3}$$

Hence, xy-plane divides externally in the ratio 2 : 3.

Aliter We know that the xy-plane divides the line segment joining $P(x_1, y_1, z_1)$ and $Q(x_2, y_2, z_2)$ in the ratio $-z_1 : z_2$. Therefore, xy-plane divides the segment joining $(-1, 3, 4)$ and $(2, -5, 6)$ in the ratio $-4 : 6$ i.e. 2 : 3 externally.

2. Suppose zx-plane divides the join of (1, 2, 3) and (4, 2, 1) in the ratio $\lambda : 1$. Then, the co-ordinates of the point of division are

$$\left(\frac{4\lambda+1}{\lambda+1}, \frac{2\lambda+2}{\lambda+1}, \frac{\lambda+3}{\lambda+1}\right)$$

This point lies on zx-plane

$$\therefore \quad y\text{-coordinate} = 0 \Rightarrow \frac{2\lambda+2}{\lambda+1} = 0 \Rightarrow \lambda = -1$$

Hence, zx-plane divides the join of (1, 2, 3) and (4, 2, 1) externally in the ratio 1 : 1.

Aliter We know that the zx-plane divides the segment joining $P(x_1, y_1, z_1)$ and $Q(x_2, y_2, z_2)$ in the ratio $-y_1 : y_2$.

$\therefore zx$-plane divides the join of (1, 2, 3) and (4, 2, 1) in the ratio $-2 : 2$ i.e. 1 : 1 externally.

3. Suppose R divides PQ in the ratio $\lambda : 1$. Then, the coordinates of R are

$$\left(\frac{5\lambda+3}{\lambda+1}, \frac{4\lambda+2}{\lambda+1}, \frac{-6\lambda-4}{\lambda+1}\right)$$

But, the coordinates of R are given as $(9, 8, -10)$.

$$\therefore \quad \frac{5\lambda+3}{\lambda+1} = 9, \quad \frac{4\lambda+2}{\lambda+1} = 8$$

$$\text{and} \quad \frac{-6\lambda-4}{\lambda+1} = -10 \;\Rightarrow\; \lambda = -\frac{3}{2}$$

Hence, R divides PQ externally in the ratio 3 : 2.

4. D divides BC in the ratio $AB : AC$ i.e. 3 : 13. Therefore, coordinates of D are

$$\left(\frac{3\times -9+13\times 5}{3+13}, \frac{3\times 6+13\times 3}{3+13}, \frac{3\times -3+13\times 2}{3+13}\right) \text{ or } \left(\frac{19}{8}, \frac{57}{16}, \frac{17}{16}\right).$$

5. Let l, m, n be the direction cosines of the given line. Then, as it makes an acute angle with x-axis. Therefore, $l > 0$. The line passes through $(6, -7, -1)$ and $(2, -3, 1)$. Therefore, its direction ratios are

$$6-2, -7+3, -1-1 \text{ or } 4, -4, -2 \text{ or } 2, -2, -1$$

Hence, direction cosines of the given line are $\frac{2}{3}, -\frac{2}{3}, -\frac{1}{3}$

6. We have, $\alpha = 45°$ and $\beta = 60°$

Suppose **OP** makes angle γ with OZ. Then,

$$\cos^2\alpha + \cos^2\beta + \cos^2\gamma = 1$$

$$\Rightarrow \quad \left(\frac{1}{\sqrt{2}}\right)^2 + \left(\frac{1}{2}\right)^2 + \cos^2\gamma = 1$$

$$\Rightarrow \quad \cos^2\gamma = \frac{1}{4} \Rightarrow \cos\gamma = \pm\frac{1}{2}$$

$$\Rightarrow \quad y = 60°, 120°$$

7. Let OA and OB be two lines with direction l_1, m_1, n_1 and l_2, m_2, n_2.

Let $OA = OB = 1$. Then, the coordinates of A and B are (l_1, m_1, n_1) and (l_2, m_2, n_2) respectively. Let OC be the bisector of $\angle AOB$. Then, C is the mid point of AB and so its coordinates are

$$\left(\frac{l_1+l_2}{2}, \frac{m_1+m_2}{2}, \frac{n_1+n_2}{2}\right)$$

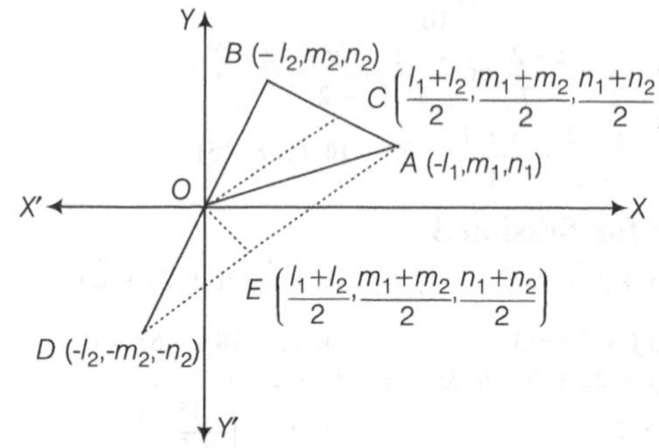

$\therefore$ Direction ratios of OC are $\frac{l_1+l_2}{2}, \frac{m_1+m_2}{2}, \frac{n_1+n_2}{2}$

$$\text{Now,} \quad OC = \sqrt{\left(\frac{l_1+l_2}{2}\right)^2 + \left(\frac{m_1+m_2}{2}\right)^2 + \left(\frac{n_1+n_2}{2}\right)^2}$$

$$OC = \frac{1}{2}\sqrt{(l_1^2+m_1^2+n_1^2)+(l_2^2+m_2^2+n_2^2)+2(l_1l_2+m_1m_2+n_1n_2)}$$

$$\Rightarrow \quad OC = \frac{1}{2}\sqrt{2+2\cos\theta} \quad [\because \cos\theta = l_1l_2+m_1m_2+n_1n_2]$$

$$\Rightarrow \quad OC = \frac{1}{2}\sqrt{2+2\cos\theta} = \cos\left(\frac{\theta}{2}\right)$$

$\therefore$ Direction cosines of **OC** are

$$\frac{l_1+l_2}{2(OC)}, \frac{m_1+m_2}{2(OC)}, \frac{n_1+n_2}{2(OC)}$$

$$\text{or,} \quad \frac{l_1+l_2}{2(\cos\frac{\theta}{2})}, \frac{m_1+m_2}{2(\cos\frac{\theta}{2})}, \frac{n_1+n_2}{(2\cos\frac{\theta}{2})}$$

8. The given line is parallel to the vector $\mathbf{n} = \hat{\mathbf{i}} - \hat{\mathbf{j}} + 2\hat{\mathbf{k}}$. The required plane passes through the point (2, 3, 1) i.e., $2\hat{\mathbf{i}} + 3\hat{\mathbf{j}} + \hat{\mathbf{k}}$ and is perpendicular to the vector $\mathbf{n} = \hat{\mathbf{i}} - \hat{\mathbf{j}} + 2\hat{\mathbf{k}}$. So, its equation is

$$\{\mathbf{r} - (2\hat{\mathbf{i}} + 3\hat{\mathbf{j}} + \hat{\mathbf{k}})\} \cdot (\hat{\mathbf{i}} - \hat{\mathbf{j}} + 2\hat{\mathbf{k}}) = 00000$$

$$\Rightarrow \quad \mathbf{r} \cdot (\hat{\mathbf{i}} - \hat{\mathbf{j}} + 2\hat{\mathbf{k}}) = 1$$

9. Le the position vectors of the given points A and B be $\mathbf{a}$ and $\mathbf{b}$ respectively and that of the variable point P be $\mathbf{r}$. It is given that

$$PA^2 - PB^2 = k \quad \text{(constant)}$$

$$\Rightarrow \quad |\mathbf{AP}|^2 - |\mathbf{BP}|^2 = k$$

$$\Rightarrow \quad |\mathbf{r}-\mathbf{a}|^2 - |\mathbf{r}-\mathbf{b}|^2 = k$$

$$\Rightarrow \{|\mathbf{r}|^2 + |\mathbf{a}|^2 - 2\mathbf{r}\cdot\mathbf{a}\} - \{|\mathbf{r}|^2 + |\mathbf{b}|^2 - 2\mathbf{r}\cdot\mathbf{b}\} = k$$

$$\Rightarrow \quad 2\mathbf{r}\cdot(\mathbf{b}-\mathbf{a}) = k + |\mathbf{b}|^2 - |\mathbf{a}|^2$$

$$\Rightarrow \quad \mathbf{r}\cdot(\mathbf{b}-\mathbf{a}) = \lambda, \text{ where } \lambda = \frac{1}{2}\{k + |\mathbf{b}|^2 - |\mathbf{a}|^2\}$$

Clearly, it represents a plane.

10. The position vectors of two given points are $\mathbf{a} = \hat{\mathbf{i}} - \hat{\mathbf{j}} + 3\hat{\mathbf{k}}$ and $\mathbf{b} = 3\hat{\mathbf{i}} + 3\hat{\mathbf{j}} + 3\hat{\mathbf{k}}$ and the equation of the given plane is

$$\mathbf{r} = (5\hat{\mathbf{i}} + 2\hat{\mathbf{j}} - 7\hat{\mathbf{k}}) + 9 = 0$$

or $\quad \mathbf{r}.\mathbf{n} + d = 0$

We have $\quad \mathbf{a}.\mathbf{n} + d = (\hat{\mathbf{i}} - \hat{\mathbf{j}} + 3\hat{\mathbf{k}})\cdot(5\hat{\mathbf{i}} + 2\hat{\mathbf{j}} - 7\hat{\mathbf{k}}) + 9$

$$= 5 - 2 - 21 + 9 < 0$$

and $\quad \mathbf{b}.\mathbf{n} + d = (3\hat{\mathbf{i}} + 3\hat{\mathbf{j}} + 3\hat{\mathbf{k}})\cdot(5\hat{\mathbf{i}} + 2\hat{\mathbf{j}} - 7\hat{\mathbf{k}}) + 9$

$$= 15 + 6 - 21 + 9 > 0$$

So, the points $\mathbf{a}$ and $\mathbf{b}$ are on the opposite sides of the plane.

11. The equation of a plane parallel to the plane

$$\mathbf{r}\cdot(4\hat{\mathbf{i}} - 12\hat{\mathbf{j}} - 3\hat{\mathbf{k}}) - 7 = 0 \text{ is,}$$

$$\mathbf{r}\cdot(4\hat{\mathbf{i}} - 12\hat{\mathbf{j}} - 3\hat{\mathbf{k}}) + \lambda = 0$$

This passes through $2\hat{\mathbf{i}} - \hat{\mathbf{j}} - 4\hat{\mathbf{k}}$

$$\therefore \quad (2\hat{\mathbf{i}} - \hat{\mathbf{j}} - 4\hat{\mathbf{k}})\cdot(4\hat{\mathbf{i}} - 12\hat{\mathbf{j}} - 3\hat{\mathbf{k}}) + \lambda = 0$$

$$\Rightarrow \quad 8 + 12 + 12 + \lambda = 0$$

$$\Rightarrow \quad \lambda = -32$$

So, the required plane is $\mathbf{r}\cdot(4\hat{\mathbf{i}} - 12\hat{\mathbf{j}} - 3\hat{\mathbf{k}}) - 32 = 0$

12. Equation of the plane containing L_1,

$$A(x-2) + B(y-1) + C(z+1) = 0$$

where $\quad A + 2C = 0;\ A + B - C = 0$

$$\Rightarrow \quad A = -2C,\ B = 3C,\ C = C$$

$\Rightarrow$ Plane is $-2(x-2) + 3(y-1) + z + 1 = 0$

or $\quad 2x - 3y - z - 2 = 0$

Hence, $\quad p = \left|\dfrac{-2}{\sqrt{14}}\right| = \sqrt{\dfrac{2}{7}}$

13. (1, 2, 3) satisfies the plane $x - 2y + z = 0$ and also

$$(\hat{\mathbf{i}} + 2\hat{\mathbf{j}} + 3\hat{\mathbf{k}})\cdot(\hat{\mathbf{i}} - 2\hat{\mathbf{j}} + \hat{\mathbf{k}}) = 0$$

Since the lines $\dfrac{x-1}{1} = \dfrac{y-2}{2} = \dfrac{z-3}{3}$ and $\dfrac{x}{1} = \dfrac{y}{2} = \dfrac{z}{3}$ both satisfy $(0, 0, 0)$ and (1, 2, 3), both are same. Given line is obviously parallel to the plane $x - 2y + z = 6$

14. Vector $(3\hat{\mathbf{i}} - 2\hat{\mathbf{j}} + \hat{\mathbf{k}}) \times (4\hat{\mathbf{i}} - 3\hat{\mathbf{j}} + 4\hat{\mathbf{k}})$ is perpendicular to $2\hat{\mathbf{i}} - \hat{\mathbf{j}} + m\hat{\mathbf{k}}$

$$\Rightarrow \quad \begin{vmatrix} 3 & -2 & 1 \\ 4 & -3 & 4 \\ 2 & -1 & m \end{vmatrix} = 0 \Rightarrow m = -2$$

15. Let $A(1, 0, -1), B(-1, 2, 2)$

Direction ratios of segment AB are $< 2, -2, -3 >$.

$$\cos\theta = \frac{|2\times1 + 3(-2) - 5(-3)|}{\sqrt{1+9+25}\sqrt{4+4+9}} = \frac{11}{\sqrt{17}\sqrt{35}} = \frac{11}{\sqrt{595}}$$

Length of projection $= (AB)\sin\theta$

$$= \sqrt{(2)^2 + (2)^2 + (3)^2} \times \sqrt{1 - \frac{121}{595}}$$

$$= \sqrt{17}\frac{\sqrt{474}}{\sqrt{17}\sqrt{35}} = \sqrt{\frac{474}{35}} \text{ units}$$

16. Let the point be A, B, C and D

The number of planes which have three points on one side and the fourth point on other side is 4. The number of planes which have two points on each side of the plane is 3.

$\Rightarrow$ Number of planes is 7.

17. Point A is $(a, b, c) \Rightarrow$ Points P, Q, R are $(a, b, -c), (-a, b, c)$ and $(a, -b, c)$ respectively.

$\Rightarrow$ Centroid of triangle PQR is $\left(\dfrac{a}{3}, \dfrac{b}{3}, \dfrac{c}{3}\right) \Rightarrow G \equiv \left(\dfrac{a}{3}, \dfrac{b}{3}, \dfrac{c}{3}\right)$

$\Rightarrow A, O, G$ are collinear $\Rightarrow$ area of triangle AOG is zero.

18. Let the equation of the plane be $\dfrac{x}{a} + \dfrac{y}{b} + \dfrac{z}{c} = 1$

$$\Rightarrow \quad \frac{1}{a} + \frac{1}{b} + \frac{1}{c} = 1$$

$\Rightarrow$ Volume of tetrahedron $OABC = V = \dfrac{1}{6}(a\,b\,c)$

Now, $\quad (abc)^{1/3} \geq \dfrac{3}{\dfrac{1}{a} + \dfrac{1}{b} + \dfrac{1}{c}} \geq 3$ (G.M $\geq$ H.M)

$$\Rightarrow \quad a\,b\,c \geq 27 \Rightarrow V \geq \frac{9}{2}$$

19.

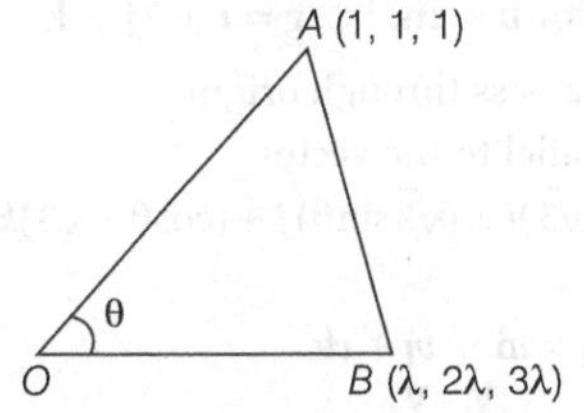

Let any point of second line be $(\lambda, 2\lambda, 3\lambda)$

$$\cos\theta = \frac{6}{\sqrt{42}},\ \sin\theta = \frac{\sqrt{6}}{\sqrt{42}}$$

$$\Delta_{OAB} = \frac{1}{2}(OA)OB\sin\theta$$

$$= \frac{1}{2}\sqrt{3}\,\lambda\sqrt{14} \times \frac{\sqrt{6}}{\sqrt{42}} = \sqrt{6} \Rightarrow \lambda = 2$$

So, B is (2, 4, 6).

20. Equation of line $x + 2y + z - 1 + \lambda(-x + y - 2z - 2) = 0 \quad \text{...(i)}$

$$x + y - 2 + \mu(x + z - 2) = 0 \quad \text{...(ii)}$$

(0, 0, 1) lies on it

$$\Rightarrow \quad \lambda = 0, \mu = -2$$

For point of intersection, $z = 0$ and solve (i) and (ii).

21. The planes are $\frac{x}{a}+\frac{y}{b}+\frac{z}{c}=1$ and $\frac{x}{a'}+\frac{y}{b'}+\frac{z}{c'}=1$

Since the perpendicular distance of the origin on the planes is same, therefore

$$\left|\frac{-1}{\sqrt{\frac{1}{a^2}+\frac{1}{b^2}+\frac{1}{c^2}}}\right|=\left|\frac{-1}{\sqrt{\frac{1}{a'^2}+\frac{1}{b'^2}+\frac{1}{c'^2}}}\right|$$

$$\Rightarrow \quad \frac{1}{a^2}+\frac{1}{b^2}+\frac{1}{c^2}-\frac{1}{a'^2}-\frac{1}{b'^2}-\frac{1}{c'^2}=0$$

22. Given one vertex $A(7, 2, 4)$ and line

$$\frac{x+6}{5}=\frac{y+10}{3}=\frac{z+14}{8}$$

General point on above line, $B \equiv (5\lambda-6, 3\lambda-10, 8\lambda-14)$

Direction ratios of line AB are $<5\lambda-13, 3\lambda-12, 8\lambda-18>$

Direction ratios of line BC are $<5, 3, 8>$

Since, angle between AB and BC is $\frac{\pi}{4}$.

$$\cos\frac{\pi}{4}=\frac{(5\lambda-3)5+3(3\lambda-12)+8(8\lambda-18)}{\sqrt{5^2+3^2+8^2}\cdot\sqrt{(5\lambda-13)^2+(3\lambda-12)^2+(8\lambda-18)^2}}$$

Squaring and solving, we have $\lambda=3, 2$

Hence, equation of lines are $\frac{x-7}{2}=\frac{y-2}{-3}=\frac{z-4}{6}$

and $\frac{x-7}{3}=\frac{y-2}{6}=\frac{z-4}{2}$

23. L_1L_2 intersecting; L_2L_3 parallel; L_3L_1 skew.

24. $\lambda=\mu=1$ (point of intersection of two lines)

$\Rightarrow$ $\mathbf{r}=\mathbf{a}+\mathbf{l}$ or $\mathbf{b}+\mathbf{m}$, i.e., $\mathbf{r}=\mathbf{i}+2\mathbf{j}+\mathbf{k}$

25. Both the lines pass through origin.

Line L_1 is parallel to the vector

$\mathbf{V}_1=(\cos\theta+\sqrt{3})i+(\sqrt{2}\sin\theta)j+(\cos\theta-\sqrt{3})k$ and L_2 is parallel to the vector

$$\mathbf{V}_2=a\mathbf{i}+b\mathbf{j}+c\mathbf{k}$$

$$\therefore \quad \cos\alpha=\frac{\mathbf{V}_1\cdot\mathbf{V}_2}{|\mathbf{V}_1||\mathbf{V}_2|}$$

$$=\frac{a(\cos\theta+\sqrt{3})+(b\sqrt{2})\sin\theta+c(\cos\theta-\sqrt{3})}{\sqrt{a^2+b^2+c^2}\sqrt{(\cos\theta+\sqrt{3})^2+2\sin^2\theta+(\cos\theta-\sqrt{3})^2}}$$

$$=\frac{(a+c)\cos\theta+b\sqrt{2}\sin\theta+(a-c)\sqrt{3}}{\sqrt{a^2+b^2+c^2}\sqrt{2+6}}$$

In order that $\cos\alpha$ in independent of θ

$$a+c=0$$

and $$b=0$$

$$\therefore \quad \cos\alpha=\frac{2a\sqrt{3}}{a\sqrt{2}\cdot2\sqrt{2}}=\frac{\sqrt{3}}{2}$$

$$\Rightarrow \quad \alpha=\frac{\pi}{6}$$

26. Given lines are skew lines and angle between them

$$=\cos^{-1}\left[\frac{12+3+0}{\sqrt{35}\sqrt{25}}\right]$$

$$=\cos^{-1}\frac{5}{\sqrt{35}}$$

27. Equation of plane

$$\begin{vmatrix} x-1 & y-1 & z-1 \\ 1 & 3 & -2 \\ 0 & 0 & 2 \end{vmatrix}=0$$

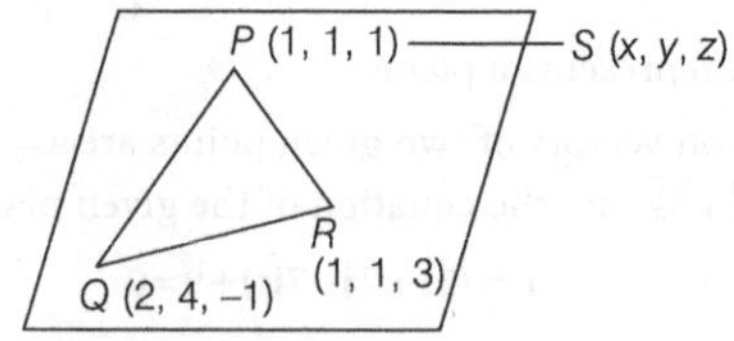

$$\Rightarrow \quad 2(3x-3-y+1)=0$$

$$\Rightarrow \quad 3x-y=2$$

28. x-intercept $=\frac{q}{\hat{\mathbf{i}}\cdot\mathbf{n}}$

$$\therefore \quad x_1\mathbf{i}\cdot\mathbf{n}=q \Rightarrow x_1=\frac{q}{\mathbf{i}\cdot\mathbf{n}}$$

29.

$$P_1=4x+6y-7z-1=0$$

$$P_2=4x+6y-7z-2=0$$

$$d=\frac{1}{\sqrt{16+36+49}}=\frac{1}{\sqrt{101}}$$

Hence, $\frac{101\times102}{2}=5151$

30. x and z-intercept of the plane is 4 and it is parallel to y-axis, hence equation of the plane is $x+z=4$.

Its distance from $(0, 0, 0)$ is $2\sqrt{2}$.

31. Coordinate of $L(0, g, h)$ and $M(f, 0, h)$. Now, to find the equation of OLM.

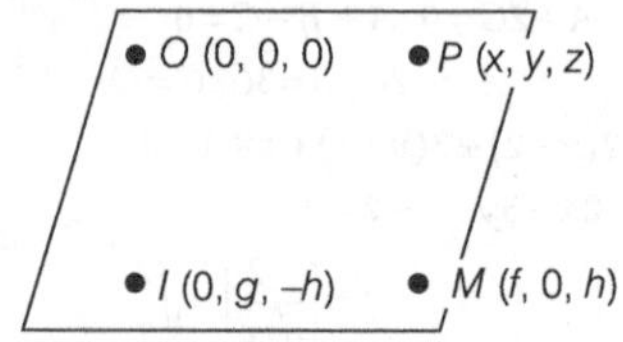

$$\Rightarrow \quad \begin{vmatrix} x & y & z \\ 0 & g & h \\ f & 0 & h \end{vmatrix}=0$$

$$\Rightarrow (gh)x+(fh)y-(gf)z=0$$

or $$\frac{x}{f}+\frac{y}{g}-\frac{z}{h}=0$$

32. y-coordinate of P is zero.

$$\Rightarrow \quad \frac{3\lambda+(-1)}{\lambda+1}=0$$

$$\Rightarrow \quad \lambda=\frac{1}{3}$$

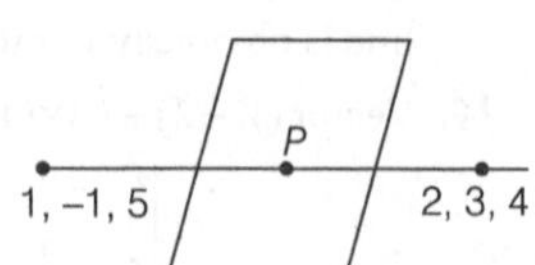

33. $\frac{x_1}{4} = u, \frac{y_1}{4} = v, \frac{z_1}{4} = w$

$$x_1 = 4u, y_1 = 4v, z_1 = 4w$$

$$V = \frac{1}{6}\begin{vmatrix} 4u & 0 & 0 \\ 0 & 4v & 0 \\ 0 & 0 & 4w \end{vmatrix} = \left(\frac{64}{6}\right)uvw$$

$$\therefore \quad 64\left(\frac{uvw}{6}\right) = 64k^3$$

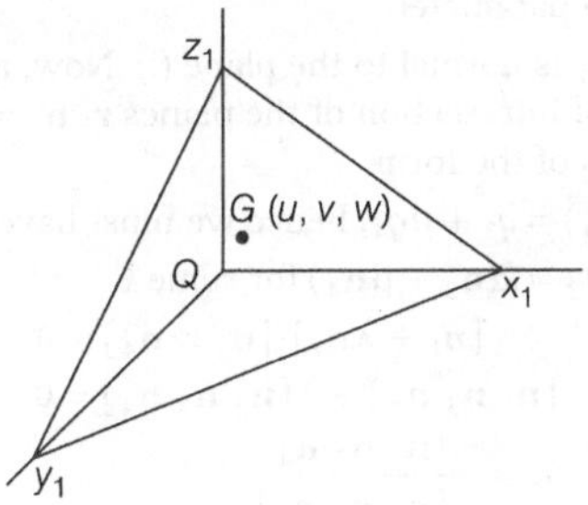

$$xyz = 6k^3$$

34.

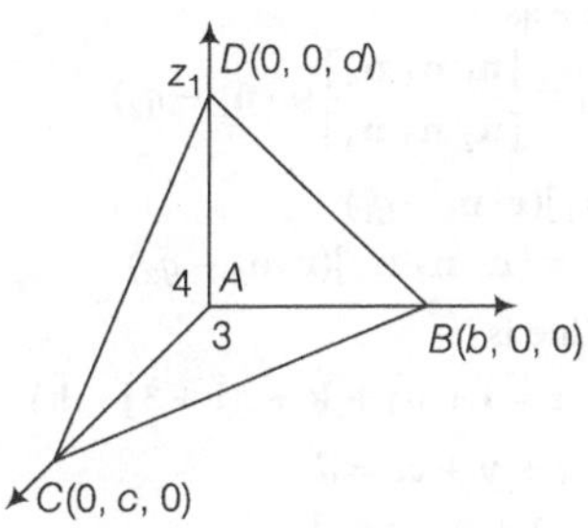

Area of $\Delta BCD = \frac{1}{2}|\mathbf{BC} \times \mathbf{BD}|$

$$= \frac{1}{2}|(b\hat{\mathbf{i}} - c\hat{\mathbf{j}}) \times (b\hat{\mathbf{i}} - d\hat{\mathbf{k}})| = \frac{1}{2}|bd\,\hat{\mathbf{j}} + bc\,\hat{\mathbf{k}} + dc\,\hat{\mathbf{i}}|$$

$$= \frac{1}{2}\sqrt{b^2c^2 + c^2d^2 + d^2b^2} \quad \text{...(i)}$$

Now, $6 = bc, 8 = cd, 10 = bd$

$$b^2c^2 + c^2d^2 + d^2b^2 = 200$$

On substituting the value in Eq. (i), we get

$$A = \frac{1}{2}\sqrt{200} = 5\sqrt{2}$$

35. $\mathbf{r} = 2\hat{\mathbf{i}} + \hat{\mathbf{j}} + 0\hat{\mathbf{k}} + t(\hat{\mathbf{i}} + \hat{\mathbf{j}}) \times (\hat{\mathbf{j}} + \hat{\mathbf{k}})$

$$= (2, 1, 0) + t(\hat{\mathbf{k}} - \hat{\mathbf{j}} + \hat{\mathbf{i}}) = (2, 1, 0) + t(1, -1, 1)$$

36. Option (a), $-\frac{2}{3} = -\frac{2}{3} \neq \frac{2}{9}$

Option (b), $2 = 2 = 2 = 2$ identical

Option (c), $\frac{-2}{3} = \frac{-2}{3} = \frac{-2}{3} \neq \frac{3}{5}$

Option (d), $2 = 2 \neq -6$

37. $\left.\begin{array}{l} x = 9; x = 1 \\ y = 8; y = 2 \\ z = 5; z = 3 \end{array}\right\}$

Edges of the cuboid are 8, 6 and 2.

38. Plane through a and parallel to two non-collinear vector

$(\mathbf{r} - \mathbf{a}) \cdot (\mathbf{b} \times \mathbf{c}) = 0$ (takes dot with $\mathbf{b} \times \mathbf{c}$ both sides)

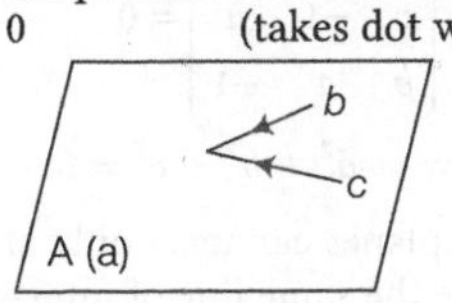

i.e., $(\mathbf{r} - (\hat{\mathbf{i}} - \hat{\mathbf{j}}) \cdot (5\hat{\mathbf{i}} - 2\hat{\mathbf{j}} - 3\hat{\mathbf{k}}) = 0$

$\Rightarrow \quad \mathbf{r} \cdot (5\hat{\mathbf{i}} - 2\hat{\mathbf{j}} - 3\hat{\mathbf{k}}) = 7$

39. Intersecting, if

$$\begin{vmatrix} 5 & 2 & 13-p \\ 1 & 2 & 3 \\ -1 & 2 & -3 \end{vmatrix} = \begin{vmatrix} 5 & 2 & 13-p \\ 0 & 2 & 0 \\ -1 & 2 & -3 \end{vmatrix}$$

$$-4(-15 + 13 - p) = 0$$

$$p = -2$$

Aliter

$$(\lambda + 2) = -(\mu + 3) \quad \text{...(i)}$$

$$2\lambda + p = 2\mu + 7 \quad \text{...(ii)}$$

$$3\lambda + 13 = p - 3\mu \quad \text{...(iii)}$$

From Eq. (i) $\mu = (-\lambda + 5)$

On putting in Eq. (ii), $2\lambda + 9 = -2(\lambda + 5) + 7$

$$\lambda = -3$$

Now, from Eq. (iii), $-9 + 13 = p + 6$

$$p = -2$$

40. $\mathbf{r} = \mathbf{a} + \gamma\mathbf{b} + \mu\mathbf{c}$

Taking dot with $\mathbf{b} \times \mathbf{c}$

$[\mathbf{r}\ \mathbf{b}\ \mathbf{c}] = [\mathbf{a}\ \mathbf{b}\ \mathbf{c}]$ [where, $\mathbf{a} = (0, 1, 1)$]

$\mathbf{b} = (1, -1, 1)$ and $\mathbf{c} = (2, -1, 0)$

$$[\mathbf{a}\ \mathbf{b}\ \mathbf{c}] = \begin{vmatrix} 0 & 1 & 1 \\ 1 & -1 & 1 \\ 2 & -1 & 0 \end{vmatrix} = 0 - (0 - 2) + 1(-1 + 2) = 3$$

and $[\mathbf{r}\ \mathbf{b}\ \mathbf{c}] = \begin{vmatrix} x & y & z \\ 1 & -1 & 1 \\ 2 & -1 & 0 \end{vmatrix} = x(0 + 1) - y(0 - 2) + z(-1 + 2)$

$$= x + 2y + z$$

Hence, equation of plane is $x + 2y + z = 3$

$$\therefore \quad p = \left|\frac{-3}{\sqrt{6}}\right| = \sqrt{\frac{3}{2}}$$

41. $\begin{vmatrix} 2-a & 9-7 & 13-(-2) \\ 1 & 2 & 3 \\ -1 & 2 & -3 \end{vmatrix} = 0$

$\Rightarrow \quad a = -3$

42. On (1, 2, 3) satisfies the plane $x - 2y + z = 0$ and also $(\hat{\mathbf{i}} + 2\hat{\mathbf{j}} + 3\hat{\mathbf{k}}) \cdot (\hat{\mathbf{i}} - 2\hat{\mathbf{j}} + \hat{\mathbf{k}}) = 0$

Since, the lines $\frac{x-1}{1} = \frac{y-2}{2} = \frac{z-3}{3}$

and $\frac{x}{1} = \frac{y}{2} = \frac{z}{3}$ both satisfy $t + 1$ and $3t + 3$.

Hence, both are same.

Given line is obviously parallel to the plane $x - 2y + z = 6$

43. Infinite solution $\begin{vmatrix} 1 & -c & -b \\ c & -1 & a \\ b & a & -1 \end{vmatrix} = 0$

$\Rightarrow \quad a^2 + b^2 + c^2 = 1$

Note that 3 such planes can meet only at one point i.e. (0, 0, 0) or they may have the same line of intersection i.e., at infinite solution.

44. The given lines are coplanar, if

$$\begin{vmatrix} 2-1 & 3-4 & 4-5 \\ 1 & 1 & -k \\ k & 2 & 1 \end{vmatrix} = 0 \Rightarrow \begin{vmatrix} 1 & -1 & -1 \\ 1 & 1 & -k \\ k & 2 & 1 \end{vmatrix} = 0$$

$$\Rightarrow \begin{vmatrix} 1 & 0 & 0 \\ 1 & 1 & 1-k \\ k & k+2 & 1+k \end{vmatrix} = 0$$

$\Rightarrow \quad 2(1+k) - (k+2)(1-k) = 0$

if $\quad k^2 + 3k = 0 \Rightarrow k = 0$ or -3

45. Put $z = 0$ in the line given $x = 5$ and $y = 1$

$\Rightarrow \quad 5 \cdot 1 = c^2$

46. Equation of the line is $\dfrac{x-2}{1} = \dfrac{y+2}{-3} = \dfrac{z-5}{2} = \lambda$ ...(i)

Hence, any point on the line (i) can be taken as

$$x = \lambda + 2$$
$$y = -(3\lambda + 2)$$
$$z = (2\lambda + 5)$$

From some λ point lies on the plane

$$2x - 3y + 4z = 163 \quad \text{...(ii)}$$
$$2(\lambda + 2) + 3(3\lambda + 2) + 4(2\lambda + 5) = 163$$
$$19\lambda = 133$$

$\Rightarrow \quad \lambda = 7$

Hence, $\quad P \equiv (9, -23, 19)$

Also, Eq. (i) intersect YZ-plane i.e., $x = 0$

$\lambda + 2 = 0$

Hence, $\quad \lambda = -2$

$\therefore \quad Q(0, 4, 1)$

So, $\quad PQ = \sqrt{9^2 + 27^2 + 18^2}$

$= 9\sqrt{1 + 3^2 + 2^2} = 9\sqrt{14}$

$\Rightarrow \quad a = 9$ and $b = 14$

Hence, $a + b = 9 + 14 = 23$

47. Equation of the plane passing through the line of intersection of the first two planes is $\mathbf{r} \cdot (\mathbf{n}_1 + \lambda \mathbf{n}_2), = p_1 + \lambda p_2$. where λ is a parameter

Since, three planes have a common line of intersection the above equation should be identical with $\mathbf{r} \cdot \mathbf{n}_3 = p_3$ for some λ. That is for some λ,

$$\mathbf{n}_1 + \lambda \mathbf{n}_2 = k\mathbf{n}_3 \quad \text{...(i)}$$

and $\quad p_1 + \lambda p_2 = kp_3$ ...(ii)

From Eq. (i)

$$\mathbf{n}_1 \times \mathbf{n}_3 + \lambda \mathbf{n}_2 \times \mathbf{n}_3 = 0$$

and $\quad \mathbf{n}_1 \times \mathbf{n}_2 = k\mathbf{n}_3 \times \mathbf{n}_2$ ...(iii)

From Eq. (ii)

$$(p_1 + \lambda p_2)(\mathbf{n}_2 \times \mathbf{n}_3) = kp_3(\mathbf{n}_2 \times \mathbf{n}_3)$$
$$= p_3(\mathbf{n}_2 \times \mathbf{n}_1) \quad \text{...(iv)}$$
$$p_1(\mathbf{n}_2 \times \mathbf{n}_3) + p_2\lambda(\mathbf{n}_2 \times \mathbf{n}_3) + p_3(\mathbf{n}_1 + \mathbf{n}_2) = 0$$
$$\Rightarrow p_1(\mathbf{n}_2 \times \mathbf{n}_3) + p_2(\mathbf{n}_3 \times \mathbf{n}_1) + p_3(\mathbf{n}_1 \times \mathbf{n}_2) = 0$$

[Using Eq. (iii)]

48. Equation of any plane through the intersection of $\mathbf{r} \cdot \mathbf{n}_1 = q_1$ and $\mathbf{r} \cdot \mathbf{n}_2 = q_2$ is of the form

$$\mathbf{r} \cdot \mathbf{n}_1 + \lambda \mathbf{r} \cdot \mathbf{n}_2 = q_1 + \lambda q_2 \quad \text{...(i)}$$

where λ is a parameter.

So, $\mathbf{n}_1 + \lambda \mathbf{n}_2$ is normal to the plane (i). Now, any plane parallel to the line of intersection of the planes $\mathbf{r} \cdot \mathbf{n}_3 = q_3$ and $\mathbf{r} \cdot \mathbf{n}_4 = q_4$ is of the form.

$\mathbf{r} \cdot (\mathbf{n}_3 + \mu \mathbf{n}_4) = q_3 + \mu q_4$, hence we must have

$$\mathbf{n}_1 + \lambda \mathbf{n}_2 = k(\mathbf{n}_3 + \mu \mathbf{n}_4) \text{ for some } k$$
$$\Rightarrow [\mathbf{n}_1 + \lambda \mathbf{n}_2] \cdot [\mathbf{n}_3 \times \mathbf{n}_4] = 0$$
$$\Rightarrow [\mathbf{n}_1\ \mathbf{n}_3\ \mathbf{n}_4] + \lambda[\mathbf{n}_2\ \mathbf{n}_3\ \mathbf{n}_4] = 0$$
$$\Rightarrow \lambda = -\frac{[\mathbf{n}_1\ \mathbf{n}_3\ \mathbf{n}_4]}{[\mathbf{n}_2\ \mathbf{n}_3\ \mathbf{n}_4]}$$

On putting this value in Eq. (i), we have the equation of required plane as

$$\mathbf{r} \cdot \mathbf{n}_1 - q_1 = \frac{[\mathbf{n}_1\ \mathbf{n}_3\ \mathbf{n}_4]}{[\mathbf{n}_2\ \mathbf{n}_3\ \mathbf{n}_4]}(r \cdot \mathbf{n}_2 - q_2)$$
$$\Rightarrow [\mathbf{n}_2\ \mathbf{n}_3\ \mathbf{n}_4](\mathbf{r} \cdot \mathbf{n}_1 - q_1) = [\mathbf{n}_1\ \mathbf{n}_3\ \mathbf{n}_4](\mathbf{r} \cdot \mathbf{n}_2 - q_2)$$

49. Equation of line is

$$\mathbf{r} = \hat{\mathbf{i}} + 0\hat{\mathbf{j}} + \hat{\mathbf{k}} + t(\hat{\mathbf{i}} + 3\hat{\mathbf{j}} - \hat{\mathbf{k}}) \quad \text{...(i)}$$

Eq. (i) lies in $x + y + cz = d$

$\therefore \quad 1 + 0 + c = d$

$1 + c = d$

Also, $\quad 1 \cdot 1 + 1 \cdot 3 + c(-1) = 0$

$c = 4$

$\therefore \quad 1 + 4 = d \Rightarrow d = 5$

$\Rightarrow \quad (c + d) = 4 + 5 = 9$

50. Any point on $\dfrac{x-2}{2} = \dfrac{y+1}{4} = \dfrac{z-2}{12}$ can be

$(2r + 2, 4r - 1, 12r + 2)$

which lies on $x - y + z = 5$

$\therefore \quad (2r + 2) - (4r - 1) + 12r + 2 = 5$

$r = 0$

$\therefore$ Point on the plane $\equiv (2, -1, 2)$

Distance between $(2, -1, 2)$ and $(-1, -5, -10)$

$$= \sqrt{(2+1)^2 + (-1+5)^2 + (2+10)^2}$$
$$= 13$$

51. $R(\mathbf{r})$ moves on $\mathbf{PQ}$,

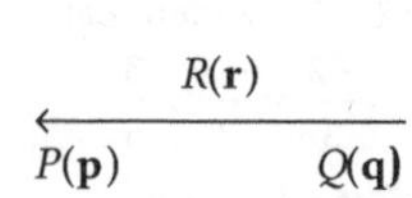

52. $(\hat{\mathbf{i}} + \hat{\mathbf{j}}) \times (\hat{\mathbf{j}} + \hat{\mathbf{k}}) = \hat{\mathbf{i}} - \hat{\mathbf{j}} + \hat{\mathbf{k}} \Rightarrow$ Unit vector perpendicular as to the plane of $\hat{\mathbf{i}} + \hat{\mathbf{j}}$ and $\hat{\mathbf{j}} + \hat{\mathbf{k}}$ is $\dfrac{1}{\sqrt{3}}(\hat{\mathbf{i}} - \hat{\mathbf{j}} + \hat{\mathbf{k}})$.

Similarly, other two unit vectors are

$\frac{1}{\sqrt{3}(\hat{\mathbf{i}}+\hat{\mathbf{j}}-\hat{\mathbf{k}})}$ and $\frac{1}{\sqrt{3}}(-\hat{\mathbf{i}}+\hat{\mathbf{j}}+\hat{\mathbf{k}})$.

$$V=[\hat{\mathbf{n}}_1, \hat{\mathbf{n}}_2, \hat{\mathbf{n}}_3]=\frac{1}{3\sqrt{3}}\begin{vmatrix}1 & -1 & 1\\ 1 & 1 & -1\\ -1 & 1 & 1\end{vmatrix}=\frac{4}{3\sqrt{3}}$$

Aliter

Let $\mathbf{a}=\hat{\mathbf{i}}+\hat{\mathbf{j}}$, $\mathbf{b}=\hat{\mathbf{j}}+\hat{\mathbf{k}}$ and $\mathbf{c}=\hat{\mathbf{k}}+\hat{\mathbf{i}}$.

Now, $[\mathbf{a}\times\mathbf{b}, \mathbf{b}\times\mathbf{c}, \mathbf{c}\times\mathbf{a}]=[\mathbf{a}\ \mathbf{b}\ \mathbf{c}]^2$

$$=\begin{vmatrix}1 & 1 & 0\\ 0 & 1 & 1\\ 1 & 0 & 1\end{vmatrix}^2=[1(1)-1(0-1)]^2=4$$

Hence, actual volume with unit vectors

$$=\frac{4}{|\mathbf{a}\times\mathbf{b}|\,|\mathbf{b}\times\mathbf{c}|\,|\mathbf{c}\times\mathbf{a}|}$$

Now, $|\mathbf{a}\times\mathbf{b}|=\sqrt{a^2b^2-(\mathbf{a}\cdot\mathbf{b})^2}=\sqrt{4-1}=\sqrt{3}$ etc.

$$V_{\text{actual}}=\frac{4}{3\sqrt{3}}$$

53. $\mathbf{n}=3\hat{\mathbf{i}}-\hat{\mathbf{j}}+4\hat{\mathbf{k}}$

Line through A are parallel to $\mathbf{n}$ is

$$\mathbf{r}=\hat{\mathbf{i}}+2\hat{\mathbf{j}}+3\hat{\mathbf{k}}+\lambda(3\hat{\mathbf{i}}-\hat{\mathbf{j}}+4\hat{\mathbf{k}})$$

$$=3\lambda+1, 2-\lambda, 3+4\lambda \quad \text{...(i)}$$

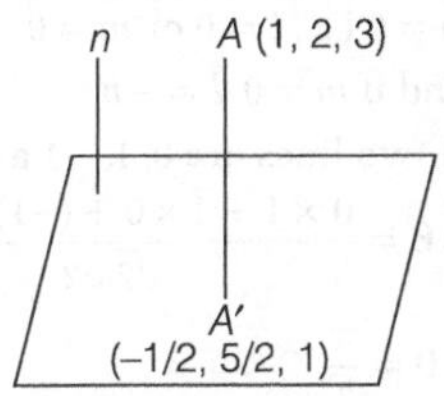

Hence, Eq. (i) must satisfy the plane

$$3x-y+4z=0$$
$$3(3\lambda+1)-(2-\lambda)+4(3+4\lambda)=0$$
$$26\lambda+13=0$$
$$\lambda=-\frac{1}{2}$$

Hence, A' is $\left(-\frac{1}{2}, \frac{5}{2}, 1\right)$ which is the foot of the perpendicular from A on the given plane.

54. On solving $x+2y-4z=0$ and $2x-y+2z=0$, we get

$$\frac{x}{0}=\frac{y}{-10}=\frac{z}{-5}$$

One point P on line is $(0, -10t, -5t)$ and $Q\equiv(1, 1, 1)$

Direction ratio of $PQ\equiv(1, 1+10t, 1+5t)$

$$0-10-100t-5-25t=0$$

$$\Rightarrow \quad t=-\frac{3}{25}$$

$$\Rightarrow \quad P\equiv\left(0, \frac{6}{5}, \frac{3}{5}\right)$$

Hence, required equation $\frac{x-1}{5}=\frac{1-y}{1}=\frac{z-1}{2}$.

55. Let $\frac{x}{a}+\frac{y}{b}+\frac{z}{c}=1$ be the variable.

So that, $\left|\frac{1}{\sqrt{\frac{1}{a^2}+\frac{1}{b^2}+\frac{1}{c^2}}}\right|=1$

Then, the coordinates of ΔABC are $A(a, 0, 0)$, $B(0, b, 0)$ and $C(0, 0, c)$.

The centroid of ΔABC is $\left(\frac{a}{3}, \frac{b}{3}, \frac{c}{3}\right)$

$$x=\frac{a}{3}, y=\frac{b}{3}, z=\frac{c}{3}$$

$$\frac{1}{x^2}+\frac{1}{y^2}+\frac{1}{z^2}=9$$

56. Direction ratios of AB are $1:1:1$.

Direction ratios of CD are $1:2:1$.

Angle between AB and CD,

$$\cos\theta=\frac{1\times1+1\times2+1\times1}{\sqrt{3}\sqrt{6}}=\frac{4}{3\sqrt{2}}$$

57. Equation of plane is $\frac{x}{1}+\frac{y}{2}+\frac{z}{3}=1$

$$\therefore \text{ Required distance }\left|\frac{\frac{1}{1}+\frac{2}{2}-\frac{3}{3}-1}{\sqrt{\left(\frac{1}{1}\right)^2+\left(\frac{1}{2}\right)^2+\left(\frac{1}{3}\right)^2}}\right|=0$$

58. Angle between the faces = Angle between the normals

$\mathbf{n}_1$ = Vector normal to face OAB

$$=\mathbf{OA}\times\mathbf{OB}=\begin{vmatrix}\hat{\mathbf{i}} & \hat{\mathbf{j}} & \hat{\mathbf{k}}\\ 1 & 2 & 1\\ 2 & 1 & 3\end{vmatrix}=5\hat{\mathbf{i}}-\hat{\mathbf{j}}-3\hat{\mathbf{k}}$$

$\mathbf{n}_2$ = Vector normal to face ABC

$$=\mathbf{AB}\times\mathbf{AC}=\begin{vmatrix}\hat{\mathbf{i}} & \hat{\mathbf{j}} & \hat{\mathbf{k}}\\ 1 & -1 & 2\\ -2 & -1 & 1\end{vmatrix}$$

$$=\hat{\mathbf{i}}-5\hat{\mathbf{j}}-3\hat{\mathbf{k}}$$

Angle between faces $=\cos^{-1}\frac{\mathbf{n}_1\cdot\mathbf{n}_2}{[\mathbf{n}_1][\mathbf{n}_2]}=\cos^{-1}\left(\frac{19}{35}\right)$

59. $Q\equiv(1+2\lambda, 2+3\lambda, 3+4\lambda)$

Direction ratio of $PQ=2\lambda, 3\lambda-1, 4\lambda-1$

Now, $(2\lambda)2+(3\kappa-1)3+(4\lambda-1)4=0)$

$$29\lambda=7$$

$$\lambda=\frac{7}{29}$$

Direction ratio of line PQ is $(14, -8, -1)$.

60. Equation of the plane passing through three points A, B and C with position vector $\mathbf{a}$, $\mathbf{b}$ and $\mathbf{c}$ is

$$\mathbf{r}\cdot(\mathbf{a}\times\mathbf{b}+\mathbf{b}\times\mathbf{c}+\mathbf{c}\times\mathbf{a})=\mathbf{a}\cdot\mathbf{b}\times\mathbf{c}$$

So that, if **a**, **b**, **c** represent the given vectors, then

$$\mathbf{a}\times\mathbf{b}+\mathbf{b}\times\mathbf{c}+\mathbf{c}\times\mathbf{a}=\begin{vmatrix}\hat{\mathbf{i}} & \hat{\mathbf{j}} & \hat{\mathbf{k}}\\ -6 & 3 & 2\\ 3 & -2 & 4\end{vmatrix}+\begin{vmatrix}\hat{\mathbf{i}} & \hat{\mathbf{j}} & \hat{\mathbf{k}}\\ 3 & -2 & 4\\ 5 & 7 & 3\end{vmatrix}+\begin{vmatrix}\hat{\mathbf{i}} & \hat{\mathbf{j}} & \hat{\mathbf{k}}\\ 5 & 7 & 3\\ -6 & 3 & 2\end{vmatrix}$$

$$=-13\hat{\mathbf{i}}+13\hat{\mathbf{j}}-912\hat{\mathbf{k}}$$

and $\mathbf{a}\cdot\mathbf{a}\times\mathbf{c}=\begin{vmatrix}-6 & 3 & 2\\ 3 & -2 & 4\\ 5 & 7 & 3\end{vmatrix}=299$

So, the required equation of the plane is
$\mathbf{r}.(-13\hat{\mathbf{i}}+13\hat{\mathbf{j}}-91\hat{\mathbf{k}})=299$ or $\mathbf{r}\cdot(\hat{\mathbf{i}}-\hat{\mathbf{j}}+7\hat{\mathbf{k}})+23=0$

61. The volume of tetrahedron

$$=\frac{1}{6}(\mathbf{OA}\ \mathbf{OB}\ \mathbf{OC})=\frac{1}{6}\begin{vmatrix}1 & 0 & 0\\ 1 & 1 & 0\\ 0 & 1 & 1\end{vmatrix}=\frac{1}{6}\text{ units}$$

Area of the base $=\frac{1}{2}|(\hat{\mathbf{i}}+\hat{\mathbf{j}}-\hat{\mathbf{i}})\times(\hat{\mathbf{j}}+\hat{\mathbf{k}}-\hat{\mathbf{i}})|$

$$=\frac{1}{2}|\hat{\mathbf{i}}+\hat{\mathbf{k}}|=\frac{1}{\sqrt{2}}$$

$$\text{Height}=\frac{3\times\text{Volume}}{\text{Area of base}}=\frac{3\sqrt{2}}{6}=\frac{1}{\sqrt{2}}$$

62. $x-y-z-4=0,\ x+y+2z-4=0$

Required plane is of the form

$(x-y-z-4)+\lambda(x+y+2z-4)=0$

Since, this plane is perpendicular to the plane $x-y-z-4=0$

$\therefore\quad 1+\lambda+(\lambda-1)(-1)+(2\lambda-1)(-1)=0,\ \lambda=\frac{3}{2}$

$\therefore$ Required plane $5x+y+4z=20$

63. Let l, m, n be the direction cosines of the normal to the plane on which lies the plane area A.

Then, $A_{xy}=$ projection of A on the xy-plane
$=A\cos\alpha$, where α is the angle between the plane and xy-plane.

$\therefore\quad \cos\alpha=\dfrac{l.0+m.0+n.1}{1}$

Since, the normal to the xy-plane has direction cosines $(0, 0, 1)$

$\therefore\quad A_{xy}=A_n$

Similarly, $\quad A_{yxz}=A_l$

$A_{zx}=A_m$

$\therefore A_{xy}^2+A_{yz}^2+A_{zx}^2=A^2$

64. Equation of the plane through $P(h, k, l)$ perpendicular to OP is $xh+yk+zl=h^2+k^2+l^2=p^2$

where, $\quad p^2=h^2+k^2+l^2$

$$\Rightarrow\quad \frac{x}{\frac{p^2}{h}}+\frac{y}{\frac{p^2}{k}}+\frac{z}{\frac{p^2}{l}}=1$$

$$A_{xy}=\frac{1}{2}\cdot\frac{p^2}{h}\cdot\frac{p^2}{k},\ A_{yz}=\frac{1}{2}\cdot\frac{p^2}{k}\cdot\frac{p^2}{l},$$

$$A_{zx}=\frac{1}{2}\cdot\frac{p^2}{l}\cdot\frac{p^2}{h}$$

$$A=\sqrt{A_{xy}^2+A_{yz}^2+A_{zx}^2}$$

$$=\frac{p^4}{4}\sqrt{\frac{l^2+h^2+k^2}{h^2k^2l^2}}=\frac{p^4}{2}\sqrt{\frac{p^2}{h^2k^2l^2}}=\frac{p^5}{2hkl}$$

Hence $Ar(DABC)=\dfrac{p^5}{2hkl}$

65. Equation of the given plane can be written as

$$\frac{x}{20}+\frac{y}{15}+\frac{z}{-12}=1$$

which meets the co-ordinates axes in points $A(20, 0, 0)$, $B(0, 15, 0)$ and $C(0, 0, -12)$ and the co-ordinates of the origin are $(0, 0, 0)$.

$\therefore$ The volume of the tetrahedron $OABC$ is

$$\left|\frac{1}{6}\begin{vmatrix}20 & 0 & 0\\ 0 & 15 & 0\\ 0 & 0 & -12\end{vmatrix}\right|=\left|\frac{1}{6}\times 20\times 15\times(-12)\right|=600$$

66. $l+m+n=0,\ l^2+m^2-n^2=0$

$\Rightarrow\quad l^2+m^2-(-l-m)^2=0$

$\Rightarrow\quad 2lm=0$ i.e., $l=0$ or $m=0$

If $l=0$, $m=-n$ and if $m=0$, $l=-n$

Since d.r.'s of the two lines are $0, 1, -1$ and $1, 0, -1$

$$\cos\theta=\frac{0\times 1+1\times 0+(-1)\times(-1)}{\sqrt{2}\sqrt{2}}=\frac{1}{2}$$

$$\Rightarrow\quad \theta=\frac{\pi}{3}$$

67. $(\hat{\mathbf{i}}-\hat{\mathbf{j}}+4\hat{\mathbf{k}})\cdot(\hat{\mathbf{i}}+5\hat{\mathbf{j}}+\hat{\mathbf{k}})=0$

Therefore the line and the plane are parallel. A point on the line is $(2, -2, 3)$. Required distance is equal to distance of $(2, -2, 3)$ from the given plane $=\dfrac{|2+5(-2)+3-5|}{\sqrt{1^2+5^2+1^2}}=\dfrac{10}{\sqrt{27}}$

68. $\because$ Plane is perpendicular to the line

$\therefore$ Equation of plane is of the form $2x-y+2z+k=0$

$\because$ If passes through origin $\therefore k=0$

$\therefore\quad 2x-y+2z=0$

69. $\mathbf{PQ}=\hat{\mathbf{i}}(-2-3\mu)+\hat{\mathbf{j}}(\mu-3)+\hat{\mathbf{k}}(5\mu-4)$

PQ is parallel to $x-4y+3z=1$

$\Rightarrow\ 1(-2-3\mu)-4(\mu-3)+3(5\mu-4)=0$

$\Rightarrow\quad \mu=\frac{1}{4}$

70. Plane meets axes at $A(a, 0, 0)$, $B(0, b, 0)$ and $C(0, 0, c)$.

Then area of $\Delta ABC=\frac{1}{2}|\mathbf{AB}\times\mathbf{AC}|$

$$=\frac{1}{2}|(-a\hat{i}+b\hat{j})\times(-a\hat{i}\times c\hat{k})|$$

$$=\frac{1}{2}\sqrt{(a^2b^2+b^2c^2+c^2a^2}$$

71. Centre of the sphere is $(-1, 1, 2)$ and its radius

$$= \sqrt{1 + 1 + 4 + 19} = 5$$

CL, perpendicular distance of C from plane, is

$$\left| \frac{-1 + 2 + 4 + 7}{\sqrt{1 + 4 + 4}} \right| = 4$$

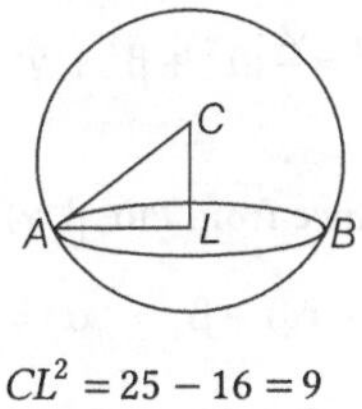

Now, $AL^2 = CA^2 - CL^2 = 25 - 16 = 9$

Hence, radius of the circle $= \sqrt{9} = 3$

72. Let $\quad \mathbf{r} \times \mathbf{a} = \mathbf{b} \times \mathbf{a}$

$\Rightarrow \quad (\mathbf{r} - \mathbf{b}) \times \mathbf{a} = 0$

$\Rightarrow \quad \mathbf{r} = \mathbf{b} + t\mathbf{a}$

Similarly, other line $\mathbf{r} = \mathbf{a} + k\mathbf{b}$, where t and k are scalars.

Now $\quad \mathbf{a} + k\mathbf{b} = \mathbf{b} + t\mathbf{a}$

$\Rightarrow \quad t = 1, k = 1$ (equation the coefficients of $\mathbf{a}$ and $\mathbf{b}$)

$\therefore \quad \mathbf{r} = \mathbf{a} + \mathbf{b} = \hat{\mathbf{i}} + \hat{\mathbf{j}} + 2\hat{\mathbf{i}} - \hat{\mathbf{k}}$

$\quad = 3\hat{\mathbf{i}} + \hat{\mathbf{j}} - \hat{\mathbf{k}}$

i.e. $(3, 1, -1)$

73. Let the point P be (x, y, z), then the vector $x\hat{\mathbf{i}} + y\hat{\mathbf{j}} + z\hat{\mathbf{k}}$ will lie on the line

$\Rightarrow \quad (x - 1)\hat{\mathbf{i}} + (y - 1)\hat{\mathbf{j}} + (z - 1)\hat{\mathbf{k}}$

$\quad = -\lambda\hat{\mathbf{i}} + \lambda\hat{\mathbf{j}} - \lambda\hat{\mathbf{k}}$

$\Rightarrow \quad x = 1 - \lambda, y = 1 + \lambda$ and $z = 1 - \lambda$

Now point P in nearest to the origin.

$\Rightarrow \quad D = (1 - \lambda)^2 + (1 + \lambda)^2 + (1 - \lambda)^2$

$\Rightarrow \quad \dfrac{dD}{d\lambda} = -4(1 - \lambda) + 2(1 + \lambda) = 0$

$\Rightarrow \quad \lambda = \dfrac{1}{3}$

$\Rightarrow$ The point is $\left(\dfrac{2}{3}, \dfrac{4}{3}, \dfrac{2}{3}\right)$.

74. We have, $|\mathbf{v}_1| = 2, |\mathbf{v}_2| = \sqrt{2}$ and $|\mathbf{v}_3| = \sqrt{29}$

If θ is the angle between $\mathbf{v}_1$ and $\mathbf{v}_2$, then

$$2\sqrt{2}\cos\theta = -2$$

$\Rightarrow \quad \cos\theta = -\dfrac{1}{\sqrt{2}} \Rightarrow \theta = 135°$

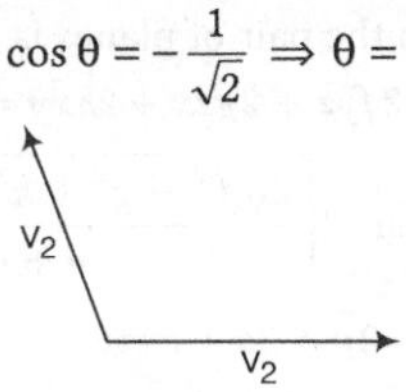

Let $\mathbf{v}_1 = 2\mathbf{i}, \mathbf{v}_2 = -\hat{\mathbf{i}} + \hat{\mathbf{j}}$ and $\mathbf{v}_3 = \alpha\hat{\mathbf{i}} + \beta\hat{\mathbf{j}} + \gamma\hat{\mathbf{k}}$

Since $\quad \mathbf{v}_3 \cdot \mathbf{v}_1 = 6 = 2\alpha$

$\Rightarrow \quad \alpha = 3$

Also, $\quad \mathbf{v}_3 \cdot \mathbf{v}_2 = -5 = -\alpha + \beta$

$\Rightarrow \quad \beta = -2$

and $\quad \mathbf{v}_3 \cdot \mathbf{v}_2 = 29 = \alpha^2 + \beta^2 + \gamma^2$

$\Rightarrow \quad \gamma = \pm 4$

$\therefore \quad \mathbf{v}_3 = 3\hat{\mathbf{i}} - 2\hat{\mathbf{j}} \pm 4\hat{\mathbf{k}}$

75. The given plane is $\mathbf{r} \cdot (5\mathbf{i} + 2\mathbf{j} - 7\mathbf{k}) = -9$

Length of the perpendicular from $\hat{\mathbf{i}} - \hat{\mathbf{j}} + 3\hat{\mathbf{k}}$ to it is

$$\frac{-9 - (\hat{\mathbf{i}} - \hat{\mathbf{j}} + 3\hat{\mathbf{k}}) \cdot (5\hat{\mathbf{i}} + 2\hat{\mathbf{j}} - 7\hat{\mathbf{k}})}{\sqrt{5 + 4 + 49}} = \frac{-9 - 5 + 2 + 21}{\sqrt{78}} = \frac{9}{\sqrt{78}}$$

Length of the perpendicular from $3\hat{\mathbf{i}} + 3\hat{\mathbf{j}} + 3\hat{\mathbf{k}}$

$$\frac{-9 - (3\hat{\mathbf{i}} + 3\hat{\mathbf{j}} + 3\hat{\mathbf{k}}) \cdot (5\hat{\mathbf{i}} + 2\hat{\mathbf{j}} - 7\hat{\mathbf{k}})}{\sqrt{78}} = \frac{-9 - 15 - 6 + 21}{\sqrt{78}} = -\frac{9}{\sqrt{78}}$$

Thus, the length of the two perpendiculars are equal in magnitude but opposite in sign. Hence, they are located on opposite side of the plane.

76. Let the position vector of A, B, C, D be $\mathbf{a}, \mathbf{b}, \mathbf{c}$ and $\mathbf{d}$ respectively.

Then, $AC^2 + BD^2 + AD^2 + BC^2$

$$= (\mathbf{c} - \mathbf{a}) \cdot (\mathbf{c} - \mathbf{a}) + (\mathbf{d} - \mathbf{b}) \cdot (\mathbf{d} - \mathbf{b}) + (\mathbf{d} - \mathbf{a}) \cdot (\mathbf{d} - \mathbf{a}) + (\mathbf{c} - \mathbf{b}) \cdot (\mathbf{c} - \mathbf{b})$$

$$= |\mathbf{c}|^2 + |\mathbf{a}|^2 - 2\mathbf{a} \cdot \mathbf{c} + |\mathbf{d}|^2 + |\mathbf{b}|^2 - 2\mathbf{d} \cdot \mathbf{b} + |\mathbf{d}|^2 + |\mathbf{a}|^2 - 2\mathbf{a} \cdot \mathbf{d} + |\mathbf{c}|^2 + |\mathbf{b}|^2 - 2\mathbf{b} \cdot \mathbf{c}$$

$$= |\mathbf{a}|^2 + |\mathbf{b}|^2 - 2\mathbf{a} \cdot \mathbf{b} + |\mathbf{c}|^2 + |\mathbf{d}|^2 - 2\mathbf{c} \cdot \mathbf{d} + |\mathbf{a}|^2 + |\mathbf{b}|^2 + |\mathbf{c}|^2 + |\mathbf{d}|^2 + 2\mathbf{a} \cdot \mathbf{b} + 2\mathbf{c} \cdot \mathbf{d} - 2\mathbf{a} \cdot \mathbf{c} - 2\mathbf{b} \cdot \mathbf{d} - 2\mathbf{a} \cdot \mathbf{d} - 2\mathbf{d} \cdot \mathbf{c}$$

$$= (\mathbf{a} - \mathbf{b}) \cdot (\mathbf{a} - \mathbf{d}) + (\mathbf{c} - \mathbf{d}) \cdot (\mathbf{c} - \mathbf{d}) + (\mathbf{a} + \mathbf{b} - \mathbf{c} - \mathbf{d}) \cdot (\mathbf{a} + \mathbf{b} - \mathbf{c} - \mathbf{d})$$

$$= AB^2 + CD^2 + (\mathbf{a} + \mathbf{b} - \mathbf{c} - \mathbf{d}) \cdot (\mathbf{a} + \mathbf{b} - \mathbf{c} - \mathbf{d})$$

$$\geq AB^2 + CD^2$$

77. If the given vectors are coplanar, then $\begin{vmatrix} x_1 & y_1 & z_1 \\ x_2 & y_2 & z_2 \\ x_3 & y_3 & z_3 \end{vmatrix} = 0$

The set of equation

$$x_1x + y_1y + z_1z = 0$$
$$x_2x + y_2y + z_2z = 0$$
$$x_3x + y_3y + z_3z = 0$$

has a non-trivial solution.

Let the given set has a non-trivial solution x, y, z without loss of generality, we can assume that $x \geq y \geq z$.

For the given equation $x_1x + y_1y + z_1z = 0$, we have

$$x_1x = -y_1y - z_1z$$

$\Rightarrow \quad |x_1x| = |y_1y + z_1z| \leq |y_1y| + |z_1z|$

$\Rightarrow \quad |x_1x| \leq |y_1x| + |z_1x|$

$\Rightarrow \quad |x_1| < |y_1| + |z_1|$

Which is a contradiction to the given inequality.

$$|x_1| > |y_1| + |z_1|$$

Similarly, the other inequalities rule out the possibility of a non-trivial solution.

Therefore, the given equations have only a trivial solution.

So, the given vectors are non-coplanar.

78. The vectors $\mathbf{n}_1 \times \mathbf{n}_2$, $\mathbf{n}_2 \times \mathbf{n}_3$ and $\mathbf{n}_3 \times \mathbf{n}_1$ are non-coplanar vectors, so every vector can be written as

$$\mathbf{r} = a(\mathbf{n}_1 \times \mathbf{n}_2) + b(\mathbf{n}_2 \times \mathbf{n}_3) + c(\mathbf{n}_3 \times \mathbf{n}_1)$$

Substituting this value in $\mathbf{r} \cdot \mathbf{n}_1 = q_1$, $\mathbf{a}, \mathbf{b}$, we get

$$a(\mathbf{n}_1 \times \mathbf{n}_2) \cdot \mathbf{n}_1 + b(\mathbf{n}_2 \times \mathbf{n}_3) \cdot \mathbf{n}_1 + c(\mathbf{n}_3 \times \mathbf{n}_1) \cdot \mathbf{n}_1 = q_1$$

$$b(\mathbf{n}_1\ \mathbf{n}_2\ \mathbf{n}_3) = q_1 \Rightarrow b = \frac{q_1}{\mathbf{n}_1\ \mathbf{n}_2\ \mathbf{n}_3}$$

Since, the required point of intersection will have the position vector,

$$\mathbf{r} = \frac{1}{(\mathbf{n}_1\ \mathbf{n}_2\ \mathbf{n}_3)}[q_3(\mathbf{n}_1 \times \mathbf{n}_2) + q_1(\mathbf{n}_2 \times \mathbf{n}_3) + q_2(\mathbf{n}_3 \times \mathbf{n}_1)]$$

79. Since $r^2 + s^2 = e^2$

$\Rightarrow e = 31$ or $e = 19$ is not possible.

Therefore, e equals 13, 20 or 25.

The possibility for triplet $\{r, s, e\}$ are $\{5, 12, 13\}$, $\{12, 16, 20\}$, $\{15, 20, 25\}$, $\{7, 24, 25\}$.

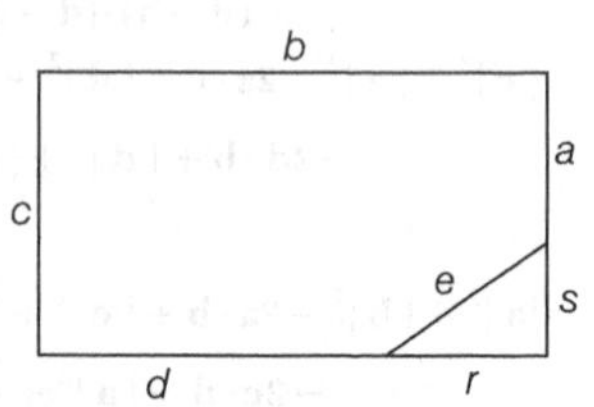

Since 16, 15 and 24 do not appear among any of pair wise differences of 13, 19, 20, 25, 31

$\Rightarrow a = 19, b = 25, c = 31, d = 20, e = 13$

Hence, required area = 745 sq units.

80. Point A is (a, b, c)

$\Rightarrow$ Points P, Q, R are $(a, b, -c), (-a, b, c)$ and $(a, -b, c)$ respectively.

$\Rightarrow$ Centroid of triangle PQR is $\left(\frac{a}{3}, \frac{b}{3}, \frac{c}{3}\right)$

$\Rightarrow G \equiv \left(\frac{a}{3}, \frac{b}{3}, \frac{c}{3}\right)$

$\Rightarrow A, O, G$ are collinear $\Rightarrow$ Area of triangle AOG is zero.

81. Let line joining AB meet plane $2x + 3y + 5z = 1$ at P.

Let $$P = \left(\frac{\lambda + 1}{\lambda + 1}, \frac{-5\lambda}{\lambda + 1}, \frac{7\lambda - 3}{\lambda + 1}\right) \quad \left[\frac{AP}{PB} = \lambda\right]$$

$$2.1 + 3\left(\frac{-5\lambda}{\lambda + 1}\right) + 5\left(\frac{7\lambda - 3}{\lambda + 1}\right) = 1$$

$$2(+1) - 15\lambda + 35\lambda - 15 = \lambda + 1$$

$$\Rightarrow \quad \lambda = \frac{2}{3}$$

$$\Rightarrow \quad P = (1, -2, 1)$$

$$\Rightarrow \quad AP = 2\sqrt{5}$$

82. $$Q = (r, r, -r)$$

PQ perpendicular $\frac{x}{1} = \frac{y}{1} = \frac{z}{-1}$

$$\therefore \quad (\alpha - r) \cdot 1 + (\beta - r) \cdot 1 + (\gamma + r)(-1) = 0$$

$$r = \frac{\alpha + \beta - \gamma}{3}$$

$$\therefore \quad PQ^2 = \frac{2}{3}\{\alpha^2 + \beta^2 + \gamma^2 - \alpha\beta + \beta\gamma + \gamma\}$$

But $PQ = 2$

{perpendicular distance from $P(\alpha, \beta, \gamma)$ to plane $x + y + z = 0$}

$$\therefore \quad \frac{2}{3}(\alpha^2 + \beta^2 + \gamma^2 - \alpha\beta + \beta\gamma + \gamma\alpha) = 4\left\{\frac{\alpha + \beta + \gamma}{\sqrt{3}}\right\}^2$$

$$\therefore \quad \alpha^2 + \beta^2 + \gamma^2 + 5\alpha\beta + 3\beta\gamma + 3\gamma\alpha = 0$$

83. The cut $x = y$ separates the cube into points with $x < y$ and those with $x > y$.

$\therefore$ So, number of pieces equals to the number of ways of arrangements of x, y and z which is $3! = 6$.

Aliter Since in each coordinate there is inequality $x > y > z$.

So, number of pieces = number of ways of arranging $x, y, z = 6$

84. Here P and Q lie on the same side of XY plane

Image $P(1, 2, 3)$ on the XY plane is $P'(1, 2, -3)$

Reflected ray is $P'Q \Rightarrow \frac{x-3}{2} = \frac{y-2}{0} = \frac{z-5}{8}$

$$\Rightarrow \quad \frac{x-3}{1} = \frac{y-2}{0} = \frac{z-5}{4}$$

85. Let $\frac{x}{a} + \frac{y}{b} + \frac{z}{c} = 1$ be required plane.

Let the sphere be $x^2 + y^2 + z^2 + 2ux + 2vy + 2wz + d = 0$

$d = 0$ if it passes through origin.

Also, $a = -2u, b = -2v, c = -2w$

and $$\frac{\alpha - \beta}{-2u} + \frac{\beta - \gamma}{-2v} + \frac{\gamma - \alpha}{-2w} = 1$$

Locus of centre $(-u, -v, -w)$ is $\Sigma(\alpha - \beta)yz = 2xyz$

86. On solving the given planes, the vertices are $O(0, 0, 0)$, $A(-a, a, a)$, $B(a, -a, a)$, $C(a, a, -a)$.

Consider the edges OA, BC whose equations are $\frac{x}{-1} = \frac{y}{1} = \frac{z}{1}$;

$$\frac{x-a}{0} = \frac{y+a}{1} = \frac{z-a}{-1}$$

Now, find S.D. between the lines.

87. The angle between the pair of planes is $ax^2 + by^2 + cz^2 + 2fyz + 2gzx + 2hxy = 0$ is

$$\theta = \tan^{-1}\left(\frac{2\sqrt{f^2 + g^2 + h^2 - ab - bc - ca}}{a + b + c}\right)$$

88. $$2p + 2q + r = 6$$

$$\Rightarrow (2\hat{\mathbf{i}} + 2\hat{\mathbf{j}} + \hat{\mathbf{k}}) \cdot (p\hat{\mathbf{i}} + q\hat{\mathbf{j}} + r\hat{\mathbf{k}}) = 6$$

$$(\mathbf{a} \cdot \mathbf{b})^2 \le |\mathbf{a}|^2 |\mathbf{b}|^2$$

$$\Rightarrow \quad 6^2 \le 9(p^2 + q^2 + r^2)$$

$$p^2 + q^2 + r^2 \le 4$$

89. Let $A(x_1, y_1, z_1), B(x_2, y_2, z_2), C(x_3, y_3, z_3), D(x_4, y_4, z_4)$ be the vertices of tetrahedron. If E is the centroid of face BCD and G is the centroid of $ABCD$ then $AG = 3/4(AE)$

$\therefore \quad K = 3/4.$

90. $$y(x+y) + z(x+y) = 0$$

$x + y = 0 \Rightarrow$ dr's are $\mathbf{b}_1 = (1, 1, 0)$

and $y + z = 0 \Rightarrow$ dr's are $\mathbf{b}_2 = (0, 1, 1)$

Now, $$\mathbf{b}_1 \times \mathbf{b}_2 = \begin{vmatrix} \hat{\mathbf{i}} & \hat{\mathbf{j}} & \hat{\mathbf{k}} \\ 1 & 1 & 0 \\ 0 & 1 & 1 \end{vmatrix} = (1, -1, 1)$$

and $\mathbf{a}_1 - \mathbf{a}_1 = (1, 1, 1) - (0, 0, 0) = (1, 1, 1)$

$$\therefore \quad \text{SD} = \frac{|(1, 1, 1) \times (1, -1, 1)|}{|(1, -1, 1)|} = \frac{|(-2, 0, 2)|}{|(1, -1, 1)|}$$

$$= \frac{\sqrt{4+0+4}}{\sqrt{1+1+1}} = \sqrt{\frac{8}{3}}$$

91. $L_1 \| L_2$ Then required distance = distance between $(k_1, k_2, 0)$, $(k_3, k_4, 0)$

$$= \sqrt{(k_1 - k_3)^2 + (k_2 - k_4)^2}$$

92. Let $\mathbf{a} = l_1\hat{\mathbf{i}} + m_1\hat{\mathbf{j}} + n_1\hat{\mathbf{k}}, \mathbf{b} = l_2\hat{\mathbf{i}} + m_2\hat{\mathbf{j}} + n_2\hat{\mathbf{k}}$

and $\mathbf{c} = l_3\hat{\mathbf{i}} + m_3\hat{\mathbf{j}} + n_3\hat{\mathbf{k}}$

Given that **a**, **b**, **c** are three mutually perpendicular unit vectors.

Then, $p_1\hat{\mathbf{i}} + q_1\hat{\mathbf{j}} + r_1\hat{\mathbf{k}} = \mathbf{b} \times \mathbf{c} = \mathbf{a}$

($\because$ $\mathbf{b} \times \mathbf{c}$ parallel to **a** and $\mathbf{b} \times \mathbf{c}$, **a** are unit vectors)

Similarly, $p_2\hat{\mathbf{i}} + q_2\hat{\mathbf{j}} + r_2\hat{\mathbf{k}} = \mathbf{c} \times \mathbf{a} = \mathbf{b}$

and $p_3\hat{\mathbf{i}} + q_3\hat{\mathbf{j}} + r_3\hat{\mathbf{k}} = \mathbf{a} \times \mathbf{b} = \mathbf{c}$

These vectors also mutually perpendicular unit vectors.

93. Let us suppose A be origin.

$$\text{ar}(\Delta ABC)^2 + \text{ar}(\Delta ACD)^2 + \text{ar}(\Delta ABD)^2 = \text{ar}(\Delta BCD)^2$$

Hence the result follows.

94. **a**, **b**, **c** be P.V. of $A, B, C, |\mathbf{a}| = |\mathbf{b}| = |\mathbf{c}| = K$

$$\mathbf{OD} = \frac{\mathbf{a}}{2} \left| |\mathbf{OE}| = \frac{\mathbf{b}+\mathbf{c}}{2} \right|$$

Given, $$|\mathbf{DE}| = 1 \Rightarrow \left| \frac{\mathbf{b}+\mathbf{c}-\mathbf{a}}{2} \right| = 1 \Rightarrow |\mathbf{b}+\mathbf{c}-\mathbf{a}| = 2$$

$$\Rightarrow \quad 3k^2 + 2k^2\left(\frac{1}{2} - \frac{1}{2} - \frac{1}{2}\right) = 4$$

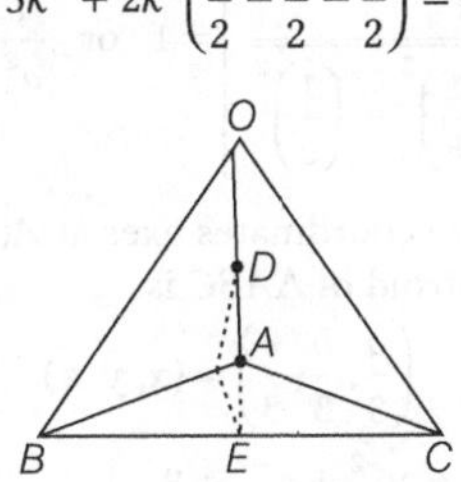

$$\Rightarrow \quad K = \sqrt{2}$$

$$\text{Volume} = \frac{1}{3}\,(\text{base area} \times \text{height}) = \frac{1}{3}\left(\frac{\sqrt{3}}{4}(k)^2 \times \sqrt{\frac{2}{3}}k\right) = \frac{1}{3}$$

95. The plane equation in the intercept forms is $\frac{x}{a} + \frac{y}{b} + \frac{z}{c} = 1.$

Volume of tetrahedron $OABC$ is

$$v = \frac{abc}{6} = 64 \Rightarrow abc = 384$$

Foot of perpendicular from (0, 0, 0) on this plane is

$$\frac{x}{1/a} = \frac{y}{1/b} = \frac{z}{1/c} = \frac{1}{\frac{1}{a^2} + \frac{1}{b^2} + \frac{1}{c^2}} = k$$

$$\Rightarrow \quad x = \frac{k}{a}, y = \frac{k}{b}, z = \frac{k}{c}$$

and $$\frac{1}{k} = \frac{1}{a^2} + \frac{1}{b^2} + \frac{1}{c^2}$$

$$\Rightarrow \quad \frac{1}{k} = \frac{x^2 + y^2 + z^2}{k^2}$$

$$\Rightarrow x^2 + y^2 + z^2 = k$$

$\therefore (x^2 + y^2 + z^2)^3 = k^3 = abc\,xyz = 384\,xyz$ is the required locus.

96. Let $A(x_1, y_1, z_1), B(x_2, y_2, z_2)$

$C(x_3, y_3, z_3), D(x_4, y_4, z_4)$

and the equation of the plane containing P, Q, R and S is $ax + by + cz + d = 0$ and $k_R = ax_r + by_r + cz_r + d$

$$\therefore \quad \frac{AP}{PB} \cdot \frac{BQ}{QC} \cdot \frac{CR}{RD} \cdot \frac{DS}{SA} = \frac{-K_1}{K_2} \cdot \frac{-K_2}{K_3} \cdot \frac{-K_3}{K_4} \cdot \frac{-K_4}{K_1} = 1$$

97. Let a, b and c be the DR's of the given line. Then,

we have $$3a - b + c = 0$$

$$5a + b + 3c = 0$$

On solving, we get $\frac{a}{1} = \frac{b}{1} = \frac{c}{-2}$

Again, suppose the given line intersect the plane $z = 0$ at $(x_1, y_1, 0)$, then $3x_1 - y_1 + 1 = 0$ and $5x_1 + y_1 = 0$

On solving, we get $x_1 = -\frac{1}{8}, y_1 = \frac{5}{8}$

Hence, the symmetrical form of the line is

$$\frac{x + \frac{1}{8}}{1} = \frac{y - \frac{5}{8}}{1} = \frac{z}{-2}$$

Equation of plane through (2, 1, 4) is

$$a(x-2) + b(y-1) + c(z-4) = 0$$

when $a = 1, b = 1$ and $c = -21$

$$x - 2 + y - 1 - 2(z - 4) = 0$$

$$x + y - 2z + 5 = 0$$

98. Vector normal to the plane

$$\mathbf{n} = \hat{\mathbf{i}} + \hat{\mathbf{j}} + \hat{\mathbf{k}}$$

$$\mathbf{V}_x = \hat{\mathbf{i}}, \mathbf{V}_y = \hat{\mathbf{j}}, \mathbf{V}_z = \hat{\mathbf{k}}$$

$$\cos(90° - \alpha) = \frac{\mathbf{V}_x \cdot \mathbf{n}}{|\mathbf{n}|}$$

$$\Rightarrow \quad \sin\alpha = \frac{1}{\sqrt{3}}$$

Similarly, $\sin\beta = \frac{1}{\sqrt{3}}$ and $\sin\gamma = \frac{1}{\sqrt{3}}$

Hence, $\sum \sin^2\alpha = 1$

and $\sum \cos^2\alpha = 2$

Also, plane is equally inclined with the coordinate axes.

Also, $A = \frac{1}{2}\sqrt{9^2 + 9^2 + 9^2} = \frac{9\sqrt{3}}{2}$

99. $2x - y - z - 2 + \lambda(x + y + z - 1) = 0$ satisfies (1, 1, 1)

$$2 - 1 - 1 - 2 + \lambda(3 - 1) = 0$$
$$\lambda = 1$$
$$x = 1 \qquad \text{...(i)}$$
$$x - y - z - 3 + \mu(2x + 4y - z - 4) = 0$$
$$-4 + \mu(1) = 0$$
$$\mu = 4$$
$$x - y - z - 3 + 4(2x + 4y - z - 4) = 0$$
$$9x + 15y - 5z - 19 = 0 \qquad \text{...(ii)}$$

From Eqs. (i) and (ii), we get

$$x - 1 = 0, 9x + 15y - 5z - 19 = 0$$

$a = 0$ and $c = 3b$

$$x - 1 = 0, \frac{y-1}{1} = \frac{z-1}{3}$$

100. $OP = \sqrt{h^2 + k^2 + l^2}$

Direction ratios of OP are $\left(\frac{h}{p}, \frac{k}{p}, \frac{l}{p}\right)$

Equation of plane is $\frac{hx}{p} + \frac{ky}{p} + \frac{lz}{p} = p$

$$A\left(\frac{p^2}{h}, 0, 0\right), B\left(0, \frac{p^2}{k}, 0\right), C\left(0, 0, \frac{p^2}{l}\right)$$

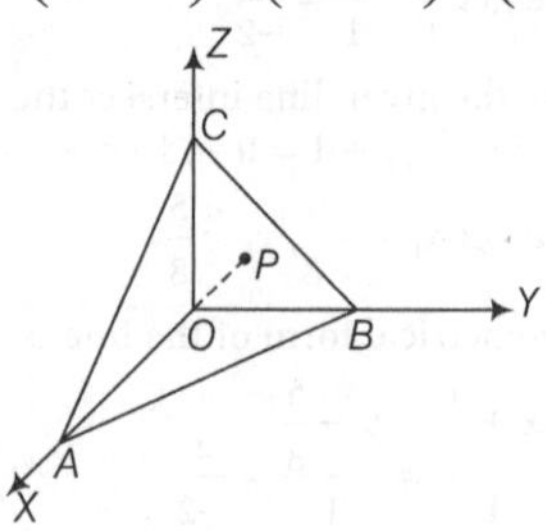

101. (a) Since, $\mathbf{n} \cdot \mathbf{a} = \mathbf{n} \cdot \mathbf{b} = \mathbf{n} \cdot \mathbf{c} = 0$

$\therefore$ **a**, **b** and **c** are coplanar

$\therefore \qquad [\mathbf{a}, \mathbf{b}, \mathbf{c}] = 0$

(b) $\cos^3 30° + \cos^2 45° + \cos^2\gamma = 1$

$\therefore \sin^2 r = \frac{3}{4} + \frac{1}{2} = \frac{5}{4}$ which is not possible.

(c) Obvious

(d) $\mathbf{AB} \times \mathbf{BC}$ is perpendicular to the plane ABC.

$$\mathbf{AB} \times \mathbf{BC} = (\mathbf{OB} - \mathbf{OA}) \times (\mathbf{OC} - \mathbf{OB})$$
$$= \mathbf{OB} \times \mathbf{OC} - \mathbf{OA} \times \mathbf{OC} + \mathbf{OA} \times \mathbf{OB}$$
$$= \mathbf{OA} \times \mathbf{OB} + \mathbf{OB} \times \mathbf{OC} + \mathbf{OA} \times \mathbf{OB}$$

i.e. $\mathbf{a} \times \mathbf{b} + \mathbf{b} \times \mathbf{c} + \mathbf{c} \times \mathbf{a}$ is perpendicular the plane ABC.

$(\mathbf{a} + \mathbf{b} + \mathbf{c}) \times (\mathbf{a} \times \mathbf{b} + \mathbf{b} \times \mathbf{c} + \mathbf{c} \times \mathbf{a})$

$$= \{(\mathbf{a} + \mathbf{b} + \mathbf{c}) \cdot \mathbf{c}\}\mathbf{b} - \{(\mathbf{a} + \mathbf{b} + \mathbf{c}) \cdot \mathbf{a}\}\mathbf{b}$$
$$+ \{(\mathbf{a} + \mathbf{b} + \mathbf{c}) \cdot \mathbf{c}\}\,\mathbf{b} - \{(\mathbf{a} + \mathbf{b} + \mathbf{c}) \cdot \mathbf{b}\} \times \mathbf{c}$$
$$+ \{(\mathbf{a} + \mathbf{b} + \mathbf{c}) \cdot \mathbf{a}\}\mathbf{c} - \{(\mathbf{a} + \mathbf{b} + \mathbf{c}) \cdot \mathbf{a}\}\mathbf{a}$$
$$= (\mathbf{a} + \mathbf{b} + \mathbf{c}) \cdot (\mathbf{b} - \mathbf{c})\mathbf{a} + (\mathbf{a} + \mathbf{b} + \mathbf{c})$$
$$\cdot(\mathbf{c} - \mathbf{a})\mathbf{b} + \{(\mathbf{a} + \mathbf{b} + \mathbf{c}) \cdot (\mathbf{a} - \mathbf{b})\}\mathbf{c}$$
$$= (\mathbf{a} \cdot \mathbf{b} - \mathbf{a} \cdot \mathbf{c} + \mathbf{b}^2 - \mathbf{c}^2)\mathbf{a} + (\mathbf{b} \cdot \mathbf{c} - \mathbf{b} \cdot \mathbf{a}$$
$$+ (\mathbf{c}^2 - \mathbf{a}^2)\mathbf{b} + (\mathbf{a}^2 - \mathbf{b}^2 + \mathbf{c} \cdot \mathbf{a} - \mathbf{c} - \mathbf{b})\mathbf{c} = \mathbf{0}$$

Thus, the statement is true.

102. Let $A(\mathbf{a})$, $B(\mathbf{d})$, $C(\mathbf{c})$ and $D(\mathbf{d})$ be the vertices of a tetrahedron, then centroid of the tetrahedron is

$$\frac{\mathbf{a} + \mathbf{b} + \mathbf{c} + \mathbf{d}}{4}$$

centroid G_1 of the face BCD is $\frac{\mathbf{b} + \mathbf{c} + \mathbf{d}}{3}$

Now, centroid of the tetrahedron G_1 divides AG_1 in the ratio 3 : 1.

i.e. $$\frac{3(\mathbf{b} + \mathbf{c} + \mathbf{d}) + \mathbf{a}}{3 + 1} = \frac{\mathbf{a} + \mathbf{b} + \mathbf{c} + \mathbf{a}}{4}$$

$\therefore$ C lies on AC_1.

(b) The edges AB and CD. Let E be the mid point of AB and F be the mid point of CD

$\therefore$Positive vector of E is $\frac{\mathbf{a} + \mathbf{b}}{2}$

Positive vector of F is $\frac{\mathbf{c} + \mathbf{a}}{2}$

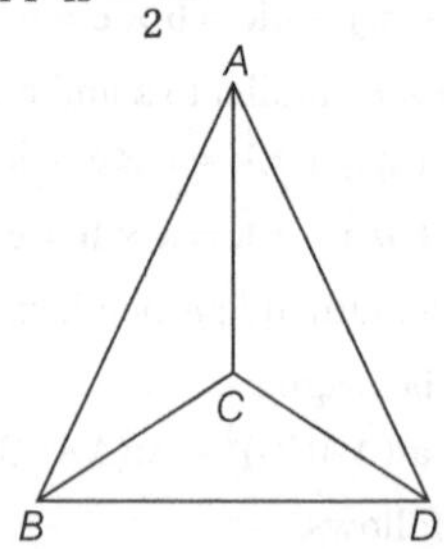

Mid-point of EF is $\frac{\mathbf{a} + \mathbf{b} + \mathbf{c} + \mathbf{d}}{4}$ which is the centroid of the tetrahedron $ABCD$.

103. Let the plane be $\frac{x}{a} + \frac{y}{b} + \frac{z}{c} = 1$

$$\left|\frac{1}{\sqrt{\left(\frac{1}{a}\right)^2 + \left(\frac{1}{b}\right)^2 + \left(\frac{1}{c}\right)^2}}\right| = 1 \text{ or } \frac{1}{a^2} + \frac{1}{b^2} + \frac{1}{c^2} = 1$$

The plane cuts the coordinates axes at $A(a, 0, 0)$, $B(0, b, 0)$, $c(0, 0, c)$. The centroid of ΔABC is

$$\left(\frac{a}{3}, \frac{b}{3}, \frac{c}{3}\right) = (x, y, z) \qquad \text{(let)}$$

$$x^{-2} + y^{-2} + z^{-2} = 9$$

or $$\frac{1}{9}\left\{\frac{1}{x^2} + \frac{1}{y^2} + \frac{1}{z^2}\right\}^{-1} = 0$$

104. When a line lies in a plane, then it is at right angles to the normal to the plane. Here, d.r's of the line are $< a, b, c >$ and altitude numbers of the plane are being taken as $< A, B, C >$. So, we must $aA + bB + cC = 0$.

105. For the given curve $z = 0$, therefore, the line and the curve meet where $\frac{x-2}{3} = \frac{y+1}{2} = \frac{0-1}{-1}$

i.e. where $\frac{x-2}{3} = 1, \frac{y+1}{z} = 1$ i.e., where $x = 5, y = 1$

So, the given line and the given curve meet in the point (5, 1, 0). Since, this point lies on the curve also, therefore, $5^2 + 1^2 = r^2$

$$\Rightarrow \quad r^2 = (\sqrt{26})^2$$

$$\Rightarrow \quad r = \pm\sqrt{26}$$

106. A vector coplanar with given vectors is $(1+\lambda)\hat{\mathbf{i}} + (\lambda - 1)\hat{\mathbf{j}} + (1-\lambda)\hat{\mathbf{k}}$. Since it is equally inclined to the two given vectors

$$\frac{(1+\lambda)\hat{\mathbf{i}} + (\lambda-1)\hat{\mathbf{j}} + (1-\lambda)\hat{\mathbf{k}}}{|(1+\lambda)\hat{\mathbf{i}} + (\lambda-1)\hat{\mathbf{j}} + (1-\lambda)\hat{\mathbf{k}}|} \cdot \frac{(\hat{\mathbf{i}} - \hat{\mathbf{j}} + \hat{\mathbf{k}})}{\sqrt{3}}$$

$$= \frac{(1+\lambda)\hat{\mathbf{i}} + (\lambda-1)\hat{\mathbf{j}} + (1-\lambda)\hat{\mathbf{k}}}{|(1+\lambda)\hat{\mathbf{i}} + (\lambda-1)\hat{\mathbf{j}} + (1-\lambda)\hat{\mathbf{k}}|} \cdot \frac{(\hat{\mathbf{i}} + \hat{\mathbf{j}} - \hat{\mathbf{k}})}{\sqrt{3}}$$

$$\therefore \quad \lambda = 1$$

Required vector is $2\hat{\mathbf{i}}$ or $\hat{\mathbf{i}}$

107. The equation of the plane through (2, 3, −1) and perpendicular to the vector $3\hat{\mathbf{i}} - 4\hat{\mathbf{j}} + 7\hat{\mathbf{k}}$ is

$$3(x-2) + (-4)(y-3) + 7(z-(-1)) = 0$$

or $\quad 3x - 4y + 7z + 13 = 0$

Distance of this plane from the origin

$$= \frac{|3 \times 0 - 4 \times 0 + 7 \times 0 + 13|}{\sqrt{3^2 + (-4)^2 + 7^2}} = \frac{12}{\sqrt{74}}$$

108. Let A, B, C be $(\alpha, 0, 0)$, $(0, \beta, 0)$ and $(0, 0, \gamma)$ then the plane ABC is $\frac{x}{\alpha} + \frac{y}{\beta} + \frac{z}{\gamma} = 1$

Since it always passes through a, b, c

$$\frac{a}{\alpha} + \frac{b}{\beta} + \frac{c}{\gamma} = 1 \quad \text{...(i)}$$

If p is (u, v, w) then $OP^2 = AP^2 = BP^2 = CP^2$

$$\Rightarrow \quad u^2 + v^2 + w^2 = (u-\alpha)^2 + v^2 + w^2$$

$$= \dots \Rightarrow \alpha = \frac{u}{2}, \beta = \frac{v}{2}, \gamma = \frac{w}{2}$$

On putting α, β, γ in (i), we get

$$\frac{a}{u} + \frac{b}{v} + \frac{c}{w} = 2$$

$\Rightarrow$ Locus of (u, v, w) is $\frac{a}{x} + \frac{b}{y} + \frac{c}{z} = 2$

109. Normal of plane P_1 is

$$\mathbf{n}_1 = (2\hat{\mathbf{j}} + 3\hat{\mathbf{k}}) \times (4\hat{\mathbf{j}} - 3\hat{\mathbf{k}}) = -18\hat{\mathbf{i}}$$

Normal to plane P_2 is

$$\mathbf{n}_2 = (\hat{\mathbf{j}} - \hat{\mathbf{k}}) \times (3\hat{\mathbf{i}} + 3\hat{\mathbf{j}}) = 3\hat{\mathbf{i}} - 3\hat{\mathbf{j}} - 3\hat{\mathbf{k}}$$

$\therefore$ **A** is parallel to $(\mathbf{n}_1 < \mathbf{n}_2) = \pm(-54\hat{\mathbf{j}} - 54\hat{\mathbf{k}})$

$\therefore$ Angle between **A** and $2\hat{\mathbf{i}} + \hat{\mathbf{j}} - 2\hat{\mathbf{k}}$ is

$$\cos\theta = \pm\frac{(-54\hat{\mathbf{j}} + 54\hat{\mathbf{k}})\cdot(2\hat{\mathbf{i}} + \hat{\mathbf{j}} - 2\hat{\mathbf{k}})}{54\sqrt{2}3} = \pm\frac{1}{\sqrt{2}}$$

$$\theta = \frac{\pi}{4} \text{ or } \frac{3\pi}{4}$$

110. Any plane through the second line is

$$2x + y + z - 1 + k(3x + y + 2z - 2) = 0$$

If this is parallel to the first line, then

$$(2 + 3\mathbf{k}) + (1 + \mathbf{k}) + (1 + 2\mathbf{k}) = 0$$

$$\Rightarrow \quad \mathbf{k} = -\frac{2}{3}$$

$\Rightarrow$ Plane is $2x + y + z - 1 - \frac{2}{3}(3x + y + 2z - 2) = 0$

or $y - z + 1 = 0$. The required SD must be distance of this plane from any point on the line $x = y = z$ say (1, 1, 1)

$$\Rightarrow \quad \text{SD} = \frac{1 - 1 + 1}{\sqrt{0^2 + 1^2 + (-1)^2}} = \frac{1}{\sqrt{2}}$$

111. $p_1 = \frac{4}{\sqrt{29}}, p_2 = \frac{1/2}{\sqrt{29}} = \frac{1}{2\sqrt{29}}, p_3 = \frac{8}{\sqrt{29}}$

For these values all the choices are easily verified.

112. Let the components of the line segment vector be a, b, c, then

$$a^2 + b^2 + c^2 = (63)^2 \quad \text{...(i)}$$

also $\quad \frac{a}{3} = \frac{b}{-2} = \frac{c}{6} = \lambda$ (say) then

$$a = 3\lambda, b = -2\lambda \text{ and } c = 6\lambda$$

and from (i), we have

$$9\lambda^2 + 4\lambda^2 + 36\lambda^2 = (63)^2$$

$$\Rightarrow \quad 49\lambda^2 = (63)^2$$

$$\Rightarrow \quad \lambda = \pm\frac{63}{7} = \pm 9.$$

The required components are 27, −18, 54 or −27, 18, −54

113. The given lines are coplanar if

$$0 = \begin{vmatrix} 2-1 & 3-4 & 4-5 \\ 1 & 1 & -k \\ k & 2 & 1 \end{vmatrix}$$

$$= \begin{vmatrix} 1 & -1 & -1 \\ 1 & 1 & -k \\ k & 2 & 1 \end{vmatrix} = \begin{vmatrix} 1 & 0 & 0 \\ 1 & 2 & 1-k \\ k & k+2 & 1+k \end{vmatrix}$$

or if $2(1+k) - (k+2)(1-k) = 0$

or if $k^2 + 3k = 0$ or if $k = 0, -3$.

114. Direction ratios of AB are $4-2, 5-3, 10-4$ or 1, 1, 3. So AB in parallel to the vector $\hat{\mathbf{i}} + \hat{\mathbf{j}} + 3\hat{\mathbf{k}}$ and passes through B (2, 3, 4), the vector $2\hat{\mathbf{i}} + 3\hat{\mathbf{j}} + 4\hat{\mathbf{k}}$, its equation is

$2\hat{\mathbf{i}} + 3\hat{\mathbf{j}} + 4\hat{\mathbf{k}} + \lambda(\hat{\mathbf{i}} + \hat{\mathbf{j}} + 3\hat{\mathbf{k}})$

Similarly, BC passes through the points B (2, 3, 4) and its direction ratios are $2-1, 3-2, 4+1$ or 1, 1, 5.

So its cartesian equation is

$$\frac{x-2}{1} = \frac{y-3}{1} = \frac{z-4}{5}$$

Next, if D is (a, b, c), then since $ABCD$ is a parallelogram mid point of AC and BD is same. (diagonals of a parallelogram bisect each other)

$$\Rightarrow \left(\frac{a+2}{2}, \frac{b+3}{2}, \frac{c+4}{2}\right) = \left(\frac{5}{2}, \frac{7}{2}, \frac{9}{2}\right)$$

$$\Rightarrow \qquad (a, b, c) = (3, 4. 5)$$

AB is not perpendicular to BC because

$$1 \times 1 + 1 \times 1 + 3 \times 5 \neq 0.$$

$ABCD$ is not a rectangle.

115. The coordinates of P where the line $x = y = z$ meets the plane $x + y + z = 1$ are $(1/3, 1/3, 1/3)$ and the co-ordinates of R and S where the line meets the sphere $x^2 + y^2 + z^2 = 1$ are $(1/\sqrt{3}, 1/\sqrt{3}, 1/\sqrt{3})$ and $(-1/\sqrt{3}, -1/\sqrt{3}, -1/\sqrt{3})$

So that $\qquad PR = \sqrt{3}\left|\left(\frac{1}{3} - \frac{1}{\sqrt{3}}\right)\right| = \sqrt{3}\left(\frac{1}{\sqrt{3}} - \frac{1}{3}\right)$

and $\qquad PS = \sqrt{3}\left(\frac{1}{3} + \frac{1}{\sqrt{3}}\right)$

$$\Rightarrow \qquad PR \cdot PS = 3\left(\frac{1}{3} - \frac{1}{9}\right) = \frac{2}{3}$$

$$PR + PS = 2$$

and $\qquad RS = \sqrt{(1/\sqrt{3} + 1/\sqrt{3})^2 \times 3} = 2$

So that, $\quad PR + PS = RS$

116. The rod sweeps out the figure which is a cone.

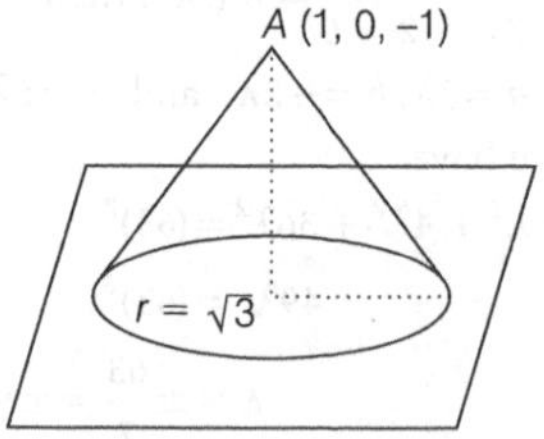

The distance of point $A(1, 0, -1)$ from the plane is $\frac{|1 - 2 + 4|}{\sqrt{9}} = 1$ unit.

The slant height l of the cone is 2 units.

Then the radius of the base of the cone is

$$\sqrt{l^2 - 1} = \sqrt{4 - 1} = \sqrt{3}.$$

Hence, the volume of the cone is $\frac{\pi}{3}(\sqrt{3})^2 \cdot 1 = \pi$ cubic units.

Area of the circle on the plane which the rod traces is 3π.

Also, the centre of the circle is $Q(x, y, z)$.

Then $\qquad \frac{x-1}{1} = \frac{y-0}{-2} = \frac{z+1}{2}$

$$= \frac{-(1 - 0 - 2 + 4)}{1^2 + (-2)^2 + 2^2}$$

or $\qquad Q(x, y, z) \equiv \left(\frac{2}{3}, \frac{2}{3}, \frac{-5}{3}\right).$

117. Observe that the lines L_1, L_2 and L_3 are parallel to the vector $(1, -1, -1)$.

Also, $\Delta = 0 = \Delta_1$ and $b_1c_2 - b_2c_1 \neq 0$

118. Volume = Area of base × height

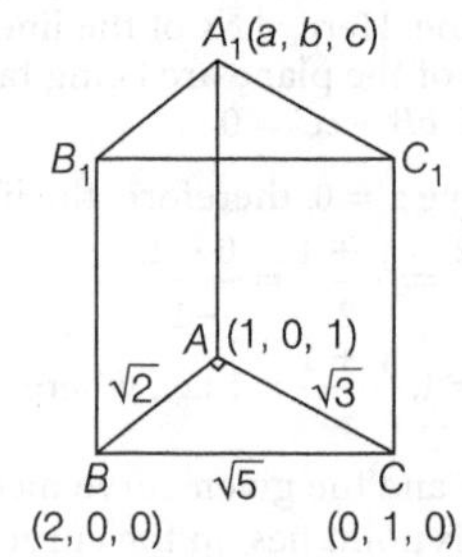

$$3 = \frac{1}{2} \times \sqrt{2} \times \sqrt{3} \times h$$

$$h = \sqrt{6}$$

$$(A_1A)^2 = h^2 = 6$$

$$\mathbf{A_1A} \cdot \mathbf{AB} = 0$$

$$\mathbf{A_1A} \cdot \mathbf{AC} = 0$$

$$\mathbf{AA_1} \cdot \mathbf{BC} = 0$$

On solving, we get position vector of A_1 are $(0, -2, 0)$ or $(2, 2, 2)$.

119. Let the equation of plane be $lx + my + nz = 0$, where l, m, n be d.c's $\Rightarrow l^2 + m^2 + n^2 = 1 \rightarrow$ (i)

Given line $\frac{x-1}{2} = \frac{y+3}{-1} = \frac{z+1}{-2} \Rightarrow 2l - m - 2n = 0 \rightarrow$ (ii);

Also, $\qquad \frac{l - 3m - n}{\sqrt{l^2 + m^2 + n^2}} = \frac{5}{3}$

$$\Rightarrow \qquad l - 3m - n = \frac{5}{3} \rightarrow \text{(iii)}$$

Solving (i), (ii) and (iii), we get equation of plane as

$$x - 2y + 2z = 0 \text{ or } 2x + 2y + z = 0.$$

120. (a) Height $= h = \sqrt{1 - \frac{1}{3}} = \sqrt{\frac{2}{3}}$

(b) Required distance $= \frac{1}{2}\left(\sqrt{\frac{2}{3}}\right) = \frac{1}{\sqrt{6}}$

(c) Angle $= \frac{\pi}{2}$

(d) Required distance $= \frac{3}{4}(h) = \frac{3}{4}\left(\sqrt{\frac{2}{3}}\right) = \sqrt{\frac{3}{8}}$

121. Let $OA = a$, $OB = b$, $OC = c$ then

$$a \cdot a + (b - c) \cdot (b - c) = b \cdot b + (c - a) \cdot (c - a)$$

$$\Rightarrow \qquad -2b \cdot c = -2c \cdot a \Rightarrow (a \cdot b) \cdot c = 0$$

or $\qquad BA \cdot OC = 0$

Hence, $AB \perp OC$ similarly $BC \perp OA$ and $CA \perp OB$.

122. Intersection of line with both the planes are the same

$$\Rightarrow \qquad \frac{3}{3\beta^2 + 6(1 - 2\alpha) + 3} = \frac{-6}{6\alpha^2 + 6(1 - 2\beta) + 6}$$

$$\Rightarrow \qquad 2(\beta - 1)^2 + 3(\alpha - 2)^2 = 0 \Rightarrow \alpha = 2, \beta = 1$$

123. If P be (x, y, z) then from the figure,

$$x = r\sin\theta\cos\phi, \; y = r\sin\theta\sin\phi, \; z = r\cos\theta$$

$$\Rightarrow \quad 1 = r\sin\theta\cos\phi, \; 2 = r\sin\theta\sin\phi, \; 3 = r\cos\theta$$

$$\Rightarrow 1^2 + 2^2 + 3^2 = r^2 \Rightarrow r = \pm \sqrt{14}$$

$$\therefore \quad \sin\theta\cos\phi = \frac{1}{\sqrt{14}}, \ \sin\theta\sin\phi = \frac{2}{\sqrt{14}}, \ \cos\theta = \frac{3}{\sqrt{14}}$$

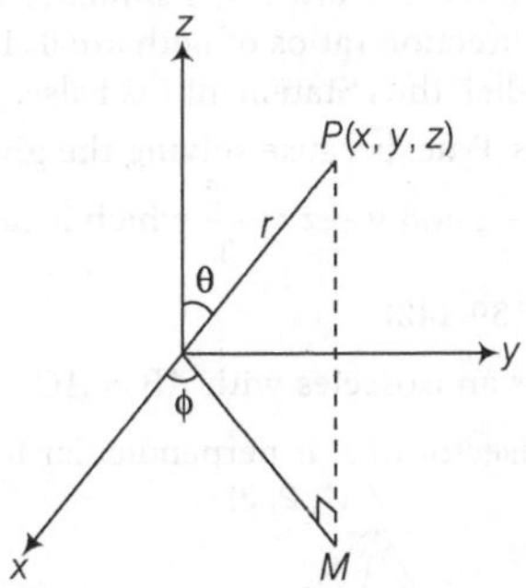

(neglecting negative sign as θ and ϕ are acute)

$$\therefore \quad \frac{\sin\theta\sin\phi}{\sin\theta\cos\phi} = \frac{2}{1} \Rightarrow \tan\phi = 2$$

Also, $\tan\theta = \dfrac{\sqrt{5}}{3}$

124. Let (l, m, n) be the direction cosines of the line perpendicular to the plane.

$\Rightarrow$ Equation of the plane $lx + my + nz = p$

$$\frac{x}{\left(\frac{p}{l}\right)} + \frac{y}{\left(\frac{p}{m}\right)} + \frac{z}{\left(\frac{p}{n}\right)} = 1$$

$A(p/l, 0, 0), B(0, p/m, 0), C(0, 0, p/n)$

Centroid of tetrahedron $OABC$ is

$$(x, y, z) = \left(\frac{p}{4l}, \frac{p}{4m}, \frac{p}{4n}\right)$$

Using, $l^2 + m^2 + n^2 = 1$

$$x^2y^2 + y^2z^2 + z^2x^2 = \frac{16}{p^2}x^2y^2z^2$$

$$\Rightarrow \quad \frac{1}{x^2} + \frac{1}{y^2} + \frac{1}{z^2} = \frac{16}{p^2}$$

Put $x = \dfrac{p}{4}\sec\alpha\sec\beta, y = \dfrac{p}{4}\sec\alpha \operatorname{cosec}\beta, z = \dfrac{p}{4}\operatorname{cosec}\alpha$

$$\frac{1}{x} = \frac{4}{p}\cos\alpha\cos\beta, \frac{1}{y} = \frac{4}{p}\cos\alpha\sin\beta, \frac{1}{z} = \frac{4}{p}\sin\alpha$$

$$\left(\frac{1}{x}\right)^2 + \left(\frac{1}{y}\right)^2 + \left(\frac{1}{z}\right)^2 = \frac{16}{p^2}$$

$$[\cos^2\alpha\cos^2\beta + \cos^2\alpha\sin^2\beta + \sin^2\alpha]$$

$$= \frac{16}{p^2}[\cos^2\alpha + \sin^2\alpha] = \frac{16}{p^2}$$

125. Statement I $\mathbf{PA} \cdot \mathbf{PB} = 9 > 0$

$\therefore P$ is exterior to the sphere

Statement II is true (standard result)

126. **Statement II** $\mathbf{r} \times (\mathbf{i} + 2\mathbf{j} - 3\mathbf{k}) = \begin{vmatrix} \hat{\mathbf{i}} & \hat{\mathbf{j}} & \hat{\mathbf{k}} \\ x & y & z \\ 1 & 2 & -3 \end{vmatrix}$

$$\hat{\mathbf{i}}(-3y - 2z) - \hat{\mathbf{j}}(-3x - z) + \hat{\mathbf{k}}(2x - y)$$

$\therefore \quad -3y - 2z = 2, 3x + z = -1, 2x - y = 0$

i.e. $-6x - 2z = 2, 3x + z = 1.$

$\therefore$Straight line $2x - y = 0, 3x + z = -1$

Statement I

$$\mathbf{r} \times (2\mathbf{i} - \hat{\mathbf{j}} + 3\hat{\mathbf{k}}) = \begin{vmatrix} \hat{\mathbf{i}} & \hat{\mathbf{j}} & \hat{\mathbf{k}} \\ x & y & z \\ 2 & -1 & 3 \end{vmatrix}$$

$$= \hat{\mathbf{i}}(3y + z) - \hat{\mathbf{j}}(3x - 2z) + \hat{\mathbf{k}}(-x - 2y)$$

$\therefore \quad 3y + z = 3, 3x - 2z = 0, -x - 2y = 1$

$$3x - 2(3 - 3y) = 0$$

$$3x + 6y = 6$$

$$\Rightarrow \quad x + 2y = 2$$

Now, $x + 2y = -1, x + 2y = 2$ are parallel planes.

$\therefore \mathbf{r} \times (2\hat{\mathbf{i}} - \hat{\mathbf{j}} + 3\hat{\mathbf{k}}) = 3\hat{\mathbf{i}} + \hat{\mathbf{k}}$ is not a straight line.

127. $\sin\theta = \left|\dfrac{2 - 3 + 2}{\sqrt{4 + 9 + 4\sqrt{3}}}\right| = \dfrac{1}{\sqrt{51}}$

$\therefore$Statement I is true, Statement II is true by definition.

128. **Statement I**

$$3y - 4z = 5 - 2k$$

$$-2y + 4z = 7 - 3k$$

$\therefore x = k, y = 12 - 5k, z = \dfrac{31 - 13k}{4}$ is a point on the line for all real value of k.

Statement I is true.

Statement II Direction ratios of the straight line are $< bc' - kbc, kac - ac', 0 >$ direction ratios of normal to be plane $< 0, 0, 1 >$

Now, $0 \times (bc' - kbc) + 0 \times (kac - ac') + 1 \times 0 = 0$

$\therefore$The straight line is parallel to the plane.

Statement II is the true but does not explain Statement I.

129. Equation of the plane, perpendicular to the plane P and containing line L is $8x + y - 7z = 4$.

130. L_1 and L_2 are obviously not parallel.

Consider the determinent

$$D = \begin{vmatrix} 2 & -4 & 1 \\ 2 & 4 & -3 \\ 1 & 3 & 2 \end{vmatrix}$$

$$= 2(8 + 9) + 4(4 + 3) + 1(6 - 4)$$

$$= 34 + 28 + 2$$

$$D \neq 0 \Rightarrow \text{Skew}$$

Hence, Statement I is false.

131. $\mathbf{n} = \mathbf{a} \times \mathbf{b}$. Equation of the plane

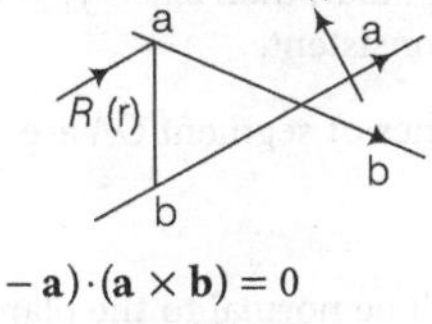

$$(\mathbf{r} - \mathbf{a}) \cdot (\mathbf{a} \times \mathbf{b}) = 0$$

$$[\mathbf{r}\ \mathbf{a}\ \mathbf{b}] = 0$$

132. Statement II is not true because image of P in a plane is a point M such that PM is perpendicular to the plane and the mid-point of PM lies on the plane.

The point A, B, C are respectively $(-a, b, c), (a, -b, c)$ and $(a, b, -c)$

which lie on the plane $\frac{x}{a}+\frac{y}{b}+\frac{z}{c}=1$ and thus Statement I is true.

133. In Statement II, let $\mathbf{r}$ be the position vector of the point on the locus, then

$$|\mathbf{r}-\mathbf{a}|=|\mathbf{r}-\mathbf{b}| \Rightarrow (\mathbf{r}-\mathbf{a})^2=(\mathbf{r}-\mathbf{b})^2$$

$$\Rightarrow |\mathbf{r}|^2+|\mathbf{a}|^2-2\mathbf{r}\cdot\mathbf{a}=|\mathbf{r}|^2+|\mathbf{b}|^2-2\mathbf{r}\cdot\mathbf{b}$$

$$\Rightarrow 2\mathbf{r}\cdot(\mathbf{a}-\mathbf{b})+|\mathbf{b}|^2-|\mathbf{a}|^2=0$$

$$\Rightarrow 2\mathbf{r}\cdot(\mathbf{a}-\mathbf{b})+(\mathbf{b}-\mathbf{a})\cdot(\mathbf{b}+\mathbf{a})=0$$

$$\Rightarrow \left(\mathbf{r}-\frac{\mathbf{a}+\mathbf{b}}{2}\right)\cdot(\mathbf{a}-\mathbf{b})=0$$

Showing that Statement II is true using it for Statement I.

we get the required locus as

$$\left[\mathbf{r}-\frac{3\hat{\mathbf{i}}-2\hat{\mathbf{j}}+5\hat{\mathbf{k}}+\hat{\mathbf{i}}+2\hat{\mathbf{j}}+\hat{\mathbf{k}}}{2}\right]$$

$$\cdot(3\hat{\mathbf{i}}-2\hat{\mathbf{j}}+5\hat{\mathbf{k}}-(\hat{\mathbf{i}}+2\hat{\mathbf{j}}-\hat{\mathbf{k}})=0$$

$$\Rightarrow [\mathbf{r}-(2\hat{\mathbf{i}}+2\hat{\mathbf{k}})\cdot(2\hat{\mathbf{i}}-4\hat{\mathbf{j}}+6\hat{\mathbf{k}})]=0$$

$$\Rightarrow \mathbf{r}\cdot(\hat{\mathbf{i}}-2\hat{\mathbf{j}}+3\hat{\mathbf{k}})=2\times1-0\times2+2\times3=8$$

and thus Statement I is also true.

134. Since $\mathbf{a}$ and $\mathbf{c}$ are non-collinear. Equating the coefficients of $\mathbf{a}$ and $\mathbf{c}$ in the two values of $\mathbf{r}$ we get.

$$6-\lambda=1+\mu, 2\lambda-1=3\mu-1 \Rightarrow \lambda=3, \mu=2$$

So there exist values for λ and μ such that the two values of $\mathbf{r}$ are same showing that the lines intersect and hence they are coplanar. Thus, statement I and statement II both are true and the first follows from the second.

135. Since, $\begin{vmatrix}1-2 & 0+1 & -1-0\\ 1 & -1 & 1\\ 1 & 2 & 3\end{vmatrix}=\begin{vmatrix}-1 & 1 & -1\\ 1 & -1 & 1\\ 1 & 2 & 3\end{vmatrix}=0$

The lines in Statement I are coplanar and equation of the plane containing them is

$$\begin{vmatrix}x-1 & y & z+1\\ 1 & -1 & 1\\ 1 & 2 & 3\end{vmatrix}=-(5x+2y-3z-8)=0$$

So Statement I is true.

Also, Statement II is true because $\frac{1}{3}=\frac{2}{6}=\frac{3}{9}$ and $1+2-3=0$

But does not lead to Statement I.

136. Any point on the first line is $(2x_1+1, x_1-3, -3x_1+2)$.

Any point on the second line is $(y_1+2, -3y_1+1, 2y_1-3)$

If two lines are coplanar, then $2x_1-y_1=1$, $x_1+3y_1=4$ and $3x_1+2y_1=5$ are consistent.

137. The direction cosines of segment OA are $\frac{2}{\sqrt{14}}, \frac{1}{\sqrt{14}}$ and $\frac{-3}{\sqrt{14}}$.

$OA=\sqrt{14}$

This means OA will be normal to the plane and the equation of the plane is $2x+y-3z=14$.

138. If l, m, n denote the direction ratios of L_1 and $l+m-n=0$ and $l-3m+3n=0 \Rightarrow l=0, m=n$

$\Rightarrow$ direction ratios of L_1 are 0, 1, 1 similarly for L_2 and L_3, we find that the direction ratios of both are 0, 1, 1 showing that L_1, L_2, L_3 are parallel, thus Statement I is False.

Statement II is True, because solving the given equation we get $x=0$, $y-z=-1$ and $y-z=\frac{-2}{3}$ which is not possible.

Solution (Q. Nos. 139-142)

139. Here, ΔABC is an isosceles with $AB=AC$

So, internal bisector of A is perpendicular to BC.

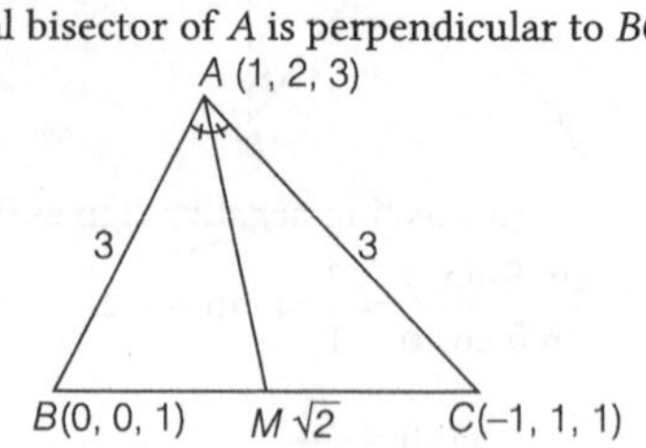

$$\Delta AMB \cong \Delta AMC \text{ (RHS rule)}$$

M is mid-point of BC.

So, $M \equiv \left(\frac{-1}{2}, \frac{-1}{2}, 1\right)$

$\therefore$ Equation of internal bisector through A to side BC is

$$\mathbf{r}=(\hat{\mathbf{i}}+2\hat{\mathbf{j}}+3\hat{\mathbf{k}})+\mu\left(\frac{3}{2}\hat{\mathbf{i}}+\frac{3}{2}\hat{\mathbf{j}}+2\hat{\mathbf{k}}\right)$$

$$\Rightarrow \mathbf{r}=(\hat{\mathbf{i}}+2\hat{\mathbf{j}}+3\hat{\mathbf{k}})+\mu(3\hat{\mathbf{i}}+3\hat{\mathbf{j}}+4\hat{\mathbf{k}})$$

Aliter Equation of BC is $\mathbf{r}=\hat{\mathbf{k}}+\lambda(\hat{\mathbf{i}}-\hat{\mathbf{j}})$

Let position vector of M on BC be $\mathbf{r}$.

Now, $\mathbf{AM}=$ Position vector of $M-$ Position vector of A

$$=(\lambda\hat{\mathbf{i}}-\lambda\hat{\mathbf{j}}+\hat{\mathbf{k}})-(\hat{\mathbf{i}}+2\hat{\mathbf{j}}+3\hat{\mathbf{k}})$$

$$=(\lambda-1)\hat{\mathbf{i}}-(\lambda+2)\hat{\mathbf{j}}-2\hat{\mathbf{k}}$$

Since, $\mathbf{AM}\cdot(\hat{\mathbf{i}}-\hat{\mathbf{j}})=0 \Rightarrow \lambda=\frac{-1}{2}$

Position vector of point, $M=\frac{-1}{2}\hat{\mathbf{i}}+\frac{1}{2}\hat{\mathbf{j}}+\hat{\mathbf{k}}$

Equation of internal bisector through A to the side BC is

$$\mathbf{r}=(\hat{\mathbf{i}}+2\hat{\mathbf{j}}+3\hat{\mathbf{k}})+\mu\left(\frac{3}{2}\hat{\mathbf{i}}+\frac{3}{2}\hat{\mathbf{j}}+2\hat{\mathbf{k}}\right)$$

$$\Rightarrow \mathbf{r}=(\hat{\mathbf{i}}+2\hat{\mathbf{j}}+3\hat{\mathbf{k}})+\mu(3\hat{\mathbf{i}}+3\hat{\mathbf{j}}+4\hat{\mathbf{k}})$$

140. Now, equation of AC is

$$\mathbf{r}=(\hat{\mathbf{i}}+2\hat{\mathbf{j}}+3\hat{\mathbf{k}})+\lambda(2\hat{\mathbf{i}}+\hat{\mathbf{j}}+2\hat{\mathbf{k}})$$

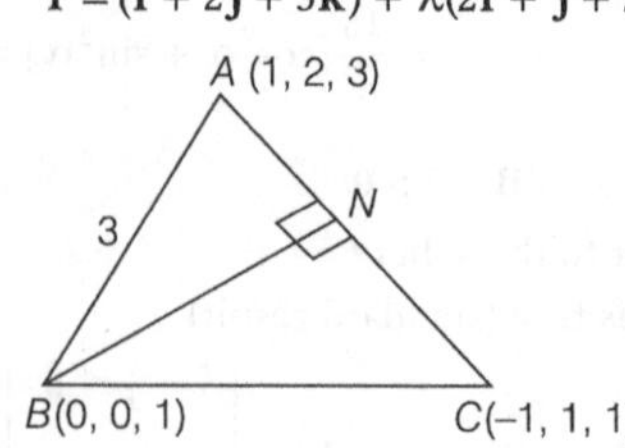

Also, $\mathbf{BM}=(1+2\lambda)\hat{\mathbf{i}}+(2+\lambda)\hat{\mathbf{j}}+2(1+\lambda)\hat{\mathbf{k}}$

$$\mathbf{BM}\cdot(2\hat{\mathbf{i}}+\hat{\mathbf{j}}+2\hat{\mathbf{k}})=0$$

$\Rightarrow \quad 2(1 + 2\lambda) + (2 + \lambda) + 4(1 + \lambda) = 0$

$\Rightarrow \quad \lambda = \frac{-8}{9}$

Position vector of $N = \frac{-7\hat{i} + 10\hat{j} + 11\hat{k}}{9}$

Equation of altitude through B to side AC is

$$\mathbf{r} = \hat{k} + t\left(-\frac{7}{\theta}\hat{i} + \frac{10}{\theta}\hat{j} + \frac{11}{\theta}\hat{k} - \hat{k}\right)$$

$$\mathbf{r} = \hat{k} + t(-7\hat{i} + 10\hat{j} + 2\hat{k})$$

141. Clearly, mid-point L of AB is $\left(\frac{1}{2}, 1, 2\right)$.

Equation of median through C to AB is

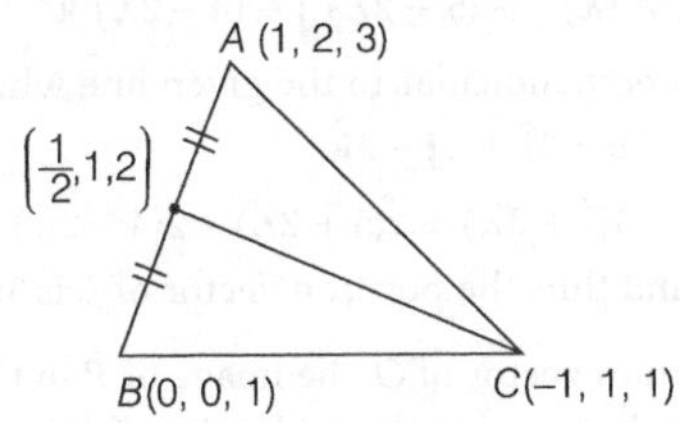

$$\mathbf{r} = (-\hat{i} + \hat{j} + \hat{k}) + p\left(\frac{3}{2}\hat{i} + \hat{k}\right)$$

$\Rightarrow \quad \mathbf{r} = (-\hat{i} + \hat{j} + \hat{k}) + p(3\hat{i} + 2\hat{k})$

142. We have, $\cos A = \frac{3^2 + 3^2 - (\sqrt{2})^2}{2(3)(3)}$

$\therefore \quad \cos A = \frac{16}{18} = \frac{8}{9}$

Now, area $(\Delta\ ABC) = \frac{1}{2}(3)(3)$

$$\sin A = \frac{9}{2}\sqrt{1 - \frac{64}{81}}$$

$$= \frac{9}{2} \times \frac{\sqrt{17}}{9} = \frac{\sqrt{17}}{2} \text{ sq units}$$

Solution (Q. Nos. 143-144)

143. Line $\frac{x-1}{3} = \frac{y-2}{-1} = \frac{z-3}{4} = r$

Any point $B \equiv 3r + 1, 2 - r, 3 + 4r$ (on the line L)

$\mathbf{AB} = 3r, -r, 4r + 6$

Hence, **AB** is parallel to $x + y - z = 1$.

Hence, $\quad 3r - r - 4r - 6 = 0$

$2r = -6;\ r = -3$

Hence, B is $(-8, 5, -9)$

144. Equation of plane containing the line L is

$$A(x-1) + B(y-2) + C(z-3) = 0$$

where, $\quad 3A - B + 4C = 0 \quad$ ...(ii)

$\therefore$ Eq. (i) also contains the point $A(1, 2, -1)$

Hence, $\quad C = 0, 3A = B$

Equation of plane $\ x - 1 + 3(y - 2) = 9$

$$x + 3y - 7 = 0$$

Solution (Q. Nos. 145-148)

145. $|\mathbf{AG}|^2 = \left(\frac{5}{3}\right)^2 + \frac{1}{9} + \left(\frac{5}{3}\right)^2 = \frac{51}{9}$

$$|\mathbf{AG}| = \frac{\sqrt{51}}{3}$$

146. $\mathbf{AB} = -4\hat{i} + 4\hat{j} + 0\hat{k}$

$\mathbf{AC} = 2\hat{i} + 2\hat{j} + 2\hat{k}$

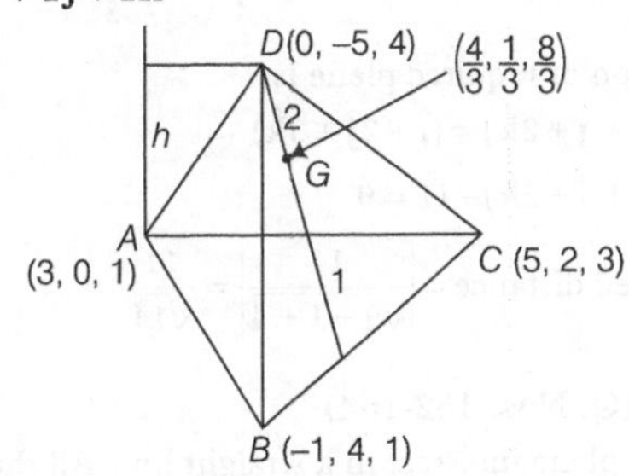

$$\therefore \quad \mathbf{AB} \times \mathbf{AC} = -8\begin{vmatrix} \hat{i} & \hat{j} & \hat{k} \\ 1 & -1 & 0 \\ 1 & 1 & 1 \end{vmatrix}$$

$$= -8(-\hat{i} - \hat{j} + 2\hat{k}) = 8(\hat{i} + \hat{j} - 2\hat{k}) = \mathbf{n}$$

$\therefore$ Area of $\Delta ABC = \frac{1}{2}|\mathbf{AB} \times \mathbf{AC}| = 4\sqrt{6}$

147. $h = |$ Projection of **AD** on $\mathbf{n}\,|$

$\mathbf{AD} = -3\hat{i} - 5\hat{j} + 3\hat{k}$

$$= \left|\frac{\mathbf{AD} \cdot \mathbf{n}}{|\mathbf{n}|}\right| = \left|\frac{(-3\hat{i} - 5\hat{j} + 3\hat{k})(\hat{i} + \hat{j} - 2\hat{k})}{\sqrt{6}}\right|$$

$$= \left|\frac{-3 - 5 - 6}{\sqrt{6}}\right| = \frac{14}{\sqrt{6}}$$

148. Equation of the plane ABC

$$A(x-3) + By + (z-1) = 0$$

where, $\quad A = 1, B = 1, C = -2$

$\therefore \quad x - 3 + y - 2z + 2 = 0$

$$x + y - 2z = 1$$

Solution (Q. Nos. 149-151)

149. Line L_1 is parallel to $\mathbf{a} = \hat{i} + 2\hat{j} + 3\hat{k}$

Line L_2 is parallel to $\mathbf{b} = 3\hat{i} + \hat{j} + 2\hat{k}$

Normal to the plane perpendicular to line L_1 and L_2 is

$$\mathbf{a} \times \mathbf{b} = (\hat{i} + 7\hat{j} - 5\hat{k})$$

and plane passes through the point with positive vector

$$= \frac{3}{2}\hat{i} + \frac{5}{2}\hat{j} + 2\hat{k}$$

Equation of plane is $\mathbf{r} \cdot (\hat{i} + 7\hat{j} - 5\hat{k}) = 9$

150. Angle bisector of vector **a** and **b** is,

$$\mathbf{r}_1 = \frac{1}{\sqrt{14}}(2\hat{i} - \hat{j} - \hat{k})$$

and $\quad \mathbf{r}_2 = \frac{1}{\sqrt{14}}(4\hat{i} + 3\hat{j} + 5\hat{k})$

Hence, the plane with either (2, −1, −1) or (4, 3, 5) as the direction ratio of normal and passing through (2, −3, 2) is the required plane.

$\therefore$ Equation of line is $\frac{x-2}{2}=\frac{y+3}{-1}=\frac{3-2}{-1}$

and $\frac{x-2}{4}=\frac{y+3}{3}=\frac{3-2}{5}$

$$\frac{x-2}{2}=y+3=z-2 \text{ or } \frac{x-2}{4}=\frac{y+3}{3}=\frac{z-2}{5}$$

151. $\therefore$ Equation of required plane is

$$\mathbf{r}\cdot(3\hat{\mathbf{i}}+\hat{\mathbf{j}}+2\hat{\mathbf{k}})=(\hat{\mathbf{i}}+2\hat{\mathbf{j}}+3\hat{\mathbf{k}})$$

$$\Rightarrow \mathbf{r}\cdot(3\hat{\mathbf{i}}+\hat{\mathbf{j}}+2\hat{\mathbf{k}})-11=0$$

$\therefore$ Required distance $=\left|\frac{11}{\sqrt{9+1+4}}\right|=\frac{11}{\sqrt{14}}$

■ **Solutions** (Q. Nos. 152-154)

The three plane intersect in a straight line. All three plane pass through origin (clearly).

$$\begin{vmatrix} 1 & -n & -m \\ n & -1 & l \\ m & l & -1 \end{vmatrix}=1(1-l^2)+n(-n-lm)-m(nl+m)$$

$$1=l^2+m^2+n^2+2lmn$$

Let $l=\cos\theta_1$, $m=\cos\theta_2$, $n=\cos\theta_3$ [since, $l, m, n \in (0,1)$]

$$\cos^2\theta_1+\cos^2\theta_2+\cos^2\theta_3+2\cos\theta_1\cos\theta_2\cos\theta_3=1$$

$$\cos^2\theta_1+(2\cos\theta_2\cos\theta_3)\cos\theta_1+\cos^2\theta_2+\cos^2\theta_3-1=0$$

$$\cos\theta_1=\frac{-2\cos\theta_2\cos\theta_3\pm\sqrt{4\cos^2\theta_2\cos^2\theta_3-4\cos^2\theta_2-4\cos^2\theta_3+4}}{2}$$

$$=-\cos\theta_2\cos\theta_3\pm\sqrt{1-\cos\theta_2}\sqrt{1-\cos^2\theta_3}$$

$$\Rightarrow \cos\theta_1=(\cos\theta_2\cos\theta_3-\sin\theta_2\sin\theta_3)$$

$$\Rightarrow \cos\theta_1=-\cos(\theta_2+\theta_3)$$

$$\theta_1+\theta_2+\theta_3=\pi$$

$$ny+mz=\frac{y-lz}{n}$$

$$n^2y+mnz=y-lz$$

$$(1+mn)z=(1-n^2)y$$

$$\frac{z}{y}=\frac{1-n^2}{1+mn}$$

$$\frac{x-ny}{m}=\frac{y-nx}{1}$$

$$lx-nly=my-mnx$$

$$(l+mn)x=(m+nl)y$$

$$\frac{y}{x}=\frac{l+mn}{m+nl}$$

$$\frac{y-lz}{n}=\frac{z-ly}{m}$$

$$my-mlz=nz-nly \Rightarrow (m+nl)y=(n+ml)z$$

$$\frac{z}{y}=\frac{m+nl}{n+ml}=\frac{\sin\theta_1\sin\theta_3}{\sin\theta_2\sin\theta_1}=\frac{\sin\theta_3}{\sin\theta_2}$$

$$\frac{x}{y}=\frac{m+nl}{l+mn}=\frac{\sin\theta_1}{\sin\theta_2}$$

$$\frac{x}{\sin\theta_1}=\frac{y}{\sin\theta_2}=\frac{z}{\sin\theta_3}$$

$$\frac{x}{\sqrt{1-l^2}}=\frac{y}{\sqrt{l-m^2}}=\frac{z}{\sqrt{1-n^2}}$$

152. (b) **153.** (c) **154.** (a)

Solution (Q. Nos. 155-157)

155. Let the position vector of L be $\mathbf{a}+\lambda\mathbf{b}$

$$=(6+3\lambda)\hat{\mathbf{i}}+(7+2\lambda)\hat{\mathbf{j}}+(7-2\lambda)\hat{\mathbf{k}}$$

So, $\mathbf{PL}=(6+3\lambda)\hat{\mathbf{i}}+(7+2\lambda)\hat{\mathbf{j}}+(7-2\lambda)\hat{\mathbf{k}}-(\hat{\mathbf{i}}+2\hat{\mathbf{j}}+3\hat{\mathbf{k}})$

$$=(5+3\lambda)\hat{\mathbf{i}}+(5+2\lambda)\hat{\mathbf{j}}+(4-2\lambda)\hat{\mathbf{k}}$$

Since, **PL** is perpendicular to the given line which is parallel to

$$\mathbf{b}=3\hat{\mathbf{i}}+2\hat{\mathbf{j}}-2\hat{\mathbf{k}}$$

$$\Rightarrow 3(5+3\lambda)+2(5+2\lambda)-2(4-2\lambda)=0$$

$\Rightarrow \lambda=-1$ and thus the position vector of L is $3\hat{\mathbf{i}}+5\hat{\mathbf{j}}+9\hat{\mathbf{k}}$

156. Let the position vector of Q, the image of P in the given line be $x_1\hat{\mathbf{i}}+y_1\hat{\mathbf{j}}+z_1\hat{\mathbf{k}}$, then L is the mid-point of PQ.

$$\Rightarrow 3\hat{\mathbf{i}}+5\hat{\mathbf{j}}+9\hat{\mathbf{k}}=\frac{\hat{\mathbf{i}}+2\hat{\mathbf{j}}+3\hat{\mathbf{k}}+x_1\hat{\mathbf{i}}+y_1\hat{\mathbf{j}}+z_1\hat{\mathbf{k}}}{2}$$

$$\Rightarrow \frac{x_1+1}{2}=3,\ \frac{y_1+2}{2}=5,\ \frac{z_1+3}{2}=9$$

$$\Rightarrow x_1=5, y_1=8, z_1=15$$

$\Rightarrow$ Image of P in the line is (5, 8, 15)

157. Area of the $\Delta PLA=\frac{1}{2}|\mathbf{PL}||\mathbf{AL}|$

$$=\frac{1}{2}|2\hat{\mathbf{i}}+3\hat{\mathbf{j}}+6\hat{\mathbf{k}}||-3\hat{\mathbf{i}}-2\hat{\mathbf{j}}+2\hat{\mathbf{k}}|$$

$$=\frac{1}{2}\sqrt{4+9+36}\sqrt{9+4+4}=\frac{7\sqrt{17}}{2}\text{ sq units.}$$

Solution (Q. Nos. 158-160)

158. Let $P(x, y, z)$ be any point on the locus then $3PA=2PB$

$$\Rightarrow 9(PA)^2=4(PB)^2$$

$$\Rightarrow 9[(x+2)^2+(y-2)^2+(z-3)^2]$$

$$=4[(x-13)^2+(y+3)^2+(z-13)^2]$$

$$\Rightarrow 5(x^2+y^2+z^2)+140x-60y+50z-1235=0$$

$$\Rightarrow x^2+y^2+z^2+28x-12y+10z-247=0$$

159. The required coordinates are

$$\left(\frac{2\times13+3(-2)}{2+3},\frac{2\times(-3)+3(2)}{2+3},\frac{2\times13+3\times3}{2+3}\right)=(4,0,7)$$

160. Direction ratios of AB are $13+2, -3-2, 13-3$

i.e. 15, −5, 10

Let the equation of the required line L be

$$\frac{x+2}{l}=\frac{y-2}{m}=\frac{z-3}{n}$$

then $15l-5m+10n=0$

which satisfied by (c)

Solution (Q. Nos. 161-163)

161. Equation $\mathbf{r} = \mathbf{a} + t\hat{\mathbf{n}}$ is line passing through $\mathbf{a}$ and parallel to $\hat{\mathbf{n}}$. This will meet the plane $\mathbf{r}\cdot\hat{\mathbf{n}} = d$ at point for which

$$(\mathbf{a} + t\hat{\mathbf{n}})\,.\,\hat{\mathbf{n}} = d \Rightarrow t = d - \mathbf{a}\,.\,\hat{\mathbf{n}}$$

Required distance $= |\mathbf{a} + (d - \mathbf{a}.\hat{\mathbf{n}})\hat{\mathbf{n}} - \mathbf{a}| = |d - \mathbf{a}.\hat{\mathbf{n}}|$

162. Foot of the perpendicular from the point A to plane $\mathbf{r}.\hat{\mathbf{n}} = d$

$$= \mathbf{a} + (d - \mathbf{a}.\hat{\mathbf{n}})\,\hat{\mathbf{n}}$$

163. Let $\mathbf{b}$ be position vector of image of $\mathbf{a}$

$$\therefore \quad \frac{\mathbf{b} + \mathbf{a}}{2} = \mathbf{a} + (d - \mathbf{a}.\hat{\mathbf{n}})\,\hat{\mathbf{n}}$$

$$\mathbf{b} = \mathbf{a} + 2(d - \mathbf{a}.\hat{\mathbf{n}})\,\hat{\mathbf{n}}$$

Solution (Q. Nos. 164-166)

164. The centre of the sphere is at the mid-point of the extremities of a diameter $\Rightarrow$ the centre $\left(-\frac{3}{2}, \frac{3}{2}, -\frac{3}{2}\right)$

and hence the radius $= \sqrt{\left(\frac{7}{2}\right)^2 + \left(\frac{9}{2}\right)^2 + \left(\frac{11}{2}\right)^2}$

165. Equation of the circle can be written as

$$x^2 - 16x + y^2 - 9 + z^2 = 0$$

or $$x^2 + y^2 + z^2 = 25$$

166. Distance of the point $(3, 6, -4)$ from the given plane is equal to the radius of the sphere $\Rightarrow$ the radius of the sphere

$$= \left|\frac{(3\hat{\mathbf{i}} + 6\hat{\mathbf{j}} - 4\hat{\mathbf{k}})\cdot(2\hat{\mathbf{i}} - 2\hat{\mathbf{j}} - \hat{\mathbf{k}}) - 10}{\sqrt{4 + 4 + 1}}\right| = \left|\frac{6 - 12 + 4 - 10}{3}\right| = 4$$

Solution (Q. Nos. 167-168)

167. Mid-point of BC $= \left(\frac{\lambda - 1}{2}, \frac{\mu + 2}{2}\right)$

DR's $\quad AD = \left(\frac{\lambda - 5}{2}, 1, \frac{\mu - 8}{2}\right)$

AD is equally inclined to axes $\Rightarrow \lambda = 7, \mu = 10, 2\lambda - \mu = 4$

168. $A(2, 3, 5)\ B(-1, 2, 3)\ C(7, 5, 10)$

Projection of $AB = -3\hat{\mathbf{i}} - 3\hat{\mathbf{k}}$ on $BC = 8\hat{\mathbf{i}} + 2\hat{\mathbf{j}} + 8\hat{\mathbf{k}}$

$$\frac{AB \cdot BC}{|BC|} = \frac{-8\sqrt{3}}{11}$$

Solution (Q. Nos. 169-171)

169. Horizontal plane P_1 is of the form

$$\mathbf{r}\cdot\mathbf{n}_1 = 0, \quad \text{where } \mathbf{n}_1 = (4, -3, 7)$$

Plane P_2 is of the form $\mathbf{r}\cdot\mathbf{n}_2 = 0$, where $\overline{n}_2 = (2, 1, -5)$

The vector $\mathbf{b}$ along the line of interaction

$$= \mathbf{n}_1 \times \mathbf{n}_2 = (4, 17, 5) = n_3 \text{ (say)}$$

Since the line of greatest slope is perpendicular to $\mathbf{n}_2$ and $\mathbf{n}_3$, the vector along the line of greatest slope

$$= \mathbf{n}_2 \times \mathbf{n}_3 = (3, -1, 1) = \mathbf{n}_4$$

and $$\mathbf{n}_4 = \left(\frac{3}{\sqrt{11}}, \frac{-1}{\sqrt{11}}, \frac{1}{\sqrt{11}}\right)$$

170. Since $(0, 0, 0)$ is a point on both the planes, it is a point on the line of intersection and hence the equation of a line of greatest slope is

$$\frac{x}{\frac{3}{\sqrt{11}}} = \frac{y}{\frac{-1}{\sqrt{11}}} = \frac{z}{\frac{1}{\sqrt{11}}}$$

171. The point on the line at a distance $\sqrt{11}$ from the origin is the required point and it is $(3, -1, 1)$

Solution (Q. Nos. 172-174)

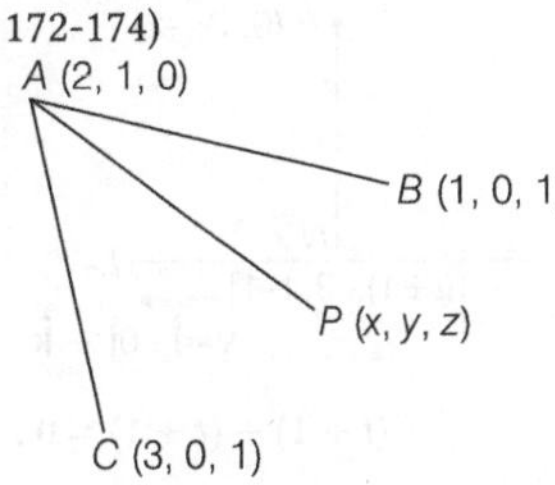

$$\begin{vmatrix} x-2 & y-1 & z \\ -1 & -1 & 1 \\ 1 & -1 & 1 \end{vmatrix} = 0$$

$$(x-2)[(-1)-(1)] - (y-1)[(-1)-1] + z[1+1] = 0$$

$$2(y-1) + 2z = 0$$

$$\Rightarrow \quad y + z - 1 = 0$$

The vector normal to the plane is $\mathbf{r} = 0\hat{\mathbf{i}} + \hat{\mathbf{j}} + \hat{\mathbf{k}}$

The equation of the line through $(0, 0, 2)$ and parallel to $\mathbf{n}$ is

$$\mathbf{r} = 2\hat{\mathbf{k}} + \lambda(\hat{\mathbf{j}} + \hat{\mathbf{k}})$$

The perpendicular distance of $D(0, 0, 2)$ from plane.

172. (b) **173.** (c) **174.** (d)

175. (A) $L_1 : \frac{x-1}{1} = \frac{y-0}{1} = \frac{z-2}{-5}$; $\mathbf{V}_1 = \hat{\mathbf{i}} + \hat{\mathbf{j}} - 5\hat{\mathbf{k}}$;

$L_2 : \frac{x-2}{2} = \frac{y-1}{2} = \frac{z+3}{-10}$; $\mathbf{V}_2 = 2(\hat{\mathbf{i}} + \hat{\mathbf{j}} - 5\hat{\mathbf{k}})$

Hence, lines are parallel and both contains the points $(1, 0, 2)$ and $(2, 1, -3)$ Coincident lines both L_1 and L_2 may lie in an infinite number of planes.

(B) $\left.\begin{matrix}\mathbf{V}_1 = 2\hat{\mathbf{i}} + 2\hat{\mathbf{j}} - \hat{\mathbf{k}} \\ \mathbf{V}_2 = \hat{\mathbf{i}} - 2\hat{\mathbf{j}} + 3\hat{\mathbf{k}}\end{matrix}\right] \Rightarrow$ Lines not parallel

Also, both intersect at $(3, 5, 1)$

Hence, lines are intersecting, hence they lie on a unique plane.

(C) $L_1 : \frac{x-0}{-6} = \frac{y-1}{9} = \frac{z-0}{-3} = t$

$L_2 : \frac{x-1}{2} = \frac{y-4}{-3} = \frac{z-0}{1} = s$

$\left.\begin{matrix}L_1 \text{ is parallel to } -6\hat{\mathbf{i}} + 9\hat{\mathbf{j}} - 3\hat{\mathbf{k}} \\ L_2 \text{ is parallel to } 2\hat{\mathbf{i}} - 3\hat{\mathbf{j}} + \hat{\mathbf{k}}\end{matrix}\right]$

$\Rightarrow$ Lines parallel but not coincident.

Since, $(0, 1, 0)$ does not lie on L_2, not intersecting.

Hence L_1, L_2 lies in unique planes.

(D) Lines are skew can be verified.

176. $L_1 : \frac{x}{3} = \frac{y-3}{4} = \frac{z+2}{1}$ ...(i) (passing through P and Q)

$L_2 : \frac{x-1}{1} = \frac{y-3}{0} = \frac{z+1}{1}$ ...(ii)

(passing through R and parallel to $\mathbf{v} = \hat{\mathbf{i}} + \hat{\mathbf{k}}$)

(A) Distance of $P(0, 3, -2)$ from L_2

$$\mathbf{PN} = (t+1)\,\hat{\mathbf{i}} - 6\hat{\mathbf{j}} + 2(t-1)\,\hat{\mathbf{k}}$$

Now, $\quad \mathbf{PN} \cdot \mathbf{V} = 0$

$$[(t+1)\hat{\mathbf{i}} - 6\hat{\mathbf{j}} + (t+1)\hat{\mathbf{k}}] \cdot (\hat{\mathbf{i}} + \hat{\mathbf{k}}) = 0$$

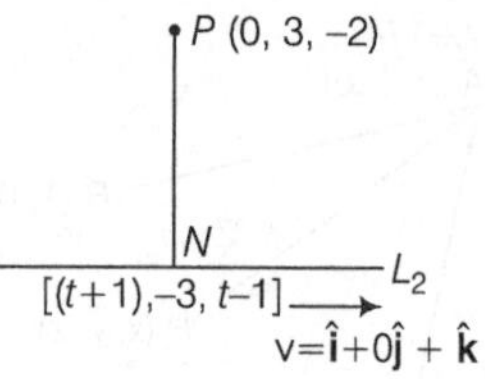

$$(t+1) + (t+1) = 0\,;\; t = -1$$

Hence, $\mathbf{PN} = 6\hat{\mathbf{j}}$

$$|\mathbf{PN}| = |-6\hat{\mathbf{j}}| = 6$$

(B) Distance between L_1 and L_2

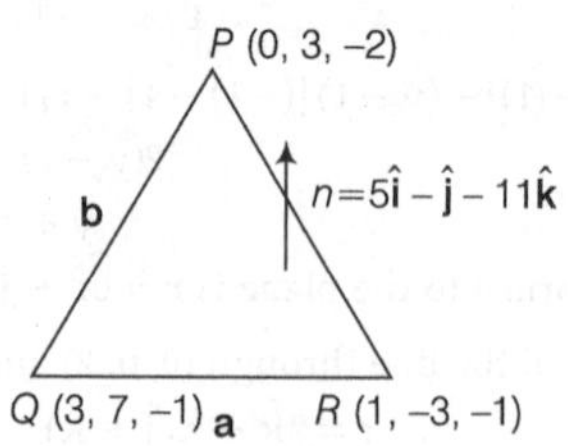

Equation of plane containing L_1 and parallel to L_2

$$Ax + B(y-3) + C(z+2) = 0$$

where $\quad 3A + 4B + C = 0$

And $\quad A + B + C = 0$

$$A + C = 0$$

$$C = \lambda, A = -\lambda, B = +\lambda/2$$

$\therefore$ Equation of plane

$$-\lambda x + \frac{\lambda}{2}(y-3) + \lambda(z+2) = 0$$

$$2x - y + 3 - 2z - 4 = 0$$

$$2x - y - 2z = 1 \qquad \text{...(i)}$$

Now, distance of the point $(1, -3, -1)$ lying on the line L_1 from the plane (i)

$$d = \left|\frac{2+3+2-1}{3}\right| = 2$$

(C) Area of ΔPQR

$$\mathbf{QR} = a = 2\hat{\mathbf{i}} + 10\hat{\mathbf{j}} + 0\hat{\mathbf{k}}$$

$$\mathbf{QP} = b = 3\hat{\mathbf{i}} + 4\hat{\mathbf{j}} + \hat{\mathbf{k}}$$

$$\mathbf{a} \times \mathbf{b} = 2\begin{vmatrix} \hat{\mathbf{i}} & \hat{\mathbf{j}} & \hat{\mathbf{k}} \\ 1 & 5 & 0 \\ 3 & 4 & 1 \end{vmatrix}$$

$$= 2[\hat{\mathbf{i}}(5) - \hat{\mathbf{j}}(1) + \hat{\mathbf{k}}(4-15)] = 2[5\hat{\mathbf{i}} - \hat{\mathbf{j}} - 11\hat{\mathbf{k}}]$$

$$\frac{|\mathbf{a} \times \mathbf{b}|}{2} = \sqrt{25+1+121} = \sqrt{147} = \sqrt{3 \cdot 49} = 7\sqrt{3}$$

(D) Distance of $(0, 0, 0)$ from PQR

Equation of plane PQR is $(r - p) \cdot n$

$$= [x\hat{\mathbf{i}} + (y-3)\hat{\mathbf{j}} + (z+2)\hat{\mathbf{k}}] \cdot [5\hat{\mathbf{i}} - \hat{\mathbf{j}} - 11\hat{\mathbf{k}}]$$

$$= 5x - (y-3) - 11(z+2) = 0$$

$$= 5x - y - 11z - 19 = 0$$

Distance from $(0, 0, 0)$ of the plane

$$d = \left|\frac{19}{\sqrt{25+1+121}}\right| = \frac{19}{\sqrt{147}}$$

177. (A) $3 \cdot 1 - 2(-2) + 5(\lambda) = 0$

$$\Rightarrow \qquad \lambda = -\frac{7}{5}$$

(B) Point $(3, \lambda, \mu)$ lies on

$$2x + y + z - 3 = 0$$

$$= x - 2y + z - 1$$

$$3 \cdot 2 + \lambda + \mu - 3 = 0 \text{ and } 3 - 2\lambda + \mu - 1 = 0$$

$$\lambda + \mu + 3 = 0 \text{ and } 2\lambda - \mu - 2 = 0$$

So, $\quad \lambda + \mu = -3$

(C) $\sin\theta = \dfrac{1 \cdot 4 + 1(-3) + 1 \cdot 5}{\sqrt{1^2+1^2+1^2}\,\sqrt{16+9+25}} = \dfrac{6}{\sqrt{3}\,\sqrt{50}}$

$$\theta = \sin^{-1}\sqrt{\frac{6}{25}}$$

(D) $\cos\theta = \dfrac{1 \cdot 3 + 1(-4) + 1 \cdot 5}{\sqrt{3}\,\sqrt{16+9+25}} = \dfrac{4}{\sqrt{3}\,\sqrt{50}}$

$$\theta = \cos^{-1}\sqrt{\frac{8}{75}}$$

178. $\begin{vmatrix} 1 & 3 & -5 \\ 3 & -k & -1 \\ 5 & 2 & -12 \end{vmatrix} = 1(12k+2) - 3(-36+5) - 5(6+5k)$

$$= 12k + 2 + 108 - 15 - 30 - 25k = 0$$

$k = 5$

L_1, L_2 and L_3 are concurrent for $k = 5$.

Slope of $L_1 = -\dfrac{1}{3}$, Slope of $L_2 = \dfrac{3}{k}$,

Slope of $L_3 = -\dfrac{5}{2}$

$$\frac{3}{k} = -\frac{1}{3} \Rightarrow k = -9$$

$$\frac{3}{k} = -\frac{5}{2} \Rightarrow k = -\frac{6}{5}$$

L_1, L_2 and L_3 form a triangle, if they are non-concurrent or any two out or three are not parallel.

$$k \neq -9, -\frac{6}{5}, 5$$

$k = \dfrac{5}{6}$ and 0 will be the values for which L_1, L_2 and L_3 form a triangle.

179. Given, $\dfrac{abc}{6} = 32$, where A, B and C are respectively, $(a, 0, 0)$, $(0, b, 0)$, $(0, 0, c)$.

(A) Centroid of tetrahedron $[\alpha, \beta, \gamma] = \left(\dfrac{a}{4}, \dfrac{b}{4}, \dfrac{c}{4}\right)$

$$a = 4\alpha, b = 4\beta, c = 4\gamma$$

$$64\alpha\beta\gamma = 32 \times 6$$

$$\alpha\beta\gamma = 3$$

(B) Equidistant point $(\alpha, \beta, \gamma) \equiv \left(\dfrac{a}{2}, \dfrac{b}{2}, \dfrac{c}{2}\right)$

$$a = 2\alpha, b = 2\beta, c = 2\gamma$$
$$8\alpha\beta\gamma = 32 \times 6$$
$$\alpha\beta\gamma = 24$$

(C) The equation of the plane is $\frac{x}{a} + \frac{y}{b} + \frac{z}{c} = 1$

$\therefore$Foot of the perpendicular from the origin

$$\equiv (\alpha, \beta, \gamma) \equiv \left(\frac{1/a}{\sum 1/a^2}, \frac{1/b}{\sum 1/b^2}, \frac{1/c}{\sum 1/c^2}\right)$$

$$\frac{1}{a\alpha} = \frac{1}{b\beta} = \frac{1}{c\gamma} = t$$

where, $t = \frac{1}{a^2} + \frac{1}{b^2} + \frac{1}{c^2} = \sum \frac{1}{a^2}$

or $t = (\alpha^2 + \beta^2 + \gamma^2)t^2$

$$t = \frac{1}{\alpha^2 + \beta^2 + \gamma^2}$$

and $\alpha = \frac{\alpha^2 + \beta^2 + \gamma^2}{\alpha}, b = \frac{\alpha^2 + \beta^2 + \gamma^2}{\beta}$,

$$c = \frac{\alpha^2 + \beta^2 + \gamma^2}{\gamma}$$

Now, $abc = 6 \times 32$

$$(\alpha^2 + \beta^2 + \gamma^2) = 192\,\alpha\beta\gamma$$

(D) Let P be (α, β, γ), then $PA \perp PB$

$\Rightarrow$ $a(\alpha - a) + \beta(\beta - b) + \gamma\gamma = 0$

$\Rightarrow$ $a\alpha + b\beta = \alpha^2 + \beta^2 + \gamma^2$

$PB \perp PC$

$\Rightarrow$ $a\alpha + b(\beta - b) + c(\gamma - c) = 0$

$\Rightarrow$ $b\beta + c\gamma = \alpha^2 + \beta^2 + \gamma^2$

$\therefore$ $\frac{a}{1/\alpha} = \frac{b}{1/\beta} = \frac{c}{1/\gamma}$

$$a = \frac{\alpha^2 + \beta^2 + \gamma^2}{2\alpha}, b = \frac{\alpha^2 + \beta^2 + \gamma^2}{2\beta}$$

and $c = \frac{\alpha^2 + \beta^2 + \gamma^2}{2\gamma}$

$\therefore$ $abc = 6 \times 32$

$\Rightarrow$ $(\alpha^2 + \beta^2 + \gamma^2) = 192 \times 8\alpha\beta\gamma$

$$= 1536\alpha\beta\gamma$$

180. Let $O(0, 0, 0)$, $A(3, 4, 7)$ and $B(5, 2, 6)$ be the given point

Area of $\Delta OAB = \frac{1}{2} OA \cdot OB \sin(\angle AOB)$

Now, $OA = \sqrt{3^2 + 4^2 + 7^2} = \sqrt{74}$

$$OB = \sqrt{5^2 + 2^2 + 6^2} = \sqrt{65}$$

Also dc's of the line OA and OB are

$$= \frac{3}{\sqrt{74}}, \frac{4}{\sqrt{74}}, \frac{7}{\sqrt{74}} \text{ and } \frac{5}{\sqrt{65}}, \frac{2}{\sqrt{65}}, \frac{6}{\sqrt{65}}$$

$\therefore$ Required area $\frac{1}{2} \times \sqrt{74} \times \sqrt{65} \times \frac{3}{\sqrt{74}} = \frac{3}{2}\sqrt{65}$

(B) Let the required sphere be

$x^2 + y^2 + z^2 + 2ux + 2vy + wz + d = 0$ ...(1) substituting given points then we get $1 + 2u + d = 0$

$$1 + 2v + d = 0 \text{ and } 1 + 2w + d = 0$$

$\Rightarrow$ $u = v = w = \frac{1 + d}{2}$

If R be the radius of the sphere,

then $R^2 = u^2 + v^2 + w^2 - d$

covert above equation in terms of d differentiate, equate to zero solve for d.

(C) Let the given points be A, B and C respectively.

Then find AB, AC, BC and then apply $AB^2 + AC^2 = BC^2$ then solve for the λ.

(D) Any point on the line is $(1 - r, r + 1, r)$

The direction ratio of the line joining $(1, 3, 4)$ & $(1 - r, r + 1, r)$ is $- r, r - 2, r - 4$

$\therefore$ $(-1)(-r) + 1 \cdot (r - 2) + (r - 4) = 0$

$$r + r - 2 + r - 4 = 0$$
$$3r = 6 \Rightarrow r = 2$$

$\therefore$Foot of the perpendicular is $(-1, 3, + 2)$

$\therefore$Distance $= \sqrt{(2)^2 + 0 + 4} = 2\sqrt{2}$

$\therefore$ $d = 2\sqrt{2}$

$$\frac{d}{2\sqrt{3}} = \frac{2\sqrt{2}}{2\sqrt{3}} = \frac{\sqrt{2}}{\sqrt{3}}$$

181. The solid diagonals may be taken as the lines join $(0, 0, 0)$, (a, a, a) and $(a, a, 0)$ and $(0, 0, a)$. The direction ratios will be $a, a, a,; a, a, -a$.

$\Rightarrow$ $\cos\theta = \frac{a^2 + a^2 - a^2}{\sqrt{3a^2} \times \sqrt{3a^2}} \frac{1}{3} \Rightarrow \theta = \cos^{-1}\frac{1}{3}$

Let us take the solid diagonal as the one joining $(0, 0, 0)$ (a, a, a) and plane diagonal as joining $(0, 0, 0)$ and $(a, a, 0)$. We easily get the angle as $\cos^{-1}\frac{2}{\sqrt{6}}$.

The third part is easily found as $\cos^{-1}\left(\frac{1}{2}\right)$

Hence, matching follows (A) $\rightarrow$ (r) ; (B) $\rightarrow$ (p) ; (C) $\rightarrow$ (q)

182. (i) Shortest distance

$$= \frac{|\mathbf{OB} \cdot \mathbf{OA} \times \mathbf{BC}|}{|\mathbf{OA} \times \mathbf{BC}|} = \frac{|(\hat{\mathbf{i}} + \hat{\mathbf{j}}) \cdot \hat{\mathbf{i}} \times (\hat{\mathbf{j}} \times \hat{\mathbf{k}})|}{\hat{\mathbf{i}} \times (\hat{\mathbf{j}} + \hat{\mathbf{k}})} = \frac{1}{\sqrt{2}}$$

$\Rightarrow$ $\sqrt{2}m = 1$

183. The length of the edges are given by $a = 5 - 2 = 3$, $b = 9 - 3 = 6$ and $c = 7 - 5 = 2$, so length of the diagonal

$$= \sqrt{a^2 + b^2 + c^2} = \sqrt{9 + 36 + 4} = 7 \text{ units}$$

184. Foot of perpendicular r from $(6, 5, 8)$ on y-axis is $(0, 5, 0)$.

Required distance

$$= \sqrt{(6 - 0)^2 + (5 - 5)^2 + (8 - 0)^2} = 10$$

$\Rightarrow$ $5\lambda = 10 \Rightarrow \lambda \Rightarrow \frac{10}{5} = 2$

185. Given lines are

$$\mathbf{r} = (3, 8, 3) + \lambda(-3, 1, 1)$$

and $\mathbf{r} = (-3, -7, 6) + \mu(-3, 2, 4)$

where, λ and μ are parameters

Shortest distance

$$= \frac{(-3-3, -7-8, 6-3)\cdot[(3, -1, 1)\times(-3, 2, 4)]}{|(3, -1, 1)\times(-3, 2, 4)|}$$

$$= \frac{(-6, -15, 3)\cdot(-6, -15, 3)}{\sqrt{36+225+9}}$$

$$= \sqrt{270} = 3\sqrt{30} \text{ units} = \lambda\sqrt{30}$$

$\therefore \quad \lambda = 3$

186. Given planes are

$$x - cy - bz = 0 \quad \ldots(i)$$
$$cx - y + az = 0 \quad \ldots(ii)$$
$$bx + acy - z = 0 \quad \ldots(iii)$$

Equation of plane passing through the line of intersection of planes (i) and (ii) may be taken as

$$(x - cy - bz) + \lambda(cx - y + az) = 0 \quad \ldots(iv)$$

Now, eliminating λ we get

$$a^2 + b^2 + c^2 + 2abc = 1$$

187. We must have $\frac{\lambda(-1) + 1\times 3}{\lambda + 1} = 0 \Rightarrow \lambda = 3$

$\therefore \quad \lambda = 3$

188. The coordinates of vertices of projected triangle will be $A'(-1, 1, 0)$, $B'(1, -1, 0)$, $C'(1, 1, 0)$

area of triangle $= \frac{1}{2}\begin{vmatrix} -1 & 1 & 1 \\ 1 & -1 & 1 \\ 1 & 1 & 1 \end{vmatrix}$ (Two dimension area formula)

$= 2$ square units.

189. Plane must pass through

$$\left(\frac{1-3}{2}, \frac{5+1}{2}, \frac{7-1}{2}\right) \text{ or } (-1, 3, 3)$$

$$\Rightarrow \quad -1 + 3 + 2\times 3 = \lambda \Rightarrow \lambda = 8$$

190. $x^2 + y^2 + z^2 =$ square of distance from origin

$$4\sin^2 t + 4\cos^2 t + 9t^2 = 4 + 9t^2$$

which is shortest at $t = 0$

$\Rightarrow$ Shortest distance $= 2$

191. The point $(-1, \lambda, -2)$ must be lie on the plane

$$2x - 2y + z + 12 = 0$$
$$-2 - 2\lambda - 2 + 12 = 0$$
$$\lambda = 4$$

We can easily show that the distance of $(-1, 4, -2)$ from centre of the sphere $(1, 2, -1)$ is equal to its radius.

192. $1 = \frac{a+1+2+0}{4}$, $2 = \frac{2+b+1+0}{4}$,

$$3 = \frac{3+2+c+0}{4}$$

$\Rightarrow \quad a = 1, b = 5, c = 7$

$\Rightarrow$ Distance of centroid from origin is

$$\sqrt{1^2 + 25 + 49} = \sqrt{75} = 5\sqrt{3} \Rightarrow \lambda = 3$$

193. Equating the distances of circumcentre $(-1, \lambda, -3)$ from $(3, 2, -5)$ and $(-3, 8, -5)$ we get

$$2^2 + (\lambda + 2)^2 + (-3+5)^2 = (-1+3)^2 + (\lambda - 8)^2 + (-3+5)^2$$

$\Rightarrow \quad \lambda = 4$

Note : Verify

(i) $(-1, \lambda, -3)$ is at the same distance from third vertex.

(ii) $(-1, \lambda, -3)$ lies on the plane containing three points $(3, 2, -5)$; $(-3, 8, -5)$ and $(-3, 2, 1)$.

194. D.R's of $P_1P_2 = (k, -1, 3)$

D.R's of $P_2P_3 = (2, k, -1)$

$\because \quad P_1P_2 \perp P_2P_3$

$\therefore \quad k(2) - k - 3 = 0 : k = 3$

195. A plane containing line of intersection of the given planes is

$$x - y - z - 4 + \lambda(x + y + 2z - 4) = 0$$

i.e., $(\lambda + 1)x + (\lambda - 1)y + (2\lambda - 1)z - 4(\lambda + 1) = 0$

vector normal to it

$$V = (\lambda + 1)\hat{\mathbf{i}} + (\lambda - 1)\hat{\mathbf{j}} + (2\lambda - 1)\hat{\mathbf{k}}$$

Now the vector along the line of intersection of the planes

$$2x + 3y + z - 1 = 0$$

and $x + 3y + 2z - 2 = 0$ is given by

$$\mathbf{n} = \begin{vmatrix} \hat{\mathbf{i}} & \hat{\mathbf{j}} & \hat{\mathbf{k}} \\ 2 & 3 & 1 \\ 1 & 3 & 2 \end{vmatrix} = 3(\hat{\mathbf{i}} - \hat{\mathbf{j}} + \hat{\mathbf{k}})$$

As $\mathbf{n}$ is parallel to the plane (i), therefore $\mathbf{n}\cdot\mathbf{V} = 0$

$$(\lambda + 1) - (\lambda - 1) + (2\lambda - 1) = 0$$

$$2 + 2\lambda - 1 = 0 \Rightarrow \lambda = \frac{-1}{2}$$

Hence, the required plane is $\frac{x}{2} - \frac{3y}{2} - 2z - 2 = 0$

$$x - 3y - 4z - 4 = 0$$

Hence $\quad |A + B + C - 4| = 7$

196. Clearly, minimum value of $a^2 + b^2 + c^2$

$$= \left(\frac{|3(0) + 2(0) + (0) - 7|}{\sqrt{(3)^2 + (2)^2 + (1)^2}}\right) = \frac{49}{14} = \frac{7}{2} \text{ units}$$

197.

$$4x + 7y + 4z + 81 = 0 \quad \ldots(i)$$
$$5x + 3y + 10z = 25 \quad \ldots(ii)$$

Equation of plane passing through their line of intersection is

$$(4x + 7y + 4z + 81) + \lambda(5x + 3y + 10z - 25) = 0$$

or $(4 + 5\lambda)x + (7 + 3\lambda)y + (4 + 10\lambda)z + 81 - 25\lambda = 0 \quad \ldots(iii)$

plane (iii) $\perp$ to (i), so

$$4(4 + 5\lambda) + 7(7 + 3\lambda) + 4(4 + 10\lambda) = 0$$

$\therefore \quad \lambda = -1$

From (iii), equation of plane is

$$-x + 4y - 6z + 106 = 0 \quad \ldots(iv)$$

Distance of (iv) from $(0, 0, 0)$

$$= \frac{106}{\sqrt{1 + 16 + 36}} = \frac{106}{\sqrt{53}}$$

198. Line through point $P(-2, 3, -4)$ and parallel to the given line

$$\frac{x+2}{3} = \frac{2y+3}{4} = \frac{3z+4}{5}$$

is $$\frac{x+2}{3} = \frac{y+\frac{3}{2}}{2} = \frac{z+\frac{4}{3}}{\frac{5}{3}} = \lambda$$

Any point on this line is $Q\left[3\lambda - 2, 2\lambda - \frac{3}{2}, \frac{5}{3}\lambda - \frac{4}{3}\right]$.

Direction ratios of PQ are $\left[3\lambda, \frac{4\lambda - 9}{2}, \frac{5\lambda + 8}{3}\right]$

Now, PQ is parallel to the given plane $4x + 12y - 3z + 1 = 0$

$\Rightarrow$ line is perpendicular to the normal to the plane

$$\Rightarrow \quad 4(3\lambda) + 12\left(\frac{4\lambda - 9}{2}\right) - 3\left(\frac{5\lambda + 8}{3}\right) = 0$$

$$\Rightarrow \quad \lambda = 2 \Rightarrow Q\left(4, \frac{5}{2}, 2\right)$$

$$\Rightarrow \quad PQ = \sqrt{(6)^2 + \left(\frac{5}{2} - 3\right)^2 + (6)^2} = \frac{17}{2}$$

199. The given points are $O(0, 0, 0)$, $A(0, 0, 0)$, $B(0, 4, 0)$ and $C(6, 0, 0)$

Here, three faces of tetrahedron are xy, yz, zx plane.

Since point P is equidistance from zx, xy and yz planes, its coordinates are $P(r, r, r)$

Equation of plane ABC is

$$2x + 3y + 6z = 12 \quad \text{(from intercept form)}$$

P is also at distance r from plane ABC

$$\Rightarrow \quad \frac{|2r + 3r + 6r - 12|}{\sqrt{4 + 9 + 36}} = r \Rightarrow |11r - 12| = 7r$$

$$\Rightarrow \quad 11r - 12 = \pm 7r \Rightarrow r = \frac{12}{18}, 3$$

$$\therefore \quad r = 2/3 \quad (\text{as } r < 2)$$

200. The equation of the given planes can be written as

$$x - y + z + 1 = 0$$
$$\lambda xz + 3y + 2z - 3 = 0$$
$$3x + \lambda y + z - 2 = 0$$

The rectangular array is

$$\left\|\begin{matrix} 1 & -1 & 1 & 1 \\ \lambda & 3 & 2 & -3 \\ 3 & \lambda & 1 & -2 \end{matrix}\right\| = 0$$

$$\therefore \quad \Delta_4 = \begin{vmatrix} 1 & -1 & 1 \\ \lambda & 3 & 2 \\ 3 & \lambda & 1 \end{vmatrix}$$

Applying $C_2 \to C_2 + C_1$ and $C_3 \to C_3 + C_1$, then

$$\Delta_4 = \begin{vmatrix} 1 & ...0 & ...0 \\ \lambda & 3+\lambda & 2-\lambda \\ 3 & 3+\lambda & -2 \end{vmatrix} = (\lambda - 4)(\lambda + 3) \quad \text{... (i)}$$

Also, $$\Delta_1 = \begin{vmatrix} -1 & 1 & 1 \\ 3 & 2 & -3 \\ \lambda & 1 & -2 \end{vmatrix}$$

Applying $C_2 \to C_2 + C_1$ and $C_3 \to C_3 + C_1$, then

$$\Delta_1 = \begin{vmatrix} -1 & 0 & 0 \\ 3 & 5 & 0 \\ \lambda & \lambda + 1 & \lambda - 2 \end{vmatrix} = -5(\lambda - 2) \quad \text{... (ii)}$$

$$\Delta_2 = \begin{vmatrix} 1 & 1 & 1 \\ \lambda & 2 & -3 \\ 3 & 1 & -2 \end{vmatrix}$$

Applying $C_2 \to C_2 - C_1$ and $C_3 \to C_3 - C_1$, then

$$\Delta_2 = \begin{vmatrix} 1... & 1... & 0 \\ \lambda & 2-\lambda & -3-\lambda \\ 3 & -2 & -5 \end{vmatrix} = 3\lambda - 16 \quad \text{... (iii)}$$

$$\Delta_3 = \begin{vmatrix} 1 & -1 & 1 \\ \lambda & 3 & -3 \\ 3 & \lambda & -2 \end{vmatrix}$$

Applying $C_2 \to C_2 + C_1$ and $C_3 \to C_3 - C_1$, then

$$\Delta_3 = \begin{vmatrix} 1... & 1... & 0 \\ \lambda & 3+\lambda & -3-\lambda \\ 3 & 3+\lambda & -5 \end{vmatrix} = (\lambda + 3)(\lambda - 2) \quad \text{... (iv)}$$

If the given planes form a triangular prism, then we know that $\Delta_4 = 0$ and none of $\Delta_1, \Delta_2, \Delta_3$ is zero. Here from Eqs. (i), (ii), (iii) and (iv) we find that if $\lambda = 4$, then $\Delta_4 = 0$ and none of Δ_1, Δ_2, Δ_3 is zero. Consequently for $\lambda = 4$, then given planes form a triangular prism.

201. $7x + 6y + 2z = 272$ and $x - y + z = 16$

$$\Rightarrow \quad 5x + 8y = 240 \Rightarrow x = 48 - \frac{8}{5}y$$

Let $y = 5\lambda, \lambda \in I \Rightarrow x = 48 - 8\lambda$

and $z = 16 + y - x = 13\lambda - 32$

But $x > 0$, $y > 0$ and $z > 0 \quad \Rightarrow 48 - 8\lambda > 0 \Rightarrow \lambda > \frac{48}{8}$

$\Rightarrow \lambda \le 5$ and $13\lambda - 32 > 0 \Rightarrow \quad \lambda > \frac{32}{13}$

$$\Rightarrow \quad \lambda \ge 3$$

$$\therefore \quad \lambda \in [3, 5]$$

$$\therefore \quad Z_{\min} = 39 - 32 = 7 \Rightarrow x = 24, y = 15$$

$$\therefore \quad x + y + z - 42 = 4$$

202. The given two lines are intersect each other, then

$$\frac{a_1\alpha + b_1\beta + c_1\gamma + d_1}{a_1 l + b_1 m + c_1 n} = \frac{a_2\alpha + b_2\beta + c_2\gamma + d_2}{a_2 l + b_2 m + c_2 n}$$

$$\Rightarrow \quad \frac{-2d^2}{\sin A + \sin B + \sin C} = \frac{-d^2}{\sin 2A + \sin 2B + \sin 2C}$$

$$\Rightarrow \quad \sin\frac{A}{2}\sin\frac{B}{2}\sin\frac{C}{2} = \frac{1}{16}$$

203. $$[\mathbf{c - a\ b\ c}] = 0 \quad \Rightarrow \begin{vmatrix} k & k-1 & 2 \\ 1 & k & -1 \\ 2k & 3k-1 & k \end{vmatrix} = 0$$

$$\Rightarrow \quad k^3 - 4k^2 + 8k - 2 = 0$$

Here $$f'(k) = 3k^2 - 8k + 8 > 0 \ \forall \ k \in R$$

($\because$ Its discriminate is negative)

$\therefore$ The equation has only one real root.

204. Taking O as the origin, let the position vectors of A, B and C be $\mathbf{a}$, $\mathbf{b}$, $\mathbf{c}$ respectively. Then the position vectors of G_1, G_2 and G_3 are

$$\frac{\mathbf{b}+\mathbf{c}}{3}, \frac{\mathbf{c}+\mathbf{a}}{3} \text{ and } \frac{\mathbf{a}+\mathbf{b}}{3}$$

$$V_1 = \frac{1}{6}[\mathbf{a}\ \mathbf{b}\ \mathbf{c}] \text{ and } V_2 = [\mathbf{OG_1}\ \mathbf{OG_2}\ \mathbf{OG_3}]$$

$$\Rightarrow \quad V_2 = \frac{1}{27}[\mathbf{b}+\mathbf{c}\ \ \mathbf{c}+\mathbf{a}\ \ \mathbf{a}+\mathbf{b}] = \frac{2}{27}[\mathbf{a}\ \mathbf{b}\ \mathbf{c}]$$

$$\Rightarrow \quad V_2 = \frac{2}{27}\times 6V_1 \Rightarrow 9V_2 = 4V_1$$

205. Let the equation of planes is $lx + my + nz = p$

$$\therefore A = \left(\frac{p}{l}, 0, 0\right) B = \left(0, \frac{p}{m}, 0\right) C = \left(0, 0, \frac{p}{n}\right) \text{ respectively}$$

$$\text{Centroid of } OABC = \left(\frac{p}{4l}, \frac{p}{4m}, \frac{p}{4n}\right) = (x_1, y_1, z_1) \quad \text{(say)}$$

$$\because \quad l^2 + m^2 + n^2 = 1$$

$$\therefore \quad \frac{p^2}{16x_1^2} + \frac{p^2}{16y_1^2} + \frac{p^2}{16z_1^2} = 1$$

$$\Rightarrow \quad x_1^2y_1^2 + y_1^2z_1^2 + z_1^2x_1^2 = \frac{16}{p^2}x_1^2y_1^2z_1^2$$

$$\therefore \quad k = 16 \Rightarrow 2k = 32 \Rightarrow \sqrt[5]{2k} = 2$$

206. $l_1^2 + m_1^2 + n_1^2 = 1,\ l_2^2 + m_2^2 + n_2^2 = 1$

$$(l_1^2 + m_1^2 + n_1^2)(l_2^2 + m_2^2 + n_2^2) - (l_1l_1 + m_1m_2 + n_1n_2)^2$$

$$= (m_1n_2 - m_2n_1)^2 + (n_1l_2 - n_2l_1)^2 + (l_1m_2 - l_2m_1)^2$$

$$(l_1m_2 - l_2m_1)^2 + (m_1n_2 - m_2n_1)^2 + (n_1l_2 - n_2l_1)^2 + (l_1l_2 + m_1m_2 + n_1n_2)^2 = 1$$

207. Coordinates of the point, $S = \left(\frac{n^5}{2}, \frac{n^4}{2}, \frac{n}{2}\right)$

$$\Rightarrow \quad 2\times\left(\frac{n^5}{2} + \frac{n^4}{2} + \frac{n}{2}\right) = -1$$

$$\Rightarrow \quad n(n^4 + n^3 + 1) = -1$$

$n = -1$ is the only solution.

208. We have, $\quad l + m + n = 0 \quad$ ...(i)

and $\quad 2l^2 + 2m^2 - n^2 = 0 \quad$ ...(ii)

Now, $\quad 2(l^2 + m^2) - n^2 = 0$

$$\Rightarrow \quad 2(1 - n^2) - n^2 = 0 \quad [\because l^2 + m^2 + n^2 = 1]$$

$$\Rightarrow \quad 3n^2 = 2$$

$$\Rightarrow \quad n = \pm\sqrt{\frac{2}{3}}$$

Again, $\quad 2(l^2 + m^2) = n^2$

$$\Rightarrow \quad 2[(l+m)^2 - 2lm] = (-(l+m))^2$$

$$\Rightarrow \quad l = m$$

$$\therefore \quad l + m = \pm\sqrt{\frac{2}{3}} \Rightarrow 2l = \pm\sqrt{\frac{2}{3}}$$

$$\Rightarrow \quad l = \pm\frac{1}{\sqrt{6}} = m$$

$\therefore$ Direction cosines are

$$\left(\frac{1}{\sqrt{6}}, \frac{1}{\sqrt{6}}, \sqrt{\frac{2}{3}}\right), \left(\frac{1}{\sqrt{6}}, \frac{1}{\sqrt{6}}, -\sqrt{\frac{2}{3}}\right)$$

or $$\left(-\frac{1}{\sqrt{6}}, -\frac{1}{\sqrt{6}}, \sqrt{\frac{2}{3}}\right)$$

and $$\left(-\frac{1}{\sqrt{6}}, -\frac{1}{\sqrt{6}}, -\sqrt{\frac{2}{3}}\right)$$

The angle between in both the cases is $\cos^{-1}\left(\frac{-1}{3}\right)$.

209. Elimination n between the given relations, we get

$$ul^2 + vm^2 + w\left(\frac{al + bm}{-c}\right)^2 = 0$$

$$\Rightarrow \quad (c^2u + a^2w)\frac{l^2}{m^2} + 2abw.\frac{l}{m} + (b^2w + c^2v) = 0 \quad \text{...(i)}$$

$$\therefore \quad \frac{l_1}{m_1}\cdot\frac{l_2}{m_2} = \text{product of roots} = \frac{b^2w + c^2v}{c^2u + a^2w}$$

$$\text{or} \quad \frac{l_1l_2}{b^2w + c^2v} = \frac{m_1m_2}{c^2u + a^2w} = \frac{n_1n_2}{a^2v + b^2u} \quad \text{(by symmetry)}$$

If lines are perpendicular, then

$$l_1l_2 + m_1m_2 + n_1n_2 = 0$$

$$\Rightarrow \quad a^2(v + w) + b^2(w + u) + c^2(u + v) = 0$$

Again, if the lines be parallel, then their $d'c$ are equal so that the roots of Eq. (i) should be equal, i.e. discriminate = 0

$$\therefore \quad 4a^2b^2w^2 - 4(c^2u + a^2w)(b^2w + c^2v) = 0$$

$$\Rightarrow \quad a^2c^2vw + b^2c^2uw + c^4uv = 0$$

$$\Rightarrow \quad \frac{a^2}{u} + \frac{b^2}{v} + \frac{c^2}{w} = 0$$

210. The coordinates of any point on the line $\frac{x+2}{3} = \frac{y+1}{2} = \frac{z-3}{2} = \lambda$ are given by

$(3\lambda - 2, 2\lambda - 1, 2\lambda + 3)$

The distance between the above point and $(1, 2, 3)$ is $3\sqrt{2}$.

$$\therefore \quad \sqrt{(3\lambda - 2 - 1)^2 + (2\lambda - 1 - 2)^2 + (2\lambda + 3 - 3)^2} = 3\sqrt{2}$$

$$\Rightarrow \lambda = \frac{30}{17}, 0$$

$\therefore$ Required points are $(-2, -1, 3)$ and $\left(\frac{56}{17}, \frac{43}{17}, \frac{111}{17}\right)$

211. The required line is perpendicular to the lines which are parallel to vectors $\mathbf{b}_1 = 2\hat{\mathbf{i}} - 2\hat{\mathbf{j}} + \hat{\mathbf{k}}$ and $\mathbf{b}_2 = \hat{\mathbf{i}} + 2\hat{\mathbf{j}} + 2\hat{\mathbf{k}}$ respectively. So, it is parallel to the vector $\mathbf{b} = \mathbf{b}_1 \times \mathbf{b}_2$.

$$\text{Now, } \mathbf{b} = \mathbf{b}_1 \times \mathbf{b}_2 = \begin{vmatrix} \hat{\mathbf{i}} & \hat{\mathbf{j}} & \hat{\mathbf{k}} \\ 2 & -2 & 1 \\ 1 & 2 & 2 \end{vmatrix} = -6\hat{\mathbf{i}} - 3\hat{\mathbf{j}} + 6\hat{\mathbf{k}}$$

Thus, the required line passes through the point $(2, -1, 3)$ and is parallel to the vector $\mathbf{b} = -6\hat{\mathbf{i}} - 3\hat{\mathbf{j}} + 6\hat{\mathbf{k}}$.

So, its vector equation is

$$\mathbf{r} = (2\hat{\mathbf{i}} - \hat{\mathbf{j}} + 3\hat{\mathbf{k}}) + \lambda(-6\hat{\mathbf{i}} - 3\hat{\mathbf{j}} + 6\hat{\mathbf{k}})$$

or $$\mathbf{r} = (2\hat{\mathbf{i}} - \hat{\mathbf{j}} + 3\hat{\mathbf{k}}) + \mu(2\hat{\mathbf{i}} + \hat{\mathbf{j}} - 2\hat{\mathbf{k}}),$$

where $\quad \mu = -3\lambda$.

212. The coordinates of any point on the line $\frac{x-3}{2}=\frac{y-3}{1}=\frac{z}{1}$ are given by $\frac{x-3}{2}=\frac{y-3}{1}=\frac{z}{1}=\lambda.$

So, let the coordinates of A be $(2\lambda+3, \lambda+3, \lambda)$.

Let the line through $O(0, 0, 0)$ and making an angle $\frac{\pi}{3}$ with the given line be along OA. Then, its $d'r$ are proportional to

$$2\lambda+3-0, \lambda+3-0, \lambda-0$$

or $$2\lambda+3, \lambda+3, \lambda$$

The direction ratios of the given line are proportional to 2, 1, 1.
It is given that the angle between the given line and the line along OA is $\frac{\pi}{3}$.

$$\therefore \quad \cos\frac{\pi}{3}=\frac{(2\lambda+3)\times 2+(\lambda+3)\times 1+\lambda\times 1}{\sqrt{(2\lambda+3)^2+(\lambda+3)^2+\lambda^2}\sqrt{2^2+1^2+1^2}}$$

$$=\frac{6\lambda+9}{\sqrt{6\lambda^2+18\lambda+18}\sqrt{6}}$$

$$\therefore \quad \lambda=-1, -2.$$

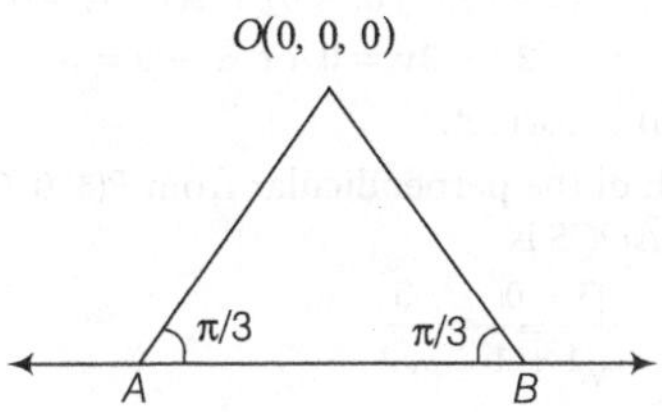

$$\frac{x-3}{2}=\frac{y-3}{1}=\frac{z}{1}$$

Putting these values of λ in the coordinates of A i.e. $(2\lambda+3, \lambda+3, \lambda)$, we find the coordinates of A and B i.e. $A(1, 2, -1)$ and $B(-1, 1, -2)$.

So, the equations of OA and OB are

$$\frac{x-0}{1-0}=\frac{y-0}{2-0}=\frac{z-0}{-1-0}$$

$$\frac{x-0}{-1-0}=\frac{y-0}{1-0}=\frac{z-0}{-2-0}$$

or $$\frac{x}{1}=\frac{y}{2}=\frac{z}{-1}$$

and $$\frac{x}{-1}=\frac{y}{1}=\frac{z}{-2}$$

213. Clearly, height h of ΔABC is the length of perpendicular from $A(1, -1, 2)$ to the line $\frac{x+2}{2}=\frac{y-1}{1}=\frac{z-0}{4}$ which passes through $P(-2, 1, 0)$ and is parallel to $\mathbf{b}=2\hat{\mathbf{i}}+\hat{\mathbf{j}}+4\hat{\mathbf{k}}$.

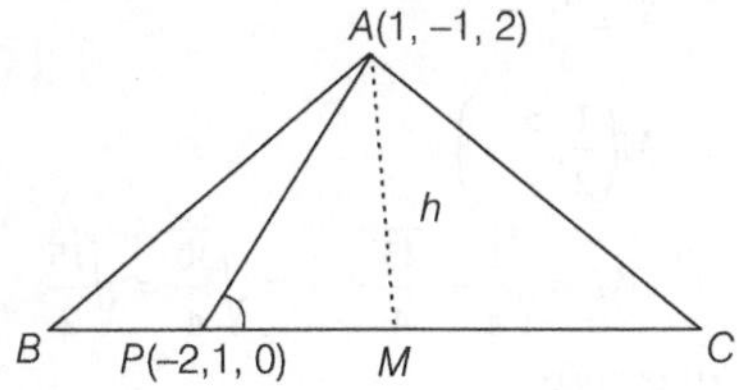

$$\therefore \quad h=\frac{|\mathbf{PA}\times\mathbf{b}|}{|\mathbf{b}|}$$

Now, $\mathbf{PA}=-3\hat{\mathbf{i}}+2\hat{\mathbf{j}}-2\hat{\mathbf{k}}$ and $\mathbf{b}=2\hat{\mathbf{i}}+\hat{\mathbf{j}}+4\hat{\mathbf{k}}$

$$\therefore \quad \mathbf{PA}\times\mathbf{b}=\begin{vmatrix}\hat{\mathbf{i}} & \hat{\mathbf{j}} & \hat{\mathbf{k}}\\ -3 & 2 & -2\\ 2 & 1 & 4\end{vmatrix}=10\hat{\mathbf{i}}+8-2\hat{\mathbf{k}}$$

$$\therefore \quad |\mathbf{PA}\times\mathbf{b}|=\sqrt{10^2+8^2+(-7)^2}=\sqrt{213}$$

and $$|\mathbf{b}|=\sqrt{2^2+1^2+4^2}=\sqrt{21}$$

$$\therefore \quad h=\frac{|\mathbf{PA}\times\mathbf{b}|}{|\mathbf{b}|}=\frac{\sqrt{213}}{\sqrt{21}}=\sqrt{\frac{71}{7}}$$

It is given that the length of BC is 5 units.

$\therefore$ Area of $\Delta ABC=\frac{1}{2}(BC\times h)$

$$=\frac{1}{2}\times 5\times\sqrt{\frac{71}{7}}=\sqrt{\frac{1775}{28}} \text{ sq units.}$$

214. If the coordinates of the point P be (α, β, γ).

Then, $$\frac{\alpha}{a}+\frac{\beta}{\beta}+\frac{\gamma}{c}=1 \quad \text{...(i)}$$

Again d′ c of OP are proportional to α, β, γ and hence these are also the d′ r of the normal to the plane which is perpendicular to OP and since it passes through P, its equation is

$$\alpha(x-\alpha)+\beta(y-\beta)+\gamma(z-\gamma)=0$$

or $$\alpha x+\beta y+\gamma z=\alpha^2+\beta^2+\gamma^2 \quad \text{...(ii)}$$

It meets the axes in A, B, C and hence the coordinates of these points are $\left(\frac{\alpha^2+\beta^2+\gamma^2}{\alpha}, 0, 0\right)$ etc.

The equation of the plane through A and parallel to the YZ plane is $x=\frac{\alpha^2+\beta^2+\gamma^2}{\alpha}$.

Similarly the equations of other planes are

$$y=\frac{\alpha^2+\beta^2+\gamma^2}{\beta} \text{ and } z=\frac{\alpha^2+\beta^2+\gamma^2}{\gamma}.$$

The locus of their point of intersection is obtained by elimination α, β, γ between the three equations of the planes and relation (i)

$$\frac{1}{x^2}+\frac{1}{y^2}+\frac{1}{z^2}=\frac{\alpha^2+\beta^2+\gamma^2}{(\alpha^2+\beta^2+\gamma^2)^2}=\frac{1}{\alpha^2+\beta^2+\gamma^2}$$

Again, $$\frac{1}{ax}+\frac{1}{by}+\frac{1}{cz}=\frac{\left(\frac{\alpha}{a}\right)+\left(\frac{\beta}{b}\right)+\left(\frac{\gamma}{c}\right)}{\alpha^2+\beta^2+\gamma^2}$$

$$=\frac{1}{\alpha^2+\beta^2+\gamma^2} \quad \text{[from Eq. (i)]}$$

$$\therefore \quad \frac{1}{x^2}+\frac{1}{y^2}+\frac{1}{z^2}=\frac{1}{ax}+\frac{1}{by}+\frac{1}{cz}$$

215. Any point on the line is $(3r+2, 4r-1, 12r+2)$.

If it lies on the plane $x-y+z=5$, then

$(3r+2)-(4r-1)+(12r+2)=5$

$\Rightarrow r=0$

Hence, point of intersection is $(2, -1, 2)$.

Its distance from $(-1, -5, -10)$ is

$$\sqrt{(2+1)^2+(-1+5)^2+(2+10)^2}$$

$$=\sqrt{9+16+144}=\sqrt{169}=13$$

216. Any plane through the intersection of given planes is

$(x+3y+6+\lambda(3x-y-4z)=0$

or $(1+3\lambda)x+(3-\lambda)y-4\lambda z+6=0$ ...(i)

Its perpendicular distance from $(0, 0, 0)$ is 1.

$$\therefore \left|\frac{6}{\sqrt{(1+3\lambda)^2+(3-\lambda)^2+(-4\lambda)^2}}\right|=1$$

$\Rightarrow \lambda=\pm 1$

$\therefore$ Required planes are $2x+y-2z+3=0$ and $x-2y-2z-3=0$.

217. The image of the plane

$$x-2y+2z-3=0 \quad \text{...(i)}$$

in the plane $x+y+z-1=0$ ...(ii)

passes through the line of intersection of the given planes.

Therefore, the equation of such a plane is

$$(x-2y+2z-3)+t(x+y+z-1)=0$$

$$\Rightarrow (1+t)x+(-2+t)y+(2+t)z-3-t=0 \quad \text{...(iii)}$$

Now, plane (ii) makes the same angle with plane (i) and image plane (iii). Thus,

$$\frac{1-2+2}{3\sqrt{3}}=\pm\frac{1+t-2+t+2+t}{\sqrt{3}\sqrt{(t+1)^2+(t-2)^2+(2+t)^2}}$$

$$\Rightarrow t=0, -\frac{2}{3}$$

For $t=0$, we get plane (i); hence for image plane, $t=-\frac{2}{3}$

The equation of the image plane

$3(x-2y+2z-3)-2(x+y+z-1)=0$

$\Rightarrow x-8y+4z-7=0.$

218. Given, square base $OP=OR=3$

$\therefore \quad P(3,0,0), R=(0,3,0)$

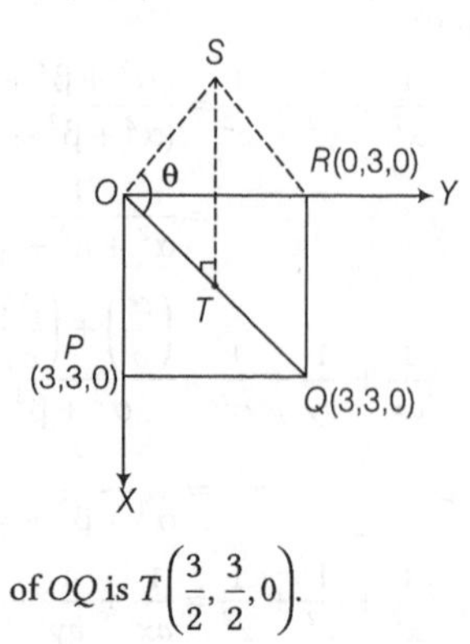

Also, mid-point of OQ is $T\left(\frac{3}{2},\frac{3}{2},0\right)$.

Since, S is directly above the mid-point T of diagonal OQ and $ST=3$.

i.e. $S\left(\frac{3}{2},\frac{3}{2},3\right)$

Here, DR's of OQ $(3, 3, 0)$ and DR's of $OS\left(\frac{3}{2},\frac{3}{2},3\right)$.

$$\therefore \quad \cos\theta=\frac{\frac{9}{2}+\frac{9}{2}}{\sqrt{9+9+0}\sqrt{\frac{9}{4}+\frac{9}{4}+9}}=\frac{9}{\sqrt{18}\cdot\sqrt{\frac{27}{2}}}=\frac{1}{\sqrt{3}}$$

$\therefore$ Option (a) is incorrect.

Now, equation of the plane containing the ΔOQS is

$$\begin{vmatrix} x & y & z \\ 3 & 3 & 0 \\ 3/2 & 3/2 & 3 \end{vmatrix}=0$$

$$\Rightarrow \begin{vmatrix} x & y & z \\ 1 & 1 & 0 \\ 1 & 1 & 2 \end{vmatrix}=0$$

$$\Rightarrow x(2-0)-y(2-0)+z(1-1)=0$$

$$\Rightarrow 2x-2y=0 \text{ or } x-y=0$$

$\therefore$ Option (b) is correct.

Now, length of the perpendicular from $P(3, 0, 0)$ to the plane containing ΔOQS is

$$\frac{|3-0|}{\sqrt{1+1}}=\frac{3}{\sqrt{2}}$$

$\therefore$ Option (c) is correct.

Here, equation of RS is

$$\frac{x-0}{3/2}=\frac{y-3}{-3/2}=\frac{z-0}{3}=\lambda$$

$$\Rightarrow x=\frac{3}{2}\lambda, y=-\frac{3}{2}\lambda+3, z=3\lambda$$

To find the distance from $O(0, 0, 0)$ to RS.

Let M be the foot of perpendicular.

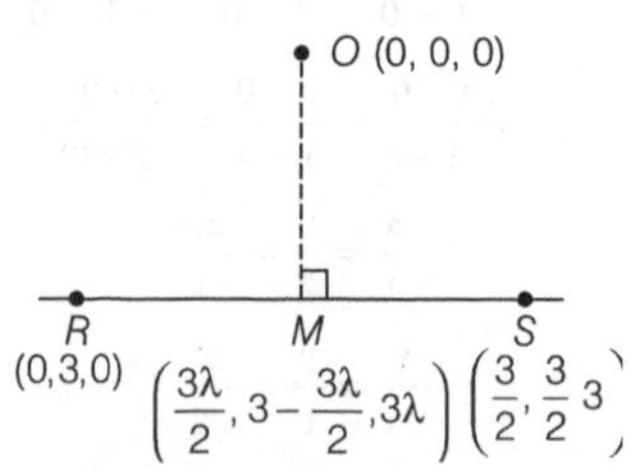

$$\because \quad \overline{OM}\perp\overline{RS} \Rightarrow \overline{OM}\cdot\overline{RS}=0$$

$$\Rightarrow \frac{9\lambda}{4}-\frac{3}{2}\left(3-\frac{3\lambda}{2}\right)+3(3\lambda)=0$$

$$\Rightarrow \lambda=\frac{1}{3}$$

$$\therefore \quad M\left(\frac{1}{2},\frac{5}{2},1\right)$$

$$\Rightarrow OM=\sqrt{\frac{1}{4}+\frac{25}{4}+1}=\sqrt{\frac{30}{4}}=\sqrt{\frac{15}{2}}$$

$\therefore$ Option (d) is correct.

219. Let image of $Q(3, 1, 7)$ w.r.t. $x - y + z = 3$ be $P(\alpha, \beta, \gamma)$.

$\therefore \quad \dfrac{\alpha - 3}{1} = \dfrac{\beta - 1}{-1} = \dfrac{\gamma - 7}{1} = \dfrac{-2(3 - 1 + 7 - 3)}{1^2 + (-1)^2 + (1)^2}$

$\Rightarrow \quad \alpha - 3 = 1 - \beta = \gamma - 7 = -4$

$\therefore \quad \alpha = -1, \beta = 5, \gamma = 3$

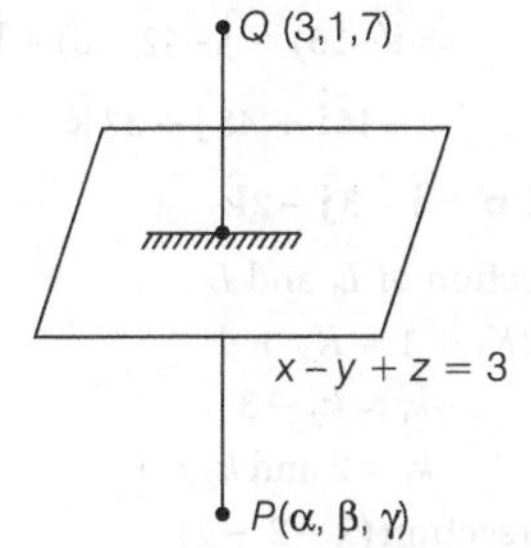

Hence, the image of $Q(3, 1, 7)$ is $P(-1, 5, 3)$.

To find equation of plane passing through $P(-1, 5, 3)$ and containing $\dfrac{x}{1} = \dfrac{y}{2} = \dfrac{z}{1}$

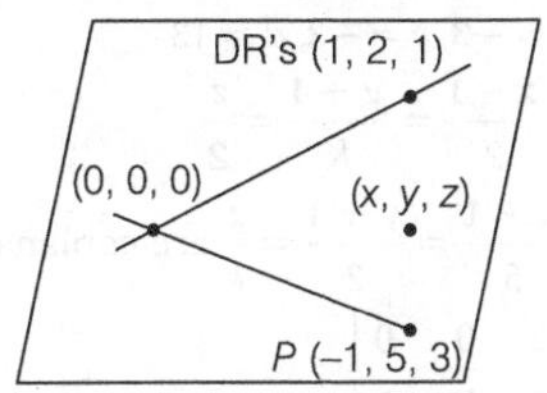

$$\Rightarrow \begin{vmatrix} x - 0 & y - 0 & z - 0 \\ 1 - 0 & 2 - 0 & 1 - 0 \\ -1 - 0 & 5 - 0 & 3 - 0 \end{vmatrix} = 0$$

$\Rightarrow x(6 - 5) - y(3 + 1) + z(5 + 2) = 0$

$\therefore \quad x - 4y + 7z = 0$

220. (i) Direction ratios of a line joining two points (x_1, y_1, z_1) and (x_2, y_2, z_2) are $x_2 - x_1, y_2 - y_1, z_2 - z_1$.

(ii) If the two lines with direction ratios $a_1, b_1, c_1; a_2, b_2, c_2$ are perpendicular, then $a_1 a_2 + b_1 b_2 + c_1 c_2 = 0$

Line L_1 is given by $y = x; z = 1$ can be expressed

$$L_1: \frac{x}{1} = \frac{y}{1} = \frac{z - 1}{0} = \alpha \qquad \text{[say]}$$

$\Rightarrow \quad x = \alpha, y = \alpha, z = 1$

Let the coordinates of Q on L_1 be $(\alpha, \alpha, 1)$.

Line L_2 given by $y = -x, z = -1$ can be expressed as

$$L_2: \frac{x}{1} = \frac{y}{-1} = \frac{z + 1}{0} = \beta \qquad \text{[say]}$$

$\Rightarrow \quad x = \beta, y = -\beta, z = -1$

Let the coordinates of R on L_2 be $(\beta, -\beta, -1)$.

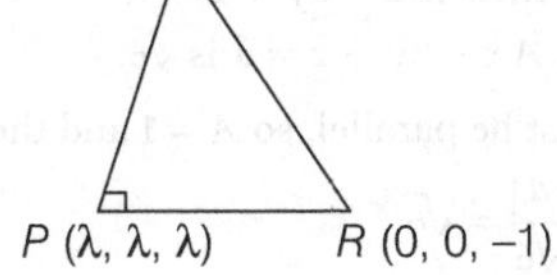

Direction ratios of PQ are $\lambda - \alpha, \lambda - \alpha, \lambda - 1$.

Now, $PQ \perp L_1$

$\therefore \quad 1(\lambda - \alpha) + 1 \cdot (\lambda - \alpha) + 0 \cdot (\lambda - 1) = 0$

$\Rightarrow \quad \lambda = \alpha$

Hence, $Q(\lambda, \lambda, 1)$

Direction ratios of PR are $\lambda - \beta, \lambda + \beta, \lambda + 1$.

Now, $PR \perp L_2$

$\therefore \quad 1(\lambda - \beta) + (-1)(\lambda + \beta) + 0(\lambda + 1) = 0$

$\lambda - \beta - \lambda - \beta = 0 \Rightarrow \beta = 0$

Hence, $R(0, 0, -1)$

Now, as $\angle QPR = 90°$

[as $a_1 a_2 + b_1 b_2 + c_1 c_2 = 0$, if two lines with DR's $a_1, b_1, c_1; a_2, b_2, c_2$ are perpendicular]

$\therefore \quad (\lambda - \lambda)(\lambda - 0) + (\lambda - \lambda)(\lambda - 0) + (\lambda - 1)(\lambda + 1) = 0$

$\Rightarrow (\lambda - 1)(\lambda + 1) = 0 \Rightarrow \lambda = 1 \quad \text{or} \quad \lambda = -1$

$\lambda = 1$, rejected as P and Q are different points.

$\Rightarrow \quad \lambda = -1$

221. If two straight lines are coplanar,

i.e. $\dfrac{x - x_1}{a_1} = \dfrac{y - y_1}{b_1} = \dfrac{z - z_1}{c_1}$

and $\dfrac{x - x_2}{a_2} = \dfrac{y - y_2}{b_2} = \dfrac{z - z_2}{c_2}$ are coplanar

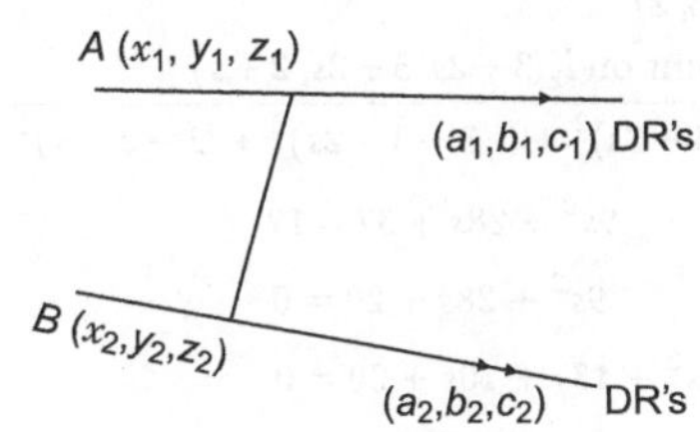

Then, $(x_2 - x_1, y_2 - y_1, z_2 - z_1)$, (a_1, b_1, c_1) and (a_2, b_2, c_2) are coplanar,

$$\text{i.e.} \quad \begin{vmatrix} x_2 - x_1 & y_2 - y_1 & z_2 - z_1 \\ a_1 & b_1 & c_1 \\ a_2 & b_2 & c_2 \end{vmatrix} = 0$$

Here, $\quad x = 5, \dfrac{y}{3 - \alpha} = \dfrac{z}{-2}$

$\Rightarrow \quad \dfrac{x - 5}{0} = \dfrac{y - 0}{-(\alpha - 3)} = \dfrac{z - 0}{-2} \qquad \text{...(i)}$

and $\quad x = \alpha, \dfrac{y}{-1} = \dfrac{z}{2 - \alpha}$

$\Rightarrow \quad \dfrac{x - \alpha}{0} = \dfrac{y - 0}{-1} = \dfrac{z - 0}{2 - \alpha} \qquad \text{...(ii)}$

$$\Rightarrow \begin{vmatrix} 5 - \alpha & 0 & 0 \\ 0 & 3 - \alpha & -2 \\ 0 & -1 & 2 - \alpha \end{vmatrix} = 0$$

$\Rightarrow \quad (5 - \alpha)[(3 - \alpha)(2 - \alpha) - 2] = 0$

$\Rightarrow \quad (5 - \alpha)[\alpha^2 - 5\alpha + 4] = 0$

$\Rightarrow \quad (5 - \alpha)(\alpha - 1)(\alpha - 4) = 0$

$\therefore \quad \alpha = 1, 4, 5$

222. Equation of straight line is $l : \frac{x - x_1}{a} = \frac{y - y_1}{b} = \frac{z - z_1}{c}$

Since, l is perpendicular to l_1 and l_2.

So, its DR's are cross-product of l_1 and l_2.

Now, to find a point on l_2 whose distance is given, assume a point and find its distance to obtain point.

Let $\quad l : \frac{x - 0}{a} = \frac{y - 0}{b} = \frac{z - 0}{c}$

which is perpendicular to

$$l_1 : (3\hat{\mathbf{i}} - \hat{\mathbf{j}} + 4\hat{\mathbf{k}}) + t(\hat{\mathbf{i}} + 2\hat{\mathbf{j}} + 2\hat{\mathbf{k}})$$

$$l_2 : (3\hat{\mathbf{i}} + 3\hat{\mathbf{j}} + 2\hat{\mathbf{k}}) + s(2\hat{\mathbf{i}} + 2\hat{\mathbf{j}} + \hat{\mathbf{k}})$$

$\therefore$ DR's of l is $\begin{vmatrix} \hat{\mathbf{i}} & \hat{\mathbf{j}} & \hat{\mathbf{k}} \\ 1 & 2 & 2 \\ 2 & 2 & 1 \end{vmatrix} = -2\hat{\mathbf{i}} + 3\hat{\mathbf{j}} - 2\hat{\mathbf{k}}$

$$l : \frac{x}{-2} = \frac{y}{3} = \frac{z}{-2} = k_1, k_2$$

Now, $A(-2k_1, 3k_1, -2k_1)$ and $B(-2k_2, 3k_2, -2k_2)$.

Since, A lies on l_1.

$\therefore \quad (-2k_1)\hat{\mathbf{i}} + (3k_1)\hat{\mathbf{j}} - (2k_1)\hat{\mathbf{k}} = (3 + t)\hat{\mathbf{i}} + (-1 + 2t)\hat{\mathbf{j}} + (4 + 2t)\hat{\mathbf{k}}$

$\Rightarrow \quad 3 + t = -2k_1, -1 + 2t = 3k_1, 4 + 2t = -2k_1$

$\therefore \quad k_1 = -1$

$\Rightarrow \quad A(2, -3, 2)$

Let any point on $l_2 (3 + 2s, 3 + 2s, 2 + s)$

$$\sqrt{(2 - 3 - 2s)^2 + (-3 - 3 - 2s)^2 + (2 - 2 - s)^2} = \sqrt{17}$$

$\Rightarrow \quad 9s^2 + 28s + 37 = 17$

$\Rightarrow \quad 9s^2 + 28s + 20 = 0$

$\Rightarrow \quad 9s^2 + 18s + 10s + 20 = 0$

$\Rightarrow \quad (9s + 10)(s + 2) = 0$

$\therefore \quad s = -2, \frac{-10}{9}$.

Hence, $(-1, -1, 0)$ and $\left(\frac{7}{9}, \frac{7}{9}, \frac{8}{9}\right)$ are required points.

223. Any point on $\frac{x + 2}{2} = \frac{y + 1}{-1} = \frac{z}{3} = \lambda$

$\Rightarrow \quad x = 2\lambda - 2, y = -\lambda - 1, z = 3\lambda$

Let foot of perpendicular from $(2\lambda - 2, -\lambda - 1, 3\lambda)$ to $x + y + z = 3$ be (x_2, y_2, z_2).

$$\therefore \quad \frac{x_2 - (2\lambda - 2)}{1} = \frac{y_2 - (-\lambda - 1)}{1} = \frac{z_2 - (3\lambda)}{1}$$

$$= -\frac{(2\lambda - 2 - \lambda - 1 + 3\lambda - 3)}{1 + 1 + 1}$$

$\Rightarrow \quad x_2 - 2\lambda + 2 = y_2 + \lambda + 1 = z_2 - 3\lambda = 2 - \frac{4\lambda}{3}$

$\therefore \quad x_2 = \frac{2\lambda}{3}, y_2 = 1 - \frac{7\lambda}{3}, z_2 = 2 + \frac{5\lambda}{3}$

$\Rightarrow \quad \lambda = \frac{x_2 - 0}{2/3} = \frac{y_2 - 1}{-7/3} = \frac{z_2 - 2}{5/3}$

Hence, foot of perpendicular lie on

$$\frac{x}{2/3} = \frac{y - 1}{-7/3} = \frac{z - 2}{5/3} \Rightarrow \frac{x}{2} = \frac{y - 1}{-7} = \frac{z - 2}{5}$$

224. $L_1 : \frac{x - 1}{2} = \frac{y - 0}{-1} = \frac{z - (-3)}{1}$

Normal of plane $P : \mathbf{n} = \begin{vmatrix} \hat{\mathbf{i}} & \hat{\mathbf{j}} & \hat{\mathbf{k}} \\ 7 & 1 & 2 \\ 3 & 5 & -6 \end{vmatrix}$

$$= \hat{\mathbf{i}}(-16) - \hat{\mathbf{j}}(-42 - 6) + \hat{\mathbf{k}}(32)$$

$$= -16\hat{\mathbf{i}} + 48\hat{\mathbf{j}} + 32\hat{\mathbf{k}}$$

DR's of normal $\mathbf{n} = \hat{\mathbf{i}} - 3\hat{\mathbf{j}} - 2\hat{\mathbf{k}}$

Point of intersection of L_1 and L_2.

$\Rightarrow \quad 2K_1 + 1 = K_2 + 4$

and $\quad -k_1 = k_2 - 3$

$\Rightarrow \quad k_1 = 2$ and $k_2 = 1$

$\therefore$ Point of intersection $(5, -2, -1)$

Now equation of plane,

$$1 \cdot (x - 5) - 3(y + 2) - 2(z + 1) = 0$$

$\Rightarrow \quad x - 3y - 2z - 13 = 0$

$\Rightarrow \quad x - 3y - 2z = 13$

$\therefore \quad a = 1, b = -3, c = -2, d = 13$

225. Since, $\quad \frac{x - 1}{2} = \frac{y + 1}{K} = \frac{z}{2}$

and $\quad \frac{x + 1}{5} = \frac{y + 1}{2} = \frac{z}{k}$ are coplanar.

$$\Rightarrow \quad \begin{vmatrix} 2 & 0 & 0 \\ 2 & K & 2 \\ 5 & 2 & K \end{vmatrix} = 0$$

$\Rightarrow \quad K^2 = 4 \Rightarrow K = \pm 2$

$\therefore \quad \mathbf{n}_1 = \mathbf{b}_1 \times \mathbf{d}_1 = 6\hat{\mathbf{j}} - 6\hat{\mathbf{k}}$, for $k = 2$

$\therefore \quad \mathbf{n}_2 = \mathbf{b}_2 \times \mathbf{d}_2 = 14\hat{\mathbf{j}} + 14\hat{\mathbf{k}}$, for $k = -2$

So, equation of planes are $(\mathbf{r} - \mathbf{a}) \cdot \mathbf{n}_1 = 0$

$\Rightarrow \quad y - z = -1$ and $(\mathbf{r} - \mathbf{a}) \cdot \mathbf{n}_2 = 0$

$\Rightarrow \quad y + z = -1$

226. Equation of the plane containing the lines

$$\frac{x - 2}{3} = \frac{y - 3}{4} = \frac{z - 4}{5}$$

and $\quad \frac{x - 1}{2} = \frac{y - 2}{3} = \frac{z - 3}{4}$

is $a(x - 2) + b(y - 3) + c(z - 4) = 0$ ...(i)

where, $\quad 3a + 4b + 5c = 0$ ...(ii)

$\quad 2a + 3b + 4c = 0$ ...(iii)

and $a(1 - 2) + b(2 - 3) + c(2 - 3) = 0$

i.e. $\quad a + b + c = 0$ ...(iv)

From Eqs. (ii) and (iii), $\frac{a}{1} = \frac{b}{-2} = \frac{c}{1}$, which satisfy Eq. (iv).

Plane through lines is $x - 2y + z = 0$.

Given plane is $Ax - 2y + z = d$ is $\sqrt{6}$.

$\therefore$ Planes must be parallel, so $A = 1$ and then

$$\frac{|d|}{\sqrt{6}} = \sqrt{6}$$

$\Rightarrow \quad |d| = 6$

227. The equation of the plane passing through the point $(-1, -2, -1)$ and whose normal is perpendicular to both the given lines L_1 and L_2 may be written as

$(x + 1) + 7(y + 2) - 5(z + 1) = 0 \Rightarrow x + 7y - 5z + 10 = 0$

The distance of the point (1, 1, 1) from the plane

$$= \left|\frac{1 + 7 - 5 + 10}{\sqrt{1 + 49 + 25}}\right| = \frac{13}{\sqrt{75}} \text{ units}$$

228. The shortest distance between L_1 and L_2 is

$$\left|\frac{\{(2-(-1))\,\hat{\mathbf{i}} + (2-2)\,\hat{\mathbf{j}} + (3-(-1))\,\hat{\mathbf{k}}\} \cdot (-\hat{\mathbf{i}} - 7\hat{\mathbf{j}} + 5\hat{\mathbf{k}})}{5\sqrt{3}}\right|$$

$$= \left|\frac{(3\hat{\mathbf{i}} + 4\hat{\mathbf{k}}) \cdot (-\hat{\mathbf{i}} - 7\hat{\mathbf{j}} + 5\hat{\mathbf{k}})}{5\sqrt{3}}\right|$$

$$= \frac{17}{5\sqrt{3}} \text{ units}$$

229. The equations of given lines in vector form may be written as

$$L_1 : \vec{\mathbf{r}} = (-\hat{\mathbf{i}} - 2\hat{\mathbf{j}} - \hat{\mathbf{k}}) + \lambda\,(3\hat{\mathbf{i}} + \hat{\mathbf{j}} + 2\hat{\mathbf{k}})$$

and $L_2 : \vec{\mathbf{r}} = (2\hat{\mathbf{i}} - 2\hat{\mathbf{j}} + 3\hat{\mathbf{k}}) + \mu\,(\hat{\mathbf{i}} + 2\hat{\mathbf{j}} + 3\hat{\mathbf{k}})$

Since, the vector is perpendicular to both L_1 and L_2.

$$\begin{vmatrix} \hat{\mathbf{i}} & \hat{\mathbf{j}} & \hat{\mathbf{k}} \\ 3 & 1 & 2 \\ 1 & 2 & 3 \end{vmatrix} = -\hat{\mathbf{i}} - 7\hat{\mathbf{j}} + 5\hat{\mathbf{k}}$$

$\therefore$ Required unit vector

$$= \frac{(-\hat{\mathbf{i}} - 7\hat{\mathbf{j}} + 5\hat{\mathbf{k}})}{\sqrt{(-1)^2 + (-7)^2 + (5)^2}}$$

$$= \frac{1}{5\sqrt{3}}(-\hat{\mathbf{i}} - 7\hat{\mathbf{j}} + 5\hat{\mathbf{k}})$$

230. Given three planes are

$$P_1 : x - y + z = 1 \quad \text{...(i)}$$
$$P_2 : x + y - z = -1 \quad \text{...(ii)}$$
and $$P_3 : x - 3y + 3z = 2 \quad \text{...(iii)}$$

On solving Eqs. (i) and (ii), we get

$$x = 0, z = 1 + y$$

which does not satisfy Eq. (iii).

As $\quad x - 3y + 3z = 0 - 3y + 3(1 + y) = 3\,(\neq 2)$

So, Statement II is true.

Next, since we know that direction ratios of line of intersection of planes $a_1x + b_1y + c_1z + d_1 = 0$

and $\quad a_2x + b_2y + c_2z + d_2 = 0$ is

$b_1c_2 - b_2c_1, c_1a_2 - a_1c_2, a_1b_2 - a_2b_1$

Using above result,

Direction ratios of lines L_1, L_2 and L_3 are

$$0, 2, 2\,;\, 0, -4, -4\,;\, 0, -2, -2$$

Since, all the three lines L_1, L_2 and L_3 are parallel pairwise.

Hence, Statement I is false.

231. Given planes are $3x - 6y - 2z = 15$ and $2x + y - 2z = 5$.

For $z = 0$, we get $x = 3, y = -1$

Since, direction ratios of planes are

$$< 3, -6, -2 > \text{ and } < 2, 1, -2 >$$

Then the DR's of line of intersection of planes is $< 14, 2, 15 >$ and line is

$$\frac{x-3}{14} = \frac{y+1}{2} = \frac{z-0}{15} = \lambda \quad \text{[say]}$$

$$\Rightarrow \quad x = 14\lambda + 3,\ y = 2\lambda - 1,\ z = 15\lambda$$

Hence, Statement I is false.

But Statement II is true.

232. Let $\Delta = \begin{vmatrix} a & b & c \\ b & c & a \\ c & a & b \end{vmatrix}$

$$= -\frac{1}{2}(a + b + c)[(a-b)^2 + (b-c)^2 + (c-a)^2]$$

(A) If $a + b + c \neq 0$ and $a^2 + b^2 + c^2 = ab + bc + ca$

$\Rightarrow \Delta = 0$ and $a = b = c \neq 0$

$\Rightarrow$ The equations represent identical planes.

(B) $a + b + c = 0$ and $a^2 + b^2 + c^2 \neq ab + bc + ca$

$\Rightarrow \Delta = 0$

$\Rightarrow$ The equations have infinitely many solutions.

$ax + by = (a + b)z, \quad bx + cy = (b + c)z$

$\Rightarrow (b^2 - ac)y = (b^2 - ac)z \Rightarrow y = z$

$\Rightarrow ax + by + cy = 0 \Rightarrow ax = ay \Rightarrow x = y = z$

(C) $a + b + c \neq 0$ and $a^2 + b^2 + c^2 \neq ab + bc + ca$

$\Rightarrow \quad \Delta \neq 0$

The equations represent planes meeting at only one point.

(D) $a + b + c = 0$ and $a^2 + b^2 + c^2 = ab + bc + ca$

$\Rightarrow a = b = c = 0$

$\Rightarrow$ The equations represent whole of the three-dimensional space.

233. Any line parallel to $\frac{x}{1} = \frac{y}{4} = \frac{z}{5}$ and passing through $P(1, -2, 3)$ is

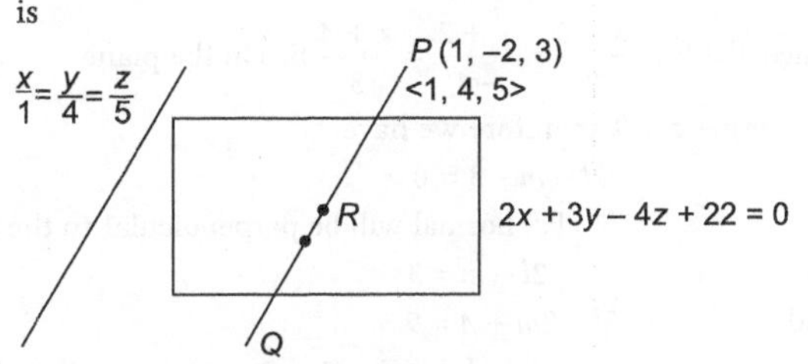

$$\frac{x-1}{1} = \frac{y+2}{4} = \frac{z-3}{5} = \lambda \quad \text{(say)}$$

Any point on above line can be written as $(\lambda + 1, 4\lambda - 2, 5\lambda + 3)$.

$\therefore$ Coordinates of R are $(\lambda + 1, 4\lambda - 2, 5\lambda + 3)$.

Since, point R lies on the above plane.

$\therefore \quad 2(\lambda + 1) + 3(4\lambda - 2) - 4(5\lambda + 3) + 22 = 0 \Rightarrow \lambda = 1$

So, point R is (2, 2, 8).

Now, $\quad PR = \sqrt{(2-1)^2 + (2+2)^2 + (8-3)^2} = \sqrt{42}$

$\therefore \quad PQ = 2PR = 2\sqrt{42}$

234. Given, equations of lines are

$$\frac{x-1}{1}=\frac{y+2}{-2}=\frac{z-4}{3} \text{ and } \frac{x-2}{2}=\frac{y+1}{-1}=\frac{z+7}{-1}$$

Let $\mathbf{n}_1 = \hat{\mathbf{i}} - 2\hat{\mathbf{j}} + 3\hat{\mathbf{k}}$ and $\mathbf{n}_2 = 2\hat{\mathbf{i}} - \hat{\mathbf{j}} - \hat{\mathbf{k}}$

$\therefore$ Any vector $\mathbf{n}$ perpendicular to both $\mathbf{n}_1, \mathbf{n}_2$ is given by

$$\mathbf{n} = \mathbf{n}_1 \times \mathbf{n}_2$$

$$\Rightarrow \quad \mathbf{n} = \begin{vmatrix} \hat{\mathbf{i}} & \hat{\mathbf{j}} & \hat{\mathbf{k}} \\ 1 & -2 & 3 \\ 2 & -1 & -1 \end{vmatrix} = 5\hat{\mathbf{i}} + 7\hat{\mathbf{j}} + 3\hat{\mathbf{k}}$$

$\therefore$ Equation of a plane passing through $(1, -1, -1)$ and perpendicular to $\mathbf{n}$ is given by

$$5(x-1) + 7(y+1) + 3(z+1) = 0$$

$$\Rightarrow \quad 5x + 7y + 3z + 5 = 0$$

$$\therefore \text{Required distance} = \left| \frac{5 + 21 - 21 + 5}{\sqrt{5^2 + 7^2 + 3^2}} \right| = \frac{10}{\sqrt{83}} \text{ units}$$

235. Equation of line passing through $(1, -5, 9)$ and parallel to $x = y = z$ is

$$\frac{x-1}{1} = \frac{y+5}{1} = \frac{z-9}{1} = \lambda \quad \text{(say)}$$

Thus, any point on this line is of the form $(\lambda + 1, \lambda - 5, \lambda + 9)$.

Now, if $P(\lambda + 1, \lambda - 5, \lambda + 9)$ is the point of intersection of line and plane, then

$$(\lambda + 1) - (\lambda - 5) + \lambda + 9 = 5$$

$$\Rightarrow \quad \lambda + 15 = 5$$

$$\Rightarrow \quad \lambda = -10$$

$\therefore$ Coordinates of point P are $(-9, -15, -1)$.

Hence, the required distance

$$= \sqrt{(1+9)^2 + (-5+15)^2 + (9+1)^2}$$

$$= \sqrt{10^2 + 10^2 + 10^2} = 10\sqrt{3}$$

236. Since, the line $\frac{x-3}{2} = \frac{y+2}{-1} = \frac{z+4}{3}$ lies in the plane $lx + my - z = 9$, therefore we have

$$2l - m - 3 = 0$$

[$\because$ normal will be perpendicular to the line]

$$\Rightarrow \quad 2l - m = 3 \quad \text{...(i)}$$

and $\quad 3l - 2m + 4 = 9$

[$\because$ point $(3, -2, -4)$ lies on the plane]

$$\Rightarrow \quad 3l - 2m = 5 \quad \text{...(ii)}$$

On solving Eqs. (i) and (ii), we get

$$l = 1 \text{ and } m = -1$$

$$\therefore \quad l^2 + m^2 = 2$$

237. Given equation of line is

$$\frac{x-2}{3} = \frac{y+1}{4} = \frac{z-2}{12} = \lambda \quad \text{(say) ...(i)}$$

and equation of plane is

$$x - y + z = 16 \quad \text{...(ii)}$$

Any point on the line (i) is

$$(3\lambda + 2, 4\lambda - 1, 12\lambda + 2)$$

Let this point be point of intersection of the line and plane.

$$\therefore \quad (3\lambda + 2) - (4\lambda - 1) + (12\lambda + 2) = 16$$

$$\Rightarrow \quad 11\lambda + 5 = 16$$

$$\Rightarrow \quad 11\lambda = 11$$

$$\Rightarrow \quad \lambda = 1$$

$\therefore$ Point of intersection is $(5, 3, 14)$.

Now, distance between the points $(1, 0, 2)$ and $(5, 3, 14)$

$$= \sqrt{(5-1)^2 + (3-0)^2 + (14-2)^2}$$

$$= \sqrt{16 + 9 + 144}$$

$$= \sqrt{169}$$

$$= 13$$

238. Let equation of plane containing the lines $2x - 5y + z = 3$ and $x + y + 4z = 5$ be

$$(2x - 5y + z - 3) + \lambda(x + y + 4z - 5) = 0$$

$$\Rightarrow (2 + \lambda)x + (\lambda - 5)y + (4\lambda + 1)z - 3 - 5\lambda = 0 \quad \text{...(i)}$$

This plane is parallel to the plane $x + 3y + 6z = 1$.

$$\therefore \quad \frac{2+\lambda}{1} = \frac{\lambda - 5}{3} = \frac{4\lambda + 1}{6}$$

On taking first two equalities, we get

$$6 + 3\lambda = \lambda - 5$$

$$\Rightarrow \quad 2\lambda = -11$$

$$\Rightarrow \quad \lambda = -\frac{11}{2}$$

On taking last two equalities, we get

$$6\lambda - 30 = 3 + 12\lambda$$

$$\Rightarrow \quad -6\lambda = 33$$

$$\Rightarrow \quad \lambda = -\frac{11}{2}$$

So, the equation of required plane is

$$\left(2 - \frac{11}{2}\right)x + \left(\frac{-11}{2} - 5\right)y + \left(-\frac{44}{2} + 1\right)z - 3 + 5 \times \frac{11}{2} = 0$$

$$\Rightarrow \quad -\frac{7}{2}x - \frac{21}{2}y - \frac{42}{2}z + \frac{49}{2} = 0$$

$$\Rightarrow \quad x + 3y + 6z - 7 = 0$$

239. Given, $l + m + n = 0 \Rightarrow l = -(m + n)$

$$\Rightarrow \quad (m+n)^2 = l^2$$

$$\Rightarrow \quad m^2 + n^2 + 2mn = m^2 + n^2 \quad [\because l^2 = m^2 + n^2, \text{ given}]$$

$$\Rightarrow \quad 2mn = 0$$

Case I When $m = 0$, then

$$l = -n$$

Hence, (l, m, n) is $(1, 0, -1)$.

Case II When $n = 0$, then

$$l = -m$$

Hence, (l, m, n) is $(1, 0, -1)$.

$$\therefore \quad \cos\theta = \frac{1 + 0 + 0}{\sqrt{2} \times \sqrt{2}} = \frac{1}{2}$$

$$\Rightarrow \quad \theta = \frac{\pi}{3}$$

240. Plane and line are parallel to each other. Equation of normal to the plane through the point (1, 3, 4) is

$$\frac{x-1}{2}=\frac{y-3}{-1}=\frac{z-4}{1}=k \quad \text{[Say]}$$

Any point in this normal is

$(2k+1, -k+3, 4+k)$.

$\Rightarrow \left(\frac{2k+1+1}{2}, \frac{3-k+3}{2}, \frac{4+k+4}{2}\right)$ lies on plane.

$$\Rightarrow \quad 2(k+1)-\left(\frac{6-k}{2}\right)+\left(\frac{8+k}{2}\right)+3=0$$

$$\Rightarrow \quad k=-2$$

Hence, point through which this image pass is

$(2k+1, 3-k, 4+k)$

i.e. $(2(-2)+1, 3+2, 4-2)=(-3, 5, 2)$

Hence, equation of image line is

$$\frac{x+3}{3}=\frac{y-5}{1}=\frac{z-2}{-5}$$

241. Given planes are

$$2x+y+2z-8=0$$

and $2x+y+2z+\frac{5}{2}=0$

Distance between two parallel planes

$$=\frac{|d_1-d_2|}{\sqrt{a^2+b^2+c^2}}=\left|\frac{-8-\frac{5}{2}}{\sqrt{2^2+1^2+2^2}}\right|$$

$$=\frac{\frac{21}{2}}{3}=\frac{7}{2}$$

242. The given line are

$$\frac{x-2}{1}=\frac{y-3}{1}=\frac{z-4}{-k} \quad \text{...(i)}$$

and $\frac{x-1}{k}=\frac{y-k}{2}=\frac{z-5}{1}$ ...(ii)

Condition for two lines are coplanar.

$$\begin{vmatrix} x_1-x_2 & y_1-y_2 & z_1-z_2 \\ l_1 & m_1 & n_1 \\ l_2 & m_2 & n_2 \end{vmatrix}=0$$

wher,e (x_1, y_1, z_1) and $(x_2, y_2 z_2)$ are any points on the lines (i) and (ii), respectively and $< l_1, m_1, n_1 >$ and $< l_2, m_2, n_2 >$ are direction cosines of lines (i) and (ii), respectively.

$$\therefore \quad \begin{vmatrix} 2-1 & 3-4 & 4-5 \\ 1 & 1 & -k \\ k & 2 & 1 \end{vmatrix}=0$$

$$\Rightarrow \quad \begin{vmatrix} 1 & -1 & -1 \\ 1 & 1 & -k \\ k & 2 & 1 \end{vmatrix}=0$$

$\Rightarrow 1(1+2k)+1(1+k^2)-(2-k)=0$

$\Rightarrow \quad k^2+2k+k=0$

$\Rightarrow \quad k^2+3k=0$

$\Rightarrow \quad k=0, -3$

Note : If 0 appears in the denominator, then the correct way of representing the equation of straight line is

$$\frac{x-2}{1}=\frac{y-3}{1}; z=4 \text{ and } x=l; \frac{y-4}{2}=\frac{z-5}{1}$$

243. Given A plane $P: x-2y+2z-5=0$

To find The equation of a plane parallel to given plane P and at a distance of 1 unit from origin. Equation of family of planes parallel to the given plane P is

$$Q: x-2y+2z+d=0$$

Also, perpendicular distance of Q from origin is 1 unit.

$$\Rightarrow \quad \left|\frac{0-2(0)+2(0)+d}{\sqrt{1^2+2^2+2^2}}\right|=1$$

$$\Rightarrow \quad \left|\frac{d}{3}\right|=1 \Rightarrow d=\pm 3$$

Hence, the required equation of the plane parallel to P and at unit distance from origin is

$$x-2y+2z\pm 3=0$$

Hence, out of the given equations, option (a) is the only correct option.

244. Given Two lines $L_1: \frac{x-1}{2}=\frac{y+1}{3}=\frac{z-1}{4}$

and $L_2: \frac{x-3}{1}=\frac{y-k}{2}=\frac{z-0}{1}$

To find The value of 'k' of the given lines L_1 and L_2 are intersecting each other.

Let $L_1: \frac{x-1}{2}=\frac{y+1}{3}=\frac{z-1}{4}=p$

and $L_2: \frac{x-3}{1}=\frac{y-k}{2}=\frac{z-0}{1}=q$

$\Rightarrow$ Any point P on line L_1 is of type $P(2p+1, 3p-1, 4p+1)$ and any point Q on line L_2 is of type $Q(q+3, 2q+k, q)$.

Since, L_1 and L_2 are intersecting each other, hence both points P and Q should coincide at the point of intersection, *i.e.*, corresponding coordinates of P and Q should be same.

$$2p+1=q+3,\ 3p-1=2q+k \text{ and } 4p+1=q$$

On solving $2p+1=q+3$ and $4p+1=q$, we get the values of p and q as

$$p=\frac{-3}{2} \text{ and } q=-5$$

On substituting the values of p and q in the third equation $3p-1=2q+k$, we get

$$3\left(\frac{-3}{2}\right)-1=2(-5)+k$$

$$\Rightarrow \quad k=\frac{9}{2}$$

245. Angle between straight line $\mathbf{r}=\mathbf{a}+\lambda\mathbf{b}$ and plene $\mathbf{r}\cdot\hat{\mathbf{n}}=\mathbf{d}$ is

$$\sin\theta=\frac{\mathbf{b}\cdot\hat{\mathbf{n}}}{|\mathbf{b}||\hat{\mathbf{n}}|}$$

$$\therefore \quad \sin\theta=\frac{(\hat{\mathbf{i}}+2\hat{\mathbf{j}}+\lambda\hat{\mathbf{k}})\cdot(\hat{\mathbf{i}}+2\hat{\mathbf{j}}+3\hat{\mathbf{k}})}{\sqrt{1+4+\lambda^2}\sqrt{1+4+9}}$$

$$= \frac{5+3\lambda}{\sqrt{\lambda^2+5}\,\sqrt{14}}$$

Given, $\cos\theta = \sqrt{\frac{5}{14}}$

$\therefore$ $\sin\theta = \frac{3}{\sqrt{14}} \Rightarrow \frac{3}{\sqrt{14}} = \frac{5+3\lambda}{\sqrt{\lambda^2+5}\cdot\sqrt{14}}$

$\Rightarrow$ $9(\lambda^2+5) = 9\lambda^2 + 30\lambda + 25$

$\Rightarrow$ $9\lambda^2 + 45 = 9\lambda^2 + 30\lambda + 25$

$\Rightarrow$ $30\lambda = 20 \Rightarrow \lambda = \frac{2}{3}$

246. Mid-point of AB is $M(1, 3, 5)$.

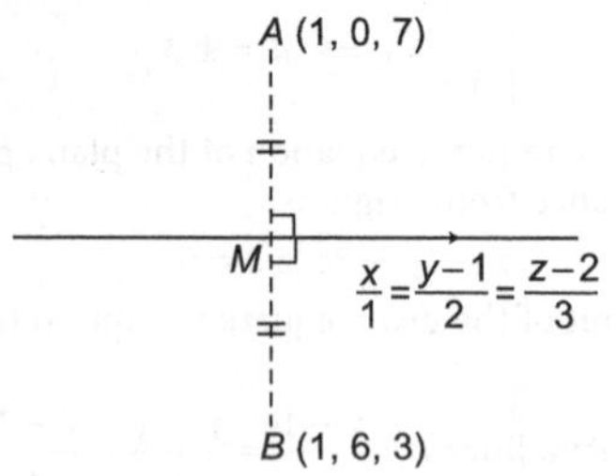

which lies on $\frac{x}{1} = \frac{y-1}{2} = \frac{z-2}{3}$

as $\frac{1}{1} = \frac{3-1}{2} = \frac{5-2}{3} \Rightarrow 1 = 1 = 1$

Hence, Statement II is true.

Also, directions ratios of AB is

$(1-1, 6-0, 3-7)$

i.e. $(0, 6, -4)$ ...(i)

and direction ratios of straight line is

$(1, 2, 3)$ ...(ii)

The two lines are perpendicular, if

$0(1) + 6(2) - 4(3) = 12 - 12 = 0$

Hence, Statement I is true and statement II is a correct explanations of statement II.

247. Let the coordinates of Q be $(2\lambda, 3\lambda+2, 4\lambda+3)$ which is any point on the straight line AB.

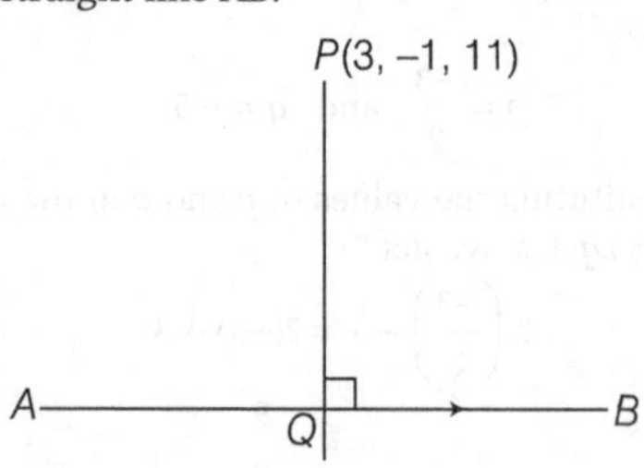

$\therefore$ DR's of PQ is $(2\lambda-3, 3\lambda+3, 4\lambda-8)$

Also, perpendicular to straight line AB $\frac{x}{2} = \frac{y-2}{3} = \frac{z-3}{4} = \lambda$

having DR's (2, 3, 4).

Thus, $2(2\lambda-3) + 3(3\lambda+3) + 4(4\lambda-8) = 0$

$\Rightarrow$ $4\lambda - 6 + 9\lambda + 9 + 16\lambda - 32 = 0$

$\Rightarrow$ $29\lambda - 29 = 0$

$\therefore$ $\lambda = 1$

Hence, coordinates of Q are (2, 5, 7)

$\therefore$ $|PQ| = \sqrt{(3-2)^2 + (-1-5)^2 + (7-11)^2}$

$= \sqrt{1+36+16} = \sqrt{53}$

248. Let Q be any point on the plane.

Then equation of PQ is

$$\frac{x-1}{1} = \frac{y+5}{1} = \frac{z-9}{1} = \lambda$$

where $P = (1, -5, 9)$

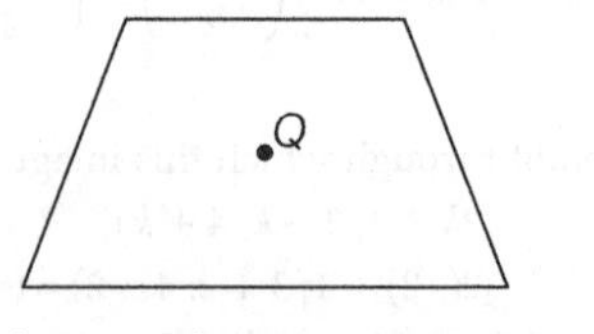

$\therefore$ $x = \lambda+1, y = \lambda-5, z = \lambda+9$ lies on the plane

$x - y + z = 5$

$\Rightarrow \lambda + 1 - \lambda + 5 + \lambda + 9 = 5$

$\therefore$ $\lambda = -10$

Hence, coordinate of Q is $Q(-9, -15, -1)$

$\therefore$ $|PQ| = \sqrt{(10)^2 + (10)^2 + (10)^2} = 10\sqrt{3}$

249. We know that, $\cos^2 45° + \cos^2 120° + \cos^2\theta = 1$

$\Rightarrow$ $\frac{1}{2} + \frac{1}{4} + \cos^2\theta = 1 \Rightarrow \cos^2\theta = \frac{1}{4}$

$\Rightarrow$ $\cos\theta = \pm\frac{1}{2} \Rightarrow \theta = 60°$ or $120°$

250. The image of the point (3, 1, 6) with respect to the plane $x - y + z = 5$ is

$$\frac{x-3}{1} = \frac{y-1}{-1} = \frac{z-6}{1} = \frac{-2(3-1+6-5)}{1+1+1}$$

$$\left[\because \frac{x-x_1}{a} = \frac{y-y_1}{b} = \frac{z-z_1}{c} = \frac{-2(ax_1+by_1+cz_1+d)}{a^2+b^2+c^2}\right]$$

$\Rightarrow$ $\frac{x-3}{1} = \frac{y-1}{-1} = \frac{z-6}{1} = -2$

$\Rightarrow$ $x = 3 - 2 = 1$

$y = 1 + 2 = 3$

and $z = 6 - 2 = 4$

which shows that Statement I is true.

We observe that the line segment joining the points $A(3, 1, 6)$ and $B(1, 3, 4)$ has direction ratios $2, -2, 2$ which are proportional to $1, -1, 1$. The direction ratios of the normal to the plane. Hence, Statements II is true. Thus, the Statements I and II are true and Statement II is correct explanation of Statement I.

251. Dr's of given line are $(3, -5, 2)$.

Dr's of normal to the plane $= (1, 3, -\alpha)$

$\therefore$ Line is perpendicular to the normal.

$\Rightarrow$ $3(1) - 5(3) + 2(-\alpha) = 0$

$\Rightarrow$ $3 - 15 - 2\alpha = 0$

$\Rightarrow$ $2\alpha = -12 \Rightarrow \alpha = -6$

Also, point $(2, 1, -2)$ lies on the plane.

$\therefore$ $2 + 3 + 6(-2) + \beta = 0 \Rightarrow \beta = 7$

$\Rightarrow$ $(\alpha, \beta) = (-6, 7)$

252. Projection of a vector on coordinate axes are

$$x_2 - x_1, y_2 - y_1, z_2 - z_1$$

$$x_2 - x_1 = 6,$$

$$y_2 - y_1 = -3,$$

$$z_2 - z_1 = 2$$

Now, $\sqrt{(x_2 - x_1)^2 + (y_2 - y_1)^2 + (z_2 - z_1)^2}$

$$= \sqrt{36 + 9 + 4} = 7$$

So, the DC's of the vector are $\frac{6}{7}, -\frac{3}{7}, \frac{2}{7}$.

253. Equation of line passing through (5, 1, a) and (3, b, 1) is

$$\frac{x-3}{5-3} = \frac{y-b}{1-b} = \frac{z-1}{a-1} \quad \text{... (i)}$$

$$\left[\because \frac{x - x_1}{x_2 - x_1} = \frac{y - y_1}{y_2 - y_1} = \frac{z - z_1}{z_2 - z_1}\right]$$

Point $\left(0, \frac{17}{2}, -\frac{13}{2}\right)$ satisfies Eq. (i), we get

$$-\frac{3}{2} = \frac{\frac{17}{2} - b}{1-b} = \frac{-\frac{13}{2} - 1}{a-1}$$

$$\Rightarrow \quad a - 1 = \frac{\left(-\frac{15}{2}\right)}{\left(-\frac{3}{2}\right)} = 5 \Rightarrow a = 6$$

Also, $\quad -3(1-b) = 2\left(\frac{17}{2} - b\right)$

$$\Rightarrow \quad 3b - 3 = 17 - 2b$$

$$\Rightarrow \quad 5b = 20 \Rightarrow b = 4$$

254. Given, $\quad \frac{x-1}{k} = \frac{y-2}{2} = \frac{z-3}{3} \quad \text{... (i)}$

and $\quad \frac{x-2}{3} = \frac{y-3}{k} = \frac{z-1}{2} \quad \text{... (ii)}$

Since, lines intersect at a point. Then, shortest distance between them is zero.

$$\therefore \quad \begin{vmatrix} k & 2 & 3 \\ 3 & k & 2 \\ 1 & 1 & -2 \end{vmatrix} = 0$$

$$\Rightarrow k(-2k - 2) - 2(-6 - 2) + 3(3 - k) = 0$$

$$\Rightarrow \quad -2k^2 - 5k + 25 = 0$$

$$\Rightarrow \quad 2k^2 + 5k - 25 = 0$$

$$\Rightarrow 2k^2 + 10k - 5k - 25 = 0$$

$$\Rightarrow 2k(k + 5) - 5(k + 5) = 0$$

$$\Rightarrow \quad k = \frac{5}{2}, -5$$

Hence, integer value of k is -5.

255. Let the drection cosines of line L be l, m and n. Since, the line intersect the given planes, then the normal to the planes are perpendicular to the line L.

$$\therefore \quad 2l + 3m + n = 0 \quad \text{...(i)}$$

and $\quad l + 3m + 2n = 0 \quad \text{...(ii)}$

From Eqs. (i) and (ii), we get

$$\frac{l}{3} = \frac{m}{-3} = \frac{n}{3} = k \quad \text{[say]}$$

We know that, $l^2 + m^2 + n^2 = 1$

$$\therefore \quad (3k)^2 + (-3k)^2 + (3k)^2 = 1$$

$$\Rightarrow \quad 27k^2 = 1 \Rightarrow k = \frac{1}{3\sqrt{3}}$$

$$\therefore \quad l = \frac{1}{\sqrt{3}} \Rightarrow \cos\alpha = \frac{1}{\sqrt{3}}$$

256. Since, a line makes an angle of $\frac{\pi}{4}$ with positive direction of each of X-axis and Y-axis, therefore

$$\alpha = \frac{\pi}{4}, \beta = \frac{\pi}{4}$$

We know that, $\cos^2\alpha + \cos^2\beta + \cos^2\gamma = 1$

$$\Rightarrow \quad \cos^2\frac{\pi}{4} + \cos^2\frac{\pi}{4} + \cos^2\gamma = 1$$

$$\Rightarrow \quad \frac{1}{2} + \frac{1}{2} + \cos^2\gamma = 1$$

$$\Rightarrow \quad \cos^2\gamma = 0 \Rightarrow \gamma = 90°$$

257. Given, equation of sphere is

$$x^2 + y^2 + z^2 - 6x - 12y - 2z + 20 = 0$$

whose coordinates of centre are (3, 6, 1).

Since, one end of diameter are (2, 3, 5) and the other end of diameter be (α, β, γ),

then $\quad \frac{\alpha+2}{2} = 3, \frac{\beta+3}{2} = 6, \frac{\gamma+5}{2} = 1$

$$\Rightarrow \quad \alpha = 4, \beta = 9$$

and $\quad \gamma = -3.$

Hence, the coordinates of other point are (4, 9, − 3).

258. Given equations of lines are

$$x = ay + b, z = cy + d$$

and $\quad x = a'y + b', z = c'y + d'$

These equations can be rewritten as

$$\frac{x-b}{a} = \frac{y-0}{1} = \frac{z-d}{c}$$

and $\quad \frac{x-b'}{a'} = \frac{y-0}{1} = \frac{z-d'}{c'}$

These lines will perpendicular, if $aa' + 1 + cc' = 0$

$$\because \quad l_1 l_2 + m_1 m_2 + n_1 n_2 = 6$$

259. We know that, the image (x, y, z) of a point (x_1, y_1, z_1) in a plane $ax + by + cz + d = 0$ is given by

$$\frac{x - x_1}{a} = \frac{y - y_1}{b} = \frac{z - z_1}{c}$$

$$= \frac{-2(ax_1 + by_1 + cz_1 + d)}{a^2 + b^2 + c^2}.$$

Thus, the image of point (−1, 3, 4) in a plane $x - 2y = 0$ is given by

$$\frac{x+1}{1} = \frac{y-3}{-2} = \frac{z-4}{0}$$

$$= \frac{-2\,[1\times(-1)+(-2)\times 3+0\times 4]}{1+4}$$

$\Rightarrow \quad \frac{x+1}{1} = \frac{y-3}{-2} = \frac{z-4}{0} = \frac{-2\,(-7)}{5}$

$\Rightarrow \quad x = \frac{14}{5} - 1 = \frac{9}{5},\ y = -\frac{28}{5} + 3 = -\frac{13}{5}$

and $\quad z = 4$

Hence, the image of point $(-1, 3, 4)$ is $\left(\frac{9}{5}, -\frac{13}{5}, 4\right)$.

260. Centre of sphere $x^2 + y^2 + z^2 + 2ux + 2vy + 2wz + d = 0$ is $(-u, -v, -w)$.

Given equation of first sphere is

$$x^2 + y^2 + z^2 + 6x - 8y - 2z = 13 \quad \text{...(i)}$$

whose centre is $(-3, 4, 1)$

and equation of second sphere is

$$x^2 + y^2 + z^2 - 10x + 4y - 2z = 8 \quad \text{...(ii)}$$

whose centre is $(5, -2, 1)$.

Mid-point of $(-3, 4, 1)$ and $(5, -2, 1)$ is $(1, 1, 1)$.

Since, the plane passes through $(1, 1, 1)$.

$\therefore \quad 2a - 3a + 4a + 6 = 0$

$\Rightarrow \quad 3a = -6 \ \Rightarrow a = -2$

261. Direction ratios of line normal are

$(a_1, b_1, c_1) = (1, 2, 2)$

and direction ratios of a plane are

$(a_2, b_2, c_2) = (2, -1, \sqrt{\lambda})$

Since, $\sin\theta = \frac{a_1a_2 + b_1b_2 + c_1c_2}{\sqrt{a_1^2 + b_1^2 + c_1^2}\ \sqrt{a_2^2 + b_2^2 + c_2^2}}$

$$= \frac{1\times 2 + 2(-1) + 2\times\sqrt{\lambda}}{\sqrt{(1) + (2)^2 + (2)^2}\sqrt{(2)^2 + (1)^2 + (\sqrt{\lambda})^2}}$$

$\Rightarrow \quad \frac{1}{3} = \frac{2\sqrt{\lambda}}{3\sqrt{5+\lambda}} \Rightarrow 5 + \lambda = 4\lambda$

$\Rightarrow \quad \lambda = \frac{5}{3}$

262. If a_1, b_1, c_1 and a_2, b_2, c_2 are DR's of two lines, then the angle between them is given by

$$\cos\theta = \frac{a_1a_2 + b_1b_2 + c_1c_2}{\sqrt{a_1^2 + b_1^2 + c_1^2}\ \sqrt{a_2^2 + b_2^2 + c_2^2}}$$

The given equations can be rewritten as

$$\frac{x}{3} = \frac{y}{2} = \frac{z}{-6} \text{ and } \frac{x}{2} = \frac{y}{-12} = \frac{z}{-3}$$

$\therefore$ Angle between the lines is given by

$$\cos\theta = \frac{6 - 24 + 18}{\sqrt{9+4+36}\sqrt{4+144+9}}$$

$$= \frac{0}{\sqrt{49}\sqrt{157}} = 0$$

$\Rightarrow \quad \theta = 90°$

263. Since, the centre of sphere

$x^2 + y^2 + z^2 - x + z - 2 = 0$ is $\left(\frac{1}{2}, 0, -\frac{1}{2}\right)$ and radius of sphere

$$= \sqrt{\frac{1}{4} + \frac{1}{4} + 2} = \frac{\sqrt{10}}{2}$$

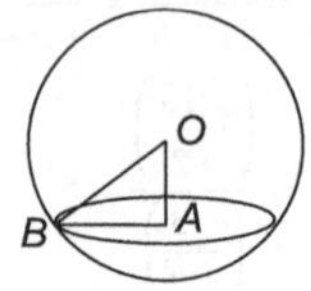

Distance of plane from centre of sphere

$$= \left|\frac{\frac{1}{2} + \frac{1}{2} - 4}{\sqrt{1+4+1}}\right| = \frac{3}{\sqrt{6}}$$

So, radius of circle $= \sqrt{\frac{10}{4} - \frac{9}{6}}$

$$= \sqrt{\frac{30-18}{12}} = \sqrt{\frac{12}{12}} = 1$$

Previous Years' Questions

JEE Main and Advanced (2021-18)

Vector Algebra

1. Let $P_1, P_2, \ldots\ldots, P_{15}$ be 15 points on a circle. The number of distinct triangles formed by points $P_{\hat{i}}, P_{\hat{j}}$ and $P_{\hat{k}}$, such that

$\hat{i} + \hat{j} + \hat{k} \neq 15$is **[JEE Main 2021]**

(a) 12 (b) 419 (c) 443 (d) 455

2. A vector a has components $3p$ and 1 with respect to rectangular cartesian system. This system is rotated through a certain angle about the origin in the counter clockwise sense. If, with respect to new system, a has components $p+1$ and $\sqrt{10}$, then a value of p is equal to **[JEE Main 2021]**

(a) 1 (b) $-\frac{5}{4}$ (c) $\frac{4}{5}$ (d) -1

3. Let a vector $\alpha\hat{\mathbf{i}} + \beta\hat{\mathbf{j}}$ be obtained by rotating the vector $\sqrt{3}\,\hat{\mathbf{i}} + \hat{\mathbf{j}}$ by an angle 45° about the origin in counterclockwise direction in the first quadrant. Then the area of triangle having vertices (α, β), $(0, \beta)$ and $(0, 0)$ is equal to **[JEE Main 2021]**

(a) $\frac{1}{2}$ (b) 1 (c) $\frac{1}{\sqrt{2}}$ (d) $2\sqrt{2}$

4. If vectors $\mathbf{a}_1 = x\hat{\mathbf{i}} - \hat{\mathbf{j}} + \hat{\mathbf{k}}$ and $\mathbf{a}_2 = \hat{\mathbf{i}} + y\hat{\mathbf{j}} + z\hat{\mathbf{k}}$ are collinear, then a possible unit vector parallel to the vector $x\hat{\mathbf{i}} + y\hat{\mathbf{j}} + z\hat{\mathbf{k}}$ is **[JEE Main 2021]**

(a) $\frac{1}{\sqrt{2}}(-\hat{\mathbf{j}} + \hat{\mathbf{k}})$ (b) $\frac{1}{\sqrt{2}}(\hat{\mathbf{i}} - \hat{\mathbf{j}})$

(c) $\frac{1}{\sqrt{3}}(\hat{\mathbf{i}} + \hat{\mathbf{j}} - \hat{\mathbf{k}})$ (d) $\frac{1}{\sqrt{3}}(\hat{\mathbf{i}} - \hat{\mathbf{j}} + \hat{\mathbf{k}})$

5. Let $S_1 = \{(\hat{\mathbf{i}}, \hat{\mathbf{j}}, \hat{\mathbf{k}}) : \hat{\mathbf{i}}, \hat{\mathbf{j}}, \hat{\mathbf{k}} \in \{1, 2, \ldots, 10\}\}$,

$S_2 = \{(\hat{\mathbf{i}}, \hat{\mathbf{j}}) : 1 \le \hat{\mathbf{i}} < \hat{\mathbf{j}} + 2 \le 10,\ \hat{\mathbf{i}}, \hat{\mathbf{j}} \in \{1, 2, \ldots, 10\}\}$

$S_3 = \{(\hat{\mathbf{i}}, \hat{\mathbf{j}}, \hat{\mathbf{k}}, l) : 1 \le \hat{\mathbf{i}} < \hat{\mathbf{j}} < \hat{\mathbf{k}} < l,\ \hat{\mathbf{i}}, \hat{\mathbf{j}}, \hat{\mathbf{k}}, l \in \{1, 2, \ldots, 10\}\}$

and $S_4 = \{(\hat{\mathbf{i}}, \hat{\mathbf{j}}, \hat{\mathbf{k}}, l) : \hat{\mathbf{i}}, \hat{\mathbf{j}}, \hat{\mathbf{k}}$ and l are distinct elements in $\{1, 2, \ldots, 10\}\}$.

If the total number of elements in the set S_r is n_r, $r = 1, 2, 3, 4$, then which of the following statements is (are) TRUE? **[JEE Advanced 2021]**

(a) $n_1 = 1000$ (b) $n_2 = 44$

(c) $n_3 = 220$ (d) $\frac{n_4}{12} = 420$

6. If a unit vector $\mathbf{a}$ makes angles $\frac{\pi}{3}$ with $\hat{\mathbf{i}}$, $\frac{\pi}{4}$ with $\hat{\mathbf{j}}$ and $\theta \in (0, \pi)$ with $\hat{\mathbf{k}}$, then a value of θ is **[JEE Main 2019]**

(a) $\frac{5\pi}{6}$ (b) $\frac{\pi}{4}$

(c) $\frac{5\pi}{12}$ (d) $\frac{2\pi}{3}$

7. Let $\alpha = (\lambda - 2)\mathbf{a} + \mathbf{b}$ and $\beta = (4\lambda - 2)\mathbf{a} + 3\mathbf{b}$ be two given vectors where vectors $\mathbf{a}$ and $\mathbf{b}$ are non- collinear. The value of λ for which vectors α and β are collinear, is **[JEE Main 2019]**

(a) 4 (b) -3

(c) 3 (d) -4

8. Let $\sqrt{3}\,\hat{\mathbf{i}} + \hat{\mathbf{j}}$, $\hat{\mathbf{i}} + \sqrt{3}\,\hat{\mathbf{j}}$ and $\beta\hat{\mathbf{i}} + (1-\beta)\hat{\mathbf{j}}$ respectively be the position vectors of the points A, B and C with respect to the origin O. If the distance of C from the bisector of the acute angle between OA and OB is $\frac{3}{\sqrt{2}}$, then the sum of all possible values of β is **[JEE Main 2019]**

(a) 1 (b) 3 (c) 4 (d) 2

9. Let a and b be two unit vectors such that $a \cdot b = 0$. For some $x, y \in R$, let $c = xa + yb + (a \times b)$. If $|c| = 2$ and the vector c is inclined at the same angle α to both a and b, then the value of $8\cos^2\alpha$ is .... **[JEE Advanced 2018]**

10. Consider the cube in the first octant with sides OP, OQ and OR of length 1, along the X-axis, Y-axis and Z-axis, respectively, where $O(0, 0, 0)$ is the origin. Let $S\left(\frac{1}{2}, \frac{1}{2}, \frac{1}{2}\right)$ be the centre of the cube and T be the vertex of the cube opposite to the origin O such that S lies on the diagonal OT. If $\mathbf{p} = \mathbf{SP}$, $\mathbf{q} = \mathbf{SQ}$, $\mathbf{r} = \mathbf{SR}$ and $\mathbf{t} = \mathbf{ST}$, then the value of $|(\mathbf{p} \times \mathbf{q}) \times (\mathbf{r} \times \mathbf{t})|$ is ..... . **[JEE Advanced 2018]**

Answers

1. (c) **2.** (d) **3.** (a) **4.** (d) **5.** (a, b, and d) **6.** (d) **7.** (d) **8.** (a) **9.** (3) **10.** (0.5)

Product of Vectors

1. Let **a** and **b** be two vectors such that $|2\mathbf{a}+3\mathbf{b}| = |3\mathbf{a}+\mathbf{b}|$ and the angle between **a** and **b** is 60°. If $\frac{1}{8}\mathbf{a}$ is a unit vector, then $|\mathbf{b}|$ is equal to **[JEE Main 2021]**

(a) 4 (b) 6 (c) 5 (d) 8

2. A hall has a square floor of dimension 10m × 10m (see the figure) and vertical walls. If the $\angle GPH$ between the diagonals AG and BH is $\cos^{-1}\frac{1}{5}$, then the height of the hall (in m) is **[JEE Main 2021]**

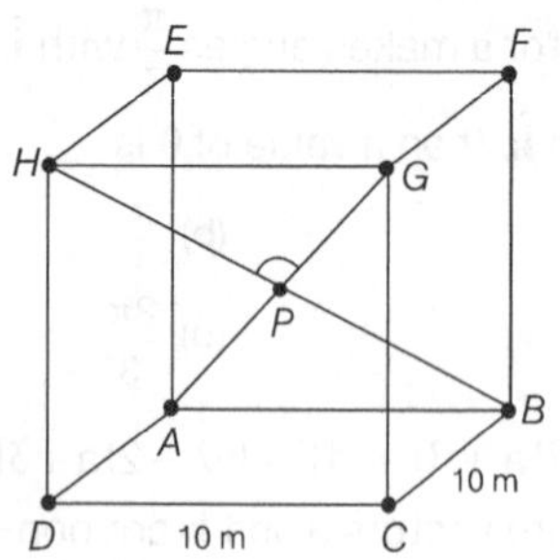

(a) 5 (b) $2\sqrt{10}$ (c) $5\sqrt{3}$ (d) $5\sqrt{2}$

3. If the projection of the vector $\hat{i}+2\hat{j}+\hat{k}$ on the sum of the two vectors $2\hat{i}+4\hat{j}-5\hat{k}$ and $-\lambda\hat{i}+2\hat{j}+3\hat{k}$ is 1, then λ is equal to **[JEE Main 2021]**

4. If $(\mathbf{a}+3\mathbf{b})$ is perpendicular to $(7\mathbf{a}-5\mathbf{b})$ and $(\mathbf{a}-4\mathbf{b})$ is perpendicular to $(7\mathbf{a}-2\mathbf{b})$, then the angle between **a** and **b** (in degrees) is .... **[JEE Main 2021]**

5. Let **a, b, c** be three mutually perpendicular vectors of the same magnitude and equally inclined at an angle θ, with the vector $\mathbf{a}+\mathbf{b}+\mathbf{c}$. Then, $36\cos^2 2\theta$ is equal to .... **[JEE Main 2021]**

6. For $p>0$, a vector $\mathbf{V}_2 = 2\hat{i}+(p+1)\hat{j}$ is obtained by rotating the vector $\mathbf{v}_1 = \sqrt{3}p\,\hat{i}+\hat{j}$ by an angle θ about origin in counter clockwise direction. If $\tan\theta = \frac{(\alpha\sqrt{3}-2)}{(4\sqrt{3}+3)}$, then the value of α is equal to **[JEE Main 2021]**

7. In ΔABC, if $|\mathbf{BC}|=3$, $|\mathbf{CA}|=5$ and $|\mathbf{BA}|=7$, then the projection of the vector **BA** on **BC** is equal to **[JEE Main 2021]**

(a) $\frac{19}{2}$ (b) $\frac{13}{2}$ (c) $\frac{11}{2}$ (d) $\frac{15}{2}$

8. In a triangle ABC, if $|\mathbf{BC}| = 8$, $|\mathbf{CA}| = 7$, $|\mathbf{AB}| = 10$, then the projection of the vector **AB** on **AC** is equal to **[JEE Main 2021]**

(a) $\frac{25}{4}$ (b) $\frac{85}{14}$ (c) $\frac{127}{20}$ (d) $\frac{115}{16}$

9. Let **x** be a vector in the plane containing vectors $\mathbf{a}=2\hat{i}-\hat{j}+\hat{k}$ and $\mathbf{b}=\hat{i}+2\hat{j}-\hat{k}$. If the vector **x** is perpendicular to $(3\hat{i}+2\hat{j}-\hat{k})$ and its projection on **a** is $\frac{17\sqrt{6}}{2}$, then the value of $|\mathbf{x}|^2$ is equal to .... **[JEE Main 2021]**

10. Let $\mathbf{a}=\hat{i}+5\hat{j}+\alpha\hat{k}$, $\mathbf{b}=\hat{i}+3\hat{j}+\beta\hat{k}$ and $\mathbf{c}=-\hat{i}+2\hat{j}-3\hat{k}$ be three vectors such that, $|\mathbf{b}\times\mathbf{c}|=5\sqrt{3}$ and **a** is perpendicular to **b**. Then, the greatest amongst the values of $|\mathbf{a}|^2$ is **[JEE Main 2021]**

11. Let $\mathbf{p}=2\hat{i}+3\hat{j}+\hat{k}$ and $\mathbf{q}=\hat{i}+2\hat{j}+\hat{k}$ be two vectors. If a vector $\mathbf{r}=(\alpha\hat{j}+\beta\hat{j}+\gamma\hat{k})$ is perpendicular to each of the vectors $(\mathbf{p}+\mathbf{q})$ and $(\mathbf{p}-\mathbf{q})$, and $|\mathbf{r}|=\sqrt{3}$, then $|\alpha|+|\beta|+|\gamma|$ is equal to .... **[JEE Main 2021]**

12. If $|\mathbf{a}|=2, |\mathbf{b}|=5$ and $|\mathbf{a}\times\mathbf{b}|=8$, then $|\mathbf{a}.\mathbf{b}|$ is equal to **[JEE Main 2021]**

(a) 6 (b) 4 (c) 3 (d) 5

13. Let **a** and **b** be two non-zero vectors perpendicular to each other and $|\mathbf{a}|=|\mathbf{b}|$. If $|\mathbf{a}\times\mathbf{b}|=|\mathbf{a}|$, then the angle between the vectors $[\mathbf{a}+\mathbf{b}+(\mathbf{a}\times\mathbf{b})]$ and **a** is equal to **[JEE Main 2021]**

(a) $\sin^{-1}\left(\frac{1}{\sqrt{3}}\right)$ (b) $\cos^{-1}\left(\frac{1}{\sqrt{3}}\right)$ (c) $\cos^{-1}\left(\frac{1}{\sqrt{2}}\right)$ (d) $\sin^{-1}\left(\frac{1}{\sqrt{6}}\right)$

14. Let $\mathbf{a}=2\hat{i}-3\hat{j}+4\hat{k}$ and $\mathbf{b}=7\hat{i}+\hat{j}-6\hat{k}$. If $\mathbf{r}\times\mathbf{a}=\mathbf{r}\times\mathbf{b}, \mathbf{r}.(\hat{i}+2\hat{j}+\hat{k})=-3$, then. $\mathbf{r}(2\hat{i}-3\hat{j}+\hat{k})$ is equal to **[JEE Main 2021]**

(a) 12 (b) 8 (c) 13 (d) 10

15. Let $\mathbf{a}=\hat{i}+2\hat{j}-3\hat{k}$ and $\mathbf{b}=2\hat{i}-3\hat{j}+5\hat{k}$. If $\mathbf{r}\times\mathbf{a}=\mathbf{b}\times\mathbf{r}$, $\mathbf{r}.(\alpha\hat{i}+2\hat{j}+\hat{k})=3$ and $\mathbf{r}.(2\hat{i}+5\hat{j}-a\hat{k})=-1, \alpha\in R$, then the value of $\alpha+|\mathbf{r}|^2$ is equal to **[JEE Main 2021]**

(a) 9 (b) 15 (c) 13 (d) 11

16. Let **c** be a vector perpendicular to the vectors $\mathbf{a}=\hat{i}+\hat{j}-\hat{k}$ and $\mathbf{b}=\hat{i}+2\hat{j}+\hat{k}$

If $\mathbf{c}\cdot(\hat{i}+\hat{j}+3\hat{k})=8$ then the value of $\mathbf{c}\cdot(\mathbf{a}\times\mathbf{b})$ is equal to .... **[JEE Main 2021]**

17. Let $\mathbf{a}=\hat{i}+\alpha\hat{j}+3\hat{k}$ and $\mathbf{b}=3\hat{i}-\alpha\hat{j}+\hat{k}$. If the area of the parallelogram whose adjacent sides are represented by the vectors **a** and **b** is $8\sqrt{3}$ square units, then $\mathbf{a}\cdot\mathbf{b}$ is equal to .... **[JEE Main 2021]**

18. Let $\mathbf{a}=\hat{i}+2\hat{j}-\hat{k}$, $\mathbf{b}=\hat{i}-\hat{j}$ and $\mathbf{c}=\hat{i}-\hat{j}-\hat{k}$ be three given vectors. If **r** is a vector such that $\mathbf{r}\times\mathbf{a}=\mathbf{c}\times\mathbf{a}$ and $\mathbf{r}\cdot\mathbf{b}=0$, then $\mathbf{r}\cdot\mathbf{a}$ is equal to .... **[JEE Main 2021]**

19. Let **a**, **b** and **c** be three vectors mutually perpendicular to each other and have same magnitude. If a vector **r** satisfies.

$\mathbf{a}\times\{(\mathbf{r}-\mathbf{b})\times\mathbf{a}\}+\mathbf{b}\times\{(\mathbf{r}-\mathbf{c})\times b\}+\mathbf{c}\times\{(\mathbf{r}-\mathbf{a})\times\mathbf{c}\}=\mathbf{0}$, then **r** is equal to **[JEE Main 2021]**

(a) $\frac{1}{3}(\mathbf{a}+\mathbf{b}+\mathbf{c})$ (b) $\frac{1}{3}(2\mathbf{a}+\mathbf{b}-\mathbf{c})$
(c) $\frac{1}{2}(\mathbf{a}+\mathbf{b}+\mathbf{c})$ (d) $\frac{1}{2}(\mathbf{a}+\mathbf{b}+2\mathbf{c})$

20. Let $\mathbf{a}=\hat{\mathbf{i}}+\hat{\mathbf{j}}+\hat{\mathbf{k}}$ and $\mathbf{b}=\hat{\mathbf{j}}-\hat{\mathbf{k}}$. If **c** is a vector such that $\mathbf{a}\times\mathbf{c}=\mathbf{b}$ and $\mathbf{a}\cdot\mathbf{c}=3$, then $\mathbf{a}\cdot(\mathbf{b}\times\mathbf{c})$ is equal to **[JEE Main 2021]**

(a) −2 (b) −6 (c) 6 (d) 2

21. Let $\mathbf{a}=\hat{\mathbf{i}}+\hat{\mathbf{j}}+2\hat{\mathbf{k}}$ and $\mathbf{b}=-\hat{\mathbf{i}}+2\hat{\mathbf{j}}+3\hat{\mathbf{k}}$. Then the vector product $(\mathbf{a}+\mathbf{b})\times[(\mathbf{a}\times\{(\mathbf{a}-\mathbf{b})\times\mathbf{b})\}\times\mathbf{b}]$ is equal to **[JEE Main 2021]**

(a) $5(34\hat{\mathbf{i}}-5\hat{\mathbf{j}}+3\hat{\mathbf{k}})$ (b) $7(34\hat{\mathbf{i}}-5\hat{\mathbf{j}}+3\hat{\mathbf{k}})$
(c) $7(30\hat{\mathbf{i}}-5\hat{\mathbf{j}}+7\hat{\mathbf{k}})$ (d) $5(30\hat{\mathbf{i}}-5\hat{\mathbf{j}}+7\hat{\mathbf{k}})$

22. Let $\mathbf{a}=\hat{\mathbf{i}}+\hat{\mathbf{j}}+\hat{\mathbf{k}}$, **b** and $\mathbf{c}=\hat{\mathbf{j}}-\hat{\mathbf{k}}$ be three vectors such that $\mathbf{a}\times\mathbf{b}=\mathbf{c}$ and $\mathbf{a}\cdot\mathbf{b}=1$. If the length of projection vector of the vector **b** on the vector $\mathbf{a}\times\mathbf{c}$ is l, then the value of $3l^2$ is equal to .... **[JEE Main 2021]**

23. Let **a**, **b** and **c** be three vectors such that $\mathbf{a}=\mathbf{b}\times(\mathbf{b}\times\mathbf{c})$. If magnitudes of the vectors **a**, **b** and **c** are $\sqrt{2}$, 1 and 2, respectively and the angle between **b** and **c** is $\theta\left(0<\theta<\frac{\pi}{2}\right)$, then the value of $(1+\tan\theta)$ is equal to **[JEE Main 2021]**

(a) $\sqrt{3}+1$ (b) 2 (c) 1 (d) $\frac{\sqrt{3}+1}{\sqrt{3}}$

24. Let $\mathbf{a}=\hat{\mathbf{i}}-\alpha\hat{\mathbf{j}}+\beta\hat{\mathbf{k}}$, $\mathbf{b}=3\hat{\mathbf{i}}+\beta\hat{\mathbf{j}}-\alpha\hat{\mathbf{k}}$ and $\mathbf{c}=-\alpha\hat{\mathbf{i}}-2\hat{\mathbf{j}}+\hat{\mathbf{k}}$, where α and β are integers. If $\mathbf{a}\cdot\mathbf{b}=-1$ and $\mathbf{b}\cdot\mathbf{c}=10$, then $(\mathbf{a}\times\mathbf{b})\cdot\mathbf{c}$ is equal to .... **[JEE Main 2021]**

25. Let the vectors

$(2+a+b)\hat{\mathbf{i}}+(a+2b+c)\hat{\mathbf{j}}-(b+c)\hat{\mathbf{k}}$, $(1+b)\hat{\mathbf{i}}+2b\hat{\mathbf{j}}-b\hat{\mathbf{k}}$ and $(2+b)\hat{\mathbf{i}}+2b\hat{\mathbf{j}}+(1-b)\hat{\mathbf{k}}$, $a,b,c\in R$ be coplanar. **[JEE Main 2021]**

Then, which of the following is true?

(a) $2b=a+c$ (b) $3c=a+b$
(c) $a=b+2c$ (d) $2a=b+c$

26. Let three vectors **a**, **b** and **c** be such that $\mathbf{a}\times\mathbf{b}=\mathbf{c}$, $\mathbf{b}\times\mathbf{c}=\mathbf{a}$ and $|\mathbf{a}|=2$. Then, which one of the following is not true ? **[JEE Main 2021]**

(a) $\mathbf{a}\times((\mathbf{b}+\mathbf{c})\times(\mathbf{b}-\mathbf{c}))=\mathbf{0}$ (b) Projection of **a** on $(\mathbf{b}\times\mathbf{c})$ is 2
(c) $[\mathbf{a}\ \mathbf{b}\ \mathbf{c}]+[\mathbf{c}\ \mathbf{a}\ \mathbf{b}]=8$ (d) $|3\mathbf{a}+\mathbf{b}-2\mathbf{c}|^2=51$

27. Let a vector **a** be coplanar with vectors $\mathbf{b}=2\hat{\mathbf{i}}+\hat{\mathbf{j}}+\hat{\mathbf{k}}$ and $\mathbf{c}=\hat{\mathbf{i}}-\hat{\mathbf{j}}+\hat{\mathbf{k}}$. If **a** is perpendicular to $\mathbf{d}=3\hat{\mathbf{i}}+2\hat{\mathbf{j}}+6\hat{\mathbf{k}}$ and $|\mathbf{a}|=\sqrt{10}$. Then a possible value of $[\mathbf{a}\ \mathbf{b}\ \mathbf{c}]+[\mathbf{a}\ \mathbf{b}\ \mathbf{d}]+[\mathbf{a}\ \mathbf{c}\ \mathbf{d}]$ is equal to **[JEE Main 2021]**

(a) −42 (b) −40 (c) −29 (d) −38

28. Let $\mathbf{a}=2\hat{\mathbf{i}}+\hat{\mathbf{j}}-2\hat{\mathbf{k}}$ and $\mathbf{b}=\hat{\mathbf{i}}+\hat{\mathbf{j}}$. If **c** is a vector such that $\mathbf{a}\cdot\mathbf{c}=|\mathbf{c}|$, $|\mathbf{c}-\mathbf{a}|=2\sqrt{2}$ and the angle between $(\mathbf{a}\times\mathbf{b})$ and **c** is $\frac{\pi}{6}$, then the value of $|(\mathbf{a}\times\mathbf{b})\times\mathbf{c}|$ is **[JEE Main 2021]**

(a) 2 / 3 (b) 4 (c) 3 (d) 3/2

29. If $\mathbf{a}=\alpha\hat{\mathbf{i}}+\beta\hat{\mathbf{j}}+3\hat{\mathbf{k}}$, $\mathbf{b}=-\beta\hat{\mathbf{i}}-\alpha\hat{\mathbf{j}}-\hat{\mathbf{j}}$ and $\mathbf{c}=\hat{\mathbf{i}}-2\hat{\mathbf{j}}-\hat{\mathbf{k}}$, such that $\mathbf{a}\cdot\mathbf{b}=1$ and $\mathbf{b}\cdot\mathbf{c}=-3$, then $\frac{1}{3}[(\mathbf{a}\times\mathbf{b})\cdot\mathbf{c}]$ is equal to .... **[JEE Main 2021]**

30. Let O be the origin. Let $\mathbf{OP}=x\hat{\mathbf{i}}+y\hat{\mathbf{j}}-\hat{\mathbf{k}}$ and $\mathbf{OQ}=-\hat{\mathbf{i}}+2\hat{\mathbf{j}}+3x\hat{\mathbf{k}}$, $x,y\in R, x>0$, be such that $|\mathbf{PQ}|=\sqrt{20}$ and the vector **OP** is perpendicular to **OQ**. If $\mathbf{OR}=3\hat{\mathbf{i}}+z\hat{\mathbf{j}}-7\hat{\mathbf{k}}$, $z\in R$, is coplanar with **OP** and **OQ**, then the value of $x^2+y^2+z^2$ is equal to **[JEE Main 2021]**

(a) 7 (b) 9 (c) 2 (d) 1

31. If (1, 5, 35), (7, 5, 5), (1, λ, 7) and (2λ, 1, 2) are coplanar, then the sum of all possible values of λ is **[JEE Main 2021]**

(a) $\frac{39}{5}$ (b) $-\frac{39}{5}$ (c) $\frac{44}{5}$ (d) $-\frac{44}{5}$

32. If **a** and **b** are perpendicular, then $\mathbf{a}\times(\mathbf{a}\times(\mathbf{a}\times(\mathbf{a}\times\mathbf{b})))$ is equal to **[JEE Main 2021]**

(a) **0** (c) $\frac{1}{2}|\mathbf{a}|^4\mathbf{b}$ (c) $\mathbf{a}\times\mathbf{b}$ (d) $|\mathbf{a}|^4\mathbf{b}$

33. Let three vectors **a**, **b** and **c** be such that **c** is coplanar with **a** and **b**, $\mathbf{a}\cdot\mathbf{c}=7$ and **b** is perpendicular to **c**, where $\mathbf{a}=-\hat{\mathbf{i}}+\hat{\mathbf{j}}+\hat{\mathbf{k}}$ and $\mathbf{b}=2\hat{\mathbf{i}}+\hat{\mathbf{k}}$, then the value of $2|\mathbf{a}+\mathbf{b}+\mathbf{c}|^2$ is .... **[JEE Main 2021]**

34. Let **u**, **v** and **w** be vectors in three-dimensional space, where **u** and **v** are unit vectors which are not perpendicular to each other and $\mathbf{u}\cdot\mathbf{w}=1$, $\mathbf{v}\cdot\mathbf{w}=1$, $\mathbf{w}\cdot\mathbf{w}=4$

If the volume of the parallelopiped, whose adjacent sides are represented by the vectors **u**, **v** and **w**, is $\sqrt{2}$, then the value of $|3\mathbf{u}+5\mathbf{v}|$ is **[JEE Advnaced 2021]**

35. Let O be the origin and $\mathbf{OA}=2\hat{\mathbf{i}}+2\hat{\mathbf{j}}+\hat{\mathbf{k}}$, $\mathbf{OB}=\hat{\mathbf{i}}-2\hat{\mathbf{j}}+2\hat{\mathbf{k}}$ and $\mathbf{OC}=\frac{1}{2}(\mathbf{OB}-\lambda\mathbf{OA})$, for some $\lambda>0$. If $|\mathbf{OB}\times\mathbf{OC}|=\frac{9}{2}$, then which of the following statements is (are) TRUE? **[JEE Advnaced 2021]**

(a) Projection of **OC** on **OA** is $-3/2$.
(b) Area of the ΔOAB is 9/2.
(c) Area of the ΔABC is 9/2.
(d) The acute angle between the diagonals of the parallelogram with adjacent sides **OA** and **OC** is $\frac{\pi}{3}$.

36. **a**, **b** and **c** be three unit vectors such that $|\mathbf{a}-\mathbf{b}|^2+|\mathbf{a}-\mathbf{c}|^2=8$. Then $|\mathbf{a}+2\mathbf{b}|^2+|\mathbf{a}-2\mathbf{c}|^2$ is equal to .... **[JEE Main 2020]**

37. Let the position vectors of points 'A' and 'B' be $\hat{\mathbf{i}}+\hat{\mathbf{j}}+\hat{\mathbf{k}}$ and $2\hat{\mathbf{i}}+\hat{\mathbf{j}}+3\hat{\mathbf{k}}$, respectively. A point 'P' divides the line segment AB internally in the ratio $\lambda:1(\lambda>0)$. If O is the origin and $\mathbf{OB}\cdot\mathbf{OP}-3|\mathbf{OA}\times\mathbf{OP}|^2=6$, then λ is equal to .... **[JEE Main 2020]**

38. Let $a,b,c\in R$ be such that $a^2+b^2+c^2=1$. If $a\cos\theta=b\cos\left(\theta+\frac{2\pi}{3}\right)=c\cos\left(\theta+\frac{4\pi}{3}\right)$, where $\theta=\frac{\pi}{9}$, then the angle between the vectors $a\hat{\mathbf{i}}+b\hat{\mathbf{j}}+c\hat{\mathbf{k}}$ and $b\hat{\mathbf{i}}+c\hat{\mathbf{j}}+a\hat{\mathbf{k}}$ is **[JEE Main 2020]**

(a) $\frac{\pi}{2}$ (b) 0
(c) $\frac{2\pi}{3}$ (d) $\frac{\pi}{9}$

39. Let the vectors $\mathbf{a}$, $\mathbf{b}$, $\mathbf{c}$ be such that $|\mathbf{a}|=2$, $|\mathbf{b}|=4$ and $|\mathbf{c}|=4$. If the projection of $\mathbf{b}$ on $\mathbf{a}$ is equal to the projection of $\mathbf{c}$ on $\mathbf{a}$ and $\mathbf{b}$ is perpendicular to $\mathbf{c}$, then the value of $|\mathbf{a}+\mathbf{b}-\mathbf{c}|$ is .... **[JEE Main 2020]**

40. If $\mathbf{a}$ and $\mathbf{b}$ are unit vectors, then the greatest value of $\sqrt{3}|\mathbf{a}+\mathbf{b}|+|\mathbf{a}-\mathbf{b}|$ is .... **[JEE Main 2020]**

41. If $\mathbf{x}$ and $\mathbf{y}$ be two non-zero vectors such that $|\mathbf{x}+\mathbf{y}|=|\mathbf{x}|$ and $2\mathbf{x}+\lambda\mathbf{y}$ is perpendicular to $\mathbf{y}$, then the value of λ is **[JEE Main 2020]**

42. A vector $\mathbf{a}=\alpha\hat{\mathbf{i}}+2\hat{\mathbf{j}}+\beta\hat{\mathbf{k}}$ $(\alpha,\beta\in\mathbf{R})$ lies in the plane of the vectors, $\mathbf{b}=\hat{\mathbf{i}}+\hat{\mathbf{j}}$ and $\mathbf{c}=\hat{\mathbf{i}}-\hat{\mathbf{j}}+4\hat{\mathbf{k}}$. If $\mathbf{a}$ bisects the angle between $\mathbf{b}$ and $\mathbf{c}$, then **[JEE Main 2020]**

(a) $\mathbf{a}\cdot\hat{\mathbf{i}}+3=0$ (b) $\mathbf{a}\cdot\hat{\mathbf{k}}+2=0$
(c) $\mathbf{a}\cdot\hat{\mathbf{i}}+1=0$ (d) $\mathbf{a}\cdot\hat{\mathbf{k}}+4=0$

43. Let $\mathbf{a}$, $\mathbf{b}$ and $\mathbf{c}$ be three unit vectors such that $\mathbf{a}+\mathbf{b}+\mathbf{c}=\mathbf{0}$. If $\lambda=\mathbf{a}\cdot\mathbf{b}+\mathbf{b}\cdot\mathbf{c}+\mathbf{c}\cdot\mathbf{a}$ and $\mathbf{d}=\mathbf{a}\times\mathbf{b}+\mathbf{b}\times\mathbf{c}+\mathbf{c}\times\mathbf{a}$, then the ordered pair, $(\lambda,\mathbf{d})$ is equal to **[JEE Main 2020]**

(a) $\left(\frac{3}{2},3\mathbf{b}\times\mathbf{c}\right)$ (b) $\left(-\frac{3}{2},3\mathbf{c}\times\mathbf{b}\right)$
(c) $\left(\frac{3}{2},3\mathbf{a}\times\mathbf{c}\right)$ (d) $\left(-\frac{3}{2},3\mathbf{a}\times\mathbf{b}\right)$

44. Let $\mathbf{a}=\hat{\mathbf{i}}-2\hat{\mathbf{j}}+\hat{\mathbf{k}}$ and $\mathbf{b}=\hat{\mathbf{i}}-\hat{\mathbf{j}}+\hat{\mathbf{k}}$ be two vectors. If $\mathbf{c}$ is a vector such that $\mathbf{b}\times\mathbf{c}=\mathbf{b}\times\mathbf{a}$ and $\mathbf{c}\cdot\mathbf{a}=0$, then $\mathbf{c}\cdot\mathbf{b}$ is equal to **[JEE Main 2020]**

(a) $\frac{1}{2}$ (b) $-\frac{3}{2}$ (c) $-\frac{1}{2}$ (d) -1

45. Let x_0 be the point of local maxima of $f(x)=\mathbf{a}\cdot(\mathbf{b}\times\mathbf{c})$, where and $\mathbf{c}=7\hat{\mathbf{i}}-2\hat{\mathbf{j}}+x\hat{\mathbf{k}}$. Then the value of $\mathbf{a}\cdot\mathbf{b}+\mathbf{b}\cdot\mathbf{c}+\mathbf{c}\cdot\mathbf{a}$ at $x=x_0$ is **[JEE Main 2020]**

(a) 14 (b) -4 (c) -22 (d) -30

46. If $\mathbf{a}=2\hat{\mathbf{i}}+\hat{\mathbf{j}}+2\hat{\mathbf{k}}$, then the value of $|\hat{\mathbf{i}}\times(\mathbf{a}\times\hat{\mathbf{i}})|^2+|\hat{\mathbf{j}}\times(\mathbf{a}\times\hat{\mathbf{j}})|^2+|\hat{\mathbf{k}}\times(\mathbf{a}\times\hat{\mathbf{k}})|^2$ is equal to .... **[JEE Main 2020]**

47. If the volume of a parallelopiped, whose coterminus edges are given by the vectors $\mathbf{a}=\hat{\mathbf{i}}+\hat{\mathbf{j}}+n\hat{\mathbf{k}}$, $\mathbf{b}=2\hat{\mathbf{i}}+4\hat{\mathbf{j}}-n\hat{\mathbf{k}}$ and $\mathbf{c}=\hat{\mathbf{i}}+n\hat{\mathbf{j}}+3\hat{\mathbf{k}}$ $(n\geq 0)$, is 158 cu units, then **[JEE Main 2020]**

(a) $n=9$ (b) $\mathbf{b}\cdot\mathbf{c}=10$
(c) $\mathbf{a}\cdot\mathbf{c}=17$ (d) $n=7$

48. Let the volume of a parallelopiped whose conterminous edges are given by $\mathbf{u}=\hat{\mathbf{i}}+\hat{\mathbf{j}}+\lambda\hat{\mathbf{k}}$, $\mathbf{v}=\hat{\mathbf{i}}+\hat{\mathbf{j}}+3\hat{\mathbf{k}}$ and $\mathbf{w}=2\hat{\mathbf{i}}+\hat{\mathbf{j}}+\hat{\mathbf{k}}$ be 1 cu. unit. If θ be the angle between the edges $\mathbf{u}$ and $\mathbf{w}$, then $\cos\theta$ can be **[JEE Main 2020]**

(a) $\frac{5}{3\sqrt{3}}$ (b) $\frac{7}{6\sqrt{3}}$ (c) $\frac{7}{6\sqrt{6}}$ (d) $\frac{5}{7}$

49. If the vectors $\mathbf{p}=(a+1)\hat{i}+a\hat{j}+a\hat{k}$, $\mathbf{q}=a\hat{i}+(a+1)\hat{j}+a\hat{k}$ and $\mathbf{r}=a\hat{i}+a\hat{j}+(a+1)\hat{k}(a\in R)$ are coplanar and $3(\mathbf{p}.\mathbf{q})^2-\lambda|\mathbf{r}\times\mathbf{q}|^2=0$, then the value of λ is .... **[JEE Main 2020]**

50. Let $\mathbf{a}$, $\mathbf{b}$ and $\mathbf{c}$ be three vectors such that $|\mathbf{a}|=\sqrt{3}$, $|\mathbf{b}|=5$, $\mathbf{b}\cdot\mathbf{c}=10$ and the angle between $\mathbf{b}$ and $\mathbf{c}$ is $\frac{\pi}{3}$. If $\mathbf{a}$ is perpendicular to the vector $\mathbf{b}\times\mathbf{c}$, then $|\mathbf{a}\times(\mathbf{b}\times\mathbf{c})|$ is equal to .... **[JEE Main 2020]**

51. In a triangle PQR, let $\mathbf{a}=\mathbf{QR}$, $\mathbf{b}=\mathbf{RP}$ and $\mathbf{c}=\mathbf{PQ}$. If $|\mathbf{a}|=3$, $|\mathbf{b}|=4$ and $\frac{\mathbf{a}.(\mathbf{c}-\mathbf{b})}{\mathbf{c}.(\mathbf{a}-\mathbf{b})}=\frac{|\mathbf{a}|}{|\mathbf{a}|+|\mathbf{b}|}$, then the value of $|\mathbf{a}\times\mathbf{b}|^2$ is .... **[JEE Advanced 2020]**

52. Let a and b be positive real numbers. Suppose $\mathbf{PQ}=a\hat{i}+b\hat{j}$ and $\mathbf{PS}=a\hat{\mathbf{i}}-b\hat{\mathbf{j}}$ are adjacent sides of a parallelogram $PQRS$. Let $\mathbf{u}$ and $\mathbf{v}$ be the projection vectors of $\mathbf{w}=\hat{\mathbf{i}}+\hat{\mathbf{j}}$ along $\mathbf{PQ}$ and $\mathbf{PS}$, respectively. If $|\mathbf{u}|+|\mathbf{v}|=|\mathbf{w}|$ and if the area of the parallelogram $PQRS$ is 8, then which of the following statements is/are TRUE? **[JEE Advanced 2020]**

(a) $a+b=4$
(b) $a-b=2$
(c) The length of the diagonal PR of the parallelogram $PQRS$ is 4
(d) $\mathbf{w}$ is an angle bisector of the vectors $\mathbf{PQ}$ and $\mathbf{PS}$

53. Let $A(3,0,-1)$, $B(2,10,6)$ and $C(1,2,1)$ be the vertices of a triangle and M be the mid-point of AC. If G divides BM in the ratio $2:1$, then $\cos(\angle GOA)$ (O being the origin) is equal to **[JEE Main 2019]**

(a) $\frac{1}{\sqrt{15}}$ (b) $\frac{1}{2\sqrt{15}}$
(c) $\frac{1}{\sqrt{30}}$ (d) $\frac{1}{6\sqrt{10}}$

54. Let $\mathbf{a}=\hat{\mathbf{i}}+\hat{\mathbf{j}}+\sqrt{2}\,\hat{\mathbf{k}}$, $\mathbf{b}=b_1\,\hat{\mathbf{i}}+b_2\,\hat{\mathbf{j}}+\sqrt{2}\,\hat{\mathbf{k}}$ and $\mathbf{c}=5\,\hat{\mathbf{i}}+\hat{\mathbf{j}}+\sqrt{2}\,\hat{\mathbf{k}}$ be three vectors such that the projection vector of $\mathbf{b}$ on $\mathbf{a}$ is $\mathbf{a}$. If $\mathbf{a}+\mathbf{b}$ is perpendicular to $\mathbf{c}$, then $|\mathbf{b}|$ is equal to **[JEE Main 2019]**

(a) 6 (b) 4
(c) $\sqrt{22}$ (d) $\sqrt{32}$

55. Let $\mathbf{a}=2\hat{\mathbf{i}}+\lambda_1\hat{\mathbf{j}}+3\hat{\mathbf{k}}$, $\mathbf{b}=4\hat{\mathbf{i}}+(3-\lambda_2)\hat{\mathbf{j}}+6\hat{\mathbf{k}}$ and $\mathbf{c}=3\hat{\mathbf{i}}+6\hat{\mathbf{j}}+(\lambda_3-1)\hat{\mathbf{k}}$ be three vectors such that $\mathbf{b}=2\mathbf{a}$ and $\mathbf{a}$ is perpendicular to $\mathbf{c}$. Then a possible value of $(\lambda_1,\lambda_2,\lambda_3)$ is **[JEE Main 2019]**

(a) (1, 3, 1) (b) (1, 5, 1)
(c) $\left(-\frac{1}{2}, 4, 0\right)$ (d) $\left(\frac{1}{2}, 4, -2\right)$

56. Let $\mathbf{a}=3\hat{\mathbf{i}}+2\hat{\mathbf{j}}+x\hat{\mathbf{k}}$ and $\mathbf{b}=\hat{\mathbf{i}}-\hat{\mathbf{j}}+\hat{\mathbf{k}}$, for some real x. Then $|\mathbf{a}\times\mathbf{b}|=r$ is possible if **[JEE Main 2019]**

(a) $0<r\le\sqrt{\frac{3}{2}}$

(b) $\sqrt{\frac{3}{2}}<r\le 3\sqrt{\frac{3}{2}}$

(c) $3\sqrt{\frac{3}{2}}<r<5\sqrt{\frac{3}{2}}$

(d) $r\ge 5\sqrt{\frac{3}{2}}$

57. Let $\vec{\alpha}=3\hat{\mathbf{i}}+\hat{\mathbf{j}}$ and $\vec{\beta}=2\hat{\mathbf{i}}-\hat{\mathbf{j}}+3\hat{\mathbf{k}}$. If $\vec{\beta}=\vec{\beta}_1-\vec{\beta}_2$, where $\vec{\beta}_1$ is parallel to $\vec{\alpha}$ and $\vec{\beta}_2$ is perpendicular to $\vec{\alpha}$, then $\vec{\beta}_1\times\vec{\beta}_2$ is equal to **[JEE Main 2019]**

(a) $\frac{1}{2}(3\hat{\mathbf{i}}-9\hat{\mathbf{j}}+5\hat{\mathbf{k}})$ (b) $\frac{1}{2}(-3\hat{\mathbf{i}}+9\hat{\mathbf{j}}+5\hat{\mathbf{k}})$
(c) $-3\hat{\mathbf{i}}+9\hat{\mathbf{j}}+5\hat{\mathbf{k}}$ (d) $3\hat{\mathbf{i}}-9\hat{\mathbf{j}}-5\hat{\mathbf{k}}$

58. The distance of the point having position vector $-\hat{\mathbf{i}}+2\hat{\mathbf{j}}+6\hat{\mathbf{k}}$ from the straight line passing through the point $(2,3,-4)$ and parallel to the vector, $6\hat{\mathbf{i}}+3\hat{\mathbf{j}}-4\hat{\mathbf{k}}$ is **[JEE Main 2019]**

(a) $2\sqrt{13}$ (b) $4\sqrt{3}$
(c) 6 (d) 7

59. Let $\mathbf{a}=3\hat{\mathbf{i}}+2\hat{\mathbf{j}}+2\hat{\mathbf{k}}$ and $\mathbf{b}=\hat{\mathbf{i}}+2\hat{\mathbf{j}}-2\hat{\mathbf{k}}$ be two vectors. If a vector perpendicular to both the vectors $\mathbf{a}+\mathbf{b}$ and $\mathbf{a}-\mathbf{b}$ has the magnitude 12, then one such vector is **[JEE Main 2019]**

(a) $4(2\hat{\mathbf{i}}+2\hat{\mathbf{j}}+\hat{\mathbf{k}})$ (b) $4(2\hat{\mathbf{i}}-2\hat{\mathbf{j}}-\hat{\mathbf{k}})$
(c) $4(2\hat{\mathbf{i}}+2\hat{\mathbf{j}}-\hat{\mathbf{k}})$ (d) $4(-2\hat{\mathbf{i}}-2\hat{\mathbf{j}}+\hat{\mathbf{k}})$

60. If the volume of parallelopiped formed by the vectors $\hat{\mathbf{i}}+\lambda\hat{\mathbf{j}}+\hat{\mathbf{k}}$, $\hat{\mathbf{j}}+\lambda\hat{\mathbf{k}}$ and $\lambda\hat{\mathbf{i}}+\hat{\mathbf{k}}$ is minimum, then λ is equal to **[JEE Main 2019]**

(a) $-\frac{1}{\sqrt{3}}$ (b) $\frac{1}{\sqrt{3}}$
(c) $\sqrt{3}$ (d) $-\sqrt{3}$

61. Let $\alpha\in R$ and the three vectors

$$\mathbf{a}=\alpha\hat{\mathbf{i}}+\hat{\mathbf{j}}+3\hat{\mathbf{k}},\ \mathbf{b}=2\hat{\mathbf{i}}+\hat{\mathbf{j}}-\alpha\hat{\mathbf{k}}$$

and $\mathbf{c}=\alpha\hat{\mathbf{i}}-2\hat{\mathbf{j}}+3\hat{\mathbf{k}}$. Then, the set

$S=\{\alpha:\mathbf{a},\mathbf{b}$ and $\mathbf{c}$ are coplanar$\}$ **[JEE Main 2019]**

(a) is singleton.
(b) is empty.
(c) contains exactly two positive numbers.
(d) contains exactly two numbers only one of which is positive.

62. Let $\mathbf{a}=\hat{\mathbf{i}}-\hat{\mathbf{j}}$, $\mathbf{b}=\hat{\mathbf{i}}+\hat{\mathbf{j}}+\hat{\mathbf{k}}$ and $\mathbf{c}$ be a vector such that $\mathbf{a}\times\mathbf{c}+\mathbf{b}=\mathbf{0}$ and $\mathbf{a}\cdot\mathbf{c}=4$, then $|\mathbf{c}|^2$ is equal to **[JEE Main 2019]**

(a) 8 (b) $\frac{19}{2}$ (c) 9 (d) $\frac{17}{2}$

63. Let $\mathbf{a}=\hat{\mathbf{i}}+2\hat{\mathbf{j}}+4\hat{\mathbf{k}}$, $\mathbf{b}=\hat{\mathbf{i}}+\lambda\hat{\mathbf{j}}+4\hat{\mathbf{k}}$ and $\mathbf{c}=2\hat{\mathbf{i}}+4\hat{\mathbf{j}}+(\lambda^2-1)\hat{\mathbf{k}}$ be coplanar vectors. Then, the non-zero vector $\mathbf{a}\times\mathbf{c}$ is **[JEE Main 2019]**

(a) $-10\hat{\mathbf{i}}+5\hat{\mathbf{j}}$ (b) $-10\hat{\mathbf{i}}-5\hat{\mathbf{j}}$
(c) $-14\hat{\mathbf{i}}-5\hat{\mathbf{j}}$ (d) $-14\hat{\mathbf{i}}+5\hat{\mathbf{j}}$

64. The sum of the distinct real values of μ, for which the vectors, $\mu\hat{\mathbf{i}}+\hat{\mathbf{j}}+\hat{\mathbf{k}}$, $\hat{\mathbf{i}}+\mu\hat{\mathbf{j}}+\hat{\mathbf{k}}$, $\hat{\mathbf{i}}+\hat{\mathbf{j}}+\mu\hat{\mathbf{k}}$ are coplanar, is **[JEE Main 2019]**

(a) 2 (b) 0 (c) 1 (d) -1

65. Let $\mathbf{a}$, $\mathbf{b}$ and $\mathbf{c}$ be three unit vectors, out of which vectors $\mathbf{b}$ and $\mathbf{c}$ are non-parallel. If α and β are the angles which vector $\mathbf{a}$ makes with vectors $\mathbf{b}$ and $\mathbf{c}$ respectively and $\mathbf{a}\times(\mathbf{b}\times\mathbf{c})=\frac{1}{2}\mathbf{b}$, then $|\alpha-\beta|$ is equal to **[JEE Main 2019]**

(a) 30° (b) 45° (c) 90° (d) 60°

66. Let $\mathbf{a}=2\hat{\mathbf{i}}+\hat{\mathbf{j}}-\hat{\mathbf{k}}$ and $\mathbf{b}=\hat{\mathbf{i}}+2\hat{\mathbf{j}}+\hat{\mathbf{k}}$ be two vectors. Consider a vector $\mathbf{c}=\alpha\mathbf{a}+\beta\mathbf{b}$, $\alpha,\beta\in R$. If the projection of $\mathbf{c}$ on the vector $(\mathbf{a}+\mathbf{b})$ is $3\sqrt{2}$, then the minimum value of $\mathbf{c}-(\mathbf{a}\times\mathbf{b}))\cdot\mathbf{c}$ equals ……… **[JEE Advanced 2019]**

67. Let $\mathbf{u}$ be a vector coplanar with the vectors $\mathbf{a}=2\hat{\mathbf{i}}+3\hat{\mathbf{j}}-\hat{\mathbf{k}}$ and $\mathbf{b}=\hat{\mathbf{j}}+\hat{\mathbf{k}}$. If $\mathbf{u}$ is perpendicular to $\mathbf{a}$ and $\mathbf{u}\cdot\mathbf{b}=24$, then $|\mathbf{u}|^2$ is equal to **[JEE Main 2018]**

(a) 336 (b) 315 (c) 256 (d) 84

Answers

1. (c)	**2.** (d)	**3.** (5)	**4.** (60)	**5.** (4)	**6.** (6)	**7.** (c)	**8.** (b)	**9.** (486)	**10.** (90)
11. (3)	**12.** (a)	**13.** (b)	**14.** (a)	**15.** (b)	**16.** (28)	**17.** (2)	**18.** (12)	**19.** (c)	**20.** (a)
21. (b)	**22.** (2)	**23.** (b)	**24.** (9)	**25.** (a)	**26.** (d)	**27.** (a)	**28.** (d)	**29.** (2)	**30.** (b)
31. (c)	**32.** (d)	**33.** (75)	**34.** (7)	**35.** (a, b, c)	**36.** (2)	**37.** (0.80)	**38.** (a)	**39.** (6.00)	**40.** (4.00)
41. (1)	**42.** (b)	**43.** (d)	**44.** (c)	**45.** (c)	**46.** (18)	**47.** (b)	**48.** (b)	**49.** (1)	**50.** (30)
51. (108.00)	**52.** (a,c)	**53.** (a)	**54.** (a)	**55.** (c)	**56.** (d)	**57.** (b)	**58.** (d)	**59.** (b)	**60.** (b)
61 (b)	**62.** (b)	**63.** (a)	**64.** (d)	**65.** (a)	**66.** (18)	**67.** (a)			

Three Dimensional Coordinate System

1. The angle between the straight lines, whose direction cosines are given by the equations $2l+2m-n=0$ and $mn+nl+lm=0$, is **[JEE Main 2021]**

(a) $\frac{\pi}{2}$ (b) $\pi-\cos^{-1}\left(\frac{4}{9}\right)$

(c) $\cos^{-1}\left(\frac{8}{9}\right)$ (d) $\frac{\pi}{3}$

2. Let α be the angle between the lines whose direction cosines satisfy the equations $l+m-n=0$ and $l^2+m^2-n^2=0$. Then, the value of $\sin^4\alpha+\cos^4\alpha$ is **[JEE Main 2021]**

(a) $\frac{3}{4}$ (b) $\frac{3}{8}$ (c) $\frac{5}{8}$ (d) $\frac{1}{2}$

3. The distance of line $3y-2z-1=0=3x-z+4$ from the piont $(2,-1,6)$ is **[JEE Main 2021]**

(a) $\sqrt{26}$ (b) $2\sqrt{5}$ (c) $2\sqrt{6}$ (d) $4\sqrt{2}$

4. If the lines $\frac{x-k}{1}=\frac{y-2}{2}=\frac{z-3}{3}$ and $\frac{x+1}{3}=\frac{y+2}{2}=\frac{z+3}{1}$ are co-planar, then the value of k is .... **[JEE Main 2021]**

5. If the shortest distance between the straight lines $3(x-1)=6(y-2)=2(z-1)$ and $4(x-2)=2(y-\lambda)=(z-3)$, $\lambda\in R$ is $\frac{1}{\sqrt{38}}$, then the integral value of λ is equal to **[JEE Main 2021]**

(a) 3 (b) 2 (c) 5 (d) -1

6. If the shortest distance between the lines $\mathbf{r}_1=\alpha\hat{\mathbf{i}}+2\hat{\mathbf{j}}+2\hat{\mathbf{k}}+\lambda(\hat{\mathbf{i}}-2\hat{\mathbf{j}}+2\hat{\mathbf{k}})$, $\lambda\in R$, $\alpha>0$ and $\mathbf{r}_2=-4\hat{\mathbf{i}}-\hat{\mathbf{k}}+\mu(3\hat{\mathbf{i}}-2\hat{\mathbf{j}}-2\hat{\mathbf{k}})$, $\mu\in R$ is 9, then α is equal to .... **[JEE Main 2021]**

7. The lines $x=ay-1=z-2$ and $x=3y-2=bz-2$, $(ab\neq 0)$ are coplanar, if **[JEE Main 2021]**

(a) $b=1$, $a\in R-\{0\}$ (b) $a=1$, $b\in R-\{0\}$

(c) $a=2$, $b=2$ (d) $a=2$, $b=3$

8. Let the position vectors of two points P and Q be $3\hat{\mathbf{i}}-\hat{\mathbf{j}}+2\hat{\mathbf{k}}$ and $\hat{\mathbf{i}}+2\hat{\mathbf{j}}-4\hat{\mathbf{k}}$, respectively. Let R and S be two points such that the direction ratios of lines PR and QS are $(4,-1,2)$ and $(-2,1,-2)$, respectively. Let lines PR and
QS intersect at T. If the vector **TA** is perpendicular to both **PR** and **QS** and the length of vector **TA** is $\sqrt{5}$ units, then the modulus of a position vector of A is **[JEE Main 2021]**

(a) $\sqrt{482}$ (b) $\sqrt{171}$ (c) $\sqrt{5}$ (d) $\sqrt{227}$

9. If the foot of the perpendicular from point $(4,3,8)$ on the line $L_1\Rightarrow\frac{x-a}{l}=\frac{y-2}{3}=\frac{z-b}{4}$, $l\neq 0$ is $(3,5,7)$, then the shortest distance between the line L_1 and line $L_2\Rightarrow\frac{x-2}{3}=\frac{y-4}{4}=\frac{z-5}{5}$ is equal to **[JEE Main 2021]**

(a) $\frac{1}{2}$ (b) $\frac{1}{\sqrt{6}}$ (c) $\sqrt{\frac{2}{3}}$ (d) $\frac{1}{\sqrt{3}}$

10. A line 'l' passing through origin is perpendicular to the lines

$l_1:\mathbf{r}=(3+t)\hat{\mathbf{i}}+(-1+2t)\hat{\mathbf{j}}+(4+2t)\hat{\mathbf{k}}$

$l_2:\mathbf{r}\,(3+2s)\hat{\mathbf{i}}+(3+2s)\hat{\mathbf{j}}+(2+s)\hat{\mathbf{k}}$

If the coordinates of the point in the first octant on 'l_2' at a distance of $\sqrt{17}$ from the point of intersection of 'l' and 'l_1' are (a,b,c), then $18(a+b+c)$ is equal to .... **[JEE Main 2021]**

11. The equation of the line through the point (0,1,2) and perpendicular to the line $\frac{x-1}{2}=\frac{y+1}{3}=\frac{z-1}{-2}$ is **[JEE Main 2021]**

(a) $\frac{x}{3}=\frac{y-1}{4}=\frac{z-2}{3}$ (b) $\frac{x}{3}=\frac{y-1}{-4}=\frac{z-2}{3}$

(c) $\frac{x}{3}=\frac{y-1}{4}=\frac{z-2}{-3}$ (d) $\frac{x}{-3}=\frac{y-1}{4}=\frac{z-2}{3}$

12. Let λ be an integer. If the shortest distance between the lines $x-\lambda=2y-1=-2z$ and $x=y+2\lambda=z-\lambda$ is $\frac{\sqrt{7}}{2\sqrt{2}}$, then the value of $|\lambda|$ is .... **[JEE Main 2021]**

13. Let $a, b \in R$. If the mirror image of the point $P(a, 6, 9)$ with respect to the line $\frac{x-3}{7}=\frac{y-2}{5}=\frac{z-1}{-9}$ is $(20, b, -a-9)$, then $|a+b|$ is equal to **[JEE Main 2021]**

(a) 88 (b) 86 (c) 84 (d) 90

14. Let the acute angle bisector of the two planes $x-2y-2z+1=0$ and $2x-3y-6z+1=0$ be the plane P. Then, which of the following points lies on P? **[JEE Main 2021]**

(a) $\left(3, 1, -\frac{1}{2}\right)$ (b) $\left(-2, 0, -\frac{1}{2}\right)$
(c) $(0, 2, -4)$ (d) $(4, 0, -2)$

15. Let the plane passing through the point $(-1, 0, -2)$ and perpendicular to each of the planes $2x+y-z=2$ and $x-y-z=3$ be $ax+by+cz+8=0$. Then the value of $a+b+c$ is equal to **[JEE Main 2021]**

(a) 3 (b) 8 (c) 5 (d) 4

16. Let a, b and c be distinct positive numbers. If the vectors $a\hat{\mathbf{i}}+a\hat{\mathbf{j}}+c\hat{\mathbf{k}}$, $\hat{\mathbf{i}}+\hat{\mathbf{k}}$ and $c\hat{\mathbf{i}}+c\hat{\mathbf{j}}+b\hat{\mathbf{k}}$ are coplanar, then c is equal to **[JEE Main 2021]**

(a) $\frac{2}{\frac{1}{a}+\frac{1}{b}}$ (b) $\frac{a+b}{2}$
(c) $\frac{1}{a}+\frac{1}{b}$ (d) $\sqrt{ab}$

17. Let P be an arbitrary point having sum of the squares of the distance from the planes $x+y+z=0$, $lx-nz=0$ and $x-2y+z=0$, equal to 9. If the locus of the point P is $x^2+y^2+z^2=9$, then the value of $l-n$ is equal to .... **[JEE Main 2021]**

18. Let the plane, $ax+by+cz+d=0$ bisect the line joining the points $(4, -3, 1)$ and $(2, 3, -5)$ at the right angles. If a, b, c, d are integers, then the minimum value of $(a^2+b^2+c^2+d^2)$ is .... **[JEE Main 2021]**

19. The equation of the planes parallel to the plane $x-2y+2z-3=0$ which are at unit distance from the point $(1, 2, 3)$ is $ax+by+cz+d=0$. If $(b-d)=K(c-a)$, then the positive value of K is .... **[JEE Main 2021]**

20. The equation of the plane which contains the Y-axis and passes through the point $(1, 2, 3)$ is **[JEE Main 2021]**

(a) $x+3z=10$ (b) $x+3z=0$
(c) $3x+z=6$ (d) $3x-z=0$

21. If for $a>0$, the feet of perpendiculars from the points $A(a, -2a, 3)$ and $B(0, 4, 5)$ on the plane $lx+my+nz=0$ are points $C(0, -a, -1)$ and D respectively, then, the length of line segment CD is equal to **[JEE Main 2021]**

(a) $\sqrt{31}$ (b) $\sqrt{41}$ (c) $\sqrt{55}$ (d) $\sqrt{66}$

22. If (x, y, z) be an arbitrary point lying on a plane P, which passes through the point $(42, 0, 0)$, $(0, 42, 0)$ and $(0, 0, 42)$, then the value of expression

$$3+\frac{x-11}{(y-19)^2(z-12)^2}+\frac{y-19}{(x-11)^2(z-12)^2}+\frac{z-12}{(x-11)^2(y-19)^2}-\frac{x+y+z}{14(x-11)(y-19)(z-12)}$$

[JEE Main 2021]

(a) 0 (b) 3 (c) 39 (d) –45

23. The equation of the plane passing through the point $(1, 2, -3)$ and perpendicular to the planes $3x+y-2z=5$ and $2x-5y-z=7$, is **[JEE Main 2021]**

(a) $6x-5y+2z+10=0$ (b) $11x+y+17z+38=0$
(c) $6x-5y-2z-2=0$ (d) $3x-10y+2z+11=0$

24. Consider the three planes

$P_1: 3x+15y+21z=9$,
$P_2: x-3y-z=5$ and
$P_3: 2x+10y+14z=5$

Then, which one of the following is true? **[JEE Main 2021]**

(a) P_1 and P_2 are parallel (b) P_1 and P_3 are parallel
(c) P_2 and P_3 are parallel (d) P_1, P_2 and P_3 all are parallel

25. The square of the distance of the point of intersection of the line $\frac{x-1}{2}=\frac{y-2}{3}=\frac{z+1}{6}$ and the plane $2x-y+z=6$ from the point $(-1, -1, 2)$ is **[JEE Main 2021]**

26. The distance of the point $(-1, 2, -2)$ from the line of intersection of the planes $2x+3y+2z=0$ and $x-2y+z=0$ is **[JEE Main 2021]**

(a) $\frac{1}{\sqrt{2}}$ (b) $\frac{5}{2}$ (c) $\frac{\sqrt{42}}{2}$ (d) $\frac{\sqrt{34}}{2}$

27. Suppose the line $\frac{x-2}{\alpha}+\frac{y-2}{-5}=\frac{z+2}{2}$ lies on the plane $x+3y-2z+\beta=0$. Then, $(\alpha+\beta)$ is equal to **[JEE Main 2021]**

28. Let the equation of the plane, that passes through the point $(1, 4, -3)$ and contains the line of intersection of the planes $3x-2y+4z-7=0$ and $x+5y-2z+9=0$ be $\alpha x+\beta y+\gamma z+3=0$, then $\alpha+\beta+\gamma$ is equal to **[JEE Main 2021]**

(a) – 23 (b) – 15 (c) 23 (d) 15

29. The distance of the point (1, –2, 3) from the plane $x-y+z=5$ measured parallel to a line, whose direction ratios are 2, 3, –6 is **[JEE Main 2021]**

(a) 3 (b) 5 (c) 2 (d) 1

30. Equation of a plane at a distance $\sqrt{2/21}$ from the origin, which contains the line of intersection of the planes $x-y-z-1=0$ and $2x+y-3z+4=0$, is **[JEE Main 2021]**

(a) $3x-y-5z+2=0$ (b) $3x-4z+3=0$
(c) $-x+2y+2z-3=0$ (d) $4x-y-5z+2=0$

31. The equation of the plane passing through the line of intersection of the planes $\mathbf{r}\cdot(\hat{\mathbf{i}}+\hat{\mathbf{j}}+\hat{\mathbf{k}})=1$ and $\mathbf{r}\cdot(2\hat{\mathbf{i}}+3\hat{\mathbf{j}}-\hat{\mathbf{k}})+4=0$ and parallel to the X-axis is **[JEE Main 2021]**

(a) $\mathbf{r}\cdot(\hat{\mathbf{j}}-3\hat{\mathbf{k}})+6=0$ (b) $\mathbf{r}\cdot(\hat{\mathbf{i}}+3\hat{\mathbf{k}})+6=0$
(c) $\mathbf{r}\cdot(\hat{\mathbf{i}}-3\hat{\mathbf{k}})+6=0$ (d) $\mathbf{r}\cdot(\hat{\mathbf{j}}-3\hat{\mathbf{k}})-6=0$

32. Let S be the mirror image of the point $Q(1, 3, 4)$ with respect to the plane $2x-y+z+3=0$ and let $R(3, 5, \gamma)$ be a point of this plane. Then the square of the length of the line segment SR is **[JEE Main 2021]**

33. A plane P contains the line $x+2y+3z+1=0=x-y-z-6$ and is perpendicular to the plane $-2x+y+z+8=0$. Then which of the following points lies on P? **[JEE Main 2021]**

(a) (–1, 1, 2) (b) (0, 1, 1) (c) (1, 0, 1) (d) (2, –1, 1)

34. Let P be the plane passing through the point (1,2,3) and the line of intersection of the planes $\mathbf{r}\cdot(\hat{\mathbf{i}}+\hat{\mathbf{j}}+4\hat{\mathbf{k}})=16$ and $\mathbf{r}\cdot(-\hat{\mathbf{i}}+\hat{\mathbf{j}}+\hat{\mathbf{k}})=6$. Then which of the following points does not lie on P? **[JEE Main 2021]**

(a) (3, 3, 2) (b) (6, – 6, 2) (c) (4, 2, 2) (d) (– 8, 8, 6)

35. Let the line L be the projection of the line $\frac{x-1}{2}=\frac{y-3}{1}=\frac{z-4}{2}$ in the plane $x-2y-z=3$. If d is the distance of the point (0, 0, 6) from L, then d^2 is equal to **[JEE Main 2021]**

36. Let Q be the foot of the perpendicular from the point $P(7,-2,13)$ on the plane containing the lines $\frac{x+1}{6}=\frac{y-1}{7}=\frac{z-3}{8}$ and $\frac{x+1}{3}=\frac{y-2}{5}=\frac{z-3}{7}$. Then, $(PQ)^2$ is equal to **[JEE Main 2021]**

37. Let a plane P pass through the point $(3,7,-7)$ and contain the line, $\frac{x-2}{-3}=\frac{y-3}{2}=\frac{z+2}{1}$. If distance of the plane P from the origin is d, then d^2 is equal to .... **[JEE Main 2021]**

38. For real numbers α and $\beta\neq 0$, if the point of intersection of the straight lines

$$\frac{x-\alpha}{1}=\frac{y-1}{2}=\frac{z-1}{3}$$

and $\frac{x-4}{\beta}=\frac{y-6}{3}=\frac{z-7}{3}$ lies on the plane $x+2y-z=8$, then $\alpha-\beta$ is equal to **[JEE Main 2021]**

(a) 5 (b) 9 (c) 3 (d) 7

39. The distance of the point $P(3,4,4)$ from the point of intersection of the line joining the points. $Q(3,-4,-5)$ and $R(2,-3,1)$ and the plane $2x+y+z=7$, is equal to .... **[JEE Main 2021]**

40. Let the foot of perpendicular from a point $P(1,2,-1)$ to the straight line $L \Rightarrow \frac{x}{1}=\frac{y}{0}=\frac{z}{-1}$ be N. Let a line be drawn from P parallel to the plane $x+y+2z=0$ which meets L at point Q. If α is the acute angle between the lines PN and PQ, then $\cos\alpha$ is equal to **[JEE Main 2021]**

(a) $\frac{1}{\sqrt{5}}$ (b) $\frac{\sqrt{3}}{2}$
(c) $\frac{1}{\sqrt{3}}$ (d) $\frac{1}{2\sqrt{3}}$

41. Let L be the line of intersection of planes $\mathbf{r}\cdot(\hat{\mathbf{i}}-\hat{\mathbf{j}}+2\hat{\mathbf{k}})=2$ and $\mathbf{r}\cdot(2\hat{\mathbf{i}}+\hat{\mathbf{j}}-\hat{\mathbf{k}})=2$. If $P(\alpha,\beta,\gamma)$ is the foot of perpendicular on L from the point (1, 2, 0), then the value of $35(\alpha+\beta+\gamma)$ is equal to **[JEE Main 2021]**

(a) 101 (b) 119 (c) 143 (d) 134

42. Let P be a plane passing through the points (1, 0, 1), (1, –2, 1) and (0, 1, –2). Let a vector $\mathbf{a}=\alpha\hat{\mathbf{i}}+\beta\hat{\mathbf{j}}+\gamma\hat{\mathbf{k}}$ be such that $\mathbf{a}$ is parallel to the plane P, perpendicular to $(\hat{\mathbf{i}}+2\hat{\mathbf{j}}+3\hat{\mathbf{k}})$ and $\mathbf{a}\cdot(\hat{\mathbf{i}}+\hat{\mathbf{j}}+2\hat{\mathbf{k}})=2$, then $(\alpha-\beta+\gamma)^2$ equals .... **[JEE Main 2021]**

43. Consider the line L given by the equation $\frac{x-3}{2}=\frac{y-1}{1}=\frac{z-2}{1}$.

Let Q be the mirror image of the point (2, 3, –1) with respect to L. Let a plane P be such that it passes through Q, and the line L is perpendicular to P. Then which of the following points is on the plane P? **[JEE Main 2021]**

(a) (–1, 1, 2) (b) (1, 1, 1)
(c) (1, 1, 2) (d) (1, 2, 2)

44. If the equation of the plane passing through the line of intersection of the planes $2x-7y+4z-3=0$, $3x-5y+4z+11=0$ and the point (–2, 1, 3) is $ax+by+cz-7=0$, then the value of $2a+b+c-7$ is .... **[JEE Main 2021]**

45. If the equation of plane passing through the mirror image of a point (2, 3, 1) with respect to line $\frac{x+1}{2}=\frac{y-3}{1}=\frac{z+2}{-1}$ and containing the line $\frac{x-2}{3}=\frac{1-y}{2}=\frac{z+1}{1}$ is $\alpha x+\beta y+\gamma z=24$, then $\alpha+\beta+\gamma$ is equal to **[JEE Main 2021]**

(a) 20 (b) 19 (c) 18 (d) 21

46. Let the mirror image of the point $(1,3,a)$ with respect to the plane $\mathbf{r}\cdot(2\hat{i}-\hat{j}+\hat{k})-b=0$ be $(-3, 5, 2)$. Then the value of $|a+b|$ is equal to .... **[JEE Main 2021]**

47. Let P be a plane containing the line $\frac{x-1}{3}=\frac{y+6}{4}=\frac{z+5}{2}$ and parallel to the line $\frac{x-3}{4}=\frac{y-2}{-3}=\frac{z+5}{7}$. If the point $(1,-1,\alpha)$ lies on the plane P, then the value of $|5\alpha|$ is equal to .... **[JEE Main 2021]**

48. Let P be a plane $lx+my+nz=0$ containing the line, $\frac{1-x}{1}=\frac{y+4}{2}=\frac{z+2}{3}$. If plane P divides the line segment AB joining points $A(-3,-6,1)$ and $B(2,4,-3)$ in ratio $k:1$, then the value of k is equal to **[JEE Main 2021]**

(a) 1.5 (b) 3 (c) 2 (d) 4

49. If the distance of the point $(1,-2,3)$ from the plane $x+2y-3z+10=0$ measured parallel to the line, $\frac{x-1}{3}=\frac{2-y}{m}=\frac{z+3}{1}$ is $\sqrt{\frac{7}{2}}$, then the value of $|m|$ is equal to......... . **[JEE Main 2021]**

50. If the mirror image of the point (1, 3, 5) with respect to the plane $4x-5y+2z=8$ is (α,β,γ), then $5(\alpha+\beta+\gamma)$ equals **[JEE Main 2021]**

(a) 47 (b) 43 (c) 39 (d) 41

51. Let L be a line obtained from the intersection of two planes $x+2y+z=6$ and $y+2z=4$. If point $P(\alpha,\beta,\gamma)$ is the foot of perpendicular from $(3,2,1)$ on L, then the value of $21(\alpha+\beta+\gamma)$ equals **[JEE Main 2021]**

(a) 142 (b) 68 (c) 136 (d) 102

52. Let $(\lambda, 2, 1)$ be a point on the plane which passes through the point $(4, -2, 2)$. If the plane is perpendicular to the line joining the points $(-2, -21, 29)$ and $(-1, -16, 23)$, then $\left(\frac{\lambda}{11}\right)^2-\frac{4\lambda}{11}-4$ is .... **[JEE Main 2021]**

53. A plane passes through the points $A(1,2,3)$, $B(2,3,1)$ and $C(2,4,2)$. If O is the origin and P is $(2,-1,1)$, then the projection of **OP** on this plane is of length **[JEE Main 2021]**

(a) $\sqrt{\frac{2}{3}}$ (b) $\sqrt{\frac{2}{11}}$ (c) $\sqrt{\frac{2}{7}}$ (d) $\sqrt{\frac{2}{5}}$

54. The distance of the point (1, 1, 9) from the point of intersection of the line $\frac{x-3}{1}=\frac{y-4}{2}=\frac{z-5}{2}$ and the plane $x+y+z=17$ is **[JEE Main 2021]**

(a) $2\sqrt{19}$ (b) $19\sqrt{2}$ (c) $\sqrt{38}$ (d) 38

55. The vector equation of the plane passing through the intersection of the planes $\mathbf{r}\cdot(\hat{\mathbf{i}}+\hat{\mathbf{j}}+\hat{\mathbf{k}})=1$ and $\mathbf{r}\cdot(\hat{\mathbf{i}}-2\hat{\mathbf{j}})=-2$ and the point (1, 0, 2) is **[JEE Main 2021]**

(a) $\mathbf{r}\cdot(\hat{\mathbf{i}}+7\hat{\mathbf{j}}+3\hat{\mathbf{k}})=\frac{7}{3}$ (b) $\mathbf{r}\cdot(3\hat{\mathbf{i}}+7\hat{\mathbf{j}}+3\hat{\mathbf{k}})=7$

(c) $\mathbf{r}\cdot(\hat{\mathbf{i}}+7\hat{\mathbf{j}}+3\hat{\mathbf{k}})=7$ (d) $\mathbf{r}\cdot(\hat{\mathbf{i}}-7\hat{\mathbf{j}}+3\hat{\mathbf{k}})=\frac{7}{3}$

56. The projection of the line segment joining the points $(1,-1,3)$ and $(2,-4,11)$ on the line joining the points $(-1,2,3)$ and $(3,-2,10)$ is......... . **[JEE Main 2020]**

57. If the variance of the terms in an increasing AP, $b_1, b_2, b_3, \ldots\ldots, b_{11}$ is 90, then the common difference of this AP is .... **[JEE Main 2020]**

58. If (a, b, c) is the image of the point $(1, 2, -3)$ in the line $\frac{x+1}{2}=\frac{y-3}{-2}=\frac{z}{-1}$, then $a+b+c$ is equal to **[JEE Main 2020]**

(a) 3 (b) 2 (c) −1 (d) 1

59. If for some $\alpha\in R$, the lines

$L_1:\frac{x+1}{2}=\frac{y-2}{-1}=\frac{z-1}{1}$ and

$L_2:\frac{x+2}{\alpha}=\frac{y+1}{5-\alpha}=\frac{z+1}{1}$ are coplanar, then the line L_2 passes through the point **[JEE Main 2020]**

(a) (10, 2, 2) (b) (2, −10, −2)
(c) (10, −2, −2) (d) (−2, 10, 2)

60. If the foot of the perpendicular drawn from the point (1, 0, 3) on a line passing through $(\alpha, 7, 1)$ is $\left(\frac{5}{3},\frac{7}{3},\frac{17}{3}\right)$, then α is equal to .... . **[JEE Main 2020]**

61. The shortest distance between the lines $\frac{x-3}{3}=\frac{y-8}{-1}=\frac{z-3}{1}$ and $\frac{x+3}{-3}=\frac{y+7}{2}=\frac{z-6}{4}$ is **[JEE Main 2020]**

(a) $3\sqrt{30}$ (b) $2\sqrt{30}$ (c) $\frac{7}{2}\sqrt{30}$ (d) 3

62. The lines $\mathbf{r}=(\hat{\mathbf{i}}-\hat{\mathbf{j}})+l(2\hat{\mathbf{i}}+\hat{\mathbf{k}})$ and $\mathbf{r}=(2\hat{\mathbf{i}}-\hat{\mathbf{j}})+m(\hat{\mathbf{i}}+\hat{\mathbf{j}}-\hat{\mathbf{k}})$ **[JEE Main 2020]**

(a) do not intersect for any values of l and m.
(b) intersect for all values of l and m.
(c) intersect when $l=2$ and $m=\frac{1}{2}$.
(d) intersect when $l=1$ and $m=2$.

10 Vectors and 3D Geometry

63. The mirror image of the point (1, 2, 3) in a plane is $\left(-\frac{7}{3},-\frac{4}{3},-\frac{1}{3}\right)$. Which of the following points lies on this plane? **[JEE Main 2020]**

(a) (1, – 1, 1) (b) (– 1, – 1, 1)
(c) (1, 1, 1) (d) (– 1, – 1, – 1)

64. If the distance between the plane, $23x-10y-2z+48=0$ and the plane containing the lines $\frac{x+1}{2}=\frac{y-3}{4}=\frac{z+1}{3}$ and $\frac{x+3}{2}=\frac{y+2}{6}=\frac{z-1}{\lambda}(\lambda\in R)$ is equal to $\frac{k}{\sqrt{633}}$, then k is equal to ..... . **[JEE Main 2020]**

65. A plane passing through the point (3, 1, 1) contains two lines whose direction ratios are 1, –2, 2 and 2, 3, –1 respectively. If this plane also passes through the point $(\alpha, -3, 5)$, then α is equal to **[JEE Main 2020]**

(a) 5 (b) – 10 (c) 10 (d) – 5

66. The plane which bisects the line joining the points (4,–2, 3) and (2, 4, – 1) at right angles also passes through the point **[JEE Main 2020]**

(a) (0, – 1, 1) (b) (4, 0, – 1)
(c) (4, 0, 1) (d) (0, 1, – 1)

67. If the equation of a plane P, passing through the intersection of the planes, $x+4y-z+7=0$ and $3x+y+5z=8$ is $ax+by+6z=15$ for some $a,b\in\mathbf{R}$, then the distance of the point $(3,2,-1)$ from the plane P is .... **[JEE Main 2020]**

68. The plane passing through the points (1, 2, 1), (2, 1, 2) and parallel to the line, $2x=3y, z=1$ also passes through the point **[JEE Main 2020]**

(a) (– 2, 0, 1) (b) (0,6,–2)
(c) (0,–6, 2) (d) (2,0,–1)

69. The foot of the perpendicular drawn from the point (4, 2, 3) to the line joining the points (1, –2, 3) and (1, 1, 0) lies on the plane **[JEE Main 2020]**

(a) $2x+y-z=1$ (b) $x-y-2z=1$
(c) $x-2y+z=1$ (d) $x+2y-z=1$

70. The distance of the point (1, –2, 3) from the plane $x-y+z=5$ measured parallel to line $\frac{x}{2}=\frac{y}{3}=\frac{z}{-6}$ is **[JEE Main 2020]**

(a) $\frac{1}{7}$ (b) 1 (c) 7 (d) $\frac{7}{5}$

71. The shortest distance between the lines $\frac{x-1}{0}=\frac{y+1}{-1}=\frac{z}{1}$ and $x+y+z+1=0, 2x-y+z+3=0$ is **[JEE Main 2020]**

(a) 1 (b) $\frac{1}{\sqrt{3}}$ (c) $\frac{1}{\sqrt{2}}$ (d) $\frac{1}{2}$

72. Let a plane P contain two lines $\hat{\mathbf{i}}+\lambda(\hat{\mathbf{i}}+\hat{\mathbf{j}}),\lambda\in\mathbf{R}$ and $\mathbf{r}=-\hat{\mathbf{j}}+\mu(\hat{\mathbf{j}}-\hat{\mathbf{k}}),\mu\in\mathbf{R}$. If $Q(\alpha,\beta,\gamma)$ is the foot of the perpendicular drawn from the point $M(1, 0, 1)$ to P, then $3(\alpha+\beta+\gamma)$ equals .... **[JEE Main 2020]**

73. Let L_1 and L_2 be the following straight lines.

$L_1:\frac{x-1}{1}=\frac{y}{-1}=\frac{z-1}{3}$ and $L_2:\frac{x-1}{-3}=\frac{y}{-1}=\frac{z-1}{1}$.

Suppose the straight line

$$L:\frac{x-\alpha}{l}=\frac{y-1}{m}=\frac{z-\gamma}{-2}$$

lies in the plane containing L_1 and L_2 and passes through the point of intersection of L_1 and L_2. If the line L bisects the acute angle between the lines L_1 and L_2, then which of the following statements is/are TRUE? **[JEE Advanced 2020]**

(a) $\alpha-\gamma=3$ (b) $l+m=2$
(c) $\alpha-\gamma=1$ (d) $l+m=0$

74. Let $\alpha,\beta,\gamma,\delta$ be real numbers such that $\alpha^2+\beta^2+\gamma^2\neq 0$ and $\alpha+\gamma=1$. Suppose the point (3, 2, – 1) is the mirror image of the point (1, 0, – 1) with respect to the plane $\alpha x+\beta y+\gamma z=\delta$. Then which of the following statements is/are TRUE? **[JEE Advanced 2020]**

(a) $\alpha+\beta=2$ (b) $\delta-\gamma=3$
(c) $\delta+\beta=4$ (d) $\alpha+\beta+\gamma=\delta$

75. A tetrahedron has vertices $P(1,2,1)$, $Q(2,1,3)$, $R(-1,1,2)$ and $O(0,0,0)$. The angle between faces OPQ and PQR is **[JEE Main 2019]**

(a) $\cos^{-1}\left(\frac{7}{31}\right)$ (b) $\cos^{-1}\left(\frac{9}{35}\right)$
(c) $\cos^{-1}\left(\frac{19}{35}\right)$ (d) $\cos^{-1}\left(\frac{17}{31}\right)$

76. The length of the perpendicular from the point $(2,-1,4)$ on the straight line, $\frac{x+3}{10}=\frac{y-2}{-7}=\frac{z}{1}$ is **[JEE Main 2019]**

(a) greater than 3 but less than 4.
(b) less than 2.
(c) greater than 2 but less than 3.
(d) greater than 4.

77. The vertices B and C of a ΔABC lie on the line, $\frac{x+2}{3}=\frac{y-1}{0}=\frac{z}{4}$ such that $BC=5$ units. Then, the area (in sq units) of this triangle, given that point $A(1,-1,2)$ is **[JEE Main 2019]**

(a) $\sqrt{34}$ (b) $2\sqrt{34}$ (c) $5\sqrt{17}$ (d) 6

78. If a point $R(4, y, z)$ lies on the line segment joining the points $P(2,-3,4)$ and $Q(8,0,10)$, then the distance of R from the origin is **[JEE Main 2019]**

(a) $2\sqrt{21}$ (b) $\sqrt{53}$ (c) $2\sqrt{14}$ (d) 6

79. If the length of the perpendicular from the point $(\beta, 0, \beta)$ $(\beta \neq 0)$ to the line, $\frac{x}{1} = \frac{y-1}{0} = \frac{z+1}{-1}$ is $\sqrt{\frac{3}{2}}$, then β is equal to **[JEE Main 2019]**

(a) 2 (b) -2 (c) -1 (d) 1

80. The magnitude of the projection of the vector $2\hat{\mathbf{i}}+3\hat{\mathbf{j}}+\hat{\mathbf{k}}$ on the vector perpendicular to the plane containing the vectors $\hat{\mathbf{i}}+\hat{\mathbf{j}}+\hat{\mathbf{k}}$ and $\hat{\mathbf{i}}+2\hat{\mathbf{j}}+3\hat{\mathbf{k}}$, is **[JEE Main 2019]**

(a) $\frac{\sqrt{3}}{2}$ (b) $\sqrt{6}$ (c) $3\sqrt{6}$ (d) $\sqrt{\frac{3}{2}}$

81. If the line $\frac{x-1}{2} = \frac{y+1}{3} = \frac{z-2}{4}$ meets the plane, $x+2y+3z=15$ at a point P, then the distance of P from origin is **[JEE Main 2019]**

(a) $7/2$ (b) $9/2$ (c) $\sqrt{5}/2$ (d) $2\sqrt{5}$

82. A plane passing through the points $(0,-1,0)$ and $(0,0,1)$ and making an angle $\frac{\pi}{4}$ with the plane $y-z+5=0$, also passes through the point **[JEE Main 2019]**

(a) $(\sqrt{2}, 1, 4)$ (b) $(-\sqrt{2}, 1, -4)$
(c) $(-\sqrt{2}, -1, -4)$ (d) $(\sqrt{2}, -1, 4)$

83. If the plane $2x-y+2z+3=0$ has the distances $\frac{1}{3}$ and $\frac{2}{3}$ units from the planes $4x-2y+4z+\lambda=0$ and $2x-y+2z+\mu=0$, respectively, then the maximum value of $\lambda+\mu$ is equal to **[JEE Main 2019]**

(a) 13 (b) 15 (c) 5 (d) 9

84. A plane which bisects the angle between the two given planes $2x-y+2z-4=0$ and $x+2y+2z-2=0$, passes through the point **[JEE Main 2019]**

(a) $(1,-4,1)$ (b) $(1,4,-1)$ (c) $(2,4,1)$ (d) $(2,-4,1)$

85. Let A be a point on the line $\mathbf{r}=(1-3\mu)\hat{\mathbf{i}}+(\mu-1)\hat{\mathbf{j}}+(2+5\mu)\hat{\mathbf{k}}$ and $B(3,2,6)$ be a point in the space. Then, the value of μ for which the vector **AB** is parallel to the plane $x-4y+3z=1$ is **[JEE Main 2019]**

(a) $\frac{1}{4}$ (b) $-\frac{1}{4}$ (c) $\frac{1}{8}$ (d) $\frac{1}{2}$

86. The plane passing through the point $(4,-1,2)$ and parallel to the lines $\frac{x+2}{3} = \frac{y-2}{-1} = \frac{z+1}{2}$ and $\frac{x-2}{1} = \frac{y-3}{2} = \frac{z-4}{3}$ also passes through the point **[JEE Main 2019]**

(a) $(-1,-1,-1)$ (b) $(1,1,-1)$
(c) $(1,1,1)$ (d) $(-1,-1,1)$

87. Let S be the set of all real values of λ such that a plane passing through the points $(-\lambda^2,1,1)$, $(1,-\lambda^2,1)$ and $(1,1,-\lambda^2)$ also passes through the point $(-1,-1,1)$. Then, S is equal to **[JEE Main 2019]**

(a) $\{\sqrt{3},-\sqrt{3}\}$ (b) $\{3,-3\}$
(c) $\{1,-1\}$ (d) $\{\sqrt{3}\}$

88. The direction ratios of normal to the plane through the points $(0,-1,0)$ and $(0,0,1)$ and making an angle $\frac{\pi}{4}$ with the plane $y-z+5=0$ are **[JEE Main 2019]**

(a) $2,-1,1$ (b) $\sqrt{2},1,-1$
(c) $2,\sqrt{2},-\sqrt{2}$ (d) $2\sqrt{3},1,-1$

89. If the point $(2,\alpha,\beta)$ lies on the plane which passes through the points $(3,4,2)$ and $(7,0,6)$ and is perpendicular to the plane $2x-5y=15$, then $2\alpha-3\beta$ is equal to **[JEE Main 2019]**

(a) 17 (b) 7 (c) 5 (d) 12

90. The equation of a plane containing the line of intersection of the planes $2x-y-4=0$ and $y+2z-4=0$ and passing through the point $(1,1,0)$ is **[JEE Main 2019]**

(a) $x-3y-2z=-2$ (b) $2x-z=2$
(c) $x-y-z=0$ (d) $x+3y+z=4$

91. The vector equation of the plane through line of intersection of the planes $x+y+z=1$ and $2x+3y+4z=5$, which is perpendicular to the plane $x-y+z=0$ is **[JEE Main 2019]**

(a) $\mathbf{r}\cdot(\hat{\mathbf{i}}-\hat{\mathbf{k}})-2=0$ (b) $\mathbf{r}\times(\hat{\mathbf{i}}+\hat{\mathbf{k}})+2=0$
(c) $\mathbf{r}\times(\hat{\mathbf{i}}-\hat{\mathbf{k}})+2=0$ (d) $\mathbf{r}\cdot(\hat{\mathbf{i}}-\hat{\mathbf{k}})+2=0$

92. Let P be the plane, which contains the line of intersection of the planes, $x+y+z-6=0$ and $2x+3y+z+5=0$ and it is perpendicular to the XY-plane. Then, the distance of the point $(0,0,256)$ from P is equal to **[JEE Main 2019]**

(a) $63\sqrt{5}$ (b) $205\sqrt{5}$
(c) $\frac{11}{\sqrt{5}}$ (d) $\frac{17}{\sqrt{5}}$

93. If $Q(0,-1,-3)$ is the image of the point P in the plane $3x-y+4z=2$ and R is the point $(3,-1,-2)$, then the area (in sq units) of ΔPQR is **[JEE Main 2019]**

(a) $\frac{\sqrt{91}}{2}$ (b) $2\sqrt{13}$ (c) $\frac{\sqrt{91}}{4}$ (d) $\frac{\sqrt{65}}{2}$

94. A perpendicular is drawn from a point on the line $\frac{x-1}{2} = \frac{y+1}{-1} = \frac{z}{1}$ to the plane $x+y+z=3$ such that the foot of the perpendicular Q also lies on the plane $x-y+z=3$. Then, the coordinates of Q are **[JEE Main 2019]**

(a) $(-1,0,4)$ (b) $(4,0,-1)$
(c) $(2,0,1)$ (d) $(1,0,2)$

95. If the line $\frac{x-2}{3} = \frac{y+1}{2} = \frac{z-1}{-1}$ intersects the plane $2x+3y-z+13=0$ at a point P and the plane $3x+y+4z=16$ at a point Q, then PQ is equal to **[JEE Main 2019]**

(a) 14 (b) $\sqrt{14}$
(c) $2\sqrt{7}$ (d) $2\sqrt{14}$

96. If lines $x=ay+b, z=cy+d$ and $x=a'z+b', y=c'z+d'$ are perpendicular, then **[JEE Main 2019]**

(a) $ab'+bc'+1=0$ (b) $bb'+cc'+1=0$
(c) $aa'+c+c'=0$ (d) $cc'+a+a'=0$

97. The equation of the plane containing the straight line $\frac{x}{2}=\frac{y}{3}=\frac{z}{4}$ and perpendicular to the plane containing the straight lines $\frac{x}{3}=\frac{y}{4}=\frac{z}{2}$ and $\frac{x}{4}=\frac{y}{2}=\frac{z}{3}$ is **[JEE Main 2019]**

(a) $5x+2y-4z=0$ (b) $x+2y-2z=0$
(c) $3x+2y-3z=0$ (d) $x-2y+z=0$

98. The length of the perpendicular drawn from the point (2, 1, 4) to the plane containing the lines $\mathbf{r}=(\hat{\mathbf{i}}+\hat{\mathbf{j}})+\lambda(\hat{\mathbf{i}}+2\hat{\mathbf{j}}-\hat{\mathbf{k}})$ and $\mathbf{r}=(\hat{\mathbf{i}}+\hat{\mathbf{j}})+\mu(-\hat{\mathbf{i}}+\hat{\mathbf{j}}-2\hat{\mathbf{k}})$ is **[JEE Main 2019]**

(a) 3 (b) $\frac{1}{3}$
(c) $\sqrt{3}$ (d) $\frac{1}{\sqrt{3}}$

99. The equation of the line passing through $(-4, 3, 1)$, parallel to the plane $x+2y-z-5=0$ and intersecting the line

$\frac{x+1}{-3}=\frac{y-3}{2}=\frac{z-2}{-1}$ is **[JEE Main 2019]**

(a) $\frac{x+4}{3}=\frac{y-3}{-1}=\frac{z-1}{1}$ (b) $\frac{x+4}{-1}=\frac{y-3}{1}=\frac{z-1}{1}$
(c) $\frac{x+4}{1}=\frac{y-3}{1}=\frac{z-1}{3}$ (d) $\frac{x-4}{2}=\frac{y+3}{1}=\frac{z+1}{4}$

100. The plane through the intersection of the planes $x+y+z=1$ and $2x+3y-z+4=0$ and parallel to Y-axis also passes through the point **[JEE Main 2019]**

(a) $(3, 3, -1)$ (b) $(-3, 1, 1)$
(c) $(3, 2, 1)$ (d) $(-3, 0, -1)$

101. The plane which bisects the line segment joining the points $(-3,-3,4)$ and $(3,7,6)$ at right angles, passes through which one of the following points? **[JEE Main 2019]**

(a) $(4, -1, 7)$ (b) $(2, 1, 3)$
(c) $(-2, 3, 5)$ (d) $(4, 1, -2)$

102. On which of the following lines lies the point of intersection of the line, $\frac{x-4}{2}=\frac{y-5}{2}=\frac{z-3}{1}$ and the plane, $x+y+z=2$? **[JEE Main 2019]**

(a) $\frac{x-4}{1}=\frac{y-5}{1}=\frac{z-5}{-1}$
(b) $\frac{x+3}{3}=\frac{4-y}{3}=\frac{z+1}{-2}$
(c) $\frac{x-2}{2}=\frac{y-3}{2}=\frac{z+3}{3}$
(d) $\frac{x-1}{1}=\frac{y-3}{2}=\frac{z+4}{-5}$

103. The plane containing the line $\frac{x-3}{2}=\frac{y+2}{-1}=\frac{z-1}{3}$ and also containing its projection on the plane $2x+3y-z=5$, contains which one of the following points? **[JEE Main 2019]**

(a) $(-2, 2, 2)$ (b) $(2, 2, 0)$
(c) $(2, 0, -2)$ (d) $(0, -2, 2)$

104. Two lines $\frac{x-3}{1}=\frac{y+1}{3}=\frac{z-6}{-1}$ and $\frac{x+5}{7}=\frac{y-2}{-6}=\frac{z-3}{4}$ intersect at the point R. The reflection of R in the xy-plane has coordinates **[JEE Main 2019]**

(a) $(2, -4, -7)$ (b) $(2, -4, 7)$
(c) $(-2, 4, 7)$ (d) $(2, 4, 7)$

105. The perpendicular distance from the origin to the plane containing the two lines,

$\frac{x+2}{3}=\frac{y-2}{5}=\frac{z+5}{7}$

and $\frac{x-1}{1}=\frac{y-4}{4}=\frac{z+4}{7}$, is **[JEE Main 2019]**

(a) $11\sqrt{6}$ (b) $\frac{11}{\sqrt{6}}$ (c) 11 (d) $6\sqrt{11}$

106. If an angle between the line, $\frac{x+1}{2}=\frac{y-2}{1}=\frac{z-3}{-2}$ and the plane, $x-2y-kz=3$ is $\cos^{-1}\left(\frac{2\sqrt{2}}{3}\right)$, then a value of k is **[JEE Main 2019]**

(a) $\sqrt{\frac{5}{3}}$ (b) $\sqrt{\frac{3}{5}}$ (c) $-\frac{3}{5}$ (d) $-\frac{5}{3}$

107. Let L_1 and L_2 denote the lines

$$\mathbf{r}=\hat{\mathbf{i}}+\lambda(-\hat{\mathbf{i}}+2\hat{\mathbf{j}}+2\hat{\mathbf{k}}), \lambda\in R$$

and $\mathbf{r}=\mu(2\hat{\mathbf{i}}-\hat{\mathbf{j}}+2\hat{\mathbf{k}}), \mu\in R$

respectively. If L_3 is a line which is perpendicular to both L_1 and L_2 and cuts both of them, then which of the following options describe(s) L_3? **[JEE Advanced 2019]**

(a) $\mathbf{r}=\frac{2}{9}(2\hat{\mathbf{i}}-\hat{\mathbf{j}}+2\hat{\mathbf{k}})+t(2\hat{\mathbf{i}}+2\hat{\mathbf{j}}-\hat{\mathbf{k}}), t\in R$
(b) $\mathbf{r}=\frac{1}{3}(2\hat{\mathbf{i}}+\hat{\mathbf{k}})+t(2\hat{\mathbf{i}}+2\hat{\mathbf{j}}-\hat{\mathbf{k}}), t\in R$
(c) $\mathbf{r}=\frac{2}{9}(4\hat{\mathbf{i}}+\hat{\mathbf{j}}+\hat{\mathbf{k}})+t(2\hat{\mathbf{i}}+2\hat{\mathbf{j}}-\hat{\mathbf{k}}), t\in R$
(d) $\mathbf{r}=t(2\hat{\mathbf{i}}+2\hat{\mathbf{j}}-\hat{\mathbf{k}}), t\in R$

108. Three lines are given by

$\mathbf{r}=\lambda\hat{\mathbf{i}}, \lambda\in R$

$\mathbf{r}=\mu(\hat{\mathbf{i}}+\hat{\mathbf{j}}), \mu\in R$

and $\mathbf{r}=\nu(\hat{\mathbf{i}}+\hat{\mathbf{j}}+\hat{\mathbf{k}}), \nu\in R$

Let the lines cut the plane $x+y+z=1$ at the points A, B and C respectively. If the area of the triangle ABC is Δ then the value of $(6\Delta)^2$ equals .... **[JEE Advanced 2019]**

109. Three lines

$L_1 : \mathbf{r} = \lambda\hat{\mathbf{i}}, \lambda \in R,$ $\quad L_2 : \mathbf{r} = \hat{\mathbf{k}} + \mu\hat{\mathbf{j}}, \mu \in R$ and

$L_3 : \mathbf{r} = \hat{\mathbf{i}} + \hat{\mathbf{j}} + \nu\hat{\mathbf{k}}, \nu \in R$

are given. For which point(s) Q on L_2 can we find a point P on L_1 and a point R on L_3 so that P, Q and R are collinear? **[JEE Advanced 2019]**

(a) $\hat{\mathbf{k}}$ (b) $\hat{\mathbf{k}} + \hat{\mathbf{j}}$ (c) $\hat{\mathbf{k}} + \frac{1}{2}\hat{\mathbf{j}}$ (d) $\hat{\mathbf{k}} - \frac{1}{2}\hat{\mathbf{j}}$

110. If L_1 is the line of intersection of the planes $2x - 2y + 3z - 2 = 0$, $x - y + z + 1 = 0$ and L_2 is the line of intersection of the planes $x + 2y - z - 3 = 0$, $3x - y + 2z - 1 = 0$, then the distance of the origin from the plane, containing the lines L_1 and L_2 is **[JEE Main 2018]**

(a) $\frac{1}{4\sqrt{2}}$ (b) $\frac{1}{3\sqrt{2}}$

(c) $\frac{1}{2\sqrt{2}}$ (d) $\frac{1}{\sqrt{2}}$

111. The length of the projection of the line segment joining the points $(5, -1, 4)$ and $(4, -1, 3)$ on the plane, $x + y + z = 7$ is **[JEE Main 2018]**

(a) $\frac{2}{\sqrt{3}}$ (b) $\frac{2}{3}$

(c) $\frac{1}{3}$ (d) $\sqrt{\frac{2}{3}}$

112. Let $P_1 : 2x + y - z = 3$ and $P_2 : x + 2y + z = 2$ be two planes. Then, which of the following statement(s) is (are) TRUE? **[JEE Advanced 2018]**

(a) The line of intersection of P_1 and P_2 has direction ratios $1, 2, -1$

(b) The line $\frac{3x - 4}{9} = \frac{1 - 3y}{9} = \frac{z}{3}$ is perpendicular to the line of intersection of P_1 and P_2

(c) The acute angle between P_1 and P_2 is $60°$

(d) If P_3 is the plane passing through the point $(4, 2, -2)$ and perpendicular to the line of intersection of P_1 and P_2, then the distance of the point $(2, 1, 1)$ from the plane P_3 is $\frac{2}{\sqrt{3}}$

113. Let P be a point in the first octant, whose image Q in the plane $x + y = 3$ (that is, the line segment PQ is perpendicular to the plane $x + y = 3$ and the mid-point of PQ lies in the plane $x + y = 3$) lies on the Z-axis. Let the distance of P from the X-axis be 5. If R is the image of P in the XY-plane, then the length of PR is .... **[JEE Advanced 2018]**

Answers

1. (a)	**2.** (c)	**3.** (c)	**4.** (1)	**5.** (a)	**6.** (6)	**7.** (a)	**8.** (b)	**9.** (b)	**10.** (44)
11. (d)	**12.** (1)	**13.** (a)	**14.** (b)	**15.** (d)	**16.** (d)	**17.** (0)	**18.** (28)	**19.** (4)	**20.** (d)
21. (d)	**22.** (b)	**23.** (b)	**24.** (b)	**25.** (61)	**26.** (d)	**27.** (7)	**28.** (a)	**29.** (d)	**30.** (d)
31. (a)	**32.** (72)	**33.** (b)	**34.** (c)	**35.** (26)	**36.** (96)	**37.** (3)	**38.** (d)	**39.** (7)	**40.** (c)
41. (b)	**42.** (81)	**43.** (d)	**44.** (4)	**45.** (b)	**46.** (1)	**47.** (38)	**48.** (c)	**49.** (2)	**50.** (a)
51. (d)	**52.** (8)	**53.** (b)	**54.** (c)	**55.** (c)	**56.** (8)	**57.** (3)	**58.** (b)	**59.** (b)	**60.** (4)
61. (a)	**62.** (a)	**63.** (a)	**64.** (3)	**65.** (a)	**66.** (b)	**67.** (3)	**68.** (a)	**69.** (a)	**70.** (b)
71. (b)	**72.** (5)	**73.** (a, b)	**74.** (a, b, c)	**75.** (c)	**76.** (a)	**77.** (a)	**78.** (c)	**79.** (c)	**80.** (d)
81. (b)	**82.** (a)	**83.** (a)	**84.** (d)	**85.** (a)	**86.** (c)	**87.** (a)	**88.** (b, c)	**89.** (b)	**90.** (c)
91. (d)	**92.** (c)	**93.** (a)	**94.** (c)	**95.** (d)	**96.** (c)	**97.** (d)	**98.** (c)	**99.** (a)	**100.** (c)
101. (d)	**102.** (d)	**103.** (c)	**104.** (a)	**105.** (b)	**106.** (a)	**107.** (a,b,c)	**108.** (0.75)	**109.** (c,d)	**110.** (b)
111 (d)	**112.** (c,d)	**113.** (8)							

For Solutions
Scan the QR Code